THE SHARKS OF NORTH AMERICA

THE SHARKS OF NORTH AMERICA

JOSÉ I. CASTRO

COLOR ILLUSTRATIONS BY
DIANE ROME PEEBLES

OXFORD
UNIVERSITY PRESS

OXFORD
UNIVERSITY PRESS

Oxford University Press, Inc., publishes works that further
Oxford University's objective of excellence
in research, scholarship, and education.

Oxford New York
Auckland Cape Town Dar es Salaam Hong Kong Karachi
Kuala Lumpur Madrid Melbourne Mexico City Nairobi
New Delhi Shanghai Taipei Toronto

With offices in
Argentina Austria Brazil Chile Czech Republic France Greece
Guatemala Hungary Italy Japan Poland Portugal Singapore
South Korea Switzerland Thailand Turkey Ukraine Vietnam

Published by Oxford University Press, Inc.
198 Madison Avenue, New York, New York 10016

www.oup.com

Library of Congress Cataloging-in-Publication Data
Castro, José I. (José Ignacio), 1948–
The sharks of North America / Jose I. Castro.
p. cm.
Includes bibliographical references and index.
ISBN 978-0-19-539294-4 (cloth : alk. paper)
1. Sharks—North America. I. Title.
QL638.9.C346 2011
597.3097—dc22 2009032858

1 3 5 7 9 8 6 4 2

Printed in the United States of America
on acid-free paper

FOREWORD

In 1980 I met a young man who was visiting second-hand bookstores in Bethesda, Maryland, where I lived. He came to see me and told me he was writing a book on sharks. He had not published any papers on sharks but was writing a book on all the known sharks of North American waters! I was skeptical at first, but as we spoke I realized there was an awesome amount of knowledge about sharks, butterflies, rare books, Persian carpets, airplanes, and hunting guns in the cranial cavity behind his long, black, curly eyelashes.

When his first book came out in 1983, it was amazingly good and well received. *The Sharks of North American Waters* became a must-have item on the bookshelves of every student and specialist in sharks, including shark fishermen. José Castro even correctly predicted that the then new "Hawaiian megamouth" shark and the goblin shark would be caught in North American waters and so had included them in his book. Because most North American sharks are found in both hemispheres, this handy volume, 180 pages, 6 × 9 inches, could be taken anywhere in one's briefcase. It was crammed with information on more than 100 sharks. It was readable and accurate and had black-and-white drawings of each species.

Since then José and I have examined and dissected many rare and large sharks together, including the first female megamouth (we even shared a taste of its muscle: raw, fried, and baked by a Japanese gourmet cook). We dissected a huge basking shark in the pouring, freezing rain near Hatteras Inlet, North Carolina. We viewed live sixgill sharks from submersibles in the Cayman Islands and Bermuda. Here at Mote Marine Laboratory, I watched him elegantly dissect specimens of rare, great sharks and then take his gorgeous, colorful photographs of the whole shark, especially the reproductive system—all truly works of art. His photograph of the reproductive tract of a bigeye thresher shark—with two oophagous embryos and the eggs they were in the process of eating—is framed and hanging on my living room wall. Happily, his photo artwork is presented in more than 20 color plates in this new big book on North American sharks. These photographs are equal in beauty and accuracy to the 160+ color paintings by Diane Rome Peebles. These remarkable color plates of each shark, its teeth, and ventral views of the head were meticulously reviewed by José. It is my hope that, after a century passes, this first edition will not be taken apart by art connoisseurs as has been done with the color plates in Bleeker's *Atlas Ichthyologique*, and the text and plates, including the gorgeous, gutsy dissections, will remain intact.

At the banquet sessions of our American Elasmobranch Society meetings, the members love to affectionately kid José. We've had contests to see who could give the best imitation of José's charmingly accented voice dismissing his own artwork after he had wowed us with his color renditions of embryos *in situ* in a female shark. At our most recent meeting in Tampa in 2005, they raffled off, for a tidy sum in support of student travel, a José Castro doll in hunting clothes, carrying a camera and an address booklet. This fun doll now resides in my office.

José has expanded and updated the tremendous and pertinent information we now have on sharks, but you will need an extra suitcase to take this encyclopedic, 600-page, 10-pound treasure when you travel! I predict that there will never be another book on sharks as informative and beautiful as this one.

Genie Clark

PREFACE

This book is a comprehensive guide to North American sharks, summarizing the knowledge of these species through 2008. This book contains descriptions and life histories of all the species of sharks reported within 500 nautical miles of North American shores, as well as selected species from adjacent areas. The area covered includes the Arctic Ocean, the eastern Pacific from Alaska to southern Mexico, and the western Atlantic from Greenland to the Gulf of Mexico and the northern Caribbean Sea. There are 135 species accounts, each beginning with the etymology of the common and scientific names and a reference to the original description of each species. The accounts provide summaries of the biology, range, length and weight, size at maturity, mode of reproduction, brood size, size at birth, nurseries, age and growth, and relation to humans. Each species is illustrated by one or more color drawings, snout and tooth outlines, and an scanning microphotograph of the skin denticles.

This book started as a revision to my field guide, *The Sharks of North American Waters*, shortly after its publication in 1983 and was a long-term project carried out in spare time over many years. In the years after publication of the field guide, I often received two requests from users of the book. Many students and biologists requested references to source publications, so that they could obtain additional details while researching and writing papers; others simply requested that I give more details about given species. The present book is an effort to satisfy those requests. My attempt to produce a more comprehensive book, and the large body of knowledge accumulated in the last two decades, has resulted in a much larger book than I envisioned.

I started by asking, "What species of sharks inhabit North America, and what do we *really* know about them?" My goal was to publish a comprehensive and accurate book that would answer those questions. Early in the development of the book, I realized that it was not possible to rely on the literature to determine the species found in North America. Many specimens taken in our fisheries were not identified or were misidentified, and there were both unreported and undescribed species. I concluded that it would be necessary to examine the sharks actually collected in many of our fisheries and research operations throughout the continent, although the lack of deep-water fisheries in North America would greatly complicate the task. The literature would have to be examined critically, and the facts would have to be verified. I also decided that I would not copy existing illustrations, to avoid replicating errors, and found it necessary to create original illustrations based on actual specimens. Of course, this required obtaining fresh specimens of all the species covered, or at least preserved specimens in excellent condition.

Although I work for the National Oceanic and Atmospheric Administration, one of the great scientific organizations, I purposely did not seek institutional or financial support for this project. My reason for this decision was that I simply did not know how long the project would take. In all institutions, large and small, all projects must be supervised and must adhere to some schedule. I knew that if this project was subject to managerial or bureaucratic control, time limits would be imposed, and I simply did not have any idea of how long it was going to take. My concern was that a schedule, which may seem reasonable at the onset of a project, is often demonstrated by hindsight to have been overly optimistic. In this case there were too many questions that could be solved only by looking at shark specimens, and I could not predict how long it would take to obtain them. Making the illustrations alone would take years. There were myriad questions and logistical problems to be solved before the book could be finished. I did not want to be forced to publish an incomplete or rushed work just to satisfy someone else's schedule. Similarly, I decided to do the project alone, to avoid the problems of imposing my ideas on others and to avoid the dissentions of multiauthor projects. Thus, it was best to do the project by myself and on my own schedule. As it turned out, the project took well over a decade.

Along the way, the Mote Scientific Foundation offered to underwrite the color plates, and I gratefully accepted their generous offer. In the end, they covered the cost of most of the color illustrations and provided great assistance. In the long process of creating the book and illustrations, there were moments of despair, when I thought that obtaining and examining fresh specimens of all the species involved was an impossible task, or when financial considerations threatened the project. Amazingly, we obtained nearly all the needed specimens (how some rare specimens were obtained probably deserves to be told someday). The project was possible through the assistance of dozens of friends: biologists, commercial fishermen, aquarists, and sportsmen who helped me to collect specimens and data or who sent me much-needed specimens often at great trouble or financial loss. If there was ever a book that required the direct help of many people, this is it!

Since the publication of my 1983 field guide, a terrible onslaught has been unleashed upon sharks. The high price of shark fins destined for Asian markets and the depletion of bony fish stocks have combined to make sharks a highly profitable product, resulting in the rapid proliferation of shark fisheries throughout the world. These generally unrestrained fisheries have greatly reduced the populations of sharks and threaten the survival of many species. It is my hope that education, enlightenment, and appreciation of these interesting predators will induce the public to demand an end to the destruction of these great fishes. I hope that this book will contribute to that endeavor.

José I. Castro

ACKNOWLEDGMENTS

Diane Rome Peebles did the color illustrations. Diane and I had talked about the possibility of her illustrating this book in the early 1990s. Having admired her elegant work for many years, I was delighted when she agreed to undertake this project with me. Working with Diane, a consummate professional, was both an honor and a pleasure. Taryn Baacke Estell and Faith B. Keller drew most of the snout and teeth illustrations. Bryan Stone and Paul Vecsi also contributed some snout and teeth illustrations. The illustrations were prepared under my guidance, and I am responsible for any inaccuracies that may have escaped me.

Most of the illustrations were funded through a generous grant from the Mote Scientific Foundation to Mote Marine Laboratory, and for that I am grateful to Kumar Mahadevan and the late Perry W. Gilbert. It was Perry's idea that the foundation should sponsor this work, and Kumar did the rest, solving many financial problems and allowing me to proceed with the color illustrations. At Mote Marine Laboratory, Genie Clark allowed me the free use of her extensive field notes on Florida sharks and Cape Haze Marine Laboratory records. She generously shared her knowledge and friendship and encouraged me throughout the project. Bob Hueter provided two valuable assistants who helped me at different times; he also provided useful comments on the illustration of the whale shark. Peter Hull helped by securing many specimens.

In making this book I received help from many people. The late Perry Gilbert's help was significant. He was aware that I had lost my library (among other greater losses) in Miami during hurricane Andrew in 1992. Perry's generous gift of a wonderful reprint collection and numerous ichthyological books from his library allowed me to continue the work. Over the years, other friends helped me to replace my library or gave much needed references: Jim Atz, Eugene Balon, Eugenie Clark, Howard Evans, Bob Hueter, Glen Loates, Larry Page, and Bill Richards. Their help with books and reprints was invaluable and saved me considerable time. Maria Elena Ibarra, Consuelo "Coqui" Aguilar, and Gaspar Gonzales kindly provided me with a set of Felipe Poey's complete works that was most useful and enlightened me greatly.

I am also most grateful to all the friends and colleagues who helped me collect specimens and data, often at a great deal of trouble to them, and who encouraged me throughout the project: Hugo Aguirre, Alberto Amorim, the late Shelton Applegate, Jim Atz, Scott Bachman, Eugene Balon, Christine Balon, Henry "Hank" Bart, Larry Beerkircher, Maria Bello, Sally Boynton, David Brindle, Cheryl Brown, Michelle Bruni, George Burgess and the staff of the Florida Museum of Natural History, Bob Burhans, Greg Cailliet, Steve Campana, John Carlson, Jeff Carrier, Nicole Castagna, Leonardo Castillo, José Luis Castro Aguirre, Eugenie Clark, Graeme Charter, all my Chiapaneco fishermen friends, Geremy Cliff, Tristram Colkett IV, April Cook, Joao Correia, Chip Cotton, Clark and Diana Crabbe, Jerry Crow, Tobey Curtis, Dean Dougherty, Bob Davis, Cyndi Dawson, the late René de Dios, William "Trey" Driggers, Sheldon Dudley, Héctor Espinosa, Howard Evans, Manny Ezcurra, Brooke Flammang, Sarah Fretzer, John Galbraith, Jeffrey Gallant, Felipe Galván-Magaña, Jimmy Gelsleichter, R. Grant Gilmore, Ken Goldman, Dean Grubbs, Sonny Gruber, Peter Hall, Karsten Hartel, Chris Harvey-Clark, Michael and Linda Heithaus, Eric Hoffmeyer, Mauricio Hoyos, Gordon Hubbell, Hua-Hsun Hsu, Steve Kajiura, Glenda Kelly, Duncan and Carol Kniseley, Nancy Kohler, Hera Konstantinou, Jay Lamee, Jeff Landesman, Christine Light, Henry Luciano, Carl Luer, Dennis Lee, Anabela Maia, Meredith Marchioni, Jennifer Martin, Fernando Márquez, Yosuke Matsumoto, Matthew Miller, Masaki Miya, John Morrissey, Ken "Curly" Moran, John A. Moore, Jeff Morris, John A. "Jack" Musick, David McGowan, Matt McLeod, Kazuhiro Nakaya, Lisa Natanson, Cheryl Nicholson, Neil Overstrom, Ted Otis, José Luis Oviedo, Reinaldo "Ray" Pérez, Gus Pérez-Abreu, Juan Carlos Pérez-Jiménez, Andrew Piercy, Dan Pondella, Steve Poston, Gregg Poulakis, Harold "Wes" Pratt, Manny Puig, Mark Quartiano, Christopher Rackley, Mark Rackley, Sandra Raredon, Nelson E. Rios, Claudia Lorena Ruiz, Mark Sampson, Jorge Sánchez, Eric Sander, Keiichi Sato, Mike Schaadt, Ron Schatman, Ivan Schultz, Jason Seitz, Katie Shade, Robert Siders, Candance Silva, Guylaine Simard, Buck Snelson, Sandra R. Soriano, Oscar Sosa Nishizaki, Charlott Stenberg, Rick Stringer,

Adam Summers, Chris Tanaka, Hisao Teshima, Taketeru Tomita, Antonio Sandes Torres, John Tyminski, Rick Waites, H. J. Walker, Tom Watts-FitzGerald, Nick Whitney, Tonya Renee Wiley, Jeff Williams, Makio Yanagisawa, Kara Yopak, and Forrest Young. I thank the Florida Institute of Oceanography, Dean Milliken, Randy Maxon, Rob Walker, and the captain and crew of RV *Suncoaster* for their support and use of their vessel, which allowed me to obtain many deep-water specimens. I thank Larry Page, George Burgess, Robert H. Robins, and the staff at the Florida Museum of Natural History for their splendid assistance with all the specimen loans, and their great camaraderie. I thank Henry "Hank" L. Bart for his assistance with specimens at the Tulane University Museum of Natural History and for being a gracious and splendid host. I thank Jorge and Lázaro Sánchez for their many courtesies at their facility, Casablanca Seafood, where we processed some very large specimens. Seven of my friends contributed a great number of specimens: George Burgess, Chip Cotton, William B. "Trey" Driggers III, John Galbraith, Rey Pérez, Juan Carlos Pérez-Jiménez, and Ron Schatman. They were the "knights in shining armor" who came to my rescue many times with the "impossible-to-get" specimens. John Galbraith, ichthyologist and naturalist extraordinaire, and Professor Kazuhiro Nakaya helped me to discriminate among the species of *Apristurus* catsharks in North America, which would have been impossible without their combined help. Given the nature of the project and the need for fresh specimens, I would have never been able to accomplish this book without the generous help of all these folks.

The use of the excellent scanning electron microscope at the Okinawa Churaumi Aquarium, through the courtesy of Senzo Uchida and Keiichi Sato, allowed me to obtain many of the images of the dermal denticles. I thank Tom Beasley and Barbara Maloney for their guidance in using the electron microscope at Florida International University. Their help was invaluable in obtaining many of the images presented here.

Beverly McMillan, Michael Heithaus, Stephen Spotte, Gregg Poulakis, Juan Carlos Pérez-Jiménez, and William B. "Trey" Driggers, III critically read the manuscript and made many valuable suggestions. Cheryl Nicholson, Katherine Shade, and Meredith Marchioni proofread the manuscript, pointing out many errors that I had missed. April Cook performed many tasks in the preparation of specimens during part of the project, and her help was invaluable. Elizabeth Simonson assisted (as a volunteer) with specimen preparation and many other organizational tasks. She was the catalyst that made many tasks possible at the end of the project. She alone was not intimidated by the prospect of going into the shark freezers looking for a long-lost specimen. The onerous and tedious task of preparing the initial bibliography fell to Sally Boynton, Esq., a long-time volunteer, who would come into my office chortling with mirth after finding some small transgression in the manuscript. Her help, support, and friendship meant a lot to me. Juan Carlos Pérez-Jiménez helped me immensely with the sharks of the Gulf of California and proofread the final bibliography. Meredith Marchioni took time from writing her dissertation to locate and organize the pen-and-ink drawings, coming to my help at various times. I also received splendid help from Mote Marine Laboratory librarian Susan Stover and the IAMSLIC[1] librarians who provided me with so many requested references. To all these great folks I express my sincere gratitude, while retaining all responsibility for all errors of fact or omission.

I thank my Japanese friends Senzo Uchida, Keiichi Sato, and Masaki Miya, who made it possible for me examine fresh goblin and frill sharks, as well as many other deep-water sharks in Japan. Senzo Uchida made it possible for me to visit Japan several times, and I learned a great deal from him and his staff at the splendid facilities of the Okinawa Churaumi Aquarium. My good friend Keiichi Sato shared his great knowledge of deep-water sharks and was always a splendid host.

I thank my friends Bob and Ida Fowler for their wonderful hospitality at their home, my favorite hideaway in Albany, Georgia, where I accomplished parts of the work. I thank Clark and Diana Crabbe for their friendship, hospitality, and splendid support at their home in Rancho El Barril, Baja California, where I had the opportunity to observe many species of sharks of the Sea of Cortez under most pleasant conditions. My friend Gerald (Gerry) Dratch was a pleasant cheerleader who always encouraged me by reminding me that we were both mortals and that I had to finish the book before one of us died. Jim Bohnsack encouraged me and did much to facilitate the project. Divina Grossman encouraged me throughout the project. Her advice and moral support meant a great deal to me.

Finally, I thank those who gave me the tools to undertake this project—first, my parents and family, who, in just a few years, planted the seed of education, which was to survive many years of separation and adversity, to germinate and bear fruit many years later. A series of gifted science and biology teachers influenced and helped me greatly from primary school (when there was no one else to help) to postgraduate training: Mr. Tabor at Citrus Grove Junior High, Mary Ellen Chestnut at Miami Senior High, E. Morton Miller and Taylor R. Alexander at the University of Miami, and, much later in graduate school, Christopher C. Koenig. John P. Wourms, and A. "Budd" Bodine were both mentors and friends at Clemson University during my doctorate there. I owe so much to all of them.

1. The International Association of Aquatic and Marine Science Libraries and Information Centers (see http://www.iamslic.org/).

CONTENTS

THE SHARKS OF NORTH AMERICA

INTRODUCTION

HOW TO USE THIS BOOK

This book uses standard dichotomous keys to help readers identify families and species of sharks. These keys are aids in shark identification that use easily observable anatomical features (e.g., fin position and shape of the teeth) or geographic range to identify families or species of sharks. Each key consists of a series of alternatives presented in numbered couplets. Starting with the first couplet, the reader should select the alternative that best matches the specimen being identified, and then follow it to the next indicated couplet, continuing the process until the family or species is identified. Once a species is reached, additional characteristics of the species may be added in parentheses to confirm the identification. These keys are not based on phylogenetic or taxonomic relationships; their design follows diagnostic convenience.

To identify and learn about a shark, first turn to the Key to the Families of North American Sharks (p. 11) to determine which family a given shark belongs to; the identification can be assisted using the figure on the right margin. Then use the key to that family to determine the species of shark. Finally, go to the indicated pages to confirm the identification and read about the species.

Description of Species Accounts This book covers 135 species of sharks, with species accounts that summarize our present knowledge of North American species. I have also included some species from adjacent areas. These are clearly marked as "extralimital species" and are included because they have been reported as occurring in North American waters due to identification errors, or are included to complete families of sharks and extend the area where this book can be useful for identifying sharks. Some accounts are brief because of our limited knowledge of the species. Statements or observations that are not the author's are followed by a bibliographic citation in parentheses. All accounts are written in the same format to facilitate comparisons, and contain the following sections:

Common name: The species accounts begin with the common English names of the species and the etymology of the name. The common names used are those designated by the American Fisheries Society (A.F.S.) in Special Publication No. 29, Common and Scientific Names of Fishes from the United States, Canada, and Mexico, sixth edition (Nelson et al. 2004). When a common name has not been designated, but a vernacular name exists, the latter is used. When no common names are in use, the name given to the species by the Food and Agriculture Organization of the United Nations (F.A.O.) is used.

Spanish name: The Spanish names given are the native names I have encountered in usage or in the literature of the Spanish Caribbean and Mexico. Some useful references for Spanish names are "*Guía para los tiburones de aguas cubanas,*" by Guitart Manday (1968); *Lista de nombres científicos y comunes de peces marinos cubanos (Nomenclator),* by Rodríguez et al. (1984); *Guía de campo para la identificación de especies de tiburones y cazones en la Sonda de Campeche,* by Uribe Martínez (1990); *Guía para la identificación de las especies de tiburones de importancia comercial del Golfo de México* by this author (Castro 2001a); and *Guía para la identificación de las especies de tiburones de importancia comercial del Océano Pacífico* (Castro 2001b). I also provide the Spanish names given by the American Fisheries Society (Nelson et al. 2004), although these names often differ widely from the names actually used in the Spanish Americas.

To placate editors who thought this work was already too long, I have explained only a few of the Spanish etymologies, primarily when a colorful name may have been misunderstood by nonnative Spanish speakers, or when I just could not resist explaining it.

Scientific name: The scientific name given is that currently used in the scientific literature. Each scientific name consists of two words in Latin or in latinized form. The first word is the genus to which the species belongs, or generic name; the second word is the species or specific name. A scientific name includes the last name of the author or authors that formally described the species, and the year that the scientific name was published. Some authors' names are enclosed in parentheses, indicating that the author originally placed the species in a different genus than the one to which it is presently assigned. The scientific name is followed by a reference to the publication where it was first published. The etymology of each scientific name is given in an attempt to explain the name. Many different sources were used in deciphering scientific names; the most useful of these is Jaeger's *A Source-Book of Biological Names and Terms,* third edition (1962).

Synonyms: Most sharks have several obsolete scientific names, published after the original, accepted description of the shark. These names, known as synonyms, are invalid as scientific names. Synonyms are often encountered in older literature, and a complete list would cover many pages. These synonyms are of interest to specialists and students and are already available in publications by Bigelow and Schroeder (1948) and Compagno (1984 a, b). Thus, I have not attempted to repeat the extensive synonymies found in those works. I have merely listed the more common synonyms that may be encountered by the reader.

Identification: The identification section includes the most important, diagnostic, anatomical characteristics used to recognize the species, with the key identifying marks in italics. The shape and number of teeth are given because these are often diagnostic. The shape of the teeth is also useful in elucidating the feeding habits of a species. The number of teeth in each side of the upper jaw is written before the number of teeth in each side of the lower jaw. For example, U: 12–1–12, L: 15 to 17–2–15 to 17 indicates 12 teeth on the animal's left side of the upper jaw, 1 tooth in the center of the jaw (symphysis), and 12 teeth on the right side of the upper jaw; 15 to 17 teeth on the left side of the lower jaw, 2 teeth in the center, and 15 to 17 on the right side. In cases where the center of the jaw cannot be distinguished readily, the total number of teeth in that jaw is indicated. For example, U: 24 to 30, L: 28 to 32 indicates that there are 24 to 30 teeth in the upper jaw and 28 to 32 in the lower jaw. The shape of the dermal denticles is given because it is often useful to distinguish among species. The dermal denticles pictured are from the flanks just below the first dorsal fin. The shape of the denticles varies all over the body. The denticles protect the shark with a tough skin and have other important properties. Their primary function appears to be, in many cases, to provide laminar flow of the water over the body, minimizing drag and allowing the shark to move effortlessly and quietly through the water. The shape of the denticles can be species specific and in some cases can be a great aid to identification. In some cases sharks of a given genus can have similar denticles. The denticles are covered by a thin layer of skin and mucus, but that layer is sometimes lost upon preservation. The denticles of neonates and juveniles are often different from those of adults. The denticles shown here are from the illustrated specimens.

The color of the illustrations is that of live or freshly caught specimens, since the coloration of most sharks will fade or dull shortly after death. I have attempted to illustrate sharks in their fresh coloration, often going to great lengths to verify these colors. However, the coloration of most sharks is variable, often changing with light conditions and angle of view, and often varying from one specimen to another. In the cases of some deep-water sharks, it is possible, or even likely, that we may have illustrated sharks that I considered fresh but were actually faded specimens.

The colors of sharks (or other animals) can be caused by either structural colors or pigments. Structural colors result from the interaction of light and thin surface tissues, and are responsible for the iridescence[1] of some sharks when fresh out of the water. These colors change with the angle of view or light and can be easily destroyed by simply scratching the skin or by a jet of water. Different colors may be reflected in different directions by surface structures.[2] As tissues dry or shrivel up, iridescence disappears. By contrast, colors due to pigments do not change with the direction of viewing and fade only slowly as the pigment decays. Pigments are organic compounds found in or among the animal's cells and can serve many different purposes, such as camouflage, protection against ultraviolet rays, or adding strength to structures. Pigments can be endogenous (produced by the animal) or exogenous (acquired, usually through the diet). The golden hammerhead is an example of a species with both structural and pigmentary colors. When the shark is fresh out of the water, it has an overall iridescence that changes with the viewing angle. This is due to interference or diffraction of light in the thin outer skin. The golden hammerhead also has yellow and orange pigments in and among its skin cells. These pigments are acquired through the diet (see p. 526) and cause its bright yellow or orange colors. These pigmentary colors look the same regardless of angle of view and cannot be scratched or washed off. Some pigmentary colors endure for a long time and can even survive freezing for many days.

1. Iridescence (from the Greek *iris,* the rainbow) "is the phenomenon of glittering with different colors that change according to the angle from which the object is seen" (Fox and Vevers 1960: 5).

2. A discussion of this phenomenon is beyond the scope of this book; the reader may refer to texts such as *The Nature of Animal Colours* by H. Munro Fox and G. Vevers (1960) or to *Animal Biochromes and Structural Colors* by D. L. Fox (1976).

Similar species: Similar species occurring in the same area are listed with their distinguishing characteristics. The most similar species is listed first, and the least similar is listed last.

Range: The range given for a species is its worldwide distribution, the known or approximate limits of its North American distribution, and areas where the species may be locally common. A species can be referred to as polar, subpolar, temperate, subtropical, or tropical, depending on the waters it inhabits. Polar waters are those where the surface temperature is usually below 5°C. In polar areas, or in seas influenced by polar currents, bottom temperatures as low as –1°C can occur, due to the presence of salt in seawater, which depresses the freezing point. Subpolar waters range from 5 to 10°C. Temperate waters include a wide zone, ranging from 10°C on the polar side to 20°C on the side toward the equator. Those waters in the warmer part of the temperate zone are called warm-temperate; those on the polar side are referred to as cold-temperate. Subtropical waters range from 15 to 30°C. Tropical waters form two wide zones, one on each side of the Equator, where the surface temperature is 25°C or higher.

The range of a species is a three-dimensional space determined by its tolerance limits to physical factors. Temperature is one of the most important factors affecting the distribution of sharks; it usually decreases rapidly with depth, from maxima of 25–30°C at the surface to 2°C at depths greater than 2,000 m. The average temperatures at different depths for the oceans as a whole, according to Murray and Hjort (1912), are as follows:

Meters	Degrees Centigrade
183	15.95
366	10.05
549	7.05
732	5.44
1,097	3.89
2,012	2.28
2,743	1.83
4,023	1.78

Whereas the temperature of deep waters is generally uniform and stable regardless of latitude or season, the temperature of surface waters is highly variable, changing with latitude and season. Thus, species inhabiting surface waters are often migratory and move about seasonally seeking favorable conditions, while species inhabiting deep waters live under stable conditions with few barriers to their distribution and thus are often widely distributed. In temperate latitudes of the Northern Hemisphere, the warmest temperatures are found close to shore during summer. In winter, coastal waters can cool off rapidly, while offshore waters remain warmer and more stable. It follows that a species adapted to warm waters may be found close to shore in the northern parts of its range during the summer; when the temperature begins to fall in autumn, it may seek warmer offshore waters or migrate southward. A cool-water-adapted species may be found close to the surface in the northern parts of its range, while in the southern parts it may inhabit much deeper waters where correspondingly cooler temperatures are found.

Size and weight: This section gives representative sizes and weights for the species and its maximum known size. The length given is total length, which is measured in a straight line between perpendiculars, from the tip of the snout to the tip of the tail at its maximum extension (see figure X). Length measurements are in centimeters (cm), or in millimeters (mm) for very small structures or small embryos. Occasionally, I have used meters as measurement units for the largest sharks. Weights are usually given in kilograms (kg). In a few cases, I have given the fork length or the precaudal length of specimens when quoting works using those measurements, and when the measurement could not be accurately converted to total length.

Biology: The general section on biology gives a summary of habitat, including depth and temperature preferences, habits, diet, and migrations.

Size at maturity: This section summarizes the size at which males and females become sexually mature, giving the methods used to determine maturity.

Reproduction: The section on reproduction gives mode of reproduction, mating and birth seasons, length of the gestation period, duration of the reproductive cycle, size of the young at birth, and brood[3] size. When the size of the young at birth is unknown, the size of the smallest free-swimming specimen known is given. See Appendix 1 for more on shark reproduction.

3. Throughout this book, I use the terms "young" and "brood," purposely avoiding the terms "pups" and "litter" that prevail throughout the elasmobranch literature. Pups are the young of certain mammals, such as canines and pinnipeds, and litter refers to the offspring produced at one birth by a multiparous mammal. Brood is the traditional term for the progeny of fishes and birds. Those who insist on using definitions other than those provided by dictionaries can claim that the terms "pup" and "litter" are well entrenched in the elasmobranch literature or can advocate the Humpy Dumpy rule: "'When *I* use a word,' Humpty Dumpy said, in a rather scornful tone, 'it means just what I choose it to mean—neither more nor less'" (Carroll 1936, p. 214).

Nurseries: Nursery areas, or simply nurseries, are geographically discrete parts of the species range where the gravid females deliver their young or deposit their eggs and where the young spend their first weeks, months, or years (Castro 1993b). Nursery areas are often in shallow coastal areas of high productivity, such as *Spartina* marshes and mangroves, where prey is abundant and where large predators are scarce.

The use of nursery areas by elasmobranchs has been known since ancient times. Aristotle first recorded, in his *Historia Animalium* (probably written around 343 B.C.) that "The Selachia come in from the high seas and out of deep water towards land and produce their young there; this is for the sake of the warmth and because they are concerned for the safety of their young" (Aristotle 1970: 265). Little else was said about the subject for the next 2,000 years.

Age and growth: The life span of sharks can be determined by keeping them in captivity, by tagging, or, more commonly, by counting growth rings in their vertebrae or spines and estimating the age. Only a few species of sharks have been kept in captivity for long periods of time. When sharks have been captured young and have lived in captivity for prolonged periods, a rough idea of their potential life span can be obtained. Although keeping sharks in captivity for many years is a difficult feat, some sharks have lived for decades in captivity, indicating that they have long life spans. Unfortunately, most aquaria do not keep accurate records of the life span of their captive sharks, and valuable data on the longevity of sharks is lost. Tagging sharks with long-lasting tags that are not easily shed or rejected by the shark has produced estimates of longevity for some species. When a shark is tagged, if the tag is not shed, and if the shark is recaptured many years later, a rough estimate of its life span can be obtained. When a large number of tagged sharks of a given species are recaptured, one begins to get an idea of the possible life span, assuming that the tag does not decrease the life span. For example, more than 60,000 blue sharks have been tagged by the National Marine Fisheries Service's Cooperative Shark Tagging Program over the last few decades (Kohler et al. 1998). The longest period between tagging and recovery is 8.5 years. Thus, if the sharks were tagged when only a few years of age, it is likely that the maximum life span of the blue shark is about 15–20 years.

The most commonly used method for aging sharks is to count the number of growth bands that sharks periodically deposit in their vertebrae or spines. These rings can be visually counted, usually after the rings have been enhanced with dyes, or viewed under X-rays. If the rate at which the shark produces the growth rings is determined, one can then estimate the age. When a shark is tagged and at the same time injected with the antibiotic oxytetracycline, a fluorescent ring is deposited on the vertebrae. If the shark is recaptured at a later time and its vertebrae are recovered, the number of rings formed after the deposition of the tetracycline ring can be used to confirm the rate of ring deposition and the age of the shark. When the rate of ring deposition is confirmed by injection with tetracycline or any other method, the age estimate is said to be "validated." Although there are many age estimates for many species of sharks, few estimates have been validated. To further complicate the situation, it is possible, and even likely, that some sharks may deposit growth rings regularly during their juvenile stage and then stop depositing such rings after maturation, or deposit them at irregular periods. Age determination and validation of age estimates are dynamic areas of shark research, and much remains to be learned.

Relation to humans: This section summarizes the interactions of humans with the species, such as commercial exploitation, uses of the species, and protected status.

Illustrations Each shark species is illustrated by one or more profile figures and outlines of the snout, upper and lower teeth, and dermal denticles. The left side of the shark is illustrated, as is customary in the ichthyological literature, with the pectoral and pelvic fins depressed to show their shape and relationship to the other fins. Diane Rome Peebles prepared all of the color illustrations. The illustrations were generally prepared from fresh specimens. It was our goal to show the colors of live or fresh specimens. Whenever possible, each specimen was photographed soon after capture. In the early part of the project, I used Kodachrome 64 film to ensure color fidelity. As this film became difficult to obtain, I switched to Fujichrome Velvia and eventually to digital Nikon cameras. Each specimen was photographed in profile and in anatomical detail. Some 60 standard measurements were taken on each specimen, and proportions were calculated. The specimen was then forwarded to the illustrator for her examination. She then prepared a profile outline based on the specimen, the photographs, and the proportional measurements. The outline was then sent to me for checking against the photographs and proportional measurements. The proportions in the outline had to be within 1% of the proportions previously calculated for the specimen. After the profile outline was accepted, we agreed on a color scheme, and then the illustration could be prepared.

Some sharks become severely distorted when taken out of the supporting water. When taking profile photographs, with the shark on its right side, every attempt was made to support the shark being photographed to avoid a lateral flattening. This flattening was more pronounced in large or heavy sharks and in deep-water species. Although we attempted to avoid the resulting distortion, some illustrations may show a shark as having a deeper body than would be observed when alive in the water. All the illustrations were prepared under my direction or supervision, and I am responsible for any errors or inaccuracies.

The outline drawings of snouts, teeth, and denticles were prepared from the actual snouts and teeth. Tooth sets were prepared from the jaws of the specimens illustrated. The teeth were separated and placed on a cardboard backing. Their outlines were then traced or drawn free-hand. The dermal denticles illustrated were taken from an area just below the first dorsal fin and were photographed using scanning electron microscopes.

Glossary A glossary at the end of the book explains selected words that may not be familiar to the casual reader.

Bibliography The references cited in the text are listed in the back of the book. Entries are listed alphabetically under the first author's last name(s). Please note that, for Spanish authors, it is customary to use two last names, the first last name being the patronymic. For example, in Juan Carlos Perez Delgado, Perez is the patronymic. In English literature, this person's name would often be indexed as Delgado, omitting the patronymic, thus corrupting the name. To preserve the correct names, I have treated such Spanish last names as if they were hyphenated last names.

THE KNOWLEDGE AND STUDY OF SHARKS

The knowledge of sharks has lagged considerably behind that of bony fishes. There are several reasons for this. First, fishes were studied because of their commercial value as food or sport, and it was desirable to know their life histories to know when and where to find them. The impetus of commercial value ensured that funds were allocated to scientific studies on commercially important species. Sharks usually lacked commercial value because they were considered unappetizing due to their often unpleasant smell, caused by the presence of urea and trimethylamine in their blood. Consequently, most ichthyologists studied commercially important bony fishes and generally ignored elasmobranch fishes. Classical books on fishes devoted only a few pages to the study of sharks. For example, the great book *Fishes* by David Starr Jordan (1907), used to train ichthyologists for decades, covers all the elasmobranchs in only 37 of its 789 pages. Thus, the study of sharks was often left to the self-trained.

Numerous logistical problems make the study of sharks difficult. The large size attained by many species of sharks often makes it very difficult for biologists to study them. Large sharks are usually difficult for fishermen to bring back to port, and many rare or unusual specimens are not boated or are discarded overboard. Even when specimens are obtained by scientists, their large size often prevents them from being preserved, even in museum collections. In commercial fisheries, it is difficult for scientists to examine sharks because they are typically gutted, finned, or discarded quickly after capture. Thus, biologists wishing to study sharks must often examine them at sea, because the gutted carcasses brought back to port yield little biological data. Today, a multitude of fishing regulations prevent or discourage fishermen from bringing back rare or unusual sharks found dead on their lines. Thus, these specimens are discarded back into the sea and the opportunity to learn about them is lost.

Sharks are often fast-moving and wide-ranging fishes that cannot easily be studied in their environment. The status of our underwater technologies prevents us from observing and studying sharks in their environments, as we do with land animals or even some marine bony fishes. In most cases we are able to obtain only brief glances of these fishes or at best a few minutes of observation. Observations in captivity are also difficult because only a relatively few species of sharks can be maintained for long periods of time in present-day aquaria. Although much has been learned about the behavior of small bony fishes in captivity, attempting to study the behavior of most sharks in aquaria is akin to trying to study gorilla behavior by watching gorillas confined to small cages. Because of the problems outlined above, much of what is learned about sharks today is still inferred from dissection of dead specimens. Although some small sharks could be studied in aquaria, little has been published on sharks based on captive studies. The interesting behavioral study of captive bonnetheads by Myrberg and Gruber (1974) seems to be the only one of its kind.

In the late 1970s and early 1980s, shark fishing tournaments became very popular along the eastern United States, probably influenced by Peter Benchley's 1974 popular book and movies about a man-eating shark. At about the same time, commercial shark fisheries started to develop in the southeastern United States and throughout the world, engendered by the rising price of shark fins and declining stocks of other fishes. Since 1990 the commercial value of sharks has increased dramatically because of the reduction in catches of bony fishes and the increasing demand for shark fins in Asia. (The fins are,an essential ingredient in the traditional Chinese shark fin soup, and their value exceeds the value of the flesh of a shark). With the impetus of commercial shark fisheries and the popular interest in sharks, the desire and need for information on sharks grew. As shark fisheries grew unrestrained, the need for regulation became evident, which in turn demanded more knowledge of sharks. These factors engendered a new interest in sharks, and more biologists started to study sharks. At present there is great commercial and scientific interest in sharks.

THE SHARK LITERATURE

I have reviewed the relevant scientific literature to the best of my ability, making every effort to review all the articles on each North American species. The overall quality of the shark literature is poor. Confusion about species or specific names was common until the middle of the twentieth century. The publication of Bigelow and Schroeder's (1948) monumental work on sharks of the western North Atlantic solved many of the existing identification problems. However, considerable confusion remained about the identity and species names of hammerheads, requiem sharks, and squaloid sharks until relatively recent times. The taxonomy of the hammerheads fluctuated until the publication of Gilbert's (1967) revision of the family, although questions about the identity of some hammerheads still remain. The taxonomy of the requiem sharks of the genus *Carcharhinus* was not settled until the publication of Garrick's (1982, 1985) excellent reviews. Thus, the older literature must be used with caution unless the species identity can be verified. Considerable confusion still exists today about the species in many genera, such as *Centrophorus, Mustelus,* and *Apristurus.*

The shark literature up to the 1980s suffered from a perennial lack of qualified "reviewers."[4] Because there were few biologists who had studied sharks, manuscripts were often reviewed by ichthyologists who knew little about sharks. Consequently, many flawed articles have been published, containing much misinformation and misleading concepts that became established lore. The reader should also be aware that, because of the small number of scientists involved and the logistical difficulties in working with sharks, there is little verification of published works, so erroneous information, once published, tends to survive for a long time.

Even today, the lack of reviewers qualified to critically read articles on sharks continues to bedevil journal editors. Even the modern literature must be read with caution, as I have seen numerous recent articles and theses where common species have been misidentified, judging by the accompanying illustrations. The problems of the current literature are not limited to the usual identification problems or lack of qualified reviewers. Today there is steep competition for manuscripts among the numerous journals, and some editors are compelled to fill a journal issue with many flawed articles. I know of several recently published articles that had been rejected by various independent reviewers but were published nevertheless, probably due to the editorial need to fill a journal issue. Another unfortunate trend of the current shark literature is the publication of much "armchair" research. The current interest in sharks, combined with the increasing logistical and legal difficulties in obtaining the necessary study specimens, has engendered numerous "armchair" studies by authors lacking firsthand knowledge of sharks, who did not examine any specimens, or who used questionable data collected by others in the distant past. Thus, even the current literature must be examined critically by the reader before it is accepted. I have attempted to rectify the numerous errors I found in the literature, providing the necessary evidence to justify my emendations. I do not claim to have corrected all the misinformation in the shark literature.

In this book I have cited references and justified, qualified, or explained nearly every statement in the species accounts. I have attempted to give evidence for most of my statements, writing in such a way that the reader can follow the logic behind the facts or conclusions stated. This may cause the writing to sometimes appear verbose and longwinded, and has resulted in a book much larger than I had envisioned.

4. When articles are submitted to a scientific journal, it is customary to send them for review to two or more qualified "reviewers" or "referees," who check it for scientific merit and accuracy. The reviewers recommend to the editor to accept the article for publication as it is, to accept it conditionally pending a revision, or to reject it entirely, based on their evaluations of its merit. Journals that publish articles so reviewed are said to be "peer reviewed," and this is the standard of most scientific disciplines. This system works well only when the editors make an effort to find qualified, unbiased, and knowledgeable reviewers, and when the reviewers take the time to carefully evaluate a manuscript. Lazy editors, who ignore reviewer's recommendations just to fill the journal issue, or ignorant and biased reviewers often cause the system to fail and cause bad science and misinformation to be published or may keep worthy articles from being published. The peer-review system often works poorly with articles that challenge accepted dogmas (which history sometimes demonstrates to be wrong!), causing articles that push the frontiers of knowledge to be rejected. Nevertheless, this system is necessary and valuable in the dissemination of knowledge.

Articles that are not "peer reviewed," such as project reports, "working documents," or articles published in newsletters and magazines, are often referred to as the "gray literature." I have generally avoided using these.

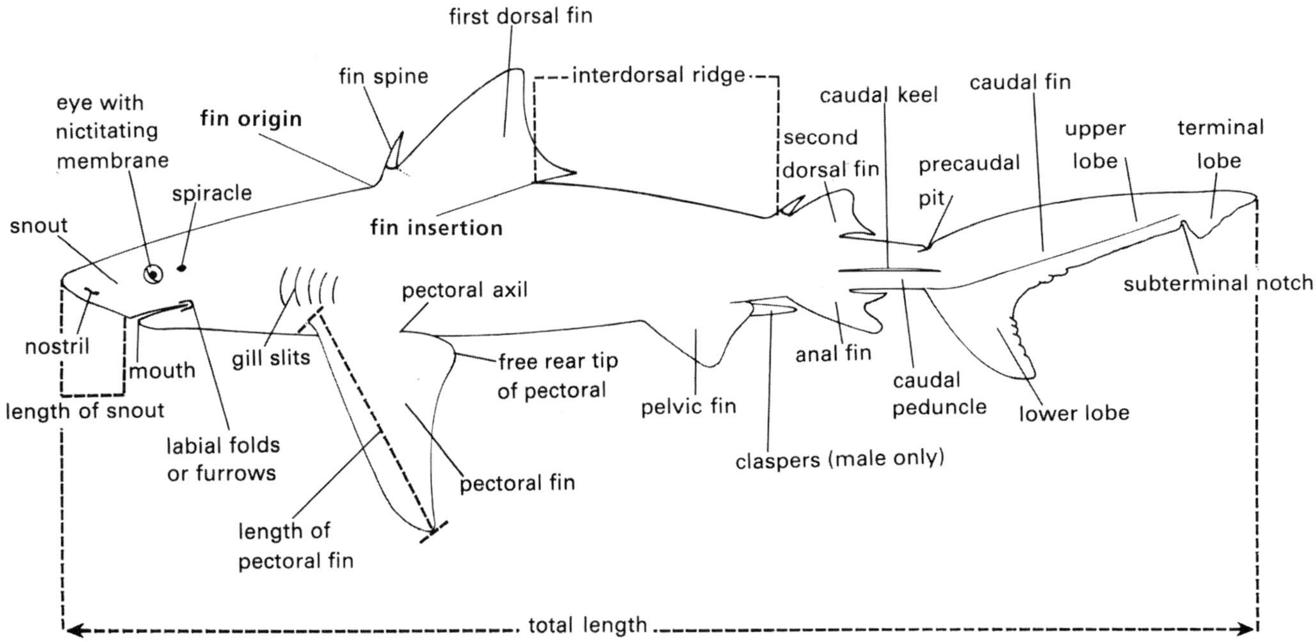

Parts and measurements of a shark

A KEY TO THE FAMILIES OF NORTH AMERICAN SHARKS

1a. Snout elongated into a flat, sawlike blade with teeth and two long barbels on the ventral surface... **PRISTIOPHORIDAE**[1] Sawsharks (p. 159)

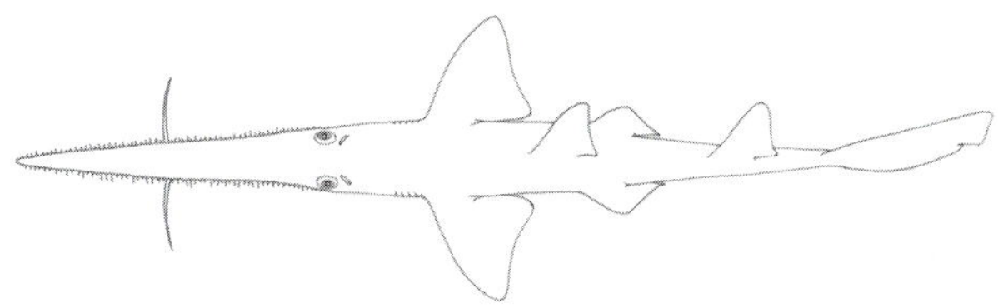

1b. Snout without a flat, sawlike blade 2

2a. Body flattened and skatelike, eyes on top of the head and two large spiracles behind the eyes.. **SQUATINIDAE,** Angel sharks (p. 162)

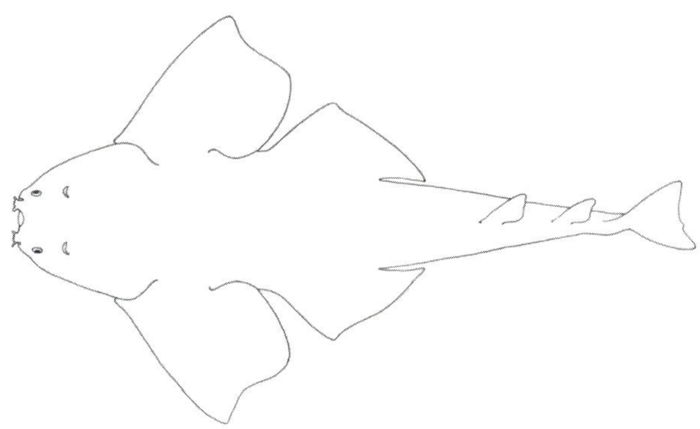

2b. Body not flattened, eyes on the side of the head 3

3a. Gill slits 6 or 7, only one dorsal fin 4

3b. Gill slits 5, two dorsal fins 5

1. The sawsharks are small deep-water sharks and should not be confused with the shallow-water sawfishes. The sawfishes are sharklike rays and are among the largest elasmobranchs. The sawfishes have ventral gill slits and pectoral fins that arise from the head far anterior to the dorsal fins. They also lack the nasal barbels on the underside of the snout. Sawfishes are endangered species and may not be taken.

4a. Mouth terminal, first gill slit continuous under throat................ **CHLAMYDOSELACHIDAE,** Frill shark (p. 21)

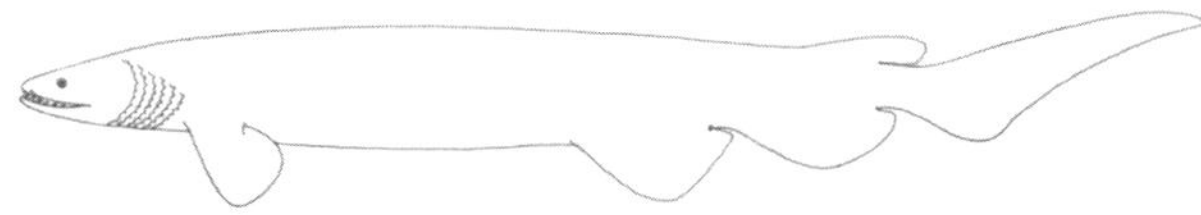

4b. Snout projecting ahead of mouth, first pair of gill slits not continuous under throat.. .... **HEXANCHIDAE,** Cowsharks (p. 26)

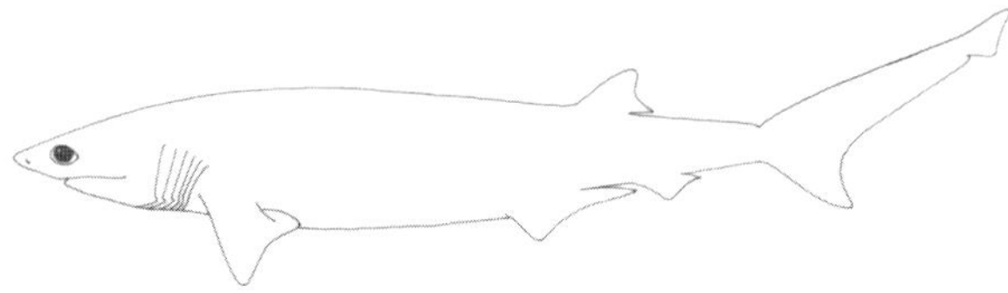

5a. Anal fin absent.. 6

5b. Anal fin present 13

6a. First dorsal fin originating well behind the origin of the pelvic fins, skin covered with large, thornlike denticles (dorsal fins without spines) .... **ECHINORHINIDAE,** Bramble sharks (p. 42)

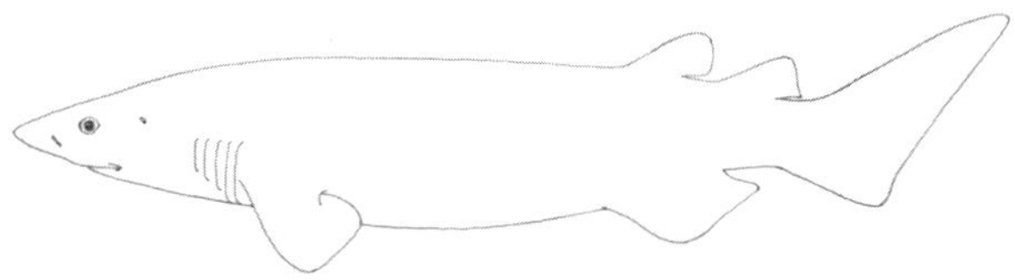

6b. First dorsal fin originating ahead of origin of pelvic fins; skin lacking large, thornlike denticles 7

7a. Trunk triangular in cross section; first dorsal fin originating over pectoral anterior base, first dorsal fin larger and longer than pectoral fin, with strongly falcate rear margins (large dorsal spines embedded in the skin, with only the tips showing). . ... **OXYNOTIDAE,** Rough sharks (p. 137)

7b. Trunk rounded or oval in cross section, first dorsal fin originating behind pectoral fin axil, first dorsal fin smaller than pectoral fin, with straight, rounded, or weakly falcate margins... 8

8a. Teeth similar in both jaws, but may be larger in lower jaw, with a single, bladelike, oblique cusp .... **SQUALIDAE,** Dogfishes (p. 50)

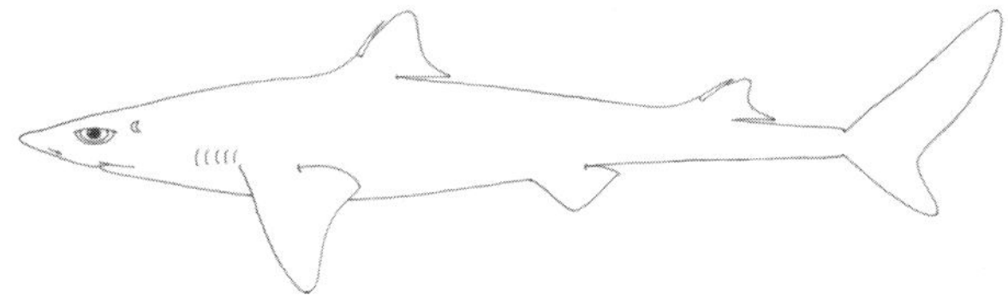

8b. Teeth dissimilar in both jaws or with several cusps (a large central cusp with lateral cusplets) 9

9a. Upper teeth with several cusps (a large central cusp with lateral cusplets), small shark less than 1 m, coloration uniformly black or with black undersides .... **ETMOPTERIDAE,** Black dogfishes and Lanternsharks (p. 91)

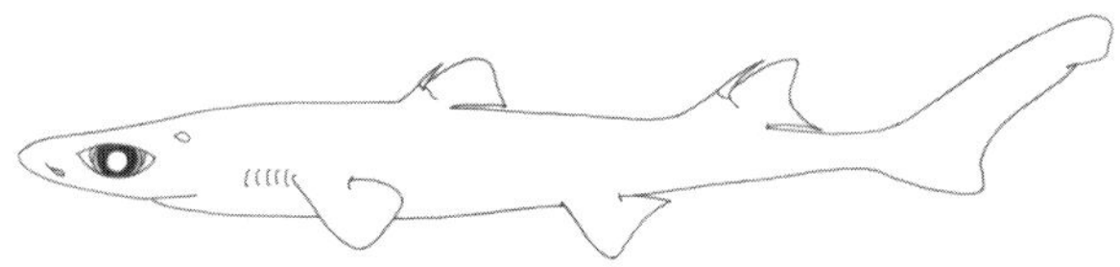

9b. Upper teeth with a single cusp 10

10a. Pectoral fin free, rear tip pointed (often drawn in to a long pointed lobe); a series of grooves on throat behind mouth; OR snout very long, its length greater than distance from mouth to pectoral fin origin .... **CENTROPHORIDAE,** Gulpers and birdbeak dogfishes (p. 67)

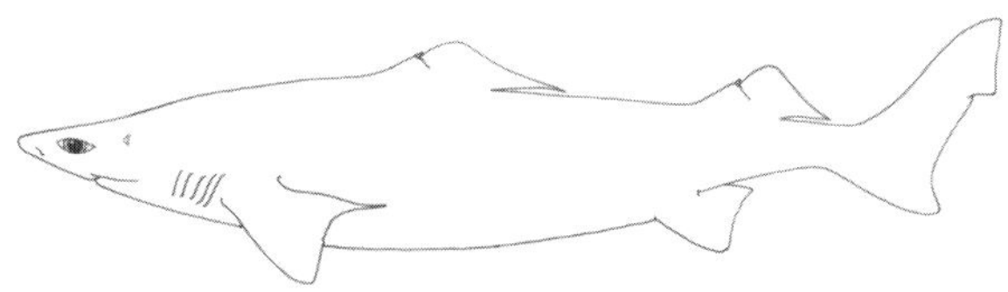

10b. Pectoral fin free, rear tip rounded; throat area smooth, lacking grooves behind mouth; snout short, its length shorter than distance from mouth to pectoral fin origin 11

11a. Second dorsal fin with a spine.............. **SOMNIOSIDAE** (in part), Genera *Scymnodon* and *Centroscymnus* (p. 115)

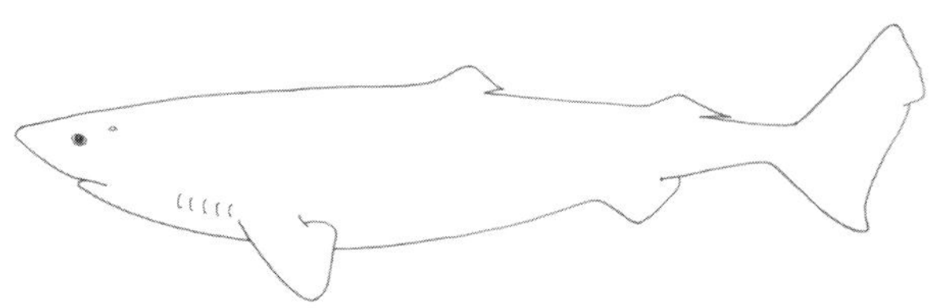

11b. Second dorsal fin spineless 12

12a. Snout-to-eye distance equal to 1–2 eye diameters or less, length of first dorsal fin base smaller than second dorsal fin base (usually very small sharks except for *Dalatias*)................**DALATIIDAE,** kitefin, cookiecutter, and pygmy sharks (p. 140)

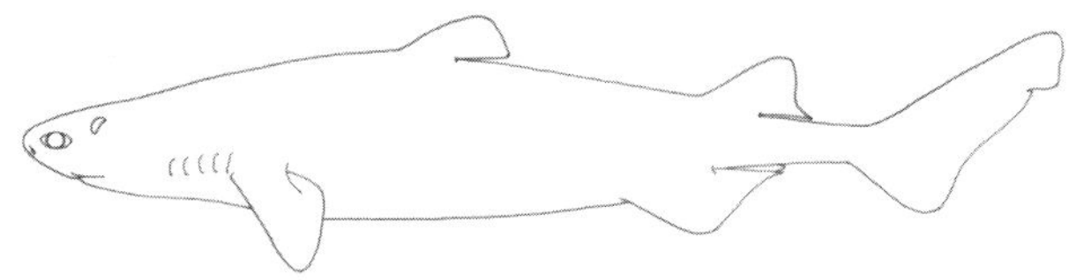

12b. Snout-to-eye distance at least 3 eye diameters, length of dorsal fin bases about equal or first dorsal fin base greater than second dorsal fin base (often giant sharks, with one exception)...**SOMNIOSIDAE** (in part), Genus *Somniosus,* sleeper sharks (p. 115)

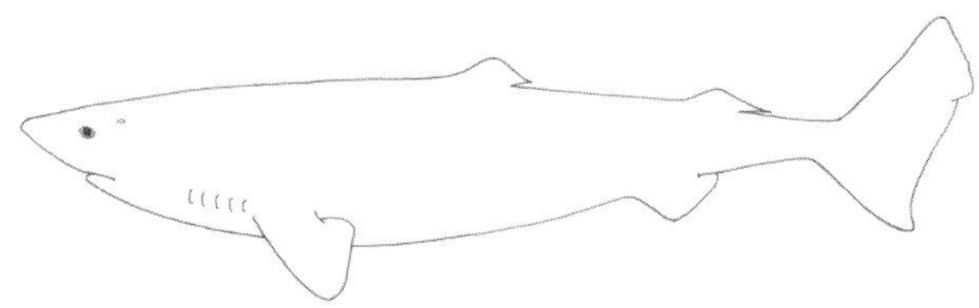

13a. Both dorsal fins with a spine...**HETERODONTIDAE,** Bullhead sharks (p. 174)

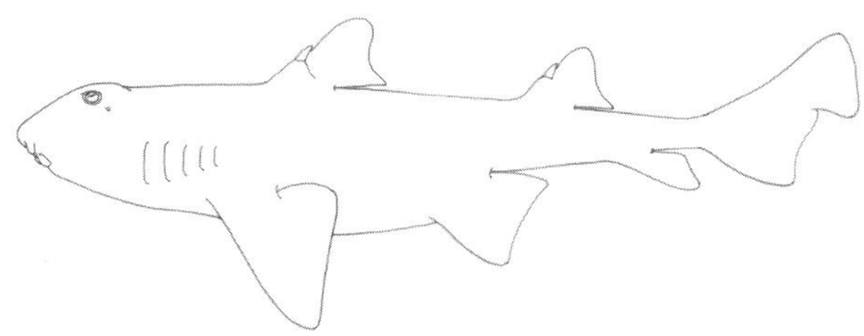

13b. Dorsal fins without spines.. 14

14a. Mouth terminal, nearly even with tip of snout.. 15
14b. Mouth ventral, snout projecting well ahead of mouth.. 16

15a. Body covered with white spots...**RHINCODONTIDAE,** Whale shark (p. 193)

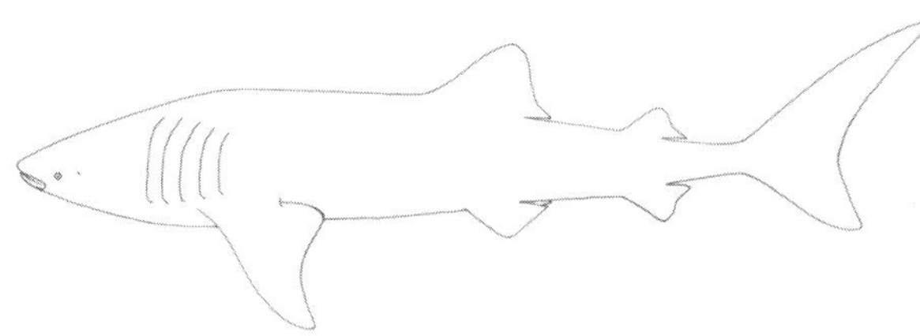

15b. Body not covered with white spots (white band above upper lip, minute teeth) .... **MEGACHASMIDAE,** Megamouth shark (p. 225)

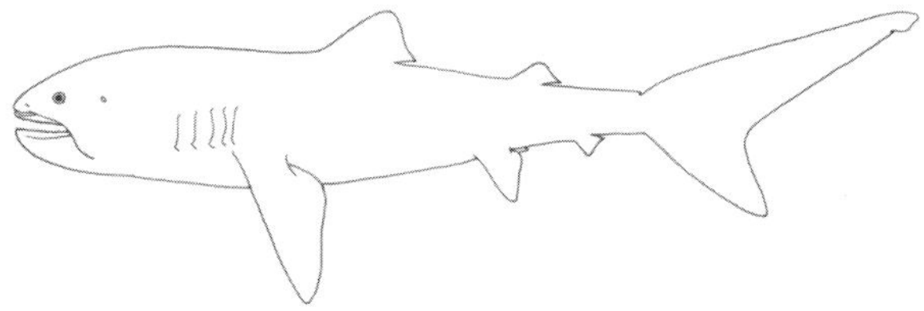

16a Nasal barbels present (first dorsal fin over pelvic fin) .... **GINGLYMOSTOMATIDAE,** Nurse sharks (p. 183)

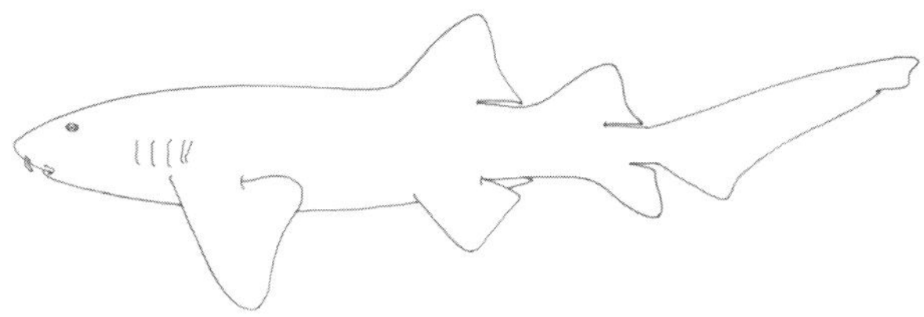

16b. Nasal barbels absent 17

17a. Snout elongated into long, flat blade; anal fin much larger than first dorsal fin .... **MITSUKURINIDAE,** Goblin shark (p. 201)

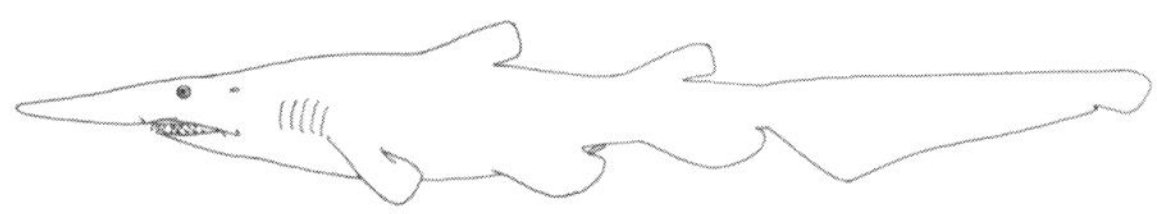

17b. Snout not flattened into a flat blade, anal fin smaller than first dorsal fin 18

18a. Caudal fin as long as the rest of the body .... **ALOPIIDAE,** Thresher sharks (p. 231)

18b. Caudal fin shorter than half the body length 19

19a. Head flattened, shovel or hammer shaped (eyes at ends of lobes) ...... **SPHYRNIDAE,** Hammerhead sharks (p. 504)

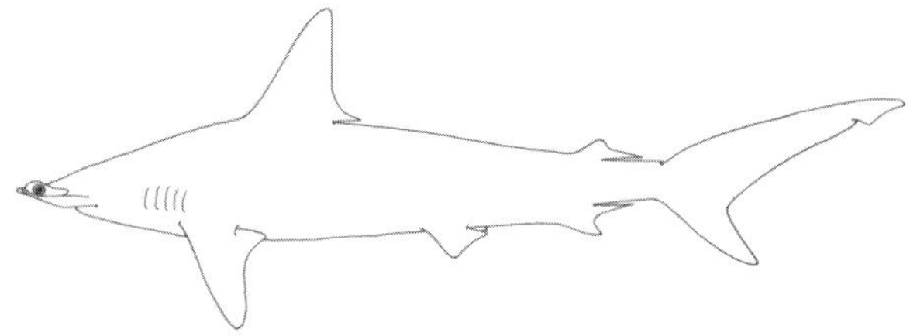

19b. Head not flattened, not hammer or shovel shaped 20

20a. Caudal peduncle with pronounced lateral keels, lunate tail with nearly equal lobes.. 21

20b. Caudal peduncle without lateral keels, heterocercal tail with one lobe much larger than the other............................ 22

21a. Teeth large and few in number, gill slits not extending to the upper surface of head, no gill rakers on the internal gill arches .... **LAMNIDAE,** Mackerel sharks (p. 256)

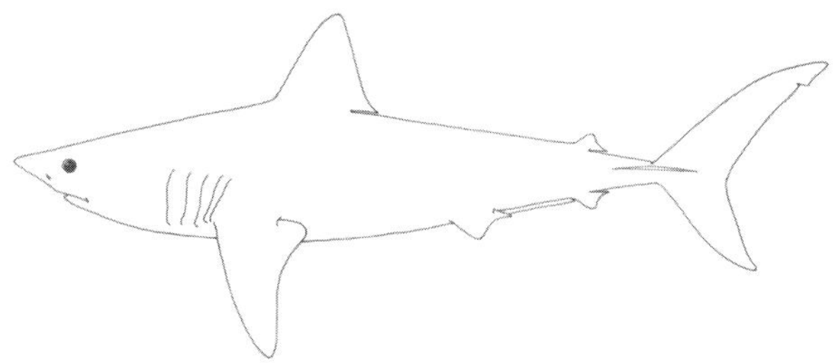

21b. Teeth minute and numerous, gill slits extending on to the upper surface of head (black gill rakers usually present on the internal gill arches) .... **CETORHINIDAE,** Basking shark (p. 248)

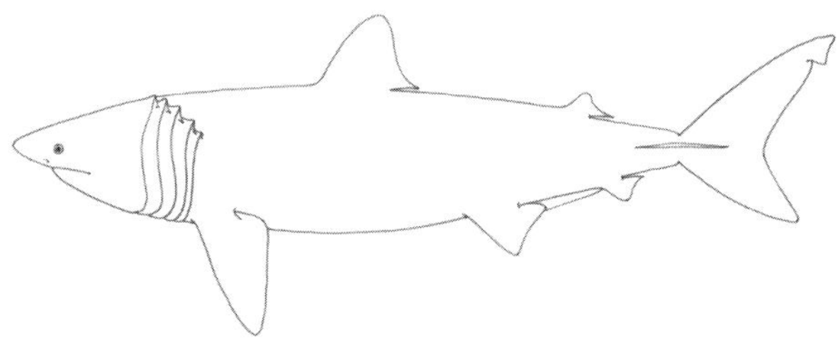

22a. First dorsal fin much longer (six or seven times) than high............ **PSEUDOTRIAKIDAE,** False catshark (p. 352)

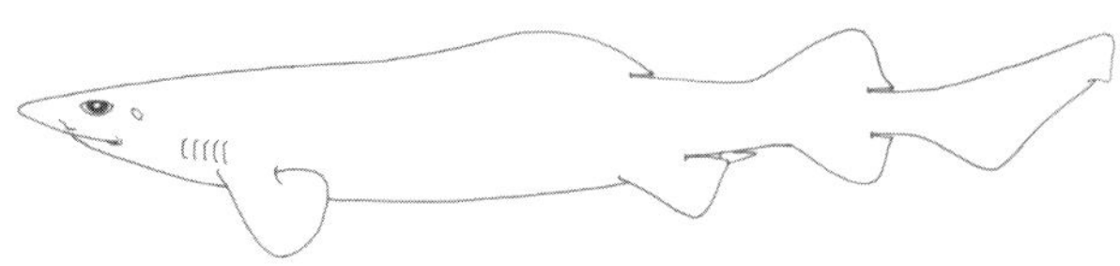

22b. First dorsal fin not longer than high.. 23

23a. Fifth gill slit anterior to pectoral fin base, eyes without a nictitating eyelid 24

23b. Fifth gill slit over or posterior to pectoral fin base, eye with a nictitating eyelid 25

24a. Eyes very large (horizontal diameter about half the length of the first gill slit), lower precaudal pit present .... **PSEUDOCARCHARIIDAE,** Crocodile shark (p. 221)

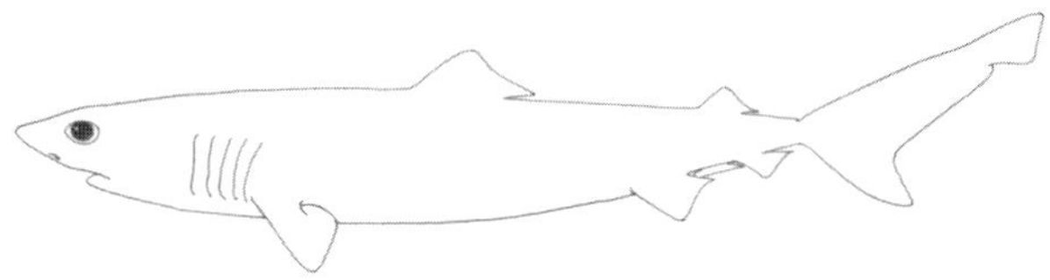

24b. Eyes small (horizontal diameter of eye less than one half the length of the first gill slit), lower precaudal pit absent .... **ODONTASPIDIDAE,** Sand tiger and ragged tooth sharks (p. 206)

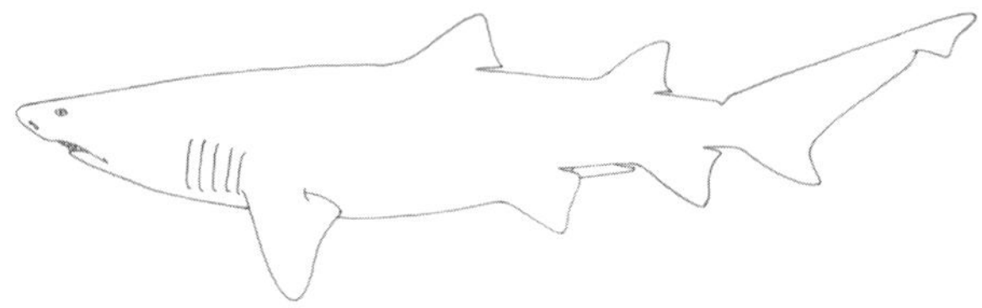

25a. First dorsal fin origin over or behind pelvic fin base .... **SCYLIORHINIDAE,** Catsharks (p. 290)

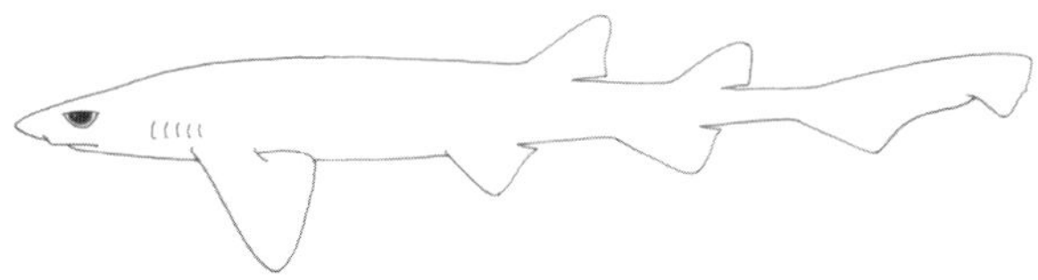

25b. First dorsal fin origin anterior to pelvic fin base 26

26a. First dorsal fin height less than second dorsal fin height, the midpoint of first dorsal fin base closer to pelvic fins origin than to pectoral fin axil (a small shark less than 50 cm) .... **PROSCYLLIDAE,** Ribbontail catsharks (p. 349)

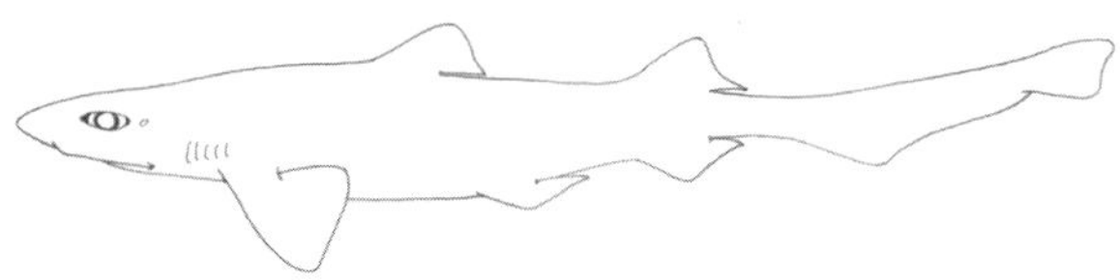

26b First dorsal fin height greater than second dorsal fin height.. 27

27a. Precaudal pit absent.. **TRIAKIDAE,** Hound sharks (p. 357)

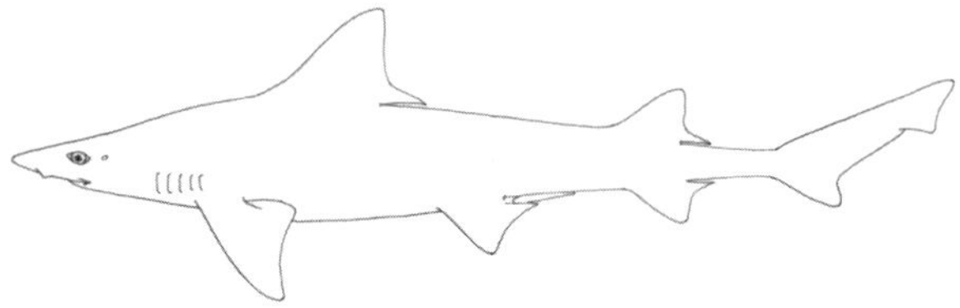

27b. Precaudal pit present .... **CARCHARHINIDAE,** Requiem sharks (p. 388)

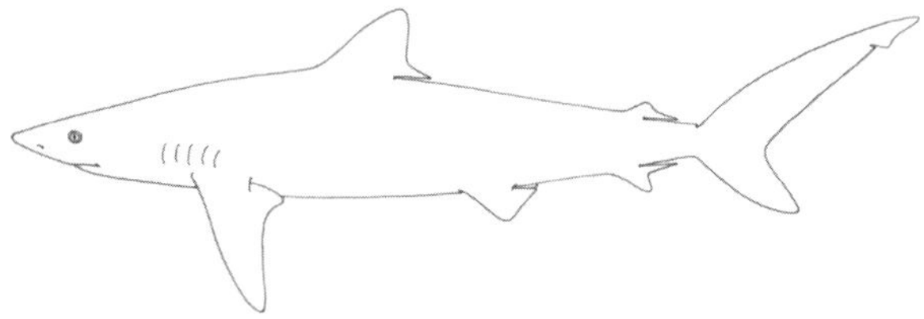

SPECIES ACCOUNTS

FAMILY CHLAMYDOSELACHIDAE

The Frill Shark

The family Chlamydoselachidae includes a single species, the frill shark, the most primitive living shark. It is an eel-like shark with a large terminal mouth, a very flexible, uncalcified skeleton, and a single dorsal fin set far back on the body. The frill shark is one of the deepest dwelling sharks and, although widely distributed, is a rare catch. The existence of a second species of frill shark inhabiting South African waters has been suggested (Compagno et al. 2005) but has not been described.

FRILL SHARK

Chlamydoselachus anguineus Garman, 1884

Fig. 1a. *Chlamydoselachus anguineus*, frill shark, female, 179 cm, from southern New England (39°57′ N, 71°00′ W; Walter Tate, collector; MCZ # 153745, courtesy of Karsten Hartel).

Common name Frill shark, an allusion to the appearance of the head and its large gill slits, which encircle the head, forming a frill.

Spanish name The species lacks a vernacular name in the Spanish Caribbean. The A.F.S. name is *tiburón anguila.*

Scientific name *Chlamydoselachus anguineus* Garman, 1884, in *Bulletin of the Essex Institute* 16: 47–55, fig. 10. The original description is brief but was soon followed by a more comprehensive version (Garman 1885). According to Garman, *Chlamydoselachus* means "the snake-like frilled shark," from the Greek *chlamydos,* cloak or mantle, an allusion to the gill covers that fit like a cloak or frill around the animal's throat, and the Greek *selachus,* shark; *anguineus:* snakelike, from the Latin *anguis,* a snake.

Synonyms None.

Identification The *eel-like* frill shark has a *terminal mouth* with long jaws, six large gill slits with large gill covers, and a single dorsal fin set far back, posterior to the pelvic fins. The lower jaw extends almost to the tip of the snout, creating an enormous gape, and *the first pair of gills is continuous across the throat.* The teeth have *three large, fanglike, inwardly directed cusps,* with an additional minute cusp on each side of the central cusp. The teeth form interlocking, longitudinal rows and are alike in both jaws. Tooth number varies, averaging U: 13–13 and L: 13–1–13. The first six or seven rows of teeth on each side are much larger than those in the corners of the mouth. The dermal denticles are the most primitive of any shark. They are flattened, arrowhead-shaped with a central grooved ridge, and sparse. Frill sharks are dark to light brown with darker fin edges. Some specimens have paler undersides.

Similar species The sixgill shark and the bigeye sixgill shark also have six gill slits and a single dorsal fin, but none of their gill covers is continuous across the throat, and the lower jaws extend only to beneath the eye.

Range The frill shark is probably cosmopolitan in deep waters, although its distribution appears to be patchy. Whether this reflects a narrow niche, the difficulty of capturing such a deep-dwelling fish, or the lack of deep-water fisheries in some areas is uncertain. Trunov (1968) noted that all four captures by AtlantNIRO[1] vessels in 1965 were at depths of 260–363 m in the area of the southwestern African shelf between 18°53′ and 21°22′ S latitude, and he mentioned that the distribution of frill sharks was highly localized.

In the Atlantic Ocean, the frill shark has been reported mainly in the eastern Atlantic off Arctic Norway (Varanger Fjord, Barents Sea; Collett 1897), the British Isles (four records; Wheeler 1962), the Iberian Peninsula (Roule 1912; Bolívar 1907; Gudger and Smith 1933), Madeira (seven records: one from Roule 1912; six from Cadenat and Blache 1981), and the North African coast (off Cap Blanc; Domanevskiy 1975). There are three reports from the western North Atlantic: two from the eastern United States (see below) and the other from Suriname and French Guiana, where three specimens were caught at 754–810 m (Uyeno et al. 1983). In the South Atlantic, the species has been reported from Namibia (Smith, J. L. B. 1967; Bass et al. 1975a).

In the western Pacific, the frill shark is recorded from Japan (Nishikawa 1898; Gudger 1940; Tanaka et al. 1990), Australia (Last and Stevens 1994), and New Zealand (Nakaya and Bass 1978). Most records, however, are from Japanese waters (mainly Suruga and Sagami bays), where several hundred specimens have been recorded since 1888. In the eastern Pacific the species has been reported from California (see below) and the western coast of South America.

1. Atlantic Research Institute of Marine Fisheries and Oceanography of the State Fisheries Committee of the Russian Federation.

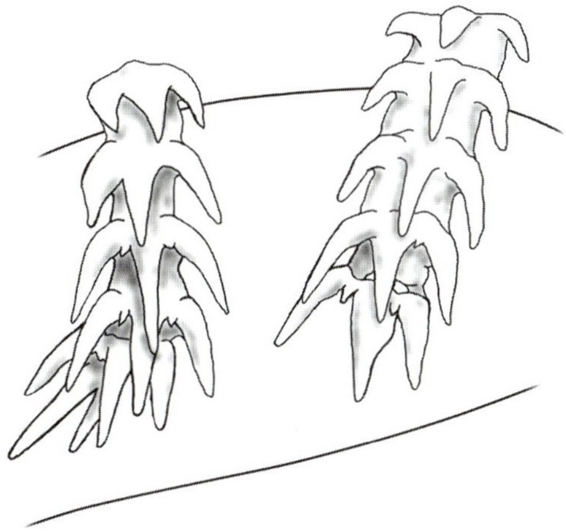

Fig. 1b. Teeth.

Records also exist for the Northwestern and Hawaiian submarine ranges (Borets 1986).

The frill shark has also been reported from the Indian Ocean. According to Tumokhin (1980), 10 frill sharks—seven females and three males—were caught in August 1976 and April 1977 by bottom trawls in open waters of the southwestern Indian Ocean at 1,200–1,440 m.

There are three North American records of frill sharks as of February 2007. In June 1948, a 172-cm female was captured by drift net at 17–18 m, 33 km southwest of Point Arguello, Santa Barbara County, California (Noble 1948). In 2003, a 186-cm female was caught on the continental slope off Martha's Vineyard, Massachusetts, between Block and Alvin canyons (39°57′ N, 71°00′ W) at 400 m (Moore et al. 2003). In 2004, a frill shark was observed and photographed at 874 m on the outer Blake Plateau east of Georgia (Sedberry et al. 2007).

Size and weight Most specimens captured measured 130–150 cm, adult females being slightly larger. The largest specimen on record is a 196-cm female obtained by Bashford Dean in Japan in the early 1900s (Gudger and Smith 1933). A 191-cm female captured in 1896 at 180–275 m in Varanger Fjord, on the Arctic coast of Norway, is apparently the second largest specimen on record (Collett 1897). A 147-cm adult male I examined weighed 5.4 kg, whereas a 182-cm, ovigerous female weighed 15.4 kg.

Biology The frill shark is the most primitive living shark, a relict surviving in the deep oceans since the time when sharks had eel-like bodies, terminal mouths, multicuspid teeth, incompletely segmented vertebral columns, and poorly calcified vertebrae. Based on catch records, the frill shark usually inhabits deep cold waters (240–1,500 m). It is rarely caught near the surface, and little is known of its biology.

Despite the highly localized distribution of the frill shark and its relative scarcity, its anatomy and reproductive processes are better known than those of many common species. Its discovery in 1888 caught the attention of ichthyologist Bashford Dean at the American Museum of Natural History in New York City. In the early 1900s, Dean traveled to Japan and secured 39 specimens (Gudger and Smith 1933). Dean, a scholar with diverse interests ranging from archaic fishes to medieval body armor, studied the frill shark for many years, making exquisite drawings of developing embryos. He also provided other scholars with anatomical

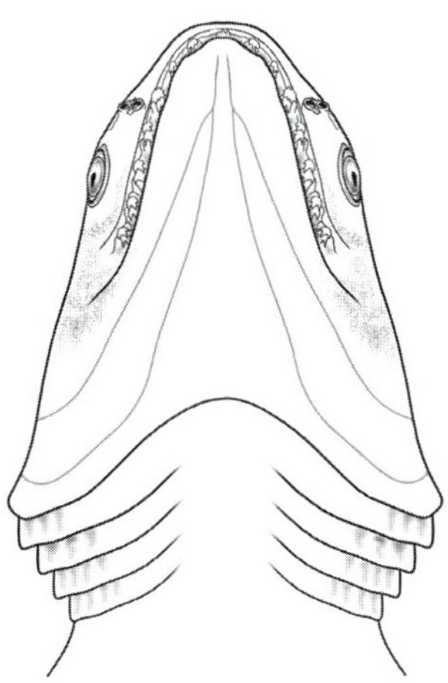

Fig. 1c. Snout.

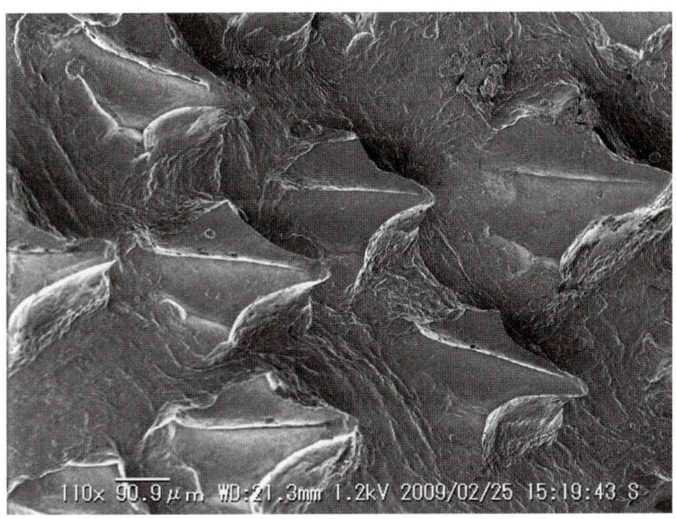

Fig. 1d. Dermal denticles.

material that resulted in several published works. Unfortunately, Dean died in 1928 before finishing his studies of the frill shark. Later, Eugene W. Gudger, also of the American Museum of Natural History, and Bertram G. Smith, professor of anatomy at New York University, used Dean's materials and notes to prepare a series of monographs on the frill shark and the Japanese horn shark. The resulting monographs are splendid, comprehensive works seldom equaled in the study of sharks. Those dedicated to the frill shark are *The Cranial Anatomy of Chlamydoselachus anguineus* (Allis 1923), *The natural history of the frilled shark Chlamydoselachus anguineus* (Gudger and Smith 1933), *The anatomy of the frilled shark Chlamydoselachus anguineus* (Smith, B. G. 1937), and *The breeding habits, reproductive organs and external embryonic development of Chlamydoselachus, based on notes and drawings by Bashford Dean* (Gudger 1940). Little has since been learned about the frill shark, and all that we know about it is derived from Japanese-caught fishes in Suruga and Sagami bays; its ecology and behavior remain mysteries.

Frill sharks have been taken over a wide depth range. Tanaka et al. (1990) collected 264 specimens with midwater and bottom trawl nets set at depths of 60–240 m in the inner part of Suruga Bay from 1981 to 1988. According to those authors, few frill sharks were caught between August and November, when the temperature of the 100 m layer rises above 15°C. This increase in temperature apparently prevents the sharks from moving upward, and they might actually migrate to deeper waters or colder latitudes. In the African shelf, specimens have been taken at 260–500 m (Trunov 1968; Domanevskiy 1975). The greatest known depth at which this shark has been caught is 1,340–1,570 m (Golovan 1976). The specimen observed off Georgia by Sedberry et al. (2007) was in 874 m and 4.3°C temperature.

Diet: The diet of the frill shark was unknown for many years because nearly all captured specimens had empty stomachs (Wheeler 1962; Tumokhin 1980). Its small, slender teeth evolved for seizing thin-skinned prey such as squids. Its distensible jaws, enormous gape, and inward-directed teeth suggest that it feeds on relatively large prey. Frill sharks have an interesting longitudinal fold along the ventral midline. It appears to serve as an expansion fold, allowing the stomach and body wall to expand after a large meal.

The only recent report on the frill shark diet is that of Kubota et al. (1991), who analyzed stomach contents of 139 specimens caught in Suruga Bay in the 1980s. Squid was the most common food item, present in 23 stomachs. Bony fishes were found in four stomachs, and unidentifiable remains in 11 stomachs. The squids included two intact specimens of *Todarodes pacificus* measuring 16–20 cm mantle length. The fishes were not identified because only vertebrae remained.

The empty stomachs of such a large percentage of captured sharks led to the speculation that a gorged frill shark might lie quiescent until its large meal has been digested, indifferent to prey and baited hooks (Smith, J. L. B. 1967). The reasons for the high percentage of empty stomachs remain unknown.

According to Gudger (1935), this shark has a fold of tissue in the mouth believed to act as a breathing valve. The structure would allow it to breathe while motionless or resting on the bottom (Gudger 1935). However, I have not been able to discern such structure in the mouth of the fresh frill sharks I have examined.

Size at maturity

Males: Tanaka et al. (1990) reported that males mature at about 110 cm, based on males of 117.8–154.8 cm having elongated, hard claspers and appreciable amounts of sperm in the seminal vesicles and sperm sacs.

Females: Females apparently mature at about 135 cm. B. G. Smith (1937) refers to a 155-cm female as "nearly mature," although the methods he used to determine maturity are unclear. Given that Smith also mentioned two fully mature females of 135 cm and 148.5 cm, it is likely that the specimen was a mature female in the resting stage of the ovarian cycle. Tanaka et al. (1990) used the size of ovarian eggs (oocytes) and "condition of the uterus" to assess the maturity of females. Those having small oocytes, threadlike uteri, and oviducal (shell) glands of less than 15 mm were classified as immature; other females were deemed mature. The oviducal gland criterion was not explained. Tanaka et al. reported that their largest immature female measured 155.9 cm and the smallest gravid female was 137.7 cm, concluding that females mature at 140–150 cm.

Reproduction Despite the restricted availability of gravid frill sharks, their reproductive processes have been well described by Gudger (1940), based on specimens obtained and studied by Bashford Dean. The frill shark is an aplacental viviparous species. Like most primitive sharks, it has two functional ovaries of equal size. The ripe oocytes and fertilized eggs are huge, measuring 90 × 96 mm (Gudger 1940). As many as 12 large, ripe oocytes have been reported in one shark (Nishikawa 1898). Interestingly, only the right uterus is typically functional, the left one remaining rudimentary. Presumably, there is no space in the abdominal cavity of this long, slender shark for two gravid uteri, given the large size of the eggs and embryos.

Development: The ovulated egg passes through the shell gland, acquiring a thin, pale brown, transparent capsule. The encapsulated egg is ellipsoidal, measuring 65–75 mm × 102–124 mm. The eggs were illustrated by Nishikawa (1898) and Garman (1913). Later, Gudger and Smith (1933) and Gudger (1940) reproduced their illustrations. The blastodisc is yellowish-red, as in other sharks, and is circular, 1.3 mm in diameter. The embryos have been described by Ziegler (1908) and Gudger (1940).

Gestation: The duration of gestation in frill sharks is unknown. Two estimates have been proposed, both speculative. Gudger and Smith (1933: 302) speculated that gestation would last "for a year or more" because of the huge size of the eggs, the great size of the embryos, and the low temperature at depths of 450–760 m where these sharks live. They concluded that "taking into consideration the large size of the yolk-sac in the second year, it seems possible that hatching may not take place until at least the second summer following impregnation, and indeed that it may not take place till after the little shark is quite two years old" (Gudger and Smith 1933: 303). Tanaka et al. (1990) suggested an even longer gestation period of "at least three and a half years." This conclusion was derived from the growth rate of early encapsulated embryos, which ranged from 10.2 to 16.6 mm per month (average 14 mm per month). Tanaka et al. estimated that if the growth rate remained continuous, gestation would last more than 39 months. As they admitted, their estimate is based on several assumptions, the most critical (and, to me, unlikely) is that the rate of embryonic growth of 14 mm per month is continuous throughout development.

Size at birth: The largest reported embryo is 54.9 cm long (Tanaka et al. 1990); the smallest free-swimming specimen, captured in Suruga Bay, measured 53.5 cm in length (Yasuhara et al. 1983). These data suggest that birth occurs at 53–55 cm. Neonate specimens are seldom captured. The only other reference to a neonate is that of Collett (1890), who reported a 61-cm free-swimming female obtained at Funchal, Madeira.

Brood size: Tanaka et al. (1990) reported that frill shark females from Suruga Bay, Japan, carried 2–10 embryos still inside the egg capsules, whereas females with "free embryos" (hatched from the egg capsules) carried two to eight embryos per female.

Nurseries Unknown.

Age and growth Frill shark specimens have not been aged because of their uncalcified skeletons.

Relation to humans Although numerous specimens have appeared in Japanese fish markets over the years, the frill shark is too difficult to capture to be of much economic importance. It is a rare catch outside Japan.

FAMILY HEXANCHIDAE

The Cowsharks

The family Hexanchidae, or cowsharks, includes the sixgill and sevengill sharks, a small group of widely distributed deep-water sharks. The family is easily recognized by the six or seven gill slits, subterminal mouth, and single dorsal fin set posterior to the pelvic fins. The only other sharks with six gill slits are the frill shark and one species of sawshark; all others have five gill slits. Cowsharks have dissimilar teeth in the upper and lower jaws: the upper teeth are fanglike, and the lower teeth sawlike and rectangular. Cowsharks are aplacental viviparous. Their skeletons are poorly calcified, which is considered a primitive characteristic. Four species are presently recognized: three are deep-water, bottom-dwelling species found over the outer continental shelves, and one is a shallow-water, coastal species. These species are classified in three different genera: *Heptranchias, Hexanchus,* and *Notorynchus.* Sometimes these genera are elevated to families. Sharks of the genus *Hexanchus* have six gill slits, while those in *Heptranchias* and *Notorynchus* have seven gills slits.

KEY TO THE HEXANCHIDAE

1a. Six pairs of gill slits 2

1b. Seven pairs of gill slits 3

2a. Five large, broad sawlike teeth on each side of lower jaw (maximum size about 180 cm) ....
.... **Bigeye sixgill,** *Hexanchus nakamurai* (p. 36).

2b. Six large, broad sawlike teeth on each side of lower jaw .... **Sixgill shark,** *Hexanchus griseus* (p. 31).

3a. Snout broadly rounded, dorsal surface with numerous dark spots (a shark of the North Pacific) ....
.... **Broadnose sevengill shark,** *Notorynchus cepedianus* (p. 38).

3b. Snout narrowly tapering, dorsal surfaces uniformly light brown or gray, spots ....
.... **Sharpnose sevengill shark,** *Heptranchias perlo* (p. 28).

SHARPNOSE SEVENGILL, OR PERLON SHARK

Heptranchias perlo (Bonnaterre, 1788)

Fig. 2a. *Heptranchias perlo*, sharpnose sevengill shark, female, 66.2 cm, 0.7 kg; from Ft. Pierce, Florida (Scott Bachman, collector).

Common name Sharpnose sevengill, sevengill, seven-gilled, or perlon shark. The name sharpnose sevengill has been used to distinguish this species from the broadnose sevengill shark; perlon shark is derived from the Old French name (see below).

Spanish name *Tiburón de siete branquias* (Cuba, A.F.S.).

Scientific name *Heptranchias perlo* (Bonnaterre, 1788), originally described as *Squalus perlo* in *Tableau Encyclopédique et Méthodique des Trois Règnes de la Nature, Ichthyologie:*10. *Heptranchias:* from the Greek *hept,* seven, and *branchos,* gill; *perlo:* from *le perlon,* an old French name for the species, first described by Broussonet (1780), from *perle,* pearl. The allusion is not clear; it probably refers to the skin, which Broussonet calls smooth and grayish ("*lisse & grisâtre*").

Synonyms *Squalus cinereus* Gmelin, 1788; *Heptranchias cinereus* Rafinesque, 1810; *Heptranchias angio* Costa, 1837; *Heptanchus cinereus* Müller and Henle, 1839; *Heptranchias deani* Jordan and Starks, 1901; *Heptranchias dakini* Whitley, 1931 [a].

Identification The sharpnose sevengill shark is recognized by a *single dorsal fin* with its origin behind the pelvic fins, *seven gill slits,* and *a narrow tapering snout.* The upper teeth are fanglike with long tapering cusps, and the side teeth have small lateral cusplets. The lower teeth are broad and sawlike, with one larger cusp followed by four or five smaller ones, except for the small symmetrical central tooth. Teeth number U: 9 to 11–9 to 11, L: 5–1–5. The dermal denticles have three points, with the central point the largest, and are closely overlapping. Coloration is brownish gray above with paler undersides. The tips of the dorsal and caudal fins are black.

Similar species The broadnose sevengill shark also has seven gill slits, but its snout is broadly rounded (see illustration), and it has *dark dorsal spots.* The broadnose sevengill shark, abundant off western North America, has not been reported from eastern North America, although it inhabits the South Atlantic off Brazil (Sadowsky 1970).

Range The sharpnose sevengill shark is cosmopolitan in deep subtropical and warm-temperate waters. It has long been known in the Mediterranean (Risso 1810; Garman 1913). It has also been reported from Japan (Garman 1913), Australia (as *Heptranchias dakini* by Whitley 1931a), and New Zealand (as *H. perlo* by Garrick and Paul 1971). Garrick and Paul (1971) have shown that the Australian *H. dakini* is a synonym of *H. perlo.* The species has also been found in the eastern Atlantic from southern Ireland to southern England and in the Gulf of Gascogne (Cappetta 1985), the Great Meteor Seamount in the central eastern Atlantic (Frentzel-Beyme and Köster 2002), equatorial Africa (Poll 1951), and Natal and southern Mozambique (Bass et al. 1975a).

Although the species is common in moderately deep water (100–200 m) along southeastern North America, there are few records of it in the literature. The North American specimens examined by Bigelow and Schroeder (1948) were limited to two animals from Cuba, and they knew of no other material from the western North Atlantic. Nevertheless, the sharpnose sevengill is found from Cape Cod (where it is rare) to Florida and the Gulf of Mexico. I have seen dozens of specimens caught by tilefish fishermen off South Carolina, as well as some from the Gulf of Mexico.

Size and weight The sharpnose sevengill is a small shark; most specimens caught measure less than 100 cm. The largest specimen that I have examined was a 114.1-cm female weighing 2.5 kg, obtained off Georgetown, South Carolina, in 1984. Forster et al. (1970) reported that their largest specimen from the Indian Ocean was a male measuring 101 cm. Garrick and Paul (1971) wrote that the largest female from New Zealand measured 137 cm. Capapé (1980) gives 118 cm for the largest male and 139 cm for the largest

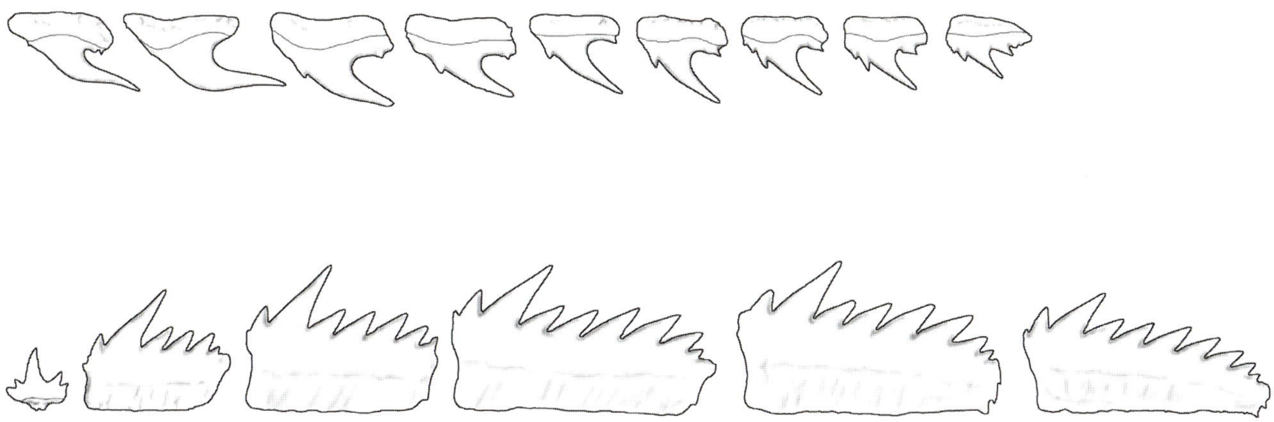

Fig. 2b. Teeth.

female from Tunisia. These last two are the largest specimens I have been able to document.

Larger sizes published for this species are likely caused by confusion with the broadnose sevengill shark or the sixgill shark. Bigelow and Schroeder (1948) quoted a maximum size of 2.1 m based on Günther (1870). According to Tortonese (1956) it is said to reach 3 m. Bini (1967) wrote that the largest sevengill shark reliably measured was slightly less than 2 m and that a specimen captured off Israel measured 2.5 m. There is, however, no evidence to validate these sizes.

I have recorded the following lengths and weights for specimens caught off the southeastern United States:

Male: 56 cm, 0.54 kg; 62 cm, 0.68 kg.

Female: 66 cm, 0.74 kg; 87.5 cm, 1.45 kg; 98.2 cm, 2.7 kg.

Biology This is a warm-water species common along the edges of the continental shelves, where it appears to be most abundant at depths of 180–450 m (Castro 1983). Despite its abundance from the Carolinas to the Gulf of Mexico, its life history is poorly known. There are no American publications on this species, and it is necessary to turn to works published in other areas to complement our meager knowledge of American specimens. Capapé (1980) examined a total of 154 individuals taken off the northern coast of Tunisia, concluding that they showed no differences from Mediterranean specimens or those from both sides of the Atlantic. Garrick and Paul (1971) found no differences between the Australian form (described as *H. dakini*) and the

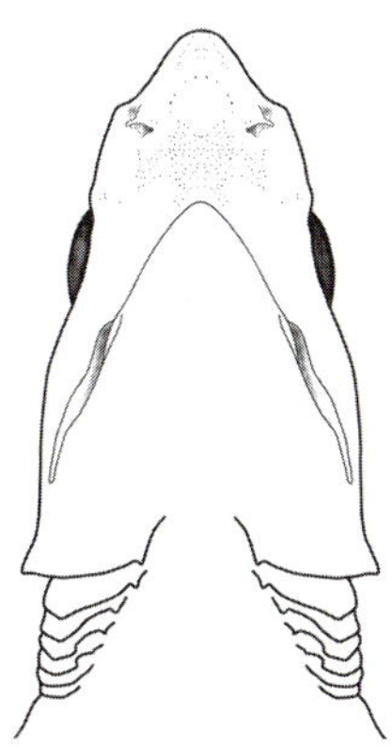

Fig. 2c. Snout.

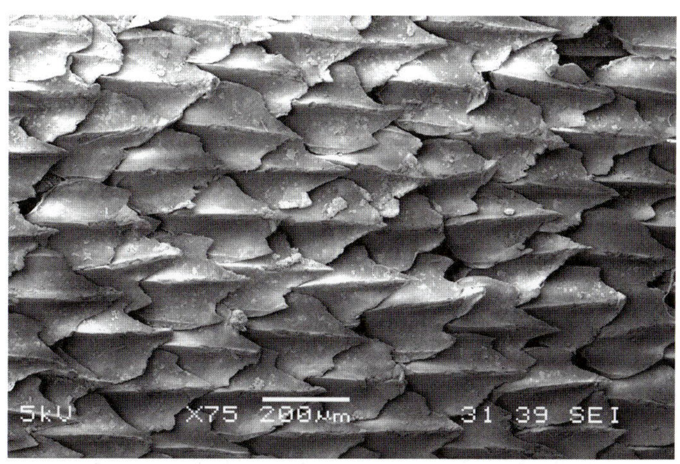

Fig. 2d. Dermal denticles.

worldwide *H. perlo.* The differences between the American population and those of other areas appear to be minor.

The sharpnose sevengill shark is occasionally caught on longlines set for tilefish along the southeastern coast of the United States (North Carolina to Georgia) and in the Gulf of Mexico, usually at depths of 180–450 m. Captures elsewhere indicate a similar depth range. In the central eastern Atlantic, the species has been caught in traps at 297–435 m (Frentzel-Beyme and Köster 2002). Australian captures indicate a depth range of 100–400 m, with some captures in shallower waters. Captures off New Zealand usually occur at 235–275 m, with some at 50 m (Garrick and Paul 1971). In Japan the species has been reported in 150–250 m (Tanaka et al. 1975). This shark feeds on squids and small fishes, but little else is known about its habits.

Size at maturity

Males: Most published information on size of males at maturity is too imprecise to accurately determine this characteristic. Bigelow and Schroeder (1948) wrote that males "may mature" at 60–75 cm, without stating the basis for their claim. Capapé (1980) used the relative clasper length to determine maturity, but he did not associate the length of the clasper with calcification or spermatogenesis. He recognized three growth phases or groups of individuals: juvenile phase, or individuals less than 80 cm; maturation phase, or individuals 81–92 cm; and an adult phase comprising individuals longer than 92 cm. From his graphs it can be determined that the male adult phase must begin at about 95 cm. Tanaka et al. (1975) examined 191 specimens from the Japanese coast off Kyushu. They used relative clasper length and relative testis weight to determine maturity, but they, too, failed to correlate the clasper and testis growth with clasper calcification, spermatogenesis, or any other functional indicator of maturity. These authors concluded that males reach maturity at 70–85 cm.

Females: Capapé (1980) found that vitellogenesis (formation of yolk) begins when females reach 85 cm. The largest oocytes (45 mm in diameter and weighing more than 37 g) are not found in females of less than 100 cm. The smallest gravid females were larger than 105 cm. Capapé concluded that females mature when they attain a total length of 85–100 cm and are invariably mature at lengths greater than 100 cm.

Reproduction The sharpnose sevengill shark is aplacental viviparous. Little else is known of its reproductive processes.

Size at birth: Parturition probably occurs when the embryos reach about 25 cm. Bigelow and Schroeder (1948) reported that a 26-cm specimen from Japan had a faint yolksac scar, indicating its recent birth. Bass et al. (1975a) mentioned that the smallest sharpnose sevengill shark caught in the Indian Ocean was a 27-cm female. Capapé (1980) gives the birth size at around 30 cm.

Brood size: Bigelow and Schroeder (1948) reported a brood of nine young from a 93-cm female (Museum of Comparative Zoology no. 36897); they also cited an unverified report of a brood of 18, but the report is questionable.

Nurseries Unknown.

Age and growth Unknown.

Relation to humans In North American waters the sharpnose sevengill shark is too small and scarce to be of economic importance. The flesh is said to be mildly poisonous (Halstead et al. 1990), but I have not found firsthand evidence of this effect.

SIXGILL SHARK

Hexanchus griseus (Bonnaterre, 1788)

Fig. 3a. *Hexanchus griseus*, sixgill shark, female, 375 cm, 354.5 kg; from Key West, Florida (Jay Lamee, collector).

Common name Sixgill shark, reflecting its distinctive number of gill slits.

Spanish name *Cañabota, marrajo* (Cuba). The A.F.S. name is *tiburón de seis branquias.*

Scientific name *Hexanchus griseus* (Bonaterre, 1788), originally described as *Squalus griseus* in *Tableau Encyclopédique et Méthodique des Trois Règnes de la Nature, Ichthyologie:* 9. *Hexanchus:* one of Rafinesque's words, apparently a mistake for *Hexancus,* from the Greek *hex,* six, and *ankos,* a bend or hollow; *griseus:* Latin, gray.

Synonyms *Squalus griseus* Gmelin, 1788; *Squalus vacca* Bloch and Schneider, 1801; *Notidanus griseus* Günther, 1870; *Hexanchus corinus* Jordan and Gilbert, 1880.

The sixgill shark of the Pacific coast of North America was described as *Hexanchus corinus* by Jordan and Gilbert, in 1880. Supposedly, *H. corinus* differs from *H. griseus* primarily in that its lower teeth (other than the median tooth) have fine serrations along their inner edges. Bigelow and Schroeder (1948) noted that differences in dentition and other characteristics were not significant and referred the Pacific form to *H. griseus.* To my knowledge, there are no recent comprehensive comparisons of the two forms, probably because comparative material is lacking. The large size of these sharks usually prevents the preservation of entire specimens, which are necessary for such comparative work.

Identification The sixgill shark can be recognized by its *single dorsal fin, six gill slits,* and *six large, broad, sawlike teeth on each side of the lower jaw.* The frontal upper teeth have one large curved cusp; other upper teeth have oblique cusps with an increasing number of additional side cusplets. The lower teeth are much broader and more sawlike than the upper teeth. There is one small, broad, and symmetrical tooth at the symphysis. Teeth number U: 9 or 10–9 or 10, L: 6–1–6, disregarding the minute, budlike teeth at the corners of the jaws. Compared to females, males have lower teeth with a much larger first cusp, a feature that presumably helps them to grasp females securely during copulation. The dermal denticles have three points with a prominent central ridge terminating in the central point. There are two color morphs (forms) of the sixgill shark. One is chocolate brown with little, if any, countershading. The other has distinct countershading, being pale brown above and lighter below. I have seen both forms together in Bermuda, where Eugenie Clark and National Geographic photographer Emory Kristof photographed them together (Clark and Kristof 1990). These color morphs appear to be variations of the same species, although the possibility that they might be separate species cannot be discounted.

Similar species The bigeye sixgill shark has five large teeth on each side of the lower jaw. The frill shark has a terminal mouth and fanglike teeth.

Range The sixgill shark is cosmopolitan in deep temperate, subtropical, and tropical waters. In North America, it has been reported from Nova Scotia to Florida and the Gulf of Mexico and from British Columbia to California; the species is common in deep water along the southeastern coast of the United States and in shallow waters of the Pacific Northwest. Records of sixgill sharks from the Atlantic coast of North America are few because of its deep-water habitat. The first record for North America was a 310-cm specimen caught in March 1886 in North Carolina, at the Currituck Inlet lifesaving station. This specimen was forwarded to the U.S. National Museum, where a plaster cast was made and exhibited for several years (Smith, H. M. 1907). In 1963 a "13-foot, one-ton" sixgill shark was reported from 400 m, 72 km off the Mississippi River delta (Sport Fishing Institute 1963), with the comment that the species had not been taken in North American waters since

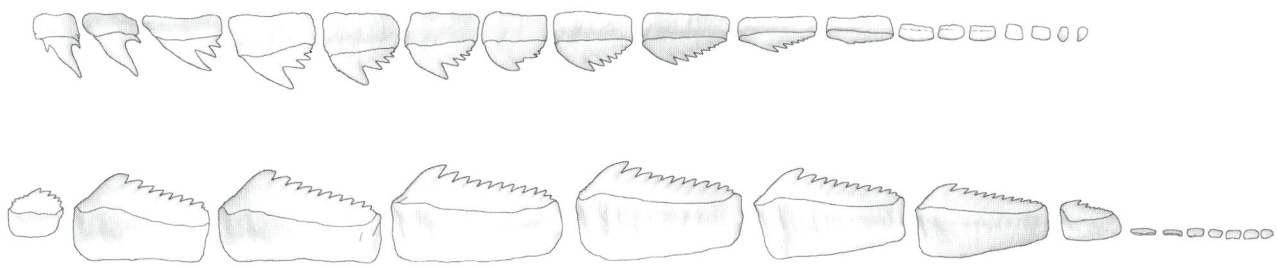

Fig. 3b. Teeth.

1886. Sixgill sharks are seldom seen north of Cape Hatteras, although the northernmost records are two small juveniles caught at 155–183 m off Nova Scotia (Gilhen 1989), and I have examined a 195-cm juvenile was taken off New England by John Galbraith in 2005. All other current records for eastern North America are from North Carolina southward. Branstetter and McEachran (1986a) reported a 325-cm male caught 108 km off Port Isabel, Texas. I examined specimens from Bermuda and Florida. Sixgill sharks have been seen from submersibles on many occasions off the southeastern coast of the United States, but most of these observations have not been published because they are considered commonplace.

Sixgill sharks appear in shallower water (40 m) off Vancouver Island, British Columbia, from June to September (Dunbrack and Zielinski 2003). Small juveniles have been reported from shallower water in San Francisco Bay (Herald and Ripley 1951) and Puget Sound (my observation).

Size and weight Few measurements exist of the large sixgill sharks. The largest specimen that appears to have been measured reliably and illustrated is a 482-cm female caught off La Coruña, Spain, on 26 October 1906 (Bolívar 1907). Some incredible sizes have been stated for this species, both in the literature and anecdotally. There are reports in the older literature of a sixgill shark "26 feet 5 inches" (8.05 m) long, based on an account by Day (1880–1884). Bigelow and Schroeder (1948: 83) called this fish "a giant of its kind if its size was stated correctly." Years later, Lineweaver and Backus (1970) traced the error with an explanation that shows how such mistakes are often created and perpetuated. In *The Zoologist,* Jonathan Couch stated that "On the 19th of February of the present year [1846] there was caught by a fisherman of Polperro, and immediately brought to me, a specimen of a fish, which I recognized as the Six-branchial, or Gray Shark: a species new to the Fauna of Cornwall. . . . The length of the specimen was 2 feet, 2 1/2 inches" (Couch 1846: 1337–1338). Later, Day (1880–1884: 308) in his *Fishes of Great Britain and Ireland,* in the account of *Hexanchus griseus,* reported that "February 19th, 1846, one 26 feet 5 inches in length [was] captured at Polperro, in Cornwall. . . . It is said to grow to a large size." Couch, in his major work *A History of the Fishes of the British Islands* (1868: 21), makes no mention of the gigantic specimen mentioned by Day, referring again to his small 1846 specimen: "The example from which the description is taken, measured in length no more than two feet two inches and a half; but it has been caught at a length of eleven or twelve feet." Polperro fishermen commonly brought specimens to Couch, so it is unlikely that a giant sixgill shark captured the same day as the specimen 2 feet 2 1/2 inches in length would have escaped his notice. As Lineweaver and Backus (1970) suggested, Day probably converted 2 feet 2 1/2 inches into 26.5 inches. This figure was subsequently misinterpreted by the printer or editor and printed as "26 feet 5 inches," thus engendering an error that would live long in the shark literature.

Convincing anecdotal evidence suggests that sixgill sharks do perhaps exceed 550 cm. I occasionally hear reports of huge sixgill sharks from fishermen and scientists alike, but so far no one has actually measured one of those elusive giants.

Few accurate weight records exist for sixgill sharks from the Atlantic coast of North America. Branstetter and McEachran (1986a) reported a 325-cm, 211-kg male from the Gulf of Mexico. A 195-cm female taken off New England weighed 45.9 kg. The large female illustrated here was caught off Key West, Florida; it measured 381 cm and weighed 354 kg, while a 452-cm female taken off Bimini, Bahamas, in 1993 weighed 558 kg (Ron Schatman, pers. comm., December 2002). Ebert (1986a) gave the following sizes for Pacific specimens: 150 cm, 18.2 kg; 168 cm, 27.3 kg; 208 cm, 60.0 kg; 210–242 cm, 54.1 kg; 273 cm, 107.3 kg.

Biology The sixgill shark is a common, bottom-dwelling, deep-water species, usually reported from depths of 300–1,000 m. Most specimens caught or observed at these depths are juveniles measuring less than 400 cm. Carey and Clark

(1995) tracked two female sixgill sharks off northwestern Bermuda in 1987. One shark, estimated to be 310–350 cm, was captured and outfitted with a transmitter. After release it moved offshore and descended to 1,000 m. That evening it moved farther offshore, making an excursion to 1,500 m, the maximum depth recorded. At dawn, it rose from 1,000 m and spent the day between 700 and 900 m. Tracking was discontinued that evening and resumed 24 hours later. The shark continued swimming at 600–1,000 m for the following 36 hours, ordinarily following the bottom contour at 914 m. A second female, estimated to be 380 cm, swam at 700–1,000 m, rising a few times to 600 m. At 800 m the temperature was 9.5°C, and at 600 m it was 14–16°C.

Adult sixgill sharks exceed 450 cm in length. They seldom are captured or seen, presumably because few people fish or dive in the deep water they inhabit or because their large size allows them to break away from most fishing gear. Perhaps their poorly calcified jaws tear easily and hooked animals are freed. Although records are scant, adult sixgill sharks apparently live at depths exceeding 1,200 m. According to Bigelow and Schroeder (1948), sixgill sharks have been caught in Portuguese waters at 800–1,875 m. In the Indian Ocean, Forster et al. (1970) collected sixgill sharks at 200–950 m. Most of Forster et al.'s specimens were caught at night at 200–499 m. The two specimens captured during daylight came from the deeper part of the range, at 455 and 950 m. All specimens examined were juvenile females. Their only large specimen, estimated at 450 cm, was caught at about 950 m, but, unfortunately, it was not examined.

If adult sixgill sharks dwell at depths greater than 1,000 m, what is their source of food? Such large animals must require large prey and great amounts of energy, even if they are slow moving and inhabit cold water. The large, flat, saw-like teeth of the lower jaw form a continuous cutting edge. The fact that the cutting teeth are on the lower jaws, and the thinness of the teeth suggest a diet of relatively soft prey. I suspect that sixgill sharks scavenge dead whales as sleeper sharks do at higher latitudes. Sunken whale carcasses can persist for many months (Smith and Baco 2003) at depths few sharks are thought to inhabit. Perhaps they also feed on giant squid, as sleeper sharks do in the Antarctic (Cherel and Duhamel 2004), but just how these sharks catch giant squid is an interesting mystery.

Despite living most of their lives in darkness, sixgill sharks have the largest pineal window that I have seen. The pineal window is a light-colored spot on top of the head between the eyes, and is common in many deep-water sharks. When compared to the surrounding tissue, this structure allows up to seven times more light to enter the brain cavity and impinge on the pineal organ (Gruber et al. 1975). The pineal organ (*epiphysis cerebri*) is an area of the brain (diencephalon) known to be as photosensitive as the retina, and its threshold of light reception is below that of moonlight (Hamasaki and Streck 1971). However, we can only speculate on its function in deep-water sharks.

The differences between Atlantic and Pacific forms of the sixgill shark are not well studied, and the possibility that they could be separate species cannot be ruled out. I treat the two forms separately here.

Atlantic form: I have observed several Atlantic sixgill sharks from submersibles off the Bahamas and Grand Cayman at depths of 300–850 m. This slow-moving shark gives the impression of gliding effortlessly a meter or two above the bottom. Its movements are slow when searching for food or even while scavenging. This is probably an energy-saving

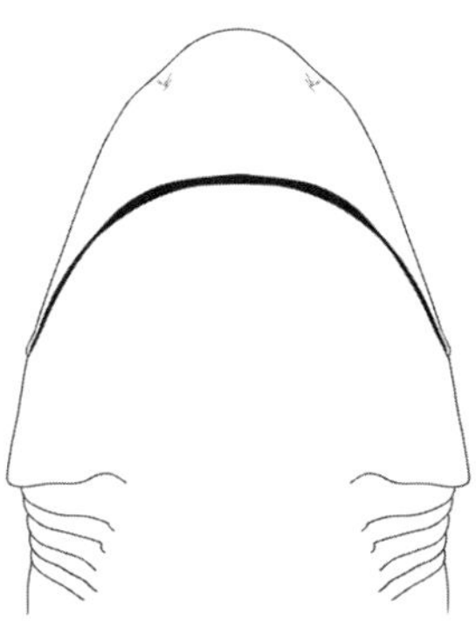

Fig. 3c. Snout.

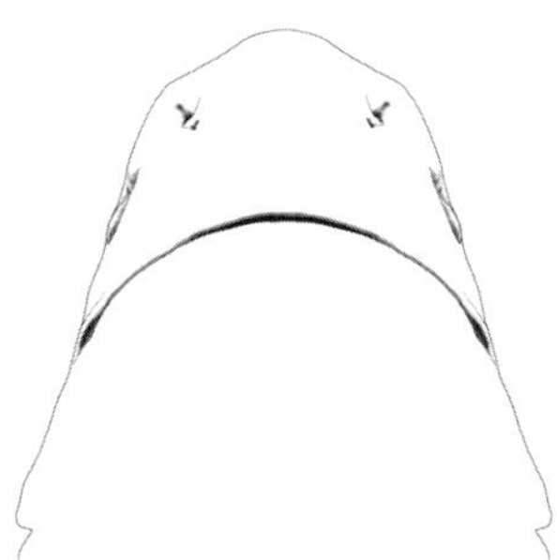

Fig 3d. Snout (neonate)

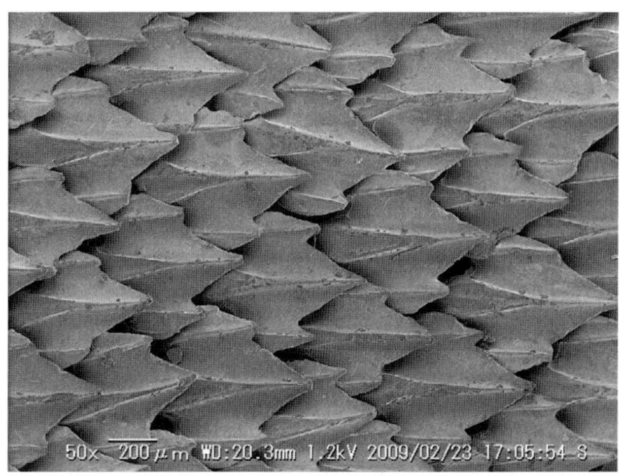

Fig. 3e. Dermal denticles.

Fig. 3f. *Hexanchus griseus*, sixgill shark, immature female, 93.4 cm; from upper Gulf of California, (Juan Carlos Pérez-Jiménez, collector).

adaptation for an animal that must search vast areas for a meal, whether by predation or scavenging. Obviously, it is capable of rapid movement when attacking prey or otherwise stimulated. I saw a large specimen (~360 cm) veer off quickly with powerful tail strokes when encountering the cone of illumination in front of the submersible. The sixgill shark also has the remarkable ability to stand vertically on its head with little horizontal movement, remaining over a given spot in this position while feeding. I watched a 330-cm female use suction feeding to pick up bait we had scattered over the bottom. There are few reports of the diet of Atlantic sixgill sharks. One 300-cm specimen I examined in Bermuda contained the remains of a small whale or a dolphin (vertebrae and blubber). Glenn Ulrich (South Carolina Marine Resources Division) observed and filmed a sixgill shark from a submersible while it made several unsuccessful attempts to capture a very nimble crab (*Gerion* sp.).

Pacific form: Unlike the Atlantic form, the Pacific sixgill shark is often found in shallow water along the colder northwestern coast of North America. Herald and Ripley (1951) reported three sixgill sharks taken inside San Francisco Bay, California. One was a female (129 cm, 9 kg) caught on hook and line at a depth of 11 m between Hunter Point and Coyote Point. The other was a 67-cm male caught on 19 March 1945 "at the intake of the Pacific Gas and Electric plant." The third was a 335-cm, 210-kg female captured in July 1928, about 1 km inside the Golden Gate Bridge near Sausalito. Miller and Greenfield (1965) reported a 75-cm juvenile sixgill shark caught in October 1964 at 27 m in East Sound, Orcas Island, Washington, where the bottom temperature was 9.5°C. This shark bore a yolk-sac scar between the two pectoral fins. Knaggs et al. (1975) reported the capture of an 85-cm specimen in a midwater trawl fished at 14–38 m in Todos Santos Bay, Mexico, on 6 October 1970. Ebert (1986a) recorded six specimens (150–242 cm) from 30 m inside San Francisco Bay, California. Dunbrack and Zielinski (2003) wrote that sixgill sharks occur regularly at 20–40 m on a shallow reef in the Strait of Georgia, British Columbia, between June and September.

Diet: Ebert (1986a) examined the stomach contents of 19 sixgill sharks from California, finding prey items in nine of them. All prey items occurred only once, except for spiny dogfish (*Squalus acanthias*), which were found in three stomachs. Other items included whale blubber, pinniped remains, prickly sharks (*Echinorhinus cookei*), ratfish (*Hydrolagus collei*), hake (*Merluccius productus*), Pacific lamprey (*Lampetra tridentata*), Pacific hagfish (*Eptatretus stoutii*), and unidentified bony fishes.

Size at maturity

Atlantic form: Size of *males* at maturity is not known because of previous confusion with the bigeye sixgill shark. Eugenie Clark and I examined a 282-cm immature male caught 7 July 1986 off Bermuda. Branstetter and McEachran (1986a) reported that a 325-cm male from the Gulf of Mexico had calcified claspers. These observations suggest that Atlantic males mature at about 300 cm.

Females appear to reach maturity at about 450 cm. A 432-cm female that I examined, caught off Marathon Key, Florida, in February 1992, was immature. Its two ovaries were immature, measuring 56 × 5 cm, and contained whitish oocytes 5–10 mm in diameter, most measuring 5–7 mm. Desbrosses (1938) reported that the female sixgill sharks from coastal France were immature to lengths of 350 cm and that gravid females had been reported with lengths of 452, 482, 450, and 465 cm. Vaillant (1901) related the capture of a 452-cm gravid female off Arcachon, France, in November 1900.

Pacific form: Males of the Pacific form appear to reach maturity at about 310 cm. Crow et al. (1996) reported that males captured off Hawaii measuring 273–308 cm had uncalcified claspers, and those 309–331 cm had calcified claspers.

The size at maturity of Pacific *females* is not precisely known. Springer and Waller (1969) stated that no mature sixgill shark had ever been reported from the eastern Pacific. Ebert (1986a) reported a 421-cm gravid female caught at 160 m in May 1975 off Church Rock, San Luis Obispo

County, California. According to Ebert this is apparently the only mature sixgill shark reported from the eastern North Pacific.

Reproduction The sixgill shark is an aplacental viviparous species like other members of the family. Few gravid females have been examined by biologists, and little is known about the reproductive processes of the species.

Size at birth: Few data exist on size of the young at birth. Based on the female reported by Vaillant (1901), which carried embryos of 65–68 cm, and the specimen reported by Ebert (1986a) that carried embryos measuring 68–74 cm, it appears that sixgill sharks are born at 65–74 cm. However, this needs verification.

Brood size: There is little information on the brood size of the sixgill shark because pregnant specimens are seldom caught, probably because of their large size and deep-water habitat. The few available reports indicate very large broods. Vaillant (1901) reported the capture of a 452-cm female off Arcachon, France, in November 1900 that, according to the fishermen, contained 108 young. Bolívar (1907) mentioned that a 482-m female caught off La Coruña, Spain, carried 42 young and was observed to abort five others during capture. The female reported by Ebert (1986a) carried 51 young.

Nurseries There is no information on the location of the nurseries of the sixgill shark along the eastern coast of North America. Small juveniles (80–120 cm) are common all along the Pacific coast from Puget Sound to the Gulf of California, but reports of neonates are rare. One such report is of a 67-cm specimen from San Francisco Bay (Herald and Ripley 1951). I have examined a few small sixgill sharks (79–83 cm) collected at 200–220 m in the northern Gulf of California in March by Oscar Sosa and J. C. Pérez Jiménez (one of these neonates is illustrated on page 34). These observations suggest that the nurseries along western North America are distributed over a large area.

Three gravid females have been taken in the eastern Atlantic, off the French and Spanish coasts. In addition to the two females mentioned above, a gravid female said to be about 400 cm and weighing some 400 kg, taken near St. Raphaël (near Cannes), France, at 850 m in August 2003. Nine "near-term" embryos averaging 61 cm in length were obtained from this shark (Michèle Bruni, curator of the Musée Océanographique de Monaco, pers. comm., January 2006).

Age and growth The sixgill shark has not been aged because of its uncalcified vertebrae.

Relation to humans The sixgill shark is too rarely caught in North American waters to be of economic importance. A few specimens are caught seasonally in deep-water grouper and snapper fisheries off Florida. The sixgill shark has been on the list of federally protected species on the eastern coast of the United States since 1992. Given that there are no fisheries for it in this region, and that sixgill sharks are seldom encountered in pelagic fisheries because their deep-water habitat, the reasons for protecting the species are precautionary, the intention evidently to prevent future fisheries from developing.

In the late 1970s, divers in coastal British Columbia, Canada, began seeing sixgill sharks with regularity during the summer at Tyler Rock in Barkley Sound, on the west coast of Vancouver Island and off Flora Island in the Strait of Georgia. By 1993 shark ecotourism had developed in the area, with several companies regularly taking scuba divers to see sixgill sharks (Harvey-Clark 1995).

BIGEYE SIXGILL SHARK

Hexanchus nakamurai Teng, 1962

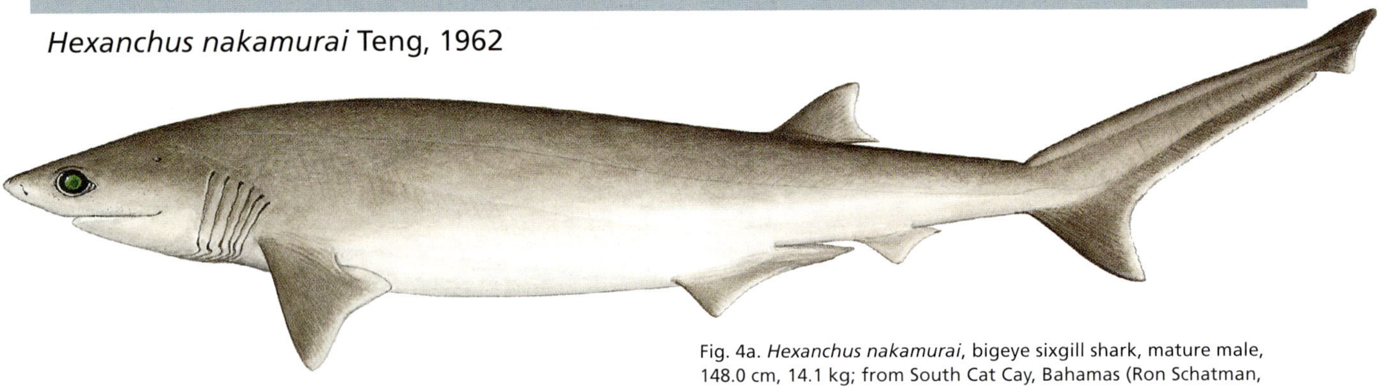

Fig. 4a. *Hexanchus nakamurai*, bigeye sixgill shark, mature male, 148.0 cm, 14.1 kg; from South Cat Cay, Bahamas (Ron Schatman, collector).

Common name Bigeye sixgill shark or small sixgill shark. The name alludes to having six gills and eyes proportionally larger than those of the much bigger sixgill shark (*Hexanchus griseus*).

Spanish name *Cazón de fondo* (Cuba). The A.F.S. name is *cazón de seis branquias.*

Scientific name *Hexanchus nakamurai* Teng, 1962, in *Classification and Distribution of the Chondrichthyes of Taiwan:* 30–33, fig. 5. *Hexanchus:* one of Rafinesque's words, apparently a mistake for *Hexancus,* from the Greek *hex,* six, and *ankos,* a bend or hollow; *nakamurai:* named after H. Nakamura (1936), who first illustrated the species.

Synonyms *Hexanchus vitulus* Springer and Waller, 1969. The bigeye sixgill shark was known by this scientific name until recently. Teng (1962) described this species as *Hexanchus griseus nakamurai,* a subspecies of the sixgill shark, in his unpublished Ph.D. dissertation. According to Compagno (1984a), doubts as to whether Teng's unpublished dissertation constituted a formal "publication" led to the wide use of the name *vitulus.* The name *H. nakamurai* has been adopted by the A.F.S. (Nelson et al. 2004).

Identification The bigeye sixgill shark is recognized by its *single dorsal fin, six gill slits,* and *five large, broad, sawlike teeth on each side of the lower jaw.* The upper teeth are long and pointed; the lower teeth are broad and sawlike, with a small central tooth at the symphysis. The teeth number U: 9–9, L: 5–1–5, disregarding the minute, budlike teeth at the sides of the mouth. The dermal denticles have three ridges that end in three points, the central one being the largest. Coloration is pale grayish brown above with lighter undersides.

Similar species The sixgill shark has *six* large, broad teeth at each side of the lower jaw. The frill shark has a terminal mouth and fanglike teeth.

Range The distribution of the bigeye sixgill shark is patchy and poorly understood. The species has been reported from the Gulf of Mexico, Florida, and West Indies region (Springer and Waller 1969), the southwest Indian Ocean (Forster et al. 1970; Bass et al. 1975a), and Taiwan (Nakamura 1935;

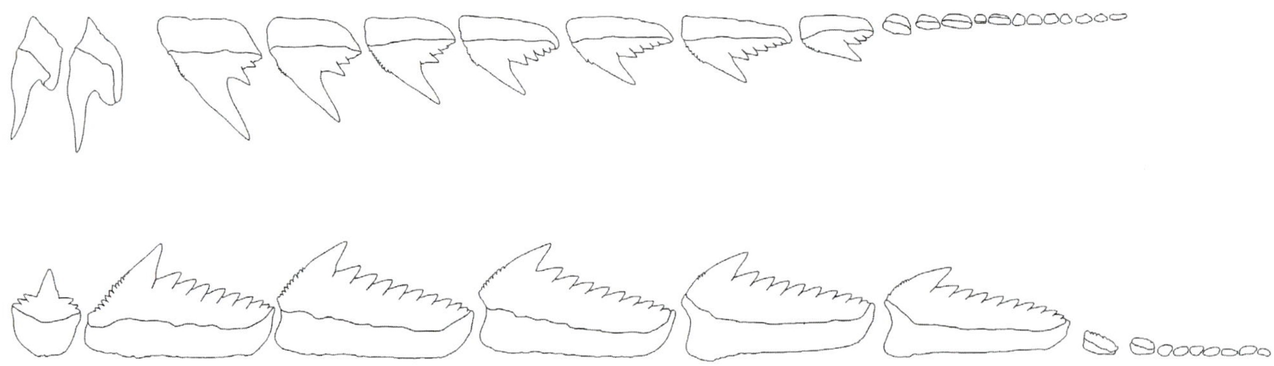

Fig. 4b. Teeth.

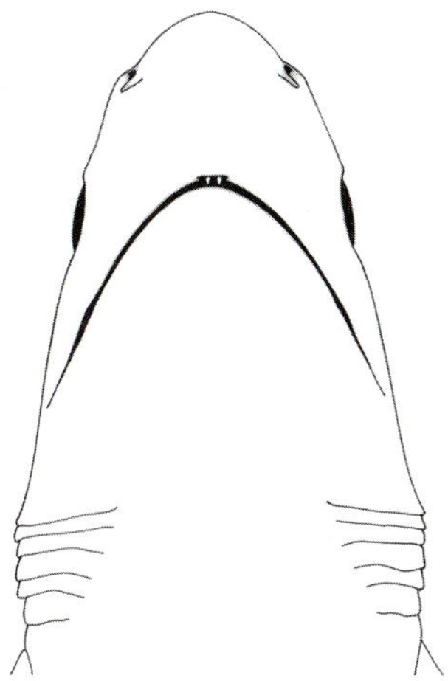

Fig. 4c. Snout.

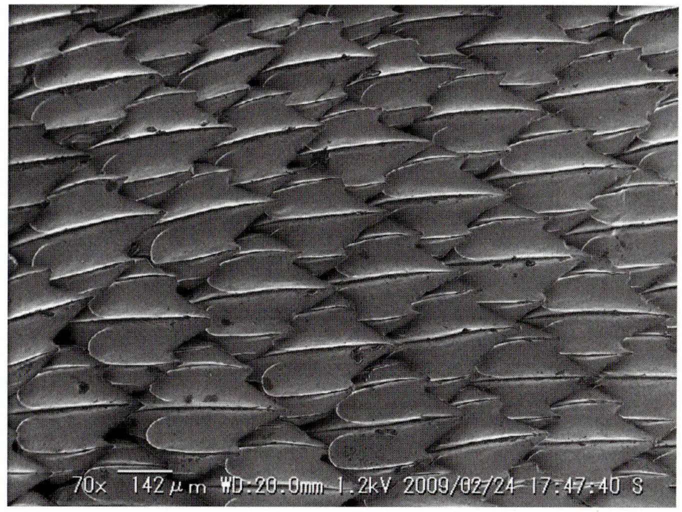

Fig. 4d. Dermal denticles.

Teng 1962). I have found it to be abundant in deep waters of the northern Gulf of Mexico, off the Bahamas, and off Grand Cayman, suggesting that the bigeye sixgill is common throughout the region. Like many other deep-water sharks, it is probably widely distributed in deep tropical, subtropical, and warm-temperate waters, there being few barriers to its movements.

Size and weight The bigeye sixgill shark probably grows to about 180 cm. Representative sizes and weights of specimens that I have examined are as follows:

Male: 144 cm, 12.2 kg; 157 cm, 15 kg.

Female: 155 cm, 16.4 kg; 164 cm, 18.4 kg; 165 cm, 20.6 kg; 169 cm, 23.6 kg.

Biology The bigeye sixgill is a shark of the deep continental slopes. This species was not generally recognized as distinct from the sixgill shark until 1969, and thus is poorly known. Although locally common at depths of 200–500 m in warm waters of the Florida–Caribbean region, there have been no published works on the species since 1969.

I have observed bigeye sixgill sharks from submersibles off the Bahamas at 465 m and off Grand Cayman at 305 m. The shark moves slowly with gentle tail beats, appearing to glide effortlessly. Its flexible body allows it to turn quickly and, like its larger congener, the sixgill shark, the bigeye sixgill has the ability to feed while positioned head down and almost motionless. On one occasion off the Bahamas, during a dive to 465 m in the *Johnson-Sea-Link II* submersible, I watched as a bigeye sixgill, swimming just above the submersible's sphere, suddenly attacked a lane snapper (*Lutjanus synagris*) on the bottom barely three meters away. The shark dashed in front of the sphere so quickly that I was not fully aware of what had just crossed my field of view. In a few seconds I saw the shark practically standing on its head, seizing the snapper.

Size at maturity There are no published data on size at maturity for the bigeye sixgill shark. I examined a mature 144-cm male and a 155-cm female that carried 40–50 mm oocytes.

Reproduction Development is aplacental viviparous, but little else is known about its reproductive processes.

Size at birth: Young are born at about 40–43 cm. Forster et al. (1970) reported embryos 39–42 cm from females in the Indian Ocean. Bigelow and Schroeder (1948) recorded a 43-cm newborn female from Cuba, indicating that the young measure about 40 cm at birth.

Brood size: A brood of 13 was reported from the Indian Ocean (Forster et al. 1970). Females that I examined, taken off Bimini, Bahamas, from May to July, carried from 7 to 20 ripe oocytes in each ovary, providing evidence of large broods. The largest oocytes measured 50 mm in diameter.

Nurseries I have examined a few small juveniles of 68–82 cm from deep water off Panama City, Florida, which indicates that some of the bigeye sixgill nurseries are in the northern Gulf of Mexico.

Age and growth The bigeye sixgill has not been aged.

Relation to humans None.

BROADNOSE SEVENGILL SHARK

Notorynchus cepedianus Péron, 1807

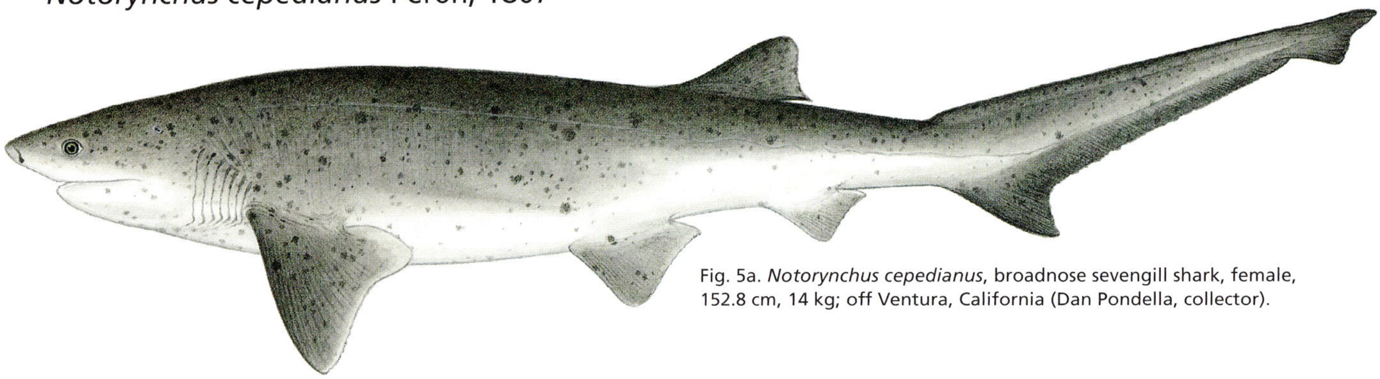

Fig. 5a. *Notorynchus cepedianus*, broadnose sevengill shark, female, 152.8 cm, 14 kg; off Ventura, California (Dan Pondella, collector).

Common name Broadnose sevengill shark, a recent name intended to distinguish it from the sharpnose sevengill, or perlon shark. The species is usually called the sevengill shark or simply "sevengill." In the past it was also called cow shark and mud shark (Walford 1935).

Spanish name *Tiburón manchado* (Mexico). The A.F.S. name is *tiburón pinto.*

Scientific name *Notorynchus cepedianus* Péron, 1807, in *Voyages Australes* 1: 337. *Notorynchus:* from the Greek *notos,* back, and *rhynchus,* from the Greek *rhynchos,* snout (the etymology or allusion is not clear); *cepedianus* honors the French naturalist B. G. E. de la Ville, Comte de Lacépède (1756–1825).

Synonyms *Heptanchus indicus* Müller and Henle, 1839; *Notidanus indicus* Agassiz, 1835; *Notorynchus maculatus* Ayres, 1855; *Notorhynchus borealis* Gill, 1864 [a]; *Heptranchias pectorosus* Garman, 1884; *Notorynchus macdonaldi* Whitley, 1931 [b]; *Notorynchus platycephalus* Fowler, 1925; among others. This species has been described under numerous names from diverse localities throughout the world. Bass et al. (1975a) considered it a single species with worldwide distribution, a view generally accepted today. In the recent past the name *Notorhynchus maculatus* Ayres, 1855 was widely used.

Identification The broadnose sevengill shark is recognized by its *seven gill slits, a flattened head with a broadly rounded snout* (see illustration), *a single dorsal fin,* and *numerous dark spots on the dorsal surface.* The upper teeth are long and pointed; the lower teeth are broad and sawlike. Counting the large teeth only, teeth number U: 7–1–7, L: 6–1–6. The dermal denticles have three points and a prominent central ridge that terminates in a long point. Sevengills are sandy gray to reddish brown above with dark spots, and whitish below. Albinos and partial albinos have been reported. Herald (1953) described a partial albino sevengill caught in the 1952 Coyote Point Derby, California. This specimen (now at the California Academy of Sciences, cataloged as CAS no. 20623) was white with numerous dark specks on its dorsal surface. The eye had an unpigmented iris and a dark blue pupil. Ebert (1985) described a "piebald" specimen caught in San Francisco Bay. It had a white background dorsally with unusually large spots 3–18 mm in diameter that covered the dorsal surface and extended onto the sides and upper surfaces of the fins. Its lower side was the normal olive brown. Ebert described the shark's eye color as "black consistent with normal eye coloration."

Similar species The sharpnose sevengill shark also has seven gill slits, but it has a pointed snout and lacks spots on its dorsal surface.

Range The broadnose sevengill shark is widely distributed in temperate waters. It has been reported from both sides of the Pacific Ocean, the South Atlantic, and the Indian Ocean. In the eastern Pacific, it has been reported from British Columbia to southern California (Walford 1935; Hart 1973), Colombia (Franke and Acero 1991), and Chile (Pequeño 1979). The report from Colombia of two specimens caught off the beach at Playa Blanca (Franke and Acero 1991) is interesting because the distribution of this species has been considered antitropical. In the western Pacific, it has been reported from southern Japan (Lindberg and Legeza 1967), Australia (Whitley 1940), and New Zealand (Phillipps 1935). In the South Atlantic, the sevengill has been reported from the Strait of Magellan (Guzman and Campodonico 1976), Argentina (Lahille 1928), Brazil (Sadowsky 1970), and Namibia (Bass et al. 1975a). In the Indian Ocean, it has been reported from South Africa (Bass et al. 1975a).

Fig. 5b. Teeth.

In North America the broadnose sevengill shark ranges from northern British Columbia (off Butedale and Bonilla Island; Hart 1973) to southern California (San Diego; Walford 1935). Bonham (1942) reported a 260-cm female found on the beach at Grayland, Grays Harbor County, Washington, and mentioned that the species is not commonly recorded from the state. Barnhart (1936) noted that it was only occasionally taken off San Diego, but rather commonly from Monterey to Washington. Roedel (1953) stated that the sevengill shark was common in San Francisco, Tomales, and Monterey bays.

Size and weight Sevengill sharks reach about 300 cm. Most sevengills caught are juveniles of 70–125 cm and 2–9 kg. A 197-cm adult male weighed 35 kg. Herald (1968) recorded the capture of a 264-cm female weighing 107 kg in San Francisco Bay in March 1966. Van Dykhuizen et al. (1998) stated that a female caught in Humboldt Bay, California, measured 295 cm and weighed 125 kg. Ebert (1989) noted that the heaviest female sevengill shark recorded was 291 cm and 182 kg.

Biology The sevengill is a common, large shark of the shallow coastal bays of central California. It is the only member of the family Hexanchidae that lives in shallow coastal waters; all other sixgill and sevengill sharks generally inhabit deep waters. Herald (1953) relates that, at the Sixth Coyote Point Shark Derby, held 14 September 1952, with the fishing limited to South San Francisco Bay between 7 A.M. and 3 P.M., the 1,484 registered fishermen caught 1,871 sharks, of which 301 (16%) were sevengills. Despite its abundance, the biology of the sevengill shark is poorly understood, and

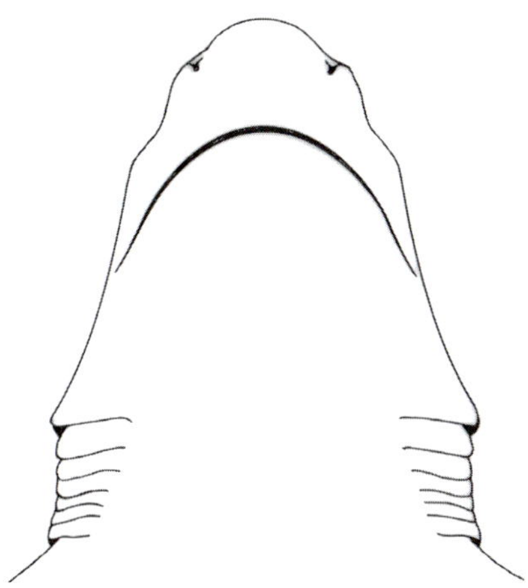

Fig. 5c. Snout.

Fig. 5d. Dermal denticles.

much of what has been written about it is anecdotal or based on questionable old data. However, its anatomy is well known thanks to the extensive work of J. Frank Daniel in the early 1900s in California. Much of Daniel's work was later published in his book *The Elasmobranch Fishes* (Daniel 1934), which remains a useful anatomical reference.

Nearly all sevengills caught in the shallow bays of California are immature. Herald and Ripley (1951) examined 37 of 59 specimens caught in a 1950 shark derby in San Francisco Bay, and all were juveniles. According to these authors, sevengill sharks apparently leave the bay by the time they reach about 170 cm and 23 kg, which might account for the rarity of gravid females in the literature.

Ebert (1991a) observed sevengill sharks foraging in the shallow mudflats of Humboldt Bay, California, in the predawn hours during spring tides. They usually moved into the deeper channels after sunrise.

Diet: According to Ebert (1989), the sevengill shark is an apex predator in the ecosystem of the northern California bays, preying on numerous other elasmobranchs, bony fishes, and marine mammals. The species preys on at least five species of sharks and rays, with the brown smoothhound (*Mustelus henlei*) being the most important. Other prey species reported by Ebert (1989, 1991a) include the Pacific lamprey (*Lampetra tridentata*), leopard shark (*Triakis semifasciata*), bat ray (*Myliobatis californica*), big skate (*Raja binoculata*), jack smelt (*Atherinopsis californiensis*), salmon (*Oncorhynchus* sp.), white surf perch (*Phanerodon furcatus*), and California sea lion (*Zalophus californianus*). Ebert (1991b) also found that South African sevengill sharks fed on numerous species of sharks and rays.

Two authors have described the homing ability of the sevengill. Ebert (1996) stated that a large specimen estimated at 300 cm was recaptured at the tagging site 730 days later. Van Dykhuizen et al. (1998) discussed a female sevengill collected in Humboldt Bay on 23 July 1990. It measured about 298 cm and was estimated at 112–135 kg. The shark was displayed at Monterey Bay Aquarium until 16 June 1994, when it was released in Monterey Bay because of abrasions caused by collisions with the aquarium windows. Tagged prior to release, the shark was recaptured 845 days later in Humboldt Bay, 503 km north of the release location.

Size at maturity

Males: Size at maturity is not precisely known because most authors have used poor or confusing criteria to define maturity. Herald (1968) stated that a specimen 197.4 cm and 34.5 kg was the youngest mature sevengill on record and gave the size of the testes as 179 mm with a diameter of 39 mm, but did not elaborate further on the criteria used to determine maturity. Compagno (1984a) reported that males mature at 150–180 cm, but the source of this information was unclear. Ebert (1989) stated that all male sevengill sharks larger than 150 cm that he had examined were mature. However, he judged maturity based on the abrupt increase in clasper length that occurs at that body length and on development of the "clasper sac mechanism" but did not demonstrate the correlation of clasper size with calcification or maturity. Although clasper elongation provides an indication of approaching maturity, it is not a definitive indicator. This is because the claspers usually elongate well before the animal reaches maturity (Castro 1996). Likewise, the presence of sperm alone is not a good criterion for maturity, because in many species sperm production precedes clasper calcification (Castro 1996), and sperm production is also dependent on the stage of the reproductive cycle. Ebert later (1996) adopted the criterion of calcification of the "terminal cartilage elements" of the clasper, revising the size at maturity to 155 cm.

Females: Size at maturity is not known precisely because gravid females have seldom been encountered or reported. Bass et al. (1975a) mentioned having traced records of two mature females of 209 and about 192 cm, the condition of maturity being perhaps anecdotal. Compagno (1984a) wrote that females mature at 192–208 cm, probably based on the literature. However, Ebert (1986a) stated that the smallest mature female was 268 cm and weighed 127.3 kg, based on oocyte diameter. Later, Ebert (1989) stated that females mature at about 250 cm and weigh in excess of 91 kg, predicted on seeing a female with 75-mm oocytes and uterine eggs (D. Ebert, pers. comm., July 2003).

Reproduction The sevengill shark is an aplacental viviparous species. As in most primitive sharks, females have two functional ovaries of equal size (Daniel 1934). Beyond this, there is little information on its reproductive processes, despite the abundance of sevengills in coastal California waters. Bass et al. (1975a) were unable to find any records of embryos, and more than a quarter-century later, I have found only one vague report. Ebert (1989) mentioned that a female examined in May contained 82 near-term young and had no developed oocytes in the ovaries. From this observation he speculated that female sevengill sharks give birth every 18–24 months. This conjecture was based on "Herald and Ripley's (1951) speculation that sevengills enter San Francisco Bay for breeding is correct, then the time from parturition to fertilization may be from 6 to 12 months. Additionally, based on Holden's (1974) calculations for *Hexanchus griseus,* the time from fertilization to parturition (gestation) would extend another year. Therefore, after first parturition adult female sevengill sharks would give birth every 18 to 24 months."(Ebert 1989:107) This is all highly speculative. All that can really be said about the reproductive cycle of the sevengill is that the ovarian cycle and gestation are not concurrent, and that the cycle is possibly biennial, considering that many other species have biennial cycles. The gestation period must be determined empirically; it cannot be extrapolated from Holden's (1974) calculations of the length of gestation for the sixgill shark (*H. griseus*), a different species living in a different environment.

Size at birth: According to Ebert (1989) the young are born at a length of about 45 cm.

Brood size: Broods appear to be large. Herald (1968) reported that a 264-cm gravid female of 107 kg contained 83 oocytes each about 51 mm in diameter. Another female of 295 cm and 125 kg carried about 100 oocytes about 10 mm in diameter (Van Dykhuizen et al. 1998). Ebert (1989) refers to a female that contained 82 near-term embryos in May.

Nurseries It is not known where females give birth to their young. Juveniles are found in the shallow bays of central California.

Age and growth The sevengill shark has not been aged because of its uncalcified skeleton.

Relation to humans The sevengill is considered one of the most palatable sharks, and small amounts of meat are found in California markets. Sevengills are often exhibited in marine aquariums, where they can live for several months, but often require force-feeding (Herald 1968). Several authors have described the aggressiveness of sevengill sharks. Roedel and Ripley (1950: 41) observed that a sevengill "is a pugnacious shark that will attempt to bite its captors," and Herald and Ripley (1951: 326) stated that "the belligerent disposition of this shark as well as its tendency to use its jaws and sharp teeth on anything in the vicinity makes it a dangerous liability." Herald (1968) described how a sevengill attacked a diver at San Francisco's Steinhart Aquarium after the diver attempted to capture the shark by hand. The shark slipped from his grasp and swam away, then circled back toward the diver, who threw up his arm in front of his face. The sevengill bit his arm, removing a section of flesh. Herald (1968) also related other cases of divers and fishermen who experienced the business end of sevengill sharks they had provoked.

FAMILY ECHINORHINIDAE

The Bramble Sharks

The family Echinorhinidae includes two widely distributed species, the bramble shark and the prickly shark. These large, heavy-bodied sharks have two small, spineless dorsal fins, the first originating behind the pelvic fin origins; thin, bladelike multicuspid teeth (except juveniles) that are alike on both jaws (they are diagnostic for the genus); and a caudal fin without a subterminal notch. In both species the skin is covered with characteristic dermal denticles that are large and spinelike. Both species dwell in deep waters but regularly enter shallow waters. They have often been included in the Squaloidae because they lack an anal fin; however, most authors now place them in a separate family because of their many unique characteristics, such as the shape of their teeth and their peculiar skin denticles. The family consists of a single genus, *Echinorhinus.*

KEY TO THE ECHINORHINIDAE

1a. Very large, sparse, and irregularly distributed spinelike dermal denticles, some measuring up to 1.5 cm in diameter at the base and larger when two or more are joined together; denticle bases with a scalloped edge, and the spines are finely ridged.... .... **Bramble shark,** *Echinorhinus brucus* (p. 44).

1b. Dermal denticles less than 4 mm in diameter and never joined together, with stellate bases and heavily ridged spines.... .... **Prickly shark,** *Echinorhinus cookei* (p. 47).

BRAMBLE SHARK

Echinorhinus brucus (Bonnaterre, 1788)

Fig. 6a. *Echinorhinus brucus,* bramble shark, female, 261.4 cm, 81.36 kg; from Gulf of Mexico, edge of Mississippi Canyon, 27 km south of Grand Isle, Louisiana (James E. Flanner, collector).

Common name Bramble shark, an allusion to the enlarged, spinelike dermal denticles that cover its body. In the older English literature (e.g., Yarrell 1841; Couch 1868), it was known as the spinous or spiny shark.

Spanish name There are no native common names for it in the Spanish Caribbean or Mexico because of its rarity.

Scientific name *Echinorhinus brucus* (Bonnaterre, 1788), originally described as *Squalus brucus* in *Tableau Encyclopédique et Méthodique des Trois Règnes de la Nature, Ichthyologie:* 11. *Echinorhinus: echino* means spiny or prickly, from the Greek *echinos* (Latin *echinus*), a sea urchin, and *rhinus* is from the Greek *rhine,* the name of a shark with rough skin (also a file or rasp), both allusions to its spiny skin; *brucus:* New Latin from the Greek *brux,* the depths of the sea, or *bruchios,* of the depths of the sea, referring to the bramble shark's deep habitat.

Synonyms *Squalus spinosus* Gmelin, 1788; *Echinorhinus obesus* Smith, 1849; *Echinorhinus* (*Rubusqualus*) *mccoyi* Whitley, 1931 [a].

Identification The bramble shark has two small dorsal fins located far back on the body, a *first dorsal fin located over or behind the pelvic fin origin;* it *lacks an anal fin,* and the caudal fin lacks a subterminal notch. The teeth have several smooth-edged cusps, with the largest cusp curved toward the corners of the mouth; the teeth number U: 10 to 13–10 to 13, L: 11 to 13–11 to 13 and are similar in both jaws. It has *conspicuous, large, spinelike dermal denticles that are sparsely and irregularly distributed.* The denticles measure up to 1.5 cm in diameter at the base and larger when two or more are joined together; their bases have a scalloped edge, and the spine bases are finely ridged. In adults, large denticles cover the underside of the snout. Color varies from brown to black above, with purplish metallic hues; the undersides are paler. A fisherman who caught a specimen off Louisiana reported that when fresh it had a greenish glow that disappeared sometime after death. Some specimens have been reported with dark blotches.

Similar species The more common prickly shark has smaller, uniformly distributed dermal denticles, less than 4 mm in diameter and never joined together; the bases of its denticles are stellate, and the spines are heavily ridged.

Range The bramble shark is widely distributed in deep temperate and tropical waters. It has been reported numerous times from the eastern Atlantic and the western Indian Ocean (Smith, J. L. B. 1949; Nair and Lal Mohan 1971), but it is rare in the western Atlantic, and there are no records of its presence in the eastern Pacific. It is also present in the South Atlantic, but there are few records. There is one report from Brazil (Soto et al. 1995), one from Patagonia (Caille and Olsen 2000), and an unpublished account of a large specimen caught off Tobago in 1986. Unfortunately, only the tail of the Tobago specimen was brought to me for identification after the specimen had been discarded. However, its characteristic skin denticles made the identification obvious.

The bramble shark is a rare catch in North American waters. Only a handful of specimens have been reported in the western North Atlantic in the last 100 years. The first known North American specimen washed ashore at Provincetown, Massachusetts, in December 1878 (Goode and Bean 1879). The second report is of a 216-cm, 78-kg female caught some 120 km northeast of Cape Hatteras, North Carolina, at a depth of 187 m, and a bottom temperature of 10.6°C (Musick and McEachran 1969). The third reported specimen was a female 280 cm and 200 kg, caught some 69 km northeast of Cape Hatteras at 111 m, in March

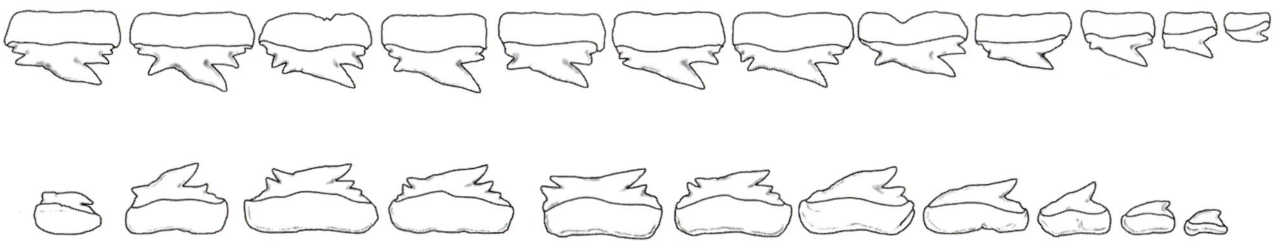

Fig. 6b. Teeth.

1992 (Schwartz 1993). Three recent unpublished records from the Gulf of Mexico (Louisiana) also exist. Two of these specimens are preserved at the Florida Museum of Natural History in Gainesville. The third was a 261-cm, 81-kg female captured at about 200 m, at the edge of Mississippi Canyon, 27 km south of Grand Isle, Louisiana, in December 1994. I examined this specimen while fresh, and it is illustrated in this work. It is preserved in the Tulane University Collections (cat. no. 172379).

Reports of bramble sharks from the western coast of North America are doubtful because of earlier confusion with the prickly shark, and early reports of bramble sharks there are about prickly sharks. There are no confirmed reports of bramble sharks in the eastern Pacific Ocean.

Size and weight The bramble shark is a heavy-bodied shark. The largest on record is a 280-cm female from North Carolina (Schwartz 1993), but the species is said to grow to 3.1 m (Compagno 1984a). Reported weights for females are 228 cm, 72 kg (gravid female, Ivory Coast); 261 cm, 78 kg; and 280 cm, 200 kg.

Biology The bramble shark is a widely distributed, but scarce, deep-water species that occasionally enters shallow coastal waters. The species has been known for centuries, but because of its scarcity, its life history is still poorly known. The first accurate depiction of the species, with a good accompanying anatomical description, is that of Turner (1875). Until the early 1960s, prickly sharks of the Pacific Ocean were mistaken for bramble sharks. Hence, pre-1960 publications referring to bramble sharks off the western coast of North America (e.g., Hubbs and Clark 1945; Collyer 1953) are actually about prickly sharks. Garrick, in 1960, published an excellent redescription of the prickly shark, clarifying distinctions between the two species (Garrick 1960a).

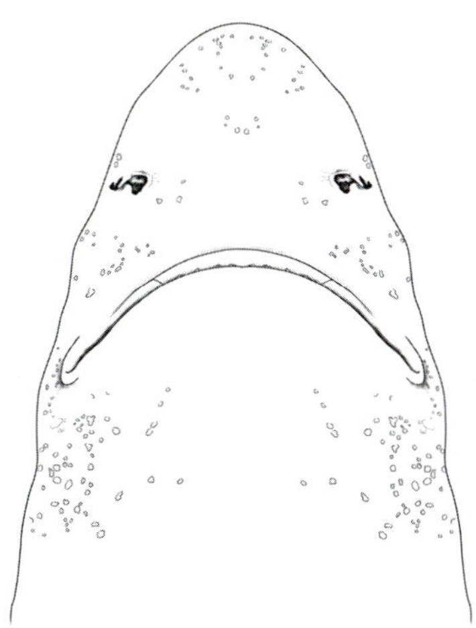

Fig. 6c. Snout.

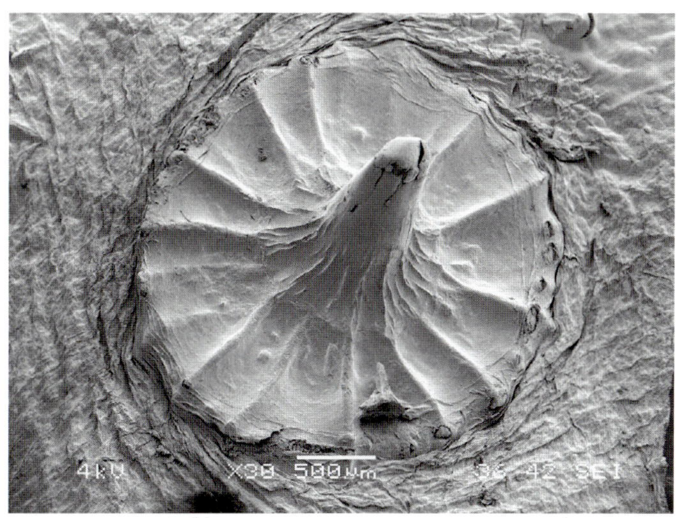

Fig. 6d. Dermal denticle.

Based on European catches, it is believed that bramble sharks usually dwell near the bottom at depths of 400–900 m, undertaking a summer migration to shallow waters of 20–200 m (Bauchot and Pras 1980). Little is known of the diet. The stomach of a Virginia specimen (Musick and McEachran 1969) contained one spiny dogfish about 54 cm long, a segment of the vertebral column of another spiny dogfish, and a hake (*Urophycis tenuis*) about 35 cm long.

There is evidence that the bramble shark was abundant in the eastern Atlantic in the eighteenth and nineteenth centuries, becoming scarcer afterward. Yarrell (1841) mentioned five captures on the English coast between 1830 and 1838, plus a dead specimen cast ashore in the Moray Firth, Scotland. Rappé (1984) listed 10 captures in the North Sea, all between 1830 and 1893, all during the summer. Rappé also mentioned that there were very few records of bramble sharks taken around the British Isles in the twentieth century and none in the North Sea, and that in the 1960s and 1970s, the species had been captured only once, near Cornwall in 1969. Wheeler (1969) in his work on fishes of the British Isles and northwestern Europe stated that, comparing the nineteenth-century records to those of the twentieth century, the bramble shark appeared to have become much rarer in northern waters. This scarcity has also been noted on the continental coasts. Québro and Emonnet (1993) quoted references showing that that bramble sharks were regularly fished along the Arcachon coast of France during the eighteenth century and that they started to become rare there in the second half of the nineteenth century, although they were still common between Bidassoa and Adour. Quéro and Emonnet (1993) noted that in 25 years of observations, they had seen only one bramble shark in 1968, and knew of two others, one caught in 1981 and another landed at Santander, Spain. It is suspected that the decline in catches resulted from fishing mortality. It appears that the bramble shark is highly susceptible to overfishing, like many other sharks.

Size at maturity Unknown. The Ivory Coast female was pregnant at 228 cm.

Reproduction The bramble shark is an aplacental viviparous species. One female I examined had two active and well-developed ovaries. Silas and Selvaraj (1972) illustrated a 30-cm embryo but gave no details of the mother or its capture, other than that it was captured in April off India.

Size at birth: The young are probably born at about 40–50 cm.

Brood size: A female caught off southern Natal, South Africa, in April 1973 carried 24 embryos averaging 16.5 cm (Bass et al. 1976). A 228-cm female from Ivory Coast contained 15 young 62–68 mm, weighing about 250 g each (Cadenat and Blache 1981).

Nurseries Unknown.

Age and growth The species has not been aged because of its rarity. It is thought to be a long-lived species as a result of its deep-water habitat, large size, and apparent susceptibility to overfishing.

Relation to humans None. Given its rarity, catches should be preserved for scientific study.

PRICKLY SHARK

Echinorhinus cookei Pietschmann, 1928

Fig. 7a. *Echinorhinus cookei,* prickly shark, female, 235 cm, 103.4 kg; from Monterey Canyon, California (Cyndi Dawson, collector).

Common name Prickly shark, a reference to the thornlike dermal denticles that cover its body.

Spanish name *Tiburón espinoso* (Mexico). The A.F.S. name is *tiburón espinoso negro.* The color reference is without meaning.

Scientific name *Echinorhinus cookei* Pietschmann, 1928, in *Anzeiger der Akademie der Wissenschaften in Wien* (*Mathematisch-Naturwissenchaftliche Klasse*) 27: 36. Pietschmann actually wrote two accounts of the prickly shark, labeling both as new species descriptions. The original 1928 description in German is brief and lacks illustrations. It was followed in 1930 by a more extensive description in English of "*Echinorhinus cookei,* new species" in an article titled "Remarks on Pacific Fishes" (Pietschmann 1930) that includes illustrations of the teeth, dermal denticles, and the entire specimen. *Echinorhinus: echino* means spiny or prickly, from the Greek *echinos* (Latin *echinus*), a sea urchin, and *rhinus* is from the Greek *rhine,* the name of a shark with rough skin (also a file or rasp), both allusions to the spiny skin; *cookei:* latinized form of Cooke, after Dr. C. Montague Cooke Jr., a conchologist at the Bishop Museum.

Synonyms None.

Identification The prickly shark has two small dorsal fins located far back on the body, the *first dorsal fin located over or behind the pelvic fin origin.* It *lacks an anal fin,* and the tail lacks a subterminal notch. A conspicuous lateral line extends like an open furrow from just over the gills to the tail. The teeth have three to seven smooth-edged cusps, the longest cusp curved toward the corners of the mouth; they number U: 10 to 12–10 to 12, L: 11 to 14–11 to 14 and are similar in both jaws. The juveniles have single-cusped teeth. The body is *covered by very large dermal denticles,* with *stellate bases and strongly ridged spines.* These denticles measure up to 0.4 cm in diameter at the base; they are *uniformly distributed* over the body. In adults the underside of the snout has very small dermal denticles and is almost smooth. Color in life is brown to grayish brown, often with purplish hues above and paler below.

Similar species The bramble shark has larger, irregularly distributed denticles with rounded bases, scalloped edges, and finely ridged spinelike cusps.

Range The prickly shark inhabits tropical and temperate waters of the Pacific and Indian oceans. Numerous records indicate that it is distributed throughout the Pacific basin: New Zealand (Garrick 1960a), Australia (Last and Stevens 1994), Taiwan (Compagno 1984a), Japan (Taniuchi and Yanagisawa 1983), Hawaiian Islands (Pietschmann 1928, 1930; Borets 1986; Crow et al. 1996), California (Miller and Lea 1972), Mexico (Collyer 1953), Peru (Chirichigno 1963), and Chile (Flores and Rojas 1979).

In North America it has been reported from southern California to central Mexico (Hubbs and Clark 1945, as *E. brucus;* Miller and Lea 1972; Chávez-Ramos and Castro-Aguirre 1974; Aguirre et al. 2002). The prickly shark once was considered uncommon along the California coast. However, the exploration of Monterey Canyon has revealed that the species is at least locally abundant in summer and early fall (Crane and Heine 1992).

Size and weight The prickly shark attains a large size, possibly about 4 m. Garrick (1960a) mentioned a New Zealand (Moeraki) specimen believed to be about 4 m long. Later, Garrick and Moreland (1968) recorded a 298.5-cm specimen from Cook Strait, New Zealand. There are few available weights for prickly sharks: 267 cm, 192 kg; 295 cm, 222 kg; 305 cm, 266 kg. As Garrick (1960a) has pointed

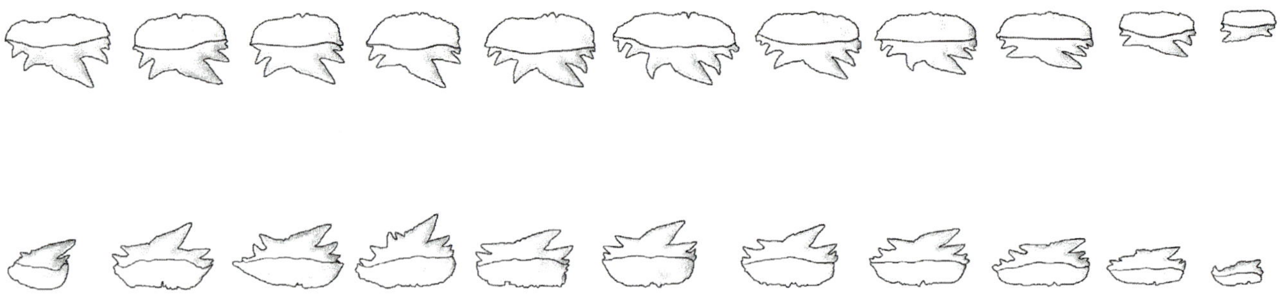

Fig. 7b. Teeth.

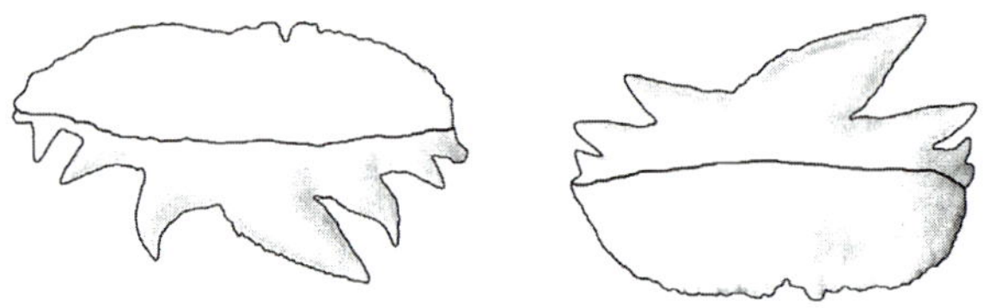

Fig. 7c. Teeth (detail).

out, the juveniles are quite slender when compared to the heavy-bodied adults.

Biology Although the prickly shark is fairly common off California, its habits are poorly known. Hubbs and Clark (1945: 67) first reported the species from California as a "bramble shark," presciently noting that "probably *Echinorhinus* will prove to be much less rare in California than the available data would seem to indicate." Prickly sharks continued to be mistaken for bramble sharks until the early 1960s. For this reason, publications prior to 1960 referring to bramble sharks off the western coast of North America (e.g., Hubbs and Clark 1945; Collyer 1953) are actually about the prickly shark. In 1960, Garrick published an excellent redescription of the prickly shark, clarifying the distinctions between the two species (Garrick 1960a).

Almost nothing is known of the habits of the prickly shark. It appears to be a deep-water species that frequently visits surface waters, perhaps seasonally. Prickly sharks have been observed by scuba divers in Monterey Canyon, Cali-

Fig. 7d. Snout.

Fig. 7e. Dermal denticles.

fornia, at depths of 15–35 m. Crane and Heine (1992) reported encountering from one to more than 30 prickly sharks on 13 of 16 dives in the canyon. Even in conditions of poor visibility, up to four sharks could be seen at a time. The sharks did not appear disturbed by the divers and even tolerated contact. Prickly sharks have also been caught in shallow water at other locations. Collyer (1953) reported capturing two females, 267 and 295 cm in length, while bottom fishing at 37 m off Guadalupe Island, Mexico. Other reported catches have occurred at much greater depths. Two specimens were caught at 400–424 m in the southern Gulf of California (Chávez-Ramos and Castro-Aguirre 1974). Off Hawaii, Tester (1969) reported prickly sharks to be particularly abundant at about 320–400 m. Prickly sharks off Japan have been captured at 550–650 m (Kobayashi 1986).

Few firsthand accounts of the diet of prickly sharks exist. It appears to feed on a wide variety of bony fishes, sharks, squid, and octopuses. J. C. Pérez Jiménez (pers. comm., February 2009) informed me that he had found the beaks of Humboldt squid (*Dosidicus gigas*) in the stomachs of prickly sharks from the northern Gulf of California.

Size at maturity Size at maturity is not precisely known. According to Last and Stevens (1994: 46), "Some males mature by 198 cm; females mature between 250 and 300 cm."

Reproduction The prickly shark is an aplacental viviparous species and produces very large broods. Nothing else is known of its reproductive processes.

Size at birth: A free-swimming male of 37.7 cm and 268 g was captured off the Michoacan coast, Mexico, at a depth of 137 m (Leonardo Castillo, pers. comm., November 2005), suggesting that birth size may be less than 40 cm.

Brood size: The broods are apparently very large. One brood that I examined at the U.S. National Museum (USNM no. 179768–93) consisted of 114 embryos (16–24 cm), taken from a 305-cm female weighing 266 kg. The shark was caught off Ala Moana Beach, Oahu, Hawaii, at a depth of 266 m in early November 1959. This is the only prickly shark brood on record that I know of (and one of the largest shark broods).

Nurseries No information on nurseries is available other than the location of the 37.7-cm specimen noted above.

Age and growth Prickly sharks have not been aged.

Relation to humans None.

FAMILY SQUALIDAE

The Dogfishes

The Squalidae is a large family of sharks, known as the dogfishes, its members characterized by having bladelike teeth on both jaws that are imbricated or interlocked with each other and forming a continuous cutting edge, two dorsal fins with conspicuous spines, the absence of an anal fin, and a tail that lacks a subterminal notch. The family comprises two genera, *Cirrhigaleus* and *Squalus.* The genus *Cirrhigaleus* is characterized by the presence of a nasal barbel on the nostril. It contains two species, one of which is found in North America, the roughskin spiny dogfish, *Cirrhigaleus asper.* The genus *Squalus* is noted for a supraethmoidal process on the skull and by a precaudal pit at the base of the tail (Shirai 1992). There are about 12 species in the genus, three of which inhabit North America.

KEY TO THE SQUALIDAE

1a. Dorsal fins with white tips and a wide white band extending along the rear margin (first dorsal fin originating well behind the free rear tips of the pectoral fins, having long, stout dorsal fin spines that often reach beyond the fin tip) .... **Roughskin spiny dogfish,** *Cirrhigaleus asper* (p. 52).

1b. Dorsal fins with black or dark tips and lacking a wide white band along the rear margin 2

2a. First dorsal fin originating well behind the free rear tip of the pectoral fins, a series of small white dots along the flanks. .... **Spiny dogfish,** *Squalus acanthias* (p. 55).

2b. First dorsal fin originating from the level of the pectoral fin axil to the free rear tip of the pectoral fins, lacking white dots along the flanks 3

3a. Pectoral fins with rounded free rear tips, three pointed dermal denticles (a species reaching 110 cm)............................ .... **Shortspine dogfish,** *Squalus mitsukurii* (p. 65).

3b. Pectoral fins with a pointed free rear tip, lanceolate dermal denticles (a small species seldom reaching 60 cm)............... .... **Cuban dogfish,** *Squalus cubensis* (p. 63).

ROUGHSKIN SPINY DOGFISH

Cirrhigaleus asper (Merrett, 1973)

Fig. 8a. *Cirrihgaleus asper*, roughskin spiny dogfish, mature male, 98.5 cm, 5.0 kg; from Ft. Pierce, Florida (Scott Bachman, collector).

Common name Roughskin spiny dogfish, allusions to the large skin denticles that produce a rough-textured skin and the presence of dorsal fin spines (see the origin of "dogfish" under spiny dogfish, p. 55).

Spanish name The species lacks vernacular names in the Spanish Caribbean.

Scientific name *Cirrhigaleus asper* (Merrett, 1973), originally described as *Squalus asper* in *Journal of Zoology, London* 171: 93–110. *Cirrhigaleus:* from the Latin *cirrus,* a curl, an allusion to the nasal barbels (which this species does not have) characteristic of the genus, and *galeus,* from the Greek *galeos,* a shark; *asper:* from the Latin *asper,* rough, an allusion to its skin texture. The spelling *Cirrhigaleus* comes from *cirrh,* "a wrong but very common from of *cirr,* from the mistaken notion that the Latin *cirrus,* a curl of hair, was represented by the Greek kirrhos" (Jaeger 1962: 58).

Synonyms None. The species was known as *Squalus asper* until recently.

Identification The roughskin spiny dogfish is characterized by two dorsal fins, each with a long, stout spine, a first dorsal fin originating *posterior* to the free rear tips of the pectoral fins, a second dorsal fin *almost as large* as the first, very broad pectoral fins with rounded free rear tips, and a *broad anterior nasal flap.* The teeth have strongly oblique, smooth-edged cusps with a pronounced notch on their outer margins, forming a nearly continuous cutting edge. Teeth number U: 12 to 14–12 to 14, L: 11 or 12–11 or 12 and are similar in both jaws. The dermal denticles are *very large,*

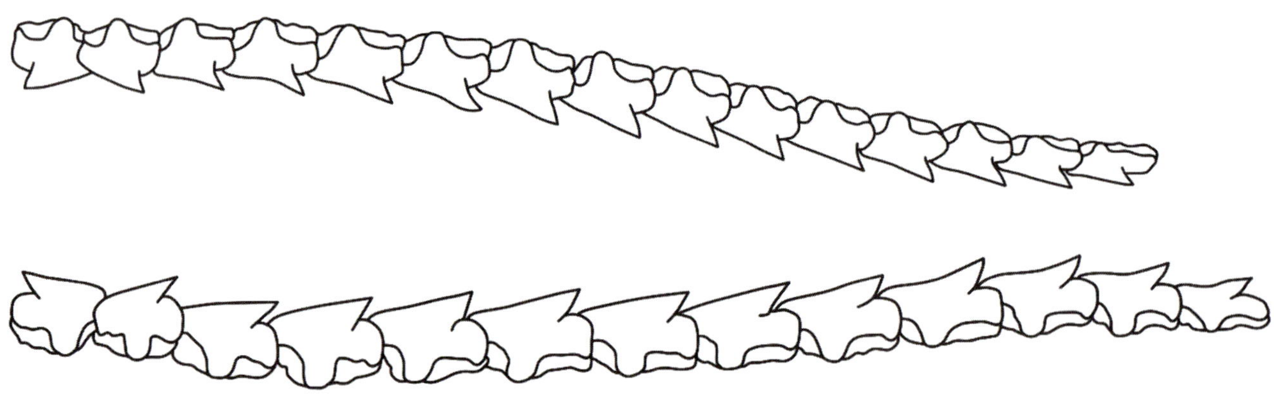

Fig. 8b. Teeth.

Fig. 8c. Snout.

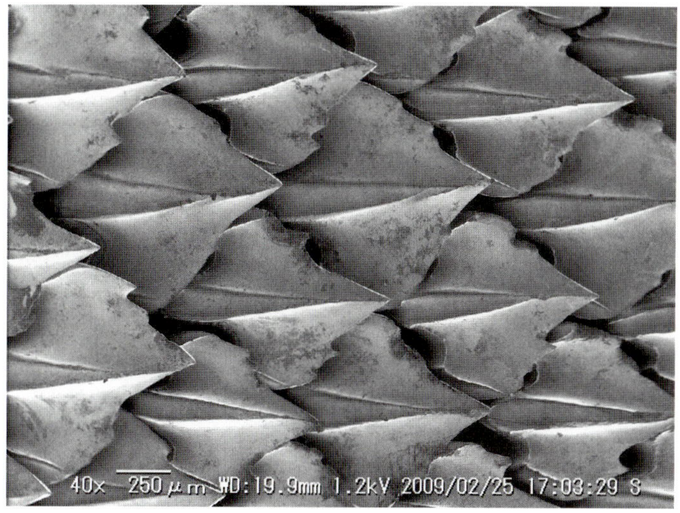

Fig. 8d. Dermal denticles.

usually 0.08–0.09% of the total length (about three times larger than those of comparably sized specimens of similar species), or 1–2 mm wide in grown specimens. The dermal denticles of juveniles are leaf shaped with a strong central ridge; those of the adults have a strong central ridge terminating in a point and flanked on either side by winglike extensions and weaker ridges. The upper sides are brown with bronze hues in life, fading to brownish gray after death, with paler or whitish undersides. *The rear margins of the dorsal fins are white.* Neonates and small young are beautifully marked; their dorsal and caudal fins have brownish bases, an inner black band, and white edges.

Similar species Other dogfishes in North America lack the white dorsal fin margins. The spiny dogfish has smaller dermal denticles, a second dorsal fin smaller than the first, and, usually, small white spots on the dorsal surface. The shortspine dogfish and the Cuban dogfish have a more anterior first dorsal fin, originating over the pectoral fin and smaller dermal denticles (about 0.3 mm in adult specimens).

Range The roughskin spiny dogfish appears to be very widely distributed in deep waters. It has been reported from the western Indian Ocean (Merrett 1973), Hawaiian Islands (Compagno 1984a), western North Atlantic (Castro 1983), and Gulf of Mexico. It is abundant from South Carolina to Florida and the Gulf of Mexico at depths of 200–300 m.

Size and weight Most specimens seen are about 90 cm, but the species grows to at least 118 cm (Merrett 1973). Representative sizes and weights are as follows:

Male: 71 cm, 1.6 kg.

Female: 69 cm, 1.3 kg; and 109 cm, 7 kg.

Biology This is a poorly known species, which was not described until 1973. Most specimens have been isolated captures at 110–600 m over the continental slopes (Merrett 1973; Bass et al. 1976; this report). It is abundant off South Carolina at depths of 200–300 m. Most of the specimens I have examined had empty stomachs. The 109-cm female mentioned above had a 30-cm squid in its stomach. Nothing else is known of its habits.

Size at maturity

Males: Males mature between 71 and 85 cm.

Females: Females mature at about 88–90 cm. Females that I have examined were immature up to 86 cm and carried oocytes of 1–2 mm in diameter, while a female 100 cm carried a large number of uterine eggs. These observations indicate a length at maturity of about 88–90 cm for specimens off the eastern coast of North America. Merrett (1973) reported an 88.8-cm female from the Indian Ocean as mature. Given that the 86-cm female that I examined had oocytes of only 2 mm in diameter, it is possible that the North American population may mature at a slightly larger size.

Reproduction The roughskin spiny dogfish is aplacental viviparous. I examined a 109-cm, gravid female caught by E. Jayroe in 220 m off Georgetown, South Carolina, on 4 August 1984. It carried four 25 to 28-cm embryos in the left uterus, apparently having aborted the young on the right side. The young had fully absorbed yolk sacs and appeared to be ready for birth. I also examined a preserved brood of 21 midterm young caught off Florida in summer. If the capture date of that brood is correct, embryos of two different sizes in summer would indicate a two-year gestation period. This requires confirmation.

Size at birth: The embryos appear to be born at about 25–28 cm, based on the well-formed 25–28-cm embryos mentioned above, and a 33-cm free-swimming neonate reported by Bass et al. (1976).

Brood size: Broods usually consist of 21–23 embryos. This is based on a small number of gravid females reported. I have examined a brood of 21 caught off Florida. Bass et al. (1976) reported broods of 21 and 22 embryos.

Nurseries Unknown.

Age and growth The species has not been aged.

Relation to humans It is a frequent bycatch in tilefish operations along the southeastern coast of the United States, where it is often discarded. It has no commercial importance.

SPINY DOGFISH

Squalus acanthias Linnaeus, 1758

Fig. 9a. *Squalus acanthias*, spiny dogfish, female, 95 cm, 3.6 kg; from McClellanville, South Carolina (Steve Poston, collector).

Common name Spiny dogfish, picked dogfish, spiked dogfish, bone dog, often just dogfish, or simply dog. The term "spiny" was added to distinguish it from the smooth dogfish. In Great Britain, it is also called spurdog or piked dogfish. The species was often labeled grayfish or flake in government or industry publications that sought to enhance its public appeal as an edible product. According to Ketchen (1986), the name dogfish has its origin in ancient Europe, where the prefix "dog" was used to denote some plant or animal worthless or unfit for human consumption. This is only partially correct. The prefix dog was applied to plants unfit for human food, and we have many examples of this in the *Oxford English Dictionary:* dog's-apple, dog cabbage, dog-thistle, dog's-wheat, and so on. Animal names were seldom modified with the prefix dog to indicate lack of suitability for human consumption. Dog-meat was not meat unfit for humans or fit only for dogs, but dog's flesh used as food. The only animal name modified by the prefix dog (that I could find) is the dog salmon (or chum salmon, *Oncorhynchus keta*) that used to be considered of poor quality in earlier times. Spawning dog salmon deteriorate quickly and "are not prime fish for human consumption, but native people in Canada and Alaska dry and smoke them to feed to their dogs" (Behnke 2002: 52). In any case, the name dogfish can be traced to the translation of the Latin *canes in mari* ("dogs in the sea") of medieval bestiaries. An example of this can be found in the Bodleian Library version of the *Bestiary* (Oxford M.S. Bodley 764), produced during the first half of the thirteenth century. A recent English translation of the manuscript by Barber (1992: 202) reads: "As men learnt little by little about the kinds of fishes they were given names, often from their similarity to creatures on land, such as frogs, calves, lions, blackbirds, and peacocks. . . . Other fishes were so called because their ways were like those of creatures on land: dogfish were called after dogs, because they bite."

Spanish name The species does not reach the Spanish Caribbean, is generally unknown to fishermen along western Mexico, and has no vernacular names there. In Spain it is known as *mielga.* The A.F.S. calls it *cazón espinoso común.*

Scientific name *Squalus acanthias* Linneus, 1758, in *Systema Naturae* (10th ed., 1758, I: 233). *Squalus:* from the Latin *squalus,* some sea-fish, later a shark; *acanthias:* thorny or prickly, from the Greek *akantha,* a thorn or prickle, a reference to its dorsal spines.

Synonyms There are numerous synonyms for *Squalus acanthias,* and a complete list would cover many pages, most placing it in the genera *Spinax* or *Acanthias.* Girard (1856) described the Pacific Ocean form of the spiny dogfish as a separate species, *Spinax* (*Acanthias*) *suckleyi.* Girard's description is imprecise, but Garman (1913) stated that *S. suckleyi* differed from *S. acanthias* by the more anterior position of its first dorsal fin spine in relation to the pectoral fin (over or a little behind the angle of the pectoral fin in *S. suckleyi,* and behind and remote from the angle of the pectoral in *S. acanthias*). The name *suckleyi* remained in use for the Pacific population of the spiny dogfish during the remainder of the nineteenth century and the first half of the twentieth century, some authors considering the Pacific Ocean population as belonging to a subspecies, *S. acanthias suckleyi.* By the middle of the twentieth century, most authors agreed that the differences cited by Girard (1856) did not hold and that there is only one form of *Squalus acanthias* in North

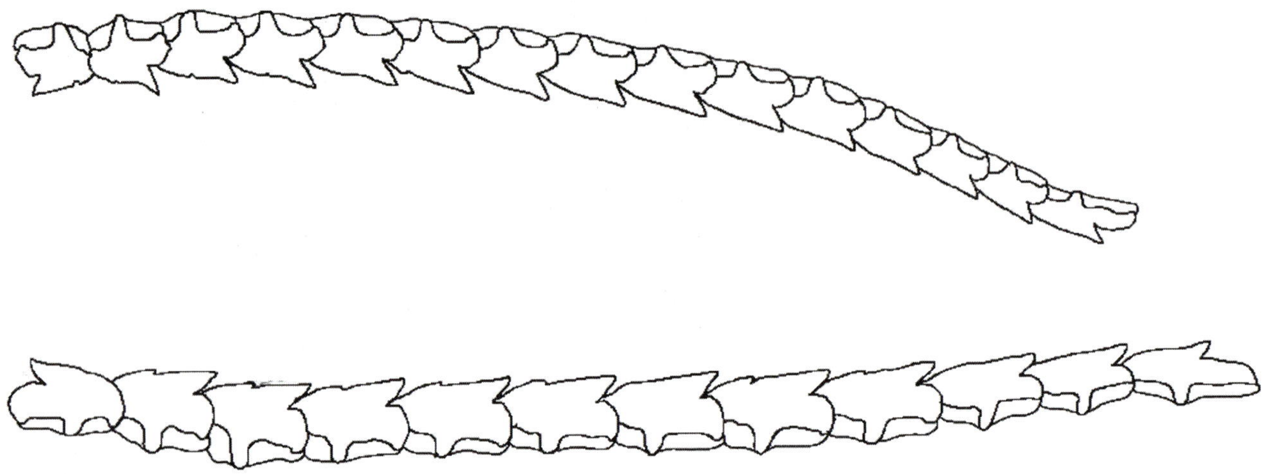

Fig. 9b. Teeth.

America (Bigelow and Schroeder 1948; Jones and Geen 1976c). Those desiring more taxonomic details should consult the account in Bigelow and Schroeder (1948).

Identification The spiny dogfish is characterized by two dorsal fins, each with a spine: a *first dorsal fin originating posterior to the free rear tips of the pectoral fins* (occasionally over), and a second dorsal fin considerably *smaller* than the first; pectoral fins with curved rear margins and rounded free rear tips; the *midpoint of the pelvic fin bases much closer to the second dorsal fin rather than the first;* and a *narrow anterior nasal flap (not bilobed),* and *no anal fin.* The teeth have strongly oblique, smooth-edged cusps with a strong notch on their outer margin, forming a continuous cutting edge. Teeth number U: 14–14, L: 11 or 12–11 or 12 and are similar in both jaws. The dermal denticles have three points with a strong central ridge and broad, winglike extensions on either side of the central ridge. The dorsal fin spines are said to have venom glands associated with them (Halstead 1970), but there are few reports on the effects of the poison. Coloration is slate gray or brownish gray above, *usually with small white spots scattered over the body,* and pale gray, grayish white, or pure white below. Albino or partial albino specimens have been reported from Norwegian waters (Frøiland 1975) and the Bay of Fundy, Nova Scotia (Coad and Gilhen 2002). I have also obtained an albino from Puget Sound.

Similar species The roughskin spiny dogfish has a second dorsal fin almost as large as the first, dorsal fins with white patches along the rear margins, and very large dermal denticles (>1 mm wide) and lacks the white spots on the body. In the Cuban dogfish and the shortspine dogfish, the origin of the first dorsal fin is above the midpoint of the inner margin of the pectoral fin; these species also lack white spots on the body.

Range The spiny dogfish inhabits the temperate and subpolar latitudes of the North Atlantic and North Pacific oceans. It has also been reported from the Mediterranean and Black seas. Along the eastern coast of North America, it ranges from Greenland to Georgia, being rare in Greenland and present only in summer (Jensen 1914). Hansen (1949) stated that the spiny dogfish was a great rarity in Greenland but that several had been caught off Sukkertoppen (southwestern Greenland). Backus (1957) reported it present at Grady Harbor (about 53°50′ N) and stated that the species was unknown from Indian Harbor on the north side of Hamilton Inlet, Labrador. Spiny dogfish are seasonally common from Newfoundland to Georgia (Dahlberg 1975; see "Migrations," below). Evermann and Bean (1898: 239) wrote in error that the spiny dogfish "is probably the most abundant shark in the Indian River [central Florida]. Said to be a permanent resident." This obvious mistake must have been caused by confusion with another species.

Off the West Coast of North America, the spiny dogfish ranges from the Bering Sea to Baja California (Miller and Lea 1972). The northernmost records are from the Chukchi Sea near Kotzebue, Alaska (Mecklenburg et al. 2002). According to Alverson and Stansby (1963), large concentrations of spiny dogfish are present in Puget Sound, the Strait of Georgia, and continental shelf waters from northern Washington to Hecate Strait. Starks (1917b) mentioned that the spiny dogfish was extremely abundant off California, occurring in great numbers in the shallow bays north of Point Conception, but southward occurring only in deep

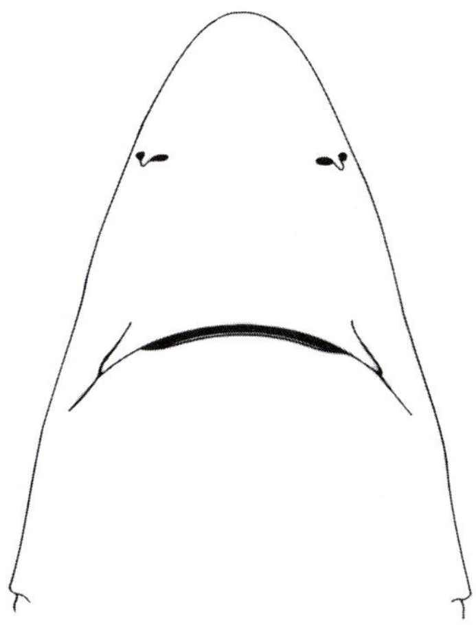

Fig. 9c. Snout.

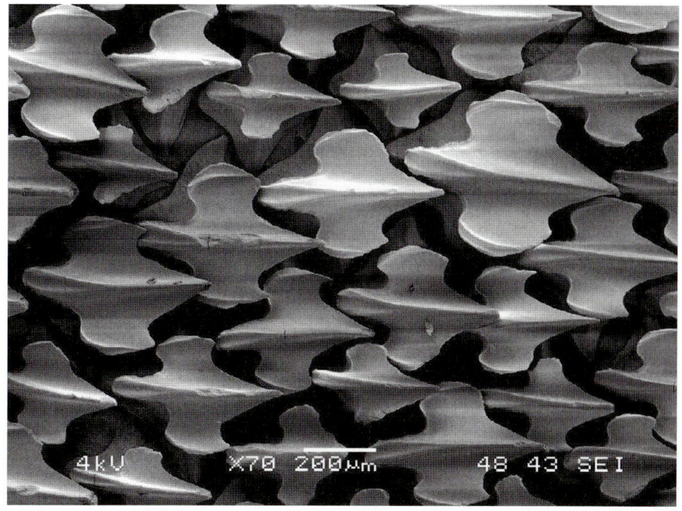

Fig. 9d. Dermal denticles.

water. The southernmost specimens that I have examined were three gravid females caught in June 2004 near Laguna San Ignacio, Baja California Sur, Mexico. These were the first spiny dogfish that local fishermen and biologists recalled encountering.

Size and weight Spiny dogfish males reach about 110 cm and 4 kg; females can reach 130 cm and 7 kg. The following sizes and weights have been recorded for Newfoundland dogfish: for males, 60 cm, 0.77 kg; 75 cm, 1.5 kg; 83 cm, 2.0 kg; for females, 75 cm, 1.7 kg; 83 cm, 2.3 kg; 96 cm, 3.8 kg. The heaviest male was 2.26 kg at 86 cm, and the heaviest female weighed 5.0 kg at 99 cm (Templeman 1944). In the western North Atlantic, the largest females can reach 122 cm and 7 kg. In the Pacific Northwest, males have been reported to reach 107 cm, while females can reach 130 cm (Ketchen 1972).

Biology The spiny dogfish is one of the most abundant sharks in the world, and it is also one of the few species that survives in captivity for long periods. Because of its abundance and small size, the spiny dogfish is the convenient "typical shark" dissected each year by thousands of biology and anatomy students in the United States and elsewhere. The spiny dogfish has also been the "laboratory white rat" of shark species, having been used in innumerable physiological experiments, because of its hardiness and prolonged survival in captivity. Consequently, it is the most studied shark in the world, and we have extensive knowledge of its anatomy and highly specific knowledge of its physiology. Therefore, the scientific literature on the spiny dogfish is the most extensive of any shark species. There must be at least 2,000 articles on the species, many highly specialized and of narrow scope. There are also some elegant monographs on the spiny dogfish, such as Scammon's (1911) *Normal Plates of the Development of Squalus acanthias,* a standard reference on the developmental stages of sharks (although the book is quite scarce today), and Norris and Hughes's (1920) "The cranial, occipital, and anterior spinal nerves of the dogfish, *Squalus acanthias.*" There is also a monograph on the development of the spiny dogfish nervous system by Neal (1898), as well as dozens of dissection manuals for the species. There are some good regional life histories, notably Hickling (1930, Ireland), Ford (1921, England), Templeman (1944, Newfoundland), Bigelow and Schroeder (1948, western North Atlantic), Alverson and Stansby (1963, northeastern Pacific), and Jensen (1966, North America). There is also a comprehensive history of the use of spiny dogfish in the Pacific Northwest by Ketchen (1986), and even a useful bibliography by Jones and Geen (1976a) with 1,009 entries. However, despite this extensive literature, some of the information on the spiny dogfish remains contradictory or ambiguous, and to this day, many aspects of its basic biology are poorly understood.

The spiny dogfish is a highly gregarious fish that forms very large, highly localized schools,[1] usually composed of hundreds or thousands of individuals of uniform size or sex. These schools exhibit north–south coastal movements and onshore–offshore migrations that are controlled primarily by temperature, although these movements are not fully

1. In earlier times dogfish were said to travel in packs or troops, while other fishes, such as herring and pilchards, were said to travel in shoals. Here I use "schools" to conform to modern usage.

understood. According to Jensen (1966), spiny dogfish in the northwest Atlantic prefer bottom water temperatures between 7.2 and 12.8°C (average 9.8°C). The average temperature at which 100 or more dogfish per haul were caught during January–June was 8.3°C; for the period July–December, it was 11.4°C. The preferred range of temperatures on the offshore wintering grounds seems to be 6–11°C (Jensen 1966). According to Bigelow and Schroeder (1948), spiny dogfish disappear from inshore waters of Cape Cod when the surface waters have warmed to about 15°C, moving northward or offshore.

The spiny dogfish is our only squaloid shark that routinely enters shallow coastal waters and is found from the surface to depths of 700 m or more. A comprehensive sampling for spiny dogfish has been carried out for many years by the NOAA's Northeast Fisheries Science Center along the northeastern states. This long-term survey revealed that adult spiny dogfish were captured in the spring at 7–439 m and bottom temperatures of 3–17°C, although most individuals were caught between 50 and 149 m, at bottom temperatures of 7–11°C (McMillan and Morse 1999).

Diet: The spiny dogfish is an opportunistic feeder, preying on numerous invertebrates and small fishes and taking whatever is locally abundant. Juveniles feed mainly on invertebrates, but with growth, the diet changes to fish (Jones and Geen 1977b). Reported invertebrate prey includes squids, shrimps, crabs, amphipods, snails, and worms (Bigelow and Schroeder 1948). When food is scarce, spiny dogfish will even eat such low-calorie prey as jellyfish and ctenophores (comb jellies). According to Bigelow and Schroeder (1948), off Woods Hole, Massachusetts, spiny dogfish are full of ctenophores when they appear in the spring.

In the western North Atlantic, the spiny dogfish diet includes numerous fishes such as capelin, herring, menhaden, mackerel, scup (*Stenotomus*), silver hake, cod, haddock, pollock, blennies, and croakers (Templeman 1944; Bigelow and Schroeder 1948). Although adult spiny dogfish are primarily fish eaters, they frequently rise in the water column, even to the surface, to gorge on krill (euphausiids) (Ketchen 1986).

Several studies have described the diet of the spiny dogfish in the northeastern Pacific. In one of the earliest reports, Hubbs (1917) examined specimens from the southern end of Monterey Bay, California. Stomach contents consisted mainly of squid (*Loligo opalescens*), Pacific sardine (*Sardinops sagax*), squid eggs, and the remains of hake (*Merluccius productus*). Interestingly, in some of Hubbs's specimens, lanternfishes (Myctophidae, probably *Stenobrachius leucopsarus*) formed the bulk of the diet. Jones and Geen (1977b) carried out the most comprehensive study, examining stomach contents of 14,796 spiny dogfish caught in a variety of gear off British Columbia. The major dietary components (by occurrence) were teleosts (55%), crustaceans (35%), and mollusks (5%). Krill were the major prey of juveniles at all times of the year and were important in the diet of adult spiny dogfish during spring and summer. Small dogfish also consumed plankton, amphipods, and small fishes such as herring and eulachon (*Thaleichthys pacificus*). This study revealed that as spiny dogfish mature the diet shifts from invertebrates to fishes, and when they reach 78 cm, herring becomes the principal prey. Adult spiny dogfish consume mainly herring, shrimp, hake, and eulachon. In a study by Jones and Geen (1977b), these prey constituted more than 42% of the diet, while salmon was less than 0.3%. Chatwin and Forrester (1953) studied the spring diet of the spiny dogfish in the Fraser River estuary and found them to be feeding primarily on eulachon that were entering the river. Of 249 dogfish stomachs examined, 229 contained food, and all stomachs containing food contained some eulachon. Bonham (1954) recorded at least 60 different food items and reported that the three more prominent items in the diet of the spiny dogfish caught off Washington State were ratfish (20%), herring (18%), and krill.

There are few records of species that prey on spiny dogfish. Ketchen (1986) mentions the occasional presence of young dogfish in the stomachs of lingcod (*Ophiodon elongatus*) and sablefish (*Anoplopoma fimbria*) and notes an unconfirmed report of an adult in the stomach of a white shark. Off Virginia, spiny dogfish have been found in the stomachs of swordfish (Scott and Tibbo 1968). Both seals and sea lions are known to prey on dogfish. Mansfield (1963) reported that gray seals (*Halichoerus grypus*) of the Canadian North Atlantic occasionally feed on dogfish, and Spalding (1964) reported the northern sea lion (*Eumetopias jubata*) also prey on dogfish off British Columbia.

Migrations, East Coast: The population of spiny dogfish along the eastern coast of North America moves north of Cape Cod in summer and southward or into deep water in winter. The schools usually winter from Chesapeake Bay southward to off the Carolinas and Georgia, remaining in the wintering grounds until early spring, when they begin their northward migration.

Spiny dogfish appear from Georges Bank to the New Jersey coast in March or April after the temperature has risen above 6°C. When the waters south of Cape Cod warm to 15°C, spiny dogfish schools continue moving north to the cooler waters of Newfoundland and the Labrador coast (Bigelow and Schroeder 1948). These sharks first appear in Newfoundland in June, off the western and southern ends of the island (Templeman 1944). The largest fish, probably pregnant females, appear first. Mature males arrive in late autumn. As the season progresses, spiny dogfish appear farther north along the coast and are off Labrador by September. In general, spiny dogfish are plentiful around Newfoundland from June through November or December, although a few can be caught there all year. Newfoundland and the southeastern Labrador coast are the northern limit of the migration. Although a few spiny dogfish reach the Greenland coast (Jensen 1914; Hansen 1949), it is possible or likely that they come from Iceland or the North Sea. The very cold Labrador Current might keep the spiny dogfish from regularly crossing the Labrador Sea to Greenland.

According to Jensen (1969), tagging evidence suggests that spiny dogfish found off New England in June and July move south in autumn into Nantucket Shoals and off Long Island, New York, and to Delaware in December. From there they may continue southward, depending on water temperature. Spiny dogfish have been reported as common in the lower Chesapeake Bay (south of the mouth of the Potomac River) from late fall to early spring (Murdy et al. 1997). Schools of spiny dogfish have been reported off South Carolina from mid-December to early March, when bottom temperatures ranged from 7.5 to 12°C (Bearden 1965a), and in St. Simons Sound, Georgia, in February when the temperature was below 15°C (Dahlberg and Heard 1969).

Spiny dogfish that are found off Newfoundland in July and August perhaps winter off Virginia and North Carolina. A specimen tagged near St. John's, Newfoundland, in July 1942 was recovered off Cape Henry, Virginia, in February 1947 (Templeman 1954). Dogfish found off New England in summer spend the winter much farther south off South Carolina and Georgia. However, these movements are still not well understood, and many aspects of the migrations cannot be explained.

A few tagged dogfish have undertaken transatlantic movements. One spiny dogfish tagged on the Grand Banks was recovered 10 years later off Iceland (Templeman 1958). Another dogfish tagged on St. Pierre Bank, Newfoundland, in June 1963, was recovered 11 years later (September 1974) in the Shetlands–North Sea area (Templeman 1976). The scarcity of transatlantic records indicates that only small numbers of dogfish undertake these movements.

Migrations, West Coast: The migratory movements of spiny dogfish along the western coast of North America are complex and poorly understood. Of 10,000 spiny dogfish tagged off Washington and British Columbia in the 1940s, 655 (6.7%) were recovered by 1953 (Holland 1957). Tag returns demonstrated a northerly migration in spring and summer and a southward, coastal migration in autumn and winter. The northward migration extended as far north as the Gulf of Alaska and the Bering Sea, with dogfish occurring there in July and August, when the surface water temperature is 7–13.3°C (Neave and Hanavan 1960). According to Holland (1957), the tagging data suggested that there are resident populations of dogfish in Puget Sound and the Strait of Georgia (British Columbia) and a migratory population off the Pacific coast ranging from (at least) Baja California north and west to Japan. Some long-distance movements have been recorded in this population. One dogfish tagged off Willapa Bay, Washington, in 1944 was recovered off the northern end of Honshu, Japan, in 1952 (Kauffman 1955). This is a great circle route of some 4,000 nautical miles, although Kauffman noted that the shark probably followed a much longer route along the coastal shelf. Another dogfish tagged off the Queen Charlotte Islands, British Columbia, traveled 1,125 miles (1,770 km) in 171 days to Santa Cruz, California (Manzer 1946).

Size at maturity

Males: Depending on population, male spiny dogfish attain maturity at 64–72 cm. According to Templeman (1944), 50% of males in the northwestern Atlantic reach maturity at 64 cm, and all specimens were mature by 70 cm. Bonham et al. (1949) reported that 50% of male dogfish in the northeastern Pacific reached maturity at about 72 cm and 1.6 kg, maturity being determined primarily by the presence of sperm. Jones and Geen (1977c) reported that, in the Strait of Georgia, 50% of male dogfish reached maturity at 78.5 cm. Their smallest mature male was 72 cm, and their largest immature male was 93 cm.

Females: The size at which female spiny dogfish reach maturity ranges from 84 to 94 cm, again depending on the population. Templeman (1944) reported that off Newfoundland, females 84.5 cm carried tiny embryos while the largest immature female was 88 cm. Ketchen (1972) reported that the size at which 50% of the females in the waters of the western coast of Canada were mature was 93.5 cm and that the minimum size at maturity was 76 cm. This study sampled more than 4,300 female dogfish, which were considered mature if they carried uterine eggs or embryos or if they had empty uteri but carried oocytes larger than 2.5 cm in diameter. Jones and Geen (1977c) also reported that 50% of the females in the Strait of Georgia reach maturity at 93.5 cm.

Reproduction Spiny dogfish development is aplacental viviparous. They mate during the winter months. Off British Columbia, Ketchen (1972) estimated the mating period as December to February, while Jones and Geen (1977c) estimated it to occur between October and January, peaking in December. Females have two well-developed and functional ovaries. Mature oocytes are released from the ovaries when they reach a diameter of 4–4.5 cm. Embryonic development begins shortly after mating. As the eggs pass through the oviducal gland, they are fertilized and acquire a thin, horny, transparent shell. This shell usually encloses two to four eggs and is referred to as a "candle." The blastodisc embryo is visible as a small (1–2 mm) orange spot on the caudal portion of the egg. The shell ruptures when the embryos are about four months old (Jones and Geen 1977c). After hatching, the embryos are free in the uterus for the rest of gestation and are nourished by the contents of the yolk sac. By the end of the first year of gestation, the embryos average 14–15 cm long (Ketchen 1972). Hisaw and Albert (1947) stated that it appeared that the developing young derive all the necessary inorganic matter from the egg and that they required only additional water from the mother. This was later demonstrated by experiments in which 15-cm embryos were able to survive outside the uterus while confined to finger bowls or plastic containers, being nourished by their yolk-sac reserves alone, for up to 20 months and into a feeding stage (Gilbert 1959; Jones et al. 1983). During the final weeks of gestation, the embryos are bathed in a uterine solution that resembles seawater in ionic composition but that is high in ammonia (Kormanik 1989), and the contents of the external yolk sac are absorbed and moved to an internal yolk sac. The contents of this internal yolk sac may reach 5 cc about 1.5 months prior to birth (Jones and Geen 1977c), serving as a source of nourishment for the neonate shark before it finds its first meal. The latest estimates of gestation indicate that it extends for about 23 months, from December of the first year to November two years later (Ketchen 1972; Jones and Geen 1977c). This is the longest gestation period known in vertebrates.

The spiny dogfish has a biennial reproductive cycle with concurrent vitellogenesis (ovarian cycle) and gestation. That is, females carry developing oocytes in the ovaries while they carry developing embryos in the uteri. Parturition occurs when the embryos measure about 28–30 cm, at which time the oocytes have reached their maximum development and measure about 4 cm in diameter. The female mates and ovulates soon after parturition and both the ovarian cycle and gestation begin anew.

Size at birth: Spiny dogfish are born at an average length of 27 cm. There are few actual reports of the size at birth for spiny dogfish. According to Bigelow and Schroeder (1948), most spiny dogfish are born at lengths of 22–33 cm, although occasionally some may be born at 16.5–17.9 cm. According to Bonham et al. (1949), birth in the northeastern Pacific occurs at an average length of 27 cm. The few large embryos that I have seen off South Carolina measured 28–30 cm in February.

Brood size: Spiny dogfish broods range from 2 to 16 embryos, with the average being six or seven embryos per female. Smaller females have the smaller broods.

Templeman (1944) examined a sample of 389 gravid females from Newfoundland. He reported an average of 3.72 large embryos, with the larger females having the larger broods. Other more recent studies have found a higher number of young per female. Off the northeastern United States, Nammack et al. (1985) found broods ranging from 2 to 15 embryos, with an average of 6.6 embryos per female, in a sample of 439 gravid females. Off British Columbia, Ketchen (1972) found that the number of embryos in a sample of 440 females from British Columbia ranged from 2 to 16, with an average of 6.6 embryos, while Jones and Geen (1977c) found that the average brood consisted of 7.3 embryos.

Nurseries The nursery areas of the spiny dogfish along the eastern coast of North America are not well defined but are believed to be in deep water in the wintering grounds. Few data are available, although I have seen a few females with 28–30-cm young ready for birth off the Carolinas in January. According to Bigelow and Schroeder (1948), females with young "nearly ready for birth" are taken off Newfoundland in October, in the Gulf of Maine in late summer, and off Virginia and North Carolina in January to February. However, it seems unlikely that the dogfish would have such a wide birthing area or season. It is more likely that the young assumed to be "nearly ready for birth" by Bigelow and Schroeder (1948) actually still had about five more months of development to complete. In most sharks, the young may appear ready for birth many months before they are born, with growth in these last few months being primarily in girth. It is worth noting that small dogfish are seldom seen along the eastern coast of North America. Hisaw and Albert (1947) speculated that birth occurred "in late fall south of the Woods Hole region." Bigelow and Schroeder (1948) also suggested that most of the young are born in late fall and winter in offshore wintering grounds, probably repeating Hisaw and Albert's speculation. Thus, from what few data are available and from my personal observations, it appears that the spiny dogfish of the East Coast give birth in offshore wintering grounds in February and March.

Off the West Coast, parturition takes place in midwater layers during the winter. According to Jones and Geen (1977c), parturition occurs between September and January, with a peak in November. Other authors have reported neonates at subsequent times. Beamish and Smith (1976) observed that neonates were found from 10 to 140 m off the bottom and 290–300 m from the surface during January in the Strait of Georgia. Ketchen (1986) reported finding newborn young while midwater trawling in March at the entrance of Juan de Fuca Strait, at depths of 145–155 m overlaying depths of 240–250 m.

Age and growth The spiny dogfish is one of the slowest growing and longest living sharks known. The rings in dogfish spines and vertebrae have both been used for aging the species. Jones and Geen (1977a) used X-ray spectrometry to count the second dorsal fin annuli (rings), reporting that the age at 50% maturity is 19 years for males and 29 years for females. Ketchen (1972, 1975) used the second dorsal fin spine to age spiny dogfish off British Columbia. He estimated that males mature at an average age of 14 years, reach a maximum length of 107 cm, and live more than 40 years. Ketchen estimated that females mature at 94 cm, at an average age of 24 years (range, 16–35), and may live for 54 years before attaining the maximum length of 130 cm. How long spiny dogfish actually live can only be guessed. Given the available data, one can speculate that they may live 60–80 years, perhaps longer.

Relation to humans Despite its abundance and palatability, the spiny dogfish was little utilized in North America during the eighteenth, nineteenth, and twentieth centuries because the general prejudice of the population against eating sharks. That prejudice was exemplified by John Lawson's 1709 account of animals of the Carolinas: "The Dog-fish are a small sort of the Shark kind; and are caught with Hook and Line, fishing for Drums. They say, they are good Meat; but we have so many other sorts of delicate Fish, that I shall hardly ever make Trial what they are" (Lawson 1709: 155). The attitude has prevailed ever since.

The spiny dogfish was considered the bane of commercial fishermen because of the great damage it caused to fisheries. Large numbers of spiny dogfish often became entangled in fishing nets, and their removal was (and is) a tedious process that resulted in great loss of fishing time and excessive wear and tear on nets. When catches were large enough (2,000–4,500 kg of spiny dogfish per tow were not unusual), nets could rupture or be lost. Hook-and-line fisheries for groundfish, such as cod and haddock, were also devastated by the sudden appearance of large schools of spiny dogfish. Once such schools appeared, fishing for other species practically came to a halt, as the spiny dogfish either drove other fishes away or beat them to the bait. The spiny dogfish was also blamed for loss of fishing stock through predation. As a result of all these factors, numerous elaborate schemes for the elimination of spiny dogfish were devised and sometimes implemented in areas affected by its presence. These schemes,[2] ranging from bounties to the blasting of schools, as well as utilization plans, have invariably failed because they were shortsighted or naive, or did not take into account the prejudices of the population concerning sharks as food.

Along the Pacific coast of North America, the dogfish was first used by the natives, who extracted clear oil from the liver by heat and pressure (Lord 1866). The oil was used for tanning animal skins, finishing wood, and preparing paints. Early European settlers of the Pacific Northwest used the liver oil in lamps and as a lubricant. Barraclough (1953) reported that large quantities of dogfish oil were used for lubrication and lighting in sawmills in British Columbia, and Ketchen (1986) stated that, in British Columbia, from 1870 to 1916 there was extensive use of dogfish oil for lubrication and lighting purposes. Clemens and Wilby (1946) mentioned that, prior to 1880 there was a reduction plant in Skidegate Inlet, British Columbia, for the extraction of oil from dogfish livers. Fishmeal made from dogfish was also a particularly valuable fertilizer due to its high nitrogen content (Clemens and Wilby 1946).

A fishery for spiny dogfish existed along the Pacific coast of North America from 1937 to 1950, when the species was valued for the vitamin A extracted from its liver. From 1941 until 1950 the fishery was intensive because the interruption of European supplies of vitamin A due to World War II and its effects. Some 13.5 million pounds of dogfish livers were landed during 1944, the peak year of the fishery (Holland 1957). After 1950 other sources of vitamin A sharply reduced demand for dogfish livers, and by 1955 the fishery had ceased for lack of markets (Holland 1957). Some 20 years later, in the mid-1970s, small food fisheries reappeared in the Pacific Northwest (Ketchen 1986).

In 1969, A. Jensen wrote that the spiny dogfish in the northwestern Atlantic had little economic value and that commercial fishermen considered it a nuisance. This situation continued until the mid-1990s, when other fish stocks collapsed, and there was a revival of the spiny dogfish trade in the Pacific Northwest. Spiny dogfish fillets were exported to Europe for "fish and chips" in England and *schillerlocken* jerky in Germany, or to Japan for the production of *kamaboko* fish cakes (Vovcsko 1995). From 1987 to 1993, commercial landings of spiny dogfish in the United States increased fivefold to more than 22,000 metric tons. Minimum estimates for 1993 suggested an additional 25,000 metric

2. The Massachusetts Commissioners on Fisheries and Game (1906) explain the effects of schools of dogfish on fishing communities, and their report makes very interesting reading. Some harebrained schemes designed to increase the killing of dogfish are proposed in this publication; a partial list is quoted below:

1. "Liberate alive some hundreds of dogfish having securely fastened outside their bodies (by means of hooks, wires, etc) glittering and gaudy streamers or jingling chains or bells, calculated to terrify and frighten the schools of dogfish, on the old principle of setting at liberty a rat with a bell hung around its neck.
2. "Inoculate a number of dogfish with some fatal or contagious disease, thus securing the infection and death of all the schools of dogfish that may hover near, on the principle adopted in reducing the pest of rabbits in Australia some years ago.
3. "Dynamite the great schools of dogfish when they appear.
4. "Employ the government cruisers and their men in capturing these pests, or let the government employ special vessels for the purpose until the plague is reduced.
5. "Pay a bounty of 1 cent for every five tails of dogfish ($2 per 1000) brought to a fishing officer."

tons of dogfish were discarded, of which 14,000 metric tons were killed (Rago et al. 1998). More than 95% of landings consist of mature females larger than 80 cm. The landings of dogfish in the western North Atlantic have continued to increase.

In 1986, K. S. Ketchen wrote, "It would not be a gross exaggeration to say that to speak seriously of the need for conservation of spiny dogfish is to incur suspicions of dementia" (Ketchen 1986: 1). This old attitude that the dogfish is a nuisance fish still prevails in fishery management circles and has hampered the rational management and conservation of the species, even though it is obvious that a shark with the life history characteristics of the spiny dogfish cannot support the current intensive levels of fishing. So, the destruction of spiny dogfish stocks continues into the twenty-first century with little concern from scientists or the public.

CUBAN DOGFISH

Squalus cubensis Howell Rivero, 1936

Fig. 10a. *Squalus cubensis*, Cuban dogfish, female, 64 cm, 1.2 kg; from northern Bimini, Bahamas (Ron Schatman, collector).

Common name Cuban dogfish, an allusion to Cuba, where the species was first described.

Spanish name *Galludo* or *galludo cubano* (Cuba); *tiburón bagre, cazón espinoso* (Mexico). The A.F.S. name is *cazón aguijón cubano,* a name that Spanish speakers would never use.

Scientific name *Squalus cubensis* Howell Rivero, 1936, in *Proceedings of the Boston Society of Natural History* 41(4): 45–47, plates 10 and 11. *Squalus:* from the Latin *squalus,* some sea-fish, later a shark; *cubensis:* latinized form, "from Cuba."

Synonyms None. However, any old reports of the spiny dogfish, *Squalus acanthias,* from Cuba or the Caribbean Sea, are undoubtedly *Squalus cubensis.*

Identification The Cuban dogfish is characterized by two dorsal fins, each with a spine; a first dorsal fin spine as long as the base of the fin and originating over the midpoint of the inner margin of the pectoral fin; a second dorsal fin smaller than the first; and pectoral fins with *curved posterior margins* and *pointed free rear tips.* The teeth have strongly oblique, smooth-edged cusps with a strong notch on their outer margins forming a nearly continuous cutting edge. Teeth number U: 13–13, L: 13–13 and are similar in both jaws. The dermal denticles are irregularly shaped, with a strong central ridge that divides anteriorly into two to three ridges and a broad winglike extension on either side of the central ridge (I think of them as an arrowhead with a long, wide base). Coloration is dark gray above and pale gray to white below. The apexes of both dorsal fins are black; the other fins are white edged. The color of the iris is bright gold.

Similar species The shortspine dogfish has pectoral fins with rounded free rear tips, its caudal fin is entirely grey (lacks the white edge of the Cuban dogfish), and its dermal denticles are three-pointed. The spiny dogfish and roughskin dogfish have the first dorsal fin originating above or posterior to the free rear tips of the pectoral fin.

Range The Cuban dogfish inhabits the waters of the western North Atlantic from North Carolina (Schwartz 2003) to Florida, the Gulf of Mexico (Springer and Bullis 1956), and the Caribbean (Bigelow and Schroeder 1948). It has also been reported from the western Atlantic south to Brazil. It is common in deep water in the Gulf of Mexico and off eastern Florida and the Bahamas.

Size and weight The Cuban dogfish reaches about 65 cm in length. I have recorded the following weights for males: 59 cm, 0.77 kg; 60 cm, 0.83 kg; 61 cm, 0.89 kg; and 63 cm, 1.0 kg.

Biology This is a little-known, bottom-dwelling species that inhabits deep warm-temperate and tropical waters and forms dense schools. Most catches are made in 110–500 m. It forms large schools of same-sex and size individuals. Cortés (1999) reported that it feeds on benthic invertebrates, squid, cuttlefish, and bony fishes.

Size at maturity There are few published data on the size at maturity of the Cuban dogfish.

Males: Males probably mature at about 45 cm.

Females: The Cuban dogfish females mature at about 49–50 cm. Cruz Pacheco (1997) reported that the smallest

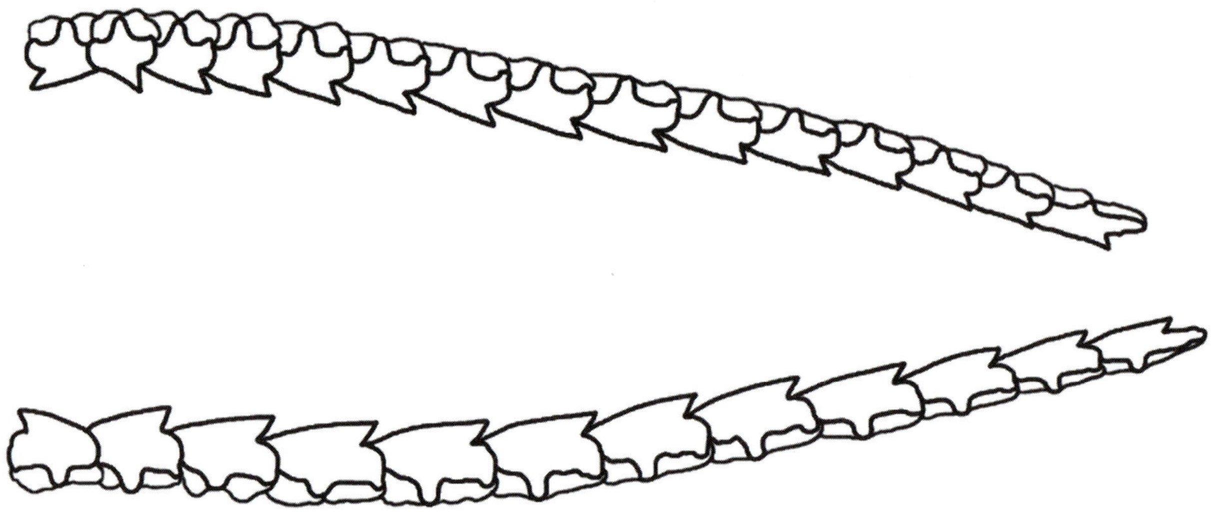

Fig. 10b. Teeth.

of 25 gravid females caught off Veracruz, Mexico, measured 49 cm.

Reproduction The Cuban dogfish is aplacental viviparous. There are few published data on its reproductive processes. Cruz Pacheco (1997) reported that the largest embryos (22 cm) were found off Veracruz, Mexico, in February, suggesting a birth season in February and March, just like the spiny dogfish.

Size at birth: The young are apparently born at about 25–27 cm.

Brood size: Cruz Pacheco (1997) reported broods of 1–26.

Nurseries Unknown.

Age and growth The species has not been aged.

Relation to humans None.

Fig. 10c. Snout.

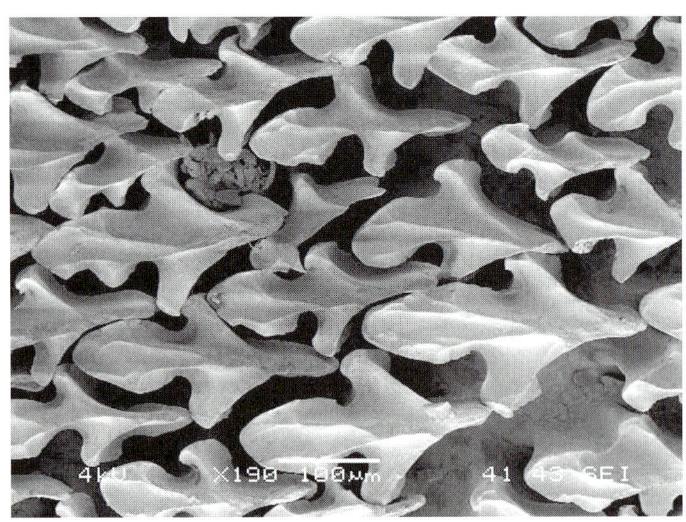

Fig. 10d. Dermal denticles.

SHORTSPINE DOGFISH

Squalus mitsukurii Jordan and Snyder, 1903

Fig. 11a. *Squalus mitsukurii*, shortspine dogfish, female, 74.7 cm, 1.6 kg; off Ft. Pierce, Florida (Scott Bachman, collector).

Common name Shortspine dogfish, an allusion to the dorsal fin spines that are somewhat shorter than those of some other dogfishes (an F.A.O. name).

Spanish name The species lacks a vernacular name in the Spanish Americas. The A.F.S. has given it the cumbersome Spanish name of *cazón aguijón galludo,* which no Spanish speaker is likely to use. The F.A.O. name, *galludo espinilla,* is a slightly better rendition of the English name.

Scientific name *Squalus mitsukurii* Jordan and Snyder, 1903, in *Proceedings of the United States National Museum* 26 (No. 1324): 629–630, but not fig. 3, which appears to be *Squalus acanthias.* The name *S. mitsukurii* had appeared earlier in a 1901 checklist by the same authors, but without a description (a *nomen nudum*). Later, the species was described in Jordan and Fowler (1903). *Squalus:* from the Latin *squalus,* some sea-fish, later a shark; *mitsukurii:* named after Kakichi Mitsukuri (1858–1909), professor of zoology, dean of the Imperial University of Tokyo and, in his day, the foremost zoologist in Japan. David S. Jordan admired him and considered him a valued friend, naming two species of sharks after him.

Synonyms *Squalus fernandinus* Molina, 1782; *Squalus blainville* (Risso, 1826). This last name, as *Squalus blainvillei,* was in wide use until recently for the southeastern U.S. population. The name was changed after the most recent revision by Chen et al. (1979) of this species and related dogfishes.

Identification The shortspine dogfish is characterized by two dorsal fins, each with a spine, a first dorsal fin originating over the *midpoint* of the inner margin of the pectoral fin, a second dorsal fin smaller than the first, and pectoral fins with slightly curved rear margins and *rounded free rear tips.* The teeth have strongly oblique, smooth-edged cusps with a strong notch on their outer sides, forming a nearly continuous cutting edge. Teeth number U: 13–1–13, L: 11–1–11 and are similar in both jaws. The dermal denticles are small and *broad,* with *three points,* and have a strong central ridge. Coloration is gray-brown above and white below. The tips of the dorsal fins may be black or dusky. The posterior edges of the fins are white edged in juveniles, but adults lack the white edge.

Similar species The Cuban dogfish has pectoral fins with curved posterior margins and pointed free rear tips, a caudal fin with a white rear margin, and dermal denticles with a single narrow point. The spiny dogfish and the roughskin spiny dogfish have the origin of the first dorsal fin posterior to the free rear tip of the pectoral fins.

Range If reports are to be believed, and if the identification of the species is correct, the shortspine dogfish has an almost cosmopolitan distribution. Compagno (1984a) lists reports for the western North Pacific, western South Pacific, central Pacific (including the Hawaiian Islands), Indian Ocean, and both sides of the Atlantic. Whether the species is this widely distributed, or whether several undescribed species are being confused, is unknown. In any case, in North America the shortspine dogfish ranges from North Carolina to the Gulf of Mexico, being abundant throughout the range.

Size and weight Chen et al. (1979) listed a maximum length of 110 cm for their specimens. A 77-cm female I examined weighed 1.7 kg.

Biology The shortspine dogfish inhabits the continental slopes of warm-temperate waters at depths of 180–400 m.

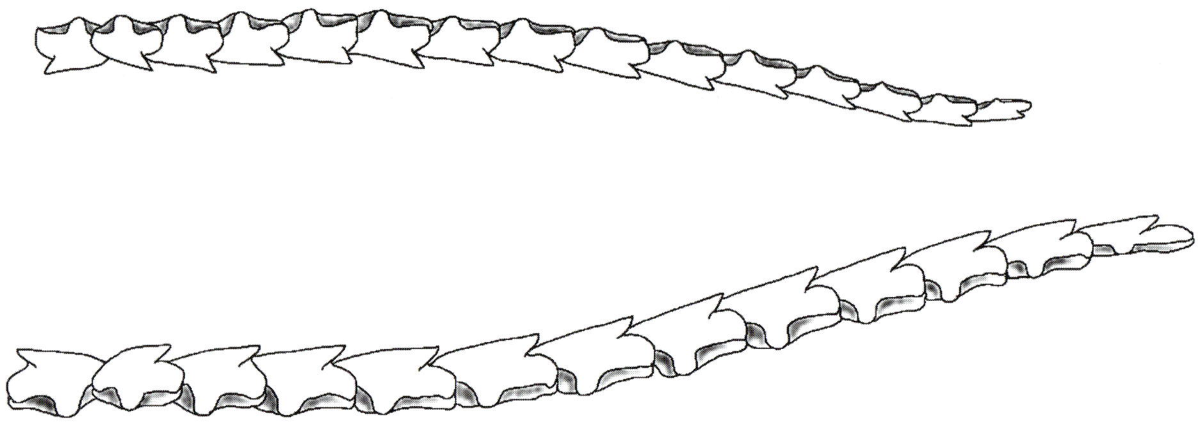

Fig. 11b. Teeth.

Although frequently caught in tilefish operations at depths of 240–250 off the Carolinas, its habits are poorly known. I have taken it in slightly deeper waters (300–370 m) in the Gulf of Mexico.

Size at maturity The American population has not been studied. Bass et al. (1976) reported that South African females mature at about 70 cm and males at about 58–60 cm.

Reproduction The shortspine dogfish is aplacental viviparous, like other dogfishes. It appears to have a two-year gestation period, like its congener, the spiny dogfish. Embryos of 15.5–16.8 cm that I examined in January had yolk sacs 57 mm long, suggesting that they were at midterm and would not be born until the next year. Little else is known about its reproductive processes.

Size at birth: There are no published data on the birth size of the American population. The young are probably born at about 24 cm. Bass et al. (1976) reported that the species (as *S. blainvillei*) was born at 22–26 cm off South Africa.

Brood size: The only gravid female I have examined carried eight young.

Nurseries Unknown.

Age and growth The shortspine dogfish has not been aged.

Relation to humans None. It is often taken as bycatch in tilefish operations in the Carolinas and usually discarded at sea.

Fig. 11c. Snout.

Fig. 11d. Dermal denticles.

FAMILY CENTROPHORIDAE

The Gulper Sharks and Birdbeak Dogfishes

The family Centrophoridae was redefined by Shirai (1992) based on similarities of teeth and skeletal characteristics and includes the gulper sharks and birdbeak dogfishes. These are all deep-dwelling squaloid sharks characterized by two dorsal fins each with a spine, the second dorsal fin originating just behind the base of the pelvic fin or, at least, only partly over it; the free rear tip of the pectoral fin often prolonged into a long, tapering point; teeth dissimilar in both jaws; upper teeth smaller and narrower than lower teeth, with erect to oblique cusps linearly arranged, their bases not imbricated or overlapping; and lower teeth with large, bladelike, oblique cusps with imbricated or overlapping bases. Like all the squaloid sharks, they lack an anal fin.

The family Centrophoridae consists of two genera, *Centrophorus,* the gulper sharks, and *Deania,* the birdbeak dogfishes. The gulpers take their name from their ability to swallow large prey. They have numerous ridges on the throat that presumably allow it to distend for swallowing large prey. Gulpers have snouts shorter than the distance from the mouth to the origin of the pectoral fin and characteristic ridges across the throat, and most species have dermal denticles with low, flat, ridged crowns without pedicels.

Centrophorus is one of the most confusing genera of squaloid sharks, and it is in need of revision. Many of the species were described long ago, using characteristics that are not specific or that change ontogenetically, and there are no extant type specimens for many of these species. There are few modern specimens in museums due to their deep-water habitat and size, and it is almost certain that there are undescribed species of gulpers. Thus, there is considerable confusion about the identity of many gulper sharks. Compagno et al. (2005) included 10 species of *Centrophorus* in their field guide to the sharks of the world. Last et al. (2008) described two additional species of the genus. Until recently it was believed that there were only two species of *Centrophorus* along the eastern United States. However, my survey of this group demonstrates that there are at least seven species in eastern North America, and all are poorly known. *The keys that follow have been designed for identifying adult specimens.* Juveniles may be difficult to identify because their body proportions differ from those of adults due to allometric growth.

Sharks of the genus *Deania* have enormously elongated snouts (the snout length being greater than the distance from the mouth to the origin of the pectoral fin) and curious dermal denticles resembling minute pitchforks (they have high pedicels, cusps with three ridges and three points, and stellate bases). Like *Centrophorus, Deania* are deep-dwelling sharks, and, with the exception of the birdbeak dogfish (*D. calcea*), they are seldom encountered. Four nominal species are presently recognized, and at least two are found in North America.

KEY TO THE CENTROPHORIDAE

1a. Snout length shorter than distance from the mouth to the origin of pectoral fins, dermal denticles not pitchfork shaped 2

1b. Snout length longer than distance from the mouth to the origin of pectoral fins, dermal denticles along the back with long pedicels and pitchfork-shaped crowns 8

2a. Free rear tips of pectoral fins short, not reaching past the origin of the first dorsal fin; dermal denticles with very large (1–2 mm) leaf-shaped crowns on tall pedicels (body coloration uniformly dark chocolate brown or black) .... **Leafscale gulper shark,** *Centrophorus squamosus* (p. 76).

2b. Free rear tips of pectoral fins long, reaching to the origin of the first dorsal fin (and past the origin of the first dorsal fin spine in most cases) 3

3a. Dorsal fin base more than half the interdorsal space (from the first dorsal fin insertion to the origin of the second dorsal fin); dorsal fin spines short, usually about 1 cm long in adults; free rear tip of pectoral fin not reaching the origin of the first dorsal fin spine (adults only) (pectoral fin anterior margin longer than distance from the tip of the snout to the spiracle, corners of mouth and labial furrows dark, a large species reaching 180 cm) .... **Taiwan gulper shark,** *Centrophorus niaukang* (p. 73).

3b. Dorsal fin base equal to or less than half the interdorsal space (from the first dorsal fin insertion to the origin of the second dorsal fin), free rear tip of pectoral fin reaching past the origin of the first dorsal fin spine 4

4a. Pectoral fin anterior margin shorter than distance from the tip of the snout to the spiracle 5

4b. Pectoral fin anterior margin equal to or longer than distance from the tip of the snout to the spiracle 6

5a. Snout longer than width of the mouth, corners of mouth and labial furrows tan or light brown, not white (a small species perhaps reaching 85 cm) .... **Minigulper shark,** *Centrophorus* sp. A (p. 84).

5b. Snout shorter than width of mouth, corners of mouth and labial furrows white (a small species reaching about 100 cm) .... **Little gulper shark,** *Centrophorus uyato* (p. 81).

6a. Interdorsal distance shorter than head length (from tip of the snout to the origin of the pectoral fin), caudal fin with a wide black rear margin (a small species reaching a length of about 95 cm) .... **Blackfin gulper shark,** *Centrophorus isodon* (p. 71).

6b. Interdorsal distance greater than head length (from tip of the snout to the origin of the pectoral fin), snout length less than half of the head length 7

7a. Length of anterior margin of pectoral fin equal to distance from tip of snout to the spiracle (teeth similar in both jaws, with small pointed cusps and coarse serrations on both sides) .... **Mosaic gulper shark,** *Centrophorus tesselatus* (p. 79).

7b. Length of anterior margin of pectoral fin longer than distance from the tip of the snout to the spiracle .... **Slender gulper shark,** *Centrophorus* sp. B (p. 85).

8a. A noticeable keel on the ventral surface between the pelvic and caudal fins (a slender and small shark less than 70 cm in length).. **Arrowhead dogfish,** *Deania profundorum* (p. 89).

8b. Lacking a noticeable keel on the ventral surface between the pelvic and caudal fins (dermal denticles along the back tall and large, their crowns more than 0.5 mm long) .... **Birdbeak dogfish,** *Deania calcea* (p. 86).

BLACKFIN GULPER SHARK

Centrophorus isodon Chu, Meng, and Liu, 1981

Fig. 12a. *Centrophorus isodon*, blacktail gulper shark, female, 87 cm, 3.6 kg; off Cat Cay, Bahamas (Ron Schatman, collector).

This is one of the species of gulpers encountered in the making of this book. The identification of the species is tentative; it is based on the illustration and description published by Compagno et al. (2005). The species was originally described as *Pseudocentrophorus isodon* Chu, Meng, and Liu, 1981 in *Acta Zootaxonomica Sinica* (1981, 6(1): 100–103). The holotype is from the South China Sea. The species is included here based on two specimens taken by Ron Schatman off Cat Island, Bahamas, in February 2002. The two sharks, an 89-cm female (illustrated) and an 87-cm male, were taken on the bottom at 652 m. Whether this Atlantic species is the same as the similar Pacific gulper remains to be determined.

This is a small gulper shark, perhaps reaching 85 cm. It has a long free rear tip of the pectoral fin that reaches to about the midpoint of the first dorsal fin base. Its interdorsal space is much shorter than the head length. One of its distinguishing characteristics is the black coloration of its dorsal fin tips and tail.

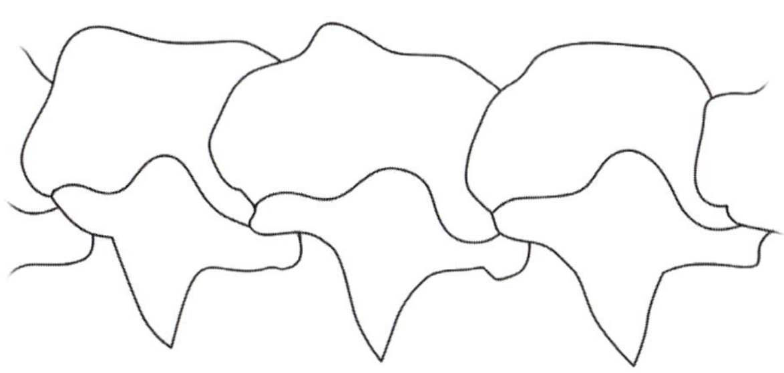

Fig. 12b. Teeth (upper).

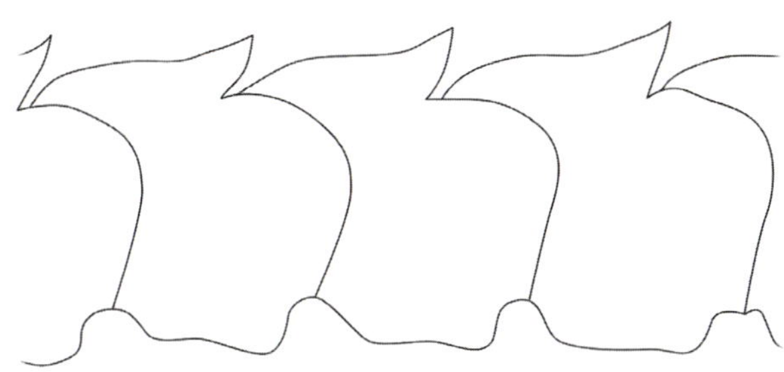

Fig. 12c. Teeth (lower).

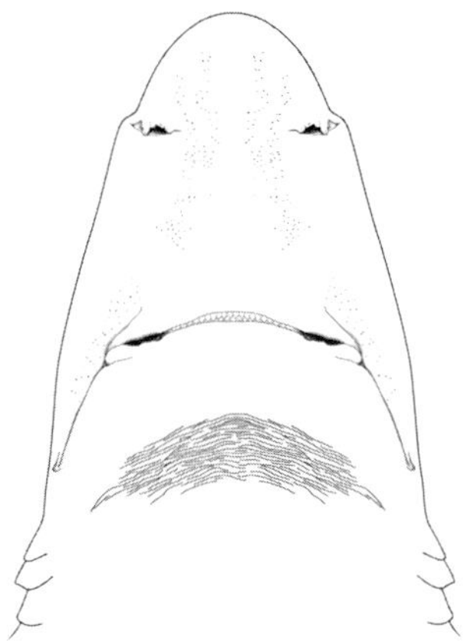

Fig. 12d. Snout.

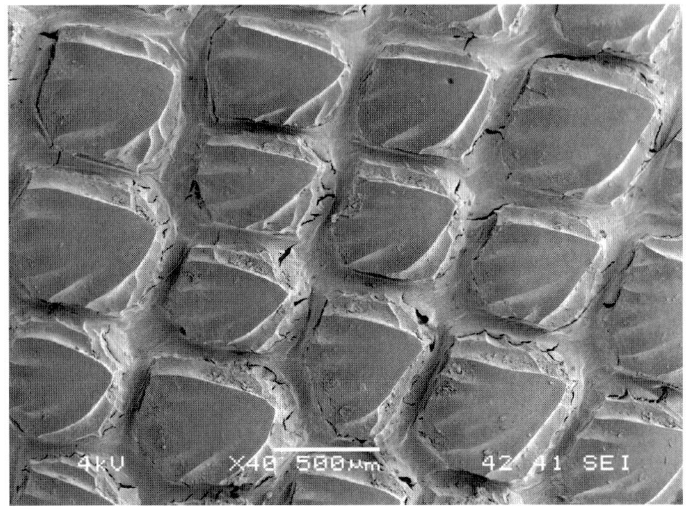

Fig. 12e. Dermal denticles.

TAIWAN GULPER SHARK

Centrophorus niaukang Teng, 1959

Fig. 13a. *Centrophorus naiukang*, Taiwan gulper shark, gravid female, 176 cm, 35.5 kg; from the Gulf of Mexico (Collected by the author).

Common name Taiwan gulper shark.

Spanish name *Tiburón quelvacho* (A.F.S.).

Scientific name *Centrophorus niaukang* Teng, 1959 [b], in *Taiwan Fisheries Research Institute (Keelung), Report-Laboratory of Fishery Biology* 9: 1–6, plate 1. *Centrophorus:* from the Greek *kemtron,* spine, and *phorus,* from *phoreus,* bearer, hence "spine-bearer," an allusion to the spines of the dorsal fins; *niaukang:* from *niaukang sua,* a local name for the species in the Tungchiang area (Taiwan) meaning "the male catshark."

The large gulper shark of the western North Atlantic has been considered to be *Centrophorus granulosus* Bloch and Schneider, 1801 until recently (Bigelow and Schroeder 1957). The original description was based on a "five-foot" stuffed specimen of unknown origin that, according to Krefft and Tortonese (1973), has long since been lost. This description, as *Squalus granulosus* by Bloch and Schneider (1801) in Latin, is so vague that, according to Bigelow and Schroeder (1957: 73–74):

> Obviously [it] is not enough to place Bloch and Schneider's five-foot specimen, whether as to genus among the Squalinae, or as to species, for it does not mention either the shape of the pectoral fins or the nature of the teeth. It is therefore fortunate for the sake of nomenclatural stability that while Müller and Henle's (1839: 89, plate 33) description and illustrations of their *granulosus* were based on an alcoholic specimen from near Sicily, they did have for comparison, a dried specimen in Berlin from Bloch's collection, probably the one on which Bloch and Schneider's 1801 account was based, for they mention only one. Müller and Henle's identification of their "*granulosus*" with the "*granulosus*" of Bloch and Schneider 1801 is therefore to be accepted—unless re-examination of the type specimen (if still in existence) should eventually prove otherwise.

Unfortunately, Müller and Henle's 1839 description shows a long-spined gulper with a short-based first dorsal fin and pectoral fin free rear tip reaching well behind the level of the first dorsal fin spine. Thus, their specimen cannot be one of the larger gulpers of eastern North America. The large gulpers have long-based first dorsal fins with very small fin spines and the pectoral fin free rear tip that does not reach past the level of the first dorsal fin spine. The next available name for a large gulper is *Centrophorus niaukang,* Teng 1959. Most writers have accepted this name for the largest member of the genus (e.g., Muñoz-Chapuli and Ramos 1989; Moore et al. 2003; Compagno et al. 2005). I am provisionally using this name until a comparison between the holotype of *C. niaukang* and the American gulpers is performed.

Large gulpers (adults > 150 cm) believed to be *Centrophorus niaukang* are taken routinely off the eastern United States using deep trawls or longlines. Large adult female gulpers are caught off New England at depths of 150–660 m (Moore et al. 2003). Similarly, adult female gulpers are taken in the Gulf of Mexico at depths of about 450 m. There are only slight proportional, color, and denticle differences between the New England specimens and those from the Gulf of Mexico. For example, northern specimens are usually of a dark reddish brown, while Gulf of Mexico specimens are of a duller grayish brown color. It is likely that both northern and southern forms belong to the same species, but this needs to be verified.

Synonyms *Centrophorus granulosus* Bloch and Schneider, 1801?

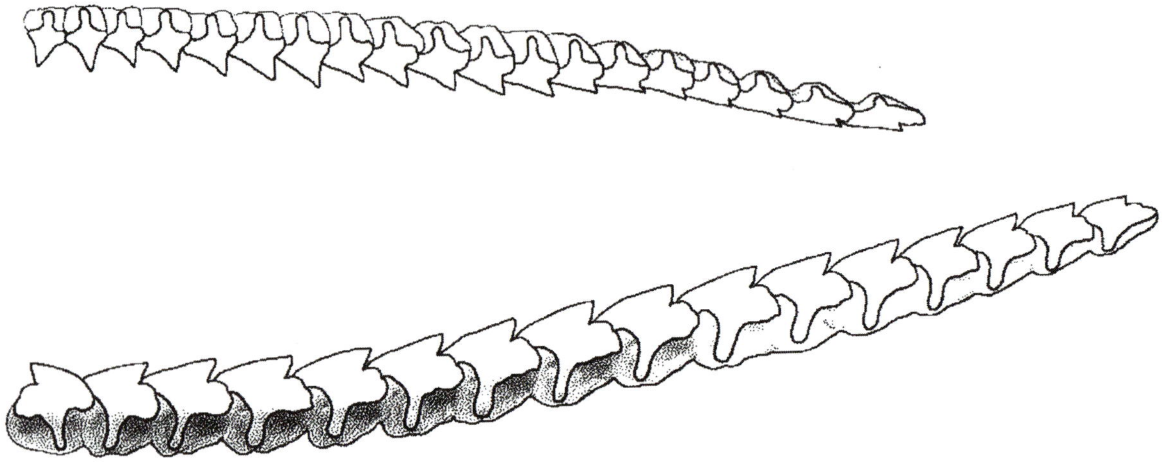

Fig. 13b. Teeth.

Identification The Taiwan gulper is characterized by two dorsal fins with short spines (usually < 1 cm long in adult specimens), *a first dorsal fin base that is 14–17% of the total length,* a second dorsal fin higher than the first, and *pectoral fins with long and pointed free rear tips* that do not reach past the level of the spine of the first dorsal fin. The corners of the mouth are dark. *There are numerous creases on the ventral side of the throat.* It lacks an anal fin. The upper teeth are triangular, pointed, smooth edged, *symmetrical all along the central part of the jaw,* and slightly oblique at the corners. The lower teeth have strongly oblique cusps with microscopically fine serrations, a notch on their outer margins, and overlapping bases. Teeth number U: 33 to 40, L: 14 or 15–1–14 or 15. The dermal denticles are low, squarish, and widely separated; their anterior edges have five to seven weak ridges converging rearward. Coloration is light grayish brown to cinnamon brown above, lighter below. The young have a bluish gill area, a black-tipped second dorsal fin, and a black tail.

Similar species The other gulpers in the area have the free rear tip of the pectoral fin reaching past the level of the spine of the first dorsal fin, higher first dorsal fins (6–7% of total length), and shorter first dorsal fin bases (12–14% of total length). Other squaloid sharks in the area lack the pectoral fins with long, pointed free rear tips.

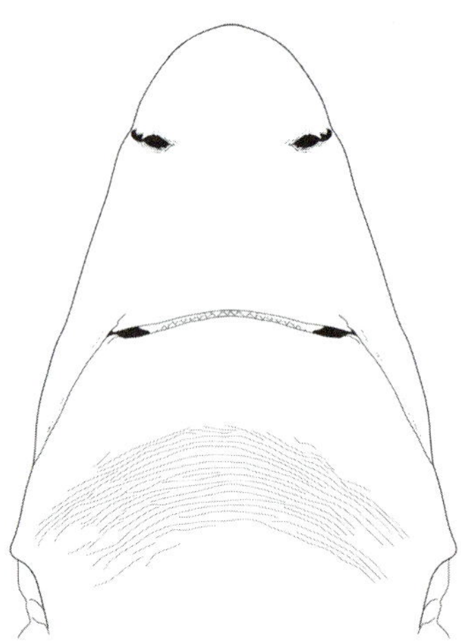

Fig 13c. Snout.

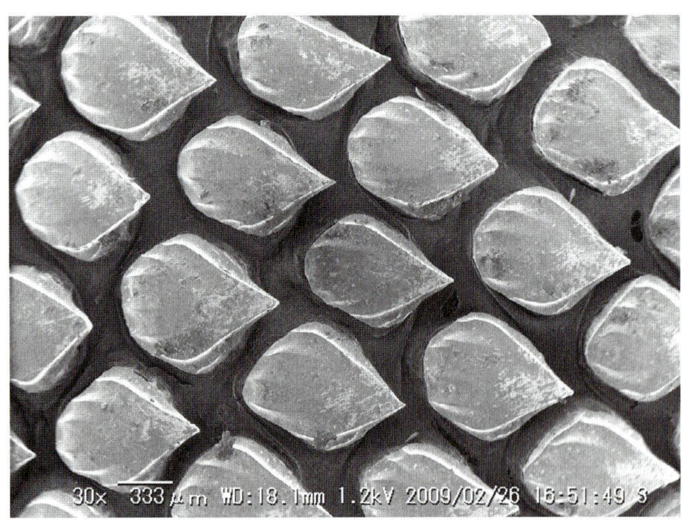

Fig 13d. Dermal denticles.

Fig. 13e. *Centrophorus naiukang*, Taiwan gulper shark, gravid female, 153 cm, 25.8 kg; from the western North Atlantic at 36°11 N, 74° 46′ W (John Galbraith, collector).

Range The Taiwan gulper is widely distributed in deep waters around the world, including the North Atlantic. It has been reported in the eastern Atlantic from Portugal to North Africa and the Mediterranean Sea. In North America it ranges from off New England to Florida, the Bahamas, and the Gulf of Mexico, although it is rarely seen because of its habitat and the absence of deep-water fisheries. It is seasonally common in the deep waters throughout its North American range.

Size and weight A 122-cm adult male from the Bahamas weighed 9.5 kg. A 111-cm immature female from the Gulf of Mexico weighed 7.5 kg. A 148-cm adult female from New England weighed 23.5 kg, while a 161-cm gravid female from the Gulf of Mexico weighed 28 kg. This last specimen is the largest I have encountered.

Biology The Taiwan gulper inhabits continental slopes at 150–600 m, being most abundant at 400–500 m and bottom temperatures of 9–12°C. Its biology is poorly known because of confusion with other species of gulper sharks.

Nearly all the specimens caught off the eastern United States that I have examined, or know of, have been females, indicating that the species segregates by sex and size, like so many other species. Large schools of female gulpers appear off New England every March at depths of 440 m (J. Galbraith, pers. comm., March 2002). Moore et al. (2003) wrote that on two occasions more than 20 large females were caught in single 30-minute trawls on the continental shelf at 150–200 m. All the specimens that I have caught or examined from the Gulf of Mexico have also been females, all caught at depths of 446–460 m. The single adult male that I have examined was taken at 836 m off Cat Cay, Bahamas.

Little is known about the Taiwan gulper shark's diet. Most of the specimens that I have examined that had food in their stomachs contained squids of 30–45 cm.

Size at maturity Unknown. A 122-cm male was mature. The only gravid female that I have encountered measured 161 cm and weighed 28.8 kg.

Reproduction Like all squaloids, the Taiwan gulper is an aplacental viviparous species. Females carry developing oocytes and developing young at the same time; that is, vitellogenesis and gestation are concurrent. Given that females taken at the same time carry either developing young or candled eggs (Moore et al. 2003), it is obvious that the Taiwan gulper shark has a gestation period longer than a year. The Taiwan gulper shark produces proportionately large oocytes that can reach more than 86 mm in diameter before being ovulated. Based on concurrent oogenesis and gestation, gestation longer than a year, and large size of the oocytes, it is likely that the Taiwan gulper has a biennial reproductive cycle similar to that of the spiny dogfish.

Size at birth: Young are born at about 40–42 cm and 0.3–0.4 kg. Based on the examination of a single aborting female, the newborn young apparently measure about 26% of adult size.

Brood size: The usual brood size appears to be four young, based on very limited data. A 153-cm female caught off New England in March carried four oocytes 75–80 mm in diameter, apparently ready for ovulation. A gravid female I examined aborted three young 40–41.5 cm on deck and appeared to have aborted a fourth in the water.

Nurseries Unknown.

Age and growth Unknown.

Relation to humans The Taiwan gulper has no economic importance in North America. In Portugal the species is often marketed for its liver oil.

LEAFSCALE GULPER SHARK

Centrophorus squamosus (Bonaterre, 1788)

Fig. 14a. *Centrophorus squamosus*, leafscale gulper shark, ovigerous female, 136 cm, 15.9 kg; from 260 km NE of the Bahamas (Reinaldo "Rey" Pérez, collector). The gravid specimen appears somewhat laterally flattened in this illustration.

Common name Leafscale gulper, allusions to the large and conspicuous leaf-shaped skin denticles and to its ability to swallow large prey due to its expanding throat.

Spanish name There are no names in use for the species in he Spanish Caribbean. The A.F.S. has not designated a name for it. In Spain it is known as *quelvacho negro.*

Scientific name *Centrophorus squamosus* (Bonaterre, 1788), originally described as *Squalus squamosus* in *Tableau Encyclopédique et Méthodique des Trois Règnes de la Nature, Ichthyologie:* 12. *Centrophorus: Centro,* from the Greek *kemtron,* spine, and *phorus,* from the Greek *phoreus,* bearer, hence "spine-bearer," an allusion to the spines on the dorsal fins; *squamosus:* from the Latin *squama,* scale, + *osus,* a termination suggesting full of, thus "covered with scales."

Synonyms *Centrophorus foliaceus* Günther, 1877.

Identification This squaloid shark has two dorsal fins each with a spine, a snout as long as the mouth width, a pectoral fin free rear tip that tapers to a short point and does not reach the level of the first dorsal fin spine, and numerous creases across the throat and lacks an anal fin. The best diagnostic character is its *exceptionally large, broad, leaflike dermal denticles,* which may be 1.3–1.5 mm wide and 2–2.5 mm long and are mounted on long pedicels. These denticles make the shark look as if covered in a thin chain mail and give it a peculiar feel to the touch. Color is chocolate brown to blackish brown with a slight purplish tinge above, and slightly lighter below, fading to black after death.

Similar species The dark, chocolate brown coloration, the short free rear tip of the pectoral fin that does not reach

Fig.14b. Teeth (upper).

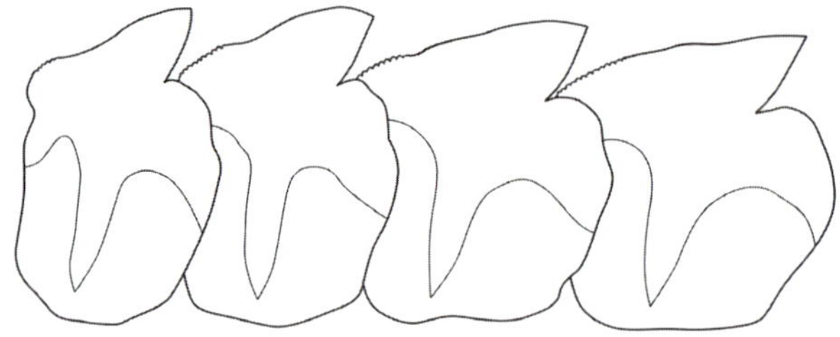

Fig.14c. Teeth (lower).

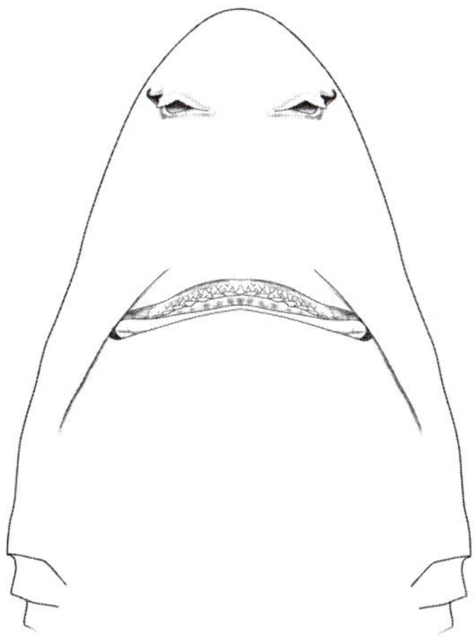

Fig. 14d. Snout.

Fig.14e. Dermal denticles.

the level of the first dorsal fin spine, and the very large denticles on pedicels distinguish it from the other gulper sharks.

Range The leafscale gulper is probably cosmopolitan in deep waters, having been reported from many localities in the Pacific and Indian oceans (Garrick 1959, Penrith 1969). In the eastern Atlantic, it has been reported from numerous locations from Scandinavia to Senegal (Cadenat and Blache 1981). There are no published reports of its presence in the western Atlantic, although it is present there. I have examined two fresh specimens taken in the western Atlantic some 250 km northeast of the Bahamas by Rey Pérez.

Size and weight A 136-cm female carrying large oocytes weighed 15.8 kg.

Biology This is a deep-water species of the continental slopes, usually found at depths of 400–1,400 m. Although it is common in the eastern Atlantic, it is a rare catch in the western Atlantic, probably due to the lack of deep-water fisheries. The species is common in the continental slopes at depths of 600–1,400 m west of the British Isles (Girard and Du Buit 1999).

The only western Atlantic specimens that I have examined are the two taken off the Bahamas by Rey Pérez (see above). Both of these sharks were taken at a depth of about 1,100 m, on the two separate (and rare) occasions when a portion of his swordfish longline was sunk to the bottom by large dead bluefin tunas. Given that one specimen was taken each time that the longline was sunk to the bottom, it may be that the leafscale gulper is abundant in such habitat.

Size at maturity There is only one report on size at maturity (Girard and Du Buit 1999), which uses very confusing terminology and vague criteria of maturity (e.g., "other mature males had underdeveloped claspers" [324]).

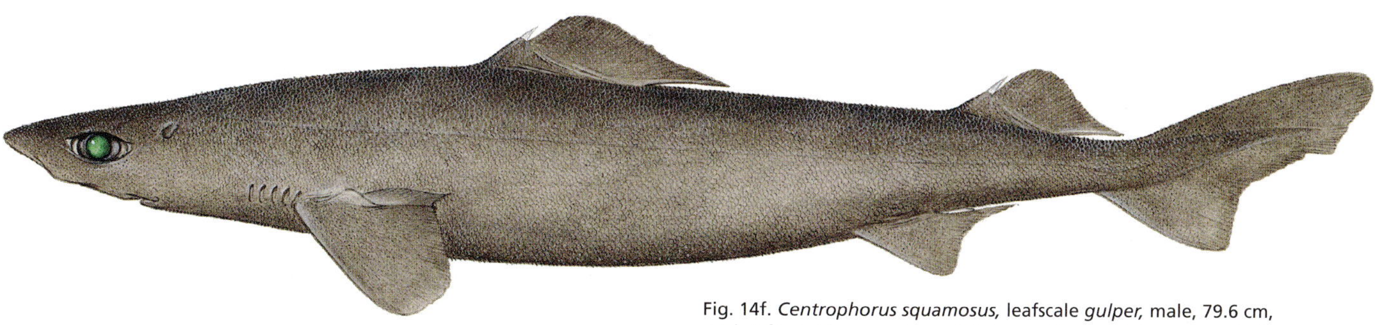

Fig. 14f. *Centrophorus squamosus,* leafscale *gulper,* male, 79.6 cm, 2.1 kg; from Okinawa, Japan (Okinawa Churaumi Aquarium staff, collectors).

Males: According to Girard and Du Buit (1999), males mature at about 98 cm, although they listed a 94-cm male as "the smallest male showing some signs of maturity" (325).

Females: It appears that females reach maturity at about 124 cm. Girard and Du Buit (1999: 926) stated, "Size at first maturity is reached at about 124 cm. . . . The smallest female ready to ovulate measured 108 cm."

Reproduction The leafscale gulper is an aplacental viviparous species. Little is known of its reproductive processes. The mature oocytes are proportionately very large. The 136-cm female that I examined in February carried 10 large oocytes averaging 80 × 70 × 50 mm. These large oocytes suggest a long reproductive cycle, probably two years, or perhaps longer. I speculate this based on the spiny dogfish model and on the assumption that, metabolically, it must take a long time for such a relatively small fish to acquire the necessary energy and store it into enough yolk to make the large oocytes. Similarly, it must also take a long time for the embryo to take up and consume the large amount of yolk. All this remains to be studied and demonstrated.

Size at birth: Unknown.

Brood size: It appears that females carry an average of 10 young per brood. Girard and Du Buit (1999) reported a mean of 10 ova (embryos?) per female. The ripe female that I examined carried 10 large oocytes.

Nurseries Unknown.

Age and growth The leafscale gulper has not been aged.

Relation to humans None in North America. In Portugal it is used for oil production.

MOSAIC GULPER SHARK

Centrophorus tesselatus Garman, 1906

Fig. 15a. *Centrophorus tesselatus,* mosaic gulper shark, 90 cm, from the southern Caribbean, off Panama (9°17′ N, 81°09′ W (RV *Oregon,* collector).

Common name Mosaic gulper, a reference to the skin denticles, which appear as small squares of a mosaic.

Spanish name *Galludito* (F.A.O.).

Scientific name *Centrophorus tesselatus* Garman, 1906, in *Bulletin of the Museum of Comparative Zoölogy at Harvard College* 46(11): 205. *Centrophorus: Centro:* from the Greek *kemtron,* spine, and *phorus,* from the Greek *phoreus,* bearer, hence "spine-bearer," an allusion to the spines on the dorsal fins; *tesselatus:* Latin, made of small square stones, from *tessella,* a small cube.

Synonyms None.

Identification This is a small and slender gulper shark with characteristic teeth. *It has an interdorsal space much longer*

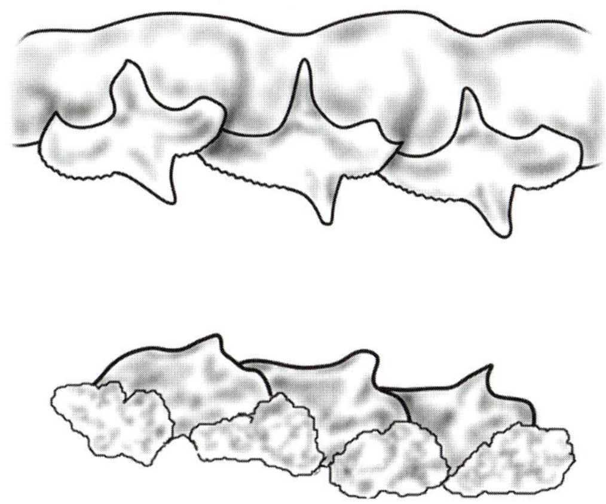

Fig. 15b.Teeth.

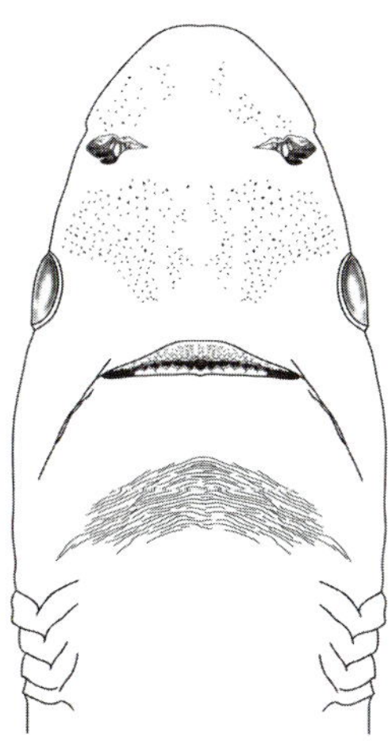

Fig. 15c. Snout.

Fig. 15d. Dermal denticles.

than the head length, a first dorsal fin base that measures about 40% of the interdorsal space, a pectoral fin with a long free rear tip that reaches well past the midpoint of the first dorsal fin base, and *the anterior margin of the pectoral fin is much longer than the distance from the tip of the snout to the spiracle.* The skin denticles have flat and squarish crowns with barely noticeable converging ridges and are arranged in irregular rows. *The teeth are characteristic: they have wide crowns with a short central pointed cusp and low and weak serrations on each side of the cusp and are similar in both jaws.* The coloration is brownish above and whitish below. The color of the preserved specimen is yellowish tan.

Similar species The slender gulper is very similar in profile to the mosaic gulper, but it has a more anterior second dorsal fin, its origin over the rear base or the free tip of the pelvic fin. The blackfin gulper has an interdorsal distance shorter than the head length.

Range The type specimen was taken off Japan. Since then specimens have been reported from Hawaii and North America (see below). The species is probably widely distributed in deep waters.

Size and weight The specimen illustrated here measured 90.3 cm. The species probably reaches 120 cm.

Biology This is a little-known gulper of tropical and warm temperate waters. Its biology is basically unknown. Only two specimens have been reported from North America, and the illustration here is based on the only specimen I have seen. It was taken at 640 m in the southern Caribbean, off Panama, by the RV *Oregon II* in 1970 (Station 11229). It is a 90.3-cm male that appears to be mature. The specimen is currently in the collection of the Florida Museum of Natural History (no. 3161).

Size at maturity Unknown.

Reproduction No data available.

Size at birth: No data available. Probably around 30 cm.

Brood size: Unknown.

Nurseries Unknown.

Age and growth Unknown.

Relation to humans None.

LITTLE GULPER SHARK

Centrophorus uyato (Rafinesque 1810)

Fig. 16a. *Centrophorus uyato*, little gulper shark, mature female, 95.5 cm, 5.7 kg; off Chub Cay, Bahamas (Tracey Sutton, collector).

Common name Little gulper shark, a reference to being smaller than the Taiwan gulper shark. For many years it was believed there were only two species of gulper sharks along the eastern coast of the United States, and this species was the "small" one. Now we know that there are several species of small gulpers in the area. The other species remain to be identified and have not yet acquired common names.

Spanish name None in use in the Spanish Caribbean. In Spanish-speaking countries the name *tiburón quelvacho* is applied to several species of gulper sharks.

Scientific name *Centrophorus uyato* (Rafinesque 1810), in *Caratteri di alcuni nouvi generi e nuove specie di animali e piante della Sicilia* 13, plate 14. *Centrophorus:* from the Greek *kemtron,* spine, and *phorus,* from *phoreus,* bearer, hence "spine-bearer," an allusion to the spines of the dorsal fins; *uyato:* from the common name *uyat* or *uyato* (derived from *agullats* or *ahullats*) given to dogfish sharks in the Mediterranean (de Buen 1926).

Synonyms It is difficult to determine which species was described by Rafinesque under the name *Squalus uyato* and which subsequent descriptions are based on it. Poor species descriptions, the lack of type specimens for most species, and our limited knowledge of the genus all combine to make species identification uncertain.

The name *Centrophorus uyato* is problematic. Muñoz-Chapuli and Ramos (1989) gave a long list of reasons why the name *uyato* should be considered a junior synonym of *granulosus.* This change was adopted by Compagno et al. (2005) in their field guide to sharks of the world, calling the small gulper of the Florida Bahamas region *Centrophorus granulosus.* This creates an even more confusing situation, because the name *granulosus* has been applied to a *large* species of *Centrophorus* for many years. Although there are at least four other species of small gulper sharks in the Florida–Bahamas region, I have retained the name *Centrophorus uyato* for the most common small gulper reaching about one meter in length (see description below) because it is a well-established name. This name may not endure, even if a lectotype of *C. uyato* is designated, because *C. uyato* was based on a European species, which probably differs from the American species. The situation will remain complicated until the genus is revised, a task that is beyond the scope of this book. Compagno (1984a) lists *Squalus infernus* Blainville, 1825; *Acanthias nigrescens* Nardo, 1860; and *Centrophorus armatus barbatus* Teng, 1962 as possible synonyms.

Identification This is a *small, stout shark* (probably not exceeding 100 cm) with two dorsal fins, each with a large spine (the spine of the first dorsal fin is about one-third the height of the fin, whereas the spine of the second dorsal is about half the height of the fin), a second dorsal fin smaller than the first, and pectoral fins with *long and pointed free rear tips that reach to about the midpoint of the base of the first dorsal fin.* The *anterior margin of the pectoral fin is shorter than the distance from the tip of the snout to the spiracle.* The *snout is slightly longer than the mouth width.* It has *thick and conspicuous white mouth corners and labial furrows.* The upper teeth have narrow, *triangular, smooth-edged cusps* and squarish bases; the cusps become increasingly oblique toward the corners of the mouth. The lower teeth are larger than the upper and *have oblique, smooth-edged cusps notched on their outer margins and interlocking bases.* Teeth number U: 18 or 19–1–18 or 19, L: 16–0 or 1–16. The dermal denticles are ovoid, narrowing rearward to a point, with one strong central ridge flanked by two weaker lateral ridges. Coloration is grayish brown above, paler or whitish on the

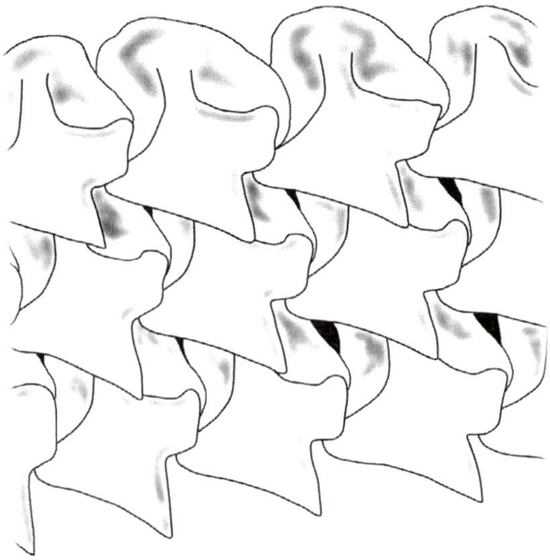

Fig. 16b. Teeth (upper).

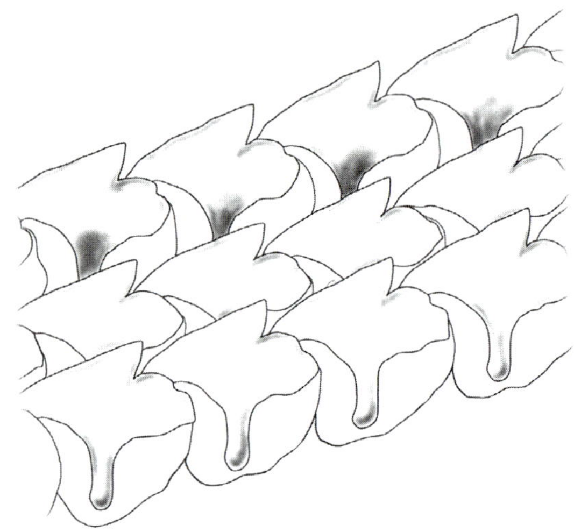

Fig. 16c. Teeth (lower).

sides and below. The *lining of the mouth is very dark grayish blue.*

Similar species The minigulper lacks the white corners of the mouth. The other gulper sharks in the area have a pectoral fin anterior margin that is as long as or longer than the distance from the tip of the snout to the spiracle. The Taiwan gulper, generally much larger, has pectoral fins with much shorter free rear tips that do not reach the midpoint of the base of the first dorsal fin, its dermal denticles have five to seven ridges diverging rearward, and it lacks the white labial corners. The mosaic gulper has characteristic teeth that are similar in both jaws. The leafscale gulper is chocolate colored and has a pectoral fin free rear tip that does not reach the level of the first dorsal fin spine.

Range The little gulper is common around the Bahamas and the Gulf of Mexico. The limits of the range are not precisely known because of confusion with other small gulper species found in the region.

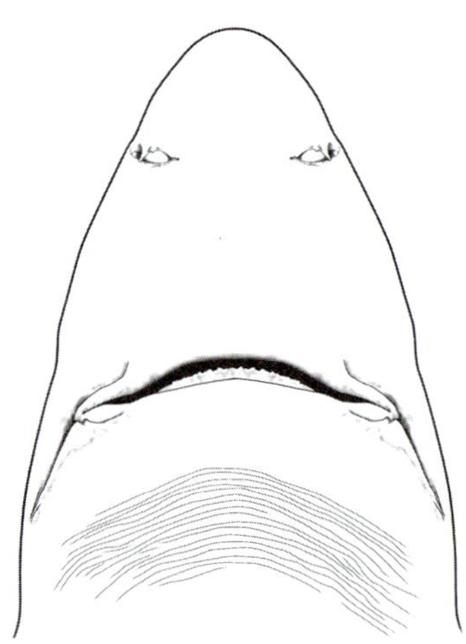

Fig. 16d. Snout.

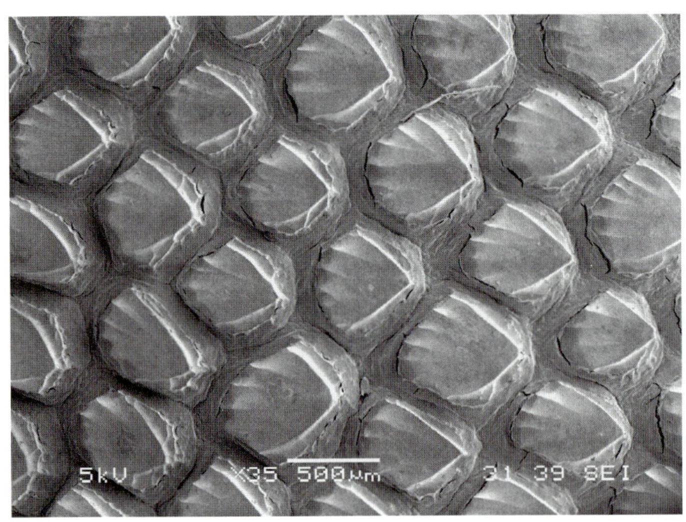

Fig. 16e. Dermal denticles.

Size and weight The 95-cm female illustrated here weighed 5.7 kg.

Biology Unknown.

Size at maturity Unknown.

Reproduction The little gulper is an aplacental viviparous species. Females usually carry one or two embryos at a time, one in each uterus. Females carry developing embryos and developing ova at the same time, indicating that gestation and vitellogenesis are concurrent and suggesting a 2-year reproductive cycle, similar to that of the spiny dogfish. The ripe oocytes are huge and may reach 80–90 mm in diameter just before ovulation.

Size at birth: Unknown.

Brood size: Generally two young are produced, one in each uterus. Smaller females often carry only one young.

Nurseries Unknown.

Age and growth The species has not been aged.

Relation to humans The little gulper shark is seldom caught in North America because of its deep-water habitat. It has no commercial importance in North America.

THE CONFUSING GULPERS

The following species of North American gulper sharks were encountered during the production of this book. Their identities are uncertain.

Minigulper, *Centrophorus* sp. A

Fig. 17a. *Centrophorus* sp. A., minigulper, female, 71 cm, 1.5 kg; from the northern Gulf of Mexico (Wm B. Driggers III, collector).

This is the smallest of the gulper sharks I have encountered in North America. The identity of this species remains to be determined. The two specimens that I have examined are a 71-cm immature female (illustrated) and a 79-cm mature male. The species probably reaches 85–90 cm. The female was taken in the Gulf of Mexico in 2003 by Trey Driggers. The male was taken off the Bahamas by Ron Schatman in 2005. At first I believed these specimens to be small *C. uyato,* but the mature male specimen at hand is too small for that species. The minigulper is a *small gulper shark with a pectoral fin anterior margin shorter than the distance from the snout to the spiracle and tan (not white) corners of the mouth.* It differs from the larger *C. uyato* in lacking the distinctive white corners of the mouth.

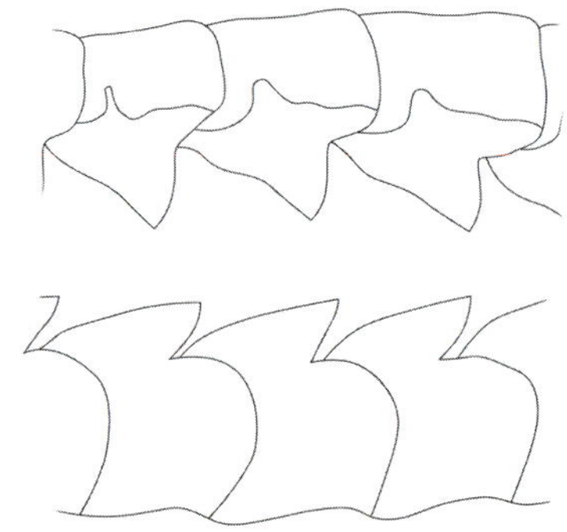

Fig. 17b. Teeth.

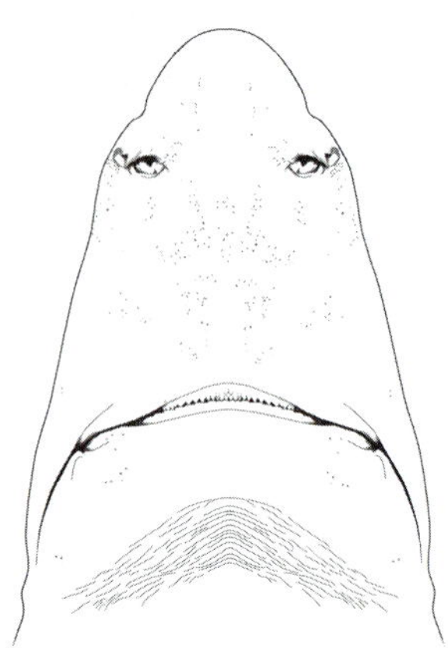

Fig. 17c. Snout.

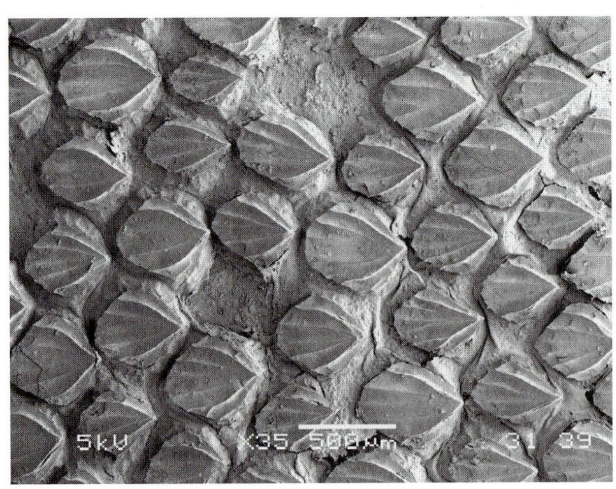

Fig. 17d. Dermal denticles.

Slender gulper, *Centrophorus* sp. B

Fig. 18a. *Centrophorus* sp. B., slender gulper, mature female, 100.0 cm, 4.7 kg., off Cat Island, Bahamas (Ron Schatman, collector).

This gulper shark resembles *Centrophorus lusitanicus* Barboza du Bocage and de Brito Capello, 1864, a common species of the eastern Atlantic that was originally described from Portugal (*lusitanicus:* Latin "of Portugal," from *Lusitania,* an ancient region and Roman province of the Iberian Peninsula that corresponds roughly to modern Portugal). The description appeared in *Proceedings of the Scientific Meetings of the Zoological Society of London* (1864: 260–263).

This is a medium-size, slender gulper shark, reaching more than 100 cm, characterized by an *interdorsal space much longer than the length of the head,* an *anterior margin of the pectoral fin that is as long as the distance from the tip of the snout to the spiracle,* and a *second dorsal fin originating over the rear margins of the pelvic fins.* The upper teeth have narrow, triangular, smooth-edged cusps that become increasingly oblique toward the corners of the mouth and squarish bases; the lower teeth are larger than the upper and have oblique, smooth-edged cusps notched on their outer margins and interlocking bases. The largest specimen that I have seen is a mature female of 105 cm and weighing 7.4 kg. All five specimens that I have seen were taken off the Bahamas by Ron Schatman.

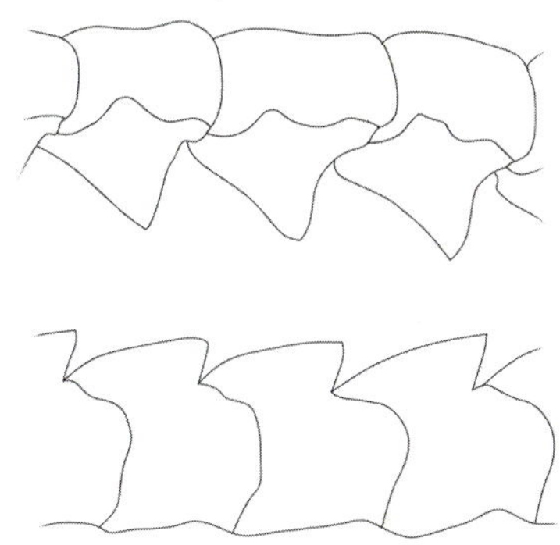

Fig. 18b. Teeth.

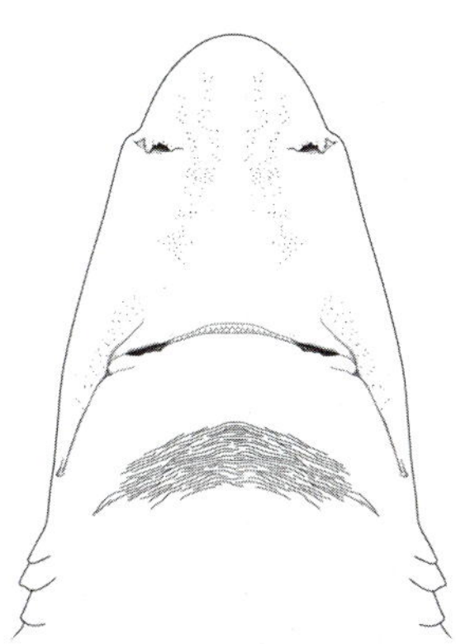

Fig. 18c. Snout.

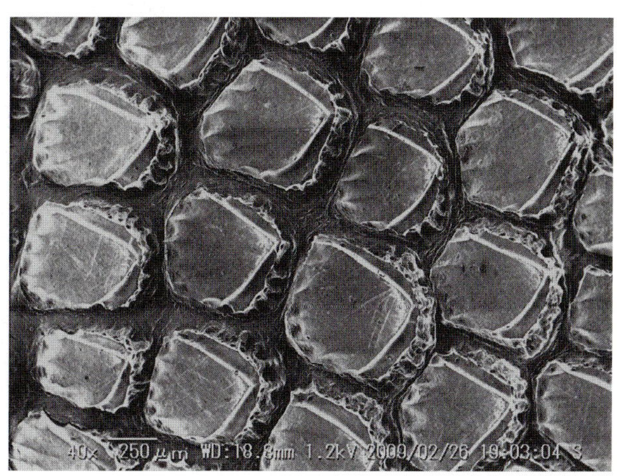

Fig. 18d. Dermal denticles.

BIRDBEAK DOGFISH

Deania calcea (Lowe, 1839)

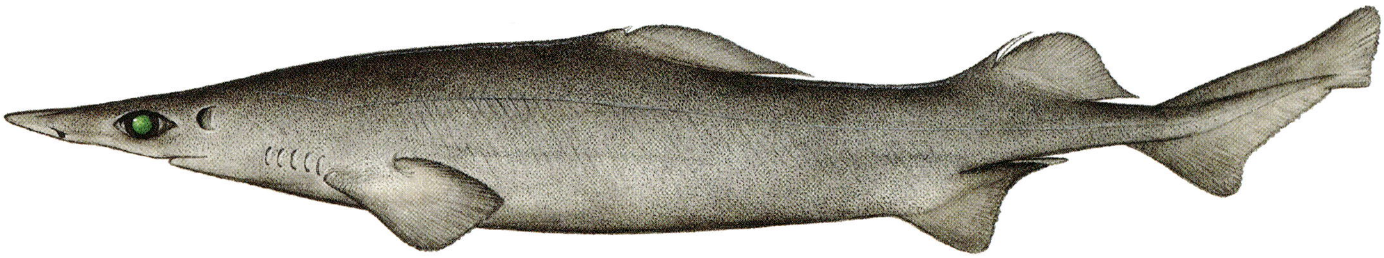

Fig. 19a *Deania calcea*, Birdbeak dogfish, mature male, 85 cm, from the fish market, Sesimbra, Portugal (Joao Correia, collector).

Common name Birdbeak dogfish, allusions to its very long snout and to its dogfish-like dorsal spines.

Spanish name The species lacks vernacular names in the Spanish Caribbean.

Scientific name: *Deania calcea* (Lowe, 1839), originally described as *Acanthidium calceus* in *Transactions of the Zoological Society, London* 3(1): 19. *Deania:* a name created by Jordan and Snyder (1902) honoring Prof. Bashford Dean (1867–1928), in recognition of his research on sharks; *calcea:* from the Latin *calceus,* a shoe or slipper, based on the old Portuguese name for the species, *sapata branca* = white shoe, allusions to the long snout that resembles a shoe and its coloration that is lighter than that of the other species.

Synonyms *Centrophorus crepidalbus* Barboza du Bocage and de Brito Capello,[1] 1864; *Deania eglantina* Jordan and Snyder, 1902; *Acanthidium rostratum* Garman, 1906; *Acanthidium aciculatum* Garman, 1906; and *Centrophorus kaikourae* Whitley, 1934.

Identification This small shark is characterized by a *broad, flat, very long snout,* its length greater than the distance from the center of the mouth to the origin of the pectoral fins; and two dorsal fins each with a spine, the second one larger than the first. The teeth vary markedly with size and sex. In small specimens the teeth of both jaws have oblique cusps. Adult males have teeth with erect, slender, smooth-edged cusps and squarish bases on both jaws; adult females have lower teeth with oblique cusps. Teeth number U: 26 to 31, L: 26 to 30. The dermal denticles have three points, are *pitchfork shaped,* and stand nearly erect on the skin *on a tall pedicel.* Coloration is dark brown above, fading to dark gray after death, with slightly lighter undersides.

1. The names of these authors have been alternately listed as "Bocage and Capello" and "du Bocage and de Capello." Here I follow latest usage trend.

Similar species Other squaloid sharks in the area have shorter snouts. The arrowhead dogfish has a noticeable keel between the pelvic and caudal fins.

Range The birdbeak dogfish appears to be cosmopolitan in deep waters, having been reported from numerous locations around the world. In the western Pacific it has been listed from Japan (as *Deania eglantina;* Jordan and Snyder 1902), New Zealand (Garrick 1960b), and Australia (Last and Stevens 1994). In the Indian Ocean it has been reported from South Africa (Penrith 1969; Bass et al. 1976). It has been reported from several locations in the eastern Atlantic, including Portugal (Barboza du Bocage and de Brito Capello 1864), Madeira (Cadenat and Blache 1981), and Namibia (Yano 1991). Birdbeak dogfishes are often taken as bycatch in the scabbard fish fishery in Portugal, where I have seen dozens landed daily at Sesimbra. The species is a rare catch in the western North Atlantic, perhaps due to the general lack of deep-water fisheries. A few specimens were taken some years ago off Virginia and are now in the Virginia Institute of Marine Science collection (J. A. Musick, pers. comm., April 2004). A specimen caught in deep water off Florida by Ron Schatman in June 1980 was subsequently obtained and preserved by David McClellan, who kindly made it available to me along with photographs. To my knowledge, this is the only specimen from southeastern United States waters.

Size and weight Most specimens I have examined measured 90–110 cm and weighed 4–7 kg. The species reaches at least 120 cm.

Biology This is a deep-water species found over temperate continental slopes throughout the world. Its habits are poorly known despite the wide distribution. Studies on

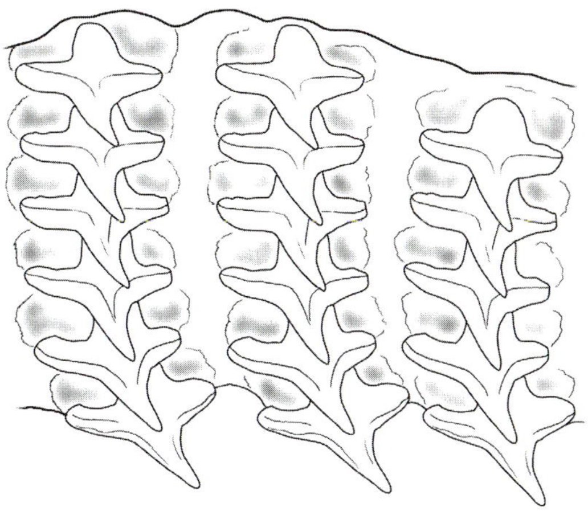

Fig. 19b. Teeth (upper).

Fig. 19c. Teeth (lower).

the species have not been carried out in North America, undoubtedly because of its scarcity here. In the eastern Atlantic it is commonly taken at 400–900 m (Bordalo Machado and Figueiredo 2000). In the northeastern Atlantic, larger individuals have been reported to feed primarily on rockling (*Antonogadus macrophthalmus*), whereas the smaller sharks feed on blue whiting (*Micromesistius poutassou*) (Mauchline and Gordon 1983). Myctophids have been reported in the stomach contents by several authors (Mauchline and Gordon 1983; Marshall and Merrett 1977), indicating that the species feeds at some height above the bottom.

Size at maturity

Males: Males 80.2–86.5 cm long had calcified claspers, while males 47.5–82.0 cm had uncalcified claspers. According to Yano (1991), males in the southeastern Atlantic off Namibia mature at about 80 cm, based on clasper calcification.

Females: Females are believed to mature at about 100 cm (Yano 1991).

Reproduction Reproductive processes remain undescribed, but the species is undoubtedly aplacental viviparous, like other dogfishes.

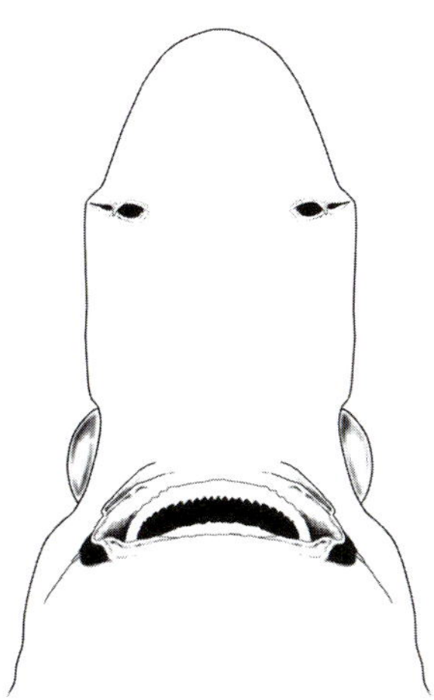

Fig. 19d. Snout.

Fig. 19e. Dermal denticles.

Size at birth: Unknown.

Brood size: Compagno (1984) gave the number of embryos as 6–12. Yano (1991) reported that Japanese specimens carried 12–23 embryos per brood.

Nurseries Unknown.

Age and growth The species has not been aged. Bordalo Machado and Figueiredo (2000) suggested that growth bands on the second dorsal spine might be discernible after decalcification and staining with Mayer's hematoxylin. Whether these bands are periodic, and whether they can be used for aging, remains to be determined.

Relation to humans It has no economic importance in North America. In Portugal, many specimens, taken in scabbard fisheries, are sold in the Sesimbra fish market, probably for their oil.

ARROWHEAD DOGFISH

Deania profundorum (Smith and Radcliffe, 1912)

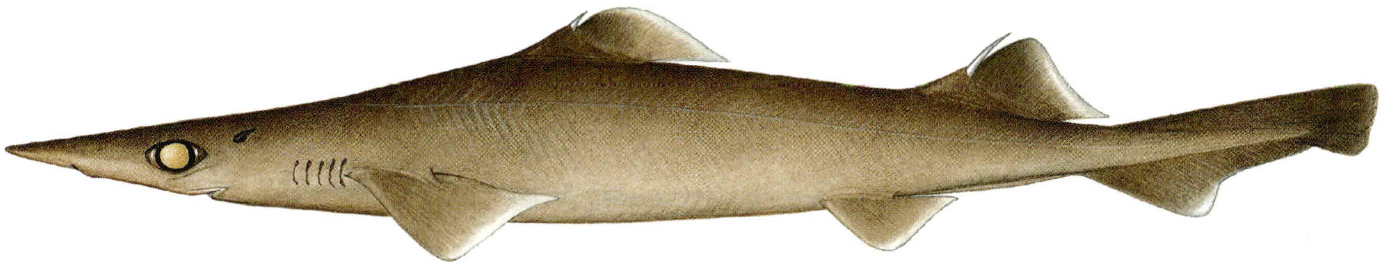

Fig. 20a. *Deania profundorum*, arrowhead dogfish, female, 48.5 cm, from the Gulf of Mexico (29°29′ N, 86°55′ W; Wm. B Driggers III, collector).

Common name Arrowhead dogfish, an allusion to the shape of the head.

Spanish name The species lacks a vernacular name in the Spanish Caribbean. However, the names *tollo flecha* and *visera flecha* have been used in some Spanish publications, obviously literal translations of the English name.

Scientific name *Deania profundorum* (Smith and Radcliffe, 1912), originally described as *Nasisqualus profundorum* in *Proceedings of the United States National Museum* 41(No. 1877): 681–683, fig. 3, plate 53. *Deania:* named by Jordan and Snyder (1902) honoring Prof. Bashford Dean (1867–1928) in recognition of his research on sharks; *profundorum:* Latin, of the depths.

Synonyms *Acanthidium natalense* Gilchrist, 1922; *Deania elegans* Springer, 1959; *Deania cremouxi* Cadenat, 1960.

Identification This small dogfish is characterized by a *broad, flat, very long snout,* its length greater than the distance from the center of the mouth to the origin of the pectoral fins; two dorsal fins each with a spine, the second spine larger than the first; and a noticeable, *low keel or flap of skin between the pelvic and caudal fins.* This keel, or ridge, termed a "postanal median keel"[2] by H. M. Smith (1912), is considered diagnostic of the species *at this time.* However, it is likely that future research may reveal that there are other species of *Deania* having this anatomical feature. Like all squaloid sharks, it lacks an anal fin. The teeth vary markedly with size and sex. In small specimens the teeth of both jaws have oblique cusps. Adult males have teeth with slender, smooth-edged cusps and wide, squarish bases on both jaws; the central upper teeth are erect, but they become increasingly oblique toward the corners of the mouth; the lower teeth are erect. Adult females have lower teeth with oblique cusps. Teeth number U: 26 to 31, L: 26 to 30. The dermal denticles have three points, are *pitchfork shaped,* and stand nearly erect *on a tall pedicel.* North American specimens that I examined were chocolate brown, with darker undersides. However, photographs of other North Atlantic specimens reveal a variable coloration: some are gray above and whitish below, while others are chocolate brown all over.

2. This keel or ridge has been called a "pseudoanal fin" by Penrith (1969), and a "pseudanal flap" by Bass et al. (1976). I have chosen to retain the original term of postanal median keel, because it best describes this structure.

Similar species The birdbeak dogfish lacks the low keel between the pelvic and anal fins. Other squaloid sharks in the area have shorter snouts and lack the low keel between pelvic and caudal fins.

Range The arrowhead dogfish is probably widely distributed in deep waters, but its small size and deep-water habitat make it a rare catch. The species was originally reported from the Philippine Sea (Smith, H. M. 1912). It has also been reported from Africa (Senegal, as *Deania cremouxi,* by Cadenat 1960; South Africa by Gilchrist 1922 and Bass et al. 1976). In North America, it has been caught off North Carolina (Springer 1959), and there are unreported captures in the Gulf of Mexico (George Burgess, pers. comm., July 2004). I have examined three fresh specimens caught in the northern Gulf of Mexico.

Size and weight Average size is about 50 cm; maximum size is about 76 cm.

Biology The specimens taken in the northern Gulf of Mexico (by W. Driggers III, one of which is illustrated here) were taken at 457 m. Nothing else is known of its habits.

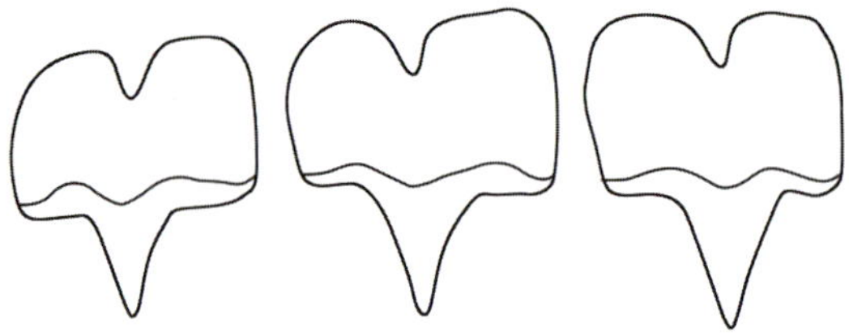

Fig. 20b. Teeth (upper).

Fig. 20c. Teeth (lower).

Size at maturity Unknown.

Reproduction The species is certainly aplacental viviparous, like all other dogfishes. Its reproductive processes have not been described.

Size at birth: Unknown.

Brood size: Unknown.

Nurseries Unknown.

Age and growth The species has not been aged.

Relation to humans None.

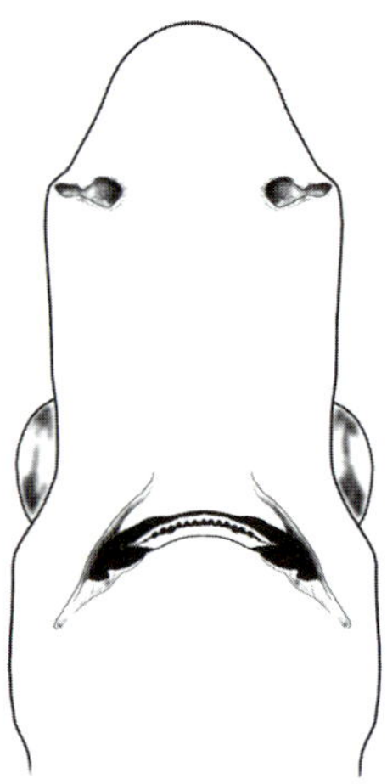

Fig. 20d. Snout.

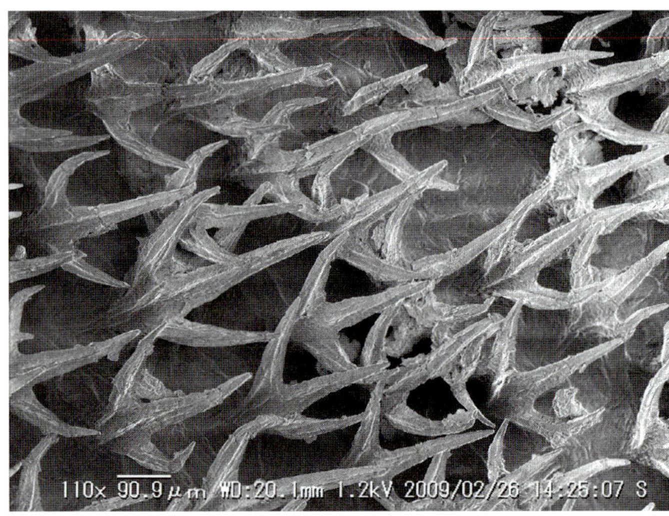

Fig. 20e. Dermal denticles.

FAMILY ETMOPTERIDAE

The Black Dogfishes and Lanternsharks

The Etmopteridae are small, deep-dwelling sharks having two dorsal fins with well-developed spines and lacking an anal fin. The spines have two longitudinal grooves on each side, and the second dorsal fin spine is larger than the first. The family Etmopteridae includes the genera *Centroscyllium,* the black dogfishes, and *Etmopterus,* the lanternsharks (Shirai 1992).

The genus *Centroscyllium* is characterized by similar teeth in both jaws, the teeth having three to five cusps, the central cusp being largest. Two species occur in North America: the black dogfish (*Centroscyllium fabricii*) of the Atlantic and the combtooth dogfish (*Centroscyllium nigrum*) of the Pacific Ocean.

The genus *Etmopterus* includes the small and beautiful lanternsharks. Their upper and lower teeth are dissimilar. The upper teeth are erect and have three to five cusps, the median cusp usually being largest. The lower teeth resemble those of *Squalus.* They have a single oblique cusp with a notch on its outward margin, and interlocking bases forming a continuous cutting edge. Lanternsharks usually have dark markings and bioluminescent skin organs (photophores). Some lanternsharks have a broad bioluminescent band along the sides and scattered photophores on their ventral surfaces. Others have bioluminescent undersides. The bioluminescent sides probably serve to conceal the shark against the ambient light in side view, while the bioluminescent undersides serve to prevent these small to tiny mesopelagic sharks from being silhouetted against the brighter surface by predators swimming below. This process, known as counterillumination, allows the animal to blend in with the downwelling light. Unfortunately, live lanternsharks are seldom seen. There are seven species known in North America, but it is likely that there are undescribed species as well.

Traditionally, authors have used body proportions or ratios to describe species of lanternsharks. Some of these proportions change with growth (allometric growth). Thus, one must not rely on single proportions to identify lanternsharks; all the diagnostic characteristics must be considered.

KEY TO THE ETMOPTERIDAE

1a. Teeth similar in both jaws 2

1b. Teeth dissimilar in both jaws 3

2a. First dorsal fin origin well behind pectoral fin rear margins when laid back (by a distance equal to the pectoral fin base, a shark of the western North Atlantic) .... **Black dogfish,** *Centroscyllium fabricii* (p. 93).

2b. First dorsal fin origin over pectoral fin rear margins when laid back (a shark inhabiting the eastern Pacific) .... **Pacific black dogfish,** *Centroscyllium nigrum* (p. 96).

3a. First dorsal fin origin over or close to a perpendicular over pectoral fin rear margins (skin denticles on the flanks arranged in longitudinal rows) .... **Lined lanternshark,** *Etmopterus bullisi* (p. 101).

3b. Origin of first dorsal fin well behind pectoral fin rear margins 4

4a. Distance between the dorsal fin bases (interdorsal space) longer than head length (from the tip of the snout to the origin of the pectoral fin) 5

4b. Distance between the dorsal fin bases (interdorsal space) shorter than head length (from the tip of the snout to the origin of the pectoral fin) 7

5a. Caudal fin upper margin much shorter than the distance from the tip of the snout to the pectoral axil (its length about equal to the distance from the tip of the snout to the second gill slit; skin denticles with flat or concave crowns and four-pointed bases; margins of pectoral fins usually intact, not showing its naked rays (ceratotrichia) .... **Emerald lanternshark,** *Etmopterus bigelowi* (p. 98)

5b. Caudal fin upper margin about as long as distance from the tip of the snout to the pectoral fin axil 6

6a. Pectoral fin with rounded corners, its margins frayed and forming a fringe of naked rays (ceratotrichia), underside of snout covered with skin denticles and rough to the touch .... **Fringefin lanternshark,** *Etmopterus schultzi* (p. 110).

6b. Pectoral fin with angular corners and with intact borders, not showing the naked fin fibers, underside of snout naked, lacking denticles and smooth to the touch .... **Green lanternshark,** *Etmopterus virens* (p. 112).

7a. Midpoint of first dorsal fin much closer to pectoral fin axil than to the origin of the pelvic fins 8

7b. Midpoint of first dorsal fin base equidistant from pectoral axil and pelvic fin origin (a light spot on the top of the head between the eyes) .... **Broadband lanternshark,** *Etmopterus gracilispinis* (p. 103).

8a. Distance from snout to pectoral fin axil nearly equal to distance from pectoral axil to origin of pelvic fins; skin denticles conical and heavy, visible to the naked eye as raised white spots giving the skin a pebbly appearance; coloration uniformly blackish brown or black (a shark of the cold waters of the North Atlantic, reaching a length of 80 cm) .... **Great lanternshark,** *Etmopterus princeps* (p. 107).

8b. Distance from snout to pectoral fin axil longer than distance from the pectoral fin axil to the origin of the pelvic fins; skin denticles on the flanks below first dorsal fin slender and bristlelike; coloration of ventral surfaces black, much darker than upper surfaces (a small Caribbean shark of less than 30 cm) .... **Caribbean lanternshark,** *Etmopterus hillianus* (p. 105).

BLACK DOGFISH

Centroscyllium fabricii (Reinhardt, 1825)

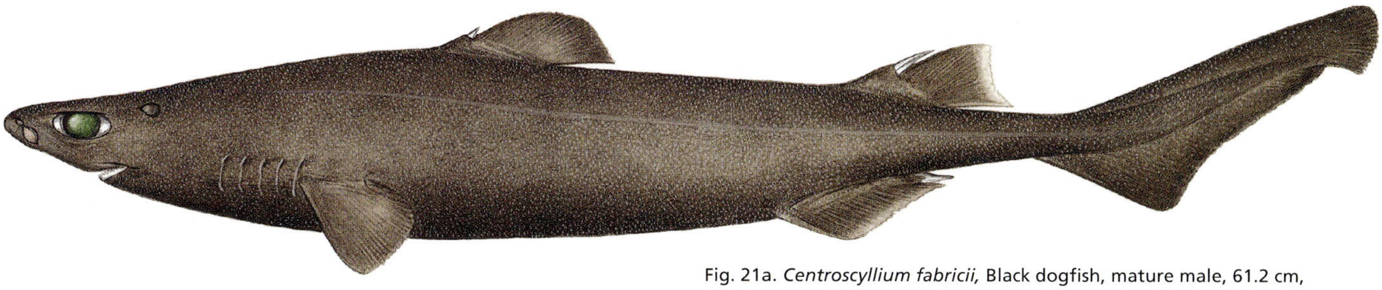

Fig. 21a. *Centroscyllium fabricii*, Black dogfish, mature male, 61.2 cm, 0.96 kg; from Oceanographer Canyon, Georges Bank (John Galbraith, collector).

Common name Black dogfish. See explanation of "dogfish" under spiny dogfish (p. 55).

Spanish name The species lacks a vernacular name in the Spanish Caribbean.

Scientific name *Centroscyllium fabricii* (Reinhardt, 1825), originally named *Spinax fabricii,* in *Danske Videnskabernes Selskabs Forhandlinger* 1824–1825, 111, XVI. *Centroscyllium:* from the Greek *kentron,* spine or point, and *scyllium,* from the Greek *skylion,* a dogfish; thus, a spined dogfish; *fabricii:* latinized form of Fabricius, after Otto Fabricius (1744–1822), a Danish naturalist who first studied this species and the fishes of Greenland.

Synonyms None.

Identification The black dogfish is characterized by two dorsal fins each with a prominent spine, and *a second dorsal fin originating over the pelvic fins.* The *teeth have three (sometimes four or five) sharp, smooth-edged cusps, the central cusp being largest, and are similar in both jaws.* The teeth number U: 34–34, L: 34–34. The dermal denticles are sparse, minute, thornlike, and with ridged sides and stellate bases. Juveniles are inky black below and slightly lighter or chocolate brown above, with the margins of the dorsal and pectoral fins being white, the pelvic fins tipped with white, and white fin spines. Half-grown and *adult specimens are dark brown all over,* with white spines. This is likely a bioluminescent shark, although I found no reports to that effect. Jensen (1948: 8) reported that "this shark must be considered luminous like other species of *Spinacidae*" and that the skin between the spines, when viewed under a lens, showed minute dark points that must be luminous organs.

Similar species Other Atlantic coast dogfishes and lanternsharks lack the tricuspid teeth in the lower jaw, and their second dorsal fins originate behind the pelvic fin bases.

Range The black dogfish inhabits the fringes of the polar waters of the North Atlantic from Davis Strait in the west to Iceland, the Faroes, and the Shetlands in the east. In North American waters it ranges from Davis Strait south to Georges Bank. It is common in deep water from Davis Strait to Nova Scotia (Jensen 1948; Templeman 1963). The species has not been reported from Ungava Bay (Dunbar and Hildebrand 1952; Templeman 1963) or Hudson Bay (Vladykov 1933).

Size and weight According to Templeman (1963), most of the Nova Scotian specimens are 60–75 cm, only a few are larger than 80 cm, and 84 cm was the greatest length observed. Jakobsdóttir (2001) reported Icelandic specimens up to 106 cm. She also gave length–weight relationships for both males and females. According to these, a 60-cm male weighs about 1 kg, the largest 80-cm males weigh about 2.3 kg, and the larger females reach 90 cm and 4 kg. Most of the specimens I have examined, taken off Georges Bank by John Galbraith, were less than 66 cm.

Biology The black dogfish inhabits the coldest waters of the North Atlantic, but it does not enter polar waters. It appears to remain within a narrow temperature range: most captures have occurred where bottom water temperatures are 3.1–4.5°C. In the western Atlantic the northern limit of black dogfish distribution appears to be the submarine ridge across Davis Strait, which separates the colder waters of Baffin Bay from the warmer Labrador Sea. Jensen (1948) reported the capture of 42 individuals in a single trawl at

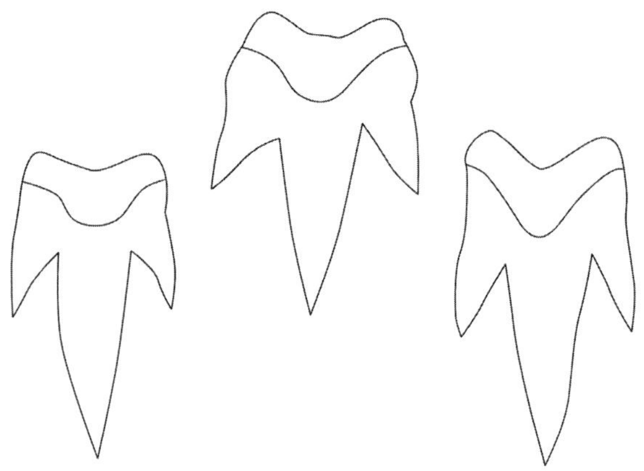

Fig. 21b. Teeth (upper).

Fig. 21c. Teeth (lower).

66°37′ N, 56°37′ W, on the south slope of the Davis Strait ridge, in water 460 m deep and 3.12°C, while another set of the net just north of the ridge took no black dogfish in water of 2.47°C (Bigelow and Schroeder 1948). Templeman (1963) likewise reported that no black dogfish were obtained in half-hour sets at 567, 538, and 700 m on the western side of the Davis Strait ridge between 66°04′ N and 66°48′ N, and 59°00′ W and 59°49′ W. Bottom temperatures during these sets were 1.0°C,–1.1°C, and –1.2°C, respectively. Templeman (1963) reported that on the slope of the northeastern Newfoundland Shelf, the greatest number of black dogfish captured in a half-hour was 68 specimens at 550 m and 3.6°C and that no specimens were obtained in three deep-water sets in Ungava Bay at 550–730 m where the temperature was 1.46–1.75°C. Off Iceland, Jakobsdóttir (2001) found the species most abundant at 800–1,200 m and temperatures of 4.1–4.4°C.

The black dogfish has a wide depth range. Captures off Greenland range from close to the surface through the ice in winter (Fabricius 1780) to 1235 m (Yano 1995b), the species probably staying within its narrow temperature tolerance or preference. Off the southwest coast of Iceland, Saemundsson (1949) reported it as abundant at depths of 500–950 m.

Localized catches of large numbers of black dogfish indicate that it forms large schools. Templeman (1963) mentioned that in St. Pierre Bank, south of Newfoundland, a trawler caught 1,400 kg of black dogfish in a single 1.5-hour set, while other tows in the same area immediately before and after produced only a few.

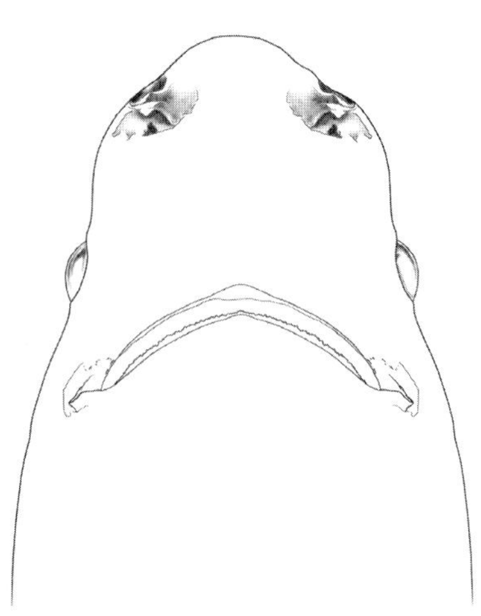

Fig. 21d. Snout.

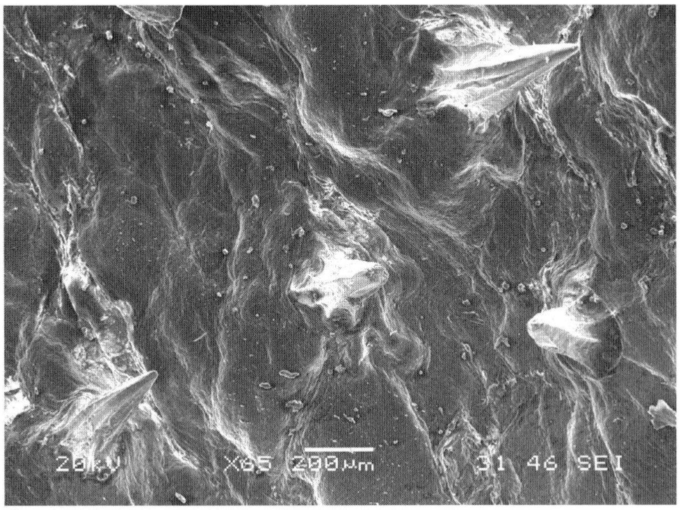

Fig. 21e. Dermal denticles.

Reports of the black dogfish diet are scarce. Jakobsdóttir (2001) examined the stomachs of 316 Icelandic specimens, but 236 (75%) were empty. In the 80 that contained food, teleosts were found in 47.4% of the stomachs, crustaceans in 44.7%, and cephalopods in 28.9%.

Despite the abundance of the black dogfish, almost nothing is known of its life history. Bigelow and Schroeder (1948) mentioned that the black dogfish skin has minute, deeply pigmented papillae resembling the luminous organs of the brightly bioluminescent cookiecutter shark (*Isistius brasiliensis*), but such luminescence has not been reported in the literature.

Size at maturity Yano (1995b) wrote that 97% of females less than 64.9 cm were immature and that the smallest gravid females were 64.4 cm. He also reported that 52% of males 55–59 cm were mature with calcified claspers, and 87% of specimens larger than 60 cm had full sperm sacs.

Reproduction Development is aplacental viviparous. The only comprehensive report of the reproductive processes is that of Yano (1995b), who examined some 2,500 specimens caught off Greenland. He reported that the black dogfish does not appear to have a well-defined breeding season, based on finding females with mature oocytes from June to November and females with large embryos from August to November, and that the black dogfish differs from most squaloid sharks in that its oocytes do not develop during gestation.

Size at birth: Yano (1995b) observed near-term embryos with fully absorbed external yolk sacs ranging from 15.2 to 19.2 cm and free-swimming young 16.5–17.5 cm. Thus, it appears that black dogfish are born at 16–19 cm, or at 21–25% of the adult size.

Brood size: The brood size has been reported as ranging from 4 to 40, with a mean of 16.4 (Yano 1995b).

Nurseries The nurseries have not been characterized. However, Yano (1995b) reported catching neonates of 16.5–17.5 cm at depths of 575–668 m off western Greenland.

Age and growth The black dogfish has not been aged.

Relation to humans None. It is often taken as bycatch in the halibut fisheries from the Grand Banks to Iceland and discarded.

PACIFIC BLACK DOGFISH

Centroscyllium nigrum Garman, 1899

Fig. 22a. *Centroscyllium nigrum,* Pacific black dogfish, mature male, from the eastern Pacific (RV *Oregon*, collector; courtesy of David Ebert).

Common name Pacific black dogfish, obvious allusions to the area where it is found and its coloration.

Spanish name None.

Scientific name *Centroscyllium nigrum* Garman, 1899, in *Memoirs of the Museum of Comparative Zoology at Harvard College* 24: 28, plates 1, 4, 5, 69. *Centroscyllium:* from the Greek *kentron,* spine or point, and *scyllium,* from the Greek *skylion,* a dogfish, thus a spined dogfish; *nigrum:* from the Latin *niger,* dark or black.

Synonyms *Centroscyllium ruscosum* Gilbert, 1905.

Identification This small shark *of the eastern Pacific Ocean* has *two dorsal fins of equal size, each with a long spine.* The *teeth have three (occasionally four or five) sharp, smooth-edged cusps, the median cusp being largest, and are similar in both jaws.* The dermal denticles are thornlike and erect, with stellate bases, and set far apart. Coloration is *blackish brown on the sides and black below,* with white dorsal spines and *a narrow white margin on the dorsal, pectoral, and pelvic fins.*

Similar species Other Pacific coast dogfishes lack the tricuspid teeth in the lower jaw.

Range The Pacific black dogfish inhabits the central and eastern Pacific Oceans. It has been reported from numerous locations in the eastern Pacific, from California (Garman 1899) to the tip of South America (Galapagos and Cocos Island, Kato et al. 1967; Chile, de Buen 1960; Strait of Magellan, Long 1994); it has also been reported from Hawaii

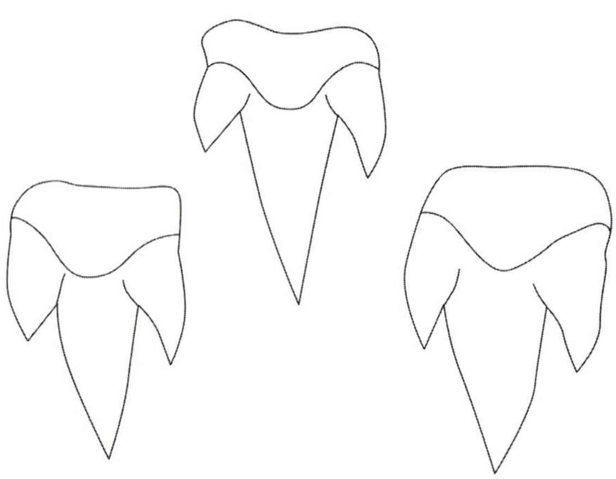

Fig. 22b. Teeth (upper).

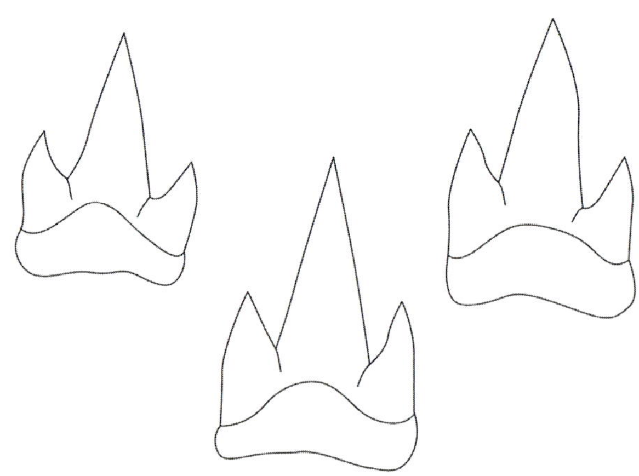

Fig. 22c. Teeth (lower).

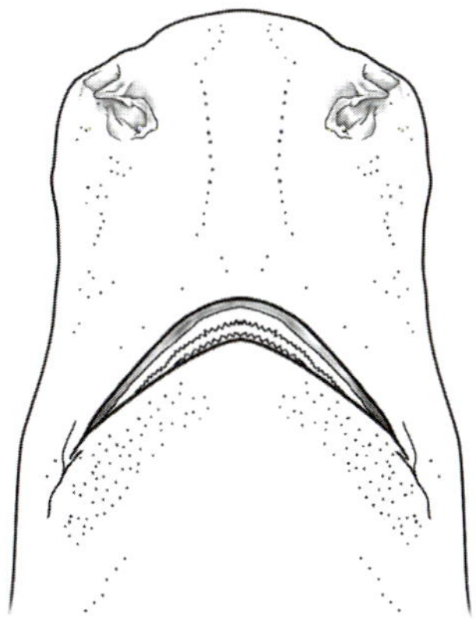

Fig. 22d. Snout.

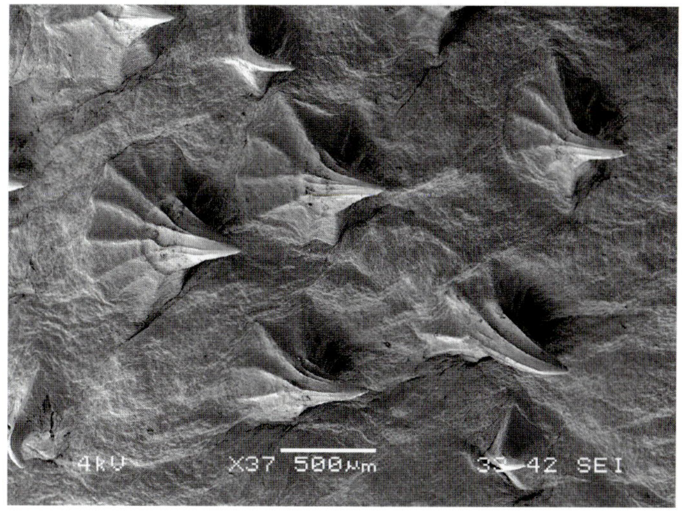

Fig. 22e. Dermal denticles.

(Gilbert 1905). It is an unusual catch in North America because of the absence of deep-water fisheries.

Size and weight Most specimens seen are about 30 cm, but it is said to grow to 50 cm (Compagno 1984a).

Biology This is a poorly known deep-water species; most specimens are caught at 730–1,140 m. It has been reported to feed on shrimp, but nothing else is known of its habits (Beebe and Tee-Van 1941).

Size at maturity Little published data are available on the size at maturity. Gilbert (1905) described a 40-cm male from Kauai, Hawaiian Islands, as adult with fully developed claspers.

Reproduction Development is undoubtedly aplacental viviparous, like the other dogfishes. Nothing else is known of its reproductive processes.

Size at birth: Unknown.

Brood size: Unknown.

Nurseries Unknown.

Age and growth The Pacific black dogfish has not been aged.

Relation to humans None.

EMERALD OR BLURRED LANTERNSHARK

Etmopterus bigelowi Shirai and Tachikawa, 1993

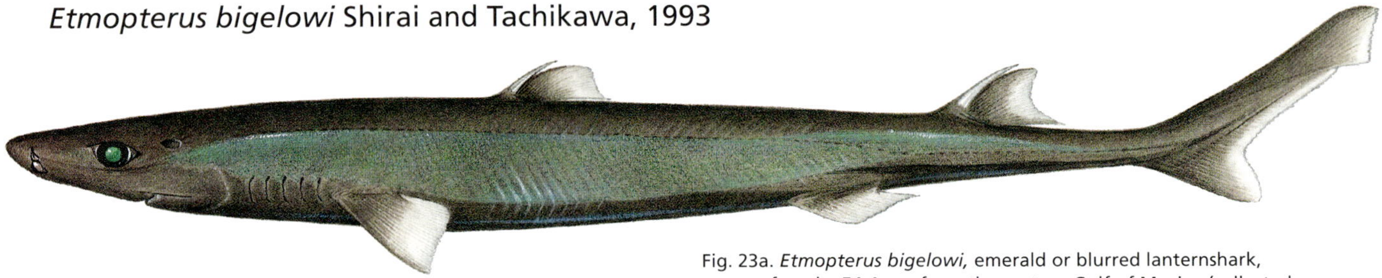

Fig. 23a. *Etmopterus bigelowi,* emerald or blurred lanternshark, mature female, 54.4 cm, from the eastern Gulf of Mexico (collected by the author on RV *Suncoaster*).

Common name Emerald lanternshark, or blurred lanternshark. In their description of the species, Shirai and Tachikawa (1993) named it "blurred smooth-dogfish." The American Fisheries Society later adopted the name "blurred lanternshark," following the name proposed by Shirai and Tachikawa (Nelson et al. 2004). Both these names are cumbersome and confusing for general use, and certainly inappropriate for the most spectacular of our lanternsharks. Therefore, I suggest the more descriptive name of emerald lanternshark.

Spanish name There are no vernacular names for this species in the Spanish Caribbean.

Scientific name *Etmopterus bigelowi* Shirai and Tachikawa, 1993, in *Copeia* 1993: 483–495. *Etmopterus: Etm,* from the Greek *etmagen,* to cut, and Greek *pteron,* diminutive of *pteridion,* wing or fin, thus "having cut or torn fins," a reference by Rafinesque to the frayed fins of the original type; *bigelowi,* Latinized form of Bigelow, after Henry B. Bigelow (1869–1967), eminent scientist of Harvard's Museum of Comparative Zoology, who published major treatises on oceanography, plankton, and fishes, including the monumental volumes on sharks and rays with W. C. Schroeder, in the series *Fishes of the Western North Atlantic.*

Synonyms *Etmopterus frontimaculatus* Pietschmann, 1907.

Until recently, all the North Atlantic *Etmopterus* having flattened, concave denticles with four-pointed bases were believed to belong to one species, the smooth lanternshark, *Etmopterus pusillus* (Lowe, 1839). Shirai and Tachikawa (1993) indicated that in fact there are two species of *Etmopterus* with such denticles: the smooth lanternshark, *Etmopterus pusillus* (Lowe, 1839), found in the eastern Atlantic, and the new species *Etmopterus bigelowi* found in the western Atlantic. Both species are widely distributed and nearly cosmopolitan. Although the specimens I have collected in the Gulf of Mexico do not conform fully with Shirai and Tachikawa's description of *E. bigelowi,* I have adopted that new name for the Gulf of Mexico *Etmopterus* having truncate dermal denticles with four-pointed bases. The specimens that I have collected have much larger fin spines and narrower dorsal fin bases, and their distance from the first dorsal fin to a perpendicular through the pectoral axil is about one-fourth the length of the interdorsal space.

Identification This small shark has two dorsal fins, each with a large spine. The interdorsal space is almost as long as the distance from the tip of the snout to the pectoral axil, the upper margin of the caudal fin is almost as long as the distance from the tip of the snout to the level of the second gill slit, and the posterior end of the first dorsal fin base is much nearer to the pectoral fin axil than to the origin of the pelvic fins. This species shows profound morphometric changes with growth; in both sexes the length of the trunk increases allometrically from 18% to 32% of the total length (Shirai and Tachikawa 1993). The upper teeth have three (usually) to five triangular, smooth-edged cusps, the median cusp being much larger; the lower teeth have one strongly oblique, smooth-edged cusp with a deep notch on its outer margin. Females usually have upper teeth with only two cusps flanking a larger central cusp; adult males have upper teeth with multiple side cusps, presumably to help hold the female during copulation. Teeth number U: 22 to 26, L: 17 to 19–1–17 to 19. The dermal denticles have *low concave crowns* and *four-pointed bases,* and they are randomly distributed. Color in life is black above, with a wide band of bright metallic green on the sides extending from head to tail, and brownish black undersides. The green side band is highly visible in fresh specimens in bright daylight; it fades a few hours after death. The undersides bear numerous minute blue photophores. In live or freshly caught sharks in daytime, the ventral photophores appear as round, pearly, bright blue structures scattered among the much larger skin denticles (the photophores are about 0.1 mm or less, or about half the size of the denticles). The photophores are seen only in dim light or with the help of a 10× magnifying lens. They can remain bright blue for several days after death if

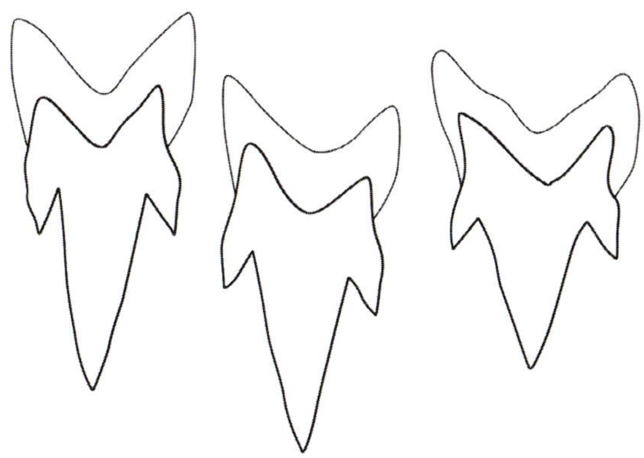

Fig. 23b. Teeth (upper).

Fig. 23c. Teeth (lower).

the specimen is kept on ice; they are even visible after a few days of formalin storage. They turn dark brown after preservation. The anterior borders of the fins are black while the body of the fins is translucent. The lateral line near the tip of the tail is open and shows as a thin black streak. There is a pale yellowish spot on top of the head (pineal window).

Similar species The other squaloid sharks in the western Atlantic lack the flat, four-pointed denticles. The smaller smooth lanternshark (*E. pusillus*) of the eastern Atlantic differs from Bigelow's lanternshark in having a smaller number of intestinal valve turns (10–13 in *E. pusillus,* 16–19 in *E. bigelowi*) and in having more teeth (U: 22–31, L: 30–53 in *E. pusillus,* U: 19–24, L: 25–39 in *E. bigelowi*).

Range This shark appears to be a very widely distributed dogfish, nearly cosmopolitan in warm temperate waters. Prior confusion with the smooth lanternshark prevents a precise description of its range. Shirai and Tachikawa (1993) identified specimens from the Gulf of Mexico and the Caribbean Sea, the South Atlantic, the Indian Ocean, and the central and western Pacific Ocean. In North America, Bigelow's lanternshark has only been reported from the Gulf of Mexico (Bigelow et al. 1955). I have found it to be common in the eastern Gulf, having caught several in the area of Vernon Bank.

Size and weight Maximum known size is 67 cm for males and 65 cm for females (Shirai and Tachikawa 1993). A

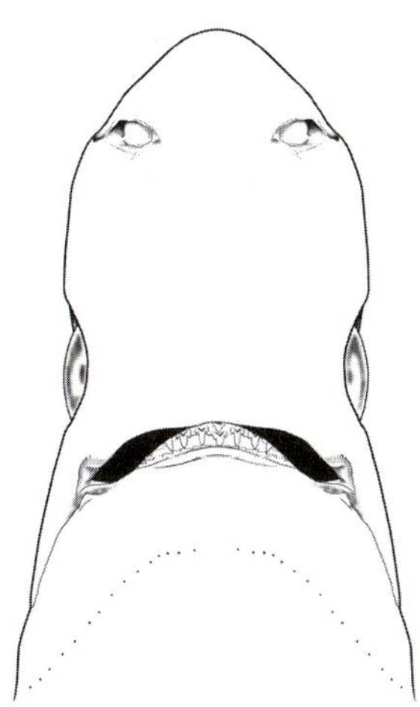

Fig. 23d. Snout.

Fig. 23e. Dermal denticles.

49-cm female that I examined weighed 0.46 kg; a 42-cm male weighed 0.14 kg.

Biology This is one of the largest and most beautiful of the lanternsharks. As with many other lanternsharks, the emerald lanternshark is bioluminescent, having a bright green side stripe and blue ventral photophores. Based on my observations, the luminescence is emitted only in daytime. None of the live specimens caught at night that I have observed exhibited any luminescence; in these specimens the green band along the flanks was visible only as a band of a different shade of dark brown, and the ventral photophores were not visible. These observations suggest that the purpose of the luminescence is to break the shark's outline and prevent it from being seen when silhouetted against the daytime surface glare or when viewed from the side in the deep but clear waters it inhabits.

The emerald lanternshark is a deep-water species, having been recorded from depths of 163–1,000 m (Shirai and Tachikawa 1993). The specimens that I caught in the Gulf of Mexico came from depths of 250–1,000 m; most were taken at 550–750 m.

The emerald lanternshark appears to feed on a variety of small (5–15 cm) deep-dwelling fishes and squid. I have found barracudina (Paralepididae), lanternfish (Myctophidae), and squid, all swallowed whole, in the stomachs of specimens taken in the Gulf of Mexico. Nothing else is known of its habits.

Size at maturity According to Shirai and Tachikawa (1993), males and females are both believed to mature at a size greater than 42 cm.

Reproduction This species is aplacental viviparous, like other squaloid sharks.

Size at birth: Unknown.

Brood size: Females that I examined carried 8–10 mature oocytes in the ovary, suggesting broods of such range.

Nurseries Unknown.

Age and growth Unknown.

Relation to humans None.

LINED LANTERNSHARK

Etmopterus bullisi Bigelow and Schroeder, 1957

Fig. 24a. *Etmopterus bullisi*, lined lanternshark, mature male, 30.3 cm, from the western Atlantic (31°24′ N, 77°54′ W; UF # 96292, RV *Gillis*, collector).

Common name Lined lanternshark, a reference to the longitudinal lines of skin denticles on its back and sides, and to the bioluminescence of the genus.

Spanish name The species lacks a vernacular name in the Spanish Caribbean.

Scientific name *Etmopterus bullisi* Bigelow and Schroeder, 1957, in *Bulletin of the Museum of Comparative Zoology* 117(1): 50–54, fig. 5 A–D). *Etmopterus: Etm,* from the Greek *etmagen,* to cut, and Greek *pteron,* diminutive of *pteridion,* wing or fin, thus "having cut or torn fins," a reference by Rafinesque to the frayed fins of the original type; *bullisi:* latinized form of Bullis, after Harvey R. Bullis, Jr., of the Bureau of Commercial Fisheries, who was active in exploratory fishing in the Gulf of Mexico in the mid-twentieth century.

Synonyms None.

Identification This small shark is characterized by two dorsal fins, each with a spine; a *first dorsal fin originating over the free rear tips of the pectoral fins,* and a *space between the dorsal fins that is shorter than the distance from the tip of the snout to the first gill slit.* The upper teeth have three smooth-edged cusps, the central cusp being much larger; the lower teeth have a single cusp with a nearly horizontal cutting edge and a notch on its outer margin. Teeth number U: 18–20, L: 27–31. The dermal denticles are long, spinelike, and erect; those on the back and sides are *arranged in regular longitudinal rows.* The coloration of freshly dead specimens has not been reported. The species is probably bioluminescent like many other lanternsharks. The coloration of preserved specimens is brown above and darker or blackish below, with distinct flank markings. The dark ventral coloration extends upward, just posterior to the cloaca, into a wide band that bifurcates into two thin dark lines, one extending forward for a few centimeters and another extending to the rear. Several lines of dark spots extend along the dorsal surface. One line of dark spots runs along the dorsal midline; a double row of spots extends from the spiracle to the base of the tail, and another double row extends from the pectoral axil to the pelvic fin region. The fins are light colored. There is a pale band along the dorsal midline with a series of black dots along its center; these dots are presumably bioluminescent organs. There is a pale yellowish spot on top of the head.

Similar species The broadband lanternshark has bristlelike dermal denticles. In other similar lanternsharks, the space between the dorsal fins is longer than the distance from the tip of the snout to the first gill slit.

Range This little shark inhabits the western North Atlantic from North Carolina to the Caribbean Sea, and it is common throughout its range.

Size and weight Most specimens captured have been juveniles of 18–24 cm. The size of adults is not known, but it probably reaches 40–50 cm.

Biology This is a deep-water species usually found at 350–650 m. It feeds on small crustaceans and squids. It is capable of swallowing relatively large prey whole, probably by distending the jaws. One 15-cm specimen I examined had an intact 5- to 6-cm squid in its stomach, a truly astounding sight.

Size at maturity Unknown.

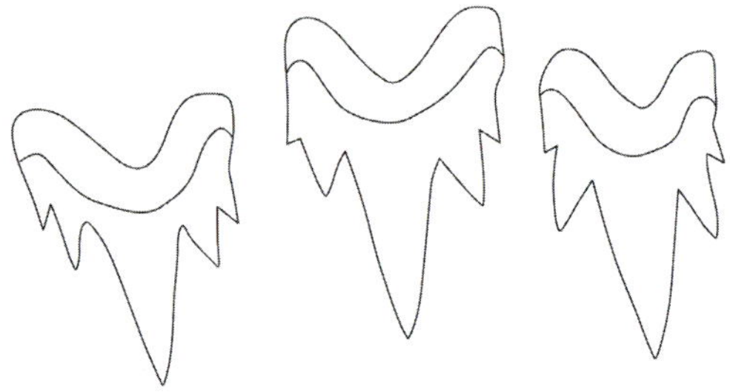

Fig. 24b. Teeth (upper).

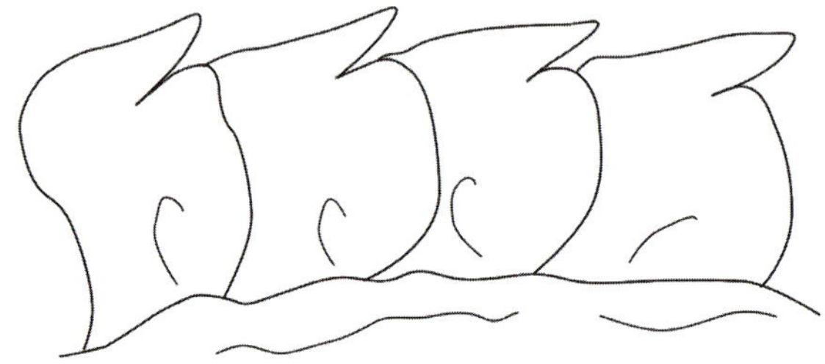

Fig. 24c. Teeth (lower).

Reproduction The lined lanternshark is probably aplacental viviparous like other lanternsharks.

Size at birth: Unknown.

Brood size: Unknown.

Nurseries Unknown.

Age and growth The species has not been aged.

Relation to humans None.

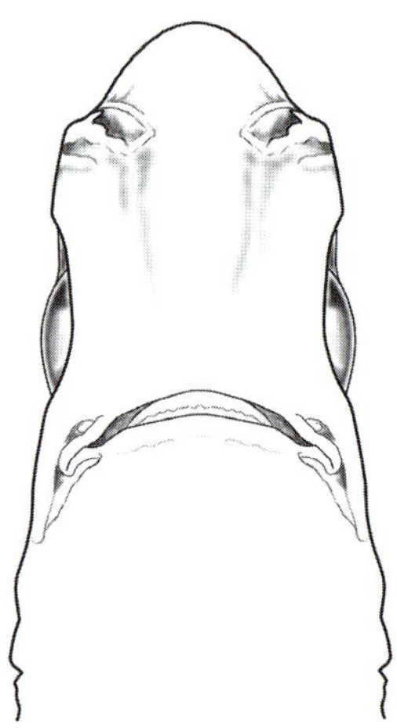

Fig. 24d. Snout.

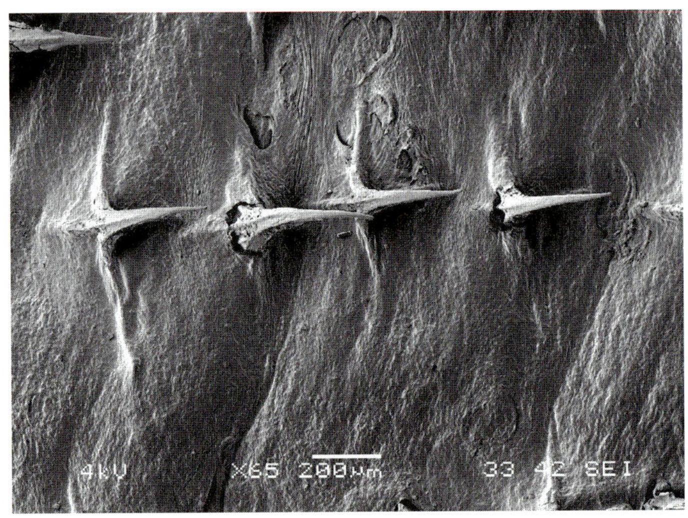

Fig. 24e. Dermal denticles.

BROADBAND LANTERNSHARK

Etmopterus gracilispinis Krefft, 1968

Fig. 25a. *Etmopterus gracilispinis*, broadband lanternshark, female, 16.3 cm, 0.02 kg; off Cape Hatteras, North Carolina (John Galbraith, collector).

Common name Broadband lanternshark, a reference to the markings on the flanks and to the bioluminescence of members of the family.

Spanish name The species lacks a vernacular name in the Spanish Caribbean.

Scientific name *Etmopterus gracilispinis* Krefft, 1968, in *Archiv fuer Fischerewissenschaft* 19 (1): 3–9. *Etmopterus: Etm,* from the Greek *etmagen,* to cut, and Greek *pteron,* diminutive of *pteridion,* wing or fin, thus "having cut or torn fins," a reference by Rafinesque to the frayed fins of the original type; *gracilispinis:* from the Latin *gracilis,* slender or thin, and *spinis,* from *spina,* thorn or spine, an allusion to its slender dorsal fin spines.

Synonyms None.

Identification This small shark has two dorsal fins, each with a spine; an *interdorsal space shorter than the head length* (the distance from the tip of the snout to the origin of the pectoral fins); and *pectoral fins ending far short of the dorsal fin origin.* The *midpoint of the first dorsal fin is equidistant from the pectoral fin axil and the origin of the pelvic fins.* The upper teeth have five cusps, with the median being the largest. The lower teeth have one outwardly pointing cusp. Teeth number about U: 27, L: 28. The dermal denticles are high and bristlelike and have four-pointed bases. The underside of the snout is rough with denticles. Coloration is dark brown above with a reddish band along the dorsal surface and black winglike markings on the flanks. The undersides are black. The head has a pale yellowish spot on top.

Similar species The lined lanternshark has pectoral fins that reach to the level of the first dorsal fin origin. Other small squaloids have a space between the dorsal fins longer than the distance from the tip of the snout to the first gill slit.

Range Although the range limits of the broadband lanternshark are not well known, it appears that the species is widely distributed in the temperate Atlantic. In the western North Atlantic, the broadband lanternshark has been found off New England (J. Galbraith, pers. comm., June 2004), New Jersey (Hudson Canyon, collected by J. A. Musick in 1991, Virginia Institute of Marine Science no. 08332), and North Carolina (Schwartz and Burgess 1975). In the South Atlantic the species has been reported from Uruguay and Argentina (Krefft 1968).

Size and weight This is a very small shark that apparently does not exceed 40 cm. I have recorded the following weights:

Male: 23 cm, 30 g; 26 cm, 60 g.

Female: 24 cm, 60 g; 28 cm, 80 g.

Biology This is a deep-water species that has been caught at 450–800 m in temperate zones. It is believed to ascend in the water column at night. Nothing else is known of its habits.

Size at maturity The broadband lanternshark probably matures at 26–28 cm. The 26-cm male holotype was reported as immature (Krefft 1968). However, a 26-cm, 60-g specimen that I examined had calcified, mature claspers. Krefft also reported a 33-cm female as immature. A 24.8-cm female that I examined was immature and had oocytes less that 1-mm in diameter, while a 28-cm female carried eight oocytes that were 10 mm in diameter, suggesting that she was ready to ovulate. These differences between Krefft's material and the numerous specimens that I have examined suggest that there may be population differences, or even different species involved.

Reproduction All other species of *Etmopterus* are aplacental viviparous, and the broadband lanternshark should be

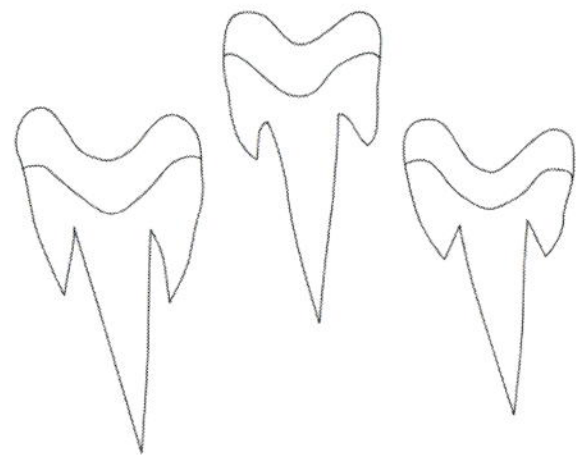

Fig. 25b. Teeth (upper).

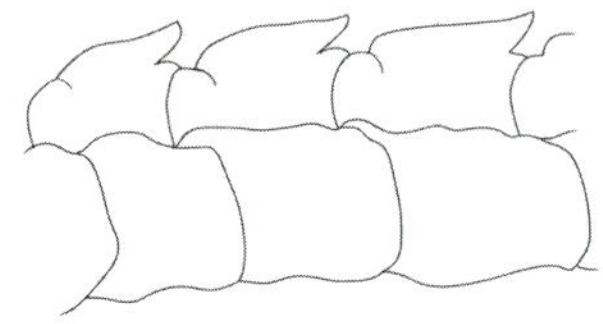

Fig. 25c. Teeth (lower).

no exception. The oocytes reach a proportionally huge size. The apparently ripe, 10-mm-diameter oocytes I observed in a 28-cm, 80-g female must require considerable energy investment for such a small creature, suggesting long reproductive cycles (probably two-year cycles), similar to many other squaloids.

Size at birth: Krefft (1968) reported a 13-cm specimen as a neonate.

Brood size: The females I have examined carried six to nine very large oocytes that appeared mature. It is reasonable to assume that all or most of these would produce embryos.

Nurseries Unknown.

Age and growth The species has not been aged.

Relation to humans None.

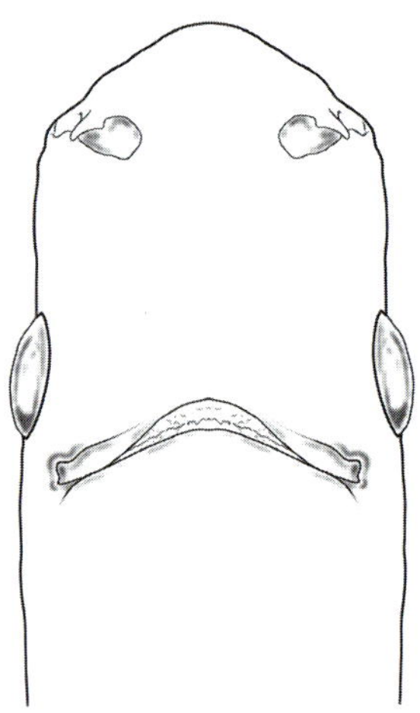

Fig. 25d. Snout.

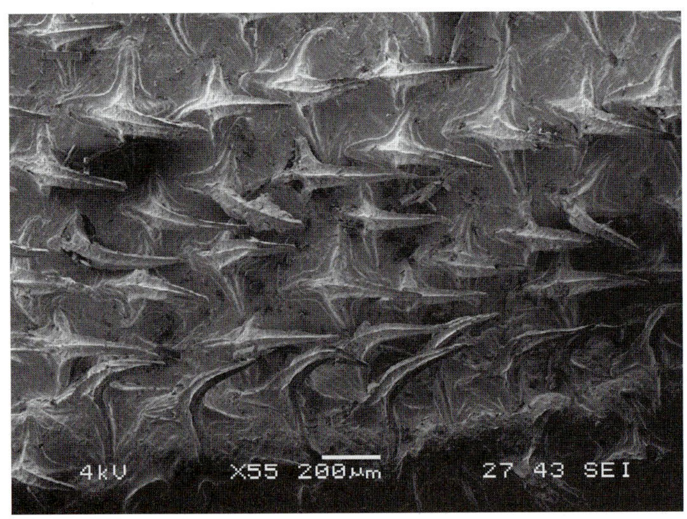

Fig. 25e. Dermal denticles.

CARIBBEAN LANTERNSHARK

Etmopterus hillianus (Poey, 1861)

Fig. 26a. *Etmopterus hillianus*, Caribbean lanternshark, female, 22 cm, 0.03 kg; from off Jamaica (George Burgess, collector).

Common name Caribbean lanternshark, allusions to its habitat and luminescence.

Spanish name The species lacks vernacular or common names in the Spanish Caribbean.

Scientific name *Etmopterus hillianus* (Poey, 1861), originally described as *Spinax hillianus* in *Memorias sobre la Historia Natural de la Isla de Cuba* 2(203): 340–342, plate 19, figs. 13, 14. *Etmopterus: Etm,* from the Greek *etmagen,* to cut, and Greek *pteron,* diminutive of *pteridion,* wing or fin, thus "having cut or torn fins," a reference by Rafinesque to the frayed fins of the original type; *hillianus:* latinized form of the name Hill, or "of Hill," named after Richard Hill (1795–1872), a Jamaican antislavery agitator, judge, and naturalist who corresponded with Poey.

Synonyms None.

Identification This beautiful little shark is characterized by two dorsal fins, each with a spine; a distance from the spiracle to the tip of the snout that is slightly longer than the distance from the spiracle to the pectoral fin origin; and a *space between the pelvic fins and the caudal fin that is as long as the distance from the pectoral axil to the origin of the pelvic fins.* The upper teeth usually have five cusps (rarely three or seven), the central cusp being much larger. The lower teeth have one cusp with a nearly horizontal cutting edge and a notch on its outer margin. Teeth number about U: 12–12, L: 18–18. The dermal denticles are bristlelike, with long tapering points and four-pointed bases concealed in the skin. *The underside of the snout is rough with dermal denticles.* Coloration in life is chocolate brown above, with brilliant purplish sides. The undersides are black, with a band of the black color extending upward from the vent and continuing both anteriorly and posteriorly. The tail is black except for the anterior margin, which is purple. These colors fade after death to a dark chocolate brown. The head has a pale yellowish spot on top (a pineal window). Small black dots are scattered over the top of the head and continue posteriorly into a single row along the midline; two to four rows of narrow black dashes extend along the sides. These black dots and dashes are presumably bioluminescent.

Similar species The green lanternshark lacks dermal denticles on the underside of the snout. The lined lanternshark and the broadband lanternshark have a space between the dorsal fins that is much shorter than the distance from the tip of the snout to the first gill slit. The emerald lanternshark has a longer interdorsal space, almost as long as the distance from the tip of the snout to the pectoral fin axil. In the greater lanternshark and the fringefin lanternshark, the space between the pelvic fins and the caudal fin is much shorter than the distance from the rear tips of the pectoral fins to the origin of the pelvic fins.

Range This small shark is found from Virginia to Florida and into the Gulf of Mexico and Caribbean basin. It was first reported from waters off Havana by Poey (1861), although it was not included in Guitart Manday's (1968) guide to Cuban sharks.

Size and weight This is a tiny shark, averaging about 25 cm. It grows to at least 32 cm.

Biology It appears to be confined to deep water and has been reported from 380 to 716 m. Nothing else is known of its habits.

Size at maturity Females mature at about 29 cm, and males at about 25 cm (Bigelow and Schroeder 1948).

Reproduction The Caribbean lanternshark is an aplacental viviparous species.

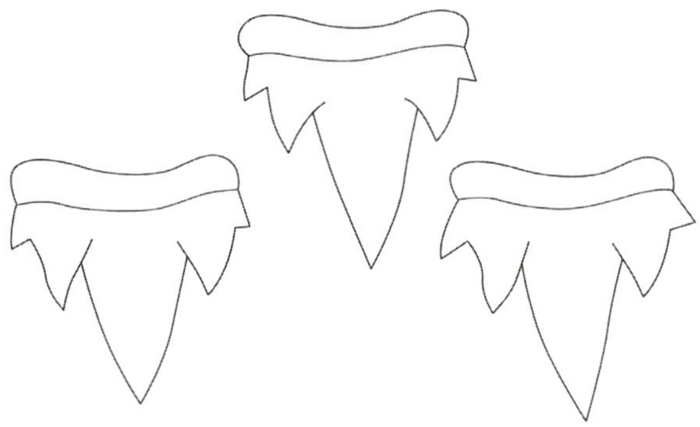

Fig. 26b. Teeth (upper).

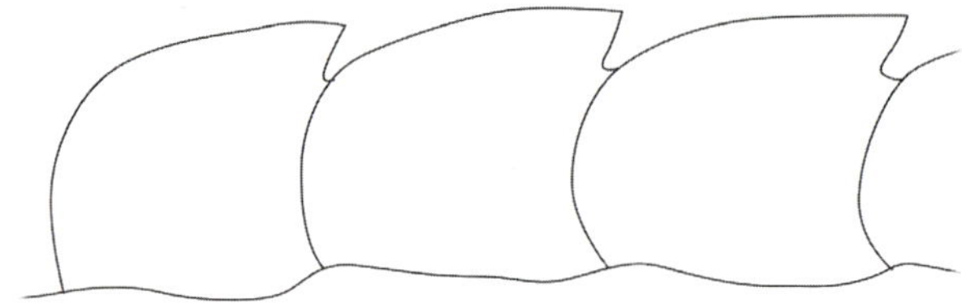

Fig. 26c. Teeth (lower).

Size at birth: Bigelow and Schroeder (1948) reported that a 29.5-cm female contained four young 8–8.5 cm and suggested that length at birth is slightly more than 9 cm.

Brood size: Bigelow and Schroeder (1948) reported broods of four to five.

Nurseries Unknown.

Age and growth The species has not been aged.

Relation to humans None.

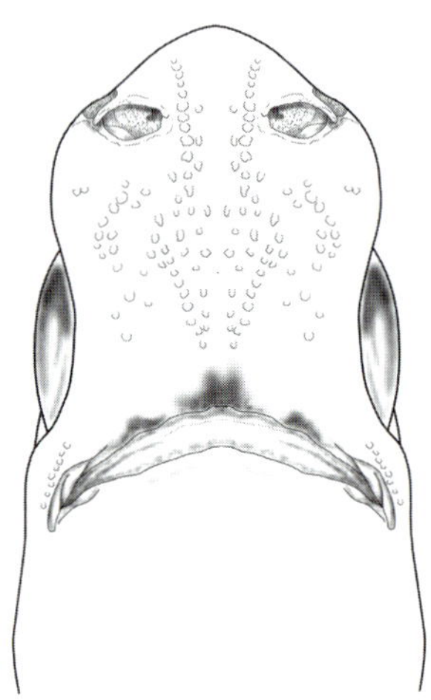

Fig. 26d. Snout.

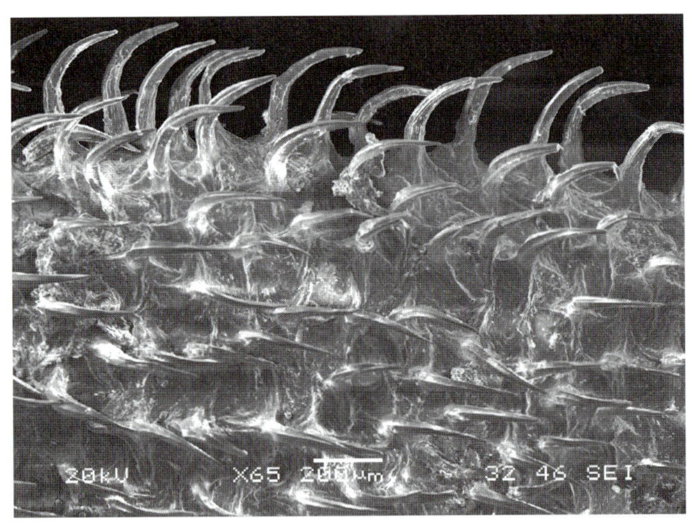

Fig. 26e. Dermal denticles.

GREAT LANTERNSHARK

Etmopterus princeps Collett, 1904

Fig. 27a. *Etmopterus princeps*, great lanternshark, immature male, 32.0 cm, 0.15 kg; off Georges Bank (John Galbraith, collector).

Common name Great lanternshark, a reference to the species being one of the largest lanternsharks, although, despite its name, it is not known to be bioluminescent.

Spanish name The species lacks vernacular names in the Spanish Caribbean. The A.F.S. has not designated a name for it. In Spain it is known as *tollo raspa.*

Scientific name *Etmopterus princeps* Collett, 1904, in *Christiania Videnskabs-Selskabs Forhandlinger for 1904* 9: 3–4. *Etmopterus: Etm,* from the Greek *etmagen,* to cut, and Greek *pteron,* diminutive of *pteridion,* wing or fin, thus "having cut or torn fins," a reference by Rafinesque to the frayed fins of the original type; *princeps:* Latin for first or chief, probably an allusion to its large size or being the largest of the genus.

Synonyms None.

Identification This is one of the larger and stouter of the American lanternsharks. It has two dorsal fins each with a spine, the second dorsal fin and spine being much larger than the first. The spine of the second dorsal fin arises above or behind the level of the pelvic fin bases. The midpoint of the first dorsal fin is much closer to the pectoral fin axil than to the origin of the pelvic fins. The space between the dorsal fins is greater than the distance from the tip of the snout to the first gill slit. The upper teeth have five smooth-edged cusps, the central cusp being much larger. Small specimens may have upper teeth with only three cusps. *The lower teeth have a single, strongly oblique cusp with a notch on its outer margin.* Teeth number about U: 29–32, L: 40–50, the number varying with age. The dermal denticles are nearly erect, with prominent ridges, squarish bases, and the point turned rearward. *The denticles are large enough to be seen with the naked eye and give the skin a fuzzy or sandy appearance.* The denticles from the first dorsal fin to the tail are arranged in rows that converge at the caudal peduncle. *The underside of the snout is free of denticles and feels smooth to the touch.* Small juveniles up to 30 cm are black; larger specimens are blackish brown. Both have darker undersides. The edges and rear tips of the dorsal fins are usually white or whitish. This species, unlike other lanternsharks, *lacks the dots or dashes on the flanks and the pale yellowish spot on top of the head.* It does not appear to be bioluminescent.

Similar species The black dogfish has three-cusped teeth on both jaws. The emerald lanternshark has a yellowish spot on top of the head and has flat dermal denticles. The lined lanternshark has conspicuous body markings and a yellowish spot on top of the head, the interspace between the dorsal fins is shorter than the distance from the tip of the snout to the first gill slit, and its dermal denticles are arranged in rows. The Caribbean lanternshark and the green lanternshark have conspicuous body markings. The broadband lanternshark has distinct winglike flank markings, the interspace between its dorsal fins is shorter than the distance from the snout to the first gill slit, and the midpoint of its first dorsal fin is equidistant from the pectoral fin axil and the pelvic fin origin.

Range The great lanternshark is known from both sides of the North Atlantic. In the eastern North Atlantic, it has been reported from Iceland (Jakobsdóttir 2001), the Faroes (Collett 1904), the European coast south to the Strait of Gibraltar, and the Moroccan coast (Koefoed 1927; Dollfus 1955). In North America it occurs from southern Nova Scotia to southern New England (Bigelow et al. 1953; Bigelow and Schroeder 1954). Templeman (1963) mentioned that it had not been reported for Newfoundland, but that some small specimens had been caught on the Scotian shelf seaward from La Have Bank. Other authors have also reported catching the species from the Scotian shelf to New England (Bigelow et al. 1953; Scott and Scott 1988). It is abundant in deep waters of the Scotian shelf off New England, where a single trawl can produce dozens of specimens.

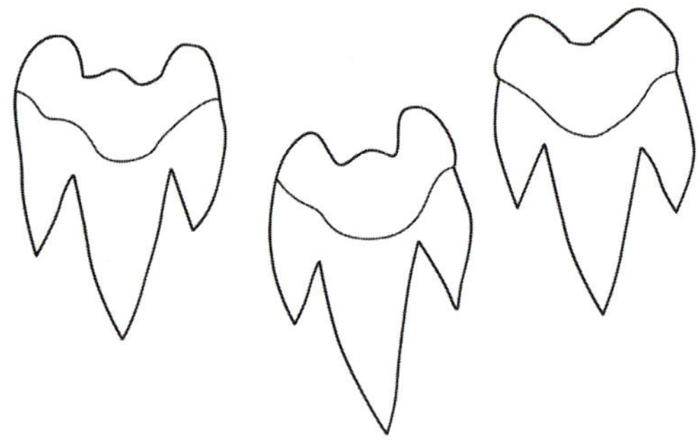

Fig. 27b. Teeth (upper).

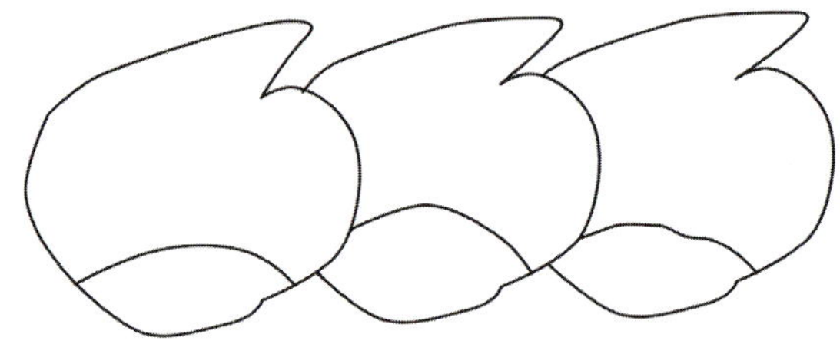

Fig. 27c. Teeth (lower).

Size and weight Males reach 70 cm and 1.8 kg; large females can attain 80 cm and weigh 3.2 kg (Jakobsdóttir 2001). Representative sizes and weights of specimens I have examined are as follows:

Male: 51.6 cm, 0.64 kg; 56.3 cm, 0.8 kg.

Female: 40.7 cm, 0.32 kg; 56.4 cm, 0.91 kg; 60.6 cm, 0.95 kg; 61.9 cm, 1.14 kg.

Biology This is an abundant shark of the deep continental slopes of the North Atlantic. Bigelow et al. (1953) reported that specimens had been caught at 566–950 m from southern Nova Scotia to southern New England. Gordon and Swan (1997) reported catching it at 689–1,861 m, in the Rockall Trough on the Hebridean slope between 55 and 57° N. Jakobsdóttir (2001) noted a depth range of 436–1,653 m for the species off Iceland and stated that it occurred at 800–1,000 m. She also observed that the species was caught at bottom temperatures of 2–6.4°C and that it was most abundant at 3–4.5°C.

Dietary information is scanty. Mauchline and Gordon (1983) reported that seven specimens, taken in the Rockall Trough west of Scotland and Iceland, had consumed decapod crustaceans, fishes, and cephalopods. Jakobsdóttir (2001) examined the stomach contents of 74 specimens caught in Icelandic waters. Of these, 32 (43%) were empty; teleosts had the highest frequency of occurrence (42%), with myctophids being the most abundant group of those identified. Cephalopods and crustaceans had similar frequency of oc-

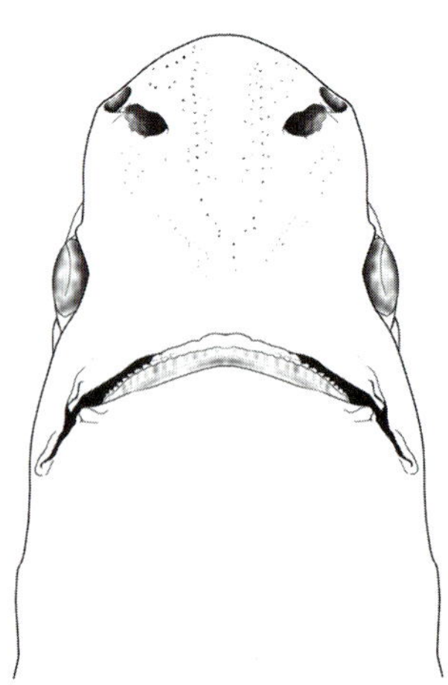

Fig. 27d. Snout.

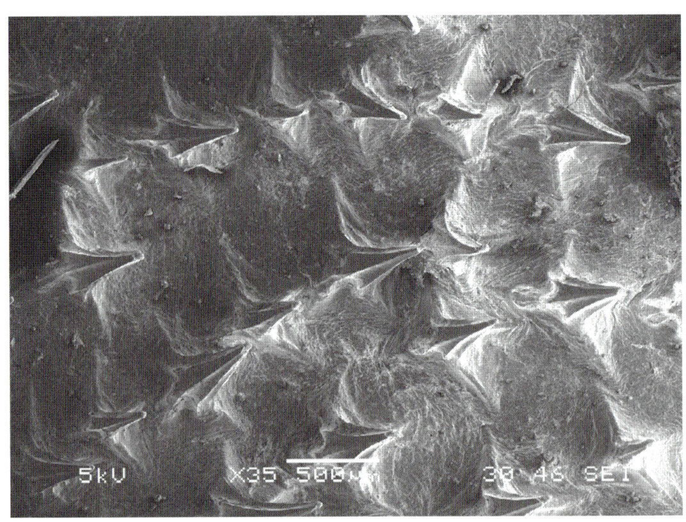

Fig. 27e. Dermal denticles.

currence, both at 26%. North American specimens I have examined had consumed small squids and shrimps. These food items suggest a bottom-feeding predator.

Size at maturity

Males: Males apparently mature at about 55 cm. Bigelow and Schroeder (1953) reported that a 55-cm male appeared to be mature. The largest immature male I have examined measured 52 cm, while the smallest mature male measured 56.3 cm.

Females: Females mature at about 62 cm. A 61.4-cm, 1.3-kg female I examined in late May carried 11 large oocytes 28 mm in diameter. Their relatively large size suggests that the female was close to maturity and ovulation. The values presented here agree well with those for the Icelandic population described by Jakobsdóttir (2001), who wrote that the length at which 50% of the population reached maturity is 62.2 cm for females and 57.3 cm for males.

Age and growth The species has not been aged.

Relation to humans None. The species is a frequent bycatch in the commercial fishery for Greenland halibut (*Reinhardtius hippoglossoides*) off western Iceland, and large quantities of great lanternsharks are discarded by the fishery (Jakobsdóttir 2001).

FRINGEFIN LANTERNSHARK

Etmopterus schultzi Bigelow, Schroeder, and Springer, 1953

Fig. 28a. *Etmopterus schultzi*, fringefin lanternshark, female, 31.5 cm; from about 100 km south of Mobile Bay, Alabama (UF # 81745; RV *Oregon*, collector).

Common name Fringefin lanternshark, or fuzzy dogfish. The name fuzzy dogfish is an older name for the species (Springer and Bullis 1956). More recently, the name fringefin lanternshark has been used by the F.A.O. (Compagno 1984a). Both names allude to the edges of its pectoral fins, which appear as a thin, whitish fringe. The names "dogfish" and "lanternshark" refer, respectively, to the presence of dorsal fin spines and the bioluminescence of the genus.

Spanish name The species lacks vernacular names in the Spanish Caribbean.

Scientific name *Etmopterus schultzi* Bigelow, Schroeder, and Springer, 1953, in *Bulletin of the Museum of Comparative Zoology* 109(3): 252–257. *Etmopterus: Etm,* from the Greek *etmagen,* to cut, and Greek *pteron,* diminutive of *pteridion,* wing or fin, thus "having cut or torn fins," a reference by Rafinesque to the frayed fins of the original type; *schultzi:* latinized form of Schultz, after Leonard P. Schultz, curator of fishes (1938–1965) at the Smithsonian Institution, U.S. National Museum.

Synonyms None.

Identification This small shark is characterized by two dorsal fins, each with a spine; *pectoral fins with a broad fringe of naked, hornlike rays* (ceratotrichia); a space between the dorsal fins that is as long as the distance from the tip of the snout to the origin of the pectoral fin; and a *caudal fin upper lobe about as long as the distance from the tip of the snout to the pectoral fin axil.* The upper teeth have five to eight (usually seven) pointed, smooth-edged cusps, with the median cusp being the much larger. The lower teeth have one cusp with a nearly horizontal cutting edge and a notch on its outer margin. Teeth number U: 32 to 38, L: 32. The dermal denticles are *minute and bristlelike, curve strongly backward,* and have four-pointed bases. The denticles are closely spaced and randomly distributed (not in any linear arrangement). The denticles are often covered with mucus or mud and may be difficult to see in sharks preserved in alcohol. Coloration is dark sooty gray above and black below. There are two irregular lines of short, narrow, black markings along each side, with a single line along the dorsal midline. These black markings can be difficult to discern. The head has a pale yellowish spot on top. Lanternsharks are usually bioluminescent, but luminescence has not been reported in this species.

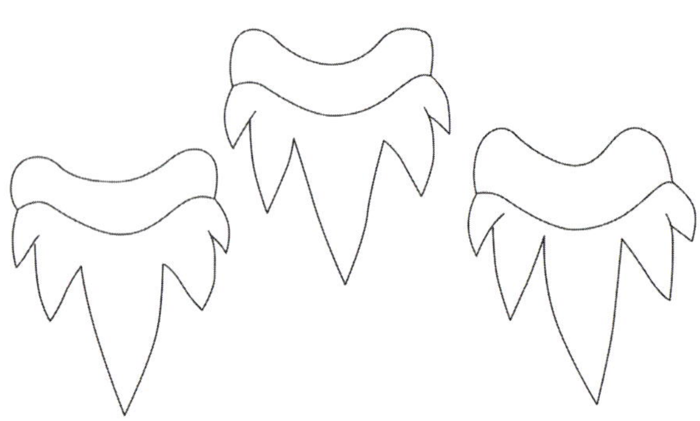

Fig. 28b. Teeth (upper).

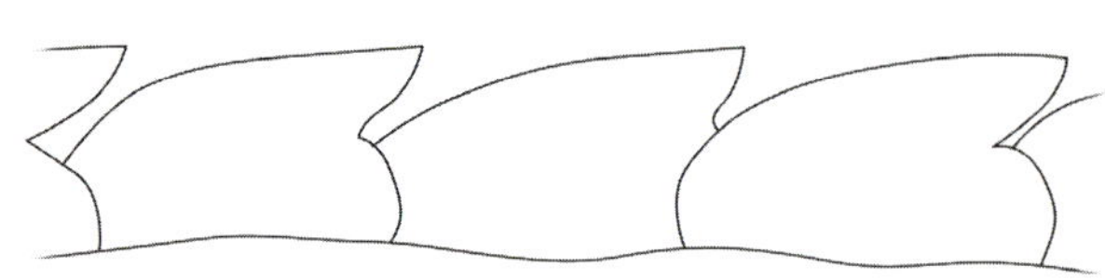

Fig. 28c. Teeth (lower).

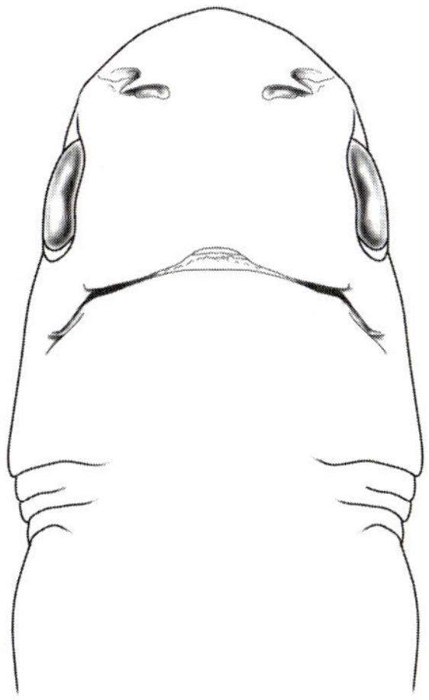

Fig. 28d. Snout.

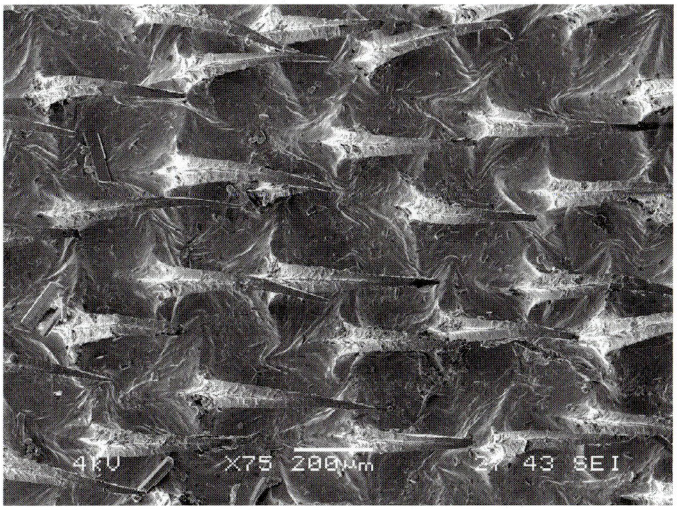

Fig. 28e. Dermal denticles.

Similar species Other lanternsharks in North America lack the pectoral fins with a conspicuous fringe of ceratotrichia on the rear margins.

Range The fringefin lanternshark was first described from specimens caught in trawls in the northern Gulf of Mexico, where the species appears to be common. Because of its small size, deep-water habitat, and the obvious difficulties of capturing specimens, the limits of the range are unknown.

Size and weight A 27.5-cm female carried "several large eggs, apparently not fertilized," whereas the claspers of the 27-cm type specimen appear full grown, suggesting a maximum length around 30 cm (Bigelow et al. 1953).

Biology This is a deep-water species, as most captures have been reported from 380–730 m (Bigelow et al. 1953). It is known to feed on squid.

Size at maturity It apparently matures at less than 27 cm, but the size at maturity is not known precisely.

Reproduction The reproductive processes have not been described. The species is almost surely aplacental viviparous, like other squaloid sharks.

Size at birth: Unknown.

Brood size: Unknown.

Nurseries Unknown.

Age and growth The species has not been aged.

Relation to humans None.

GREEN LANTERNSHARK

Etmopterus virens Bigelow, Schroeder, and Springer, 1953

Fig. 29a. *Etmopterus virens*, green lanternshark, female, 23.3 cm, 0.05 kg; from south-west of Loggerhead Key, Florida (George Burgess, collector).

Common name Green lanternshark, referring to its coloration and bioluminescence in life.

Spanish name The species lacks vernacular names in the Spanish Caribbean.

Scientific name *Etmopterus virens* Bigelow, Schroeder, and Springer, 1953, in *Bulletin of the Museum of Comparative Zoology* 109(3): 257–263. *Etmopterus: Etm,* from the Greek *etmagen,* to cut, and Greek *pteron,* diminutive of *pteridion,* wing or fin, thus "having cut or torn fins," an allusion by Rafinesque to the frayed fins of the original type; *virens:* Latin for green.

Synonyms None.

Identification This beautiful small shark is characterized by two dorsal fins, each with a spine, and a distance from the spiracle to the pectoral fin origin that is about two-thirds the distance from the spiracle to the tip of the snout. The head length (distance from the tip of the snout to the origin of the pectoral fin) is about 25% of the length of the trunk (from the tip of the snout to the caudal fin origin).[1] The upper teeth have five cusps (occasionally four), with the median cusp being much larger than the others; the lower teeth have one cusp with a nearly horizontal cutting edge and a notch on its outer margin. Teeth number about U: 29 to 34, L: 24 to 32. The dermal denticles are long and pointed, with ridged sides and squarish bases concealed in the skin. The *skin on the lower surface of the snout up to the mouth lacks dermal denticles and is smooth to the touch.* Coloration is dark sooty brown with metallic purple hues on the upper surfaces, and black with green luminescence on the ventral surfaces. The black coloration extends upward along the flanks, and just below the second dorsal fin it continues upward into a narrow black band that divides and extends both anteriorly and posteriorly. The tips of upper and lower lobes of the caudal fin are black tipped. The ventral surfaces of a freshly caught specimen shine with a green bioluminescence and conspicuous green photophores; this bioluminescence is lost upon preservation. There are three conspicuous lines of photophores on each flank, and two narrow, pale bluish gray *longitudinal stripes along the sides.* The stripes merge together on the rear portion of the trunk, forming a conspicuous pattern of *dark and pale flank markings.* The head has usually a pale yellowish spot on top.

1. In their description of the green lanternshark, Bigelow et al. (1953) mentioned that the caudal fin is as long as the head to the origin of the pectoral fins. The caudal fin origin may be difficult to discern, and consequently, different observers may get highly variable results. In any case, in the specimens that I have examined, the length of the caudal fin was longer than the distance from the snout tip to the pectoral fin origin (from 5.4–5.6% of the total length, whereas the head length was 4.6–5.4%).

Similar species The Caribbean lanternshark has a longer head length (about 30% of the length of the trunk) and bristlelike dermal denticles; the lower surface of its snout is rough with denticles, and it has different markings on the sides and the rear portion of the trunk. The lined lanternshark has a first dorsal fin originating over the rear margins of the pectoral fin. The fringefin lanternshark has an interdorsal space longer than the head length. The broadband lanternshark has bristlelike dermal denticles on the underside of the snout. The great lanternshark and the emerald lanternshark have uniform dark backgound color and lack the markings on the rear portion of the trunk.

Range The green lanternshark is known from the northern Gulf of Mexico and Caribbean Sea.

Size and weight This is a very small species that probably reaches 30 cm.

Biology This tiny shark inhabits moderately deep waters of the northern Gulf of Mexico, most commonly at depths

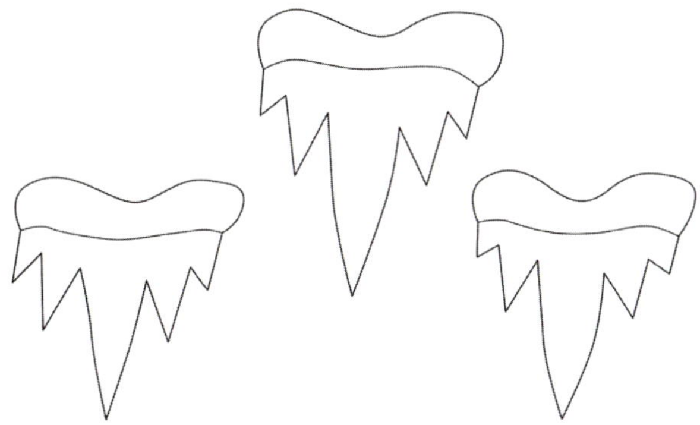

Fig. 29b. Teeth (upper).

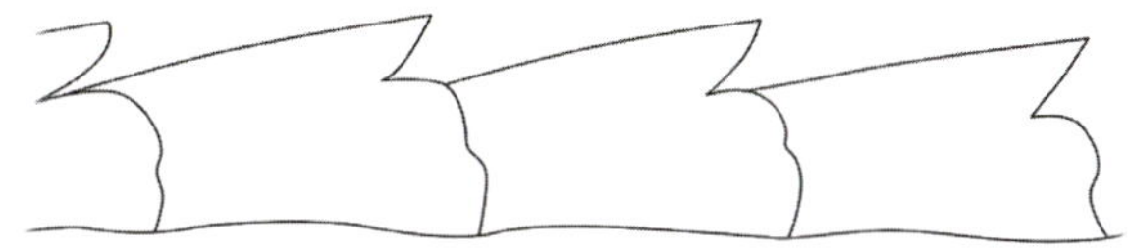

Fig. 29c. Teeth (lower).

of 320–530 m (Bigelow et al. 1953; Eisert 1969). Although the species has been recognized since the 1950s, what little is known about it is based on specimens captured by Springer in the 1950s and 1960s. The green lanternshark seems to be quite common in the northern Gulf, Bigelow et al. (1953) reporting having seen more than 100 specimens taken in a single trawl. Springer (1967) suggested that the species lives in rather dense aggregations, basing this on the fact that repeated trawls at the same depth and habitat usually catch many or none.

Green lanternsharks are known to feed on squids. According to Springer (1967), analysis of hundreds of stomachs revealed that more than half their food is squid or octopus. The cephalopod beaks and eyes commonly found in the stomach contents are so large that the shark's jaws must have been greatly stretched at the time of swallowing. Springer (1967: 162) speculated that green lanternsharks hunt and feed in large aggregations and "may literally swarm over squid or octopus much larger than they are, biting chunks with the razor sharp band of lower-jaw teeth."

Size at maturity

Males: Males appear to mature at less than 20 cm. According to Bigelow et al. (1953), the claspers of the 20.3-cm type specimen and those of another male 22.5 cm appeared to be mature.

Females: Females mature at about 18 cm. Eisert (1969) mentioned an 18.6-cm gravid female; Bigelow et al. (1953) reported a 23-cm female that carried an embryo.

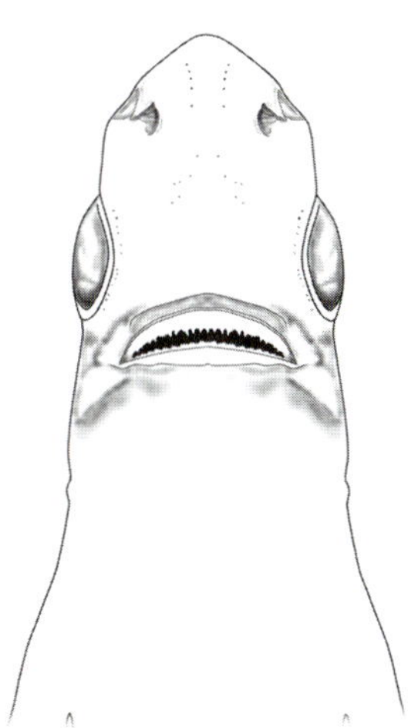

Fig. 29d. Snout.

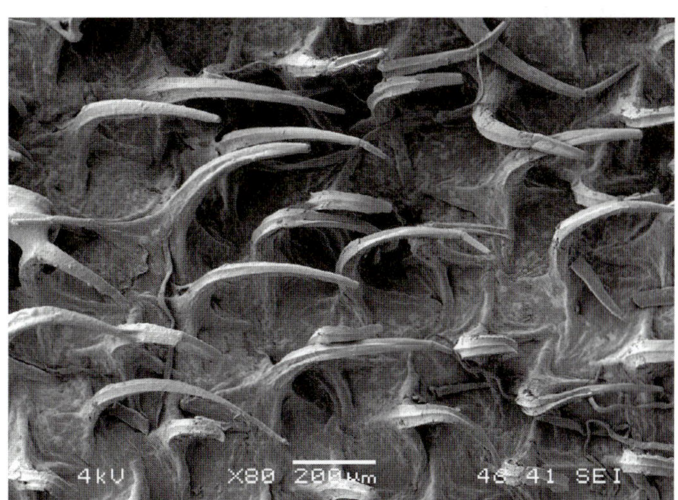

Fig. 29e. Dermal denticles.

Reproduction The green lanternshark is an aplacental viviparous species. Very little is known about its reproductive processes.

Size at birth: Green lanternshark young are probably born at about 9 cm. Eisert (1969) reported embryos 7.9–8.5 cm with "umbilical scars and apparently ready for birth." She also reported a 9.9-cm free-swimming specimen, which she suspected of having been "aborted prematurely." Given the sizes of the largest embryos and the adult size, there is no reason to suspect that a 9- to 10-cm specimen would be aborted (note that the specimen was not measured in a straight line; it probably measured about 9.5 cm).

Brood size: Broods consist of one to three young.

Nurseries Unknown.

Age and growth The species has not been aged.

Relation to humans None.

FAMILY SOMNIOSIDAE
The Sleeper Sharks

According to the latest taxonomy of the squaloid sharks (Shirai 1992), the Somniosidae is a family comprising the genera *Centroscymnus, Scymnodon, Somniosus,* and *Zameus.* These squaloids share two common characteristics of the skeleton: a postorbital fenestra and anteriorly elongated prehypochordal cartilages. The postorbital fenestra is a large opening at the postorbital process of the skull; the prehypochordal cartilages are a series of cartilages at the origin of the lower caudal lobe, which extend forward along the hypaxial musculature. These characteristics cannot be observed externally and are visible only upon careful dissection.

The genera *Centroscymnus* and *Scymnodon* are very similar, and according to Compagno (1984a: 54), the two genera might need to be synonymized. They are small, deepwater squaloid sharks with small, inconspicuous dorsal fin spines that are often covered by the skin or barely protrude through it. These two genera have been separated based on the shape of the lower teeth. *Centroscymnus* has lanceolate or awl-like upper teeth in the center of the mouth, and the lower teeth with low oblique cusps. *Scymnodon* has lanceolate upper teeth that are much longer midway along each side of the jaw than toward either the center of the mouth or its outer corners, and lower teeth with erect, large cusps and interlocking bases. The genus *Zameus* is distinguished from the two genera above by having dermal denticles with a peculiar microstructure of transverse ridges.

Sharks of the genus *Somniosus* have lanceolate upper teeth without cusplets and bladelike lower teeth with oblique cusps and high, interlocking bases. These sharks have uncalcified vertebral centra and lack luminous organs. Some species are very large, reaching more than 4–5 m in length. Although sharks of the genus *Somniosus* are widely distributed, they are very poorly known. Except in high latitudes, *Somniosus* sharks generally inhabit very deep waters, and they are infrequently caught. Specimens can seldom be preserved whole because of their usually large size, and small specimens that can be preserved are rare in museum collections. Thus, series of preserved specimens needed to make comparisons are not available anywhere, and the identification of the large species of the genus is problematic.

The number of species in the genus *Somniosus* has been the subject of debate for many years. In the latest review of the genus Yano et al. (2004) concluded that there were four species: the Greenland shark (*Somniosus microcephalus*) of the North Atlantic and Arctic oceans, the Pacific sleeper shark (*Somniosus pacificus*) of the Pacific and Atlantic oceans, the Antarctic sleeper shark (*Somniosus antarcticus*) of the southern Pacific Ocean, and the little sleeper shark (*Somniosus rostratus*) currently known from the North Atlantic, the Mediterranean Sea, and Japan, although it is probably widely distributed in deep waters.

The identification problems of the large *Somniosus* result from the lack of large series of comparative materials and from workers selecting diagnostic characteristics that are quite variable within a given species, overlapping among species, or difficult to measure accurately. For example, the location of the origin of the first dorsal fin has been used as a diagnostic characteristic to separate the Greenland shark from the Pacific sleeper shark. After I measured only a few specimens of both species, it became obvious that there was considerable variability and that the measurements were difficult to replicate. This fin arises so slowly on the mid-dorsal line that two people measuring the same specimen will invariably obtain different measurements. Benz et al. (2007) have pointed out some of these problems.

As of 2010, there is no reliable or effective way of separating species of the larger *Somniosus,* except by geographic range. The Pacific sleeper and the Greenland shark are common in the North Pacific and in the North Atlantic, respectively, and usually there is no difficulty in identifying specimens in those areas. However, identifying specimens in Arctic waters where their ranges may overlap, in the southern latitudes of the Northern Hemisphere, or in the Southern Hemisphere is problematic. The confusion will continue until modern methods of species separation, such as DNA analysis, are applied to this group and a series of representative specimens become available; then, it will be possible to select suitable morphological characteristics for species separation.

PORTUGUESE SHARK

Centroscymnus coelolepis Barboza du Bocage and de Brito Capello, 1864

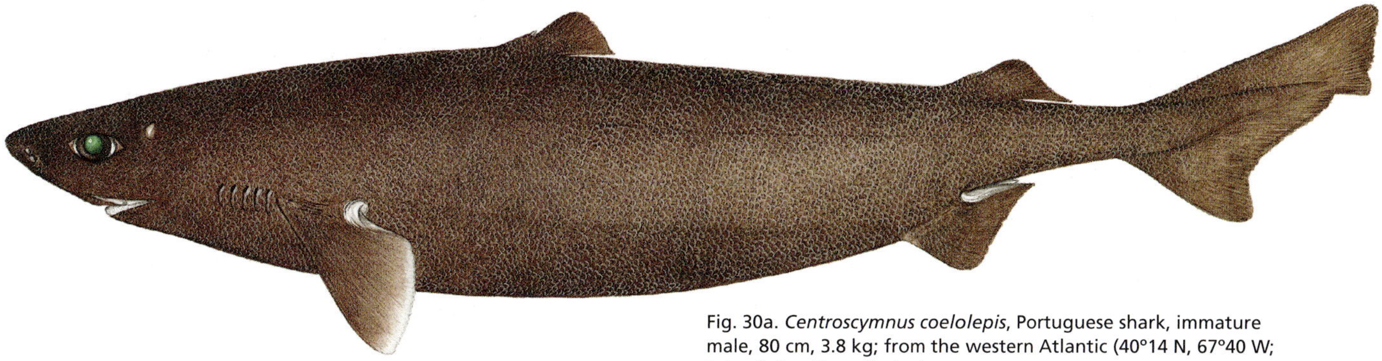

Fig. 30a. *Centroscymnus coelolepis*, Portuguese shark, immature male, 80 cm, 3.8 kg; from the western Atlantic (40°14 N, 67°40 W; John Galbraith, collector).

Common name Portuguese shark. The species was first described from Portugal, hence its name.

Spanish name The species lacks a vernacular name in the Spanish Americas. In Portugal and Spain it is known as *pailona.*

Scientific name *Centroscymnus coelolepis* Barboza du Bocage and de Brito Capello, 1864, in *Zoological Society of London, Proceedings of the Scientific Meetings 1864:* 260–263. *Centroscymnus: Centro,* from the Greek *kemtron,* spine, and *skymnos,* the ancient name for some shark, thus a spined shark; *coelolepis:* from the Greek *koilos,* hollow, and *lepis,* scale, thus hollow scales, an allusion to the concave skin denticles of the species.

Synonyms *Scymnodon melas* Bigelow, Schroeder, and Springer, 1953.

Identification The Portuguese shark is characterized by *a snout shorter than the mouth width; two dorsal fins with very small, inconspicuous spines at their origins;* a second dorsal fin slightly larger than the first; and the absence of an anal fin. The *upper teeth are lanceolate and smooth edged.* The *lower teeth have short, broad, strongly oblique cusps and large, overlapping bases.* Teeth number about U: 58, L: 40. The dermal denticles of the young and juveniles have three points and are widely spaced. The *adult denticles are very large, 2–3 mm long, and clearly visible to the naked eye.* The *denticles are somewhat rounded with a point on their rear edges and a central depression.* They have short pedicels and overlap each other in scalelike fashion. *To the naked eye, the denticles resemble bony fish scales.* Juveniles are black above and below; adults are dark brown, almost black, above and below.

Similar species The roughskin dogfish is very similar and *presumably* distinguished from the Portuguese shark by a longer snout. Its snout is as long as the mouth width or as long as the distance from the mouth to the level of the gill slits. The Greenland shark has spineless dorsal fins, and its denticles are thorny. The kitefin shark has spineless dorsal fins and lower teeth that are erect and triangular, with pointed cusps and serrated edges.

Range The Portuguese shark inhabits both sides of the North Atlantic, from off Newfoundland to off New England in the west, and from Iceland and the Faroe Islands south to the Azores and Cape Verde Islands in the east (Bigelow and Schroeder 1948, 1953). It has also been reported from the western Mediterranean (Torchio and Michelangeli 1971) and Japan (Yano and Tanaka 1983b). It is a common shark of the deeper continental slopes, although it is infrequently caught in North America because of the lack of deep-water fisheries. Bean (1881) gave records for eastern Canada, from the Grand Banks, Sable Island Bank, and La Have Bank. Templeman (1963) mentioned that there have been few records of Portuguese sharks in the Canadian Atlantic since those of Bean (1881), probably because of the 730 m limit of their scientific otter trawling. Scott and Scott (1988) later reported catching two specimens on the Scotian shelf at a depth of 950 m and temperature of 5°C, reinforcing Templeman's opinion that the Portuguese shark is found at depths greater than 730 m off Newfoundland and the Scotian shelf. Similarly, there are few reports of Portuguese sharks from the United States, obviously due to the great depths it inhabits. Specimens have been reported from the continental shelf edge south of Nantucket (Bigelow and Schroeder 1948), and off New York and Delaware Bay at depths of 887–1,234 m (Bigelow and Schroeder 1954).

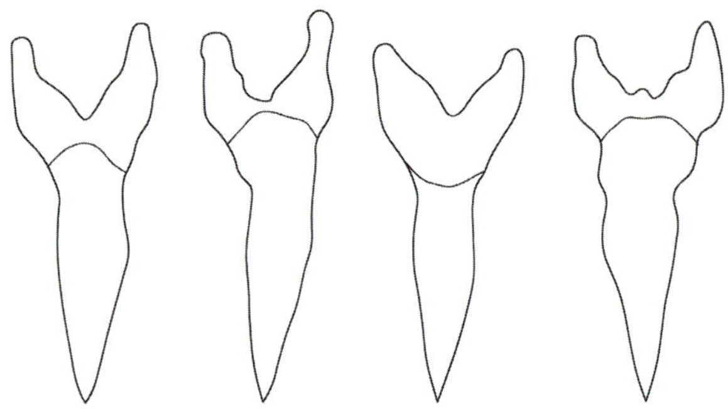

Fig. 30b. Teeth (upper).

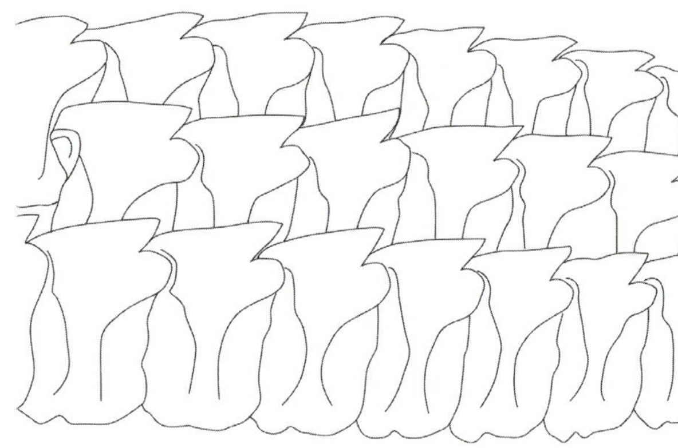

Fig. 30c. Teeth (lower).

The specimen illustrated here was taken while trawling at 820–1,307 m, off New York (40°14 N, 67°41 W) in January 2004. The species probably ranges farther south in deep water. However, a specimen, taken just north of Cape Hatteras and reported by Moore et al. (2003) as a southern record, is actually a gulper shark, *Centrophorus* sp. (Karsten Hartel and John Galbraith, pers. comm., January 2005).

Size and weight According to Last and Stevens (1994) the Portuguese shark attains 120 cm. Bigelow and Schroeder (1948) reported a 112-cm female. The 78-cm specimen illustrated here weighed 3.8 kg.

Biology The Portuguese shark is one of the deeper dwelling sharks. In the Atlantic, it has been taken at depths of 340–2,718 m (Bigelow and Schroeder 1948), with most captures at more than 900 m and at temperatures of 5–6°C. Carrassón et al. (1992) reported that in the western Mediterranean, the species is found at 1,419–2,251 m and that it was rare at depths of less than 1,600 m. In Suruga Bay, Japan, Yano and Tanaka (1984b, 1988) reported that Portuguese sharks occurred mainly at 600–1,300 m.

Although this is a common shark of the deeper continental slopes, its habits are nonetheless poorly known. Its diet includes squid and small fishes, but nothing else is known of its habits.

Curiously, the Portuguese shark has very large erythrocytes (red blood cells). Sherburne (1973) reported the erythrocytes of the Portuguese shark to have a mean size of 33.4 × 23.6 μm, larger than any previously recorded size for other

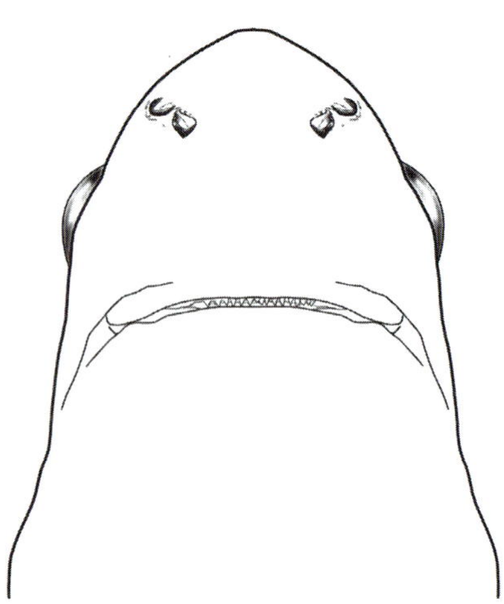

Fig. 30d. Snout.

Fig. 30e. Dermal denticles.

chondrichthyes. Whether this is an adaptation for its very deep habitat or serves some other function is unknown.

Size at maturity There are no North American data available on the Portuguese shark because of its scarcity here.

Males: According to Yano and Tanaka (1988), off Japan males mature at about 70 cm, based on spermatozoa in the seminal vesicles and calcification of the claspers. Veríssimo et al. (2003) suggested that males off Portugal reach maturity at about 90 cm.

Females: Females appear to mature at about 100 cm, based on oocytes larger than 20 mm or ova or embryos in the uteri (Yano and Tanaka 1988). Veríssimo et al. (2003) estimated that in females from Portugal the length at 50% maturity is 98.5 cm. These authors also reported that that the smallest "mature" female measured 80 cm and the largest immature specimen measured 119 cm. However, their criteria for maturity are ambiguous, which probably accounts for the proportionally wide gap between the smallest reportedly adult specimen and the largest juvenile.

Reproduction The Portuguese shark is aplacental viviparous. There is little published information about its reproductive processes. The Portuguese shark apparently does not have a well-defined breeding season, because broods in various stages of development are encountered at any given time throughout the year (Yano and Tanaka 1988; Veríssimo et al. 2003).

Size at birth: The young are born at 23–30 cm. This is based on Bigelow and Schroeder's (1954) report of a female giving birth prematurely on deck to embryos of that size, and on Veríssimo et al. (2003) reporting 23–30-cm embryos with fully absorbed yolk-sacs.

Brood size: The brood of Atlantic Portuguese sharks ranges from 10 to 16. The only brood reported from the western North Atlantic consisted of 15 young (Bigelow and Schroeder 1954), and broods of 10–16 have been reported for females off Portugal (Bigelow and Schroeder 1953; Veríssimo et al. 2003). In Suruga Bay, Japan, Yano and Tanaka (1988) found that females carried 15–29 embryos.

Nurseries The nurseries of the Portuguese shark remain undiscovered. However, Bigelow and Schroeder (1948) reported examining a 33-cm specimen, obviously a neonate, from the continental slope south of Nantucket. In the eastern North Atlantic, Vaillant (1888) reported that females caught near Setubal, Portugal, were very close to parturition.

Age and growth The Portuguese shark has not been aged.

Relation to humans The Portuguese shark is of no commercial importance in North American waters. Portugal is the only country in the North Atlantic where the species is taken in commercial fisheries. The landings are usually reported as "siki," consisting of both Portuguese sharks and leafscale gulper sharks (*Centrophorus squamosus*). This fishery landed 302 metric tons in 1991, reached a peak of 3,284 metric tons in 1996, and then declined to 1,939 metric tons in 1999, raising concerns about sustainability of the resource (Veríssimo et al. 2003).

ROUGHSKIN DOGFISH

Centroscymnus owstoni Garman, 1906

Fig. 31a. *Centroscymnus owstoni,* roughskin dogfish, mature male, 80.5 cm: from Suruga Bay, Japan (paratype, Museum of Comparative Zoology # 1126; Alan Owston, collector; courtesy of Karsten Hartel).

Common name Roughskin dogfish, an allusion to roughness of the skin denticles. See the origin of the term dogfish on p. 55.

Spanish name None.

Scientific name *Centroscymnus owstoni* Garman, 1906, in *Bulletin of the Museum of Comparative Zoology at Harvard College* 46(11): 207–208. *Centroscymnus:* from the Greek *kemtron,* spine, and *skymnos,* the ancient name for some shark; *owstoni:* after Alan Owston, a shipmaster and naturalist of Yokohama, Japan, who first obtained the type specimen.

Synonyms None.

Identification This small squaloid shark has *a snout about as long as the mouth width* and *two dorsal fins with small inconspicuous spines that barely protrude through the skin, and lacks an anal fin.* The first dorsal fin starts as a ridge almost at the level of the pectoral fins. The second dorsal fin is larger and higher than the first, with its origin over the midpoint of the pelvic fin base. *The upper teeth are lanceolate* in the center of the mouth. The lower teeth have low, oblique, bladelike cusps. *The skin denticles of the lateral trunk are round or oval with a central depression.* Coloration is dark blackish brown or black above and below.

Similar species The Portuguese shark has a shorter snout (it is shorter than the mouth width). The velvet dogfish has lower teeth with erect cusps.

Range The roughskin dogfish appears to be cosmopolitan, as it has been reported from the Pacific, Indian, and Atlantic Oceans. In the Pacific Ocean, it has been reported from Japan (Garman 1906; Yano and Tanaka 1988), Australia

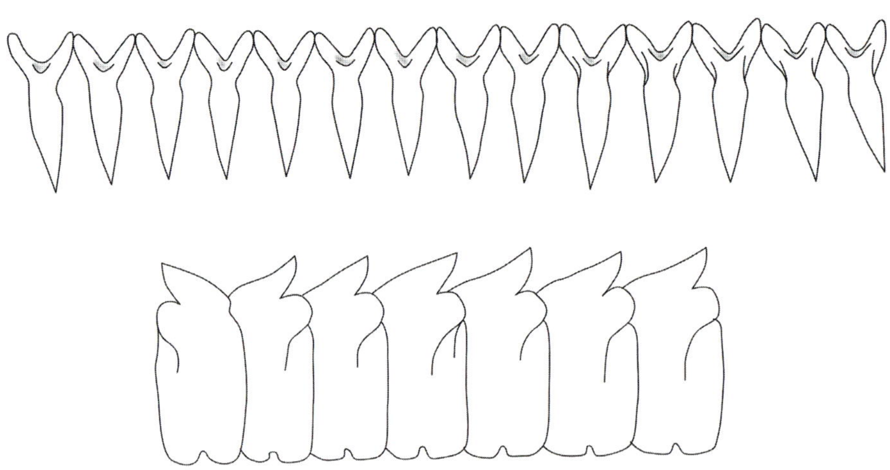

Fig. 31b. Teeth.

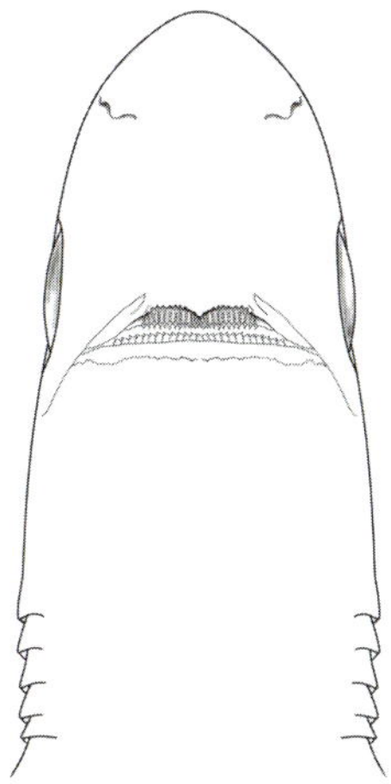

Fig. 31c. Snout.

Fig. 31d. Dermal denticles.

(Bass 1979), and New Zealand (Garrick 1959). In the Indian Ocean, it has been reported from Madagascar (Forster et al. 1970). In the Atlantic, it has been reported from the Gulf of Mexico (Compagno 1984a). I have never encountered it in North America; the specimen illustrated here is a paratype taken off Japan.

Size and weight It grows to at least 117 cm (Yano and Tanaka 1983b).

Biology This is a poorly known, small squaloid shark inhabiting the continental slopes at depths of 500–1,080 m (Garrick 1959; Forster et al. 1970). Nothing has been published on the habits of the species in the Gulf of Mexico. What little is known of the species is primarily based on the studies of Yano and Tanaka (1983b, 1988) on specimens caught in Suruga Bay, Japan.

Size at maturity Yano and Tanaka (1988) reported that, off Japan, males mature at about 70 cm and females mature at about 100 cm.

Reproduction The roughskin dogfish is aplacental viviparous. Oocytes may reach 50–60 mm at ovulation (Yano and Tanaka 1987). According to Yano and Tanaka (1988), females carrying mature oocytes 50–60 mm in diameter were found from February to November, and females carrying fertilized eggs in the uteri were caught from February to August, indicating that the species does not have a well-defined breeding cycle off Japan. Ripe males were also found throughout the year. Consequently, the duration of gestation could not be determined. Interestingly, Yano and Tanaka reported that oocytes do not develop during embryonic development, although they observed no full-term females. If their observation is correct, it would indicate that the gestation and oogenesis are not concurrent and therefore must be consecutive. This would be unusual for a dogfish.

Size at birth: The size at birth is unknown because gravid females or newborn specimens have not been reported. Based on the largest embryos and smallest specimens caught (Yano and Tanaka 1988), it appears that roughskin dogfish are born at 30–35 cm.

Brood size: The brood size ranges from 6 to 28 (Yano and Tanaka 1988).

Nurseries Unknown.

Age and growth The species has not been aged.

Relation to humans None.

VELVET DOGFISH

Zameus squamulosus (Günther, 1877)

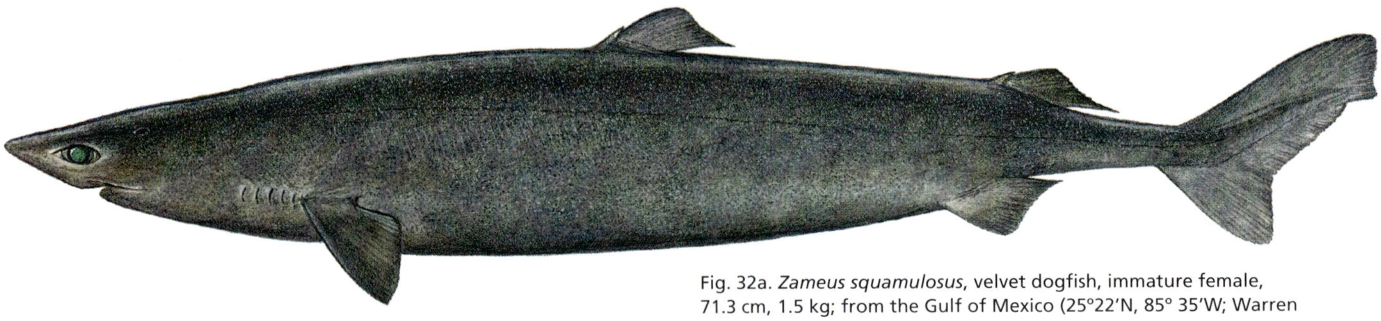

Fig. 32a. *Zameus squamulosus*, velvet dogfish, immature female, 71.3 cm, 1.5 kg; from the Gulf of Mexico (25°22′N, 85° 35′W; Warren Mitchell, collector).

Common name Velvet dogfish. The name refers to the appearance and feel of fresh specimens. See origin of the term dogfish on page 55. The F.A.O. name is smallmouth velvet dogfish.

Spanish name None.

Scientific name *Zameus squamulosus* (Günther 1877), originally described as *Centrophorus squamulosus* in *Annals and Magazine of Natural History,* Series 4, 20(56): 433. *Zameus:* from the Japanese *zame,* shark; *squamulosus:* from the Latin *squama,* scale, diminutive, *squamula,* and *osus,* an ending suggesting full of, thus "covered with small scales."

This species was long known as *Scymnodon squamulosus* (Günther, 1877) and was considered distinct from *Scymnodon obscurus* (Vaillant 1888). Most recent authors (Krefft 1980; Compagno 1984a; Yano and Tanaka 1984a; McEachran and Fechhelm 1998) have considered *Scymnodon obscurus* to be a junior synonym of *S. squamulosus,* the type specimen of which was caught off Japan during the *Challenger* expedition. Taniuchi and Garrick (1986) placed *Scymnodon squamulosus* in the genus *Zameus* Jordan and Fowler, 1903.

Synonyms *Scymnodon obscurus* (Vaillant, 1888).

Identification This small shark has a snout longer than the mouth width; *two dorsal fins with small spines that barely protrude through the skin,* with the second dorsal fin originating over the base of the pelvic fin; and no anal fin. The upper teeth have lanceolate, smooth-edged cusps. The *upper teeth are much longer midway along each side of the jaw than toward either the center of the mouth or its outer corners.* The lower teeth have bladelike triangular cusps and interlocking bases; the cusps are erect at the center of the mouth and become increasingly oblique toward the corners. The dermal denticles are overlapping and leaflike with three points and have *characteristic traverse microridges over their surface.* Coloration is dull *black above and below without any fin markings.* The iris is blue in life.

Similar species The roughskin dogfish has a shorter snout, its length about equal to the mouth width, its frontal upper teeth are of similar size, and its frontal lower teeth have strongly oblique cusps. The Portuguese shark has a snout that is much shorter than the width of its mouth and much

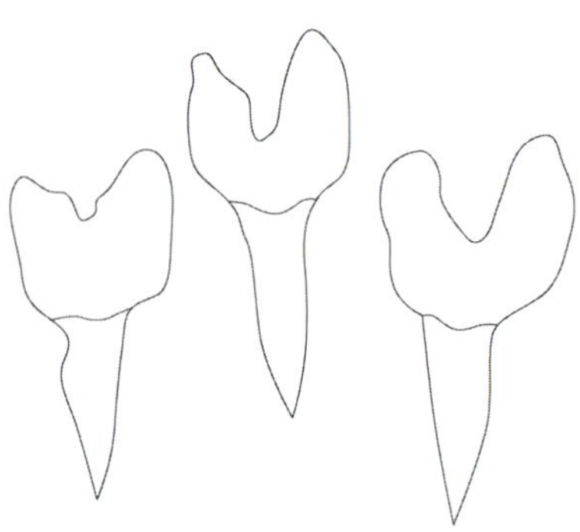

Fig. 32b. Teeth (upper).

Fig. 32c. Teeth (lower).

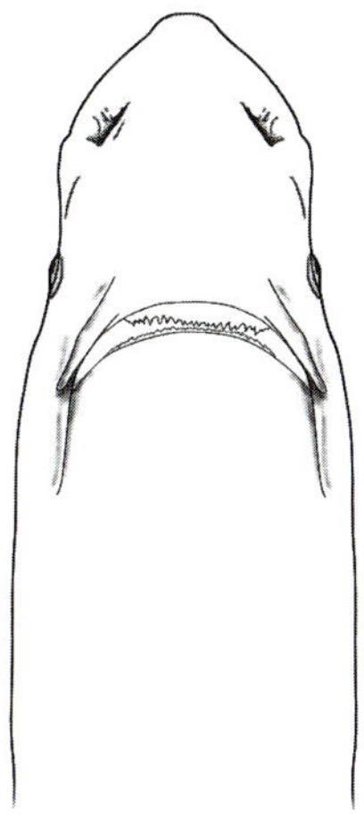

Fig. 32d. Snout.

Fig. 32e. Dermal denticles.

larger and conspicuous skin denticles that resemble the scales of bony fishes. Both the roughskin dogfish and the Portuguese shark can be distinguished by their much larger skin denticles that are scalelike and lack the thin ridge lines that cover the denticles of the velvet dogfish.

Range The velvet dogfish is apparently cosmopolitan in deep waters. It has been reported from the Pacific and Indian oceans (Japan, Günther 1877; Australia, Last and Stevens 1994; New Zealand, Taniuchi and Garrick 1986; South Africa, Bass et al. 1976). It is widespread in the Atlantic, having been reported from numerous locations on both sides of the ocean: off Sudan (Vaillant 1888), Madeira and Ivory Coast (Cadenat and Blache 1981), French Guiana (Uyeno et al. 1983), and the Gulf of Mexico (McEachran and Branstetter 1984). It appears to be common in deep waters off the Florida Keys and in the Gulf of Mexico, based on occasional captures by longliners.

Size and weight According to Last and Stevens (1994), it reaches at least 84 cm. A 71-cm specimen weighed 1.5 kg.

Biology This is a poorly known deep-water species. Most catches have been recorded at 500–1,435 m (Bigelow and Schroeder 1957; Yano and Tanaka 1984a). Some specimens I have examined were obtained from swordfish vessels fishing in the Straits of Florida, indicating an epipelagic existence. Nothing else is known of its habits.

Size at maturity Unknown.

Reproduction The velvet dogfish is presumably aplacental viviparous.

Size at birth: Unknown.

Brood size: Unknown.

Nurseries Unknown.

Age and growth The species has not been aged.

Relation to humans None.

GREENLAND SHARK

Somniosus microcephalus (Bloch and Schneider, 1801)

Fig. 33a. *Somniosus microcephalus*, Greenland shark, immature male, 283.0 cm, 229 kg; Saguenay River, Quebec, Canada (courtesy of Musée du Fjord).

Common name Greenland or sleeper shark, a sluggish species and the only shark found in Greenland waters. The name "sleeper shark" was first used by Le Sueur (1818), who deduced that this shark must be slow and sluggish because its fins are relatively small. In the past, it has also been called the nurse, ground, or gurry shark.

Scientific name *Somniosus microcephalus* Bloch and Schneider (1801) originally described as *Squalus microcephalus* in *Systema Ichthyoligiae iconibus ex illustratum:* 135. *Somniosus:* Latin, sleepy, from *somnus,* sleep; *microcephalus:* Greek for small-headed, from *mikros,* small, and *kephale,* head.

Synonyms *Squalus carcharias* Gunnerus, 1766; *Squalus* or *Somniosus brevipinna* Le Sueur,[1] 1818; *Squalus borealis* Scoresby, 1820; *Laemargus borealis* Bonaparte, 1846.

Identification The Greenland shark is *usually a large stout shark lacking an anal fin* and has *two spineless dorsal fins of nearly equal size.* The first dorsal fin originates about midway on the trunk, the distance from the tip of the snout to the origin of the first dorsal fin usually being less than 44% of total length. The upper teeth have long, pointed, smooth-edged cusps; the lower teeth are large with strongly oblique smooth-edged cusps and overlapping bases. Although several rows of upper teeth are functional at one time, only one row of lower teeth is functional at a time. Apparently, an entire row of teeth moves forward as a unit over the previous row. The older row of lower teeth then is found in front and below the cutting edge of the new row. The number of teeth is variable, usually about 40–52 teeth in the upper jaw (the symphysis, or center of the jaw, is not always discernible) and the lower teeth usually numbering 23 to 27. The dermal denticles are short, thornlike, ridged, and closely apposed. Coloration is variable, ranging from mottled gray to uniform brown throughout. Jensen (1914) reported two chalk-white specimens, possibly albinos.

1. The name Le Sueur also appears as Lesueur in different issues of *Journal of the Academy of Natural Sciences of Philadelphia* from 1818 through 1822. I retain the original spelling as it appears in the relevant publication.

Similar species Adult Greenland sharks are so large that they can be confused with few other species. The equally large sixgill shark has a single dorsal fin. The prickly and bramble sharks have the first dorsal fin originating above or posterior to the origin of the pelvic fins. The Portuguese shark resembles a small Greenland shark, but it has dorsal fins with small, inconspicuous spines and overlapping scale-like dermal denticles instead of the thornlike denticles of the Greenland shark.

The Greenland shark is replaced in the Pacific Ocean by the Pacific sleeper shark. These two species, or forms, are very similar and difficult to separate, other than by geographic area. Some observers have stated that the Pacific sleeper shark has a more posterior first dorsal fin, with an origin almost as close to the tip of the snout as to the tip of the tail, and that the space between the two dorsal fins is about two-thirds as long as from the tip of the snout to the first gill slit. These differences might not be sufficient to separate the species (see section on the Pacific sleeper shark, p. 131).

Range The Greenland shark inhabits the North Atlantic from polar latitudes south to at least the southeastern United States in the west and the North Sea in the east. In the eastern Atlantic, it occurs fairly regularly in the northern North Sea and Scottish waters but becomes rare to the south (Wheeler 1969). In the western North Atlantic, it has been reported from Ellesmere Island and Baffin Bay (Templeman 1963) to off Savannah, Georgia (Herdendorf and Berra

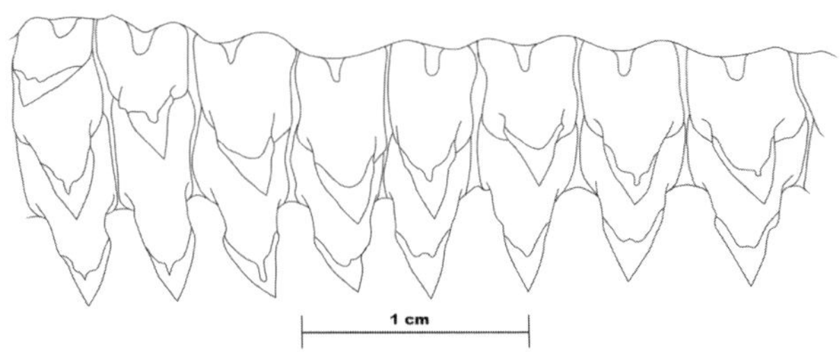

Fig. 33b. Teeth (upper).

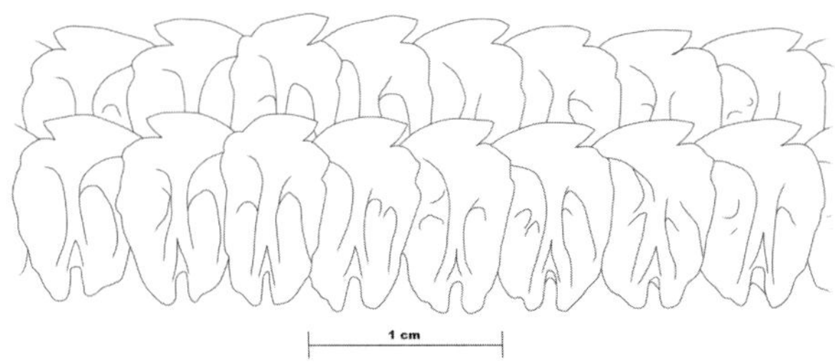

Fig. 33c. Teeth (lower).

1995). The Greenland shark is abundant above 76° N latitude and on both western and eastern sides of Baffin Bay (Templeman 1963). The species is doubtless abundant throughout the polar seas, and the lack of reports north of 76° N is simply attributable to the hindrances the ice presents to fishing activities. Greenland sharks are common from Ellesmere Island to the Gulf of St. Lawrence. It penetrates far into the St. Lawrence estuary, entering the Saguenay Fjord (Drainville and Brassard 1960).

The Greenland shark is occasionally caught in the Gulf of Maine (Bigelow and Schroeder 1948) and rarely in deep waters south of Cape Hatteras, North Carolina. While the Greenland shark is generally considered a denizen of polar and high latitudes, it might be far more widely distributed in deep cold waters in lower latitudes than is currently believed. Recent reports suggest that it may enter warm temperate latitudes and even the tropics, submerging into the colder deep layers of southern waters. Based on video images and photographs, Herdendorf and Berra (1995) reported a large Greenland shark cruising over the wreck of the SS *Central America* at 2,200 m on the inner side of the Blake Ridge, about 370 km east of Savannah, Georgia. In October 2003, James Ray, of Shell Oil Company, sent me a video recording of a sleeper shark around an oil platform in the Gulf of Mexico, taken from a remotely operated vehicle at 2,700 m. This specimen was later reported by Benz et al. (2007). The scarcity of specimens south of Cape Hatteras perhaps reflects the absence of deep-water fisheries and activities in the area, and the difficulties of deploying conventional gear to depths inhabited by Greenland sharks.

Some evidence exists that the Greenland shark might have a bipolar distribution and a much wider range than previously thought. Gushchin et al. (1987) reported catching a shark of the genus *Somniosus* on the continental slope of Patagonia (42°24 S, 58°19 W) at 800 m. The specimen, a 4.16-m male, was compared with descriptions of *Somniosus microcephalus* and considered to be this species. However, Francis et al. (1988) and Cione (1998) believe that the

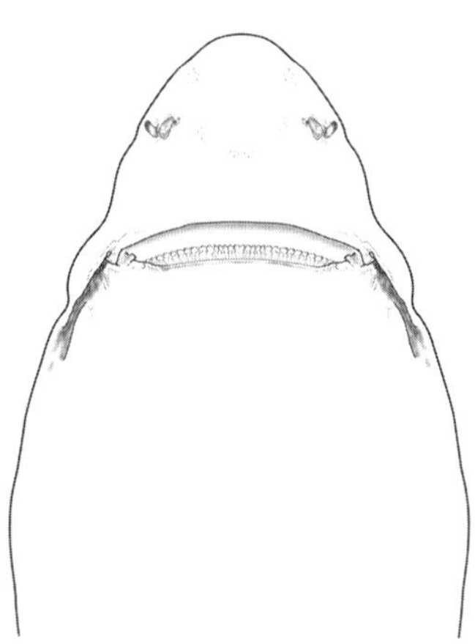

Fig. 33d. Snout.

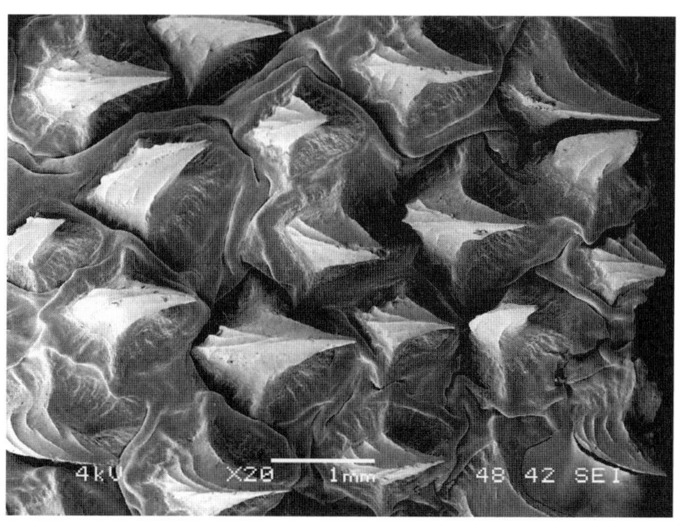

Fig. 33e. Dermal denticles.

Greenland shark occurs only in the northern Atlantic and that it is the Pacific sleeper shark that has a bipolar distribution. Given the present confusion about the number of species of large *Somniosus* and how to distinguish them, the identity of large sleeper sharks in the Gulf of Mexico and other southern waters cannot be established at this time. Much more comparative material is needed before this problem can be solved.

Size and weight The Greenland shark is one of the largest sharks. However, documentation on the biggest specimens is scanty because when one of these large sharks becomes entangled in fishing gear, it is quickly cut adrift or discarded overboard. There are rumors of Greenland sharks measuring 7 m, but they all are old reports or lacking credible evidence. Jacobson (1885) noted that in Greenland the species rarely exceeds 3.65 m ("12 feet"), but that "Lieut. C. Trolle states that he caught one measuring 23 feet [7 m] in length." Crozier et al. (2002) reported a 7.3-m female without confirming evidence. Bigelow and Schroeder (1948) stated that the longest specimen on record was 6.40 m (Jenkins 1925) and that specimens of 4.88–5.49 m were unusual but taken occasionally. They also reported several specimens from the western North Atlantic, ranging from 4.57 to 5.03 m. The largest measured by Beck and Mansfield (1969) was 3.11 m. The largest of several hundred Greenland specimens reported by Hansen (1963) was 4.0 m (Hansen tagged 411 sharks and reported that mortality in the tagging process was a little more than 50%). There are very few recorded weights of Greenland sharks: 200 cm, 68 kg (MacCoy 1939); 236 cm, 125 kg (Kneeland 1847); and 283 cm, 230 kg (specimen illustrated in this book). Koefoed (1957) mentioned a 5-m female weighing 1,000–1,200 kg.

Biology Sleeper sharks are the largest polar fishes and the only sharks regularly inhabiting polar areas. The Greenland shark is a cold-water species of the North Atlantic that can be found from shallow cold estuaries to deep oceanic waters. It has been taken in waters as cold as –0.6°C and as warm as 10–12°C; however, according to Bigelow and Schroeder (1948), most have been caught when temperatures were 2–7°C at the bottom.

The Greenland shark has the widest depth range of any known shark. At polar latitudes, it can be caught at 250–500 m during the warmer months (June–September), although the highest catches occur at 300–400 m (Hansen 1963). At lower latitudes, specimens have been observed at 2,200 m off the Georgia coast (Herdendorf and Berra 1995) and captured as deep as 3,000 m off the Canary Islands (J. A. Musick, pers. comm., December 2003). The Greenland shark is also found in shallow waters just a few meters deep in cold river mouths, estuaries, and fjords. There are reports of Greenland sharks stranding at the mouths of rivers in Ungava Bay (Dunbar and Hildebrand 1952) and of others ascending the Saguenay River of Quebec and becoming stranded at low tides (Drainville and Brassard 1960). A 274-cm specimen stranded at low tide in June 1944 at Baie des Îles, on the Saguenay River, 81 km from its mouth at Tadoussac. In another instance, in September 1955 a 320-cm specimen stranded after a spring tide near the village of Saint-Fulgence, on the north shore of the Saguenay River 95 km upriver from its mouth (Drainville and Brassard 1960). The shark was alive when found, erasing any doubt about its origin (i.e., being discarded dead from a boat). Since those early reports, Greenland sharks have been found to be common in the St. Lawrence River estuary at temperatures of 4–6°C (Stokesbury et al. 2005). Three Greenland sharks were tagged with acoustic and satellite tags by scuba divers in Baie St. Pancrace in the St. Lawrence estuary. One of these sharks remained at depths of 2–67 m during a 47-day tracking period, while a second one ranged from the surface to 72 m. The third shark moved into the main channel of the St. Lawrence estuary and traveled 115 km upstream in 66 days, remaining at depths of 132–352 m and temperatures of 1–5.4°C (Stokesbury et al. 2005). Since 2003, Chris Harvey-Clark and Jeffrey Gallant (pers. comm., 3 May 2006) have been diving in the Saguenay fjord and in the St. Lawrence estuary to observe Greenland sharks and have filmed them on many occasions. Their films reveal sharks moving extremely slowly just above the bottom at depths of 10–20 m.

We can only speculate what these deep-water sharks achieve by entering deep into cold estuaries and their brackish waters. There is little available prey in those environments, so it appears unlikely that they are in search of food. Although these estuaries are highly stratified, the sharks are often found in shallow water where conditions of decreased salinities prevail (C. Harvey-Clark, pers. comm., 4 June 2006).

Most Greenland sharks have one or more large copepods (*Ommatokoita elongata*) attached to each cornea (Grant 1827; Dunbar and Hildebrand 1952). Of 1,505 specimens inspected in Greenland waters, 84% were infected in both eyes, 15% were infected on one eye only, and 1% were not infected, although it is possible that the copepods could have been lost during the haul or have been too small to be noticed. One copepod on each eye is the rule, but in some cases two or (rarely) more specimens were attached to one eye (Berland 1961). Borucinska et al. (1998) examined the ocular lesions caused by this parasite and concluded that attachment of the parasite to the cornea resulted in lesions and corneal opacity that appeared to cause severe vision impairment or even blindness. The few Greenland sharks I have examined had such opaque corneas that it was difficult to imagine that the shark could see at all. However, a review of close-up films of Greenland sharks taken by Gallant and

Harvey-Clark reveals that even in conditions of low visibility, the sharks appeared to follow the divers with their eyes. It is interesting to note that in all these films and photographs of Greenland sharks of the St. Lawrence and Saguenay waters, none of the sharks (mainly large juveniles) had any eye parasites. Perhaps the sharks' sojourn into brackish water is related to parasite removal. It has been speculated that these copepods may be bioluminescent and may act as a lure, attracting prey to the shark (Berland 1961). However, no evidence supports this putative case of mutualism.

Eyewitnesses agree that the Greenland shark is one of the most sluggish sharks, offering no resistance whatsoever when hooked or even when drawn out of the water (Bigelow and Schroeder 1948). Bøje (1939: 3) wrote that in Greenland

> the Eskimo makes a hole in the ice and, from a groove at the edge of this hole, allows blood to ooze out into the water; sharks that are lured by the blood are harpooned or simply caught by the eyes and dragged out of the water—a form of sport devoid of all danger, for this shark is incredibly apathetic. I myself have caught a live, unwounded shark from a boat, by simply seizing its tail.

Similarly, A. S. Jensen (1914: 11) wrote: "The dullness or stupidity of the shark, namely, is such that this huge animal allows itself to be hauled up on a line no thicker than stout twine, so that is not rarely taken by chance on the same lines that are used to catch the Greenland halibut." Scoresby (1820, 1: 539) wrote:

> The sailors imagine this shark is blind, because it pays not the least attention to the presence of a man; and is, indeed, so apparently stupid, that it never draws back when a blow is aimed at it with a knife or lance. . . . It is so insensible of pain, that though it has been run through the body with a knife and escaped; yet after a while, I have seen it return to banquet again on the whale [being butchered], at the very spot where it received its wounds.

Note that this perceived sluggishness must be evaluated against the evidence that these sharks feed on fast-moving seals and squids (see below).

Greenland sharks are often attracted to surface waters during the summer when offal from sealing or whaling operations is disposed of (Hansen 1953; Templeman 1963; Beck and Mansfield 1969), and they often attack netted or harpooned whales. According to Beck and Mansfield (1969), Greenland sharks cleanly removed large semicircular chunks of skin and blubber off netted narwhals, despite their unimpressive set of teeth. However, at this writing, there is no evidence of attacks on free-swimming whales.

Diet: The diet of the Greenland shark includes a very long list of prey, ranging from slow-moving snails to fast-moving seals and squids. Raven (1926) reported that five specimens examined in Greenland had consumed mainly Greenland cod (*Gadus ogac*) and Greenland sculpins (*Myoxocephalus scorpius*), as well as numerous invertebrates. The stomach of one shark held the remains of a small harbor seal (*Phoca vitulina*) that had been bitten into pieces 17–25 cm long (Raven 1926). Beck and Mansfield (1969) reported that specimens of 217–311 cm taken in Koluktoo Bay, northern Baffin Island, contained amphipods, sea urchins, gastropods, egg cases of the arctic skate (*Raja hyperborea*), chars (*Salvelinus alpinus*), seals, and whale offal. A 236-cm specimen caught off the coast of Massachusetts contained a 76-cm wolf-fish (*Anarhichas lupus*) and many fragments of sea urchins and shells (Kneeland 1847). A 200-cm specimen caught off New England contained skin and blubber from a pilot whale or porpoise, nine large redfish (*Sebastes marinus*), one gray sole, and several pieces of jellyfish (MacCoy 1939). Williamson (1963) found a neonate or stillborn harbor porpoise (*Phocoena phocoena*) in the stomach of a 4.15-m female Greenland shark captured in June 1962 on the northwestern slope of the Grand Bank of Newfoundland. Dunbar and Hildebrand (1952) also recorded a small seal in one of the stomachs they examined. Fisk et al. (2002) found Greenland halibut (*Reinhardtius hippoglossoides*) to be the most common prey in the stomach contents of 14 Greenland sharks from Cumberland Sound but considered that this might have been an artifact of the longline sampling method, which provided a large number of hooked halibut. Marine snails followed halibut in abundance, and one specimen contained a ringed seal (*Phoca hispida*). The 283-cm Greenland shark illustrated in this work, taken in the Saguenay River in February, contained remains of a juvenile gray seal (*Halichoerus grypus*), two skates, a 50-cm Greenland halibut, and crab remains.

How such a seemingly sluggish and apparently nearly blind shark can catch intelligent, alert, and agile prey such as seals is not known. It is likely that these sharks detect prey using smell or sound and then use a very stealthy approach to get within biting range, with electroreception guiding the bite. These sharks have the ability of remaining almost motionless, and perhaps the sharks wait undetected close to seal breathing holes, attacking suddenly when the seals enter the water. Greenland sharks are suspected of being the cause of spiral-shaped wounds on seal carcasses found on Sable Island, but this has yet to be demonstrated.

What the largest Greenland sharks feed on is an interesting puzzle. The largest specimens live in deep water below 1,000 m, an environment considered energy poor regardless of latitude. What could those large creatures feed upon effectively in such an energy-poor environment? Large marine organisms (i.e., the larger whales) have adopted two general feeding strategies: plankton eating (the baleen whales) and squid eating (the sperm whale). The few sharks known to grow longer than 6 m (the whale shark, the basking shark, and megamouth) are plankton feeders. These behemoths obtain their food with a minimal expenditure of energy, just as baleen whales do. Adult Greenland sharks approach the size at which pursuing active prey may cease to be energy effective. What are their possible feeding strategies? Two recent papers provide partial answers. C. R. Smith and Baco (2003) observed, from submarines and with remotely operated vehicles, the fate of large whale carcasses that fall to the ocean floor ("whale falls"). When one of these massive giants (30–160 metric tons adult body weight) dies and eventually sinks to the bottom, it provides a massive food fall to the normally nutrient-poor deep-sea floor. The carcass is first attacked by such "mobile scavengers" as hagfish, sleeper sharks, and amphipods. A large whale fall can so saturate the local scavenging capacity of bathyal systems that it lasts for months or years. Smith and Baco (2003) calculated that the mobile-scavenger stage for an adult blue whale carcass of 100,000 kg at bathyal depths off southern California can last about five years, and much longer in the abyssal North Atlantic! It is possible that such whale falls (along with other massive creatures, e.g., whale sharks and the larger teleosts) could provide food for sleeper sharks in warm temperate or tropical regions. Thus, Greenland sharks could have evolved as scavengers at the greatest depths, where there are few other competing sharks, spending little energy in their search for food, just as other giant sharks do.

Another food source for sleeper sharks are large, deep-water squids, including the giant squid (*Architeuthis dux*) and the even larger colossal squid (*Mesonychoteuthis hamiltoni*). Cherel and Duhamel (2004) examined the stomach contents of 36 sleeper sharks[2] caught in the upper slope waters of the Kerguelen Archipelago in the Southern Ocean. All the stomachs contained cephalopods, including 19 species of squid and one species of an unknown large octopod. Four species of squids dominated the cephalopod diet, accounting for 81% of the total number of beaks and 99% of the reconstituted mass. The squid *Kondakovia longimana* (mantle length of up to 90 cm) was the main prey by number (50% by number, 21% by mass). The colossal squid (the largest invertebrate on earth, a squid of mantle length up to 2.3 m and a total length exceeding 12 m) was the main prey by mass (16% by number, 52% by mass). The squid *Taningia danae* (mantle length to 1.7 m) constituted 13% by number and 11% by mass. The giant squid, *Architeuthis dux* (mantle length of 2.2 m), constituted 3% by number and 15% by mass (Cherel and Duhamel 2004). The presence of squid in all the samples indicates that squids are of primary importance to sleeper sharks around Kerguelen, and perhaps in other areas of the world. The abundance of beaks (553) in the stomachs of the sharks also suggests that the squid were captured by the sharks and not scavenged. How such seemingly sluggish sharks as sleeper sharks catch alert and fast-moving creatures such as squid was not explained. A diet based on large squid is similar to that of the sperm whale. The mean estimated size of cephalopods eaten by the Kerguelen sleeper sharks was 68 cm mantle length and 8.2 kg (Cherel and Duhamel 2004), which is slightly larger than the average weight (7.2 kg) estimated for cephalopods eaten by sperm whales in the Antarctic (Clarke 1980: 41). Thus, some sleeper sharks (perhaps including the adult Greenland sharks) have developed a diet similar to that of the sperm whale, although their method of obtaining squid may be quite different.

2. Cherel and Duhamel (2004) identified the sleeper sharks as *Somniosus* cf. *microcephalus*. The number of species of *Somniosus* is uncertain, as is their distribution. Thus, the identity of the specimens examined by Cherel and Duhamel must await further understanding of these sharks. The information is presented here because it is relevant to sleeper sharks in general.

Reproduction Although incalculable numbers of Greenland sharks have been caught in northern European and Arctic waters in past centuries, the reproductive processes of the species are basically unknown. Lack of knowledge and the lack of gravid specimens engendered early absurd ideas about Greenland shark reproduction. Bigelow and Schroeder (1948: 520), repeating Collett's (1904) notions, stated,

> Adult females have been found repeatedly containing great numbers of soft eggs without horny capsules, with up to as much as 1 1/2 barrels of them in large specimens, these eggs ranging in size up to that of a goose egg. This, combined with the fact that none of the many examined have ever been found with embryos, supports the general belief that this shark, unlike other squalids is oviparous. If so, it seems that the eggs are deposited on the bottom in mud.

Jensen (1948) reported that an ovary removed from a female caught near Disko, Greenland, in May 1925 was 37 cm long and contained about 110 eggs, 90% of which were large, measuring up to 35 mm in diameter. Koefoed (1957) ended the confusion when he reported a 37-cm embryo from a female 5 m long and weighing 1,000–1,200 kg that was caught off the Faroe Islands in August 1954. The right uterus contained 10 embryos and the left was said to be empty. Koefoed considered this embryo to be close to birth because it was fully formed and uniformly dark colored and no trace of the yolk sac could be observed. If the small embryo examined by Koefoed was really close to birth, and given the large numbers of oocytes reported in ovigerous females and the huge size of adult females, then it is likely that Greenland shark broods must be among the largest.

Size at birth: We know little of the birth size of Greenland sharks. Jensen (1914) stated that he saw no Greenland sharks smaller than 1.8 m. In August 1981, two Greenland sharks of about 1 m were caught in the eastern region of Jan Mayen Island (Norwegian Sea) by midwater trawl, at a depth of 250–300 m over water of more than 2,000 m deep. Traces of the yolk sac and yolk residues in the digestive tract suggested that the specimens had been born recently. The catch of these two fish suggested that Greenland sharks are born in the Norwegian Sea in July–August and that the young measure about 1 m at birth (Kondyurin and Myagkov 1983). The discrepancy between this size and that estimated by Koefoed (1957) remains to be resolved.

Brood size: Unknown (see "Reproduction," above).

Nurseries Unknown.

Age and growth Tagging studies have revealed the Greenland shark to grow slowly; medium-size specimens seem to grow about 1 cm or less per year. According to Hansen (1963), a tagged Greenland shark grew from 262 to 270 cm in 16 years. Another recaptured shark had grown 1 cm after two years in the sea (from 271 to 272 cm). A third specimen, recaptured 14 years after tagging, grew from 285 to 300 cm. These measurements suggest that the Greenland shark may be very long-lived.

Relation to humans Norway had a commercial fishery for Greenland sharks for centuries. Most of the catch came from the Arctic, where the fishery was combined with sealing. The Greenland fishery started in the early nineteenth century. According to Jensen (1914), it grew from 2,000–3,000 sharks yearly in 1857 to 11,500–15,000 sharks a year in the early 1890s and to 32,000 sharks yearly by 1914. This fishery ended in 1960 because the oil was no longer profitable (International Council for the Exploration of the Sea 1995).

Jensen (1914) gave an excellent account of the capture and use of the species for oil and meat in Greenland in the early 1900s. This fishery was conducted just out from the houses, through holes in the ice, or from primitive boats. The sharks were attracted to the surface at night by means of a light, and then a simple bent hook was stuck into them, or a simple baited hook attached to 4–6 feet of iron chain was suspended in the water. The principal product of the Greenland shark was liver oil, which was used in lamps, because petroleum was not permitted in houses owing to the danger of fire. The oil, then known as "self-running shark oil," was obtained by beating the frozen shark to a mash with a hammer. The mash was then placed in a tub. When the weather became warmer, the oil would run out of the mash by itself (Jensen 1914). A second product from the Greenland shark was dog food. However, in earlier times of scarcity, the Greenland Eskimo also ate shark meat, cooked or dried, but "especially after it had been made halfway to putrefy" (Jensen 1914).

In northern Greenland, where dogs played an important role as draft animals for the sleds, the shark fishery provided an important source of food for the dogs (Jensen 1914). The flesh was considered excellent dog food when dried but toxic and dangerous when fresh. The flesh was detoxified by drying or cooking. The usual practice was to cut the flesh into thin strips, which were hung to dry in the sun and air (Jensen 1914). Boiling in three or four different changes of water could also detoxify it. Symptoms of poisoning are the same as those of intoxication. In dogs it causes stiff walking, vomiting, and spurting watery, fetid diarrhea. According to Bøje (1939), the symptoms and the violence of the intoxication vary widely, depending on the quantity of meat ingested and its condition, and on individual susceptibility of the animals. Bøje also gave interesting descriptions of the symptoms of the flesh on experimental dogs and correctly deduced that the substance responsible for the intoxication was trimethylamine oxide. Much later, Anthoni et al. (1991) confirmed that the symptoms appear to be caused by acute trimethylamine poisoning arising from intestinal reduction of trimethylamine oxide, which is present in large amounts in the flesh.

Jacobson (1885, p. 301) gave an account of an Icelandic fishery for the Greenland shark. He stated that the Greenland shark was caught principally for its oil but that in Iceland the meat was frequently used for food, after it had been allowed to hang for some time or having been put in the ground and undergone a process of fermentation: "When fresh the meat is indigestible and unwholesome; when dried is has a peculiar but by no means disagreeable flavor, somewhat resembling old cheese. When fermented the meat is slimy and jelly-like, but it is stated that at present is not much used for food when in that condition" (Jacobson 1885).

When locally abundant, the Greenland shark often becomes a nuisance to fishermen. Bigelow and Schroeder (1948) reported that at least in some years the Greenland shark is sufficiently numerous to be a nuisance in the Straits of Belle Isle and inward along the north shore of the Gulf of St. Lawrence. In early 1970 there was a subsidized fishery off western Norway for reducing the stocks of Greenland sharks because they had become a problem for other fisheries. Currently, few fisheries target the Greenland shark. According to F.A.O. statistics (1994), Iceland is the only country landing some 40 metric tons of Greenland sharks annually.

Old tales of Greenland sharks attacking the in their kayaks appear to be mythical (Bigelow and Schroeder 1948). Scoresby (1820) stated that although whalers frequently slipped off the ships and fell into the water where sharks abound, he knew of no whalers ever being attacked by sharks. Caloyianis (2000) dived around Greenland sharks off Baffin Island in the summer without eliciting an attack. In 2006 I met a few hardy divers who had taken up the sport of diving to see Greenland sharks in the frigid, murky waters of Baie St. Pancrace, St. Lawrence Estuary, Canada. Chris Harvey-Clark and Jeffrey Gallant, who have been diving with these sharks repeatedly since 2003, have recounted to me several instances of coming face to face with them in conditions of very poor visibility. In each case, the sharks rushed to the divers, and quickly veered away after approaching within 50 cm. Given the large size of these sharks, and knowing that they routinely feed on large seals, the possibility of a shark mistaking a human for a seal should not be dismissed entirely.

PACIFIC SLEEPER SHARK

Somniosus pacificus Bigelow and Schroeder, 1944

Fig. 34a. *Somniosus pacificus*, Pacific sleeper shark, female, 232 cm, 72.1 kg; from western Prince William Sound, Alaska (Ted Otis, collector).

Common name Pacific sleeper shark; also mud shark (Alaska), a reference to its being covered in mucus, which appears mudlike when out of the water.

Scientific name *Somniosus pacificus* Bigelow and Schroeder, 1944, in *Proceeding of the New England Zoölogical Club* 23: 35. *Somniosus:* Latin for sleepy; *pacificus:* from the Latin *pacificus,* peace maker or peaceable, a reference to being from the Pacific Ocean.

Synonyms None.

Identification The Pacific sleeper shark is usually a large stout shark *of the northeastern Pacific Ocean,* with *two spineless dorsal fins of nearly equal size,* the *first originating midway on the trunk,* and *lacking an anal fin.* The distance from the tip of the snout to the origin of the first dorsal fin is usually more than 44% of total length. The upper teeth have long, pointed, smooth-edged cusps; the *lower teeth have strongly oblique, smooth-edged cusps* and large overlapping bases. The number of teeth is variable, U: 46–50, L: 26 to 30–26 to 30. The dermal denticles are bladelike with one central thick ridge and curve to the rear. They have stout stellate bases and are closely apposed. Coloration is black or blackish brown above and below. Abraded or washed-off specimens are dark gray.

Range The Pacific sleeper shark is widely distributed in the North Pacific Ocean. In the western North Pacific it ranges from the Bering Sea and along the Kamchatka Peninsula (Orlov 1999) south to Japan (Suruga Bay, Tanaka et al. 1982); in the eastern North Pacific, it occurs from the Alaskan coast south to California (see below). The southern limits of its distribution are not well known because the identity of large sleeper sharks of the southern waters of the Northern and Southern Hemispheres is not clear.

Large sleeper sharks in the Atlantic Ocean are sometimes identified as Pacific sleeper sharks. A 4.4-m female trawled from a depth of 677 m off Cape Columbine, South Africa, in 1963 was tentatively identified as "*S. microcephalus* or a closely allied species" by Bass et al. (1976). Gushchin et al. (1987) referred to a 4.2-m sleeper shark from the continental slope of Patagonia (42°24′ S, 58°19′ W) as *S. microcephalus.* Francis et al. (1988) referred both of these specimens to *S. pacificus* on the basis of length from snout to first dorsal fin. Díaz de Astarloa et al. (1999) collected 10 sleeper sharks, 3.85–6.0 m in length, in the continental shelf off Argentina between latitudes 35° S and 55° S. After examination these were identified as *S. pacificus.* Moreno and Pol (1992) reported the capture of three large sleeper sharks, 3.81–4.02 m, from western Cuba at a depth of 500 m, identifying them as *S. pacificus.* These identifications are unreliable and will continue to be until the taxonomic problems of the genus are solved. In any case, it appears that several species of large *Somniosus* exist in the Atlantic Ocean. Although the reports of *S. pacificus* in the Atlantic are all recent, it had been suspected for many years that there were three species of *Somniosus* in the North Atlantic. Maul (1955) suggested that the Portuguese shark described and pictured under the name *S. rostratus* by de Brito Capello (1870) actually represented a third Atlantic species (still unnamed). Bigelow and Schroeder (1957) also concluded that the Capello specimen may well represent a third species more nearly related to *S. microcephalus* than to *S. rostratus.*

In North America the Pacific sleeper shark ranges from the Bering Sea to at least Baja California, having been reported from the Bering Sea (Wilimovsky 1954), Alaska (Bright 1959), British Columbia (Hart 1973), Washington (DeLacy et al. 1972), and California (Scofield 1920; Roedel and Ripley 1950; Gotshall and Jow 1965). Anderson et al. (1979) referred to a specimen photographed by a baited

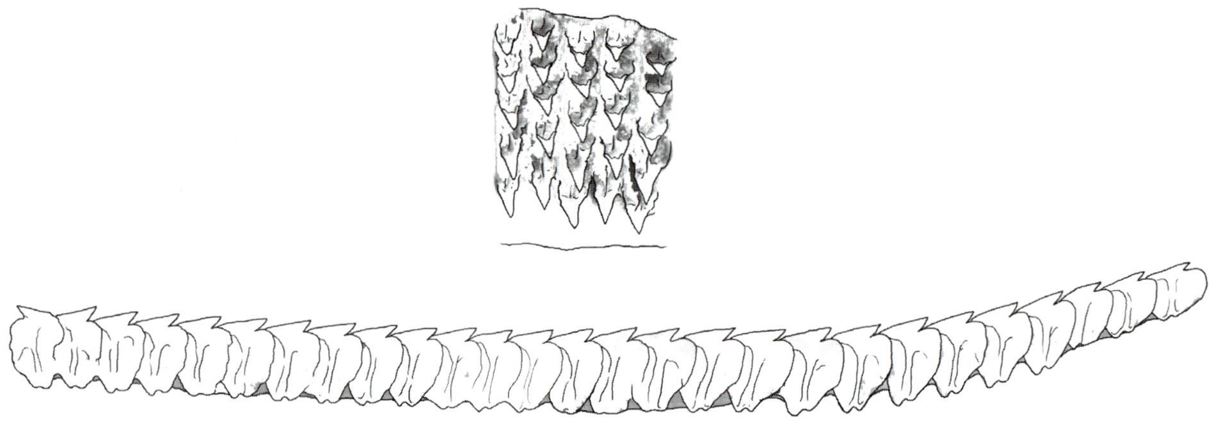

Fig. 34b. Teeth.

camera at a depth of 2,008 m off Baja California as the southernmost record in the eastern Pacific, although the species must certainly be found in deep waters farther south. It is abundant from southeastern Alaska to Washington.

Similar species The Pacific sleeper and the Greenland shark are common in the North Pacific and in the North Atlantic, respectively, and usually there is no difficulty in identifying specimens in those regions. However, identifying specimens in Arctic waters where their ranges may overlap, in the southern latitudes of the northern Hemisphere, or in the Southern Hemisphere has been problematic because the anatomical characteristics that workers have selected as diagnostic features have turned out to be quite variable.

For example, Compagno (1984a) used tooth form and counts, size, and the presence of lateral keels to separate the little sleeper shark (*S. rostratus*) from both the Pacific sleeper shark and the Greenland shark. According to Compagno, the Pacific sleeper and the Greenland shark lack caudal keels. However, other authors have found caudal keels in Pacific sleeper sharks from Suruga Bay (Tanaka et al. 1982), Chile (Pequeño et al. 1991), and Australia (Last and Stevens 1994). Okamura et al. (1995) also reported caudal keels in Greenland sharks caught around Greenland. My own observations confirm that both species have caudal keels. Compagno (1984a) also used the distance between the dorsal fins in relation to prebranchial length as a diagnostic feature to distinguish between the Greenland shark and the Pacific sleeper. According to Díaz de Astarloa et al. (1999), this character is variable. In two of three Pacific sleeper specimens from Australia, the interdorsal space was just slightly shorter than the prebranchial length. A Uruguayan specimen had an interdorsal space 1.5 times longer than the prebranchial length, whereas that of a Patagonian (Argentina) shark was 1.4 times longer than the prebranchial length.

In a revised key to the genus *Somniosus,* Francis et al. (1988) used tooth counts and length from snout to first dorsal fin to discriminate between Pacific sleepers and Greenland sharks. According to these authors, the tooth count for the Greenland shark is U:45–52 L; 48–53, while that for the Pacific sleeper is U: 22–48, L: 53–60. However, Díaz de Astarloa et al. (1999) reported a 385-cm male Pacific sleeper shark from the southwest Atlantic as having tooth counts of U: 64, L: 59. A specimen from Cape Columbine (southeastern Atlantic) seen by Bass et al. (1976) had tooth counts of U: 48, L: 53, agreeing with the counts of both the Pacific sleeper and the Greenland shark. Díaz de Astarloa concluded that of all the diagnostic features used to separate the species, the distance from the tip of the snout to the origin of the first dorsal fin is the most useful for separating the two species. All Southern Hemisphere specimens (reported as Greenland sharks or Pacific sleeper sharks) have a more posterior first dorsal fin (the distance from the tip of the snout to the origin of the first dorsal fin is greater than 43% of total length) than in the Greenland shark. This is in agreement with the original description of the Pacific sleeper shark by Bigelow and Schroeder (1944), who stated that the Pacific sleeper had a first dorsal fin that originates considerably farther posteriorly.

A recent analysis by Benz et al. (2007) reveals that the anatomical characteristics used to separate these species overlap to such an extent that species cannot be separated reliably. This agrees with my own observations. The identification problems of the large species of *Somniosus* will not be solved until sufficient numbers of specimens are collected and separated by modern methods. Once these specimens are separated, then diagnostic anatomical characteristics can be determined.

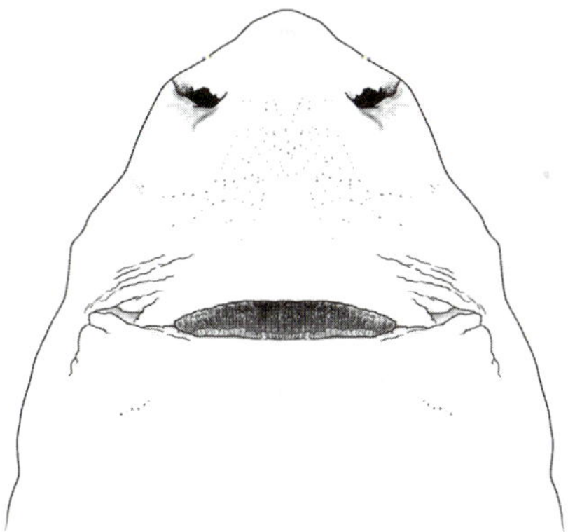

Fig. 34c. Snout.

Fig. 34d. Dermal denticles.

Other squaloid sharks can be differentiated more easily. The little sleeper shark has teeth with more erect cusps that are higher than wide, and it does not exceed 1.5 m. The kitefin shark has lower teeth with triangular cusps with serrated edges. The Portuguese shark has small inconspicuous dorsal fin spines and scalelike dermal denticles.

Size and weight The Pacific sleeper shark is one of the largest sharks. Specimens up to 420–430 cm have been reported (Ebert et al. 1987; Orlov 1999). Other authors claim that the species reaches 500 cm (Soldatov and Lindberg 1930). There are few reliable weights for the Pacific sleeper shark because of its large size. A 114-cm female from California weighed 9 kg (Gotshall and Jow 1965), while a 232-cm specimen weighed 72 kg. A 430-cm female weighed 545.5 kg, whereas one of 401 cm weighed more than 888 kg, the maximum capacity of the available scale (Ebert et al. 1987).

Biology The Pacific sleeper shark is adapted to cold waters and has a very broad depth range. It is often found close to the surface in polar and subpolar latitudes, and in deeper colder waters in temperate and tropical latitudes. Its deepwater habitat and large size protect it from most fisheries, except in Alaska, where many are taken as bycatch in trawls and longline fisheries. Specimens that are hooked or entangled in nets are usually released or discarded because of their large size and lack of commercial value. Thus, they are seldom seen by biologists.

Off the California coast, most specimens are caught at 200–500 m (Phillips 1952; Gotshall and Jow 1965). In the Bering Sea and Kuril Islands, sleeper sharks occurred at 85–717 m (mean 451 m), with no evidence of a deeper range in summer (Orlov 1999). In the Kurils, the bottom temperatures where the sharks were taken was –0.16°C to 4.1°C, and more than half the sharks were taken where the temperature was 2–3°C. Off Alaska, Mecklenburg et al. (2002) reported that the species is frequently caught on the Commander shelf northwest of Bering Island on longline hooks at depths below 300 m. Hulbert et al. (2006) found that sharks carrying electronic tags in the Gulf of Alaska spent most of their time at 150–450 m, often ascending above 100 m. These authors reported that tagged sharks carried out nearly continuous vertical movements traveling below the photic zone during the day and approaching the surface at night.

In some areas, Pacific sleeper sharks appear to be invariably infected with a large parasitic copepod, *Ommatokoita elongata,* which attaches to the cornea. All eight sleeper sharks captured by Benz et al. (1998) in Prince William Sound, Alaska, had these parasites or showed signs of previous infection. Four of the eight sharks had a copepod attached to each cornea; three sharks had a copepod attached only to one cornea, and one shark showed signs of previous infection (G. Benz, pers. comm., October 2005). I have examined six sharks from Prince William Sound and two from the Oregon coast; all had a copepod attached to each cornea. This parasite is also found on the Greenland shark.

Diet: The few reports of the Pacific sleeper shark's diet give a glimpse of a voracious predator that consumes a variety of prey. The smallest specimen for which the stomach contents have been reported was a 114-cm female that contained five sets of squid beaks (*Moroteuthis robusta*), one unidentified squid, and an unidentified octopus (Gotshall and Jow 1965). Larger specimens have a more varied diet. Phillips (1952) reported that a 396-cm male caught off Fort Bragg, California, disgorged several large rockfish (*Sebastes alascanus*), a species found in deeper water than other most rockfishes. Bright (1959) wrote that a 345-cm specimen from Kachemak Bay, Cook Inlet, Alaska, in June 1958 contained at least three harbor seals (*Phoca vitulina*), three complete octopuses (*Octopus* sp.), tanner crabs (*Chionoecetes bairdi*), and triton snails (*Fusitriton oregonensis*). A 366-cm female caught off Trinidad, California, held about 136 kg of fish in its stomach, mostly rex sole (*Glyptocephalus zachirus*) and Dover sole (*Microstomus pacificus*), three king salmon (*Oncorhynchus tshawytscha*), and a 4.5-kg Pacific halibut (*Hippoglossus stenolepis*) (Gotshall and Jow 1965). Yang and Page (1999) analyzed the stomach contents of 13 sleeper sharks 218–295 cm in length from the Gulf of Alaska. Two had empty stomachs, but in the others the arrowtooth flounder (*Atheresthes stomias*) was the most common prey species. Other prey items included walleye pollock (*Theragra chalcogramma*), rockfish (*Sebastes* sp.), salmon (*Oncorhynchus* spp.), and flathead sole (*Hippoglossoides elassodon*). The most important invertebrate species was Pacific giant octopus (*Octopus dofleini*), which ranged from 5 to 29 kg. Other invertebrate prey included squids, snails (*Fusitriton* spp.), hermit crabs, and gammarid amphipods. A 423-cm sleeper shark taken off the Kuril Islands contained six Alaskan pollock (*Theragra chalcogramma*), seven pink salmon (*Oncorhynchus gorbuscha*), one eelpout (*Lycodes* sp.), two sculpins (*Malacocottus zonurus*), about 10 pairs of large cephalopod beaks (probably Pacific giant octopus, or the robust clubfoot squid, *Moroteuthis robusta*), and large gammarid amphipods (Orlov 1999).

Size at maturity There is very little information on the size at maturity of Pacific sleeper sharks. A 366-cm female reported by Gotshall and Jow (1965) carried 300 large oocytes measuring 45–58 mm in diameter (mean 52 mm). Ebert et al. (1987) estimated that maturity occurs at about 370 cm, based on a length–weight curve of 15 specimens and assuming that the 366-cm specimen of Gotshall and Jow (1965) was mature. Although that specimen *might* have been close to maturity, the size of the oocytes is not evidence of maturity, unless the maximum size of the oocytes at ovulation is known. This size, in turn, can only be determined by examining ovulating females. In the absence of such data, maturity can be determined only by observing ovulation or pregnancy. Ebert et al. (1987) reported that a 430-cm female contained one large egg in each uterus. Given these limited data, we can merely state that Pacific sleeper sharks reach maturity at a size between 3.7 m and 4.3 m.

Reproduction Gravid Pacific sleeper sharks have never been reported, and almost nothing is known about its reproductive processes. The species can be assumed to be aplacental viviparous, like other squaloid sharks. The very limited evidence available suggests that the embryos are relatively small at birth and that the broods are large. The 366-cm female reported by Gotshall and Jow (1965) contained 300 large oocytes measuring 45–58 mm in diameter. The relatively uniform size of the vitellogenic oocytes suggests that they would all be ovulated (and fertilized) in a given cycle. Some evidence exists that sleeper sharks are relatively small at birth, in relation to the size of the female. Francis et al. (1988) reported a 41.7-cm free-swimming male from the Challenger Plateau caught in a bottom trawl at 938–941 m. It was newly born, having an unhealed "umbilical" [yolk-sac] oval scar on the ventral surface between the gill slits. Given the very large size of mature sleeper shark females and the very small size of the young, one can conjecture that the species produces broods of up to 200–300.

Relation to humans The Pacific sleeper shark has little economic importance today, although the species has been accused of causing considerable damage to sablefish traps in California (Anderson et al. 1979). During World War I there was a small fishery for the Pacific sleeper shark in Alaskan waters. A pamphlet titled *Shark Hides for Leather*, circulated circa 1918 by the Marine Products Company of Seattle (an affiliate of the Ocean Leather Company of New York), explained how sharks are captured and their hides cured. This pamphlet states:

> For two years the waters of Southeastern and Southwestern Alaska have been quietly prospected for sharks, and over 1000 skins have been taken and tanned, and, by comparison with other varieties secured from other points, the Alaska mud shark [Pacific sleeper shark] has been proven to be on the average the largest and the skin the best for tanning.

A sample of tanned sleeper shark leather, with its characteristic denticles, was attached to the pamphlet. The leather was still supple and in fine condition when I examined it in 2002, 84 years after tanning.

LITTLE SLEEPER SHARK

Somniosus rostratus (Risso, 1826)

Fig. 35a. *Somniosus rostratus*, little sleeper shark, mature male, 96.5 cm, 3.87 kg; Station 3 MAR-ECO Cruise 42°-34"N 28°-09"W (Chip Cotton, collector).

Common name Little sleeper shark, an apt reference to being the smallest, by far, of the known sleeper sharks.

Spanish name None.

Scientific name *Somniosus rostratus* (Risso, 1826), originally described as *Scymnus rostratus* in *Histoire naturelle des principales productions de l'Europe méridionale . . . Poissons* 3: 138. The species has been redescribed by Cigala Fulgosi and Gandolfi (1983). Recently, Yano et al. (2004) suggested placing this species in the subgenus *Rhinoscymnus,* based on size and denticle differences. *Somniosus:* Latin for sleepy; *rostratus:* Latin beaked or hooked, from the Latin *rostrum,* snout or beak.

Synonyms *Heteroscymnus longus* Tanaka, 1912; *Laemargus rostratus* Helbing, 1904; *Somniosus bauchotae* Quéro, 1975.

Identification The little sleeper shark is a *small squaloid 150 cm or less,* with two spineless dorsal fins, and a *small lateral keel on the caudal peduncle at the beginning of the caudal fin.* The upper teeth have smooth-edged, erect cusps. The lower teeth have very finely serrated oblique cusps and overlapping bases. The teeth number U: 57 to 63, L: 15 to 18–15 to 18. The skin denticles are nearly flat and arrowheadlike, with sharp margins. Coloration is dark brown above and below with darker chocolate brown fins.

Similar species The Greenland shark and the Pacific sleeper have a greater number of lower teeth (46–60). The kitefin shark lacks the caudal keel, and its lower teeth have erect, triangular cusps. The Portuguese shark has dorsal fins with small, inconspicuous spines, and overlapping, scalelike dermal denticles.

Range The little sleeper shark has been reported from various locations in the North Atlantic Ocean and the Mediterranean Sea (Maul 1955; Quéro 1975); it has also been reported from Japan (Tanaka 1998). If the North Atlantic and Japanese specimens are indeed conspecific, then the species must be more widely distributed than the few records indicate. The only records from the western North Atlantic are from the northern coast of Cuba (Moreno and Pol 1992). It is probably found in the deep waters of the Gulf of Mexico.

Size and weight The largest specimen on record measured 130 cm (Quéro 1975).

Biology The little sleeper shark is a rarely encountered, deep-dwelling species, probably of wide distribution. Most captures have occurred at depths of 300–1,000 m. However, the capture of a specimen at 2,582–2,734 m in the North Atlantic during the Mar-Eco cruise of 2004 shows this to be one of the deeper dwelling sharks. Due to its deep habitat, few specimens are caught, and little is known of its habits. The only specimen I have examined contained a large number of squid.

The anatomy and embryos of the little sleeper (as *Laemargus rostratus*) were described in a dissertation by Helbing (1904), which also contains an excellent list of the older literature on sleeper sharks. The little sleeper shark was reported to be bioluminescent by Bigelow and Schroeder (1957) and by Hubbs et al. (1967), who misread Burckhardt (1900). The supposed light organs were actually pit organs (Cigala Fulgosi and Gandolfi 1983).

Size at maturity The size at maturity is not known precisely, but it is probably around 80 cm, given that an 82.2-cm

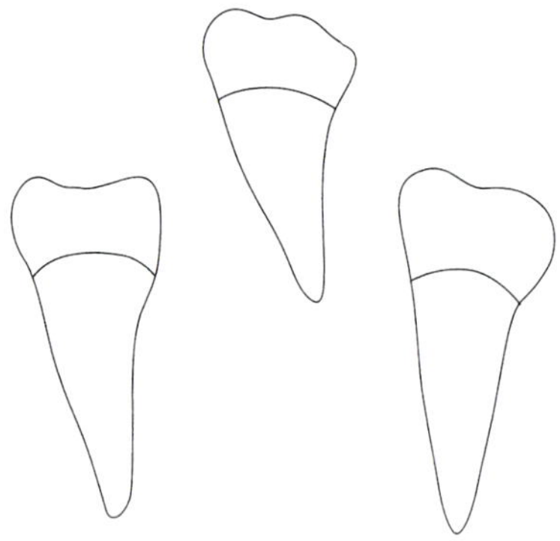

Fig. 35b. Teeth (upper).

Fig. 35c. Teeth (lower).

female from the Ligurian Sea was reported to be carrying nine young (Cigala Fulgosi and Gandolfi 1983).

Reproduction The little sleeper shark is aplacental viviparous. Helbing (1904) illustrated embryos of 3.5–16 cm. Nothing else is known of its reproductive processes.

Size at birth: Unknown.

Brood size: Broods of 9 and 17 have been reported (Cigala Fulgosi and Gandolfi 1983; Tortonese 1956).

Nurseries Unknown.

Age and growth The species has not been aged.

Relation to humans None.

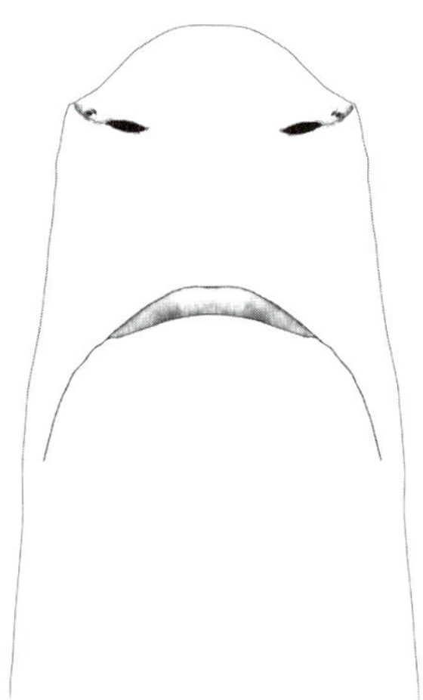

Fig. 35d. Snout.

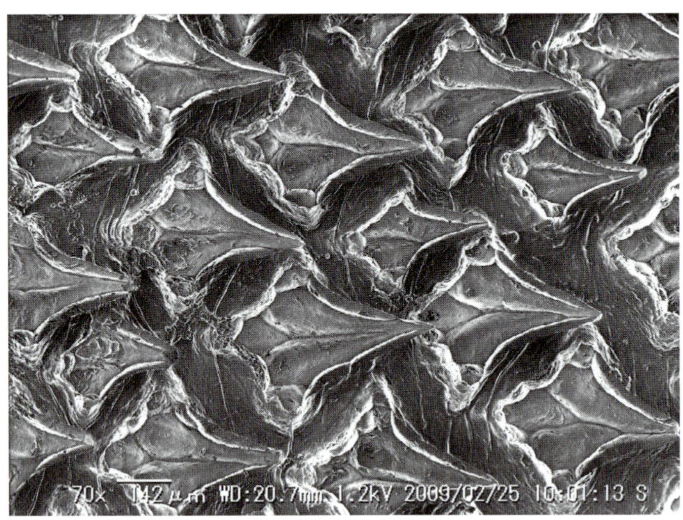

Fig. 35e. Dermal denticles.

FAMILY OXYNOTIDAE

The Rough Sharks

The Oxynotidae is a family of peculiar and unmistakable sharks with laterally compressed bodies that are triangular in cross section, with strong abdominal ridges and two very large dorsal fins with inconspicuous spines (only a small tip protrudes through the skin). These sail-like fins serve as antiroll devices to counteract the rolling tendency of a triangular body. There is a single genus in the family, *Oxynotus,* with four known species in the world. Only one species is found in North America.

CARIBBEAN ROUGHSHARK

Oxynotus caribbaeus Cervigón, 1961

Fig. 36a. *Oxynotus caribbaeus,* Caribbean roughshark, immature female, 52 cm, from Grand Caille Point, St. Lucia, (John Reed, collector; courtesy of R. Grant Gilmore).

Common name Caribbean roughshark, a reference to its range and skin texture.

Spanish name None known in use in the Caribbean.

Scientific name *Oxynotus caribbaeus* Cervigón, 1961, in *Novedades Científicas, Fundación La Salle de Ciencias Naturales,* contribución #9: 3–10. *Oxynotus: oxy,* from the Greek *oxys,* sharp, and *notus,* from the Greek *noteus,* one who carries on his back, an allusion to carrying a sharp spine on its back; *caribbaeus:* Latinized form of Caribbean, plus *eus,* suffix used to indicate provenance, thus "from the Caribbean Sea."

Synonyms None.

Identification The Caribbean roughshark is unique in having a *body triangular in cross section,* with two large *dorsal fins with spines,* the spines being quite large but with only their tips protruding through the skin, and *lacking an anal fin.* The upper teeth are conical and arranged into several functional rows. The lower teeth have broadly triangular cusps with weak serrations and overlapping squarish bases and are arranged into a single functional row. The coloration consists of *a whitish gray background with numerous dark brown or black blotches and spots, with nearly black undersides.*

Similar species Its body form and coloration distinguish the Caribbean roughshark from the other squaloid sharks of North America.

Range This species is known only from the Caribbean Sea. The type specimen was caught approximately 140 km north of Isla La Blanquilla, Venezuela, in 1960. Other captures include a specimen collected off Grand Caille Point, St. Lucia, West Indies (*Johnson-Johnson-Sea-Link* Dive 1763, 4 May 1989), and a 22-cm female caught 20 km south of Isla La Blanquilla, Venezuela (*Oregon* Station 4478, 21 October 1963). Charles Messing filmed a live Caribbean roughshark off Grand Bahama and kindly sent me a copy of the video. The species appears to be widely distributed in the Caribbean Sea, and it is likely found in the Gulf of Mexico.

Size and weight The Caribbean roughshark reaches at least 60 cm.

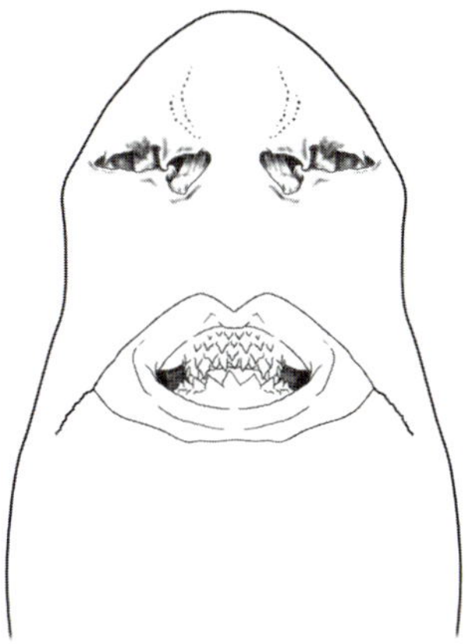

Fig. 36b. Snout.

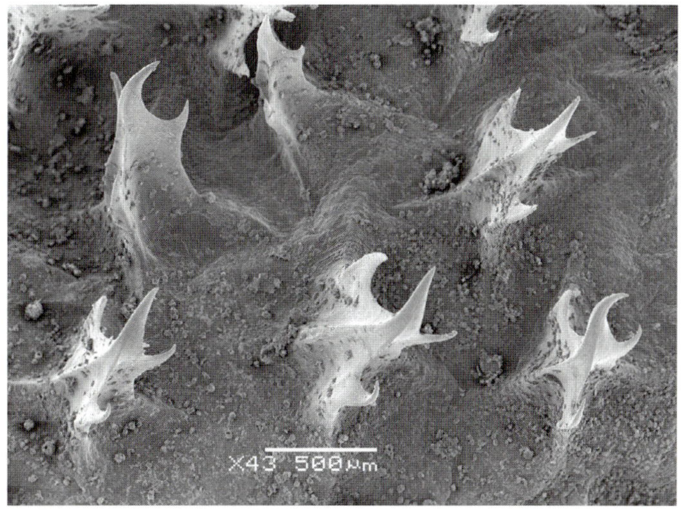

Fig. 36c. Dermal denticles.

Biology This is a poorly known, deep-water shark of the Caribbean Sea. Slow moving, and small, the Caribbean roughshark dwells at depths of 250–500 m around the islands of the Caribbean Sea. This is one of the rarest sharks in collections. I know of only three specimens in collections. A few live sharks have been observed on the seafloor from the *Johnson-Sea-Link* submersible (R. Grant Gilmore, pers. comm., July 2003), suggesting greater abundance in the deep Caribbean Sea than indicated by the small number of specimens in collections. The 52-cm specimen illustrated in this work is the shark from Grand Caille Point, St. Lucia, mentioned above. The bottom temperature at the capture site was 14°C. Videotapes taken of these sharks show them to be sluggish swimmers; the pilot of the *Johnson-Sea-Link* submersible was able to catch the specimen using the manipulator arm! Apparently, the shark was dazzled by the submarine lights. Through the courtesy of Grant Gilmore, I examined the fresh specimen shortly after its capture. The illustration here is based on my photographs and those of Richard Robins taken immediately after capture. The 22-cm female from La Blanquilla was caught at 430 m, the deepest record for the species.

Size at maturity Unknown. The 52-cm specimen I examined was immature.

Reproduction The species is probably aplacental viviparous, like other squaloid sharks.

Size at birth: No data. The smallest specimen I have examined is the 22-cm female caught off La Blanquilla.

Brood size: Unknown.

Nurseries Unknown.

Age and growth The species has not been aged.

Relation to humans None.

FAMILY DALATIIDAE

The Kitefin Shark, the Cookiecutter Sharks, and Dwarf Sharks

The Dalatiidae is a family of dwarf to medium-size squaloid sharks characterized by dorsal fins without spines (or with a small first dorsal fin spine in the case of *Squaliolus*). This family includes the genera *Dalatias, Isistius, Euprotomicrus,* and *Squaliolus.* According to Shirai (1992), these genera have in common nasal capsules almost united with each other but separated by a narrow internasal septum. Another common characteristic is the type of insertion of the suborbitalis muscle into the mandibular. Within this family, two separate lineages are evident: *Dalatias* + *Isistius* and *Euprotomicrus* + *Squaliolus.*

The genera *Dalatias* and *Isistius* share several skeletal characteristics (see Shirai 1992 for details). They also share similar dentitions: their upper teeth are thornlike and increasingly curved toward the corners of the mouth; their lower teeth are much larger than the uppers and have triangular erect cusps with finely serrated edges. Their teeth are perfectly suited for carving plugs out of large prey. *Dalatias* consists of a single species, the kitefin shark (*Dalatias licha*), and *Isistius* encompasses at least two species, the well-known cookiecutter shark (*Isistius brasiliensis*) and the poorly known largetooth cookiecutter shark (*Isistius plutodus*).

The genera *Euprotomicrus* and *Squaliolus* consist of dwarf sharks that do not exceed 30 cm when fully grown. In addition to their tiny size, these sharks share a similar body shape with spineless, low, elongated second dorsal fins and nearly triangular tails. Shirai (1992) has pointed out other minor skeletal similarities that show a common origin. *Euprotomicrus* is a monotypic genus containing only the pygmy shark (*Euprotomicrus bispinatus*), whereas *Squaliolus* comprises two species, one of which, the spined pygmy shark (*Squaliolus laticaudus*), is found off North America.

KEY TO THE DALATIIDAE

1a. First dorsal fin with a small spine, either exposed or buried in the skin (fin originating over the rear portion of the pectoral fin, second dorsal fin base about twice as long as first dorsal fin base) .... **Spined pygmy shark,** *Squaliolus laticaudus* (p. 156).

1b. Dorsal fins without spines 2

2a. First dorsal fin origin much closer to pectoral fin axil than to the origin of the pelvic fins (lips thick and fleshy, white or whitish, with numerous folds; lower teeth erect, triangular, and with serrated edges .... **Kitefin shark,** *Dalatias licha* (p. 142).

2b. First dorsal fin origin much closer to the origin of the pelvic fin than to the pectoral fin axil 3

3a. Second dorsal fin base about three to four times longer than base of first dorsal fin, first dorsal fin free rear tip well anterior to origin of pelvic fin, coloration uniformly black.................................. **Pygmy shark,** *Euprotomicrus bispinatus* (p. 153).

3b. Second dorsal fin base about equal or slightly longer than first dorsal fin base, first dorsal fin free rear tip over or behind origin of pelvic fin, coloration brown or brownish 4

4a. A black or dark collar encircling the gill area, tail with broad black or dark band, second dorsal fin base only slightly larger than first dorsal fin base, 12–15 lower teeth on each side of the jaw.. **Cookiecutter shark,** *Isistius brasiliensis* (p. 145).

4b. Lacking a dark collar encircling the gill area, tail lacking a dark band over tail, second dorsal fin base more than twice the width of the first dorsal fin, about nine lower teeth on each side of the jaw .... **Largetooth cookiecutter shark,** *Isistius plutodus* (p. 151)

KITEFIN SHARK

Dalatias licha (Bonnaterre, 1788)

Fig. 37a. *Dalatias licha*, kitefin shark, mature male, 109.8 cm, 6.56 kg; from off the Azores (courtesy of João Correia).

Common name Kitefin shark. The allusion is unclear, and it is difficult to understand which fin makes it resemble a kite. This name apparently is a relatively recent designation by the A.F.S. (Robins et al. 1980) because I cannot find any prior usage. Other English names include darkie charlie (England), black shark (Australia, New Zealand), and seal shark (South Africa).

Spanish name The species lacks vernacular names in the Spanish Caribbean. In Spain it is known as *carocho* or *negra.*

Scientific name *Dalatias licha* (Bonnaterre, 1788), originally described as *Squalus licha,* in *Tableau Encyclopédique et Méthodique des Trois Règnes de la Nature, Ichthyologie:* 12. *Dalatias:* A generic name of uncertain origin and meaning created by Rafinesque; *licha:* a latinized version of the French name for the species, *la liche.*

Synonyms *Squalus americanus* Gmelin, 1788; *Dalatias sparophagus* Rafinesque, 1810; *Squalus nicaensis* Risso, 1810; *Scymnorhinus phillippsi* Whitley, 1931 [a]; *Pseudoscymnus boshuensis* Herre, 1935; *Scymnorhginus brevipinnis,* J. L. B. Smith, 1936; *Dalatias tachiensis* Shen and Ting, 1972.

Identification The kitefin shark has *two spineless dorsal fins,* the first set forward on the trunk, its origin much closer to the origin of the pectoral fins than to origin of the pelvic fins, and it lacks an anal fin. The upper teeth have long, smooth-edged, pointed cusps; *the lower teeth have triangular erect cusps with serrated edges and have, on their outer side, a strong notch at the base and a small rounded cusplet that overlaps the adjacent tooth.* Teeth number U: 9 or 10–9 or 10, L: 9–1–9. The denticles are broad and pointed, with three weak ridges united posteriorly in a point of variable length, and with scalloped rear margins. The denticles are of variable sizes. Coloration is uniformly dark brown or black above and below (occasionally light brown). *The lips are fleshy and have white or whitish corners.*

Similar species Cookiecutter sharks have two spineless dorsal fins set much farther back on the trunk, the rear tip of the first dorsal fin reaching the level of the pelvic fins. The spineless dorsal fins and the triangular and pointed lower teeth with serrated edges distinguish the kitefin shark from other squaloid sharks in the region.

Range The kitefin shark appears to be almost cosmopolitan in deep waters. It has been reported from the western North Atlantic (Nichols and Firth 1939); the eastern Atlantic from the Shetlands, the North Sea, and the English Channel (Wheeler 1969; De Groot and Nijssen 1976) to Madeira and tropical western Africa (Senegal and Ivory Coast, Cadenat and Blache 1981); the Mediterranean (Bini 1967); the western Pacific (New Zealand, Parker, T. J. 1882; Garrick 1960b); and the southwestern Indian Ocean (Mozambique and South Africa, Bass et al. 1976). It has not been reported from the eastern Pacific, the South Atlantic, or the northwestern Indian Ocean. Although the kitefin shark is common in deep water in the eastern Atlantic and along the east coast of southern Africa, it appears to be scarce off the North American coasts. The only specimens from North America came from Georges Bank (Nichols and Firth 1939) and the Gulf of Mexico (Bigelow et al. 1955; Springer and Bullis 1956). Clark and Kristof (1990) saw and photographed specimens in Bermuda and Grand Cayman. The scarcity of North American records probably reflects the lack of deep-water fisheries rather than the rarity of the shark.

Size and weight The kitefin shark reaches 160–170 cm.

Males: Males reach 150 cm and 18 kg (Marques da Silva 1988; Correia da Costa 1994).

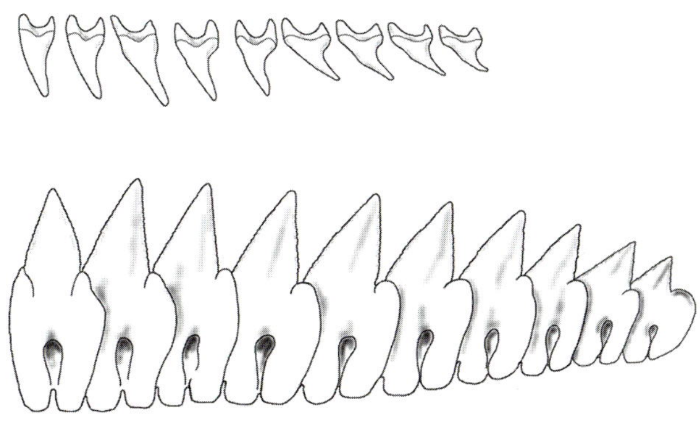

Fig. 37b. Teeth.

Females: Females reach 162 cm and may reach 25–27 kg. Marques da Silva (1983) gave the following weights for females: 125 cm, 15.5 kg; 141 cm, 17 kg; 150 cm, 20.3 kg; and 161 cm, 19.1 kg. Bigelow and Schroeder (1948) listed the maximum size as 182 cm based on an old report by Duméril (1865). De Groot and Nijssen (1976) reported a 170-cm specimen from the Netherlands (see below). Most modern reports (Marques da Silva 1988; Bass et al. 1976; Correia da Costa 1994) all give a maximum size of about 162 cm.

Biology Although the kitefin shark is abundant in European waters, its habits are poorly known. Most captures have been reported from 300–1,000 m (Bigelow and Schroeder 1948; Marques da Silva 1988). However, there are records of its occasional wanderings into shallow coastal waters. A 170-cm specimen was reportedly caught alive in a trawl at a depth of 20 m in the coastal waters of the Netherlands (De Groot and Nijssen 1976). Another specimen was caught in a trawl at 71 m in the English Channel (Wheeler and Blacker 1972). One of the few North American specimens was taken at 91 m on Georges Bank.

The diet of the kitefin shark includes numerous bony fishes, sharks, crustaceans, and cephalopods (Matallanas 1982). Marques da Silva (1983) found squid (*Loligo forbesi*) and scabbard fish (*Lepidopus caudatus*) in the stomachs of two Azorean kitefin sharks. Large quantities of oil are often found in the stomachs of these sharks (Heller et al. 1957), believed to be a dietary residue (Lewis 1969). Both the stomach and liver oils of this shark are rich in squalene, containing 45% and 90% squalene, respectively (Heller et al. 1957). The liver squalene also appears to be of dietary origin (Lewis 1969).

The kitefin shark is an intriguing shark. It has unusually thick, fleshy lips, labial cartilages, and teeth that are very similar to those of its close relative, the cookiecutter shark. Given these similarities, one can only wonder if the kitefin shark has similar "cookiecutting" habits. There are hints of this habit in the literature. Clark and Kristof (1990: 278) wrote that "the strong teeth and tongue and fleshy lips of *Dalatias* indicate that it may at times have similar feeding behavior to *Isistius*." Wheeler (1969: 71) commented that "only small fish are eaten whole, more usually large chunks are bitten out of the back of the prey." Bass et al. (1976: 43) reported the presence of bonito (*Sarda sarda*) in the stomachs of at least three specimens they examined, writing: "It is remarkable that this rather sluggish shark can capture such swift prey, and the method by which it does so is a

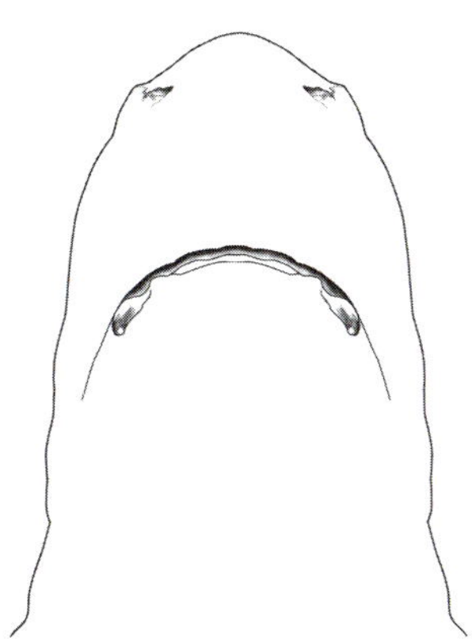

Fig. 37c. Snout.

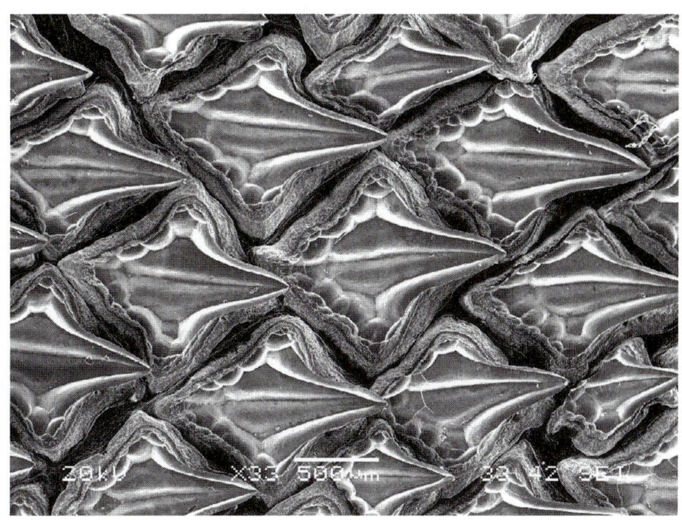

Fig. 37d. Dermal denticles.

complete mystery." This is the same question that has been asked about the cookiecutter shark. Perhaps the kitefin shark and the cookiecutter shark share more than anatomical similarities.

Size at maturity There are few data on the kitefin shark's size at maturity.

Males: Males probably mature at about 115 cm. A 111.6-cm male from Okinawa that I examined had long, developing claspers that were poorly calcified, suggesting that it was approaching maturity. Bass et al. (1976) reported that males 115–121 cm were mature.

Females: Bass et al. (1976) reported that 117-cm female from the Indian Ocean was immature, whereas others 134–159 cm were mature. Marques da Silva (1983) wrote that the smallest gravid female in his sample measured 140 cm.

Reproduction The kitefin shark is aplacental viviparous. T. J. Parker (1882) described the early embryos, while Ranzi (1934) described the uterine wall of the gravid female. Bass et al. (1976) mentioned that Indian Ocean females usually carried 10–20 ripe oocytes 7–9 cm in diameter. Marques da Silva (1988) suggested that the gestation period lasts about two years. Because embryos of different sizes were found in the same month (i.e., the females are not reproductively synchronous), Marques da Silva had to find the best fit for the available data, which yielded a gestation period of 25 months. A gestation period of two years appears to be common in deep-water squaloid sharks.

Size at birth: The kitefin shark is probably born at 36–40 cm. Bigelow and Schroeder (1948) suggested that birth occurs at about 30 cm, but also mentioned a 27-cm embryo with a large yolk sac. Bass et al. (1976) reported that a 36-cm female had an unhealed "umbilical scar" (yolk-sac scar). Marques da Silva (1988) estimated that, off the Azores, the young are born at about 42 cm. I have collected a 40-cm neonate off Japan.

Brood size: Broods reportedly consist of 10–16 young. However, there are few firsthand accounts of gravid females and the brood size. One of these is an extensive report on the anatomy and embryology of the kitefin shark by T. J. Parker (1882), who reported a brood of 10 embryos. Marques da Silva (1988) noted that Azorean females carry 10–14 young.

Nurseries The nurseries of the kitefin shark have not been located. A 38-cm neonate was captured in the northern Gulf of Mexico (Bigelow and Schroeder 1957). Marques da Silva (1988) stated that although hundreds of adult kitefin sharks are caught in the Azorean fisheries, small kitefin sharks have never been taken there. He suggested that the nurseries are in deep unexploited waters or in very restricted areas.

Age and growth Marques da Silva (1988) estimated that a 123-cm kitefin shark would be six years old. Given that sharks and other fishes dwelling in cold deep waters are usually slow growing, it appears unlikely that such a shark could mature in six years.

Relation to humans The kitefin shark is fished in the Azores and dried and exported to Portugal, where the flesh is much appreciated. Its liver, about 20% of total body weight, yields thin oil with a high content of squalene, which is exported (Marques da Silva 1983). The species has no economic value in North America due to its scarcity.

COOKIECUTTER SHARK

Isistius brasiliensis (Quoy and Gaimard, 1824)

Fig. 38a. *Isistius brasiliensis*, cookiecutter shark, mature female, 48.5 cm, 0.48 kg; from the northern Gulf of Mexico (Cheryl Brown, collector).

Common name Cookiecutter or cigar shark. The name "cookiecutter" refers to its characteristic bites, which are uniformly round or oval and of similar size, "as if made with a cookie cutter." More appropriate names might be "ice cream scoop shark" or "watermelon baller shark," as the concave bites appear to have been made with such tools and are not flat, as would result from a cookie cutter. The name "cookiecutter" is fairly recent in the popular and scientific literatures; prior to the mid-1970s the species was simply referred to as *Isistius* because the species and its curious habits were poorly known. Ellis (1975) mentions the name cookiecutter in his popular book on sharks but attributes it to *Etmopterus.* The species has also been called the cigar shark (Le Boeuf et al. 1987) because small specimens resemble a cigar in size, proportion, and coloration.

Spanish name *Tiburón cigarro* (A.F.S.).

Scientific name *Isistius brasiliensis* (Quoy and Gaimard, 1824), originally described as *Scymnus brasiliensis* in *Voyage autour du monde . . . exécuté sur les corvettes de S. M. L'Uranie et La Physicienne, pendant les années 1817–20* 198. The species was later assigned to the genus *Isistius* (Gill 1864b). *Isistius:* the origin of the name is unclear, perhaps alluding to the Egyptian goddess Isis, who was represented in statuary with her head veiled, a symbol of mystery, or to the dark collar of *Isistius,* which could also suggest a veil over the head; *brasiliensis:* from Brazil, location of the type specimen.

Identification This small shark has *two spineless dorsal fins of equal height set far back on the trunk,* and the rear tip of the first dorsal over the pelvic fins. The *base of the second dorsal fin is only slightly longer than the base of the first.* The upper teeth are small, slender, thornlike, and increasingly curved toward the corners of the mouth. The lower teeth are much larger with erect, triangular, smooth-edged cusps and overlapping bases; unlike most other sharks, the lower teeth are shed as a unit, ensuring that this shark has a fully functional cutting edge at all times. Teeth number U: 15 to 19–1–15 to 18, L: 12 to 15–1–12 to 15, their number increasing with age, the larger specimen having the larger numbers. The denticles are roughly square, with a square depression in the center and raised points at the corners.

Coloration is dark brown above and lighter below *with a dark collar encircling the throat;* the median fins have pale, nearly translucent margins, and the *caudal fin has dark brown markings.* Along the upper lobe of the tail the light margin is very narrow and does not extend to the caudal notch. The light margin becomes wider and very distinct from the caudal notch to the tip of the lower caudal lobe. The eyes are proportionally very large; the iris is dusky brown or black and the pupil is green.

The undersides are often bioluminescent when viewed in the dark. F. D. Bennett (1840: 257) was the first to report the bioluminescence of the cookiecutter. His excellent and vivid description has been quoted by others (e.g., Garman 1899; Fowler 1941; Parin 1966) and deserves to be quoted again:

> When the larger specimen, taken at night, was removed to a dark apartment, it afforded a very extraordinary spectacle. The entire inferior surface of the body and head emitted a vivid and greenish phosphorescent gleam, imparting to the creature, by its own light, a truly ghastly and terrible appearance. The luminous effect was constant, and not perceptibly increased by agitation or friction. I thought at one time, that it shone brighter when the fish struggled, but I was not satisfied that such was the fact. When the shark expired, (which was not until it had been out of the water more than three hours,) the luminous appearance faded entirely from the abdomen, and more gradually from other parts; lingering the longest around the jaws and on the fins.
>
> The only part of the under surface of the animal that which was free from luminosity was the black

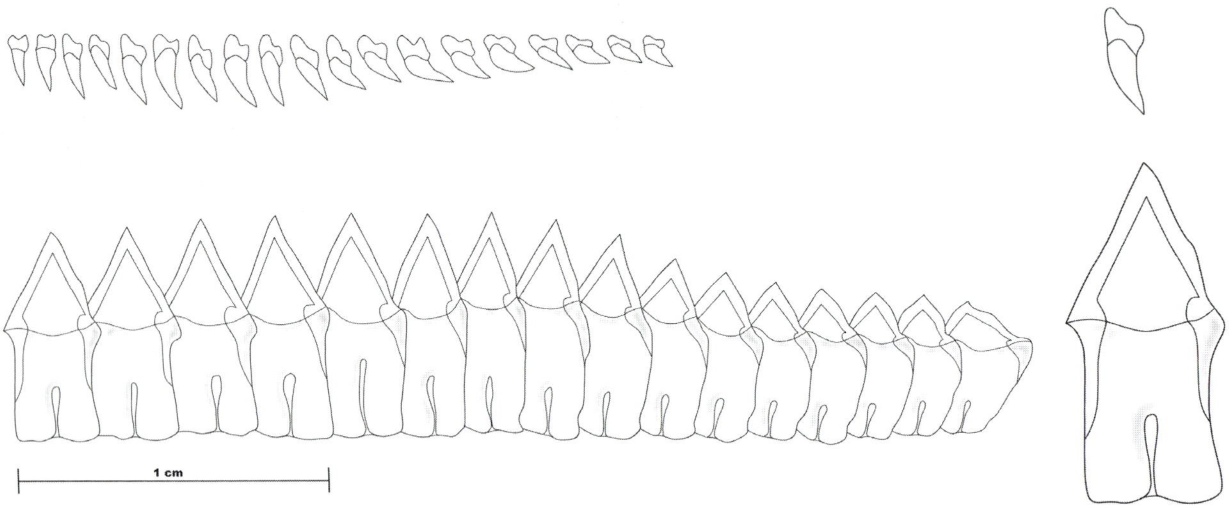

Fig. 38b. Teeth.

collar around the throat; and while the inferior surface of the pectoral, anal, and caudal fins shone with splendour, their superior surface (including the upper lobe of the tail-fin) was in darkness, as also were the dorsal fins, back, and summit of the head.

Other authors have confirmed the bioluminescence of the cookiecutter shark. Strasburg (1963) wrote that it was difficult to observe the well-known bioluminescence of this shark on a brightly lit deck, but that one of the specimens emitted a light blue glow whereas the eyes of two others gleamed with reflected light. Parin (1966) wrote that the entire ventral surface (except the dark collar at the throat) radiated a slight greenish luminescence that faded intermittently. However, others who have observed living specimens, reported a lack of luminescence (Duncker and Mohr 1929). There are at least two possible explanations for this discrepancy. According to Parin, the number and distribution of photophores are quite variable, which could account for the lack of luminescence in some specimens. A more likely explanation is that the luminescence depends on environmental light conditions. Generally, bioluminescent undersides serve to render the shark invisible when viewed from below against the surface light or glare during daylight or bright moonlight. Thus, specimens captured during a moonless night may not luminesce.

Similar species The largetooth cookiecutter shark, *Isistius plutodus,* has fewer and larger teeth (U: 14–1–14, L: 9–1–9); it lacks the dark collar and dark tips on the caudal fin. The pygmy shark has a very long second dorsal fin, its base being more than twice as long as the base of the first dorsal.

Size and weight Cookiecutter males reach about 41 cm, and females reach about 56 cm. The largest cookiecutter shark on record is a 56-cm female reported by Nakano and Tabuchi (1990). I examined three freshly dead specimens from the Gulf of Mexico and recorded the following sizes and weights: male, 41 cm, 0.21 kg; female, 45 cm, 0.48 kg; 48.5 cm, 0.5 kg (carrying large 27-mm oocytes).

Range The cookiecutter shark is cosmopolitan in the deep waters of the tropical and subtropical belts of the Atlantic, Pacific, and Indian oceans. Most reports are from latitudes from 20° N to 20° S; catches from higher latitudes are usually from warm oceanic currents (Bigelow and Schroeder 1957; Strasburg 1963; Jones 1971). Nakano and Tabuchi (1990) reported catching cookiecutter sharks in the Pacific Ocean from latitude 23° to 38° 30′ N and from longitude 146° E to 131° W. Reports of cookiecutter sharks at high latitudes by Bennett (1840) and Seigel (1978) are now considered to be erroneous (Jahn and Haedrich 1987).

In North America the cookiecutter shark has been reported from north of the Bahamas (Bigelow and Schroeder 1948) and the Gulf of Mexico (Garrick and Springer 1964; Castro-Aguirre and Garcia-Dominguez 1988; McEachran and Fechhelm 1998). I have examined numerous specimens from the Gulf. It is interesting, as Jahn and Haedrich (1987) remarked, that there are no reports of cookiecutter sharks in the Gulf Stream, despite the considerable sampling done there, given the presence of the species in other warm currents such as the Agulhas, Kuroshio, and East Australia currents.

The species might be abundant in the South Atlantic. Beaked whale (*Mesoplodon europaeus*) carcasses I examined

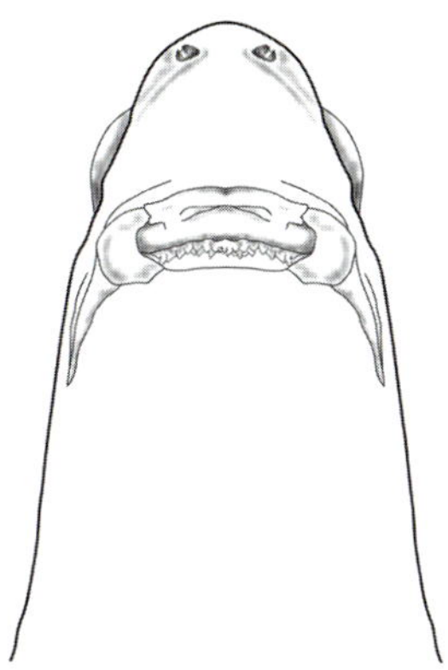

Fig. 38c. Snout.

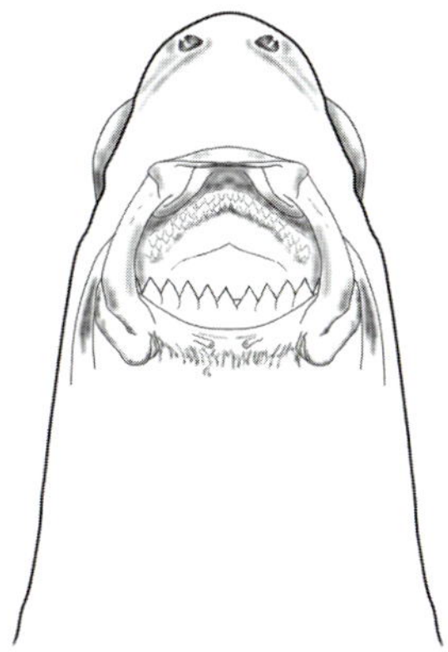

Fig. 38d. Snout.

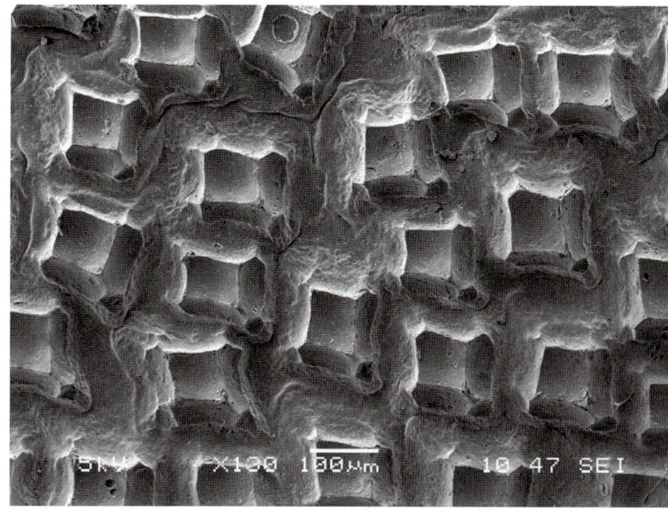

Fig. 38e. Dermal denticles.

on Ascension Island were covered by fresh and healed wounds caused by cookiecutter sharks, suggesting that the species (or another with similar feeding habits) may be common in that area. Also, most of the swordfish carcasses I observed off Santos, Brazil, bore the characteristic bites of these sharks.

Biology This small shark inhabits the upper layers of warm, tropical, oceanic waters. Most catches occur at night from the surface to 550 m (e.g., King and Ikehara 1956; Strasburg 1963: Jones 1971). Strasburg (1963) reported catching cookiecutter sharks only at night, while trawling at depths of 65–357 m in the Pacific Ocean, although trawling and plankton collections were done at all hours. This implies that the species makes large vertical migrations, spending the daylight hours beyond reach at depths greater than 350 m. Jahn and Haedrich (1987) examined capture data for 90 specimens taken by the *Dana* expeditions and concluded that the modal depth of capture was about 150 m. Parin (1966: 180) reported that *Isistius* "inhabits the upper layers of the bathypelagial, for this is also proved bycatches carried out at the surface and at limited depths of the order of 100–500." However, reports of cookiecutter sharks actually on the surface are rare, and it is not generally attracted to lights. Nevertheless, one specimen was dip-netted under a light at the Equator, where upwelling had reduced the surface temperature to 23.6°C (SIO 52–413–5A, Hubbs et al. 1967).

Parin (1966) reported that, based on observations made from the Russian RV *Vitiaz,* cookiecutter sharks swim in schools. Parin cited one occasion when three specimens were taken in a net and another when an entire school was captured, although all but one slipped out by biting through the mesh. This schooling behavior needs to be confirmed, as all other reports indicate catches of single specimens only (Strasburg 1963; Jones 1971). Evidence of bite marks suggests that cookiecutter sharks are solitary. If cookiecutter sharks traveled and fed in schools, the unfortunate animal encountering a school would be attacked by several sharks, and if lucky enough to escape, would bear multiple scars. However, most observed victims bear single bites (Iversen 1959 and others). Le Boeuf et al. (1987) observed fresh crater wounds on 20 juvenile northern elephant seals (*Mirounga angustirostris*) in the islands west of Baja California. Most animals had one wound, two had two wounds, and one had three. Similarly, swordfish carcasses I examined in Santos, Brazil, had one to four bites, most having only one. These hooked swordfish would be highly vulnerable to a passing school of cookiecutter sharks, and if they lived in schools, the swordfish carcasses would be covered by bites. Therefore, it appears unlikely that cookiecutter sharks are found in schools.

According to Parin (1966), temperature data taken at the locations of *Isistius* catches in the Pacific Ocean were 23.5–29.6°C at the surface, 21.5–25.7°C at a depth of 100 m, 13.9–20.5°C at 200 m, and 7.0–8.8°C at 500 m. Nakano and Tabuchi (1990) reported taking cookiecutter sharks when surface water temperatures were 18–26°C in the Pacific Ocean.

Diet: Despite the abundance and ubiquity of cookiecutter sharks, there are few analyses of stomach contents. Squids apparently are common prey of cookiecutter sharks. Strasburg (1963) found squid remains in six of eight stomachs opened. These remains consisted of beaks, pens, tentacles, and pieces of squid estimated at 15–30 cm mantle length (excluding tentacles). Strasburg remarked that this was surprisingly large prey for a small shark, and that it was difficult to understand how such small sharks could capture relatively large, live squid. He ruled out scavenging based on the high proportion of squid in stomachs. Strasburg also reported fibrous tissue believed to be of vertebrate origin. Jahn and Haedrich (1987) examined 130 cookiecutter sharks and reported that 20 contained cephalopod remains, 15 contained plugs of flesh bitten off large fishes, and 6 contained pieces of cetacean skin and blubber. I examined a 41-cm male from the Gulf of Mexico that contained three beaks from small squids.

In contrast to the few reports of stomach contents, there are numerous reports of cookiecutter shark bites on diverse large marine organisms: yellowfin tuna (*Thunnus albacares*) and dolphinfish (*Coryphaena hippurus*) (Nemoto 1955, Iversen 1959), swordfish (*Xiphias gladius*) (Jones 1971), opah (*Lampris regius*) (Klawe 1966), megamouth shark (*Megachasma pelagios*) (Taylor et al. 1983; Yamaguchi and Nakaya 1997), northern elephant seal (*Mirounga angustirostris*) (Le Boeuf et al. 1987), pygmy right whale (*Caperea marginata*) (Baker 1985), dense-beaked whales (*Mesoplodon densirostris*) (Paterson et al. 1993), and spinner dolphin (*Stenella longirostris*) (Norris and Dohl 1980).

The cookiecutter shark's razor-sharp cutting teeth allow it to effectively remove a plug of flesh from large prey, leaving a characteristic "crater wound." These are the concave and crescent-shaped wounds and scars commonly found in the skin of marine mammals and large pelagic fishes. These wounds are usually 3–5 cm in diameter. The smallest wounds reported were 1.2 cm × 2 cm (Jones 1971); the largest recorded were 4–5 cm × 7 cm (Mackintosh and Wheeler 1929). The cause of these bites or sores was a mystery for many decades. Over the years, these bites, or scars, were attributed to spearfishes, lampreys, copepods, sea lice, and even bacteria (Pike 1951; Nemoto 1955; Iversen 1959). Nemoto (1955) was the first to recognize cookiecutter shark bites as distinct from lamprey marks but did not identify their cause. Strasburg (1963) provided the first clues when he reported that the entire mandibular dentition of *Isistius* was shed as a unit and that two stomachs he examined contained fibrous matter believed to be vertebrate connective tissue. Later, crediting Strasburg for the idea, Jones (1971) demonstrated that cookiecutter sharks could inflict hemispherical and crescent-shaped wounds and reported finding hemispherical plugs of flesh in their stomachs. Jones also observed that, in fresh specimens, the tongue is quite mobile and capable of fitting closely against the roof of the mouth. He also observed that when the tongue is retracted and in a vertical position, the lips protrude around the mouth, forming a complete ovoid of labial margins on a single plane. These observations led Jones to hypothesize that cookiecutter sharks could achieve a vacuum with its mouth on a smooth surface when the spiracles are closed and the tongue is retracted. Shirai and Nakaya (1992) performed an in-depth analysis of the functional morphology of the biting apparatus of the cookiecutter shark. They found that a cookiecutter shark stabs the prey with the sharp lower teeth. Immediately afterward, the ventral muscles of the shark contract, resulting in an oral vacuum that fastens the mouth to the prey. The contraction of right and left body muscles results in a simultaneous twisting of the body, with or without help of the drag caused by the water. A half-turn of the body probably makes a complete crater wound.

The shedding of the cookiecutter shark's mandibular teeth as a unit allows the shark to maintain a sharp continuous cutting edge at all times. This is necessary when removing a plug of flesh from vertebrates with tough, fibrous skin. If teeth were shed singly, gaps in the dentition would occur. A gap would allow a strip of skin to remain uncut when the shark bites, perhaps preventing the shark from removing a clean plug of flesh.

Behavior: How this slow-moving dwarf shark can approach and attack fast-moving prey such as squids and pelagic fishes is an interesting mystery. According to Iversen (1959), the wounds on yellowfin tuna were always close to the anal fin. Jones (1971) stated that crescentic wounds that he examined in tunas were inflicted from a frontal attack, when the shark and prey were going in opposite directions. Jones (1971) suggested that the shark is approached by predators, which reject it as food and veer off; at that instant, the shark might be able to make a quick dash and effect a bite.

The idea of the cookiecutter shark using its bioluminescence to attract prey is an old one. Describing the cookiecutter shark as a new species, F. D. Bennett (1840: 258) wrote, "The phosphorescent power it possesses is of use to attract its prey." Widder (1998) attempted to explain how these sharks manage to attack larger and swifter prey, suggesting that the nonbioluminescent collar produces the silhouette of a small fish, when viewed from below, luring prey into striking distance. I have found no evidence to support this hypothesis, and it is difficult to follow the logic behind it. First, cookiecutter sharks often bite very large organisms such as baleen whales and sperm whales that would not be attracted to such prey. Second, the luminescence is directed entirely downward as in many deep-water sharks, and some cookiecutter sharks have photophores only around the anal region. Any larger predator attracted from below and behind would be undetectable or invisible to the shark, with obvious consequences. If cookiecutter sharks were attempting to attract such large prey with their luminescence, there would be no reason to diminish their silhouettes because larger silhouettes would be more effective attractants. Third, the Widder hypothesis ignores that the presence of photophores and production of bioluminescence on the ventral surface are a common adaptation of many mesopelagic fishes. Other small sharks, such as the closely related dwarf shark (*Euprotomicrus bispinatus*) and the spined pygmy shark (*Squaliolus laticaudus*), as well as the small lanternsharks (*Etmopterus*), also have numerous photophores covering the ventral and lower sides of the body and head, although their diets and habits are quite different from the cookiecutter shark.

A more credible hypothesis of the function of bioluminescence in the cookiecutter shark and other mesopelagic organisms is that advanced by W. D. Clarke (1963), who suggests that the light is used as a form of counter illumination, to aid in concealment from predators. Most mesopelagic predators hunt using vision to locate their prey from below. They often have upward-directed eyes to pick up the silhouettes of their prey outlined against the brighter surface. Thus, the ventral position and downward orientation of the photophores on the body surfaces of many pelagic organisms function to eliminate the shadowed area that would form on the ventral surfaces in directional light from above. By blending against ambient light, the organisms become difficult to see from below. Bioluminescent mesopelagic organisms that have been investigated produce light of the same color as the solar light reaching the depths where they dwell—blue light peaking at 478 nm—thus matching their bioluminescence to the ambient light. Organisms in the scattering layers move actively seeking a light level that matches their own luminescence and affords concealment from predators. For further discussion on the function of bioluminescence of mesopelagic organisms, the interested reader can turn to W. D. Clarke's (1963) elegant essay on this subject.

Cookiecutter shark attacks are not confined to living organisms. Cookiecutter sharks have been reported to attack the soft neoprene rubber that covers parts of nuclear submarines. These attacks were reported only on submarines operating in the Pacific Ocean (Gray and Johnson, 1979; Johnson 1978). Obviously, the sharks mistake the submarines for large whales. Cookiecutter sharks also attack the plastic tubing that covers towed arrays used by surface ships and submarines for underwater detection of submarines. I have received reports and evidence (cable sheathing samples bearing bite marks) of cookiecutter sharks damaging magnetic cables off the Bahamas and off Brazil. These magnetic cables were covered with polyurethane rubber sheathing 2–3 mm thick. Previous nonmagnetic cables had not elicited shark bites (C. Berglund and T. Leary, pers. comm., 16 December 1991).

Cookiecutter shark attacks are not confined to large organisms or objects. I once found a rubber sandal (a black flip-flop) on a beach in Tobago that bore the unmistakable bite of a cookiecutter shark with the teeth marks neatly preserved in the rubber. A shark must have bitten it while it floated in the ocean.

Bennett (1840) indicated that a cookiecutter shark survived out of water for nearly three hours. If this is correct (and there is no reason to doubt it, for some squaloid sharks survive remarkably long periods out of the water), and given the warm-water habitat of the cookiecutter shark, it might be possible to keep the species in captivity. It would make a most interesting aquarium fish, and its interesting biting mechanism could be observed.

Almost nothing is known about predators on cookiecutter sharks. Isouchi (1970) reported collecting a live 16.6-cm cookiecutter shark from the stomach of a mackerel (*Scomberomorus* sp.).

Size at maturity

Males: Parin (1966) reported that a 31.4-cm male was immature, whereas males 36.9 and 37.1 cm long were mature. Jahn and Haedrich (1987: 298) gave the size at maturity for males as 36 cm. Their determination of maturity was based on testis size: "Individuals with testes larger that the collapsed condition found in the smallest males were considered mature." The determination of maturity by these methods is very subjective, as the authors admitted, although it is in general agreement with the sizes reported by other authors.

Females: Bigelow and Schroeder (1948) stated that females mature at about 46 cm, but they provided no evidence. However, a pregnant 46.5-cm female taken off Brazil in 1995 confirms that they are mature at that size (Gadig and Gomes 2002). Parin (1966) reported that a 44-cm female carried seven ellipsoidal eggs (presumably in the uteri), suggesting that it was mature. Jahn and Haedrich (1987: 298) stated that females mature at 39 cm. They considered a female as mature if the uteri stood out from the body wall and had differentiated shell glands. However, these workers did not present their data or measurements and encountered no gravid females.

Reproduction The cookiecutter shark is aplacental viviparous, but nothing else is known of its reproductive processes because only one gravid female has been reported (Gadig and Gomes 2002). As in many small squaloid sharks, the oocytes are proportionally large. I have examined a 49-cm female taken in the Gulf of Mexico on 4 September 1995 that contained oocytes 27 mm in diameter.

Size at birth: The only embryos reported measured 12.4 to 13.7 cm (Gadig and Gomes 2002). These embryos still had large yolk sacs attached and obviously were far from term. The smallest free-swimming specimens captured measured

15–17 cm (Jahn and Haedrich 1987). This suggests that the young are born at a length of 14–15 cm.

Brood size: Only one gravid female, carrying nine young, has been reported (Gadig and Gomes 2002).

Nurseries The nursery areas of the cookiecutter shark are unknown. Strasburg (1963) mentioned that nurseries might be near oceanic islands, but this needs confirmation.

Relation to humans Although the cookiecutter shark is of no commercial value, it causes cosmetic damage to hooked tunas and swordfish. It also causes damage to nuclear submarines and to sound or magnetic anomaly detecting cables towed by warships. There is evidence that cookiecutter sharks have attacked shipwrecked sailors in the tropical Atlantic. On the night of 12 September 1942, the German U-boat *U-156* torpedoed and sank the British passenger liner *Laconia* in the South Atlantic. The *Laconia,* en route from Suez to England via the Cape of Good Hope, carried about 3,000 passengers, including 1,800 Italian prisoners of war. It sank at approximately 4°50′ S, 11°30′ W. Summoned by the U-boat captain,[1] several ships attempted to rescue survivors. A French ship, the *Dumont-d'Urville,* rescued 42 of the Italians and "when dressing wounds on arms and legs, the ship's doctor noticed that these were neat round holes as if made with a punch, from three and a half to four inches in diameter—the bites of small sharks" (Peillard 1963: 211). The descriptions of the bites are typical of those caused by cookiecutter sharks, although, if the measurements were correct, those bites are slightly larger than the sizes reported in most cases of cookiecutter shark bites. The shape of the bites makes it unlikely that they could be caused by small carcharhinid sharks, and it is difficult to imagine that the survivors were attacked *only* by very small carcharhinid sharks (there are no accounts of attacks by larger sharks in the narratives).

Another series of attacks on shipwrecked sailors occurred just north of the Equator at 26° W:

> Men on the [ship's] raft were greatly bothered by the attacks of an extremely ferocious small fish less than a foot long and blunt-nosed. Sharks were around but were no bother. These small fish swam in schools and were very persistent. During the first and each successive night many men sustained deep bites. The bites were clean cut and upward of an inch or more deep, mostly in the soles of the feet, although in some cases the fish sprang out the water and bit the men's arms. Frequently the bites resulted in severed arteries and many men died from this cause. The gratings and nettings of the rafts did not prevent the fish from attacking. (Llano 1955: 75)

Llano attributed these attacks to bluefish (*Pomatomus saltatrix*). However, the description of the fish, behavior, and wounds are consistent with cookiecutter sharks and their bites. Whether the fish actually swam in schools or they simply aggregated around the sailors cannot be determined.

Both of these attacks occurred in equatorial waters of the South Atlantic, where cookiecutter sharks are very abundant. If the speculations above are correct, these would be the only documented attacks by cookiecutter sharks on humans. Yet, given the habits and ferocity of cookiecutter sharks, it is not difficult to imagine that shipwrecked sailors, floating in the warm waters of the South Atlantic for many hours or days, would be attacked by these little monsters.

1. After the sinking, the U-boat cruised into the wreckage in an attempt to find the ship's master for intelligence purposes. Expecting to find British soldiers in the water, the U-boat commander was surprised to find the sea full of floating Italians—German allies in World War II. Attempts by the young U-boat commander to rescue his allies and other victims resulted in one of the more interesting and far-reaching episodes of World War II. Unfortunately, the story is beyond the reach of this book. Interested readers can pursue the story in *The Laconia Affair* by L. Peillard (1963) and in the numerous accounts about Admiral Dönitz and the Nuremberg trials.

LARGETOOTH COOKIECUTTER SHARK

Isistius plutodus Garrick and Springer, 1964

Fig. 39a. *Isistius plutodus,* largetooth cookiecutter shark, 42.3 cm female, from the Gulf of Mexico off Alabama (holotype, USNM # 188386, collected by RV *Oregon).*

Common name Largetooth cookiecutter shark, a reference to having larger teeth than those of its congener, the cookiecutter shark (*Isistius brasiliensis*).

Spanish name None.

Scientific name *Isistius plutodus* Garrick and Springer, 1964, in *Copeia* 1964: 678–682. *Isistius:* the origin is unclear, perhaps alluding to the Egyptian goddess Isis, who is represented in statuary with her head veiled, a symbol of mystery, or to the dark collar of *Isistius,* which could also suggest a veil covering the head; *plutodus:* from the Greek *ploutos,* wealth, riches, or abundance, plus *odus,* Greek *odous,* tooth, thus having an abundance of teeth.

Synonyms None.

Identification This small shark has *two spineless dorsal fins set far back on the trunk,* the first having its rear tip above the pelvic fins; *the second dorsal is higher than the first,* and its base is only slightly larger than the base of the first dorsal fin. The upper teeth are small, narrowly triangular, smooth edged, erect near the center of the mouth, and slightly oblique at the sides. The lower teeth are very large with broadly triangular cusps and high rectangular bases; those at the center of the jaw are erect and symmetrical, and those at the sides having slightly oblique cusps with their points angled toward the center of the mouth. Teeth number U: 14–1–14, L: 9–1–9. The skin denticles are low and squarish, with a depression at the center. Coloration is uniform dark brown above and below, except for a band of palcr brown on the underside of the head from the mouth to the gill slits. The type specimen was described as lacking the dark collar present in the cookiecutter shark (*Isistius brasiliensis*). The second specimen reported by Parin (1975) possessed a dark collar but lacked the spots on the tail. The ventral surfaces are probably bioluminescent.

Similar species The cookiecutter shark has dorsal fins of equal size, its lower teeth are smaller and number 12 to 15–1–12 to 15, and its caudal fin is black tipped. The pygmy shark has a long second dorsal fin, the base of which is more than twice the length of the base of the first dorsal.

Range The type specimen was captured in the Gulf of Mexico (Garrick and Springer 1964). The second specimen reported was captured off Okinawa, Japan (Parin 1975). A third specimen was taken off New South Wales, Australia, in 1988 (McGrouther 2001). The species probably has a cosmopolitan distribution like the cookiecutter shark.

Size and weight The type specimen measured 42 cm.

Biology Its habits are probably similar to those of the cookiecutter shark, based on its dentition and similarity to its congener.

Size at maturity No data available.

Reproduction No data are available, but the species is undoubtedly aplacental viviparous, like other squaloid sharks.

Size at birth: Unknown.

Brood size: Unknown.

Nurseries Unknown.

Age and growth The species has not been aged.

Relation to humans None.

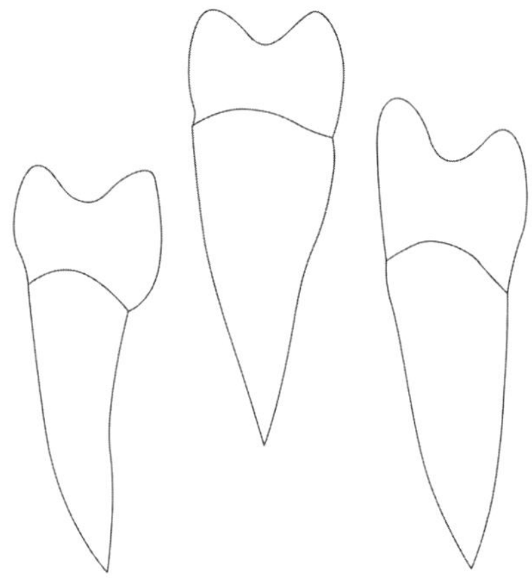

Fig.39b. Teeth (upper).

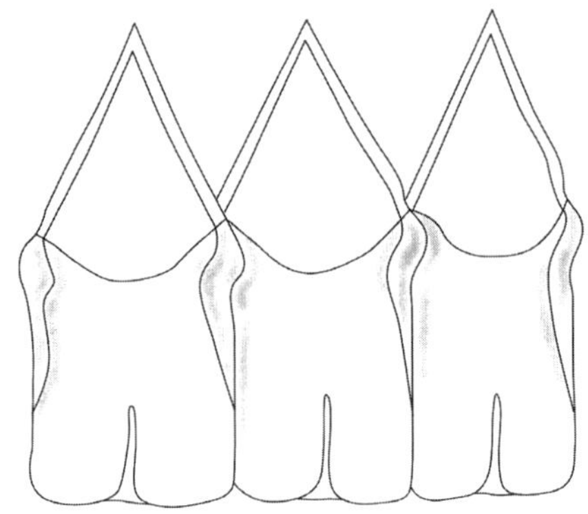

Fig.39c. Teeth (lower).

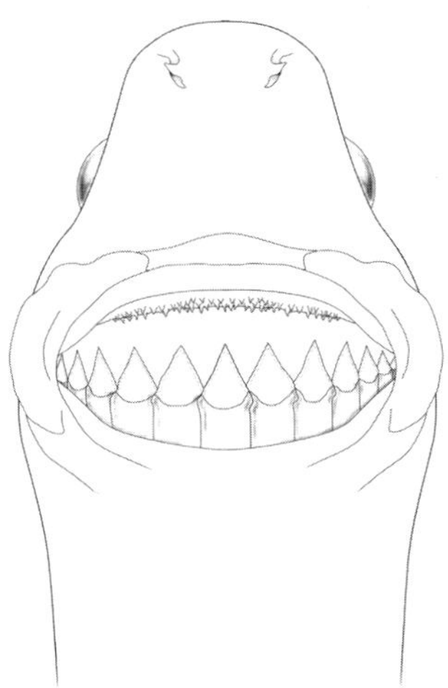

Fig. 39d. Snout (based on a photograph).

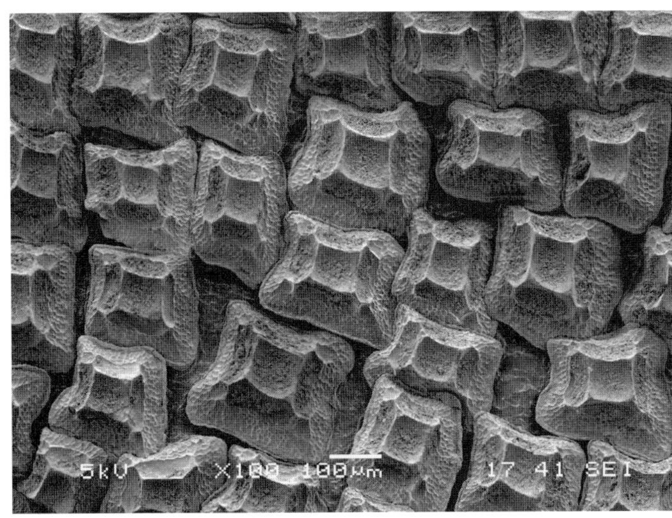

Fig. 39e. Dermal denticles. The denticles are whitish four-sided pyramids with truncated, concave tops. However, many viewers report seeing an optical illusion of "fuzzy black square holes" when looking at this picture. If you get the optical illusion, try looking at the picture upside down.

PYGMY SHARK

Euprotomicrus bispinatus (Quoy and Gaimard, 1824)

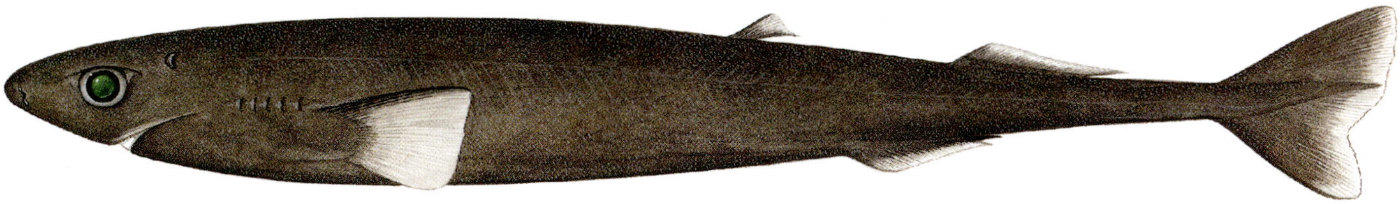

Fig. 40a. *Euprotomicrus bispinatus*, pygmy shark, male 17.3 cm, 0.023 kg; from 4.5 km off Waianae, Oahu, Hawaii, (Jerry Crow, collector).

Common name Pygmy shark.

Spanish name *Tiburón pigmeo* (A.F.S.).

Scientific name *Euprotomicrus bispinatus* (Quoy and Gaimard, 1824), originally described as *Scymnus bispinatus* in *Voyage autour du monde . . . exécuté sur les corvettes de S. M. L'Uranie et La Physicienne, pendant les années 1817–20*: 197, pl. 44, figs. 1–2. *Euprotomicrus:* Greek *eu,* true, plus *proto* (*protos*), first or primary, and *micrus* (*mikros*), small or little, perhaps an allusion to being the first truly small shark described; *bispinatus:* having two spines, a reference to the spiny processes on each pelvic fins (claspers) of the male holotype and not to any dorsal fin spines (which it lacks).

Synonyms *Scymnus* (*Laemargus*) *labordii* Müller and Henle, 1839; *Scymnus labordii* Gill, 1865; *Euprotomicrus hyalinus* Eigenmann, 1890.

Identification The pygmy shark is a very small shark less than 30 cm long. It has a soft, slender body with a tiny spineless *first dorsal fin set far back on the trunk, its origin much nearer the insertion of the pelvic fins than to the pectoral fin axil;* and a *very long spineless second dorsal fin,* its base more than twice as long as the base of the first dorsal fin. The upper teeth are narrow, triangular, smooth edged, and needle pointed, with teeth from adjacent rows set in alternate arrangement. The lower teeth have strongly oblique, pointed cusps and overlapping bases, forming a continuous cutting edge. Teeth number U: 10–1–10, L: 9 to 11–1–9 to 11. The denticles are low and square, with a depression in the center, and of two sizes, the smaller ones being the more numerous. Color varies from light brown to brownish black above, slightly lighter below, with clear fin margins. The ventral side is bioluminescent.

Similar species The spined pygmy shark has a first dorsal fin with a spine, and the fin originates above the pectoral fin. In the cookiecutter and the largetooth cookiecutter sharks the base of the second dorsal fin is only slightly longer than the base of the first dorsal fin.

Size and weight Average size of pygmy sharks is about 20 cm. The largest specimen on record measured almost 27 cm and weighed 70 g (Hubbs et al. 1967).

Range The pygmy shark inhabits subtropical and warm-temperate central waters of the North Pacific (King and Ikehara 1956), South Pacific, Southern Indian Ocean (Mauritius, original locality, Quoy and Gaimard 1824; Last and Stevens 1994), and South Atlantic oceans (800 km west of Cape Town, Bass et al. 1976). On two occasions strays have been caught not far north of the Antarctic Convergence. North American records are from waters several hundred kilometers off the California coast (Hubbs and McHugh 1951).

Biology As Hubbs et al. (1967: 1) noted: "The pygmy shark is a strange creature, defying most concepts of a shark." It is a tiny bioluminescent creature that inhabits the surface layers of the central waters of warm-temperate oceans. The species is infrequently captured and rare in collections. However, it is the best known of the dwarf pelagic sharks because of the excellent monograph by Hubbs et al. (1967). At the time of their study, these authors found only 37 specimens in museums throughout the world. Little else has been learned since.

The pygmy shark is a pelagic species usually found where the surface temperature is 16–27°C and the depth exceeds 1,800 m. Nearly all specimens for which capture data exist were taken in surface waters at night, often while attracted to lights. Hubbs and McHugh (1951) reported dip-netting a pygmy shark under a surface light nearly 750 km off the California coast, where depths were 4,390–5,120 m. Otsu and Yoshida (1957) likewise reported dip-netting a 24.2-cm female at a night-light station over water 4,754 m deep some 480 km north of the Midway Islands. Parin (1966) also reported that all the pygmy sharks cap-

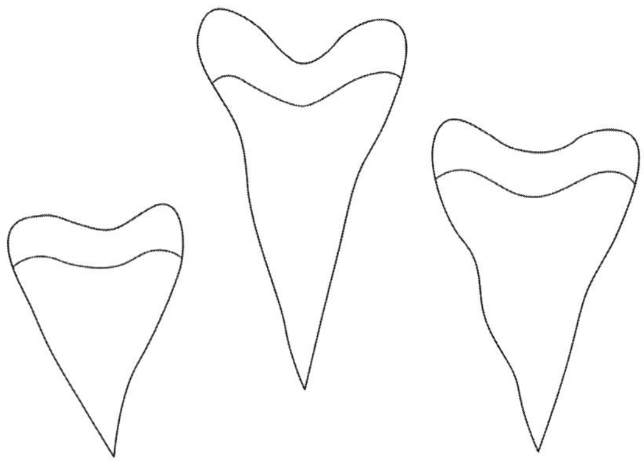

Fig. 40b. Teeth (upper).

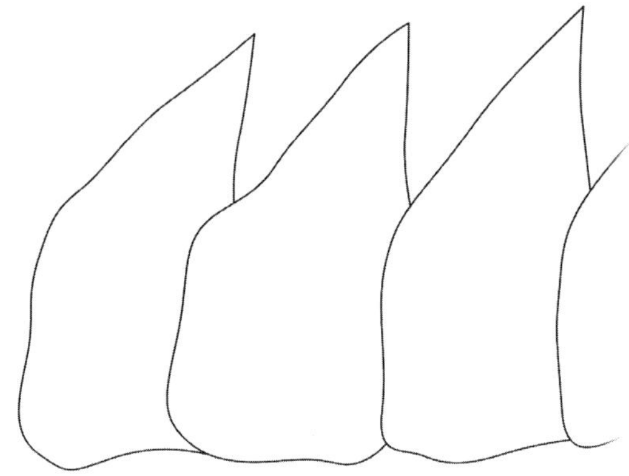

Fig. 40c. Teeth (lower).

tured during the RV *Vitiaz* expeditions in the Indian and Pacific Oceans were caught at the surface during the night using lights.

Although there are no records of pygmy sharks ever being captured in midwater trawls, pygmy shark stomach contents of bathypelagic squids and fishes indicate that the pygmy shark makes considerable diel vertical migrations, coming to the surface to feed during the night hours. One could speculate that it spends the day out of reach of most midwater trawls.

Surface water temperatures at or near the place and time of capture of 21 specimens, for which data are available, were 16–27°C, while the temperature at 200 m were 13–21°C (Hubbs et al. 1967). Parin (1966) gave water temperatures at the locations where pygmy sharks were caught as 16–22.3°C (once at 26.2°C) at the surface, 12.4–21.6°C (once at 26.2°C) at 100 m, 10.3–17.4°C (once at 20.0°C) at 200 m, and 6.5–11.0°C at 500 m. The female mentioned by Otsu and Yoshida (1957) was caught off Midway Islands where the surface water temperature was 16.9°C.

Diet: Little is known about the diet of the pygmy shark. Parin (1966: 186) reported finding "well-preserved pieces of tentacles of an obviously rather large squid" in the stomach of a 21.6-cm specimen. However, it was not clear whether these pieces belonged to a dead animal mutilated by a large predator, or whether the pygmy shark (or more likely, a school of pygmy sharks) had attacked a living squid. Hubbs et al. (1967) surveyed the stomach and intestines of 12 specimens. Nearly all contained at least a trace of much digested blackish to jet-black remains of flesh and skin believed to belong to squids or pelagic octopuses. This survey revealed that the pygmy shark also feeds on a variety of small deep-water fishes, such as lanternfish (*Myctophum nitidulum*), lightfish (*Vinciguerria nimbaria*), and hatchetfish (*Argyropelecus aculeatus*).

Bioluminescence: The pygmy shark's ventral surfaces are covered with thousands of tiny photophores, or light organs, which combine to produce a bluish green glow. Burckhardt (1900) was first to correctly interpret the numerous black specks on the ventral surfaces of the species (as *Euprotomicrus labordii,* a synonym) as light-emitting organs. The bioluminescence of *Euprotomicrus* was first reported by Dickens (1956), who described the greenish glow as being sufficiently strong to be noticed at a distance of 15 m. Parin (1966) reported that three of eight specimens obtained in the Pacific Ocean had been described as bioluminescent, and quoted field notes from logs of the RV *Vitiaz* describing the phenomenon:

> The entire ventral surface and the lower borders of the sides radiated an even, pale greenish light, which flared up with the sudden movements of the shark and faded when the fish became quiet again. . . . One specimen from st. 5199 had an especially intensive luminescence along the midline of the abdomen. Actually the entire lower surface of the body was radiating, from the end of the snout to the peduncle, except for the tips of the ventral fins (pterigiopodia). First the luminescence appeared only at irritation, but before death the luminescence became constant and its traces were maintained for quite a while after death. (Parin 1966: 186)

As with most pelagic fishes, the light is directed almost entirely downward. No luminous organs can be seen on the dorsal surface, and the bioluminescent areas along the trunk region scarcely reach to the lateral line; in the head to the line from the eye to the gill openings (Hubbs et al. 1967).

Hubbs et al. (1967: 17) described the photophores of the pygmy shark as forming a uniform network over the entire ventral surface and lower sides of body and head.

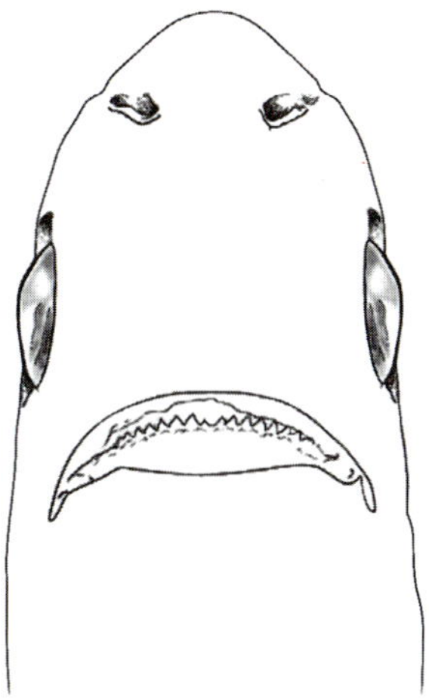

Fig. 40d. Snout.

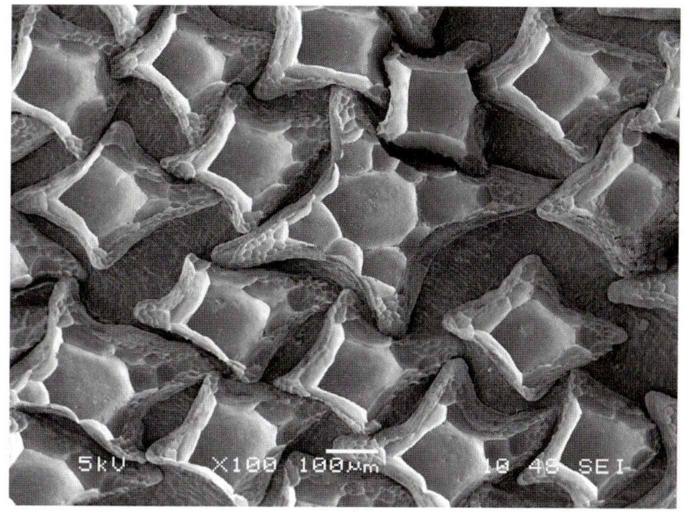

Fig. 40e. Dermal denticles.

> Each unit of the network is a roughly V-shaped aggregation of photophores paralleling the two posterior oblique sides of each denticle. Where they are densely distributed, as on the belly, more than 25 make up each single V-shaped cluster, and no fewer than 60 can be seen in an area of 1 mm^2. Each organ is so tiny, 50–90 μm in diameter, as to be scarcely visible to the unaided eye.

Under low magnification, each organ appears as a black spot no larger than an erythrocyte. According to Hubbs et al. (1967), the photophores of the pygmy shark generally resemble those of the cookiecutter shark but differ notably from those of *Etmopterus lucifer* and *Centroscyllium ritteri.* These observations suggest that bioluminescence in sharks has evolved independently at least twice.

Size at maturity The size at maturity for pygmy sharks is not known precisely.

Males: Hubbs et al. (1967) referred to males 17–23 cm as mature.

Females: Gravid specimens of 23–26 cm have been reported (see below).

Reproduction The pygmy shark is an aplacental viviparous species, but little is known about its reproductive processes. Males probably mature at about 17 cm and females at about 23 cm or less (Hubbs et al. 1967), although size at maturity is not known with precision. Hubbs and coworkers noted that females 23.3, 24.0, 24.2, and 26.5 cm contained "well-formed ova" (oocytes). The first two of these contained four oocytes in each ovary, while the other had oocytes in successive stages of development. Only two gravid females have been reported. The smallest was a 23.4-cm female reported to have given birth to six embryos (Dickens 1956). Marshall (1956) later found two more embryos in this specimen's uteri; hence, the total number was eight. Parin (1966) illustrated, but did not describe, one of eight embryos of 5.4–6.0 cm, four to each uterus, which he found in a 25.6-cm specimen from the Pacific Ocean. This 6.0-cm embryo had external gill filaments, a long yolk stalk, and a large yolk sac, indicating that the species is aplacental viviparous like most squaloid sharks.

Size at birth: Pygmy shark young are probably born at 7.0–9.5 cm (Hubbs et al. 1967).

Brood size: The presence of eight well-developed oocytes in each of two females, and eight embryos in both gravid females reported, led Hubbs et al. (1967) to hypothesize that the usual brood is eight.

Age and growth Unknown.

Nurseries Unknown.

Relation to humans None.

SPINED PYGMY SHARK

Squaliolus laticaudus Smith and Radcliffe, 1912)

Fig. 41a. *Squaliolus laticaudus,* spined pygmy shark, female, 14.4 cm, 0.022 kg; from off Santos, Brazil (courtesy of Alberto Amorim).

Common name Spined pygmy shark (F.A.O.); also dwarf shark (Sadowsky et al. 1985).

Spanish name *Tollo pigmeo espinudo* (F.A.O.).

Scientific name *Squaliolus laticaudus* Smith and Radcliffe, 1912, in *Proceedings of the U. S. National Museum* 41 (No. 1877): 684–685; note that Smith is the sole author of the paper, the name of the species is credited to Smith and Radcliffe). *Squaliolus:* from Latin *squalus,* shark, and *olus,* a diminutive ending, thus a little shark; *laticaudus:* from Latin *lati* (*latus*), broad, and *caudus,* tailed, having a broad tail.

Synonyms Two other species of *Squaliolus* have been described: *Squaliolus sarmenti* di Noronha, 1926, from Madeira, which is now considered a synonym of *S. laticaudus,* and *Squaliolus alii* Teng, 1959 [a], from Taiwan. Seigel et al. (1977) considered *S. alii* conspecific with *S. laticaudus,* but Sasaki and Uyeno (1987) recognized it as a valid species confined to the waters of Japan, Taiwan, and the Philippines.

Identification The spined pygmy shark is a *tiny* shark, usually less than 25 cm in length. It has a cigar-shaped body; a *long and pointed snout,* its length about equal as the distance from the mouth to the origin of the pectoral fin; *a first dorsal fin with a minute spine at its origin* (often difficult to see or feel, and may require the use of a magnifying glass), located anteriorly on the trunk, its origin at about the level of the rear margin of the pectoral fin; and *a very long, spineless second dorsal fin,* its base more than twice as long as that of the first dorsal fin. The upper lip is smooth and lacks any protruding papillae. The upper teeth are long, pointed, and smooth edged. The lower teeth have oblique cusps (except for a median erect tooth) and overlapping bases. The number of teeth in the upper jaw is variable, with two or more series of teeth being functional at a time; the lower teeth number 8 to 10–1–8 to 10. The dermal denticles have four-pointed bases and a central depression. The center of the depression appears to be filled with a substance. Coloration is uniform blackish brown or jet black throughout, with white or white-margined fins. There are numerous light organs on the ventral surface, but their luminescence has not been observed (Seigel 1978).

Similar species According to Sasaki and Uyeno (1987), the smalleye spined pygmy shark (*Squaliolus alii*) has a chevron-shaped upper eye orbit margin (instead of a straight upper eye orbit margin in the spined pygmy shark, *S. laticaudus*), and the upper lip has two papillae. It also has a smaller horizontal eye diameter (43–66% of snout length, measured from the tip of the snout to the anterior eye orbit, compared to 61–82% in the spined pygmy shark). Its eye is 46–70% of the interorbital distance, compared to 73–86% in the spined pygmy shark. The putative differences of the upper eye orbit margin are subtle, variable, and difficult to distinguish.

Specimens of the spined pygmy shark (*S. laticaudus*) taken in the Gulf of Mexico and off Brazil that I have examined conform in the relationship between eye diameter and snout length but not in the eye diameter–interorbital length relationship established by Sasaki and Uyeno (1987). Whether the Pacific and Atlantic spined pygmy sharks are conspecific remains to be determined.

The pygmy shark, cookiecutter shark, and largetooth cookiecutter shark all have spineless dorsal fins set far back on the trunk, their origins closer to the origin of the pelvic fins than to the pectoral fins.

Range The spined pygmy shark is a difficult shark to capture because of its minute size and deep habitat. It is probably cosmopolitan in deep warm waters and more widely distributed and abundant than the few catch records indicate. In the North Atlantic, it has been reported from off France (Belloc 1937; Sigalas 1939–1940), the Azores (Aires da Silva et al. 1998), Madeira (di Noronha 1926), and Ber-

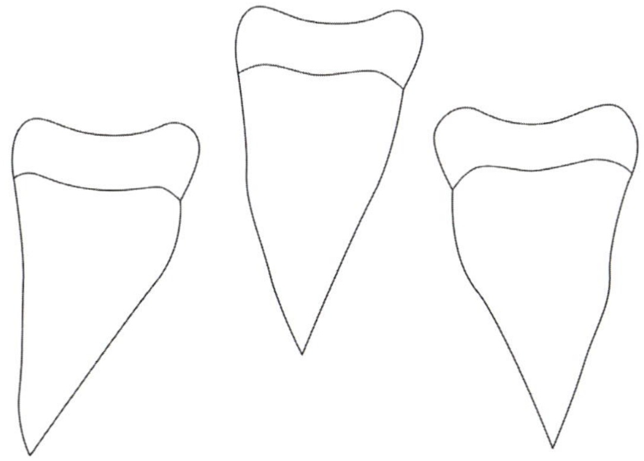

Fig. 41b. Teeth (upper).

Fig. 41c. Teeth (lower).

muda and the Cape Verde Islands (Seigel 1978). The northernmost Atlantic record was taken at 49°37′ N, 28°52′ W (Glukhov and Kuz'michev 1984), probably in the North Atlantic Current. The spined pygmy shark has been taken in open waters of the South Atlantic (Krefft 1980) and off Brazil (Sadowsky et al. 1985). In the western Indian Ocean, it has been reported off Somalia (Seigel 1978). In the western Pacific Ocean, it has been reported from the Philippines (type locality, Smith 1912) and from Suruga and Sagami Bays, Japan (Abe 1962; Seigel 1978). I have recently examined a 19-cm female (courtesy of J. L. Oviedo) taken at 284 m on the Campeche Bank, Gulf of Mexico, in 1998.

Size and weight This is one of the smallest sharks, similar in size to the pygmy shark and the smaller lanternsharks (*Etmopterus*). A specimen 23.1 cm weighed 61.5 gm (Glukhov and Kuz'michev 1984). Its maximum length is about 25 cm.

Biology This tiny shark inhabits deep warm waters near continents or islands. Although it has been reported from depths of 120–1,800 m, it is usually taken at 200–500 m. The type specimen, a 15-cm male, was trawled from 310 m in Batangas Bay, Luzon, Philippines, in 1912 (Smith, H. M. 1912). The type specimen of *Squaliolus sarmenti* (= *S. laticaudus*) was taken on a longline, presumably fishing as deep as 900 m. Off the Azores, it has reportedly been caught at 700–750 m (Aires da Silva et al. 1998). Cadenat and Blache (1981) wrote that off Madeira it was taken in deep longlines set at 600–1,800 m. Spiny pygmy sharks have not been

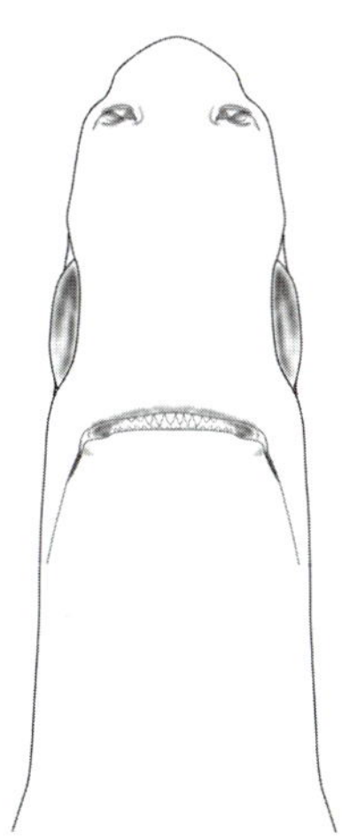

Fig. 41d. Snout.

Fig. 41e. Dermal denticles.

caught at the surface, except for one interesting capture from shoal water near Arcachon, France (Belloc 1937).

The spined pygmy shark is locally abundant in areas of high productivity. Sadowsky et al. (1985) reported 45 specimens removed from the stomach of a bigeye tuna caught off Brazil at depths below 200 m, which, according to the fishermen catching the tuna, contained more spined pygmy sharks than those given to the researchers. The specimens were in different stages of digestion, indicating that they had been eaten over time. Sadowsky et al. remarked that most captures of spined pygmy sharks are single individuals and that the discovery of such large number in the stomach of a tuna was unusual. They also recovered two additional specimens from the stomachs of swordfishes caught off Brazil.

Diet: The spined pygmy shark feeds on squid and small deep-water fishes, such as lanternfishes (Myctophidae). There are few actual reports of its stomach contents. Seigel (1978) reported the stomach contents of 14 spined pygmy sharks from Suruga Bay, Japan, the Cape Verde Islands, and Bermuda. These specimens contained squids (*Chiroteuthis imperator* and *Histioteuthis dofleini*), black dragonfish (*Idiacanthus* sp.), lanternfish (*Diaphus dumerili*), lightfish (*Gonostoma elongatum*), and unidentified bony fishes. A specimen from Brazil that I examined contained squid remains in the stomach.

The spined pygmy shark is one of the rarest sharks in collections. Seigel (1978) listed only 35 museum specimens in his revision of the genus *Squaliolus*. The scarcity in collections reflects the difficulties in capturing such a small, deep-dwelling shark, and not its abundance.

Size at maturity

Males: A 19-cm Brazilian male that I examined was apparently mature and had claspers that protruded 15 mm from the pelvic fins. According to Cadenat and Blache (1981), males of 21.2–24.6 observed off Madeira were mature, based on the size of their claspers.

Females: Females appear to mature at about 21 cm. Cadenat and Blache (1981) also noted that females 21.9–24.3 cm were mature, but did not specify how they had determined maturity. One of their females, 22.5 cm in length, was gravid and carried three eggs in each oviduct. William Driggers (pers. comm., February 2008) examined a 21-cm gravid female from the Gulf of Mexico.

Reproduction The spined pygmy shark is aplacental viviparous (Cadenat and Blache 1981). A 26-cm female taken off Brazil carried four young 9.0–9.7 cm, which had already absorbed their yolk sacs and were fully pigmented (black), suggesting that they were very close to term (Magenta Cunha and Bueno Gonzalez 2006).

Size at birth: Spined pygmy sharks are probably born at a length of 9–10 cm.

Brood size: The limited data available suggest broods of 4–12. Compagno (1984a) reported that a female carried 12 mature oocytes. The two gravid females reported carried four and six young (Cadenat and Blache 1981; Magenta Cunha and Bueno Gonzalez 2006).

Nurseries Unknown.

Relation to humans None.

FAMILY PRISTIOPHORIDAE

The Sawsharks

The sawsharks are characterized by a snout, or rostrum, elongated into a flat blade with teeth along its lateral margin that resembles a saw. Another unique characteristic is the presence of two long barbels on the ventral side of the snout. Sawsharks have five to six pairs of gill slits, large spiracles just behind the eyes, pectoral fins well separated from the head, two spineless dorsal fins, and they lack an anal fin. The family comprises two genera, *Pliotrema* (the sixgill sawshark of the Indian Ocean), and *Pristiophorus* (the fivegill sawsharks of the Indo-Pacific and Atlantic oceans). There are five described species of *Pristiophorus* and probably several still undescribed. Only one species is known from North America.

AMERICAN SAWSHARK

Pristiophorus schroederi Springer and Bullis, 1960

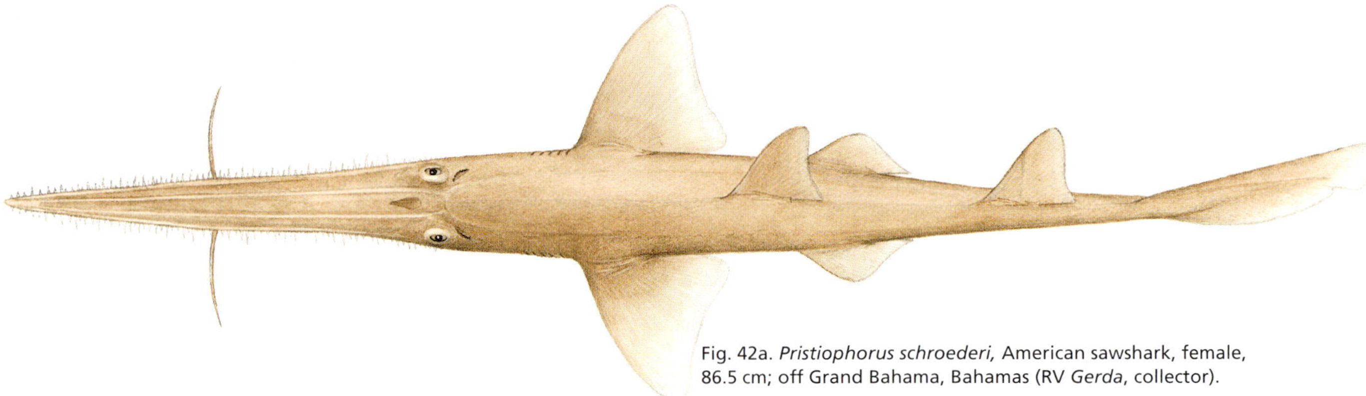

Fig. 42a. *Pristiophorus schroederi,* American sawshark, female, 86.5 cm; off Grand Bahama, Bahamas (RV *Gerda*, collector).

Common name American sawshark: the only sawshark known to occur in the western hemisphere. The teeth on its rostrum (snout) resemble those of a saw.

Spanish name None in the Spanish Caribbean. The A.F.S. has not designated a name for it. I suggest *tiburón serrucho.*

Scientific name *Pristiophorus schroederi* Springer and Bullis, 1960, in *Bulletin of Marine Science of the Gulf and Caribbean* 10(2): 241–252. *Pristiophorus:* from Greek *pristes,* a saw or file, and Greek *phoreus,* a bearer or carrier, thus "sawbearer"; *schroederi:* latinized form honoring the American ichthyologist William C. Schroeder (1895–1977), coauthor of the well-known volume on sharks in *Fishes of the Western North Atlantic* (Bigelow and Schroeder 1948).

Synonyms None.

Identification The American sawshark is recognized by its elongated bladelike snout or "saw," which is *armed with "teeth" of varying sizes along its lateral margin,* and by the presence of *two long barbels about midway on the underside of the snout.* In addition to the "teeth" along the lateral margins of the snout, there are two longitudinal rows of widely spaced, small, thin "teeth," on the *ventral* side of the snout, one row near on each border. The snout (rostral) "teeth" are actually enlarged dermal denticles and are replaced when lost, unlike those of the sawfishes. The sawshark has large spiracles behind the eyes, five *lateral gill slits,* and two equally large dorsal fins. The teeth in the mouth are quite small, having one smooth conical cusp. They number U: 36, L: 32 and are alike in both jaws. The dermal denticles are closely overlapping, have a marked central ridge that terminates in a long point, and have two smaller lateral points. Coloration is light gray to brownish gray above and whitish below.

Size and weight Springer and Bullis (1960) reported an 81-cm specimen. The specimen illustrated here measured 86.5 cm and is the largest known. It was taken in the Northwest Providence Channel, Bahamas, by RV *Gerda* in 1965.

Similar species This is the only shark with a sawlike snout known in the western hemisphere. The only other animals having a similar snout are the sawfishes. Sawfishes are sharklike rays having pectoral fins joined to the head at their an-

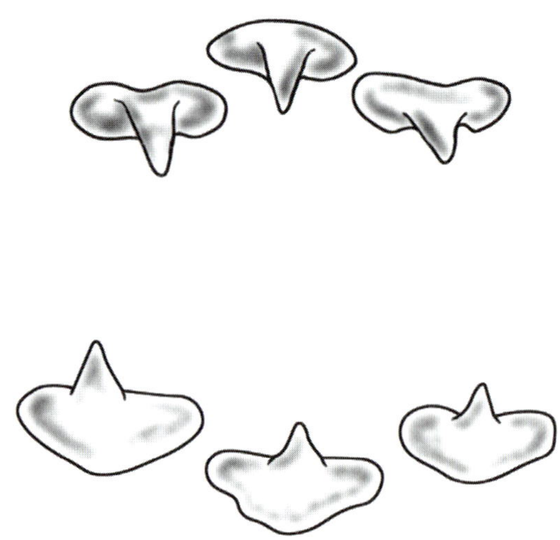

Fig. 42b. Teeth.

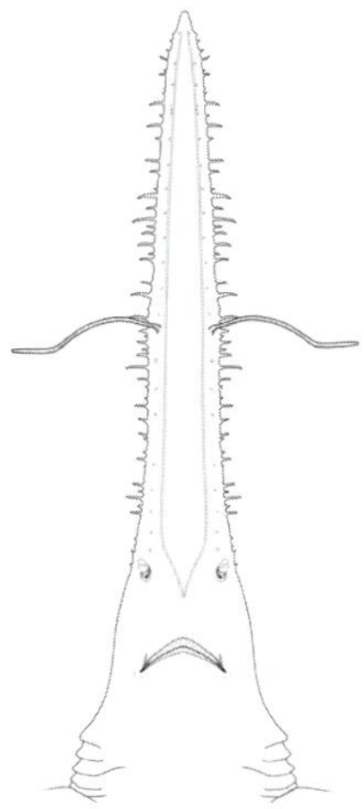

Fig. 42c. Snout.

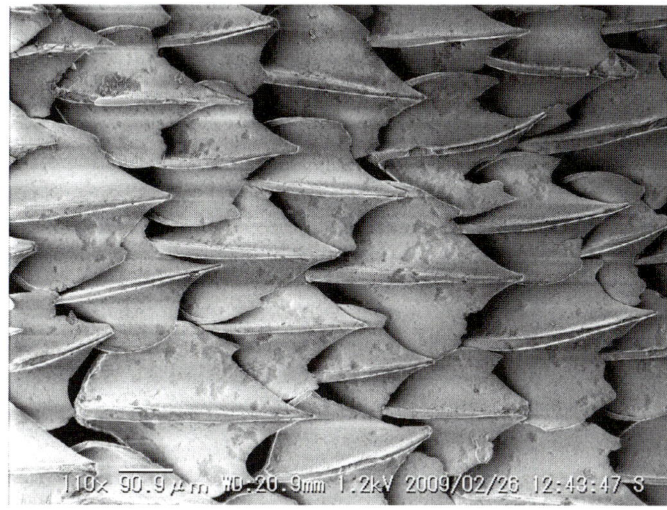

Fig. 42d. Dermal denticles.

terior margins, ventral gill slits, and the absence of barbels on the underside of the snout. However, despite the similar snouts, it should be difficult to confuse the two species. The sawfishes attain very large sizes and inhabit shallow coastal waters (they are usually found at depths less than 100 m); the sawsharks are small and live in very deep waters.

Range The American sawshark is found in deep waters around the Bahamas and probably as far south as Cuba. The only known specimens are from Bahamian waters, where it appears to be locally common on Cay Sal Bank. The limits of its distribution are unknown.

Biology This is a deep-water, bottom-dwelling shark found in the tropical waters of the Bahamas. The species is extremely rare in collections, and nothing is known of its habits. It probably uses its saw (rostrum) to stun and disable prey, like sawfishes do. The long barbels are probably used to locate buried prey, such as shrimp.

The holotype is a female of 38.3 cm, caught about 15 nautical miles east of Dog Rocks, Cay Sal Bank, at 640 m in June 1957. Two other larger specimens were collected north of Little Bahama Bank at 914–950 m, in June 1958 (Springer and Bullis 1960).

Size at maturity Unknown.

Reproduction The sawshark is an aplacental viviparous species. Nothing else is known of its reproductive processes.

Size at birth: Unknown (the young probably measure about 30 cm at birth).

Brood size: Unknown.

Nurseries Unknown.

Age and growth Unknown.

Relation to humans None. The American sawshark received protected status in Florida in 1999, although the species has never been collected there. Apparently, legislators feared that sawsharks would be confused with the protected sawfishes or that someone would claim that a protected sawfish was a nonprotected sawshark. Later, the mistake was realized and the species was removed from the protected list. The deep-water habitat of the sawshark protects it better than any legislation could.

FAMILY SQUATINIDAE

The Angel Sharks

Angel sharks, family Squatinidae, are characterized by flattened bodies, a terminal mouth, eyes on top of the head, large spiracles behind the eyes, greatly expanded pectoral fins, two equally small dorsal fins near the tail, and no anal fin. Angel sharks greatly resemble skates and can be considered intermediate forms between sharks and the batoid fishes. They differ from the batoids in that their pectoral fins are not attached anteriorly to the head, their five gill slits are on the side of the head instead of on the ventral surface, and they have movable eyelids. Angel sharks are small to medium-size bottom dwellers found in shallow coastal waters throughout tropical and warm-temperate seas. Development is aplacental viviparous. At present, some 12 or 13 species are recognized, all of which closely resemble each other. Until very recently, it was considered that there were only two species of angel sharks in North America, the Pacific angel shark and the Atlantic angel shark. In September 2006, Castro-Aguirre et al. published a revision of the genus and described two new species, both from the Gulf of Mexico, making it a total of four species in North America, one in the Pacific and three in the Atlantic. The characters used for separation of the Atlantic species are sometimes difficult to see.

PACIFIC ANGEL SHARK

Squatina californica Ayres, 1859

Fig. 43a. *Squatina californica*, Pacific angel shark, female, 84.5 cm, 5.8 kg; from El Barril, Baja California, Mexico (Cruz Aguilar Arce, collector).

Common name Angel shark; also monkfish, angelfish (Jordan and Evermann 1896). Angel fish is an archaic name for the species, bestowed to it in medieval times because, according to some authors, the pectoral fins resemble angel wings (Pennant 1776).

Spanish name *Angelito* (Mexico), diminutive of angel, probably after the English name; *perro* (Sonora, Mexico), "dog," probably due to its biting tendencies. The A.F.S. name is *angelote del Pacífico.*

Scientific name *Squatina californica* Ayres, 1859, in *Proceedings of the California Academy of Sciences* 2: 29. *Squatina:* an ancient Latin name for a skate; *californica:* an obvious reference to its being found in California.

Synonyms *Rhina (Squatina) armata* Philippi, 1887.

Identification *The Pacific angel shark is a raylike shark. It has a flattened body, terminal mouth, eyes on top of the head, large spiracles behind the eyes, and greatly expanded pectoral fins.* The pectoral fins are clearly separated from the side of the head and end anteriorly on a pointed projection. *Its gill slits are on the side of the head.* It has two equally small dorsal fins near the tail and lacks an anal fin. The internal nasal lobe is wide and spatulate. The teeth are conical, pointed, smooth edged, broad based, and similar in both jaws; they number U: 9–9, L: 10–10, with a broad gap at the symphysis. The dermal denticles on the dorsal side are flat, lanceolate points set on round bases.

The dermal denticles on the mid-dorsal line are conical and much larger than those on the sides, forming a distinctive band along the mid-dorsal line extending from the nuchal area to the caudal peduncle. Coloration is sandy gray or tan speckled with darker spots above; the undersides are white.

Similar species The flattened body and the terminal mouth distinguish this species readily from other sharks in the area. The skates have the pectoral fins joined to the head and have ventral gill slits.

Range The Pacific angel shark inhabits the Pacific coast of North America from southeastern Alaska to Baja California.

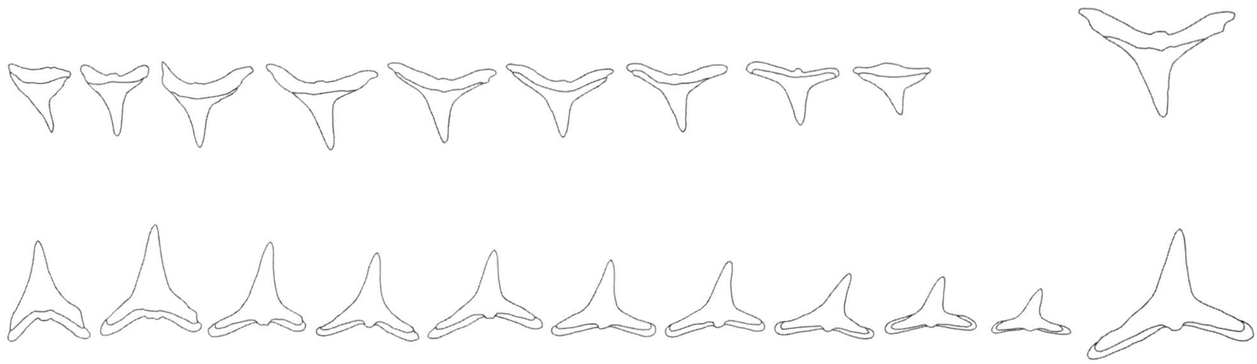

Fig. 43b. Teeth.

The northernmost record is for the area south of Yakutat Bay, Alaska (Wilimovsky 1954). It is abundant from southern California southward to Mexico. I have found it to be extremely abundant off El Barril, Baja California, Mexico, on the western side of the central Gulf of California.

The southern limit of its range is uncertain due to confusion with a similar form, the angelote (Squatina armata), which is found off the western coast of South America. Kato et al. (1967) suggested that the angelote may be a synonym of the California angel shark, but the relationship or status of these two forms remains to be elucidated.

Size and weight According to Kato et al. (1967) the species reaches 150 cm, but all the specimens that I have seen have been smaller than 120 cm. Feder et al. (1974) reported that a 113-cm female weighed 14 kg. An 85-cm male and an 85-cm female from Baja California both weighed 5.8 kg.

Biology The Pacific angel shark is a bottom-dwelling ambush predator. It is often found lying partly buried in the sand or mud, near rocks or ledges, with only the eyes and upper parts exposed, but well camouflaged against the bottom by its spotted dorsal surface. Angel sharks can fold or depress their dorsal fins against the body so that when they are buried in the sand their fins do not betray their presence. A partly buried angel shark waits for prey to come within striking distance. When a fish swims over the shark, the shark suddenly presses its pectoral fins against the substrate and rapidly lunges upward, with its head at a 90° angle to the body and its mouth open, quickly sucking in the prey. Fouts and Nelson (1999) recorded the attack sequences and estimated that the attack lasted 30–100 ms.

Angel sharks are nocturnal predators, becoming active soon after sunset. According to Standora and Nelson (1977), angel sharks began their nighttime activity when light levels decreased to 1–16 lux. These authors also found that angel sharks at Ship Rock, California, stayed within specific home ranges. Using telemetry, the researchers found that tagged sharks moved an average of 4 km during the nightly tracking sessions, but nearly all movements were in an area of approximately 150 ha (1 × 1.5 km). Tagged sharks usually swam close to the bottom, although on occasion they were estimated to be between 15 and 91 m off the bottom. During the tracking, sharks were monitored at depths ranging from 27 to 100 m and temperatures ranging from 9.0° to 16.5°C. Their activity ceases at about an hour before sunrise (Pittenger 1984).

Pittenger (1984) also equipped angel sharks with transmitters, finding that all their detectable movements occurred at night. Pittenger stated that, in 116 dives at Ship Rock, almost all the angel sharks observed in daylight were resting in the sand, unresponsive, and could be easily approached. Only two angel sharks were seen swimming during the day without having been disturbed previously by the divers. Although Pittenger interpreted the motionless sharks as resting, the lack of movement in an ambush predator may indicate hunting and not resting. Telemetry tracking and visual sighting of tagged sharks showed that angel sharks may go for several days without detectable movement. According to Pittenger (1984), sharks did not move every night, and it was common to find a transmitter-equipped or tagged shark in the morning in the same location where it was the evening before. One of the tracked sharks was found in the same place for 10 consecutive days, after which it resumed moving. Pittenger stated that, while it was possible that the sharks were leaving their resting sites after dark and returning to the exact same site before dawn the next morning, he considered this unlikely because the behavior was not observed during five continuous night tracking sessions. Subsequent research has shown that angel sharks sometimes do return to their previous resting places (Fouts and Nelson 1999). In each case the shark's temporary absence was visually confirmed during a scuba search.

Diet: Pittenger (1984) found identifiable prey in 27 of 65 stomachs sampled. During the summer the prey were queen-

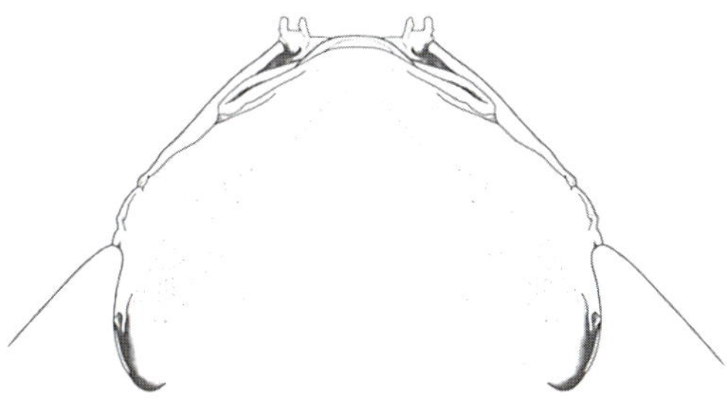

Fig. 43c. Snout.

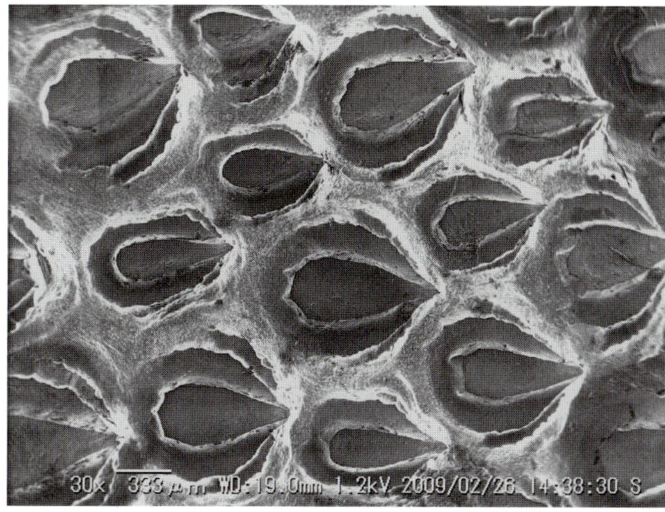

Fig. 43d. Dermal denticles.

fish (Seriphus politus), blacksmith (Chromis punctipinnis), juvenile seabass (Cynoscion nobilis), and blackeye gobies (Rhinogobiops nicholsii), along with occasional sea cucumbers (Parastichopus parvimensis). Queenfish and blacksmith were the two most common summer prey. During winter all identifiable stomach contents consisted of market squid (Loligo opalescens), which are present around Catalina Island from late October through mid-March.

Migrations: The long-term patterns of tracked sharks, and resightings or captures of tagged sharks, indicate that the angel sharks of Catalina Island constitute a single resident population distributed around the island (Pittenger 1984). Tracked sharks moved away from Ship Rock but remained near the island in a narrow band from 12 to 100 m. During several months of tracking, four of the sharks (33%) moved at least one-third of the way around the island. One moved about 80% around the island. Pittenger believed that if tracking had continued longer (i.e., one year), some of the sharks would have circled the island one or more times. Gaida (1997) used electrophoresis to examine the genetic variability of the angel sharks around three of the California Channel Islands. Significant genetic divergence was found between samples from northern and southern islands, while samples of two northern islands were of the same population. This suggests that deep-water channels around the islands may constitute effective barriers to the movements of angel sharks.

Size at maturity

Males: Male angel sharks apparently reach maturity at about 90 cm. Pittenger (1984) examined 35 males caught off Catalina Island and determined their maturity by the presence or absence of spermatophores or spermatozoa in the testes or epididymides. The smallest male found to "have active sperm or spermatophores" was 103 cm, while "active products" were found in all sharks over 108 cm. Pittenger also used visual inspection to assess the size of the claspers relative to body length. The minimum length of a male determined to have large claspers relative to its body length was 102 cm, whereas the largest male seen with small claspers was 107 cm. Although this method does not consider clasper calcification, it furnishes a rough evaluation of maturity. Natanson and Cailliet (1986) estimated that males mature between 90 and 100 cm. These authors determined maturity in the male using three methods: (1) plotting clasper length versus total length, and assuming maturity where the clasper length suddenly increases relative to total length; (2) evaluating the amount of coiling of the epididymis; and (3) detecting the presence of viable sperm in smears taken from the epididymis and examined under a microscope. Unfortunately, these criteria may include large juveniles and do not yield great precision; viable sperm were found in individuals as small as 69 cm, while another specimen 86.5 cm had a coiled epididymis. Totally coiled epididymides were found in 19 of 20 specimens larger than 100 cm. The smallest mature male that I have examined measured 91 cm. It had fully developed, calcified claspers.

Females: The available data and observations suggest that females reach maturity at about 107 cm. Pittenger (1984) used the presence of "mature ova and fully developed ovaries" as a criterion for maturity but did not define these terms and gave no sizes for the mature oocytes. He stated that all females larger than 109 cm (n = 10) had developed left ovaries and enlarged ova. Natanson and Cailliet (1986: 988) estimated that females mature at 90–100 cm. They considered females to be mature when they "had ovaries containing large, heavily yolked eggs [oocytes] 50–60 mm in diameter." This method tends to underestimate the size at maturity by including large juveniles. The fact that the smallest gravid female examined by Natanson and Cailliet measured 107.9 cm suggests that female maturity occurs at sizes larger than 100 cm.

Reproduction The Pacific angel shark is an aplacental viviparous species. In most specimens only the left ovary is developed; some have both ovaries developed. Gestation is believed to last about 10 months and parturition occurs between March and June (Natanson and Cailliet 1986).

Size at birth: The Pacific angel shark is born at 24–26 cm. The smallest neonate I have examined measured 25.8 cm. It had an open yolk-sac scar 3 × 10 mm and was caught in mid-June off El Barril, Baja California, Mexico.

Brood size: Brood size ranges from 1 to 11 (Natanson and Cailliet 1986).

Nurseries The nurseries of the Pacific Angel shark along the California coast are undescribed. I have encountered a few neonates in the Gulf of California, indicating that a part of the population gives birth there.

Age and growth The Pacific angel shark has not been aged because, as Natanson and Cailliet (1990) have demonstrated, the vertebral rings are not deposited annually but are related to somatic growth. Thus, these rings do not provide a reliable record of age.

Relation to humans The Pacific angel shark used to be considered a "trash fish," to be discarded overboard or used as crab bait, or a dangerous nuisance to be killed before it could be removed from a net. In the late 1970s, a commercial demand was created in California through astute marketing, and the angel shark was transformed into the most highly sought-after shark species in the Santa Barbara Channel (Richards 1987). The fishery grew from a total of 149 kg in 1977 to more than 454,000 kg in 1985 and more than 517,000 kg in 1986 (Richards 1987, Holts 1988). The fishery peaked in the 1985–1986 season, when more than 79,000 individual angel sharks were taken during the March to April season (Richards 1992). After 1986, landings declined due to overfishing. Subsequent minimum size restrictions and the banning of gillnets in California waters have further limited the catch. By 1991, the catches had returned to pre-1981 levels. Those desiring more information on the history of this fishery should consult the excellent accounts by J. B. Richards (1987, 1992).

ATLANTIC ANGEL SHARK

Squatina dumerili Le Sueur, 1818

Fig. 44a. *Squatina dumerili*, Atlantic angel shark, immature male, 102 cm, 9.3 kg; off Cape Hatteras, North Carolina (John Galbraith, collector).

Common name Atlantic angel shark (A.F.S.); also monkfish, angel fish (Jordan and Evermann 1896). Angel fish is an archaic English name for the species, bestowed to it in medieval times because, according to some authors, the pectoral fins resemble angel wings (Pennant 1776). Later it was called the American angel-fish by De Kay (1842).

Spanish name *Angelito, tiburón angelito* (Mexico). The A.F.S. name is *angelote del Atlántico.*

Scientific name *Squatina dumerili* Le Sueur, 1818, in *Journal of the Academy of Natural Sciences of Philadelphia* 1(2): 225–226, plate 10. *Squatina:* an ancient Latin name for a skate, an allusion to its resembling a skate; *dumerili:* named after Auguste Duméril (1812–1870), physician and professor of herpetology and ichthyology at the Museum National d'Histoire Naturelle in Paris. In the older literature, this species was considered synonymous with the larger European angel shark (*S. squatina*), from which it differs by numerous characters, including the shape of the head folds and the conformation of the nasal flaps and barbels (Bigelow and Schroeder 1948).

Synonyms None.

Identification The Atlantic angel shark is a raylike shark, characterized by its *flattened body, terminal mouth, eyes on top of the head,* large spiracles behind the eyes, *greatly expanded pectoral fins that are clearly separated from the head,* and two small dorsal fins of equal size located near the tail. *The internal nasal lobe is thin and pointed.* The teeth are conical, smooth edged, broad based, and alike in both jaws and number U: 10–10, L: 9–9, with a broad gap at the symphysis. The dermal denticles on the dorsal side are flat, lanceolate points set on large round bases. *The dermal denticles on the mid-dorsal line are conical and much larger than those on the sides, forming a distinctive band along the mid-dorsal line extending from the nuchal area to the caudal peduncle.*

The coloration of the angel shark is variable. All the specimens that I have seen were greenish or bluish gray above, with numerous scattered, small black spots, and sometimes with irregular dark splotches, and had whitish undersides. However, specimens have been described as being reddish brown above with brownish spots and splotches and having white undersides.

Similar species The two recently described species of angel sharks from the Gulf of Mexico lack the band of distinctive

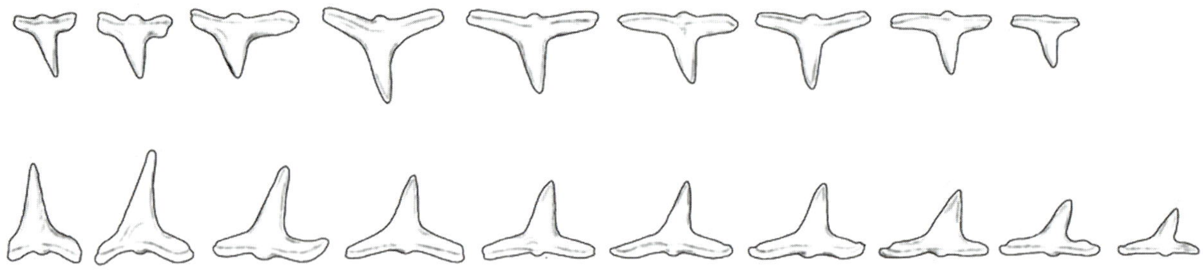

Fig. 44b. Teeth.

larger dermal denticles along the mid-dorsal line. The skates have the pectoral fins joined to the head and have ventral gill slits.

Range The Atlantic angel shark inhabits the Atlantic coast of the United States from Massachusetts to the Florida Keys. The southern limits of its distribution and whether it is present in the Gulf of Mexico are uncertain because of the recent descriptions of two other species of angel sharks inhabiting of the Gulf of Mexico. It is locally common from southern New England to North Carolina, although reports in the literature are few. Bigelow and Schroeder (1953) did not mention it in their Fishes of the Gulf of Maine. H. M. Smith (1922) reported a 109-cm specimen caught in a fish trap off Manemsha Bight, Martha's Vineyard, Massachusetts, in September, mentioning that it was only the second known occurrence in the area. Gordon (1956) reported a 91-cm specimen caught at a depth of 100 m off Rhode Island in January. Hildebrand and Schroeder (1928) mentioned that the angel shark was very rare in Chesapeake Bay and that none had been seen during extensive collecting in 1921–1922, although the species was said to have been rather common off Maryland in past years. More recently, Murdy et al. (1997) described it as a seasonal visitor to the Chesapeake Bay in summer and fall, found occasionally in the lower bay and rarely in the upper bay. John Galbraith (pers. comm., March 2003) found it to be quite abundant from Delaware Bay to Cape Lookout. It is a rare catch south of Cape Lookout.

Size and weight The Atlantic angel shark reaches at least 128 cm. Sizes and weights of Atlantic specimens are as follows: 42 cm, 0.5 kg; 75 cm, 4.5 kg; 100 cm, 7.8 kg; 107 cm, 11.3 kg; and 126 cm, 16.1 kg.

Biology This is a common, bottom-dwelling shark found in cool waters. In the northern parts of the range, it is often found buried in the sand or mud in shallow water during

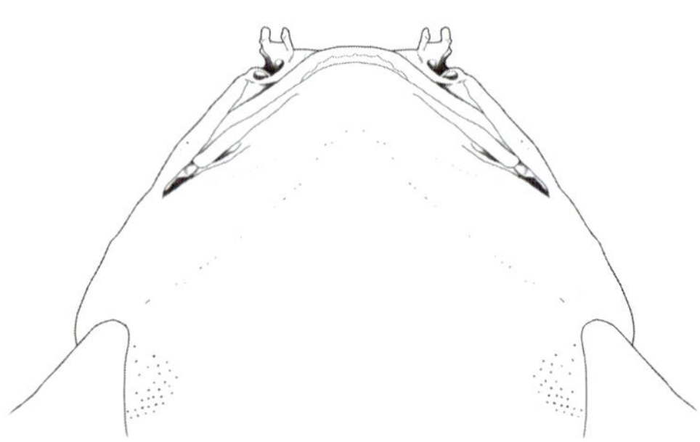

Fig. 44c. Snout.

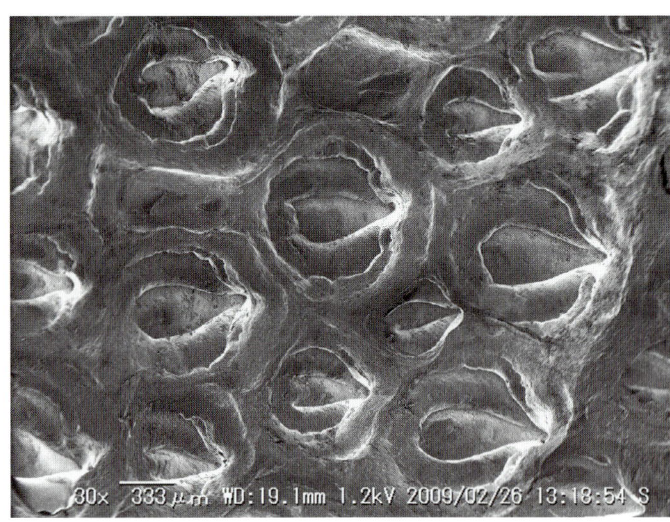

Fig. 44d. Dermal denticles.

the summer, while in the southern parts of the range it is found in the deeper (>100 m) and cooler waters of the continental shelf. Its habits are poorly known.

The angel shark migrates seasonally from shallow to deep water. Trawl survey data (J. Galbraith, pers. comm., October 2006) shows that, off the northeastern United States, the angel shark is usually found close to shore at depths less than 35 m in summer and from shore to the 90 m (50 fathom) isobath in fall. According to Bigelow and Schroeder (1948), most of the specimens at the time of their writing had been caught at depths of "only a few feet." Winter and spring trawl catches reveal that most specimens are found between the 90 m isobath and the edge of the continental shelf, and there are numerous reports of angel sharks caught at great depths. One was caught in September over the continental shelf some 140 km from the shore where the depth was 1,289 m (Albatross Dredging Station 2749, Bigelow and Schroeder 1948). Several specimens were trawled from 200 m by the U.S. fisheries Steamer Albatross about 135 km off Long Island, New York, in February 1920, showing that it is found in deep offshore waters in all seasons (Bigelow and Schroeder 1948).

The Atlantic angel shark feeds on mollusks, crustaceans, flounders, skates, and other bottom fishes. Its feeding habits or mechanisms have not been reported; they are probably similar to those of the better known Pacific angel shark.

Size at maturity Maturity is probably reached at a length of 90–105 cm.

Reproduction Development is aplacental viviparous.

Size at birth: The young measure 28–30 cm at birth.

Brood size: Up to 16 young in one brood have been observed, and it is said that gravid females can contain up to 25 young (Bigelow and Schroeder 1948).

Nurseries Birth apparently occurs at depths of 18–27 m during the spring or early summer months.

Relation to humans Its flesh is edible but seldom marketed on the East Coast. This shark can inflict a nasty bite with its sharp teeth and should be handled with caution. A live angel shark on deck can suddenly spring upward and bite an unwary observer. I know of one unlucky fisherman who was "accosted" by one as he stepped over it on the boat deck. The species is usually an accidental catch in trawls.

GULF ANGEL SHARK

Squatina heteroptera Castro-Aguirre, Pérez, and Campos, 2006

Fig. 45a. *Squatina heteroptera,* Gulf angel shark, immature male, 34.8 cm; off Tamaulipas, western Gulf of Mexico (UNAM # 12720, courtesy of Hector Espinosa).

Common name: Gulf angel shark.

Spanish name *Angelito del Golfo.*

Scientific name *Squatina heteroptera* Castro-Aguirre, Pérez, and Campos, 2006; in Revista de Biologia Tropical 54(3): 1031–1040, figs. 5–8. *Squatina:* an ancient Latin name for a skate, an allusion to its resembling a skate; *heteroptera:* from the Greek heteros, other or different, and pteron, fin or wing, hence "different fins," a reference to the species having dorsal fins of different size.

Synonyms None.

Identification The Gulf angel shark is a raylike shark; it has *a flattened body, a terminal mouth, eyes on top of the head,* large spiracles behind the eyes, and *greatly expanded pectoral fins that are clearly separated from the head* and lacks and anal fin. *The dorsal fins are of different sizes, the first one being higher and wider at the base than the second.* The first dorsal fin has a more rounded rear margin, while the second dorsal has a rear margin with a squarish edge. The endolymphatic pores appear as two small openings on the dorsal midline behind the level of the spiracles. In front of each pore there is a visible canal, *both canals converging anteriorly forming a U-shape.* The dermal denticles on the dorsal side are flat, lanceolate points set on broad bases that taper to rearward point. The mid-dorsal line lacks thorns or enlarged, sharp, pointed dermal denticles. The mid-dorsal dermal denticles have *four ridges.* The teeth are triangular in shape, with narrow, erect, pointed, smooth-edged cusps. The teeth number U: 8–8, L: 8–8. There are two functional rows in the upper jaw and one functional row in the lower jaw. The dermal denticles on the dorsal side are flat, lanceolate points set on large broad and pointed bases. Coloration is dark brown above, with two conspicuous dark blotches on the anterior margin of the pectoral fins.

Similar species The Atlantic and Mexican angel sharks have dorsal fins of similar size and shape, and the canals anterior to the endolymphatic pores converge forward in a "V" shape, not a "U."

Range The holotype was taken in 1989 in the continental shelf waters of the state of Tamaulipas, Mexico, at 22°44′ N, 97°24′ W. Other catches in waters of the continental shelf from Tamaulipas to Tabasco suggest that the species is widespread in the western Gulf of Mexico. The limits of its range are unknown due to confusion with other angel sharks.

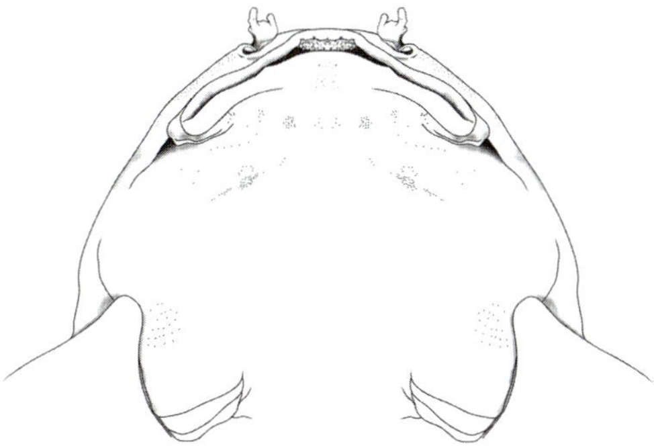

Fig. 45b. Snout.

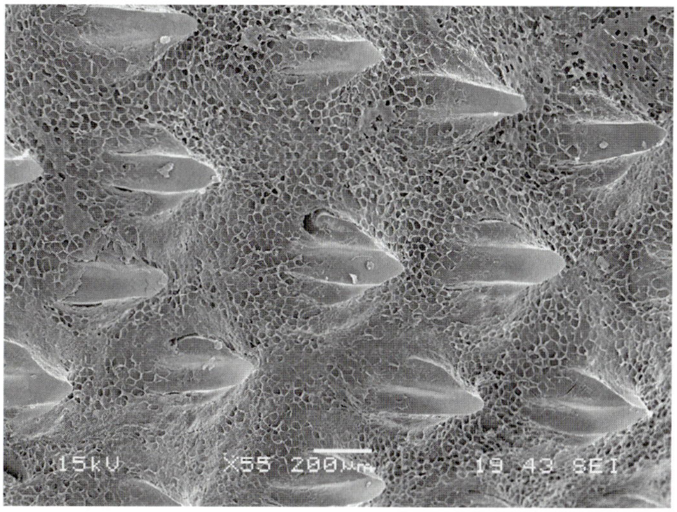

Fig. 45c. Dermal denticles.

Size and weight A 99.6-cm male weighed 7.03 kg, while a 99.0-cm female weighed 7.65 kg.

Biology This is recently described (September 2006) species, and nothing is known of its habits. The type series specimens were collected at 101–164 m.

Size at maturity Unknown.

Reproduction Angel sharks are aplacental viviparous. A female caught in the Gulf of Mexico in May contained eight large oocytes 50–54 mm in diameter.

Size at birth: Unknown.

Brood size: Unknown.

Nurseries Unknown.

Age and growth Unknown.

Relation to humans None.

MEXICAN ANGEL SHARK

Squatina mexicana Castro Aguirre, Pérez, and Campos, 2006

Fig. 46a. *Squatina mexicana*, Mexican angel shark, mature male, 99.6cm, 7 kg; from the Gulf of Mexico, southwest of Destin, Florida (John Carlson, collector).

Common name Mexican angel shark, a reference to its having been first found off Mexico.

Spanish name *Angelito mexicano.*

Scientific name *Squatina mexicana* Castro-Aguirre, Pérez, and Campos, 2006, in *Revista de Biología Tropical* 54(3): 1031–1040, figs. 1–4. *Squatina:* an ancient Latin name for a skate, an allusion to its resembling a skate; *mexicana:* latinized form, from Mexico.

Synonyms None.

Identification The Mexican angel shark is a raylike shark; it has a *flattened body, a terminal mouth, eyes on top of the head,* large spiracles behind the eyes, and *greatly expanded pectoral fins that are clearly separated from the head,* and lacks and anal fin. *The dorsal fins are of similar shape and size.* The anterior nasal lobe is longer and thinner than the posterior lobe. The central nasal lobe is squarish. The endolymphatic pores appear as two small openings between the spiracles; in front of each pore there is a visible canal, *both canals forming a "V" shape. The mid-dorsal line lacks thorns or enlarged, sharp, pointed dermal denticles. The mid-dorsal dermal denticles have three ridges.* The teeth are triangular in shape, with narrow, erect, pointed, smooth-edged cusps. Teeth number U: 10–10; L: 10–10. The dermal denticles on the dorsal side are flat, lanceolate points set on large broad and tapering bases. Coloration is dark brown above with four dark brown or black saddle blotches on the caudal region, equidistant from one another. The pectoral fins have a diffuse, round, black spot anterior to the pectoral axil. There is another round black spot on the posterior portion of the pectoral fins.

Similar species The Atlantic angel shark has a band of large, distinctive, dermal denticles along the mid-dorsal line.

Range The type specimen series was taken in the continental shelf waters off Tamaulipas, Mexico. The limits of its range are unknown due to confusion with other angel sharks.

Size and weight No data available.

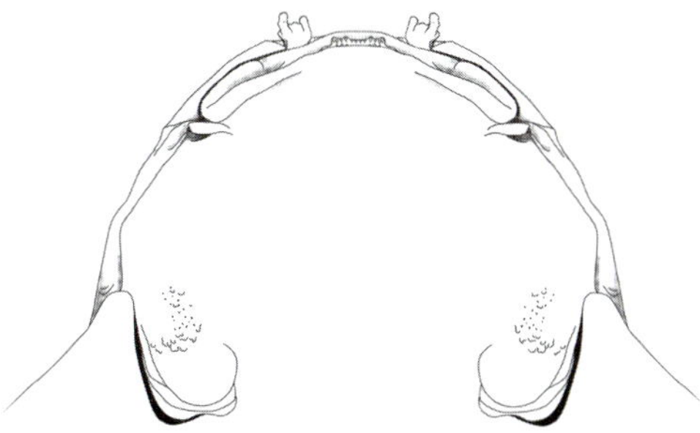

Fig. 46b. Snout.

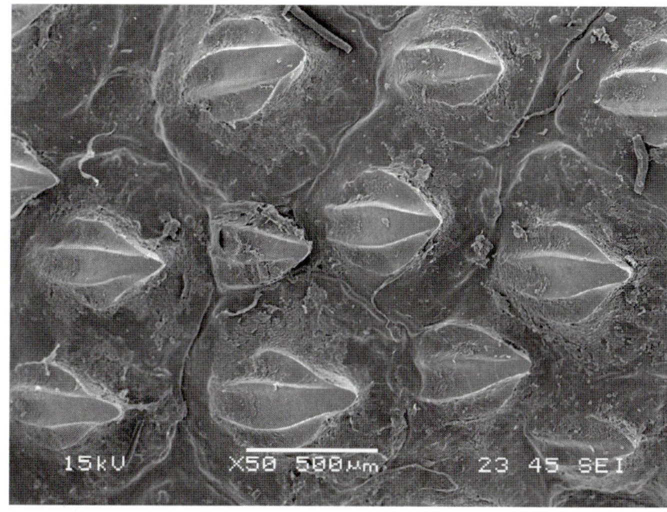

Fig. 46c. Dermal denticles.

Biology This is a recently described (September 2006) species, and nothing is known of its habits. Specimens have been taken at 96–180 m.

Size at maturity Unknown.

Reproduction Angel sharks are aplacental viviparous. There are no data on this recently described species.

Size at birth: Unknown.

Brood size: Unknown.

Nurseries Unknown.

Age and growth Unknown.

Relation to humans None.

FAMILY HETERODONTIDAE

The Bullhead Sharks

Bullhead sharks, also known as horn sharks or Port Jackson sharks, are distinguished by the presence of two dorsal fins, each with a strong spine at the origin; a first dorsal fin originating over the pectoral fin base; and the presence of an anal fin. Other characteristics include short, blunt snouts, ridges above the eyes, and conspicuous upper and lower labial furrows. They possess two types of teeth (*Heterodontus* = "different tooth"): frontal teeth that have sharp cusps for seizing prey, and rear teeth that have round blunt cusps ("molariform" teeth) that are used to crush the hard shells of their invertebrate prey. The bullhead sharks are considered to be primitive because of their hybodont dentition and the presence of spines on the dorsal fins. The family is confined to the Pacific and Indian oceans. Although several genera have been described, currently all species are placed in the genus *Heterodontus.* Presently eight species are recognized; there are two described North American species and perhaps one undescribed species. For a review of the family, see Taylor (1972b).

HORN SHARK

Heterodontus francisci (Girard, 1855)

Fig. 47a. *Heterodontus francisci*, horn shark, immature male, 68 cm, 2.3 kg; from Catalina Island, California (Adam Summers, collector).

Common name Horn shark, or horned shark (Roedel and Ripley 1950), allusions to the two spines at the origins of the dorsal fins; also bullhead shark (Jordan and Evermann 1896), a reference to its unusual head shape, which may resemble a bull's head when seen in profile after a long day in the sun and sufficient alcohol.

Spanish name *Gata* or *puerco* (Mexico). The A.F.S. name is *tiburón puerco.*

Scientific name *Heterodontus francisci* (Girard, 1855), originally described as *Cestracion francisci* in *Proceedings of the Academy of Natural Sciences of Philadelphia* 7: 196 (the date of this publication is variously given as 1855 or 1856. The pages of Girard's description are marked November 1854, but these apparently did not appear until 1856). *Heterodontus:* from the Greek *heteros,* different, and *odontos,* tooth, an allusion to its having two different types of teeth; *francisci:* a latinized form of Francisco, after the city of San Francisco, although the species range may not extend so far to the north.

Synonyms None.

Identification The horn shark is characterized by a short, blunt head with high ridges above the eyes, *two dorsal fins each with a spine at its origin,* and the *presence of an anal fin.* The spines of young horn sharks are relatively long and sharp; those of adults are usually worn and rounded probably due to abrasion against rocks. The frontal teeth are tricuspid, with the central cusp being the largest; lateral teeth are modified into flat molars. Teeth number U: 20 to 29, L: 17 to 29 and are similar in both jaws. In young specimens 20 cm or less, all teeth are similar and bear up to seven cusps. The dermal denticles are cross shaped. Coloration is light tan to light brown above, with small, scattered black spots over the body and fins. All or most of the spots are *smaller than one-third the eye diameter* and may be faded or absent in large adults. The undersides are pale yellowish tan to dirty white. Melanistic (dark) individuals are common. They are dark tan to brown and usually have very few spots or lack them entirely. In some areas, some specimens have purple-tinted teeth and spines (see "Diet," below).

Size and weight The species is said to reach 120 cm (Roedel and Ripley 1950). A 98-cm specimen weighed 10 kg.

Similar species The Mexican horn shark has spots larger than one-half the eye diameter and larger dermal denticles (about 125 denticles/cm^2 in the anterior area of the first dorsal fin, compared with 200 denticles/cm^2 in the horn shark). The cryptic horn shark (if a valid species) lacks the numerous dark spots and is much lighter in color than melanistic horn sharks (which may also lack spots). The dogfish sharks have dorsal fins with spines but lack an anal fin.

Range The horn shark inhabits shallow waters from central California to at least the Gulf of California (Taylor and Castro-Aguirre 1972). Compagno (1984a) mentions that the range extends "probably to Ecuador and Peru" but did not explain the basis of his statement. The type specimen was reported from Monterey Bay, apparently having reached that far north during a warm water event known to have occurred at the time, but since then horn sharks have not

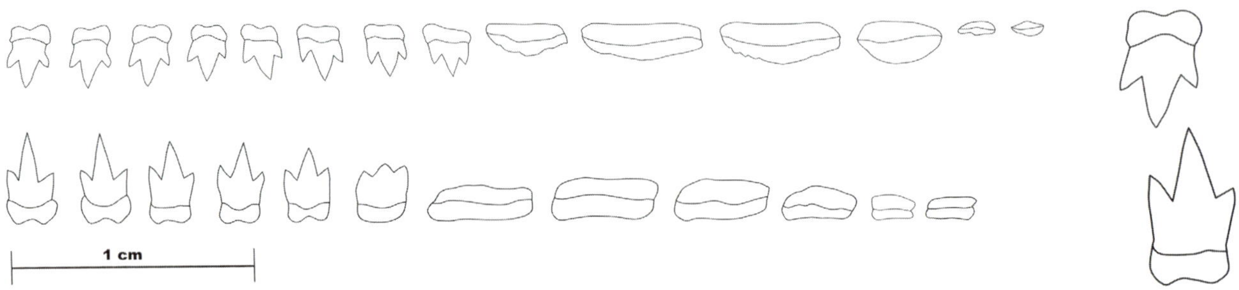

Fig. 47b. Teeth.

been reported north of Point Conception (Starks and Morris 1907; Taylor 1972a). The species is common south of Point Conception (Jordan and Evermann 1896), and it is often taken in San Diego Bay (Eigenmann 1893; Starks and Morris 1907). The southern boundary of the distribution of the horn shark is not clear due to confusion with the Mexican horn shark and perhaps other undescribed species. Taylor and Castro-Aguirre (1972) stated that the ranges of the horn sharks and Mexican horns sharks overlap in the Gulf of California and in Bahia Magdalena, Mexico, the two species being taken together in 4 of 22 collections in the area. Reports of the presence of the horn shark south of Baja California published before 1972, when the Mexican horn shark was described, must be interpreted with caution. Such publications (Beebe and Tee-Van 1941; Kato et al. 1967) would have reported any horn shark as *Heterodontus francisci.*

Biology The horn shark is a sluggish, solitary bottom dweller usually found among rocks or in kelp beds in shallow coastal areas. According to Limbaugh (1955), who studied the species throughout most of its range from central California to Cabo San Lucas, Mexico, horn sharks were observed in shallow tide pools to depths of 17 m, but most of the population was concentrated between depths of 2 to 10 m and at an average depth of 5.5 m. However, horn sharks have been caught as deep as 150 m and are suspected of migrating seasonally from shallow to deep water (Roedel and Ripley 1950).

Horn sharks are nocturnal, spending the daylight hours resting among rocks or algae, with the heads in a crevice or shaded in some way. Finstad and Nelson (1975) found that horn sharks became active 76 minutes after sunset at 10 m depth, where the light level was approximately 0.03 lux. At dusk the sharks roamed over adjacent open areas to feed.

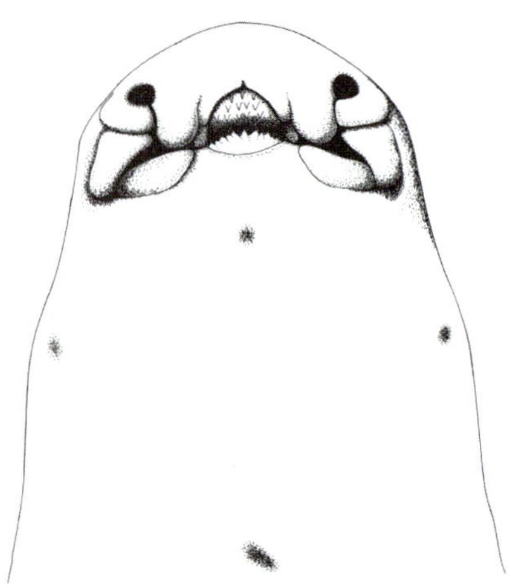

Fig. 47c. Snout.

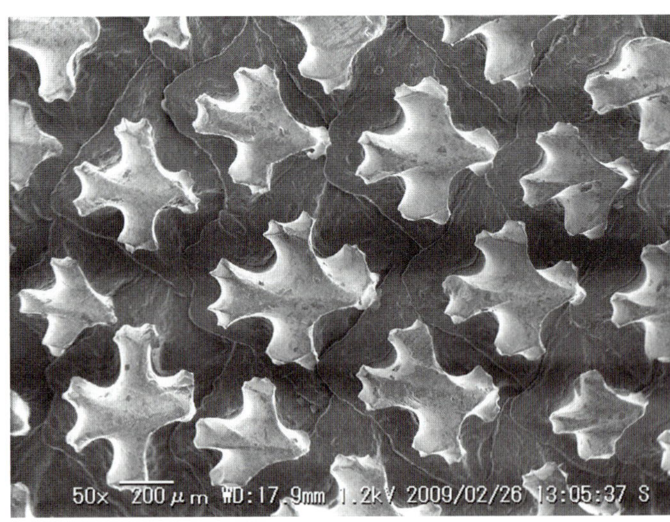

Fig. 47d. Dermal denticles.

Fig. 47e. *Heterodontus francisci*, horn shark, melanistic female, 57.4 cm, 1.3 kg; from Catalina Island, California (Adam Summers, collector).

Swimming activity usually ceased within an hour after sunrise. The linear distance covered by individual sharks from the point of origin of nightly activity to the point of cessation averaged 39.3 m ± 32 m (Strong 1989).

Diet: Horn sharks are said to feed on mollusks, crabs, and other hard-shelled invertebrates (Roedel and Ripley 1950). Despite the abundance of horn sharks, there are few reports of its diet actually based on analyses of stomach contents. Juvenile horn sharks are believed to feed on polychaete worms based on laboratory food preference studies (Strong 1989). Finstad and Nelson (1975) reported that one horn shark examined had consumed two blacksmith (*Chromis punctipinnis*) and one squid (*Loligo opalescens*).

Off Santa Catalina Island, California, Strong (1989) found that about 10% of horn sharks larger than 64 cm had purple teeth or spines. This pigmentation is believed to result from eating purple sea urchins (*Strongylocentrotus purpuratus*). Taylor (1972a) stated that naphthoquinone pigments are responsible for the purple coloration and that the pigments are believed to be deposited primarily in calcified tissues (teeth, spines, and cartilage). The mechanism of incorporation is believed to be related to the calcification process.

Little is known about predators of the horn shark. In feeding experiments, Strong (1989) presented California angel sharks with a variety of prey. Small horn sharks that swam within 10 cm of the angel shark's mouth were readily seized and quickly rejected, unlike other prey items that were consumed. Strong also presented the angel sharks with small horn sharks that had their spines removed. The angel sharks seized and retained these horn sharks for longer periods of time, repeatedly biting the horn sharks and reorienting them for swallowing. However, these horn sharks were also eventually rejected. The despined group suffered greater physical damage than the intact specimens, but none were fatally injured. This experiment demonstrated that the dorsal spines of the horn shark afford some protection against predators.

Because of its availability and small size, the horn shark has been used in anatomical studies. Daniel (1914, 1915) described the endoskeleton and the exoskeleton. Huber et al. (2005) studied the feeding mechanism of the horn shark.

Size at maturity

Male: Despite the abundance and availability of horn sharks, there are few data on the size at maturity. According to Compagno (1984a), males mature at 58–71 cm, an unusually wide size range for such a small species. Strong (1989) stated that males mature between 56 and 61 cm, based on the ratio of clasper size to total length. However, he did not ascertain whether animals had calcified claspers. Given that his largest immature male was 61 cm, it is likely that he underestimated the size at maturity, as some of the immature males may have had long claspers that had not calcified yet.

Female: There are almost no data on the size at maturity for the female. According to Compagno (1984a), females mature at a size larger than 58 cm.

Reproduction The horn shark is oviparous. Dempster and Herald (1961) observed mating in the Steinhart Aquarium, San Francisco, in December; egg cases appeared two to four weeks after mating. The horn shark egg case is conical with two broad flanges spirally twisted around it, giving it a screw shape. The egg case measures about 12 cm in length by 6.4 cm at the wider end; it is pale brown when freshly laid but turns dark after a few days. The egg case wall is tough and pliable. Egg cases are scattered among rocks. Strong (1989) reported finding egg cases in shallow water (2–13 m deep) and noted that they were difficult to locate among the rocks, some being so firmly implanted that they were difficult to extract. Strong believed that direct implantation by cloacal apposition against the rocks or crevices would have been impossible, because some egg cases had been pushed into tight spaces in gradually narrowing depressions. In the related Port Jackson shark, McLaughlin and O'Gower (1971) have suggested that eggs may be deposited in the open and secondarily inserted orally into suitable anchorages. Whether horn sharks actually do this is unknown. Dempster and Herald (1961) reported that one female horn shark that picked up an egg case of unknown origin and carried it around the aquarium for some five minutes, and that the shark eventually bit the egg case, squeezing out its contents. Thus, the report cannot be considered as evidence of females orally carrying their eggs to suitable locations.

Size at birth: In captivity, young have hatched 8–10 months after deposition of the egg case (Dempster and Herald 1961). The incubation period in the wild is not known. Horn shark hatchlings measure 15–17 cm at birth.

Brood size: According to Ebert (2003) a female may deposit up to 24 egg cases in one breeding season.

Nurseries Although horn sharks are known to lay their eggs in shallow waters of southern California, the nurseries of the horn shark have not been delimited.

Age and growth Although horn sharks have been kept in aquaria since the mid-1950s, little is known about their growth and longevity because most aquaria do not keep records on individual fishes or do not publish their findings. They are believed to be slow-growing, long-lived sharks. Hinton (1962) reported one surviving for 12 years.

Relation to humans The horn shark is often used as an experimental shark or exhibited in aquaria. The species has no other economic importance.

MEXICAN HORN SHARK

Heterodontus mexicanus Taylor and Castro-Aguirre, 1972

Fig. 48a. *Heterodontus mexicanus*, Mexican horn shark, female, 61.0 cm, 1.8 kg; from El Sahuímaro, Sonora, Mexico (Fernando Márquez, collector).

Common name Mexican horn shark, allusions to the two spines at the origin of the dorsal fins, and to its original locality.

Spanish name *Gata* or *perro* (Mexico). The A.F.S. name is *tiburón perro.*

Scientific name *Heterodontus mexicanus* Taylor and Castro-Aguirre, 1972, in *Anales de la Escuela Nacional de Ciencias Biológicas* (Mexico) 19:123–143. *Heterodontus:* from the Greek *heteros,* different, and *odontus,* tooth, an allusion to its having two different types of teeth; *mexicanus:* a Latinized form meaning "from Mexico."

Synonyms None.

Identification The Mexican horn shark is characterized by *two dorsal fins each with a spine, an anal fin,* and *large dark spots scattered over the body.* Only the tips of dorsal fin spines are exposed, and the tips are well separated from the anterior margins of the dorsal fins. The anterior teeth are tricuspid, with the central cusp being the largest. The lateral teeth are modified into flat molars; they have wide, blocklike bases and low cusps that form a thin ridge. There is one transitional tooth between the anterior tricuspic teeth and the flattened molariform teeth; it has low tricuspid cusps and wide blocklike base. Teeth number U: 19 to 25, L: 15 to 21 and are similar in both jaws. In specimens of less than 20 cm, all the teeth are similar, each having eight to nine cusps. The dermal denticles are very large (1–1.5 mm wide and visible to the naked eye) and broadly cross shaped with scalloped edges. Coloration is light golden brown with a slight bronze iridescence above and grayish white to yellowish brown below, with a *pale bar across the top of the head* and many scattered dark spots over the body. The spots are *about one-half the eye diameter or larger.*

Similar species The horn shark has a broader snout, higher ridges above the eyes, and smaller body spots (less than one-third the eye diameter), and the tips of the dorsal fins spines rest closely against the margins of the dorsal fins. The cyptic horn shark lacks the spots or has only a few spots. The dogfish sharks have dorsal fins with spines but lack an anal fin.

Range The Mexican horn shark inhabits the Gulf of California and the Mexican coast south to at least Oaxaca (Taylor and Castro-Aguirre 1972; Espinosa Pérez et al. 2004). It has also been reported from Peru (Chirichigno 1998). However, the limits of its distribution are not well known, and the identity of specimens from the Southern Hemisphere is questionable due to confusion with other species. I have examined specimens taken off Guaymas and El Barril, Mexico, in the Gulf of California.

Size and weight The largest specimen reported by Taylor and Castro-Aguirre (1972) measured 70 cm. I recorded the following sizes and weights for specimens taken in the Gulf of California.

Male: 37 cm, 0.4 kg; 59.6 cm, 1.6 kg.

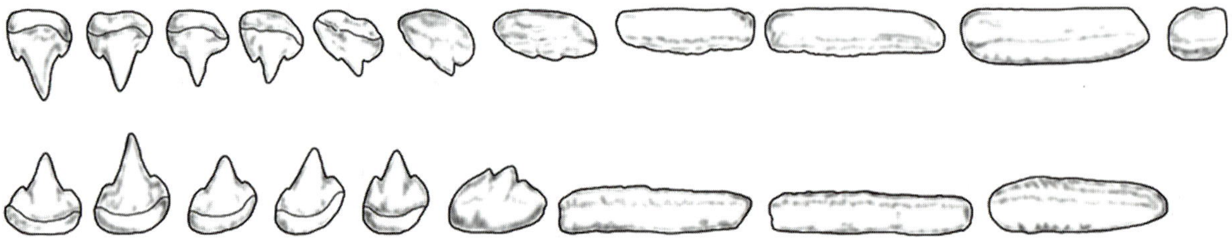

Fig. 48b. Teeth.

Female: 44 cm, 0.7 kg; 59 cm, 1.67 kg; and 67 cm, 2.61 kg.

Biology This is a poorly known species. Its diet and habits have not been described; they are probably similar to those of the horn shark. At El Barril, Baja California, Mexico, I examined specimens caught at 50–60 m, a greater depth than the other horn sharks in the area.

Size at maturity Males reach maturity at lengths of 55 cm or less. Females measuring 66–67 cm that I examined were mature.

Reproduction The Mexican horn shark is oviparous. The egg case is screw shaped, with a thick ridge spirally twisted around it. This ridge bears wide flanges, which lie parallel to the axis of the case. The egg case measures 6–8 cm in length and bears a pair of short tendrils at its tip. Females measuring 66–67 cm that I examined in Guaymas, Mexico, on 30 May 2003 carried six to eight six turgid oocytes 27–31 mm in diameter that appeared ready for ovulation.

Size at birth: The hatchlings measure about 14 cm at birth.

Brood size: The number of eggs or young produced by a given female in a season is unknown.

Nurseries The shallow waters and lagoons around Bahia Magdalena are known nursery areas of the Mexican horn shark.

Age and growth The species has not been aged.

Relation to humans None.

Fig. 48c. Snout.

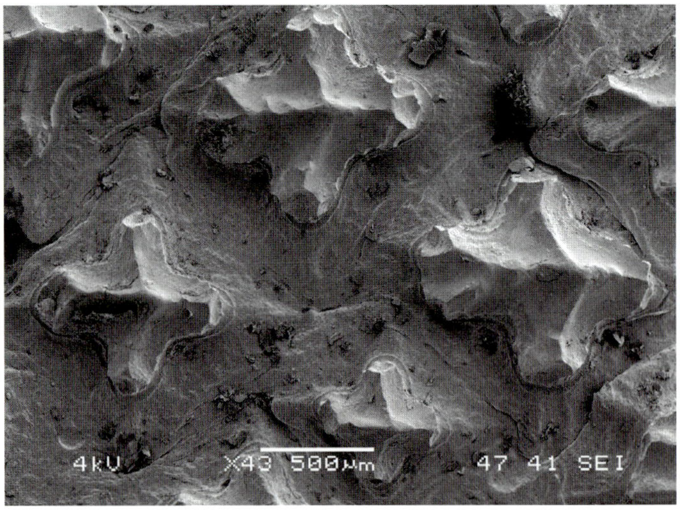

Fig. 48d. Dermal denticles.

Fig. 48e. *Heterodontus mexicanus*, Mexican horn shark, immature female, 44 cm, 0.7 kg; from El Barril, Baja California, Mexico (Clark and Diana Crabbe, collectors)

CRYPTIC HORN SHARK

***Heterodontus** sp. X*

Fig. 49a. *Heterodontus sp.,* mature female, 80.2 cm, 4.5 kg; from El Barril, Baja California, Mexico (Gonzalo the Fisherman, collector).

This horn shark, which I call the cryptic horn shark, inhabits the Gulf of California, where I have examined several specimens taken off El Barril, Baja California. This shark differs both from the horn shark (*Heterodontus francisci*) and from the Mexican horn shark (*H. mexicanus*) in lacking the markings and spots of those species. Morphologically this shark is very similar to the horn shark, and at first I thought that these specimens were adult horn sharks (*H. francisci*). The dermal denticles are similar to those of the horn shark. However, genetic analysis by Gavin Naylor suggests that it may be a separate species.

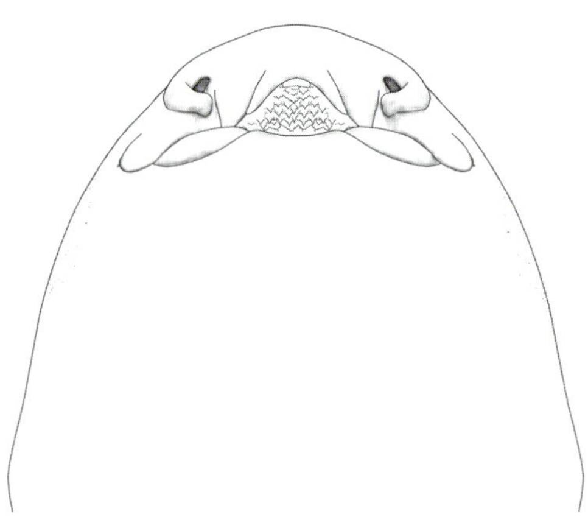

Fig. 49b. Snout.

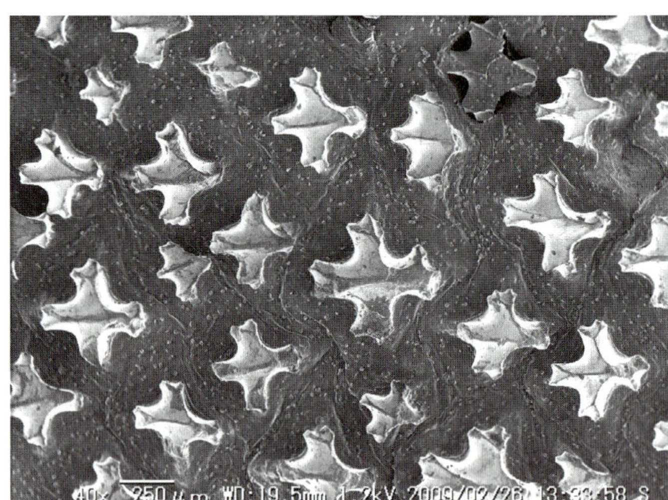

Fig. 49c. Dermal denticles.

FAMILY GINGLYMOSTOMATIDAE

The Nurse Sharks

The nurse sharks are sluggish, bottom-dwelling species of shallow tropical and subtropical waters. They are characterized by fleshy nasal barbels just anterior to the mouth and a deep groove that connects the nostril and the mouth. Other characteristics include a short snout with a rectangular mouth and small eyes; a slightly dorsoventrally flattened body that is broadest in the pectoral region but quickly tapers to a tail in line with the body; five gill slits, the fifth one smaller and almost covered by the fourth; spiracles just behind the eye; and multicuspid teeth. Nurse sharks eat small fishes and invertebrates. This family includes oviparous and aplacental viviparous species. The family includes the genera *Ginglymostoma, Nebrius,* and *Pseudoginglymostoma.* The number of recognized species varies between 12 and 24, most of which inhabit Indo-Pacific waters or the Red Sea. Only one genus, *Ginglymostoma,* is represented in North America; one species is found in the Atlantic, and another in the eastern Pacific. The Pacific species is presently undescribed.

NURSE SHARK

Ginglymostoma cirratum (Bonaterre, 1788)

Fig. 50a. *Ginglymostoma cirratum*, nurse shark, gravid female, 256 cm, 98.6 kg; from Bahia Honda Bridge, Florida Keys (Mike Heithaus and Jeff Carrier, collectors).

Common name Nurse shark. The name "nurse" comes from "nusse," meaning fish, which appears in print as early as 1440 in the *Promptorium Parvulorum,* an early English/Latin dictionary. The term was applied to large fish, probably sharks. Richard Hakluyt, who explored the waters around Novaya Zemlya, Russia, in the 1500s wrote, "and there we gate a great Nuse, which Nuses were there so plentie, that they would scarcely suffer any other fish to come neere the hookes: the said Nuses carried away sundrie of our hookes and leads" (Hakluyt 1589 [1907]: 352). According to the Oxford English Dictionary, "nusse" is derived from the earlier "huss," the ancient name for the catsharks *Scyliorhinus stellaris* and *S. caniculus,* common in English waters, the skins of which were much used by fletchers for smoothing and polishing arrows. Even today *S. stellaris* is known as the nurse hound, bull huss, or huss (Wheeler 1969). When English explorers came to the American tropics in the seventeenth century, they quickly distinguished the nurse shark from the requiem sharks and dubbed it "nurse," apparently because of its superficial morphological and behavioral similarity to the catsharks that English sailors knew as "nurses."

Spanish name *Gata* (Spanish Caribbean); also *tiburón gata* (Mexico, A.F.S.). The name *gata,* the feminine of *gato,* or cat, presumably because of its "whiskers," or nasal barbels, is centuries old. Parra (1787) used it in one of the earliest descriptions of the nurse shark.

Scientific name *Ginglymostoma cirratum* (Bonaterre, 1788). Originally described as *Squalus cirratus* in *Tableau Encyclopédique et Méthodique des Trois Règnes de la Nature, Ichthyologie:* 7. *Ginglymostoma:* from *ginglymos,* Greek for hinge or hinge joint, and *stoma,* mouth, hence "hinge mouth," an allusion to the peculiar mouth of the nurse shark; *cirratum:* from *cirratus,* Latin for bearing cirri (barbels).

Synonyms *Squalus punctatus* Bloch and Schneider, 1801; *Ginglymostoma fulvum* Poey, 1861; *Nebrius cirratum* Fowler, 1945.

Identification The nurse shark has conspicuous *nasal barbels* on the anterior margins of the nostrils and a first dorsal fin base midpoint *above or posterior* to the origin of the pelvic fin. It has a wide head that gives it a tadpolelike appearance from above. The mouth is full of minute teeth, numbering U: 30 to 36, L: 28 to 31, and are similar in both jaws. The teeth have one large cusp flanked on each side by two or three cusplets. The dermal denticles have a pebblelike appearance; they are large and closely apposed with little overlapping and have three to five weak ridges. Coloration is highly variable, usually ranging from light yellowish tan to dark brown, most specimens being light brown. There are several reports of xanthic (yellowish) specimens. Wright (1981)[1] reported a brilliant yellow 213-cm specimen from Ft. Pierce Inlet, Florida, and Baughman and Springer (1950) also saw yellowish specimens. I described and illustrated a milk-white specimen that was captured, photographed, and released by Captain Dana Banks off Key Largo, Florida, in July 1992 (Castro 2000). This animal, estimated at 210 cm, might have been a partial albino. Its body was white with scattered small brown spots, with

1. Wright, V. D. (1981). *Some observations on the biology of sharks of the Florida Keys and adjacent waters.* This is an unpublished manuscript prepared as a master's thesis at Florida Atlantic University but never completed. It contains a considerable amount of useful data on Florida sharks.

Fig. 50b. Teeth.

more intense pigmentation on the flanks and first dorsal fin. The color of the iris was not noted.

Neonate nurse sharks and juveniles up to 45 cm have alternating bands of light and dark pigmentation along the dorsal surfaces and are covered by small black spots, each spot surrounded by an area of lighter pigmentation. Juvenile nurse sharks 70–120 cm long are capable of limited color changes caused by direct exposure to sunlight (Castro 2000).

Similar species The nurse shark is the only Atlantic shark that has long, tapering nasal barbels. Its first dorsal fin base midpoint over or behind the origin of the pelvic fins, the nasal barbels, and its coloration distinguish it from other Atlantic sharks. The nurse shark of the Pacific Ocean is a different (undescribed) species.

Range The nurse shark is widely distributed in littoral waters on both sides of the tropical and subtropical Atlantic. It ranges from tropical West Africa to the Cape Verde Islands in the eastern Atlantic, and from southern Brazil to North Carolina in the western Atlantic. The species is abundant all year in the shallow waters of tropical Florida and the Caribbean. Bigelow and Schroeder (1948) noted that the nurse shark is common in southern Florida and the Florida Keys and is a year-round resident of the west coast of Florida north to Tampa, and for some distance north of Florida on the east coast. The nurse shark expands its range in the warm months, reaching the northern coast of the Gulf of Mexico, and, in the Atlantic, to Charleston, South Carolina (Bigelow and Schroeder 1948), and occasionally to Cape Hatteras, North Carolina. It is an uncommon summer visitor to the South Carolina coast, where I have seen 90- to

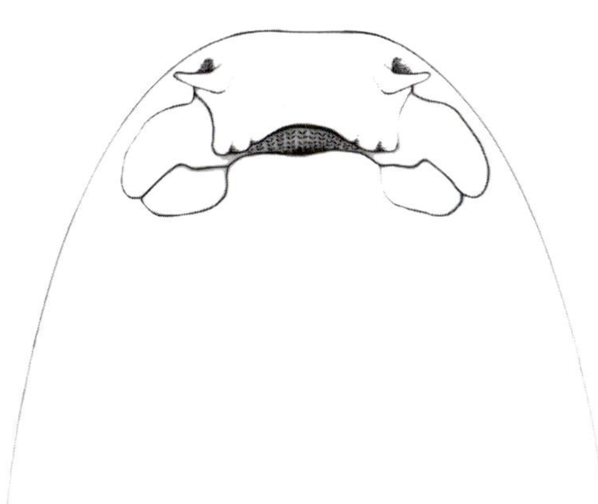

Fig. 50c. Snout.

Fig. 50d. Dermal denticles.

191-cm juveniles from late August to late September (Castro 2000). According to Glenn Ulrich (South Carolina Department of Natural Resources, pers. comm., June 2005), numerous large juveniles and some adults have been caught in St. Helena Sound, South Carolina, from June to October. Coles (1915) reported a large school of nurse sharks in the breakers of the Cape Lookout Shoals, North Carolina, in the summer of 1913. Because schooling nurse sharks have never again been reported from that area, it is likely that these were sand tiger sharks, which are abundant in the area. Schwartz and Burgess (1975) wrote that nurse sharks were only sporadic summer visitors in inshore waters or inlets south of Cape Hatteras. Nurse sharks may stray north of Cape Hatteras, but most reports of strays are suspect. In 1877, a nurse shark was reported in "the southern part of Chesapeake Bay," but none have been seen since (Murdy et al. 1997). Bigelow and Schroeder (1948) reported one stray from Rhode Island; however, the Narragansett Laboratory of the National Marine Fisheries Service has no records of any other Rhode Island specimens (N. Kohler, pers. comm., 1998).

Size and weight Nurse sharks in North America have been measured up to 275 cm. The maximum size and weight of the nurse shark have been greatly exaggerated in the scientific literature, with values that have been repeated so often that they have become an established canon of nurse shark lore. Bigelow and Schroeder (1948) stated that specimens of 213–335 cm were frequently reported, with lengths of 335–365 cm not being unusual, and that the reported maximum was about 427 cm, although, in their study, they had examined only specimens 27–65 cm long. The larger-than-life maximum length appears to have originated with Fowler (1906a), who is cited in Bigelow and Schroeder's nurse shark references with the comment "large size." Fowler (1906a: 79) observed mating nurse sharks in the shallows of the Marquesas (Florida) and stated, "All I observed were large, nearly twelve or fourteen feet in length, and were not at all shy." Fowler's exaggerated estimates could have been influenced by an earlier reference by Storer (1846), who wrote that nurse sharks reach 457 cm. People who have actually measured specimens invariably report much smaller sizes. For example, Poey (1868) stated that the species reached 244 cm. The largest specimen seen by Rivera-López (1970) in Puerto Rico was a 249-cm, 85.7-kg gravid female. From 1965 to 1970, Eugenie Clark and her staff at the Cape Haze Marine Laboratory in Placida, Florida, examined at least 200 nurse sharks, according to her records now stored at Mote Marine Laboratory. These records reveal that adults 240–256 cm were common off southwest Florida and that the largest sharks examined were a 258-cm male, caught on 14 July 1966, and a 270-cm female of undetermined weight, examined on 1 June 1968. The largest specimen reported by Randall (1967) in Puerto Rico and the Virgin Islands measured 274 cm. The largest specimen seen by Dodrill (1977) off Melbourne Beach, Florida, was 251 cm. The largest specimen I have examined was an ovigerous female of 265 cm and 114.5 kg caught off Miami in January 1999 (Castro 2000). H. Wes Pratt (pers. comm., March 2005) captured and measured a 275-cm female at Dry Tortugas, Florida, in June 2000. Nurse sharks up to 305 cm have been reported from the South Atlantic (Brazil, Castro, A. L. F., and Rosa 2005).

The weights attained by nurse sharks have also been greatly exaggerated in the scientific literature. Bell and Nichols (1921) gave the weight of a 260-cm female as 168 kg, and that of a 254-cm male as 151 kg. Bigelow and Schroeder (1948) noted the weight of a specimen of "about 8 1/2 feet" (259 cm) as 150–168 kg, evidently repeating the Bell and Nichols figures. Other writers have quoted Bigelow and Schroeder, establishing these weights in nurse shark lore. However, Cadenat and Blache (1981) stated that the Bigelow and Schroeder (1948) data seemed erroneous. These authors gave the size of the largest specimen they actually measured in the eastern Atlantic as 251 cm but indicated an anecdotal report of a 280-cm specimen that weighed 106 kg. The heaviest specimen I examined was the 114.5-kg female noted above. I have been unable to authenticate a heavier specimen, but given the sizes of the largest specimens actually measured and weighed, it is possible, and even likely, that some unusual North American specimens could reach around 300 cm and 125 kg.

Representative weights of Florida nurse sharks are as follows (Castro 2000).

Male: 38 cm, 0.45 kg; 48 cm, 0.6 kg; 57 cm, 1.0 kg; 70 cm, 2.0 kg; 107 cm, 6.8 kg; 142 cm, 16.8 kg; 160 cm, 21.8 kg; 184 cm, 34.1 kg; 200 cm, 42.7 kg; 228 cm, 58.9 kg; 241 cm, 83.6 kg; 251 cm, 90.0 kg; and 257 cm, 103.1 kg.

Female: 43 cm, 0.39 kg; 57 cm, 1.26 kg; 76 cm, 3.6 kg; 116 cm, 8.18 kg; 121 cm, 8.86 kg; 140 cm, 13.6 kg; 164 cm, 27.2 kg; 184 cm, 35.2 kg; 202 cm, 35.9 kg; 224 cm, 74.1 kg; 248 cm, 81.8 kg; 251 cm, 91.8 kg; and 256 cm, 109.4 kg.

Biology The nurse shark is a common inhabitant of shallow coastal waters in warm tropical seas. Juveniles of 50–120 cm inhabit coral reefs and sea grass flats in water 1–3 m deep. Several small juveniles are often found concealed together in holes or under rocks or ledges. Juveniles 120–170 cm occur on shallow reefs and around mangrove islands at depths of 1–4 m. Larger juveniles and adults frequent deeper reefs, rocky areas, and bridge passes at 3–75 m during daytime and in shallower waters of 3–20 m at night. From May to July, larger adults are present in waters 1–6 m

Fig. 50e. *Ginglymostoma cirratum, n*urse shark, neonate male, 35.7 cm from Marathon, Florida (Forrest Young, collector).

deep; they move to deeper waters 3–75 m during the rest of the year. Individuals of all sizes can be found in small groups during daytime, concealed under ledges, or among boulders and rocks. These sharks are often close together and sometimes almost on top of each other. In some cases, the same shark can be observed resting in the same place over several days (Castro 2000). The nurse shark is primarily a nocturnal predator, becoming active at dusk.

There are few data about predators on the nurse shark, other than man. Wright (unpublished ms. 1981) reported finding two nurse sharks of unstated lengths in the stomach of a 224-cm tiger shark. I retrieved a 45-cm nurse shark from the stomach of a 234-cm lemon shark, caught during a shark fishing tournament off Long Key, Florida. I also have two well-observed and reliable accounts of attacks on adult nurse sharks by great hammerheads and bull sharks during a shark attracting and feeding operation in the Florida Keys. In each case, the attackers failed to cut through the skins of the nurse sharks, and the nurse sharks quickly escaped.

Diet: The nurse shark feeds on numerous species of small fishes. Well-documented analyses of stomach contents of nurse sharks are scarce. Bigelow and Schroeder (1948) stated that the nurse shark feeds chiefly on invertebrates (squids, shrimps, crabs, spiny lobsters, sea urchins) and small fishes, probably based on literature sources. Randall (1967) reported that of 24 nurse shark stomachs examined in Puerto Rico and the Virgin Islands, only nine (38%) contained food. Eight had eaten fishes, and the ninth contained a cephalopod beak. Rivera-López (1970) examined 127 nurse shark stomachs. Of 27 (13%) with food, 48% contained fish and 39% had invertebrate remains (cephalopod beaks and gastropod opercula). Cruz et al. (1986) reported that the nurse shark is an important predator of juvenile spiny lobsters (*Panulirus argus*) at Isla de la Juventud (formerly Isla de Pinos), Cuba. Of 21 nurse sharks 68–132 cm, 60% had food in their stomachs. Of the stomachs containing food, lobsters made up 50% of the prey items. This is the only study to have reported that nurse sharks feed on lobsters. I have examined the stomach contents of 91 nurse sharks 45–265 cm (Castro 2000). Of the food items identified, teleosts comprised 89% of the total prey items, mollusks constituted 4.5%, and crustaceans 2%. Numerous species of bony fishes were identified from the stomach contents, but the majority of these were grunts (Haemulidae). Analysis of prey size showed that most of the prey items were small, 7–22 cm standard length. Occasionally, prey animals were surprisingly small in proportion to the shark (e.g., a 7-cm wrasse consumed by a 239-cm shark; Castro 2000).

Size at maturity

Male: Development of the claspers is gradual until a length of about 180 cm. At that size, the claspers begin to elongate rapidly and begin to calcify, and males reach maturity at about 215 cm (Castro 2000).

Female: Bigelow and Schroeder (1948) stated that a 152-cm female had been found to be carrying embryos, based on an uncertain anecdotal reference by Beebe (1941). This was apparently an error because female nurse sharks mature at a much larger size (223–231 cm). Immature females have ovaries containing undeveloped, nonvitellogenic oocytes about 1–3 mm in diameter, poorly differentiated oviducal glands less than 15 mm wide, and tubelike oviducts 5–10 mm in diameter (Castro 2000). Female nurse shark start to mature at about 210 cm. Maturation involves abrupt enlargement of the oviducal glands to a width of 48 mm or more, enlargement of the uteri to a width of 100 mm, and the growth of the oocytes to their maximum diameter of about 60 mm. Nearly all females over 235 cm have oviducal glands wider than 48 mm. Expansion of the uteri begins at the distal (cloacal) end, and each uterus expands from a narrow tube 7–9 mm in diameter into a wide sac more than 100 mm wide. Adults have uteri 100–275 mm wide, depending on the stage of the reproductive cycle. Vitellogenesis and enlargement of the oocytes begin in the fall after a female reaches 210 cm. Only the right ovary is functional;

the left ovary is absent. Both oocytes and ovary become very large in ovigerous females. Ripe oocytes measure 52–56 mm in diameter and weigh an average of 100 g each, and the ovary can reach 45 × 30 × 10 cm and weigh 5.7 kg (Castro 2000). Fifty percent of the females I measured attained maturity at 223–231 cm.

Reproduction The nurse shark is an aplacental viviparous species. Despite the abundance and ease of capture of nurse sharks, most of what we know about its reproductive processes has been learned only in the last decade.

Mating: The nurse shark is the only large shark whose mating behavior has been observed repeatedly, and much of what we know has been contributed by Jeff Carrier of Albion College and Harold Wes Pratt, formerly with the National Marine Fisheries Service.

Mating aggregations of nurse sharks in the lower Florida Keys, Marquesas, and Dry Tortugas have been known since the early 1900s. Fowler (1906a) visited the Marquesas in June 1904 and reported that copulating nurse sharks were common in the warm shallows around the Marquesas and that copulating pairs could be located in very shallow water by their dorsal and caudal fins, which projected out of the water. Fowler also stated that these aggregations occurred only at that time of the year, the species being absent from the area at other times. Gudger (1912) studied the breeding habits of the nurse shark off Bush Key in the Dry Tortugas. He reported that during copulation a male grabbed one of the female's pectoral fins, threw her on her back, and inserted a clasper into her cloaca to transmit the seminal fluid. Gudger stated that females would often break away from the males before copulation ensued, and that consequently the pectoral fins of females often were torn and scarred. Carrier et al. (1994) described cases in which several males attempted to mate simultaneously with a single female. Pratt and Carrier (1995) witnessed some 165 mating events off Dry Tortugas, noting that fewer than 10% of the observed mating attempts by males resulted in copulation. According to these authors, females often avoid males by retreating into shallow water and digging their pectoral fins into the bottom, and that only the most aggressive and persistent males succeed in towing a female into deeper water to copulate with her. A female can copulate with several males (Anonymous 1992; Saville et al. 2002).

Copulation occurs in Florida and the Bahamas in June and July. Klimley (1980) reported mating activity observed by other investigators in mid and late June at Bush Key, Dry Tortugas; in early July at Long Key, Dry Tortugas; and in June at Chub Cay, Bahamas. John Morrissey (pers. comm., 1988) saw 13 copulating pairs of nurse sharks along the west shoreline of North Sound, Bimini, Bahamas, from 20 June to 2 July and one copulating pair off the southwest tip of North Bimini on 7 June. Carrier et al. (1994) observed 50 mating attempts at Dry Tortugas from 16 to 24 June. I calculated that mating activity peaks about 15 June in Florida and the Bahamas, based on mating observations and on embryo sizes at specific times (Castro 2000). Off La Parguera, Puerto Rico, Rivera-López (1970) observed nurse sharks mating from late June to early August.

Development: Embryos are lecithotrophic, feeding solely on nutrients stored in the yolk sac (Castro 2000). Ovulation is a prolonged process probably lasting at least two to three weeks. For the first 12–14 weeks of gestation, the embryos are enclosed in their horny egg cases, hatching when they reach a length of 218–233 mm. Because of their different ages, embryos in a brood are at different developmental stages through the first months of gestation. For example, a female I examined in October contained embryos measuring 215–233 mm that were still inside the egg cases and embryos measuring 270–278 mm that were free in the uterus and had fully absorbed yolk sacs and open yolk-sac scars.

The empty egg cases somehow find their way to the cloaca, become folded lengthwise, and are expelled. By the end of October all embryos have hatched and the egg cases have been expelled. By mid-November, the embryos have generally absorbed their yolk sacs. Embryos are born at 280–305 mm in late November and early December, after a gestation period of five to six months. Parturition is spread out over a few weeks (Castro 2000), because the embryos are not ready for birth at the same time. Embryos are released singly or a few at a time as they become fully developed. Females that I sampled carried fewer and fewer young as the second week of December approached.

Reproductive cycle: Male nurse sharks show a clear reproductive cycle. The testes are streaklike and embedded in the dorsal surfaces of the epigonal organs from August to January. From February to late August, the testes of ripe adult males become sausage shaped and emerge from the epigonal organs.

The reproductive cycle of the female nurse shark is biennial with consecutive vitellogenesis and gestation. Vitellogenesis and growth of oocytes probably begin in June (beginning of the first year), although oocyte growth is not obvious until early September. By late winter and early spring, females are carrying large vitellogenic oocytes 45–59 mm in diameter. The oocytes reach their maximum development and a diameter of 55–60 mm in late May and early June (end of the first year). Females then mate and ovulate, and their eggs are fertilized in June, at the start of the second year of the cycle. They gestate for about five to six months, giving birth in late November or early December. Throughout gestation and parturition, females carry small oocytes 7–14 mm in diameter. They continue to have undeveloped oocytes through the following spring and summer, and thus females cannot mate in June (end of year 2). This period has been referred to as a "resting interval," although it is a time of active energy storage in the liver. Then the vitellogenic cycle begins again, and the female will mate and ovulate again a year later in June, two years from its previous mating and ovulation. Thus, the reproductive cycle is biennial, with female nurse sharks producing broods every two years (Castro 2000).

Brood size: The literature contains few data on brood size in nurse sharks. I have observed nine broods ranging from 21 to 50 young, with a mean of 34 (Castro 2000). Because birth of the entire brood may be spread out over several weeks, females observed in late November and early December may appear to have small broods, having released some of their progeny.

Age and growth Little is known about age and growth of the nurse shark. Carrier and Luer (1990) showed that wild juvenile nurse sharks (mean total length 127 cm, 9.8 kg) grew at a rate of 13.1 ± 9.5 cm/year and 2.3 kg ±1.3 kg/year. The nurse shark is a long-lived species (see aquarium records below), but little is known about its potential life span in the wild.

Relation to humans The nurse shark has been used for its liver oil, hide, and meat, as well as for display and experimentation. In the past, nurse shark liver oil was used for various purposes. Gosse (1851) stated that, in Jamaica, the nurse shark was fished solely for its liver, each shark yielding about a gallon of liver oil that was used as lamp fuel. In the Florida sponge fishery of the 1880s, fishermen used nurse shark liver oil to calm the water surface so that they could scan the bottom. A teaspoon of oil was said to produce a smooth surface for as long a time as a small boat cared to work in one spot. Residents of Key West killed nurse sharks in summer and extracted the oil, which at the time sold for $1 per gallon (Rathbun 1887).

Nurse shark hides were the most valuable hides in the Florida shark fishery of the 1940s. In 1943, shark hides were classified into only two categories: nurse shark hides and the hides of "other species." Nurse shark hides were bought by the shark leather industry at prices about 25% higher than those of other species. The price of the hide from a 230 cm nurse shark was about US $3.10 in 1943 (Bigelow and Schroeder 1945). In the year 2000, I received a belt made of (Atlantic) nurse shark leather from my friend Enrique de la Torre, an expert leather tanner in Guaymas, Mexico. This belt is made of the toughest and best-looking shark leather I have ever seen, and after seven years of hard wear, the leather remains pristine.

The hardiness of the nurse shark and its ability to survive in confinement have made it one of the most popular aquarium and laboratory sharks. It is certainly the most commonly displayed shark in public aquariums throughout the Americas. According to E. Clark (1963a), one survived 25 years at the Shedd Aquarium in Chicago, and another lived for 24 years at the Government Aquarium in Bermuda. The nurse shark is one of the most important species in shark research and has been used in many physiological, sensory, immunological, and behavioral studies. Captive nurse sharks have been trained using operant conditioning to retrieve hoops in the manner of trained bottlenose dolphins (McManus et al. 1984). I saw films of nurse sharks being trained in the 1980s, the sharks retrieving small hoops by inserting their heads through the hoops and then lifting the hoops out of the water to the trainer, just as dolphins are trained to do.

Nurse sharks today have little commercial value in Florida. Some are fished for crab bait and sold for about $0.20/kg. Nurse shark fins are worthless in today's markets. Although the nurse shark is edible, its meat is seldom found in United States markets. Small numbers of neonates are caught for the aquarium trade. Unfortunately, this limited use affords only partial protection to the species because fishermen occasionally kill nurse sharks caught on their longlines, considering them a nuisance. The species is included in the Fishery Management Plan for Sharks of the Atlantic Ocean (National Marine Fisheries Service 1992), which has regulated the shark fisheries along the eastern coast of the United States since 1993. The nurse shark was placed in the "large coastal sharks" category, thus benefiting from the closed seasons and catch limits applied to that category. However, such protection is tenuous, because most captured nurse sharks are not landed. They are sometimes killed and discarded at sea, so usually they are not counted toward the commercial quotas.

The greatest resource value of the nurse shark probably lies in ecotourism. This is the shark seen most often by recreational divers in Florida and the Caribbean. Nurse sharks are large, and seeing one unexpectedly or at close range can be a thrilling experience for divers. In recent years, numerous shark-watching operations have emerged in Florida and the Bahamas. At some locations, nurse sharks have become habituated to being fed by divers. Although the long-term consequences and risks of these operations are still unclear, one can only hope that public awareness and concern may result in some form of protection for these interesting animals.

Nurse sharks have been involved in provoked and unprovoked attacks on divers. Unwary divers often attempt to capture small nurse sharks by picking them up by their tails as they lie motionless on the bottom; others foolishly pick up these sharks to pose for a picture. These sharks can quickly turn on their tormentors and are capable of inflicting serious injury with their tenacious bite.

I have seen more anglers and biologists injured by hooked nurse sharks than by any other species of shark. Injuries usually occurred when people handled the shark carelessly, were surprised by the quick twisting motions of the shark, or underestimated the shark's strength. Unlike carcharhinid or lamnid sharks, hooked nurse sharks offer little resistance on the line and can be easily brought to a boat. However, when one attempts to boat them or remove the hook, nurse sharks explode into a twisting action, and sometimes they repay their tormentors with a bite or catch them in the twisted lines or hooks.

PACIFIC NURSE SHARK

Ginglymostoma (sp.)

Fig. 51a. *Ginglymostoma sp.*, immature male 191.2 cm, 42.8 kg; from El Sahuímaro, Sonora, Mexico (Fernando Márquez, collector).

Common name Pacific nurse shark, an allusion to differentiate it from the nurse shark of the Atlantic. See discussion of the name "nurse" under nurse shark (p. 184).

Spanish name *Gata* (western Mexico). See discussion on p. 184.

Scientific name *Ginglymostoma:* from *ginglymus,* Greek for hinge or hinge joint, and *stoma,* Greek mouth, hence "hinge mouth," an allusion to its peculiar mouth. This is an undescribed species.

Synonyms None. Authors have generally assumed this species to be the same as the Caribbean nurse shark (*Ginglymostoma cirratum*), so they have used the names given to that species (*G. cirratum* or *G. fulvum,* a name given to a Caribbean nurse shark specimen by F. Poey).

Identification The Pacific nurse shark has conspicuous *nasal barbels* on the anterior margins of the nostrils and a first dorsal fin originating *above or posterior* to the origin of the pelvic fins, the trailing lobe of the first dorsal fin reaching almost to the origin of the second dorsal fin (see plate 4), the trailing lobe of the second dorsal fin reaching to the upper lobe of the caudal fin, and the trailing lobe of the anal fin extending to the lower lobe of the caudal fin; and has separate and nonoverlapping fourth and fifth gill slits. It has a very wide head that gives it a tadpole appearance from above. The teeth have one large cusp flanked on each side by with four or five cusplets on each side of the main cusp. The teeth toward the sides of the jaw may have up to six small cusplets on each side of the central cusp. The teeth number U: 17–17, L: 14–14 and are similar in both jaws. The dermal denticles are large and visible to the naked eye and have a pavement-like appearance. They are squarish and have several weak ridges. Coloration is light brown above and tan below.

Similar species The Pacific nurse shark is the only shark with long, tapering nasal barbels along the western coast of North America.

Range The Pacific nurse shark inhabits the eastern Pacific Ocean from the Gulf of California to Peru (Kato et. al. 1967). Beebe and Tee-van (1941) collected or observed numerous specimens from Chamela Bay, Mexico, to Bahia Honda, Panama.

Size and weight There are very few data on the species. The largest specimen measured in the Pacific by Beebe and Tee-Van (1941) was 280 cm long. A 191.2-cm male weighed 42.75 kg.

Biology The species has long been considered to be the Atlantic nurse shark (*Ginglymostoma cirratum*), from which it differs anatomically and genetically after a separation of more than four million years. Because it was assumed that it was the same species as its Atlantic congener, and because of its current scarcity, biologists have paid little attention to the species. The species is only briefly mentioned in faunistic reports of the nineteenth century and the early part of the twentieth century. Presumably its habits are similar to those of the Atlantic species.

Size at maturity Two males, 228 and 229 cm respectively, examined by J. C. Pérez Jiménez in Islas Marias, Mexico, were both mature based on clasper calcification.

Fig. 51b. Teeth.

Reproduction The species is most likely aplacental viviparous like its Atlantic congener.

Size at birth: Unknown, but probably 28–30 cm at birth.

Brood size: Unknown.

Nurseries Unknown.

Age and growth The species has not been aged.

Relation to humans The Pacific nurse shark is currently too scarce along the western coast of Mexico to be of any economic importance. The species appears to have been overfished in the Gulf of California in the last few decades. Small numbers of Pacific nurse sharks used to be routinely processed at the shark leather factory of Guaymas, Mexico, in the 1960s and early 1970s, but the species has become very scarce in the last few decades (Enrique de la Torre, pers. comm., 2003). The species was described as rare in the artisanal fisheries off Guaymas, Mexico, in 1974 (Compagno 2001). Today the species is seldom encountered in the fisheries of the Gulf of California. In a recent survey of shark catches in the area, conducted by Mote Marine Laboratory in collaboration with several other institutions, in the states of Baja California, Baja California Sur, Sonora, and Sinaloa, observers recorded more than 61,000 sharks, none a nurse shark (J. Bizarro, pers. comm., August 2006).

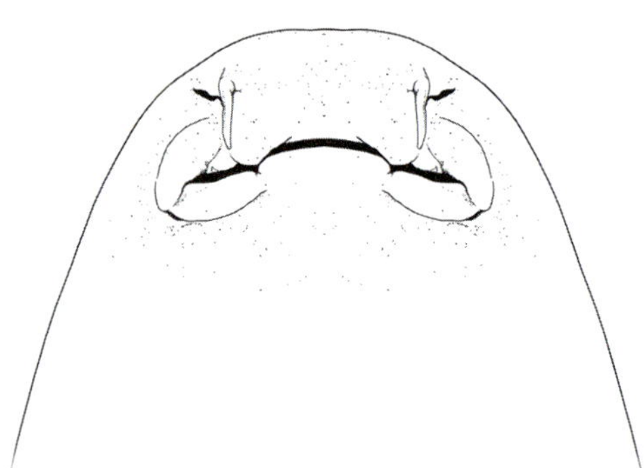

Fig. 51c. Snout.

Fig. 51d. Dermal denticles.

FAMILY RHINCODONTIDAE

The Whale Shark

This family consists of only one species, the whale shark, the largest fish in the world.

WHALE SHARK

Rhincodon typus Smith, 1828

Fig. 52a. *Rhincodon typus*, whale shark immature male, 790 cm; off Okinawa, Japan. Drawing based on a captive male held at the Okinawa Churaumi Aquarium in February 2009, accomplished with the assistance of Yosuke Matsumoto, Keiichi Sato, and Senzo Uchida).

Common name Whale shark, an obvious allusion to its size and feeding habits, which are similar to those of baleen whales.

Spanish name *Tiburón ballena* (Mexico, A.F.S.), *dómino* (eastern Mexico), and *pez dama* or *damero* (Cuba), Spanish for "checkers" or "checkerboard fish," not, as some have said, "lady fish"—*dama* comes from *juego de damas,* literally "game of ladies," the game of checkers. The best explanation of the Cuban name is given by Gudger and Hoffmann (1931: 33–36):

> The intersection of keels and [white or yellow] bars gives a checkerboardlike appearance to the fish. This is accentuated by the large white or yellow spots set in the middle of each square. The local Cuban name for the fish is *pez dama,* in allusion to this. Now *pez* means fish and *dama* lady, but the Cubans do not mean "lady fish" by this appellation, for *dama* also means checkerboard. Hence, when it is locally called *pez dama,* this means "checkerboard fish" to the hearers. The blocked-off sections of skin with their enclosed large round spots certainly do resemble a checkerboard with the checkers in the center of each square.

The checkerboard pattern is well illustrated by plate 5 of that article.

Scientific name *Rhincodon typus* Smith, 1828; in *South African Commercial Advertiser* 3(145): 2. *Rhincodon:* a misspelling of *Rhineodon,* from the Greek *rhine,* file, and *odont,* tooth, or file-toothed shark, a reference to its teeth being aligned like rows on a file; *typus:* Greek, type. Smith almost certainly wrote *Rhineodon typus,* but the printer mistook Smith's "e" for a "c" and so printed it (Gudger 1934).

Synonyms There have been numerous versions (synonyms) of the generic name, all of which are simply emended spellings of the 1829 name. An earlier description of the whale shark using the name *Rhiniodon typus* was published in 1828 in a South African paper but was overlooked until 1972 (Penrith 1972). Robins and Lea (1975) gave a good account of the history of the name and its numerous spellings. In 1984, the International Commission on Zoological Nomenclature (1984) placed the name *Rhincodon* in the Official List of Generic Names in Zoology, suppressing the other spellings. The A.F.S. uses the 1828 date.

Identification The whale shark is an unmistakable fish. It is a *huge (usually 5–10 m!), docile shark covered with white or yellowish spots and irregular light and dark bars.* Its snout is short, with a *huge terminal mouth.* Its nostrils have short, blunt nasal barbels. *Three pronounced longitudinal ridges extend along each side of the trunk,* the lowermost becoming strong keels near the tail. The whale shark's teeth are minute: the cusps measure only 3 mm in an adult shark. Each tooth has a single backward-curving, sharp cusp that overlaps the following tooth. The teeth are arranged linearly, from front to back, in files of 12–15 active teeth. There are about 300 such files in the entire jaw. The teeth are similar in both jaws. The dermal denticles are three pointed with three ridges, the central ridge being the largest and terminating in a long point. In life it has dark greenish or brownish gray dorsal surfaces (purplish blue in sun-exposed specimens), with numerous small white or yellowish spots over

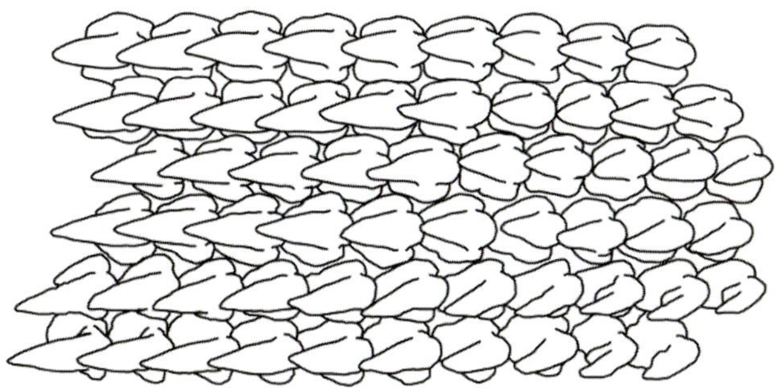

Fig. 52b. Teeth.

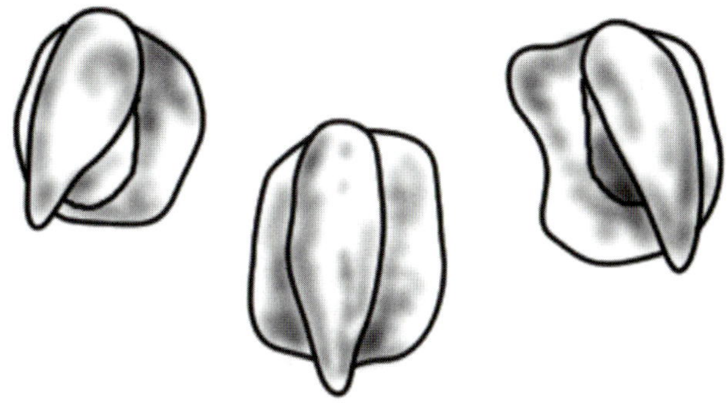

Fig. 52c. Teeth (detail).

the head region and the dorsal surfaces of the pectoral fins, and a complex pattern of narrow whitish bands and white spots over the body. Behind the gill region, vertical rows of bright white or yellowish spots alternate with rows of fainter spots that tend to coalesce into lighter vertical bars of irregular width. The spots, vertical bars, and longitudinal ridges along the flanks often create a checkerboard appearance. The undersides are whitish.

The markings are quite variable and unique to each shark, and they appear to remain constant, at least for several years. Taylor (1994) and Stevens et al. (1998) used a photo identification system based on major scarring and each whale shark's unique spot pattern to recognize individuals. Using such a system, Taylor (1994) identified 44 whale sharks at Ningaloo Reef, Western Australia, in 1992, and Stevens et al. (1998) were able to distinguish 52 individuals from 1995 through 1997.

Similar species The whale shark's striking markings and usual huge size easily distinguish it from all other sharks.

Range The whale shark is cosmopolitan in tropical and subtropical coastal and oceanic waters. According to Wolfson (1986), its range extends mainly from 30° N to 35° S latitudes. In eastern North America, the whale shark ranges from the Bay of Fundy to Florida and the Gulf of Mexico. For many years the northernmost record was a 9.6-m specimen caught off Fire Island, New York (42° N), in August 1935 (Gudger 1936). Recently, Turnbull and Randell (2006) reported sighting a whale shark in the Bay of Fundy (44° N)

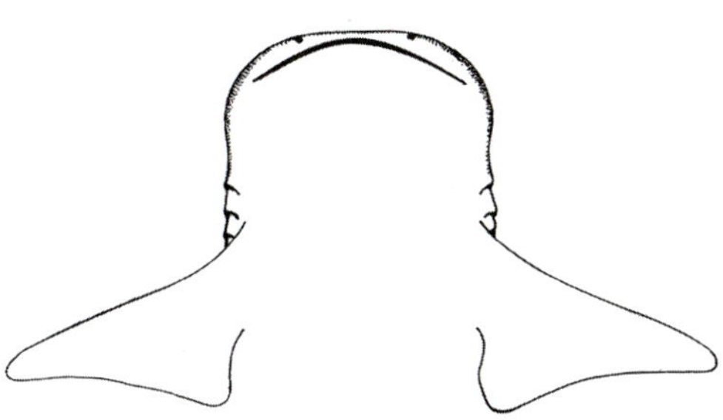

Fig. 52d. Snout.

Fig. 52e. Dermal denticles.

in August 1997 when the sea surface temperature was 15°C. The whale shark is most frequently encountered off Florida and in the Gulf of Mexico. It is seasonally common off the Yucatan Peninsula.

In western North America, the species ranges from southern California and the Gulf of California southward to the tropics. Records of whale sharks off California and off the United States–Mexico boundary are only sight records (Fitch 1951; Miller and Lea 1972; Ebert et al. 2004); it appears that whale sharks have not been landed in California. Fitch (1951) first reported a sighting of a whale shark some 45 km south of San Diego, just south of the United States–Mexico boundary, and 1,125 km farther north than previous records. The whale shark is often encountered in the Gulf of California. Clark and Nelson (1997) reported finding it usually from Bahía de la Paz to Cabo San Lucas. Nelson and Eckert (2007) reported that its presence in Bahia de Los Angeles was seasonal, occurring from June to November and with the highest abundance from August to October.

Size and weight The whale shark is the largest fish known; specimens 5.50–10.0 m are often reported. Its maximum size is not known with certainty because actual measurements of large specimens are few. The greatest length quoted most often in the literature is 13.7 m ("45 feet"). This length is based on two reports, one anecdotal and the other erroneous. The earlier report is that of E. P. Wright, who spent time in the Seychelles and mentions that Mr. Ward, the Civil Commissioner of the islands, had "measured one that somewhat exceeded forty-five feet in length" (Wright 1870: 64). Wright also reported having seen specimens he believed "to have exceeded fifty feet in length" (65). In the other document (Garman 1913), a length of 13.7 m is attributed to a specimen captured off Knight's Key, Florida, in May 1912. This specimen was first exhibited fresh on a flatbed railroad car. Later, after the shark had been mounted, it was exhibited throughout the southeastern states atop a large cart and later on a barge. Postcards and photographs of this shark were once common. A postcard of the fresh specimen bears the inscriptions "Copyright by Chas. Thompson 1912" and "The largest fish in the world, weight over 30,000 lbs. Captured off Miami, Fla. by Capt. Chas. H. Thompson." A postcard of the mounted specimen bears the inscriptions "Copyrighted by Chas. Thompson Weight 30,000 lbs. Length 45 ft. Photo by Hand Co. Miami, Florida. 1913." These cards engendered the belief that the live or fresh specimen had been measured at 13.7 m (45 ft.), although it is possible that the stretched, mounted skin might have been 13.7 m. Thus, this shark was incorrectly reported by Garman (1913) to measure 13.7 m. Later, Gudger (1915a) wrote that the specimen was 11.6 m ("38 ft."), based on a statement from a Mr. Brooks, one of its captors, who had measured the shark while it was in the water (Gudger 1915a). Gudger never saw this fish whole, having seen only the skin while he was in Miami, although it would have been possible for him to measure the skin.

Much later, Bigelow and Schroeder (1948) also quoted the 13.7 m length given by Wright (1870) without actually having seen Wright's account, stating that the whale shark is credibly reported to reach lengths of 18.2 m ("60 ft") or even more, but without providing any evidence. Chen et al. (1997b) stated that in March 1987, two of the authors had recorded one specimen in the Lotung, Taiwan, fish market that was "approximately 20 m in length" but again offered no evidence of actual measurements or how the length was estimated. They believed this specimen to be the largest whale shark ever caught in Taiwan.

Authors who have actually measured large whale sharks have reported smaller sizes. Kaikini et al. (1959) measured a 12.1-m female that stranded off Mangalore, India. Karbhari and Josekutty (1986) reported a 12.18-m male from Bombay, India, caught on 21 November 1983.

There are few reliable weights for adult specimens, given the limited availability of scales suitable for weighing such large animals. The weight of the 1912 Knight's Key specimen given in the literature has no validity. Garman (1913) reported the weight as 13,608 kg ("30,000 pounds"). This is the figure given on Capt. Thompson's postcards in circulation at the time. Later Gudger (1915a: 353) quoted a Mr. Brooks involved in the capture as writing, "His weight was pure estimate, but we thought he would weight something over five tons." Gudger (1915a) estimated the animal's weight at 12,088 kg (26,594 lbs) using "an ancient fisherman's formula" (the length in inches multiplied by the square of the girth in inches, and the product when divided by 800, gives the weight in pounds). Later, Bigelow and Schroeder (1948) repeated the 26,594 pounds weight, based on Gudger's article. Joung et al. (1996) gave the weight of an approximately 10.6-m whale shark as 16,000 kg, as weighed by a construction crane. These authors also gave weights of 15,220 kg and 36,000 kg for sharks landed in Taiwan in March 1994 but without any other data or evidence. Chen et al. (1997b) stated that the "approximately 20 m long specimen" of the Lotung fish market in 1987 weighed 34,000 kg ("34 tons"). How the weight was arrived at was not stated.

At the Okinawa Churaumi Aquarium, personnel under Senzo Uchida have been weighing whale sharks at capture or shortly afterward. These accurate weights relegate the weights reported above to the realm of fantasy: males, 3.7 m, 290 kg; 4.25 m, 550 kg; 4.5 m, 632 kg; 5.1 m, 1,400 kg; female, 4.35 m, 577 kg. The weights of two larger specimens were 6.33 m, 1,750 kg (after 46 days in captivity); 9.1 m, 6,360 kg (after 21 days in captivity). In the case of the 9.1-m specimen, even if one assumes a weight loss of 20% in the 21 days in captivity when the animal did not feed, its weight at capture could not have exceeded 7,632 kg.

Biology The whale shark is a huge, pelagic, filter/suction feeder, usually seen swimming slowly near the surface in tropical waters. Iwasaki (1970) stated that the species prefers upwelling areas where cold seawater less than 17°C surges upward from the depths, surface temperatures are 21–25°C, and salinities of 34–34.5. Hoffman et al. (1981) reported that seven sightings of whale sharks about 100 km east of South Padre Island all occurred in water with surface temperatures of 29°C. Rowat and Gore (2007) reported that whale sharks in the Indian Ocean off Seychelles spent 53% of their time at depths of less than 10 m and generally at temperatures of 25–30°C. Although whale sharks are generally seen on or near the surface, they regularly dive to great depths. Hsu et al. (2007) reported that in the northwestern Pacific Ocean, whale sharks generally inhabited areas where the water temperature was between 23 and 32°C but that individuals would dive deep, into waters where the temperature was 6°C. Graham et al. (2005) reported that whale sharks equipped with satellite tags dived to depths greater than 980 m and temperatures less than 7.6°C and that it was likely that some dives exceeded the tag limitations (1,500 m). Rowat and Gore (2007) reported whale sharks in the Indian Ocean diving to depths of 750–1,000 m. The reasons for the deep dives are unknown.

The whale shark is the subject of an extensive literature, the older literature consisting mostly of catch or sighting reports, and the recent literature consisting mostly of tracking or ecotourism reports, with little information on its natural history. Many of the older articles were contributed by E. W. Gudger (1866–1956), who published 47 articles about whale sharks over 40 years. There are even two bibliographies, one by Gudger (1934) and an annotated one by Wolfson and di Sciara (1981). In the last few years the whale shark has become a very popular attraction for skin-diving tourists in several places and a popular target for researchers' satellite tags, resulting in a flurry of articles. Despite these numerous articles, the biology of the whale shark remains poorly understood, and only recently have we cleared up the mysteries and misconceptions about its reproduction.

Whale sharks feed on a wide variety of planktonic invertebrates, fish eggs, and small fishes (see below). Documented reports of whale shark stomach contents are scarce. Gudger (1941) examined the stomach contents of a whale shark captured off Fire Island, New York. Based on its red color, he believed the material to be broken up crustacean exoskeletons. Likewise, the stomach contents of a South African specimen were described as masses of small crustaceans (Bass et al. 1975b). Silas and Rajagopalan (1963) presented the most extensive analysis of the food of a whale shark. These authors found that crustacean zooplankton comprised most of the stomach contents, which also included fishes, mollusks, and small quantities of seaweed and other algae. Springer (1957b) described whale sharks feeding at the surface in an area, where mantis shrimp (stomatopod) larvae, small fishes, and small blackfin tuna (*Thunnus atlanticus*) were abundant. According to Springer, a feeding shark would have its head at or above the surface and the tail straight down. In a rhythmic sequence, the shark would rise until its head was about a meter above the surface and then slowly sink back down, stopping as its open mouth reached the surface. At this point, water would pour into its mouth, and as the mouth filled with water, the shark would swim upward until the water had exited out the gill slits. The shark would then repeat the cycle. Springer believed that the whale sharks he observed ingested a number of tuna, although he did not catch any of them or inspect their stomach contents. Clark (1992) watched whale sharks feeding at Ningaloo Reef, Western Australia, and she observed them actively seeking dense patches of plankton. Swimming slowly at the surface toward a dense mass of plankton or anchovies, a shark would accelerate and open its mouth as it neared the prey. Swimming through the massed prey, a shark would move its head from side to side, sucking in all or part of the food mass. Clark and Nelson (1997) observed small whale sharks (3.2–5.2 m) suction feeding on dense concentrations of copepods identified as *Acartia clausi* in Bahía de la Paz, Mexico. A shark would concentrate on a patch by turning and gulping until it had consumed the densest concentrations. When a shark emerged from a patch of plankton, it would circle back sharply and continue feeding. During active feeding, the upper part of the head would be out of the water and turning from side to side, the mouth opening and closing 7–28 times per minute.

Little is known about the migrations of whale sharks, although significant numbers are known to congregate in a few localities at specific times. These aggregations are often correlated with increased local productivity (e.g., coral or fish spawning, which is often tuned to lunar cycles) or are associated with schools of pelagic fishes. Gudger (1939) reported schools of whale sharks apparently feeding on small jellyfish (15 mm in diameter) in the Bahama Islands. Springer (1957b) reported aggregations of whale sharks in the Gulf of Mexico from May to September. Taylor (1994) noted that, off Ningaloo Reef, Western Australia, the greatest numbers of sharks were seen two to three weeks after coral spawnings, which usually occurs seven to nine nights after the March and April full moons. Heyman et al. (2001) reported that whale sharks aggregated to feed on fish spawn off Gladden Spit, Belize. According to these authors, the aggregations of whale sharks coincided with spawning aggregations of cubera snappers (*Lutjanus cyanopterus*) and dog snappers (*L. jocu*), which occurred each year at specific locations at sunset, during the full and last-quarter moon periods of April and May. Graham and Roberts (2007) reported other seasonal aggregations off Isla Contoy, Mexico, and Utila, Honduras. According to Last and Stevens

(1994), small aggregations occur each winter near the coast of central Western Australia.

Based on concentrations of whale sharks that appear at widely scattered locations, it is likely that whale sharks have vast ranges and cover great distances in their migrations. However, at this time, there are few data to prove this hypothesis. Few whale sharks have been tracked, and a few have undertaken long distance movements. One shark tagged in the Sea of Cortez traveled to the western North Pacific, covering more than 13,000 km in 37 months (Eckert and Stewart 2001). One juvenile carrying a satellite transmitter was tracked for 4,567 km as it traveled from the Philippines to the South China Sea (Eckert et al. 2002).

Remoras, cobias, and small jacks often accompany whale sharks. The remoras may enter the mouth, spiracle, and cloaca, and sometimes can be seen peering out of the anus (J. I. Castro and D. Perrine, unpublished observations). Cobias swimming alongside are sometimes mistaken for whale shark offspring by uninformed observers.

The whale shark is a sluggish and docile fish that often allows divers to hang onto its fins. It is so sluggish that there are numerous old reports of its being impaled on the bows of steamships (e.g., Gudger 1923, 1927, 1937; Deraniyagala 1936). All the reported collisions occurred before 1940. Whether the sharks are scarcer now, and the collisions therefore less frequent, or people simply stopped reporting them is uncertain. Whale sharks occasionally strand on shallow reefs or in lagoons due to falling tides.

Size at maturity There are few data on the size at maturity of the whale shark. It appears that males reach maturity at around 9.2 m. This is based on two South African males measuring 9.1 and 9.2 m reported as immature (Beckley et al. 1997) and on four males 9.0–9.5 m reported as mature (Beckley et al. 1997; Wintner 2000). Females can be expected to mature at a larger size, probably at around 9.7 m.

Reproduction The whale shark is an aplacental, viviparous species. It was long believed to be oviparous, based on two reports. Southwell (1913) reported that 16 eggs had been found in a specimen from Ceylon (Sri Lanka) and that they were the "same form as dog fish." In July 1953 a shrimp trawler hauled up a large, unopened egg case from a depth of 57 m in the Gulf of Mexico, about 195 km south of Port Isabel, Texas. The egg case measured 30 × 14 × 9 cm and contained a live, fully formed whale shark 35 cm long (Baughman 1955). Reid (1957) suggested that this specimen was an advanced embryo, but the subsequent illustration by Garrick (1964) shows a large external yolk sac, suggesting that the animal was not ready for birth.

Not everyone assumed the whale shark to be oviparous. Gudger (1915a) believed the species to be viviparous. Later, he stated that it seemed inconceivable that such a giant would lay eggs (Gudger 1952). Bigelow and Schroeder (1948) also assumed that whale sharks were "probably ovoviviparous" (aplacental viviparous). The trawled egg case was considered by some to have been aborted because of its thin shell and rudimentary tendrils. Then, in July 1995, a pregnant female was harpooned off the east coast of Taiwan. This female, estimated at 10.6 m and said to weigh 16 metric tons, carried about 300 embryos in her uteri. Joung et al. (1996) measured 237 of the embryos, which ranged in length from 42 to 64 cm. These embryos fell into three developmental classes of increasing length: those measuring 42–52 cm, those of 53–58 cm, and those of 59–64 cm.[1] Embryos in the last two classes were free from the egg cases and no longer had yolk sacs, but rather exhibited vitelline scars. The largest embryos (59–64 cm) were probably ready for birth (Joung et al. 1996). This mode of reproduction is similar to that of the nurse shark (Castro 2000), in which lecithotrophic embryos are found in different stages of development and hatch out of their egg cases over many days.

Size at birth: Whale sharks are born at 55–64 cm. This estimate is based on sizes of the largest embryos encountered by Joung et al. (1996), which measured 59–64 cm and presumably were ready for birth, and on the sizes of the smallest free-swimming neonates. Wolfson (1983) reported neonates measuring 55, 56, 57, 62, and 63 cm, while Kukuyev (1996) wrote of others measuring 55.7 and 59 cm.

Brood size: The whale shark brood of 300 reported by Joung et al. (1996) is by far the largest known for any elasmobranch. It is likely that much larger broods can be expected, because this specimen is believed to have aborted some young before being hauled out of the water (E. Clark, pers. comm., 1995). Also, much larger whale sharks have been seen, and if brood size correlates to the size of the mother, as in most sharks, larger whale sharks should have even larger broods.

Nurseries The nurseries of the whale shark are poorly defined because, remarkably, few small whale sharks have been captured. Wolfson (1986: 211) stated that "all of the 11 juveniles known to date were taken pelagically by by tuna purse-seiners , five in the eastern Pacific and six in the Gulf of Guinea.." These catches are probably more indicative of the distribution of tuna purse seiners than of neonate whale sharks. Kukuyev (1996) reported a 59-cm female neonate long caught off Sierra Leone in waters more than 2,000 m

1. Joung et al. (1996: 220) gave three sizes of embryos: those "42 to 52" cm, those "52 to 58" cm, and those "58 to 64" cm. Because the last two size classes overlap, I have amended the last class to be 59 to 64 cm.

deep and a 55.7-cm male from the stomach of a blue shark caught in the "central tropical Atlantic." The dates and locations of the catches are unknown. All of these neonates came from tropical waters within 12° from the Equator.

If Wolfson's (1983) capture dates are correct, newborn whale sharks can be found both in summer and winter, suggesting year-round reproduction. Such reproductive cycles are not unusual for tropical sharks. However, two of these two dates may be wrong. Capture dates for sharks caught by tuna vessels are often lost or recorded in error many months after the event. If this is the case for the two winter captures, the rest of the meager data available suggest a summer birth season in tropical waters. The live 35-cm embryo from the egg case trawled on 2 July, had it lived, probably would have been born in August. The pregnant female mentioned above was caught on 15 July and only a third of its embryos were not ready for birth. It probably would have given birth in August also. The 55-cm and 63-cm females listed by Wolfson (1983) were born in August and June, respectively.

Age and growth The whale shark is one of the fastest-growing elasmobranchs known. The meager growth data available are derived from only a few young specimens held in captivity. Two embryos were birthed from the dead female reported by Joung et al. (1996) and were placed in aquariums. One grew from 60 cm and 1 kg to 139 cm and 20.4 kg in 120 days (Chang et al. 1997). Later, Yang and Horike (1998) reported that the surviving young at the Oita Ecological Aquarium, Japan, grew from 70 cm and 0.8 kg to 3.69 m and 151.25 kg in 3 years and 68 days (from the time of removal from the female on 15 July 1995, to the time of its death on 21 September 1998). Kitafuji and Yamamoto (1998) maintained two whale sharks at the Osaka Aquarium, Japan. One of them, a female measuring 407 cm at capture, grew 226 cm in 1,825 days, or an average of 45.2 cm annually. This growth occurred while the animal was kept at 23°C and consumed 4–5% of its body weight each week. Although growth rates in captivity are generally faster than under natural conditions, this growth rate is astounding. Interestingly, long before any growth data on whale sharks were available, Gudger (1934) shrewdly anticipated that whale sharks grow rapidly. He estimated that "a youthful looking" 549-cm specimen that came ashore in Ormond Beach, Florida, in January 1902 (the first U.S. record) was "possibly not more than a year or two old."

How long whale sharks live can only be guessed. Wintner (2000) aged specimens from South Africa by counting their vertebral growth rings, which have been shown to be deposited annually during the early part of the animal's life. The number of growth rings counted in four female whale sharks 4.2–7.5 m (precaudal length) ranged from 19 to 27, while three mature males of 6.7–7.55 m (precaudal length) had 20–27 growth rings. Longevity estimates must wait until larger specimens are aged and for validation of the annual deposition of growth rings in adulthood.

Relation to humans Historically, there have been few fisheries for the whale shark because of its large size and unpredictable presence at most locations. There have been a few small fisheries in India, the Philippines, and Taiwan, but the species is of little commercial importance elsewhere (Castro et al. 1999: 20). The whale shark was generally caught for its flesh, but presently the fins and oil are also used. According to Ramachandran and Sankar (1990), no organized fishery exists in India, except in Veraval, where fishing is conducted exclusively for the oil for a short period from February to May. According to Silas (1986), a small harpoon fishery occurs in certain years when whale sharks are abundant in Gujarat waters in northwest India. This fishery caught about 40 whale sharks in 1982, of which 22 were taken to Veraval for liver removal in the period of 12–15 April. Silas (1986) stated that captures were sporadic or even fortuitous.

The Philippines fishery is poorly known. It appears to be centered in the Bohol Sea, according to an unpublished report by the WWF-Philippines Program (Trono 1996). According to this source, fishermen in Talisayan (one of four sites visited in the Bohol Sea) reported catching 100 sharks in 1994, 80 sharks in 1995, and only 30 for the 1996 season. Based on interviews with the fishermen from the four sites, a total of 95 sharks were landed during the 1996 season. This fishery was closed in 1998 (Shark News 1998).

In the late twentieth century, Taiwan had the largest fishery for whale sharks. Chen et al. (1997b) reported that whale sharks are caught accidentally by set net or opportunistically by harpoon and that catches by gillnet or longlines are less common. These authors estimated the 1996 annual whale shark catch for Taiwan at 272, a figure that agrees with an estimate of 250 sharks caught yearly, given to them by a Suao wholesaler who specializes in whale shark meat. According to Joung et al. (1996), Taiwanese fishermen in An-Ping Harbor caught more than 70 whale sharks in 1992, only 2 in 1993, and 14 in 1994. They reported that fishermen from the southern coast of Taiwan used to catch 30–100 whale sharks in a season but that by the late 1980s fewer than 10 sharks were being taken each season.

In the 1990s the whale shark became economically important to ecotourism off Western Australia. According to Stevens et al. (1998), eight boats were engaged in the whale shark industry each season from March to May at Ningaloo Reef. Sharks swimming at the surface are located from the boat or by a small aircraft in early morning. The boat is then maneuvered into the path of an advancing shark, and parties of six to eight snorkelers are dropped off. The snorkelers then swim with the shark (which usually moves at 1–2 knots) until they become tired and the boat picks them up.

Whale sharks are now being used for ecotourism purposes in other locations where one can have a reasonable chance of seeing them on a given day during a specific season. In Mexico, a small whale shark watching operation exists seasonally near La Paz, in the Gulf of California, and another one has developed off Contoy Island, near Isla Mujeres, in the Gulf of Mexico. For those lucky enough to see a whale shark up close, the sight of this huge fish is exhilarating.

Whale sharks are the most dazzling attractions in the few aquariums lucky enough to possess them. At this time (January 2007), only a few facilities display whale sharks. Whale sharks have been displayed in the Okinawa aquarium since 1980, where Senzo Uchida spent many years developing procedures for their maintenance. At the superb Okinawa Churaumi Aquarium, three specimens are currently on display, while three other specimens are kept in reserve in an ocean pen, and sometimes the display specimens are rotated to the sea pens. Although the first specimen lived only 10 days in captivity, present-day husbandry techniques allow whale sharks to live in captivity for decades; a 7.9-m male has lived over 13 years at the Okinawa Churaumi Aquarium, and it is expected that their whale sharks will reproduce when the largest female reaches maturity in a few years (Keiichi Sato and Minoru Toda, pers. comm., August 2008). Other Japanese and Taiwanese aquariums have also kept whale sharks. In June 2005, two male whale sharks were first imported from Taiwan into the United States for exhibit at a Georgia aquarium, and two females were added in June 2006.

The whale shark received protected status in the United States in 1999, when the National Marine Fisheries Service (1999) declared it a protected species. In October 2001 the whale shark was listed as nationally threatened in Australia and protected in Commonwealth and state waters. In December 2002, the species was placed in Appendix II of the Convention on International Trade in Endangered Species (CITES), along with the basking shark.

The protected status is primarily due to the whale shark being part of what I call the "black and white charismatic megafauna,"[2] rather than based on known population declines or threats. The protected status can be considered precautionary. Although there are many aesthetic and political reasons to protect whale sharks, our knowledge of its biology is so meager at present that it is difficult to specify any unique vulnerabilities of this fast-growing and most prolific of sharks, for which there are only a few fisheries in the world and none in the western hemisphere.

2. In recent times, many conservation organizations have selected large animals for their logos. These include the killer whale or orca, the whale shark, the white shark, the panda bear, and the zebra. These are invariably black-and-white animals. I refer to them as the "black-and-white charismatic megafauna," if one can call "charismatic" such awesome predators as the orca and the white shark. Although the precarious status of the panda is well known, the status of the other animals is not so critical. There are certainly more endangered marine mammals than the widely distributed orca, and certainly there are shark species far more endangered than the incredibly prolific whale shark or the ubiquitous white shark. Once selected for a logo, the creatures soon gain protection over more threatened animals.

It appears that the reason for the selection of these species as symbols for conservation is probably because they produce showy, recognizable letterheads in black and white, which can be reproduced in photocopy machines. If one chose a highly endangered multicolored creature such as birdwing butterfly, desert pupfish, or a bird of paradise as an organizational symbol, it would be necessary to print far more expensive multicolored letterhead, for these lovely and colorful creatures may not be recognizable in black and white. Furthermore, such colored images do not reproduce well in most photocopy machines.

White polka dots on a dark background are a certain ticket toward being selected for a logo and protected status. For example, the whale shark and the spotted eagle ray were among the first protected species in the United States (the former at the federal level and the latter in the State of Florida), although there were no fisheries targeting them.

FAMILY MITSUKURINIDAE

The Goblin Shark

The Mitsukurinidae is a monotypic family. Its sole member is the goblin shark, a deep-water species characterized by a snout elongated into a long flat blade.

GOBLIN SHARK

Mitsukurina owstoni Jordan, 1898

Fig. 53a. *Mitsukurina owstoni,* goblin shark, immature male, 114 cm, 2.4 kg; off Chiba, Japan (H. Teshima, collector; courtesy of Senzo Uchida).

Common name Goblin shark, an interpretation of its early Japanese name *tenguzame:* from *tengu,* a red-faced, long-nosed, mythical half-bird and half-human creature, and *zame,* shark, derived from the shark's unusual protruding nose and pinkish color.

Spanish name *Tiburón duende,* a literal translation of goblin shark. The A.F.S. has not designated a Spanish name.

Scientific name *Mitsukurina owstoni* Jordan, 1898; in *Proceedings of the California Academy of Sciences, Third Series, Zoology* 1(6): 199–202. *Mitsukurina:* named in honor of Kakichi Mitsukuri (1858–1909), professor of zoology at the Imperial University of Tokyo and, in his day, Japan's foremost zoologist; *owstoni:* latinized surname of Alan Owston, a Yokohama shipmaster and naturalist who obtained the type specimen and presented it to the University of Tokyo. In 1897, Prof. Mitsukuri sent the specimen to David S. Jordan for identification and description, requesting that it be named after Mr. Owston for his discovery. Jordan honored both men by naming the shark after them. The specimen was later returned to the museum of the University of Tokyo (and, apparently, it has been lost).

Synonyms Some authors use the name *Scapanorhynchus owstoni* (Jordan, 1898) (see below). Other synonyms are *Odontaspis nasutus* de Braganza, 1904; *Scapanorhynchus jordani* Hussakof, 1909; and *Scapanorhynchus dofleini* Engelhardt, 1912.

The scientific name of the goblin shark has been controversial almost since Jordan named it *Mitsukurina owstoni* in 1898. The generic name *Scapanorhynchus* Woodward, 1889 had been applied to the teeth of an extinct goblin shark from the Cretaceous found in many parts of the world. Woodward considered *Mitsukurina* synonymous with *Scapanorhynchus* and even wrote an article titled "Note on *Scapanorhynchus,* a Cretaceous shark apparently surviving in Japanese Seas" (Woodward 1899). Modern authors (Signeux 1949; Cappetta 1980; Compagno 1984a) have compared more complete fossils and found differences primarily in fin structure between the fossil and extant forms, thus concluding in favor of retaining the name *Mitsukurina.* See Stevens and Paxton (1985) and Compagno (2001) for historical details of this controversy, and Cappetta (1987) for details of *Scapanorhynchus.*

Identification This is an unmistakable deep-water shark, having a *snout elongated into a long, flat blade,* greatly protrusible jaws, the *last gill slit located above the pectoral fin,* and the *long caudal fin lacking a ventral lobe.* Its body is soft and flabby, and its skeleton is extremely flexible. The eyes are small and lack a nictitating eyelid. *The teeth have very long, narrow, smooth-edged cusps;* the *side teeth usually have lateral cusplets.* There are *numerous fine longitudinal grooves covering the entire lingual surfaces of the teeth, extending from the base of the crown to the tip* (a magnifying lens is useful to see these grooves clearly). The shape of the teeth is highly variable among specimens: some are very long and thin, while others are shorter and wider; some specimens have small cusplets on all the teeth. Some specimens have symphysial teeth, and some have intermediate teeth between the large teeth and the smaller ones in the corners of the jaw. Teeth number about U: 13–13, L: 12–12, not counting the minute teeth in the back of the jaws. The dermal denticles are short, spinelike, and heavily ridged.

Fresh juveniles are white to pinkish light gray above and white below, with gray fin tips; adults are pink to light tan, and preserved specimens are gray. The skin is semitransparent in places, and the underlying blood vessels give the animal a pinkish or reddish tone. The iris is black with light blue streaks.

Similar species The goblin shark's elongated, bladelike snout and coloration distinguish it from all other sharks.

Range The goblin shark appears to be cosmopolitan in deep waters, as it has been reported from nearly every ocean. Most records are of single specimens caught accidentally on longlines or in deep trawls. Its very deep habitat makes it a rare catch. In the western Atlantic Ocean, it has been reported from the Gulf of Mexico (Parsons et al. 2002) and off Surinam and French Guiana (Uyeno et al. 1983). In the Eastern Atlantic, it has been reported from France (Quéro 1972). In the Pacific Ocean, it has been found off Japan (Jordan 1898), Australia (Last and Stevens 1994), New Zealand (Duffy 1997), and California (Ugoretz and Seigel 1999). It has been reported in the western Indian Ocean (Piotrovskiy and Prut'ko 1980), off Cape Town, South Africa, and off southern Australia (Glover 1976).

Only a few locations with deep-water fisheries regularly catch goblin sharks: Suruga and Sagami bays, Japan; and Sesimbra, Portugal. The abrupt bottom topography of the Tokyo submarine canyon is attractive or suitable to juvenile goblin sharks, and many are caught in the crab and prawn fisheries off Chiba, Sagami Bay, where I examined (courtesy of Senzo Uchida, Keiichi Sato, and Masaki Miya) three goblin sharks taken by a fisherman over two days of fishing in early December 2006. Sesimbra produces two or three specimens each year, taken in the scabbard fish fishery; I was able to see only frozen specimens there.

In May 2003, I examined (through the courtesy of Gordon Hubbell) more than 100 goblin shark jaws from specimens taken off Taiwan. These sharks, previously unknown to Taiwanese fishermen, began to be taken in large numbers as fisheries expanded into deeper waters following depletion of nearshore fish stocks. However, either the sharks or the fishermen soon moved from the area, or the sharks were fished out, as the catches were not repeated.

At this writing, there are only two North American records, one from each coast, a consequence of the deep habitat of the goblin shark and the absence of deep-water fisheries. A large female was captured in the Gulf of Mexico 190 km due south of Pascagoula, Mississippi, in July 2000. The shark was taken by commercial fishermen at 919–1,099 m after becoming entangled by a crab trap line, apparently attracted to the bait (Parsons et al. 2002). Unfortunately, the specimen was not measured, and only its jaw was retained. Based on photographs taken while the shark was suspended from lines, Parsons et al. (2002) estimated its length at 540–617 cm. The sole report of a goblin shark off western North America is that of a male caught in October 1998 in a drift gillnet set 13 km off San Clemente Island, California, by a commercial swordfish and shark-fishing vessel. Only the head of the shark was saved, and the fishermen estimated its total length at 270 cm (Ugoretz and Seigel 1999).

Size and weight This is one of the larger species, but there are few data on its maximum size or weight. A 114-cm juvenile male (illustrated in this book) weighed 2.35 kg, while a 118-cm female weighed 2.7 kg. Most specimens taken are small juveniles, but specimens of 3–4 m are occasionally reported. Masai et al. (1973) recorded a 3.5-m male caught in Suruga Bay, Japan. Piotrovskiy and Prut'ko (1980) reported a 322-cm and 160-kg male from the Western Indian Ocean. Kobayashi et al. (1982) mentioned a 373-cm female caught off Japan. Based on the estimates of Parsons et al. (2002), it appears that the goblin shark may exceed 5 m in length.

Biology Capture records indicate that this interesting creature is one of the deepest dwelling sharks; it is usually reported from 550–900 m. However, in Japan, juveniles are often taken at shallower depths in areas of abrupt bottom topography. Yano et al. (2003) reported that, in the submarine canyons off Tokyo, Japan, juvenile goblin sharks were taken at 100–350 m and were abundant at 250 m. The adults are seldom taken, presumably because they inhabit much deeper and colder waters and do not ascend up into shallower waters as the juveniles do.

Almost nothing is known of its habits, which must be inferred from its morphological characteristics. Its relatively small eyes and light coloration are typical of sharks living in the lower mesopelagic zone (1,000–2,000 m) or deeper, where no light penetrates.[1] The poorly calcified, flexible body is also typical of fishes inhabiting very deep waters. The jaw is capable of incredible protrusion and can be thrust forward to its maximum extension and retracted with incredible speed (based on remarkable footage of a live goblin shark obtained by the Japanese television station NHK shown to me in April 2008). The protrusible mouth probably aids in picking up relatively small prey near the bottom, after detection at a very close range. The slender, pointed teeth are designed for grasping small, thin-skinned small fishes or squid. The long snout is densely covered by pores (the openings of canals leading to the electroreceptors known as ampullae of Lorenzini). It may serve to enlarge the area or number of receptors and may enhance electroreception by moving the receptors away from the body. Woodward (1942: 562) considered that the shovel-shaped snout was "evidently adapted for stirring up the mud in search of food." However, there is no evidence of such habit.

In its completely dark environment, the goblin shark must depend on its hearing, olfaction, and electroreceptors for locating prey. Obviously, these sensory inputs must be integrated in the brain into a picture of the proximal envi-

1. By contrast, sharks living in the upper mesopelagic zone, at depths of 500–1,000 m, where a small amount of light penetrates, have large eyes adapted for vision at very low light levels. Their dorsal surfaces and flanks are usually black or dark brown to avoid reflecting any light. Their undersides are often lighter or emit a weak bioluminescence to avoid being silhouetted against the weak surface light that might penetrate during daylight hours or during full moon.

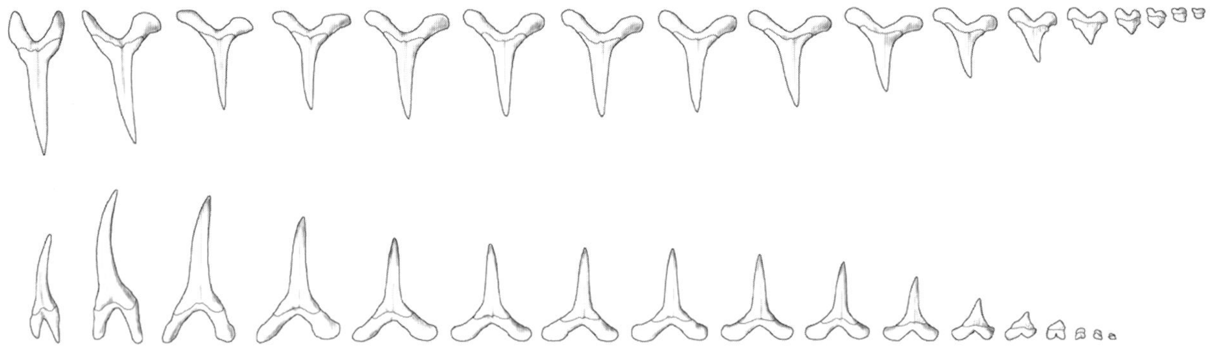

Fig. 53b. Teeth.

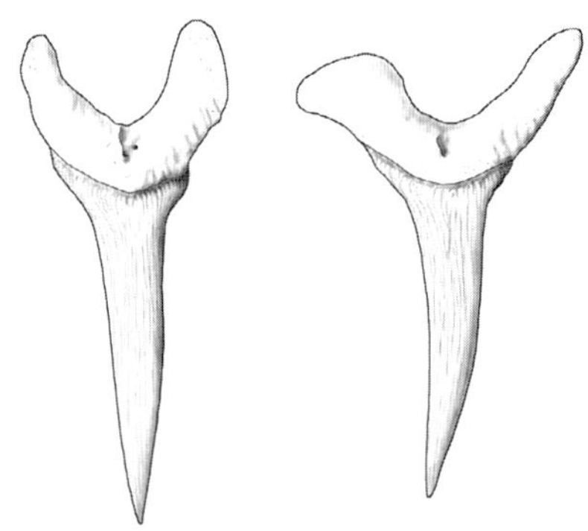

Fig. 53c. Teeth (detail).

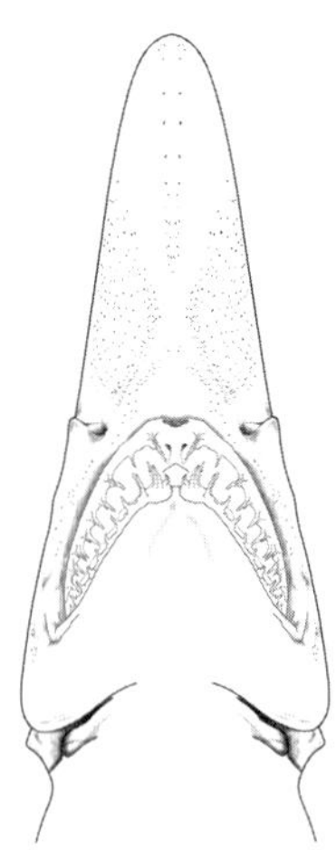

Fig. 53d. Snout.

Fig. 53e. Dermal denticles.

ronment that is sufficiently clear to locate and grasp small prey. Based on their analysis of brain development, Masai et al. (1973) concluded that the goblin shark is slow swimming and sensitive to vibration. These authors also reported that the optic tectum is atrophied (as would be expected of an animal that pursues prey by olfaction, vibration, etc., rather than by vision). Thus, the available evidence suggests that the goblin shark cruises slowly near the bottom at great depths, detecting its small prey by sound, olfaction, or electroreception.

Diet: Little is known about the diet of the goblin shark. Yano et al. (2003) examined 121 specimens from the Tokyo Submarine Canyon; prey items included bony fishes (Macrouridae and Stomiidae spp.), squid, and decapod crustaceans. The California specimen was said to be regurgitating squid when hauled aboard (Ugoretz and Seigel 1999).

Size at maturity Unknown. A 209-cm male and a 196-cm female were reported to be immature (Yano et al. 2003).

Reproduction Gravid females have not been reported, and there is no information on the goblin shark's mode of reproduction. The embryos are probably oophagous, like most lamniform sharks.

Size at birth: Unknown. For well over a century, the smallest known specimen was the 107-cm holotype (Jordan 1898). Recently, Yano et al. (2003) reported specimens of 81.7 and 92.8 cm, suggesting a birth size of 81–90 cm.

Brood size: Unknown.

Nurseries Unknown.

Age and growth The species has not been aged.

Relation to humans None.

FAMILY ODONTASPIDIDAE

The Sand Tigers

Sand tigers are large sharks of warm and temperate waters throughout the world. The four species in this family are characterized by a conical snout, eyes lacking a nictitating eyelid, five gill slits that do not extend unto the upper surface of the head and are located anterior to the origin of the pectoral fins, two large, spineless dorsal fins, and a well-marked upper precaudal pit. The teeth are long, slender, awl-like, and smooth-edged and have small lateral cusplets. The family includes the genera *Carcharias* and *Odontaspis.* The genus *Carcharias* is characterized by a two dorsal fins of about equal size, the first dorsal fin origin closer to the pelvic fins than to the pectoral axil, and three rows of large upper anterior teeth on each side of the symphysis. The genus *Odontaspis* is characterized by a first dorsal fin closer to the pectoral fin axil than to the pelvic fin origin and two rows of large, anterior upper teeth on each side of the symphysis. Three species are found in North American waters.

KEY TO THE ODONTASPIDIDAE

1a. First dorsal fin origin over pectoral fin rear margin or near free rear tip, first dorsal fin much larger than second dorsal, two rows of large anterior teeth in each side of the symphysis followed by much smaller intermediate teeth)........................ 2

1b. First dorsal fin origin well behind pectoral fin free rear tip (its midpoint closer to the pelvic fin origin than to pectoral axil), second dorsal fin almost as large as first dorsal fin, three rows of large anterior teeth on each side of the symphysis followed by a much smaller tooth.. **Sand tiger,** *Carcharias taurus* (p. 208).

2a. Teeth usually with two or three lateral cusplets on each side of the cusp, three or four rows of small teeth following the large anterior teeth ....**Ragged-tooth shark,** *Odontaspis ferox* (p. 215)

2b. Teeth with only one lateral cusplet on each side of the cusp, one small third tooth following the large anterior teeth...... .... **Bigeye sand tiger,** *Odontaspis noronhai* (p. 218).

SAND TIGER

Carcharias taurus Rafinesque, 1810

Fig. 54a. *Carcharias taurus*, sand tiger shark, mature male, 229.8 cm, 69.5 kg; from off False Cape, Virginia (John A. Musick, collector).

Common name American anglers and ichthyologists have known the sand tiger for many years under different names. These include ash-colored shark (Le Sueur 1818), ground shark (De Kay 1842), and sandbar shark (Smith, H. M. 1907). More recent and frequently used names are sand shark (e.g., Jordan and Evermann 1896; Radcliffe 1916; Breder 1929; Young and Mazet 1934; Bigelow and Schroeder 1948, 1953) and sand tiger (Springer 1938; Bigelow and Schroeder 1948). "Sand shark" likely derives from its frequent capture by surf anglers on the sandy outer beaches of Cape Cod. According to Bigelow and Schroeder (1953), sand tigers were so abundant off Nantucket that fishing for them was a popular sport. Springer (1938) appears to have first used the name "sand tiger," a reference to its supposed voraciousness; Springer noted that two adult females caught off Englewood, Florida, had enormously distended stomachs, each containing about 50 kg of fish. In Australia it is known as the gray nurse shark.

Spanish name The A.F.S. name is *tiburón arenero tigre,* a literal translation of the English name but so cumbersome that no Spanish speaker is likely to use it.

Scientific name *Carcharias taurus* Rafinesque, 1810, in *Caratteri di alcuni nuovi generi di animali e piante della Sicilia:* 10, plate 14, fig. 1. *Carcharias:* Greek, a rough shark; *taurus:* Latin, bull.

Synonyms Over the years the scientific name of the sand tiger has changed often, almost whimsically. Some earlier names include *Squalus americanus* Mitchill, 1815; *Squalus littoralis* Le Sueur, 1818; *Odontaspis taurus* Müller and Henle, 1839; *Carcharias littoralis* De Kay, 1842; *Eugomphodus griseus* Gill, 1861; *Carcharias owstoni* Garman, 1913; and *Eugomphodus taurus* Compagno, 1984a. These are only a few of the synonyms in the literature. In 1987 the International Commission of Zoological Nomenclature stabilized the confusion, restoring the name *Carcharias taurus.* Readers interested in the details of the nomenclature changes can consult the synonymies in Bigelow and Schroeder (1948) and Compagno (1984a).

Identification The sand tiger is characterized by a dorsally flattened head with a *bluntly conical snout;* five *gill openings anterior to the pectoral fin origin;* a first dorsal fin placed far back on the trunk, its origin closer to the pelvic fin origin than to pectoral fin origin; and *a second dorsal fin nearly as large as the first.* The teeth have long, *narrow, smooth-edged, sharp-pointed cusps with a small lateral cusplet on each side of the main cusp.* The first three teeth on each side of the upper jaw are large and erect, the fourth is markedly smaller, about one-third to one-half the size of the front teeth, and there is a noticeable gap between the fourth and fifth teeth. The teeth are similar in both jaws. The teeth number U: 43 to 48; L: 41 to 48; those at the corners of the jaw are minute and numerous. The dermal denticles are widely separated and have a strong central ridge and two winglike extensions on the sides. Coloration is light greenish gray or light yellowish brown above and grayish white below. *Juveniles are pale brown or tan with irregular darker brownish spots over the upper body surfaces.* The spots fade with age and disappear by the time the animals reach 180–200 cm. The fins of small juveniles have a black border 2–3 mm wide.

Similar species The ragged-tooth shark has teeth with two or more lateral cusplets on each side of the main cusp; its first upper tooth is much smaller than the second, and there is no gap between the fourth and fifth upper teeth. The lemon shark is similar in coloration and also has two equally large dorsal fins, but its snout is broadly rounded instead of

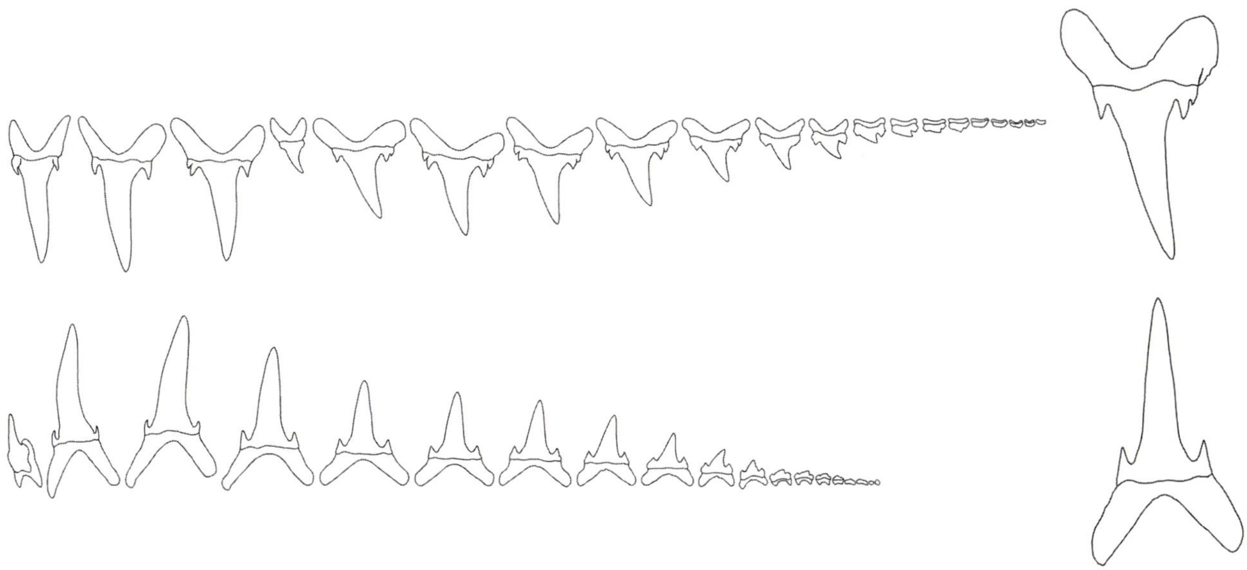

Fig. 54b. Teeth.

bluntly conical, and its pectoral fins originate at about the level of the third gill slit.

Range The sand tiger has been reported from many localities around the world: the western North Atlantic (Bigelow and Schroeder 1948), South Atlantic (Argentina, Menni 1986), Bermuda (Bigelow and Schroeder 1948; Smith-Vaniz et al. 1999), eastern Atlantic and the Mediterranean (Cadenat and Blache 1981), western Indian Ocean (South Africa; Bass et al. 1975b), and Australia (Whitley 1940; Last and Stevens 1994). Although different names have been given to local forms, it is now considered to be one species of wide distribution.

In eastern North America, the sand tiger has been reported from Passamaquoddy Bay, New Brunswick, to Salerno, Florida, and from the northern Gulf of Mexico. The northernmost record is a specimen caught in a weir in Passamaquoddy Bay near St. Andrews, New Brunswick, in 1913 (Huntsman 1922). The record of a 79-cm neonate male from the Minas Basin of the Bay of Fundy, reported by

Fig. 54c. Snout.

Fig. 54d. Dermal denticles.

Scott and Scott (1988), is obviously an error, because sand tigers are born at a much larger size and neonates are found only in the southern parts of their range. There are few valid records north of Cape Ann, Massachusetts, although the species has been reported twice from Casco Bay, Maine (Bigelow and Schroeder 1953; Moulton 1954), where it is apparently rare. South of Cape Ann, however, the sand tiger becomes common. Bigelow and Schroeder (1948) stated that next to the spiny dogfish, the sand tiger was the most abundant shark in season from Cape Cod to Delaware Bay and that it was more plentiful in this region than anywhere else in the western North Atlantic. The species is also abundant from Cape Hatteras south to Jupiter Inlet, Florida. It becomes rare south of Salerno, Florida, where I have seen a few in September. Wright (1981; see p. 184) saw none in the Florida Keys.

The sand tiger is less abundant in the Gulf of Mexico than along the Atlantic states. Clark and von Schmidt (1965) surveyed the sharks of the west-central Florida coast from January 1955 to April 1964, catching 762 sharks of which only one was a sand tiger. Springer (1960) reported seven sand tigers in a catch of 502 large sharks off the mouth of the Mississippi River from 25 June to 29 July 1947. V. G. Springer and Woodburn (1960) mentioned two specimens caught off Tampa and Clearwater, Florida. Baughman and Springer (1950) did not include the species in their survey of Texas sharks. However, Hoese and Moore (1977) called it "not uncommon" in the Gulf of Mexico.

Size and weight The sand tiger reaches about 300 cm and more than 230 kg. The largest sand tiger on record appears to be that mentioned by Clark and von Schmidt (1965), a 296-cm, 233.6-kg female caught near Gasparilla Pass, Florida, in February 1960 (I have examined the jaws of this specimen, and they leave no doubt about its large size). There is an International Game Fish Association (2006) record of a specimen 227 cm, 158.8 kg caught in 1993 off the jetties in Charleston, South Carolina. Representative weights for specimens I examined in South Carolina are male, 239 cm, 66 kg; female, 229 cm, 85.4 kg; 231 cm, 82.7 kg; 246 cm, 98 kg; 262 cm, 128 kg; and 264 cm, 118.5 kg.

Biology The sand tiger is a shallow-water shark often found close to the bottom. It has the unusual ability of remaining motionless on the bottom for extended periods. This ability has been attributed to neutral buoyancy achieved by swallowing air. This mechanism has been known for quite some time. McCormick et al. (1963: 222), quoting unnamed sources, mentioned that the sand tiger was "said to have developed" a substitute for a swim bladder by swallowing air and keeping an "air pocket" in its stomach but did not mention the source of the observations. It is likely that the source of those observations were Springer and coauthors (in Baldridge 1970), who suggested that the sand tiger gulped air for buoyancy control. Bass and Ballard (1972) stated that captive sand tigers had notably good buoyancy control and could often be seen resting nearly motionless 1–2 m above the bottom of the aquarium. These authors also described how a 210-cm female occasionally would swim at the surface with her head raised and the body axis at about 30° to the horizontal. In this position, her snout was out of the water with the mouth level with the surface. When submerging after this action, bubbles could be seen escaping from her mouth and gills. Hussain (1989) calculated that the weight of a sand tiger in water increased by 3 kg after air in the stomach had been removed.

Diet: Sand tigers are voracious fish eaters, feeding on whatever small or medium-size fishes are locally available. In reference to the sand tiger's prey, Bigelow and Schroeder (1948: 104) stated: "No doubt a complete list for any given locality would include practically all the local species that were not too large." Reported stomach contents include alewives (*Alosa*), black drum (*Pogonias*), bluefish (*Pomatomus*), bonito (*Sarda*), butterfish (*Poronotus*), catfish (*Bagre*), cunner (*Tautogolabrus*), eels (*Anguilla*), menhaden (*Brevoortia*), mullet (*Mugil*), shark suckers (*Echeneis*), seabass (*Centropristis*), silver hake (*Merluccius*), spotted seatrout (*Cynoscion*), tautog (*Tautoga*), spadefish (*Chaetodipterus*), and kingfish (*Menticirrhus*) (Springer 1938; Bigelow and Schroeder 1948; Clark and von Schmidt 1965). With teeth clearly adapted for grasping relatively small prey, there is no reason to suspect that this species is capable of attacking large prey. However, a credible witness has reported a sand tiger attacking a loggerhead turtle and biting off a flipper (Ivan Schultz, pers. comm., 30 January 2002).

Migrations: Sand tiger migrations are complex: some are annual or seasonal and tuned to water temperature, while other movements are determined by the stage of maturity. The young are born in the southern parts of the range and quickly migrate to the northern range. Neonates and small juveniles of 91–130 cm are often caught off South Carolina from February to May (the latest I have seen a small juvenile off South Carolina was 8 May 1987, when a 107.8-cm specimen was caught off McClellanville by Steve Poston.) I never observed large juveniles of 140–180 cm in South Carolina. These larger juveniles are found in the northern part of the range, from Delaware Bay to Cape Cod. According to Bigelow and Schroeder (1948), the sand tiger is most abundant from Cape Cod to New York, from June to early October. Sand tigers withdraw from the New York area in autumn, when the water temperature falls below 19–20°C, and depart the coasts of southern New England and New Jersey by November. Where the juveniles winter is unknown, but they probably move southward and offshore.

After reaching maturity, sand tigers migrate south of Cape Hatteras. Large adults of more than 220 cm are occasionally reported from scattered locations along the coasts of Connecticut, New York, and New Jersey (Bigelow and Schroeder 1948) but are common only from Cape Hatteras to Florida. I have observed adults off South Carolina in all months. They inhabit shallow water 2–3 m deep from May to October, moving to depths of 30 m and deeper in late fall and winter. Dodrill (1977), in his survey of the nearshore sharks of Melbourne Beach, Florida, caught only seven large females (236.6–263.2 cm). These sharks were captured 200–450 m from shore between 5 June and 5 August. According to Gilmore (1993) gravid females are seen only from south of Cape Hatteras to Jupiter Inlet. This agrees with my own observations; I have encountered adult, gravid females only from South Carolina to Florida.

Size at maturity

Male: Sand tiger males off the eastern coast of North America mature at about 190–195 cm (Gilmore et al. 1983; Branstetter and Musick 1994).

Female: According to Branstetter and Musick (1994), females mature at lengths greater than 220 cm. The largest immature female I observed off South Carolina was 225 cm, and the smallest gravid female was 229 cm, suggesting that maturity is reached at 225–229 cm.

Reproduction The sand tiger is an aplacental viviparous species. As with most sharks, the embryos are initially nourished by yolk stored in the egg or yolk sac. Sand tiger eggs are 7–10 mm in diameter and have only enough yolk to nourish the embryo through the first few weeks of development. Embryos hatch from their egg cases when they are 40–50 mm long. By the time they reach 100 mm, the yolk has been consumed, and the tiny yolk sac, less than 10 mm in diameter at the onset of development, has disappeared. Thereafter, the embryos eat ovulated eggs in the uterus. In the first six months of gestation, the embryos consumed large quantities of eggs in excess of their metabolic requirements and store much yolk in their highly distended yolk stomachs. By late term the ovary has released most of its oocytes to feed the embryos, and by the time of birth the ovary is small and flaccid. This mode of embryonic nutrition is known as oophagy, and it has been known in sharks since the early 1900s. Springer (1948) was the first to document the process in sand tigers. He called the egg eating embryos "oviphagous."[1]

In the southeastern United States, sand tiger females are synchronous (Gilmore et al. 1983; Gilmore 1993). In the early stages of pregnancy as many as seven embryos of different sizes can be found in each uterus (Gilmore 1983; my personal observations). Embryos develop a precocious dentition by the time they reach 50 mm. When the largest embryo reaches 80–100 mm, armed with its precocious dentition, it seeks out and kills the smaller embryos. Later, it consumes its dead siblings along with the contents of other egg cases. Bass et al. (1975b) first documented this embryonic cannibalism upon recovering an intact 40-mm embryo from the stomach of an embryo of 170 mm. Later, Wourms (1981) dubbed this phenomenon "adelphophagy" (literally "brother-eating"). However, I prefer the clearer and more colorful "embryonic cannibalism." Although the embryos of many species of sharks are known to be oophagous, embryonic cannibalism has been documented only in the sand tiger.[2] Only one embryo survives to full term in each uterus, and the brood consists invariably of two young.

Mating: Sand tigers are known to form large, mixed-sex aggregations at specific sites, presumably for mating. Such sites have been reported off North Carolina and Florida (Coles 1915; Gilmore 1993). According to Gilmore (1993), mating occurs at depths of less than 30 m off the east coast of Florida (from 27 to 29° N) from late February to April. I have seen numerous ripe males off South Carolina during May and early June, indicating that mating also occurs there.

Although the testes of the sand tiger are not larger than those of other sharks, they produce prodigious amounts of sperm, which is stored in their sperm sacs and epididymides. Sperm is exuded copiously whenever a ripe sand tiger male is removed from the water and is placed on deck. The unusually large quantities of sperm produced, the fact that sand tigers congregate in large numbers for mating, and the possibility of multiple matings by females are indirect evidence that the species might engage in sperm competition. This hypothesis remains to be investigated.

In May 2004, Cheryl Nicholson showed me a film of sand tigers mating at the Blue Planet Aquarium, Ellesmere Port, England. At the outset, the male followed the much larger female with his snout close to her cloaca. The male bit the female's flank and moved its bite anteriorly to the pectoral fin. Seizing the pectoral fin, the male then flipped the female upside down and mated with her. The female remained on her back and motionless for some 20 minutes after mating before resuming swimming.

Development: Along the coast of the southeastern United States, females begin ovulating in June. Gilmore et al. (1983), working off the east coast of Florida in May, found no embryos in three females with numerous egg cases in their uteri.

1. Springer incorrectly joined a Latin root with a Greek verb. The correct terms are oophagy and oophagous.

2. Note added in proof: A type of embryonic cannibalism has been documented in the shortfin mako (see p. 272).

These investigators reported finding five females carrying embryos of 27–42 mm in June and a single female with 13–18-mm embryos on 8 July. I examined several gravid females in the Georgetown–Charleston area of South Carolina. A 240-cm female caught off Charleston on 11 June 1992 had just begun to ovulate; her ovary weighed 4.42 kg. This shark carried seven wind eggs (cases without eggs) and five cases with eggs in one uterus, and seven wind eggs and six cases with eggs in the other uterus. At the beginning of ovulation, sand tiger egg cases contain a single fertilized egg about 10 mm in diameter. Subsequent egg cases contain 7–15 smaller eggs per case. It appears that these eggs in these smaller egg cases (destined to feed the embryos) may be fertilized, because Gilmore et al. (1983) reported blastodiscs in several eggs in a case. A female I examined in mid-July 1979 contained 35 egg capsules in each oviduct. Two capsules in each oviduct contained a single embryo 30 mm long, but no embryos were visible in the other capsules.

In late June and early July, sand tiger females can carry as many as seven embryos of different sizes in each uterus. I examined three females caught off Charleston in mid-July, all of which contained small embryos of less than 10 cm, and most had emerged from the egg cases. The first female measured 231 cm and contained embryos of 55, 65, and 85 mm and 43 egg cases in one uterus. The 55-mm embryo had been damaged and bore numerous puncture wounds along the body. Her other uterus contained embryos 40, 42, 55, 58, and 102 mm and 46 egg cases. The 55-mm embryo was just emerging from the egg case but already had been damaged, presumably by its larger siblings. The second female measured 229 cm. Her left uterus contained seven embryos (20, 30, 40 45, 50, 65, and 95 mm in length) and 57 egg cases. The right uterus contained four embryos (50, 60, 60, and 85 mm in length) and an undetermined number of egg cases. A sample of five of the egg cases revealed 7, 7, 9, 11, and 15 eggs per case. The third female, which measured 264 cm long, carried a 155-mm embryo, eight empty egg cases, and 61 full egg cases in the left uterus. The yolk sac of this embryo had already been absorbed. The right uterus contained three embryos (150, 35, and 35 mm in length) and 67 full egg cases. One of the 35-mm embryos was still inside its egg case. By late July only one embryo survives in each uterus, having killed its smaller siblings.

The embryos grow rapidly, based on females I examined off South Carolina. A female captured off Cape Romain, South Carolina, on 20 July contained two male embryos, one in each uterus; they measured 29.0 and 28.7 cm. Another female taken at the same locality on 11 August carried embryos of 22.1 and 24.5 cm. A female caught off Charleston on 7 October contained two males, 75.6 cm and 4.6 kg, and 73.7 cm and 4.4 kg. Another female I examined on 7 November contained two embryos 82.2 and 80.2 cm. These embryos had enormously distended bellies. A 273-cm female examined by Springer (1948) on November 24 carried two female embryos 96 and 97 cm long, one in each uterus. A second female seen by Springer on that date carried a single 83-cm male embryo and appeared to have aborted a second embryo.

Birth season: Few observations have been made of full-term sand tiger females in nature, because these animals occur during the late winter when there is little fishing activity. Although several females have given birth in American aquariums, almost nothing has been reported on these events. Consequently, the length of gestation and time of birth are not known with any precision. Available evidence indicates that gestation is about 11 months, with parturition occurring from December to early April.

Springer (1948) reported catching gravid females off Salerno, Florida, on 22 January. He also wrote that a 272-cm female, captured 10 April 1946, had died at Marine Studios (later known as Marineland), in St. Augustine, Florida, in March 1947 and that it contained two full-term young, one in each oviduct. Gilmore et al. (1983) reported that a female captured in November 1958 was observed to give birth to two young in February 1959 in an aquarium at Marineland. These investigators also reported that another birth had occurred on 30 December. The embryos are born out of the mother head first (Springer 1948; Gilmore et al. 1983), in contrast to carcharhinid embryos, which emerge tail first.

Reproductive cycle: The length of the reproductive cycle of the sand tiger is a subject of debate. Branstetter and Musick (1994) suggested that the reproductive cycle is biennial but provided no evidence of the capture of mature non-gravid females taken during the breeding period of the species. It is possible that females referred to as "in a resting stage with small ovarian egg follicles" (Branstetter and Musick 1994: 248) were immature, given that the oocytes in lamnoid sharks do not differ greatly in size in ovulatory and "resting" females. Lucifora et al. (2002) also suggested a biennial cycle for the sand tiger off Argentina but provided no convincing evidence (the authors had no direct evidence of mating and encountered no gravid females). Recent research by Australian workers also indicates a biennial cycle (Mike Bennett, pers. comm., July 2008).

R. Grant Gilmore, who examined a good number of sand tiger females for his 1983 paper, argues that sand tigers have an annual reproductive cycle, based on the fact that all adult females sampled from June to February (the time of gestation for the species, based on our observations), or that he has obtained records for, were pregnant in summer (pers. comm., October 2001). If sand tigers had a biennial cycle, it would be necessary for all the "resting" females to leave the southeastern United States waters after parturition and spend their off year elsewhere, just to reappear the following year at mating time. I concur with the observation of all the adult females being gravid from July to January—all the adult females that I have seen during those months were gravid. However, the flaccid condition of the ovary of term females makes it unlikely that a female could give birth in February or March and then sequester enough nutrients to start vitellogenesis and be ready to ovulate, conceive, and nourish its embryos by June, only three months later. Thus, based on the condition of the ovary at term, and on recent Australian evidence for the species, I suggest that sand tigers in North America are biennial and that the population of nongravid females is probably highly localized and has not been located. However, until such population is located during the months of July to January, the biennial reproductive cycle of the sand tiger cannot be considered demonstrated and further research is needed.

Size at birth: Young are born at 92–106 cm. The 272-cm female reported by Springer (1948) carried two young, one embryo 105 cm and 6.14 kg, and another 103 cm and 6.36 kg. One of the young born on 15 February 1959 at Marineland measured 106 cm and weighed 6.2 kg at birth (Gilmore et al. 1983). I examined a free-swimming newborn 92 cm and 4.5 kg, caught off South Carolina on 28 March.

Brood size: The brood consists of two young, one per uterus, the largest embryo in each uterus having consumed its competing siblings.

Nurseries Sand tigers are born at the southern parts of the range, off the Carolinas, primarily in February and March. Given the brief period (February–May) when neonates and small juveniles are present in South Carolina waters, it is evident that the young soon migrate northward to the nurseries of Delaware Bay, Chesapeake Bay, and southern New England. Why gravid females are found only in the southern parts of the range, and why the young migrate such long distances to their northern nurseries, is not fully understood. One can speculate that competition with the numerous species of carcharhinid sharks that use the Carolinas for nurseries forces sand tiger young to look for less crowded habitats.

Age and growth Sand tigers are long-lived sharks, although the scanty literature does not reflect the longevity of the species. Clark (1963a) reported that sand tigers had lived six years at the Fairmont Park Aquarium in Philadelphia, and Hinton (1962) recorded that one lived nine years at the New York Aquarium. The oldest fish in Branstetter and Musick's (1994) sample of 55 sharks was estimated to be 10.5 years old. According to Hans Walters of the New York Aquarium, their oldest sand tiger, a female affectionately called "Bertha," is 39–41 years old, having come to the aquarium as a yearling (H. Walters, pers. comm., 16 March 2004). There is anecdotal evidence that other sand tigers have survived in public aquariums in the United States for decades, but unfortunately, few institutions maintain or publish longevity records.

Relation to humans There were commercial fisheries for the sand tiger around Nantucket, Massachusetts, during the first quarter of the twentieth century, but they were short-lived, reportedly because the local stock became exhausted (Bigelow and Schroeder 1953). A catch of 1,900 sharks, primarily sand tigers, by three boats on Horseshoe Shoal in Nantucket Sound from June to September 1918 explains the reason for the demise of the fishery.

Sand tigers have been of little economic importance in recent times. During the 1980s and 1990s, sand tigers were generally an incidental catch of little economic importance in the shark fishery of the southeastern United States. Nonetheless, the species is extremely vulnerable to overfishing because large numbers of individuals congregate in coastal areas during the mating season. These aggregations attract fishermen because they can catch large numbers of these sharks with minimal effort, albeit with devastating effect on the local population in a short time.

In the early 1990s the populations of sand tigers off Florida, North Carolina, and the Chesapeake Bay region declined severely. According to R. Grant Gilmore (pers. comm., 15 December 1995), sand tigers practically disappeared from North Carolina and Florida waters because of commercial fishing on mating aggregations. In the Chesapeake Bay region, Musick et al. (1993) documented a decrease in catch per unit effort from an average of 1 shark per 100 hooks in 1974–1979 to 0.2 sharks per 100 hooks in 1991. In April 1997, in response to population declines and public concerns, the National Marine Fisheries Service (1997) prohibited all directed commercial and recreational fishing for sand tigers along the Atlantic coast.

The sand tiger has also become rare in Australia, where it is known as the gray nurse shark. It was one of the first sharks to receive fully protected status anywhere in the world, achieving that status in New South Wales in 1984 (Pollard et al. 1996). Severe population declines started in Australian waters during the 1960s and 1970s. Part of the decline was attributed to divers who considered it fashionable to kill sand tigers with powerheads (explosive devices attached to spearguns), even though these sharks are so sluggish that there is little sport in killing them. According to Stevens (1993), the numbers of gray nurse sharks caught in New South Wales by beach nets decreased from an average of 30 per year in 1951–1954, to one or two annually in 1988–1990, to none in 1993.

The sand tiger has been a favorite large shark of the aquarium trade since the 1930s, because it survives well in captivity, unlike most other sharks. Its toothy, ferocious countenance, size, and hardiness make it ideal for public display. Specimens have survived several decades in captivity (see "Age and growth," above). Wild-caught females have given birth in captivity numerous times, and their offspring have been raised to maturity (Henningsen et al. 2004). Small numbers of sand tigers are caught every year under special permits for aquarium displays (in the aquarium trade a large specimen can fetch up to $4,000). These small numbers of removals are unlikely to affect population sizes.

From Cape Cod to Georgia, sand tigers are abundant along shallow beaches where people swim or fish. These sharks seldom bother bathers, although they can be a nuisance or thrill to surf fishermen, depending on their sporting outlook. Nevertheless, in the past, a number of attacks on bathers have been attributed to sand tigers in North America, although few, if any, have been fatal.

RAGGED-TOOTH SHARK

Odontaspis ferox (Risso, 1810)

Fig. 55a. *Odontaspis ferox*, ragged-tooth shark, mature male, 313.5 cm, 190.5 kg; from Cat Cay, Bahamas (Ron Schatman, collector).

Common name Ragged-tooth shark, a reference to its multicuspid teeth. It is also called small-tooth sand tiger (Compagno 1984a).

Spanish name The A.F.S. name is *tiburón dientes de perro,* an awkward name.

Scientific name *Odontaspis ferox* (Risso, 1810), originally described as *Squalus ferox* in *Ichthyologie de Nice* 38. *Odontaspis:* from the Greek genitive *odontos,* tooth, and *aspis,* viper, an allusion to its slender curved awl-like teeth (i.e., having viperlike teeth); *ferox:* Latin fierce, an allusion to the toothy and ferocious countenance.

Synonyms *Odontaspis herbsti* Whitley, 1950 [a].

Identification The ragged-tooth shark is usually a heavy-bodied shark with a conical, slightly flattened snout and the *fifth gill slit set anterior to the origin of the pectoral fins.* Its first dorsal fin forms an almost perfect equilateral triangle, its origin closer to the pectoral fin origin than to the pelvic fin origin; *a large second dorsal fin,* about three-fourths the height of the first, and a well-marked precaudal pit. The teeth have long, narrow, *smooth-edged cusps with two or more lateral cusplets on each side.* The first tooth in the upper jaw is much smaller than the second and third. These large teeth are followed by four smaller teeth, with no gap between the fourth and fifth teeth. Teeth number U: 23 to 27–23 to 27, L: 18 to 24–18 to 24 and are similar in both jaws. The dermal denticles are overlapping, have three ridges, and terminate on a point. Ragged-tooth sharks vary in coloration. In the original description, Risso (1810) described a Mediterranean specimen as dark red with large irregular black spots on the back and sides and a reddish gray belly. Risso reported that the species grew to 300 cm and presumably would have been aware of any differences between juveniles and large specimens (Garrick 1974). Tortonese (1956), who illustrated a 360-cm specimen from the Gulf of Genoa, upheld Risso's color description.

Whitley (1950b) described *Odontaspis herbsti* from Australia, separating it from *O. ferox* on the basis of a slightly different dental formula, smaller pectoral fins, and coloration described as uniform gray above and lighter below. Daugherty (1964) reported the coloration of two California specimens as uniform light gray above and slightly lighter below, speculating that the dark markings could be a juvenile feature. Garrick (1974) accepted Whitley's specific name of *O. herbsti* for the gray form, retaining the *O. ferox* for the reddish form with black spots. Subsequent authors (Compagno 1984a; Last and Stevens 1994) have relegated *O. herbsti* to a synonym of *O. ferox.*

If the foregoing descriptions were not confusing enough, in March 1992 I examined a 314-cm male from the Bahamas that was light brown above with light tan or whitish undersides. This shark had numerous small white spots 3–10 mm diameter, with diffuse edges on the dorsal surface, including the first dorsal fin, the spots being most numerous along the dorsal midline. These spots were most noticeable when the shark was alive and slowly faded after death. They were still visible after 12 hours in transport. A second male from the Bahamas was light reddish brown and lacked any spots when I examined it about 24 hours after capture. Ron Schatman of Miami Beach, Florida, caught both of these specimens.

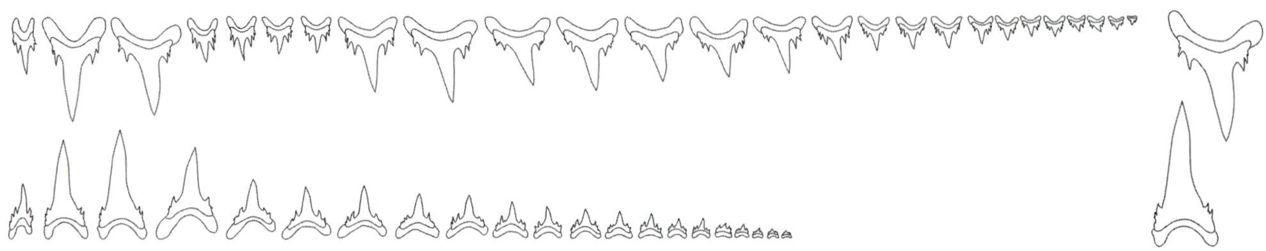

Fig. 55b. Teeth.

Similar species The sand tiger has teeth with one (or rarely two) lateral cusplets on each side of the main cusp and a very small, fourth upper tooth followed by a gap. The bigeye sand tiger's teeth have a single lateral cusplet on each side of the main cusp. In requiem sharks the fifth gill slit is located behind the origin of the pectoral fin.

Range The ragged-tooth shark is cosmopolitan in warm and temperate seas, but its distribution is poorly understood. It has been reported in the Mediterranean Sea (Risso 1810), North Atlantic (Bonfil 1995; Sheehan 1998), Pacific (New Zealand, Garrick 1974; Japan, Abe et al. 1968; New South Wales, Last and Stevens 1994; California, Daugherty 1964), and Indian Ocean (northwestern Australia, Last and Stevens 1994). The species appears to be common in Japanese waters. Abe et al. (1968) reported that large individuals are occasionally found in the Central Wholesale Market of Tokyo.

Several specimens have been captured off California. A 169-cm male was caught off San Onofre in January 1963 (Daugherty 1964). A 280-cm female was caught 1 km off Newport Beach in February 1985. A female 367 cm and 323 kg (eviscerated) was caught about 5 km off Santa Barbara Island in February 1979, and a male 274.6 cm was caught off San Clemente Island in November 1968 (Seigel and Compagno 1986). Farther south, in the Gulf of California, a 364-cm female was caught near Isla Tiburón in May 1994, and a jaw of this species was obtained from fishermen in Isla Cerralvo (Villavicencio-Garayzar 1996).

There are few reports of ragged-tooth sharks along the Atlantic coast of North America, although the species is more common than the few reports indicate. The species simply lives outside the reach of most fishing gear. Individuals are occasionally encountered by tilefish fishermen off the Carolinas but are seldom retrieved or reported. A 340-cm female was captured some 46 km south-southeast of Cape Hatteras, North Carolina, at 173 m where the temperature was 17.8°C (Sheehan 1998). A 366-cm female was caught at 130 m on the Yucatan Shelf, Mexico (Bonfil 1995).

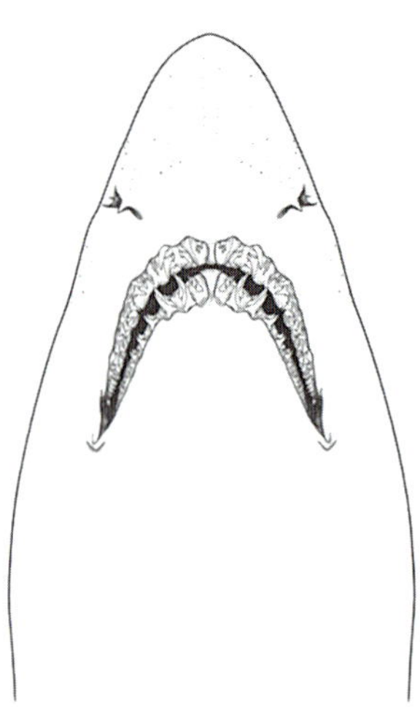

Fig. 55c. Snout.

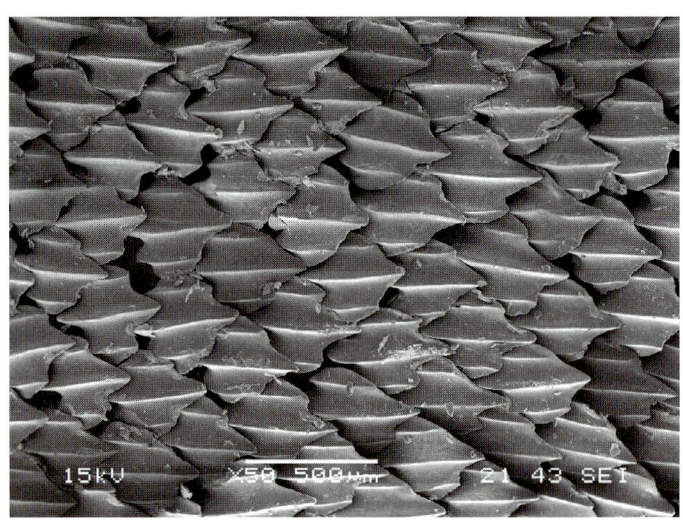

Fig. 55d. Dermal denticles.

I have examined two males caught by Ron Schatman in March 1992 and March 1993, both from the north end of South Cat Cay, Bahamas.

Size and weight Reported sizes and weights for males are 169.3 cm, 27.7 kg (Daugherty 1964), and 274.6 cm, 159 kg (Seigel and Compagno 1986); for females, 280 cm, 149 kg (Seigel and Compagno 1986), and 364 cm, 330 kg (Villavicencio-Garayzar 1996). I examined two adult males, one of 314 cm and 191 kg, and another of 309 cm and 165 kg.

Biology This is a deep-dwelling species found in warm temperate, subtropical, and tropical waters. Three Bahamian catches by Ron Schatman came from 365–655 m. The species has been filmed in shallower waters off Malpelo Island, Colombia, indicating that it enters shallow water at specific times or locations. Nothing else is known about its habits.

Diet: The stomach of a 314-cm adult male from Cat Cay that I examined contained one partly digested snapper, about 50 partly digested shark vertebrae, quantities of globular oily material, and a great quantity of clear oil.

Size at maturity Unknown.

Reproduction Gravid females have not been reported. Ragged-tooth shark embryos are presumably oophagous like those of other lamnoid sharks.

Size at birth: Unknown.

Brood size: Unknown.

Age and growth Unknown.

Relation to humans The ragged-tooth shark is too rare in North America to be of economic importance. In Japan, its oil is used, but its flesh is considered of poor quality (Abe et al. 1968).

BIGEYE SAND TIGER

Odontaspis noronhai (Maul, 1955)

Fig. 56a. *Odontaspis noronhai*, bigeye sand tiger shark, immature male, 231.5 cm, 49 kg; from the equatorial Atlantic (Tony Geisman, collector; courtesy of Lisa Natanson and Nancy Kohler).

Common name Bigeye sand tiger, a reference to its eye, which is larger than those of the other species of sand tigers.

Spanish name The species lacks a vernacular name in the Spanish Caribbean because of its rarity. The A.F.S. has not designated a Spanish name.

Scientific name *Odontaspis noronhai* (Maul, 1955), originally described as *Carcharias noronhai* in *Notulae Naturae* (Academy of Natural Sciences of Philadelphia) 279: 3, figs. 1–4. *Odontaspis:* from the Greek genitive *odontos,* tooth, and *aspis,* viper, an allusion to its slender curved awl-like teeth (i.e., having viperlike teeth); *noronhai:* honors Adolfo Cesar de Noronha, a director of the Funchal Museum in Madeira.

Synonyms None.

Identification The bigeye sand tiger has a *conical snout,* the *fifth gill slit anterior to the origin of the pectoral fin,* a *first dorsal fin with its origin closer to pectoral fin origin than to the pelvic fins,* and a *well marked upper precaudal pit.* The teeth have *one long slender central cusp with one small lateral cusplet on each side.* The teeth are similar in both jaws. The upper teeth consist of two larger frontal teeth followed by a much smaller third tooth. There is *a large gap between the third and fourth upper teeth,* and then seven intermediate teeth follow. These are followed in turn by eight smaller side teeth. In the lower jaw, there is one small tooth in the symphysis, followed by eight progressively smaller teeth. These are followed by some 12 smaller side teeth at the corners of the mouth. The dermal denticles are overlapping, have three ridges, and terminate on a point. Fresh specimens are dark brown above and below. The color changes to blackish brown or black in long dead and frozen specimens.

Similar species The ragged-tooth shark has teeth with two lateral cusplets on each side of the main cusp, three large frontal teeth followed by four small ones, and no gap between the third and fourth teeth. In the sand tiger, the origin of the first dorsal fin is closer to the origin of the pelvic fin than to the pectoral axil.

Range The bigeye sand tiger is probably cosmopolitan in deep tropical and subtropical waters, based on the few available records. The original locality was off Madeira (Maul 1955). It has also been reported from off Brazil (Sadowsky et al. 1984), in the Gulf of Mexico (Branstetter and McEachran 1986a), and off eastern Florida (Kerstetter and Taylor 2008). Compagno (1984a) indicates its possible capture in the Seychelles. I have a record of a specimen from the equatorial Atlantic.

Size and weight The 238-cm specimen from the central Atlantic specimen that I examined weighed 49 kg after many months in the freezer. Branstetter and McEachran (1986a) mentioned that a 321-cm female was immature. Sadowsky et al. (1984) examined the head of a specimen that he estimated at 360 cm. In his collection of shark jaws, Gordon Hubbell has several specimens said to come from Taiwan. These jaws indicate that this is a very large species, perhaps reaching 4 m in length.

Biology The bigeye sand tiger is very rarely caught, and almost nothing is known of its habits. Most of the catches come from tuna/swordfish longlines. However, the rarity of the catches suggests that it normally lives in deeper waters than those where gear for tunas and swordfish are set. The original description (Maul 1955) was based on a mounted specimen at the Funchal Museum in Madeira. That specimen, a 171-cm female from Câmara de Lobos, Madeira,

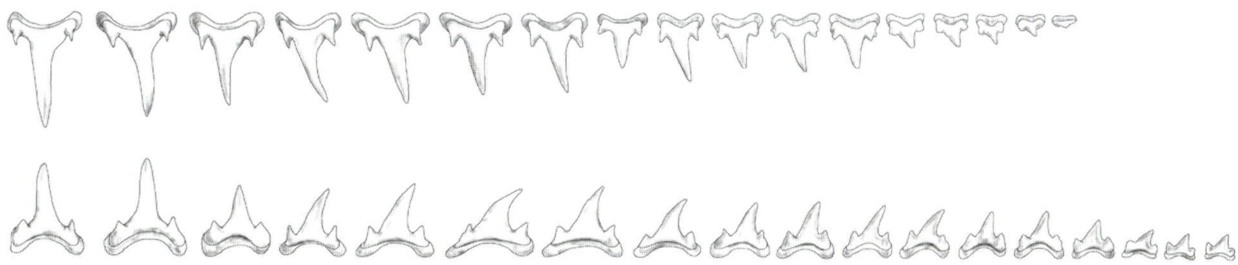

Fig. 56b. Teeth.

was caught on tackle set for black scabbard fish (*Aphanopus carbo*) at 600–1,600 m, in April 1941. Most likely, the capture occurred at 800–1,000 m (Maul 1955). The species was not reported again until 1984, when Sadowsky et al. (1984) recounted that a longliner based in Santos, Brazil, had caught eight bigeye sand tigers in early October 1981 while fishing off São Sebastião Island (24°30 S, 44°10 W). The fishing gear was baited with squid and fished at 60–120 m. These sharks were processed for commercial use, and only a head was saved. The head was given to Sadowsky, who estimated the specimen at 360 cm. In late October, the same vessel caught another specimen at 24°40 S, 44°16 W, while fishing at a depth of 600–700 m. This shark, an adult male 342 cm, was preserved at the Instituto de Pesca in Santos, Brazil.

The first reported bigeye sand tiger from North America was a 321-cm immature female captured in March 1984 in the Gulf of Mexico, some 126 km east of Port Isabel, Texas (ca. 26°30 N, 96°15 W). The specimen was deposited in the Texas Cooperative Wildlife Collections at Texas A&M University (Branstetter and McEachran 1986a). Another specimen was caught and released off eastern Florida (Kerstetter and Taylor 2008).

I examined a 238-cm male caught by Capt. Tony Geisman in the central Atlantic (at ~5°00 N, 35°00 W) in May 1997. The depth in that area was 2,700–3,660 m, and it is likely that the shark was caught at a depth of about 90 m. That specimen was used as the model for the illustration in this book.

Diet: The bigeye sand tiger has slender, awl-like teeth that are relatively small for the size of the shark. Obviously, their function is to grasp small, soft-skinned prey such as squids. The Gulf of Mexico specimen reported by Branstetter and McEachran (1986a) contained squid beaks and otoliths.

Size at maturity Unknown.

Reproduction We know very little about the reproductive processes. According to Branstetter and McEachran (1986a),

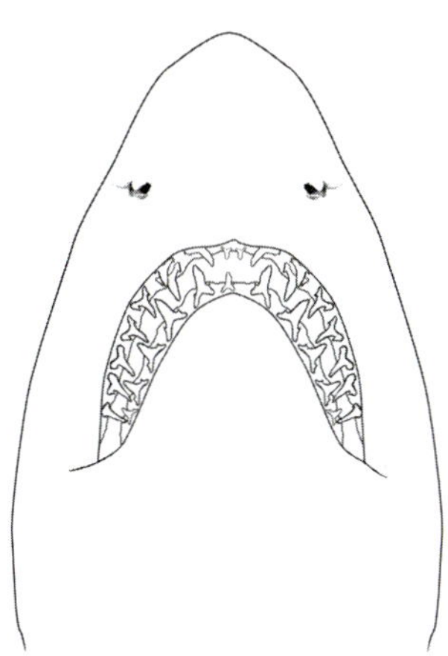

Fig. 56c. Snout.

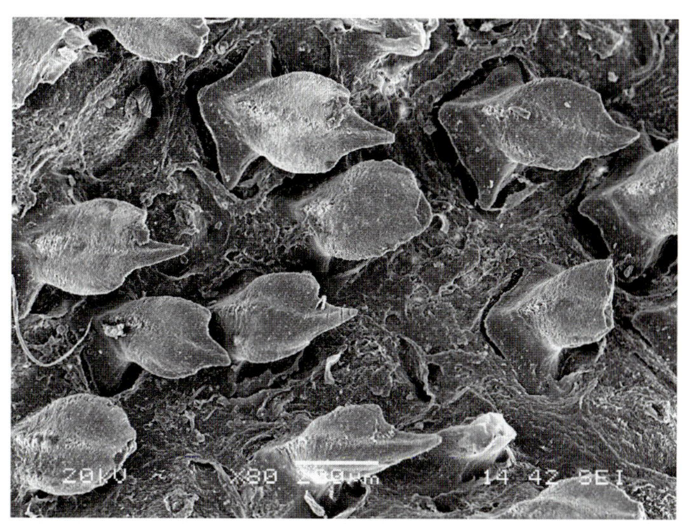

Fig. 56d. Dermal denticles.

their 321-cm immature female had a poorly developed reproductive tract. The functional right ovary was approximately 5 × 10 cm, and no follicles were discerned. The oviducts were thin and tubular, 1 cm in diameter. The uteri were short (<50 cm) and 3 cm in diameter. A hymen covered the distal end of the uteri.

Size at birth: Unknown.

Brood size: Unknown.

Nurseries Unknown.

Age and growth Unknown.

Relation to humans The bigeye sand tiger has no economic importance. Interestingly, the species was placed on the protected list in the United States in 1997, although there have been only two recorded catches of the species in the U.S. waters, and fewer than 18 worldwide. The concerns about the species are difficult to understand because the bigeye sand tiger's deep-water habitat or gear-avoiding habits protect it more effectively than any legislation could. Perhaps those drafting the protective legislation assumed that the bigeye sand tiger was reproductively similar to the sand tiger, and that, consequently, it was a "highly vulnerable species" and "highly susceptible to overexploitation." In reality, we know absolutely nothing about its reproduction. The fishery managers may have been concerned about the potential development of a fishery on bigeye sand tigers, being unaware of the difficulty of catching the species, or perhaps confusing rarity in catches with vulnerability or scarcity in nature. Some have suggested that it was included in the protected list to increase the number of protected species and appease the conservation lobby.

In 2006 and 2007, there were two instances when bigeye sand tigers were brought dead to the surface in fishing vessels off the eastern coast of the United States. In both cases, although there were official fishery observers on board, the sharks were discarded because they were protected species and the observers were reluctant to save the specimens. Thus, very rare specimens and data were lost. Unfortunately, the unnecessary protective legislation does not prevent the killing of the animals, but prevents scientists from ever learning anything about this shark. My attempts at correcting this situation have been futile.

FAMILY PSEUDOCARCHARIIDAE

The Crocodile Shark

This family consists of one species, the crocodile shark.

CROCODILE SHARK

Pseudocarcharias kamoharai (Matsubara, 1936)

Fig. 57a. *Pseudocarcharias kamoharai*, crocodile shark, mature male 87.2 cm, 2.38 kg; from Hawaii (Steven Kajiura, collector).

Common name Crocodile shark: a translation of its Japanese name *mizu-wani* (crocodile or alligator), which alludes to its toothy appearance.

Spanish name The species lacks vernacular names in the Spanish Caribbean and Mexico. The A.F.S. has not designated a Spanish name for it.

Scientific name *Pseudocarcharias kamoharai* (Matsubara, 1936), originally described as *Carcharias kamoharai* in the *Zoological Magazine* (Japan) 48(7): 380–382. *Pseudocarcharias:* from the Greek *pseudes,* false, and *carcharias,* from *karcharios,* the name for man-eating sharks; *kamoharai:* Latinized surname of Prof. Toshiji Kamohara, who obtained the type specimen at the Kôti Fish Market and gave it to Matsubara to describe.

Synonyms *Pseudocarcharias pelagicus* Cadenat, 1963.

Identification This is a *small shark,* usually less than 120 cm. It has *two spineless dorsal fins,* a first dorsal fin that originates well behind the base of the pectoral fins, about midway on the trunk, and a second dorsal fin originating over or behind the base of the pelvic fin. It has an *anal fin* that reaches nearly to the base of the tail. The *teeth are slender and smooth edged and curve backward.* The frontal teeth are much larger than the rest and protrude from the mouth. There are usually three rows of functional teeth at the front of the mouth. The teeth are similar in both jaws and number U: 13 or 14–13 or 14, L: 13–13. The dermal denticles are arrowhead shaped, with a strong central ridge terminating on a point. Coloration is light to dark brown above with purplish hues in life, fading to dark brown after death. The undersides are whitish, and the fins have a narrow whitish border. *There is often an irregular white blotch between the angle of the mouth and the gills.* The size of the white blotch is highly variable. Some specimens have a very large blotch two to three times the size of the eye, while in others the blotch is minute or absent.

Similar species The bigeye sand tiger has teeth with small basal cusplets on each side of the main cusp. The presence of an anal fin distinguishes crocodile sharks from the squaloid sharks such as *Squaliolus.*

Range The crocodile shark appears to be circumtropical in deep, warm, oceanic waters, but its distribution is poorly defined. It has been reported from the eastern Atlantic off Guinea and Angola (Cadenat 1963), the eastern Pacific (Abe 1973), the western Indian Ocean (D'Aubrey 1964; Bass et al. 1975b), and the Equatorial Indian Ocean (Romanov and Samorov 1994). However, its distribution is wider than the few reports indicate.

In North America, the crocodile shark is found in the Caribbean Sea and rarely in the Gulf Stream. The northernmost report of a crocodile shark that I know of was an 85-cm fork length male taken on a pelagic longline at 36°37′ N, 74°37′ W (off Virginia, apparently in the Gulf Stream), in December 2002, while fishing at a depth of 42 m in water 1,000–1,370 m deep (Larry R. Beerkircher, pers. comm., December 2002).

Size and weight The crocodile shark is a small shark. A 99-cm male that I examined weighed 4.09 kg. A 107.7-cm female was reported to weigh 6.0 kg. The species probably reaches 110 cm.

Biology This small shark inhabits deep, warm, tropical waters throughout the world. Little is known about its biology. Although its large eye and coloration suggest a mesopelagic existence, crocodile sharks are sometimes caught in tuna and swordfish lines, indicating that the species ascends to, or even lives in, the epipelagic zone. The specimens captured in the equatorial Atlantic that I have examined were

Fig. 57b. Teeth.

caught in surface water temperatures of 28–29°C. I have examined the stomach contents of three Atlantic specimens, all of which contained squid beaks.

Size at maturity The size at which the crocodile shark matures is not known precisely.

Male: D'Aubrey (1964) stated that the 73.5-cm male type specimen had well-developed claspers that reached beyond the origin of the anal fin, and thus she assumed that male crocodile sharks mature at that size. Bass et al. (1975b) stated that males mature at 74 cm, quoting Matsubara (1936). However, Matsubara's (1936) description mentions neither maturity nor condition of the type, nor are the claspers mentioned. As neither of these authors claimed to have examined the type specimen, its condition or state of maturity must remain in question until the type is reexamined.

Female: The smallest gravid female that I have examined measured 89 cm.

Reproduction The crocodile shark is an aplacental viviparous species. The embryos nourish themselves by eating eggs in the uterus (oophagy). I have examined six pregnant

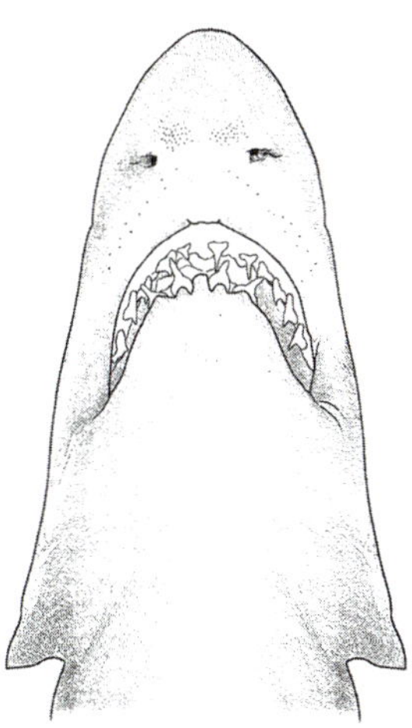

Fig. 57c. Snout.

Fig. 57d. Dermal denticles.

crocodile sharks, each of which contained four embryos, two per uterus. One of these was a 102-cm female caught in the equatorial Atlantic (4°00′ N, 38°12′ W) in March 1999, at a depth of 70 m. It carried four embryos measuring 327–335 mm, each weighing 140 g. Although the embryos had stomachs full of yolk that could be seen through the semitransparent body wall, the stomachs were not distended like those of mako or sand tiger embryos. The mother's ovary was proportionally quite large (20 × 10 × 6 cm) and contained hundreds of 3- to 5-mm oocytes. Two large feeding egg capsules were visible just below the shell gland.

Fujita (1981) reported that a 98-cm female captured in the central Pacific in June 1979 contained small embryos and egg capsules in the uteri. The left uterus contained two embryos, 38.2 and 40.0 mm, and 21 egg capsules. The right uterus contained two embryos, 38.2 and 40.8 mm, and 23 egg capsules. Another 98.5-cm female contained four large embryos, two per uterus, measuring 403, 401, 404, and 428 mm. These embryos were well formed, and their coloration was dark brown on the upper surfaces and white on the ventral surface, from the mouth to the base of the caudal fin. White blotches were present on the cheek and behind each gill slit. These embryos had no trace of a yolk stomach scar. One of the embryos, 401 mm and 307 g, contained about 75 g of yolk in its stomach, or about 24% of the weight of the embryo. These observations suggest that the crocodile shark may not have a definite or seasonal breeding cycle, as is the case with many tropical sharks.

Size at birth: There are few data on size at birth of the crocodile shark. The very limited evidence suggests birth at around 42–45 cm, based on the largest embryos reported, which measured 40.1–42.8 cm (Fujita 1981). These embryos were well pigmented and appeared outwardly to be close to birth. No neonates have been reported. The largest free-swimming young reported measured 59 cm (Abe et al. 1969).

Brood size: All the broods reported in the literature, and the six that I have examined, consisted of four young, two per uterus.

Nurseries Unknown.

Age and growth The species has not been aged.

Relation to humans None. The species is usually caught by tuna and swordfish longliners operating in the tropics, and specimens are generally discarded dead.

FAMILY MEGACHASMIDAE

The Megamouth Shark

The megamouth shark is the sole member of this recently described family. It is a gigantic, filter-feeding, pelagic shark that moves vertically in the water column with the deep-scattering layer. It is easily identified by its extremely short snout and huge terminal mouth that extends behind the level of the eyes. It is rarely encountered due to its pelagic, filter-feeding habits; many of the specimens seen have been stranded in shallow water.

MEGAMOUTH SHARK

Megachasma pelagios Taylor, Compagno, and Struhsaker, 1983

Fig. 58a. *Megachasma pelagios*, megamouth shark, immature male, 190 cm, from Brazil (courtesy of Alberto Amorim).

Common name Megamouth shark, or simply megamouth, an obvious reference to its huge mouth.

Spanish name The A.F.S. name is *tiburón bocón.*

Scientific name *Megachasma pelagios* Taylor, Compagno, and Struhsaker, 1983, in *Proceedings of the California Academy of Sciences* 43(8): 87–110. *Megachasma:* from the Greek *megas,* great, and *chasma,* the open mouth; *pelagios,* Greek "of the sea."

Synonyms None.

Identification The megamouth is characterized by a short, broadly rounded snout with a *huge terminal mouth that extends beyond the level of the eyes,* gill slits that measure 3–5% of total length and do not reach the dorsal side of the head, and two relatively small dorsal fins (~6% of total length). The teeth are very small (4–8 mm) and numerous; the Fukuoka megamouth had a dentition of U: 42–41, L: 48–49 (Yabumoto et al. 1997). The teeth have broad bases and smooth-edged cusps that taper quickly to a rear-facing point. The dermal denticles have three points with a strong central ridge terminating in a long point. *Coloration is black or dark brown above and white below. The dorsal side is uniformly colored and lacks any white or yellow spots. It has a white band along the upper margin of the mouth.*

Similar species The whale shark has a similar terminal mouth but is strikingly marked with white spots and bars.

Range As of December 2006, some 38 megamouth sharks have been reported (including an anecdotal account). The distribution of the reported specimens suggests that megamouths are widespread in the tropical, subtropical, and warm temperate waters of the Pacific and Indian oceans, and probably in the Atlantic as well. Most specimens have been caught off Japan, Taiwan, and the Philippines. Single specimens have been reported from Hawaii, Australia, Brazil, Indonesia, and South Africa. Only seven megamouth sharks have been caught in North America. Five have been caught off southern California and Baja California during October and November.

Although the presence of megamouths in the Atlantic was long suspected (Castro 1983), there is only one confirmed Atlantic specimen, a 190-cm juvenile caught off Brazil in 1995 (Amorim et al. 2000). Séret (1995) related the anecdotal capture of another juvenile male of about 180 cm off Dakar, Senegal, in 1995. The specimen supposedly was discarded after the vessel arrived in port and before fishery workers could examine it. However, according to Séret, the vessel captain's description agrees perfectly with the characteristics of a megamouth shark. Seret's account of the Senegalese specimen must be considered "anecdotal evidence" because it lacks any definitive evidence, such as a specimen or its parts, photographs, sketches or drawings by witnesses, and so forth. Even so, most authors writing about megamouth sharks have accepted this report and count it as "megamouth no. 8" despite the lack of evidence. Such is the desire to count megamouth sharks!

Size and weight Megamouth sharks are among the largest sharks. Two juveniles are known: a 190-cm juvenile male taken off Brazil that weighed 24.4 kg (Amorim et al. 2000) and a 215-cm, 27-kg juvenile female taken off Baja California (Leonardo Castillo, pers. comm., 28 November 2006). The Hawaiian megamouth, a 4.46-m male, weighed 750 kg (Taylor et al. 1983). The Australian specimen (no. 3), a 5.15-m male, weighed about 690 kg (Berra and Hutchins 1990). The 4.71-m Fukuoka female weighed 790 kg (Takada et al. 1997). The largest specimen actually measured and weighed was the Owase megamouth (no. 10), a 5.44-m female that weighed 1,040 kg. Other megamouth sharks have been estimated to reach 6.3 m, but none was actually measured.

Biology Megamouth is a large, pelagic shark that inhabits deep warm waters, a filter-feeder that migrates vertically in the water column with the day/night cycle. The species was discovered in 1976, the only new large shark species described since the discovery of the longfin mako in 1966. Consequently, our knowledge of it is extremely limited. The megamouth is the smallest of the three gigantic filter-feeding sharks (the others are the whale and basking sharks).

The first megamouth, a 4.46-m, 750-kg male, was caught on 15 November 1976, 42 km northeast of Kahuku Point, Oahu, Hawaii, at 165 m in water 4,600 m deep. The shark had become entangled in one of two parachutes deployed as sea anchors by a naval vessel. When the parachutes were hauled to the surface, crewmembers realized that they had caught an unusual animal and hauled it aboard with much difficulty. The shark was transported to Kaneohe Bay, where Leighton Taylor, of the Waikiki Aquarium, inspected the shark and quickly realized that it represented an undescribed species and should be preserved. The formal description (Taylor et al. 1983) would not be published until July 1983, nearly eight years after the capture.[1] It is, undoubtedly, the finest description of a shark ever produced.

Since the capture of the first megamouth, most of the 38 known specimens have been discarded or released without being examined, although one was released and tracked for two days. Fifteen megamouth sharks have been retained and were examined by scientists.

Most of what we know about the behavior of megamouth sharks was learned from the sixth specimen, a 4.9-m male captured off Dana Point, California, on 21 October 1990 (Nelson et al. 1997). The shark was caught in a drift gillnet set by a commercial vessel fishing for swordfish and sharks. Recognizing the still-living animal as unusual, the vessel's captain secured a rope around the caudal peduncle and slowly towed his catch to shore, a 5.5-hour journey. Surprisingly, the shark survived being tethered for more than a day in a shallow, murky harbor. On the afternoon of 22 October, the shark was towed back to open water and released with an acoustic transmitter attached. Although it had been restrained for more than 39 hours, the shark appeared to be in good condition at release. It was tracked continuously by Nelson et al. (1997) for 50.5 hours, during which the shark passed through two complete diel cycles. Although the shark remained in the epipelagic zone, it showed a clear pattern of swimming slowly at depths of 12–25 m (average 17 m) at night and at 120–166 m (average 149 m) during the day. At dawn and dusk the shark abruptly changed depth, diving at dawn, and ascending at dusk. This crepuscular behavior was well defined and lasted 1–2 hours, always spanning the times of sunrise and sunset. According to Nelson et al. (1997), the steepest parts of the shark's dives corresponded to the times of maximum illumination, suggesting that the shark's chosen path was determined by the light level.

The seventh megamouth specimen, known as the Fukuoka megamouth, was the first female to be examined. This shark stranded in Hakata Bay, Fukuoka, Japan, on 29 November 1994. How such a pelagic fish came to be stranded far into this generally shallow and polluted bay is not known. The shark was in excellent condition when found and was acquired by a nearby aquarium, Marine World Uminonakamichi. The specimen was quick-frozen and stored. The director of the aquarium and his staff assembled a group of Japanese and foreign scientists, including Eugenie Clark and me, to study the shark. The dissection of the specimen took place in February 1995 and resulted in the most comprehensive anatomical study of a single shark ever made. The collective results were published in a book titled *Biology of the Megamouth Shark* (Yano et al. 1997). This work complements the description of the first megamouth by Taylor et al. (1983). Consequently, we know a great deal about megamouth anatomy, despite only two of these animals having been examined in detail.

The ninth megamouth, the Brazilian specimen, was the first juvenile examined, and the only confirmed specimen from the Atlantic Ocean. It is a male measuring 190 cm and

1. The description of the megamouth by Leighton Taylor, Leonard Compagno, and Paul Struhsaker was eagerly awaited for eight years. This unusually long delay led Richard Ellis and John McCosker to write a fake description titled "*Megoura sandvicensis,* family Megouridae, a new genus, species, and family of deep-water shark from Hawaii, by John E. Randall, Katsumu Aida, Takashi Hibiya, Nobuhiro Mitsuura, Hisao Kamiya, and Yoshiro Hashimoto. Reprinted from *Uo,* No. 4, pp. 52–64, February 20, 1980." Ellis and McCosker used the Japanese text from an article about mating rhinoceroses from a mammology journal and actual photos of the first megamouth to complete the spoof. The fake paper was presented as genuine to Taylor, with predictable results.

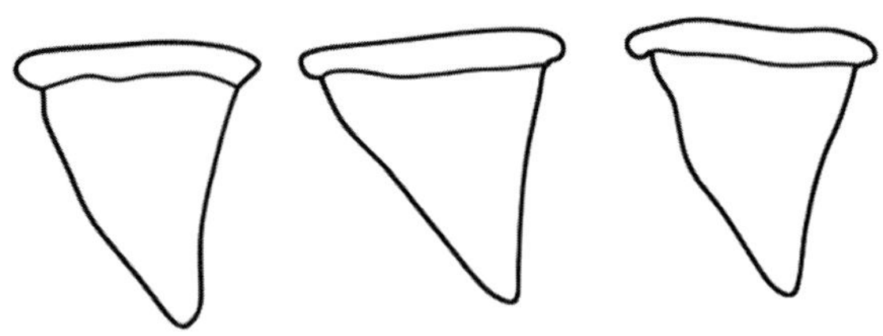

Fig. 58b. Teeth (upper).

Fig. 58c. Teeth (lower).

weighing 24.4 kg. It was accidentally caught by a longline vessel off southern Brazil (27°08 S, 43°55 W) in September 1995. The animal was hooked in the mouth at a depth of 15–40 m over water approximately 1,400 m deep. The specimen was recognized as unusual and kept on ice. It was in excellent condition when the vessel arrived in port four days later. The specimen was donated to the Instituto de Pesca in Santos, Brazil, where it was stored in formalin. When I examined it in August 1997, preparatory to writing the description (Amorim et al. 2000), it was still in excellent condition.

Most of the megamouth specimens seen have been caught in fishing nets. Four specimens were caught in swordfish gillnets off southern California, all captures taking place between 1 October and 29 November. The first California capture (megamouth no. 2) was a male caught 14 km east of Santa Catalina Island, on 29 November 1984. The specimen measured 4.5 m and apparently was not weighed. This specimen is now on display at the Los Angeles County Museum of Natural History. The second California specimen (megamouth no. 6) was the previously mentioned 4.9-m male captured off Dana Point in 1990 and tracked by Nelson et al. (1997). On 1 October 1999, a female (megamouth no. 14) estimated at 5.2 m was caught in a swordfish gillnet some 45 km west of San Diego, California. The water depth at the capture site was 350 m and the surface water temperature was 19.6°C. On 19 October 2001 another megamouth (no. 15) was caught in a swordfish gillnet some 75 km west of San Diego, California, where the depth was 841 m and the surface water temperature was 18.8°C (Don Petersen, pers. comm., October 2001). Unfortunately, these last two specimens were released or discarded and were not examined. These captures indicate that megamouth sharks are found off southern California in October and November. The area has been dubbed "Megamouth Alley."

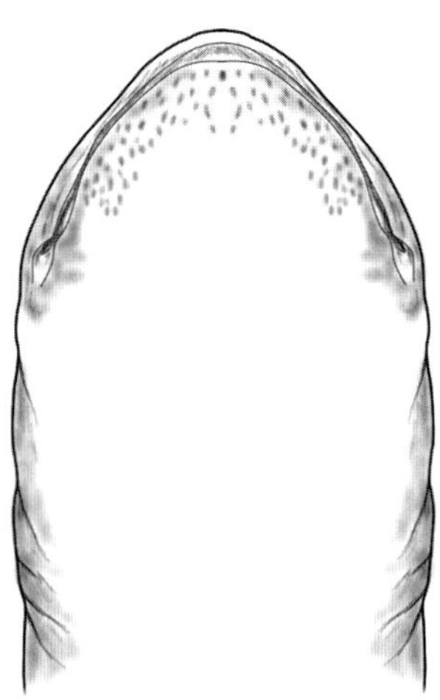

Fig. 58d. Snout.

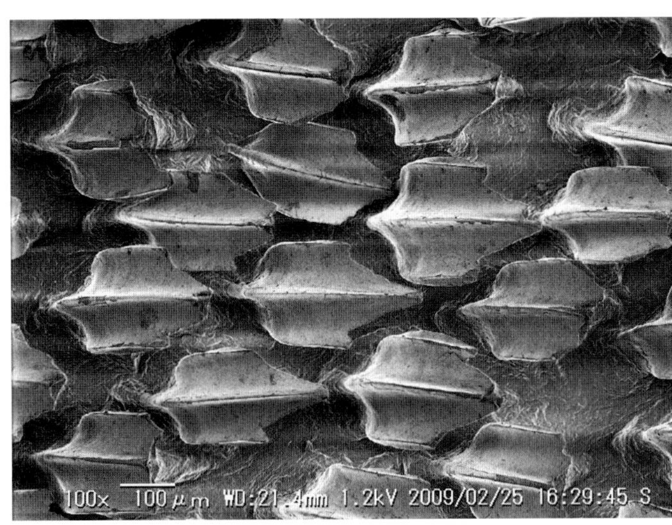

Fig. 58e. Dermal denticles.

In addition to the Fukuoka megamouth mentioned above, four other specimens have become stranded in shallow water at low tide. A 5.15-m male (megamouth no. 3) weighing about 690 kg became stranded during the receding tide at the entrance to Madurah estuary, Western Australia, on 18 August 1998 (Berra and Hutchins 1990). Another (megamouth no. 4) washed ashore at Hamamatsu, Japan, in late January 1989 (Nakaya 1989). This specimen was photographed but washed away before it could be saved. The photographs revealed it to be a male about 4 m long. Still another specimen washed ashore at a beach off Natures Valley (33°59 S, 23°34 E), Western Cape Province, South Africa, on 20 April 2002 (M. Smale, Port Elizabeth Museum, pers. comm., May 2002). Why a large pelagic fish such as megamouth would enter shallow waters and become stranded remains a mystery.

Diet: We know nothing about the habits of megamouth sharks, except for what has been revealed by acoustic tracking of one specimen and what could be deduced from the analysis of stomach contents of four preserved specimens. This analysis showed that the species is planktotrophic[2] (a plankton eater), feeding primarily on euphausiid shrimp (krill). The stomach of megamouth no. 1 (from Hawaii) contained a thick, reddish soup composed of euphausiid shrimp (*Thysanopoda pectinata*). This shrimp migrates vertically in the water column. During the day it is generally caught between 350 and 750 m depth, with some ranging up to 300 m and down to 1,000 m. At night most are found between 150 and 500 m depth. Apparently, when megamouth no. 1 was captured, it would have been in the region where these shrimp are most common at night, and quite possibly was feeding on them (Taylor et al. 1983). The second megamouth, captured off California, contained krill, copepods, and the jellyfish *Atolla vanhoeffeni* (Lavenberg and Seigel 1985). The first female megamouth (Fukuoka, Japan) also contained krill, but Yano et al. (1997) could not identify the species and suggested it was *Euphausia nana.* The mature female caught off Owase (megamouth no. 10) also contained dark-red chyme composed of many fragments of krill (Yano et al. 1999).

Interestingly, the only Atlantic specimen, a juvenile, was taken with a longline. This isolated capture is insufficient to tell us whether juvenile megamouth sharks have a different diet than adults or whether they are more prone to take hooks.

Two questions are often asked about megamouth sharks: (1) why were these large sharks not discovered sooner, and (2) why are we seeing so many of these sharks now, nearly one a year for 29 years, when none was reported before 1976? Its pelagic filter-feeding habit has probably kept megamouth away from the hooks of the ubiquitous swordfish and tuna longlines. However, many megamouth sharks must have been entangled in fishing nets before and have been released as worthless, strange creatures. Likewise, many megamouth sharks must have stranded on beaches and shallow bays in countless occasions. All these fishes simply went unnoticed, misidentified, or unpublicized. Today, fishing effort is far greater than at anytime in the past, and more and more oceanic creatures are being caught in longlines and floating gillnets. Not surprisingly, most of the known megamouth sharks have been taken in nets in areas where fishing effort is intensive. Today, preserved megamouth specimens are prominently displayed in museums and aquariums, and pictures of megamouth sharks are broadly disseminated by television and the Internet. Cameras and video recorders are everywhere, and people record anything that deserves attention. Thus, we are seeing megamouth sharks because fishermen are covering the oceans with nets and longlines, and because we know what these remarkable sharks are.

Size at maturity

Male: Males appear to mature at a smaller size than females. A 4.5-m male from Catalina Island (megamouth no. 2) was reported to have spermatophores oozing from the claspers (Lavenberg and Seigel 1985), indicating that it was mature.

Female: The 4.71-m, 790-kg Fukuoka female (megamouth no. 7) was immature, having poorly developed oviducts 12 mm wide and an ovary with small oocytes 1.2–2.7 mm. The 5.44-m, 1,040 kg Owase female (megamouth no. 10) was reported to be mature (Yano et al. 1997). It had an ovary containing oocytes 1–10 mm in diameter. Its expanded uteri measured 26 cm in width and the oviducal glands were 32 mm in width. Based on the condition of the Fukuoka shark, it appears that female megamouth sharks mature after they reach 5 m.

Reproduction The 4.71-m Fukuoka megamouth had an immature ovary of the lamnoid type containing innumerable oocytes in the same stage of development (Castro et al. 1997). This type of ovary indicates that the embryos are oophagous.

2. The word "planktivorous" has been used so often referring to the diet of the megamouth shark (e.g., Nelson et al. 1997; Yano et al. 1997) and other fishes that the word can no longer be considered a barbarism. It is now included in the 1993 *Webster's Third New International Dictionary,* but it is not found in the 1998 *Oxford English Dictionary.* Another term used in the past is "planktonophagous," used by Dean (1907) while referring to Acanthodians. I prefer the use of planktotrophic, because the word is properly constructed and is explicit.

Size at birth: Pregnant females have never been reported, so the size at birth is unknown. Oophagous species produce relatively large young; thus, newborn megamouth sharks can be expected to be large at birth, probably around 150–160 cm.

Brood size: Unknown.

Age and growth Unknown.

Relation to humans Preserved megamouth sharks are prominently displayed in a few museums and aquariums and serve as important tourist attractions.

FAMILY ALOPIIDAE

The Thresher Sharks

The Alopiidae is a small family comprising the thresher sharks, characterized by their enormously long caudal fins, which make up about half of their total length. These sharks use their long, powerful tails to stun prey with sharp blows. Live threshers exhibit beautiful metallic colors that reflect the surrounding light near the surface, camouflaging the shark and allowing it to approach schools of prey fishes undetected. Their teeth are single cusped and smooth edged. Threshers are cosmopolitan in warm and temperate waters. Although they are usually found in the open ocean, they also occur in cool in-shore waters, where they occasionally become entangled in fishing nets. Development is aplacental viviparous, and the embryos are oophagous. Three species are presently recognized, all in the genus *Alopias,* and all are widely distributed.

KEY TO THE ALOPIIDAE

1a. Midpoint of the first dorsal fin much closer to the pelvic fin origin than to the pectoral fin axil (the free rear tip of the first dorsal fin almost reaches the level of the pelvic fins), a marked groove on top of the head (from above the eyes to the gill slits) .... **Bigeye thresher,** *Alopias superciliosus* (p. 236).

1b. Midpoint of first dorsal fin closer to the pectoral fin axil than to the pelvic fin origin, lacking a marked groove on top of the head 2

2a. Light ventral coloration extending over base of pectoral fin .... **Common thresher,** *Alopias vulpinus* (p. 241).

2b. Light ventral coloration not extending over base of pectoral fin (a shark of the eastern Pacific) .... **Smalltooth thresher,** *Alopias pelagicus* (p. 233).

SMALLTOOTH THRESHER

Alopias pelagicus Nakamura, 1935

Fig. 59a. *Alopias pelagicus,* Smalltooth thresher shark, immature female, 265 cm, 54 kg; El Barril, Baja California, Mexico (courtesy of Chiapaneco fishermen).

Common name Smalltooth thresher or pelagic thresher. The term "smalltooth" refers to the size of its teeth, which are significantly smaller that those of other thresher sharks. It is also called the pelagic thresher, a simple translation of its scientific name. Thresher comes from *thresh,* to beat the stems and husks of grain or cereal plants with a flail to separate the grains from the straw; the name alludes to the shark's habit of hitting small fishes with its tail. See page XX for more details about the origin of the name.

Spanish name *Coludo* (Mexico), slang for having a long tail. The A.F.S. name is *zorro pelágico.*

Scientific name *Alopias pelagicus* Nakamura, 1935, in *Memoirs of the Faculty of Science and Agriculture, Taihoku Imperial University* 14: 1–6, plate 1, fig. 2. *Alopias:* a shortened form of Greek *alopekias,* the name of the thresher shark, derived from *alopex,* the name for the fox shark, in turn derived from *alopos,* a fox; *pelagicus:* Latin "of or pertaining to the sea," derived from the Greek *pelagios,* "of the sea."

Synonyms None.

Identification The smalltooth thresher has *the origin of the first dorsal fin much closer to the origin of the pectoral fin than to the origin of the pelvic fins;* the free rear tip of the first dorsal fin is *located far anterior* of the level of the pelvic fins and has *an enormous caudal fin,* which measures about 46–49% of the total length. The teeth have smooth-edged, curved cusps. The teeth are exceedingly small relative to the size of the adults or when compared to other species of thresher sharks. The largest frontal teeth of a 277-cm female (which I examined in Baja California) measured only 6 mm along their longest edge and 5 mm from the tip to root. The teeth of males are not noticeably longer. Teeth number U: 21 to 22–21 or 22, L: 21–21. The dermal denticles are small and closely overlapping. They have three strong ridges and three points, the central point being the largest. Coloration in life is dark indigo blue, almost black, along the dorsal midline, becoming lighter indigo toward the flanks and on the dorsal surfaces of the fins, metallic blue on the flanks, and *silvery* around the gills; the undersides are white, except for the ventral tips of the pectoral fins, which are dusky. *The area above the pectoral fin is bluish or black.* The blue and silver colors fade a few hours after death, and long-dead specimens are dark gray or nearly black.

Similar species The common thresher lacks the metallic blue coloration of the smalltooth thresher, and the area above its pectoral fin is white. The bigeye thresher has an enormous eye and a groove on top of the head. Its first dorsal fin originates much closer to the pelvic fins than to the pectoral fin origin, its free rear tip reaching to the level of the pelvic fins. It also has fewer but much larger teeth.

Range The smalltooth thresher is abundant and widely distributed in the Pacific and Indian oceans. The species was considered rare until recently, being known only from a few specimens from Taiwan (Nakamura 1935) and South Africa (Bass et al. 1975b). Confusion with the common thresher probably accounts for the perceived rarity. This large and abundant shark was not known from North America until 1973, when Castro-Aguirre and Lachica-Bonilla (1973) reported two embryos taken from a female caught off Islas Marías, Nayarit, Mexico. Along the western coast of North America, the smalltooth thresher ranges from California

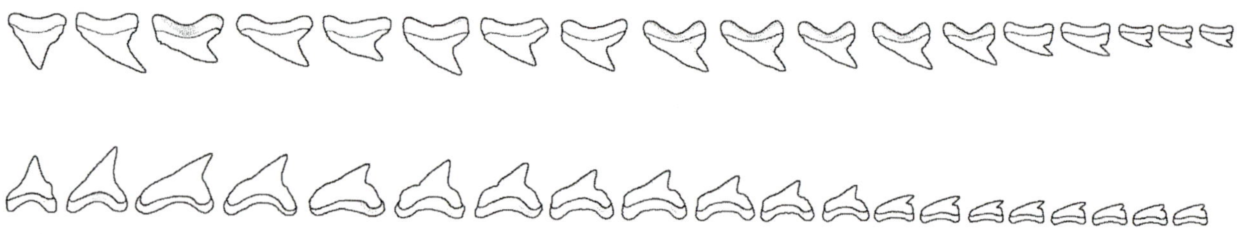

Fig. 59b. Teeth.

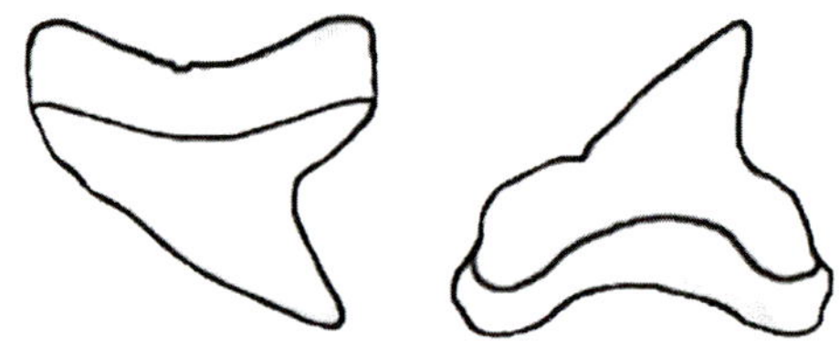

Fig. 59c. Teeth (detail).

southward to the tropics; its presence off California may be associated with warm El Niño years (Hanan et al. 1993). It is abundant off western Mexico and in the Gulf of California. I have seen dozens caught daily by the migrant fishermen (Chiapanecos) along both sides of the central Gulf of California.

Size and weight The smalltooth thresher is the smallest member of its family. The longest specimen I have encountered was a 316-cm female (see below), whereas the heaviest specimen was a 295-cm gravid female that weighed 98 kg. All the free-swimming specimens that I have examined were large juveniles or adults, all from the central Gulf of California. Representative sizes and weights are as follows:

Male: 236 cm, 34 kg; 266 cm, 57 kg; and 273 cm, 66 kg.

Female: 254 cm, 41 kg; 271 cm, 56 kg; 283 cm, 84 kg (gravid with 110-cm embryos); 299 cm, 82 kg; 310 cm, 70 kg; and 316 cm, 77 kg.

Biology The smalltooth thresher is a pelagic predator of the tropical, subtropical, and warm temperate waters of the Pacific and Indian oceans. It inhabits surface waters in oceanic

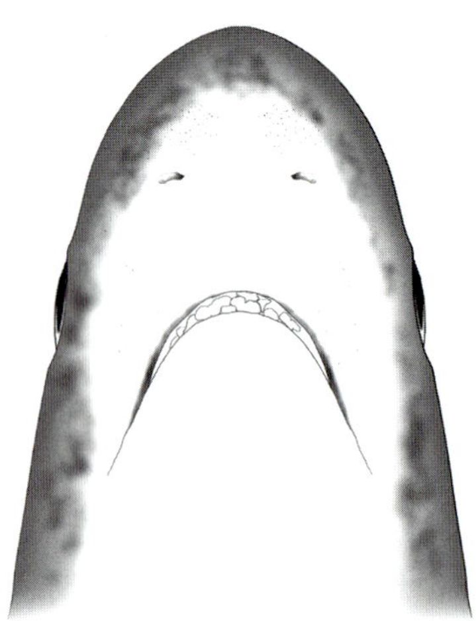

Fig. 59d. Snout.

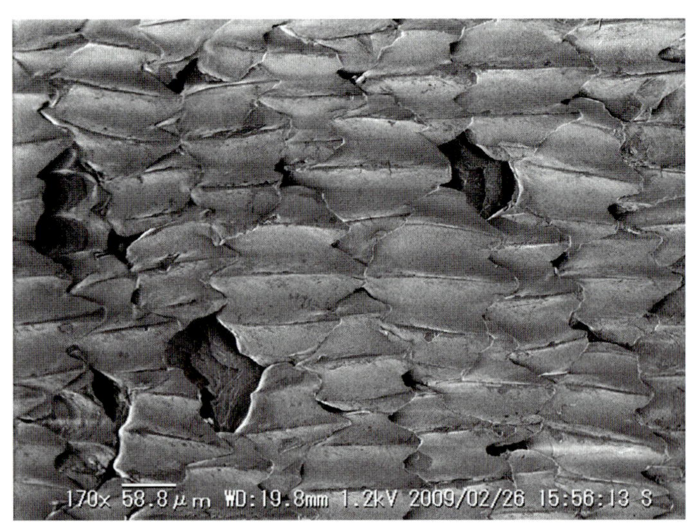

Fig. 59e. Dermal denticles.

areas as well as nearshore areas. Its biology is poorly known. In the Gulf of California, it feeds primarily on Humboldt squid (*Dosidicus gigas*) and clupeid fishes. Nearly all the specimens that I examined in the Gulf of California had squid remains or beaks in the stomach, and many had the remains of 10–15-cm clupeid fishes. The smalltooth thresher is a migratory species, but little is known about its movements along Baja California and western Mexico.

Size at maturity

Male: Males mature at about 264 cm. I have only examined about a dozen male specimens. A 262-cm, 53-kg male that I examined in July had calcified claspers that did not open well, while all males longer than 265 cm were mature. Liu et al. (1999) estimated that, off Taiwan, males mature at 267–276 cm.

Female: Based on specimens that I have examined, females mature at 272–285 cm. I considered females to be mature when they were gravid or had developed uteri and oviducal glands more than 20 mm wide. The smallest of my gravid females measured 272 cm and weighed 56 kg. The largest two juveniles measured 283 cm and 285 cm, and weighed 56 kg and 68 kg, respectively. Both these animals were seen in June. All the females that I have examined that were longer than 286 cm were mature, and nearly all were gravid. Liu et al. (1999) reported a slightly greater length at maturity, 282–292 cm, for smalltooth thresher females sampled at a Taiwan fish market.

Reproduction The smalltooth thresher is an aplacental viviparous species with oophagous embryos (Otake and Mizue 1981). Sometime after mating, one egg case containing a single fertilized egg is deposited into each uterus. This egg case is much larger than those that follow it, measuring 12 × 2 cm. It contains a single ovum of about 11 × 15 mm, its blastodisc visible as an orange spot 1.5 mm in diameter. I have not observed smalltooth embryos at hatching, but they probably emerge from the egg cases at an early age (when they measure less than 5 cm). Subsequent egg cases are smaller (9 × 2 cm) and contain numerous (30–40) small ova about 4–7 mm in diameter. The embryos feed on these egg cases after consuming their small yolk sacs; thus, they are known as "feeding egg cases." Unlike other oophagous species in which the ovulatory phase is shorter than the gestation period, smalltooth thresher females continue to ovulate eggs and produce feeding cases throughout gestation. As a consequence, the embryos feed throughout gestation and do not develop enlarged yolk stomachs as sand tiger, porbeagle, and mako embryos do. The embryos acquire bluish pigmentation at a relatively small size, 50-cm embryos being already pigmented. Consequently, from their first months of development, smalltooth thresher embryos look like miniature adults.

In rare cases where the single ovum in one of the large egg cases fails to be fertilized or develop, the uterus with the failed egg case fills up with all the feeding egg cases that were intended for the embryo. In one shark I examined, a 65-cm embryo inhabited one uterus while the other uterus contained 51 egg cases. One of these cases, much larger than the others, contained a single decomposed ovum, whereas the rest contained multiple ova. Although the egg case with the single ovum and several other egg cases with multiple ova showed signs of the yolk being slightly decomposed, most were in remarkably good condition.

All adult females that I have seen in the Gulf of California from May to September were gravid with embryos of different sizes. This indicates that the species reproduces year-round and that the reproductive period is annual, with females becoming pregnant again shortly after parturition. Liu et al. (1999) reached similar conclusions based on Taiwanese specimens, stating that the species had no specific period of parturition and no resting stage in its reproductive cycle.

Size at birth: Smalltooth thresher young are probably born at a length of 140–160 cm. The largest embryos I have seen measured 134–139 cm in September. In Taiwan, Liu et al. (1999) estimated the birth size to be 158–190 cm, based on the largest embryo they examined (158 cm) and the smallest neonate captured (190 cm).

Brood size: Smalltooth thresher broods consist of one or, usually, two young.

Nurseries Unknown.

Age and growth North American specimens have not been aged. Liu et al. (1999) examined vertebrae from smalltooth threshers from northern Taiwan, estimating that females matured at 282–292 cm and 8.0–9.2 years of age and males matured at 267–276 cm and 7–8 years of age. These authors determined that vertebral annuli form once a year up to ages of 16 and 14 years for females and males, respectively. However, their extrapolation of the growth rates for the largest sharks yielded maximum ages of 28.5 for females and 17.5 years for males.

Relation to humans Although smalltooth threshers are widely distributed through the Pacific and Indian oceans, there is little information on catches or landings. Substantial numbers of smalltooth threshers are caught annually by both artisanal and large-scale fisheries in the Gulf of California. The effects of these intensive fisheries have not been analyzed.

BIGEYE THRESHER

Alopias superciliosus (Lowe, 1839)

Fig. 60a. *Alopias superciliosus*, Bigeye thresher shark, immature female, 177.3 cm, 12.5 kg; from the Bahamas (Reinaldo "Rey" Pérez, collector).

Common name Bigeye thresher: bigeye because its eyes are larger than those of other threshers, and thresher, from the verb *to thresh,* to beat the stems and husks of grain with a flail to separate the grains from the straw. The name alludes to the use or shape of the shark's tail.

Spanish name *Zorro* or *tiburón zorro,* "fox" or "fox shark" (Cuba, Mexico), and *coludo prieto, tiburón grillo* (Mexico). The A.F.S. name is *tiburón zorro ojón,* a name not likely to be used by Spanish speakers.

Scientific name *Alopias superciliosus* (Lowe, 1839), originally described as *Alopecias superciliosus* in *Transactions of the Zoological Society of London* 3(1): 18. *Alopias:* a shortened form of the Greek *alopekias,* the name of the thresher shark, derived from *alopex,* the name of the fox shark, which in turn is derived from *alopos* (*alopex*), a fox; *superciliosus:* Latin, haughty or having a raised eyebrow, from *supercilius,* eyebrow, an allusion to the marked groove over the eyes.

Synonyms *Alopias profundus* Nakamura, 1935.

Identification The bigeye thresher has an *enormous upward-looking eye,* a *pronounced groove on top of the head,* running from above the eye to the gill slits; a *first dorsal fin free rear tip located over or slightly anterior to the pelvic fin origin;* and an *enormous caudal fin,* which makes up about 50% of the total length. Males have teeth with long, slender cusps with conical points and smooth edges. Females have teeth with narrow, bladelike cusps with smooth edges. In both sexes, the frontal tooth is erect and symmetrical; the others become increasingly curved toward the corners of the mouth. The upper teeth are larger than the lowers in both sexes. Teeth number U: 11 or 12–11 or 12, L: 10 to 12–10 to 12. The dermal denticles are simple points anchored on three-pointed bases, with little overlap. A living specimen is dark purplish brown or grayish brown along the dorsal midline, the color gradually changing to a band of metallic purple and then to metallic blue or bluish gray on the flanks to the level of the fins. The undersides are white or yellowish. At night, bigeye threshers can appear silvery brown or gray above with metallic purple and gold hues on the flanks. The colors and metallic hues are structural and not of a pigmentary nature. Thus, the colors are fragile and easily removed by touching or scratching the skin. These colors disappear shortly after death, and the upper body coloration changes to very dark purplish brown, which later turns very dark, almost black.

Similar species Both the common thresher and smalltooth thresher lack the large upward-looking eyes and the groove on top of the head. In both of these species, the free rear tip of the first dorsal fin terminates well anterior to the pelvic fins. Furthermore, these two species have twice as many teeth as the bigeye thresher.

Range The bigeye thresher is cosmopolitan in warm, deep waters. It has been reported from numerous localities throughout the world, including Cuba (Bigelow and Schroeder 1948), Florida (Springer 1943), Portugal (Antunes 1970), the eastern Pacific (Kato et al. 1967), the northwestern Pacific and the South China Sea (Ivanov 1986), east Africa (Bass et al. 1975b), southern Spain (Blasco and Muñoz-Chapuli 1981), Taiwan (Nakamura 1935), and Australia (Last and Stevens 1994).

Along the eastern coast of North America, the bigeye thresher ranges from Cape Cod to Florida, the Gulf of Mexico, and the Antilles. It can be found anywhere along the eastern coast of North America in the deep warm waters of the Gulf Stream. Schwartz and Burgess (1975) reported it for North Carolina, and Stillwell and Casey (1976) described summer concentrations of bigeye threshers off Cape Hatteras. According to Berkeley and Campos (1988), bigeye threshers made up 6% of all the sharks captured in the swordfish longline fishery along the Florida coast from south of Cape Canaveral to the Straits of Florida in 1981–1983. Branstetter and McEachran (1983) first reported it from the Gulf of Mexico. Guitart Manday (1975) noted that the bigeye threshers were found around Cuba all year, being most abundant in the fishery from August to October. I have examined about a dozen specimens caught off Miami Beach from February to April.

Off the western coast of North America, the bigeye thresher ranges from southern California southward to the tropics, although reports of bigeye threshers off the Pacific coast are few. Fitch and Craig (1964) mentioned a 375-cm male caught in a net at the bottom of a canyon off San Clemente, California, and a female captured off San Diego. Holts (1988) stated that in the California thresher shark fishery of the 1980s, the common thresher was the principal target, but bigeye threshers were also landed. He noted that in 1985, some 52,264 kg of bigeye threshers were tallied along the West Coast. I have seen a few bigeye threshers taken along with smalltooth threshers at El Barril, in the west-central Gulf of California.

Size and weight Although the bigeye thresher is a very large shark, its size has been greatly exaggerated in the literature. Bigelow and Schroeder (1948) refer to the jaws of a 548 cm ("18 foot") specimen from Cuba contributed by Louis HowellRivero who also reported the largest white shark at 640 cm and a 548-cm tiger shark from Cuba (Bigelow and Schroeder 1948). These lengths appear to be anecdotal and exceed other authenticated measurements by a wide margin. The reader has the choice of believing that everything grows bigger in Cuba or that these lengths are exaggerations.

Evidence indicates that bigeye threshers may reach 440 cm, although there are no authenticated reports of North American specimens reaching such size. Stillwell and Casey (1976) reported that their largest female measured 399 cm, while their largest male measured 352 cm. Later, Kohler et al. (1996) reported the measurements and weights of 55 bigeye threshers, including some of the same data used by Kohler's National Marine Fisheries Service predecessors (Stillwell and Casey 1976), but they gave the size of the largest as 371 cm total length (as 217 cm fork length). Kohler et al. (1996) did not include the 399-cm specimen because the datum was labeled as suspect (Pat Turner, National Marine Fisheries Service, pers. comm., December 2002). The largest bigeye threshers examined by Chen et al. (1997a) in Taiwan were a 422.8-cm female and a 357.7-cm male, while Moreno and Morón (1992) reported a 444-cm female and a 410-cm male from Spain. The largest specimen that I have examined measured 425 cm (see below).

Weights for bigeye threshers are scarce, probably because of their large size. Representative weights are as follows: males, 160 cm, 10.0 kg; 321 cm, 126 kg; 340 cm, 123 kg; females, 177 cm, 12.5 kg; 299 cm, 85 kg; 304 cm, 73 kg; 326 cm, 102 kg; 351 cm, 134 kg; 394 cm, 193 kg; 395 cm, 209 kg. Guitart Manday (1975) gave the weight of the heaviest female as 284.5 kg, without giving a corresponding length.

Biology The bigeye thresher inhabits the deeper layers of tropical and temperate oceanic waters. It was first reported from Madeira in 1839. Not until 1941 was another specimen recorded when one was captured off Florida (Springer 1943). It was not reported again until 1963 when a 375-cm, 164-kg male was caught off Newport Beach, California (Fitch and Craig 1964). Thus, the bigeye thresher was considered rare until the advent of the modern swordfish and tuna fisheries in the 1960s and 1970s. Nowadays thousands of bigeye threshers are caught annually as bycatch in the swordfish and tuna fisheries. Unfortunately, biologists have examined relatively few bigeye threshers because the species has little commercial value and most specimens that are caught are immediately discarded at sea. Thus, the species remains poorly known.

The bigeye thresher was long assumed to be a deep-water (mesopelagic) species, based on the uniform coloration of dead specimens and its large eye. However, the coloration of live specimens and night catches at depths of 35–150 m suggested a more epipelagic existence. Recent studies of bigeye threshers have demonstrated that these sharks spend the day at depths of 200–500 m and rise to 10–130 m at night. Nakano et al. (2003) used acoustic telemetry to follow two immature females in the eastern Pacific. These

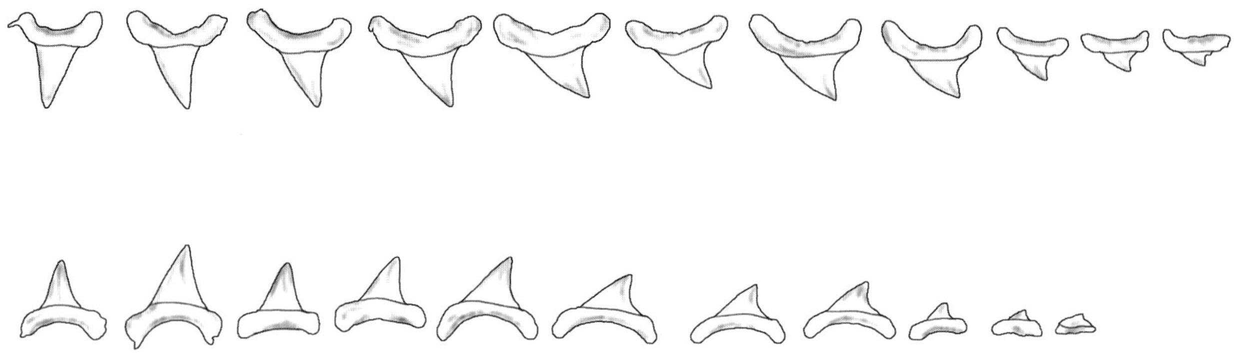

Fig. 60b. Teeth.

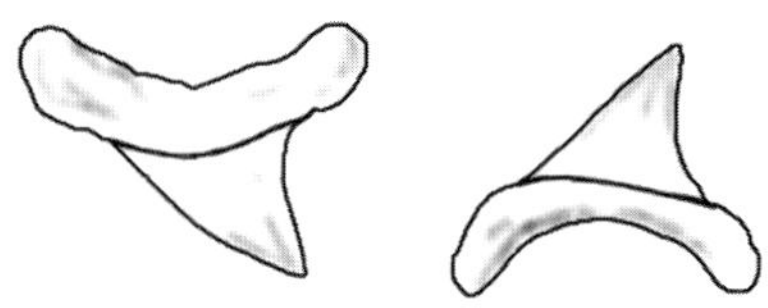

Fig. 60c. Teeth (detail).

sharks spent the day at depths of 200–500 m and temperatures of 7–11°C, ascending to 50–130 m at night. Both sharks made predictable movements within 30 minutes of sunrise or sunset, diving at sunrise and surfacing at sunset. Weng and Block (2004) used pop-up satellite archival tags to track a shark in the Gulf of Mexico and another in the Hawaiian Archipelago. The Gulf of Mexico shark spent the day below the thermocline at 300–500 m and temperatures of 6–12°C, and the night in the upper thermocline, between 10 and 100 m and temperatures of 20–26°C. The Hawaii shark spent the day at 400–500 m and the night at 10–50 m.

Fishery observer data also indicate that bigeye threshers hunt in the warm upper layers at night. Stillwell and Casey (1976) reported that the highest catches of bigeye threshers occurred in surface water temperatures of 16–25°C. However, NOAA fishery observer data reports catches of bigeye threshers off North Carolina, where water temperature was up to 27–28°C, while two females were reported in 29.5°C surface waters, some 100 km off Cape Hatteras in August (G. Burgess, pers. comm., 2003).

The reasons for the crepuscular migrations of the bigeye thresher are not fully understood. It is apparent that the bigeye thresher locates its prey from below, with its huge, upward-looking eyes, by contrasting or silhouetting the prey against the lighter background of the surface. The shark probably does this from the visibility limits, while its dark upper surface allows it to remain invisible against the dark background of the depths. Spending the day at much deeper and cooler waters could be an energy-saving behavior, although Nakano et al. (2003) speculated that the behavior may help the bigeye thresher avoid predation by other larger animals.

Little is known about the temperature regime of this species. The huge eyes (they may reach 10 cm in diameter in large specimens) have large *retia mirabilia* (see p. 256) in their associated blood vessels. Although direct temperature measurements of the eyes have not been performed, these structures are probably heat exchangers, presumably allowing the conservation of heat in the eye and serving to keep the eye at a higher temperature than the rest of the body. The countercurrent heat exchangers in the eye suggest that the bigeye thresher is warm-blooded like the lamnid sharks, but little is known on this subject. An elevated body temperature, several degrees above ambient temperature, is believed to be an adaptation for penetrating colder waters (Block et al. 1993).

Bigeye threshers occasionally approach land, even entering shallow water. Large females have been known to strand on rare occasions. Cadenat (1956) mentions a 4-m female stranded in the breakers near the mouth of the Saloum River, Senegal, in December 1955. A 348-cm female stranded on a sandbar in Ponce de Leon Inlet, Florida, in February 1992. It was gaffed and secured by a fisherman (Eric Sander, pers. comm., 1992). Another large female of 425 cm stranded in Myrtle Beach, South Carolina, in October 1981. Castro (1983) illustrated this specimen. Why strandings occur is unknown.

Diet: The bigeye thresher feeds on squid and small schooling fishes. Stillwell and Casey (1976) examined the contents of 18 bigeye thresher stomachs and found that the

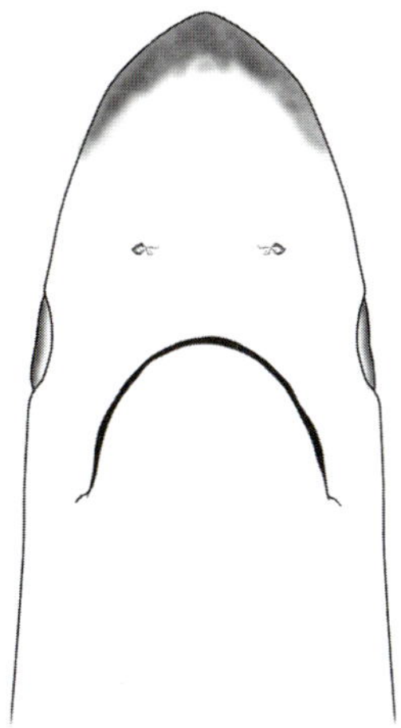

Fig. 60d. Snout.

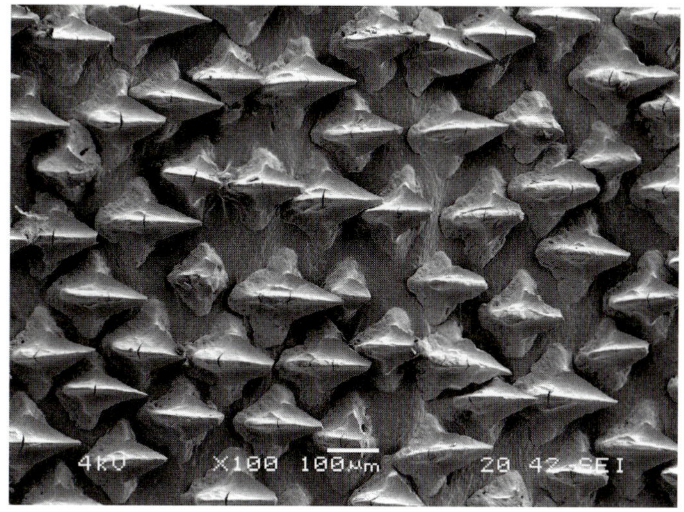

Fig. 60e. Dermal denticles.

most common items were squid (66%) and scombrid fishes. Gruber and Compagno (1981) also reported finding squid in the only stomach they examined. A 168-cm male I examined contained squid beaks and teleost vertebrae. A male caught off Newport Beach, California, contained six partially digested hake (*Merluccius productus*) (Fitch and Craig 1964).

Migrations: The bigeye thresher is probably migratory, but little is known about its movements. Stillwell and Casey (1976) reported aggregations of bigeye threshers from April to June off Cape Hatteras, although the species was caught there most of the year.

Size at maturity

Male: Stillwell and Casey (1976) examined 35 specimens and stated that males mature at 290–300 cm based on clasper calcification. Also using the criterion of clasper calcification, Chen et al. (1997a) estimated the size at 50% maturity for males off Taiwan at 270–287 cm.

Female: Chen et al. (1997a) also examined 429 females and estimated size at maturity at 332–341 cm, based only on the relative sizes of the oocytes and development of the uteri. Stillwell and Casey (1976) estimated that females do not mature until they reach 350 cm, based on a sample of 15 females. Female maturity was based only on the relative sizes of the ovary and oviducts, which were larger than those of other specimens examined.

Reproduction The bigeye thresher is an aplacental viviparous species. The embryos are oophagous (egg eaters). Unlike embryos of oophagous lamnoid sharks, bigeye thresher embryos do not acquire large distended yolk stomachs. Bigeye thresher embryos are externally well formed and partly pigmented when they are only about 60 cm. This fully formed external appearance of embryos has caused researchers to overestimate how close embryos are to birth and to underestimate the birth size. Only after examining a large number of gravid females is one likely to encounter the large term embryos.

Chen et al. (1997a) found that off Taiwan nearly all the adult females were pregnant and that in January females could be found carrying embryos 30–130 cm. Thus, these authors reported that no specific period of parturition could be noticed and the length of gestation could not be determined.

I examined a 395-cm female taken off Miami Beach in late May that carried two near-term embryos measuring 124 and 129 cm. This female had a large ovary (61 × 28 × 7 cm) that weighed 7.8 kg. Such a large ovary so close to parturition suggests that the bigeye thresher may be capable of mating again shortly after giving birth and that its reproductive cycle is annual.

Size at birth: Bigeye threshers are probably born at about 140 cm. Chen et al. (1997a) examined 35 broods and embryos up to 137 cm, leading them to estimate that the young are born at 135–140 cm. The largest embryos that I have examined measured 140 cm in May. These sizes agree with the smallest free-swimming specimens encountered. Bigelow and Schroeder (1948) reported a 130-cm young male from Cuba. I have examined a 169-cm specimen from the Bahamas. Moreno and Morón (1992) reported free-swimming specimens 156, 160, and 175.5 cm from southern Spain.

Brood size: Most of the bigeye thresher broods examined have consisted of two young, one per uterus. Guitart Manday (1975) stated that all the females examined by him carried two embryos but that there were reports of broods consisting of three and four embryos. Likewise, Moreno and Morón (1992) reported that whereas most females examined in Spain carried two embryos, one female had three and two others had four each. All other authors have reported two young per brood (Cadenat 1956; Gruber and Compagno 1981; Gilmore 1983). A 364-cm, 200-kg female that I examined was caught off Georgia on 13 January 1992, contained two embryos, a 96-cm male and a 94-cm female, each weighing about 1.75 kg. I also examined two other gravid females caught about 100 km east of Cape Hatteras on 2 August 1993. The largest measured 372 cm and carried two embryos, a male and a female, 96 and 92 cm, respectively. The smaller female measured 368 cm and carried two embryos, about 40 cm each. It is possible that broods of more than two young may occur, but these reports may be due to the vagaries of dealing with market specimens where young removed from one female may be lumped with those of another, or to confusion with the common thresher.

Nurseries Unknown.

Age and growth Unknown.

Relation to humans The bigeye thresher has no economic importance in North America because of the poor quality of its meat and fins. However, it is a local nuisance in swordfish longline fisheries, where it takes baits set for swordfish. A shark will often dislodge several baits from their hooks before impaling or hooking itself. Large numbers of bigeye threshers are caught in swordfish and tuna fisheries and are usually discarded at sea.

Bigeye threshers are occasionally caught by sport fishermen trolling deep for swordfish. A few are caught every year in the Gulf Stream off Miami Beach while fishing at depths of 150–200 m in daytime. Many are foul-hooked on their caudal fins, and they often escape capture because of their great size and strength. In the eastern United States the species has been on the list of federally protected species since 1999 and may not be landed.

COMMON THRESHER

Alopias vulpinus (Bonnaterre, 1788)

Fig. 61a. *Alopias vulpinus*, common thresher shark, immature female, 315 cm, 64 kg; from Ocean City, Maryland (Bob Davis, collector, courtesy of Mark Sampson).

Common name Since the late eighteenth century, the common thresher has been known by several English names: fox shark, sea-fox, thrasher, and thresher, to list a few. "Fox shark" seems to be the earliest. Pennant (1776: 97) gave the origin of this name: "The antients styled this fish *Alopex* and *Vulpes,* from its supposed cunning. They believed that when it had the misfortune to have taken a bait, it swallowed the hook till it got at the cord, which it bit off, and so escaped." *Alopex* was the Greek name for the fox, while *vulpes* was the Roman (Latin) name. The name thrasher comes from the English verb *to thrash,* to beat with a flail. Thresher comes from *thresh,* to beat the stems and husks of grain with a flail to separate the grains from the straw. Both "thrasher" and "thresher" are allusions to the shape of the body or to the shark's habit of using its tail like a flail to strike prey. In North America, the names "fox shark" and "thrasher" were used by J. V. C. Smith (1833) but were soon replaced by "thresher." The name "common thresher" was first used by Bigelow and Schroeder (1945) to distinguish it from the rarer bigeye thresher (*Alopias superciliosus*). In the past, before the advent of the deep-water swordfish and tuna longline fisheries, the common thresher was often caught, whereas the deeper dwelling bigeye thresher was seldom encountered. The A.F.S. currently designates the species simply as "thresher shark," but here I retain the older name "common thresher" to avoid confusion with the generic or collective "thresher sharks."

Spanish name *Zorro* or *tiburón zorro,* "fox" or "fox shark" (Cuba, Mexico), and *coludo pintado* (Mexico).

Scientific name *Alopias vulpinus* (Bonnaterre, 1788), originally described as *Squalus vulpinus* in *Tableau Encyclopédique et Méthodique des Trois Règnes de la Nature, Ichthyologie:* 9. *Alopias:* a shortened form of the Greek *alopekias,* the name of the thresher shark, derived from *alopex,* the name of the fox shark, in turn derived from *alopos,* a fox; *vulpinus:* Latin for "foxlike," from *vulpes,* a fox.

Synonyms *Squalus vulpes* Gmelin, 1788; *Alopias macrourus* Rafinesque, 1810; *Vulpecula marina* Garman, 1913; *Alopias greyi* Whitley, 1937. For many years the species was known as *Alopias vulpes.*

Identification The common thresher has *an enormous caudal fin that makes up about 50% of the total length,* and *the midpoint of the first dorsal fin base about equidistant from the pectoral fin axil and the origin of the pelvic fin.* The teeth are small, with curved, smooth-edged cusps, and are similar in both jaws. Teeth number U: 20–20, L: 21–21. The dermal denticles are small, broad, and overlapping. They have three strong ridges and three to five points. Coloration in life varies from purplish brown to black along the dorsal midline, changing to metallic bluish gray along the flanks. The metallic hues fade shortly after death, when specimens turn dark gray. The dorsal coloration extends unevenly along the sides. The undersides are white, and *the area above the pectoral fins is white.* Small areas of the fin tips and the upper surface of the tip of the tail are often white due to abrasion of the mucous layer of the skin.

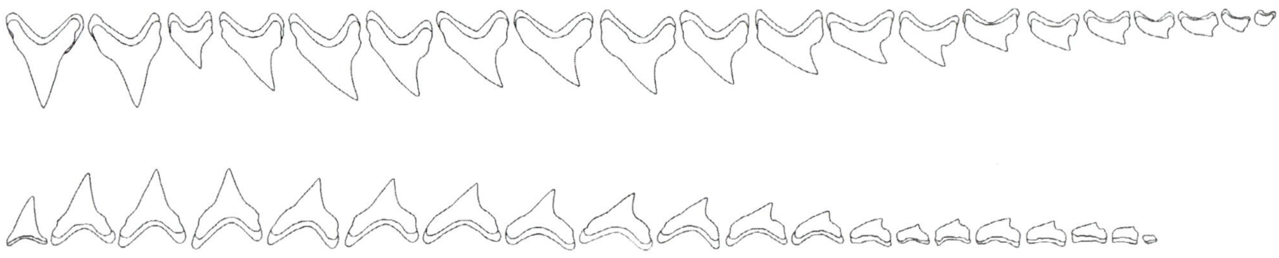

Fig. 61b. Teeth.

Similar species In the smalltooth thresher the upper metallic blue-gray coloration covers the area above the pectoral fins, and its teeth are minute, 5–6 mm in a large adult. The bigeye thresher has an enormous upward-looking eye and a pronounced groove on top of the head, and the origin of its first dorsal fin is much closer to the origin of the pelvic fins than to the origin of the pectoral fins.

Range The common thresher is cosmopolitan in warm and temperate waters. It has been reported from numerous locations in the North Atlantic, including eastern North America (Bigelow and Schroeder 1948), Portugal (Antunes 1970), and the eastern North Atlantic (Buencuerpo et al. 1998). It has also been reported from widely separated areas of the South Atlantic, including Madeira, the Moroccan coast, the Congo (Pointe-Noire), and the Angolan coast (Cadenat and Blache 1981). Tortonese (1938, 1969) reported the common thresher in the Mediterranean. Gubanov (1972) found it to be abundant in the warm temperate and subtropical waters of the Indian Ocean. Bass et al. (1975b) called it "fairly common" off the southwestern Cape of Good Hope. Strasburg (1958) reported it as uncommon in the central Pacific. It has been reported from Australia and New Zealand (Whitley 1940). It appears that there are several discontinuous populations of common threshers around the world. Little is known of the possible differences among these populations, and currently, all are considered to be one widely distributed species.

Along the eastern coast of North America, the common thresher reaches its northern limit off Newfoundland. Templeman (1963) mentions three records for eastern Newfoundland. All were large sharks (370–550 cm), with one 550-cm individual caught in Bonavista Bay (northeastern Newfoundland) in July 1954 being the northernmost record

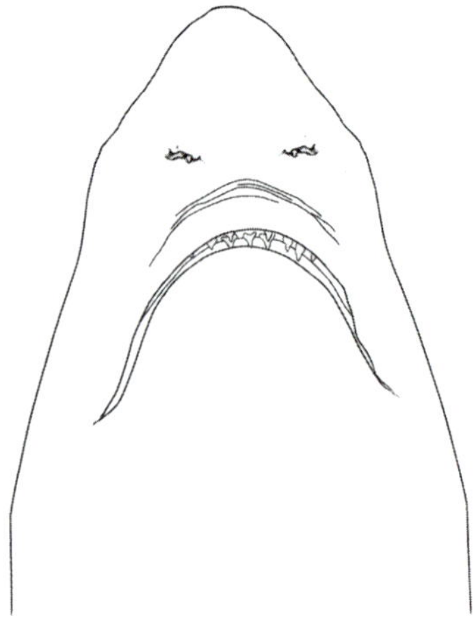

Fig. 61c. Snout.

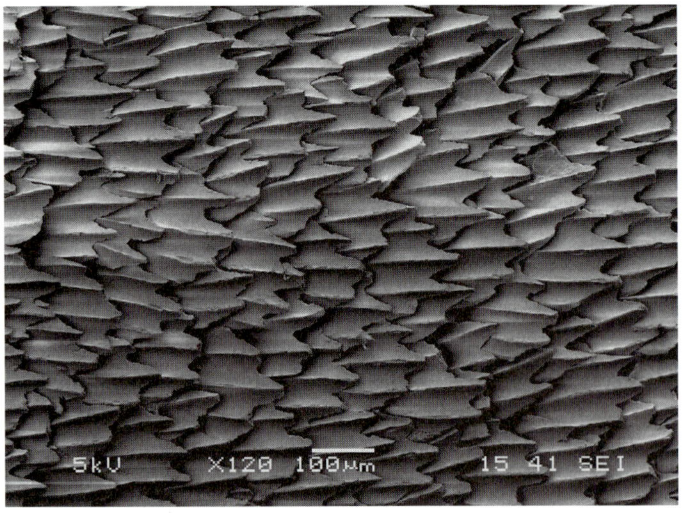

Fig. 61d. Dermal denticles.

I have found. There are several reports of the common thresher at the mouth of the Bay of Fundy and from the outer coast of Nova Scotia (McKenzie 1939; Scott and Scott 1988). It is an occasional visitor to the Gulf of Maine from May to October, becoming more abundant along the coast of southern New England (Bigelow and Schroeder 1953). It has been described as the most common shark inhabiting the continental shelf off Block Island, Rhode Island, appearing in May, becoming most plentiful in June and remaining until autumn (Nichols and Murphy 1916). Morrow (1955) reported the catch of an immature female in Long Island Sound, remarking that the species was very rare in the Sound, albeit common offshore. The common thresher is present along the continental shelf off North Carolina in July and August, and all year in offshore oceanic waters (Schwartz and Burgess 1975). Springer (1938: 34) wrote that the common thresher was "not common in Florida waters." However, based on numerous reports from fishery observers and on the fishing logs of Eric Sander, who has fished the area extensively, common threshers are abundant in coastal waters of northeastern and east-central Florida from December to April. Springer probably failed to find common threshers in the Florida fishery of the 1930s because the fishery used bottom longlines. Thresher sharks seek and strike their prey in the water column and are not likely to strike at bait on the bottom.

The common thresher is generally not found south of Cape Canaveral. Berkeley and Campos (1988) reported on 111 longline sets made both inshore and offshore off the east coast of Florida between latitudes 24°30′ and 28°00′ N. They caught a total of 613 sharks. No common threshers were caught in the area, although 10 bigeye threshers were caught. I know of only two specimens taken off Miami Beach, both taken by fishing guide Mark Quartiano; one of these measured 425 cm and was taken on 4 August 2007.

There is one report of a large female taken off Havana harbor in 1933 (Howell Rivero 1936), but this is most likely the misidentification of a bigeye thresher. When the Havana fish was taken (1933), the bigeye thresher was known only from Lowe's obscure and vague reference, and all threshers were usually considered to be one species (the common thresher). Guitart Manday (1968), who had considerable experience with Cuban sharks, stated that he had never captured a common thresher there. The bigeye thresher, however, is present in Cuban waters all year (Guitart Manday 1975).

Along the western coast of North America, the common thresher has been reported from British Columbia to the Gulf of California. The northernmost record is a small specimen (174 cm) captured near Sooke, Vancouver Island, in July 1937 (Cowan 1938). The species is abundant off southern California in summer (Roedel and Ripley 1950; Cailliet and Bedford 1983). The southern limits of the western population are along Baja California, and a few specimens have been captured in the Gulf of California.

Size and weight The common thresher is undoubtedly one of the longest sharks (although the tail makes up about 50% of the total length). However, there are few authenticated sizes for the largest specimens because few have actually been measured and weighed. Clemens and Wilby (1946) gave a maximum length of 762 cm ("to 25 ft") for the common thresher without mentioning any specimens or adducing any evidence. This length has often been repeated (e.g., Hart 1973; Cailliet and Bedford 1983) and has become established shark lore. Bigelow and Schroeder (1948) claimed that the species reaches "maximum length to 20 feet [610 cm] or more" but did not mention any particular specimens. A total length of 610 cm appears plausible given that Cailliet and Bedford (1983) reported a 573-cm specimen as the largest they examined among 167 common threshers taken in the southern California gill-net fishery. Generally, most common threshers reported are smaller. Kohler et al. (1996) measured and weighed 88 specimens caught by anglers in the northeastern United States. The largest encountered was a 450-cm, 204-kg female (P. Turner, pers. comm., December 2004). Pascoe (1986) recorded a 487-cm female from Bigbury Bay, England, that weighed 264 kg.

Bigelow and Schroeder (1948) gave weights of 136–145 kg at about 300 cm, 170–180 kg at 400 cm, and 225 kg at about 440 cm. However, at the lower size ranges, these weights do not agree with weights that I have recorded or with those that other authors (Kohler et al. 1996) have reported. Weights for some common threshers caught of the east coast of North America are as follows: 240 cm, 53 kg; 267 cm, 56 kg; 276 cm, 71 kg; 315–318 cm, 80–94 kg; 420 cm, 175 kg; and 440 cm, 191 kg.

Biology The common thresher has been known since antiquity. Aelian (Claudius Aelianus), a Roman who lived from about 170 to 230 A.D., tells us:

> The fox is a creature full of guile; this is the fox that lives on the land. But listen also to the wiles of the Fox-shark and learn the kind of things it does. Either it will not come near the hook at all, or else it swallows it and immediately turns itself inside out, reversing its body just like a garment, and in this way no doubt it gets rid of the hook. (Scholfield 1971: 231)

This is probably the earliest description of stomach eversion in sharks.

Thresher sharks use their long, massive tails to stun or kill prey with sharp, swift blows. They slap at prey with their tails with great accuracy, quickly disabling or killing their prey. The fact that the thresher uses its tail for aggressive purposes has been known for centuries, and there is a very old legend about the thresher using its tail against whales. The tale was probably old when recounted by Sir Richard Hawkins (1622: 49):

> Now of the fight betwixt the Whale and its contraries; which are the Sword-fish and the Thresher. . . . The Thresher is a greater fish, whose tayle is very broad & thick, and very waightie. They fight in this manner: The Sword-fish placeth himself under the belly of the whale and the Thresher upon the Ryme [surface] of the water, and with its tayle thresheth upon the head of the Whale, till hee force him to give way, which the Sword fish perceiving, receiveth him upon his sword, and wounding him in the belly.

There have been many versions of this story through the centuries. Perley (1852: 22) wrote:

> Its principal organ of defense, appears to be its long, broad, and flexible tail, with which it attacks, and literally *threshes* its enemies. This fish is a great enemy to the small whales, in the Gulf of Saint Lawrence. In the Bay of Chaleur, and lower part of the River Saint Lawrence, it is often seen attacking the whales that frequent those localities. In its attacks, it is most persevering; and the whale may be often seen to spring quite out of the water, and make the sea foam, from the torment it endures.

This thresher-whale tale persisted well into the twentieth century, with variations as late as 1939: "The creature [thresher shark] plays havoc with the mammoth of the deep. Whalers regard it as a special friend, for it frequently locates their prey for them" (La Gorce 1939: 212). A few writers realized that these accounts were not based on solid evidence. As early as 1833, J. V. C. Smith noted: "Perhaps no class of sea stories are more common than accounts of combats between the *thrasher* and the *whale*–the latter of which, is always said to be beaten most cruelly by the thrasher. We apprehend, however, that there is some mistake in these supposed battles" (Smith, J.V. C. 1833: 95).

Although the common thresher's use of its tail for aggressive purposes has long been known, how the shark actually killed prey was a mystery well into the early twentieth century. A common view was that it used its tail simply to herd fish into compact schools before attacking them. In 1916 Nichols and Murphy wrote that the common thresher

> uses its whiplike tail to splash the water, while it swims in narrowing circles round a school of fishes, which are kept crowded together until the moment of slaughter. Practically it whips its prey into its mouth. Sometimes a pair of threshers work together at this highly organized method of fishing. Formerly it was believed that they killed their prey by furious slashes of the flexible tail, but this theory has not been well borne out by observation, and is at least improbable. (Nichols and Murphy 1916: 21)

Later, Nichols and Breder (1927: 18) wrote: "The thresher's elongate tail is not sufficiently rigid or muscular to strike an efficient blow. These sharks are said to swim round and round a school of small fish herding them into a narrow compass where they can be readily attacked."

Direct observations of common threshers feeding were not reported until the early twentieth century. Coles (1915) reported seeing a thresher shark feeding in shallow water near Cape Lookout, North Carolina, in July 1914, describing the shark as "throwing fish to its mouth." He also reported seeing one fish being ejected from the water for a considerable distance. W. E. Allen (1923: 32) provided a better description of the process:

> While taking my plankton collections at about 7:25 A.M., April 14, 1923, I heard a splash nearby. Turning, I saw about one hundred feet distant, a swirl in the water like that made by a California sea lion. A moment later a long, slender, compressed tail (about three feet long) flashed above the surface and lashed about like a coach whip. . . . At about 7:45 . . . I saw about fifty feet from the pier what appeared *at first* to be a "soupfin shark." . . . It was coming diagonally towards the surface and swimming rapidly. Almost immediately I noticed a small fish (possibly California smelt . . . about ten inches long) frantically swimming just in front. A moment later the pursuer, a six-foot thresher shark, passed partly ahead of the victim (probably half its body length) when it turned quickly and gave the coach-whip lash with the tail which I had seen before. . . . The whip stroke was instantly repeated with very confusing speed, and then it became apparent that the victim was seriously injured. . . . It was, however, under the drip from my net, at which the shark was apparently frightened. The shark darted away and was not seen again. The victim sank, swimming feebly, and then came to the surface and lay on its side a while. . . . Finally it sank until out of sight. . . . I was much impressed with the speed and skill with which the shark worked and with the accuracy shown in its strokes at a single flying target.

The accuracy with which threshers hit their targets with their tails can be determined by examining hooked common threshers. Threshers are usually hooked on the tail after hitting the bait accurately with sufficient force to become impaled on the hook. In Gubanov's (1972) study, 97% of the threshers caught were hooked on the upper lobe of the tail, whereas only a few were hooked in the mouth area (although I have seen a few threshers hooked on their pectoral fins also).

Unlike the images presented in most textbooks, threshers carry their tails up at a high, steep angle, not in line with their bodies. Common threshers are occasionally seen leaping clear of the water and some have even been photographed while in the air. The causes or reasons for the leaping are unknown.

The common thresher is a cool-water species. Specimens captured by Eric Sander off the east coast of Florida during commercial fishing operations (1988–1998) were all taken within a narrow temperature range of 16.5–19.8°C (based on Sander's logbooks and on NOAA observer data; E. Sander and G. Burgess, pers. comm., 2001). The common thresher is usually seen in offshore waters, although it often wanders close to shore while pursuing schooling fish. Some evidence suggests that common thresher sharks are abundant only close to land. Strasburg (1958) reported that of more than 6,100 sharks taken in the open tropical and subtropical Pacific Ocean, only 127 were thresher sharks (common thresher, bigeye thresher, and smalltooth thresher combined). Bedford (1992), who studied threshers off California in the 1980s, wrote that common threshers appeared to be abundant only within 50 miles of land. Gubanov (1972) noted that during longline cruises in the Indian Ocean most catches occurred in waters about 200 m deep.

Like other lamnoid sharks, the common thresher is a warm-blooded species. Threshers achieve endothermy through a system of countercurrent heat exchangers in the circulatory system known as a *rete mirabile* (see page XX). Eschricht and Müller (1835) first described the *retia mirabilia* of the common thresher as being different from those of tunas and the porbeagle shark. More than a century would pass before these structures were recognized as adaptations for endothermy. Bone and Chubb (1983) briefly reported that the common thresher has a retial system in its red muscle mass but that the rete is not as extensive as in the lamnid sharks. Bernal and Sepulveda (2005) reported that red muscle temperatures of common threshers were as much as 5.4°C above the sea surface temperature. For an interesting view of the thresher's cartilaginous skeleton, see the excellent drawings by Molin (1860).

Little is known about the predators of the common thresher, other than humans. The large size of adults suggests that they have few natural enemies other than killer whales, white sharks, tiger sharks, and large makos. Whitley (1940) mentions a 365-cm tiger shark caught off Sydney, Australia, that had part of the 213-cm tail of a thresher shark protruding from its mouth. A 210-cm mako was reported to have two young threshers in its stomach (Bedford 1992).

Diet: The common thresher feeds on small schooling fishes. Prey reported in North American waters include anchovies, mackerel, bluefish, bonito, herring, menhaden, sardines, and squid (Nichols and Breder 1927; Herald and Ripley 1951; Bigelow and Schroeder 1953). Bedford (1992) found that common threshers caught off California contained only northern anchovies (*Engraulis mordax*), whereas blue sharks and shortfin makos taken in the same nets had gorged on the far more abundant pelagic red crabs (*Pleuroncodes planipes*) that are present in great quantities in warm-water years. Off England, common threshers have been reported to feed on scad (*Trachurus trachurus*), whiting (*Merlangius merlangus*), and mackerel (*Scomber scombrus*) (Pascoe 1986).

Migrations: Common threshers are migratory, but their routes in North American waters are poorly known. On the eastern coast of the continent, common threshers appear to winter from North Carolina to central Florida. The species is found off eastern Florida only from December to early April when the surface water is 17–18°C, based on Eric Sander's logbooks. Off North Carolina, the common thresher is present all year in offshore oceanic waters and enters continental shelf waters in July and August (Schwartz and Burgess 1975). The species appears off New York and New England in June, remaining there until October and November (Nichols and Breder 1927). Scott and Scott (1988) called it a summer visitor to the Canadian Atlantic region; some specimens have been captured there from July to November, most often during August and September, at sizes ranging from 3.3 to 5.5 m.

Off western North America, common threshers are believed to winter off Baja California, moving northward into southern California as the water warms up in the spring (Hanan et al. 1993). Both adult and immature threshers congregate in the inshore waters of southern California, with the greatest concentrations occurring in spring and summer (Cailliet and Bedford 1983). Large, mature common threshers remain in this area for up to two months. This coincides with the time of birth (Bedford 1992). The large animals continue northward to at least Vancouver Island arriving there by late summer and fall (Stick and Hreha 1988). Juveniles tend to remain in nearshore shallow areas within the Southern California Bight; they are seldom found north of Cape Mendocino, California (Holts and Bedford 1989). The reverse migration occurs in autumn. Large juveniles then appear again around the Channel Islands off southern California, but the large adults are not seen again until the following spring (Bedford 1992). There is little specific information on the migration dates. According to Cailliet and Bedford (1983), large numbers of common threshers captured off the California coast carry Japanese longline hooks, indicating movements outside the U.S. Fishery Conservation Zone.

Size at maturity

Male: Cailliet and Bedford (1983) used the length of the inner margin of the clasper versus total length to estimate that common thresher males reach maturity at about 330 cm. This nonstandard method disregards the fact that calcification follows elongation of the clasper, so it tends to consider animals as mature whose claspers have not yet calcified.

Female: Off California, female threshers reach maturity at about 390 cm, based on pregnancy as a criterion of maturity, and on the size of the smallest gravid female in a sample of 18 females (D. Bedford, California Department of Fish and Game, pers. comm., July 2005).

Other authors have proposed much different sizes at maturity. According to Gubanov (1978), common threshers in the equatorial zone of the Indian Ocean become sexually mature at a total length of 260 cm, based on the size smallest gravid female sampled. Other authors have cited Gubanov's figure of 260 cm and have speculated that different populations might reach maturity at widely differing sizes.

Reproduction The common thresher is an aplacental viviparous species with oophagous (egg eating) embryos, but its reproductive processes are poorly known. The early stages of the reproductive process following mating are undescribed, but they are probably similar to those of the smalltooth thresher. It is likely that a few fertilized eggs are packed singly into egg cases. Subsequent egg cases destined to feed the developing embryo contain several of the small eggs, and these early egg cases have been illustrated by Gubanov (1978). The embryos do not acquire the characteristic distended yolk stomachs that the embryos of porbeagles, makos, and sand tigers do. When common thresher embryos measure only 50–60 cm, they outwardly appear to be fully formed and are pigmented, even though they may require several more months to complete their development. Therefore, it is difficult to ascertain how close to birth a given brood is by inspection of the embryos. Gestation is believed to last about nine months (Cailliet and Bedford 1983).

Cailliet and Bedford (1983) found that all mature females from southern California examined during early spring were pregnant. Later, Bedford (1985) stated that females give birth every year from March to June but did not explain how he reached such a conclusion. Presumably, he deduced the yearly cycle from observing that all adult females he examined were pregnant at certain times of the year. Bedford's assumption is probably correct. However, if the females segregate, with the gravid females going to one area to give birth and the nongravid females going somewhere else to mate, a biennial cycle would not be noticed if one only samples the gravid population. In the Indian Ocean, Gubanov (1972) found widely varying sizes of embryos in a single season, suggesting year-round reproduction.

Size at birth: Common threshers are born at a length of 139–151 cm. Hixon (1979) described two of four embryos removed from an approximately 460-cm, 295-kg female taken off Solimar Beach, near Ventura, California, in June 1977. One of these embryos, a male, measured 141.7 cm and weighed 8.8 kg; the second embryo, a female, measured 138.6 cm and weighed 7.7 kg. D. Bedford (California Department of Fish and Game, pers. comm., July 2005) examined 151-cm embryos as well as a free-swimming specimen measuring 150 cm. Hanan et al. (1993), estimated the size at birth as 139 cm but did not explain their basis for the estimate.

Herald and Ripley (1951) reported a 117-cm free-swimming male weighing 3.1 kg, caught by an angler at a depth of 5–6 m near Coyote Point in San Francisco Bay, California, on 17 September 1950. The small size of this specimen, if correctly reported, suggests that it may have been a runt, or that there may be a wide variation in the birth size.

Brood size: The average brood size is four (Bedford 1985). A very large female, estimated at 5.4–5.6 m caught in a gillnet off Newport Beach, California, in May 1954 contained four young 123–137 cm that weighed 5.1–6.2 kg (Joseph 1954). Exceptional broods may consist of up to seven embryos. Moreno et al. (1989) reported a 451-cm female, apparently examined in a market in Spain that contained seven embryos 133–146 cm.

Nurseries The common thresher nurseries in the North Atlantic Ocean appear to be in coastal waters of the northeastern United States. Bigelow and Schroeder (1953) stated that threshers are probably born throughout their range along the eastern coast of the United States, because small free-living specimens have been caught off New England and Florida. However, the account of juveniles in Florida is based on a single specimen of doubtful origin, possibly a young bigeye thresher. It is highly unlikely that common threshers have nurseries in Florida, because small juveniles have not been reported again in the state. Along the western coast of North America, the nurseries appear to be in the coastal waters of southern California, but very small juveniles have been caught as far north as San Francisco Bay (Herald and Ripley 1951). According to Dan Cartamil (pers. comm., April 2007) juveniles are most abundant in shallow inshore waters between Point Conception, California, and northern Baja California, Mexico.

Age and growth Cailliet and Bedford (1983) estimated the growth of the common thresher as follows: 198 cm at 1 year of age, 323 cm at 4 years of age, 414 cm at 7 years of age, and 480 cm at 10 years of age.

Relation to humans Common threshers are seen infrequently in markets on the U.S. East Coast. In California markets, they were abundant for a few years. An intensive fishery began in California around 1977, growing from 58.8 metric tons (dressed weight) to a peak of 1,082.5 metric tons in 1982. Catches remained high through the 1985–1986 season. Beginning with the 1986–1987 season, the California legislature restricted the drift gillnet fishery to 30 days in May. As a result, mean catches of thresher sharks dropped substantially. In 1990–1991, the legislature closed the spring drift gillnet fishery within 75 miles (112 km) of shore, thus eliminating the state's directed thresher shark fishery (Hanan et al. 1993). Today, common threshers in California are taken mainly as bycatch in the gillnet fishery for swordfish. Out of concern for declining stocks, Oregon and Washington canceled experimental fisheries for coastal thresher sharks in 1989.

In the United States, the common thresher is classified as a game fish. Sports fishermen catch a few along the east coast by using trolled bait and lures, as well as drifting baits. Most specimens are hooked on the upper lobe of the caudal fin, as they impale themselves while trying to stun live bait. In commercial fisheries, common threshers are often caught on floating longlines but are unusual catches in bottom longlines.

FAMILY CETORHINIDAE

The Basking Shark

The basking shark is the only member of this family, although the southern hemisphere form is considered by some specialists to be a different species. Others place the basking shark with the Lamnidae. The basking shark is the second largest fish in the world.

BASKING SHARK

Cetorhinus maximus (Gunnerus, 1765)

Fig. 62a. *Cetorhinus maximus*, basking shark, mature male, 7.85 m, from Cape Hatteras, North Carolina (collected by personnel of USCG Hatteras Inlet Station).

Common name Basking shark. Pennant (1776) coined the name "basking shark" for this species, noting that the Irish and the Welsh called it the "sunfish" because it appears to be lying on the surface sunning itself, and that English writers of the time already used the name "sunfish" for the oceanic sunfish (*Mola mola*). Thus, the new name "basking shark" avoided confusion with the oceanic sunfish. In the United States, it was known as the bone shark in earlier times because the long, slender gill rakers were said to resemble whalebone (Goode 1884; Jordan and Evermann 1896).

Spanish name *Pez elefante* (Cuba), *peregrino* (Mexico), and *tiburón peregrino* (A.F.S.).

Scientific name *Cetorhinus maximus* (Gunnerus, 1765), originally described as *Squalus maximus* in *Skrifter-Det Kongelige Norske Videnskabers Selskab* (Trondheim): 33, plate 2. *Cetorhinus:* from the Greek *ketos,* whale, and *rhinos,* a nose; *maximus:* Latin for largest.

Synonyms *Squalus peregrinus* Blainville, 1810; *Selache maxima* Cuvier, 1817; *Squalus elephas* Le Sueur, 1822.

Identification An *enormous shark,* usually seen swimming close to the surface. It has a conical snout and *enormous gill slits* that extend from the dorsal side to the midline of the throat. When the shark swims, the mouth is usually open and the gill slits flare out, so that the head and gill area look inflated. It has *dark, bristlelike gill rakers inside the gills* (the rakers are present most of the year but are sometimes absent in winter). The gill rakers are long (10–12 cm) and slender and are located on the forward edge of each gill arch, on each side of the gill slits. They protrude into the throat and can be seen easily on the frontal edge of the gills. It has strong caudal keels on the caudal peduncle, and a lunate tail. Young specimens less than 3 m in length have curved snouts. Izawa and Shibata (1993) gave a good illustration of a young female basking shark, total length 2.6 m, captured off Wagu, Pacific coast of Honshu, Japan, in May 1977. Its snout was curved and had a ventral groove continuous with the palate. The teeth are minute (3–4 mm in a 9-m specimen) and very numerous (about 1,200 teeth in the jaws of a 9-m specimen in the Hubbell collection in Gainesville, Florida), with a single, backward-curving, conical cusp. Teeth are similar in both jaws. The dermal denticles are most unusual. They are short, ridged spines with a cavity at their core, and multiple cavities around their bases. The spines point in all directions and are fragile. The basking shark skin samples I have collected all proved to be damaged, and it was quite difficult to locate unbroken denticles to photograph. Color may be grayish brown to slate gray or black above, uniformly throughout, or with irregular lighter patches. The underside is paler, often with white patches under the snout and mouth or along the ventral side. Two albino specimens have been recorded from the North Atlantic (Frøiland 1975).

Similar species The mackerel sharks have fewer and much larger teeth and smaller gill slits without gill rakers. Large white sharks near the surface can be confused with basking

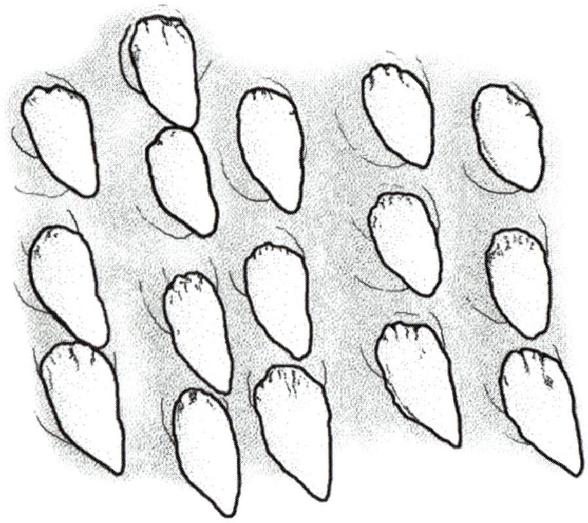

Fig. 62b. Teeth.

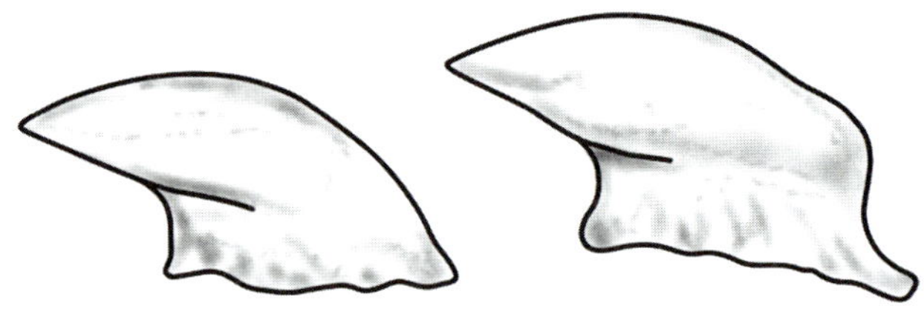

Fig. 62c. Teeth (detail).

sharks; white sharks usually appear uniformly colored, while basking sharks usually have a mottled appearance (Castro 1983). The whale shark has a terminal mouth and white spots and stripes on the dorsal surface.

Range In summer basking sharks are found in subpolar and temperate seas throughout the world; they move into warmer seas in winter. The northern limit of the distribution of the basking shark in the North Atlantic extends from Newfoundland to Iceland, to the North Cape in the northern coast of Norway, and beyond to the Kola Peninsula and the western White Sea (Lien and Fawcett 1986; Stott 1982; Konstantinov and Nizovtsev 1979). In eastern North America, the basking shark ranges from Newfoundland in summer to the southeastern coast, the Bahamas, and the Caribbean Sea in winter. There is evidence that it strays into the Gulf of Mexico (Springer and Gilbert 1976; see below). It is common from Nova Scotia to the Gulf of Maine in summer and off the North Carolina coast in winter. Along western North America, the basking shark ranges from the Gulf of Alaska to Baja California. It is common in the coastal bays of central and southern California throughout

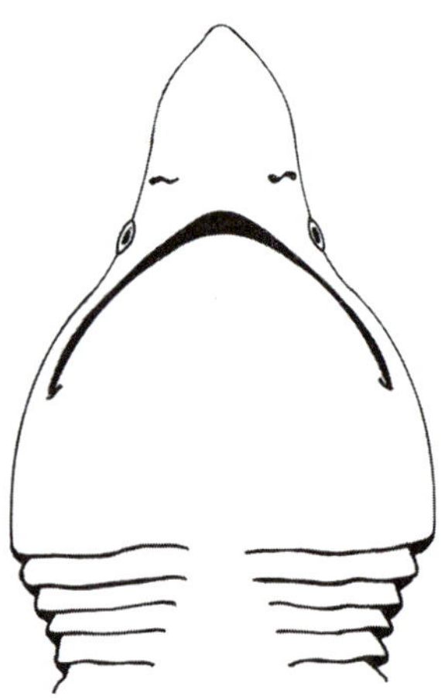

Fig. 62d. Snout.

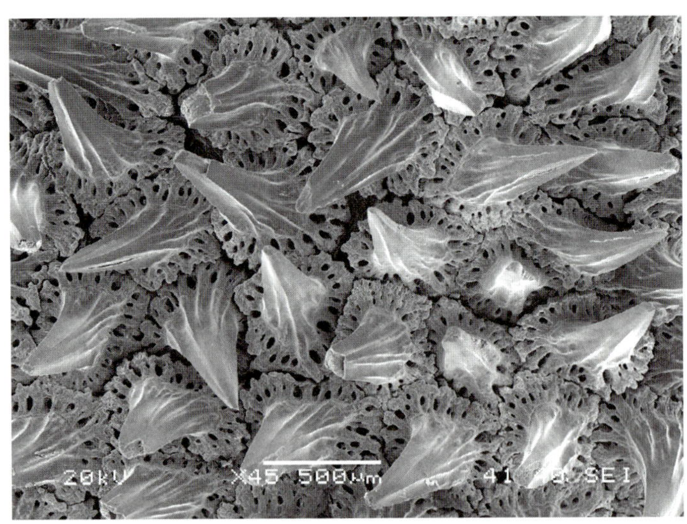

Fig. 62e. Dermal denticles.

the year, being most plentiful in December and January (MacGinitie 1931).

Size and weight The basking shark is the second largest fish in the world, surpassed in size only by the whale shark. Specimens 7–8.5 m are common, and some may reach 9–11 m. Le Sueur (1822) reported that the skin of a basking shark taken off New Jersey in 1822 measured 10.0 m. Goode (1884) reported that a 10.4-m specimen stranded in Eastport, Maine, in 1839. Bigelow and Schroeder (1948) reported that the largest basking sharks in the western Atlantic, for which they could find exact measurements, were 9.80 m, 9.75 m, 9.2 m, and 8.07 m. Yarrell (1841) reported that he examined an 11-m ("36 feet") specimen taken off Brighton, England.

Larger basking sharks (12.2–13.7 m) have been reported several times without supporting evidence. According to Collett (1905), a specimen measuring 13.7 m and three of 12.2 m were taken off the Norwegian coast in 1885–1905. Maxwell (1952) wrote that he believed that average basking sharks in the Hebrides were about 7.9 m long, that the largest shark that he measured was 9.04 m ("29 ft. 8 inches"), and that the largest ever brought to the Soay factory was 9.58 m. However, he had seen 12-m basking sharks and was convinced of their existence and that "the fact that they were always the 'the ones that got away' is due to cause and effect" (Maxwell 1952: 299).

Actual weight data for the largest basking sharks are scarce. Evermann (1919) reported that a 7.1-m specimen entangled in a net off Point Reyes, California, weighed 1,111 kg. MacGinitie (1931) gave the lengths and weights for 22 specimens brought to a processing plant in Seaside, California, between November 1930 and February 1931. The smallest was 4.57 m and 410 kg, while the largest was 8.53 m and 2,991 kg. The largest specimen ever sold to the processing plant was "just a few inches short of thirty feet [9.14 m] and weighed 8,600 pounds" (3,909 kg). MacGinitie also noted the weight of a 7.92 m specimen from Monterey as 2,812 kg.

Biology Like the whale shark, the basking shark is a huge, filter-feeding plankton eater. It is a migratory species that spends the summer at high latitudes, in subpolar and cold temperate seas throughout the world, moving into warmer water in winter. The early description by J. V. C. Smith (1833: 103) best sums up the basking shark: "Though really formidable in appearance, by the concurrent testimony of mariners, it is allowed to be a dull, unsuspecting, harmless creature."

Despite its local abundance in summer, and having been the object of many fisheries, the biology of the basking shark is poorly known. Lack of interest in sharks, and the problems of obtaining or transporting and dissecting such huge specimens, probably inhibited biologists from studying this species. Thus, there is little biological literature on the basking shark, and some of what has been written on the species is armchair biology, often by writers not directly familiar with the species or with sharks.

Basking sharks use their gill rakers to selectively strain minute copepods and other small crustaceans from the water. Hussakof (1915: 26) described the stomach contents of a 426-cm basking shark as "a bright red material. On examination under the microscope, this resolved itself into a vast multitude of minute Crustacea (species not yet determined) whose reddish bodies lent color to the entire mass." Sims and Merrett (1997) have shown that basking sharks off Plymouth, England, actively forage to locate plankton patches containing higher numbers of the larger copepods (e.g., *Calanus helgolandicus*, ~2 mm long) and fewer numbers of the smaller species (*Temora longicornis* and *Acartia clausi*, ~ 0.5–1 mm long). This suggests that the presence of feeding basking sharks at a given locality is determined by the specific characteristics of the zooplankton there.

In the early 1950s, Van Deinse and Adriani (1953) examined the gill rakers of a small number of basking sharks caught in the Netherlands. Specimens examined from April to September had gill rakers present, while those taken in October and November lacked them. A specimen examined in 20 October 1952 lacked gill rakers but had a stomach full of partly digested plankton, leading the authors to conclude that the gill rakers were not indispensable for a basking shark to feed. However, the apparent absence of sharks from northern European waters, together with the belief that they shed their gill rakers in winter, led H. W. Parker and Boeseman (1955) to postulate a novel hypothesis to explain where the basking sharks spend the winter. These authors proposed that the basking shark sheds its gill rakers in October and November, when plankton densities in northern European waters are insufficient to provide the necessary nourishment, and that it then enters a resting, nonfeeding, demersal stage during which a new set of gill rakers develops, becoming functional by February. The capture of rakerless sharks with full stomachs in winter did not deter Parker and Boeseman from developing their hypothesis; they simply said that the feeding needed not have been recent. Matthews (1962: 419) imaginatively expanded on this hypothesis, stating, "At this moment there are probably great schools of these enormous fish quietly resting on the bottom of the sea, perhaps in the heads of the canyons at the edge of the continental shelf, with their metabolism running its lowest level while they grow their new gill rakers ready for browsing on next summer's crop of plankton." This naive hypothesis that basking sharks hibernate is well established in shark lore, although it has never been confirmed. In reality, any animal, dead or alive, lying at the bottom of the continental slopes is quickly set upon by myriad scavengers (hagfish, ratfish, isopods, amphipods, shrimp, etc.) that enter through whichever orifices they can penetrate and quickly consume the soft tissues. In the deep waters of the continental slopes, I have often seen hooked sharks, dead on longlines for only a short time, be reduced to empty skins in a few hours by scavengers that entered through the gills and anus. I have actually watched the process unfold from submersibles at depths in excess of 500 m on the continental slopes, where shrimp and the "huge and ferocious" isopod *Bathynomus* alone can reduce bait and hooked fish to skeletons within hours. It is likely that the "missing" basking sharks of the eastern Atlantic have moved to warmer or southern areas, or may be simply feeding actively well below the surface in areas where they are not available to fisheries or observation (Castro 1983). Recent work using satellite tags has revealed that basking sharks travel far to their winter areas and that, once there, they go into mesopelagic depths, confirming my speculation of long ago. Gore et al. (2008) reported that a basking shark tagged off Isle of Man, British Isles, on 21 June 2007 crossed the North Atlantic, released its tag off Newfoundland, Canada, 82 days later (a horizontal distance of 9,589 km), and reached a record depth of 1,264 m. Skomal et al. (2009) demonstrated that tagged sharks moved from their feeding areas off southern New England to winter areas in the Bahamas, the Caribbean Sea, and the coast of northern South America, and that in these areas the sharks descended into mesopelagic depths (200–1,000 m) and in some cases stayed there for weeks or months.

Migrations: The seasonal movements of the basking sharks probably encompass entire ocean basins and are only partially understood. In the North Atlantic, basking sharks move to the northern parts of their range in summer while the plankton is abundant there, and then travel southward as the abundance of plankton decreases with the onset of winter. Sims et al. (1997) and Sims and Quayle (1998) have shown that basking sharks forage along thermal fronts and seek the highest densities of zooplankton, demonstrating that the migrations (or the appearance at one location) of basking sharks are probably related to zooplankton abundance.

Along the eastern coast of North America, basking sharks can be found from Newfoundland to Cape Hatteras in summer and from Virginia to Florida and the Caribbean Sea in winter. In some winters, basking sharks are common from southern Virginia to Cape Hatteras. As Skomal et al. (2009) have shown, some basking sharks travel southward in winter to the Bahamas, the Caribbean Sea, or as far as the northern coast of South America.

The general timing of the north–south migrations of the basking shark along eastern North America is not well defined but can be discerned from the following reports. Lien and Fawcett (1986) reported that, between 1980 and 1983, some 371 basking sharks were caught incidentally in inshore fishing gear, primarily salmon gillnets and cod traps, in the coastal waters of Newfoundland. Of these, 90% were taken in waters where the surface temperatures ranged between 8 and 12°C, with highest catches occurring where the average surface temperature was 9.3°C. Most sharks were caught between mid-June and mid-July. Backus (1960) reported seeing about 30 individuals during an hour's steam southeast of Nova Scotia in late June. MacCoy (1931) recorded a 380-cm female from York Harbor, Maine, in early August. Scattergood (1948) reported four specimens from Maine from 23 August to 17 September.

Kenney et al. (1985) reported that basking sharks were sighted between Cape Sable, Nova Scotia, and Cape Hatteras, North Carolina, from March to December during aerial and shipboard surveys of marine mammals, although most were seen there from May through August. Gudger (1948) reported three specimens from the North Carolina coast, all about 4 m long, caught between February and April. I examined a 7.82-m male that became entangled on a net off Cape Hatteras, North Carolina, in March 2002. Coles (1915) reported the species off Cape Lookout, North Carolina, in July 1905. Eric Sander (pers. comm., January 1987) reported two basking sharks about 22 km east of Ponce de Leon Inlet, Florida, on 28 January 1987. Choy and Adams (1995) reported a basking shark actively feeding along a distinct oceanic front off Cape Canaveral, Florida, in January 1994. The shark was feeding at the surface along the warmer and clearer eastern side of the frontal boundary. Springer and Gilbert (1976) reported on two Florida specimens. One was a 3.59-m juvenile that was caught in a shrimp trawl off Saint Augustine in February 1970; the other specimen, an adult female 8.29 m long, was caught off Sarasota, Florida, on 2 April 1969. The Sarasota specimen was the first ever—and thus far, the only—basking shark recorded in the Gulf of Mexico.

Off central California, basking sharks are present throughout the year, being most numerous from September to May and scarce in summer. MacGinitie (1931) stated that, off central California, basking sharks could be seen any day of the year, but they were most plentiful in December and January. Squire (1990) noted that basking sharks are frequently observed along the central and southern California during winter and spring months.

Basking sharks are also abundant at many northern European localities during the spring, departing in the early autumn (Matthews 1962; H. W. Parker and Stott 1965; Stott 1982). The winter whereabouts of the European population are unknown, although occasionally specimens are caught or become stranded in winter. Work by Gore et al. (2008) suggests that some specimens cross to the western Atlantic.

Basking sharks sometimes breach out of the water. This habit has been known since the days of Pennant (1776: 90), who wrote: "At certain times they were seen sporting on the waves and leaping with vast agility several feet out of the water." Matthews and Parker (1950) expressed the opinion that it was very improbable that a fish as sluggish and slow moving as the basking sharks was in the habit of breaching. However, Matthews and Parker (1951) later stated that their opinion was not based on personal experience with basking sharks and quoted the experiences of Capt. Harry Thomson off the Loch Brolum-Shiant Islands area in June 1948. Thompson described seeing basking sharks leap clear of the water, with the tail being at least 182 cm above the water. He stated that this aerobatic performance always occurred outside the 40-fathom (73 m) curve and that, after a shark leapt out of the water, a shoal would surface shortly afterward. The leaping behavior was most prevalent during mating time in May–June, but this is the time when the schools are most dense. Thus, the cause of the leaping could not be ascertained without more direct evidence. Sims et al. (2000) observed that breaching by basking sharks occurred during social interactions and speculated that breaching behavior by females could function to announce receptivity to mating. There seems to be little basis for this speculation.

An interesting study of the brain of the basking shark by Kruska (1988) concludes that the basking shark has proportionally the smallest brain of any shark investigated. The brain, which occupies only 1/16 of the brain cavity, is very small for such large shark. The cerebellum, on the other hand, is comparatively large, which may indicate a well-developed motor control center. The relatively small brain indicates that a filter-feeding plankton eater does not require a brain capable of a high level of behavioral control.

Size at maturity There are few data on when basking sharks reach maturity. Males are believed to reach maturity between 4.6 and 6.1 m (Bigelow and Schroeder 1948). Compagno (1984a) stated that females mature at 8.1–9.8 m, but the source of the information is not clear.

Reproduction Despite thousands of basking sharks caught in European fisheries, almost nothing is known about its reproductive processes because gravid females have not been examined. Matthews (1950) demonstrated that the basking shark has a lamnoid-type ovary. This type of ovary indicates oophagy (egg eating), as one would expect in a lamnoid shark. H. W. Parker and Stott (1965) suggested that gestation lasts for 3.5 years, based on vertebral ring counts. This hypothesis, in turn, led "to the expectation that, since pregnant (or gravid) females are almost or quite unknown, they must live in a hidden, perhaps demersal, life for three consecutive years after the first mating" (H. W. Parker and Stott 1965: 305). This is a far-fetched hypothesis, as no other elasmobranch (or any other animal, for that matter) has been demonstrated to have a 3.5-year gestation period. It is also difficult to explain the whereabouts and mode of living of the gravid females for three years. Stott (1984) later stated that length–frequency analysis by itself indicated a gestation period of just over a year, noting this estimate seemed a more likely possibility.

Size at birth: The basking shark is believed to give birth to young measuring about 180 cm, probably in high latitudes. Sund (1943) reported that a Norwegian fisherman catching a female basking shark, which aborted or gave birth to five live and one stillborn young, estimated to range between 1.5 and 2 m long. This capture was said to have taken place in August, although this date may not be correct, as it appears to be too late in the year for birth of a species inhabiting the high latitudes. H. W. Parker and Stott (1965) reported that the two smallest specimens on record were 168 cm (Lough Swilly, July 1849; Thompson 1856) and 180 cm (Casablanca, May 1955; Furnestin et al. 1958) and assumed that these were young-of-the-year.

Brood size: There are no modern reports on basking shark brood size.

Nurseries Unknown.

Age and growth H. W. Parker and Stott (1965) estimated that basking sharks reach maturity at an estimated age of four to five years. This unvalidated age estimate is probably far off the mark because of the usually long periods required for most temperate sharks to reach maturity.

Natanson et al. (2008) reported that vertebral band pair counts (the alternating opaque and translucent bands normally used to age sharks) varied dramatically along the length of the vertebral column and that large sharks had a difference up to 24 band pairs between the highest and lowest counts along the column. These authors concluded that vertebral band pair deposition in the basking shark may be related to growth and not age, and thus, the species cannot be directly aged using vertebral band pair counts. Therefore, the age of these large sharks can only be guessed at this time.

Relation to humans Basking shark fisheries have existed in northern Europe since the eighteenth century. The species was fished primarily for its liver oil, which was burned in lamps until replaced by petroleum products. The oil was also used for dressing wool (Went and Ó Súilleabháin 1967) and to treat bruises, burns, and rheumatic complaints (Pennant 1776). Modern fisheries of the twentieth century also pursued the basking shark for its liver oil, which was used for the production of vitamin A. A single basking shark is said to yield 200–400 gallons of oil.

A basking shark fishery already existed in Irish waters[1] in the 1750s. This fishery was characterized by large annual fluctuations in occurrence and abundance, and many authors commented on its erratic or sporadic nature. In 1836, James A. Morris wrote: "Only a few fish may be seen at one season, and a thousand at another; but one good year compensates for the disappointment of three" (Went and Ó Súilleabháin 1967: 96). In 1856, a Dr. Ball, in a lecture on the fishes of Ireland, stated: "The appearance of the fish is so uncertain, and the number so few, that the taking of one is quite a lottery" (Thompson 1856: 254). By the turn of the nineteenth century, the hunting of basking sharks was greatly reduced or had virtually been abandoned. After World War II, a fishery was revived at Achill Island, off the western coast of Ireland, lasting until 1975 (International Council for the Exploration of the Sea 1995). This fishery peaked in 1955, when 1,708 sharks were caught. It declined thereafter, and by 1965, only 47 basking sharks were caught (Went and Ó Súilleabháin 1967). After 1983, a single boat targeted basking sharks when they were available in the Clyde estuary, but that fishery has now ceased (International Council for the Exploration of the Sea 1995). H. W. Parker and Stott (1965) described this fishery during the 1950s and attributed its decline to overfishing.

In the 1760s, basking shark fisheries developed in Scotland and the Hebrides. By about 1780, small-scale fishing was established in the Firth of the Clyde and in the Sea of the Hebrides (Fairfax 1998). This fishery ended by the late 1830s, probably due to competition from whale oil and petroleum products. Nearly 100 years later, the fishery was revived on a small scale to obtain the shark liver oil, which was in great demand as a source of vitamin A. This fishery came to an abrupt end with the onset of World War II. After the war, the fishery was revived again for a few years, but it ended by 1950. There was a small operation in the Clyde from 1982 to 1994 (Fairfax 1998).

The modern basking shark fishery (1930–1950) around the British Isles is well documented in several good books (Maxwell 1952; O'Connor 1953; Watkins 1958; Geddes 1960; McNally 1976). There is also a popular book on the basking shark in Scotland (Fairfax 1998) giving a good and readable account of the fisheries. In 1933, Robert J. Flaherty made a cinematographically interesting silent film, *Man of Aran,* depicting the traditional method of fishing for basking sharks off western Ireland. This film, for the most part, had to be staged because, by then, the traditional fishing methods had been abandoned (I do not recommend searching widely for this film).

In the 1960s and 1970s, more than 30 vessels participated in the European fishery for the whole or part of the season (International Council for the Exploration of the Sea 1995). Today, the basking shark is fished for its fins, primarily in a small Norwegian fishery.

The presence of the basking shark off the western coast of North America has always been sporadic, just as elsewhere. Reporting on whaling activities off Monterey, California, A. H. Clark (1887: 58) stated that two sharks had been obtained in 1880, and that "the basking shark is rare here, sometimes not seen for 20 years." By the early part of the twentieth century, basking sharks were reported as abundant off central California. Chute (1930) noted that fishermen in Monterey Bay had reported sightings of up to 500 sharks at time. Walford (1935) stated that basking sharks appeared in schools of 20–30 individuals in Monterey and San Simeon bays during the winter months. Roedel and Ripley (1950) stated that schools of 75–100 individuals had been observed.

A basking shark harpoon fishery first developed in California in 1924, the sharks being utilized entirely for fertilizer or chicken feed, with the oil used to prepare the latter (MacGinitie 1931; Phillips 1948). From 1924 to 1938, the yearly average was 25 sharks, but in one year, 100 sharks were taken (Phillips 1948). In the late 1940s, the fishery expanded using war surplus aircraft and amphibious vehicles (Fitch 1948). Between September 1946 and May 1947, about 300 basking sharks were taken along the California coast (Phillips 1948). Fishing for the species was suspended in California in October 1950 because of the low prices paid for the oil and the low availability of basking sharks (Squire 1967).

Fisheries for the basking shark have never lasted long. The primary cause of their demise appears to be the unpredictable nature of the basking shark migrations, although local overfishing may also play a part. Stott (1982) studied basking sharks catches off Norway in 1971–1973. He found that, generally, basking sharks were first caught there in May, with last catches occurring in early August. He concluded that the migrations of the basking shark into that area were quite unpredictable and that this unpredictability resulted in its survival there (unlike the predictable bluefin tuna, which disappeared from the local fishery in the 1960s due to overfishing). Kunzlik (1988) reviewed the literature on basking sharks in Europe, and he also attributed the decline of the basking shark fisheries to the erratic nature of the seasonal appearances of the basking shark, as well as

1. For an excellent account of the Irish basking shark fishery, see the work by Went and Ó Súilleabháin (1967). The first section of this paper consists of a historical account of the fishery from published sources. The second section gives a number of accounts of fishing for basking sharks, recorded in manuscripts from the early 1930s and preserved by the Irish Folklore Commission.

to the collapse in the price paid for livers. Kunzlik (1988: 18) also noted that "there is no firm evidence to suggest that the species has been threatened, even locally, by exploitation." Thus, it appears that the unpredictable movements of the basking shark are tied to the abundance of plankton, which cannot be easily predicted. Nevertheless, the basking shark, like most elasmobranchs, is likely to be susceptible to overexploitation.

Basking sharks are often entangled in inshore fishing gear, causing great damage to nets and traps because of their size. Lien and Fawcett (1986) reported that numerous basking sharks were incidentally caught in salmon gillnets or cod traps in the coastal waters of Newfoundland. In the 1950s, basking sharks were damaging salmon nets in British Columbia waters. In response, the Canadian Fisheries Department conducted a basking shark eradication program, and a boat was equipped with a sharp ramming device for killing sharks. About 50 sharks were reported killed in the first month of operation (Squire 1990).

In 1997, the National Marine Fisheries Service placed the basking shark on the list of protected species, which may not be caught or landed along the Atlantic coast of the United States. As there have never been any basking shark fisheries in eastern North America, the move to protect basking sharks was precautionary. In December 2002, the basking shark was placed on Appendix II of the Convention on International Trade in Endangered Species (CITES), along with the whale shark. Consequently, products from the basking shark can be sold only with special permits.

FAMILY LAMNIDAE
The Mackerel Sharks

The Lamnidae is a small family of large, fast-swimming sharks also known as the isurids, or mackerel sharks. They share many adaptations for high-speed swimming: a conical snout, large gills for more efficient gas exchange, a streamlined fusiform body, a greatly reduced second dorsal fin, a caudal peduncle that is dorsoventrally flattened to form prominent keels on both sides (strengthening the tail), and a lunate tail with two nearly symmetrical lobes. These adaptations allow lamnids to be at the top of oceanic food chains by allowing them to feed on other fast-moving predators such as sharks, swordfish, tunas, salmon, seals, and sea lions. Some lamnids maintain elevated body temperatures some 7–14°C higher than the temperature of the surrounding water. The increased temperature is achieved by means of a network of highly branched blood vessels (known as the *rete mirabile* or "wonderful net") that acts as a countercurrent heat exchanger through the musculature and around the viscera, thus preventing heat from being dissipated into the external environment.[1] This adaptation allows the lamnids to enter cold waters. Future research will probably show elevated body temperature to be characteristic of all lamnids. The seasonal migrations of lamnid sharks often encompass vast oceanic areas. Some migrations may be in pursuit of prey, but few details are known. Lamnids are aplacental viviparous, and the embryos are oophagous (egg eaters). Three genera and five species are recognized. The monotypic genus *Carcharodon* is characterized by flat, triangular teeth with serrated edges. The genera *Isurus* and *Lamna* have smooth-edged teeth with narrow triangular cusps. *Isurus* is characterized by a first dorsal fin originating behind the pectoral fin free rear tip, and teeth without lateral cusplets. It comprises two species. The genus *Lamna* is characterized by the first dorsal fin originating above the pectoral fin axil and teeth with lateral cusplets. It also includes two species. All the lamnid sharks are widely distributed.

1. The *rete mirabile,* or "wonderful net," is a network of closely apposed, parallel, small veins and arteries that supply or drain the viscera and red muscle of some fishes. In this system, the venous blood leaving the viscera or muscles, after having been warmed by metabolism, gives off its heat to the cold, newly oxygenated blood coming from the fish's gills. This system, known as a "countercurrent heat exchanger," minimizes heat from being dissipated at the gills and results in a net temperature increase. This temperature increase may result in greater physiological efficiency in cold water. For excellent reviews, see Scholander (1958) and Carey (1973). For an excellent historical review of the discovery of the *rete mirabile,* see Fudge and Stevens (1996). Note that the singular term is *rete mirabile,* and the plural is *retia mirabilia.*

KEY TO THE LAMNIDAE

1a. Teeth broadly triangular, with serrated edges..**White shark,** *Carcharodon carcharias* (p. 258).

1b. Teeth long and slender, with smooth edges 2

2a. Teeth with small lateral cusplets 3

2b. Teeth lacking small lateral cusplets... 4

3a. Distance from the tip of the snout to the front of the eye less than one-half the distance from the back of the eye to the first gill slit, a shark of the North Pacific.. **Salmon shark,** *Lamna ditropis* (p. 278).

3b. Distance from the tip of the snout the front of the eye more than one-half the distance from the back of the eye to the first gill slit (a shark of the North Atlantic) ....**Porbeagle,** *Lamna nasus* (p. 284).

4a. Length of pectoral fins 16–17% of total length, ventral side of snout and area around mouth white (see illustration on p. XX) .... **Shortfin mako,** *Isurus oxyrinchus* (p. 267).

4b. Length of pectoral fins 20–28% of total length, ventral side of snout and area around mouth bluish or black, specimens less that 120 cm may have only dusky splotches (see illustration on p. XXX).. .... **Longfin mako,** *Isurus paucus* (p. 274).

WHITE SHARK

Carcharodon carcharias (Linnaeus, 1758)

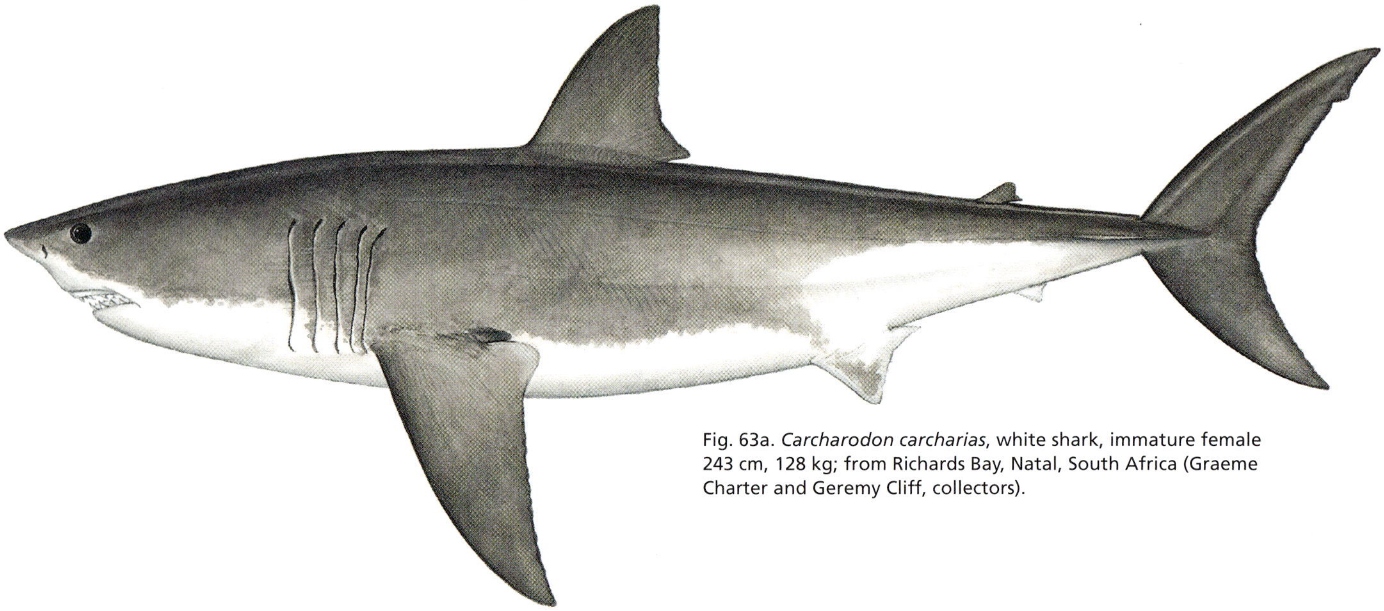

Fig. 63a. *Carcharodon carcharias*, white shark, immature female 243 cm, 128 kg; from Richards Bay, Natal, South Africa (Graeme Charter and Geremy Cliff, collectors).

Common name White shark, because it appears whitish or light gray in comparison to other sharks, when viewed in the water from above. In English-speaking countries the white shark has been known by many names: great white shark, white pointer (Australia), white death (Australia), blue pointer (South Africa), and man-eater shark, all alluding to its coloration or "man-eating" propensities.

Spanish name *Jaquetón* (Cuba) and *tiburón blanco* (Cuba, Mexico, A.F.S.).

Scientific name *Carcharodon carcharias* (Linnaeus, 1758), originally described as *Squalus carcharias* in *Systema Naturae* 10th ed., I: 235. *Carcharodon:* from the Greek *karkaros,* rough, and *odous,* tooth, an allusion to the rough edges of this shark's teeth; *carcharias:* from the Greek *karcharios,* name for man-eating sharks.

Synonyms *Carcharodon rondeletti* Müller and Henle, 1839; *Carcharodon smithii* Agassiz, 1838.

Identification The white shark is heavy bodied with a *conical snout,* prominent keels on the caudal peduncle, a lunate tail with nearly equal lobes, and *triangular teeth with serrated edges on both jaws.* The upper teeth are broadly triangular, whereas the lower teeth have narrower cusps. Small juveniles have proportionately narrower teeth with (usually) a small lateral cusplet on either side of the main cusp. In juveniles less than 200 cm, the first lower tooth lacks serrations. The teeth number U: 12–12, L: 11 or 12–11 or 12. The dermal denticles are small and closely overlapping. They have three ridges and three points, with the central ridge being the largest and terminating in a longer point. Coloration varies from lead gray to slate brown above, shading to white below. Often there is a black spot around the axil of the pectoral fin.

Similar species Makos, porbeagles, and salmon sharks all have similar stout bodies with pointed snouts and caudal keels on the caudal peduncle and tail, but their teeth have *smooth edges.* Large shortfin makos and salmon sharks swimming close to the surface can be mistaken for white sharks.

Range The white shark is widely distributed in temperate, subtropical, and tropical waters. Along eastern North America, the white shark ranges north to Labrador and Newfoundland and southward to Florida and the Gulf of Mexico. The few records for the Labrador coast and Newfoundland are all for August (Templeman 1963). The northernmost record of a white shark for the western North Atlantic appears to be a specimen estimated at 370 cm that became entangled in a cod trap leader in Hare Bay, Newfoundland, at 51° N in August 1956 (Templeman 1963). South of Newfoundland, the white shark can be found seasonally all along the coast to Florida and the Gulf of Mexico, although its presence is usually sporadic and unpredictable. Few records exist of white sharks in the Gulf of Mexico, perhaps reflecting the small numbers in the area, or more likely, the difficulties of capturing them with standard fishing gear. There is only one record of a white shark for Cuba (Guitart 1974), the fabled "Cojímar monster" that supposedly measured 6.4 m (see below). The southern boundaries, if any, of the eastern North American population are probably in the Caribbean Sea.

Off western North America, white sharks have been reported from near Craig, Alaska, at 55° N (Hart 1973), south to Mexico. Bonham (1942) reported five specimens from Willapa Bay, Washington, in August and September, suggesting that the species might be common in the area. There are numerous reports of white sharks encountered around seal and sea lion rookeries off Central California (Le Boeuf et al. 1982; Ainley et al. 1981). The white shark also enters the Gulf of California, where two specimens were landed at El Barril, Baja California, in 2001 and 2002.

Most records of white sharks are for coastal or continental waters. There is a dubious record reported by Smith-Vaniz et al. (1999) off Bermuda. Those authors relied on a photograph taken from a vessel on 15 May 1985. However, I was on that vessel when the sighting occurred and the photograph was taken; the fish swam lazily on the surface for a time and could be clearly identified as an unusually large shortfin mako. Moreover, later that day, a mile or so from the location of the sighting, angler Andrew Card captured a 299-cm (fork length), 328-kg female mako (Craig. A. Trott, Division of Fisheries, Bermuda, pers. comm., 10 September 1985). At a fish house I subsequently examined the mako's carcass, which was recorded as the largest shortfin mako ever caught in Bermudan waters. It stretches credulity to imagine that both the only white shark ever sighted off Bermuda and such a memorable mako would have been at almost the same location on the same day.

Size and weight Historical reports of white sharks have often exaggerated the animal's size. For many years it was believed that the white shark grew to 11.1 m, based on the jaws of an Australian specimen in the British Museum reported by Günther in 1870. This specimen was often referred to as a "giant" of its kind because the second largest white shark was the putative a 6.4-m shark captured off Cojímar, Cuba, in 1945 (Bigelow and Schroeder 1948; Guitart and Milera 1974). Perry Gilbert questioned the accuracy of the size attributed to the British Museum specimen after examining the jaws of a 5-m specimen from Durban, South Africa (Gilbert and Gilbert 1973). Noting that the jaws of the two animals were the same size, Gilbert concluded that the 36.5 ft (11.1 m) length attributed to the British Museum specimen was probably a printer's error, and that the correct figure should have been 16.5 ft (5 m). Randall (1973) also examined the jaws of the British Museum specimen, comparing it with the jaws of specimens of known lengths. After comparing tooth enamel height measurements, Randall also concluded that the specimen was about 5 m long.

The size of the 6.4-m (21-ft) specimen from Cojímar is also suspect. The measurement comes from Bigelow and Schroeder (1948), who had received it as a personal communication from Luis Howell-Rivero from Havana, Cuba. This measurement appears to be anecdotal. Randall (1987) analyzed the data from the 6.4-m shark from Cojímar. By studying a picture of one of the shark's teeth, figured against a millimeter scale (Guitart and Milera 1974), Randall calculated that the enamel height of 44 mm corresponded to a 4.9 m shark. In addition, Randall analyzed one of the illustrations of one of the shark's vertebrae, which measured 80 mm in diameter and 37 mm in width, using the regression for vertebral centra of white sharks given by Cailliet et al. (1985). Randall then determined that a vertebra of such size would belong to a 5-m shark.

In December 2002 I received two old photographs of the fabled Cojímar shark from Jorge Rey of Miami, Florida, who, as a boy of 12, saw the shark in Cojímar. One photograph shows the same shark illustrated by Guitart and Milera (1974) with quite a few more bystanders around it. The other photograph was taken from just behind the first dorsal fin and shows a boy—identified by Rey as himself—nearby. A handwritten note on the back of this photograph reads (translated from Spanish) that the "monster" weighed 7,125 lb, its liver weighed 1005 lb, and it measured 19.5 feet, thus two feet less than the Howell-Rivero length. The weights are the same as those given by Guitart and Milera (1974).

Several factors complicate the use of teeth and vertebrae comparisons in determining the size of white sharks. Gordon Hubbell (1996), who has probably examined more white shark jaws than anyone, stated that tooth structure is highly variable among sexually mature white sharks. Some individuals have relatively long narrow teeth, while others have short, wide teeth (G. Hubbell, pers. comm., January 2001). Vertebrae also vary greatly according to their position along the spinal column and among individuals. Even so, the Randall analysis and the Rey photographs do cast doubt on Howell-Rivero's measurement. A similar exaggeration occurred with a tiger shark reported by Howell-Rivero to Bigelow and Schroeder (1948: 269).

The largest white shark believed to have been measured reliably is a 6-m specimen from Ledge Point, Western Australia, caught on 22 March 1984. The jaws of this specimen are currently in Gordon Hubbell's collection in Gainesville, Florida. The largest upper tooth has an enamel height of 51 mm. As Randall (1987: 168) said: "Undoubtedly *Carcharodon carcharias* exceeds 6.1 m in length, but as yet there is no authenticated record of such a size."

Weights of small white sharks I have measured, or that have been reliably reported in the literature, are females

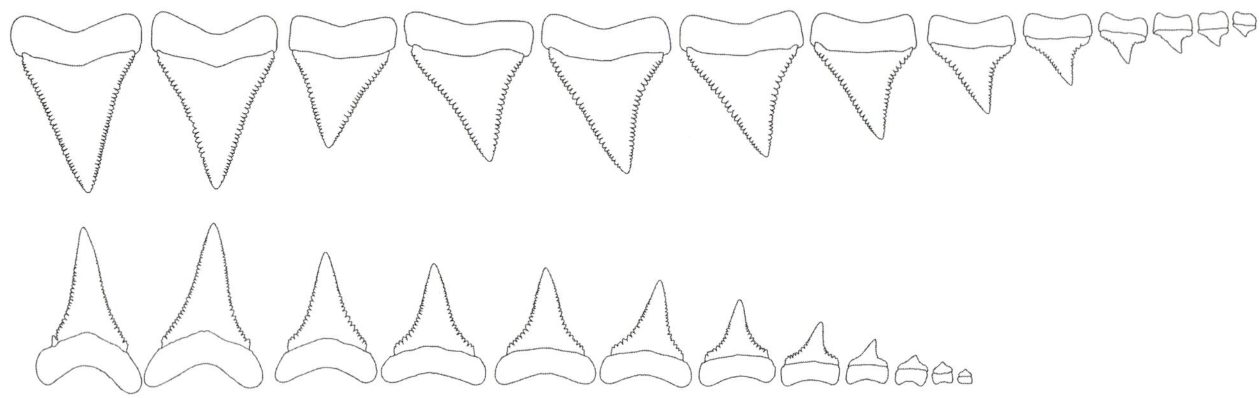

Fig. 63b. Teeth.

measuring 131 cm, 13.3 kg; 183 cm, 65.5 kg; and 243 cm, 128 kg; and a male 199.8 cm, 65.7 kg (South Africa). Reported lengths and weights for large males are 390 cm, 442 kg; 474 cm, 892 kg; 478 cm, 1,037 kg; and 528 cm, 1,279 kg (Kohler et al. 1996; Cliff et al. 1996). Few weights for the larger females can be found in the literature. Uchida et al. (1996) gave these values for females caught off Okinawa, Japan: 436 cm, 850 kg; 506 cm, 1,790 kg; and 555 cm, 1,970 kg.

Biology Despite tremendous popular and scientific interest in the white shark, the species remains poorly known. Several factors combine to make observations and research on the white shark extremely difficult. First, the white shark is a scarce apex predator with vast seasonal ranges, and its presence anywhere is difficult to predict. Second, even at locations where white sharks are known to visit seasonally, such as pinniped haul-out sites or rookeries, actually locating the specimens can be difficult. Third, the size of the white shark and its reputation as a confirmed man-eater have dissuaded most divers from attempting to observe them in the water, although the perception of white sharks is changing quickly. Fourth, in many places today, white sharks are prohibited species that cannot be landed; even when white sharks are landed, they are seldom available for anatomical studies because their large size or commercial value causes them to be butchered or discarded quickly. Fifth, only recently have aquarists succeeded in keeping white sharks in captivity for more than a few days. All these factors have not prevented laymen and scientists from writing about white sharks.

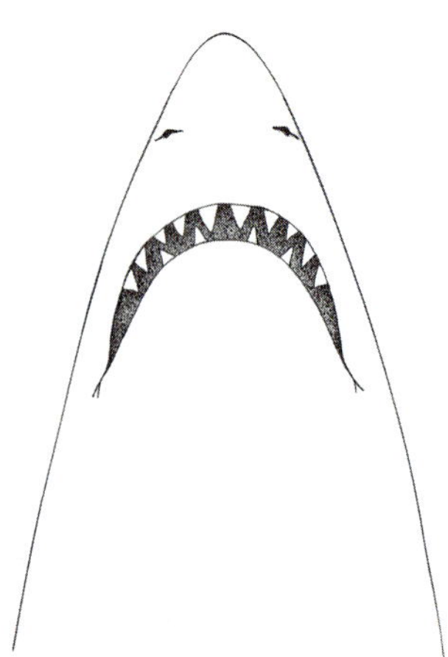

Fig. 63c. Snout.

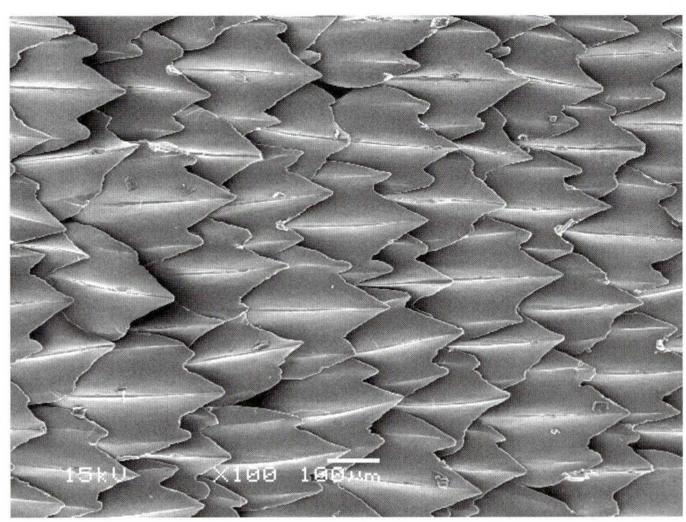

Fig. 63d. Dermal denticles.

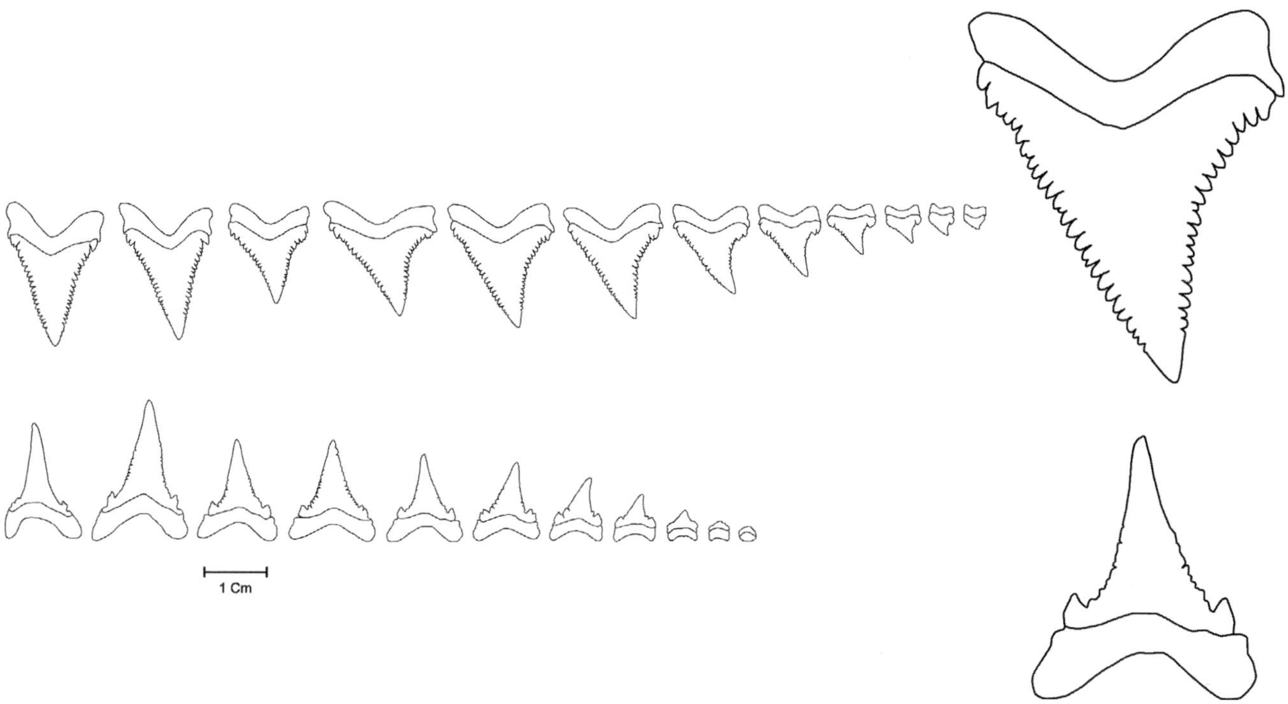

Fig. 63e. Juvenile teeth.

However, much of what has been written, both in the popular press and the scientific literature, is speculative or inaccurate.

Perhaps two dozen white sharks are caught each year in sport and commercial fisheries in the United States, and specimens are occasionally available for anatomical studies. Still, know very little about the white shark's age at maturity, growth and longevity, gestation, reproductive cycle, physiology, or behavior. What we know for certain might fill only a few pages. Therefore, we must consider white sharks throughout the world to arrive at a reasonable synopsis for the North American population.

The white shark can be found nearly anywhere, from shallow coastal waters to the deep oceans; it even wanders into the surf. While its presence is generally sporadic, small seasonal concentrations of white sharks occur in suitable habitats. Most records of white sharks are for coastal or continental waters, and there are very few oceanic records. The lack of observations in oceanic waters probably reflects the small probability of encounter. The occasional presence of white sharks around oceanic islands, a few sighting records, and an attack on a swimmer recorded in oceanic waters had long suggested that white sharks inhabited oceanic areas, but confirmation has been obtained only recently. Boustany et al. (2002: 36) tagged six sharks with satellite transmitters off California. All remained near the coast for long periods, but four later moved offshore, where they remained exclusively pelagic. One shark traveled 3,800 km to the west coast of the Hawaiian island of Kahoolawe, and the others moved to the subtropical eastern Pacific.

White sharks are found across surface temperatures ranging from 10 to 25°C. According to Casey and Pratt (1985), surface water temperatures reported in 73 cases of white shark occurrences in the western North Atlantic ranged from 11 to 24°C, with 75% of the occurrences being in waters of 15–22°C. Squire (1967) reported white sharks to occur throughout the year in Monterey Bay, California, where mean monthly temperatures range from 10.2 to 14.4°C. Bruce (1992) stated that white sharks in South Australian waters have been caught at temperatures of 12–24°C. The ability of the white shark to inhabit cold waters is probably due to its elevated body temperature, which can be 5–14°C above that of the surrounding water. Like other thermoregulating lamnids, the white shark has countercurrent heat exchangers in the circulatory system that allow it to conserve body heat. These organs, each known as a *rete mirabile,* consist of masses of parallel arteries and veins in close apposition, which minimize heat being dissipated at the body surfaces and gills (Carey et al. 1985). As a result, the shark's internal temperature is usually (at least) 4–5°C above that of the surrounding water (Carey et al. 1985) and can be even higher. Off San Francisco, Goldman et al. (1996) used acoustic telemetry to track an adult white shark for eight days at the South Farallon Islands, and they recorded a maximum body temperature of 13.7°C above the surface water temperature. Goldman (1997) recorded stomach temperatures as high as 14.3°C above ambient temperature.

Juvenile white sharks do not seem to be able to tolerate as wide a temperature range as the adults. Casey and Pratt (1985) reported that, although white sharks more than 300 cm have been reported in every region from Florida to Newfoundland, individuals less than 200 cm are common only in the "Mid-Atlantic Bight."[2] Of the 135 white sharks in their database, only four are from the Gulf of Maine, one was captured in southeastern waters, and the remaining 130 were taken in the "Mid-Atlantic Bight."

White sharks appear to spend much of their time in relatively shallow water, but they have been tracked to depths of 680 m (Boustany et al. 2002). Casey and Pratt (1985) reported catching 10 juveniles at a depth of 9 m near a popular beach. In reviewing 217 white shark records from South Australia, Bruce (1992) concluded that all were from continental waters, and the majority occurred in inshore waters less than 15 m deep. Goldman and Anderson (1999) found that four white sharks at the Farallon Islands, California, generally swam within 10 m of the bottom to depths of about 30 m, straying higher in deeper water. This pattern presumably allowed them to detect prey outlined against the surface glare while remaining concealed against the dark background.

A fully grown white shark has few predators other than man and the killer whale (*Orcinus orca*). Pyle et al. (1999) observed and filmed an adult killer whale kill and partially ingest a 3- to 4-m white shark near the Southeast Farallon Island in October 1997. It is not known how commonly this occurs. Man is probably the only significant threat to white sharks, and those animals that can avoid nets and hooks can live a very long time.

Diet: Small white sharks are piscivorous; adults prey on other sharks and marine mammals or scavenge dead whales and other large animals. Juvenile white sharks of less than 200 cm eat a variety of fishes, especially small sharks. Casey and Pratt (1985) examined the stomach contents of 54 juvenile white sharks (22–82 kg) and found sea robins, menhaden, hakes, skates, bluefish, flounder, butterfish, silver hake, and smooth dogfish, as well as invertebrates such as crabs (*Cancer* spp.) and sea stars (*Asterias* spp.). I examined a 183-cm female caught off Point Lookout, North Carolina, on 1 May 1987 that contained two 70-cm sandbar sharks. Both had been bitten in two by a single bite in front of the second dorsal fins and both halves swallowed. A 200-cm male captured at the same time and location contained the head of a 15-cm unidentified bony fish.

As white sharks grow larger than about 250–300 cm, their diet shifts to other species of sharks, sturgeons, seals, sea lions, and small cetaceans. A 249-cm, 155-kg specimen caught near the mouth of the North River, Washington, contained a 120-cm sturgeon, most likely a green sturgeon (*Acipenser medirostris*) (Bonham 1942). A 282-cm white shark harpooned near Provincetown, Massachusetts, contained a 90-cm spiny dogfish and the head of an unidentified fish (Schroeder 1938).

Along the west coast of North America, adult white sharks are the primary predators of pinnipeds (Arnold 1972; Ainley et al. 1981; Le Boeuf et al. 1982; Scholl 1983), feeding on California sea lions (*Zalophus californianus*), elephant seals (*Mirounga angustirostris*), Steller's sea lions (*Eumetopias jubatus*), and harbor seals (*Phoca vitulina*). White sharks also occasionally prey on harbor porpoises (*Phocoena phocoena*), Dall's porpoises (*Phocoenoides dalli*), Pacific white-sided dolphins (*Lagenorhynchus obliquidens*), small pygmy sperm whales (*Kogia breviceps*), and many others. However, based on stranding records, white shark predation on cetaceans along western North America seems uncommon (Long and Jones 1996).

Along the eastern coast of North America, there are also reports of predation on pinnipeds. Scattergood (1962a) noted that a 363-cm white shark caught off Mark Island, Maine, contained two harbor seals, and a 320-cm shark harpooned off Isle au Haut, Maine, contained a single seal.

Studies from other areas have also demonstrated the shift in prey size as a white shark grows. Cliff et al. (1989) examined stomach contents of 591 white sharks caught in gillnets off Natal, South Africa. They found that sharks of less than 303 cm fed most often on other elasmobranchs, these being present in 44% of the stomachs. Bony fishes were present in 35% of the stomachs, and marine mammals in 26%. In larger sharks, marine mammals were the most common category being present in 40% of the stomachs, followed by bony fishes at 32%, and elasmobranchs at 28%.

Large white sharks on both coasts of North America scavenge dead whales. Pratt et al. (1982) observed four to nine white sharks feeding on the carcass of a fin whale (*Balaenoptera physalus*), but they saw only one shark feeding at a time. McCosker (1985) reported a white shark 400–500 cm in length feeding on the floating carcass of a gray whale (*Eschrichtius robustus*) off central California. Long and Jones (1996) reported about five white sharks feeding on the carcass of a blue whale (*B. musculus*) floating off San Francisco. Again, only one individual fed at a time. Based on their examination of the relative sizes of bites on the whale carcasses, these researchers concluded that only white sharks larger than 350 cm scavenge carcasses. It has been

2. The terms "Mid-Atlantic Bight" or "Middle Atlantic Bight" usually refer to "an imprecise coastal area off the middle Atlantic states of New York, New Jersey, Delaware, Maryland, and Virginia" (Richards 1999: 56). Although these terms are often used in the fishery literature, they are problematic because they are vague and are not official geographic names.

Records of juvenile white sharks along the east coast of the United States

Location	Date	Length, cm	Source
South Carolina	15 December	~210 cm	Rick Stringer (pers. comm., 2001).
Florida	13 February	228 cm	*New-Journal,* 14 Feb. 2001, p. 7C
South Carolina	22 June	131 cm	Don Hammond (pers. comm. June 1993)
North Carolina	27 April	231 cm	J. I. Castro, unpubl. obs., 1985
North Carolina	1 May	186 cm, 200 cm	J. I. Castro, unpubl. obs., 1987
North Carolina	20 May	188 cm, 221 cm	Coles 1919
Massachusetts	25 July	213 cm	Schroeder 1938
New Jersey	August	132–198 cm*	Casey and Pratt 1985
Maine	29 August	203 cm, 216 cm	Schroeder 1938
Maine	30 August	216 cm	Schroeder 1938

*Range of 10 individuals.

suggested that the largest white sharks might maximize their energy intake by preferentially scavenging on the blubber of dead whales (Long and Jones 1996). It is possible that the largest white sharks reach a size where such passive feeding is more energy-efficient than active predation. Most other species of sharks that exceed 5–6 m, such as the whale shark, the basking shark, and the megamouth shark, are filter-feeders, whereas two other large carnivorous species, the Greenland shark and the Pacific sleeper shark, also scavenge dead whales.

To white sharks, California sea otters (*Enhydra lutris*) must present visual or auditory cues similar to those of pinnipeds because dead sea otters are sometimes washed ashore bearing bits of white shark teeth embedded in their bones (Ames and Morejohn 1980). The sharks seemingly bite and then release the otters, killing them but not consuming them (Ames et al. 1996). Perhaps white sharks find sea otters unpalatable. The attacks on sea otters demonstrate that the white shark is an ambush predator that depends on a swift and focused approach to surprise and capture unwary prey. Once it detects it prey, the shark attacks at once, attempting to close the distance and inflict a fatal or disabling wound. Such an approach might be a vital adaptation for capturing intelligent and agile animals like seals, sea lions, and porpoises (Castro 1983).

Long (1996b) reported that two leatherback sea turtles (*Dermochelys coriacea*) that washed ashore in California had wounds attributable to large white sharks. Whether these were cases of direct predation or scavenging could not be discerned. There is also evidence that white sharks sometimes feed on birds. Bass et al. (1975b) recorded a jackass penguin (*Spheniscus demersus*) and a gull (*Larus cirrocephalus*) from the stomach of a white shark, and Randall et al. (1988) wrote that injuries to jackass penguins in Algoa Bay, South Africa, indicated attack by white sharks. Most injuries revealed the involvement of only one jaw and ranged from small cuts across the feathers to the bird bitten in half. Actual attacks by white sharks were not witnessed.

Migrations: The annual migration of adult white sharks along the eastern coast of North America is not narrowly tuned to seawater temperatures like those of other sharks because of their wide temperature tolerances. However, the record of migrating juveniles yields a clearer picture.

Thus, based on this limited evidence, the migration of juvenile white sharks along the United States eastern coast can be summarized as follows. White sharks winter from South Carolina to Florida and the Caribbean. By April and early May they appear off the Carolinas. By early June the sharks are from Virginia to Delaware, reaching the Gulf of Maine in mid-July. Juveniles remain south of Cape Cod, while larger sharks reach the Gulf of St. Lawrence by late July and August. The southern migration begins in September and October. The location of the wintering areas is not well defined but appears to encompass the Florida Keys, Gulf of Mexico, and Caribbean Sea. There are few data about the migrations along the western coast of North America.

It is difficult to generalize about the migrations of adult white sharks in North America because of their broad thermal tolerance. Nevertheless, I shall try to make some broad generalizations. In summer, adult white sharks can be found just about anywhere from south Florida to Newfoundland. However, most are found from Cape Hatteras to Newfoundland (Casey and Pratt 1985). Reports of white sharks in Canadian waters are usually for April to November, although most captures occur in August (Scott and Scott 1988). Most of the Florida records that I have of adult specimens extend from January to March, with the exception of a white shark estimated at 360 cm photographed while it fed on a large, floating swordfish off Miami in May 1999. Winter records from Florida and a few from the Caribbean suggest that large white sharks winter in those areas.

White sharks also undertake transoceanic migrations. Bonfil el al. (2005) tagged a female white shark, estimated at 380 cm, with a pop-up archival satellite transmitter off the South African coast. The shark traveled from South Africa to Australia, a distance of more than 11,000 km, in

99 days. The shark then traveled back to the original tagging site in South Africa.

Behavior: Behavioral observations on most species of sharks are difficult. Behavioral research on the white shark is even more difficult because of its scarcity and other factors. However, the appeal of the white shark is so strong, and the urge to write about this great predator so compelling, that logistical problems and the lack of data have not prevented some authors from writing extensively on the subject. Nowhere is this more evident than in the descriptions of white shark behavior. Klimley (1994) even attempted to study white sharks by observing them through a telescope from a distant island hilltop. After watching sharks splash at the surface while feeding, he proposed a bizarre theory of agonistic behavior in which the sharks use their tails to splash one another. Other workers have not verified this tail slapping behavior. Goldman and Anderson (1999) observed that several sharks around a kill showed no aggression toward each other. However, Ken Goldman (pers. comm., Sept. 2001) has observed smaller individuals give wide berth to the larger ones.

It has long been known that white sharks frequent specific sites, usually close to seal and sea lion rookeries. Collier (1992) called these places "attack prone," in terms of human interactions. Klimley and Anderson (1996) studied the occurrence patterns of white sharks around the South Farallon Islands and were able to identify individuals by their distinctive scars on the body and fins. According to Klimley and Anderson, most white sharks were transient visitors, although a few were long-term residents. Three individuals returned to the area at the same time in successive years. Work by Goldman and Anderson (1999) indicated that specimens larger than 400 cm show some site fidelity. Their study also suggested an inverse relationship between body length and activity space: the larger sharks restricted their movements to *smaller* areas around the Farallon Islands than did smaller individuals. The authors doubted that white sharks in the Farallon Islands were territorial, because they often observed several sharks at kill sites without signs of aggression toward each other. However, the absence of obvious signs of aggression does not negate other dominance interactions.

Size at maturity

Male: We have only scant data on the size at maturity. Pratt (1996) analyzed the data available for males, but his sample lacked specimens at the critical lengths needed to establish maturity accurately. Specimens up to 320 cm had soft claspers. The smallest mature male (i.e., with calcified claspers) in his sample measured 379 cm, and all the other mature males known were longer than 450 cm. I examined the 379-cm, 454-kg specimen in the Florida Keys in 1992, and it had calcified claspers. The smallest mature male I have been able to authenticate is a 339.5-cm, 482-kg specimen examined by Eugenie Clark in 1967 (Cape Haze Marine Laboratory Shark Records, white shark no. 2). This shark had hardened claspers 42 cm long, and according to the record, "great quantities of sperm were present in the seminal vesicles." Based on the Florida males, we can conclude that male white sharks mature at about 340 cm. The paucity of maturing specimens may be the result of this stage being relatively brief and occurring when the animals are in their wintering areas, out of the reach of researchers. The estimated size at maturity (340 cm) is about 61% of the maximum size (560 cm) for male white sharks. This agrees well with the sizes of mature males 304–339 cm reported by Bass et al. (1975a) for South African specimens.

Female: Despite the many female white sharks that have been caught, few data are available to determine size at maturity. To my knowledge, no pregnant white shark has ever been caught in the waters of the United States or Canada. Thus, it is necessary to look elsewhere for data on maturity. Uchida et al. (1996) described the two smallest pregnant females in the literature: a female "about 470 cm" and another 480 cm, both from Japan. Ignoring possible differences in populations, it appears that female white sharks mature at about 475 cm. The estimated size at maturity (475 cm) is about 85% of the maximum size for females (560 cm).

Older literature on size at maturity of white sharks must be evaluated with some skepticism. For example, Springer (1939a: 115) reported that a 470-cm female captured off Englewood, Florida, on 1 February 1939 had small ovaries, "and the oviducts were not well developed." Subsequent authors, such as Francis (1996), have taken this to mean that the specimen was immature. However, in 1939 there was little knowledge about the reproductive biology of lamnoid sharks. It would have been easy for Springer to mistake the small ovary of a postpartum lamnoid shark, with its small oocytes, for an immature ovary if one uses a carcharhinid ovary for comparison, as is likely he did. Likewise, the upper oviducts of lamnoid sharks are quite narrow and may resemble the narrow oviducts of an immature carcharhinid shark. Springer's 1939 shark was close in size to the smallest reported gravid white sharks, so it could have been immature, but it could also have been a postpartum female. Springer's account, devoid of measurements and made at a time when knowledge of the lamnoid reproductive cycle did not exist, is insufficient to determine with certainty the reproductive condition of the specimen.

Reproduction Although white shark reproduction had long been suspected to be of the lamnoid type—aplacental viviparity with oophagous embryos—proof has been obtained only recently. A female estimated to be 470 cm long, taken off Honshu, Japan, in April 1986, carried embryos 100–110 cm with large yolk stomachs (Uchida et al. 1996).

Size at birth: It appears that no intact, gravid white shark has been examined by biologists, and most of the gravid females have come to the attention of biologists after they had been butchered or discarded. In most cases, only a few embryos or photographs remained to determine the size and number of young. Most gravid females taken carried near-term embryos. A 515-cm female white shark, caught in a net off Toyo, Japan, on 22 May 1992, was carrying 10 embryos that appeared to be near term: their yolk stomachs were absorbed, and only a yolk-stomach scar was visible. These embryos measured 135–151 cm and weighed 21.3–32.4 kg (Uchida et al. 1996). A 480-cm gravid female, captured in May 1992 off Kyushu, Japan, carried five near-term embryos, only one of which was measured at 130 cm (Uchida et al. 1996). A 1,360-kg gravid female of unknown length was captured off North Cape, New Zealand, on 13 November 1991 (Francis 1996). This animal contained seven near-term embryos, of which only two were measured and weighed, 144.9 cm, 26.1 kg, and 143.0 cm, 23.5 kg, respectively. Considering the sizes of the Japanese and New Zealand term embryos, it appears that white sharks are born at about 122–150 cm and weights of 21–32 kg, with birth occurring in May–June in the northern hemisphere and in November–December in the southern hemisphere.

The sizes of the term embryos correlate well with the sizes of the smallest known free-swimming specimens taken off New York (122–140 cm) and South Carolina (131 cm). Casey and Pratt (1985) reported that the five smallest individuals they examined from Atlantic waters were 122 cm, 12 kg; two 130-cm specimens, each 16 kg; 132 cm, 23 kg; and 136 cm, 18 kg. There is one 131-cm newborn white shark that was captured on 22 June 1993 off South Carolina at 32°18′ N, 78°58′ W, some 80 km from shore. This specimen had an unhealed yolk scar and a fleshy sheath covering its front teeth (Don Hammond, South Carolina Department of Natural Resources, pers. comm., June 1993). Casey and Pratt (1985) also reported a Pacific specimen 122 cm and 16 kg caught off Santa Catalina Island, California, on 25 August 1983. Cailliet et al. (1985) mentioned a 129-cm free-swimming specimen presumed to be from Mexico.

Brood size: Based on the Japanese specimens mentioned above, and gravid females from Australia and New Zealand that contained 4–14 embryos (Bruce 1992; Francis 1996), it appears that white shark broods consist of 7–14 young. Due to the propensity of term females to abort their young, these brood sizes should be considered minimum estimates.

Nurseries Researchers have not precisely located the nurseries of the white shark, although their general area can be deduced. Pratt (1996) suggested that the East Coast nurseries may be off New York in summer, based solely on the presence of several sharks, described as neonates 130–140 cm, captured in that area. However, the lack of gravid females caught off the New York area and the catch of the 131-cm newborn from South Carolina cast doubt on that hypothesis. The few observations of gravid females cannot be attributed solely to difficulties of capturing those formidable specimens, because large males have often been caught in that area. It is likely that gravid females inhabit areas where they are inaccessible to fisheries, or where they elude capture by the fishing gears used there. Scanty data and some anecdotal information suggest that the birthing and nursery areas for the population of eastern North America might be in tropical waters along the northern coast of the Greater Antilles in late winter or early spring. The young might then travel westward and northward along the eastern coast of the United States, reaching New York State by August or September.

The location of the West Coast nurseries is not known either, although for many years there have been anecdotal reports of very small white sharks caught along the Pacific coast of Baja California, and some evidence points to nurseries of white sharks being in Mexican waters along the Baja Peninsula. Klimley (1985: 15) stated that "females give birth to pups during late summer and early fall south of Point Conception." However, the statement does not seem to be based on evidence of gravid females and neonates, but on much larger juveniles. These one- or two-year-old white sharks are sometimes caught farther north along the California coast, but their birth place is certainly in more southern waters. Recently, John O'Sullivan, of the Monterey Bay Aquarium, informed me that he had found the remains of several small white sharks less than 200 cm in the refuse piles of the numerous fishing camps of Baja California (pers. comm., September 2004). If these specimens are neonates, this would also confirm that the nurseries are along the Baja California Peninsula.

Age and growth There is very limited information on the age and growth of white sharks. Based on vertebral centra, Cailliet et al. (1985) estimated that white sharks 366 and 427 cm were 9–10 years old. Wintner and Cliff (1999) used X-radiography to count the vertebral growth rings of South African white sharks. Assuming that a growth ring is deposited annually, these authors estimated that a 445-cm female would be 12 years old. These estimates await validation by other means but suggest impressive growth rates for a shark.

Relation to humans The white shark is one of the most dangerous sharks to humans because of its large size, aggressiveness, habit of feeding on large prey, and propensity to enter shallow waters (Castro 1983). White sharks have attacked bathers and surfers in shallow waters and divers and boats in deeper water. There are numerous cases of white sharks attacking surfers off the California and Oregon coasts, presumably because surfers resemble sea lions from below (McCosker 1985). White sharks have attacked kayaks and killed the kayakers (McCosker and Lea 1996). There is also an old record of a white shark attacking a dory in St. Pierre Bank, Newfoundland (Putnam 1874). There are also many reports of white sharks being attracted to the electromagnetic fields emitted by small boat engines and even mouthing the propellers.

Many attacks of white sharks on humans consist of a single, often fatal bite, with the shark releasing its human prey after the initial bite, or without the shark consuming the entire victim (Lea and Miller 1985; Nakaya 1996). This bite-and-release pattern supports McCosker's hypothesis that white shark attacks on humans are often the result of the victims being mistaken for pinnipeds. Most attacks in North America have occurred off California and are often fatal because of the size and severity of the bites. Burgess and Callahan (1996) reviewed the global pattern of white shark attacks on humans on a worldwide basis and noted attacks occurred on solitary individuals 84.8% of the time and that a single bite occurred one-third of the time.

As a result of numerous attacks on humans, the white shark is generally described in awe-inspiring terms: it "is the most dreadful and voracious of all animals; preys on everything that comes its way, even its own tribe, and has been known to swallow a man whole" (Linné [Linneus] 1806: 921); "of all others, the white shark is the most terrific, and therefore regarded by mariners with peculiar dread" (Smith, J. V. C. 1833: 83); "the most voracious of fish-like animals" (Jordan 1907: 195); "this swift, voracious and ferocious Shark is a terror to all who venture on or in the water" (Smith, J. L. B. 1949: 49); "huge, fearsome shark" (Compagno 1984a: 239). The species has also been the notorious subject of horrifying tales, accounts, paintings, and movies.

At one time, it was considered that being in the water with a white shark without the protection of a sturdy cage was to invite immediate attack. Leaving the shark cages was believed to be a form of suicide. However, in the 1990s, cinematographers and divers started to venture out of cages in the presence of one or more white sharks, but only in the absence of blood or food in the water. To my knowledge, not one of those brave souls has been attacked. This is probably explained either by the shark not recognizing scuba divers as potential food or, perhaps, by the general reluctance of ambush predators to attack prey that is aware of their presence.

Keeping a white shark in captivity has long been the Holy Grail of aquarists, because these sharks are a great public attraction. Unfortunately, white sharks generally die soon after capture, although juveniles occasionally survive for several months in captivity. One of these was kept at the Steinhart Aquarium for four days before being released (McCosker 1981), while the other was kept at the Monterey Bay Aquarium for a record 198 days (*USA Today* 2005). Concern for the safety of aquarium divers and public outcry after the white shark killed two soupfin sharks resulted in the shark being released back into the Pacific.

The white shark is classed as a gamefish, and it is probably the world's largest. Its pursuit was for many years the ultimate trophy of many sports fishermen (and scientists!), and such sport required highly specialized tackle and methods, and plenty of patience. Sport fishing for white sharks used to be pursued only in a few places off South Africa, South Australia, and New York, but the sport has died down in the last decade. In South Africa, anglers fished for white sharks from piers and beaches. For interesting accounts of fishing for white shark from South African shores, see L. Jones (2003, 2005).

There are markets for white shark fins, teeth, and jaws. Today, white shark jaws and teeth command the highest prices, and a prepared set of jaws from the largest specimens may bring several thousand dollars in collector circles. The white shark is becoming an important species in the lucrative business of shark ecotourism, although at present white shark watching is confined to a few places off California, Australia, Mexico, and South Africa.

Along the western coast of the United States, the white shark has been a protected species in California waters since 1994 and in federal waters since 2004. The species has been protected along the eastern United States since 1997.

Current levels of protection are effective in reducing mortality in sports fisheries in the United States, but they cannot prevent white sharks from dying in commercial nets or longlines. Fishermen in the southeastern United States have reported to me numerous cases of white sharks being found dead on their longlines. Unfortunately, there is no provision for their retrieval in the legislation; these dead specimens are discarded at sea, and valuable data are lost. The State of California has a more enlightened approach to the disposal of dead white sharks taken incidentally in commercial fisheries, and these sharks can be used for research or educational purposes.

White sharks became protected in South Africa in 1991 and in Australia in 1997. Other countries will soon give them some form of protection. The protective legislation has resulted not from any documented population declines (although these have undoubtedly occurred), but from conservationists' efforts. The white shark has become a symbol of their causes.

SHORTFIN MAKO

Isurus oxyrinchus Rafinesque, 1810

Fig. 64a. *Isurus oxyrinchus*, shortfin mako shark, immature female, 237.3 cm, 106 kg; western Atlantic specimen (Reinaldo "Rey" Perez, collector).

Common name Shortfin mako. The name mako comes from the Maori *mango-makomako* or *mako; mango,* a generic term for sharks; and *mako,* a lamnoid shark (Best 1929). For many years the species was simply called the "mako," except in California, where it used to be named the bonito shark. After the discovery of the longfin mako in 1966, "shortfin" was added to the name to avoid confusion. Most people still refer to it simply as "mako."

Spanish name *Alecrín* (Mexico), and *dientuso azul* (Cuba): *dientuso,* corruption of *dentudo,* having big teeth; *azul:* blue.

Scientific name *Isurus oxyrinchus* Rafinesque, 1810, in *Caratteri di alcuni nuovi generi e nuove specie di animali e piante della Sicilia,* p. 12, plate 13, fig. 1. *Isurus:* from the Greek *isurus,* equal tail (*isos,* equal, and *oura,* tail), a reference to the two lobes of the tail being equal in size; *oxyrinchus,* from the Greek *oxys,* sharp, and *rynchos,* snout (thus "sharp-snouted").

Synonyms *Isurus spallanzanii* Rafinesque, 1810; *Squalus cepedii* Lesson, 1830; *Oxyrhina glauca* Müller and Henle, 1839; *Lamna punctata* DeKay, 1842; *Carcharias tigris* Atwood, 1865; *Isurus glaucus* Jordan and Gilbert, 1883; *Isurus mako* Whitley, 1929; *Isurus tigris africanus* Smith, 1957.

Identification The shortfin mako has *a conical snout, moderately short pectoral fins* (usually measuring 16–17% of the total length), the *first dorsal fin origin posterior to the free rear tips of the pectoral fins,* well-developed keels on the caudal peduncle, and a lunate tail. The teeth have *long, slender, smooth-edged cusps;* they number U: 12 or 13–12 or 13, L: 12 or 13–12 or 13 and are alike in both jaws. The first two teeth on each jaw are much longer and more slender than the rest, the second lower tooth being the longest. *The outer cutting edge of the two lower frontal teeth does not extend all the way to the base of the crown;* it ends on a smooth, rounded shelf at the base of the crown. The dermal denticles are small and closely overlapping. They have three strong ridges and three points (sometimes five), the central point being slightly longer that the others. The coloration of living or fresh specimens is dark blue above with a band of metallic blue on the upper sides, and white below. The *underside of the snout and the area around the mouth are generally white,* although large specimens have some gray spots or blotches. The blue coloration of living specimens turns to gray after death or upon freezing. Consequently, dead or frozen shortfin makos are sometimes mistaken for longfin makos. Moreno and Morón (1992) described a dark form from the Azores, where they saw more than 100 specimens. In those sharks the ventral surface of the snout and the area around the mouth are dark, and the undersides are darker. The darker coloration consists of more or less dense dark dots. In the largest of these sharks, the dark coloration can cover the entire body, the white zones being reduced to isolated ventral stains or patches, similar to large longfin makos.

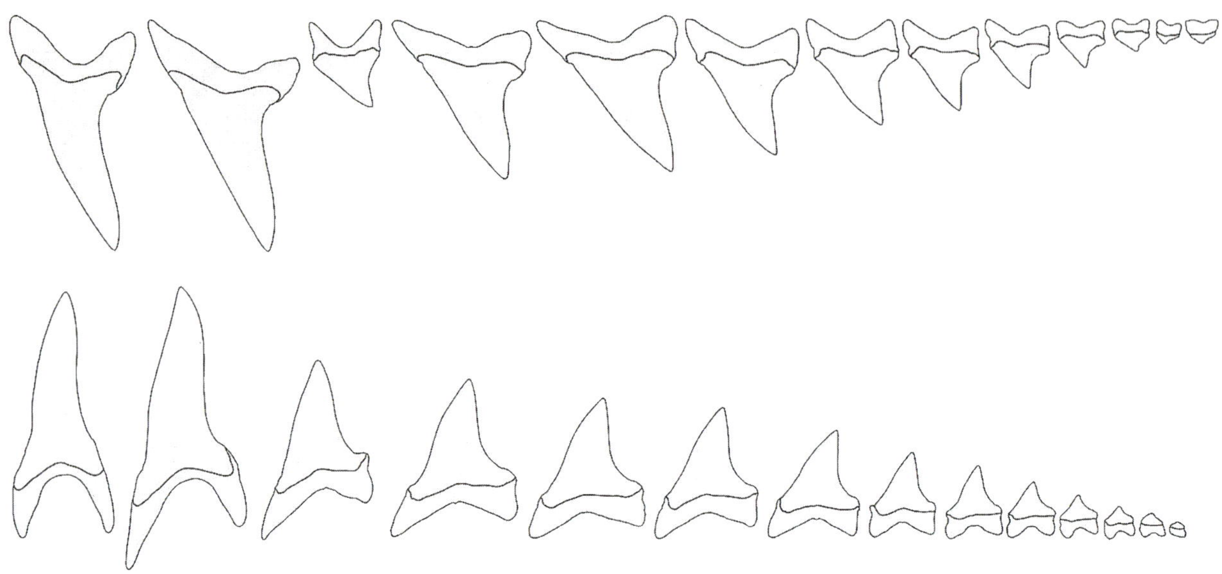

Fig. 64b. Teeth.

The coloration of most pelagic fishes closely matches their background, an adaptation for concealment. The shortfin mako is an outstanding example of this. Its dorsal coloration matches the intense blue of deep tropical and subtropical waters so well that, when viewed from above in deep water, such as in the Gulf Stream, the shortfin mako is nearly invisible against the dark blue background, even at moderate distances. The metallic light blue sides, changing from dark blue to white, reflect the brighter light near the surface, countershading the animal in lateral view. The pure white underside conceals it when viewed from below against the brightness of the surface.

Similar species The longfin mako has much longer pectoral fins (usually measuring 20–28% of the body length, with the smaller specimens having proportionally larger fins), and the underside of its snout and the area around its mouth are dusky or blue-black. The two frontal lower teeth on each side of the jaw can also be used to distinguish the longfin mako. In the longfin mako the outer cutting edge of the lower frontal teeth extends all the way to the base of the crown, whereas in the shortfin mako, this edge ends on a rounded smooth shelf at the base of the crown, just above the root. The porbeagle shark and salmon sharks have a more anterior first dorsal fin, with its origin above the rear tip of the pectoral fin, and their teeth have lateral cusplets. Both these species have small secondary keels on the tail, just below the main keel. The white shark has triangular teeth with serrated edges. When viewed from above, live large specimens of all these species may be difficult to distinguish in the water.

Range The shortfin mako is found in tropical, subtropical, and warm-temperate waters throughout all oceans. In the western Atlantic it ranges from northeast of the Grand Banks southward to the Gulf of Mexico and the tropics. In the eastern Pacific it is found from California southward to the tropics.

The northernmost records of the shortfin mako from eastern North America come from the Canadian Atlantic, where the northern boundary of the Gulf Stream limits the range of the species. According to S. Campana, of the Bedford Institute of Oceanography, shortfin makos have been observed as far north as 50° N, 40° W from July to September (pers. comm., Oct. 2004).

Shortfin makos are common in continental waters west of the Grand Banks in summer. Tibbo et al. (1963) reported catching two makos on Browns Bank in early August, along with a porbeagle. Stillwell and Kohler (1982) reported 20 specimens taken from 31 July to 2 September some 1,288 km east of Browns Bank on the Tail of the Grand Bank, at approximately 42° N, 50° W. (These specimens were reported erroneously by Scott and Scott [1988] as being captured at 43° N, 50° W.)

The shortfin mako is an occasional catch in the Gulf of Maine, although published accounts do not reflect its presence there. Bigelow and Schroeder (1953) stated that stray shortfin makos can be expected in the southern part of the Gulf of Maine but noted that they themselves had never encountered any shortfin makos there. Scattergood (1962b) reported a large specimen harpooned in July off Boon Island, Maine, and indicated that this was the first record of a mako in Maine waters.

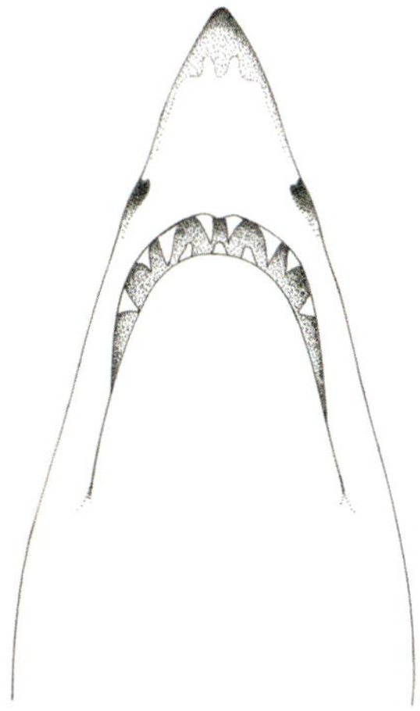

Fig. 64c. Snout.

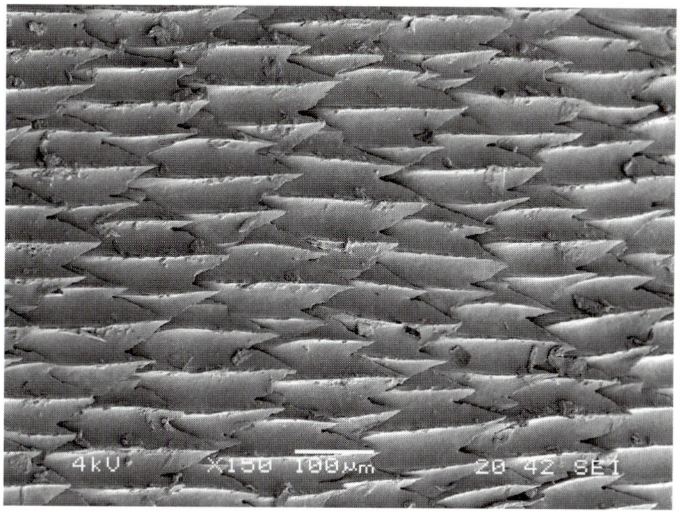

Fig. 64d. Dermal denticles.

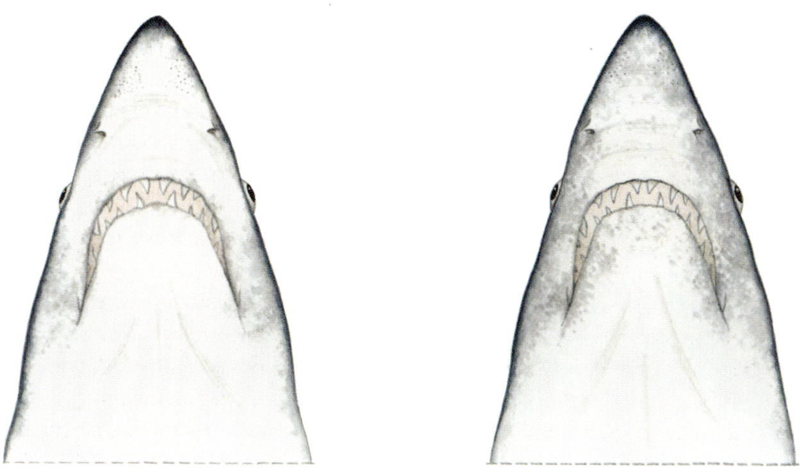

Fig. 64e. The underside of the snout of western North Atlantic shortfin makos is white, except in the largest individuals, which may show some gray spotting. Left: 335-cm mature female from Brazil; 325-cm mature female from Florida.

South of Cape Cod, the shortfin mako becomes abundant offshore and is common between Cape Cod and Cape Hatteras. South of Cape Hatteras, it is common in the Gulf Stream and in the Gulf of Mexico.

Size and weight It is not unusual for shortfin makos to measure more than 330 cm and weigh more than 400 kg; however, animals reaching this size are usually females. The largest females can reach 380 cm and weigh more than 550 kg, but sharks that big are rarely taken. Applegate (1977) recorded a 351-cm, 468-kg female from Anacapa Island, California, in 1972. Casey (1989) reported that the largest shortfin mako (335 cm) he examined weighed 568 kg. Kohler et al. (1995) reported two 366-cm females caught off Montauk, New York, that weighed 471 and 382 kg. Gubanov (1974) reported a 380-cm, 500-kg female caught in the eastern Indian Ocean (03°05′ N, 81°55′ E). Since then larger females have been taken. The current angling record is 553 kg (International Game Fish Association 2006), and according to Greg Skomal (pers. comm., June 2008) there are credible reports of harpooned specimens of 600 and 693 kg. Males reach only 260–270 cm and about 200 kg (Pratt and Casey 1983, n. 3; Kohler et al. 1996). Representative sizes and weights that I have recorded are as follows:

Male: 80 cm, 3.4 kg; 132 cm, 20 kg; 146 cm, 28 kg; 157 cm, 30.2 kg; 160 cm, 33 kg; 167 cm, 40 kg; 185 cm, 50 kg; 195 cm, 59 kg; 210 cm, 71 kg; and 228 cm, 82 kg.

Female: 72 cm, 2.6 kg; 81 cm, 3.2 kg; 165 cm, 37 kg; 214 cm, 72 kg; and 325 cm, 304 kg.

Biology The shortfin mako is an oceanic apex predator that feeds on other fast-moving predators such as sharks, swordfish, and bluefish. The species seldom approaches land except in the vicinity of oceanic islands. Its biology is poorly known, despite its abundance in warm waters, its economic importance, and its being one of the world's greatest and largest game fishes.

A few studies have been performed on shortfin makos, primarily on juveniles. The logistical problems of studying a migratory species that ranges over entire ocean basins have left many unsolved mysteries about "the mako." Little is known about its behavior or physiology. Like other lamnids, the shortfin mako can maintain a temperature 7–10°C above the ambient water. It has highly developed countercurrent heat exchangers in the vascular system, forming a thermal barrier, which minimizes heat from being carried off by the circulating blood and lost at the gills (Carey and Teal 1969). This adaptation has long been thought to increase the power available from muscles (Carey and Teal 1969), but it can also allow the species to enter cooler waters (Block et al. 1993).

Off the northeastern United States, the preferred surface water temperature of the shortfin mako is 18–21°C, although specimens have been caught in surface water temperatures of 10–28°C (Casey 1989; Loefer et al. 2005). Hoey (1983) analyzed surface temperature data from 2,700 sets of swordfish longline catches from the Gulf of Mexico to the Grand Banks and found that makos were caught between 17 and 22°C. Experiments using sonic tags have also demonstrated that this is the preferred temperature range of the shortfin mako. Carey et al. (1978) attached a temperature monitoring tag to a 160- to 180-kg shortfin mako and followed it from Florida across the Gulf Stream to the Sargasso Sea. The shark swam at depths ranging from the surface to 500 m but spent most of its time where the temperature ranged from 17 to 21°C. Off California, Holts and Bedford (1993) tracked three shortfin makos for periods of 18–25 hours. All three sharks spent 93% of the time in the mixed layer where the temperature was 18–21°C (at depths of 14 to 20 m), with only brief excursions below the thermocline. Loefer et al. (2005) used a pop-up satellite tag to track an adult male shortfin mako off South Carolina. Archived tag data indicated a depth range from 0 to 556 m in ambient temperatures 10.4–28.6°C, with the shark spending the nights in shallower warmer waters and diving to cooler and deeper waters during the day.

Diet: The shortfin mako is an opportunistic apex predator. Numerous reports (e. g., Murphy 1919) show that the bluefish (*Pomatomus saltatrix*) is its primary prey off eastern North America, although in other areas it feeds primarily on other sharks. Stillwell and Kohler (1982) examined 399 makos caught between Cape Hatteras and the Grand Banks and found bluefish in 67% of the stomachs, and squids in 15% of the stomachs. The stomach of a 337-cm female from Santa Catalina Island, California, contained the remains of three carcharhinid sharks (probably blue sharks) estimated to have been 120–150 cm long (Applegate 1966). I examined the stomach contents of a large female of about 300 cm caught off Bermuda in June and found the vertebrae of small sharks of about 90 cm. Shortfin makos often attack swordfish (*Xiphias gladius*) hooked on longlines, and there are reliable reports of attacks on free-swimming swordfish as well. Bigelow and Schroeder (1948) reported that a nearly intact 55-kg swordfish with its "sword" still attached was found in the stomach of a 330-kg shortfin mako captured near Bimini. Casey (1989) recovered about 20 kg of flesh from a swordfish (estimated at 160 kg) from the stomach of a 568-kg female. There is some evidence that shortfin makos also prey on marlins and marine mammals. In 1999, David Holts (National Marine Fisheries Service, pers. comm., October 2002) found the remains of striped marlin (*Tetrapturus audax*) in the stomach of a 183-kg male taken off Santa Monica Bay, California, and porpoise and seal flesh in the stomach of a 447-kg female caught off Redondo Beach, California, in September 1999.

Elsewhere in the world, shortfin makos prey on sharks, bony fishes, squids (Bass et al. 1975b; Cliff et al. 1990), and sea turtles. Gubanov (1974) reported that two large sea turtle heads weighing a total of 15 kg were found the stomach of a 380-cm female in the eastern Indian Ocean.

Migrations: Knowledge of the shortfin mako migrations in the western North Atlantic is limited to the time and conditions in which juveniles and males appear off the East Coast of the United States. Large females are not often found in this area, so we know little about them. Much of our knowledge of the migrations of the shortfin mako originated with Jack Casey and coworkers, who tagged thousands of specimens off the eastern coast of the United States starting in the 1970s.

In January, shortfin makos are abundant along the western edge of the Gulf Stream, with one area of high abundance (based on commercial catches) off Cape Hatteras, where the Gulf Stream flows near the continental shelf. From January to April, shortfin makos are seldom caught on the continental shelf north of Cape Hatteras. As inshore shelf waters warm up in April and May, and the axis of the Gulf Stream moves farther north, shortfin makos begin moving northward onto the continental shelf between Cape

Hatteras and southern Georges Bank (Casey and Kohler 1992). The shift inshore and northward takes place simultaneously along the Gulf Stream, from Cape Hatteras to the Grand Banks (Hoey 1983). In most years shortfin makos are taken off New Jersey in early June and off New York and southern New England in late June (Casey and Kohler 1992). By mid-July, when surface water temperatures are above 21°C, shortfin makos are more sparsely distributed over a wider inshore area (Casey 1989).

From June to October shortfin makos are caught on the continental shelf between Cape Hatteras and Cape Cod. The withdrawal to offshore waters occurs gradually in September and October and varies annually with the onset of cooler water temperatures (Casey 1989). During late fall and early winter (November and December), shortfin makos are believed to move from the area between Cape Hatteras and the Grand Banks to offshore wintering grounds in the Gulf Stream and the Sargasso Sea, although the location of the wintering grounds is not precisely known. Assuming that Sargasso Sea water temperature of 18°C represents the preferred habitat for the shortfin mako, the wintering area probably extends between latitudes of 20° and 40° N and is bordered by the Mid-Atlantic Ridge in the east and the Gulf Stream in the west (Casey and Kohler 1992).

The migration of the shortfin mako along eastern North America and back to its wintering areas might not be a simple north–south movement. Tag returns of shortfin makos tagged along the eastern seaboard suggest that some sharks move north along the East Coast and then continue northeast with the Gulf Stream along latitudes of 40–46° N to the Mid-Atlantic Ridge before moving south. If so, the migration route might be a vast circle or oval over great expanses of the western Atlantic. Such a route would begin in the northern Caribbean, perhaps in mating and nursery areas, and extend northwestward to Florida and the Gulf Stream. It would then continue with the Gulf Stream to the feeding areas of the eastern seaboard and the Grand Banks, then east to the middle latitudes of the Mid-Atlantic Ridge, and finally south to the Sargasso Sea and the northern Caribbean.

Adult females are not part of this migration. They apparently remain in the warmer waters of the Bahamas and the Caribbean or in the open Atlantic. Large females have been caught at diverse localities in the Gulf of Mexico, the northern Caribbean (e.g., the Windward Passage and off Puerto Rico), and the Bahamas. Their migratory patterns are unknown.

The movements of the shortfin mako in the Gulf of Mexico are obscure. Shortfin makos appear in the northern Gulf from mid-October to November with the onset of cooler temperatures, remaining common in the area until March. This migration coincides with the appearance and movements of yellowfin and blackfin tunas, but whether the sharks are following these potential prey fishes is not certain.

Tagging studies have shown that individual shortfin makos can cover great distances. A shortfin mako tagged on Georges Bank was recaptured 37 days later on the Grand Banks. This journey represents a straight-line distance of 2,446 km from the tagging location to the recapture locality, giving a rate of travel of 66 km/day. Another shortfin mako carrying a sonic tag had a swimming rate of 107 km/day (Casey and Kohler 1992). Tagging studies also reveal that some individuals are recaptured years later at the original tagging site, suggesting that these animals undertake cyclical travels that bring them back to the same location one or several years later. A shortfin mako estimated to be 315 cm and 275 kg was tagged off Islamorada, Florida, on 16 March 1988 and recaptured at nearly the same site four years later, on 20 April 1992. It was stated to measure 366 cm and weigh 338 kg at recapture (Miami Herald 1992). Whether the shark remained at that location during all that time or happened to return to that spot during its migrations is not known.

Size at maturity There are few studies on the size at maturity of the shortfin mako.

Male: Natanson et al. (2006), in a comprehensive age–growth study, estimated that males mature at 201 cm. In an earlier study, Stevens (1983) found that Australian males mature at about 195 cm based on clasper calcification. The estimated length at maturity of males (201 cm) is about 74% of the maximum male size (270 cm).

Female: The size at maturity for female shortfin makos is not precisely known. Female shortfin makos apparently mature at 289–296 cm. Stevens (1983) found that a 283-cm female was still immature while a female 296 cm was mature. Cliff et al. (1990) noted that their smallest gravid female measured 289 cm. The smallest graved female I have examined measured 305 cm. The estimated size at maturity (292 cm) is 77% of the maximum length for female shortfin makos (380 cm).

Reproduction The shortfin mako is an aplacental, viviparous species with oophagous embryos. Little is known about its reproduction because few gravid females have been examined by biologists and because, it appears, gravid females may abort full-term embryos during capture. Females in early stages of pregnancy are most difficult to obtain because such stages last only a short time, so they have rarely been examined. I have examined only a single female in an early stage of pregnancy. This specimen, caught off Brazil in early May, contained 47 full egg cases and five empty ones in the uteri. Five embryos of 2–3 cm had already hatched from the egg cases and were found in the uteri. These embryos had minute yolk sacs about 5 mm in diameter.

Females carrying midterm embryos are more easily encountered (I have examined four, all from the Caribbean Sea). These females reveal that embryos are oophagous (feeding on eggs in the uterus) but are not cannibalistic; thus, several embryos share each uterus. Early in development, the embryos quickly acquire huge, distended yolk stomachs. These large stomachs are absorbed before birth. Joung and Hsu (2005) reported cases of uternine cannibalism in shortfin mako embryos. These authors encountered two cases (out of 16 broods) where larger embryos had swallowed smaller embryos. In one case they found two embryos, 28 and 33 cm in length, in the stomach of a 71-cm embryo. In a second case, a 68-cm embryo contained a partially digested 20-cm embryo. Joung and Hsu attributed the cannibalism to some embryos being much smaller than others due to "unequal food." Given that shortfin mako embryos are all of similar size, it is likely that the swallowed embryos may have been dead or were runts before being ingested.

There are few estimates of the length of gestation in the shortfin mako. Pratt and Casey (1983) used length–month analysis to estimate gestation, assuming that the intrauterine growth rate was similar to the growth of young specimens. They estimated gestation to last a little more than a year. Other authors have proposed even longer periods but without providing any credible evidence. Cliff et al. (1990) examined six small shortfin makos 60–69 cm caught between July and January (three caught in November). Assuming a birth size around 57 cm, they concluded that birth would occur toward the end of the year. They then assumed that mating occurred between March and June, probably based on the capture of two "recently fertilized females" in March and June, and concluded that gestation would last for either 6 or 18 months. They selected the 18-month figure because it seemed more feasible, based on the estimate by Pratt and Casey (1983) of "a little over a year." Mollet et al. (2000) reviewed records of gravid shortfin makos from diverse localities and estimated gestation to be 15–18 months, also without providing acceptable proof or credible evidence for such an unusually long gestation. Joung and Hsu (2005) estimated the gestation period at 23–25 months, and the reproductive cycle at three years, but lacked evidence to support their claim.

Obviously, the length of the gestation period of the shortfin mako is subject to debate. As in many other sharks, the gestation period probably lasts about a year. Definitive proof of the duration of gestation will require direct measurements of embryos throughout the year in one population (if all the females are synchronous breeders), in addition to measurements of the ovary, and a demonstration of the relationship between the vitellogenic and gestation periods.

Brood size: Only a small number of gravid shortfin females have been reported, and they carried 4–16 young. The usual brood consists of 10–16 young. The smaller broods might be from females that perhaps have aborted some of their young during capture. A 229-kg female caught off Cat Cay, Bahamas, in February 1952 was reported to contain 10 young (Depperman 1953). I examined a 331-cm female caught in February 1994 in the Windward Passage that contained 13 embryos 53–59 cm (average size 57 cm). Another female, 318 cm long and weighing 312 kg (caught off Puerto Rico on 15 January 1992 and obtained by Jack Casey) carried 14 embryos measuring 50–54 cm. Stevens (1983) described four broods from Australia of 4, 6, 14, and 16 embryos. All consisted of embryos close to term (58–64 cm), so the possibility that some could have been aborted cannot be ruled out.

Size at birth: Shortfin makos are born at 65–72 cm, an estimate based on embryos reaching 72 cm in length (Costa et al. 1995) and on a 65-cm free-swimming specimen (Casey 1989). Guitart Manday (1975) illustrated the only shortfin mako embryo he was able to obtain in Cuba, a 65-cm specimen with a large yolk stomach. Costa et al. (1995) examined near-term broods from two females caught off Brazil. One consisted of 15 embryos 66–74 cm; the other, of 13 embryos 65–72 cm. Casey (1989) states that the smallest free-swimming shortfin mako that he found, in 30 years of research in the western North Atlantic, measured 65 cm. Authors working in other regions have also estimated the birth size to be about 70 cm (Bass et al. 1975b; Stevens 1983). Cliff et al. (1990) reported that their smallest free-swimming specimen was a 70-cm (57 cm precaudal length) female caught in the beach nets at Hibberdene, South Africa.

Nurseries The location of the shortfin mako nurseries in eastern North America is unknown. Limited evidence suggests that the mako nurseries are in the tropical and subtropical oceanic waters of the Gulf of Mexico and Caribbean Sea. Neonates 70–78 cm have been caught off Veracruz, Mexico; four were taken in late January, and one was taken in late April (J. L. Oviedo, pers. comm., October 2006). Small individuals appear in June over a wide area from Charleston, South Carolina, to Cape Cod. Guitart Manday (1975) reported a gravid female with near-term embryos (65 cm) from Cuba, although he doubted that the nurseries were in Cuban waters, because he had been able to obtain only one gravid female. Because this female carried just two embryos, and mako broods are known to be larger, the scarcity of gravid females in the Cuban fishery could be caused by their propensity to abort near-term embryos during the trauma of capture.

Little is known of shortfin mako nurseries along the western coast of North America. Hanan et al. (1993: 33) reported that "length–frequency and catch data from the drift gillnet fishery suggest that southern California waters are part of an important nursery area for shortfin makos in the eastern Pacific." However, whether the young makos are being born in the area or are born in more southern waters has not been determined.

Age and growth Most researchers have counted the growth rings (band pairs)[3] on the vertebrae for aging shortfin makos. Pratt and Casey (1983) stained vertebral rings with silver nitrate to reveal the age of specimens, assuming that two rings were deposited in the vertebrae per year. Based on this method, the oldest female in their sample was 11.5 years old and 328 cm; the oldest male was 4.5 years and 225 cm. Since then, other researchers have questioned the two growth rings per year hypothesis and assumed that one ring is deposited each year. Cailliet and Bedford (1983) used X-radiography to age shortfin makos and reported that their largest fish was 321 cm and had vertebrae exhibiting 17 growth rings. Campana et al. (2002b) reported that their single shortfin mako sample would have been 10 years old by the criteria of Pratt and Casey (1983) and 21 years old when using one ring (band pair) per year. Bishop et al. (2006) aged shortfin makos from New Zealand by vertebral counts, assuming one annual ring (band pair) was formed each year. They concluded that shortfin makos initially grow very fast, increasing by 39-cm fork length in their first year. Thereafter, both sexes grow at the same rate until about age 7, when the relative growth of males declines. Bishop et al. reported that males mature at 7–9 years and females at 19–21 years. The oldest male specimen in their sample was 29 years old while the oldest female was 28 years old. Recently, Natanson et al. (2006) demonstrated the validity of the one growth ring per year hypothesis, concluding that 50% of males reached maturity at 201 cm (185 cm fork length) and 8 years, and that 50% of females mature at 298 cm (275 cm fork length) and 18 years. Their longevity estimates were 29 years for males and 36 years for females.

3. In shark vertebrae, each growth ring consists of a light band and a dark band, which reflect different rates of calcium deposition. The growth rings are often referred to as "band pairs."

Relation to humans The shortfin mako is one of the great gamefishes of the world. Once hooked, they resist strenuously, often leaping high into the air, and large specimens often succeed in breaking fishing lines. Zane Grey, the famous writer and angler, first illustrated the spectacular leaps of hooked makos in an interesting photographic essay (Grey 1934). After pursuing large makos for many years, Grey (1934: 221) wrote: "The mako is a fish that often gets away." Today, given the improvements in line and terminal tackle, the landing of large shortfin makos by recreational anglers is not as difficult as it was in Grey's time. There are anecdotal reports of leaping makos landing in boats, injuring anglers and damaging equipment.

No fisheries target the shortfin mako in the United States, but the species is commonly taken as bycatch in swordfish and tuna fisheries along both coasts. West Coast landings average about 103 metric tons annually, valued at $184,000. Most of the catch is taken in the drift gillnet fishery for swordfish and thresher sharks (Holts and Bedford 1989). The flesh of the shortfin mako is flavorful, and mako steaks are sometime sold to the unwary as swordfish. It is often the only shark species kept by high seas tuna and swordfish longline fleets and is marketed everywhere.

The shortfin mako is not generally considered dangerous to divers and swimmers because of its oceanic habitat. However, Randall and Levy (1976) recorded a near-fatal attack by a shortfin mako on a swimmer at a beach in Elat, Israel, on the Red Sea. A 233-cm male repeatedly bit the swimmer, who barely survived and whose arm had to be amputated.

LONGFIN MAKO

Isurus paucus Guitart Manday, 1966

Fig. 65a. *Isurus paucus*, longfin mako shark, mature male, 242 cm, 77.8 kg; caught 270 km east of Cape Canaveral, Florida (Reinaldo "Rey" Pérez, collector).

Common name Longfin mako: longfin, having pectoral fins longer than the more common shortfin mako; mako, the Maori name for these sharks.

Spanish name *Dientuso prieto* (Cuba); *dientuso,* corruption of *dentudo,* having big teeth; *prieto,* blackish or dark in color. The A.F.S. name is *mako aletón.*

Scientific name *Isurus paucus* Guitart Manday, 1966, in *Poeyana* Ser. A 15: 3, figs. 1, 2A, 3A,C. *Isurus:* from the Greek *isurus,* from *isos,* equal, and *oura,* tail, having an equal tail, a reference to the two tail lobes being of equal size; *paucus,* Latin for scarce, an allusion to the species being scarce in comparison to the shortfin mako.

Synonyms *Isurus alatus* Garrick, 1967.

It is difficult to understand how such a large, widely distributed fish could go undescribed until 1966. Several factors probably contributed to its going unrecognized. First, it is an oceanic species seldom landed by commercial fisheries because of the poor quality of its flesh and lack of markets. Second, there are few specimens in museum collections because many of the captured specimens are of large or very large size. Third, the variability in coloration and similarities in tooth morphology of the makos often confuse laymen and biologists alike. All these factors raise the possibility that other undescribed species of makos might exist.

Identification The longfin mako has a *conical snout, pectoral fins measuring 20–28% of the total length (the smaller specimens have the proportionally larger fins),* the first dorsal fin origin posterior to the free rear tips of the pectoral fins, a caudal peduncle flattened into caudal keels, and a lunate tail with a subterminal notch on its upper lobe. The teeth have *long, slender, smooth-edged cusps,* with the first two teeth on each side of the symphysis noticeably larger than the others. *The outer cutting edge of the lower frontal teeth extends all the way to the base of the crown.* Teeth number U: 12 or 13–12 or 13, L: 12 or 13–12 or 13, and they are similar in both jaws. The dermal denticles are small and closely overlapping; they have three strong ridges and three points. Fresh specimens are an intense dark blue to bluish black with purplish hues above, becoming lighter on the sides. The upper blue color fades after death to dark brown, and to bluish black after freezing. The *underside of the snout and the area around the mouth are bluish black.* The extent of the dark coloration of the underside of the snout increases with length (see illustrations). The ventral pigmentation is variable. Some specimens are white to dirty white below, while others have dusky, blue-gray, or mottled undersides. Adults tend to have much darker undersides than juveniles.

Similar species The shortfin mako has proportionally smaller pectoral fins (their length is 16–17% of the total length); the underside of its snout and the area around the mouth are white in small and medium specimens and mottled in gray in very large specimens. In the shortfin mako, the outer cutting edge of the lower frontal teeth does not extend all the way to the base of the crown; this edge ends on a rounded smooth shelf at the base of the crown, just above the root. The white shark has triangular teeth with serrated edges. The porbeagle has teeth with lateral cusplets.

Range The longfin mako appears to be cosmopolitan in warm oceanic waters. Based on my records and those of Gordon Hubbell, it is common off central and southern Florida and off the Bahamas throughout the year. However, most collected specimens, including all the larger ones (longer than 365 cm), were caught from January to June. The longfin mako is also found in the Gulf Stream off the southeastern United States throughout much of the year. It travels north in the Gulf Stream as far as Georges Bank (Kohler et al. 1998).

There are few records of longfin makos in the eastern Pacific. Ebert (2001) reported specimens from Anacapa and Santa Barbara Islands, California, but these records are doubtful and may be misidentifications, because the waters of the California coast are generally too cold for the species. The Santa Barbara specimen illustrated by Ebert (2001) has the coloration of a shortfin mako. I attempted to verify this record by examining the jaws, which according to Ebert (2001) had been deposited at the California Academy of Sciences (CAS no. 214228). Unfortunately, the jaws were retained by the fisherman and were not deposited at the Academy (Tomio Iwamoto and David A. Ebert, pers. comm., May 2009). Leonardo Castillo (pers. comm. December 2009) sent me photographs of a longfin mako recently taken off Guadalupe Island, Mexico, by Mexican longliners.

Size and weight The longfin mako is one of the larger sharks. The largest specimen for which I have data is a 427-cm female caught off Pompano Beach, Florida, on 20 February 1984 (Gordon Hubbell, pers. comm., October 1984). This animal was missing 3–5 cm off the tip of the tail, and the total length was calculated by biologists using the fork and precaudal lengths. Its weight was estimated at 566 kg. Representative weights of specimens I have examined are 121 cm, 11.3 kg; 151 cm, 23 kg; 180.7 cm, 39.5 kg; 220 cm, 70 kg; 247.6 cm, 100 kg; 272 cm, 118 kg; and 274 cm, 139 kg. A 370-cm female was reported to weigh 351 kg (Gilmore 1983).

Biology The longfin mako is an apex predator in warm oceanic waters. It was once believed to be a deep-water species because of its coloration and relatively large eyes. However, sightings of longfin makos on the surface and analysis of their diet suggest a much broader depth range. Off northwestern Cuba, many longfin makos are hooked at 90–180 m, and infrequently at 18–50 m, off the continental shelf (Dodrill and Gilmore 1979). They are also seen at the surface when attracted to tunas caught on longlines.

Although the longfin mako is an oceanic species, on 28 December 1975 a live female missing one eye was discovered grounded in the surf at Melbourne Beach, Florida. On the evening before, an angler fishing from the beach 3 km south of the stranding site had engaged a large fish. After a brief run, the fisherman reeled in a large eye, evidently from the longfin mako that later stranded (Dodrill and Gilmore 1979).

The stomachs of six of the longfin makos I have examined contained remains of one or more porcupine fish (*Diodon hystrix*) and squid beaks. Two specimens had porcupine fish spines embedded in the leading edges of the pectoral fins, suggesting that the mako had charged at and missed the puffer, which then hit the pectoral fins, or perhaps the shark had charged into an aggregation of porcupine fish. One stomach also contained *Sargassum* weeds along with porcupine fish remains, indicating that the prey had been picked up near the surface. Other longfin mako stomachs I examined contained only squid remains, mostly from small and medium-size specimens, usually less than 40 cm mantle length. However, I retrieved the beak of a giant squid (*Architeutis*) from the stomach of a 232-cm female longfin mako caught some 260 km east of Cape Canaveral, Florida, by Rey Pérez in April 1999. This beak measured 8 cm across, suggesting that the squid may have been of about 140 cm mantle length or 400 cm total length.

Guitart Manday (1975) described the longfin mako as an irregular visitor along the northwestern coast of Cuba, although specimens could be captured nearly every month, but with peaks in abundance in April–May and August–November. Based on the capture records that I have, specimens were caught in the Gulf Stream, from Pompano Beach to Cape Canaveral, from December to May, most of them being taken from December to February. These limited observations suggest that longfin makos are in Caribbean waters from April to November and travel northward in the Gulf Stream from December to May.

Size at maturity The smallest mature male that I have data for is a 228.6-cm shark caught about 120 km east of Cape Canaveral, Florida, on 3 April 1984 and measured and examined by Gordon Hubbell. Because of the limited data on gravid females, all that can be said about size at maturity is that all the known mature females are larger than 300 cm.

Reproduction The longfin mako is an aplacental viviparous species with oophagous embryos. Little is known about its reproductive processes because biologists have examined only a few gravid females. Compagno (1984a) speculated that

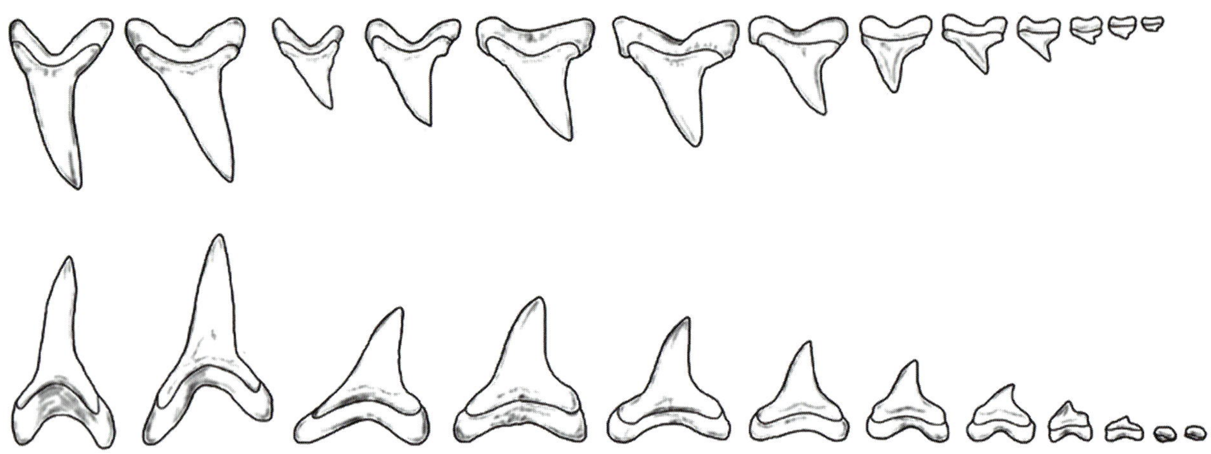

Fig. 65b. Teeth.

the longfin mako embryos are cannibalistic, probably because the first two gravid longfin mako females reported had only two young each. Subsequent broods examined have demonstrated oophagy, but they have not provided any evidence of uterine cannibalism.

Size at birth: The young measure 122–135 cm at birth. Guitart Manday (1975) reported a female with two fully formed embryos, a male and a female. He gave the size of the male as 92 cm standard length (about 116 cm total length) and a weight of 12 kg. A 303.5-cm gravid female, caught in the Florida Current near Cape Canaveral on December 1980, carried a fully formed embryo about 97 cm in one uterus and had obviously aborted another embryo from the other uterus (Gilmore 1983). A gravid female measuring 309.8 cm, taken at 18° N, 67° W, in January 1982, carried eight embryos measuring 113–122 cm and weighing 9–12.7 kg. Another 318-cm gravid female, caught north of Puerto Rico (at 20° N, 58° W) in February 1993, carried four embryos 122–134.5 cm.

Brood size: The few broods examined consisted of two to eight young.

Nurseries The location of the longfin mako nurseries is unknown. Given that females with the largest term embryos were found in the warm tropical waters of the Caribbean around 20° N latitude in late January and February, it is

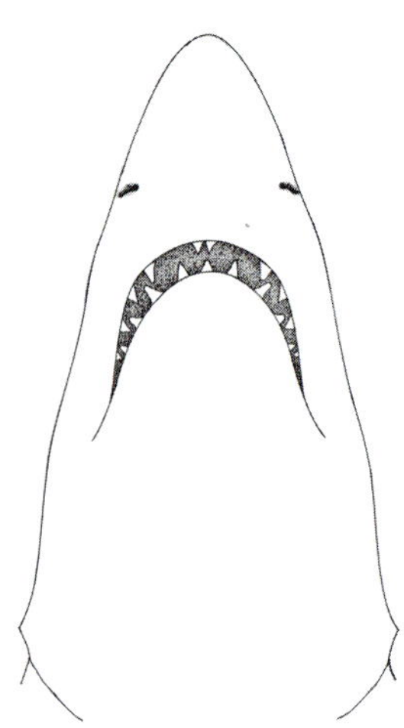

Fig. 65c. Snout.

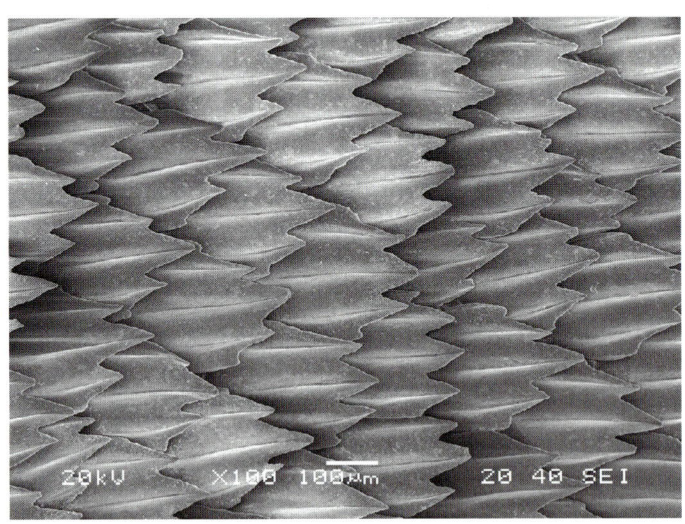

Fig. 65d. Dermal denticles.

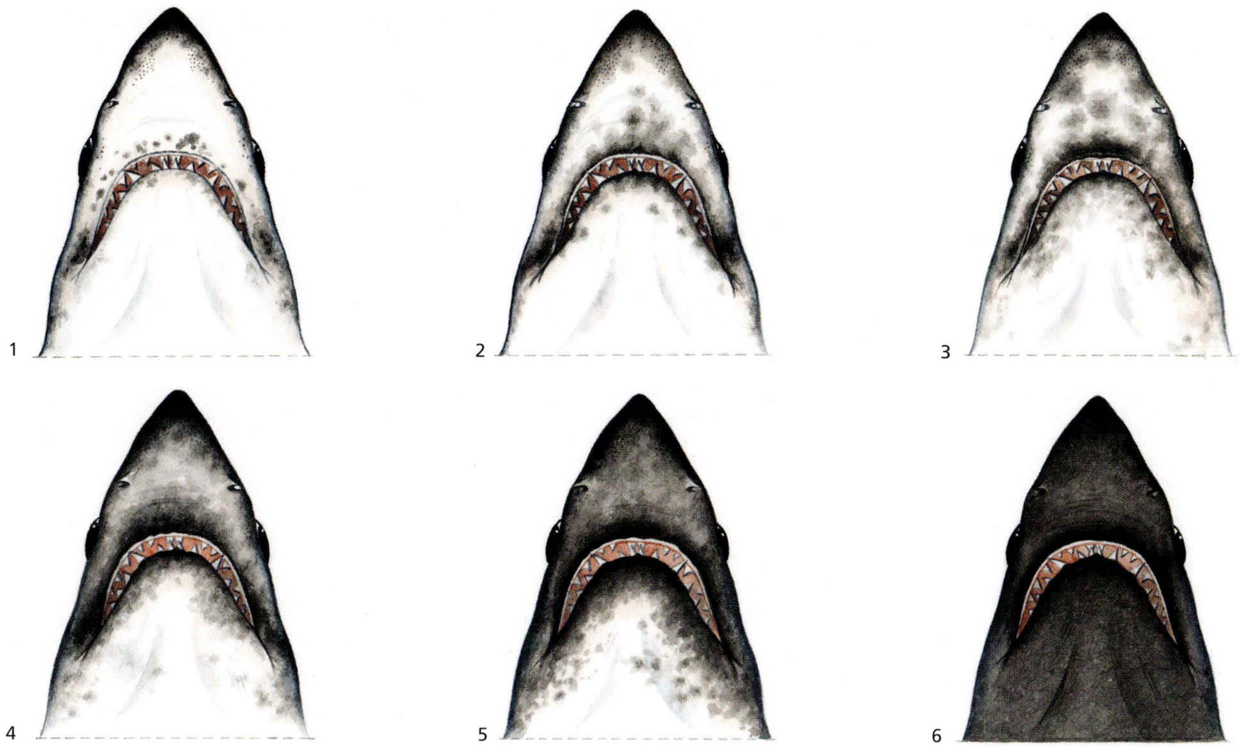

Fig. 65e. Snout color changes with size or age. Snout: 1. 121.2-cm immature female; 2. 125.0-cm immature male; 3. 156.4-cm immature female; 4. 180.7-cm immature female; 5. 232.0-cm immature female; 6. 272.0-cm immature female.

likely that the longfin mako gives birth in those areas in February and March.

Age and growth The longfin mako has not been aged.

Relation to humans The longfin mako has no commercial value in the United States because its meat is of poor quality, and there is no demand for it. Longfin makos are occasionally taken in longline fisheries off the eastern United States, but they are not normally landed. Berkeley and Campos (1988) reported that longfin makos constituted 0.3% of all sharks taken in their sampling of the swordfish fishery along the east coast of Florida in 1981–1983. Guitart Manday (1975) noted that longfin makos constituted 5.8% of all shark captures off the northwestern coast of Cuba during a two-year period (1971–1972) but that it had not been captured in 1973. According to that report, Cuban fishermen considered its meat to be watery and of inferior quality compared with that of the shortfin mako.

The longfin mako was classed as a protected species in the eastern United States in 1999 and may not be retained by fishermen. The reasons for giving it protected status were precautionary because there are no data on population declines or specific threats to the species. Nevertheless, protecting these magnificent fishes is a good idea. However, the effectiveness of the protection for longfin makos is difficult to evaluate because they are often dead on the line at retrieval. In any case, commercial fishermen quickly learn to recognize longfin makos, calling them "fakos" (from "false makos"), because they are worthless, unlike the shortfin makos. The "fakos" are quickly discarded overboard or, if alive, are cut loose. Recreational fishermen, being less familiar with longfin makos, often confuse them for shortfin makos and sometimes land them. Pride often turns to chagrin when they learn that their coveted trophy is a protected species!

SALMON SHARK

Lamna ditropis Hubbs and Follett, 1947

Fig. 66a. *Lamna ditropis*, salmon shark, mature male 240 cm, 156.9 kg mature male, from Port Gravina, Alaska (Matthew Miller, collector).

Common name Salmon shark, because it appears along the coastline at the same time as salmon schools. The species was not distinguished from the porbeagle until 1947. In the earlier literature, it was most often called the "mackerel shark" (Walford 1935; Clemens and Wilby 1946), although Walford (1935) also used the name salmon shark.

Spanish name The A.F.S. name is *tiburón salmón.*

Scientific name *Lamna ditropis* Hubbs and Follett, 1947, in *Copeia* 1947 (3): 194. *Lamna:* from the Greek *lamia,* a large shark, originally from Lamia, in Greek mythology the beautiful daughter of King Belos on whom Zeus, in acknowledgment for her favors, bestowed the singular power of plucking and replacing her eyes at will. She bore Zeus several children, most of whom were killed by Hera in a fit of jealousy. Lamia took revenge by destroying the children of others and behaved so cruelly that her face turned into a nightmarish mask (Graves 1955). The myth of the Lamia persisted well into the Renaissance; for a beautiful woodcut representation of Lamia of that epoch, see Topsell[4] (1658: 353). *Ditropis:* Greek *di,* two, and *tropis,* keel, hence "two-keeled," referring to the two caudal keels of the salmon shark.

4. Topsell (1658: 352) wrote, in his translation of Conrad Gesner's earlier work: "This word *Lamia* has many significations, being taken sometime for a Beast of Lybia, sometimes for fish, and sometimes for a Spectre or apparition of women called *Faries.* And from hence some have ignorantly affirmed, that either there were no such Beasts at all, or else that it was a compounded monster of a Beast and a Fish."

Synonyms None.

Identification This lamnid shark of the *North Pacific Ocean* is characterized by a conical snout, a first dorsal fin that originates *over or anterior* to the axil of the pectoral fin, and a *tail with a secondary keel.* The distance from the tip of the snout to the front of the eye is *less than one-half the distance from the posterior edge of the eye to the first gill slit.* The teeth have long, slender, flattened cusps with smooth edges and *lateral cusplets* that are small and are hard to see in fresh specimens because the gum tissues can partially cover them. Teeth number U: 15 to 16–15 to 16, L: 13 to 15–13 to 15 and are alike in both jaws. The teeth of the salmon shark are relatively small for its size. The largest teeth of the large male illustrated in this work (242 cm), measure only 11–11.5 mm (measuring only the crown, or enameled cusp, of the two frontal upper teeth). The side teeth of this specimen are 3–7 mm. Up to three rows of teeth can be functional at the same time. The dermal denticles are small, broad, and closely overlapping. They have three to five ridges and three to five points. Small juveniles (under 110 cm) are bluish black above with white undersides. Larger specimens are light tan to bluish black or black above, although the dorsal surfaces of most salmon sharks are black. In black specimens, the black coloration of the dorsal surface fades to gray

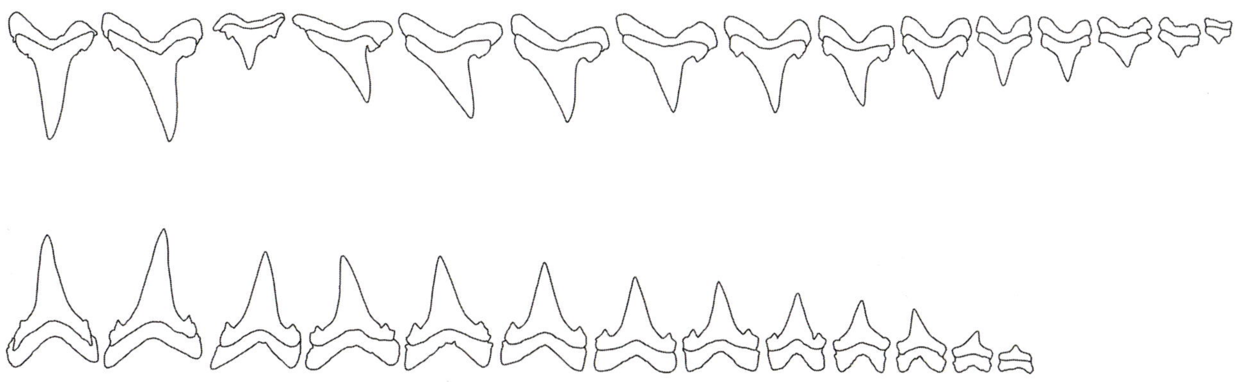

Fig. 66b. Teeth.

along the flanks. The ventral surface is white. Adults longer than 150 cm have *numerous coarse black blotches over the ventral surface.* The amount of this blotching is quite variable, with some animals having only a few black blotches where the dark upper coloration changes to white and others exhibiting numerous black blotches all over the ventral surface. The first dorsal fin has an irregular black edge and a lighter gray center.

Similar species Makos lack the secondary keel on the caudal fin and lateral cusplets on the teeth. The white shark has triangular teeth with serrated edges. The adult porbeagle shark has a longer snout, has a white patch on the free rear tip of the first dorsal fin, and lacks the dark ventral blotches.

Range The salmon shark inhabits the subpolar and cold temperate waters of the North Pacific. In the eastern Pacific, the species ranges from the Bering Sea and the Gulf of Alaska, along the western coasts of Canada and the United States, south to Baja California. For many years the southernmost record was the type specimen, caught off La Jolla (just north of San Diego), California, on 7 December 1945 (Hubbs and Follett 1947). Recently Weng et al. (2005) have published records of salmon sharks in the offshore waters of Baja California.

In the western North Pacific the salmon shark ranges from the subpolar waters of the Bering Sea southward to the Sea of Okhotsk, the Sea of Japan (Blagoderov 1994), Cheju Do, South Korea (Mori 1952), and the Ryukyu Is-

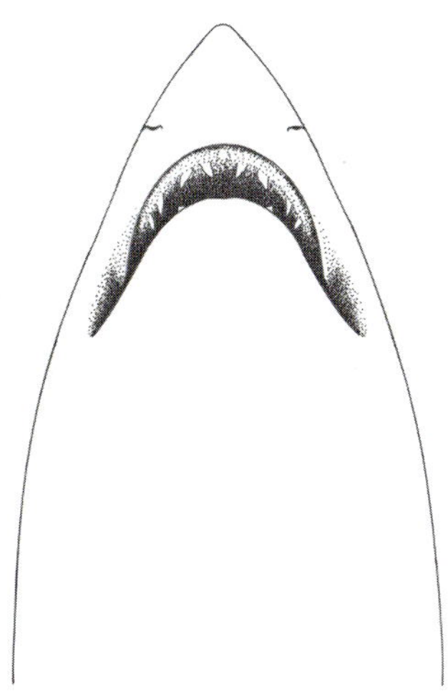

Fig. 66c. Snout.

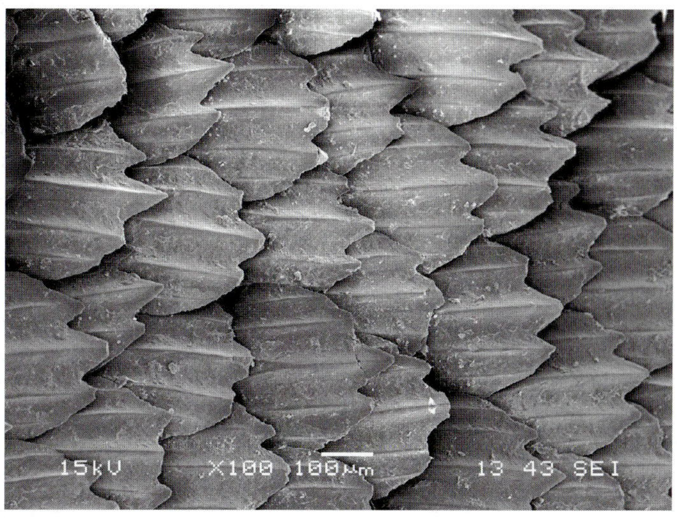

Fig. 66d. Dermal denticles.

Fig. 66e. *Lamna ditropis*, salmon shark, immature female, 225 cm, 110.2 kg; from Port Gravina, Alaska (Ken Goldman, collector). Note parasite damage near second dorsal fin.

lands (Hiyama 1951), the last reports as *Lamna cornubica.* Its southern limit is marked by the 20°C isotherm.

In the open waters of the North Pacific, the salmon shark is common from the Bering Sea south to 35° N, where the warm waters of the Kuroshio Current and its branches with 20°C surface waters create a southern boundary for the species. The Kuroshio flows northeast from Taiwan to the Ryukyus, before moving closer to the Japanese coast up to latitude 35° N. From there, the Kuroshio Extension flows directly east, becoming the North Pacific Current and reaching to 150° E longitude. Salmon sharks are abundant north of this warm water boundary. The greatest abundance of salmon sharks is found east of Japan in a region between 37° and 42° N latitude and between 140° and 160° E longitude. This is where the cold Oyashio Current and the warm Kuroshio Current converge, and the area is well known for its great abundance of salmons, sardines, mackerels, sauries, and tunas. Makihara (1980) reported that salmon sharks were abundant from the northeastern coast of Honshu along latitude 42° N to about longitude 170° W and reported catches of as many as 400 salmon sharks per gillnet set. McKinnell and Seki (1998) monitored the shark bycatch of the North Pacific Japanese high-seas squid fishery operating between 34° and 46° N latitudes and 145° W and 170° E longitudes. Of their sample of 10,496 salmon sharks, most were caught north of latitude 38° N. Catches decreased rapidly south of 38° N latitude, and no sharks were caught south of 33° N latitude. South of the warm water boundary, the salmon shark is replaced by the shortfin mako.

Neave and Hanavan (1960) found salmon sharks to be widely distributed throughout the Gulf of Alaska from May to August. Off British Columbia, Clemens and Wilby (1946) found salmon sharks to be common along the coast from Hecate Strait to the Strait of Georgia, and that individuals up to 260 cm long were abundant during summer off southeastern Vancouver Island, where sportsmen took them on light tackle using salmon as bait.

The salmon shark is not listed among species caught in the California shark fishery by Byers (1940). The first published record of the "mackerel shark" in California appears to be that of Croker (1942), who reported the capture of a specimen of 227 cm near Santa Cruz Island in February 1942. This specimen was said to be the third or fourth taken in the area that year. Miller and Lea (1972) referred to salmon sharks as "uncommon" in California. A news release from the California Department of Fish and Game, dated 21 March 1977, reports a salmon shark caught four miles south of Dana Point and mentions that the species is extremely rare in southern California waters. However, numerous reports of salmon sharks caught or stranded in central California in the 1980s and 1990s suggested that the salmon shark might be more common than previously thought. Recent studies by Weng et al. (2005) have demonstrated that salmon sharks are abundant from the Gulf of Alaska to Baja California. These authors reported that salmon sharks reach the Hawaiian Archipelago, but this needs to be verified.

Size and weight Salmon sharks are stocky and heavy sharks. The male specimen illustrated in this work (page 278) measured 242 cm and weighed 157 kg, twice the weight of the 242-cm male longfin mako illustrated on page 274. Such massive build is probably an adaptation for living in very

Fig. 66f. *Lamna ditropis*, salmon shark, neonate, 85 cm, from Santa Cruz, California (courtesy of Gilbert Van Dykhuizen and Monterey Aquarium).

cold water. The 225-cm female illustrated here weighed 110 kg. A larger female taken in Alaska measured 282 cm and weighed 162 kg (Bruce Wright, pers. comm., August 2003).

Biology The salmon shark is a fast-swimming apex predator in subpolar and cold-temperate waters. It is found in coastal waters as well as in the open ocean. It often forms large aggregations in coastal waters, probably while pursuing schools of salmon. The largest known concentrations occur at the mouths of Alaskan rivers during the salmon spawning season, and in oceanic areas where cold and warm currents converge. Along the coast and in fjords, salmon sharks are often seen at the surface chasing salmon, and there are spectacular photographs of salmon sharks leaping out of the water with salmon in their mouths. Although thousands of salmon sharks are caught every year in diverse fishing operations, the species is poorly understood.

Like other lamnid sharks, the salmon shark is warm blooded, maintaining a body temperature 8–21°C above that of the surrounding water (Smith, R. L., and Rhodes 1983; Anderson and Goldman 2001). Although early studies of homeothermy in sharks were subject to the criticism that the body temperatures had been measured in sharks taken after strenuous captures, recent studies using acoustic telemetry demonstrate homeothermy to be a normal characteristic of salmon sharks. Goldman et al. (2004) used acoustic telemetry to record swimming depth and stomach temperature of four free-swimming salmon sharks in Prince William Sound, Alaska. These sharks maintained stomach temperatures of 25.0–27.2°C in ambient temperatures of 5–16°C, with a maximum observed elevation of stomach temperature of 21.2°C. The elevated body temperature is considered to be an adaptation to allow these fishes to enter cold waters.

Temperature range: Salmon sharks prefer waters of 8–16°C. Makihara (1980) observed that salmon sharks and blue sharks predominate among the larger pelagic species of the North Pacific where the surface temperature is 10–20°C. Makihara also stated that the best fishing conditions for salmon sharks were in surface water isotherms of 10–16°C, in the convergence zone of the Kuroshio and Oyashio currents. In an extensive fishing survey for salmon sharks conducted by Japan Marine Fishery Resource Research Center (JAMARC 1981), the best catch rates were obtained off Kushiro (Hokkaido, Japan) at 42° N, 145° E, where the surface temperatures were 14–16°C. Smith and Rhodes (1983) reported catching salmon sharks near Cape Edgecumbe, about 60 miles from Sitka, Alaska, in sea surface temperatures of 8°C. Weng et al. (2005) conducted an extensive study of the temperature preferences of salmon sharks using satellite tags. They reported that salmon sharks ranged from polar to subtropical waters and temperatures from 2 to 24°C. These authors found that salmon sharks spent much of their time in water cooler than 10°C, entering waters a cold as 2–8°C in winter.

Diet: Salmon sharks consume a wide variety of bony fishes, primarily salmon. Bright (1960) listed the stomach contents of a mature male salmon shark (erroneously reported as a porbeagle) caught at Kachemak Bay, Alaska, as Pacific tomcod (*Microgadus proximus*), pink salmon (*Oncorhynchus gorbuscha*), chum salmon (*Oncorhynchus keta*), and a cottid fish (probably *Hemilepidotus* sp.). Sano (1966) analyzed the stomach contents of 176 salmon sharks collected by commercial and research boats that fished from east of Hokkaido to the Bering Sea. Thirty-nine percent of stomachs contained salmon. The species of salmon identified in the stomach contents were sockeye (*Oncorhynchus nerka;* 16.5%), chum (8%), pink (5.7%), coho (*Oncorhynchus kisutch;* 5.1%), chinook (*Oncorhynchus tshawytscha;* 0.5%), and unidentified salmon (5.7%). Cods (Gadidae) and pollock (*Theragra chalcogramma*) were found in 16.5% of the stomachs.

Migrations: The migrations of the salmon shark are complex and at present poorly understood. According to Blagoderov (1994), in the western Pacific, salmon sharks reside continually in oceanic waters off the coast of Japan, the Kuril Islands, and the mixing zone where the Kuroshio and Oyashio currents converge. In winter, salmon sharks are concentrated in the waters off Japan. The majority of the winter catches are in the coastal waters of Honshu, although individuals can be found far out at sea. In summer and early fall, larger sharks migrate to the northern Sea of Japan, the southern part of the Sea of Okhotsk, and the southern Kuril Islands. In autumn the larger sharks enter the more northern Sea of Okhotsk and the Bering Sea. Thus, the species ranges from the island of Honshu to the Bering Sea. The greatest concentrations are found in the Sea of Japan, the Tatar Straits, and the Okhotsk and Pacific waters of the southern Kuril Islands, being most abundant off Hokkaido.

Catch reports indicate that salmon sharks occur in the Gulf of Alaska and Prince William Sound all year, where they appear to undertake onshore–offshore movements. The species is abundant close to shore and inside fjords from June to September, moving offshore in winter. I have found salmon sharks still abundant in the inlets and fjords around Cordova, Alaska, as late as mid-September. Most of the specimens from coastal Alaska are reported to be females (Goldman and Musick 2006).

Salmon sharks captured in Alaskan waters (and perhaps elsewhere) bear small open sores throughout the body (see illustration of adult female on page 280). These lesions appear as raised, necrotic, open sores 1–3 cm in diameter and 1–2 cm deep. They are caused by the parasitic copepod *Anthosoma crassum,* which is buried deep in the center of the wound.

Size at maturity

Male: Goldman and Musick (2006) estimated that the median length at maturity for males was 158 cm. They used the development of the testes, a coiled ductus deferens, calcified claspers, and well-developed siphon sacs as criteria for maturity.

Female: Females mature at 205–220 cm. Goldman and Musick (2006) used the presence of a "well-developed" ovary containing various sizes of oocytes, large oviducts and oviducal glands, and expanded uteri to determine maturity. They estimated that the median length at maturity for females in the eastern Pacific was 205 cm, although they had no gravid females to verify this estimate. Sano (1966), who examined female salmon sharks caught in the western North Pacific, reported that of 14 females measuring less than 210 cm, all were immature. Of seven females of 220 cm, one was immature and six were mature. Twelve females measuring 230–250 cm were all mature.

Reproduction The salmon shark is an aplacental, viviparous species. Embryos nourish themselves by eating eggs in the uterus (oophagy). This method of embryonic nutrition was first described by Lohberger (1910). This work, perhaps the most comprehensive description of oophagy, was based on "two giant embryos of *Lamna*" obtained by Doflein (1906), presumably in Japan. Unfortunately, little else was published on salmon shark reproduction during the rest of the twentieth century.

Size at birth: Salmon sharks are born at 65–76 cm in spring. A female, caught off Monterey, California, in March 1987, contained three embryos: a female 70.5 cm and two males, 71.0 and 72.0 cm. I have one of these embryos, received through the courtesy of Prof. G. M. Cailliet, and it still shows a large yolk stomach. Fitch (1977, courtesy of R. N. Lea) mentions that he examined a female containing four 76-cm young. The smallest salmon shark in the JAMARC survey was caught in May and measured 65 cm. The survey estimated that salmon sharks measure 65–70 cm at birth (JAMARC 1981).

Brood size: Salmon shark broods usually consist of four young. In addition to the four embryos mentioned by Fitch (1977), Makihara (1980) stated that the female salmon shark gives birth to four young. JAMARC (1981) reported that seven pregnant salmon sharks captured off Sanriku, Japan, during December–March, carried three to four embryos per female.

Nurseries Strandings and captures of neonates along the central California coast suggest that the nurseries for the North American population may be in the southern parts of the range. A male, 91 cm and 8 kg, washed ashore dead at Santa Maria Beach (between Santa Cruz and Capitola, California) on 31 December 1984 (R. N. Lea, pers. comm. March 2001). Another salmon shark estimated at 90 cm was found trapped in a tidal pool at Pleasure Point California on 29 January 1995. An 85-cm neonate was snagged from a pier in Monterey in 1987 and kept for a short time at the Monterey Bay Aquarium. Strandings of small salmon sharks appear to be common, but their causes are unknown.

Blagoderov (1994) stated that, based on the encounter rate with salmon sharks less than 60 cm (fork length?), birth occurs only in three regions: the open waters of the Pacific, the waters around Japan, and in the southern Kuril regions. The young reside in those areas until they are at least 110–120 cm.

Age and growth According to Goldman and Musick (2006), males mature in 3–5 years and females in 6–9 years, based on analysis of vertebral growth rings. Their oldest shark was estimated to be 20 years old.

Relation to humans There is little, if any, commercial demand for salmon sharks in North America. Even so, thousands of salmon sharks are caught each year in salmon nets of the summer fisheries along the Alaskan coast from Bristol Bay to British Columbia and by the Japanese high-seas fleet. Consequently, tons of unwanted salmon sharks are discarded each year (Urquhart 1981).

In the mid-1980s, the Alaska Sea Grant College Program produced a voluminous report on the salmon shark (Paust and Smith 1986), apparently to aid in developing an Alaskan fishery for the species. The report is a long, eclectic compilation of unrelated shark data, but it is useful as a source of references on the salmon shark. At this writing, the fishery has not developed. However, a minute sports fishery for salmon sharks exists in Cordoba.

In Japan, the salmon shark has been a highly favored food fish since ancient times in regions north of Tokyo (except Hokkaido). In farming villages, "spring salmon shark" is eaten in May or June at rice-planting time. A fishery for salmon sharks has long existed in northeastern Japan, where from 1968 to 1978 landings ranged from 3,000 to 6,000 metric tons annually. In a fishing trip lasting 12–20 days, about 10 longline sets are made. Each longline carries between 2,000 and 2,500 hooks baited with mackerel, resulting in a catch of about 1,000 sharks, or about 40 tons (Makihara 1980). Since 1950 some salmon shark meat has been exported to Italy from the Kesennuma region of northeastern Japan. The quantity of frozen shark meat exported to Italy was only about 50 metric tons per year in the late 1950s (Makihara 1980). By 1960, exports had grown to 300 metric tons (Commercial Fisheries Review 1960), and by 1979 the annual exports had grown to 500 metric tons (Makihara 1980). F.A.O. statistics do not reflect current levels of salmon shark exports from Japan to Italy.

PORBEAGLE

Lamna nasus (Bonaterre, 1788)

Fig. 67a. *Lamna nasus*, porbeagle shark, male, 182 cm, 44 kg; from Halifax, Nova Scotia (Frank Reyno, collector, through the courtesy of Lisa Natanson and Steve Campana).

Common name Although the origin of "porbeagle" is uncertain, most sources agree that it is derived from porpoise and beagle. Yarrell (1841: 516) quotes Jonathan Couch saying of this species: "That it associates in small companies in pursuit of prey, from which circumstance, and a distant resemblance to the Porpus [porpoise], they derive their name." Day (1880–1884: 298) wrote that the "*porbeagle* apparently so termed by Borlase as a compound of *porpoise* and a *beagle* or hunting dog." According to the *Oxford English Dictionary* the name is of Cornish origin and possibly derived from the French *porc* (swine) or porpoise (*porcus pisces* = hog-fish or fish-hog) and beagle. Pennant (1776) and nineteenth century English writers sometimes called it the Beaumaris shark, from the place in North Wales where it was first caught in the United Kingdom.

Spanish name There are no vernacular names for the species in the Spanish Caribbean because the porbeagle is not found there. In Spain the species is known as *cailón* or *marrajo.* The A.F.S. has not designated a Spanish name for it.

Scientific name *Lamna nasus* (Bonaterre, 1788), originally described as *Squalus nasus* in *Tableau Encyclopédique et Méthodique des Trois Règnes de la Nature, Ichthyologie,* p. 10. *Lamna:* Greek, a large shark, derived from *Lamia,* in Greek mythology the beautiful daughter of King Belos who, after losing her children, took revenge by destroying the children of others and behaved so cruelly that her face turned into a nightmarish mask (Graves 1955; see p. 278 for more detail); *nasus:* Latin, the nose.

Synonyms *Squalus glaucus* Gunnerus, 1768; *Squalus cornubicus* Gmelin, 1788; *Lamna cornubica* Cuvier, 1817; *Lamna monensis* Yarrell, 1836; *Lamna punctata* Storer, 1839; *Lamna philippi* Perez Canto, 1886; *Isurus punctatus* Garman, 1888; *Isurus dekayi* Smith, 1898; *Isurus nasus* Garman, 1913; *Lamna whitleyi* Phillipps, 1935. In many older works, the porbeagle is referred to as *Lamna cornubica.* This name was replaced by *Lamna nasus* by priority.

Identification The porbeagle has a conical snout, a first dorsal fin originating over or anterior to the pectoral fin axil, a second dorsal fin about one-fourth as high as the first, with its origin above the anal fin origin, strong keels on the caudal peduncle, and a small secondary keel on the tail. The distance from the tip of the snout to the front of the eye is at least one-half as long as the distance from the posterior edge of the eye to the first gill slit. The teeth have long, slender, pointed cusps with smooth edges and *with one small lateral cusplet on each side of the main cusp.* Very small specimens lack the lateral cusplets. The teeth number U: 24 to 32 and L: 23 to 26 and are alike in both jaws. The dermal denticles are small and overlapping. They have a strong central ridge and two weaker ridges on the sides. Coloration is dark bluish gray to bluish brown with metallic hues above, fading to white below. After death the color changes to a deep bluish black. There is *a patch of white* at the trailing edge of the first dorsal fin.

Similar species The white shark has triangular teeth with serrated edges. The makos lack the secondary keel on the caudal fin, and their teeth lack the lateral cusplets.

Range The porbeagle inhabits the North Atlantic (Bigelow and Schroeder 1948), South Atlantic (Springer, V. G., and Garrick 1964; Nakaya 1971), South Pacific (Garrick and Schultz 1963; Last and Stevens 1994), and the western Indian Ocean (Smith, J. L. B. 1949; Bass et al. 1975b). Its distribution is basically temperate. The closely related salmon shark replaces it in the Northern Pacific. At this time, the northern and southern hemisphere populations are considered to be a single species with temperate distribution. The differences, if any, between populations are not known.

In the North Atlantic the porbeagle ranges from Newfoundland and the Grand Banks (Templeman 1963) south to New Jersey (Fowler 1906b) and eastward to Iceland (Saemundsson 1949), the Faroes (Rae 1962), western Scandinavia, and the Murmansk coast (Bigelow and Schroeder 1948), and southward to the North Sea, the Mediterranean, and the northwestern coast of Africa (Bigelow and Schroeder 1948). Cadenat and Blache (1981) wrote that they never encountered the species off the western coast of Africa, although it was known with certainty from Moroccan waters.

Off the eastern coast of North America the porbeagle ranges from Newfoundland to New Jersey. The northernmost record is from the Strait of Belle Isle (between northern Newfoundland and the Labrador coast), where a large specimen was caught in July 1929 (Jeffers 1933). There are no records of porbeagles along the Labrador coast (Templeman 1963), and Jensen (1914) did not mention the species in his *Selachians of Greenland.* Presumably, those areas, influenced by the cold Labrador or Arctic Current, are too cold for the thermal tolerances of the porbeagle. According to Templeman (1963, 1966), porbeagles are found off Newfoundland from May to December, but in the southeastern and southern parts they are most abundant from July to early September. He also reported that the species is numerous on the Southeast Shoal of the Grand Banks in July and that Newfoundland trawlers sometimes catch large, mature specimens even in winter on the southwestern Grand Banks. The southernmost records are for Cape May, New Jersey, with several specimens listed by Fowler (1906b) and by Bigelow and Schroeder (1948). Bigelow and Schroeder (1948) reported the porbeagle to be the most common large shark in summer along the coast of Nova Scotia, including Cape Breton. They also described it as "very numerous" in the Gulf of Maine in summer and the most frequently seen large shark around the Isles of Shoals and near Cape Anne.

Bigelow and Schroeder (1948: 118) stated that the range of the porbeagle shark extended "south to New Jersey, and perhaps to South Carolina." This statement has been the basis for numerous writers to extend the range of the porbeagle down to South Carolina, even though Bigelow and Schroeder called the South Carolina report "somewhat doubtful." That South Carolina record is based solely on the misidentification of a tooth fragment removed from the leg of a man who had been attacked by a shark while in waist-deep water off Folly Beach, South Carolina, in July 1924. The tooth fragment was sent for identification to the American Museum of Natural History in New York, where E. W. Gudger and J. T. Nichols examined it and stated that they were "inclined to think that it is from one of the mackerel sharks, but presumably a young specimen" (Burton 1935: 280). However, my own examination of the fragment, at the Charleston Museum in 1981, revealed it to be from a lemon shark, a common summer visitor to South Carolina.

Size and weight Although the porbeagle has been reported to reach a length of 360 cm, this size appears to be an exaggeration. A record of a 300-cm female from Monhegan Island, Maine (Hubbs 1923), appears to be a secondhand record of doubtful accuracy. Bigelow and Schroeder (1948) stated that a 240-cm specimen was positively reported from the Gulf of Maine, and that a number of specimens up to 280 cm long have been reported from European waters. Templeman (1963) discussed several secondhand reports of porbeagles 280–300 cm, but the largest he personally measured was 263 cm (with the upper lobe of the caudal fin extended directly backward). Aasen (1963) found no significant differences between the mean total lengths of males and females. Representative weights are 90 cm, 8 kg; 143 cm, 24 kg; 182 cm, 44 kg. A 229-cm female caught off Guernsey, Great Britain, weighed about 168 kg (Richard Lord, pers. comm., 17 January 2000).

Biology Although the porbeagle is abundant and widely distributed in the North Atlantic and has been the object of several fisheries in Europe and North America, little is known about its biology. This account is based primarily on North Atlantic specimens. However, because the scarcity of published material, data from specimens in other areas are also included.

The porbeagle is a warm-blooded shark with a body temperature 7–10°C above the ambient water temperature; it maintains its warm body temperature by means of a system of countercurrent heat exchangers in the circulatory system known as a *rete mirabile* or "wonderful net" (see footnote on page XX). This system has been known for a long time, although its purpose and functioning were not elucidated until recent times. Eschricht and Müller (1835) discussed the visceral *retia* of the porbeagle and stated that

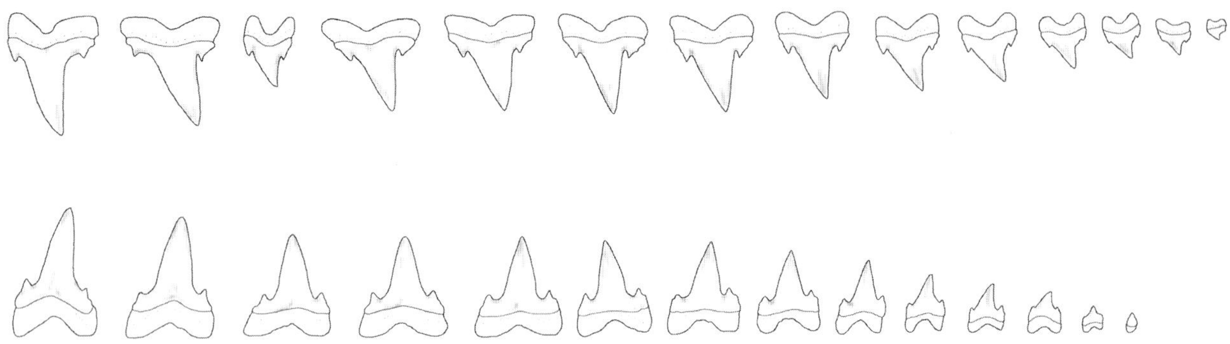

Fig. 67b. Teeth.

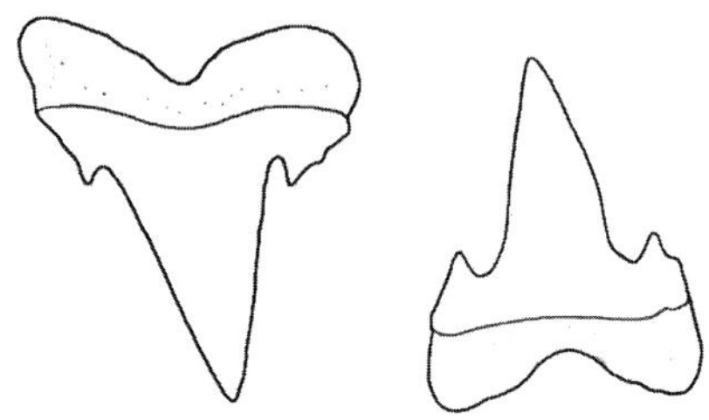

Fig. 67c. Teeth (detail).

they resembled the *retia* of tunas. Müller was aware of the warm-blooded nature of tunas through John Davy's observations. Suspecting some relationship between the *retia* and endothermy, he suggested that observing the temperature of the porbeagle would be important. Burne (1923) described and beautifully depicted the *retia* of the porbeagle but without understanding their function. Carey and Teal (1969) described the function of the *retia* in porbeagle sharks and reported maximum body temperatures of 20.2°C in 11°C water, 17.9°C in 6.6°C water, and 13.5°C in 6.5°C water. The elevated body temperature allows these animals to enter cold waters (Carey et al. 1985).

The porbeagle inhabits inshore and offshore continental waters of 4–19°C, from surface to bottom, and only rarely is it found off the continental shelf. Most porbeagles are reported from waters with surface temperatures of 8–16°C and bottom temperatures of 4–11°C. R. McKenzie (1963) stated that where porbeagles were caught on Georges Bank and Browns Bank, surface water temperatures were 11–16°C and bottom temperatures were 4.2–10.5°C. Dyer et al. (1963) reported that their best catches on Georges and Browns banks were made at the surface where temperatures were 13–16°C. More recently, Campana and Joyce (2004) reported that more than half of the porbeagle sharks from the Canadian commercial fishery were caught at temperatures of 5–10°C, inferring that this is the preferred temperature range of the species in the North Atlantic. According to Bigelow and Schroeder (1948) the isothermal belt of about 18°C limits the southern range of the porbeagle.

In the Southern Hemisphere, Stevens et al. (1983) captured porbeagles in the Tasman Sea, where surface temperatures were 13.8–15.8°C and the bottom temperature was 7.5°C. Svetlov (1978) reported a 218-cm female caught in Antarctic waters (54°28 S, 35° 29 W) at 260 m, where the surface temperature was 3.2°C and the bottom temperature was 1.7°C.

Diet: Although there are few reports of the stomach contents of porbeagles, it is clear from the reports below that the porbeagle is an opportunistic predator that feeds on a wide variety of northern fishes and squids. Fowler (1906b) reported that a porbeagle caught off New Jersey contained butterfish (*Trachinotus* sp.), menhaden (*Brevoortia*), and two or three crabs. Jeffers (1933) reported that a 259-cm specimen, caught off the Strait of Belle Isle, contained cods and capelins. Bigelow and Schroeder (1948) reported that porbeagles prey mainly on mackerel, herring, pilchards, cod, hake, cusk, flounders, and squid. Scattergood (1949) found squid (*Loligo pealei*), alewifes (*Pomolobus pseudoharengus*), herring (*Clupea harengus*), mackerel (*Scomber scombrus*), and rosefish (*Sebastes marinus*) in the stomachs of three porbeagles caught in Maine. Gauld (1989) identified the stomach contents of 86 porbeagles caught in Scotland and reported that cephalopods were the most common prey item, followed by a wide variety of bony fishes. Joyce et al. (2002) described the stomach contents of 533 porbeagles from the Canadian Atlantic and stated that the porbeagle diet was dominated by teleost fishes (91% by weight), primarily al-

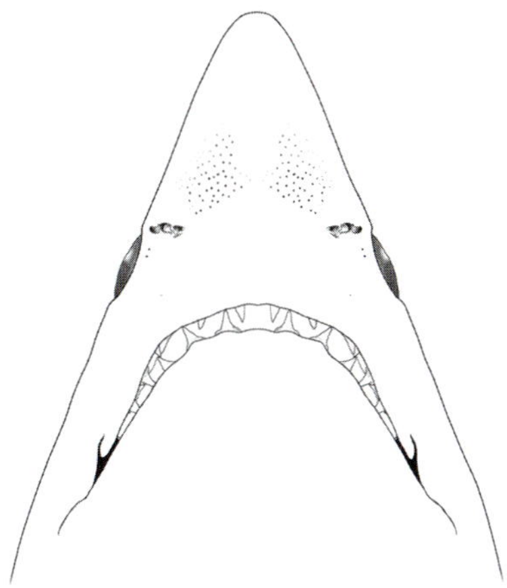

Fig. 67d. Snout.

Fig. 67e. Dermal denticles.

episaurids, gadids, pleuronectids, and clupeids. Cephalopods occurred in 12% and were the second most important food category.

Migrations: The seasonal migrations of the porbeagle in the North Atlantic are understood only vaguely. It appears that porbeagles winter in deep water in their northern range at the edges of the continental shelves. With the onset of summer, they move southward and into shallower water closer to land. In late fall, they reverse these movements.

In the eastern North Atlantic, porbeagles are caught in the Faroes most of the year at the edge of deep water, but in September they move to shallower waters where they are frequently caught by trawlers (Rae 1962). According to Rae (1962), porbeagles appear around the Shetlands in May; they then invade the North Sea, reaching peak abundance in August. Berrow (1994) reported that porbeagles were caught in the bottom-set gillnet fishery off the southern coast of Ireland in March, April, August, and October and that most were caught in October. Saemundsson (1949) said that the porbeagle approaches the coast of Iceland in the warm season, retiring to more southern waters in winter.

In the western North Atlantic, porbeagles winter in offshore areas at the edges of the continental shelf, moving southward and close to shore as summer progresses. Templeman (1963) described the porbeagle as very abundant in the southern Grand Banks in summer, stating that the species was "not uncommon" in the southwestern Grand Banks in winter (10 January to 10 February). Templeman also showed that porbeagles are present from the Flemish Cap to Sable Island, the Western Bank, and the entrance to the Gulf of Maine from March to May (based on data from the Norwegian longline fishery in 1961). Bigelow and Schroeder (1948) reported the porbeagle as the most common large shark in summer along the coast of Nova Scotia, including Cape Breton. McKenzie and Tibbo (1964b) reported that 61 specimens were caught on Browns Bank and the eastern part of Georges Bank from mid-May to mid-October. Scattergood (1949) described the porbeagle as second in abundance only to the spiny dogfish (*Squalus acanthias*) in Maine, with peak abundance from August to November, based on monthly fish landings.

The movements of porbeagles tagged off the coast of Maine show a general track to the northeast along the coast toward Newfoundland waters, with most animals traveling only relatively short distances of 185–370 km (Kohler et al. 1998). Other tagged porbeagles have traveled much greater distances. One shark tagged in Jersey, Channel Islands, Great Britain, was recaptured 13 years later off Vestfjord, Norway, 2,370 km away (Stevens 1990).

Size at maturity

Male: Males mature at 186–200 cm. Aasen (1961) used the relative size of the claspers (but not calcification) to estimate the size of males at maturity and concluded that males mature at 150–200 cm. Later, Jensen et al. (2002) used clasper calcification as a criterion for maturity and stated that males mature at 182–208 cm and that 50% of the population was mature at 196 cm.

Female: The size at which female porbeagles mature is not known precisely because the few studies available have not provided the sizes of the smallest pregnant females or have used somewhat imprecise criteria for assessing maturity. An early study by Aasen (1961) used uterus width as a criterion of maturity, concluding that females mature at 150–200 cm. However, a recent study by Jensen et al. (2002) estimated that 50% of the porbeagle population was mature at a length 245 cm (218 cm fork length).

Porbeagles in other areas have been reported to mature at smaller sizes than American specimens. Francis and Stevens (2000) estimated that off Australia and New Zealand females reached maturity at 194–211 cm, presumably based on the smallest gravid females reported to them. Duhamel and Ozouf-Costaz (1982) reported that a 201-cm gravid female caught off Kerguelen Island, in the Indian Ocean, in March 1981. Whether the differences in size at maturity from the American population are due to population differences or to methodologies is not clear.

Reproduction The porbeagle is an aplacental, viviparous species. The embryos are oophagous but not cannibalistic. The embryos are initially dependent on yolk stored in a minute yolk sac, which is depleted by the time they are 5–6 cm long. Thereafter, the embryos consume great quantities of eggs ovulated by the mother and rapidly develop huge yolk stomachs.

Early writers such as Calderwood (1888) and Shann (1911) confused the large yolk stomachs of midterm embryos with the yolk sac. Swenander (1907) was the first to describe oophagy as a mode of embryonic nutrition in the porbeagle. He noted that the yolk sac had already almost disappeared in 5–6-cm embryos, with only about 1 mm of it remaining on the ventral side. Swenander further explained that after the yolk sac is consumed, embryos ingest egg capsules and their contents, which were present around in the uterus. Later, Lohberger (1910) published a comprehensive description of the embryo, quoting much of Swenander's earlier work. Lohberger's studies were actually based on salmon shark embryos, but he identified them as *Lamna nasus* because the salmon shark had not yet been recognized as a species distinct from the porbeagle. In 1923, Shann corrected his earlier work and confirmed Lohberger's observations. Oophagous feeding continues for several months, during which time the yolk stomach continues to increase in size until it assumes gigantic proportions. The yolk stomach is absorbed before birth.

Gestation: The length of gestation in porbeagles is not clearly defined but has been estimated at eight months by Aasen (1963) and at eight to nine months by Jensen et al. (2002). Available evidence indicates a synchronous population in the North Atlantic (all females being in the same reproductive stage), with an embryonic growth period extending from about November to June. According to Gauld (1989), the average embryo size in six gravid females caught in the Shetlands in December was less than 10 cm. Shann (1923) reported that a female caught off Aberdeen, Scotland, in February containing four embryos each about 35 cm. Templeman (1963) reported that gravid porbeagles caught from 21 January to 10 February on the southwestern Grand Banks carried embryos 40–46 cm long and weighing 1.4–2.7 kg. A female caught off Guernsey, Channel Islands, Great Britain, on 16 January contained three embryos 47–52 cm and 2.7–5.2 kg (R. Lord, pers. comm., 17 January 2000). In Scotland, a female caught in May contained four embryos 60–63 cm, while one caught in June had three large embryos 66–76 cm (Gauld 1989). Aasen (1963) mentioned that fishermen also reported finding large term embryos at the Flemish Cap in late May. No embryos were present in several hundred mature porbeagles sampled by Aasen (1963) from Maine to Newfoundland during the months of July–September. These data suggest a gestation period of about eight months, agreeing with the estimates of Aasen (1963) and Jensen et al. (2002).

Reproductive cycle: Little is known about the porbeagle's reproductive cycle. Aasen (1963) suggested that the porbeagle probably reproduces every year, but he did not adduce any evidence to support that claim. Jensen et al. (2002) also suggested annual reproduction, because all the adult females they examined in December were gravid, but they added the caveat that nongravid females could reside somewhere else. Studies are needed to confirm that the population in the North Atlantic is reproductively synchronous and to confirm a yearly reproductive cycle.

There are some confusing reports that do not fit the pattern explained above. Bigelow and Schroeder (1948) cited two embryos "no doubt of this species" of about 18 cm obtained off Massachusetts in October. There is also a secondhand report by Hubbs (1923) of a "twenty pound embryo, about ready for birth" supposedly taken from a 300-cm female caught in a mackerel net three miles off Monhegan Island, Maine, on 27 August. The date of this catch appears to be in error, as it seems to be too late in the year for a female to be carrying young.

Size at birth: Porbeagles are born at 70–80 cm in late May and early June. Jensen et al. (2002) reported 59–72-cm embryos from a female caught off eastern North America in mid-April. The largest embryos reported by Gauld (1989) were 66–77 cm long in June. Shann (1923) stated that birth presumably occurs when the embryo is about 80 cm and that the smallest free-swimming specimens he observed were 82 and 87 cm. Bigelow and Schroeder (1948) stated that the smallest porbeagle on record was 74 cm. Others have reported free-swimming porbeagles of 85–87 cm from the North Atlantic (Aasen 1963; Jensen et al. 2002).

Brood size: Given how common porbeagles are, there are surprisingly few reports of its brood size. Most reports indicate that the usual brood consists of three or four young. For example, Templeman (1963) reported three broods, two consisting of four young and one of three. Gauld (1989) reported that, off Scotland, there were commonly two embryos in each horn of the uterus and that the average number of young in 12 females he examined was 3.7. Jensen et al. (2002) examined 76 gravid females from northeastern North America, finding that broods ranged from three to six embryos, with an average of four.

Nurseries Porbeagle nurseries have not been located but are believed to be in continental waters. There is an unusual record from Backus (1960) of two small specimens (112 and 98 cm) collected from the surface to 100 m over water 3,650 m deep. These records are unusual because most porbeagle catches come from continental waters. I have examined an 87-cm juvenile caught off the Grand Banks in July.

Age and growth The porbeagle is a long-lived species. Aasen (1963) estimated that porbeagles generally can live to 20 years of age and "very likely 10 years more." Forty years later, Natanson et al. (2002) analyzed vertebral rings of porbeagle sharks from the North Atlantic. Growth rings were validated as age markers up to an age of 11 years by using vertebrae from recaptured tetracycline-injected sharks of known age. This study revealed that both sexes grow at the same rate until maturity, which occurs at 8 years and 196 cm for males and 13 years and 245 cm for females. After maturity is reached, the relative growth of both sexes declines. Maximum age recorded was 25 years for males and 24 years for females. Longevity estimates by Natanson et al. suggest that porbeagles may live for 45–46 years. Campana et al. (2002) used radiocarbon assays to confirm the validity of porbeagle vertebral band counts as accurate age indicators.

Relation to humans The porbeagle is considered an excellent food fish, and Scandinavian fishermen have pursued it in European waters since the early 1800s (Gauld 1989). Directed fisheries for porbeagles in the North Sea and off the Scottish coast have been carried out during the twentieth century by Norwegian and Danish fishermen and to the south and west of England by the French (Gauld 1989). The modern Norwegian fishery started in the 1920s and has been the most intensive porbeagle fishery in the North Atlantic. Catches rose rapidly from 279 metric tons in 1926 to a high of 3,884 metric tons in 1933. Thereafter, catches declined steadily, with a period of more reduced catches caused by World War II. The fishery resumed in 1945 and reached a peak of 2,824 metric tons in 1947. Afterward, catches declined and the scarcity of porbeagles in European waters forced the Norwegians to explore other fishing areas (Rae 1962). By 1961, Norwegian vessels were fishing for porbeagles off Newfoundland and New England. This population was fished intensively in the early 1960s at catch levels of about 4,500 tons annually until the fishery collapsed (Campana et al. 2001). By 1965, plummeting catches spurred some vessels to shift to other species while others moved to African waters to fish for shortfin mako or swordfish. The fishery appeared sustainable during the 1970s and 1980s when annual landings averaged only 350 tons, and the stocks slowly recovered. Increasing annual catches of 1,000–2,000 tons throughout the 1990s appear to have reduced the stocks again. A total allowable catch of 850 tons was established for Canadian waters. However, the current porbeagle stock is so seriously depleted that recovery will depend on greatly reduced fishing mortality (Campana et al. 2001).

Other porbeagle fisheries have also declined. The Norwegian catch has declined from 207 metric tons in 1970 to 25 metric tons in 1994 (International Council for the Exploration of the Sea 1995); likewise, the Danish catch has also declined from 1954 to 1984.

FAMILY SCYLIORHINIDAE
The Catsharks

The catsharks are small sharks that inhabit deep, cold waters of the continental slopes. Catsharks are characterized by two dorsal fins (one exceptional catshark has a single dorsal fin), the origin of the first located posterior to the pelvic fin origin (over or slightly forward of the pelvic fin origin in a few exceptions), short and broad pectoral fins, an anal fin, and large oval eyes. Their teeth are small, multicuspid, and similar in both jaws. Many catsharks are strikingly marked on their dorsal surfaces with dark spots, bars, stripes, or dark blotches. Their development is usually oviparous, but a few species are viviparous. Their egg cases are distinguished from those of other cartilaginous fishes by having a posterior end with a wider diameter than that of the anterior end.

The catsharks constitute the largest family of sharks. Presently, about 100 species are recognized, most of localized distribution; at least 21 species belonging to seven genera have been reported from North America or are likely to stray into the region. Many species of catsharks live on or near the bottom, in very deep waters not exploited by fisheries; therefore, most specimens are caught in trawls or dredges operated by scientific survey ships. Consequently, many species are poorly known and are known from only one or a few specimens. As the slow exploration of the continental slopes continues, the number of catshark species will undoubtedly increase.

The genus *Apristurus* is one of the largest genera of catsharks and certainly the most problematic. Garman (1913) created the genus *Apristurus* for those catsharks that lacked a caudal crest or "filetail" (a row or crest of enlarged dermal denticles on the proximal half of the upper edge of the caudal fin). *Apristurus* are deep-water catsharks dwelling at depths greater than 800–1,000 m, generally on the bottom. They have soft bodies and slightly flattened heads with long snouts. Their snouts have enlarged pores (ampullae of Lorenzini) arranged in characteristic patterns on both dorsal and ventral surfaces; the ventral pattern is usually an oval surrounding several longitudinal rows of distinct pores. They lack a supraorbital shelf, a narrow edge of the cranium that extends above the eye orbits. *Apristurus* catsharks have prominent labial furrows that are continuous around the mouth corners, and two small dorsal fins, the first originating over or behind the pelvic fin. The number of turns in the spiral valve is variable and may be useful separating species. The number of monospondylous vertebrae has also been used as a diagnostic feature.

Most *Apristurus* are brown, black, or dark gray; their dorsal and ventral surfaces are uniformly colored, or the ventral surface may be slightly darker. They lack distinctive markings and the dark saddle blotches of other catsharks, except that they may have darker areas around the mouth, the gill slits, or the fin edges. There are profound morphological differences between juveniles and adults, and small juveniles are often difficult to identify because their body proportions differ widely from those of adults. *Apristurus* teeth are small, numerous, often multicuspid, not set in definitive rows, and are usually similar in both jaws. However, there are striking differences in tooth shape between juveniles and adults and between males and females. Mature males have teeth with conical cusps and small lateral cusplets. Females and juveniles have flat, multicuspid teeth, with one larger central cusp flanked by one or two smaller ones.

The sharks of the genus *Apristurus* are a poorly known, confusing group. There are several reasons for this. First, specimens are usually very difficult to obtain because of their deep-water habitat, often in excess of 1,000 m. Second, *Apristurus* catsharks are soft bodied; they are generally taken by trawling, and specimens are often damaged when captured. Third, even when good specimens are obtained, their soft bodies require great care in preservation (injecting), if they are to produce useful study or museum specimens. Those specimens that are just thrown in a formalin jar usually shrink into unrecognizable mummies that are of little value when identifying or comparing species. Thus, it is difficult to obtain suitable, preserved study specimens in museums. Finally, it is often difficult to determine the species, even for fresh specimens, because some of the early species descriptions were vague or based on small juveniles, or simply because of pronounced age, sex, or individual variation.

Today, *Apristurus* catsharks are seldom seen in the western North Atlantic, because most of our survey ships no longer regularly explore deep waters. With the exception of John Galbraith's recent work on the *Apristurus* of Bear Seamount, little work has been done on *Apristurus* catsharks of the western North Atlantic since Stewart Springer's days in the 1960s and 1970s. Most of the *Apristurus* species of the Gulf of Mexico have not been collected in decades.

Currently, there are more than 40 valid species of *Apristurus* in the world (Prof. Kazuhiro Nakaya, Hokkaido University, pers. comm., February 2005). Off North America, there are nine described species and several yet undescribed species. One can be certain that, as our ability to recognize species improves, more species of *Apristurus* will be identified in North American waters. A monograph on the genus by Prof. Nakaya, the recognized expert on this genus, is expected soon.

The genus *Cephaloscyllium* is characterized by robust bodies with short snouts, the presence of a supraorbital shelf on the cranium, and lack of upper and lower labial furrows. These sharks have a unique defense mechanism: they can inflate their stomachs with water (or air), becoming balloon like. This genus is represented by only one species in North America, the swell shark.

The genus *Cephalurus* is also a monotypic genus. *Cephalurus cephalus,* the lollipop shark, has a tadpole-shaped soft body, a wide, flattened head with a short, rounded snout, and a very wide mouth. It is one of the few catsharks with the first dorsal fin origin in advance of the pelvic fin origin.

The genera *Galeus* and *Parmaturus* are similar. They both lack supraorbital shelves; they both have labial furrows continuous around the mouth corners, and they both have a caudal crests consisting of enlarged skin denticles along the proximal half of the upper caudal fin. *Galeus* species have large and broad pectoral fins (the maximum width is greater than the mouth width). Their coloration is darker on the dorsal surface than the ventral surface, and the dorsal surface often has complex coloration patterns. There are three species of *Galeus* in North America. *Parmaturus* species have somewhat narrow pectoral fins; their greatest width is less than the mouth width. They are uniformly colored and lack any coloration pattern of saddle blotches or spots on their dorsal surfaces. There are at least two or three species of *Parmaturus* in North America. The Pacific species (*P. xaniurus*) is common and well known. The Atlantic *Parmaturus* (one or two species) are poorly known and rare in collections.

The genus *Schroederichthys* is easily recognized by the peculiar body with a short trunk and a very long tail; the caudal section (from the cloaca to the tip of the tail) can be as much as two-thirds of the total length. *Schroederichthys* have supraorbital shelves above the eye orbits, lack a caudal crest, and have labial furrows continuous around the mouth corners. There is a single known species of this genus in North America.

Catsharks of the genus *Scyliorhinus* have sturdy bodies with two dorsal fins, the first one being larger than the second one. They short labial furrows that are not continuous around the mouth. The upper labial furrows are absent, but the upper lip overlaps the lower lip at the corners, giving the impression that an upper furrow is present. They have supraorbital shelves over the eye orbits. Their dorsal surfaces have conspicuous markings, spots, blotches, or lines that are species specific and diagnostic. When compared to other catsharks, most of our *Scyliorhinus* live in moderately deep water at depths of 100–400 m. There are at least five species of the genus in North America.

Note: Based on our present knowledge, juveniles cannot always be identified to species; the descriptions that follow are for adult specimens.

KEY TO THE SCYLIORHINIDAE

Because of the large number of catsharks in North America, and to keep the dichotomous keys simple, it is practical to divide the species into those inhabiting the Pacific or western coast of North America, and those inhabiting the North Atlantic or eastern coast of the continent. It must be remembered that, in all probability, there are undescribed species in both areas.

Pacific Coast Species

1a. First dorsal origin over or in front of origin of pelvic fins (dorsal coloration dark, without definite markings, spots or dark blotches, head very wide, a tadpole-like small shark less than 30 cm)........ **Lollipop shark,** *Cephalurus cephalus* (p. 321).

1b Origin of first dorsal fin behind origin of pelvic fins 2

2a. Dorsal surfaces with large saddle blotches and numerous light and dark spots blotches........ **Swell shark,** *Cephaloscyllium ventriosum* (p. 318).

2b. Dorsal coloration uniform, without dark spots 3

3a. Having a row of enlarged dermal denticles on the anterior half of the upper margin of the caudal fin, dorsal fins of about equal size 4

3b. Lacking a row of enlarged dermal denticles on the anterior half of the upper margin of the caudal fin, second dorsal fin larger than first 5

4a. Gill slits in front of pectoral fin origin, lining of the mouth is dusky or black (ventral and lateral body surfaces covered with tiny, black, pepperlike spots)........ **Peppered catshark,** *Galeus piperatus* (p. 328).

4b. Last two gill slits positioned above the pectoral fin, lining of the mouth cavity white........ **Filetail catshark,** *Parmaturus xaniurus* (p. 332).

5a. Abdominal length (from the pectoral fin axil to the origin of the pelvic fins) equal to or longer than distance from tip of the snout to the first gill slit, margins of fins dark .... **Brown catshark,** *Apristurus brunneus* (p. 296).

5b. Abdominal length (from the pectoral fin axil to the origin of the pelvic fins) shorter than distance from the tip of the snout to the first gill slit, margins of fins white .... **White-edge or longnose catshark,** *Apristurus kampae* (p. 301).

Atlantic Coast Species

1a. Caudal crest present (a crest of enlarged and modified dermal denticles along the proximal half of the tail, may require magnification to see) 2

1b. Caudal crest absent........ 4

2a. Caudal crest nearly rounded across its tip, coloration uniform or dorsal surfaces lacking markings (snout widely rounded and much shorter than mouth width, labial furrows very small) .... **Campeche catshark,** *Parmaturus campechiensis* (p. 330).

2b. Caudal crest nearly flat along its top, dorsal surfaces marked with dark saddle blotches or variegated 3

3a. Dorsal surfaces with longitudinal stripes .... **Springer's catshark,** *Galeus springeri* (p. 326).

3b. Dorsal surfaces with a pattern of round or oval dark blotches or spots .... **Marbled catshark,** *Galeus arae* (p. 323).

4a. Coloration uniform (except for darker fin tips and around mouth), lacking any dark saddle markings or spots; a series of pores forming a conspicuous oval on the ventral side of snout, large and conspicuous labial furrows around the corners of the mouth, body soft or spongy, with weak dermal denticles 5

4b. Dorsal surfaces with dark or light markings, saddle blotches, or spots; lacking a series of pores forming a conspicuous oval on the ventral side of snout, lacking large and conspicuous labial furrows around the corners of the mouth, body firm with hardened dermal denticles 12

5a. Abdomen long, the distance from the pectoral fin axil to the origin of the pelvic fins is more than twice as long as the distance between dorsal fins 6

5b. Abdomen short, the distance from the pectoral fin axil to the origin of the pelvic fins less than twice the distance between dorsal fins 7

6a. Dermal denticles sparse with little overlapping, denticle sides parallel .... **Galbraith's catshark,** *Apristurus* sp. X (p. 316).

6b. Dermal denticles dense and overlapping, denticle sides divergent .... **Black roughscale catshark,** *Apristurus melanoasper* (p. 307).

7a. First dorsal fin originating over pelvic fin base 8

7b. First dorsal fin originating behind pelvic fin base 11

8a. Second dorsal fin only slightly larger than first dorsal fin (origin of first dorsal fin over or slightly behind middle of pelvic fin base, anal fin long extending from tips of pelvic fins to the origin of the tail) .... **Icelandic catshark,** *Apristurus laurussoni* (p. 303).

8b. Second dorsal fin two to four times larger than first 9

9a. First dorsal fin base slightly longer than horizontal diameter of eye (a small species of less than 50 cm, known only from the Gulf of Mexico) .... **Rivero's catshark,** *Apristurus riveri* (p. 314).

9b. First dorsal fin base over twice the horizontal diameter of eye 10

10a. Snout 7–8% of total length, dorsal fins uniformly colored and lacking dark tips, pectoral fins with squarish margins .... **Deep-water catshark,** *Apristurus profundorum* (p. 311).

10b. Snout 8–10% of total length, dorsal fins with dark edges and tips, pectoral fins with rounded margins .... **Ghost catshark,** *Apristurus manis* (p. 305).

11a. Anal fin base shorter than distance from tip of snout to first gill slit (origin of first dorsal fin at a vertical between posterior end of pelvic fin base and origin of anal fin; abdominal length, from pectoral fin axil to the origin of the pelvic fin, about 11–14% of total length, second dorsal fin two to four times larger than first) .... **Smallfin catshark,** *Apristurus parvipinnis* (p. 309).

11b. Anal fin base longer than distance from tip of snout to first gill slit (origin of first dorsal fin at a vertical between posterior end of pelvic fin base and origin of anal fin; abdominal length, from pectoral fin axil to the and the origin of the pelvic fin 7–8.5% of total length second dorsal fin two to three times larger than first, minute white spots under denticles) .... **Hoary catshark,** *Apristurus canutus* (p. 299).

12a. Distance from the tip of the snout to the cloaca less than half the distance from the cloaca to the tip of the tail; interdorsal distance much longer than head length (head length very short, about 15% of total length; a slender little shark less than 40 cm, with a very long body and covered with small white spots) .... **Narrowtail catshark,** *Schroederichthys maculatus* (p. 334).

12b. Distance from the tip of the snout to the cloaca more than half the distance from the cloaca to the tip of the tail; interdorsal length much shorter than head length (head length more than 15% of total length) 13

13a. Color pattern of black reticulating lines or dark dorsal saddle blotches, lacking white spots 14

13b. Color pattern of dark saddle blotches with black or white spots... 15

14a. Color pattern of reticulating black lines ....**Chain dogfish,** *Scyliorhinus retifer* (p. 342).

14b. Color pattern of dark saddle blotches (solid blotches on caudal area) .... **Mead's catshark,** *Scyliorhinus meadi* (p. 340).

15a. Dorsal surfaces covered with small black or dark spots .... **Boa catshark,** *Scyliorhinus boa* (p. 336).

15b. Dorsal surfaces lacking black or dark spots 16

16a. Upper body surfaces with scattered, distinct white round spots.. **Dwarf catshark,** *Scyliorhinus torrei* (p. 347).

16b. Upper body surfaces covered with closely apposed round areas of light pigmentation, not distinct white spots .... **Western catshark,** *Scyliorhinus hesperius* (p. 338).

BROWN CATSHARK

Apristurus brunneus (Gilbert, 1892)

Fig. 68a. *Apristurus brunneus*, brown catshark, mature male, 65 cm, 0.70 kg; from off Santa Cruz, California (Brooke Flammang, collector).

Common name Brown catshark.

Spanish name *Pejegato marrón* (A.F.S.).

Scientific name *Apristurus brunneus* (Gilbert, 1892), originally described as *Catulus brunneus* in *Proceedings of the United States National Museum* 14(880): 542–543. *Apristurus: A,* Greek prefix meaning the absence of something, without, and *prist,* from Greek *pristis,* a file or saw, plus *urus,* from the Greek *oura,* having a tail, thus "without a filetail" (without a caudal crest); *brunneus:* Latin brown.

Synonyms None.

Identification The brown catshark is characterized by a snout shorter than the mouth width, with a *conspicuous band of four rows of pores along its ventral surface;* a *mouth with upper labial furrows that are longer than the lower ones;* a *first dorsal fin originating posterior to the origin of the pelvic fins;* pectoral fins longer than the length of the snout; and a *long anal fin that reaches to the base of the caudal fin.* The teeth usually have five smooth-edged cusps, with the central cusp being the largest; they number U: 29 to 31, L: 30 to 34 and are similar in both jaws. Coloration is uniformly brown above and below, often with black-edged fins.

Similar species The white-edge catshark is black, has an abdominal length shorter than the distance from the tip of the snout to the first gill slit, fins with white edges, and lower labial furrows that are longer that the uppers. The filetail catshark has lower labial furrows that are more than twice the length of the uppers and lacks the enlarged pores on the underside of the snout.

Range The brown catshark inhabits the waters of the eastern Pacific from southern Alaska (Wilson and Hughes 1978) to southern California (Roedel and Ripley 1950) and the Gulf of California (Taylor 1972). Its range may extend much farther south to the tropics. The southern limits of its range are uncertain because of confusion with a very similar, perhaps identical, catshark, *Apristurus nasutus* (Springer 1979a).

The farthest-north record appears to be a specimen taken at a depth of 475 m off Icy Point, Alaska, at about 58° N, 137°40′ W (Wilson and Hughes 1978). It is abundant from British Columbia to California. According to Hart (1973), most of the British Columbia records from are from the Strait of Georgia, where Jones and Geen (1976b) reported catching 65 specimens during trawling operations during February and March 1974. The brown catshark is abundant in deep water all along the California coast. Roedel (1951) described it as abundant off northern California, while Cross (1988) reported catching 475 individuals in 50 longline sets in the upper continental slopes off southern California.

Size and weight A 65-cm male weighed 0.7 kg. The largest specimen reported by Flammang et al. (2008) measured 69.3 cm.

Biology The brown catshark inhabits the cold waters (5–8°C) of the upper continental slopes. In the northern parts of the range (British Columbia), this little shark has been reported from waters as shallow as 33 m, in Houston Passage in the Strait of Georgia, and as deep as 564 m off Vancouver Island; the mean depth of the reports is 208 m (Jones and Geen 1976b). Hart (1973) reported it from 137–360 m off British Columbia. In the southern parts of the range, the brown catshark inhabits deeper waters, some specimens having been trawled from 656–932 m off southern California (Roedel 1951).

The brown catshark feeds on shrimp and other crustaceans, squids, and small fishes. Jones and Geen (1976b) reported that they examined 65 catshark stomachs, all of which were full. Of these, 65% contained a mudlike substance they could not identify (possibly digested remains of small fishes?), 43% contained shrimp (*Pandalus* sp.), 14% contained unidentified fishes, and 2% contained squids.

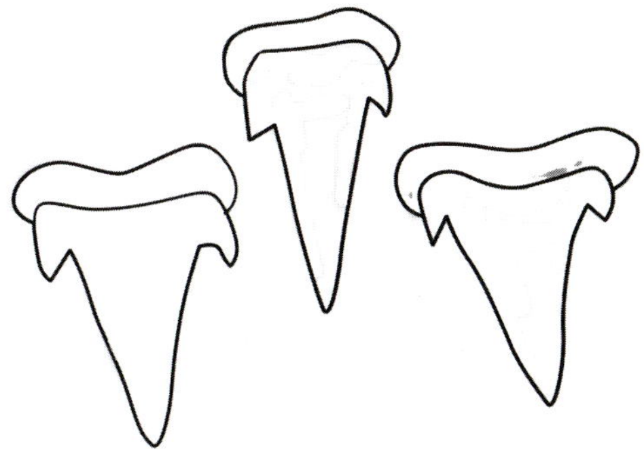

Fig.68b. Teeth (upper).

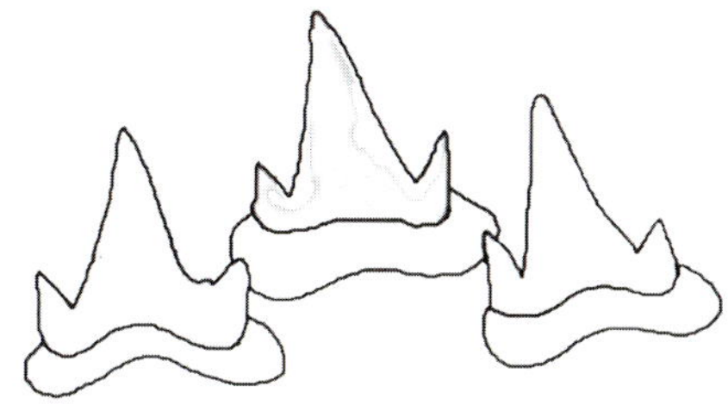

Fig.68c. Teeth (lower).

Cross (1988) examined the stomach contents of 211 brown catsharks from southern California, and crustaceans (Natantia, family Pasiphaeidae) dominated the contents.

Size at maturity

Male: It appears that males reach maturity at 49–50 cm. According to Cross (1988), males reach sexual maturity between 45 and 50 cm, based on relative clasper length. Flammang et al. (2008) also used relative clasper length as a criterion for maturity. She reported that the claspers begin to develop when the sharks reach 40 cm and that sharks 49–54 cm had the longest claspers that measured 10% of the total length.

Female: Females mature usually at about 45–48 cm. Cross (1988) reported egg cases in the oviducts of females larger than 45 cm. Flammang et al. (2008) reported egg cases in females larger than 48.5 cm. However, these authors stated that female brown catsharks matured at a smaller size in lower latitudes than at higher latitudes, and that their largest immature female was from northern California and measured 58.1 cm.

Reproduction The brown catshark is oviparous. The egg cases were described by DeLacy and Chapman (1935) and Cox (1963). The fresh egg cases are brownish and translucent; they turn darker as they age and oxidize. The egg cases have a truncated anterior end, a rounded posterior end, and a rounded flange running along their lateral edges. Egg cases measure 52–56 mm in length and 22–25 mm in width and have long tendrils at both ends. The tendrils may measure up to 37 cm. Brown catshark egg cases have been observed attached to sponges (*Aphrocallistes vastus*) and gorgonian corals (*Euplexaura marki*) at depths of 300–500 m where

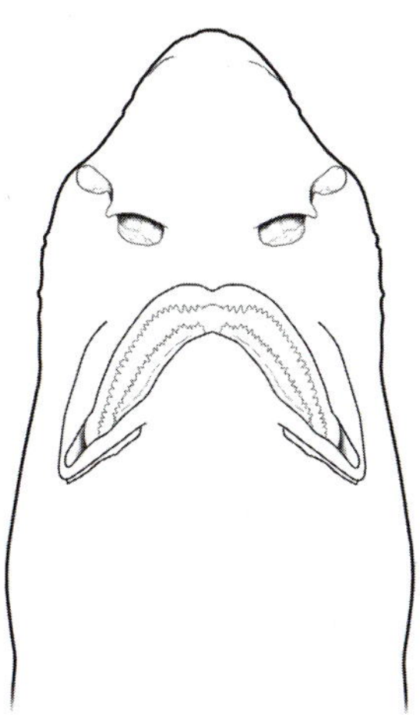

Fig. 68d. Snout.

Fig. 68e. Dermal denticles.

the average water temperature was 5°C (Flammang 2005). Jones and Geen (1976b) speculated that the incubation period in the wild lasts about a year at temperatures of 5–8°C, based on limited observations of embryo growth after 7.5 months at higher temperatures. However, incubation time in captivity at 5°C has lasted 24–27 months (Flammang 2005). According to Flammang et al. (2008), the species appears to reproduce throughout the year. Predation on egg cases by the kelp greenling (*Hexagrammos decagrammus*) has been reported (Long 1996a).

Size at birth: There are few data on the size at hatching. It is likely that the young hatch at about 7 cm, given that a free-swimming 8-cm specimen has been reported (Clemens and Wilby 1946).

Brood size: The frequency of egg laying in the wild is unknown.

Nurseries Brown catshark egg cases have been obtained by trawling at depths of 300–500 m along the coasts of California and Oregon, from 34° to 46° N latitude (Flammang 2005), indicating that the nurseries of the brown catshark extend over a wide area. In central California, brown catshark egg cases were found in rocky areas of vertical relief, where the average water temperature at 300–500 was 5°C (Flammang 2005).

Age and growth The brown catshark has not been aged.

Relation to humans None.

HOARY CATSHARK

Apristurus canutus Springer and Heemstra, 1979

Fig. 69a. *Apristurus canutus,* hoary catshark, mature male, 39.5 cm; from the Leeward Islands near Antigua (paratype, USNM 206180; RV *Oregon I*, collector).

Common name Hoary catshark (F.A.O.).

Spanish name *Pejegato cano* (F.A.O.).

Scientific name *Apristurus canutus* Springer and Heemstra, 1979, in *NOAA Technical Report NMFS Circular* 422: 16–18, figs. 10–13. *Apristurus: A,* Greek prefix meaning the absence of something, without, and *prist,* from Greek *pristis,* a file or saw, plus *urus,* from the Greek *oura,* having a tail, thus "without a filetail" (without a caudal crest); *canutus:* Latin white or gray haired, a reference to the appearance of its skin denticles.

Synonyms None.

Identification The hoary catshark is characterized by a long snout (its length about 8–10% of total length), a *first dorsal fin originating behind the pelvic fin,* a *second dorsal fin two to three times as large as the first dorsal fin,* and a *distance between the dorsal fin bases that is shorter than the distance from the tip of the snout to the eye;* the distance from the pectoral fin base to the pelvic fin base is 7–8.5% of the total length, and it has *a very long anal fin, its base is longer than the distance from the snout to the first gill slit,* or about 18–22% of total length. Both upper and lower teeth have one large, smooth-edged central cusp flank by a small cusplet on each side. The overlapping dermal denticles have rounded anterior margins, a strong central ridge, and three points. Their upper surfaces have microridges that give them a scaly look. Coloration is dark gray on the upper surfaces with darker undersides. The fins, except for the pelvic fins and the lower lobe of the caudal fin, have a narrow dark edge.

Similar species The smallfin catshark has a longer trunk segment, that is, a distance between the pectoral fin base to the pelvic fin base that is 11–14% of total length (it is only 7–8.5% in the hoary catshark), and has a shorter anal fin 16–18% of total length. Rivero's catshark has the origin of the first dorsal fin over the pelvic fin base. The Icelandic catshark has dorsal fins of nearly equal size.

Range The hoary catshark is only known from the Leeward Islands near Antigua. The limits of its distribution are unknown.

Size and weight This is a small species; the largest specimen reported measured 45.5 cm (Springer 1979a).

Biology This is little known deep-dwelling species, known only from two specimens taken by the survey ships *Oregon* and *Oregon II* in 1967 and 1969. The type specimen was taken at 687 m, while a second specimen (paratype) was taken at 750–840 m. Nothing else is known of its habits.

Size at maturity Unknown. A 42.8-cm male was said to be mature (Springer 1979a).

Reproduction The hoary catshark is presumed to be oviparous like other catsharks.

Size at birth: Unknown.

Brood size: Unknown.

Nurseries Unknown.

Age and growth The hoary catshark has not been aged.

Relation to humans None.

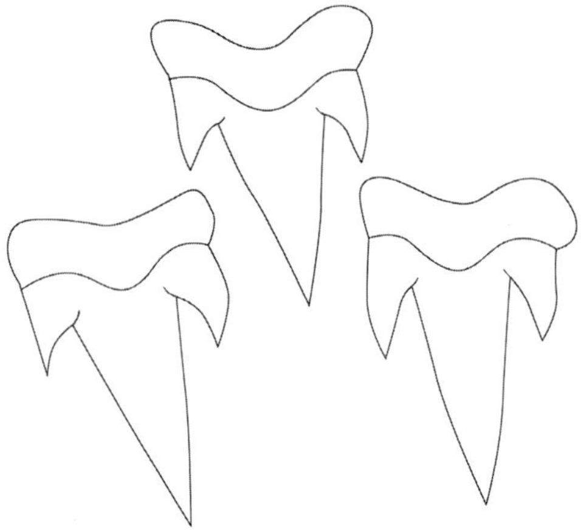

Fig. 69b. Teeth (upper).

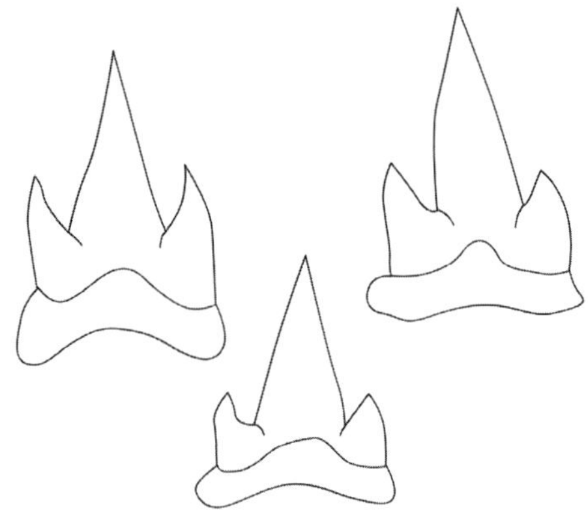

Fig. 69c. Teeth (lower).

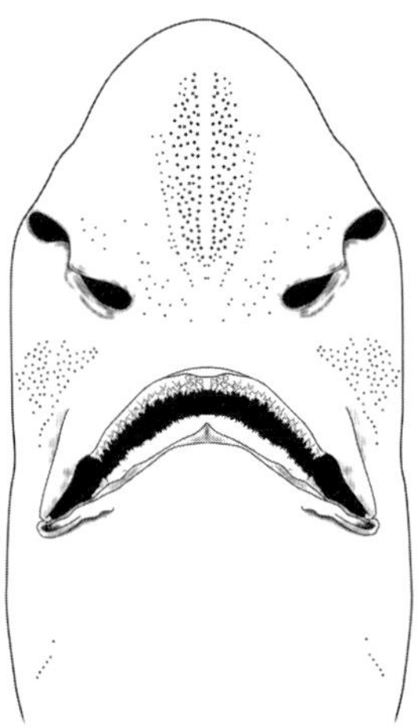

Fig. 69d. Snout.

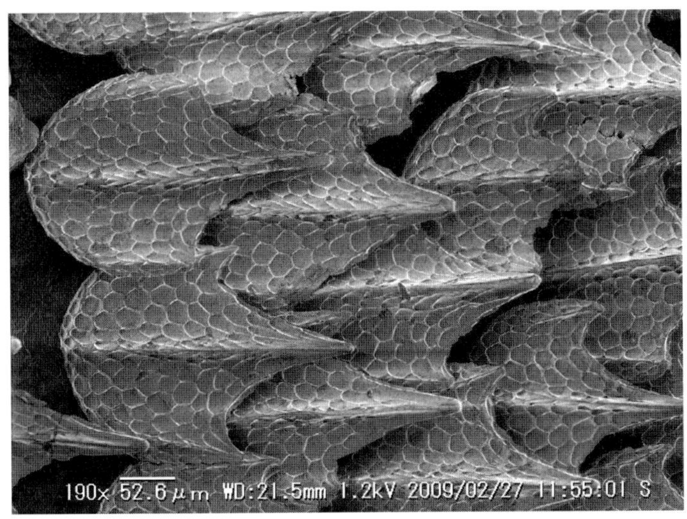

Fig. 69e. Dermal denticles.

WHITE-EDGE OR LONGNOSE CATSHARK

Apristurus kampae Taylor, 1972

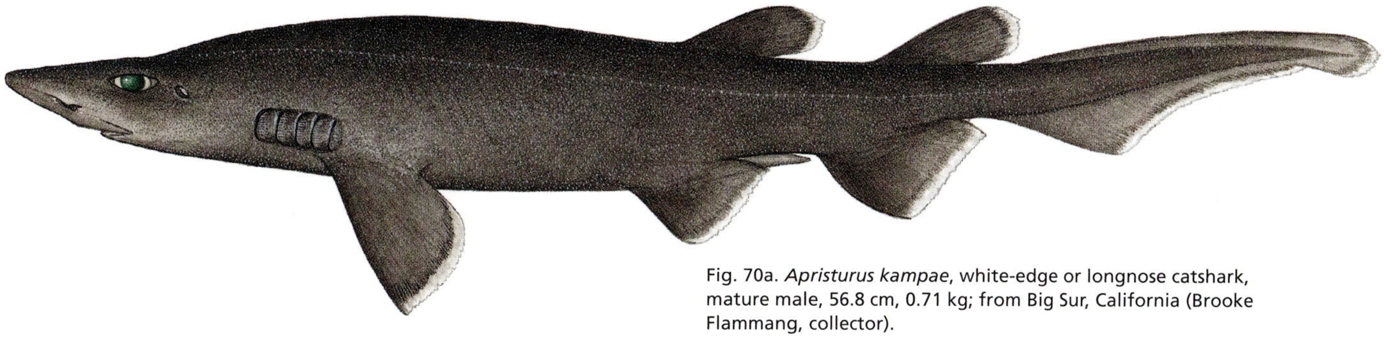

Fig. 70a. *Apristurus kampae*, white-edge or longnose catshark, mature male, 56.8 cm, 0.71 kg; from Big Sur, California (Brooke Flammang, collector).

Common name White-edge or longnose catshark. The name longnose catshark has been used for a long time (e.g., Miller and Lea 1972; Anderson et al. 1979). However, despite its name, the species' snout is shorter than the snouts of most catsharks. Thus, a more descriptive and accurate name would be "white-edge catshark," a name used by Flammang et al. (2008). Because the A.F.S. has not designated a common name for it, I have chosen to use the more accurate name of white-edge catshark.

Spanish name *Pejegato trompudo* (F.A.O.).

Scientific name *Apristurus kampae* Taylor, 1972, in *Copeia* 1972: 71–78. *Apristurus: A,* Greek prefix meaning the absence of something, without, and *prist,* from Greek *pristis,* a file or saw, plus *urus,* from the Greek *oura,* having a tail, thus "without a filetail" (without a caudal crest); *kampae,* after Elizabeth Kampa Boden, of Scripps Institution of Oceanography, chief scientist aboard RV *Argos* during the research cruise in which the type specimen was collected.

Synonyms None.

Identification This small catshark has a snout about 6–7% of the total length with a narrow, oval band of eight rows of pores along its ventral surface, *lower labial furrows longer than the upper ones,* a first dorsal fin originating posterior to the origin of the pelvic fins, and *an abdominal distance (from the pectoral fin axil to the origin of the pelvic fins) shorter than the distance from the tip of the snout to the first gill slit.* Males have smooth-edged single cusped teeth, females and juveniles have smooth-edged, multicuspid teeth with three to four cusps in front and five to seven in the sides of the jaws. Teeth are similar in both jaws. Teeth number U: 20 to 26–20 to 26, L: 18 to 25–18 to 25, depending on size of the specimen. The dermal denticles are sparse and do not overlap. They have a scalloped anterior margin, three ridges, and three points, with central ridge ending in the longest point. Their upper surfaces have the microsculpture typical of the genus *Apristurus.* Coloration is *uniformly black or dark gray. The fins have thin, white posterior margins.* The inside of the mouth is bluish black.

Similar species The brown catshark has upper labial furrows that are longer than the lowers. The filetail catshark is much lighter colored, lacks the enlarged pores on the ventral side of the snout, and has an anal fin that is much shorter, with a tip that falls quite short from the lower lobe of the caudal fin (by a distance as long as the eye length). It is likely that there are undescribed species of *Apristurus* off the California coast.

Range The white-edge catshark inhabits the eastern Pacific. It has been reported from the Oregon coast (Ebert 2003) to the Gulf of California (Taylor 1972b). The limits of its distribution are unknown.

Size and weight A 58-cm male weighed 0.75 kg. Flammang (2005) reported that the largest male in her sample measured 64.7 cm, while the largest female measured 59 cm.

Biology This is a deep-dwelling catshark of the Pacific coast of North America. Most specimens have been taken at 1,000–1,800 m. Because of its deep habitat, it is seldom taken, and nothing is known of its habits. According to Flammang (2005), juveniles and females are most often taken at 1,000–1,200 m.

Size at maturity According to Flammang (2005), both sexes mature at about 48 cm. In her sample of 52 males and 45 females, the smallest mature male was 48.6 cm, while the largest immature male was 48.4 cm; the smallest mature

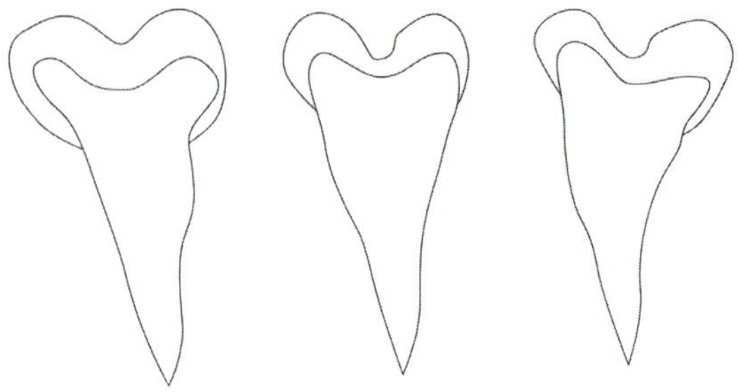

Fig. 70b. Teeth (upper).

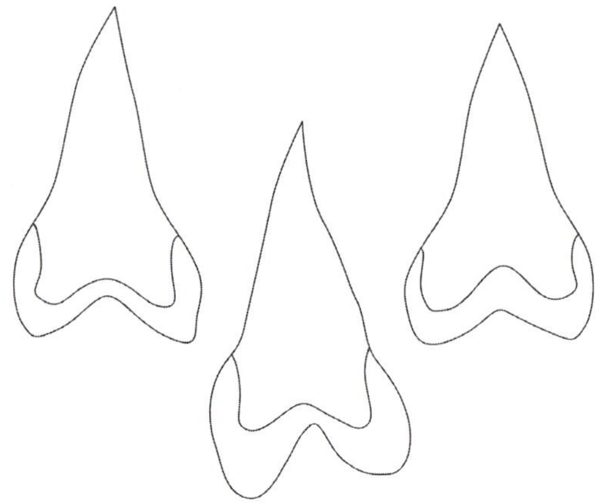

Fig. 70c. Teeth (lower).

female was 48.4 cm. Ebert (2003) reported that males mature at about 50 cm and females mature at 48–52 cm.

Reproduction The white-edge catshark is oviparous.

Size at birth: The young hatch at a length of 14 cm (Ebert 2003).

Brood size: The frequency of egg laying is unknown.

Nurseries Unknown.

Age and growth No data available.

Relation to humans None.

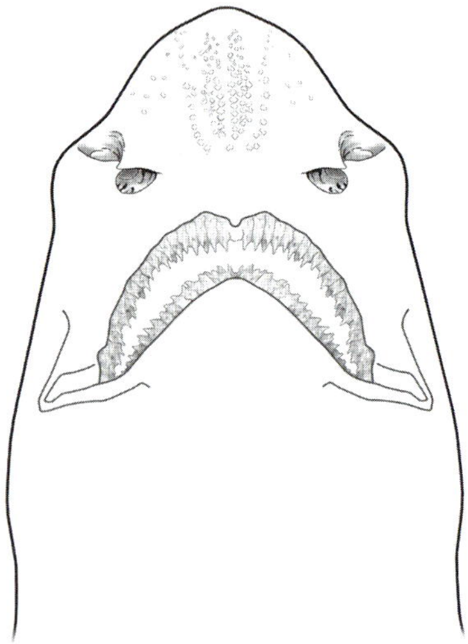

Fig. 70d. Snout.

Fig. 70e. Dermal denticles.

ICELANDIC CATSHARK

Apristurus laurussoni (Saemundsson, 1922)

Fig. 71a. *Apristurus laurussonii*, Icelandic catshark, male, 44.2 cm, 0.4 kg; Gulf of Mexico (Tulane University Collection TU# 189220; courtesy of Hank Bart).

Common name Icelandic catshark (F.A.O.), a reference to where it was described.

Spanish name *Pejegato islándico* (F.A.O.).

Scientific name *Apristurus laurussoni* (Saemundsson, 1922), originally described as *Scyllium Laurussonii* in *Videnskabelige Meddelelser fra Dansk naturhistorisk Forening i København* 74: 173, plate 4, fig. 2; plate 5, fig. 4. *Apristurus: A,* Greek prefix meaning the absence of something, without, and *prist,* from Greek *pristis,* a file or saw, plus *urus,* from the Greek *oura,* having a tail, thus "without a filetail" (without a caudal crest); *laurussoni:* name after Gísli Lárusson, a friend of Saemundsson.

Synonyms *Apristurus maderensis* Cadenat and Maul, 1966.

Identification The Icelandic catshark has a short snout (less than 8% of the total length), *two dorsal fins nearly equal in area (the second dorsal fin is usually slightly larger),* the *origin of first dorsal fin over or slightly behind middle of pelvic fin base,* a long anal fin extending from the tips of pelvic fins to the origin of the lower caudal lobe, an anal fin base as long or longer than the distance from the tip of the snout to the first gill slit and greater than the distance between the dorsal fin bases, and a *distance between the pectoral and pelvic fin bases about 15–17.8% of the total length.* A female that I examined had teeth with five to seven smooth-edged cusps, with the central cusp much larger than the others. The teeth are similar in the both jaws. The dermal denticles have rounded anterior margins, three ridges, and three points. Their surfaces have the typical microsculpture of the genus. Coloration is brown above and below.

Similar species The ghost catshark and the deep-water catshark have a greater space between the pelvic fins tips and the anal fin.

Range The Icelandic catshark is widely distributed in the North Atlantic. It has been reported from Iceland, off New England, from the north coast of Hispaniola, off Alabama, off Tamaulipas (Mexico), from the Gulf of Campeche (Springer 1979a), and from Madeira (as *Apristurus maderensis,* Cadenat and Maul 1966).

Size and weight The maximum recorded size is 68 cm (Springer 1979a).

Biology The Icelandic catshark is deep-water species. Most captures have occurred at depths of 560–675 m in the northern parts of the range and at 840–1,462 m in the Gulf of Mexico (Springer 1979a). Its habits remain undescribed.

Size at maturity Unknown.

Reproduction The Icelandic catshark is oviparous. The egg case has been described as having with prominent striations (Nakaya and Sato 1998).

Size at birth: Unknown.

Brood size: Unknown.

Nurseries Unknown.

Age and growth Unknown.

Relation to humans None.

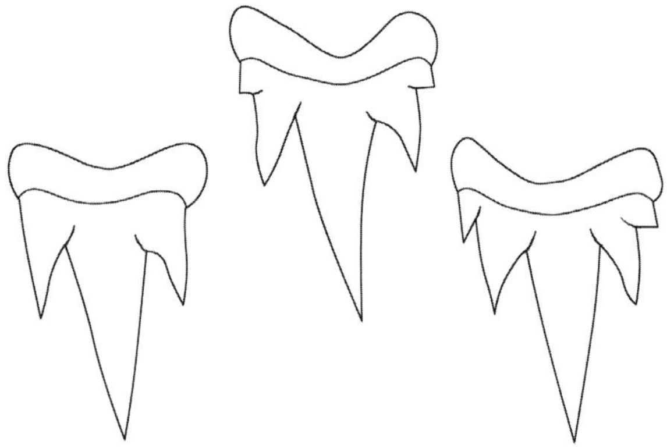

Fig. 71b. Teeth (upper).

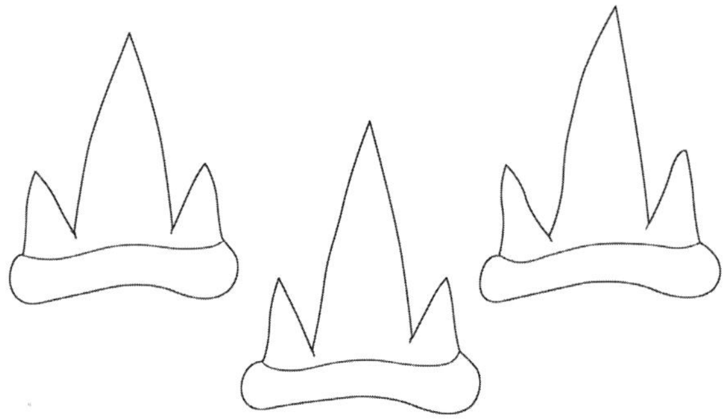

Fig. 71c. Teeth (lower).

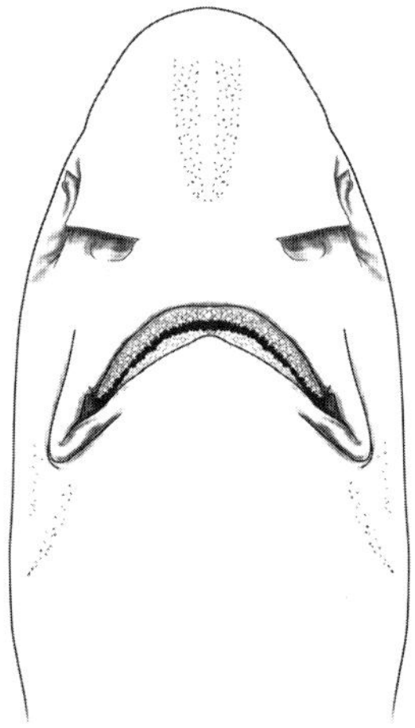

Fig. 71d. Snout.

Fig. 71e. Dermal denticles.

GHOST CATSHARK

Apristurus manis (Springer 1979)

Fig. 72a. *Apristurus manis,* ghost catshark, male, 75.2 cm, 1.5 kg; from Bear Seamount, Georges Bank (John Galbraith, collector).

Common name Ghost catshark, a reference to the pale coloration of some specimens.

Spanish name *Pejegato fantasma* (F.A.O.).

Scientific name Ghost catshark, *Apristurus manis* (Springer 1979), originally described as *Parmaturus manis* in *NOAA Technical Report NMFS Circular* 422: 102–104, figs. 60–63. *Apristurus: A,* Greek prefix meaning the absence of something, without, and *prist,* from Greek *pristis,* a file or saw, plus *urus,* from the Greek *oura,* having a tail, thus "without a filetail" (without a caudal crest); *manis:* Latin, ghost or shade of the departed.

Synonyms *Parmaturus manis* Springer, 1979.

Identification The ghost catshark has a flattened head and a *long flattened snout, its length about 8–10% of the total length.* It has two dorsal fins, the first one originating well behind the midpoint of the pelvic fin base, the *second dorsal fin originating well behind the midpoint of the anal fin base, slightly larger than the first, and reaches to the base of the tail; pectoral fins have rounded outer margins,* and the *pelvic fins stop well short of the anal fin.* The *dorsal fins are uniformly colored and lack dark tips.* Males have teeth with one long, curved, conical cusp with one small lateral cusplet on each side. The upper teeth are slightly larger than the lower teeth and have more prominent lateral cusplets than the lower teeth. The dermal denticles are sparse and widely separated. They have three points, with a wide, central ridge ending in the longest point and short side points. The upper surfaces of the denticles have the usual microsculpture of the genus. Coloration ranges from very pale brown or dark tan to dark chocolate brown, uniformly throughout. Apparently, there are two distinct color phases in the species.

Similar species The deep-water catshark has pectoral fins with squarish outer margins and dorsal fins with dark tips. The black roughscale catshark has a much darker coloration (black), and its dorsal fins have a black band on the rear edges.

Range The ghost catshark inhabits the deep waters of the western North Atlantic Ocean, but the limits of its range are unknown. It is probably found in the eastern Atlantic as well. It is abundant in slope waters and around seamounts off New England. Numerous specimens have been collected around Block Canyon (39°42′ N, 71°27′ W) and around Bear Seamount (39°52′ N, 70°50′ W) by National Marine Fisheries Service survey cruises (Moore et al. 2003; John Galbraith, pers. comm., February 2005).

Size and weight This is a small species perhaps reaching 80–85 cm: males, 66 cm, 1.0 kg; 75.2 cm, 1.5 kg; and 76 cm, 1.6 kg; females, 35 cm, 0.11 kg; 57 cm, 0.61 kg; 60 cm, 0.75 kg; and 74 cm, 1.07 kg.

Biology This is a widely distributed and abundant species of the continental slopes found at depths of 600–1,300 m. Its habits are poorly known due to its deep habitat and previous confusion with the deep-water catshark (*A. profundorum*). The stomach contents of most of the specimens that I have examined contained shrimp and squid beaks.

Size at maturity The ghost catshark matures at about 64–71 cm.

Male: Males mature at about 70 cm. A 62-cm male that I examined was immature. Two other specimens 67 and 70 cm were approaching maturity; they had elongated but uncalcified claspers. A 71-cm male was mature.

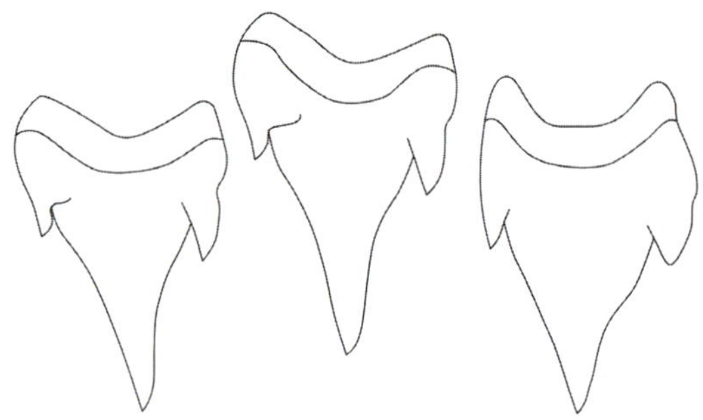

Fig. 72b. Teeth (upper).

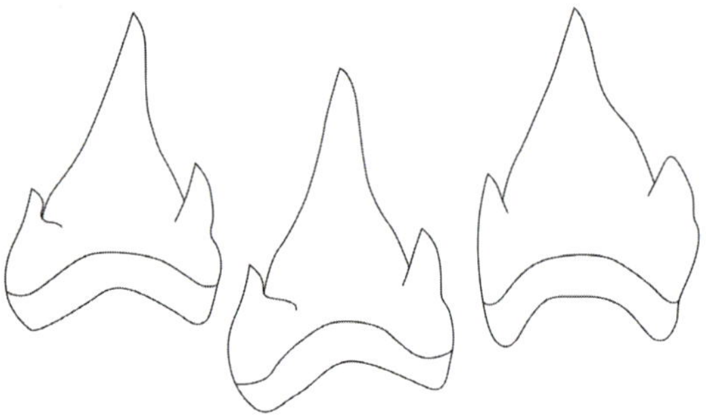

Fig. 72c. Teeth (lower).

Female: A 64-cm female carried fully formed egg cases. A 72-cm female carried oocytes 20 mm in diameter; these were the largest oocytes I have observed.

Reproduction The ghost catshark is oviparous. The egg cases are bright ochre yellow in color when fresh; they darken to yellowish brown after being laid. The egg cases are 75 mm long, 25 mm wide, and about 15 mm high. The posterior end of the case is rounded with two vestigial tendrils (about 3–5 mm long, and hard to see) and the anterior end has two short horns about 3 mm long. This suggests that the egg capsules are scattered over the bottom or perhaps placed in protected places or crevices. I have examined females carrying fully formed egg cases from March to May, all taken off Bear Seamount by John Galbraith.

Size at birth: Unknown.

Brood size: Unknown.

Nurseries Unknown.

Age and growth The species has not been aged.

Relation to humans None.

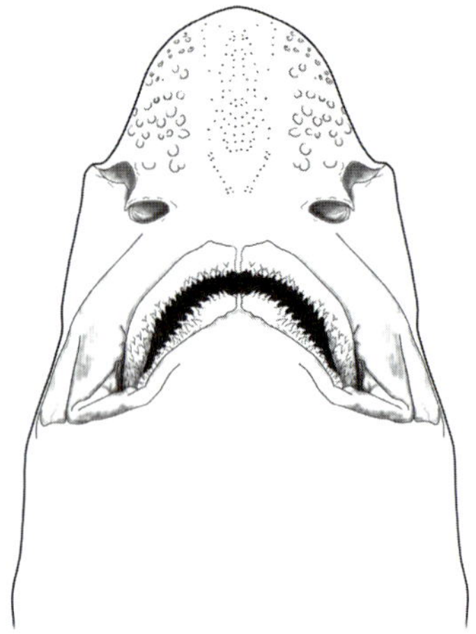

Fig. 72d. Snout.

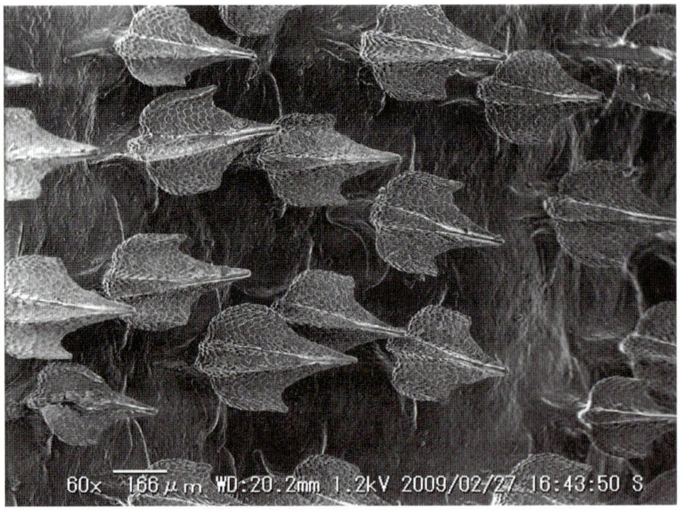

Fig. 72e. Dermal denticles.

BLACK ROUGHSCALE CATSHARK

Apristurus melanoasper Iglésias, Nakaya, and Stehmann, 2004

Fig. 73a. *Apristurus* melanoasper, black roughscale catshark, female, 47 cm female from the western Atlantic (40°13 N, 68°02 W; John Galbraith, collector).

Common name Black roughscale catshark, a name suggested by those describing the species.

Spanish name None.

Scientific name *Apristurus melanoasper* Iglésias, Nakaya, and Stehmann, 2004, in *Cybium* 28 (4): 345–356. *Apristurus: A,* Greek prefix meaning the absence of something, without, and *prist,* from Greek *pristis,* a file or saw, plus *urus,* from the Greek *oura,* having a tail, thus "without a filetail" (without a caudal crest); *melanoasper:* from the Greek *melas* or *melanos,* black, and *asper,* Latin rough, which according to Iglésias et al. references to is black coloration and rough skin, an interesting barbarism consisting of Greek and Latin adjectives.

Synonyms None.

Identification The black roughscale catsharks has two dorsal fins, the first dorsal fin originating well behind the origin of the pelvic fins and slightly smaller than the second dorsal; a snout as long as the mouth width; upper labial furrows longer than lower ones; and a *long abdomen* (its length from the pectoral fin axil to the origin of the pelvic fins is more than twice the distance between the two dorsal fins). The teeth are wide and have three to five cusps, the central being the largest. The dermal denticles are dense and widely overlapping. They have three points and a strong central ridge that ends in a long point that overlaps the adjacent denticle. The denticles have the usual microsculpture of the genus. *Coloration is black uniformly throughout.* The *rear edges of the dorsal fins are shiny black (because they lack dermal denticles).*

Similar species Galbraith's catshark (*Apristurus* sp. X) has nonoverlapping dermal denticles and a higher anal fin. Other *Apristurus* catsharks found off northeastern North America, the deep-water and ghost catsharks, are brown or lighter in color, not black, and have a short abdomen (the distance from the pectoral fin axil to the origin of the pelvic fins is less than twice the distance between dorsal fins). The ghost catshark has sparse, nonoverlapping dermal denticles.

Range The black roughscale catshark inhabits the North Atlantic, from the slopes of eastern North America to the slopes of western Europe (Ireland, the British Isles; Iglésias et al. 2004). The specimen illustrated here was taken at a depth of 718–955 m about 520 km east of New York.

Size and weight The largest male reported measured 76.1 cm and weighed 1.29 kg, while the largest female measured 73.2 cm and weighed 1.28 kg (Iglésias et al. 2004).

Biology The black roughscale catshark is a recently described species, and little is known of its habits. Much of the information here is taken from the original description, as I have only seen two specimens.

Size at maturity Males smaller than 52 cm have very short claspers. Claspers begin to elongate between 52–62 cm, and males mature between 61 and 64 cm (Iglésias et al. 2004). Females mature at about 64 cm (Iglésias et al. 2004).

Reproduction The black roughscale catshark is oviparous. The egg case is 52–67 cm long and 24–27 mm wide and yellowish khaki to light brown in color. It has fine longitudinal striations on the surface and long tendrils at the

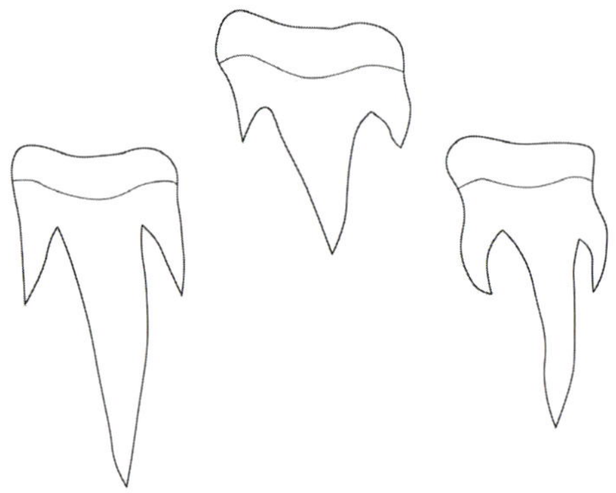

Fig. 73b. Teeth (upper).

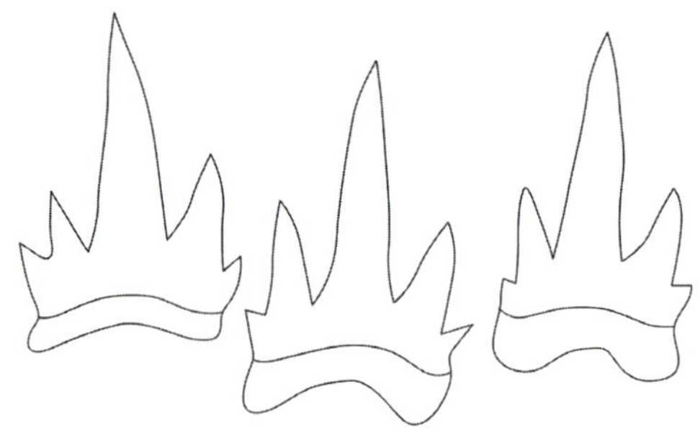

Fig. 73c. Teeth (lower).

posterior end (Iglésias et al. 2004), suggesting that the species wraps its egg capsules around upright objects as many other catsharks do.

Size at birth: Unknown.

Brood size: Unknown.

Nurseries Unknown.

Age and growth The species has not been aged.

Relation to humans None.

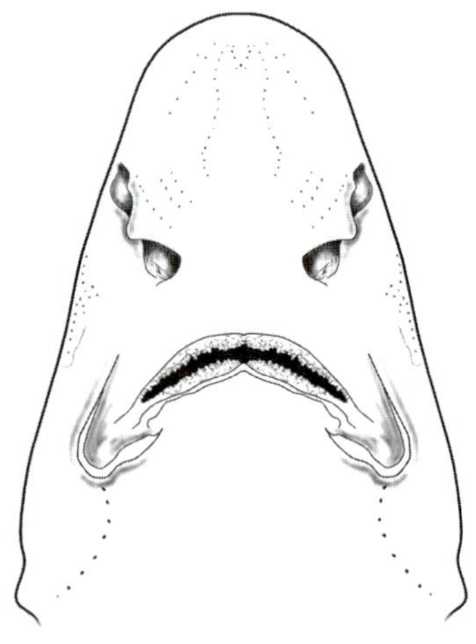

Fig. 73d. Snout.

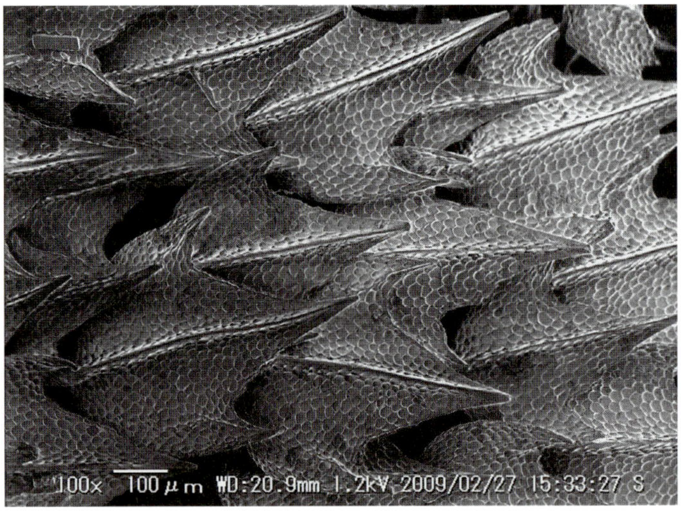

Fig. 73e. Dermal denticles.

SMALLFIN CATSHARK

Apristurus parvipinnis Springer and Heemstra, 1979

Fig. 74a. *Apristurus parvipinnis,* Smallfin catshark, 36.5 cm, 0.1 kg female; from the Gulf of Mexico (TU # 189332, courtesy of Hank Bart).

Common name Smallfin catshark (F.A.O.).

Spanish name *Mocho* (F.A.O.).

Scientific name *Apristurus parvipinnis* Springer and Heemstra, 1979, in *NOAA Technical Report NMFS Circular* 422: 25–26, fig. 15. *Apristurus: A,* Greek prefix meaning the absence of something, without, and *prist,* from Greek *pristis,* a file or saw, plus *urus,* from the Greek *oura,* having a tail, thus "without a filetail" (without a caudal crest); *parvipinnis:* from *parvus,* Latin little or small, and *pinnis,* Latin wing or fin, "having small fins."

Synonyms None.

Identification The smallfin catshark has the origin of the first dorsal fin at a vertical between the posterior end of pelvic fin base and origin of anal fin and a second dorsal fin two to four times larger than first dorsal fin, and the distance between pectoral fin base and pelvic fin base is about 11–14% of total length. The teeth are small and numerous, with more than 50 irregular rows in each jaw; they have three to six cusps each (usually four), the tooth surfaces having one to several vertical ridges extending nearly to the tooth tips (Springer 1979a). The dermal denticles have three points with a very long central ridge and point, rounded anterior margins, and the usual surface microsculpture. Coloration is black above and below.

Similar species The smallfin catshark and the hoary catshark are the only species of the genus that have a first dorsal fin origin *posterior* to the end of the pelvic fin base, and a second dorsal fin *much larger* than the first dorsal fin. The hoary catshark has a much shorter trunk space, and the distance from its pectoral fin base to the origin of the pelvic fins is 6–8.5% of total length compared to 11–14% in the smallfin catshark.

Range The smallfin catshark inhabits the western North Atlantic, from the Gulf of Mexico to the coast of northern South America (Springer 1979a).

Size and weight This is a small species. The largest specimens reported by Springer (1979a) were a 47-cm male and a 51-cm female.

Biology According to Springer (1979a), this catshark was taken more often than other *Apristurus* species of the Gulf of Mexico. Nowadays, it is seldom seen because the few survey ships do not normally trawl at the depths where *Apristurus* dwell. Most of the specimens were taken at depths of 800–1,100 m, on the bottom, although there is one record of specimens taken off French Guiana at 636 m (Springer 1979a). Its habits are unknown.

Size at maturity Unknown.

Reproduction The smallfin catshark is almost surely oviparous like all other catsharks but its reproductive processes have not been described.

Size at birth: Unknown.

Brood size: Unknown.

Nurseries Unknown.

Age and growth The smallfin catshark has not been aged.

Relation to humans None.

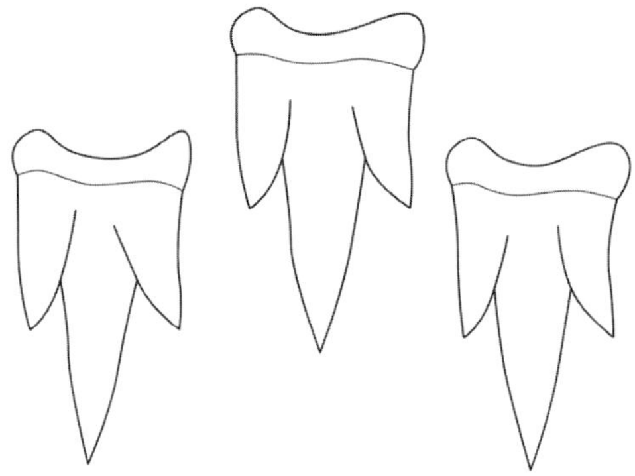

Fig. 74b. Teeth (upper).

Fig. 74c. Teeth (lower).

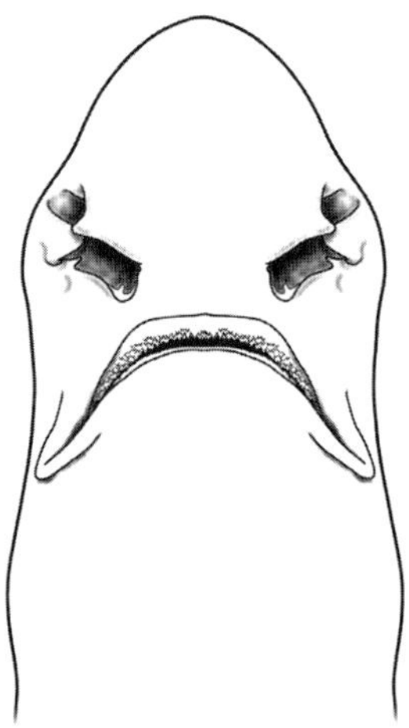

Fig. 74d. Snout.

Fig. 74e. Dermal denticles.

DEEP-WATER CATSHARK

Apristurus profundorum (Goode and Bean, 1896)

Fig. 75a. *Apristurus profundorum*, deep water catshark, light phase, mature male, 63 cm, 1.2 kg; from Bear Seamount, Georges Bank (John Galbraith, collector).

Common name Deep-water catshark.

Spanish name *Pejegato abisal* (F.A.O.).

Scientific name *Apristurus profundorum* (Goode and Bean, 1896), originally described as *Scylliorhinus profundorum* in *Oceanic Ichthyology* 17, plate 5, fig. 16. *Apristurus: A,* Greek prefix meaning the absence of something, without, and *prist,* from Greek *pristis,* a file or saw, plus *urus,* from the Greek *oura,* having a tail, thus "without a filetail" (without a caudal crest); *profundorum:* Latin, of the depths.

Synonyms In recent times, the name *Apristurus microps* Gilchrist, 1922 (South African species) has been applied to *Apristurus* living off the northeastern United States (see Compagno 1984a). Recently, Kazuhiro Nakaya, an expert on the genus *Apristurus,* and I conducted a careful comparative study of the different species of *Apristurus* found off the northeastern United States. After this study, in 2006, Prof. Nakaya examined the type specimen of *Apristurus profundorum* at the U.S. National Museum and concluded that those North American specimens of *Apristurus microps* are actually *A. profundorum.*

Identification The deep-water catshark is characterized by *two small dorsal fins* (about 7–8% of total length in adults), the *first dorsal fin originating well behind the origin of the pelvic fins,* and the second dorsal fin originating well behind the midpoint of the anal fin and just slightly larger than the first dorsal; pectoral fins with *squarish outer margins;* pelvic fins with rounded rear margins; a snout about 7–8% of the total length; and *small eyes,* about 2.4% of total length. The height of the anal fin is also variable (3.5–5.1% of total length). Juveniles have five cusped teeth, the central cusp being much larger than the others. Mature males have teeth with conical, curved cusps and may have one small lateral cusplet or cusplet on the sides. Theses cusplets are often covered by the gum line and may not always be visible. Their teeth are similar in both jaws. Females have teeth with five cusps, the central one being much larger, and their teeth are similar in both jaws. The dermal denticles are sparse with little overlap. They have three points, with a central ridge ending on a longer point. Their upper surfaces have the microsculpture characteristic of the genus. The coloration of the deep-water catshark is variable, ranging from light grayish brown to light chocolate brown above and below, *with dark or black fin edges.*

Similar species The ghost catshark has much larger snout (8–10% of total length), larger dorsal fins (about 9% of total length), and pectoral fins with rounded outer margins. Its dorsal fins are uniformly colored and lack the dark tips of the deep-water catshark. The black roughscale catshark has a much darker coloration and two dark bands on the rear edges of the dorsal fins.

Range The deep-water catshark inhabits the North Atlantic, but the limits of its distribution are unknown. It is found at least from the mid-Atlantic Ridge westward to the edges of the continental shelves of the northeastern United States. It is probably found in the eastern North Atlantic as well. Most specimens have been taken from Bear Seamount (39°52′ N, 67°20′ W) to Hudson Canyon (39°17′ N, 72°27′ W), the location where the type specimen was taken, not "off Delaware Bay" as stated in the original description (Moore et al. 2003). It is abundant around Bear Seamount, where dozens of specimens have been taken (John Galbraith, pers. comm., June 2005).

Size and weight The deep-water catshark probably reaches 75 cm. A 59-cm male weighed one kg, while a 63-cm male weighed 1.21 kg.

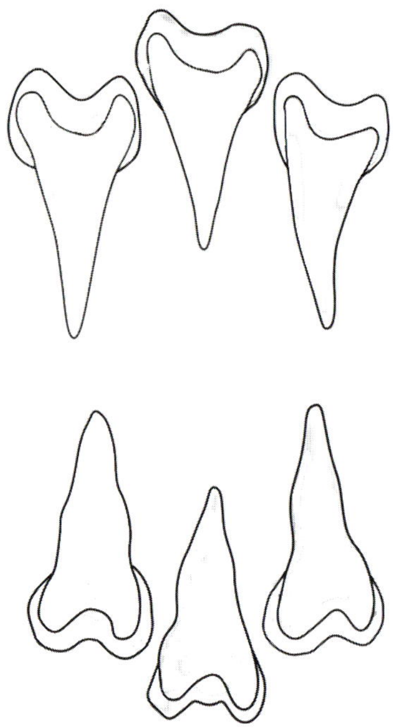

Fig. 75b.Teeth.

Biology The deep-water catshark is a bottom-dwelling species found at depths of 1,000–1,750 m along the edges of the continental shelves, in deep canyons, and around seamounts. Some of the specimens that I have examined had consumed small shrimp and small squid. Nothing else is known of its habits.

Size at maturity The deep-water catshark reaches maturity at 60–62 cm.

Male: Males reach maturity at 60–62 cm. This estimate is based on specimens I examined: a 58-cm male had elongated claspers apparently not fully calcified, a 59.2-cm male was still very immature, whereas a 63-cm male had fully calcified claspers.

Female: Females appear to reach maturity at about 62 cm. I examined a 62.5-cm female that had large (22 mm) oviducal glands and 20–22-mm oocytes in the ovary, suggesting that it had reached maturity or it was about to ovulate.

Reproduction The deep-water catshark is oviparous. Nothing is known about its egg-laying habits.

Size at birth: Unknown.

Brood size: Unknown.

Nurseries Unknown.

Age and growth Unknown.

Relation to humans None.

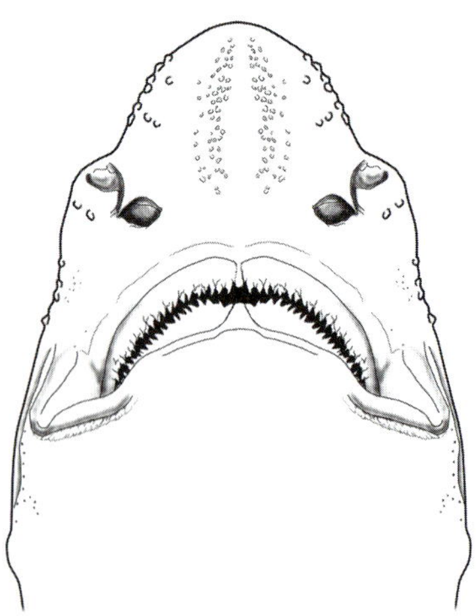

Fig. 75c. Snout.

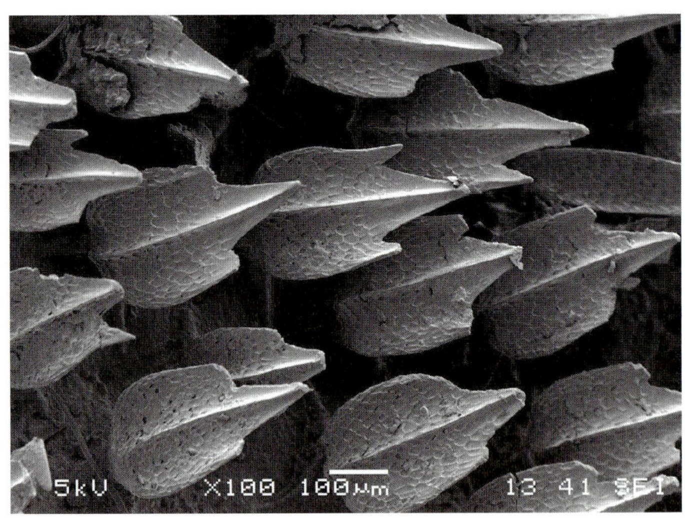

Fig. 75d. Dermal denticles.

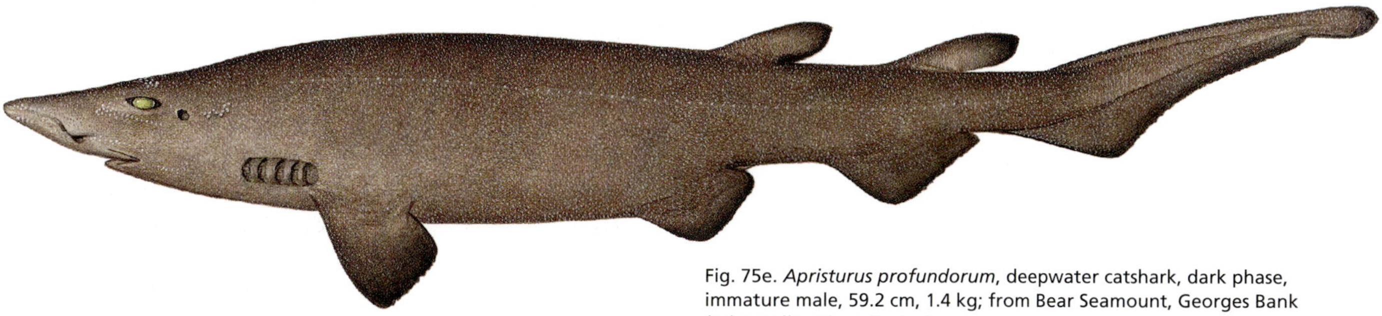

Fig. 75e. *Apristurus profundorum*, deepwater catshark, dark phase, immature male, 59.2 cm, 1.4 kg; from Bear Seamount, Georges Bank (John Galbraith, collector).

RIVERO'S CATSHARK

Apristurus riveri Bigelow and Schroeder, 1944

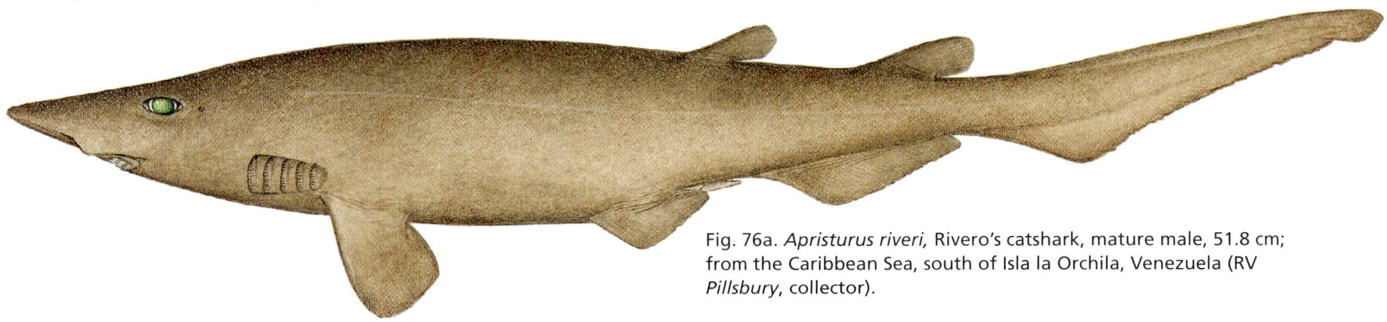

Fig. 76a. *Apristurus riveri,* Rivero's catshark, mature male, 51.8 cm; from the Caribbean Sea, south of Isla la Orchila, Venezuela (RV *Pillsbury*, collector).

Common name Rivero's catshark. The F.A.O. uses the name broadgill catshark.

Spanish name None in use in the Spanish Americas, due to its scarcity. The F.A.O. name is *pejegato agallón.*

Scientific name *Apristurus riveri* Bigelow and Schroeder, 1944, in *Proceedings of the New England Zoölogical Club* 23: 23–26, plate 7. *Apristurus: A,* Greek prefix meaning the absence of something, without, and *prist,* from Greek *pristis,* a file or saw, plus *urus,* from the Greek *oura,* having a tail, thus "without a filetail" (without a caudal crest); *riveri:* name after Luis Howell Rivero, Cuban ichthyologist and naturalist.

Synonyms None.

Identification This deep-water catshark of the Gulf of Mexico and Caribbean area has a *long snout* (its length is about 7–9% of the total length) *with a narrow band of four rows of conspicuous pores along its ventral surface;* a *first dorsal fin origin ahead of the posterior end of the pelvic fin base;* a *second dorsal fin much larger than the first dorsal fin;* and an *anal fin with a very wide base* (14–16% of total length), the fin reaching to the origin of the caudal fin. Adult males have teeth with a single, large conical cusp. Females and juveniles have multicuspid teeth, with the central cusp being the largest. The teeth are similar in both jaws. The three-pointed dermal denticles have a strong central ridge that terminates in a long point. Their upper surfaces have the typical microsculpture of the genus. Coloration is chocolate brown above and below.

Similar species The hoary and the smallfin catsharks have the origin of the first dorsal fin posterior to the end of the pelvic fin base. Other catsharks have dorsal fins of equal or nearly equal size.

Range Rivero's catshark inhabits the deep waters of the Gulf of Mexico and the Caribbean Sea. It has been reported from off Mississippi, the Dry Tortugas, the north coast of Cuba, and the Caribbean coast of Panama (Springer 1979a). The limits of its distribution are unknown.

Size and weight This is a small species. Springer (1979a) reported that specimens 39–47.5 cm were mature.

Biology This is a deep-dwelling catshark of tropical waters. Specimens have been taken at depths of 600–1,000 m (Springer 1979a). Springer (1979a) mentioned that this species had been taken by exploratory fishing vessels less frequently than other *Apristurus* catsharks. As an explanation, he suggested that Rivero's catshark may inhabit deeper waters than the other *Apristurus* in the area and that fewer trawls were made in deeper waters. Nothing else is known of its habits.

Size at maturity Unknown. Specimens 39 cm were reported as mature (Springer 1979a).

Reproduction Rivero's catshark is oviparous. The egg cases are smooth surfaced, translucent, and greenish with indistinct bands of lighter color; they measure about 5 cm long, not including the long tendrils at the ends and are about 1.5 cm wide (Springer 1966).

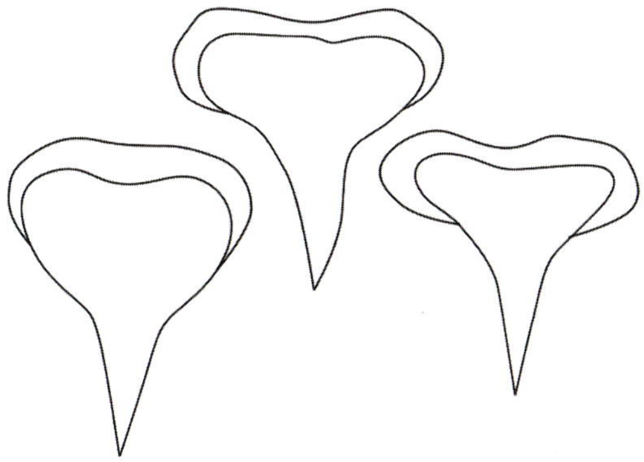

Fig. 76b. Teeth (upper).

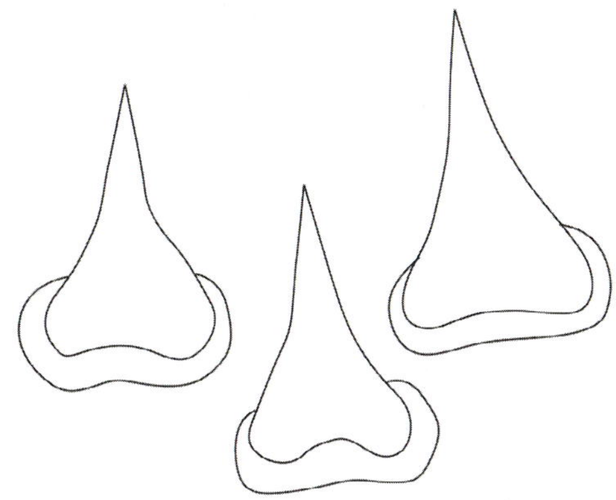

Fig. 76c. Teeth (lower).

Size at birth: Unknown, but probably less than 10 cm at hatching.

Brood size: The number of eggs produced by a female every season is unknown.

Nurseries Unknown.

Age and growth The species has not been aged.

Relation to humans None.

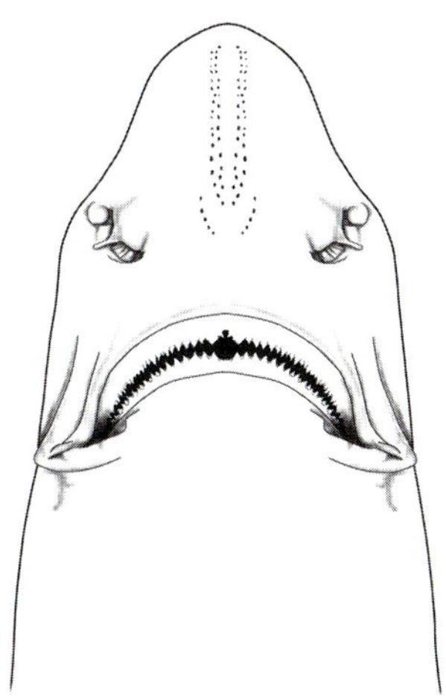

Fig. 76d. Snout.

Fig. 76e. Dermal denticles.

GALBRAITH'S CATSHARK

Apristurus sp. X

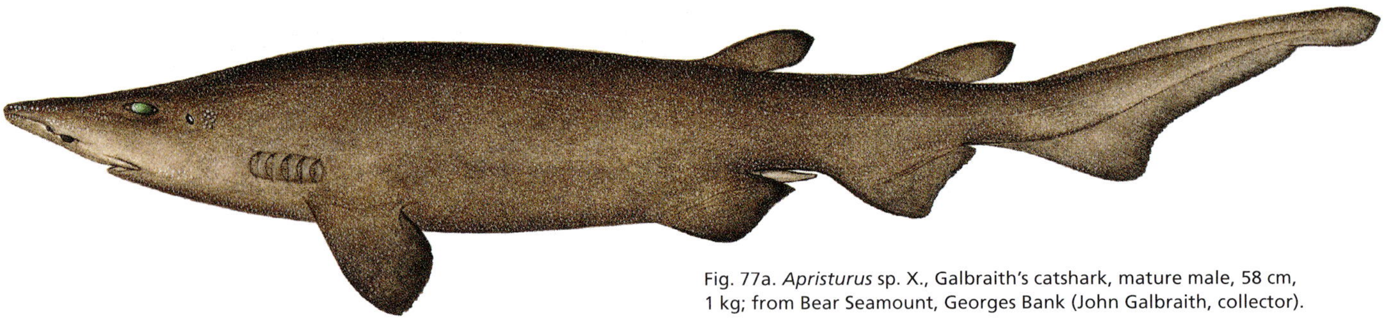

Fig. 77a. *Apristurus* sp. X., Galbraith's catshark, mature male, 58 cm, 1 kg; from Bear Seamount, Georges Bank (John Galbraith, collector).

This is a long-abdomen catshark (the distance from the pectoral axil to the origin of the pelvic fins is more than twice the distance between the dorsal fin bases). It was taken at 1,800 m off Bear Seamount (DE 2000304, Station 31) by John Galbraith in 2003. Its identity is uncertain. It is a much larger species than the other long-abdomen catshark regularly taken in the area, the black rough-scale catshark. Galbraith's catshark differs from the black rough-scale catshark in having a much shorter anal fin base; its height is about twice as long as the anal base, and its denticles are sparse with little overlapping. Its denticles are three pointed, with a strong central ridge terminating on the long central point and parallel sides. Their upper surfaces have the microsculpture of the genus. Its teeth have one large central cusp with a small cusplet on either side and are similar in both jaws.

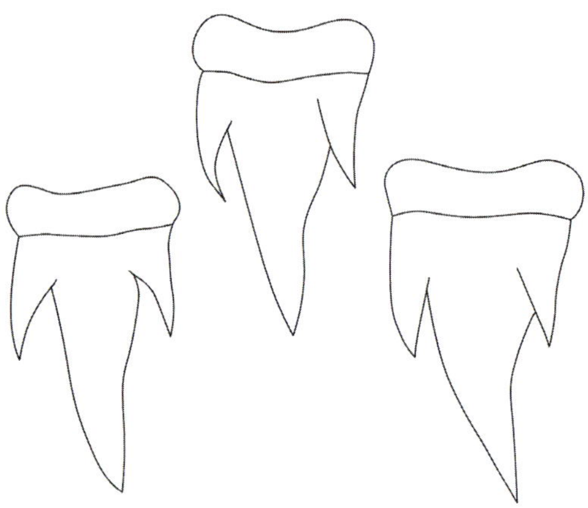

Fig. 77b. Teeth (upper).

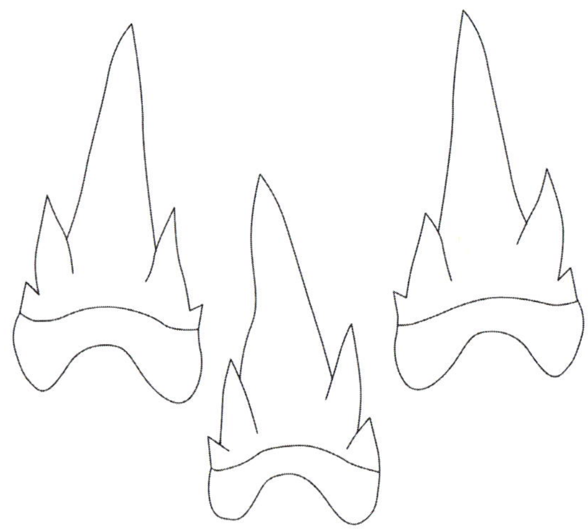

Fig. 77c. Teeth (lower).

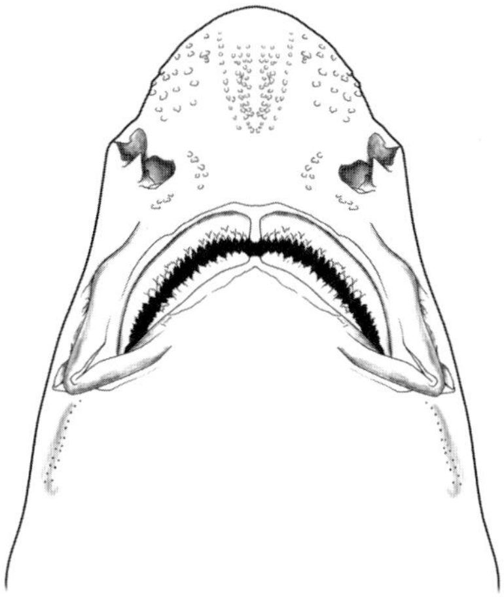

Fig. 77d. Snout.

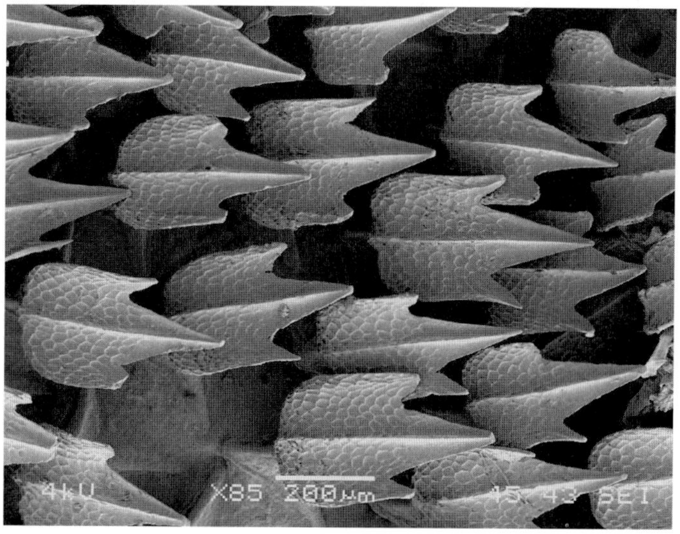

Fig. 77e. Dermal denticles.

SWELL SHARK

Cephaloscyllium ventriosum (Garman, 1880)

Fig. 78a. *Cephaloscyllium ventriosum*, swell shark, female, 71 cm, 3.7 kg; from El Barril, Baja California, Mexico (Clark Crabbe, collector).

Common name Swell shark, a reference to this shark's remarkable ability to distend or inflate its stomach by swallowing water or air, swelling its trunk to a nearly spherical shape.

Spanish name None.

Scientific name *Cephaloscyllium ventriosum* (Garman, 1880), originally described as *Scyllium ventriosum* in *Memoirs of the Museum of Comparative Zoology at Harvard College* 6(pt. 1, no. 11): 167. *Cephaloscyllium:* from the Greek *kephale,* head, and *scyllium,* from the Greek *skylion,* a dogfish, probably related to *skyllo,* to tear or mangle; *ventriosum,* from the Latin *venter,* belly, and *osum,* a Latin termination meaning full of, an allusion to its ability to inflate its stomach.

Synonyms *Catulus uter* Jordan and Gilbert in Jordan and Evermann, 1896; *Scyliorhinus ventriosum* Regan, 1908; *Cephaloscylllium uter* Roedel and Ripley, 1950.

Identification The swell shark has a stout body, a flattened head with *a very short snout (its width is about one-third the mouth width*), a very wide mouth (about 10% of total length), and a *first dorsal fin originating behind the origin of the pelvic fins.* Its nasal flaps reach to or past the upper lip. The teeth are small and numerous, usually with three smooth-edged cusps but occasionally with four or five cusps, and with the central cusp much larger than the others. Teeth number U: 55 to 60, L: 55 to 60 and are similar in both jaws. The sparse denticles are arrowhead shaped with flat surfaces. Coloration is yellowish brown above, with seven to eight dark brown *saddlelike areas over the dorsal surfaces and numerous round dark brown spots* scattered over the body; the undersides are brown with indistinct spots. The coloration is variable, northern specimens have an overall light brown background color, with darker spots and blotches. Southern specimens (Baja California) tend to have lighter or white bellies and white patches on the lower sides.

Similar species The leopard shark has saddlelike markings, but its dorsal fin originates well ahead of the pelvic fins.

Range The swell shark inhabits the tropical eastern Pacific from California to Chile. The type specimen is from Valparaiso, Chile. Although specimens have been collected both off California and Chile, none have been collected in the tropics. This distribution led Springer (1979a) to suggest that either the swell shark has an antitropical distribution, or the lack of records from the eastern tropical Pacific is due to a lack of collecting surveys in that area. Further clarification will not be possible until more specimens are collected and their genomes are analyzed.

The swell shark is common in California waters from Monterey southward to at least Acapulco, Mexico (Kato et al. 1967). I have collected it in the Gulf of California at El Barril, Baja California, where it is common.

Size and weight The largest specimen that I have examined (at the Cabrillo Marine Aquarium) measured 104 cm and weighed 7.3 kg.

Biology The swell shark is a common bottom dweller usually found at depths of 9–35 m, although it occasionally strays into depths of 200 m or more. It prefers the rocky, algae-covered areas of kelp beds. It is active only by night, spending the day hiding in crevices or caves in a lethargic state. Nelson and Johnson (1970) observed swell sharks in

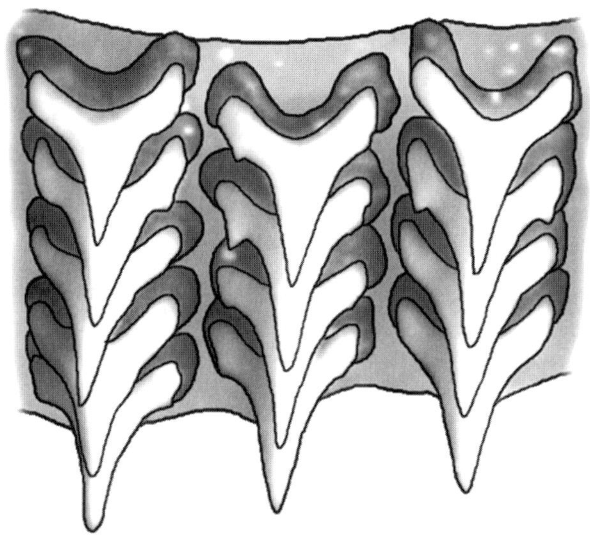

Fig. 78b. Teeth (upper).

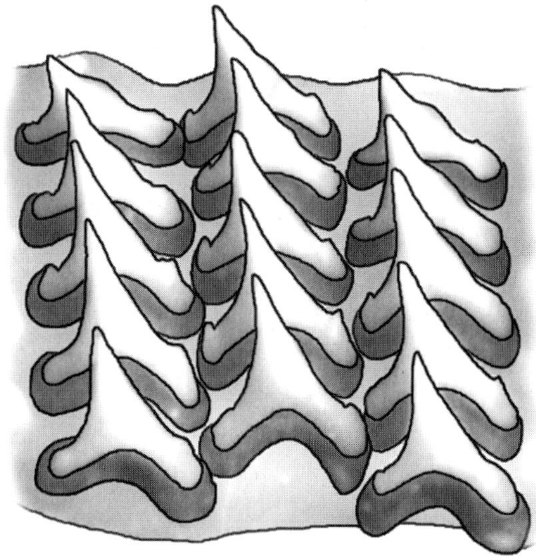

Fig. 78c. Teeth (lower).

their natural environment and reported that they become active shortly after dusk and continue until dawn, after which little activity is seen. Although the swell shark is generally solitary, many may congregate in particular areas. The swell shark generally feeds on small fishes but is capable of swallowing relatively large prey.

The swell shark has the remarkable ability to distend its cardiac stomach by swallowing water or air, swelling its trunk into a nearly spherical shape. The fluid is held in the cardiac stomach by a strong esophageal sphincter anteriorly and by the sphincter between the cardiac and pyloric stomachs posteriorly (Clark 1947). Inflation is said to be a defense mechanism to avoid being swallowed by larger predators. However, sharks that I have handled for measuring purposes, or observed while being removed from gillnets, did not inflate until after several minutes of rough handling. It is unlikely that such slow response would be of value as a defense mechanism against being swallowed by a large

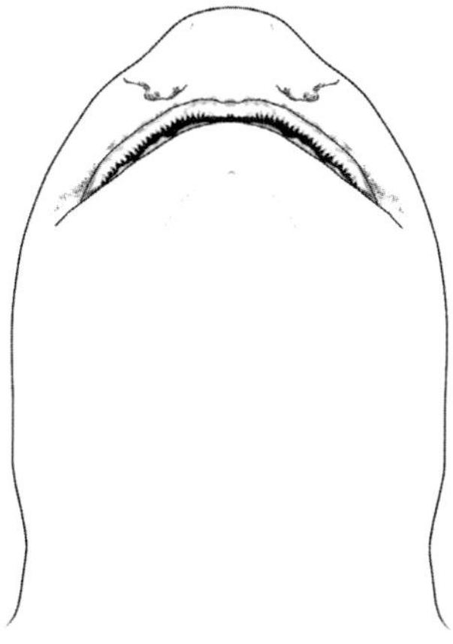

Fig. 78d. Snout.

Fig. 78e. Dermal denticles.

Fig. 78f. *Cephaloscyllium ventriosum*, swell shark, immature female, 26.1 cm, 0.1 kg; reared at Cabrillo Marine Aquarium (courtesy of Jeff Landesman).

predator. Once the shark put back in the water and is no longer threatened, it can deflate in a few minutes.

Snails (gastropods) prey heavily on swell shark egg cases. Grover (1972a) reported that of 42 egg cases found in the crevices of Ship Rock, Santa Catalina Island, California, at depths of 25–35 m, only two contained embryos. Of the remaining 40 empty egg cases, 29 had one or more holes in them. Most of the holes were circular and 2–5 mm in diameter, but one was 15 mm. Grover postulated that the damage was done by gastropods of the genus *Stenoglossa.* Whether the swell shark routinely suffers heavy predation by snails or whether these were egg cases that had been attacked because they failed to be attached to vertical structures is not known. Grover (1972b) suggested that the swell shark populations off Santa Catalina Island are reproductively isolated from those in coastal waters of California by the intervening deep water based on differences in egg case size, tendril length, and fin size.

Size at maturity Males mature at 82–85 cm (Compagno 1984a). The size of females at maturity has not been reported; it is probably around 90 cm.

Reproduction The swell shark is oviparous. The egg cases are relatively large; they measure 9–12.5 cm long and 2.8–5.5 cm wide (Edwards 1920; Cox 1963), and 2.5 cm deep. The egg cases are yellowish brown in color and have long adhesive tendrils at both ends. Some of the ones that I have measured had anterior tendrils 130–150 cm long and posterior tendrils of 60–70 cm! The egg cases may weigh 50–60 g. At the Cabrillo Marine Aquarium, California, egg cases kept at a constant 14°C temperature hatch in about 12 months (Jeff Landesman, pers. comm., June 2008).

Size at birth: The young are 14–16 cm at hatching (Edwards 1920; Cox 1963; Grover 1974).

Brood size: The frequency of egg laying has not been determined.

Nurseries The swell shark nurseries have not been described. Grover (1972a) found numerous egg cases at depths of 25–35 m off Santa Catalina Island, California (See above).

Age and growth Unknown.

Relation to humans The swell shark is a popular aquarium shark, and it lives in captivity for years with minimal care. Females lay eggs readily in captivity, and their offspring survive easily. In aquariums, the egg cases are often transilluminated and displayed so that the developing embryos can be seen by the public.

The swell shark is often a nuisance to fishermen because it enters lobster traps, attracted by decomposing fish used for bait. According to Roedel and Ripley (1950), fishermen say that eating swell shark flesh causes diarrhea.

LOLLIPOP CATSHARK

Cephalurus cephalus (Gilbert, 1892)

Fig. 79a. *Cephalurus cephalus*, lollipop catshark, mature male, 24.7 cm, from Western Central Pacific, Peru (HUMZ #180376, courtesy of K. Nakaya).

Common name Lollipop catshark or head shark, both allusions to its body shape and large head.

Spanish name *Gato renacuajo* (Mexico), a most appropriate name because small specimens resemble large tadpoles; *pejegato renacuajo* (A.F.S.).

Scientific name *Cephalurus cephalus* (Gilbert, 1892), originally described as *Catulus cephalus* in *Proceedings of the United States National Museum* 14(880): 541–542. Bigelow and Schroeder (1941) established the genus *Cephalurus* for this species because they concluded that, at the time, the species could not be referred to any existing genus. *Cephalurus:* from the Greek *kephale,* head, and *urus,* having a tail; *cephalus:* from the Greek *kephale,* head, a Latinized allusion to its huge head.

Synonyms None.

Identification The lollipop catshark is a tiny, tadpolelike shark, with a *very wide and flattened head;* a *very short, rounded snout,* about one-quarter as long as the mouth width; and *a first dorsal fin originating over or slightly anterior to the origin of the pelvic fins.* The teeth are small, usually tricuspid, smooth edged, and similar in both jaws. Lateral teeth have more cusps, up to seven per tooth. There are three to five series of functional teeth in alternate arrangement. Teeth number U: 27 to 36–27 to 36, L: 27 to 36–27 to 36. The dermal denticles are long, thin, curved spines set on four-pointed bases that resemble a bird's foot. Coloration is reddish brown, with a darker or blackish area over the gills, with slightly lighter ventral surfaces. The eye is light blue with a black iris; the pupil is green.

Similar species All other North American catsharks have the origin of the first dorsal fin behind the origin of the pelvic fins. The dogfish sharks lack the anal fin. The nurse shark has long nasal barbels.

Range The lollipop catshark has been reported from the upper Gulf of California to the Revillagigedo Islands and the western coast of Baja California Sur, at least to Cabo San Lorenzo (Castro-Aguirre 1981). It ranges southward to Peru and Chile, but the southern limits of its distribution are not clear. The species has been reported nominally from Peru (Chirichigno 1976) and from Chile (Springer 1979a). However, Springer (1979a) warned that, although he had recognized only one species of *Cephalurus,* further study could show that the specimens from Peru and Chile may represent a separate species.

Size and weight This is one of the smallest sharks. Although the maximum recorded size is 32 cm, this shark probably grows to 40 cm. Most specimens captured have been 18–26 cm long. Recorded sizes and weights are for females: 20 cm, 45 g; 23.5 cm, 70 g; and 29.5 cm, 100 g (Castro-Aguirre 1981).

Biology This is a bottom-dwelling, minute shark from deep water, usually taken at depths of 245–850 m on muddy continental slopes. The temperature at the stations where it has been caught was 5.4°C at 841 m, 7°C at 662 m, and 10°C at 250 m (Springer 1979a; Balart et al. 2000).

Large numbers of these sharks have been caught in single trawls, indicating that lollipop sharks may congregate in suitable habitat. Mathews and Ruiz (1974) reported catching 168 specimens in one trawl near Isla San Pedro Nolasco. Balart et al. (2000) captured 24 specimens at depths of 230–280 m, where the bottom temperature was 10°C. It feeds on small crustaceans and annelid worms (Castro-Aguirre 1981). Nothing else is known of its habits.

Size at maturity Based on very limited information, it appears that the species matures at 21–22 cm. A 20-cm specimen I examined was immature. Balart et al. (2000) reported 22.5-cm females carrying embryos. Gilbert (1892) reported a 24-cm male as mature.

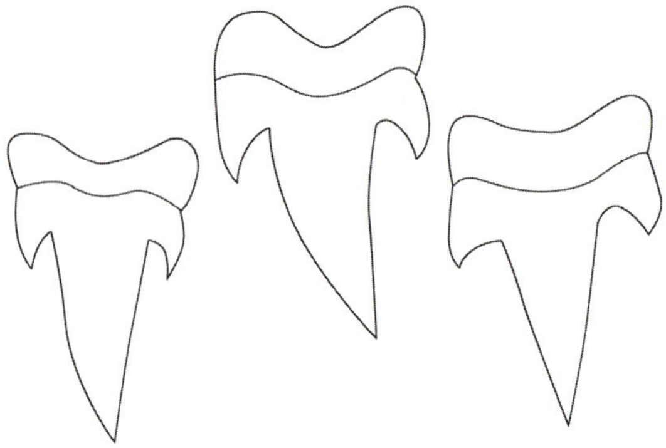

Fig. 79b. Teeth (upper).

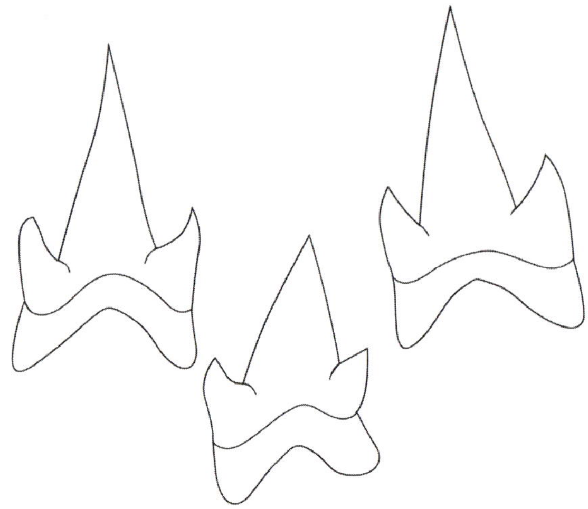

Fig. 79c. Teeth (lower).

Reproduction The species is aplacental viviparous (Springer 1979a).

Size at birth: The size at birth of the lollipop shark is not precisely known. The largest embryos reported were probably 7.5 cm. Balart et al. (2000) mentioned embryos 6.5 cm in their text, and more than 7 cm in their graphs, without describing their condition. Birth *probably* occurs when the embryos reach 8–9 cm.

Brood size: The few sketchy reports of gravid lollipop sharks neglected to mention brood size.

Nurseries Unknown.

Age and growth The species has not been aged.

Relation to humans None.

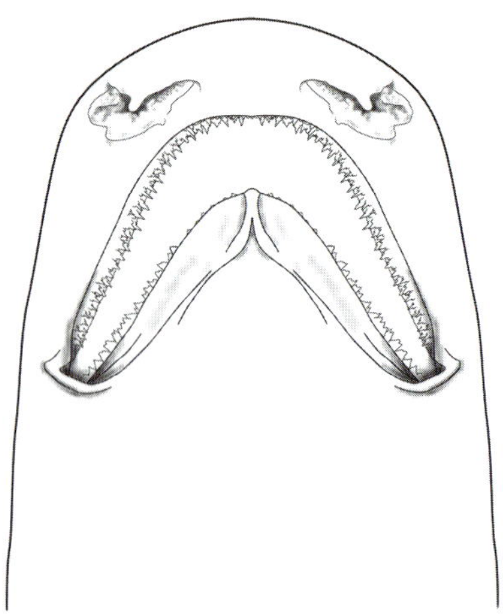

Fig. 79d. Snout.

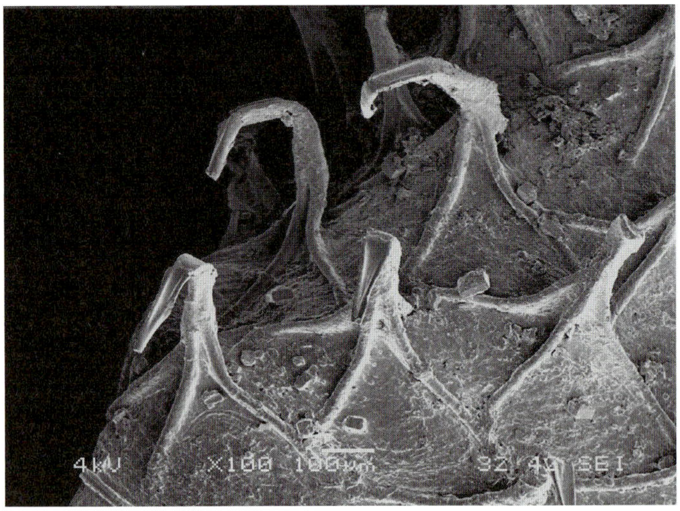

Fig. 79e. Dermal denticles.

MARBLED CATSHARK

Galeus arae (Nichols, 1927)

Fig. 80a. *Galeus arae*, marbled catshark, mature male, 25.5 cm; from the Florida Strait, southwest of Dry Tortugas (F. F. Snelson and Jason Seitz, collectors).

Common name Marbled catshark, an allusion to its coloration.

Spanish name None.

Scientific name *Galeus arae* (Nichols, 1927), originally described as *Pristiurus arae* in *American Museum Novitates* 256: 1–2. *Galeus:* from the Greek *galeos,* a shark; *arae:* named after the vessel *Ara,* which was used for ichthyological expeditions by W. K. Vanderbilt in the early twentieth century. (In Greek mythology, *Ara* was the goddess of destruction and revenge.)

Synonyms *Galeus cadenati* Springer, 1966. Springer (1966) described *Galeus cadenati* based on specimens collected off the Caribbean coasts of Panama and Colombia. Reportedly, it differs from *Galeus arae,* primarily in having a much longer anal fin, its rear tip reaching nearly to the lower origin of the caudal fin. Later, Springer (1979a) complicated the situation when he described three subspecies of the marbled catshark. The northern form, the smallest *Galeus* of the Atlantic, and the one described by Nichols (1927), he called *Galeus arae arae.* He then reduced the southern form, *Galeus cadenati,* found off Panama and Colombia to subspecies status, calling it *Galeus arae cadenati,* stating that the differences from the northern form were unclear. Springer then described a third subspecies, *Galeus arae antillensis,* to represent an island form found through the southeastern slopes of the Straits of Florida and on the slopes of the northern coasts of Cuba, Hispaniola, Puerto Rico, and the Lesser Antilles. This is a much larger form with highly variable coloration, consisting mainly of dorsal longitudinal markings or stripes, and a shorter anal fin. Konstantinou and Cozzi (1998) studied the specimens of *Galeus arae antillensis* at the Smithsonian Institute and separated them into two groups, one being a new species, Springer's catshark, or *Galeus springeri.* According to these authors, *Galeus springeri* differs from *Galeus arae antillensis* in having longitudinal stripes, more vertebrae, and a precaudal ventral crest of enlarged dermal denticles. Apparently, this crest is sometimes difficult to see.

Identification The marbled catshark is a beautiful, slender, tiny shark with a *first dorsal fin originating posterior to the pelvic fins,* and a *row of enlarged scales along the upper edge of the caudal fin.* The teeth have three to six smooth-edged cusps with the central cusp being slenderer and larger than the others; they number U: 36–36, L: 35–35 and are similar in both jaws. The sparse, three-pointed dermal denticles have curved anterior margins, a strong central ridge that forms a long point, and winglike sides, with the usual surface microsculpture. The species has a *striking color pattern of dark brown saddle marks and round or oval spots on a whitish background on the dorsal surfaces, with a bluish tinge along the flanks;* the undersides are whitish and unmarked, often with a light bluish tinge. The iris is pale silver gray.

Similar species The coloration pattern and enlarged dermal denticles in the tail distinguish the marbled catshark from other catsharks in North America, except for Springer's catshark. Springer's catshark has dark streaks on the dorsal surface and a bright blue eye.

Range The marbled catshark inhabits the slopes of the continental shelf from Georgia southward to the eastern Gulf of Mexico and Colombia. It is common or abundant throughout its range. Numerous specimens have been trawled around the Dry Tortugas. Bullis (1967) reported trawling 125 specimens there at depths of 435–510 m, while others have been trawled at depths of 200–220 m, based on the collection records at the Florida Museum of Natural History. Bullis (1967) also reported catching 316 marbled catsharks in 11 trawls off Honduras at depths of 365–585 m during a cruise of the RV *Oregon* in May 1962.

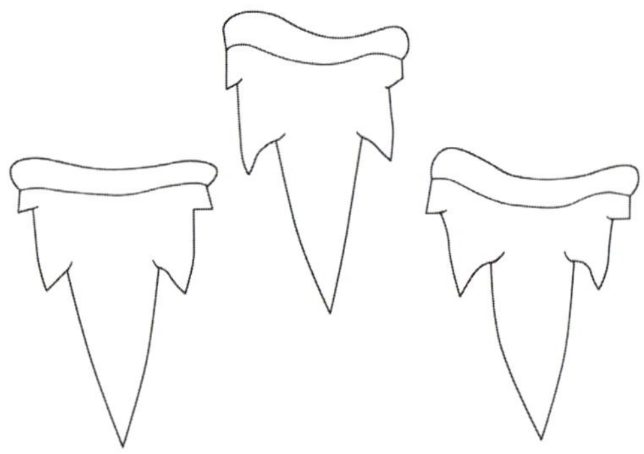

Fig. 80b. Teeth (upper).

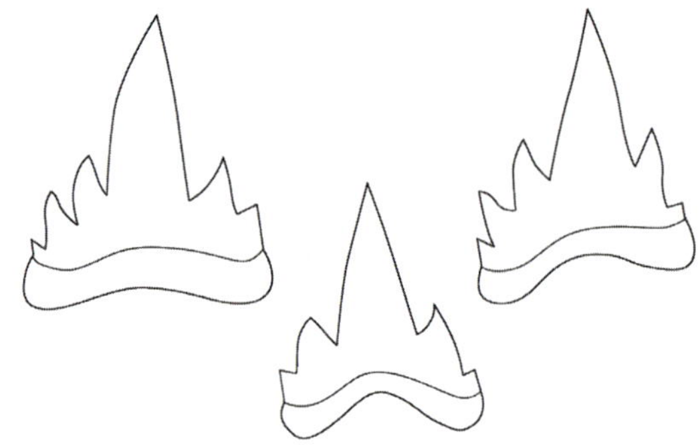

Fig. 80c. Teeth (lower).

Size and weight The marbled catshark is one of the smallest sharks. Maximum size is probably about 35 cm. The 25.5-cm, nearly adult specimen illustrated in this book weighed 42 g.

Biology This common catshark inhabits continental slopes at depths of 220–750 m and temperatures of 6–11°C. Although Bullis (1967) reported that the observed depth range of the species was 292–732 m (the mean of 169 observations was 420 m), I have seen numerous adult marbled catsharks taken at shallower depths of 200–220 m. According to Bullis, the juveniles are found in the shallower parts of the depth range, while the adults inhabit the deeper parts. It is known to feed on various species of deep-water shrimp.

Size at maturity The marble catshark matures at 26–28 cm.

Male: Males mature at about 27 cm. Bullis (1967) reported that the smallest mature males were 26.5 cm, and that all males larger than 27.5 cm had fully calcified claspers. He also reported that the largest immature males were 26 cm.

Female: Females mature at about 28 cm, based on the presence of eggs in the uterus (Bullis 1967).

Reproduction The marbled catshark is believed to be aplacental viviparous because females have been found carrying eggs that are not enclosed in a case (Bullis 1967). If correct, this observation is puzzling because other congeners and most other catsharks are oviparous. Although large numbers of females have been caught, gravid females are very seldom seen. It is likely that the reason for their rarity is that gravid females are usually found in areas where the bottom

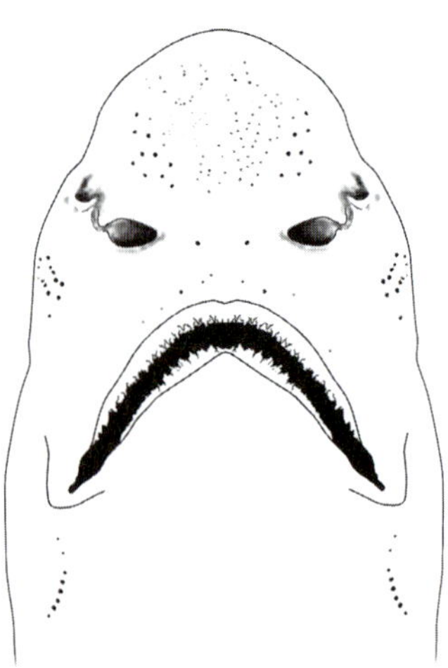

Fig. 80d. Snout.

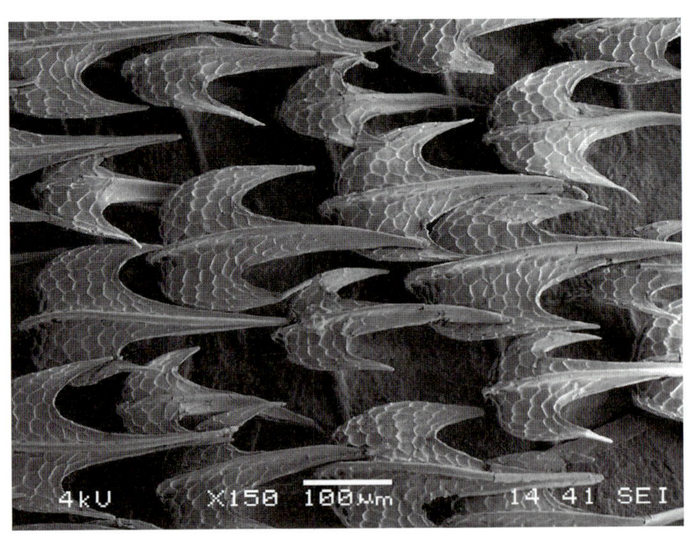

Fig. 80e. Dermal denticles.

topography is unsuitable for trawling. This is the situation with most oviparous catsharks, as the females seek areas having upright or vertical structures to which they can attach their egg cases (Castro et al. 1988).

Size at birth: The smallest free-swimming marbled catsharks on record are about 14 cm; thus, 14 cm is considered the approximate size at birth (Bullis 1967).

Brood size: Unknown.

Nurseries Unknown.

Age and growth The species has not been aged.

Relation to humans None.

SPRINGER'S CATSHARK

Galeus springeri Konstantinou and Cozzi, 1998

Fig. 81a. *Galeus springeri*, Springer's catshark, mature male, 48.1 cm, 0.33 kg; from Jamaica (John Morrissey, collector).

Common name Springer's catshark, named after Stewart Springer, who first recognized the species.

Spanish name None.

Scientific name *Galeus springeri* Konstantinou and Cozzi, 1998, in *Copeia* 1998: 151–158. *Galeus:* from the Greek *galeos,* a shark; *springeri,* Latinized form after Stewart Springer (1909–1991), shark fisherman and later biologist with the Bureau of Commercial Fisheries who first recognized the species (as *Galeus arae antillensis*). Working as a commercial fisherman in the late 1930s and 1940s, Springer saw and examined more sharks than anyone else to date and eventually wrote numerous scientific papers on the natural history of sharks.

Synonyms *Galeus arae antillensis* Springer 1979 (in part).

Identification This is a *small catshark (less than 45 cm in length),* with a *first dorsal fin originating behind the pelvic fins* and a *precaudal ventral crest of on enlarged denticles.* A hand lens or microscope is required to see the caudal crest clearly. The teeth have five cusps, with the central cusp being much larger. The upper teeth have larger central cusps. The dermal denticles are closely packed with much overlapping. The denticles are three pointed with a much longer central point. Their upper surfaces have a microsculpture that gives them a scaly look. It has *a striking coloration of dark stripes on a white background on the dorsal surfaces.* The *eye has a bright blue iris.*

Similar species The marbled catshark has round or oval spots on the dorsal surfaces and lacks the bright blue eye.

Range Springer's catshark inhabits the deep slopes of the Caribbean Sea. The limits of its distribution are not known due to previous confusion with the marbled catshark.

Size and weight Springer's catshark is a small species; the largest specimens do not exceed 45 cm.

Fig. 81b. Teeth (upper).

Fig. 81c. Teeth (lower).

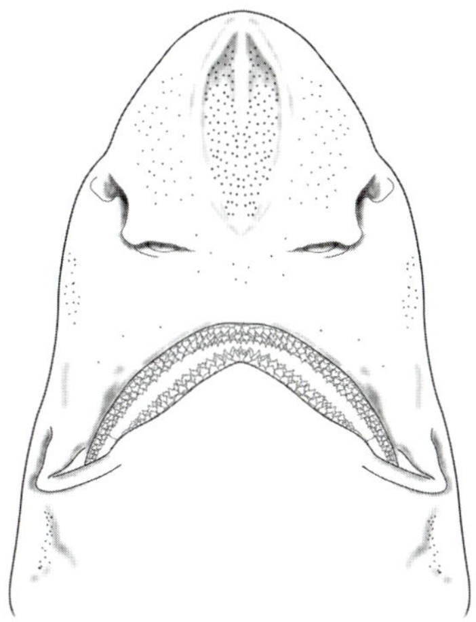

Fig. 81d. Snout.

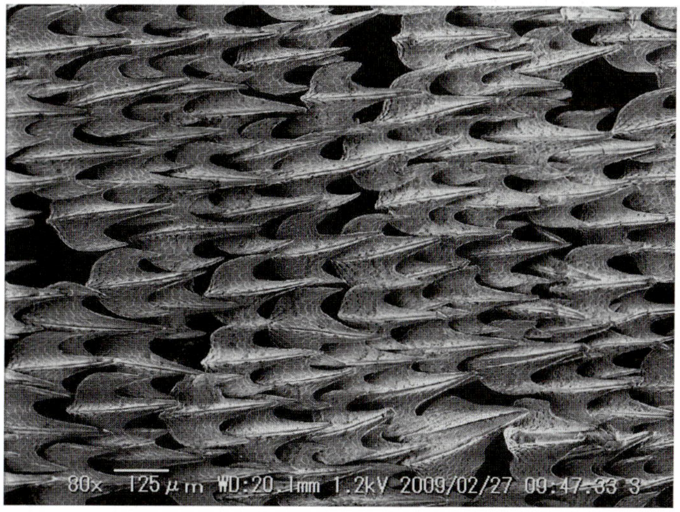

Fig. 81e. Dermal denticles.

Biology This is a shark of the Caribbean continental slopes. Springer's holotype was taken at 550–585 m where the water temperature was of 8.9°C. The specimen illustrated in this book was taken at a depth of 824 m off Jamaica. Nothing else is known of its habits.

Size at maturity Males probably mature at about 32 cm. Females mature at about 32–33 cm. A 31-cm female caught off Puerto Rico was determined to be immature, having an undeveloped ovary 2 cm long and 0.8 cm wide. A 33.5-cm female carried four large oocytes (0.8–1.4 cm in diameter), while a 37.5-cm female carried egg cases in the uteri (Parsons 1986).

Reproduction Springer's catshark is oviparous (Parsons 1986). According to Springer (1979a), one 34.5-cm female, from the Leeward Islands near Antigua, carried one egg case in each oviduct. The egg capsule measured 38 × 16.5 × 10 mm. The egg cases are golden brown and have long tendrils at the corners.

Size at birth: Unknown.

Brood size: Springer's catshark produces two egg cases at a time (Parsons 1986), as do most other catsharks.

Nurseries Unknown.

Age and growth Springer's catshark has not been aged.

Relation to humans None.

PEPPERED CATSHARK

Galeus piperatus Springer and Wagner, 1966

Fig. 82a. *Galeus piperatus*, peppered catshark, mature female, 34.5 cm, 0.14 kg; from the upper Gulf of California (Juan Carlos Pérez-Jiménez, collector).

Common name Peppered catshark, reflecting the minute pepperlike spots that cover its sides and ventral surface. The spots generally are visible only with the use of a magnifier.

Spanish name None.

Scientific name *Galeus piperatus* Springer and Wagner, 1966, in *Los Angeles County Museum Contributions in Science* 110: 1–9. *Galeus:* from the Greek *galeos,* a shark; *piperatus:* Latin for peppery, from *piper,* pepper, an allusion to the light and dark skin denticles.

Synonyms None.

Identification The peppered catshark is very small and characterized by two dorsal fins of nearly equal size, the *first dorsal fin originating posterior to the midpoint of the pelvic fins,* by *the absence of conspicuous pores on the underside of the snout,* and by the presence of *a row of enlarged dermal denticles on the upper margin of the anterior half of the upper caudal lobe.* These enlarged denticles extend from the base of the tail two-thirds of the distance to the tip. The teeth are small and numerous and have three to five smooth-edged cusps, with the central cusp being the largest. They are similar in both jaws. The three-pointed dermal denticles are large, broad, and overlapping. Coloration is grayish brown above and gray with bluish highlights below. The upper sides have a slightly darker, indistinct blotching that contrasts little with the background color. The fins and the tip of the tail are darker than the rest of the body. The body surface is covered with *tiny, black, pepperlike spots* (each spot is a melanophore) *on the ventral and lateral surfaces.* The spots are *most conspicuous on the ventral surface.* The *lining of the mouth is dusky or black.* Juveniles have brown to black mottling on the back and silvery gray undersides. Small juveniles bear dark or purplish brown saddlelike markings and have silvery undersides.

Similar species The filetail catshark also has enlarged denticles on the anterior margin of the tail, but the lining of its mouth cavity is white.

Range The peppered catshark inhabits the northern Gulf of California, where it is common. It is often found in groundfish catches, but it is usually discarded at sea. A report by Castro-Aguirre (1981: 82) mentions the surprising catch of two small juveniles, 9.7 and 10.7 cm, respectively, off Isla Maria Cleofas, Nayarit, Mexico, at a depth of 400 m. If the identification was correct, this would be the first record of the peppered catshark outside the Gulf of California.

Size and weight This is one of the smallest sharks. The largest specimens reported measured 36 cm (Mathews 1975). A gravid female 34.5 cm weighed 140 g.

Biology The peppered catshark is a poorly known small shark that inhabits the deep waters of the northern Gulf of California. Most specimens have been caught on the bottom at depths of 400–1,326 m. However, there are records suggesting that the species sometimes enters the midwater zone, perhaps at night (Hubbs and Taylor 1969; Mathews 1975). J. C. Pérez Jiménez (pers. comm., February 2009) reports that in the northern Gulf of California, the peppered catshark is taken by trawlers operating at 216–256 m.

Size at maturity There are few data on the size at maturity of the peppered catshark. Springer and Wagner (1966) mention that a 25.6-cm female showed no ovarian development, while females measuring 29.6 and 30.2 cm carried developing oocytes about 7 mm in diameter. These observations led Springer and Wagner to suggest that the species matures at about 30 cm. Mathews (1975) suggested that the species matures at about 18 cm, an estimate presumably based on observation of females larger than 18 cm carrying egg cases.

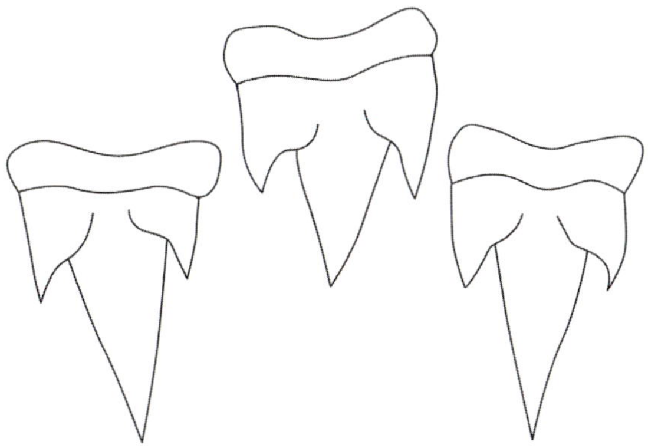

Fig. 82b. Teeth (upper).

Fig. 82c. Teeth (lower).

Reproduction The peppered catshark is oviparous. Interestingly, females carry several formed egg cases in each oviduct (multiple oviparity). The 34.5-cm female illustrated here carried four egg cases in each oviduct, and according to Mathews (1975), females may carry up to 10 eggs each. The egg cases are about 3.5 cm long and about 1.5 cm at their widest point, and have long, adhesive tendrils 20–30 cm long at each corner. The tendrils of egg cases are intertwined with each other, so that extracting one egg case from the cloaca draws out the following egg case as well. The egg cases are light reddish brown and contain pale green eggs.

Size at birth: The peppered catshark is born at about 8 cm. The smallest reported free-swimming neonate measured 8.5 cm (Hubbs and Taylor 1969).

Brood size: Given that females may carry up to 10 well-formed egg cases at the same time, and that the tendrils of adjacent egg cases are intertwined with each other, it is likely that all these egg cases are laid at once. Thus, the brood can be said to consist of 8–10 young.

Nurseries The peppered catshark nurseries are probably in areas having upright structures to which the females may attach their egg cases. Hubbs and Taylor (1969) reported recovering an egg case from a trawl at depths of 1,097–1,326 m.

Age and growth The peppered catshark has not been aged.

Relation to humans None.

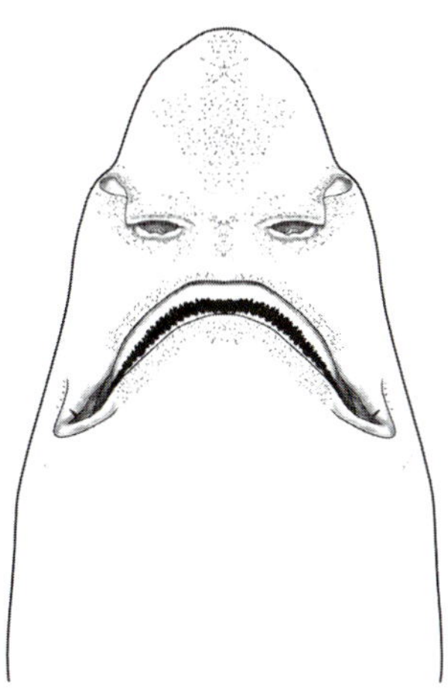

Fig. 82d. Snout.

Fig. 82e. Dermal denticles.

CAMPECHE CATSHARK

Parmaturus campechiensis Springer, 1979

Fig. 83a. *Parmaturus campechiensis*, Campeche catshark, female, 20.5 cm, 0.03 kg; off Andros Island, Bahamas (Florida Museum of Natural History, UF #101299, courtesy of George Burgess).

Common name Campeche catshark, a reference to where the species was first discovered, the Bay of Campeche, Gulf of Mexico.

Spanish name None.

Scientific name *Parmaturus campechiensis* Springer, 1979, in *NOAA Technical Report NMFS Circular* 422: 100–101. *Parmaturus:* from the Latin *parma,* a small round shield or target, and *urus,* tail, an allusion to the row of enlarged denticles (crest) on the dorsal edge of the tail; *campechiensis:* Latinized form, of or from the Bay of Campeche.

Synonyms None.

Identification This tiny shark is characterized by a *first dorsal fin originating over or slightly ahead of the pelvic fins origin,* a *second dorsal fin larger that the first, labial furrows continuous around the mouth,* and *enlarged dermal denticles on the upper edge of the caudal fin.* The teeth generally have three or four cusps, are similar in both jaws, and number about U: 60, L: 60. The dermal denticles are unique: they are sparse and have squarish sides, one very long central point, and very short side points. Coloration is grayish above with slightly darker undersides, gill slits, and outer parts of the fins. The only specimen described in the literature thus far is the type specimen, a 16-cm juvenile (Springer 1979a). It is likely that Springer's juvenile type specimen and the larger *Parmaturus* illustrated in this work are conspecific. However, this needs to be ascertained after examining more specimens.

Similar species *Apristurus* catsharks have a more posterior first dorsal fin that begins behind the origin of the pelvic fins.

Range The range of the species is not known. It is probably found throughout the Caribbean basin, given that the type specimen is from the Bay of Campeche and the probably conspecific specimen illustrated here is from the Bahamas.

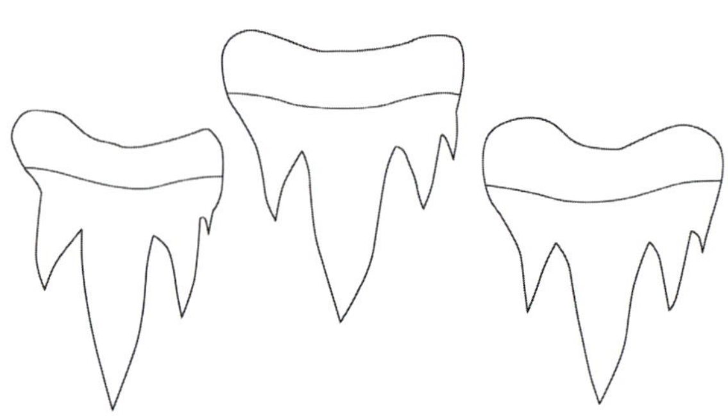

Fig. 83b. Teeth (upper).

Fig. 83c. Teeth (lower).

Fig. 83d. Snout.

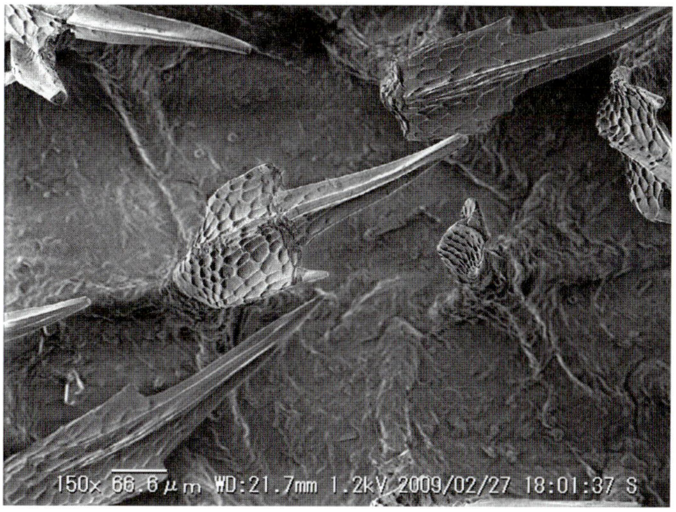

Fig. 83e. Dermal denticles.

Size and weight The species probably reaches 45 cm.

Biology This is one of the least-known sharks in North America. This shark is probably lives in midwater below 700 m. The type specimen known was caught at 1,097 m, and the Bahamian shark was taken at 1,334–1,378 m in 1974. These small deep-water sharks are extremely difficult to catch; they are probably accidental catches in bottom trawls as the equipment travels to and from the bottom.

Size at maturity Unknown.

Reproduction No data available.

Size at birth: Unknown.

Brood size: Unknown.

Nurseries Unknown.

Age and growth No data available.

Relation to humans None.

FILETAIL CATSHARK

Parmaturus xaniurus (Gilbert, 1892)

Fig. 84a. *Parmaturus xaniurus*, filetail catshark, female, 45.5 cm, 0.37 kg; off Big Sur, California (Brooke Flammang, collector).

Common name Filetail cat shark, an allusion to the rough upper edge of its tail.

Spanish name The A.F.S. name is *pejegato lima.*

Scientific name *Parmaturus xaniurus* (Gilbert, 1892), originally described as *Catulus xaniurus* in *Proceedings of the United States National Museum* 14(880): 540–541. *Parmaturus:* from the Latin *parma,* a small round shield or target, and *urus,* Latin tail, a reference to the row (crest) of enlarged denticles on the dorsal edge of the tail; *xaniurus,* from the Greek *xanion,* a comb or scraper (a reference to the comblike denticles on the upper edge of the tail), and *urus,* having a tail.

Synonyms None.

Identification The filetail catshark is characterized by a *snout much shorter than the mouth width,* a *first dorsal fin originating posterior to the origin of the pelvic fins,* the *last two gill slits positioned above the pectoral fin,* and the presence of *a row of enlarged dermal denticles in the upper margin of the anterior half of the upper caudal lobe.* The teeth have three to five smooth-edged cusps with the central cusp the largest; they number U: 41 to 45–41 to 45, L: 42–42 and are similar in both jaws. The dermal denticles have two strong ridges that meet at the tip of a long point and have wide four-pointed bases. Coloration is light grayish brown above, slightly lighter below. *The lining of the mouth cavity is white.*

Similar species The peppered shark also has enlarged dermal denticles on the tail, but the lining of its mouth cavity is dusky or black. Other North American catsharks in the area lack enlarged dermal denticles on the tail.

Range The filetail catshark is found from the Washington coast (Flammang et al. 2008) to the Gulf of California (Gilbert 1892). It is common in deep water in the Santa Barbara basin (Lee 1969).

Size and weight Specimens up to 50 cm and 0.6 kg are common. The largest specimen recorded measured 57.4 cm (Cross 1988).

Biology This is a common catshark off the western coast of North America and often found in trawl catches. It has been reported from diverse deep-water habitats. Gilbert (1892) reported catching them at depths ranging from 1,250 m. Lee (1969) reported catching juvenile filetail catsharks in midwater as much as 490 m above the bottom. The adults are believed to be benthic in habit. Springer (1979a: 109) reported that the species had been observed under low-oxygen conditions on the bottom "in the deepest part of Santa Barbara basin."

The filetail catshark feeds on pelagic prey: crustaceans, small bony fishes, and mollusks. Decapod crustaceans, particularly the galtheid *Pleuroncodes planipes,* dominate the stomach contents.

Size at maturity

Male: The size at which the filetail catshark males reach maturity is not known precisely. Cross (1988) reported that males reached maturity between 37.5 and 42.5 cm without adducing any evidence of maturity. Balart et al. (2000) suggested that males reach maturity at 34 cm, based on "the relationship between total length and clasper length," but lacked convincing evidence such as calcification of the clasper.

Female: Females mature at about 46–47 cm. Cross (1988) reported that females larger than 47 cm carried egg cases in the uteri.

Reproduction The filetail catshark is oviparous. Cox (1963) described the egg case as long and slender, tan colored, and

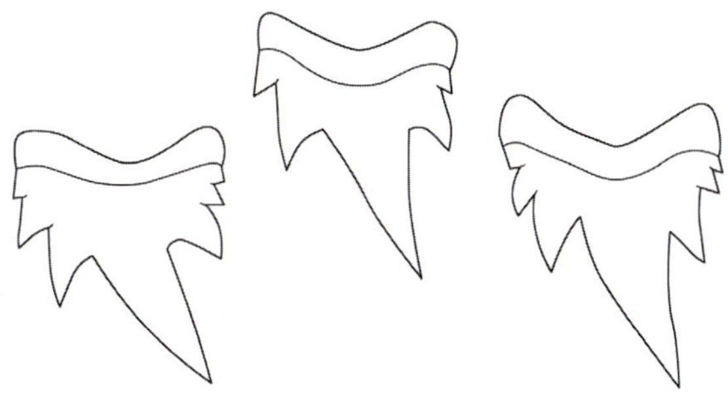

Fig. 84b. Teeth (upper).

Fig. 84c. Teeth (lower).

with long tendrils. The egg case measures 7–11 cm long and about 3 cm wide across its center, its widest point. The egg cases of the filetail catshark can be identified by the presence of a flange, T-shaped in cross section, that runs along the lateral edge.

Size at birth: According to Ebert (2003) the young are born at 7–9 cm.

Brood size: Like most oviparous catsharks, the filetail catshark is believed to produce two egg cases at a time. The frequency of egg laying is not known.

Nurseries Large masses of egg cases (some exceeding 2 m in diameter) have been found in California's Monterey Canyon and attributed to this species. These "nests" are composed of hundreds of egg cases laid one upon the other by dozens of females. These nests are the result of females seeking vertical structures for attaching their egg cases (Castro et al. 1988) and many females using the same suitable structure. Attaching the egg cases to upright structures elevates them above the bottom and away from boring snails and other predators. As the nests increase in size, they must become attractive sources of food for those predators that can reach the nest. One can only wonder what factors induce the females to lay egg cases upon other egg cases (dilution of risk or predator satiation, bad tasting eggs that would repel a predator tasting one egg case from eating others, etc.).

Age and growth The species has not been aged.

Relation to humans None.

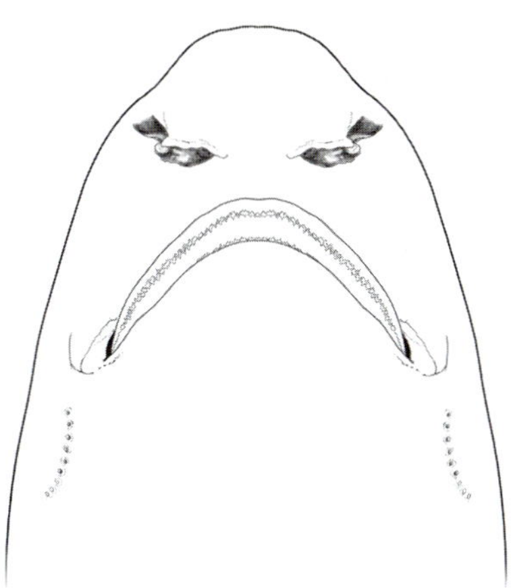

Fig. 84d. Snout.

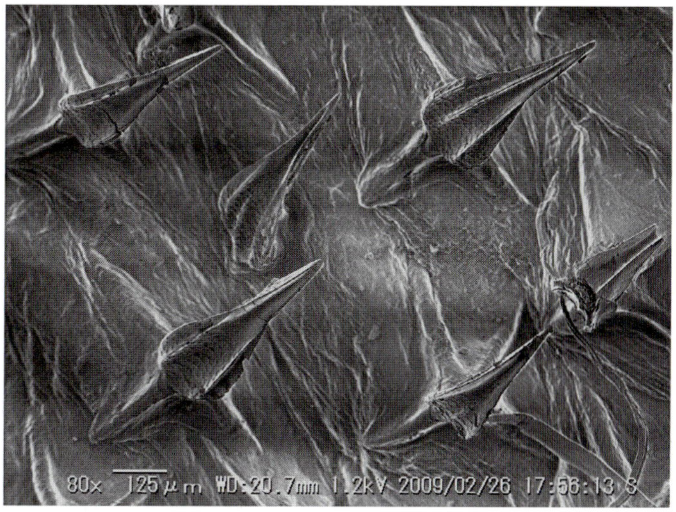

Fig. 84e. Dermal denticles.

NARROWTAIL CATSHARK

Schroederichthys maculatus Springer, 1966

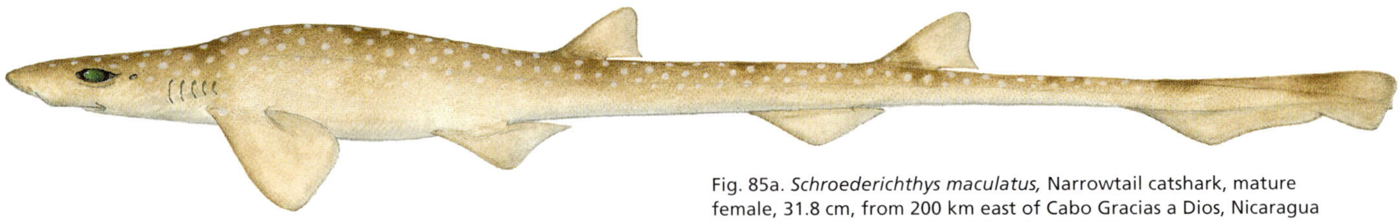

Fig. 85a. *Schroederichthys maculatus,* Narrowtail catshark, mature female, 31.8 cm, from 200 km east of Cabo Gracias a Dios, Nicaragua (RV *Pilsbry,* collector; Univ. of Florida #229707).

Common name Narrowtail catshark.

Spanish name None.

Scientific name *Schroederichthys maculatus* Springer, 1966, in *U.S. Fish and Wildlife Service Fishery Bulletin* 65(3): 605–606. *Schroederichthys: Schroeder,* after William C. Schroeder (1895–1977), who, along with Henry B. Bigelow, wrote numerous articles on cartilaginous fishes, and *ichthys,* from the Greek ichthys, fish; *maculatus:* Latin, spotted, from *macula,* spot.

Synonyms None.

Identification This small shark is characterized by a broad head with a markedly slender body, a *short trunk with an elongated caudal region* (the distance from the origin of the first dorsal fin to the tip of the tail is much longer than the rest of the body), a *first dorsal fin originating posterior to the pelvic fins,* and a *second dorsal fin larger than the first.* The teeth have one large central cusp flanked by much smaller cusps on each side; they number U: 48 to 53, L: 36 to 42 and are similar in both jaws (except that the lower teeth have shorter central cusps). The three-pointed dermal denticles have two strong ridges that form a long point. Body coloration is tan or light brown above, with *scattered white or yellowish spots 1–3 mm in diameter;* the undersides are white and unmarked. Young or half-grown specimens have seven to nine dark saddle blotches that fade or disappear in adults.

Similar species The dwarf catshark has similar white or yellowish spots, but it has a shorter caudal region, and its second dorsal fin is smaller than the first.

Range The narrowtail catshark inhabits the western Caribbean Sea, with specimens having been taken from Rosalind to Quita Sueño banks. Catches reported by Springer (1966) and subsequent catches by RV *Pillsbury* in 1971 included numerous specimens in single net hauls at those locations, suggesting schooling behavior or that the species is locally

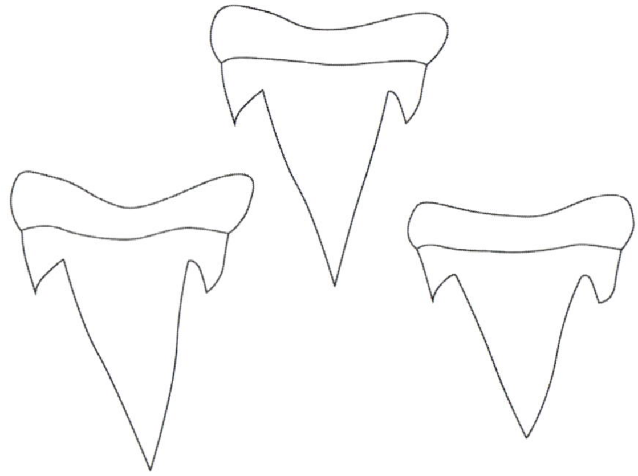

Fig. 85b. Teeth (upper).

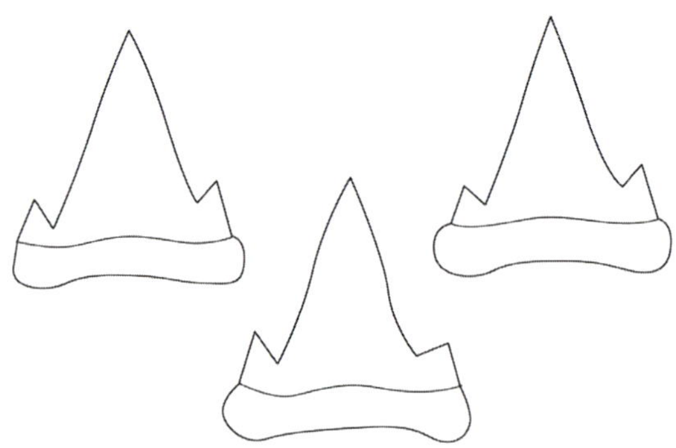

Fig. 85c. Teeth (lower).

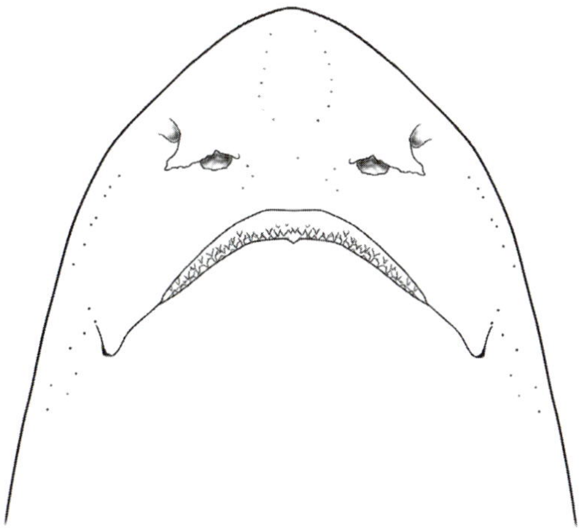

Fig. 85d. Snout.

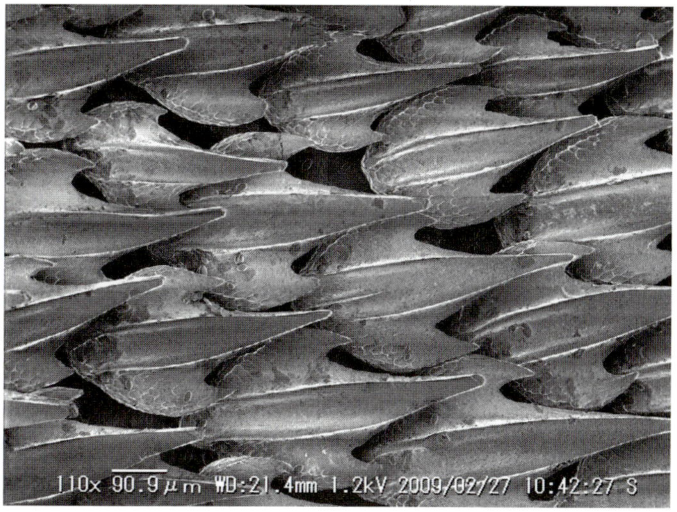

Fig. 85e. Dermal denticles.

abundant. The limits of the distribution of the narrowtail catshark are unknown.

Size and weight This is a small species; its maximum length is about 34.2 cm (Springer 1979a).

Biology The narrowtail catshark is known from only a few dozen specimens caught long ago, so we know little of its habits. I am aware of only the catches reported by Springer (1966) and those taken by RV *Pillsbury* in the early 1970s (PIL 1356, now deposited at the Florida Museum of Natural History). These specimens were trawled from depths of 190–410 m, where they appeared to favor bottoms of fine, white, calcareous sand. It feeds on octopus, squid, and small fishes (Springer 1979a). The stomach of a 32-cm female that I examined contained the remains of numerous tiny squid, their beaks measuring 1–3 mm in width.

Size at maturity Males mature at about 28 cm (Springer 1979a). The size at maturity for females is not known. A 31.8-cm female that I examined was mature.

Reproduction The narrowtail catshark is oviparous. Females produce two egg cases at a time. The egg cases measure 3–4 cm in length, about 1.3 cm in width, and have tendrils 15–22 cm long. The egg case has a narrow constriction at the caudal end and longitudinal striations on the surface. The egg cases are greenish when fresh and turn reddish brown after preservation. The duration of the incubation period is unknown.

Size at birth: Hatching size is unknown.

Brood size: Unknown.

Nurseries Unknown.

Age and growth The narrowtail catshark has not been aged.

Relation to humans None.

BOA CATSHARK

Scyliorhinus boa Goode and Bean, 1896

Fig. 86a. *Scyliorhinus boa*, Boa catshark, mature male, 46.4 cm, from Buccoo, Tobago, BWI. (R. Mounsey, collector).

Common name Boa catshark, an allusion to the markings that resemble those of a boa.

Spanish name None.

Scientific name *Scyliorhinus boa* Goode and Bean, 1896, originally described as *Scylliorhinus boa.* The description of *S. boa,* based on a 15-cm juvenile caught off Barbados, is buried in the species account of *S. retifer* in Goode and Bean's *Oceanic Ichthyology* 17. According to Goode and Bean, "It was at first our intention to describe this as a separate species under the name of *S. boa,* but it is doubtful if it can be regarded even as a color variety [of *S. retifer*] the markings being doubtless due to immaturity." The illustration by Goode and Bean (1896, plate 2, fig. 6) does not show all the markings and differs from the type in many other respects (Garman 1913). Later, upon examination of the type specimen, Garman (1913) confirmed the species status of *S. boa* (as *Catulus boa*). Much later, Bigelow and Schroeder (1948) emended the spelling to *Scyliorhinus* and correctly illustrated the species.

Scyliorhinus: from the Greek *skylion,* a dogfish, probably related to *skyllo,* to tear or mangle, and *rhinus,* from the Greek *rhinos,* nose, hence "dogfish-nose"; *boa,* from the Latin *boa,* a kind of snake.

Synonyms *Catulus boa* Garman, 1913.

Identification This small catshark has a *first dorsal fin originating posterior to the pelvic fins.* The teeth have five cusps with the central cusp being the largest and are alike in both jaws. Teeth number about U: 24–24, L: 23–23. The three-pointed dermal denticles have five to eight ridges. The dorsal surfaces of fresh specimens that I examined were pale brownish green with numerous, scattered, small brown and yellow spots 2–5 mm in diameter, seven rectangular saddle blotches over the midline of the body, and square to rectangular markings on the flanks. The undersides are very pale green to dirty white and are unmarked. The yellow spots glow even in bright daylight. The yellow spots of specimens that I examined in Tobago glowed for about an hour after death. The spots disappeared after preservation, and the

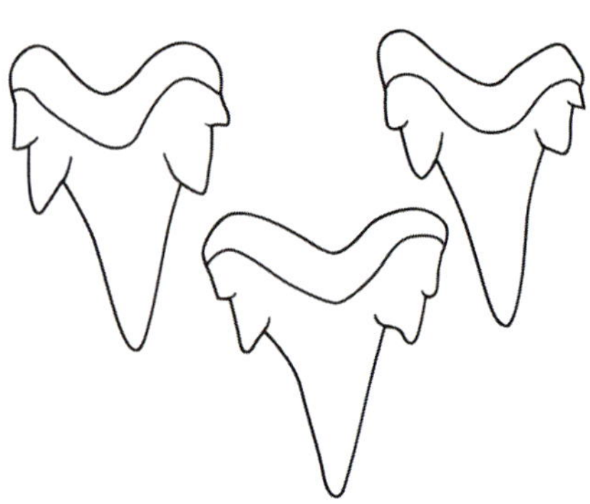

Fig. 86b. Teeth (upper).

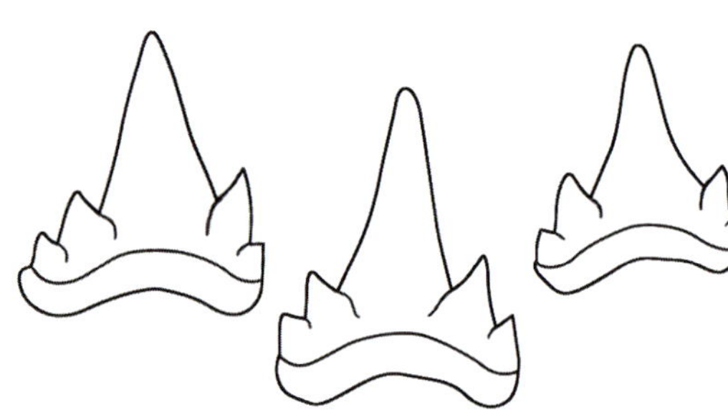

Fig. 86c. Teeth (lower).

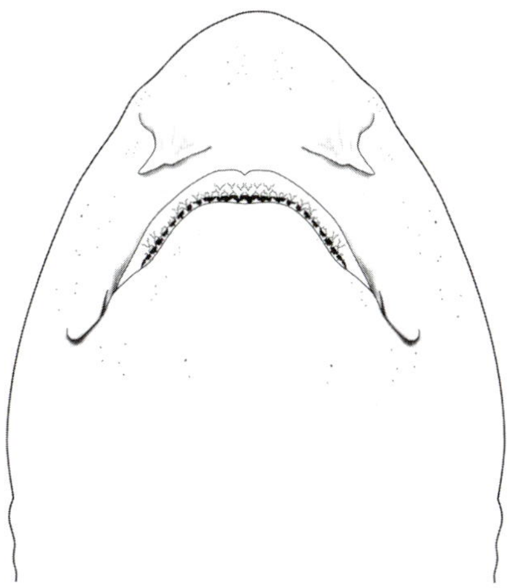

Fig. 86d. Snout.

Fig. 86e. Dermal denticles.

pale green body color changed to pale brown. The iris is pale green, of a shade similar to that of the skin color.

Similar species All other catsharks in North America lack the small scattered dark spots on the dorsal surfaces.

Range The boa catshark is found from Hispaniola, through the Lesser Antilles, to Tobago and the coast of Venezuela.

Size and weight Springer (1979a) reported an adult male 54 cm. This specimen appears to be the largest on record, but too few adults have been observed to accurately estimate of the maximum length. A 47.1-cm male that I examined weighed 0.40 kg, while a 47.6-cm male weighed 0.43 kg.

Biology This shark inhabits the continental slopes. Most specimens have been caught at depths of 274–676 m (Springer 1979a). However, the two animals that I examined were taken at 92 m off Pigeon Point, Tobago. Nothing is known of its habits.

Size at maturity Males of 47 cm that I examined in Tobago were mature. Their claspers were hard and could flex forward.

Reproduction The boa catshark is undoubtedly oviparous, but nothing is known about its reproductive processes.

Size at birth: The boa catshark probably hatches at about 10–11 cm.

Brood size: Unknown.

Nurseries Unknown.

Age and growth Unknown.

Relation to humans None.

WESTERN CATSHARK

Scyliorhinus hesperius Springer, 1966

Fig. 87a. *Scyliorhinus hesperius,* western catshark, immature male, 22.6 cm; from the southern Caribbean, off Colombia (RV *Oregon II*, Station 11253, Stewart Springer and George Burgess, collectors).

Common name Western catshark, a translation of the scientific name.

Spanish name None.

Scientific name *Scyliorhinus hesperius* Springer, 1966; in *U.S. Fish and Wildlife Service Fishery Bulletin* (1966, 65(3): 603–604, figs. 7, 9, 15, and 27). *Scyliorhinus:* from the Greek *skylion,* a dogfish, probably related to *skyllo,* to tear or mangle, and *rhinus,* from the Greek *rhinos,* nose, hence "dogfish-nose"; *hesperius:* from the Greek *hesperos* of the west or western.

Synonyms None.

Identification This small catshark is recognized by its first dorsal fin, which originates behind the pelvic fins, and its distinctive color pattern of dark, squarish, gray-brown saddle blotches and round to oval lighter markings inside the saddle blotches. The lighter markings are areas of lighter pigmentation and not white spots; they are found all over the upper body surfaces but are more visible inside the darker saddle blotches. The number of spots inside the blotches ranges from two to many. In some cases, several round or oval spots coalesce and form irregular patterns inside the darker blotches, but their outlines remain visible. The background coloration is light brown on the dorsal surfaces and slightly paler on the ventral surfaces. The teeth have five cusps with the central cusp being much larger and are similar in both jaws. The teeth number about U: 26–26, L: 21–4–21 (Springer 1979a). The dermal denticles are sparse, with three ridges that meet at the point.

Similar species The chain dogfish has a pattern of thin lines (usually 2–3 mm wide) surrounding the spots. Mead's catshark lacks small round spots within the saddle blotches. The dwarf catshark has distinct white spots on its dorsal surfaces.

Range The western catshark inhabits the western Atlantic from (at least) South Carolina to Colombia, although the limits of its distribution are not well known due its deep-water habitat and confusion with other catsharks. The type specimen was caught on the Caribbean coast of western Panama. The species has been recorded from the continental slopes from Honduras to Colombia and from Jamaica

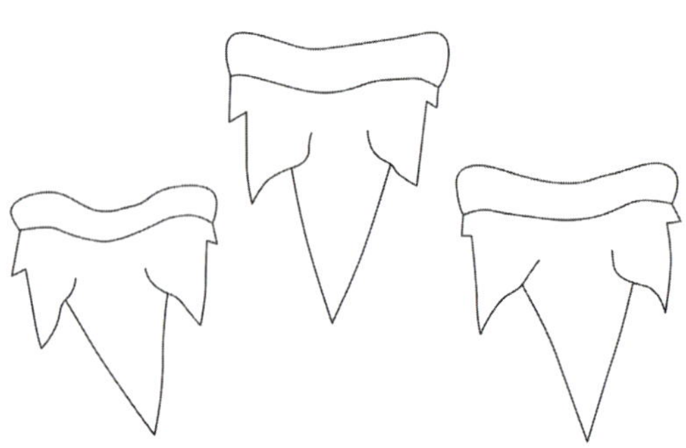

Fig. 87b. Teeth (upper).

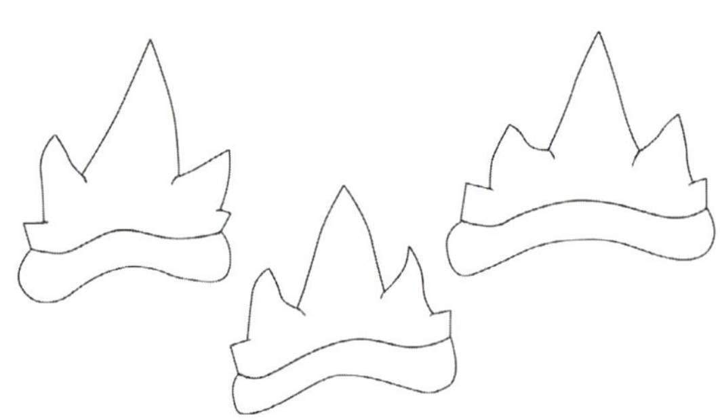

Fig. 87c. Teeth (lower).

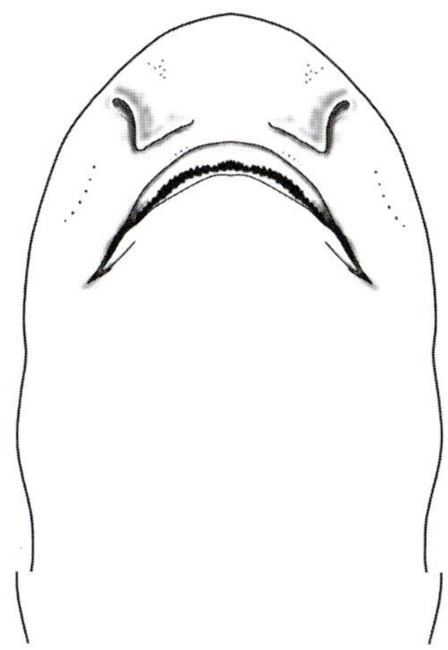

Fig. 87d. Snout.

Fig. 87e. Dermal denticles.

(Springer 1979a). Two specimens were captured off South Carolina in 1985 during experimental fishing for crabs, suggesting that the range of the western catshark extends much farther north than previously believed.

Size and weight The western catshark probably reaches about 50–52 cm, given its similarity to the chain dogfish.

Biology The western catshark inhabits the deep continental slopes. Nothing is known of its habits due to paucity of specimens and to confusion with the chain dogfish.

Size at maturity Almost nothing is known about the size of the western catshark at maturity. Springer (1979a) reported that a 47-cm female carried 20 mm oocytes that were probably ripe.

Reproduction The western catshark is undoubtedly oviparous and reproductively similar to the chain dogfish. Nothing else is known about its reproductive processes.

Size at birth: The size at birth is unknown but probably is about 10–11 cm at birth based on size similarity to the chain dogfish.

Brood size: Like its congener, the chain dogfish, female western catsharks probably produce two egg cases at a time.

Nurseries Unknown.

Age and growth The species has not been aged.

Relation to humans None.

MEAD'S CATSHARK

Scyliorhinus meadi Springer, 1966

Fig. 88a. *Scyliorhinus meadi,* Mead's catshark, female, 41.8 cm; from off Cape Canaveral, Florida (RV *Delaware II* and Glenn Parsons, collectors).

Common name Mead's catshark, named after Giles W. Mead, Jr. (1928–2003), prominent ichthyologist, philanthropist, and wine maker, at one time Curator of Fishes at the Museum of Comparative Zoology at Harvard, and later Director of the Natural History Museum of Los Angeles County.

Spanish name None.

Scientific name *Scyliorhinus meadi* Springer, 1966, in *U.S. Fish and Wildlife Service Fishery Bulletin* 65: 600–601, figs. 7, 14, and 27. *Scyliorhinus:* from the Greek *skylion,* a dogfish, probably related to *skyllo,* to tear or mangle, and *rhinus,* from the Greek *rhinos,* nose, hence "dogfish-nose"; *meadi:* named after Giles W. Mead Jr., who first called a specimen of the species to Springer's attention.

Synonyms None.

Identification This is a larger catshark, perhaps reaching 100 cm, with a stout body and a *first dorsal fin originating posterior to the pelvic fins.* The teeth have one large central cusp flanked by small side cusps; the frontal teeth have five cusps, and the others have three. Teeth number U: 25 to 28–25 to 28, L: 23–23. The dermal denticles have three ridges and a single point. Coloration is light gray above, with about *nine darker, brown saddle blotches spaced over the back and extending to the sides,* and *no dark spots in the blotches or among the blotches;* the undersides are light gray or whitish.

Similar species The color pattern of solid dark blotches without spots or lines distinguishes Mead's catshark from all other North American catsharks.

Range Mead's catshark inhabits the western Atlantic, in the Blake Plateau, from North Carolina to Florida. The limits of its distribution are not well defined.

Size and weight Mead's catshark is one of the largest *Scyliorhinus* catsharks of North America, reaching about 100 cm. The largest specimen that I examined was a male that measured 75.5 cm.

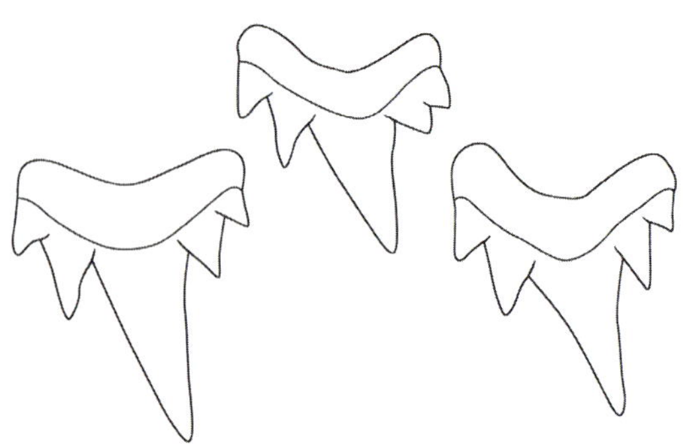

Fig. 88b. Teeth (upper).

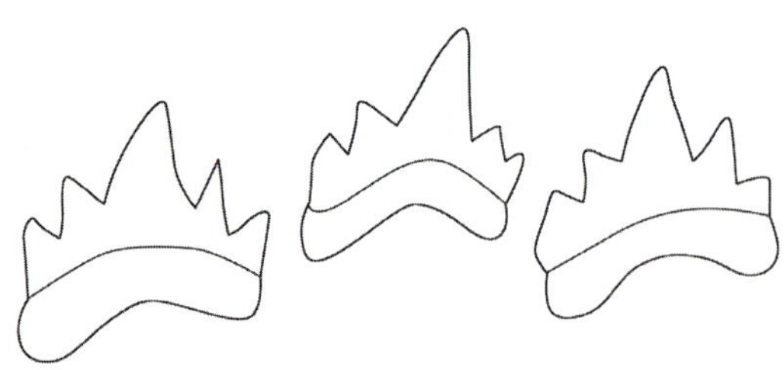

Fig. 88c. Teeth (lower).

Fig. 88d. Snout.

Fig. 88e. Dermal denticles.

Biology Nearly all reported Mead's catsharks were trawled or caught in traps at depths of 329–450 m. One specimen with cephalopod beaks and euphausid shrimp in the stomach has been reported (Burgess et al. 1979). Another specimen from South Carolina waters had crab legs and squid beaks in the stomach. Nothing else is known of its habits.

Size at maturity Unknown.

Reproduction Mead's catshark is certainly oviparous, like its congeners.

Size at birth: Unknown.

Brood size: Unknown.

Nurseries Unknown.

Age and growth Unknown.

Relation to humans None.

CHAIN DOGFISH

Scyliorhinus retifer (Garman, 1881)

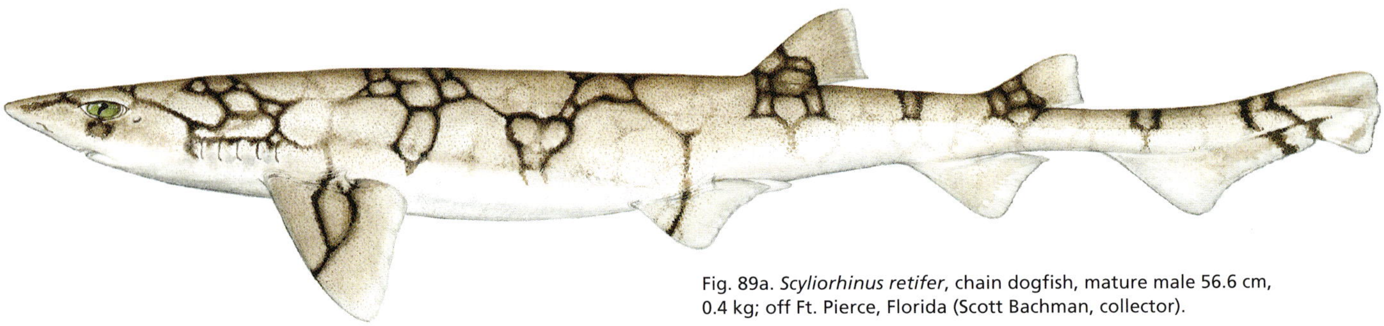

Fig. 89a. *Scyliorhinus retifer*, chain dogfish, mature male 56.6 cm, 0.4 kg; off Ft. Pierce, Florida (Scott Bachman, collector).

Common name Chain dogfish, an allusion to the chainlike or netlike markings on its dorsal surface, and dogfish, the name given to many small sharks. See the entry on the spiny dogfish (p. 55) for more on the origin of the word "dogfish."

Spanish name The A.F.S. name is *alitán mallero.*

Scientific name *Scyliorhinus retifer* (Garman, 1881), originally described as *Scyllium retiferum* in *Bulletin of the Museum of Comparative Zoology at Harvard College* 8(11): 233. *Scyliorhinus:* from the Greek *skylion,* a dogfish, probably related to *skyllo,* to tear or mangle, and *rhinus,* from the Greek *rhinos,* nose, hence "dogfish-nose"; *rotifer:* from the Latin *rete,* diminutive of *reticulum,* a net, and *fer,* from the Latin *fero,* to bear, thus "netbearer," a reference to its net or chainlike markings.

Synonyms None.

Identification The chain dogfish is a small catshark with a *first dorsal fin originating posterior to the pelvic fins* and *dark chainlike markings on the dorsal side.* The teeth are triangular and smooth edged, with one large central cusp flanked by a smaller cusp on either side, and they are alike in both jaws. Teeth number U: 21 to 26–0 to 2–21 to 26, L: 20 to 22–0 to 4–20 to 22. The dermal denticles have three ridges that taper to a single point. Some of the dermal denticles have inconspicuous side points. The body coloration is variable. Northern specimens tend to be reddish brown above and yellowish brown below. Southern specimens tend to be light tan above and whitish below. All have *distinctive black or dark brown lines forming a netlike or chainlike pattern over the back and sides.* The pattern of the markings is quite variable and differs from individual to individual. These different markings are sufficient for aquarists to recognize individual fish.

Similar species The marbled catshark has solid spots on its dorsal surface.

Size and weight This is a small shark; the largest specimens reported by Castro et al. (1988) were a 58.1-cm male weighing 682 g and a 59.0-cm female weighing 682 g. Representative sizes and weights of specimens I have examined are as follows: males, 23.2 cm, 52 g; 50.5 cm, 455 g; and 55.4 cm, 511 g; females, 11.4 cm, 4.2 g; 28.8 cm, 83 g; 42.3 cm, 227 g; 47.3 cm, 398 g; and 55.5 cm, 625 g.

Range The chain dogfish inhabits the continental slopes of the western Atlantic from La Have Bank, Nova Scotia (Gilhen et al. 2003), to Nicaragua, including the Gulf of Mexico (Springer 1979a). It is abundant in deep waters from Virginia to Florida and the Gulf of Mexico.

Biology The chain dogfish is the most abundant catshark off eastern North America. It is a bottom-dwelling species of the deep waters of the continental shelf and slopes. In the northern part of its range, the chain dogfish is usually found at 125–175 m and temperatures of 7–13.1°C, occasionally being found in shallower areas of cool water (based on several hundred National Marine Fisheries Service capture records; John Galbraith, pers. comm., October 2005). In the "northern Mid-Atlantic Bight,"[1] the chain dogfish occurs at 90–150 m, while in the southern parts they occur at 200–220 m (Able and Flescher 1991). Off the southeastern United States, the chain dogfish usually occurs at 200–450 m and bottom temperatures of 7–10°C. Springer (1979a) reported catching it in bottom temperatures of 8.5–11.3°C off the Carolinas and off Pensacola, Florida. I have examined numerous specimens caught off South Carolina at 220–256 m in tilefish operations, and at 350–440 m and temperatures of 7.4–7.5°C in experimental crab traps. I

1. Presumably off New York and New Jersey. See footnote on page 262.

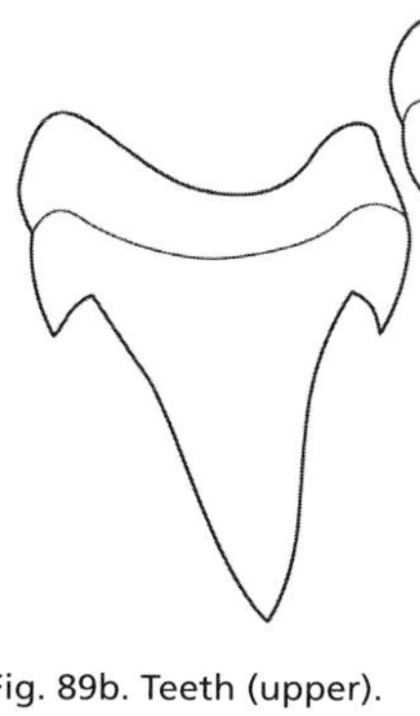

Fig. 89b. Teeth (upper).

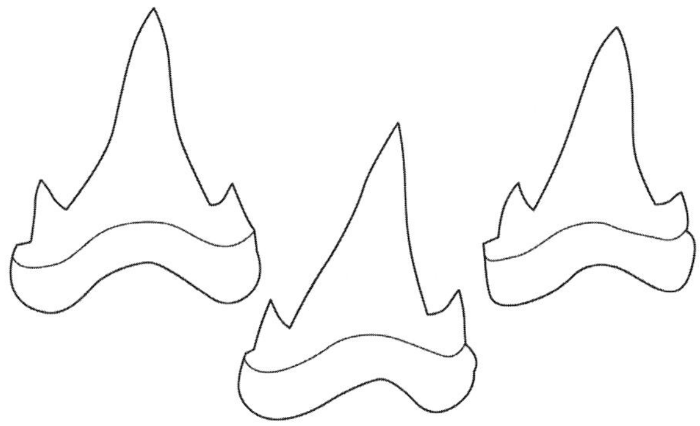

Fig. 89c. Teeth (lower).

have also examined two specimens caught in a fish trap by Mike Laudicina off Dry Tortugas at 112–122 m.

Juveniles are often caught in bottom trawls and are well represented in museum collections, but the adults are seldom taken. Springer and Sadowsky (1970) stated that all the specimens of western Atlantic *Scyliorhinus* that they could find were collected with otter trawls or dredges, and that research vessels had found fewer than two dozen mature specimens among more than a thousand taken. They correctly deduced that adult chain dogfish prefer rough bottom habitats where trawling is difficult, whereas immature sharks scatter to some extent over areas with smoother topography. Castro et al. (1988) suggested that the preference of adults for rough bottoms was due to the females seeking upright structures for egg laying. All the adults I have examined were taken in longlines set for tilefish or in fish traps.

Despite their deep and dark habitat, these sharks appear well camouflaged against the bottom rubble when seen from submersibles (see illustration in Castro et al. 1988). Why these sharks would have camouflage markings in the absence of visible light in their environment is unknown. Recently, Lisa Natanson, of the National Marine Fisheries Service Narragansett Laboratory, has discovered that chain dogfish glow under ultraviolet light. The purpose of this adaptation is unknown.

Diet: Of 81 specimens, both juveniles and adults, taken off the East Coast of the United States, 78 (96%) had food in their stomachs; 52 (64%) contained squid beaks or remains, 45 (55%) contained bony fish remains, 26 (32%) contained polychaete or other annelid worms, and 17 (21%) had crustacean remains (Castro et al. 1988). The preponderance of

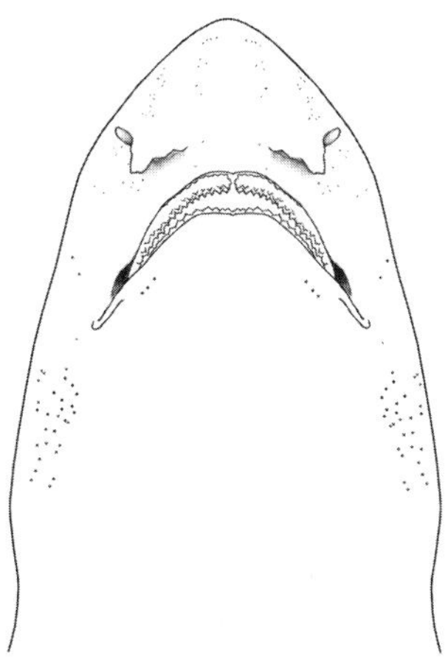

Fig. 89d. Snout.

Fig. 89e. Dermal denticles.

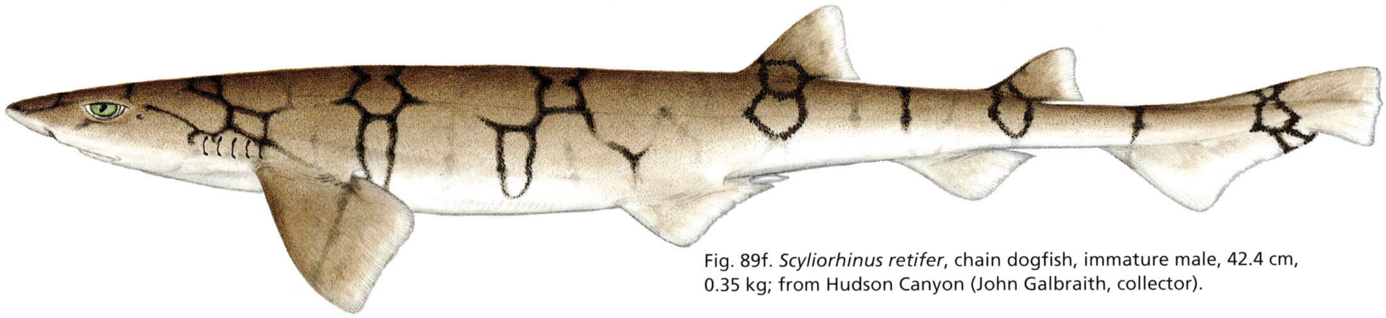

Fig. 89f. *Scyliorhinus retifer*, chain dogfish, immature male, 42.4 cm, 0.35 kg; from Hudson Canyon (John Galbraith, collector).

squid remains in the stomach contents of chain dogfish is intriguing. In aquaria the sharks spend most of their time motionless, swimming only occasionally. Likewise, chain dogfish have been observed resting on the bottom at a depth of 240 m in the Gulf of Mexico (R.G. Gilmore, pers. comm., Dec. 1994). How this seemingly sluggish shark can feed on such elusive prey as squid is unknown, for it seems unlikely that squid could always be taken by scavenging (Castro et al. 1988). Perhaps the chain dogfish is an ambush predator.

Size at maturity

Male: In the northern parts of the range, males mature at about 40 cm, based on clasper calcification (Sminkey and Tabit 1992). In the southern parts of the range (South Carolina) male chain dogfish mature at about 50 cm, based on clasper calcification and the size of the testis; mature males had hardened claspers 30 mm or more and had testis 70–100 mm in length (Castro et al. 1988).

Female: Females mature at about 39–52 cm, northern populations maturing at the smaller size while southern populations mature at the larger size (Castro et al. 1988; Sminkey and Tabit 1992). Females collected in southern New England and kept at the Mystic Marinelife Aquarium were laying eggs when they measured 41–45 cm. The smallest gravid female taken off Norfolk canyon, Virginia, measured 39 cm (Sminkey and Tabit 1992). Castro et al. (1988) used the presence of eggs in the oviduct and the size of the oviducal gland to establish maturity in the female. All specimens with eggs in the oviduct had oviducal glands at least 18 mm wide. Therefore, all specimens with oviducal glands wider than 18 mm were considered mature. At ovulation, the oviducal gland temporarily swells to about 25 mm in width, returning to its normal size of 18 mm after egg passage. The largest immature female measured 50.8 cm. A 52.0-cm specimen had a fully developed oviducal gland 18 mm wide and oocytes 16 mm in diameter; two other 52.0-cm specimens carried eggs in the oviduct. Based on these specimens, Castro et al. (1988) determined that females off South Carolina reach maturity at about 52 cm. Given that there are significant differences in size at maturity between northern and southern populations, the possibility that these are two different species needs to be investigated.

Reproduction The chain dogfish is oviparous. Castro et al. (1988) described mating in captivity. A male and a female swam actively together almost constantly, often in slow, tight circles near the bottom. The male bit the female four times in 30 minutes in the gills, flanks, or tail. At times, the female avoided the male by swimming away, but at other times it gave no indication of avoidance. Occasionally, the two sharks lay motionless on the bottom next to each other or swam with their snouts above the surface of the water with their tails down. Then, the male firmly bit the female on the tail and would not release. The female struggled violently at first but then became listless as the male progressively moved his bite anteriorly along the left flank. The male made one final bite at the left pectoral axil and then wrapped his body around the female, deflecting his left clasper to the right for copulation. Copulation ensued and the pair sank to the bottom and remained motionless for the next 30 seconds. The male then released his bite but remained coiled around the female. The female then broke away from the male, and both swam off and settled to the bottom in separate areas of the tank. The male was inactive after copulation and occasional approaches by two females failed to elicit a response. Biting behavior by the male was observed again in the following days, suggesting that mating may occur repeatedly.

Oocytes develop in pairs in the ovary, and egg cases are laid in pairs. The interval between the deposition of the first egg case of a pair and the second ranges from a few minutes to a few days, with the longest interval observed being 8 days. In captivity, females have been observed to lay one pair of egg cases roughly every two weeks. The average interval between successive pairs of eggs was measured in three undisturbed females after they had settled into an egg-laying routine. One female laid 70 pairs of eggs in 991 days, at a mean rate of one pair every 14.1 days. A second female laid 60 pairs of eggs in the same period, at a mean rate of one pair every 16.7 days. A third female laid 40 pairs in 626 days, at a mean rate of one pair per 15.2 days (Castro et al. 1988).

The eggs are enclosed in horny egg cases with long adhesive tendrils on all corners. The egg cases are 27 mm wide and 65 mm long, excluding the long tendrils, which can be as long as 35 cm. The egg cases are soft, pale yellow, and translucent when in the oviduct. After deposition, they harden and darken to a deep amber color with whitish cross bands. Each female produces egg cases of consistent size, shape, and pigmentation; thus, in captivity, egg cases can be attributed to specific females. The cross-band pigmentation tends to disappear upon preservation.

Females attach their egg cases to upright bottom structures such as corals. As egg laying approaches, the posterior pair of tendrils protrudes from the cloaca. The shark circles a vertical structure, swimming slowly and nuzzling it repeatedly. She changes direction often, thereby dragging the trailing tendrils over the structure. Adhesion of one or both of the tendrils stimulates the female to swim rapidly around the structure. As the tendrils tighten, the female pulls vigorously, causing the egg case to emerge from the cloaca. Circling continues until the egg case is secured by the anterior pair of tendrils. After the tendrils are withdrawn from the cloaca, the shark stops swimming and resumes an idle posture on the bottom (Castro et al. 1988). Females in an area tend to lay their eggs in selected structures, creating large masses of egg cases in different stages of development. Whether this is due to a scarcity of suitable structures on the bottom or whether there is some protection for the eggs in these "community nests" is unknown.

Wild females bearing egg cases have been found from March to August. However, this is simply because the availability of adult females has been limited to the months of March to September, when commercial fishing occurred in the appropriate areas and produced the specimens. Thus, the length of the egg-laying season in the wild can only be guessed.

Captive females have been shown to store sperm for extended periods. A female isolated from the male for 401 days produced an egg that hatched normally. Eggs laid 765 and 843 days after separation were developing normally after 150 and 72 days, respectively. Yet another female laid an egg 469 days after separation, although the embryo was deformed (Castro et al. 1988). Whether this capability to store sperm is functionally important in the wild is not known, as chain dogfish are quite abundant and it is difficult to envision why sperm storage would be necessary with an abundance of males.

When an egg case is deposited, the embryo is in a blastodisc stage and visible as an orange spot against the yellow yolk. Early developmental stages are similar to those of *Heterodontus* (Smith, B. G. 1942), and development is similar to that of *S. canicula* (Alluchon-Gerard and Mellinger 1971). Small slits at the anterior end of the egg cases open due to enzyme action when the embryo reaches a length of about 4.3 cm. These slits allow the exchange of water between the case interior and the surrounding seawater. The 4.3-cm embryo has greatly enlarged denticles at the base of its tail and the tip of the tail forms a small scoop. These structures, along with the embryo's almost constant fanning of the tail, serve to aerate the egg case, removing metabolites and maintaining the necessary oxygen levels. Pigmentation appears when the embryo reaches about 7 cm in length. The 7.4-cm embryo appears externally fully formed, although its yolk sac is still large. From this point onward, the yolk sac begins to be absorbed, and when absorption is complete, the embryo is ready to hatch. In the laboratory, average incubation time was 256 days at water temperatures of 11.7–12.8°C, and hatching occurred when embryos reached 10–11 cm. It is likely that development takes much longer under natural conditions, perhaps as long as one year, because the temperature in the nursery areas can be as low as 7.4°C (Castro et al. 1988).

Size at birth: The young hatch at 10–11 cm.

Brood size: Egg cases are deposited in pairs, but the numbers of egg cases produced in a season in the wild is unknown.

Nurseries The nurseries of the chain dogfish appear to extend throughout the range of the species. I have examined chain dogfish egg cases trawled from off New England to Dry Tortugas.

Age and growth Unknown.

Relation to humans The chain dogfish is often taken as bycatch in tilefish fisheries, but it has no commercial value other than as an aquarium fish. Since 1976, specimens have been displayed at Mystic Marinelife Aquarium (Mystic, Connecticut) in closed-system aquaria with chilled artificial seawater. The species reproduces easily in captivity and the offspring thrives. Its small size and easy maintenance make it an ideal laboratory shark. It would be suitable for immunological, physiological, and many other types of experiments.

OAKLEY'S CATSHARK

Scyliorhinus sp. X

Fig. 90a. *Scyliorhinus* sp X., immature female, 51.1 cm; from off Savannah, Georgia (Doug Oakley, collector).

In the late 1970s and early 1980s, I was studying reproduction of the chain dogfish, and I saw a good number of *Scyliorhinus* taken in deep water off South Carolina in research projects and tilefish operations. These specimens showed great variation in dorsal markings, ranging from light reticulating lines to wide dark lines approaching saddle blotches. The specimen illustrated here is a 51.1-cm female taken at 700 m off South Carolina (at 31°07′ N, 78°30′ W; off Savannah, Georgia) on 17 October 1985 and given to me by the late Doug Oakley along with several other chain dogfish. Although I had noticed a progression of markings in the chain dogfish that approached nearly solid markings, this specimen appeared to be different. Internal examination showed it to be very immature, at a size when southern chain dogfish females are approaching maturity. Its dermal denticles differ from those of the chain dogfish in having two parallel short ridges around the base of the main ridge. I suspect that this specimen is a juvenile of a larger species of catshark. Because I have only seen this individual, I cannot discount the possibility that it may be a chain dogfish with very dark markings. Thus, I bring it to the reader's attention, hoping that someone may obtain other specimens in the future.

Fig. 90b. Snout.

Fig. 90c. Dermal denticles.

DWARF CATSHARK

Scyliorhinus torrei Howell-Rivero, 1936

Fig. 91a. *Scyliorhinus torrei,* dwarf catshark, mature male, 29.7 cm; from the northern coast of Cuba (Museum of Comparative Zoology # 36093, courtesy of Karsten Hartel).

Common name Dwarf catshark, or Torre's catshark.

Spanish name *Gatica prieta* (Cuba).

Scientific name *Scyliorhinus torrei* Howell-Rivero, 1936, originally described as *Scylliorhinus torrei,* in *Proceedings of the Boston Society of Natural History* 41(4): 43–44, and plate 9. Later, Bigelow and Schroeder (1948) emended the spelling to *Scyliorhinus. Scyliorhinus:* from the Greek *skylion,* a dogfish, probably related to *skyllo,* to tear or mangle, and *rhinus,* from the Greek *rhinos,* nose, hence "dogfish-nose"; *torrei:* Latinized name after Carlos de la Torre (1858–1950), the foremost Cuban naturalist of his era and professor of zoology at the University of Havana.

Synonyms *Catulus boae* Sánchez Roig, 1931 (not *Scyliorhinus boa*).

Identification This small, slender shark has a *first dorsal fin that is larger than the second dorsal fin and originates posterior to the pelvic fins.* Teeth are triangular, with one larger central cusp flanked on either side by a smaller cusp, and are smooth-edged and similar in both jaws. Teeth number U: 20 to 23–20 to 23, L: 19 or 20–19 or 20. Coloration is *pale brown* above, with about seven dorsal saddle blotches of slightly darker brown spaced over the back, with *small, whitish spots evenly distributed over the back and sides,* and *lacking black spots.* The undersides are pale brown and white.

Similar species The narrowtail catshark has an identical coloration pattern, but it has only been reported from Nicaragua. It differs from the dwarf catshark in having an extremely long caudal section (the distance from the origin of its first dorsal fin to the tip of the tail is much longer than the rest of the body), and its second dorsal fin is slightly larger than the first. The other catsharks in the area lack the white spots or have black spots.

Range Specimens have been collected off the north coast of Cuba (Bigelow and Schroeder 1948), the northern Bahamas (Bullis and Thompson 1965), Jupiter Inlet on the southeast coast of Florida, and the Virgin Islands (Springer 1979a). The limits of this shark's distribution are unknown.

Size and weight Most specimens recorded measured about 25 cm (Howell-Rivero 1936; Springer 1979a), and the largest measured 29 cm (Bigelow and Schroeder 1948). The dwarf catshark probably reaches 30 cm.

Biology Dwarf catsharks have been collected in depths of 230–550 m (Bigelow and Schroeder 1948; Bullis and Thompson 1965; Springer 1979a). Reported stomach contents are squid and possibly cuttlefish. Nothing else is known about its habits.

Size at maturity No data available.

Reproduction The dwarf catshark is probably oviparous, like its congeners.

Size at birth: Unknown.

Brood size: Unknown.

Nurseries Unknown.

Age and growth The species has not been aged.

Relation to humans None.

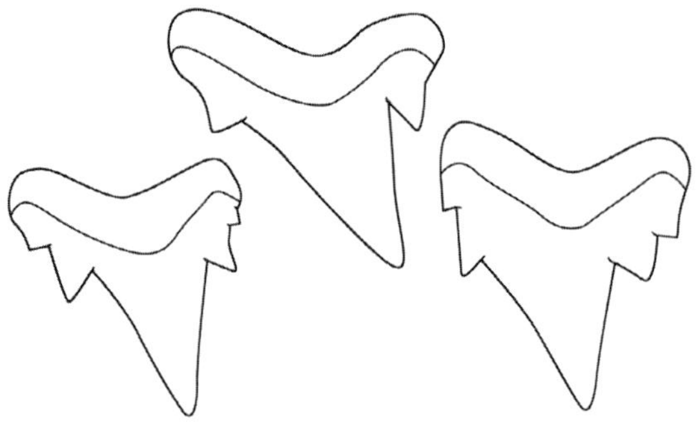

Fig. 91b. Teeth (upper).

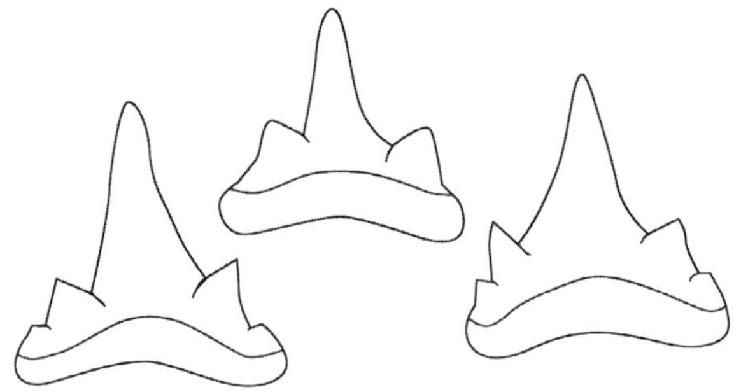

Fig. 91c. Teeth (lower).

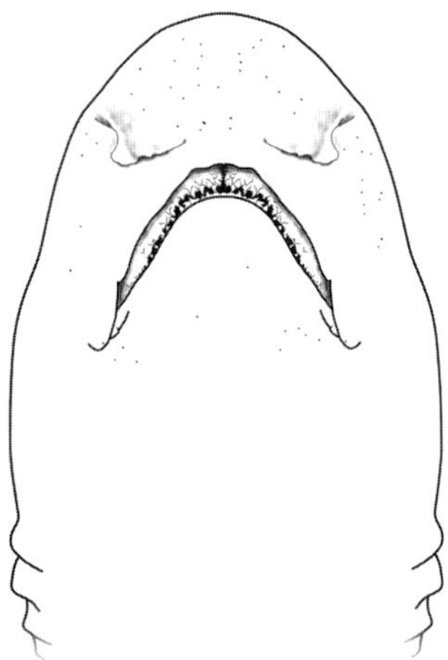

Fig. 91d. Snout.

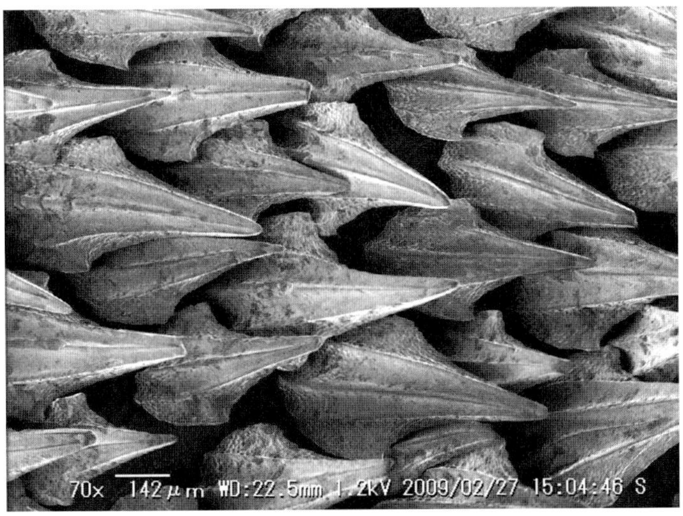

Fig. 91e. Dermal denticles.

FAMILY PROSCYLLIDAE

The Ribbontail Catsharks

This is a family of very small deep-water sharks that comprises two species. The Cuban ribbontail catshark is the only one found in North America.

CUBAN RIBBONTAIL CATSHARK

Eridacnis barbouri (Bigelow and Schroeder, 1944)

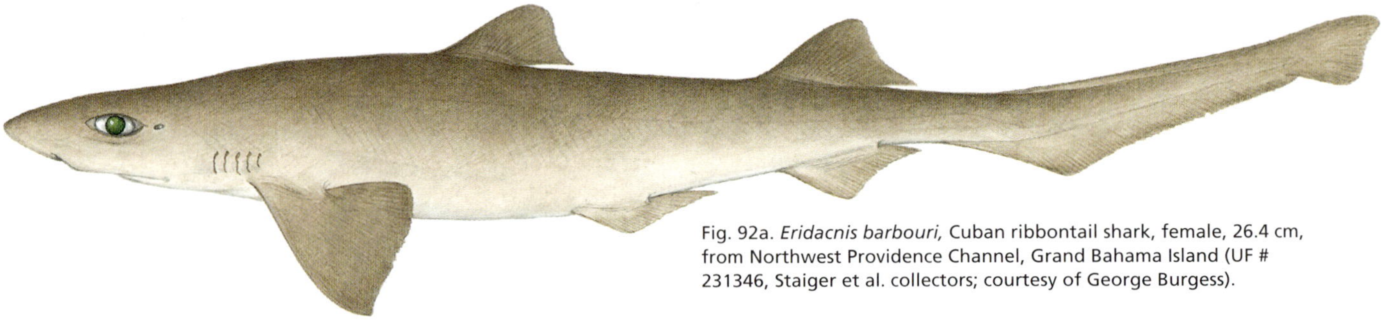

Fig. 92a. *Eridacnis barbouri,* Cuban ribbontail shark, female, 26.4 cm, from Northwest Providence Channel, Grand Bahama Island (UF # 231346, Staiger et al. collectors; courtesy of George Burgess).

Common name Cuban ribbontail catshark, named after the location where it was first found and for its elongated tail. The name catshark is a misnomer as it does not belong in the Scyliorhinidae.

Spanish name *Tiburón turquino* (Cuba).

Scientific name *Eridacnis barbouri* (Bigelow and Schroeder, 1944), originally described as *Triakis barbouri* in *Proceedings of the New England Zoölogical Club* 23: 27–30, plate 8. *Eridacnis:* from the Greek *eri,* a strengthening prefix meaning "very," and *dacnis,* from *dacno,* meaning to bite; *barbouri:* Latinized form after Thomas Barbour (1884–1946), herpetologist, wealthy patron of science, and Director of the Museum of Comparative Zoology at Harvard University (1927–1946).

Synonyms None.

Identification This little shark has a slender body, large oval eyes with a nictitating membrane, a noticeable spiracle behind the eye, short labial furrows, two equally large dorsal fins with the *first dorsal originating well ahead of the pelvic fins,* an anal fin, and an elongated caudal fin. The teeth are very small and numerous. The *front teeth have three needle-like cusps* with longitudinal striations, the central cusp being much larger than the others; the rear teeth are short and comblike. Teeth number about U: 60 to 62, L: 60 to 62. The dermal denticles have three points, with two strong ridges that converge on the point and two winglike side points. Coloration is pale gray or tan above, with darker blotches along the anterior edge of the pectoral fins and on the caudal fin, and grayish white or paler below.

Similar species The catsharks have the first dorsal fin originating over or behind the origin of the pelvic fins. The smoothhounds have short, blunt teeth and the first dorsal fin larger than the second. The requiem sharks have bladelike teeth and a well-defined caudal fin.

Range The Cuban ribbontail catshark is known only from the Straits of Florida, from south Florida to the northern coast of Cuba. It is common in deep water along the northern coast of Cuba.

Size and weight This is one of the smallest sharks. Adults measure 30–35 cm, with the largest probably reaching 40 cm and 50 g.

Fig. 92b. Teeth (upper).

Fig. 92c. Teeth (lower).

Fig. 92d. Snout.

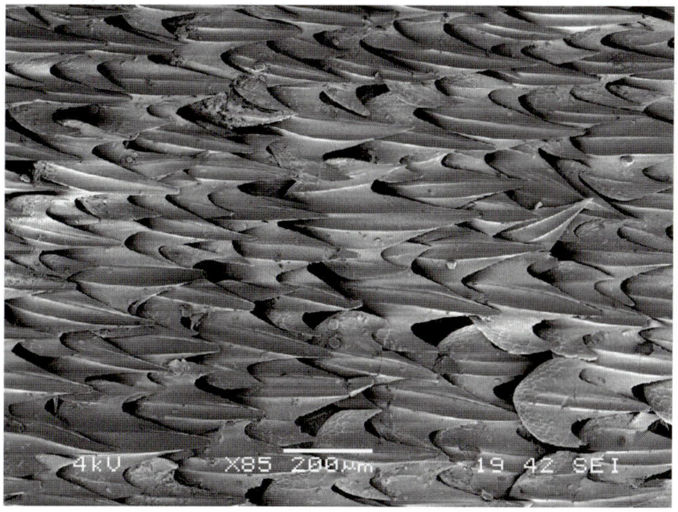

Fig. 92e. Dermal denticles.

Biology This is a poorly known species of the continental slopes. Recorded depths for the species range from 430 to 610 m, but since the trawls that yielded them were hauled to the surface open, it is not known whether the specimens were swimming at middle depths or near the bottom (Bigelow and Schroeder 1948). Nothing else is known of its habits.

Size at maturity The sizes at maturity of Cuban ribbontail catsharks are not known. The 30.3-cm female mentioned above was mature and contained two embryos, while a male 33.8 cm had well-developed claspers (Bigelow and Schroeder 1948).

Reproduction The Cuban ribbontail is aplacental viviparous. A 30.3-cm female contained two embryos 9–10 cm long, apparently ready for birth (Bigelow and Schroeder 1948).

Size at birth: The young appear to be born at about 10 cm (Bigelow and Schroeder 1948).

Brood size: The only gravid specimen reported contained two young (Bigelow and Schroeder 1948).

Nurseries Unknown.

Age and growth The species has not been aged.

Relation to humans None.

FAMILY PSEUDOTRIAKIDAE

The False Catshark

The Pseudotriakidae is a monotypic family; its sole representative is the false catshark, a rarely seen, large, deep-water shark characterized by a very long and low first dorsal fin that is as long as the caudal fin. Other characteristics include small gill slits, large spiracles, a nictitating lower eyelid, and the presence of an anal fin. In the past, this family was often included within the catsharks (Scyliorhinidae), from which it differs in the shape and position of the first dorsal fin. Development is aplacental viviparous with an interesting type of oophagy. For a long time, two species were usually recognized, *Pseudotriakis microdon* de Brito Capello, 1868, of the Atlantic Ocean, and *P. acrales* (= *acrages*) Jordan and Snyder, 1904, of the western Pacific. The validity of *acrales* was questioned shortly after its description, but the name survived many years, perhaps on Jordan's authority. Garman (1913: 105) called it "hardly distinct from *P. microdon.*" Cadenat and Blache (1981) suggested that *P. acrales* was synonymous with *P. microdon* because of the high variation among the traits used to distinguish between them. Subsequent authors (Compagno 1984; Taniuchi et al. 1984; Yano and Musick 1992) reported that the supposed differences between the two species do not hold and have agreed that the two species are synonymous, with *Pseudotriakis microdon* being the senior name.

FALSE CATSHARK

Pseudotriakis microdon de Brito Capello, 1868

Fig. 93a. *Pseudotriakis microdon,* false catshark, mature male, 224 cm, 32.5 kg; from Mid-Atlantic Ridge (*MAR-ECO Cruise* 2004, Station 11; Chip Cotton, collector).

Common name False catshark, a translation of *Pseudotriakis.* According to de Brito Capello (1868), the type specimen resembled the genus *Triakis* but differed from it in the absence of a nictitating eyelid (see below), and it could not be placed in any of the other genera in the Scylliidae (catsharks). Thus, he was forced to create a new genus for the species, choosing to call it *Pseudotriakis.*

Spanish name The species lacks a vernacular name in Spanish America due to its rarity. The name *musolón* has been used in Spain (Moreno 1995). The A.F.S. has not designated a name for it.

Scientific name *Pseudotriakis microdon* de Brito Capello, 1868, in *Jornal de Sciencias Mathematicas, Physicas e Naturaes* (Lisbon) 1: 316–323, plate 5; pp. 316–317 contain the description in Portuguese; pp. 321–323 repeat the description in French. *Pseudotriakis:* from the Greek *pseudes,* false, and *triakis,* from the Greek *triacis, tri,* three, plus *acis* (*akis*), point, an allusion to the three-pointed teeth of *Triakis,* hence a false triakid; *microdon:* from the Greek *mikros,* small, plus *odontos,* tooth, thus "small-toothed."

Synonyms *Pseudotriakis acrales* Jordan and Snyder, 1904; *Pseudotriakis acrages* Garman, 1913.

Identification The false catshark is characterized by a very short snout, its length about one-half the mouth width, and by *a very long and low first dorsal fin* (five to seven times longer than high), which is *as long as the caudal fin* and located anterior to the origin of the pelvic fins. The *teeth are minute* and extremely numerous. The upper teeth have three cusps, the central cusp being much larger; the lower teeth have three to five cusps, one cusp being only slightly larger than the others. The teeth are arranged in perpendicular files in the upper jaws and in diagonal files in the lower jaws. The dermal denticles are sparse and widely separated; they have a strong central ridge and resemble arrowheads. For a good description and drawings of the skeleton, teeth, and denticles, see Jaquet (1905). Most specimens have been described as being chocolate brown with the tips of the fins slightly darker. However, some specimens are dirty white or pale gray with darker gray mottling and numerous darker gray minute spots, presenting a mottled appearance. The female illustrated here, caught in the North Atlantic, is an example of this color pattern. It was photographed while alive, and the colors of the illustration match its live coloration. Eugenie Clark (pers. comm., November 2004) saw and filmed a similarly light-colored specimen from the submersible *Nautile* in Suruga Bay, Japan. The color changes to gray after freezing.

Contrary to de Brito Capello's (1868) description, the false catshark has a movable lower eyelid that allows it to cover the eye when biting. I recently watched a video recording, kindly sent by Chip Cotton (Virginia Institute of Marine Science), of the blinking reflex of the false catshark.

Similar species The long and low first dorsal fin distinguishes the false catshark from all other sharks. The catsharks have the first dorsal fin located over or posterior to the pelvic fin origin.

Range The false catshark appears to be cosmopolitan in very deep waters. The type specimen was caught off Setúbal, Portugal (de Brito Capello 1868), where deep-water fisheries have existed for centuries. Since then, it has been reported from both sides of the North Atlantic (e.g., New York, Bean 1884; Nova Scotia, Gilhen and Coad 1999; Cape Verde, Jaquet 1905; Iceland, Saemundsson 1949; France, Quéro et al. 1976), from the western Pacific (Sea of Kumanonada, Japan, Taniuchi et al. 1984; Okinawa and New Zealand, Yano 1992), the central Pacific (Hawaii, Tester 1969), and the Indian Ocean (Madagascar, Forster et al. 1970; Aldabra, Ellis 1975; Australia and Indonesia, Kyne

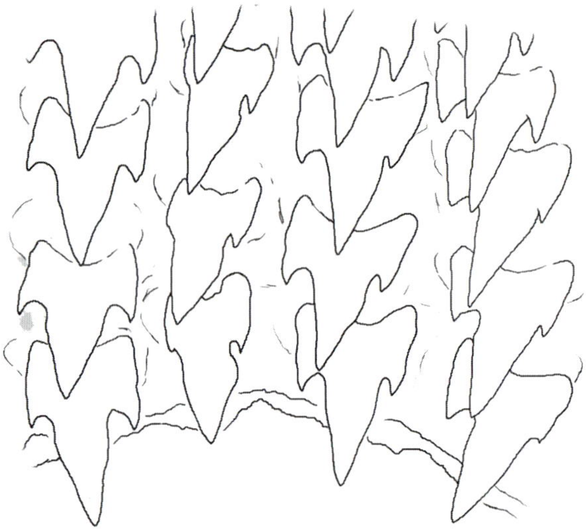

Fig. 93b.Teeth (upper).

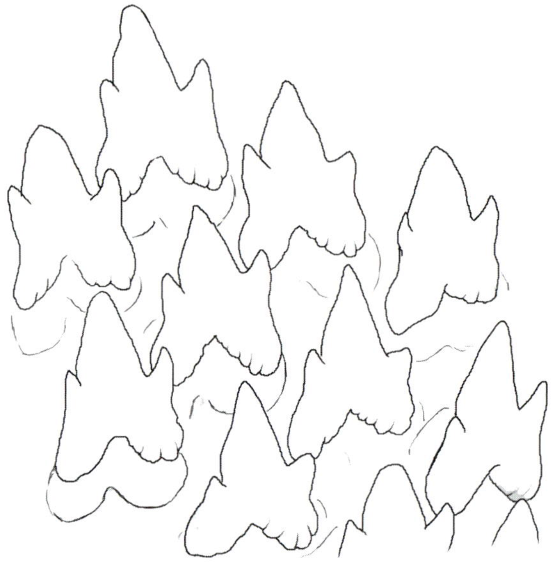

Fig. 93c.Teeth (lower).

et al. 2005). Its distribution is undoubtedly much wider than the records suggest, but its deep-water habitat usually protects it from commercial and research longlines. When specimens are taken, they may not come to the attention of biologists and are simply not recorded.

There are two published North American records. The first record dates back to February 1883, when a 295-cm specimen washed ashore in Long Island, New York; it was retrieved and taken to the U.S. National Museum (Bean 1884). The second specimen was not recorded until 115 years later (Gilhen and Coad 1999). There is an unpublished record of a specimen taken in 550 m (300 fathoms) in Corsair Canyon, some 300 km east of Nantucket in 1969 (letter of S. Springer to G. R. Case, dated 12 September 1977, courtesy of J. Seitz).

Size and weight The false catshark appears to reach a maximum size of 300 cm and 125 kg, although this is based

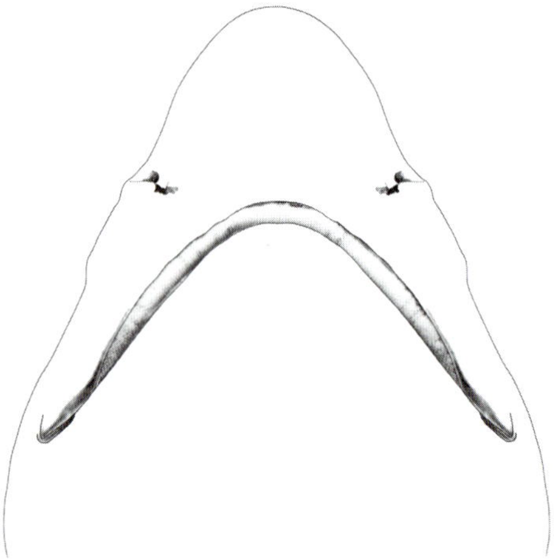

Fig. 93d. Snout.

Fig. 93e. Dermal denticles.

Fig. 93f. *Pseudotriakis microdon,* false catshark, *leucistic* female, 249 cm, 64.5 kg; from Mid-Atlantic Ridge (*MAR-ECO Cruise* 2004, Station 11; Chip Cotton, collector).

on limited data. The largest specimens reported measured slightly less than 300 cm: 295 cm (New York, Goode and Bean 1896), 295 and 296 cm (Okinawa, Yano 1992). One 295-cm specimen from Okinawa weighed 100.2 kg, while a 275-cm specimen from New Zealand weighed 121.1 kg. The 224-cm male illustrated in this work weighed 35 kg, while the 249-cm female illustrated weighed 65 kg.

Biology The false catshark is a rarely seen, bottom-dwelling, deep-water species that has been reported from 360 to 1,900 m. Most catches have occurred in 500–1,400 m, indicating its very deep habitat. It occasionally wanders into shallower water. In the Atlantic, the false catshark has been taken at 1,477 m off Cape Verde (Jaquet 1905), at 1,350 and 1,890 m between Tenerife and Gran Canaria (Yano and Musick 1992), and at depths of 883–1,358 m on the MAR-ECO cruise of 2004 in the North Atlantic. In the Pacific Ocean, most specimens have been taken in shallower waters. Two specimens were caught at 360–468 m in the Sea of Kumanonada, Japan (Taniuchi et al. 1984), and 15 were caught at 530–990 m off Okinawa and New Zealand (Yano 1992). In the Indian Ocean, two specimens were taken at 500–599 m and one at 800–899 m (Forster et al. 1970).

Although the false catshark is widely distributed, it is seldom seen or caught because of its very deep habitat. Fewer than a dozen specimens had been reported until the 1980s. One was photographed at a depth of 640 m off Aldabra atoll when it took the bait and triggered a camera (Ellis 1975). In 1975 five specimens were collected off the French Atlantic coast (Quéro et al. 1976). In the 1980s, 15 specimens were collected around the Okinawa Islands and New Zealand (Yano 1992). In 2004, four specimens were captured in the North Atlantic between the Azores and Iceland during the MAR-ECO cruise.

Almost nothing is known about the habits of the false catshark. Its chocolate brown coloration is typical of deep-dwelling sharks. The presence, or survival, of whitish or pale gray individuals in the population also indicates a deep habitat where no light penetrates.

There are few reports of its diet. Given the minute teeth of the false catshark (about 1 mm wide and 1.3 mm long in a 250-cm specimen), one can only wonder about its diet (upon seeing the teeth for the first time, one is reminded of the teeth of plankton eaters!). The two specimens I have examined contained fish and squid eye lenses in their stomachs. The stomachs of North Pacific specimens contained bony fishes, squids, and small sharks (*Etmopterus*). One Atlantic specimen had human garbage in its stomach, consisting of potatoes, a pear, a plastic bag, and a soft drink can (Yano and Musick 1992).

Size at maturity

Male: Males are believed to mature at about 255 cm. A 250-cm male was reported as immature with soft claspers, while males 261–295 cm were mature and had hard claspers (Yano 1992).

Female: Females probably mature at about 245 cm. This estimate is based on a 249-cm female that I examined that was obviously postpartum (with distended oviducts and a thin ovary). Only a few gravid females have been reported, all more than 256 cm (Indian Ocean, Forster et al. 1970; Japan, Taniuchi et al. 1984; western Pacific, Yano 1992; North Atlantic, Saemundsson 1949).

Reproduction The false catshark is an aplacental species, and its embryos are nourished by an interesting type of oophagy. Forster et al. (1970) first suggested that the false catshark embryos are oophagous based on finding an ovary with an estimated 20,000 small vitellogenic oocytes (each about 9 mm in diameter). Later, Taniuchi et al. (1984) denied the possibility of oophagy, based on their finding external and internal yolk sacs in the near-term embryo. At the time, late-term embryos of oophagous sharks, such as the sand tiger and the mako, were not known to have yolk sacs, having instead internal yolk stomachs. Some years later, Yano (1992) reported that the stomachs of two embryos examined contained yolk and egg capsules. Yano also reported that midterm, 45-cm embryos had much larger yolk sacs than the early stage embryos reported by Forster et al. (1970) and the late-term embryos reported by Taniuchi et al. (1984). Yano also observed that the embryos did not have stomachs full of yolk like other oophagous embryos, but their intestines were full of yolk. He then concluded that embryos appear to transfer ingested yolk to the external yolk sac for storage. Thus, the aplacental viviparity and oophagy of the false catshark is a different type than those previously described for lamnoid sharks. All of these observations are evolutionarily quite interesting, as oophagy had only been reported in lamnoid sharks.

Size at birth: The size at birth appears to be about 120 cm, based on the largest embryos reported. I have not been able to find reports of small free-swimming young. The 93-cm specimen illustrated in Bigelow and Schroeder (1948) appears to have been an embryo. Taniuchi et al. (1984) reported that embryos 112–113 cm still had 50 cm yolk sacs attached by a 7.5-cm yolk stalk. Yano (1992) reported "near term" embryos that measured 116 and 120.2 cm.

Brood size: The few broods reported consisted of two young, one in each uterus (Yano 1992).

Nurseries Unknown.

Age and growth The species has not been aged.

Relation to humans None. Catches are rare and should be preserved for scientific study.

FAMILY TRIAKIDAE
The Smoothhound Sharks

The Triakidae, also known as the smoothhounds (sometimes spelled "smooth hounds"), smooth dogfishes, hound sharks, or gummy sharks, is a large family of small to medium-sized carcharhiniform sharks (adults usually measuring from one to less than two meters) that inhabit shallow to moderately deep water over the continental shelves. These sharks are characterized by slender bodies, a head that is not laterally expanded like the hammerheads, large oval eyes, a variably developed nictitating eyelid, a large spiracle, teeth with low, blunt cusps, large spineless dorsal fins, a large second dorsal fin about three-quarters the size of the first dorsal fin, no precaudal pits, and a short and broad caudal fin. The smoothhounds have a spiral intestinal valve. Development is viviparous, most species being placental. The smoothhounds feed primarily on crustaceans, which they crush with their molariform teeth. Some species are of local economic importance in "artisanal" fisheries despite their small size, because, when present, they are abundant.

The Triakidae are closely related to the Carcharhinidae, from which they differ in dentition, absence of a true nictitating eyelid, lacking precaudal pits, and having a spiral intestinal valve (instead of a scroll type valve). There are sharks, such as the soupfin shark, that possess characteristics of both families and are considered transitional. There are nine genera and more than 40 species in this family. Three genera are represented in North America: *Galeorhinus, Mustelus,* and *Triakis.*

Galeorhinus is considered to be a monotypic genus; its only species is the soupfin shark or tope, a widely distributed shark of temperate coastal waters. *Galeorhinus* is distinguished by a large terminal lobe of the caudal fin, its length about one-half of the upper caudal margin; a second dorsal fin as large as the anal fin; and bladelike teeth that are similar in both jaws.

The differences between the genera *Mustelus* and *Triakis* are very subjective, and it is difficult to separate the two genera by external morphological characters, although our only species of the genus *Triakis,* the leopard shark, is easily separated from the other sympatric smoothhounds by its striking markings. According to Compagno (1988), sharks of the genus *Mustelus* have a mouth triangular or nearly triangular in shape, lower jaws with straight or nearly straight edges, and lower teeth that prominently overlap into the ventral surface of the lower jaw and are more or less enlarged at the lower symphysis, forming a knob (these jaw details are best seen in a cleaned and dried jaw). The genus *Mustelus* is by far the largest genus in the Triakidae, with some 25 species. Of these, eight described species inhabit North America; five are found in the western coast, and three along the eastern coast. It is likely that there are undescribed species of deep-water *Mustelus* in the warmer latitudes of North America (off Florida and the Gulf of Mexico, and off the Baja Peninsula and the Mexican Pacific coast).

Sharks of the genus *Triakis* have arcuate jaws, a lower jaw with convex edges, and teeth that hardly overlap onto the ventral surface of the jaw and do not form a knob at the lower symphysis (Compagno 1988). There are only five or six species in the genus, only one of which inhabits the western coast of North America.

The Atlantic weasel shark, *Paragaleus pectoralis,* inhabits the coastal waters of tropical western Africa. One specimen was reported from New England by Garman in 1906, who presumably obtained it fresh from an aquarium, the "Aquarial Gardens," "the collections for which were made off the coasts of Massachusetts and Rhode Island" (Garman 1913: 150). The specimen was depicted by Bigelow and Schroeder (1948). Whether the specimen was a stray or waif, or transported to North America in some way, cannot be determined now, more than 100 years later. It is not an American species, and no one else has encountered the species in North American waters. Thus, I have chosen not to include it in this account.

KEY THE FAMILY TRIAKIDAE

Pacific Coast Species

1a. Dorsal surfaces with numerous dark saddle barks or dark spots.. **Leopard shark,** *Triakis semifasciata* (p. 384).

1b. Dorsal surfaces unmarked, lacking any dark markings 2

2a. Second dorsal fin and anal fin equal in size (with a very wide terminal lobe on the caudal fin, about half the length of the upper caudal fin) .... **Soupfin shark,** *Galeorhinus galeus* (p. 361).

2b. Second dorsal fin much larger than anal fin 3

3a. Dorsal and caudal fins with dark and frayed trailing edges.. **Brown smoothhound,** *Mustelus henlei* (p. 374).

3b. Dorsal and caudal fins trailing edges not dark and frayed.. 4

4a. Teeth with pointed cusps; length of nostrils 2.2–2.7% total length and about 84–95% of internarial space (the space between the inner ends of the nostrils) (midpoint of first dorsal fin base about equidistant from pectoral fin axil and pelvic fin origin) .... **Sharptooth smoothhound,** *Mustelus dorsalis* (p. 372).

4b. Teeth with low, blunt, or rounded cusps, length of nostrils 1.1–2.3% total length and 34–77% of internarial space..... 5

5a. Upper and lower labial furrows about equal in length, midpoint of first dorsal fin base closer to pelvic fin origin that to pectoral fin ax.................................... **Gray smoothhound,** *Mustelus californicus* (p. 366).

5b. Upper labial furrows shorter than lower furrows, midpoint of first dorsal fin about equidistant from pectoral fin axil and pelvic fin origin 6

6a. Upper labial furrows shorter than lower furrows, internostril distance narrow about 36–44% of snout length.. **Sicklefin smoothhound,** *Mustelus lunulatus* (p. 377).

6b. Upper labial folds longer than lower folds, internostril distance wide, 45–58% of snout length (fins with white tips or whitish rear edges).................................... **Whitetip smoothhound,** *Mustelus albipinnis* (p. 364).

Atlantic Coast Species

The separation of the species of *Mustelus* inhabiting the eastern coast of North America is problematic. The diagnostic characters used in the species descriptions are variable and tend to overlap. To complicate matters, it is almost certain that there are undescribed species of *Mustelus* in the area. The keys provided here can only be considered preliminary or provisional.

1a. Internarial distance (the space between the inner ends of the nostrils) of 2.3–2.8% of total length, upper and lower labial folds about equal in length, caudal fin often with a sharp-pointed lower lobe, adults reaching less than a meter.. **Florida smoothhound,** *Mustelus norrisi* (p. 379).

1b. Internarial space 2.6–3.6% of the total length, upper labial folds longer than lower jaw folds, adults reaching more than a meter in length 2

2a. Skin denticles mainly lanceolate or weakly tridentate, precaudal vertebrae number 86–90, internarial distance of 2.7–3.6% of total length .... **Smooth dogfish,** *Mustelus canis* (p. 369).

2b. Skin denticles with three to six ridges on the surface and three distinct points on their rear margins, precaudal vertebrae number 68–71, internarial distance of 2.6–3.3% of total length.......... **Gulf smoothhound,** *Mustelus sinusmexicanus* (p. 382).

SOUPFIN SHARK

Galeorhinus galeus (Linneus, 1758)

Fig. 94a *Galeorhinus galeus*, soupfin shark, female, 163.5 cm, 20.5 kg; from off Palos Verdes, California (Dan Pondella, collector).

Common name Soupfin shark, a reference to the use of its fins for shark fin soup. The names oil shark and vitamin shark were also used in the older literature. In Australia, it is known as the school shark. The A.F.S. name used to be soupfin shark. When the scientific name of the soupfin was changed recently, the A.F.S. decided to also change the common name to "tope," a name that no one had ever used in the United States. Tope is one of the two British names for the species, the other being sweet William. The name soupfin shark is so well established in American and Canadian usage and literature that I have elected to retain it, instead of adopting the new A.F.S. name. Common names follow usage and custom, not the rules of priority of scientific nomenclature. These common names are part of a people's culture and language and cannot be changed whimsically by a committee (see below).

Spanish name *Tiburón vitamínico* (Mexico and South America). The A.F.S. name is *tiburón aceitoso,* a very novel creation ("oily shark").

Scientific name *Galeorhinus galeus* (Linneus, 1758), originally described as *Squalus galeus* in *Systema Naturae,* 10th ed. I: 234. *Galeorhinus:* from the Greek *galeos,* a shark, and *rhinus,* Greek name of a shark with a rough skin; *galeus:* from the Greek *galeos,* a shark.

The soupfin shark, *Galeorhinus zyopterus* Jordan and Gilbert, 1883, was considered to be a separate species from the more widely distributed tope (*Galeorhinus galeus*) until recently. Compagno (1984) synonymized the name *G. zyopterus* with *G. galeus.* This change was adopted by the scientific community, although there have been no studies comparing the different populations of *G. galeus.* Nowadays, the soupfin shark is believed to be one of the widely separated populations of *G. galeus,* and the name *Galeorhinus zyopterus* is considered a junior synonym of *G. galeus.*

Synonyms There are many synonyms of *G. galeus.* The following is only a partial list of the more common and recent synonyms: *Galeus australis* MacLeay, 1881; *Galeus zyopterus* Jordan and Gilbert, 1883; *Carcharhinus cyrano* Whitley, 1930; *Galeorhinus vitaminicus* de Buen, 1950.

Note: The soupfin shark possesses characteristics of both the Triakidae and the Carcharhinidae, demonstrating that these two families are closely related. It is often included in the Carcharhinidae.

Identification The soupfin shark has *a prominent spiracle* behind the eye, a *second dorsal fin origin located directly above the origin of the equal-sized anal fin,* and a very *wide caudal fin upper lobe.* The teeth have slightly *oblique, smooth-edged cusps with a deep notch on their outer margins and four to five cusplets from the notch to base;* they are similar in both jaws and number about U: 36, L: 34. The dermal denticles have three points, with three strong parallel ridges all terminating in a point. Its coloration is dark brownish gray to dark gray above, fading to white below.

Similar species In the smoothhounds, the second dorsal fin is much larger than the anal fin, and it originates well ahead of the anal fin origin. The Pacific sharpnose shark has the second dorsal fin origin well behind the anal fin origin, and its teeth lack the basal cusplets.

Range The soupfin shark is widely distributed in temperate waters throughout the world (Compagno 1984b). Interestingly, it is absent from eastern North America and eastern Asia. In North America, it ranges from northern British

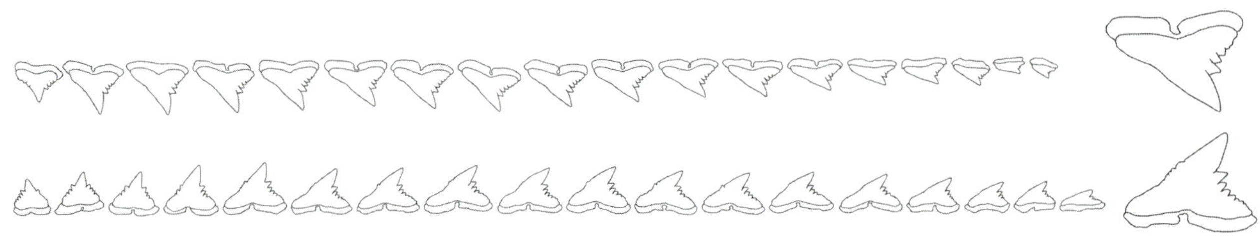

Fig. 94b. Teeth.

Columbia (Hart 1973) to at least the tip of the Baja California, where I have seen a few scattered specimens.

Size and weight Soupfin males may reach 182 cm and 27 kg; females may reach 198 cm and 45 kg (Bane and Bane 1971).

Biology The soupfin shark is a coastal species usually found at depths of less than 40 m, although it sometimes found as deep as 200–300 m. Although the soupfin shark was the object of an intensive fishery from 1938 to about 1950, we know relatively little about it. Most of the literature on the soupfin shark is old and deals primarily with the fishery and oil production and contains little biological information. Nearly all the biological information we have on North American specimens comes from Ripley's work (1946); there are no recent studies of the species in this continent.

The soupfin shark is an opportunistic feeder, often preying on sardines, flounders, midshipman, rockfish, mackerel, and squid. It migrates north in summer and south in winter, but its movements are not well defined.

The south Australian population is intensively fished, and it is the most studied. Consequently, much has been learned about that population. Olsen (1953, 1954, 1984) has published extensively on the species (as the school shark, *Galeorhinus australis*), reporting its migrations and aging.

Size at maturity Males mature at about 140–155 cm (Ripley 1946; Hart 1973). Females mature at 155–170 cm (Ripley 1946).

Reproduction The soupfin shark is aplacental viviparous. Fertilization occurs in the spring, and gestation lasts about a year (Ripley 1946). Females release their young from May to July.

Size at birth: The young measure 35–37 cm at birth (Ripley 1946).

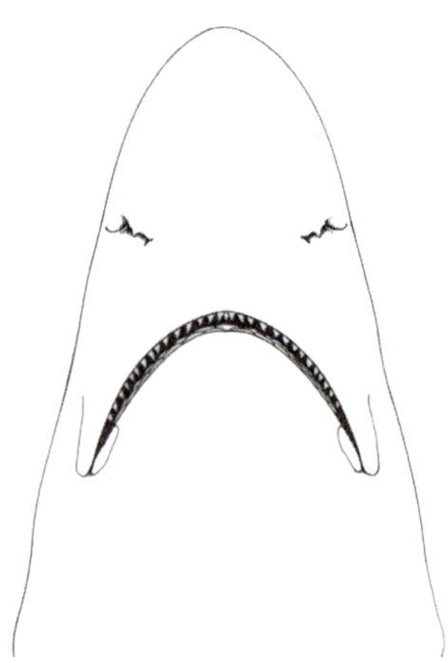

Fig. 94c Snout.

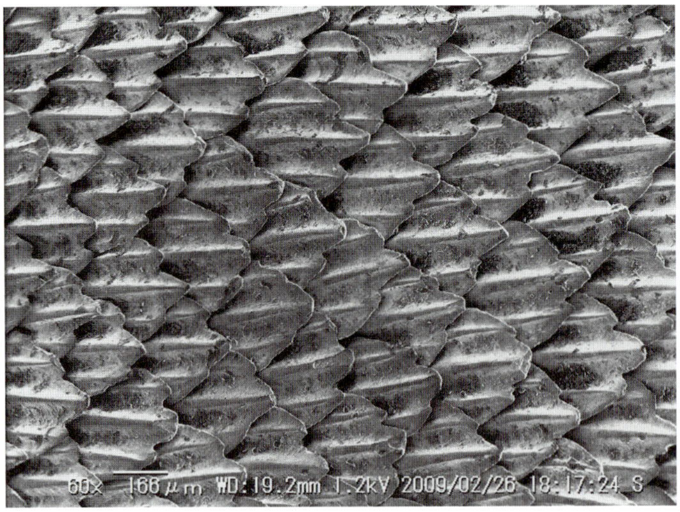

Fig. 94d. Dermal denticles.

Brood size: Soupfin shark broods range from 6 to 52, with an average of 35 (Ripley 1946).

Nurseries According to Ripley (1946), some young are found in Tomales and San Francisco bays, but most are found south of Point Conception. Ripley also reported that fishing trawlers sometimes took young soupfin sharks in considerable numbers in Ventura Flats east of Santa Barbara.

Age and growth The soupfin shark of North America has not been aged. However, Australian specimens (previously known as *Galeorhinus australis*) are known to be long-lived. Tagged specimens have been recaptured after more than 25 years at liberty, and it is believed that the species may reach 40 years of age (Grant et al. 1979).

Relation to humans The soupfin shark used to be the most economically important shark of the West Coast. In the early twentieth century, the soupfin shark was the most valuable of market sharks in California fins because the Chinese considered its fins superior to those of all other California sharks (Walford 1935). In the late 1930s, the main source of vitamin A was cod liver oil, which was imported from European sources. In 1937, it was discovered that soupfin shark liver was the richest source of high-potency vitamin A. In 1938, the outbreak of World War II in Europe, and the resulting maritime war, disrupted fishing operations and thus curtailed the supply of vitamin A. The discovery of the soupfin shark as a source of vitamin A and the curtailment of European supplies combined to create a new market for sharks and a bonanza for fishermen along the western coast of North America. (For an excellent novella about shark fishing in that era, read *The Forty Fathom Bank,* by Les Galloway [1994].) By 1940, an intensive fishery had developed from British Columbia to California for soupfin sharks and their vitamin A–rich livers, which brought as much as $13 a pound. "At one time each soupfin brought aboard was the equivalent of $50 hauled out of the sea" wrote Roedel and Ripley (1950: 24), *in 1943 dollars!* During World War II, the soupfin shark and the spiny dogfish were the two main sources of vitamin A for the United States and allied nations (Roedel and Ripley 1950). By early 1948, the average price paid per pound of liver was $10.64; by July the price of high-potency liver had risen to $14.27. The dried fins sold for $2.50 per pound.

Due to the taking of pregnant females and young in the nursery areas, the California soupfin fishery was depleted by 1946 (Ripley 1946). With the decline in the soupfin shark fishery, the industry shifted to spiny dogfish and to imports from Mexico and South America (Ripley 1946). By 1950, shark liver imports and the synthesis of vitamin A had sounded the death knell for the fishery (Nakatsu 1957).

Today, more than 60 years after the unrestricted fishing in the 1940s, the population of soupfin sharks has not recovered, and it has no economic importance in California. A recent, comprehensive publication on the marine resources of California and their utilization by Leet et al. (1992) lists several cartilaginous fishes: angel shark, leopard shark, thresher shark, shortfin mako, blue sharks, and even the cow sharks. Soupfin sharks are not mentioned.

Readers desiring more information beyond the brief summary above can turn to Ripley's (1946) excellent account of the soupfin shark and the California fishery, or to Nakatsu's (1957) brief review of the Pacific coast fishery. For an account of the soupfin fishery in Oregon in 1949, see Westrheim (1950). For a similar account for the State of Washington, see Bonham and Sanford (1949), and for the British Columbia fishery, see Barraclough (1948). There is also an account of the fishery and utilization in Uruguay (de Buen 1952).

WHITETIP SMOOTHHOUND

Mustelus albipinnis Castro-Aguirre et al., 2005

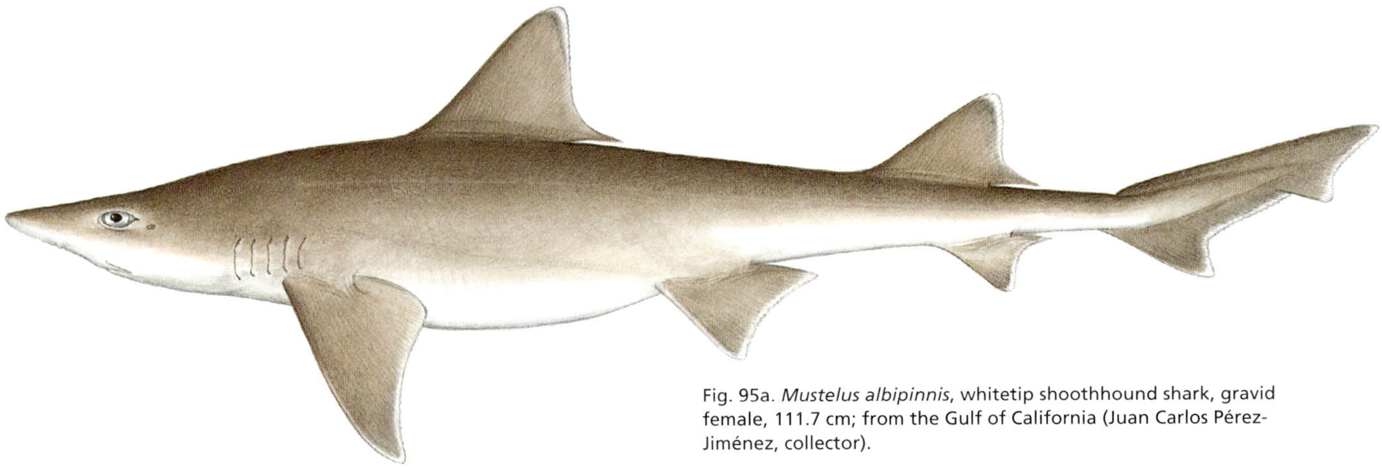

Fig. 95a. *Mustelus albipinnis*, whitetip shoothhound shark, gravid female, 111.7 cm; from the Gulf of California (Juan Carlos Pérez-Jiménez, collector).

Common name Whitetip smoothhound, a reference to the white borders of its fins.

Spanish name In their description of the species, Castro-Aguirre et al. proposed the common name "*perro lodero con aletas de bordes blancos,*" an unusually long and cumbersome name that rivals the uniquely long list of names behind the scientific name. It is unlikely that this name will be used. Names such as *mamón puntas blancas* or simply *puntas blancas* are more likely to evolve.

Scientific name *Mustelus albipinnis* Castro-Aguirre, Antuna-Mendiola, González-Acosta, and De la Cruz-Agüero, 2005, in *Hidrobiológica,* 15(2 especial): 123–130. *Mustelus:* from the Latin *mustela,* a weasel, a name used since antiquity for both mammals and fishes, a reference to the shark's long, weasel-like body; *albipinnis:* from the Latin *albus,* white, and *pinnis,* fin or wing, thus "white fin," an allusion to the white tips on the fins.

Synonyms *Mustelus hacat* Pérez Jiménez et al., 2005.

Description The whitetip smoothhound has the midpoint of the first dorsal fin equidistant or slightly closer to the pectoral axil than to the origin of the pelvic fins, and a second dorsal fin much larger than the anal fin. The fin edges are not frayed, and the fin fibers (ceratotrichia) are not visible. The upper labial folds are longer than lower jaw labial folds. The upper teeth have a low, asymmetrical, blunt-pointed cusp; the lower teeth have rounded (molariform) cusps. The dermal denticles are lanceolate and overlapping and have three to four weak ridges. Color in life is brown above and paler below. The *apex and posterior edges of the dorsal, pectoral, pelvic, anal, and lower lobe of the caudal fins are white.*

Similar species The other smoothhounds in the area lack the white fin tips and edges.

Range The whitetip smoothhound appears to be widely distributed around the Baja Peninsula, but the limits of its distribution are unknown. The type and type-series specimens were taken off Bahia Magdalena, in the Pacific coast of the Baja Peninsula (Castro-Aguirre et al. 2005). Other specimens (those described as *Mustelus hacat*) were taken off Angel de la Guardia Island, in the Gulf of California (Pérez Jiménez et al. 2005a).

Size and weight The gravid female illustrated here (the type specimen of *Mustelus hacat*) measured 113 cm and weighed 7.2 kg. Males may reach at least 109 cm and 4 kg, and females reach at least 129 cm and 10.4 kg (Pérez Jiménez 2006).

Biology This recently described species has been taken 100–281 m around the Baja Peninsula, where it appears to be the deepest dwelling of the smoothhounds (Pérez Jiménez et al. 2005a).

Size at maturity Males mature at 90–99 cm while females mature at 104–109 cm (Pérez Jiménez 2006).

Reproduction The whitetip smoothhound is a placental viviparous species.

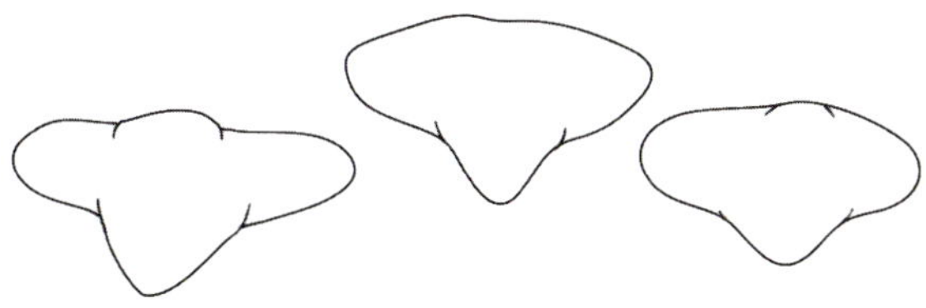

Fig. 95b. Teeth (upper).

Fig. 95c. Teeth (lower).

Size at birth: The only embryos reported measured 25–34 cm in March. The young are probably born at 30–35 cm (Pérez Jiménez et al. 2005a).

Brood size: Broods range from 3 to 23, with a mean size of 16 for nine broods. It is likely that the female with only three young had aborted some of her embryos (Pérez Jiménez et al. 2005a).

Nurseries Unknown.

Age and growth The whitetip smoothhound has not been aged.

Relation to humans The whitetip smoothhound is occasionally taken by trawlers in the northern Gulf of California, but it is of little if any economic importance.

Fig. 95d. Snout.

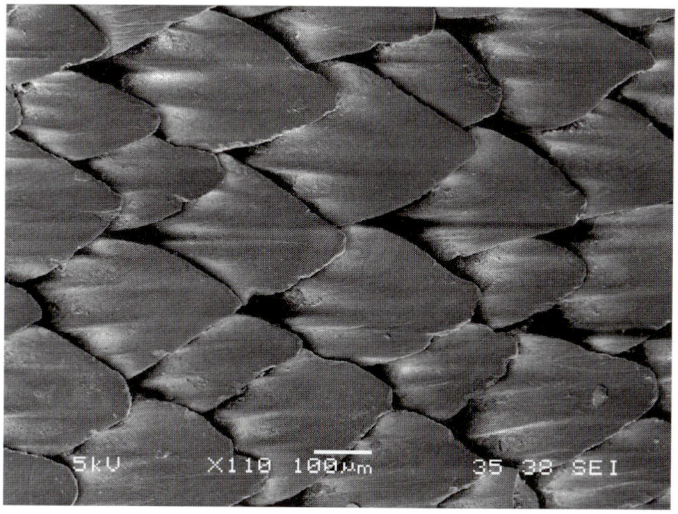

Fig. 95e. Dermal denticles.

GRAY SMOOTHHOUND

Mustelus californicus Gill, 1864

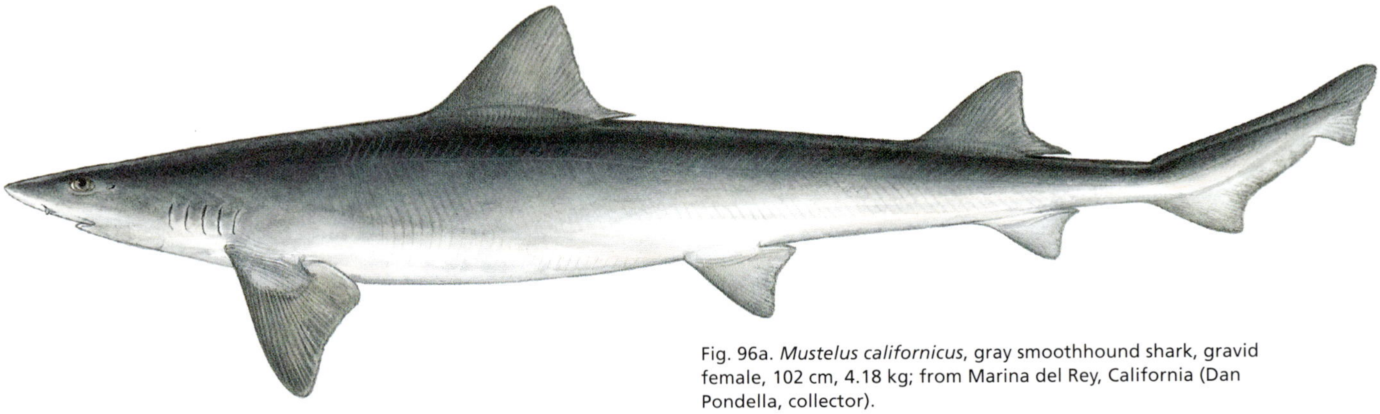

Fig. 96a. *Mustelus californicus*, gray smoothhound shark, gravid female, 102 cm, 4.18 kg; from Marina del Rey, California (Dan Pondella, collector).

Common name Gray smoothhound, a reference to its color, particularly the coloration of dead specimens.

Spanish name *Mamón, tripa* (Mexico). The A.F.S. name is *cazón mamón.*

Scientific name *Mustelus californicus* Gill, 1864, in *Proceedings of the Academy of Natural Sciences of Philadelphia* 16: 148–149. *Mustelus:* from the Latin *mustela,* a weasel, a name used since antiquity for both mammals and fishes, a reference to the shark's long, weasel-like body; *californicus:* Latinized form, "of California."

Synonyms None.

Identification The gray smoothhound has a slender body with a noticeable spiracle behind the eye. Its first dorsal fin originates *posterior* to the free rear tips of the pectoral fins, and the *midpoint of its base closer to the origin of the pelvic fins than to the pectoral axil* (very small juveniles have the first dorsal fin over or slightly behind the pectoral free rear tips). The *rear margin of the first dorsal fin is slightly falcate* and slopes gently to the rear. The second dorsal fin much larger than the anal fin. The *upper and lower labial folds are of similar length.* The teeth are small, numerous, and pavement-like, and have low, blunt cusps. The overlapping dermal denticles have single triangular points and flat surfaces. Coloration is brown or dark gray above and white below. Albino specimens from Elkhorn Slough, Monterey Bay, California, have been reported several times (see below).

Similar species The sicklefin smoothhound has the midpoint of the first dorsal fin almost equidistant to the pectoral axil and to the origin of the pelvic fins, and its rear margin is nearly vertical to the body axis. Its lower labial folds are longer than the upper ones. The brown smoothhound has a more anterior first dorsal fin, with the midpoint of its base about equidistant to the pectoral axil and the origin of the pelvic fins; the rear margins of its dorsal and caudal fins have a frayed appearance, and its teeth are pointed.

Range The gray smoothhound is found from northern California, southward along lower California and into the Gulf of California (Beebe and Tee-Van 1941; Roedel and Ripley 1950). It is rare off northern California; Bane and Bane (1971) did not include it in their *Bay Fishes of Northern California.* Herald (1953) reported that a neonate taken in San Francisco Bay in July was the northernmost record of the species at the time. It is common in inshore waters off southern California. Starks and Morris (1907) called it the most common shark in Southern California, mentioning that great numbers were taken in setlines in San Diego Bay. It is abundant in shallow waters less than 100 m in the northern Gulf of California (Pérez Jiménez et al. 2005a).

Size and weight The gray smoothhound reaches about 115 cm. Miller and Lea (1972) gave a maximum size of 163 cm, but I have not been able to verify specimens of such size. Representative sizes and weights are 80 cm, 1.8 kg; 100 cm, 3.1 kg; and 108.5 cm, 4.8 kg.

Biology The gray smoothhound is a common small shark that inhabits bays, sounds, and other coastal waters. It is generally a shallow water species; off southern California, it is found at depths of 2–45 m, usually in 5 m or less. In the Gulf of California, it has been taken at depths of 8–256 m (Pérez Jiménez 2006).

The gray smoothhound is very common in inshore waters off central California during fall and winter months

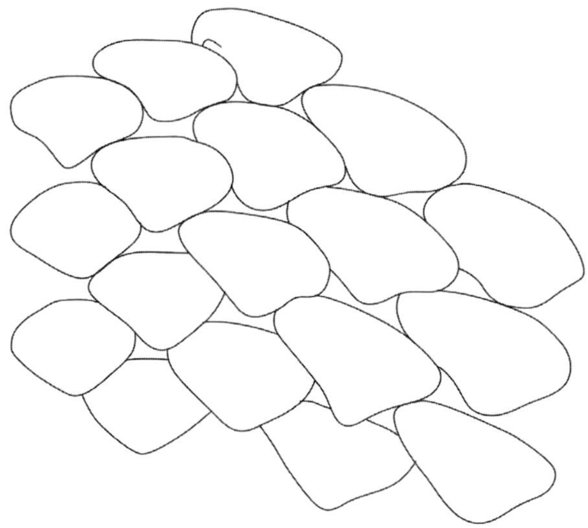

Fig. 96b. Teeth (upper).

Fig. 96c. Teeth (lower).

and is rare during the rest of the year (Talent 1973; de Wit and Scanland 1980). The extent of its migratory movements is unknown. Although it is seasonally abundant off southern California, there are no comprehensive studies on the species, and its biology has received little attention, with the exception of two studies recently carried out in the Gulf of California (Mendez Loeza 2004; Pérez Jiménez 2006).

The diet of the gray smoothhound consists primarily of crustaceans. In the Gulf of California, it feeds primarily on the stomatopod *Squilla bigelowi* and on the crab *Callinectes bellicosus* (Mendez Loeza 2004). It also feeds on various mollusks and small fishes.

Although reports of albinism in elasmobranchs are common, it is unusual to have *several* reports of albinism in a given species. The gray smoothhound is, to my knowledge, the only species in which we have several reports of albinism. All these reports are from Elkhorn Slough, Monterey Bay, California. Herald et al. (1960) reported an albino juvenile

Fig. 96d. Snout.

Fig. 96e. Dermal denticles.

male taken in Elkhorn Sough in 1959. The second case was a 26-cm neonate male taken in 1969 (Cohen 1973). The third case of albinism was a 117-cm female that carried a brood of 10 young, five of which were albinos (Talent 1973). The gray-pigmented female was obviously heterozygous for albinism. The presence of albinos in the brood would result from such female mating with an albino or heterozygous male. Either case would suggest that the albinism allele could be (or was) common in Elkhorn Slough gray smoothhounds.

Size at maturity Males mature at 72–74 cm, and females mature at 85–90 cm (Pérez Jiménez 2006).

Reproduction The gray smoothhound is a placental viviparous species. Gestation lasts for about 11 months. In the Gulf of California, it gives birth in April and May.

Size at birth: The young are born at about 28–30 cm. Pérez Jiménez (2006) reported embryos 28–30 cm in late April. Herald (1953) reported a 28 cm free-swimming neonate.

Brood size: Broods range from 7 to 16, with a mean of 11 (Pérez Jiménez 2006).

Nurseries The nurseries of the gray smoothhound are undescribed. There is some evidence of a nursery in the upper Gulf of California (Pérez Jiménez 2006).

Age and growth Yudin and Cailliet (1990) estimated that males mature at about two years of age, based on a length at maturity of 57–65 cm for males, and that females mature at an age of two to three years, based on a length at maturity of 70 cm. These lengths at maturity, based on reports by Heemstra (1973), are significantly shorter that those recently reported by Pérez Jiménez (2006). Assuming one population of smoothhounds inhabiting the California coast continuously into the Gulf of California, it is likely that the estimates of maturity at two to three years are underestimates.

Relation to humans Walford (1935) called the gray smoothhound one of the four most important sharks in the fish markets of California, but today it is of little economic importance there. It is often taken by local Mexican fisheries in the Gulf of California.

SMOOTH DOGFISH

Mustelus canis (Mitchill, 1815)

Fig. 97a. *Mustelus canis*, smooth dogfish, female, 113 cm, 5.9 kg; from off Bradenton, Florida (Rick Waites, collector).

Common name Smooth dogfish, an early name for the species, differentiating it from the spiny dogfish and implying that it lacked spines.

The name is derived from the old English names "smooth hound" and "dogfish," both of which were given to European species. The name "smooth hound" is very old. Pennant (1776, 3: 102) used it, explaining that "this species is called smooth, not that the skin is really so, but because it wants the spines on the back, which are the character of the second species, the Picked Dog" [spiny dogfish]. In the young United States, the American species was known simply as the dogfish (Mitchill 1815; Storer 1846). De Kay (1842: 356) stated that the fishermen "usually confound under the same name this and another species," likely the spiny dogfish. To avoid this confusion, De Kay called it the American hound-fish. By the early twentieth century, these names evolved into "smooth dogfish" (e.g., Hildebrand and Schroeder 1928; Breder 1929).

Spanish name *Tiburón* or *cazón mamón, mamiche* (eastern Mexico), *boca dulce* (Cuba). The A.F.S. name is *cazón dientón,* a name meaning "small shark with big teeth." This name makes no sense whatsoever, for the smooth dogfish does not have big teeth, as all the other Spanish common names allude.

Scientific name *Mustelus canis* (Mitchill, 1815), originally described as *Squalus canis* in *Transactions of the Literary and Philosophical Society* (New York) 1: 486–487. *Mustelus:* from the Latin *mustela,* a weasel, a name used since antiquity for both mammals and fishes, a reference to the shark's long, weasel-like body; *canis:* Latin, dog.

Synonyms The scientific name of the smooth dogfish changed often during the late nineteenth and twentieth centuries because many authors gave the species the scientific name of the European species, while others placed it in the genera *Galeus, Cynais, Cynias, Rhinotriacis,* or *Galeorhinus.* The confusion over the scientific name of the smooth dogfish was so great that Carl L. Hubbs (1938) found it necessary to write a 19-page article explaining the synonymy of the species. In addition, Compagno (1984b) lists *Allomycter dissutus* Guitart, 1972 as a synonym.

Identification The smooth dogfish has a slender body, a second dorsal fin much larger than the anal fin, a prominent spiracle behind the eye, upper labial furrows longer than the lower furrows, an internarial distance of 2.7–3.6% of the total length, and a caudal fin with a *rounded lower lobe (not pointed).* The skin denticles are lanceolate, with three or four weak ridges, and overlapping. Some specimens with three pointed denticles have been reported (William Driggers, pers. comm., April 2008), but this needs to be clarified. The precaudal vertebrae number 86–90. The teeth are very *small and pavement-like, with low blunt cusps,* and are very numerous and similar in both jaws. Coloration is olive gray to brown above, yellowish or grayish white below.

Similar species The Florida dogfish has upper and lower labial furrows of equal length, and a narrower internarial distance (2.3–2.8% of total length). The Gulf smoothhound has 100% tridentate skin denticles.

Range The smooth dogfish inhabits the western Atlantic from the Bay of Fundy, Canada (Huntsman 1922), to Uruguay (Heemstra 1997). It is rare in Canadian waters, having been recorded there only twice (Scott and Scott 1988). The first specimen was taken on a longline in the St. Croix River in July 1913 (Huntsman 1922). The second, and the northernmost record, is from Chance Harbor, Bay of Fundy, New

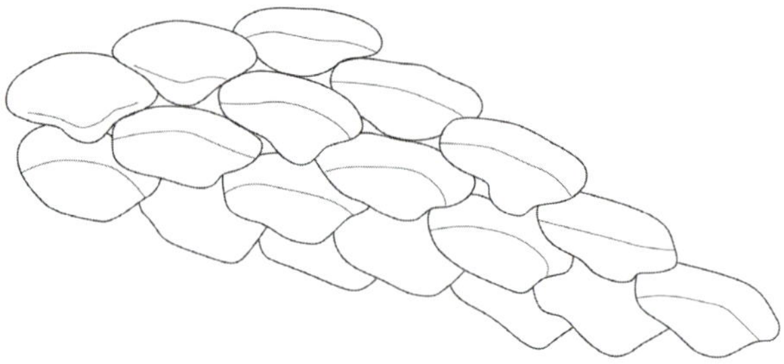

Fig. 97b. Teeth (upper).

Fig. 97c. Teeth (lower).

Brunswick, where a gravid female was taken in July 1977 (Markle et al. 1980).

The smooth dogfish is seasonally abundant from Cape Cod to South Carolina. It wanders to the central Florida coast only in colder winters, reaching the waters off Melbourne and, occasionally, off Jupiter (Dodrill 1977). Although the smooth dogfish is present in the Gulf of Mexico (Heemstra 1997), records of the smooth dogfish in the Gulf of Mexico need to be verified due to confusion with the Gulf smoothhound.

In his review of the smoothhound sharks of the western Atlantic, Heemstra (1997) described the smoothhound inhabiting Bermuda, Bahamas, Cuba, and other Caribbean islands as a subspecies, *Mustelus canis insularis*. The status of this form remains to be determined.

Size and weight Bigelow and Schroeder (1948) gave a maximum size of about 152 cm. Heemstra (1997) stated that the largest he measured was 122 cm in length. Representative sizes and weights are as follows: male, 96 cm, 2.6 kg; female, 60 cm, 0.70 kg; 87 cm, 2.0 kg; and 115 cm, 7.26 kg (gravid).

Biology The smooth dogfish is an extremely abundant shark of coastal waters, usually found at depths of less than 20 m throughout their summer range, moving to deeper waters in winter. It enters estuaries and has even been reported from tidal freshwater (Gunter 1942).

The smooth dogfish is primarily a bottom feeder, preying on crustaceans such as crabs, lobsters, and shrimp. However, it is an opportunistic feeder and a scavenger and will

Fig. 97d. Snout.

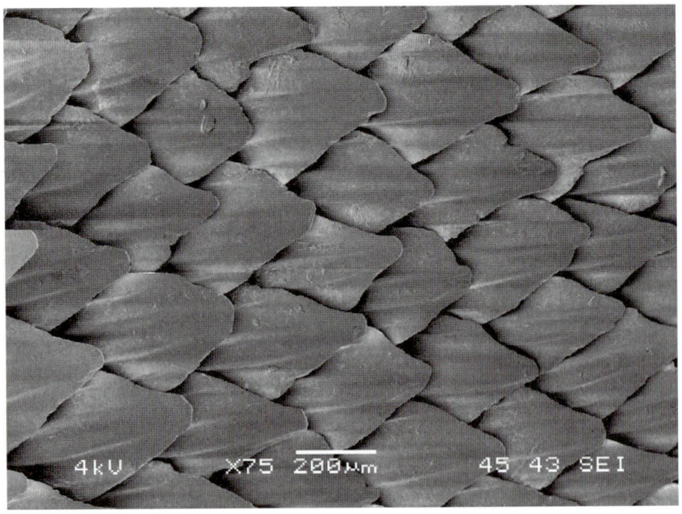

Fig. 97e. Dermal denticles.

consume whatever prey is easily available. It is primarily a nocturnal predator, which is probably related to the nocturnal habits of its crustacean prey (Casterlin and Reynolds 1979).

Bigelow and Schroeder (1948) reported that, off southern New England, lobsters may constitute as much as 16% of the smooth dogfish's diet. In the Chesapeake Bay region, Gelsleichter et al. (1999) found that the smooth dogfish eats crabs, clams, bony fishes, horseshoe crabs, and polychaete worms. The Atlantic rock crab (*Cancer irroratus*) was the primary prey, followed by lady crabs (*Ovalipes ocellatus*) and blue crabs (*Callinectes sapidus*). Razor clams (Solenidae) were the most important mollusk prey. In a New Jersey estuary, Rountree and Able (1996) found that young-of-the-year smooth dogfish preyed primarily on small shrimp (*Cragon septemspinosa* and *Palaemonetes vulgaris*), polychaete worms, and crabs.

Like many other sharks, the smooth dogfish finds its prey primarily by smell (G. H. Parker 1914). On several occasions, I have observed smoothhounds from submersibles, while they searched for bait scattered on the bottom. Whereas it took gulper sharks (*Centrophorus*) several passes over the bait for them to find and consume it, the smoothhounds found the bait on their first or second pass in the area.

The smooth dogfish is a migratory species that moves north and south seasonally. Its migrations are well tuned to the water temperature. According to Bigelow and Schroeder (1948), the temperature in their winter homes ranges from 6 to 15°C, and the species does not reappear off the Middle Atlantic states or southern New England until the temperature exceeds 6–7°C, becoming most abundant in June when the temperatures are 13–15°C.

The smooth dogfish winters from the Chesapeake Bay to South Carolina. In South Carolina, adults can be found January to May, while juveniles can be caught in low numbers throughout the summer (Castro 1993b). According to Bigelow and Schroeder (1948), the northward migration is progressive from early spring on, large numbers remaining off North Carolina waters until June, although none can be found there by July. By May, schools of smooth dogfish appear along the Chesapeake peninsula. The schools appear simultaneously along the coasts of New Jersey, New York, and southern New England by early May. The smooth dogfish summers from Cape Cod to Delaware Bay.

Since the early twentieth century, the smooth dogfish has been the "lab rat" of shark research because it survives in captivity for long periods of time and it is easily available off the northeastern United States in summer. In the early 1900s, G. H. Parker, professor of zoology at Harvard University, and R. E. Sheldon, of the U.S. Bureau of Fisheries, both working at the U.S. Fisheries Laboratory at Woods Hole, Massachusetts, were the first to use live smooth dogfish for numerous experiments on its sensory biology (G. H. Parker 1910, 1914; Sheldon 1911). Since then other researchers have used the smooth dogfish in countless experiments. Because of these experiments, much has been learned about the physiology of the smooth dogfish. G. H. Parker (1938) reported that smooth dogfish were capable of becoming fully pale or fully dark, depending on the illumination of their environment, and that the control of the melanophores was under hormonal control. Ifft and Zinn (1948) studied tooth succession in the species and reported that one tooth row was replaced every 1–12 days.

Size at maturity Males mature at 82–89 cm, and females mature at about 90 cm (Heemstra 1997).

Reproduction The smooth dogfish is a placental viviparous species. Off the northeastern United States, females ovulate between 20 June and 20 July (TeWinkel 1950). Gestation appears to last about 10 months. Off southern New England, the young are born from early May to mid-July (Bigelow and Schroeder 1948). Off South Carolina, females give birth in late April and May, just prior to the northward migration of the adults (Castro 1993b). The reproductive cycle is annual, as females carry ripe oocytes at the time of parturition.

Size at birth: The young measure 34–39 cm at birth, with an average size of 36 cm (Bigelow and Schroeder 1948).

Brood size: The number of young per brood ranges from 10 to 20, with an average of 16 (Bigelow and Schroeder 1948).

Nurseries The nursery areas of the smooth dogfish extend over most of its range, from New England to South Carolina (Castro 1993b). The nurseries are in shallow coastal waters, and the young remain there until the onset of winter.

Age and growth Conrath et al. (2002) examined the growth patterns of vertebral centra and concluded that smooth dogfish males mature at two or three years of age, and that females mature between four and five years of age. According to their age estimates, in their sample of 894 smooth dogfish, the oldest male was 10 years old and the oldest female was 16 years old.

Relation to humans The smooth dogfish is of no economic interest in the commercial shark fishery of the southeastern states because of its small size and because it is only a winter visitor to the areas where that fishery operates. Because of its abundance, exceeded only by that of the spiny dogfish, the smooth dogfish interferes with shrimp trawling operations and affects crab and lobster stocks. It is extensively used as a laboratory animal and is often displayed in aquaria.

SHARPTOOTH SMOOTHHOUND

Mustelus dorsalis Gill, 1864

Fig. 98a. *Mustelus dorsalis,* Sharptooth smoothound shark, 41.8 cm male; from Pacific Panama (LACM # 886, courtesy of Jeff Seigel).

Common name Sharptooth smoothhound.

Spanish name *Tripa, tollo,* or *cazón tripa* (A.F.S.).

Scientific name *Mustelus dorsalis* Gill, 1864, in *Proceedings of the Academy of Natural Sciences of Philadelphia* 16: 149. *Mustelus:* from the Latin *mustela,* a weasel, a name used since antiquity for both mammals and fishes, a reference to the shark's long, weasel-like body; *dorsalis,* Latin, pertaining to the back or dorsal fin—Gill used the dorsal fin shape to distinguish it from other smoothhounds.

Synonyms None.

Identification This small shark (less than 70 cm in length) has noticeable spiracle behind the eye; the *first dorsal fin origin is over the free rear tips of the pectoral fins;* and *its rear edge slopes gently to the rear (not vertically straight and perpendicular to the body).* The *midpoint of the first dorsal fin is equidistant from the pectoral axil and the origin of the pelvic fins.* The *rear edges of the dorsal fins and the caudal fin are intact and not frayed.* The *nostril length is 2.2–2.7% of the total length and about 84–95% of internarial space* (the space between the inner ends of the nostrils). The dermal denticles are lanceolate, with two to four ridges on their surfaces. The precaudal vertebrae number 99–100. The upper teeth have erect, symmetrical, smooth-edged cusps. Their tips are nearly translucent. The lower teeth have shorter cusps than the upper ones and may have one slight cusplet on each side of the main cusp. Coloration is brownish gray above and white below.

Similar species The whitetip and sicklefin smoothhounds have the upper rear edge of the first dorsal fin vertical and perpendicular to the body. The gray smoothhound has teeth with rounded cusps (molariform teeth). The brown smoothhound has dorsal and caudal fins with frayed edges.

Range The sharptooth smoothhound inhabits the eastern Pacific from southern Mexico to Panama and perhaps to northwestern South America (Beebe and Tee-Van 1941; Kato et al. 1967). Although Hildebrand (1946) reported the species in Peru, the southern limits of its distribution are unknown due to confusion with other smoothhounds. Beebe and Tee-Van (1941) cited Gulf of California records, but there are no confirmed records of the species for that locality (Kato et al. 1967; Pérez Jiménez et al. 2005a).

Size and weight The sharptooth smoothhound males reach 59 cm and 0.3 kg; females reach 66 cm and 1 kg (Rojas 2006).

Biology This is a poorly known, small tropical species that inhabits shallow coastal waters of the tropical eastern Pacific. Rojas (2006) reported that the sharptooth smoothhound feeds on crustaceans (*Squilla hancocki, S. parva, Litopenaeus* spp.), small fishes (*Anchoa* spp., *Caranx* spp., *Lutjanus* spp., and *Engraulis* spp.), and mollusks (*Loligo* spp. and *Octopus* spp.), the main prey being the mantis shrimp *Squilla hancocki.*

Size at maturity It appears that the sharptooth smoothhound matures at about 43 cm.

Male: According to Kato et al. (1967), males 43 cm were "nearly mature." Heemstra (1973) stated that the smallest male with well-calcified claspers measured 42.5 cm.

Female: The smallest mature female recorded measured 43 cm (Heemstra 1973).

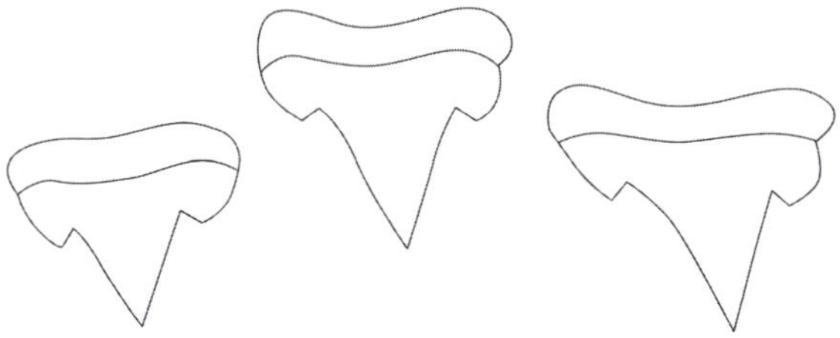

Fig. 98b. Teeth (upper).

Fig. 98c. Teeth (lower).

Reproduction The sharptooth smoothhound is a placental viviparous species (Heemstra 1973).

Size at birth: The young are born at 21–23 cm (Heemstra 1973).

Brood size: A brood of four young has been reported (Heemstra 1973).

Nurseries Rojas (2006) suggested that the shallow rocky shores of the Gulf of Nicoya, Costa Rica, are a nursery area for the sharptooth smoothhound.

Age and growth No data available.

Relation to humans None.

Fig. 98d. Snout.

Fig. 98e. Dermal denticles.

BROWN SMOOTHHOUND

Mustelus henlei (Gill, 1863)

Fig. 99a. *Mustelus henlei*, Brown smoothhound shark, mature male, 69 cm; off El Barril, Baja California, Mexico (Diana Crabbe, collector).

Common name Brown smoothhound. In the older literature, it was often called Henle's shark.

Spanish name *Tiburón mamón* or *tripa* (Mexico). The A.F.S. name is *cazón hilacho.*

Scientific name *Mustelus henlei* (Gill, 1863), originally described as *Rhinotriacis henlei* in *Proceedings of the Academy of Natural Sciences of Philadelphia,* 14: 486. *Mustelus:* from the Latin *mustela,* a weasel, a name used since antiquity for both mammals and fishes, a reference to the shark's long, weasel-like body; *henlei:* after Friedrich Gustav Jacob Henle (1807–1885), coauthor, with J. Müller, of *Systematische Beschreubung der Plagiostomen,* the greatest work on sharks of the nineteenth century.

Synonyms Until recently, the species was known as *Triakis henlei.*

Identification The brown smoothhound has a slender body with a noticeable spiracle behind the eye; the *dorsal fins and the upper margin of the caudal fin have thin and frayed rear edges,* a first dorsal fin originating over the free rear tips of the pectoral fins, with the midpoint of its base about equidistant from the pectoral axil and the pelvic fins origin (large males have the midpoint of the first dorsal fin base slightly closer to the pectoral axil than to the origin of the pelvic fins), and a second dorsal fin much larger than the anal fin. Upper labial furrows noticeably longer than lower ones. The teeth are small, numerous, and *pavement-like* and have *short, sharp points,* which may be felt by passing a finger over them, from the rear to the front. The upper teeth have a slightly oblique, median cusp flanked by a strong notch and smaller cusp; the lower teeth have more erect cusps. The dermal denticles have three points with two strong ridges that converge in a long point. Coloration is bronze brown with metallic hues above, becoming lighter on the sides, and white below.

Similar species All the other smoothhounds in the area lack the frayed edges on the rear margins of the dorsal and caudal fin. The gray smoothhound has a more posterior first dorsal fin, its origin well behind the free rear tips of the pectoral fins; its teeth are round and blunt, and it has upper and lower labial furrows of equal length. The sicklefin smoothhound has blunt teeth and upper labial furrows much shorter than the lower ones. The soupfin shark has the second dorsal fin and the anal fin of the same size, and both begin at the same level.

Range The brown smoothhound is found from at least Humboldt Bay, California, to the Gulf of California (Miller and Lea 1972). It is locally common in bays north of Monterey, with extensive populations in San Francisco, Tomales, and Humboldt bays. Roedel and Ripley (1950: 52) called it "the most abundant shark in San Francisco Bay, where it makes up over half of the total shark population."

Size and weight The brown smoothhound reaches a length of 97 cm (Bane and Bane 1971).

Biology This small shark is an abundant coastal species that inhabits shallow tidal flats and bays in the northern part of its range and deeper waters in the southern parts. In the northern parts, it is found at depths of 2–5 m, while in the southern parts (Baja Peninsula and the northern Gulf of California) most are taken below 100 m.

The brown smoothhound migrates seasonally, spending the spring and summer in shallow waters and migrating southward or to deeper water in winter. According to Flemming (1999), in San Francisco Bay, California, the brown smoothhound prefers temperatures of 14–19°C. The

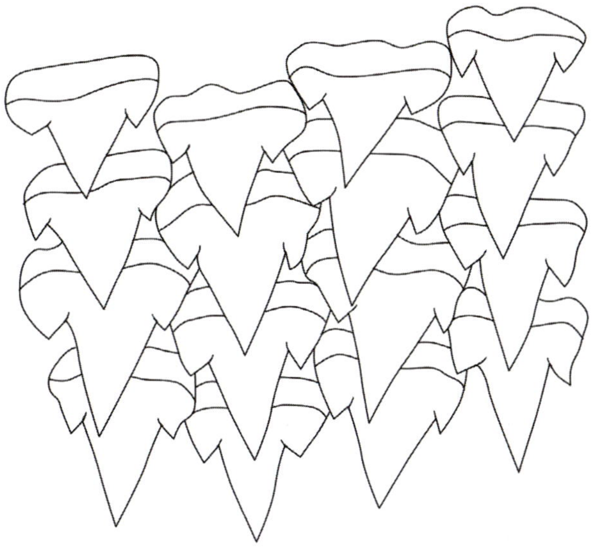

Fig. 99b. Teeth (upper).

Fig. 99c. Teeth (lower).

species enters shallow waters before parturition and leaves in autumn when the temperature falls below 14°C.

The brown smoothhound is an opportunistic crustacean predator, although it will also consume small fishes. Herald and Ripley (1951) reported that, in San Francisco Bay, brown smoothhounds fed primarily on small crabs and shrimp. Russo (1975) examined the stomach contents of 77 smoothhounds from San Francisco and Tomales Bays, California, and reported that of the sharks examined, 39% contained shrimp, 35% contained crabs, and 19% contained fishes. A subsequent study by Haeseker and Cech (1993) in Tomales Bay reported that adult sharks fed primarily on slender crabs (*Cancer gracilis*), yellow shore crabs (*Hemigrapsus oregonensis*), and bay shrimp (*Crangon stylirostris*).

Size at maturity

Male: Males mature at 55–56 cm, based on clasper calcification (Pérez Jiménez 2006).

Female: The size range at maturity for the female brown smoothhound in the literature is large (see below). Heemstra (1973) stated that the smallest gravid female he examined

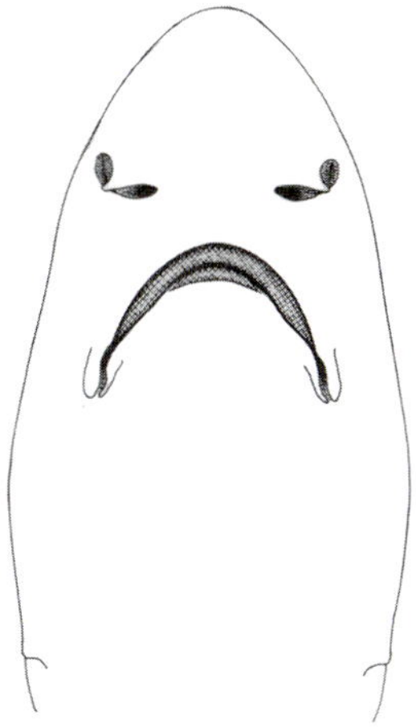

Fig. 99d. Snout.

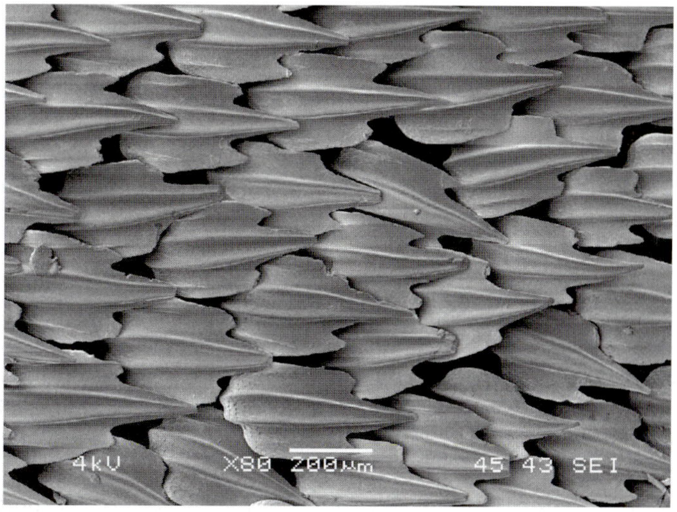

Fig. 99e. Dermal denticles.

was 50.5 cm in length. Pérez-Jiménez and Sosa-Nishizaki (2008) reported that their smallest gravid female in the northern Gulf of California measured 57 cm, whereas their largest immature female measured 66 cm.

Reproduction The brown smoothhound is placental viviparous (Starks 1917a). The reproductive cycle is annual, like other species of *Mustelus.* Gestation is believed to last about 10 months (Pérez Jiménez and Sosa-Nishizaki 2008). The embryos reach their maximum development in March (Pérez Jiménez 2006).

Size at birth: The young are born at 28–30 cm.

Brood size: According to Pérez Jiménez (2006), brood size ranges from 1 to 21, with a mean of 10.

Nurseries According to De Wit (1975), numerous brown smoothhounds, 20–23 cm in length, are taken in southern San Francisco Bay, California, throughout the summer months.

Age and growth According to Yudin and Cailliet (1990) males mature at three years of age, based on a length at maturity of 52–66 cm (Heemstra 1973; Compagno 1984b), and females mature at two or three years of age, based on a length at maturity of 51–63 cm (Compagno 1984b). The size range at maturity for both males and females is greater than that encountered by Pérez Jiménez (2006).

Relation to humans At one time, the brown smoothhound was often found in the fish markets of California (Walford 1935; Ripley 1949). Nowadays, the brown smoothhound is too small to have commercial value in the United States. However, in Mexico it is an important food species, despite its small size, because it is taken in incredibly large numbers in gillnets. I have seen daily catches of 200 brown smoothhounds per small fishing boat near La Paz, Mexico.

How these small sharks can be profitable to the Mexican fishermen is seen in their processing all along the Baja Peninsula. The returning fishermen drive their boats at high speed onto the beach, propelling the boat 4–5 m high into the sand. The catch is quickly tossed overboard and piled by species or type onto the sand. Piles of more than 100 brown smoothhounds are not unusual. Each boat is met by one or two ladies (often the fishermen's wives), who dress out the catch while the men retire for refreshments. The processing ladies take each shark by the tail with one hand, and wielding a sharp machete on the other hand, they deliver a blow that severs the head and most of the entrails. If necessary, a second blow, with the blade at a different angle, removes the rest of the entrails. The eviscerated carcass is tossed onto another pile, and they start on the next shark. The total time spent on each shark is but a few seconds. A pile of dozens of sharks is processed in 10–20 minutes. The fishermen then load and rinse the carcasses and move them to the weighing area or onto waiting trucks.

SICKLEFIN SMOOTHHOUND

Mustelus lunulatus Jordan and Gilbert, 1882

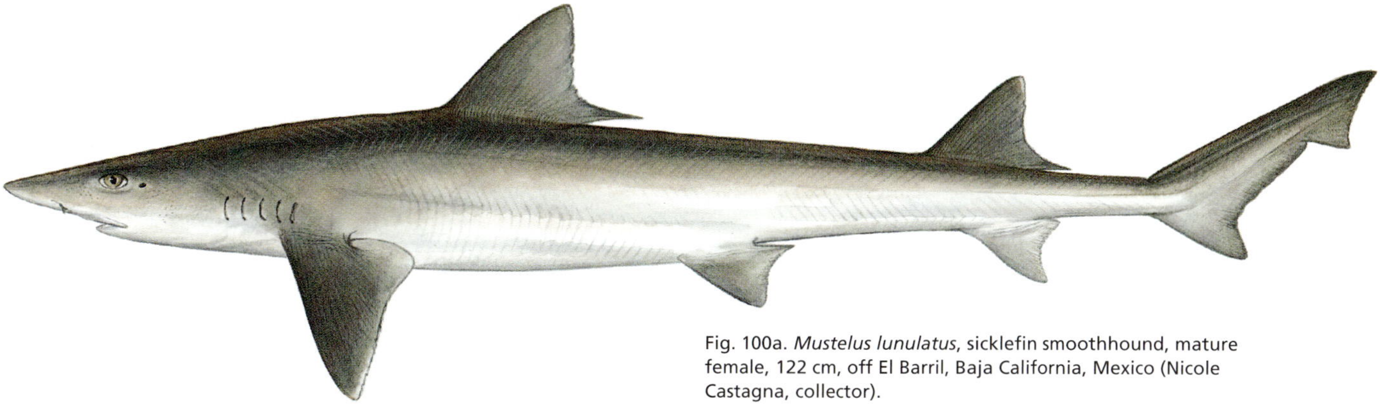

Fig. 100a. *Mustelus lunulatus*, sicklefin smoothhound, mature female, 122 cm, off El Barril, Baja California, Mexico (Nicole Castagna, collector).

Common name Sicklefin smoothhound, a reference to the concave margins of its fins.

Spanish name *Mamón,* or *tripa* (Mexico). The A.F.S. name is *cazón segador.*

Scientific name *Mustelus lunulatus* Jordan and Gilbert, 1882, in *Proceedings of the United States National Museum* 5: 108–109. *Mustelus:* from the Latin *mustela,* a weasel, a name used since antiquity for both mammals and fishes, a reference to the shark's long, weasel-like body; *lunulatus:* somewhat moon-shaped, from the Latin *luna,* the moon, a reference to the concave margins of its fins.

Synonyms None. In the past, the species has been placed in the genera *Galeus* (Jordan 1895), *Galeorhinus* (Garman 1913), and *Cynias* (Starks and Morris 1907).

Identification The sicklefin smoothhound has a slender body with a noticeable spiracle behind the eye. Its first dorsal fin originates over the free rear tips of the pectoral fins, and the midpoint of its base is about equidistant from the pectoral fin axil and the pelvic fin origin. The first dorsal fin has a nonfrayed, nearly vertical posterior edge. The second dorsal fin is much larger than the anal fin, and the lower lobe of the caudal fin has a *pointed tip.* The *upper labial folds are noticeably shorter than the lower ones.* The teeth are small and numerous, pavement-like, with rounded, blunt cusps (occasionally with a low, sharp cusp). The dermal denticles are usually lanceolate with a single point, and two weak ridges that converge on a point; some denticles may have one or two weak side points. Coloration is grayish or olive brown above, paler or yellowish below.

Similar species The gray smoothhound has upper and lower labial folds of about the same length, and the midpoint of its first dorsal fin is closer to the pelvic fins than to the pectoral axil. In doubtful cases check the number of precaudal vertebrae: 89–99 for the gray smoothhound, 74–84 for the sicklefin smoothhound. Also examine the upper jaw (palatoquadrate cartilage): the gray smoothhound has the cartilage divided near the symphysis; in the sicklefin and the other species of smoothhounds the upper jaw cartilage is undivided. The brown smoothhound has dorsal and caudal fins with frayed rear edges.

Range The sicklefin smoothhound ranges from the southern California coast (Roedel and Ripley 1950) to northern South America, off Colombia (Beebe and Tee-van 1941; Gómez et al. 2003). It is one of the two most abundant shallow water sharks in the Gulf of California (the other one is the gray smoothhound). Jordan (1895) stated that it was rather common at Mazatlan, Mexico, where the type specimen was obtained.

Size and weight The sicklefin smoothhound grows to about 174 cm (Norris 1923). Representative sizes and weights are 80 cm, 2 kg; 100 cm, 4.8 kg; 130 cm, 9 kg; and 162 cm, 16 kg.

Biology The sicklefin smoothhound is a common but little-known shark that inhabits shallow coastal waters. The

Fig. 100b. Teeth (upper).

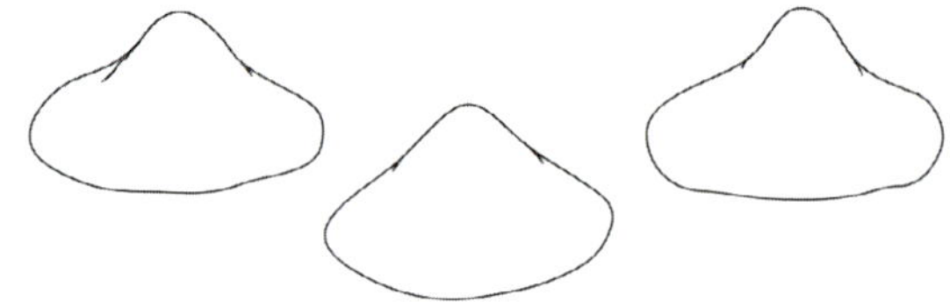

Fig. 100c. Teeth (lower).

sicklefin smoothhound is primarily a crustacean predator. Off Colombia it feeds on crabs, primarily the swimming crab *Portunus iridiscens,* stomatopods (*Squilla panamensis* and *S. mantoidea*), and numerous small fishes (Gómez et al. 2003). It is known to migrate seasonally from deep to shallow water, but its movements have not been described.

Size at maturity Males mature at 89–94 cm and females at 100–106 cm (Pérez Jiménez 2006).

Reproduction The sicklefin smoothhound is placental viviparous. Gestation appears to last 11 months. In the Gulf of California, the birth season is from mid-February to late April. The reproductive cycle is annual, like in other species of *Mustelus* (Pérez Jiménez 2006).

Size at birth: The young are born at 28–34 cm (Pérez Jiménez 2006).

Brood size: Brood size ranges from 6 to 19 young, with an average of 13 (Pérez Jiménez 2006).

Nurseries There are indications that the upper Gulf of California is a nursery for the species, but this needs confirmation.

Age and growth The sicklefin smoothhound has not been aged.

Relation to humans Large numbers of sicklefin smoothhounds are taken in local fisheries of the Gulf of California.

Fig. 100d. Snout.

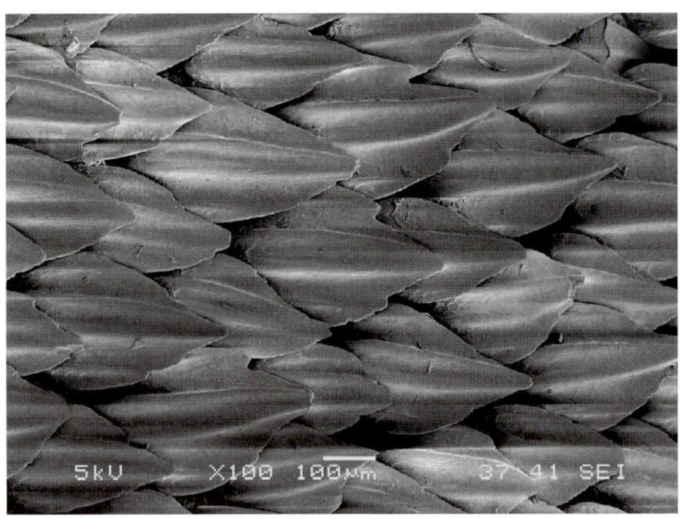

Fig. 100e. Dermal denticles.

FLORIDA SMOOTHHOUND OR FLORIDA DOGFISH

Mustelus norrisi Springer, 1939

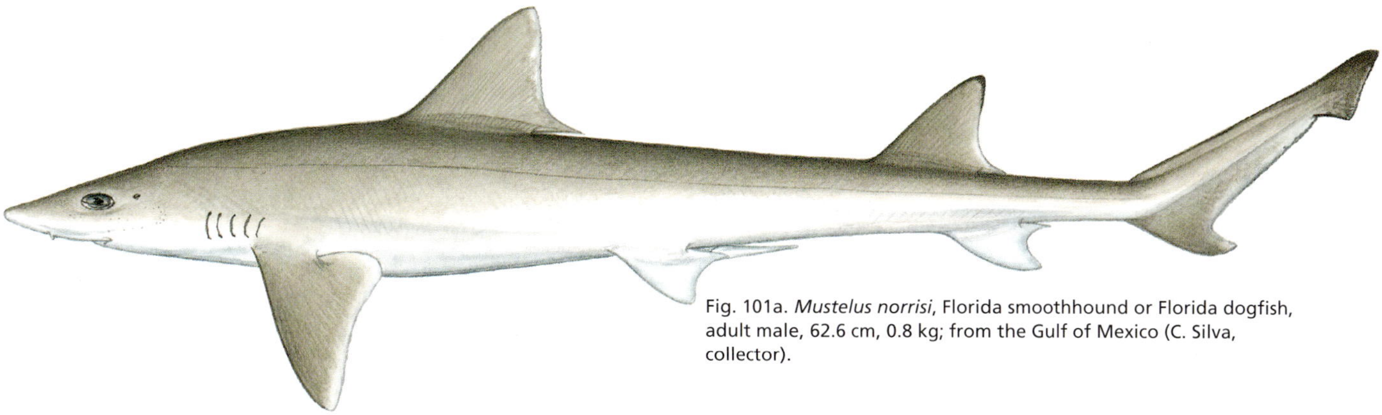

Fig. 101a. *Mustelus norrisi*, Florida smoothhound or Florida dogfish, adult male, 62.6 cm, 0.8 kg; from the Gulf of Mexico (C. Silva, collector).

Common name Florida dogfish, Florida smoothhound (A.F.S.), Gulf smoothhound, all references to its range. The name Florida dogfish is so well established in the literature (e.g., Bigelow and Schroeder 1948; Castro 1983; McEachran and Fechhelm 1998) that I have opted to use it. Obviously, it is more appropriate to call it a smoothhound.

Spanish name The A.F.S. name is *cazón viuda.*

Scientific name *Mustelus norrisi* Springer, 1939, in *Proceedings of the United States National Museum* 86(3058): 462–465. *Mustelus:* from the Latin *mustela,* a weasel, a name used since antiquity for both mammals and fishes, a reference to the shark's long, weasel-like body; *norrisi:* after Prof. Harry Waldo Norris (1862–1946), of Grinnell College, Iowa. Prof. Norris was an anatomist who published extensively on cranial nerves. His work on the cranial nerves of the spiny dogfish (Norris and Hughes 1920) is an elegant publication.

Synonyms None.

Identification The Florida dogfish is a small and slender shark (*adults reach less than a meter*), with a prominent spiracle behind the eye, a second dorsal fin much larger than the anal fin, and the caudal fin with a *sharp-pointed lower lobe* (embryos and young may have a rounded lower caudal lobe). It has an internarial distance of 2.3–2.8% of total length. The upper and lower labial furrows are about equal in length. The teeth are very *small, and pavement-like, with low blunt cusps;* they are very numerous, and similar in both jaws. The dermal denticles are lanceolate, with two prominent ridges that meet at the point. Coloration is gray or grayish brown above and paler or dirty white below; some specimens have fins with lighter or white trailing edges.

Similar species The smooth dogfish and the Gulf smoothhounds have upper labial furrows longer than the lowers.

Range The Florida dogfish inhabits the western Atlantic from the Gulf of Mexico to Brazil. It has been reported from the Florida Keys (Springer 1939b), the Florida west coast (Clark and von Schmidt 1965), Texas and the southern Caribbean (Cervigón 1966), and the western Atlantic south to Brazil (Heemstra 1997). The limits of its distribution are not well known. It is seasonally common throughout the Florida west coast.

Size and weight The Florida dogfish reaches a length of about 100 cm; Heemstra (1997) recorded a maximum size of 98 cm, but there is a Cape Haze Marine Laboratory record of a 101 cm gravid female (CHML/MML no. 122, 12 February 1968). Representative sizes and weights are as follows:

Male: 49 cm, 0.3 kg; 60 cm, 0.80 kg; 63 cm, 0.82 kg; 65 cm, 0.88 kg; and 69 cm, 1.0 kg.

Female: 57 cm, 0.45 kg; 68 cm, 1.0 kg; and 82 cm, 1.7 kg.

Biology The Florida dogfish is a common, but poorly known, small shark that inhabits shallow coastal waters. It

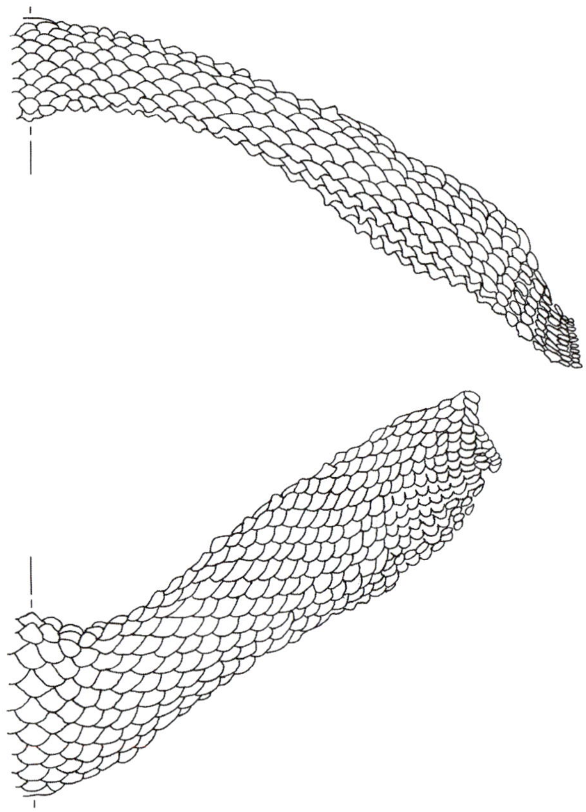

Fig. 101b. Teeth.

is usually found at depths of 5–7 m or less during the winter months. According to Heemstra (1997), the deepest record for the species is 84 m at the mouth of the Gulf of Venezuela.

The Florida dogfish is a migratory species that enters shallow waters during cool months. The species is found off the eastern Yucatan Peninsula in November and December at depths up to 20 m (D. de Anda, pers. comm., October 2006). It is a seasonal visitor to the west coast of Florida. Based on the Cape Haze Marine Laboratory records compiled by Eugenie Clark, and on my own more limited observations, it is present in shallow water off Sarasota from December through April, although most specimens are encountered from January through March. There is one single Cape Haze Marine Laboratory record for 1 July 1967, indicating that it may linger in the area into early summer in some years. Presumably it moves into deeper and cooler waters in summer.

Size at maturity According to Heemstra (1997) males are mature at 57–61 cm and females at about 65 cm.

Reproduction The Florida dogfish is a placental viviparous species. The length of the gestation period has not been determined. It gives birth in March and April.

Size at birth: The young are born at 29–30 cm (Heemstra 1997).

Brood size: Broods range from 10 to 14 young (Clark and von Schmidt 1965).

Nurseries Unknown.

Age and growth The Florida dogfish has not been aged.

Relation to humans None.

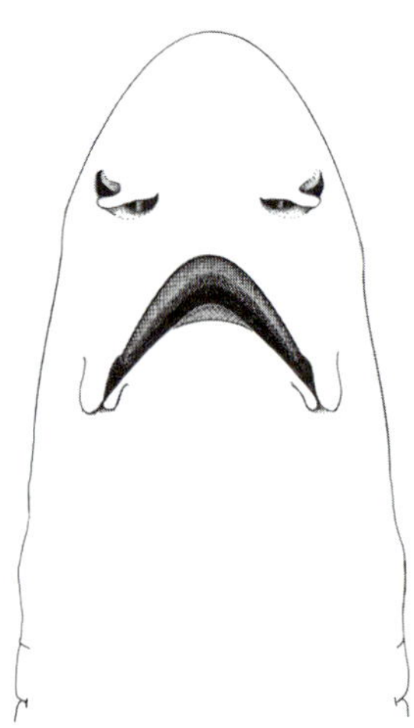

Fig. 101c. Snout.

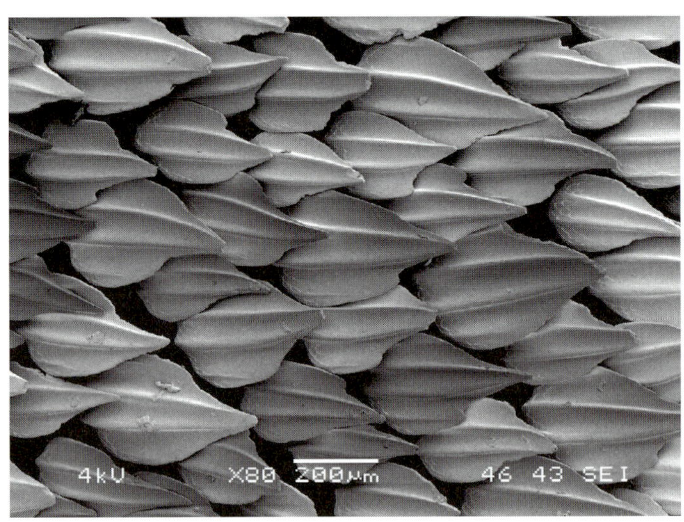

Fig. 101d. Dermal denticles.

Fig. 101e. *Mustelus norrisi*, Florida smoothhound or Florida dogfish, gravid female, 96.2 cm, 3.5 kg; Gulf of Mexico, caught March 2002 (Eric Sander collector, courtesy of Gregg Poulakis).

GULF SMOOTHHOUND

Mustelus sinusmexicanus Heemstra, 1997

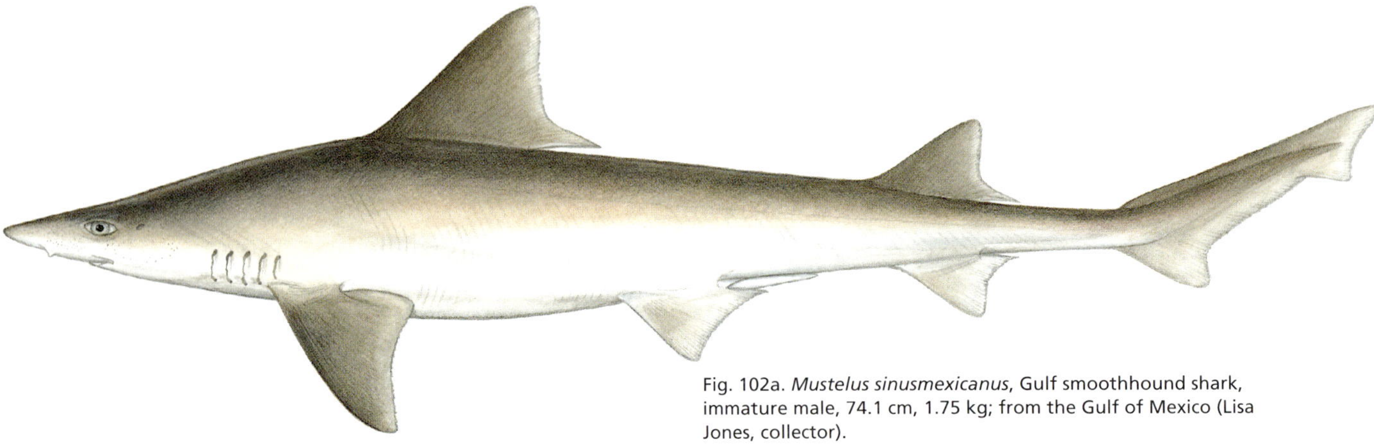

Fig. 102a. *Mustelus sinusmexicanus*, Gulf smoothhound shark, immature male, 74.1 cm, 1.75 kg; from the Gulf of Mexico (Lisa Jones, collector).

Common name Gulf smoothhound (A.F.S.), a reference to the area where it was first discovered.

Spanish name *Cazón del Golfo* (A.F.S.).

Scientific name *Mustelus sinusmexicanus* Heemstra, 1997, in *Bulletin of Marine Science* 60(3): 918–924, figs. 16b, 17, 18c. *Mustelus:* from the Latin *mustela,* a weasel, a name used since antiquity for both mammals and fishes, a reference to the shark's long, weasel-like body; *sinusmexicanus:* from the Latin *sinus,* a bay or gulf of the sea, and *mexicanus,* of Mexico, thus "of the Gulf of Mexico."

Synonyms None.

Identification The Gulf smoothhound has a prominent spiracle behind the eye, a second dorsal fin much larger than the anal fin, a caudal fin with a prominent narrowly rounded lower lobe, an internarial distance of 2.6–3.3% of total length, and *upper labial furrows longer than the lower furrows.* The teeth have low, blunt, asymmetric cusps with a smaller cusplet on the posterolateral margin of the main cusp and are similar in both jaws. Teeth number about U: 58–69; L:60–62. The precaudal vertebrae number 68–71. The *dermal denticles have three points, with two strong ridges that meet on the longer central point.* Coloration is brownish gray to gray above and lighter below.

Similar species The smooth dogfish and the Florida dogfish lack the tridentate skin denticles. The smooth dogfish has 86–90 precaudal vertebrae (Heemstra 1997). The Florida dogfish has upper and lower labial furrows of equal length.

Range The Gulf smoothhound inhabits the Gulf of Mexico. It has been taken from off Panama City to the Bay of Campeche (Heemstra 1997). The limits of its distribution are unknown.

Size and weight The Gulf smoothhound is one of the larger smoothhounds. The largest specimen reported is a 140-cm female (Heemstra 1997).

Biology The Gulf of smoothhound is a recently described species that has been taken at depths of 36–229 m (Heemstra 1997). It is likely that this is the dominant species of *Mustelus* in the Gulf of Mexico. Nothing is known of its habits.

Size at maturity Males mature at about 80 cm; the size at maturity of females is not known (Heemstra 1997).

Reproduction The Gulf smoothhound is believed to be a placental viviparous species.

Size at birth: The young are born at 39–43 cm (Heemstra 1997).

Brood size: Unknown.

Nurseries Unknown.

Age and growth The Gulf smoothhound has not been aged.

Relation to humans None.

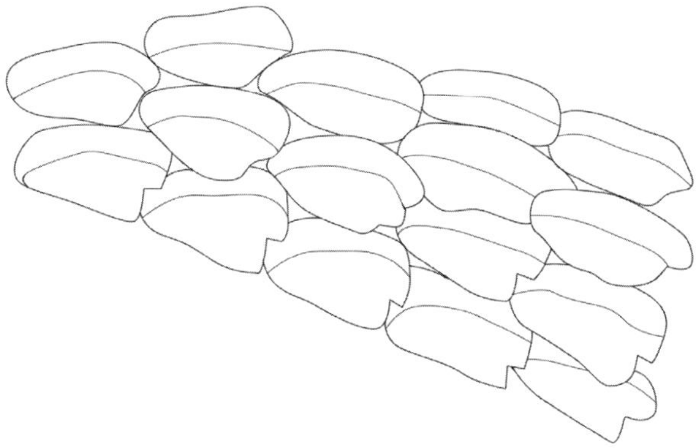

Fig. 102b. Teeth (upper).

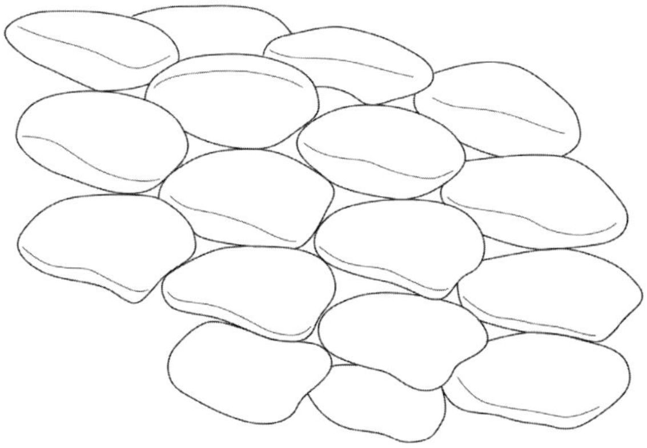

Fig. 102c. Teeth (lower).

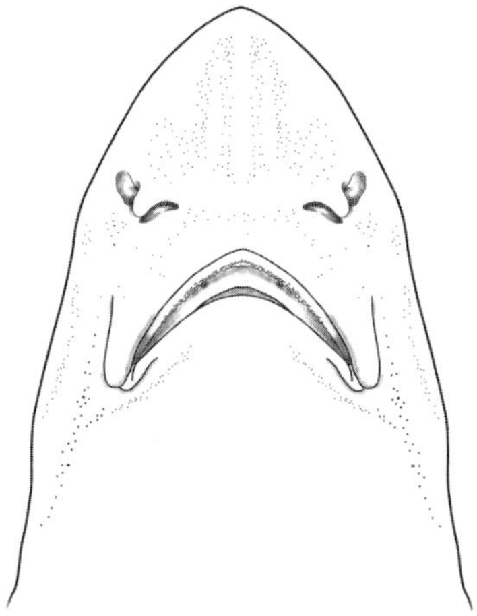

Fig. 102d. Snout.

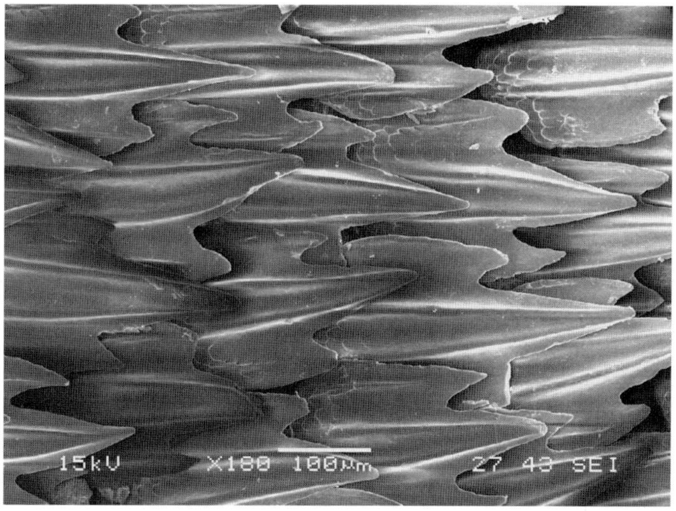

Fig. 102e. Dermal denticles.

LEOPARD SHARK

Triakis semifasciata Girard, 1855

Fig. 103a. *Triakis semifasciata*, leopard shark, immature male, 75 cm, 1.6 kg; from El Barril, Baja California, Mexico (Neal Overstrom, collector).

Common name Leopard shark, an allusion to its dark spots and markings.

Spanish name *Leopardo* (Mexico), *tiburón leopardo* (A.F.S.).

Scientific name *Triakis semifasciata* Girard, 1855, originally described as *Triakis semifasciatum* in *Proceedings of the Academy of Natural Sciences of Philadelphia* 7(6): 196; the date of this publication is variously given as 1854 or 1855; the pages of Girard's description are marked November 1854, but these apparently did not appear until 1855). *Triakis:* from the Greek *triacis, tri,* three, plus *acis* (*akis*), point, an allusion to the three-pointed teeth of the species; *semifasciata: semi,* Latin prefix *semi,* and *fasciata,* Latin, banded, thus "semibanded" or "half-banded," a reference to the dark bands or saddle blotches that cover only the dorsal half the shark's body.

Synonyms *Triakis californicus* Gray, 1851; *Mustelus felis* Ayres, 1854.

Identification The beautiful leopard shark is easily recognized by its striking color pattern of numerous, *well-defined, large dark saddle bars over the dorsal surface, interspaced with dark spots,* and a *row of rounded dark spots on the side of the body.* The pattern of the markings is quite variable (see Schott 1964 for an analysis of the patterns), and sometimes there are additional markings between the saddle bars. Occasionally individuals have irregular, dark, longitudinal streaks instead of saddle bars or, more rarely, irregular dark marks and numerous spots all over the dorsal side and flanks (see illustrations on page 386). Its *first dorsal fin originates far ahead of the pelvic fins.* The teeth are small and numerous; they have one long, oblique, central cusp flanked by one or two smaller cusps on each side, and they number about U: 25–25, L: 19–19. The dermal denticles are sparse with little overlapping. They are leaflike and variable in shape; most have a weak central ridge terminating in a point, but some have short side points. The dorsal coloration is variable, ranging from bronze brown with darker markings to silvery gray with almost black markings. The undersides are paler or tan. The coloration and metallic hues also change with the light and the angle of view. A 60-cm albino male was taken in San Pablo Bay, California, in 1972 (Follett 1976).

Similar species The swell shark has indistinct dark saddle markings, and its first dorsal fin originates behind the pelvic fin origin.

Range The leopard shark ranges from the Oregon coast (Roedel 1953) to Mazatlan, Mexico, including the Gulf of California (Miller and Lea 1972). It is common in inshore waters throughout California; it has been reported as the most common shark in Elkhorn Slough, Monterey Bay (Talent 1976), and as the second most abundant shark in San Francisco Bay (Herald and Ripley 1951).

Size and weight There is a report of a 213.4-cm specimen (Bane and Bane 1971), which is the largest leopard shark on record. A 146-cm female weighed 15.8 kg.

Biology The leopard shark is a common inhabitant of shallow, cool estuarine and coastal waters. It has been reported from bays, estuaries, mud flats, sloughs, and so on, as well as from shallow rocky shores and kelp forests. It is often encountered in schools composed of individuals of the same size and sex. Despite the abundance of the leopard shark in accessible areas, its biology is poorly understood.

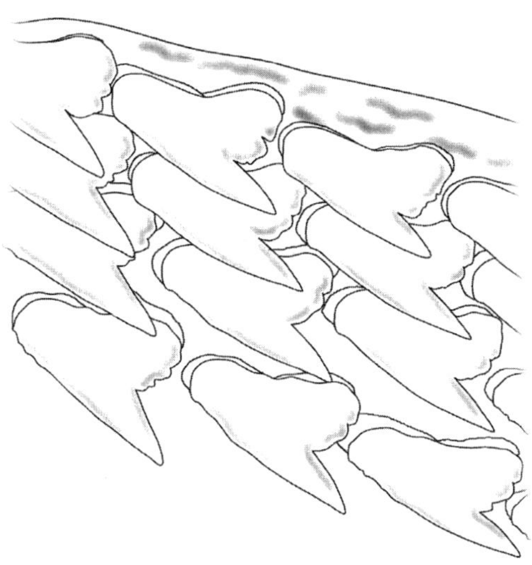

Fig. 103b. Teeth (upper).

Fig. 103c. Teeth (lower).

Diet: The leopard shark is an opportunistic benthic feeder that preys on a great variety of invertebrates and small fishes and will take whatever prey is locally abundant and easy to capture.

Herald et al. (1960) reported that leopard sharks off Monterey Bay, California, fed mainly on crabs, clams, and small fishes, primarily on midshipman (*Porichthys notatus*). Russo (1975) reported that leopard sharks fed on shrimp, crabs, fish, fish eggs (smelt, *Atherinopsis* sp.), clam necks, and worms. He also observed that, in Tomales Bay, California, in June, leopard sharks were so engorged with fish eggs that handling caused them to regurgitate. Talent (1976) undertook the most comprehensive study to date, examining 436 specimens from Elkhorn Slough, Monterey Bay, California. Of these, 367 had food in their stomachs. Leopard sharks fed mainly upon crabs, clams, fish, fish eggs, and innkeeper worms (*Urechis caupo*). The size of the sharks determined their diet. Sharks less than 90 cm in length fed mainly on the yellow shore crab (*Hemigrapsus oregonensis*), but as sharks grew larger, echiuroid worms became the most

Fig. 103d. Snout.

Fig. 103e. Dermal denticles.

Fig. 103f. *Triakis semifasciata*, leopard shark, adult female, about 170 cm, with unusual markings. Drawing based on a captive shark held at the Birch Aquarium, La Jolla, California in May 2009, accomplished with the assistance of Bob Burhans and Kara Yopak.

important item in the diet. Sharks 130–140 cm fed mainly on fishes. Webber and Cech (1998) examined 33 leopard sharks from Tomales Bay, California, and reported that at one site *Cancer* crabs were found in 87% of the stomachs containing food. At another site in the bay, innkeeper worms were found in 82% of the stomachs, while midshipman were the fishes most often encountered in the stomach contents. Ebert and Ebert (2005) reported that fish eggs (smelt, *Atherinopsis californiensis*) were the most important item in the diet of leopard sharks in Humboldt Bay, California.

The movements of leopard sharks are complex and poorly understood. A study by Ackerman et al. (2000) in Tomales Bay showed leopard sharks moved toward the inner bay with rising tides, presumably to feed on the abundant clams and worms found in the bay. The large-scale migrations of the leopard shark, if any, remain to be studied.

There is some evidence of habitat partitioning by leopard sharks in some areas. Talent (1985) reported that, in Elkhorn Slough, Monterey Bay, California, there were pronounced differences in the seasonal length–frequency distribution of leopard sharks larger than 60 cm. Leopard sharks 60–100 cm were often taken during summer and fall, but relatively few were taken in winter and spring. By contrast, leopard sharks 100–140 cm were captured throughout the year but were most abundant in winter and spring.

Size at maturity

Male: According to Kusher et al. (1992), males mature at 100–105 cm.

Female: Females mature at about 104–106 cm. This is based on the sizes of the smallest gravid females reported (106 cm, Bane and Bane 1971; 104 cm, Talent 1985; 105 cm Kusher et al. 1992).

Reproduction Development is aplacental viviparous. There are no comprehensive reproductive studies on leopard sharks, and some of the observations in the literature are difficult to reconcile with each other.

The reproductive cycle appears to be annual. Several authors (Herald et al. 1960; Talent 1985; Ebert and Ebert 2005) have commented that females examined in April and early May contained near-term young, but they failed to observe the condition of the ovary (an active ovary with ripe oocytes at the end of gestation would indicate an annual cycle, while an inactive ovary would indicate a longer

Fig. 103g. *Triakis semifasciata*, leopard shark, neonate male, 27.8 cm, 0.7 kg; from California, showing unusual markings (specimen courtesy of Tom Watts-FitzGerald).

cycle). Given that many other triakid sharks have annual reproductive cycles, it is *likely* that the leopard shark has an annual cycle also.

The gestation period apparently lasts 10–11 months. This is based on the observation of females carrying ripe oocytes or newly fertilized eggs in May and June and term young in late April and May. Herald et al. (1960: 65) stated that many small juveniles "no more than a few weeks old" were taken in June at the fishing derbies in Monterey Bay, California, and that none of the adult females carried embryos at that time, although "many of these adult females had conspicuous eggs in the ovaries." Talent (1985: 213) reported that, in Elkhorn Slough, Monterey Bay, females release their young in April and May, and stated that "all gravid females examined in April had near-term young but many of those examined in late May contained eggs with little embryonic development." Ebert (2003) estimated that gestation lasts 10–12 months.

A group of leopard sharks, apparently engaged in mating behavior, was observed, from a distance above the water, in the shallows off La Jolla, California, in early August (Smith, S. E. 2005). It is difficult to reconcile this report with the May mating assumed above. This may be a case of a wider mating season than currently assumed. It is best to leave the analysis of this observation until more evidence of mating is obtained.

Size at birth: Undoubtedly, many researchers have examined gravid, term leopard sharks, but it appears that no one has recorded the sizes of the largest embryos, nor have they recorded the sizes of the smallest free-swimming young. Most workers have reported the length of the young at birth as "less than a foot long" (Feder et al. 1974). The size of the young at birth appears to be variable. The young measure 17–20 cm at birth (Ackerman 1971).

Brood size: The broods of the leopard shark are relatively large. The brood size range has been reported as 6–24 young (Talent 1985) and 7–36 young (Ackerman 1971).

Nurseries The nurseries of the leopard shark are in the shallow coastal bays of the California coast. Monterey, San Francisco, Tomales, Bodega, and Humboldt bays have all been reported as nursery areas of the leopard shark (Bane and Bane 1971; Talent 1985; Ebert 2003). According to Ebert (2003), in Humboldt and San Francisco bays, females release their young in eel grass beds.

Age and growth The available age estimates for the leopard shark vary widely. Kusher et al. (1992) counted growth rings in vertebral centra that had been stained with silver nitrate and exposed to ultraviolet light to elucidate the growth rings. Later, they used mark–recapture studies to validate their age estimates. They estimated that males mature at 100–105 cm, at ages 7–13 years, and females at 105–135 cm and ages 10–15 years, but these estimates lack precision. Ackerman (1971) suggested that leopard sharks live 13–16 years, while S. E. Smith et al. (2003) estimated that they live 25–30 years. If the estimates of age at maturity of 7–15 years are accurate, Smith et al.'s longevity estimate appears plausible.

Relation to humans The leopard shark is often sold in California fish markets, its flesh being of excellent quality. The leopard shark is often exhibited in aquaria, where it is a popular shark that survives several years. Clark (1963a) reported that one survived 10 years in captivity. In the early twenty-first century, neonate leopard sharks became popular aquarium sharks due to their small size, showy color pattern, and hardiness. Their popularity and profitability resulted in many specimens being exported both legally and illegally from California into the eastern states and elsewhere.

The leopard shark is generally timid and harmless, usually fleeing from divers. There is one unusual incident attributed to the leopard shark, when a 90-cm shark repeatedly struck a scuba diver on the head. The diver had been bleeding through the nose and the blood-tainted, exhaled bubbles apparently attracted the shark, since it was next observed making a pass at the rising stream of bubbles and snapping at them. A few seconds later the shark made another strike made at the diver's face. The strike was warded off, and the fish disappeared (Dewitt 1955).

FAMILY CARCHARHINIDAE

The Requiem Sharks

The carcharhinids, also known as requiem sharks, whaler sharks, or gray sharks, comprise one of the largest shark families. These are small to large sharks characterized by a flattened but not laterally expanded snout, eyes with a well-developed nictitating eyelid, the fifth gill slit positioned over or behind the pectoral fin origin, the first dorsal fin origin well ahead of the pelvic fins, well-developed precaudal pits, and a caudal fin that measures less than one-third the total length, its upper lobe about twice as long as the lower lobe. Requiem shark teeth are usually bladelike; characteristically, the upper teeth are triangular with serrated edges, while the lower teeth are narrow and smooth edged. Despite the name "gray sharks," some are brown, bronze, blue, or olive. All requiem sharks are viviparous; most are placental. Most requiem sharks are voracious predators, feeding on crustaceans, mollusks, smaller sharks, rays, and bony fishes. Usually the smaller species are found close to shore, while the larger ones are found offshore. However, some large species, for example, the tiger and lemon sharks, often enter very shallow water.

Economically, this is the most important shark family. Many species are used for food, oil, leather, or fish meal. Some species cause great losses to longline fisheries through depredation of hooked fishes and to trawl fisheries by damaging nets. A few species are dangerous to humans.

Twelve genera and about 60 species are currently recognized in the Carcharhinidae; seven genera are represented in North America: *Carcharhinus, Galeocerdo, Nasolamia, Negaprion, Prionace, Rhizoprionodon,* and *Triaenodon.*

The genus *Carcharhinus* constitutes the largest genus in the Carcharhinidae. These sharks are characterized by an internal nictitating eyelid; no spiracle; bladelike upper teeth with single, triangular, or nearly triangular serrated cusps; erect lower teeth with either serrate or smooth-edged cusps; first dorsal fin origin closer to the pectoral axil (insertion) than to the pelvic fin origin; second dorsal fin less than half the height of the first; upper and lower precaudal pits present; and a caudal peduncle without lateral keels. For identification purposes, the genus is often divided into "smooth-backed" and "ridge-backed" species, depending to the presence or absence of an interdorsal ridge, an elevated dermal ridge extending between the two dorsal fins. The neonates and small young of many species have black-tipped or black-margined fins; these markings are soon lost. For additional information on the genus *Carcharhinus,* see the excellent monographs by Garrick (1982, 1985). Although there are more than 100 nominal species of *Carcharhinus,* Garrick (1982, 1985) reduced the number of species in the genus to 30; of these, 16 species inhabit North American waters. There are reports, sometimes questionable, of a few other species straying into this area.

Galeocerdo is certainly the most distinctive genus. It sole representative is the tiger shark. It is characterized by very long upper labial furrows, a relatively large spiracle, prominent keels on the caudal peduncle, and characteristic teeth that are similar in both jaws.

Nasolamia, a monotypic genus, differs from all other carcharhinid sharks in its long snout with large, transverse nostrils, which are set so closely together that the internarial distance is only slightly greater than the nostril width. Its only species is the whitenose shark.

Negaprion is distinguished by a second dorsal fin about three-quarters as high as the first, erect narrowly triangular teeth with smooth-edged cusps with weak serrations at their bases, no spiracle, the midpoint of the first dorsal fin nearer to the pectoral axil than to the pelvic fin origin, and lacking an interdorsal ridge. The only North American representative of the genus is the lemon shark.

Prionace is also a monotypic genus; its single species is the blue shark. The genus differs from all other carcharhinid sharks by the position of its first dorsal fin; its midpoint is much nearer to the origin of the pelvic fins than to the pectoral axil. Its upper teeth with strongly curved cusps are also characteristic.

Sharks of the genus *Rhizoprionodon* are small sharks of warm coastal waters and are characterized by teeth with smooth-edged, oblique cusps; second dorsal fin origin over or behind the middle of the anal fin base; long upper labial furrows; and enlarged hyomandibular pores on the outer side of each corner of the mouth. There are at least seven species in the genus (Springer, V. G. 1964), one of which is found in North America.

The genus *Triaenodon* is characterized by an internal nictitating eyelid; very short and blunt snout; tricuspid teeth with smooth edges, the central cusp being much larger than the others; first dorsal fin origin about equidistant between pectoral fin axil and the pelvic fin origin; and no interdorsal ridge. Although several nominal species of *Triaenodon* have been described, most authors consider the genus to be monotypic, its sole representative, the whitetip reef shark, being widely distributed in the Indo-Pacific region.

The genus *Isogomphodon* includes only one species, the daggernose shark. This is a peculiar shark with an extremely long snout, small eyes, large pectoral fins, and very small and numerous teeth. It inhabits the northeastern coast of South America and is included here as an extralimital species because the limits of its northern range are uncertain and to round out the accounts of American requiem sharks.

KEY TO THE CARCHARHINIDAE[1]

1a. Spiracles present, (teeth with curved cusps having serrated edges and a deep notch on the lateral margins, upper labial furrow as long as snout, prominent ridges on caudal peduncle, dorsal surface with dark blotches or bars) .... **Tiger shark,** *Galeocerdo cuvier* (p. 466).

1b. Spiracles absent 2

2a. Second dorsal fin at least three-fourths as high as first dorsal fin................................ 3

2b. Second dorsal fin less than one-half as high as first dorsal fin 4

3a. Upper and lower teeth with narrow erect cusps on broad bases (snout much shorter than mouth width, fins lacking white tips)............................**Lemon shark,** *Negaprion brevirostris* (p. 481).

3b. Upper and lower teeth with erect to oblique cusps and large lateral cusplets (snout much shorter than mouth width, darker dorsal fins with white tips).................. **Whitetip reef shark,** *Triaenodon obesus* (p. 501).

4a. Midpoint of first dorsal fin base closer to the pelvic fin origin than to the pectoral axil (*dorsal coloration bright blue in life, fading to dark gray after death, snout much longer than mouth width*).................. **Blue shark,** *Prionace glauca* (p. 486).

4b. Midpoint of first dorsal fin closer to pectoral axil than to pelvic fin origin 5

5a. Nostrils transverse to body axis and closely set, the internarial space about 1.3 times nostril width or less; snout tip with a small black spot surrounded by a white area (sharks of less than 150 cm inhabiting the eastern Pacific) .... **Whitenose shark,** *Nasolamia velox* (p. 478).

5b. Nostrils oblique to body axis and not perpendicular to it, internarial space at least three times nostril width, snout plain or with a dusky spot 5

6a. First dorsal fin only slightly tapered toward its apex, which is broadly rounded, most fin tips white mottled (specimens over 120 cm), or if black tipped (specimens less than 120 cm) there are also black dorsal saddle blotches on the caudal peduncle .... **Oceanic whitetip shark,** *Carcharhinus longimanus* (p. 438).

6b. First dorsal fin obviously tapered toward the apex, fins not white mottled, and if black tipped there are no black dorsal saddle blotches on the caudal peduncle................................ 6

7a. Interdorsal ridge absent................................ 7

7b. Interdorsal ridge present 15

8a. Labial furrows present, at least three-fourths of eye diameter in length 8

8b. Labial furrows absent or, if present, less than one-half of eye diameter in length 9

---

1. The key does not include extralimital species. The key is not designed for neonates or small juveniles which may have dark markings.

9a. Total number of teeth on outer row of upper jaw 27–30 (a small shark less than 120 cm inhabiting the eastern Pacific). .... **Pacific sharpnose shark,** *Rhizoprionodon longurio* (p. 493).

9b. Total number of teeth on outer row of upper jaw 23–27 (a small shark less than 125 cm inhabiting the western Atlantic) .... **Atlantic sharpnose shark,** *Rhizoprionodon terraenovae* (p. 496).

10a. Teeth with curved cusps, the lateral margin curved .... **Copper shark,** *Carcharhinus brachyurus* (p. 403).

10b. Teeth cusps with erect cusps, not curved, their inner margins straight or nearly straight, and lateral margins that are either straight, notched, or weekly curved.. 10

11a. Pectoral fins and lower lobe of caudal fin with black tips.. 11

11b. Pectoral fins and lower lobe of caudal fin plain or with dusky tips but not black tipped.. 12

12a. First dorsal fin origin anterior to pectoral fin free rear tip.. 13

12b. First dorsal fin origin over or posterior to pectoral fin free rear tip (lower teeth smooth edged, anal fin black tipped on sharks of more than 80 cm) .... **Spinner shark,** *Carcharhinus brevipinna* (p. 407).

---

13. Upper teeth with oblique cusps and a pronounced notch on their outer margins (inhabiting the eastern Pacific).. **Pacific smalltail shark,** *Carcharhinus cerdale* (p. 410).

13b. Upper teeth with erect symmetrical cusps (anal fin white in western Atlantic specimens, black tipped in eastern Pacific specimens).. **Blacktip shark,** *Carcharhinus limbatus* (p. 432).

14a. Snout much shorter than mouth width; teeth broadly triangular, lacking a notch on their outer margins; pectoral fins with dusky undersides (anal fin and lower lobe of caudal fin black tipped in young specimens under 100 cm).. **Bull shark,** *Carcharhinus leucas* (p. 426).

14b. Snout longer than the mouth width; teeth with either narrow and erect cusps or oblique cusps with a notch on the outer margins; dorsal, pectoral, and caudal fins unmarked.. 14

15a. Upper and lower teeth similar, with narrow, erect, smooth-edged cusps (gill slits about as long as second dorsal fin base) .... **Finetooth shark,** *Carcharhinus isodon* (p. 421).

15b. Upper and lower teeth dissimilar, upper teeth with narrow, increasingly oblique cusps with serrated margins and a deep notch on their outer margins.. 15

16a. Second dorsal fin origin over anal fin origin (snout tip usually with a dusky or black blotch).. **Blacknose shark,** *Carcharhinus acronotus* (p. 393).

16b. Second dorsal fin origin over the midpoint of anal fin base.. **Smalltail shark,** *Carcharhinus porosus* (p. 459).

17a. First dorsal fin origin anterior or over pectoral fin axil, or at least, nearer to axil than to free rear tip.. 17

17b. First dorsal fin origin slightly anterior to, over, posterior to pectoral fin free rear tip; if in front it is still nearer to the pectoral free rear tip than to the pectoral axil.. 18

18a. Usually 14 teeth on each side of the upper jaw, upper lateral teeth not longer than wide (the length of the enamel of the lateral teeth is about equal to the width of the root), first dorsal fin almost twice the snout length (adults only), precaudal vertebrae 82–97 (nonoverlapping skin denticles) .... **Sandbar shark,** *Carcharhinus plumbeus* (p. 453).

18b. Usually 15 or more teeth on each side of the upper jaw, upper teeth much longer than wide, first dorsal fin only slightly longer than snout length, precaudal vertebrae 101–110 (nonoverlapping skin denticles)...................................
.... **Bignose shark,** *Carcharhinus altimus* (p. 400).

19a. First dorsal fin origin clearly posterior to the pectoral fin free rear tip (inner margin of second dorsal fin free rear tip twice as long as height of the fin, teeth strongly notched on their lateral margins and weekly notched on their medial margins, with slightly coarser serrations at their bases) .... **Silky shark,** *Carcharhinus falciformis* (p. 413).

19b. First dorsal fin origin over or slightly posterior to the pectoral fin free rear tip 19

20a. First dorsal, pectoral, anal, and caudal fins with white or silvery tips (inhabits the Pacific Ocean) ....
.... **Silvertip shark,** *Carcharhinus albimarginatus* (p. 397).

20b. Fins tips dark, not white-tipped.. 21

21a. Not more than 13 upper teeth and 12 lower teeth on each side of the jaw (inner margin of second dorsal fin free rear tip about as long as fin height).. **Caribbean reef shark,** *Carcharhinus perezii* (p. 449).

21b. Usually at least 14 upper teeth and 13 lower teeth on each side of the jaw 20

22a. Upper teeth narrow, their lateral margins deeply concave or notched 21

22b. Upper teeth broad, their lateral margins concave but not notched, their medial margins straight or convex but not concave 22

23a. Upper teeth with concave lateral margins and finely serrated margins ....
.... **Copper shark,**[2] *Carcharhinus brachyurus* (p. 403).

23b. Upper frontal teeth with strongly oblique cusps with a marked notch on the lateral margin and two to five coarse serrations from notch to base, and smooth tips.. **Night shark,** *Carcharhinus signatus* (p. 463).

24a. Height of second dorsal fin 2.1–3.3% of total length, 103–109 precaudal vertebrae (inhabiting oceanic islands)..........
.... **Galapagos shark,** *Carcharhinus galapagensis* (p. 417).

24b. Height of second dorsal fin 1.5–2.3% of total length, 86–97 precaudal vertebrae (inhabiting continental waters) ....
.... **Dusky shark,** *Carcharhinus obscurus* (p. 444).

---

2. The copper sharks that I have examined all were smooth backed. However, other authors have reported some specimens to have weak interdorsal ridges. I have included it in both groups (smooth backed and ridge backed) in the key.

BLACKNOSE SHARK

Carcharhinus acronotus (Poey, 1860)

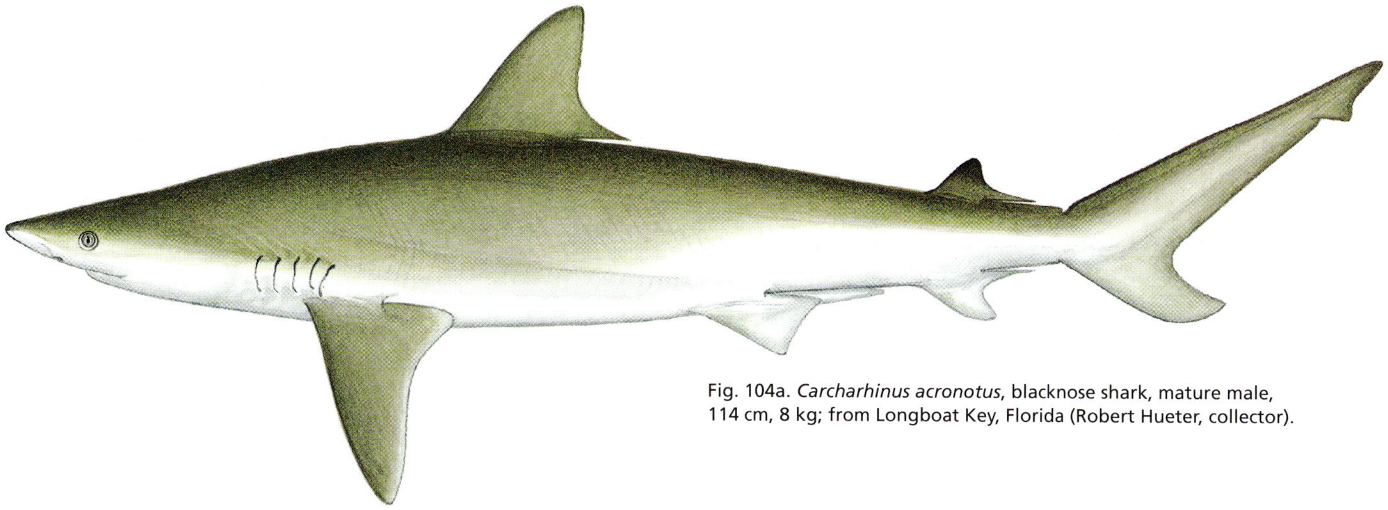

Fig. 104a. *Carcharhinus acronotus*, blacknose shark, mature male, 114 cm, 8 kg; from Longboat Key, Florida (Robert Hueter, collector).

Common name Blacknose shark, or just blacknose, a reference to the black mark on the tip of the snout.

Spanish name *Tiburón limón* (Cuba), a reference to its yellowish color; *cazón limón* or *canguay* (Mexico). The A.F.S. name is *tiburón canguay.*

Scientific name *Carcharhinus acronotus* (Poey, 1860), originally described as *Squalus acronotus* in *Memorias sobre la Historia Natural de la Isla de Cuba* 2(197): 335–336, plate 18, figs. 3 and 4. *Carcharhinus:* from the Greek *karcharos,* sharp pointed or jagged, from *karcharias,* a species of dogfish having jagged teeth, and *rhine,* Greek file or rasp, also the name of a shark, originally given to *Squatina* due to its rasplike skin; *acronotus:* from the Greek *acron,* peak or extremity, and *notus,* from the Greek *notos,* the back, thus "having a raised or humped back," a point made by Poey about the type specimen.

Synonyms *Carcharias remotus* Valenciennes in Dumeril, 1864.

Identification The blacknose shark is *a smooth-backed species, lacking an interdorsal ridge.* It has a snout as long as, or longer than, the mouth width; a first dorsal fin originating over or posterior to the free rear tips of the pectoral fin; and a second dorsal fin originating slightly posterior to the origin of the anal fin. The *upper teeth have narrow, oblique cusps strongly notched on their outer margins, with serrated edges* that become progressively coarser toward the bases. The lower teeth have narrow, erect, symmetrical cusps, with more finely serrated edges than the upper teeth. Teeth number U: 12 or 13–1 or 2–12 or 13, L: 11 or 12–1–11 or 12. The dermal denticles are dense and overlapping; they are squarish with five to seven parallel ridges. Coloration is pale greenish or yellowish gray above, although in the sun live specimens appear yellowish brown with bronze hues above; undersides are whitish, yellowish, or pale greenish gray. The tip of the second dorsal fin is dusky or black. *The undersides of the pectoral fins are unmarked.* There is a *dusky blotch at the tip of the snout,* distinct and dark in the young, diffuse and dusky in adults. The blotch tends to fade with age and may disappear in the larger individuals.

Similar species The bull shark has a shorter snout, pectoral fins with dusky ventral tips, and teeth that are triangular and unnotched, and lacks the dark blotch on the snout. The sharpnose shark lacks the spot on the nose and has marked labial furrows. The smalltail shark has a second dorsal fin originating farther back, over or slightly posterior to the midpoint of the anal fin base, and lacks the dusky blotch on the nose. Other similar requiem sharks have an interdorsal ridge or lack strongly notched teeth.

Range The blacknose shark inhabits the western North Atlantic from North Carolina to Brazil (Bigelow and Schroeder 1948). In North America, the species ranges from southern Virginia to Florida and the Gulf of Mexico. It is only a stray north of Cape Hatteras, occasionally being found as far north as Wreck Island, Virginia (J. A. Musick, pers. comm., May 2004). In summer, the blacknose shark is one of the most abundant small sharks in coastal waters, from the Carolinas to Florida and the Gulf of Mexico (Castro 1983). Dodrill (1977) reported it as the third most

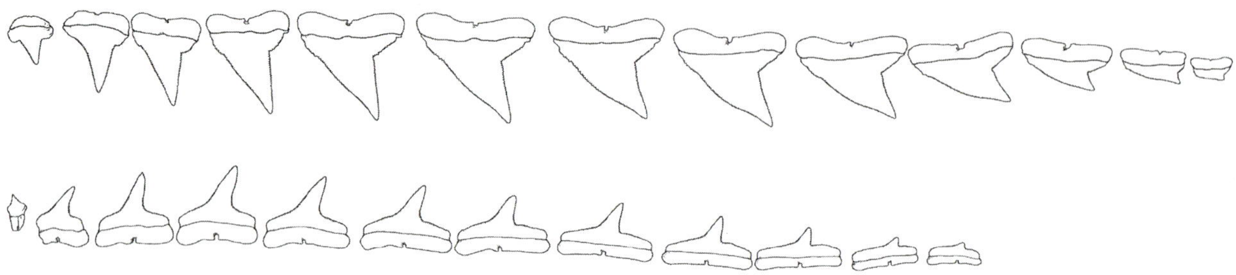

Fig. 104b. Teeth.

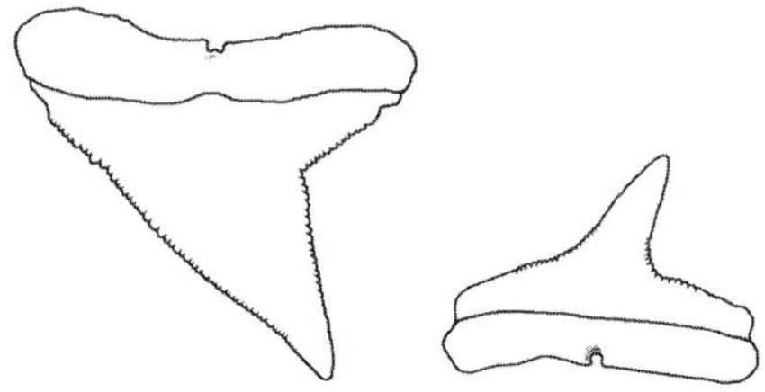

Fig. 104c. Teeth (detail).

abundant shark off Melbourne Beach, Florida, present there throughout the year. Clark and von Schmidt (1965) reported the species was found off Sarasota from March to November, with the greatest catches in May.

Size and weight The blacknose shark may grow to 150 cm and about 17 kg. Specimens from the Gulf of Mexico tend to be somewhat smaller than those taken in the Atlantic. Springer (1938) stated that the blacknose shark reaches about 137 cm, based on specimens from the Gulf coast of Florida. The largest specimen reported by Clark and von Schmidt (1965) was a 126-cm female, while the largest reported by Carlson et al. (1999) was a female 129.6 cm. Driggers (2004) reported that the largest specimen sampled off South Carolina measured 135.3 cm. The largest specimen I have seen was a female that measured 141.3 cm and weighed 14.25 kg; it was taken off South Carolina in 1992. The heaviest specimen I examined was a 140-cm female

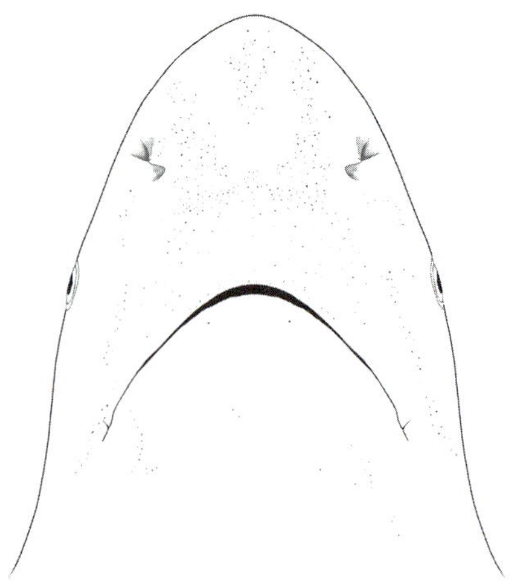

Fig. 104d. Snout.

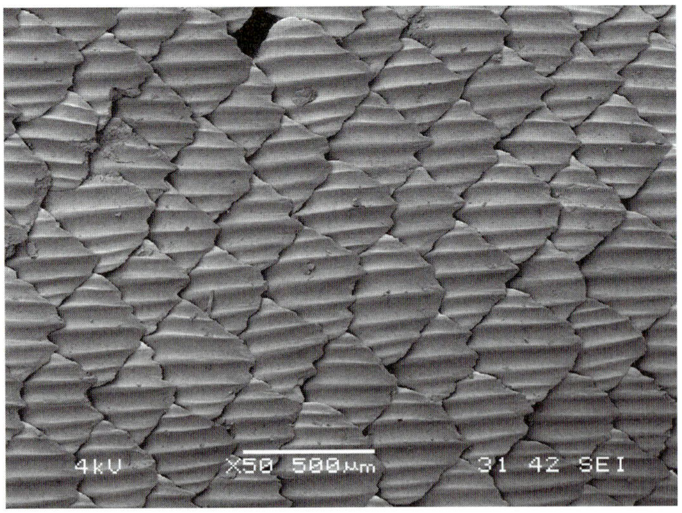

Fig. 104e. Dermal denticles.

weighing 15 kg. It was caught off South Carolina in 1979. Schwartz (2003) reports North Carolina specimens measuring 183 and 206 cm caught in 1975. These sizes are so much larger than other blacknose sharks reported that I suspect these specimens were misidentified. Representative weights of specimens that I have examined in South Carolina are as follows:

Male: 69 cm, 2.5 kg; 92.8 cm, 4.55 kg; 102 cm, 6.8 kg; 121 cm, 10.0 kg; 126.7 cm, 11.1 kg; and 128.5 cm, 11.8 kg.

Female: 102 cm, 5.9 kg; 120 cm, 8.1 kg; 131.5 cm, 10.4 kg; 136.2 cm, 14.3 kg; 140 cm, 14.97 kg; and 141.3 cm, 14.25 kg.

Biology The blacknose shark is a common, small species inhabiting shallow, warm, coastal waters, usually at depths of 10 m or less. It is quite tolerant of warm temperatures. In August and September in South Carolina, where the temperatures in shallow waters (2–4 m) exceed 28–30°C, the blacknose and the finetooth sharks become the dominant sharks in shallow water, the sharpnose sharks and other species having migrated to cooler waters farther offshore.

The blacknose shark feeds primarily on small bony fishes, usually less than 40 cm, and on small rays. I found menhaden, spot, anchovies, whiting, tonguefish, and rays in the stomachs of South Carolina specimens. According to Dodrill (1977). blacknose sharks off Melbourne Beach, Florida, feed on small mullet (*Mugil* spp.) and bluefish (*Pomatomus saltatrix*).

The blacknose shark is a migratory species but its movements are poorly known. Available data indicate that some blacknose sharks migrate along the eastern coast of North America (Kohler et al. 1998). It is possible that they may undertake offshore/onshore movements as well. The blacknose shark occurs year-round off southern Florida; I have seen gravid females caught off Long Key bridge channel almost year-round. Dodrill (1977) reported that the blacknose shark was absent off Melbourne Beach from early July to late August and suggested that the species moves either northward or offshore during that time.

Schwartz (1984) suggested that there are two populations of blacknose sharks in the southeastern United States, but these have not been characterized. Driggers (2002) analyzed the mitochondrial DNA of blacknose sharks from the Atlantic and from the Gulf of Mexico and concluded that there were significant genetic differences between the two groups.

Size at maturity

Male: According to Driggers (2002), males off South Carolina reach maturity between 108 and 112 cm, based on clasper calcification.

Female: Female blacknose sharks mature at 103–118 cm, depending on location. Springer (1938) and Clark and von Schmidt (1965) estimated that along the western coast of Florida the blacknose shark matures at about 103 cm. As Driggers (2002) has shown, Atlantic blacknose sharks grow larger than Gulf of Mexico specimens, and they also mature at a larger size. He estimated that 50% of female blacknose sharks off South Carolina reach maturity at 118 cm.

Reproduction The blacknose shark is a placental, viviparous species. I found females off South Carolina carrying ripe oocytes 26 mm in diameter from early May to mid-June, just prior to ovulation. According to Driggers (2002), mating takes place in South Carolina in early to mid-June. In South Carolina, I found embryos of the following sizes at the indicated times: 8 July, eggs in blastodisc stages, 20-mm embryos; 19 July, 40–45 mm; 18 August, 77–102 mm; 8 September, 160–177 mm; 21 April, 404–426 mm; 14 May, 489–506 mm; 28 May, 389–409 mm.

The young have smooth umbilical cords, like other species of *Carcharhinus.* Hamlett's (1986) report of blacknose embryos having appendiculae on the umbilical cord is likely a misidentification of an Atlantic sharpnose shark. Embryos become implanted by late August and early September when they measure about 120 mm. I estimate gestation at 11–12 months, based on finding blastodisc-stage embryos in late June and early July, and on the latest date on which I found gravid females (28 May). Driggers (2002), also working in South Carolina, estimated the gestation period at about 11 months. Sulikowski et al. (2007) estimated a gestation period of 9–10 months for the Gulf of Mexico population. Birth occurs in coastal waters of the Atlantic and eastern Gulf of Mexico in June.

The blacknose shark is believed to have a biennial reproductive cycle (Dodrill 1977; Driggers et al. 2004), like other species of *Carcharhinus.* However, other authors (Schwartz 1984; Hazin et al. 2002; Sulikowski et al. 2007) have suggested that vitellogenesis and gestation occur concurrently and that blacknose sharks have an annual reproductive cycle. Sulikowski et al. (2007) stated that blacknose sharks in the Gulf of Mexico have an annual reproductive cycle, their most compelling evidence being the absence of nongravid females during the study. Why there are such profound differences between the Atlantic and Gulf of Mexico populations is unknown. A partial explanation for this discrepancy is that some term females carry very large vitellogenic oocytes. Some gravid females that I have examined during the last week of May off South Carolina carried term young 45–47 cm *and* six to eight oocytes 18–19 mm in diameter. The presence of relatively large oocytes at parturition may suggest an annual reproductive cycle. However, these oocytes are far from their ripe diameter of 26 mm, and it appears unlikely that they could reach maturity by mating time in June. Obviously, more research is needed on this subject.

Size at birth: Blacknose shark young are born at 43–50 cm. This estimate is based on embryos of 48.9–50.6 cm and on free-swimming young of 43–51 cm that I have examined off South Carolina in May. These data agree well with observations by others. Clark and von Schmidt (1965) reported a 45-cm neonate from the west coast of Florida. Dodrill (1977) estimated the birth size as 41.7–49.9 cm, based on the largest embryos that he had examined and on free-swimming young measuring 45 cm.

Brood size: Blacknose shark broods range from one to eight, but the usual number is five. In a sample of 27 gravid females from South Carolina that I examined, the average was five. One female carried only one young and another carried a brood of eight.

Nurseries The nurseries of the blacknose shark are extensive. They range from, at least, the shallow coastal waters of Georgia and South Carolina, to the shallow waters of central and southern Florida. Females carrying term young appear over a wide area in late April and early May. I found gravid females off McClellanville, South Carolina, from the third week in April; they also appear at this time off Melbourne Beach, Florida (Dodrill 1977), and in the Florida Keys (my observation). Although gravid females are often caught, from the Carolinas to Florida, in late May and early June, blacknose young are often difficult to collect because of their small size and because they are often found in areas of live bottom at depths of 10–12 m. During my years in South Carolina, I saw few neonate blacknose sharks (52–58 cm), although the replacement of the previously abundant gravid females by postpartum females, and a few neonates or small young observed indicated that birth had occurred nearby. Neonate blacknose sharks do not inhabit estuaries or shallow beach areas. Driggers et al. (2004) caught no neonates despite extensive effort in the estuaries and near shore waters of South Carolina. Although I examined the bycatch of many shrimp trawlers, I collected only one neonate from shrimp trawls, a 51-cm specimen caught off Charleston, South Carolina, on 31 July 1981. I collected two small young 56–57 cm at a depth of 10–12 m, using hook and line at two separate locations in the old "Charleston drill minefield," near the mouth of the Stono River on 19 August 1983.

In Florida, Dodrill (1977) reported a female releasing a 45-cm embryo off Melbourne beach in mid-May. I have examined gravid females with 47- to 48-cm embryos caught in Long Key channel, Florida, in early April, which indicated that the area and time of parturition were not too far away. Michel (2002) found no blacknose young in the mangrove estuary of the Ten Thousand Islands, on the southwestern coast of Florida, where there are young of several other species of sharks.

In the Gulf of Mexico, a blacknose shark nursery extends from (at least) Charlotte Harbor, Florida, to Tampa Bay. Mote Marine Laboratory personnel have collected large numbers of neonates and small juveniles in those areas, at depths of 2–30 m, in June and July (R. Hueter, pers. comm., May 2004).

Age and growth Clark and von Schmidt (1965) used catch data to estimate that the species probably reaches maturity in one or two years. Schwartz (1984) used vertebral growth rings to estimate that a male 123 cm was six years old, while females 115 cm were five or six years old. Carlson et al. (1999) used vertebral centra to estimate that the oldest blacknose sharks in their sample were more than 4.5 years old and gave a theoretical longevity of 10–16 years for females and 5–9 years for males. Driggers et al. (2004a, 2004b) reported that Atlantic males mature at 111 cm and 4.3 years, and females at 119 cm and 4.5 years, and gave a theoretical longevity of and 16.4 years for males and 19.0 years for females.

Relation to humans The blacknose shark is too small to be of interest to the shark fisheries in the United States. Since 1992, the blacknose shark has been managed in the United States in the category of "small coastal sharks." This category is separate from the larger requiem sharks, so the species has increased in importance because it can be fished when the season for the larger sharks is closed. In Mexico, the blacknose shark is locally consumed in the Campeche area, where it is smoked and used to prepare a regional dish known as *pan de cazón.*

SILVERTIP SHARK

Carcharhinus albimarginatus (Rüppell, 1837)

Fig. 105a. *Carcharhinus albimarginatus*, whitetip reef shark, immature female, 136.7 cm, 22 kg; from Ishigaki Islands, Japan (Keiichi Sato, collector).

Common name Silvertip shark, a reference to the white or silvery fin tips.

Spanish name *Tiburón puntas blancas* (A.F.S.).

Scientific name *Carcharhinus albimarginatus* (Rüppell, 1837), originally described as *Carcharias albimarginatus* in *Neue Wirbelthiere zu der Fauna von Abyssinien gehorig* 64, plate 18, fig. 1. *Carcharhinus:* from the Greek *karcharos,* sharp pointed or jagged, from *karcharias,* a species of dogfish having jagged teeth, and *rhine,* Greek file or rasp, also the name of a shark, originally given to *Squatina* due to its rasplike skin; *albimarginatus:* Latin, having white edges or borders, from *albus,* white, and *margo,* edge or border, an allusion to the white-bordered fins.

Synonyms *Eulamia* (*Platypodon*) *platyrhynchus* Gilbert, 1892. Rosenblatt and Baldwin (1958) first pointed out that Gilbert's description of *C. platyrhynchus* was based on specimens of two or three different species from Clarion and Socorro Islands and from Magdalena Bay, Baja California. Of these specimens, only one, the largest, had fins "conspicuously marginated in with white" (Gilbert 1892: 544). This largest specimen, from Socorro Island, was designated as the lectotype of *C. platyrhynchus* by Rosenblatt and Baldwin. Later, Garrick (1982) stated that Gilbert's specimen from Magdalena Bay (U.S. National Museum, USNM no. 46850) is a dusky shark (*C. obscurus*); thus, there are no valid records of the silvertip shark in continental North America.

Identification This is *a ridge-backed species* with a low interdorsal ridge. The snout is shorter than the mouth width. The first dorsal fin originates above or slightly anterior to the pectoral fins free rear tips; the second dorsal fin originates over or slightly behind the origin of anal fin. The apex of the first dorsal fin is obtusely pointed. The anal fin is deeply notched. The upper teeth have wide overlapping bases that narrow abruptly after a notch into a pointed crown, with coarse serrations on the bases and fine serrations on the cusp. The cusps become increasingly oblique toward the corners of the mouth. The lower teeth have broad bases and narrower cusps and have finely serrated edges. The dental formula is usually U: 13–1–13, L: 12–1–12 but can vary by one or two teeth in each jaw. The dermal denticles are dense and closely overlapping, with five to seven parallel ridges and points. This is one of the most beautiful carcharhinid sharks. The back is dark grayish brown. The upper sides are bronze brown with metallic hues in life that becomes lighter along the flanks, with a band of the bronze color extending down into the lighter lower sides to almost the level of the pelvic fins. There is a band of metallic blue that extends from behind the gills along the lower flanks and fades toward the pelvic fins. The second dorsal fin is darker than the back and has a narrow inconspicuous white margin. The undersides are white. The *fins have white rear*

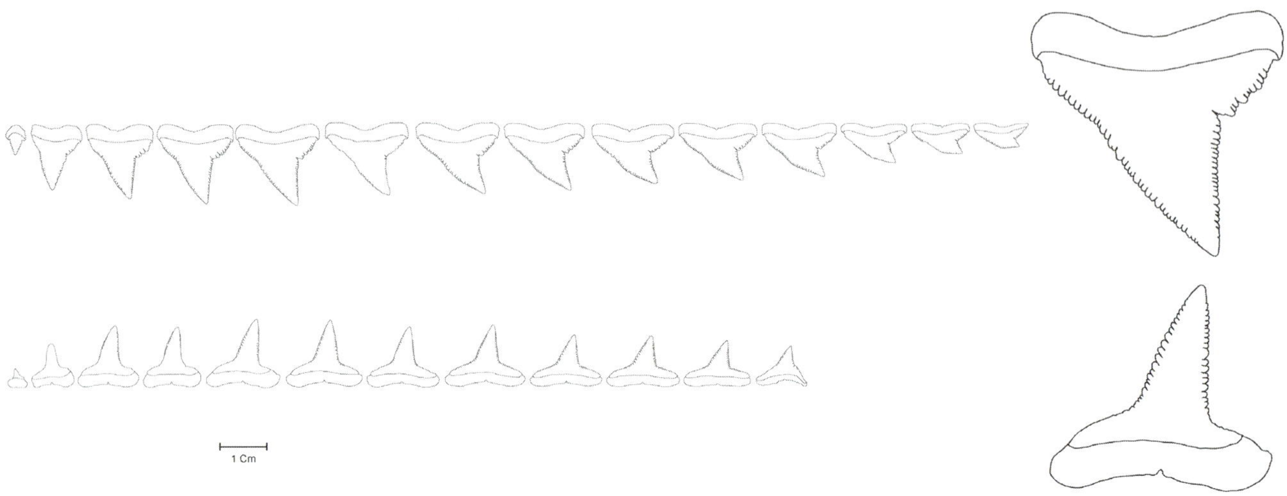

Fig. 105b. Teeth.

margins; the *first dorsal fin, pectoral fins, and tail have conspicuous white tips and rear margins;* the *second dorsal, pelvic, and anal fins have less conspicuous white margins.* The iris is a bronze color in daytime and pale light blue at night.

Similar species The only other species in the genus with white fin tips is the oceanic whitetip shark, whose fins are broadly rounded, and the white coloration of the fin tips does not extend to the trailing margins of the fins.

Range The silvertip shark is widely distributed throughout the tropical Indo-Pacific region. In the western Indian Ocean, it occurs from the Red Sea (type locality) south to Mozambique (Smith, J. L. B. 1958) and the Natal coast (Bass et al. 1973), Madagascar and the Comoros (Fourmanoir 1961), and the Mauritius and Seychelles islands (Wheeler 1962). According to Randall (1986), it is apparently absent from the Arabian Gulf and the Gulf of Oman. It has been reported from numerous western Pacific localities: northern Australia (Last and Stevens 1994), Taiwan (Chen 1963), the South China Sea (Kyushin et al. 1985), Bikini and Eniwetok Atolls in the Marshall Islands (Schultz et al. 1953), and Guam, Mariana Islands (Bryan 1973), just to name a few. Interestingly, it is not recorded from the

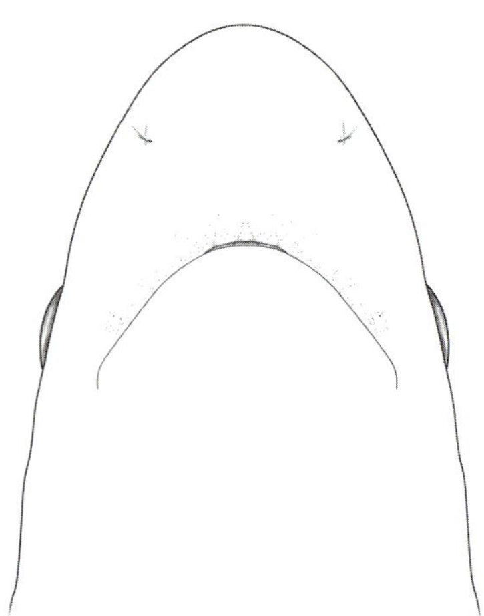

Fig. 105c. Snout.

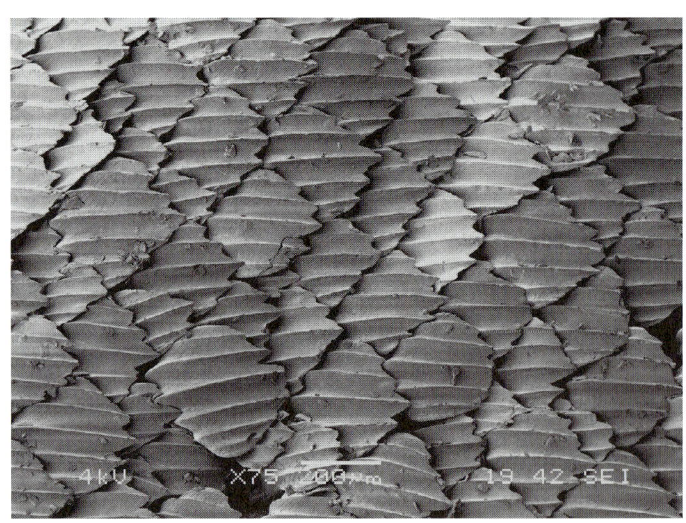

Fig. 105d. Dermal denticles. .

extensive surveys of the central Pacific and the Hawaiian Islands by Strasburg (1958) and Tester (1969), nor it is recorded in the survey of the sharks of Malaysia by Yano et al. (2005). In the eastern Pacific, it has been reported from Clipperton Island (Limbaugh 1963), and Cocos, Galapagos, and Revillagigedo islands and off Guatemala and Colombia (Kato et al. 1967). North American records are confined to the Revillagigedo Islands, some 375 km from the tip of the Baja Peninsula, where Kato and Hernandez Carvallo (1967) reported that juvenile silvertip and Galapagos sharks were the dominant species in the islands of Socorro and San Benedicto. The silvertip shark is not mentioned in faunistic records of California sharks by Miller and Lea (1972) or Ebert (2003); presumably California waters are too cold for the species.

Size and weight The silvertip shark reaches at least 275 cm. According to Wheeler (1962) sharks of up to 250 were fairly abundant in the Mauritius–Seychelles area, and his largest specimen measured 261 cm. Fourmanoir (1961) recorded a specimen of 275 cm. The few silvertip shark specimens that I have examined were all relatively stout for their sizes; the 137-cm immature female illustrated in this book weighed 22 kg.

Biology The silvertip shark is a common shark of the warm waters of the Indo-Pacific region. Despite its wide range and abundance, and its being an easily recognized species, the silvertip shark has received little attention from scientists, and there are few published reports of its biology. The silvertip shark does not enter the continental waters of North America, and I have seen only a few specimens in the Ishigaki Islands. Thus, this account is based on specimens from the western Pacific. The silvertip shark is usually reported from tropical oceanic islands, although it is also found in warm coastal waters. Randall (1986) reported that the silvertip shark inhabits the clear waters of outer-reef areas of islands and banks, particularly steep slopes generally at depths greater than 30 m. Juveniles are occasionally found in shallow lagoons, but adults are seldom seen there.

Size at maturity Males mature at about 180 cm and females at about 200 cm. Off the east coast of Africa, males start to mature at 160–170 cm and are mature at 174 cm (Bass et al. 1973). Bass et al. reported a gravid female of 199 cm. Wheeler (1962) stated that the smallest gravid female taken in his Mauritius–Seychelles Survey measured 215 cm.

Reproduction The silvertip shark is placental viviparous. According to Wheeler (1962), the silvertip shark drops its young in December or January (in the Mauritius–Seychelles area) after a gestation period of about a year.

Size at birth: Bass et al. (1973) reported that the silvertip shark is born at 70–80 cm and that four free-swimming specimens with open or recently closed umbilical scars ranged from 72 to 83 cm.

Brood size: Bass et al. (1973) reported a female with six embryos and combined this single brood with those reported by Fourmanoir (1961) and Wheeler (1962) to obtain a range of 1–10, with a mode of 6 and an average of 5.5, for 25 broods. Wheeler (1962) mentions a brood of 11.

Nurseries The nurseries have not been described, but juveniles have been reported as restricted to the shallow waters surrounding oceanic islands (Kato and Hernandez Carvallo 1967).

Age and growth Kato and Hernandez Carvallo (1967) reported that in the Revillagigedo Islands tagged juveniles grew 31–54 mm yearly.

Relation to humans Because the silvertip shark inhabits clear warm waters around islands, where good underwater photography is possible, and perhaps because of its showy markings, the silvertip shark is a favorite subject for underwater photographers. In the past, the silvertip shark was often described as potentially dangerous to divers (any large predatory shark can be dangerous to humans under some circumstances). Limbaugh (1963: 75) wrote, "They seemed vicious and unpredictable in their behavior." However, judging from the hundreds of photographs of silvertip sharks that have appeared in magazines and the Internet in the last decade, humans must be interacting with this species without any problems. Given its abundance around oceanic islands of the eastern Pacific, where dozens of illegal fishing vessels have been targeting sharks during the last decade, it is certain that large numbers of silvertip sharks have been taken for their fins. There are numerous accounts in the press of illegal vessels fishing in the protected areas of the Galapagos, Malpelo, and Cocos islands.

BIGNOSE SHARK

Carcharhinus altimus (Springer, 1950)

Fig. 106a. *Carcharhinus altimus,* bignose shark, mature male, 263 cm, 126 kg; from Bimini, Bahamas (Dean Grubbs, collector).

Common name Bignose shark, a reference to its long snout.

Spanish name *Baboso, tiburón baboso* (Cuba), *tiburón de aleta de cartón* (Mexico), *tiburón narizón* (A.F.S.).

Scientific name *Carcharhinus altimus* (Springer, 1950), originally described as *Eulamia altima* in *American Museum Novitates* 1451: 9–12. *Carcharhinus:* from the Greek *karcharos,* sharp pointed or jagged, from *karcharias,* a species of dogfish having jagged teeth, and *rhine,* Greek file or rasp, also the name of a shark, originally given to *Squatina* due to its rasplike skin; *altimus:* of the deep sea, from the Latin *altus,* deep, a reference to the species' habitat.

Synonyms *Carcharhinus radamae* Fourmanoir, 1961.

Identification The bignose shark has *a snout as long as, or longer than, the mouth width;* a *first dorsal fin usually originating over the pectoral axil, or behind it almost to the midpoint of the inner pectoral margin;* a *second dorsal fin origin anterior to the anal fin origin;* and a *marked interdorsal ridge.* The *upper teeth have broadly triangular cusps with finely serrated edges, and overlapping base;* the *front two teeth are large, symmetrical, and erect, while subsequent teeth (on the sides) have slightly curved cusps that become progressively oblique.* The lower teeth have narrow, erect cusps with finely serrated edges and broad bases; their serrations are so fine that they can only be seen with a 10× lens. Normally, there are two rows of functional lower teeth. Teeth number U: 14 to 16–0 to 2–14 to 16, L: 14 or 15–1–14 or 15. The *skin denticles are slightly oval shaped, with three parallel ridges and points, and are nonoverlapping* (or barely touch each other, and the skin can be seen in the spaces among the denticles). Coloration is bronze gray or brown above with a band of the darker coloration extending downward along the lighter flanks to the pelvic fins (Z pattern) and white or whitish below. The ventral sides of the pectoral tips are dusky or black. The young have black-tipped fins.

Similar species The sandbar shark is another ridge-backed shark with nonoverlapping skin denticles. The sandbar shark has a shorter snout (shorter than the mouth width) and a higher first dorsal fin (about twice the snout length). Sandbar sharks generally are not found in the deep waters where bignose sharks live. The Galapagos shark has a shorter snout and overlapping skin denticles. The dusky shark has the first dorsal fin origin over the free rear tips of the pectoral fins and has overlapping skin denticles. The silky shark has the first dorsal fin origin behind the free rear tips of the pectoral fin and has overlapping skin denticles.

Range The bignose shark is cosmopolitan in deep tropical and subtropical waters. In the western Atlantic Ocean, it ranges from off Delaware Bay (Kohler et al. 1998) to Brazil (Sadowsky et al. 1987). Springer (1950a) reported it from Cape Canaveral to Dry Tortugas, the Bahamas, off Nicaragua, and Costa Rica and in deep water in the Dragon's Mouth at the entrance to the Gulf of Paria. Guitart Manday (1968) reported it from Cuba. In the eastern Atlantic, it has been reported from off Senegal and the Ivory Coast

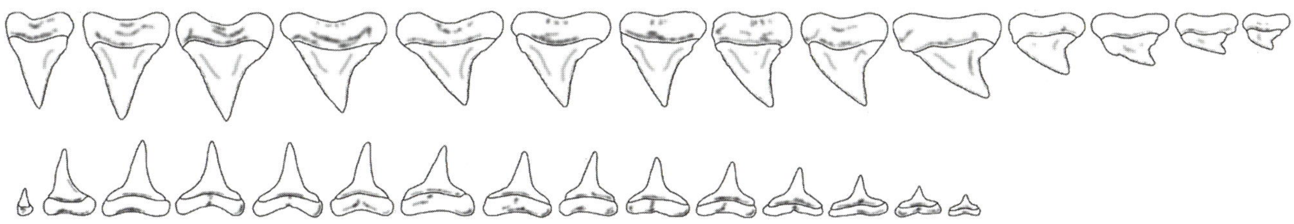

Fig. 106b. Teeth.

(Cadenat 1961). In the Pacific Ocean, it has been found off Hawaii (Tester 1969), and in the eastern Pacific, it has been found from near the mouth of the Gulf of California to Colombia, Ecuador, and the Revillagigedo Islands (Kato et al. 1967). In the Indian Ocean, Bass et al. (1973) reported it from the Natal coast, South Africa. It has also been reported from Australia (Anderson and Stevens 1996). In North America it has been reported from New York to Florida and the Gulf of Mexico (Kohler et al. 1998; Springer 1950a). The bignose shark is common in deep waters of southern Florida, the Bahamas, and the West Indies.

Size and weight Springer (1950a) gave the mean total length of a sample of 26 bignose sharks taken around Bimini and South Florida as 240 cm for males and 253 cm for females; the largest male measured 267 cm and the largest female 282 cm. Representative sizes and weights are as follows: 118 cm, 8.9 kg; 136 cm, 15.8 kg; 157 kg, 33 kg; 169 cm, 26 kg; and 263 cm, 126 kg.

Biology This is a bottom-dwelling shark found in the deeper waters of the continental shelf in tropical and subtropical areas. According to Springer (1950a) the bignose shark is the most common shark on the edges of the continental shelf of the West Indian region. It is usually caught at depths of 130–365 m, although some have been taken at 90 m off Ft. Pierce, Florida (Springer 1950a). Reports from other areas indicate that the bignose shark occasionally visits surface waters at night. Baranes and Ben-Tuvia (1978) reported catching one specimen at depths of 25–30 m at night in the Red Sea. Off Hawaii, Tester (1969) reported taking two specimens on bottom longlines set depths of 27–46 m at night. Anderson and Stevens (1996) suggested that the bignose shark undertakes periodic vertical migrations, spending the day on deep continental shelves and slopes and moving into shallower water by night, with some individuals swimming up into surface waters.

Because of its deep-water habitat, the bignose shark is poorly known, despite its abundance in tropical and sub-

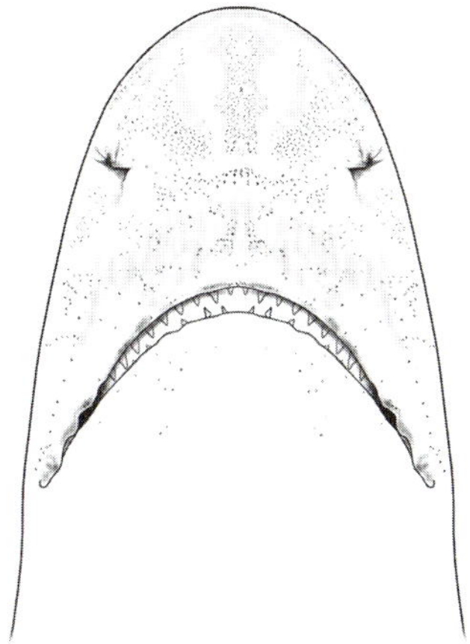

Fig. 106c. Snout.

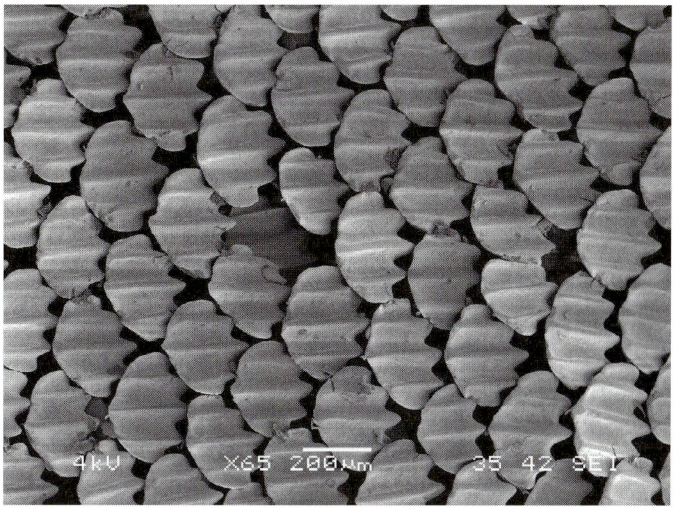

Fig. 106d. Dermal denticles.

tropical areas. It is known to feed on chimaeras, smaller sharks, squids, and octopuses.

The bignose shark undertakes north–south seasonal migrations along the East Coast of the United States, but little is known of these movements. Sharks tagged off Maryland and Delaware have been recaptured at various locations in the Gulf of Mexico and off Cuba (Kohler et al. 1998). This suggests that the bignose shark is a warm-water species that travels north from the Caribbean in the summer, migrating southward in the fall.

Size at maturity Maturity is probably reached at about 200 cm (Castro 1983).

Reproduction The bignose shark is placental viviparous. Springer (1950a) reported two females carrying 60-cm embryos in May, off Salerno, Florida.

Size at birth: The young are born in summer and can reach 75 cm at birth (Castro 1983).

Brood size: The broods range from 3 to 11, with seven being the usual number (Castro 1983).

Nurseries Unknown.

Age and growth The species has not been aged.

Relation to humans Bignose sharks are often taken in swordfish and tuna fisheries along the southeastern United States, where small specimens are sometimes mistaken for sandbar sharks. The bignose shark received protected status along the Atlantic coast of the United States in 1999 (National Marine Fisheries Service 1999). The effects of the protected status are difficult to evaluate; they are probably illusory. Bignose sharks inhabit the same waters as swordfish and tunas and take the hooks intended for these bony fishes, with usually fatal consequences. In United States waters, those sharks are released if alive or discarded dead; in other high seas fleets, only the fins are saved and the carcasses are discarded.

NARROWTOOTH OR COPPER SHARK

Carcharhinus brachyurus (Günther, 1870)

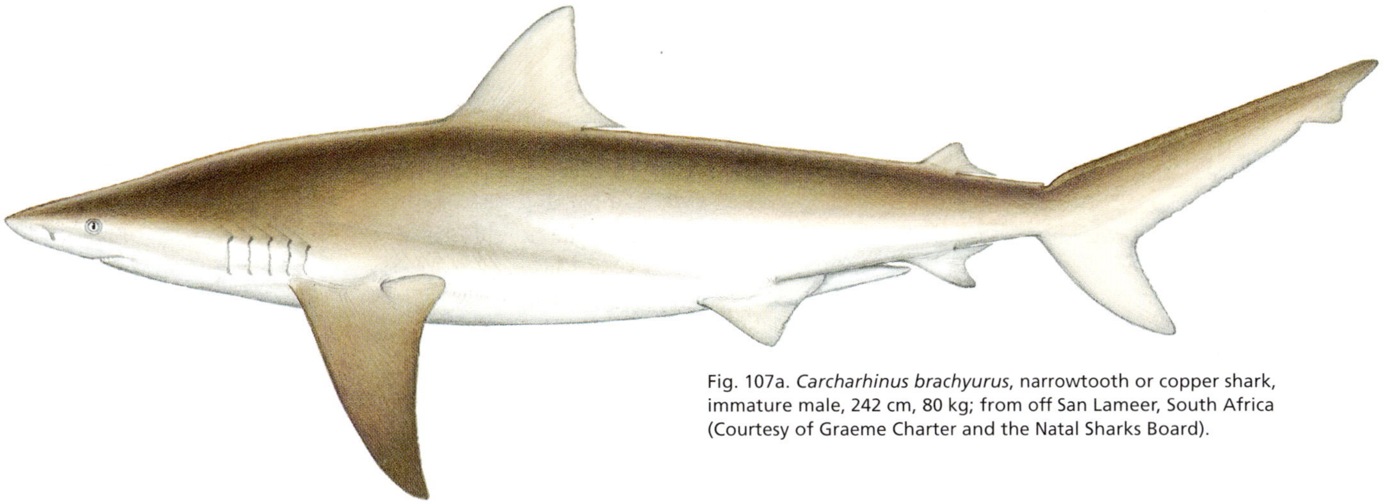

Fig. 107a. *Carcharhinus brachyurus*, narrowtooth or copper shark, immature male, 242 cm, 80 kg; from off San Lameer, South Africa (Courtesy of Graeme Charter and the Natal Sharks Board).

Common name Narrowtooth shark, a reference to the teeth of this species being somewhat narrower than those of other requiem sharks; also copper shark (South Africa) or bronze whaler (Australia), an allusion to the copper or bronze tinge of fresh specimens. The name "narrowtooth" was obviously given to it by museum workers who had never seen the coloration of a live shark. All the other vernacular names (e.g., copper shark, bronze whaler, *gambuso alazán*) refer to its conspicuous coloration. I prefer to use the shorter name copper shark.

Spanish name Among fishermen of southern Baja California, the copper shark is known as *gambuso alazán: gambuso* is the name given to several carcharhinid sharks in Mexico and the Caribbean, and *alazán* means sorrel, a brownish orange color. The A.F.S. name is *tiburón cobrizo.*

Scientific name *Carcharhinus brachyurus* (Günther, 1870), originally described as *Carcharias brachyurus* in *Catalogue of fishes in the British Museum* 8: 369. *Carcharhinus:* from the Greek *karcharos,* sharp pointed or jagged, from *karcharias,* a species of dogfish having jagged teeth, and *rhine,* Greek file or rasp, also the name of a shark, originally given to *Squatina* due to its rasplike skin; *brachyurus:* from the Greek *brachys,* short, and *urus,* from *oura,* tail, hence "having a short tail." The name *brachyurus* has a convoluted history. According to Garrick (1982), *Carcharias brachyurus* was originally described inadequately from four New Zealand type specimens. Two of these were mounted skins, since lost. The other two types are early embryos, perhaps *C. leucas.* Jordan and Gilbert (1882) later described the species as *Carcharias lamiella,* based on a 72-cm San Diego specimen. Unfortunately, subsequent interpretations of *C. lamiella* are actually dusky sharks (*C. obscurus*). To clear the confusion and stabilize the name, Garrick designated a neotype based on a New Zealand specimen. For more details of the nomenclature, see Garrick (1982).

Synonyms *Carcharias lamiella* Jordan and Gilbert, 1882 [a]; *Eulamia ahenea* Stead, 1938; *Carcharhinus improvisus* Smith, 1952; *Carcharhinus rochensis* Abella, 1972; *Carcharhinus acarenatus* Moreno and Hoyos, 1983. During most of the twentieth century, the name *Carcharhinus remotus* was used for the copper shark by Garman (1913), Bigelow and Schroeder (1948), Kato et al. (1967), and others.

Identification The copper shark is *a smooth-backed species, lacking an interdorsal ridge.* It has a snout about as long as the mouth width, a first dorsal fin originating over or slightly anterior to the pectoral fins free rear tips, and a second dorsal fin originating slightly behind the anal fin origin. Small specimens have upper teeth with erect symmetrical cusps with finely serrated edges; *large specimens have scythe-shaped upper frontal teeth with finely serrated edges.* The teeth become increasingly oblique toward the corners of the mouth. Adult males have longer, narrower, and more curved teeth than do females and juveniles. Both sexes have lower teeth with erect or slightly oblique cusps with very finely serrated edges. Teeth number U: 14 to 16–1 to 3–14 to 16, L: 14 or 15–1 to 3–14 or 15. The *teeth are diagnostic,* and this is one of the few carcharhinid sharks where the species and sex can be determined by tooth shape alone. The overlapping dermal

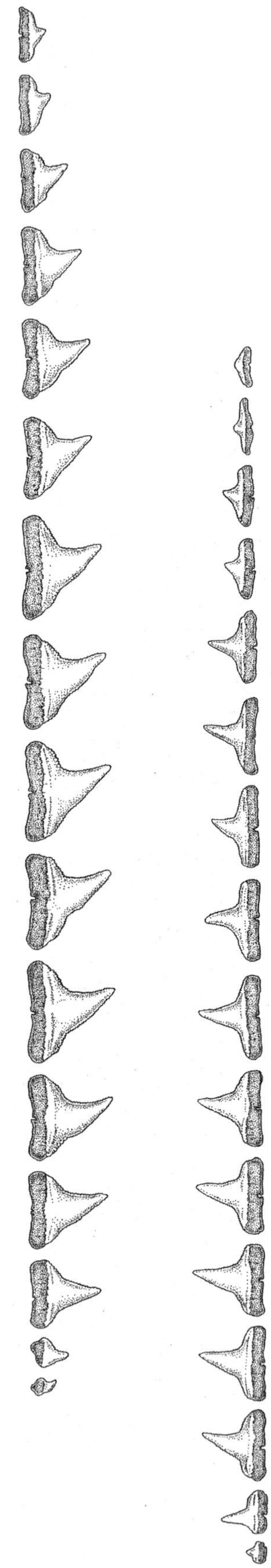

Fig. 107b. Teeth.

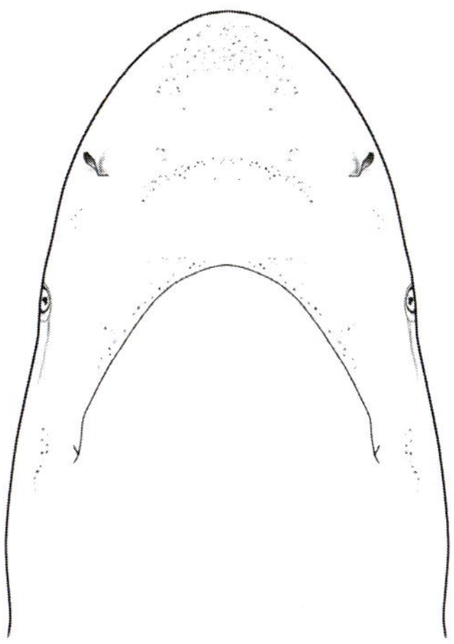

Fig. 107c. Snout.

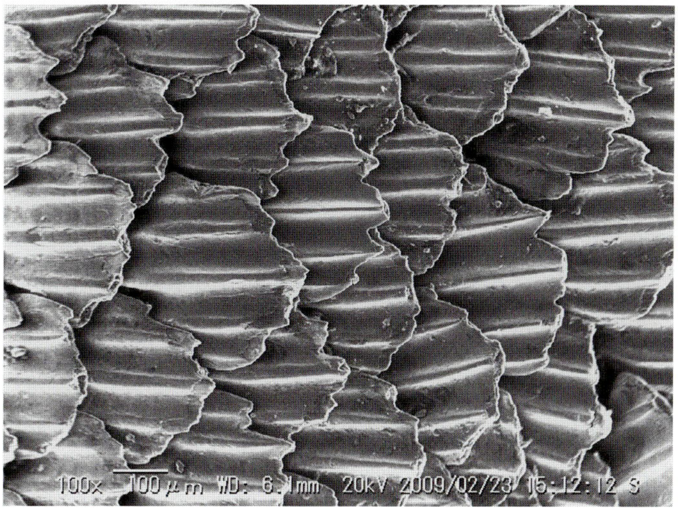

Fig. 107d. Dermal denticles.

denticles are broad with three to five ridges and points. Color in life is brownish gray with copper or bronze hues above, with a band of the upper color extending downward into the lighter sides. The undersides are white. Stead (1938: 103) described the coloration in life of Australian specimens (as *C. ahenea*) as "when fresh a bronze-brass color 'like a new penny,' almost pale golden at times; gray or cream strip along sides from eyes; creamy white below." Sadowsky (1967b: 330) described fresh Brazilian specimens as "olive-gray above, paler below, the fins with very slightly darker edge. After a few hours the back assumes a bronze color." The *fins lack conspicuous black markings,* but the pectoral fins have dusky tips on their ventral surfaces.

Similar species The blacktip shark and the spinner shark have fins with conspicuous black margins. The characteristic teeth of the copper shark distinguish it from other sharks lacking an interdorsal ridge.

Range The copper shark has worldwide distribution in subtropical to warm-temperate coastal waters. It has been reported from numerous localities in the western Pacific (New Zealand, Garrick 1982; Japan and Australia, Last and Stevens 1994), the eastern Pacific (Jordan and Gilbert 1882; Kato et al. 1967), the Indian Ocean (Bass et al. 1973, although there are many references to the species from South Africa under other names), the South Atlantic (Brazil, Sadowsky 1967a; Argentina, Chiaramonte 1998; Namibia, Bass et al. 1973), and the Mediterranean Sea (Garrick 1982). It appears to be absent from the western North Atlantic. Miller and Lea (1972) reported the species (as *C. remotus*) as being rare off southern California and in the Gulf of California. In the fishing camps of southern Baja California Sur, Mexico, I encountered a few dried jaws of the species, indicating that it is occasionally taken by migrant fishermen.

Size and weight There are no data on the sizes attained by copper sharks in North America. Bass et al. (1973) reported that the species reaches 292 cm off the Natal coast, South Africa. Later, Walter and Ebert (1991), working in the same area, reported that their largest male measured 294 cm and their largest female 288 cm. Cliff and Dudley (1992) reported that a 237-cm female weighed 203 kg and that their largest female measured 305 cm. The 242-cm male illustrated in this work weighed 81 kg. Two adult males, both measuring 286 cm, weighed 144 and 160 kg.

Biology The biology of the copper shark along the North American coast is unknown, due to its scarcity in the area and to its being confused with the dusky shark in earlier times. Most of what we know about the copper shark is based on studies carried out in South African waters (Bass et al. 1973; Smale 1991; Walter and Ebert 1991; Cliff and Dudley 1992). It is likely that habits of the copper shark in North America are similar to those observed off southern Africa. Thus, a brief summary of the life history of the copper shark in South African waters is given below. The copper shark is a coastal species, having been recorded from the surf zone to depths of 100 m (Bass et al. 1973). It preys on small schooling fishes (Clupeidae and Carangidae) and choka squid (*Loligo v. reynaudii*) (Smale 1991). Off the South African coast, copper sharks appear to follow shoals of the South African pilchard (*Sardinops ocellatus*). According to Cliff and Dudley (1992), copper sharks occur year-round off the Natal coast, but their numbers increase greatly during the winter appearance of the large pilchard shoals. This event, locally known as the sardine run, takes place in June, July, and August, when both pilchards and sharks come close to shore and even into the surf. Cliff and Dudley (1992) reported that they found the South African pilchard in 84.0% of the copper shark stomachs (containing food) sampled in winter. Copper shark catches occurred in surface temperatures of 17.3 to 25°C, but 81% of the catches occurred in temperatures of 19–21.9°C.

The copper shark is a migratory species, but little is known of its movements, other than its association with the sardine shoals. Sharks tagged off Natal have traveled 1,320 km south to the Cape of Good Hope, indicating that the species migrates long distances north–south along the African coast (Cliff and Dudley 1992).

Size at maturity

Male: Based on data presented by Cliff and Dudley (1992), it appears that most males mature around 236 cm, although their data and those of other authors (Bass et. al. 1973; Smale 1991) cover a wider size range. The smallest mature male reported by Bass et al. (1973) measured 212 cm. Smale (1991) reported that the smallest mature male he encountered was 193 cm, but most mature males were larger than 211 cm. I examined a 242-cm male, taken off Durban in July, which had elongated claspers that were not fully calcified.

Female: It appears that females mature at about 228 cm, based on the size of the smallest gravid female from a sample of 50 gravid females (Cliff and Dudley 1992).

Reproduction The copper shark is placental viviparous. Reproductive data on the species are scarce, and conclusions about the gestation period and the reproductive cycle must be considered preliminary. Cliff and Dudley (1992) noted that gestation appeared to last about a year, based on combined data of limited coverage. These authors assumed the reproductive cycle to be biennial, as in most carcharhinid sharks.

Size at birth: According to Bass et al. (1973), off South Africa, the young are born at 60–70 cm, based on observing embryos 59–63 cm and a free-swimming neonate of 72 cm. Smale (1991) recorded a 67-cm free-swimming neonate.

Brood size: Copper shark broods consist of 13–16 young. Bass et al. (1973) noted that six pregnant females averaged 16 embryos, with a range of 13–20 per brood. Smale (1991) gave a mean of 13 young and a range of 11–16 for four broods. Cliff and Dudley (1992) gave a mean of 15 and a range of 8–20 for 46 broods.

Nurseries The nurseries of the copper shark are undescribed. The location of the North American nurseries of the copper shark, if there are any, is unknown.

Age and growth Walter and Ebert (1991) counted vertebral growth rings to age copper sharks in southern Africa. They estimated females reach maturity at 19–20 years of age and males at 13–19 years, and recorded maximum ages of 30 years for a 294-cm male (their largest) and 30 years for their largest 288-cm female. Although these ages have not been validated, they indicate that the copper shark is a long-lived species.

Relation to humans The species is too rare in North America to be of any economic importance. Elsewhere, it has been blamed for unprovoked attacks on humans (Garrick and Schultz 1963).

SPINNER SHARK

Carcharhinus brevipinna (Müller and Henle, 1839)

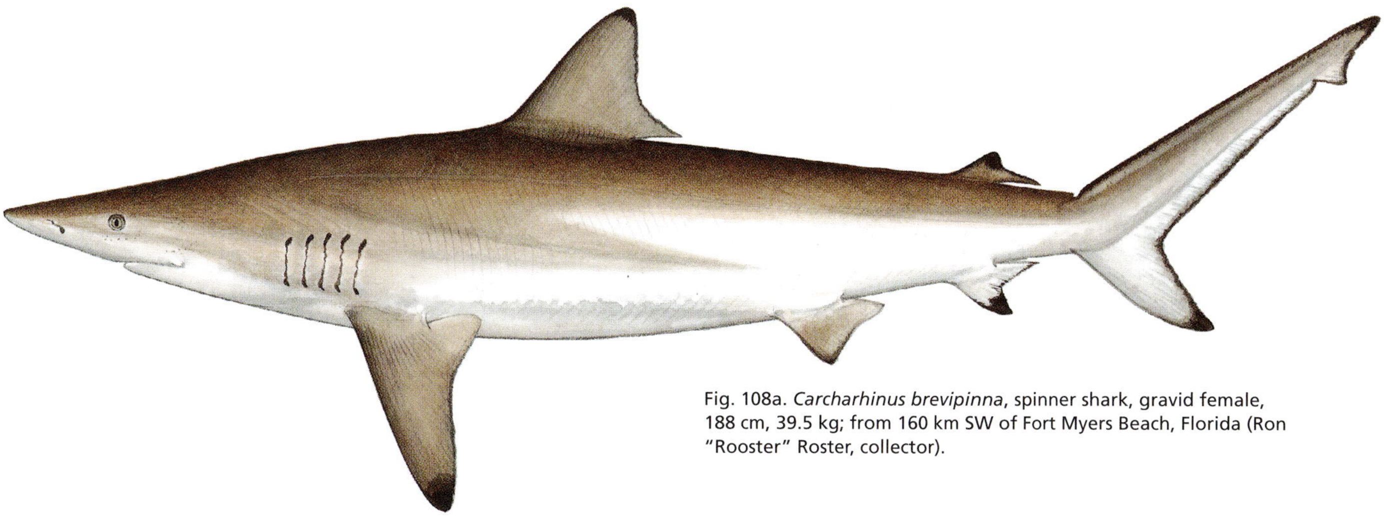

Fig. 108a. *Carcharhinus brevipinna*, spinner shark, gravid female, 188 cm, 39.5 kg; from 160 km SW of Fort Myers Beach, Florida (Ron "Rooster" Roster, collector).

Common name Spinner shark. The name "large black-tipped shark" was used until the 1960s. It appears that the name was created by Stewart Springer, who described its spinning behavior (see below).

Spanish name *Tiburón de aleta prieta, punta negra, jaquetón curro* (Mexico), *tiburón curro* (A.F.S.).

Scientific name *Carcharhinus brevipinna* (Müller and Henle, 1839), originally described as *Carcharias* (*Aprion*) *brevipinna* in *Systematische Beschreubung der Plagiostomen* 31–32, plate 9. *Carcharhinus:* from the Greek *karcharos,* sharp pointed or jagged, from *karcharias,* a species of dogfish having jagged teeth, and *rhine,* Greek file or rasp, also the name of a shark, originally given to *Squatina* due to its rasplike skin; *brevipinna:* from the Latin *brevis,* short, and *pinna,* wing or feather, hence "short-winged," an allusion to its short fins.

Synonyms *Isogomphodon maculipinnis* Poey, 1865; *Uranga nasuta* Whitley, 1943; *Longmania calamaria* Whitley, 1944; *Carcharinus johnsoni* Smith, 1951; *Aprionodon caparti* Poll, 1951. During most of the twentieth century the species was known as *Carcharhinus maculipinnis.*

Identification The spinner shark is characterized by a *lack of an interdorsal ridge;* a snout as long as, or longer than, the mouth width; the *first dorsal fin originating over or posterior to the free rear tips of the pectoral fin;* the second dorsal fin originating over or slightly behind the origin of the anal fin; and short upper labial furrows. The teeth have narrow, triangular, erect cusps and broad bases in both jaws; the *upper teeth have finely serrated edges,* and the *lower teeth have smooth edges.* Young specimens may have smooth-edged upper teeth. Teeth number U: 16–2 or 3–16, L: 15 or 16–1–15 or 16. The dermal denticles are overlapping, with five to seven ridges and weak points. They are set on an orderly rhomboidal overlapping. Coloration is brown to brownish gray with bronze above and white below, with a band of gray extending downward over the lighter undersides from the level of the pectoral fins to over the pelvic fins. Neonates or young specimens of less than 70 cm have unmarked fins. Larger specimens have the *second dorsal, anal, pectorals, and the lower lobe of the caudal fin black tipped;* the other fins may or may not be black tipped. The black markings of spinner sharks are more distinct that those of the blacktip shark, and fins have the appearance of "being dipped in ink."

Similar species The spinner shark is often confused with the blacktip shark, which has a more stout body, a more anterior first dorsal fin originating over or slightly posterior to the pectoral axil, lower teeth with finely serrated edges, and a white anal fin (Atlantic specimens only). The finetooth shark has a shorter snout, smooth-edged upper teeth, the first dorsal fin originating over the pectoral fin axil, and unmarked fins.

Range The spinner shark has an almost circumtropical distribution, having been reported form the Atlantic, Mediterranean, Red Sea, Indian Ocean, and western Pacific (Garrick 1982). It has not been reported from the eastern Pacific. In North American waters, it ranges from North Carolina (Schwartz 2003) to Florida and the Gulf of Mexico (Bigelow and Schroeder 1948; Clark and von Schmidt

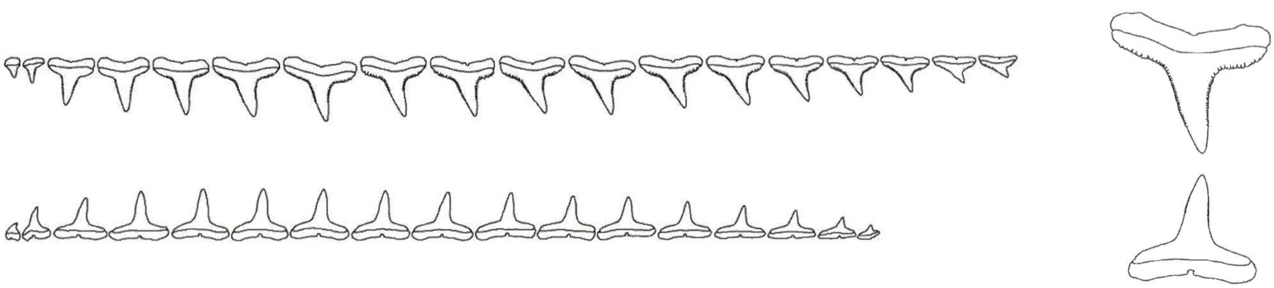

Fig. 108b. Teeth.

1965), being common throughout its range during the warmer months.

Size and weight Spinner shark males reach 229 cm and females reach 242 cm (Springer 1960). Representative weights for specimens I examined off South Carolina are as follows:

Male: 73 cm, 1.8 kg; 94 cm, 4.5 kg; 160 cm, 23 kg; 174 cm, 29 kg; and 198 cm, 48 kg.

Female: 73 cm, 1.9 kg; 100 cm, 5.4 kg; 155 cm, 20.4 kg; and 195 cm, 51 kg.

Biology The spinner shark is a common coastal species in tropical and subtropical waters. Neonates and juveniles are found in shallow coastal waters; adults are found in deeper offshore waters. I encountered neonates and juveniles of all sizes while fishing in the near shore waters 3–5 m deep off South Carolina, In several years of research off South Carolina, I encountered only a few adult females in the shallow water shark fishery or in shrimp trawl bycatch. These females were gravid with term young and about to give birth. This suggested that females enter shallow water only briefly to drop their young, and then return to offshore waters (Castro 1993b). The adult males were seen only in offshore waters.

The spinner shark feeds on small fishes. Most of the specimens that I have examined off South Carolina had consumed menhaden (*Brevoortia tyrannus*), sometimes in great numbers. One 202-cm female taken off Charleston had eaten 50 menhaden 10–23 cm each, five windowpanes (*Scophthalmus aquosus*) about 10 cm each, and a few small (7–13 cm) unidentified fishes. Bethea et al. (2004) also found that the prey of juvenile spinner sharks in Apalachicola Bay, Florida, consisted of clupeid fishes, mainly menhaden.

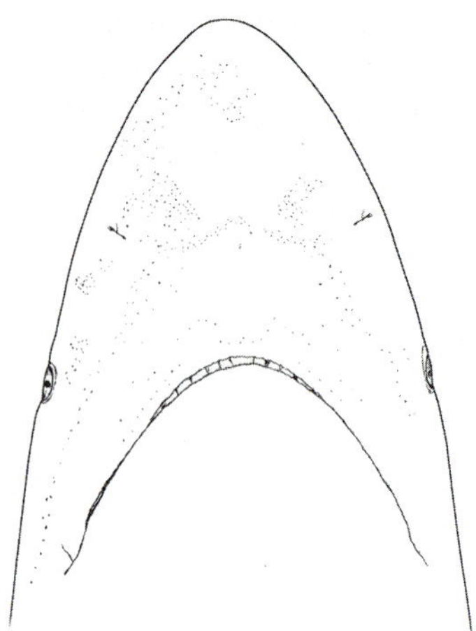

Fig. 108c. Snout.

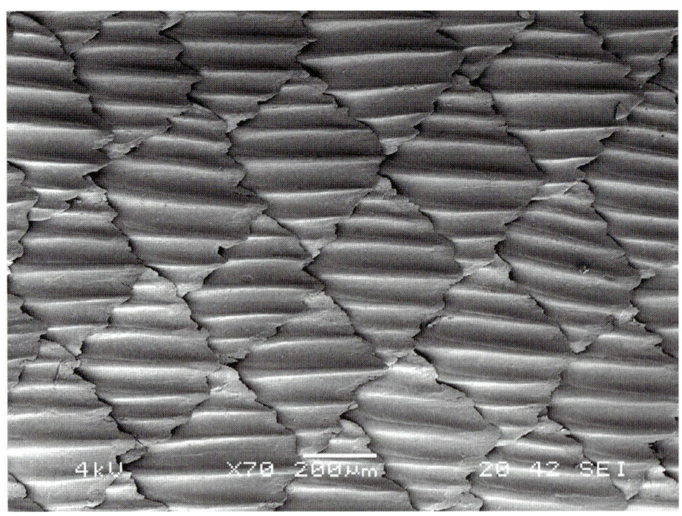

Fig. 108d. Dermal denticles.

The spinner shark is believed to leap out of the water while spinning, like blacktip sharks do. This belief is based on a report by Stewart Springer (1963: 110):

> An early-spring northward migration of spinners past Salerno, Florida, called for substantial but unsuccessful efforts to catch these sharks in quantity. In this migration the sharks make spinning leaps regularly. By following them with a boat it could be seen that the leaps sometimes started below a school of very small fish, through which the sharks passed with their mouths open while following a course toward the surface. I am confident that the relatively small numbers of spinners that I have seen from distances of less than 30 yards were *Carcharhinus maculipinnis* [= *C. brevipinna*], but spinning sharks were also rather commonly observed at greater distances and at a considerable distance from shore. I have no way of knowing whether all spinners observed were *C. maculipinnis.*

Interestingly, the report does not appear to have been confirmed by other observers. Although many authors have mentioned the leaping behavior, these accounts are based on Springer's account and are not first-hand reports. Dodrill (1977) noted that he did not see evidence of such a spring migration off Melbourne Beach, and that he noticed neither spinning sharks off the beach nor increased landings of spinner sharks at that time. Dodrill (1977: 84) stated: "When sharks spinning out of the water were seen in numbers in November at Melbourne Beach, only *C. limbatus* were being caught." I cannot confirm or deny this leaping behavior; like Dodrill, the only sharks that I have seen leaping and spinning out of the water were blacktip sharks.

The spinner shark is a migratory species, but almost nothing is known of its movements. The species appears off the Carolinas and Georgia in late spring and moves southward when waters begin to cool in September.

Size at maturity

Male: Males mature at about 170–175 cm, based on clasper calcification. Branstetter (1981) reported that the smallest mature male he examined measured 171 cm. A 175-cm male I examined was immature and had elongated but uncalcified claspers.

Female: Females appear to mature at 190–200 cm. Branstetter reported that the smallest mature female in his sample measured 191 cm and the smallest gravid female I have examined measured 202 cm. The size at maturity for males (~173 cm) is 76% of the maximum size (229 cm); for females the size at maturity (~195 cm) is 81% of the maximum size (242 cm).

Reproduction The spinner shark is a placental viviparous species. There are few published data on its reproductive processes. Branstetter (1981) estimated that parturition occurs in late May or early June after a gestation period of 11–12 months. This agrees with my observations of females carrying eggs in blastodisc stage on the last week of June and term young during the last two weeks in May.

Size at birth: Spinner sharks are born at 58–65 cm. Branstetter (1981) reported that the largest embryo that he examined measured 64.2 cm. The largest embryos that I encountered measured 62.3–63.8 cm. Free-swimming neonates of 63 and 68 cm have been reported (Dodrill 1977; Branstetter 1981).

Brood size: Observed broods have been between six and eight young. Two females that I examined carried six young each. Branstetter (1981) reported broods of seven, whereas Dodrill reported a brood of eight. Bigelow and Schroeder (1948) reported a female carrying 10 young based on a photograph from Stewart Springer.

Nurseries The nurseries of the spinner shark are poorly known but appear to extend over a wide area, because neonates have been reported from South Carolina (Castro 1993b) to Florida (Dodrill 1977).

Age and growth According to Branstetter (1987a), males mature at about 170 cm or six to seven years of age, and females mature at 180 cm or seven to eight years of age. His oldest specimen was a 208-cm female estimated to be 11.3 years old. Carlson and Baremore (2005) used stained vertebral centra to age spinner sharks caught off the southeastern United States and in the Gulf of Mexico. Their largest male measured 210 cm and was estimated to be 13.5 years old; their largest female was 217 cm and was estimated to be 17.5 years old.

Relation to humans The spinner shark is often taken in the coastal shark fisheries of the southeastern United States, but it is of minor economic importance,

PACIFIC SMALLTAIL SHARK

Carcharhinus cerdale Gilbert in Jordan and Evermann, 1898

Fig. 109a. *Carcharhinus cerdale*, Pacific smalltail shark, immature male, 103 cm, 6.9 kg; from upper Gulf of California (Juan Carlos Pérez-Jiménez, collector).

Common Name Pacific smalltail shark.

Spanish name None.

Scientific name *Carcharhinus cerdale* Gilbert in Jordan and Evermann, 1898, in *Fishes of North and Middle America* 3: 2746–2747. *Carcharhinus:* from the Greek *karcharos,* sharp pointed or jagged, from *karcharias,* a species of dogfish having jagged teeth, and *rhine,* Greek file or rasp, also the name of a shark, originally given to *Squatina* due to its rasplike skin; *cerdale:* from the Greek *kerdaleos,* of the fox, meaning wary, crafty, or foxlike.

Synonyms None. This species is not *Carcharhinus porosus* (Ranzani, 1840).

Gilbert (Jordan and Evermann 1898) described this species as *Carcharhinus cerdale,* based on a specimen from Pacific Panama. Various authors (Meek and Hildebrand 1923; Bigelow and Schroeder 1948; Garrick 1982) have considered *C. cerdale* as a junior synonym of *Carcharhinus porosus* (Ranzani, 1840), which is based on a specimen from Brazil. Garrick (1982: 69) illustrated a specimen of *C. cerdale* from Pacific Panama and considered it synonymous with the Atlantic *C. porosus,* although he cited differences in vertebral numbers. I have examined numerous of Atlantic specimens of *C. porosus* in Trinidad and a handful of specimens from the Gulf of California, and there are significant morphological differences between these two forms. The Atlantic smalltail shark (*C. porosus*) has a first dorsal fin originating over the free rear tips of the pectoral fins, a second dorsal fin originating over the midpoint of the anal fin, and unmarked fins. The two species (*C. cerdale* and *C. porosus*) have somewhat similar teeth, and if one examines only the teeth one could erroneously conclude that the two forms are one species. The two forms *are not* synonymous; they are different species. However, it is beyond the scope of this book to determine the relationship between these species. Thus, here I simply describe the major differences between the two, one species inhabiting the eastern Pacific, originally described by Gilbert as *C. cerdale,* and found in the Gulf of California; and the other inhabiting the western North Atlantic, described by Ranzani as *C. porosus.*

Identification The Pacific smalltail shark is a small species of less than 1.4 m. It is *a smooth-backed species, lacking an interdorsal ridge.* It has a long snout, its length greater than the mouth is width. Its *first dorsal fin originates behind the pectoral fin axil, at about the midpoint of the free rear tip of the pectoral fin,* and the *second dorsal fin originates slightly behind the origin of the anal fin base.* The caudal fin measures about one-fourth of the total length. The *upper teeth have* broadly triangular, increasingly oblique cusps with *serrated margins* and a notch on the outer margin. The serrations from the notch to the base are much larger than those in the rest of the cusp. The lower teeth have narrow, erect to slightly oblique, triangular cusps with more finely serrated edges. Teeth number U: 13 to 15–1 or 2–13 to 15, L: 12 to 15–0 to 2–12 to 15. The skin denticles are roughly ovoidal in shape and have minimal overlapping. The denticles have three strong ridges terminating on weak points and have a

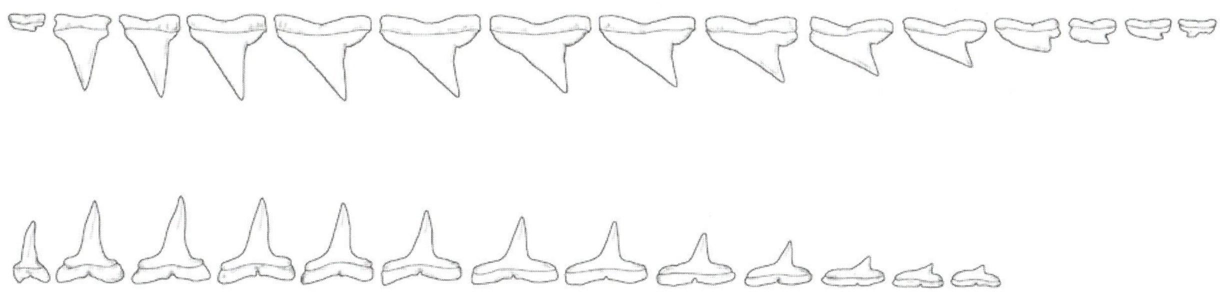

Fig. 109b. Teeth.

peculiar surface microsculpture. Coloration is dark to light gray above, with whitish undersides. The first and second dorsal fins, the pelvic fins, and the anal fin can have dark or dusky edges; the pectoral fins usually have dusky or black tips on their dorsal surfaces; the *lower caudal lobe is dusky or black tipped.* Neonates have caudal fins with a black border, and the pectoral fins have a wide, white rear margin.

Similar species The sharpnose sharks have well-marked labial furrows and smooth-edged upper teeth. The blacktip shark has a more anterior first dorsal fin, its origin over the pectoral fin axil; the origin of its second dorsal fin is anterior to the anal fin origin, and its black fin tips are more conspicuous.

Range The smalltail shark inhabits the eastern Pacific from the Gulf of California to Peru (Kato et al. 1967). According to Garrick (1982), the species is also found in Asia, off Borneo Thailand, and Vietnam. However, Compagno (1984) discounts this and suggests that Asian specimens referred by Garrick could represent an undescribed species of *Carcharhinus.* The Eastern Pacific smalltail shark is occasionally taken in the upper Gulf of California and becomes more abundant southward to the tropics.

Size and weight The largest specimen that I examined was a 116-cm, 12.2-kg gravid female that carried term young (see below).

Biology Almost nothing is known of this species due to confusion with similar sharks. In the Gulf of California, it is sometimes taken at depths of 30–40 m.

Size at maturity The Pacific smalltail shark probably reaches maturity at about 100 cm.

Male: According to Gilbert, an 85-cm male was mature. However, males 88 and 93 cm examined by J. C. Pérez

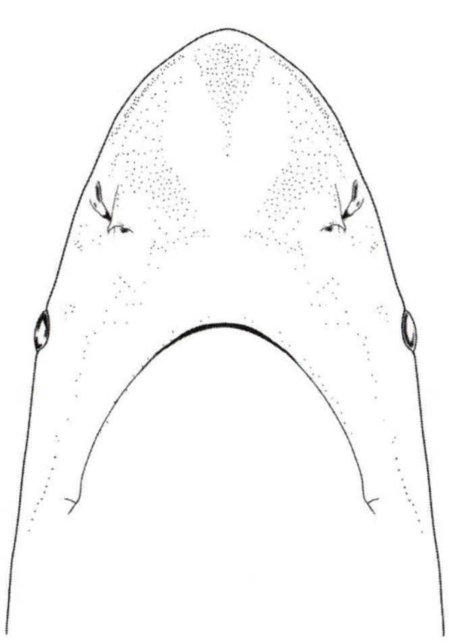

Fig. 109c. Snout.

Fig. 109d. Dermal denticles.

Jiménez (pers. comm., 21 July 2005) at Isla Isabel, Nayarit, Mexico, were immature; a 108-cm male was mature, based on clasper calcification.

Female: A 116-cm female was gravid with term young.

Reproduction The Pacific smalltail shark is a placental viviparous species. Term females carry small oocytes, suggesting a biennial reproductive cycle.

Size at birth: The young are probably born at 30–32 cm, based on a female that carried eight young 28–30 cm that appeared ready for birth in mid-April.

Brood size: I have only examined a brood of eight.

Nurseries Females with term young are sometimes encountered in the upper Gulf of California, but the extent of the nurseries is unknown.

Age and growth Unknown.

Relation to humans This species is taken in small numbers in the Gulf of California and marketed locally.

SILKY SHARK

Carcharhinus falciformis (Müller and Henle, 1839)

Fig. 110a. *Carcharhinus falciformis*, silky shark, immature male, 143 cm, 17 kg; from the northern Gulf of Mexico, off Louisiana (Wm. B. Driggers, III, collector).

Common name Silky shark, a reference to the smoothness of its skin compared to other sharks.

Spanish name Because of its wide distribution and abundance, the silky shark is well known to Caribbean and Mexican fishermen, so it is known by many different local names. In Mexico it is generally known as *tiburón prieto* or *tiburón de ley*, and regionally as *piloto* or *tiburón piloto* (Baja California) and *tutzúnxok* (Campeche). In Cuba, it is known as *tiburón de ley* or *jaquetón*. The A.F.S. names it *tiburón piloto*.

Scientific name *Carcharhinus falciformis* (Müller and Henle, 1839), originally described as *Carcharias (Prionodon) falciformis* Bibron, in Müller and Henle's *Systematische Beschreibung der Plagiostomen* 47–48. Müller and Henle credited the species to Bibron in their publication. Consequently, modern publications have given the scientific name as *Carcharhinus falciformis* (Bibron, 1841), *Carcharhinus falciformis* (Bibron in Müller and Henle, 1839), *Carcharhinus falciformis* (Müller and Henle, 1839), and so on. The publication date is given both as 1839 or 1841. Here, I follow A.F.S. usage. *Carcharhinus:* from the Greek *karcharos,* sharp pointed or jagged, from *karcharias,* a species of dogfish having jagged teeth, and *rhine,* Greek file or rasp, also the name of a shark, originally given to *Squatina* due to its rasplike skin; *falciformis:* sickle or scythe shaped, from the Latin *falx,* a sickle or scythe, probably an allusion to the shape of its pectoral fins.

Synonyms *Carcharias (Prionodon) menisorrah* Valenciennes in Müller and Henle, 1839; *Carcharhinus floridanus* Bigelow, Schroeder, and Springer, 1943; *Eulamia malpeloensis* Fowler, 1944; *Carcharhinus atrodorsus,* Deng, Xiong, and Zhan, 1981.

Note: The description of *C. falciformis* was based on a 53-cm silky shark embryo; while that of *C. floridanus* was based on two 240-cm adults. This, and confusion with dusky and night sharks, resulted in the species being misidentified for a couple of decades. Garrick et al. (1964) demonstrated that differences between the two species were due to allometric growth and synonymized the two species under the older name, *C. falciformis.*

Identification The silky shark is characterized by a snout shorter than the mouth width, a low sloping *first dorsal fin originating behind the free rear tips of the pectoral fins,* an interdorsal ridge, and a *second dorsal fin with a very long posterior free rear tip* (about 2.25 times the vertical height of the fin). The upper teeth are broadly triangular and become increasingly oblique toward the corners of the mouth. *Their edges are serrated; the serrations are coarse from the base to a weak notch about one-third of the distance from the root to the tip, and much finer from the notch to the tip.* The notch is much more noticeable on the outer margin of the tooth. The lower teeth have narrow cusps and broad bases and are symmetrical, erect, and smooth edged. Teeth number U: 14 to 16–1 to 3–14 to 16, L: 14 to 17–1–14 to 17. In small juveniles, the notch on the medial side of the upper teeth is not noticeable, and neophytes may confuse isolated teeth with those of night sharks. The dermal denticles are overlapping and have five ridges and points. Color in life is bronze-brown above and white or whitish below, fading to gray after death.

Similar species The night shark has a snout much longer than the mouth width. The dusky shark has first dorsal fin

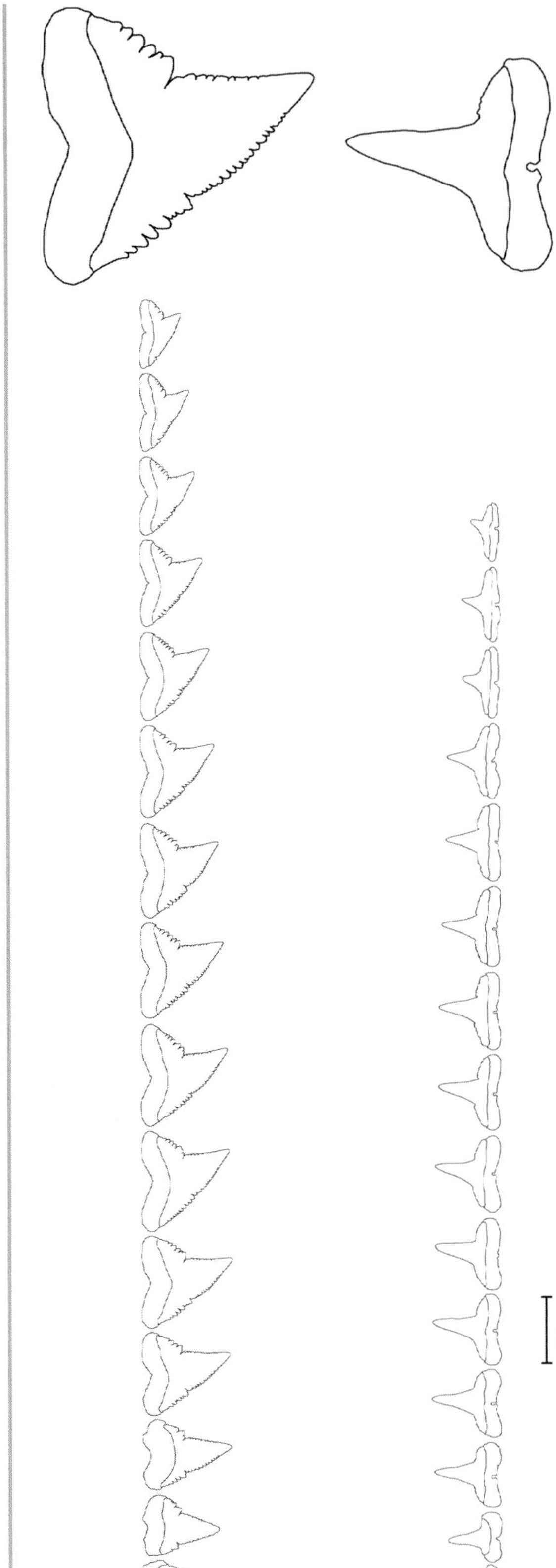

Fig. 110b. Teeth (143 cm male).

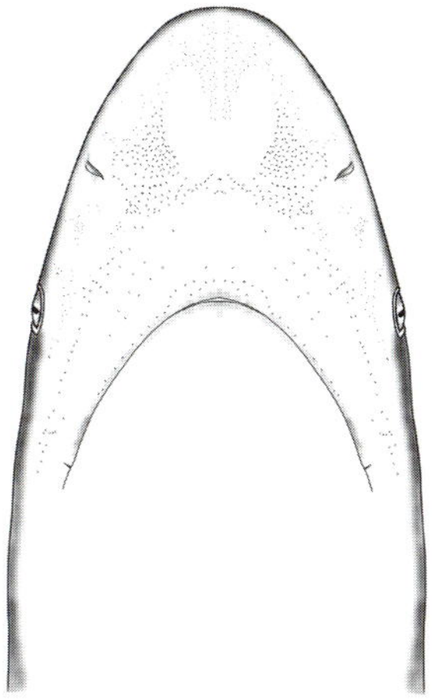

Fig. 110c. Snout.

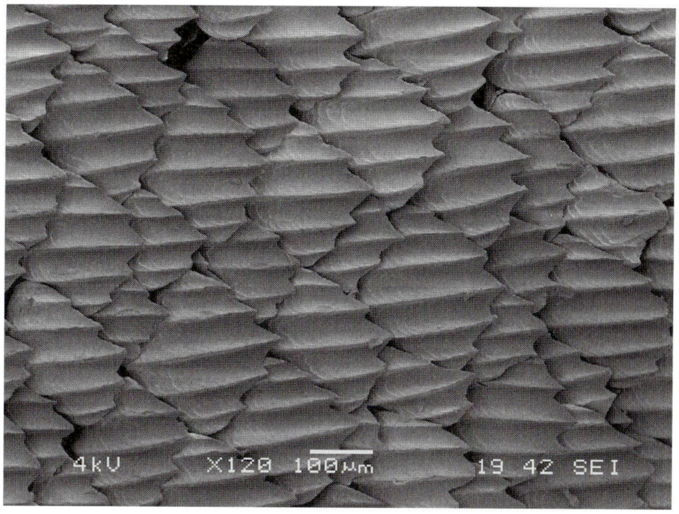

Fig. 110d. Dermal denticles.

originating over the pectoral fin rear tips and teeth that are uniformly serrated and lack any notches on their outer margins.

Range The silky shark inhabits warm-tropical and subtropical oceanic and coastal waters throughout the world. In North American waters, the silky shark ranges from Delaware to Florida and the Gulf of Mexico, and from Baja California southward. It is common in offshore waters from South Carolina to Florida and the Gulf of Mexico. Silky sharks can be found off the Florida Keys nearly year-round but are most abundant from December to February.

Size and weight According to Springer (1960), in the western North Atlantic, silky shark males reach a maximum size of 274 cm and females reach 305 cm. A 185-cm female that I examined weighed 39 kg, whereas a 203-cm gravid female weighed 55 kg.

Biology The silky shark is found in warm waters of the continental slopes and beyond, often straying into coastal waters. It is considered one of the most abundant and widely distributed sharks. Berkeley and Campos (1988) reported that silky sharks constituted 27.2% of all 613 sharks taken in their sampling of the swordfish fishery along the east coast of Florida in from 1981 to 1983. In the equatorial Pacific Ocean, Strasburg (1958) reported that the total catch south of 10° N latitude consisted of 4,157 sharks, of which 2,176 (52%) were silky sharks. Despite its abundance, its life history is poorly known, probably because of earlier confusion with night and dusky sharks.

The silky shark is a surface-dwelling species found in open, clear waters where the depth exceeds 200 m, although small juveniles often enter much shallower waters in summer. Branstetter (1981) reported taking two juveniles at depths of 36 m off Panama City, Florida. I have taken a few silky shark juveniles 89–95 cm at depths of 55 m off Charleston Harbor, South Carolina, in August 1979 and one 96-cm female in a gillnet set at a depth of 3–4 m off Morris Island, South Carolina. In the pelagic realm, silky sharks are sometimes associated with floating masses of *Sargassum* (Dooley 1972).

Small specimens are sometimes found with schools of small tuna, although the relationship is not understood. Adult silky sharks are also often associated with schools of tunas, primarily yellowfin tuna, upon which they prey (Strasburg 1958; Bane 1966). Silky sharks prey on numerous small fishes, squid, and crabs. Wright (1981, see p. 184) reported that two juvenile silky sharks had consumed 109 banded rudderfish (*Seriola zonata*). Off western Mexico, silky sharks often feed on pelagic red crabs (*Pleuroncodes planipes*), consuming large numbers of the crabs (F. Márquez-Fárias, I.N.P., Mexico, pers. comm., June 2001).

Size at maturity The estimates of size at maturity for the silky shark vary widely. According to Branstetter and McEachran (1986b), Gulf of Mexico males mature at about 210–220 cm and six to seven years of age, and females mature at 225–232 cm and about nine years of age. According to Bonfil et al. (1993), Gulf of Mexico silky shark males attain maturity at 225 cm and 10 years of age, while females attain maturity at 232–246 cm at more than 12 years of age. The smallest gravid female that I have examined was a 203 cm, 55 kg female taken in the Gulf of California. It carried four young 56–57 cm on 21 June. It is possible that silky sharks in the Gulf of Mexico may mature at a larger size than those in the Gulf of California.

Reproduction The silky shark is placental viviparous. Gilbert and Schlernitzauer (1966) described the placenta and gravid uterus of the species. Branstetter and McEachran (1986b) reported that silky shark young were born throughout the year with little seasonality. Similarly, Strasburg (1958) reported that there was no relation between embryo size and time of year in Pacific Ocean specimens. Later, Branstetter (1987b: 169) stated that "the population in the Gulf of Mexico may have a definable 12 month gestation period," based on three broods examined in March and April that "were near-term," and that "several full term litters have been observed in May." Similarly, Bonfil et al. (1993) implied a synchronous population and a gestation period of a year, based on the subjective evaluation of meager data. It is difficult to conclude that the females are synchronous breeders given the contradictory information from other areas and the limited data on the Gulf of Mexico population. Although it is reasonable to assume that the silky shark has a gestation period of about a year, as many other large *Carcharhinus* do, the gestation period of the silky shark remains to be demonstrated conclusively.

Size at birth: Branstetter and McEachran (1986b) stated that the young were born at 70–85 cm, with little seasonality. Earlier and subsequent reports support this estimate. Springer (1960) gave the maximum embryo size at about 83 cm. Bonfil et al. (1993) reported a 65-cm free-swimming neonate. Oshitani et al. (2003) gave the birth size as 65–81 cm.

Brood size: Silky shark broods consist of 4–12 young. Strasburg (1958) observed 12 silky shark broods that ranged from 2 to 11 young, with a mean of 6.5 young per female.

Nurseries The nurseries of the silky shark are in open waters of the continental shelves. There are areas off the Bahamas, over clear waters 200–300 m deep, where one can always encounter large numbers of neonate and very small silky sharks. However, the extent or limits of such areas are unknown.

Age and growth According to Branstetter and McEachran (1986b) the oldest male (267 cm) in their sample of Gulf of Mexico silky sharks, was 12 or more years old, while the largest female (276 cm) was 13 or more years old.

Relation to humans The silky shark is of limited economic importance in the United States. It is only an occasional catch in the commercial shark fishery of the southeastern United States, which targets sandbar and blacktip sharks and normally operates in waters shallower than those inhabited by silky sharks.

The silky shark is used for food in many parts of the world. In Mexico, it is the most abundant large shark in the fisheries of the Yucatan peninsula (Bonfil-Sanders 1987).

According to Strasburg (1958), silky and oceanic whitetip sharks used to cause great damage to tunas caught on longline gear in equatorial waters, sometimes involving as much as 20% of the tuna catch. Today the damage is likely not as severe as it was decades ago, as much of the tuna catch is taken by purse seiners and the number of sharks is much decreased.

GALAPAGOS SHARK

Carcharhinus galapagensis (Snodgrass and Heller, 1905)

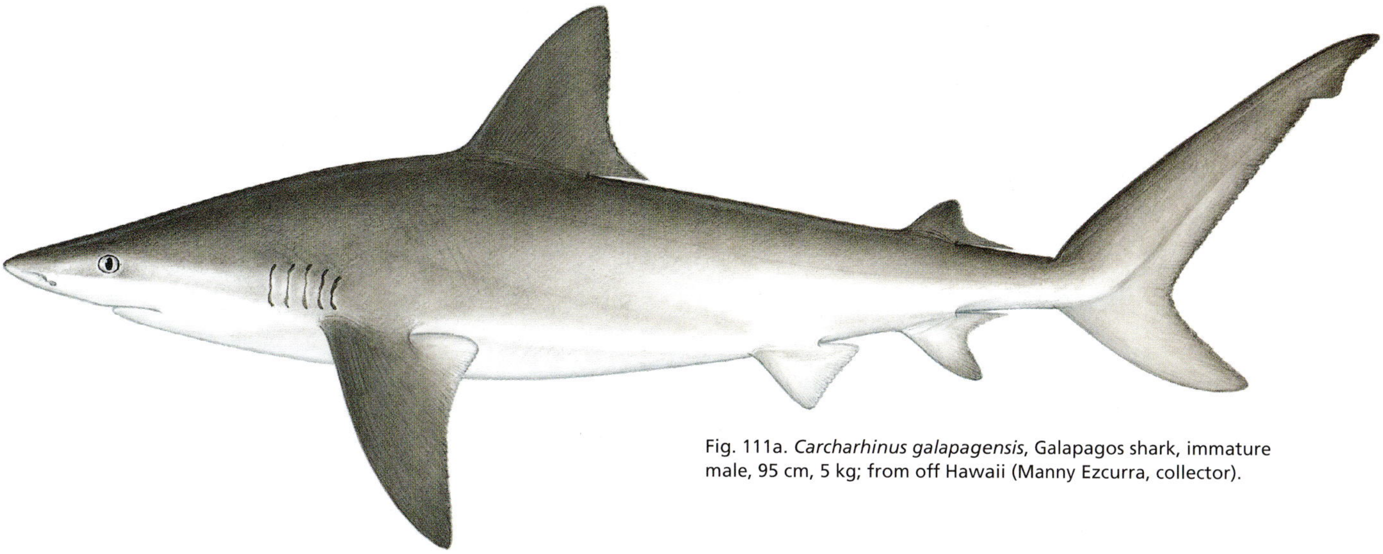

Fig. 111a. *Carcharhinus galapagensis*, Galapagos shark, immature male, 95 cm, 5 kg; from off Hawaii (Manny Ezcurra, collector).

Common name Galapagos shark, a reference to the islands where the species was first recognized.

Spanish name The A.F.S. name is *tiburón de Galápagos.*

Scientific name *Carcharhinus galapagensis* (Snodgrass and Heller, 1905), originally described as *Carcharias galapagensis* in *Proceedings of the Washington Academy of Sciences* 6: 343–344. *Carcharhinus:* from the Greek *karcharos,* sharp pointed or jagged, from *karcharias,* a species of dogfish having jagged teeth, and *rhine,* Greek file or rasp, also the name of a shark, originally given to *Squatina* due to its rasp-like skin; *galapagensis,* latinized form meaning "from the Galapagos."

Synonyms None.

Identification The Galapagos shark is a ridge-backed species with a snout shorter than the mouth width, a *first dorsal fin* that is shorter than twice the snout length and *originates over or posterior to the midpoint of the inner margin of the pectoral fins,* and a marked interdorsal ridge. The upper teeth have broadly triangular cusps with serrated edges; the lower teeth are symmetrical and erect, with microscopically fine serrations. Teeth usually number U: 14–1–14, L: 14–1–14. The precaudal vertebrae number 103–109. The overlapping dermal denticles have three to five ridges and points. Coloration is brownish gray above and white below, without fin markings.

Similar species The dusky shark has a more sloping first dorsal fin and a lower second dorsal fin 1.5–2.3% of total length (2.1–3.3% of total length in the Galapagos shark) and fewer precaudal vertebrae (86–97 in the dusky shark, 103–109 in the Galapagos shark). The silky shark has a first dorsal fin with a more rounded apex; this fin originates behind the free rear tip of the pectoral fin. The sandbar shark has a higher first dorsal fin originating farther forward over the pectoral axil. The reef shark has a more sloping first dorsal fin, and its second dorsal fin has a much shorter free rear tip. The bignose shark has a longer snout.

Range The Galapagos shark is circumtropical, having been reported from numerous oceanic islands of the tropical Pacific (e.g., Hawaii, Tester 1969; Revillagigedo, Malpelo, Cocos, and Galapagos, Kato et al. 1967; Clipperton and Alijos Rocks, Limbaugh 1963), the tropical Atlantic (Bermuda; Smith-Vaniz et al. 1999; Virgin Islands, Randall 1963; Ascension [my observation]; and Saint Helena, Compagno 1984), and the Indian Ocean (Walters Shoal, Bass et al. 1973). The limits of its distribution are not well defined due to confusion with the dusky shark. Galapagos sharks occasionally stray into the waters of the southeastern United States, but when taken they are usually confused with dusky sharks. I have examined only two specimens caught off the continental United States. One was a 177.2-cm female taken off Daytona Beach, Florida, in July 1991 by Eric Sander, a most knowledgeable commercial fisherman, who recognized it at once as distinct

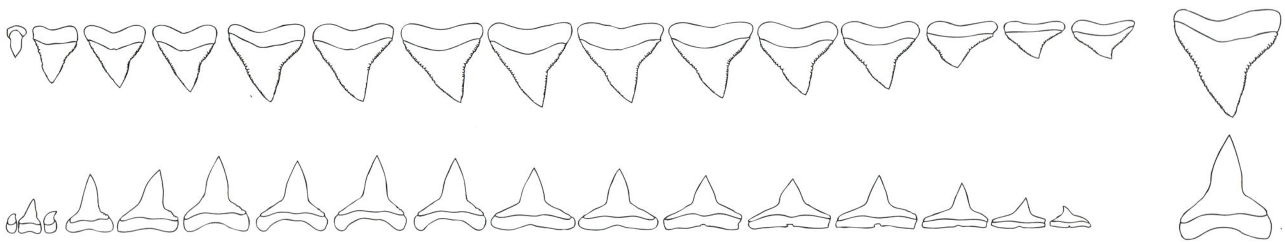

Fig. 111b. Teeth from an Ascension Island specimen estimated at 330 cm. See text (Douglas Augustus, collector).

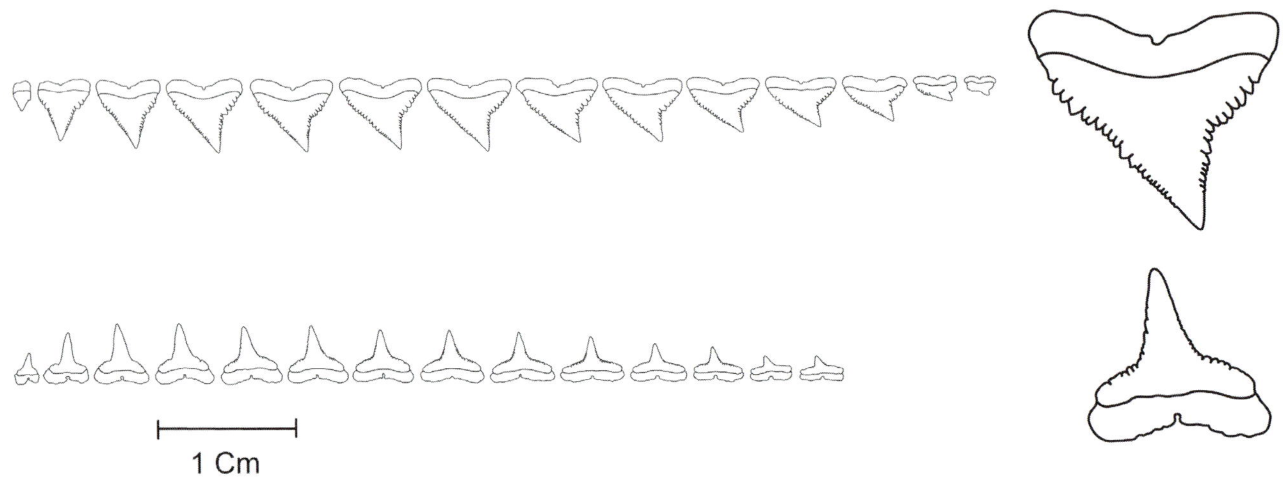

Fig. 111c. Teeth (from 95-cm male above).

from the dusky shark. I also examined a specimen from South Carolina.

Size and weight This is one of the largest sharks of the genus *Carcharhinus,* probably reaching 330 cm; only the dusky shark is larger. Kato et al. (1967) reported that the Galapagos shark grows to about 370 cm. Garrick (1982) doubted the accuracy of that measurement and noted that the largest Galapagos shark actually measured was the 292-cm specimen from the Virgin Islands (Randall 1963), if the identification was correct. Garrick gave a maximum size of about 300 cm for the Galapagos shark. However, a specimen taken off Brazil measured 323 cm (A. Amorim, pers. comm., 1997). In 1979, I received a jaw of a Galapagos shark taken in Ascension Island by Douglas Augustus the previous year. The third upper tooth in this jaw has an enamel length of 20 mm (measured along the margin of the cusp). A comparison of this tooth to those of other specimens of known lengths suggests that the species reaches at least 330 cm. I have recorded the following sizes and weights for Ascension Island specimens:

Male: 76 cm, 2.9 kg; 87 cm, 3.9 kg; 231 cm, 50 kg; and 236 cm, 45 kg.

Female: 75.8 cm, 2.9 kg; 83 cm, 3.2 kg; 130 cm, 9.1 kg; 185 cm, 40.9 kg; 231 cm, 49.9 kg; and 297 cm, 195 kg.

Biology The Galapagos shark inhabits waters around oceanic islands in tropical latitudes. Where it occurs, it is found there in very large numbers and is often the only shark species present, or at least, the dominant species. Snodgrass and Heller (1905: 343–344), who described the species, noted that it was "extremely abundant about the Galapagos Islands. . . . We examined a large number of them, several hundreds being taken aboard the schooner, and we saw probably thousands in the water." At Clipperton Island, Limbaugh (1963: 84) noted that Galapagos sharks were "extremely abundant in October 1956. . . . Juvenile Galapagos sharks were the most abundant in waters less than 200 feet deep and probably represented more than 95% of the shark population." Herre (1936, 23) wrote, "This shark swarms in the waters of the Galapagos Islands and about Cocos Island. I have never seen sharks of this genus in such abundance as in these two localities." The Galapagos shark was the only species of shark I observed off Ascension Island, juveniles being quite abundant all around the island. Despite of local abundance, the Galapagos shark is one of the least understood requiem sharks, probably due

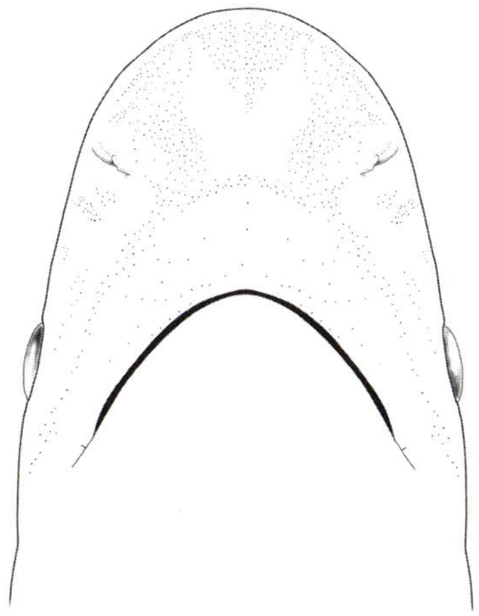

Fig. 111d. Snout.

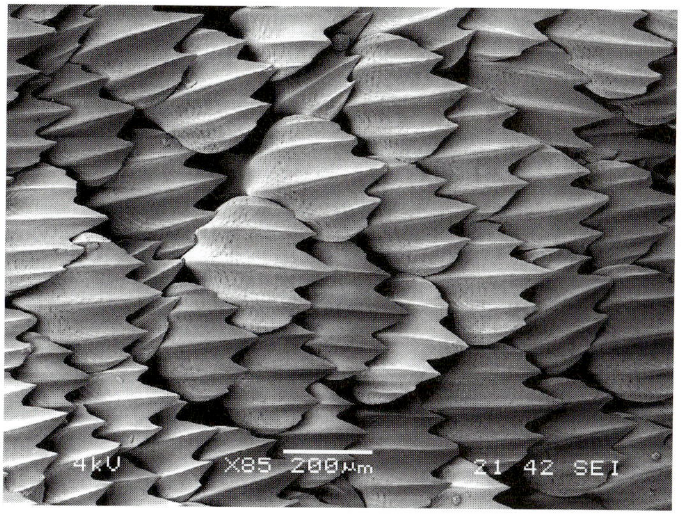

Fig. 111e. Dermal denticles.

to its oceanic habitat and confusion with the dusky shark. Consequently, there is little published information on the species.

The Galapagos shark is generally absent from continental waters, where it is replaced by the dusky shark. It strays into continental shores only rarely. Adults have been reported from the high seas (off southern Baja California, Guatemala, and Colombia by Kato et al. (1967). However, as Garrick (1982: 131) noted, "The extent to which *galapagensis* occurs in the open ocean rather than in association with oceanic islands needs further investigation."

Galapagos sharks generally stay in deep water during daytime, entering shallow rocky shores at dusk or night. However, in Ascension Island, I have observed neonates and small juveniles swimming at the water's edge during daytime. These little sharks seemed to patrol the water's edge, their dorsal fins exposed, often coming onto the beach with the breaking waves and darting back into the surf with the receding wave. In Ascension Island, juveniles could be attracted easily into shallow water by throwing rocks into the water, but the larger juveniles and adults could only be captured at a few specific sites where there was deep water against rocky cliffs and, generally, only at night. The large adults stray into shallow water only rarely. Galapagos sharks feed on groupers, triggerfishes, moray eels, flounders, octopuses, and squid.

Galapagos sharks display aggression toward conspecifics by biting them, using the upper jaw only, with a raking motion. At Ascension Island, I had the opportunity to watch three sharks competing for food. In an area of clear water in a cove, there were two 180-cm sharks and a much smaller shark about 100 cm in length. By throwing pieces of fish at the small shark I was hoping to isolate it from the other larger sharks so that I could capture it while avoiding the larger sharks. Several times the smaller and more agile shark darted in front of the larger sharks to seize a piece of fish that I had just thrown in. Shortly afterward, one of the larger sharks bit the small shark on the flank with very rapid shaking of the head, producing a half moon shaped wound that was clearly visible. The small shark quickly left the area but returned once the other sharks had been removed from the water. I captured it to examine the wound, which had been produced by raking the upper jaw teeth only over the victim. There were no lower jaw marks. I deposited the specimen in the collection at Grice Marine Laboratory.

Size at maturity Galapagos shark appear to mature at about 240–250 cm. Wetherbee et al. (1996) noted that, off Hawaii, males appear to mature between 205 and 239 cm, and females between 215 and 245 cm. I have examined immature specimens of both sexes up to 240 cm off Ascension Island. It appears, therefore, that the upper estimates given by those researchers are closer to the size at maturity.

Reproduction The Galapagos shark is a placental viviparous species. Wetherbee et al. (1996) reported that gestation appeared to last 12 months. The ovary in a female carrying term young that I examined carried 12–13 mm oocytes and appeared inactive. This suggests that reproduction could be biennial or triennial. In Ascension Island, neonates with healing or healed umbilical scars are common in July, suggesting a birth season in late May or June.

Size at birth: Galapagos sharks are born at a length of 75–89 cm. I examined two term embryos, an 89.4-cm, 5.38-kg male and an 87.8-cm, 5.0-kg female taken from a 323-cm female caught off Brazil in 1996. The smallest free-swimming neonate that I examined measured 75.8. cm and weighed 2.9 kg, taken from the Georgetown pier, Ascension Island, in July 1980. It had an open umbilicus. Other specimens 82–87 cm taken at that time had healed, but visible, umbilical scars.

Brood size: The brood size of Galapagos sharks taken off Hawaii ranged from 4 to 16, with an average of 8.7 (Wetherbee et al. 1996). The only brood I examined consisted of two young (see above).

Nurseries The Galapagos shark has its nurseries in the shallow waters of the oceanic islands where it occurs. Neonates and small juveniles can be found in water 1–3 m deep or at the water's edge, moving into deeper water as they grow.

Age and growth The species has not been aged.

Relation to humans The Galapagos shark used to have no economic importance because of its usually remote habitat. In the late twentieth century, because of the high price of shark fins, it became profitable for small boats to travel great distances to isolated oceanic islands (e.g., the Galapagos Islands, Cocos Islands) to fish for the innumerable Galapagos sharks. These fishermen would discard the carcasses and take only the easily stored, valuable fins. The flesh is of excellent quality.

One fatal attack in the Virgin Islands has been attributed to the Galapagos shark (Randall 1963). It has been described by divers in the eastern Pacific as being very aggressive. However, in Ascension Island, it is not regarded as dangerous by divers who leave the water only when large individuals appear.

FINETOOTH SHARK

Carcharhinus isodon (Müller and Henle, 1839)

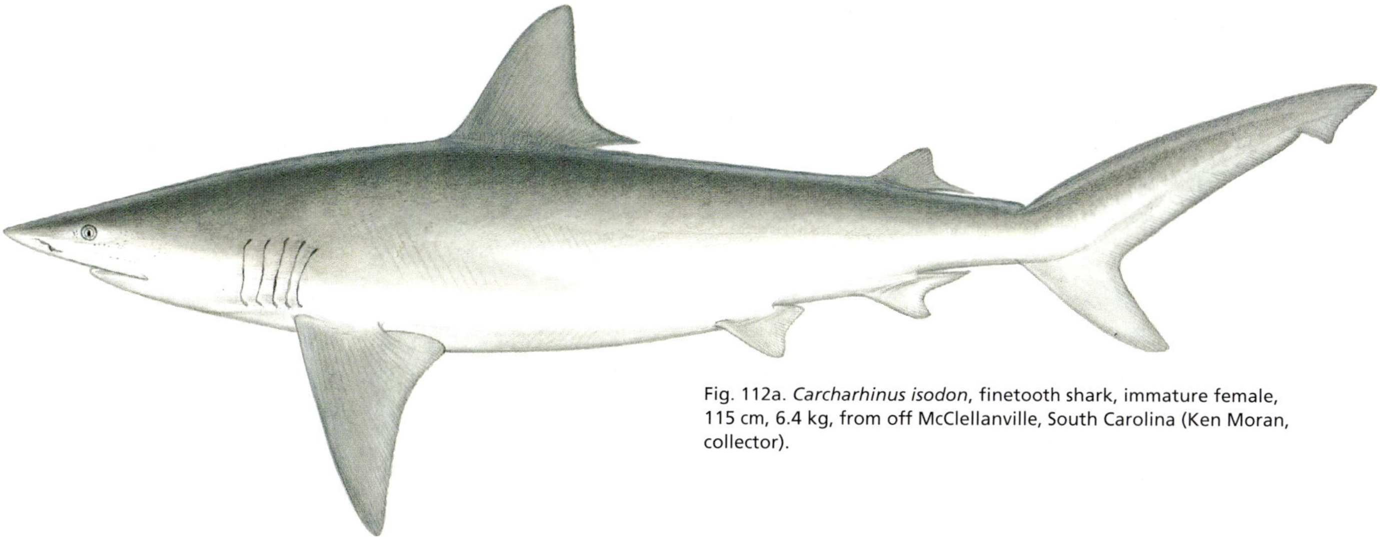

Fig. 112a. *Carcharhinus isodon*, finetooth shark, immature female, 115 cm, 6.4 kg, from off McClellanville, South Carolina (Ken Moran, collector).

Common name Finetooth shark, an allusion to its fine, smooth-edged teeth that are smaller and finer than those of other requiem sharks.

Spanish name The A.F.S. name is *tiburón dentiliso.*

Scientific name *Carcharhinus isodon* (Müller and Henle, 1839), originally described as *Carcharias* (*Aprion*) *isodon* in *Systematische Beschreubung der Plagiostomen* 32. *Carcharhinus:* from the Greek *karcharos,* sharp pointed or jagged, from *karcharias,* a species of dogfish having jagged teeth, and *rhine,* Greek file or rasp, also the name of a shark, originally given to *Squatina* due to its rasplike skin; *isodon:* from the Greek *iso,* equal, and *odontos,* tooth, thus "equal teeth," an allusion to the teeth being similar in shape and size in both jaws.

Synonyms *Aprionodon punctatus* Gill, 1861.

Identification The finetooth shark is a *smooth-backed species lacking an interdorsal ridge;* its snout is shorter than the mouth width, and the first dorsal fin originates over the free rear tips of the pectoral fin. The *teeth have slender, erect, smooth-edged cusps* and broad bases, the apex of the cusp forming an equilateral triangle with the broad bases. The *teeth are similar in both jaws,* and they number U: 13 to 15–1–13 to 15, L: 13 to 15–1 to 3–13 to 15. The dermal denticles are broad and closely overlapping. They have five to seven ridges and points, the central ones being larger. In a live animal, the upper flanks are light bluish gray with a bronze sheen, the undersides are white, and the *fins are unmarked.* The characteristic light blue-gray color, with metallic sheen, of live or fresh finetooth sharks is diagnostic and sufficient for the trained observer to identify finetooth sharks in fish catches even at a distance. The color changes to slate gray after death. An albino has been described (Jones and Shipp 2006).

Similar species The combination of a smooth back, unmarked fins, and smooth-edged, erect teeth similar in both jaws, easily distinguish the finetooth from other carcharhinid sharks. That said, a novice conceivably could mistake a very small spinner shark for a finetooth shark. The spinner shark has upper teeth with finely serrated bases and black-tipped fins. The blacktip shark has finely serrated teeth and black-tipped fins.

Range The finetooth shark inhabits coastal waters of the western Atlantic from North Carolina to Brazil. This shark appears to be a strictly American species. Records from other areas can be attributed to the misidentifications so common with carcharhinid sharks and to the perpetuation of these errors in the literature. Compagno (1984) reported that the finetooth shark inhabits the western Atlantic from North Carolina, exceptionally New York, to Florida, Cuba, the Gulf of Mexico, and southern Brazil, and that it may also occur in the Eastern Atlantic off Senegal and Guinea-Bissau. However, detailed scrutiny of the records reveals a more restricted range.

The finetooth shark is said to range as far north as New York, based on Duméril (1865–1870), who said that the type was from the New York coast (Bigelow and Schroeder

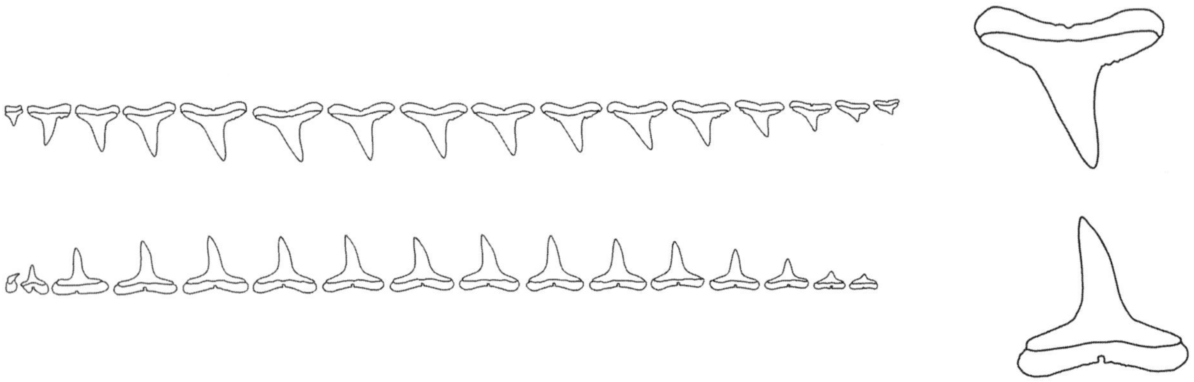

Fig. 112b. Teeth.

1948), although Müller and Henle gave no locality data for the type specimen. Jordan and Evermann (1896) noted it had been recorded in Virginia, but without mentioning specimens or source. However, the presence of the finetooth shark north of Cape Hatteras is doubtful (Castro 1993a). The finetooth shark is not mentioned in any of the faunistic works for areas north of Cape Hatteras (e.g., Nichols and Murphy 1916; Nichols and Breder 1927; Hildebrand and Schroeder 1928; Milstein 1978; Murdy et al. 1997). Other records of finetooth sharks (e.g., Bean 1905) north of Cape Hatteras are based solely on the type. Several other records note that the finetooth shark is considered rare in North Carolina (Radcliffe 1916; Schwartz and Burgess 1975), while Schwartz (1984) reported that it was found there from April to October. Frank Schwartz (pers. comm., 2000) states that the species is not found there every year.

There are few records of the finetooth shark in the Caribbean and South Atlantic. The finetooth shark is known from Cuba solely on Poey's (1868) record, which is based on some loose teeth and a caudal fin (specimen no. 356) that had been sent to him. Poey (1868: plate 4, nos. 9, 10, and 11) illustrated these teeth and attributing them to *Aprionodon* sp. *dubia.* Later, he assigned them to "*Aprionodon isodon,*" expressing doubt about the identification (Poey 1876). These three teeth, Poey's specimen no. 356, are currently in the U.S. National Museum (USNM no. 232832) in a small vial bearing an ancient label that reads "*Aprionodon* #356." I examined those teeth in 1991. The three

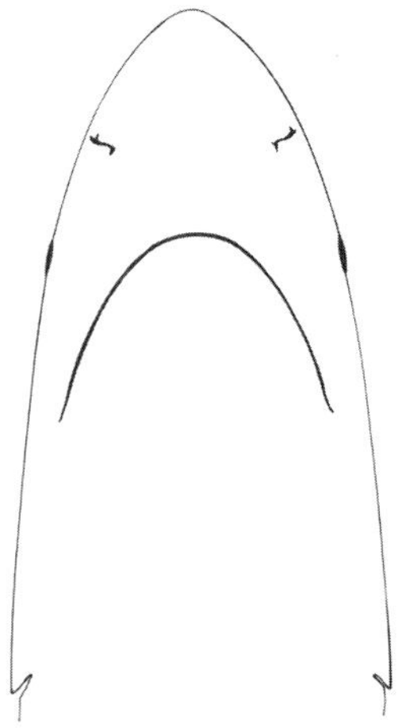

Fig. 112c. Snout.

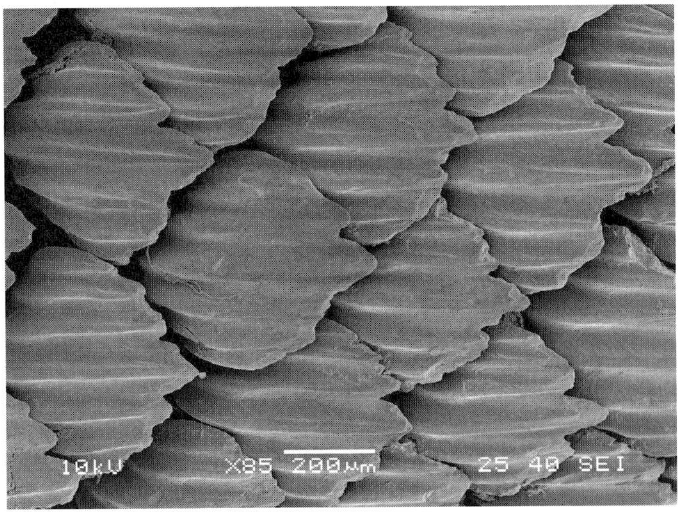

Fig. 112d. Dermal denticles.

teeth depicted by Poey (1868) cannot belong to a finetooth shark because they have bases that are too narrow and cusps that are proportionally too long. One tooth has fine serrations visible only through a 5–10× lens. The other two teeth have shorter smooth-edged cusps, and one of these has a strongly oblique cusp. The serrated tooth may belong to a blacktip shark, and the smooth-edged teeth are probably lower teeth of a scalloped hammerhead. Guitart Manday (1968) in a faunistic work on Cuban sharks noted that the finetooth shark had been recorded from Cuba only once, apparently based on Poey's record. In a 1974 work on Cuban fishes, Guitart (1974) did not include the finetooth shark. Thus, there are no valid records or specimens for Cuba (Castro 1993a). The Gulf Stream and the Straits of Florida may well be insurmountable barriers that prevent Florida finetooth sharks from migrating to Cuba.

Applegate et al. (1979) listed the finetooth shark as occurring in Mexico (Gulf of Mexico) and possibly in the Caribbean but did not list any specimens. The species is rare in Trinidad, where I did extensive sampling in 1985–1986. I saw only one finetooth shark there, a 125.4-cm mature male, out of hundreds of sharks captured there or seen in the market. There are some uncertain records for British Guiana (Lowe 1962). Sadowsky (1967a) illustrated one of two specimens from Cananeia, Brazil, at latitude 25° S, noting that these were new records for Brazil. Soto (2000) reported an immature female from Florianópolis, Brazil, as the second record for that country.

African records are obviously misidentifications. Cadenat (1950) included the species in a list of species reported from Senegal but not observed by him, stating that the species had been seen in waters around Dakar by Budker, but without offering any records or supporting evidence. Compagno (1984) states that old African records have not been confirmed and that it is possible those records are based on another species, particularly the spinner shark (*Carcharhinus brevipinna*). Garrick (1985) summarized reports of finetooth from the eastern Atlantic and concluded that those records could not be regarded as confirmed.

In summer, the finetooth shark has been reported from North Carolina to Florida and the Gulf of Mexico (North Carolina, Radcliffe 1916; South Carolina, Burton 1940 and Bearden 1965b; Georgia, Dahlberg 1975; west-central Florida, Clark and von Schmidt 1965; Texas, Baughman and Springer 1950, and others). It is seasonally abundant off South Carolina to northern Florida (Castro 1993a). There are few reports in the literature of its presence in the Gulf of Mexico, although it is certainly abundant there. I reported that during the mild weather of February 1991, unusually large schools of finetooth sharks were encountered by commercial fishermen in shallow waters off the mouth of the Mississippi River in the Gulf of Mexico, where the surface water temperature was 20°C (Castro 1993a).

Size and weight The largest male that I have seen measured 144 cm and weighed 19 kg, while the largest female, carrying four term young, measured 160 cm and weighed 28.9 kg (Castro 1993a). Compagno (1984) mentions that finetooth sharks reach a maximum size about 189 cm and possibly 200 cm; it seems unlikely that finetooth sharks reach these larger sizes. Representative sizes and weights of specimens that I have measured are as follows:

Male: 56 cm, 0.9 kg; 65 cm, 1.47 kg; 80 cm, 2.9 kg; 106 cm, 4.5 kg; 115 cm, 6.4 kg; 119 cm, 8.9 kg; 129 cm, 11.4 kg; 135 cm, 14.6 kg; 142 cm, 18.6 kg; 144 cm, 19.0 kg.

Female: 90 cm, 3.9 kg; 93 cm, 4.3 kg; 99 cm, 5.7 kg; 107 cm, 6.1 kg; 128 cm, 10.2 kg; 140 cm, 15.0 kg; 142 cm, 16.4 kg (mating female); 148 cm, 21.8 kg (term female); 151 cm, 18.1 kg; 156 cm, 28.6 kg (term female); 160 cm, 25.6 kg (postpartum); 160 cm, 28.9 kg (term female).

Biology The finetooth shark is a coastal species that inhabits waters 2–5 m deep or less and feeds on small bony fishes. Its large gill slits and small eyes suggest that the finetooth shark is a fast moving species adapted to water of poor visibility. The finetooth shark inhabits shallow waters and can be found from beyond the surf zone to depths of five meters. The species is seldom caught, even in areas where it is abundant (Castro 1993a), because it is generally too far from the beach to be caught by surf fishermen and in waters too shallow for most commercial fishing vessels. Despite its abundance in the shallow waters of South Carolina, I seldom saw the species caught in hook-and-line fishing operations or in shark tournaments, although I occasionally saw small juveniles caught in shrimp trawls. The adults appear to be too fast to be caught in shrimp trawls.

Most of the records of finetooth sharks are of juveniles caught in the surf or in very shallow water. Old reports of the finetooth shark being locally rare from southern North Carolina to Texas are due to poor sampling or to not sampling the appropriate areas. Both young and adults are easily caught in nets set at the proper depths. Gillnet catches suggest schooling behavior, at least during migration.

Diet: Finetooth sharks consume small prey; most prey fishes I saw among the stomach contents were small, usually 10–20 cm standard length. I analyzed the stomach contents of 80 specimens caught off South Carolina (Castro 1993a). Of these, 39% had empty stomachs, 38% contained menhaden (*Brevoortia tyrannus*), 14% held unidentifiable bony fish remains, 3% had juvenile spot (*Leiostomus xanthurus*), 3% had shrimp (*Penaeus* spp.), and 1% had Spanish mackerel (*Scomberomorus maculatus*), mullet (*Mugil* spp.), or juvenile Atlantic sharpnose sharks. In most cases, the menhaden consisted of digested bodies that had been swallowed whole. Late in that study, my increasing experience in identifying menhaden in advanced stages of digestion made it evident that some of the stomach contents previously recorded as "unidentified bony fish" were, in fact, well-digested menhaden. Thus, the actual percentage of menhaden in the stomachs I sampled was greater than 38%, indicating that the finetooth shark preys primarily on menhaden while in its summer areas. Bethea et al. (2004) found the same trend for juvenile finetooth sharks in Apalachicola Bay, Florida, with menhaden constituting more than 50% of the diet.

Migrations: The finetooth shark is a migratory species that moves north in summer into shallow coastal waters from northern Florida to the Carolinas and winters off Florida. The descriptions that follow are from Castro (1993a). Juvenile finetooth sharks appeared in the shallow coastal waters (2–3 m depth) of Bulls Bay, South Carolina, in early April in most years, and in late March in warm years. The adults appeared in the area in early May. The earliest records for adults in Bulls Bay are 2 May for nongravid females and 21 May for gravid females. Gravid females carrying term young and having emaciated livers are very abundant in the area in the last week of May (Castro 1993b). The surface water temperature rises from an average of 18.1°C in late April to 21.6°C in early May, so the appearance of finetooth sharks seems to coincide with the warming of the waters above 20°C. Both adults and juveniles are abundant off South Carolina during the summer. The species departs South Carolina from September to mid-October as the surface water temperature falls below 20°C. The latest record for a gravid female in the Bulls Bay area is 16 October. The latest record for the species in South Carolina is that of a 79.6-cm juvenile caught off Charleston on 26 October (Castro 1993b). After leaving South Carolina, schools of finetooth sharks migrate south to Florida, passing off Daytona Beach at depths of 15–20 m, from late September to late December, with the peak of the southward run passing off Daytona Beach area during the third week in November (Castro 1993a).

The finetooth shark winters off Florida, although the extent of the wintering areas is not well defined. Some finetooth sharks are found sporadically off Daytona Beach throughout the winter (Castro 1993a), suggesting that the wintering population is not far away. Dodrill (1977) reported the catch of six specimens on hook and line within 40 m from shore off Melbourne Beach, Florida, and stated that the species was found there from November to April. Springer (1950a) stated that relatively large numbers of finetooth shark appeared off Port Salerno, Florida, in waters less than 18 m in late December and January, and that it had not been caught there in other months. Some finetooth sharks can be found off southernmost Florida in late winter. Wiley and Simpfendorfer (2007) reported four specimens 111–130 cm, caught in the East Cape Canal (Cape Sable) and near Curry Key at depths of 1.1–1.3 m and surface temperatures of 22.2–23.5°C in February. I have examined an adult male taken by S. Spotte off Longboat Key, Florida, on 19 March 2006 when the surface temperature was 18.3°C.

The northward migration apparently begins in March. Finetooth sharks can be captured off Daytona Beach from early April to early May, with the peak of the run occurring in the first week of April. Surface water temperatures in the Daytona Beach area ranges from 20.1 to 21.3°C during both migrations (Castro 1993a).

Whether the population of finetooth sharks off the East Coast is separate from the Gulf of Mexico population is not known. Movements of East Coast finetooth sharks into the Gulf of Mexico have not been demonstrated by tagging, and the migrations of the Gulf of Mexico population have not been described. Whether the two populations actually mix in southernmost Florida in winter is unknown.

Size at maturity

Male: Finetooth shark males mature at about 133 cm. The largest immature male I have reported measured 129.8 cm (Castro 1993a). Males in a transitional stage—those having large testes 156–162 mm and large but uncalcified claspers —are found in early June in the summer range. It is likely that these males mature during the summer, but having missed the mating season, they are not likely to mate until the following summer. The smallest mature male I examined was 133 cm, and all other larger males were mature. After the mating season (by the last week in July), the testes atrophy and do not become active until the following spring (Castro 1993a).

Female: Females mature at about 135 cm. Females smaller than 125 cm have very thin, ribbonlike uteri 2–3 mm in width, oviducal glands less than 10 mm in width, and oocytes 2 mm in diameter or less. In females of about 125 cm, maturation begins in early summer, as the uteri, oviducal gland, and oocytes begin to enlarge. By early June, females of 128–133.5 cm have uteri and oviducal glands that reach a width of 24 mm, and their oocytes measure 6–8 mm in diameter. The process of female maturation is not completed until the following summer, by which time the oocytes have reached 30 mm in diameter (Castro 1993a).

Reproduction The finetooth shark is a placental, viviparous species. Mating occurs in South Carolina waters from early May to early June, based on females bearing fresh mating bites and on the presence of sperm in their uteri. Recently mated females bear one large sperm mass (spermozeugma) at the base of each uterus. The spermozeugmata are large, yellowish, almond-shaped masses consisting of individual spermatozoa embedded in a supporting matrix. After a short period of time these structures dissolve, releasing the sperm (Castro 1993a).

Recently inseminated females are found off South Carolina from early May to early June (Castro 1993a). Two females carrying spermozeugmata were caught in Bulls Bay in 2–4 May 1990. A third female, caught at the same locality on 7 May 1990, had a copious amount of sperm covering the uterine lining and carried ripe oocytes 30 mm in diameter. Another female caught off Folly Beach, South Carolina, on 1 June 1983, had fresh mating bites, ripe oocytes 30 mm in diameter, and a spermozeugma in each uterus (Castro 1993a). Ovulation occurs after mating, as demonstrated by females still having ripe oocytes in the ovary at the time of mating and insemination.

The embryos are dependent on yolk reserves (lecithotrophic) during the first 15 weeks of development. Afterward, embryos attach to the mother through a placental connection. Implantation occurs when embryos measure about 13 cm or at about the 15th week of gestation. Gestation lasts about 12 months, and gravid females carrying young 48–55 cm enter the shallow water nurseries off South Carolina in late May. Parturition occurs from late May to mid-June (Castro 1993a). After parturition, the females enter a "resting period" during which next batch of oocytes begins to develop slowly. Most oocyte development occurs in the few months prior to ovulation the following May (Castro 1993a).

The finetooth shark has a biennial reproductive cycle, with consecutive, year-long vitellogenesis and gestation, like most carcharhinid sharks. Thus, in late May the stock of adult females consists of approximately equal numbers of nongravid females and of gravid females about to give birth. The stock of nongravid females consists of females that have just attained maturity and are mating and ovulating for the first time, and of females that gave birth the previous May and are ready to mate again. All of these females carry from 9 to 11 ripe oocytes about 30 mm in diameter.

Brood size: Female finetooth sharks carry two to six embryos per brood. In a sample of 29 gravid females, the average number of embryos per female was four (Castro 1993a). Term females often abort their young upon capture (Castro 1993a).

Nurseries The finetooth shark gives birth in the shallow coastal waters of South Carolina in May and June (Castro 1993a). The nursery probably extends to southern North Carolina and Georgia. There are nurseries in the Gulf of Mexico, but these have not been delimited. The young are found in waters less than 3 m around barrier islands and beaches.

Age and growth Carlson et al. (2003) estimated the age of finetooth sharks in the Gulf of Mexico by counting vertebral rings and assuming the deposition of one ring per year. The oldest female they aged was 8.0 years old, while the oldest male was 8.1 years old. Theoretical longevity estimates were 14.4 and 8.5 years for females and males, respectively.

Relation to humans The finetooth shark is locally utilized for food. Large numbers used to be caught in gillnets set close to shore in coastal South Carolina. Some juveniles are caught accidentally in shrimp trawls in shallow water. Specimens are taken occasionally by rod and reel in coastal waters.

Under the current management plan for the shark fishery of the eastern United States, the finetooth is classed as a "small coastal shark," such as the Atlantic sharpnose shark and the bonnethead, despite having the slower growth pattern and biennial reproductive cycle of the larger sharks. Consequently, the economic importance of the finetooth shark to small-scale fisheries increases because it can be fished during times when the fishery for the larger species is closed. Given the life history and biennial life cycle of the finetooth shark, the species is vulnerable to overfishing.

BULL SHARK

Carcharhinus leucas (Müller and Henle, 1839)

Fig. 113a. *Carcharhinus leucas*, bull shark, immature male, 206 cm, from Sarasota, Florida (Peter Hull, collector).

Common name Bull shark, bullhead shark, cub shark, ground shark, or Lake Nicaragua shark. In Florida, in the early twentieth century, the species was known as the bullhead shark. Springer (1938) coined the name "bull shark" to avoid confusion with the well-established name of the bullhead sharks, family Heterodontidae. It is likely that the Floridian name bullhead shark was a reference to its large head. For many years, the sharks in Lake Nicaragua were believed to be a separate, landlocked freshwater species, the Lake Nicaragua shark. Today, the Lake Nicaragua shark is considered to be the bull shark, *C. leucas,* based on the analysis by Bigelow and Schroeder (1961). In southern Africa, the bull shark is known as the Zambezi shark, because it is often found in that river (Bass et al. 1973).

Spanish name *Cabeza de batea* (Cuba). This is one of the most colorful Cuban shark names, translating literally as "head like a wash tub"—an apt allusion to the shape and size of the snout or head of an adult bull shark. The Mexican names of *chato,* or *tiburón chato,* referring to its short nose, are equally appropriate. The A.F.S. name *tiburón toro,* a literal translation of the English name, is for the sole use of English-speaking biologists on the way to a bullfight.

Scientific name *Carcharhinus leucas* (Müller and Henle, 1839), *originally* described as *Carcharias* (*Prionodon*) *leucas* Val. in *Systematische Beschreibung der Plagiostomen* 42–43. *Carcharhinus:* from the Greek *karcharos,* sharp pointed or jagged, from *karcharias,* a species of dogfish having jagged teeth, and *rhine,* Greek file or rasp, also the name of a shark, originally given to *Squatina* due to its rasplike skin; *leucas:* from the Greek *leukos,* white, bright, or light.

Müller and Henle credited Valenciennes[3] with the new species in their publication (hence the shortened "Val." after the scientific name). Consequently, in modern publications, the scientific name of the bull shark has been given as "*Carcharhinus leucas* (Valenciennes, 1839), "*Carcharhinus leucas* (Valenciennes *in* Müller and Henle, 1839)," and *Carcharhinus leucas* (Müller and Henle, 1839). The date of the publication has been given both as 1839 and 1841, because it was published in fascicles during those years. Here, I follow the current A.F.S. usage.

Synonyms *Carcharias zambezensis* Peters, 1852; *Squalus platyodon* Poey, 1861; *Eulamia nicaraguensis* Gill and Bransford, 1877; *Carcharias azureus* Gilbert and Starks, 1904; *Galeolamna* (*Bogimba*) *bogimba* Whitley, 1943. In the literature of the first half of the twentieth century, these synonyms were often used under the genus *Carcharhinus* (e.g., *Carcharhinus nicaraguensis, Carcharhinus platyodon*).

Identification The bull shark has an *extremely short snout* (much shorter than the mouth width), a first dorsal fin origin

3. Achille Valenciennes (1794–1865), a French naturalist who spent his entire life at the Muséum National d'Histoire Naturelle in Paris. He is best known for his collaboration with Cuvier on the 22-volume *Histoire naturelle des poissons* (Cuvier 1995).

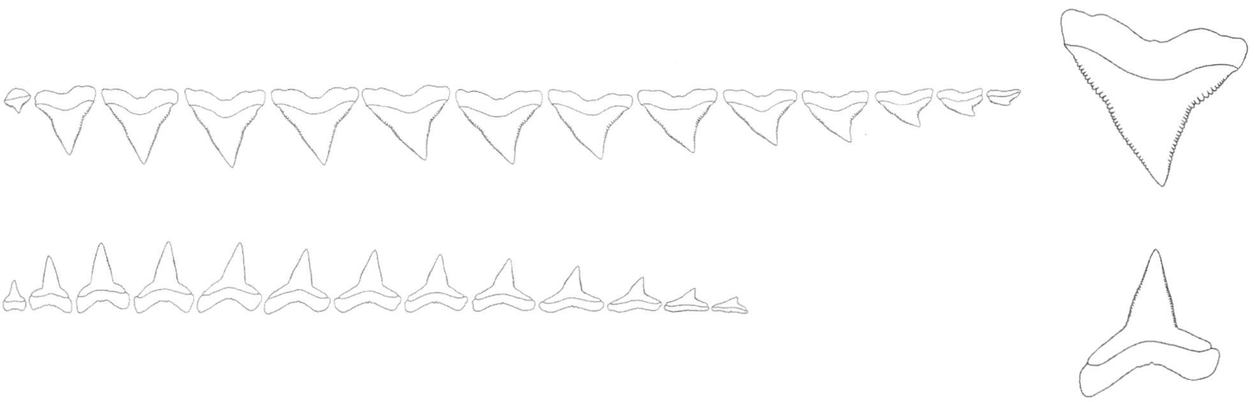

Fig. 113b. Teeth.

over or slightly anterior to the pectoral fin axil, and a second dorsal fin much smaller than the first. It is a *smooth-backed* species, lacking an interdorsal ridge. The upper teeth have *broadly triangular, heavily serrated cusps;* the lower teeth have narrow, triangular, finely serrated cusps on a broad base. The anterior teeth are erect, triangular, and symmetrical, while those on the sides are slightly oblique. Teeth number U: 12 to 14–1–12 to 14, L: 12 or 13–1 or 2–12 or 13. The denticles differ widely from those of other carcharhinid sharks in the area. The denticles are very broad with some overlapping; they have one strong central ridge that terminates on a long point, and two ridges on each side of the central point that are separated from the main point into winglike sides. Coloration is pale to dark gray above and white below. Small juveniles are brownish gray with bronze hues in life, fading to gray after death. In neonates and small juveniles, the pelvic, second dorsal, anal, and caudal fins have black tips, and the caudal fin also has a black margin.

Similar species The lemon shark has a similarly short snout, but its second dorsal fin is almost as large as the first, its teeth are slender and not broadly triangular, and it usually has a yellowish color. The tiger shark has notched teeth. The oceanic whitetip shark has a similar snout and teeth, but it is a ridge-backed shark, and the tips of the first dorsal and pectoral fins are broadly rounded and are usually white or light tan.

In other parts of the world, the Java shark, *C. amboinensis* (Müller and Henle, 1839), was for many years confused with the bull shark because of great external similarities.

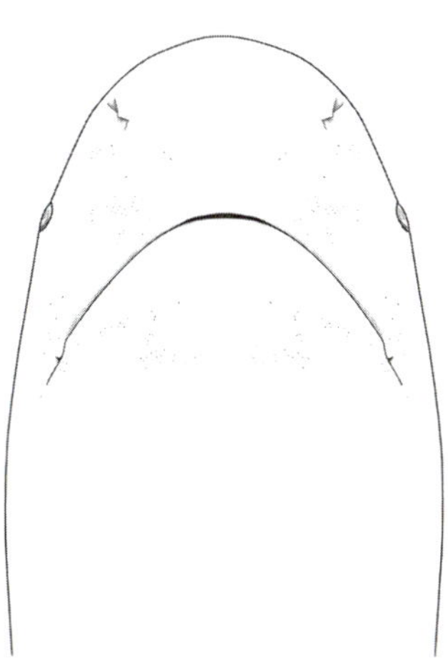

Fig. 113c. Snout.

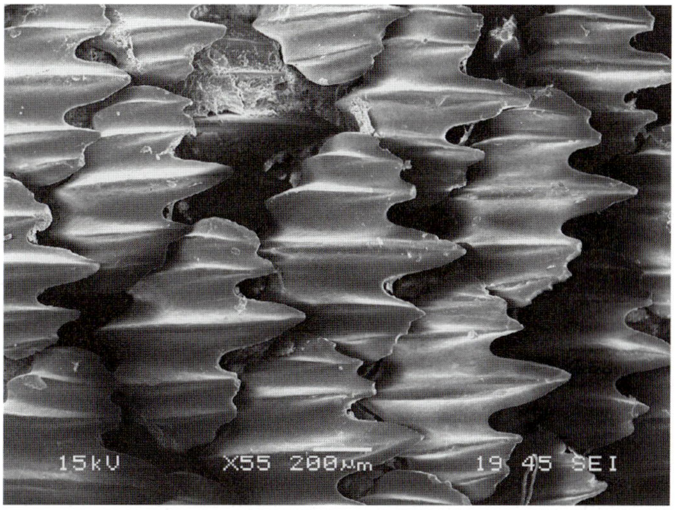

Fig. 113d. Dermal denticles.

These two species are readily differentiated on the basis of vertebral counts and by the relative sizes of the dorsal fins. The Java shark has fewer precaudal vertebrae, 93–95, and a first dorsal fin more than 3.5 times the height of the second dorsal fin. The bull shark has 101–129 precaudal vertebrae and its first dorsal fin is less than 3.5 times the height of the second dorsal fin (Bass et al. 1973; Garrick 1982).

Range The bull shark is cosmopolitan in tropical and subtropical seas. It is a coastal species that enters freshwater and can ascend far upstream into rivers and lakes. The bull shark is common along the eastern coast of North America, ranging from Chesapeake Bay southward to the Gulf of Mexico, eastern Mexico, and the tropics. It is likely that bull sharks occasionally reach New Jersey and New York, but I have been unable to verify any records. It is a rare summer visitor to Chesapeake Bay (Mansueti 1957) and is a summer shark from North Carolina to Georgia. It is common off peninsular Florida, except for the months of December through February when it moves to deeper offshore waters, and off the Keys throughout the year. Clark and von Schmidt (1965) noted that bull shark was the most common shark caught off Sarasota, Florida, accounting for 18% of the catch. Snelson et al. (1984) reported that the bull shark was the most common shark in the brackish Indian River lagoon system on the east-central coast of Florida. Steiner (2002) reported finding neonates and juveniles in the shallows of the Ten Thousand Islands estuary in Florida. There is an old record of a 38-kg bull shark being caught in the upper Mississippi River, near Alton, Illinois, on 8 September 1937, a distance then some 2,800 km from the Gulf of Mexico (Thomerson et al. 1977).

In western North America, the bull shark ranges from Baja California to the tropics. Although the species has been reported for California (albeit rare) by Miller and Lea (1972), according to Ebert (2003), there are no verifiable records of its presence there, previous records having been misidentifications of dusky sharks.

Size and weight According to Garrick (1982), bull sharks can grow to 324 cm (bull shark jaws from the western Pacific, in the Gordon Hubbell collection, confirm Garrick's estimate of maximum length). However, North American specimens actually measured are smaller. According to Springer (1960), females in Florida reach 284 cm, and males reach about 270 cm. The largest of 129 bull sharks caught by Clark and von Schmidt (1965) off Sarasota, were a 249-cm male and a 264-cm female. The largest reported by Dodrill (1977) from Melbourne Beach, Florida, measured 263 cm. The largest specimen I have measured was a 278-cm, 186-kg female caught off Charleston, South Carolina on 16 June 1981.

Caribbean specimens appear to be smaller than Florida specimens. According to Thorson and Lacy (1982), the largest of 2,742 males taken in the waters of Nicaragua and Costa Rica measured 241 cm, while the largest of 2,311 females measured 251 cm.

Representative weights are as follows:

Male: 92 cm, 5.4 kg; 180 cm, 43 kg; 188 cm, 55 kg; 241 cm, 111 kg; 47 cm, 103 kg; and 253 cm, 120 kg.

Female: 114 cm, 14 kg; 170 cm, 34 kg; 196 cm, 62 kg; 205 cm, 109 kg; 269 cm, 191 kg (gravid female); and 278 cm, 186 kg.

Biology The bull shark is a coastal species commonly found in estuaries, harbors, and creeks of tropical and subtropical America. It is one of the top predators in the coastal waters of Florida and the Caribbean. The bull shark is one of the few species of sharks that enter freshwater. Neonates and juveniles prefer salinities of 5–20, and as they grow, they leave estuaries and rivers and enter the more saline waters of coastal bays. Juveniles are common under the numerous bridges in the Florida Keys, and adults can be found farther offshore. The adults are usually found at depths of 25–100 m, but they often enter shallow waters.

Bull sharks have a remarkable tolerance to freshwater. Consequently, bull sharks can swim upstream in rivers for thousands of kilometers, sometimes entering lakes far from the sea. In North America, bull sharks have been found some 2,800 km up the Mississippi River (see "Range," above), but reports of sharks swimming up the Amazon River are more amazing. Myers (1952) reported a shark taken in October 1943 at Iquitos, Peru, some 4,000 km from the Atlantic Ocean. Thorson (1972) presented evidence of several bull sharks taken around Leticia, Colombia, some 3,500 km from the Atlantic.

The presence of bull sharks in Lake Nicaragua engendered the belief that these "lake sharks" were an endemic, strictly freshwater species. Gill and Bransford (1877) first proposed two hypotheses about the origin of the large fishes of Lake Nicaragua (the sharks, the sawfish, and the tarpon). They suggested these fishes may have evolved from saltwater ancestors that intruded into freshwaters or, more likely, from ancestors originally trapped in the lake by uplift of the land, perhaps due to volcanic activity that isolated a bay of the Pacific Ocean, forming the present lake.

The idea that the sharks of Lake Nicaragua were a distinct freshwater, landlocked species was popular until the middle of the twentieth century. Bigelow and Schroeder (1961) compared specimens from Lake Nicaragua to other bull sharks and concluded that the two forms were morphologically inseparable. Later, Thomas B. Thorson, of the

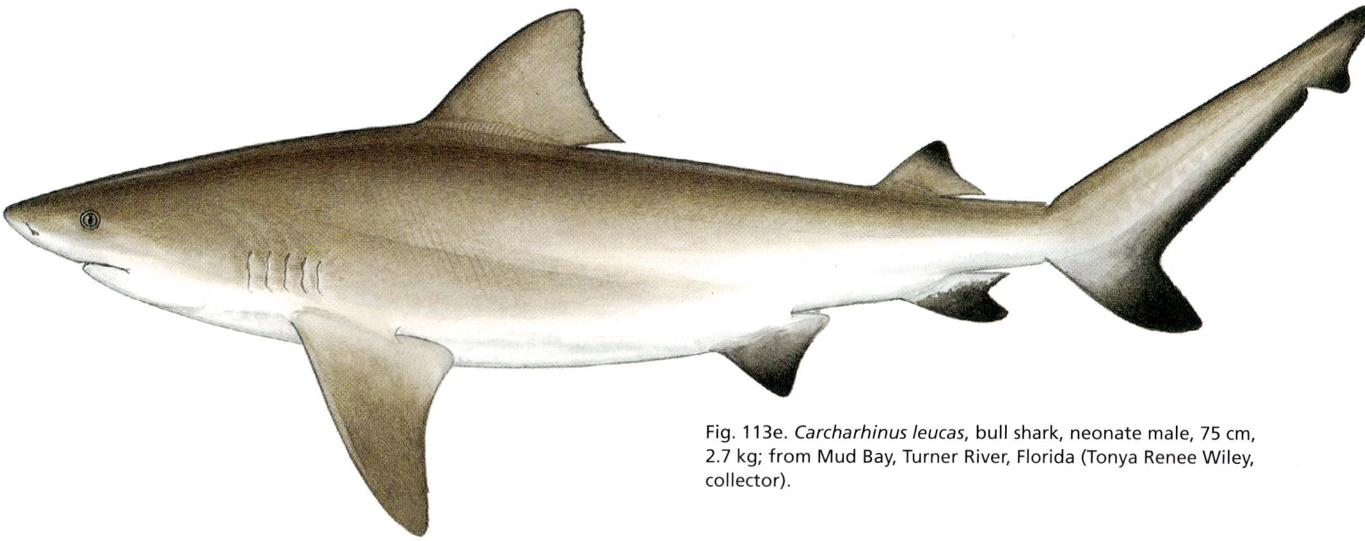

Fig. 113e. *Carcharhinus leucas*, bull shark, neonate male, 75 cm, 2.7 kg; from Mud Bay, Turner River, Florida (Tonya Renee Wiley, collector).

University of Nebraska, demonstrated by tagging that bull sharks moved from the Caribbean into Lake Nicaragua via the San Juan River and, once in the lake, also traveled back into the sea (Thorson 1971). Thorson and his colleagues tagged some 2,000 sharks at various river mouths in the Caribbean; some sharks tagged on the eastern coast of Costa Rica were recovered in Lake Nicaragua, while sharks tagged at San Carlos, on the eastern shore of Lake Nicaragua, were recovered at various locations along the Caribbean coast of Costa Rica. According to Thorson, sharks tagged at the mouth of the San Carlos River completely traversed the river in 2–25 days, while a shark tagged at San Carlos, Lake Nicaragua, was recovered the next day at San Juan del Norte, on the Caribbean shore of Nicaragua, after traveling 175 km downstream. Thorson later edited a book titled *Investigations of the Ichthyofauna of Nicaraguan Lakes* (1976) about his findings and those of his colleagues.

Today, in the early twenty-first century, the Lake Nicaragua shark and all the bull sharks of the Americas are all considered to be one species, *C. leucas.* However, given that few of these populations have been studied in detail using modern methods, it is possible that sibling species exist.

Diet: Bull sharks consume a wide variety of fishes. Snelson et al. (1984) reported that neonates and juveniles in the Indian River lagoon system, Florida, consumed mainly hardhead catfish (*Ariopsis felis*), gafftopsail catfish (*Bagre marinus*), Atlantic stingrays (*Dasyatis sabina*), and bluntnose stingray (*Dasyatis say*). The stomach of a 162-cm specimen contained the head of another bull shark estimated at 80–90 cm. Clark and von Schmidt (1965) reported finding jacks (*Caranx* spp.), tarpon (*Megalops atlanticus*), mullet (*Mugil* spp.), snook (*Centropomus undecimalis*), little tunny (*Euthynnus alletteratus*), other sharks, and mollusks. Adult bull sharks also prey on smaller sharks, including bull sharks (Springer 1960).

Migrations: The migratory patterns of the bull shark along eastern coast of North America are poorly known. From their wintering areas, some juveniles 140–170 cm in length travel north to Georgia and South Carolina in summer and return southward when the waters begin to cool in October. Juveniles up to 172 cm are found off central Florida throughout the year (Dodrill 1977).

Adult bull sharks migrate northward from their wintering areas in spring and are common from southern Florida to North Carolina from March to October. A few adults can reach Chesapeake Bay by June and July. Murdy et al. (1997) reported that during the summer months bull sharks have been recorded as far north as the Patuxent River in Maryland. Adults begin their southward migration in October. By November, the adults that had been in central Florida since April depart for deeper or more southern waters (Dodrill 1977). In winter, some bull sharks are found off southern Florida at depths greater than 100 m, but it is possible that part of the population may be in warmer southern waters. Large, pregnant females carrying term young reappear in shallow waters off southern Florida in February. They reappear off Melbourne Beach, Florida, in April (Dodrill 1977) and off the central Gulf coast of Florida in March (Clark and von Schmidt 1965). Gravid females then enter rivers and coastal lagoons to release their young (Dodrill 1977; Snelson et al. 1984).

At a shark-feeding site on the northern coast of Cuba, large female bull sharks first appear in late August. The number of sharks present at the site slowly increases in the autumn, and the greatest numbers of females are observed in November and December, when 28–30 females can be encountered at the site. The sharks disappear in January, not to be seen again until late the next summer (Elda Brizuela, pers. comm., March 2004). The periods of peak abundance of large females in Cuba coincide with the periods when they are scarce off Florida. Whether the gravid females found in Florida in autumn travel to deep water or as far as northern Cuba is unknown.

Size at maturity There is little published data on size at maturity for the bull shark, but it appears that males mature at about 220 cm and females about 230 cm. The smallest breeding male examined by Clark and von Schmidt (1965) was 225 cm. The smallest mature male that I have observed measured 220 cm. Clark and von Schmidt (1965) reported that the smallest gravid female in their sample was 233 cm.

The Caribbean population appears to differ from the bull sharks inhabiting the eastern coast of North America. Males mature at a smaller size (160–170 cm) than those in North America (Thorson 1976). N. H. Jensen (1976) reported that female bull sharks in the Lake Nicaragua–Rio San Juan system start to ovulate at a length of 170 cm and reported gravid females of 175 and 177 cm total length. These females are considerably smaller that the smallest gravid females from Florida. Furthermore, based on Jensen's limited sampling (66 gravid females), Caribbean bull shark females differ from North American specimens in that they are not reproductively synchronous and may reproduce throughout the year, although parturition is most likely to occur between April and June.

Reproduction The bull shark is a placental viviparous species. The reproductive processes of the North American population are poorly known. Gestation appears to last approximately a year. Parturition in the coastal lagoons of eastern central and southwestern Florida takes place in June and July (Snelson et al. 1984; Steiner 2002). Like most carcharhinid sharks, the bull shark has a biennial reproductive cycle (Dodrill 1977). Gravid females of the Florida population I examined are reproductively synchronous, all being in similar stages of gestation at a given time of the year.

Size at birth: Bull shark young are born at 60–80 cm, with an average of 75 cm (Clark and von Schmidt 1965; Dodrill 1977). The largest embryos I measured were 75.8–79.5 cm, from a female caught off South Carolina on 20 June. Snelson et al. (1984) reported free-swimming young 73.5–85.8 cm with fresh umbilical scars, in the Indian River, Florida, from June to August.

Brood size: Bull sharks have 7–12 young per brood, but the usual number is ten. Dodrill (1977) reported that eight females caught in the Melbourne Beach area, Florida, carried 7–12 young, both the average and the mode being 10. I have only examined six bull shark broods, three consisted of eight young and three of ten.

Nurseries Gravid bull sharks carrying term young enter river mouths, estuaries, and coastal lagoons to deliver the young. The bull shark nurseries are found primarily in the coastal lagoons and river mouths of eastern central Florida (Dodrill 1977; Snelson et al. 1984), from Daytona Beach to the southern end of the Indian River lagoon, and from the southern tip of Florida, along the southwestern coast, through the Ten Thousand Islands to at least Tampa Bay. Dodrill (1977) stated that the young-of-the-year do not spend their first winter in lagoons, apparently entering coastal waters. Juveniles 120–180 cm are present in lagoons throughout the year, and apparently reside there until the onset of sexual maturation (Snelson et al. 1984). Juveniles travel northward far along the coast in summer; I have examined numerous juveniles 120–170 cm caught off Charleston, South Carolina, in summer. Neonates and small juveniles occur in very shallow water less than 1–2 m deep in the Ten Thousand Islands area in southwestern Florida (Steiner 2002). In the same region, Simpfendorfer et al. (2005) found neonate and juvenile bull sharks in the Caloosahatchee River–Pine Island system. According to these authors, neonates spend their first year in the Caloosahatchee River, their second year near the river mouth, and thereafter enter the more saline waters of Pine Island Sound.

Age and growth The bull shark is a long-lived species, its life span probably exceeding 50 years. Most aging attempts probably reflect the growth of bull sharks in their first decade or two, and most have not been validated. It is possible that large specimens may cease to produce growth rings or lay them at irregular intervals. Thus, unvalidated estimates must be treated with some skepticism. Aquarium records, where the sharks live under unnatural conditions but are safe from human predation, provide glimpses of the potential life span, but few have been published.

Age estimates of adult bull sharks have been periodically increased. Thorson and Lacy (1982) estimated that a 228-cm female was 13.7 years old based on tagging data (7.9 years at liberty after tagging). According to Branstetter and Stiles (1987), based on vertebral ring data, their largest male (245 cm) was 21.3 years old, and the largest female (268 cm) was 24.2 years old. Cruz-Martínez et al. (2005), using vertebrae stained with a variety of techniques, estimated that the oldest male (243 cm) in their sample was 23 years old, while the oldest female (256 cm) was 28 years old. Wintner et al. (2002) used vertebral growth ring analysis to age bull sharks from the eastern coast of South Africa. Their largest female was 284 cm and 32 years old, and the largest male was 278 cm and 29 years old. These authors estimated that bull sharks live for more than 50 years and that two specimens, 269 cm and 270 cm, respectively, kept at a South African aquarium for 20 years were both more than 29 years old. A male bull shark has lived 30 years at the Okinawa Churaumi Aquarium (Senzo Uchida, pers. comm., August 2008).

Relation to humans The bull shark is often caught in commercial fisheries from South Carolina to Florida and the Gulf of Mexico. It is utilized for its flesh and fins.

This is the most dangerous shark in the southeastern United States and the Caribbean, because it is a large, aggressive species that seeks large prey in shallow coastal waters, at beaches, and in freshwater, where it may mistake humans for dolphins or other large prey. There are numerous well-authenticated cases of bull sharks attacking bathers, swimmers, and divers in beaches and coastal canals. Bull sharks become aggressive when molested. I know of a case when a juvenile bull shark repeatedly attacked a boat that was following it. The repeated attacks were filmed by videographer Mark Rackley of Little Torch Key, Florida.

Bull sharks are occasionally maintained in aquaria, where they make impressive display animals because of their large size. They are long-lived in captivity when given proper care. Many aquariums choose not to display bull sharks because they eventually kill and devour their smaller tank mates, both sharks and bony fishes.

BLACKTIP SHARK

Carcharhinus limbatus (Müller and Henle, 1839)

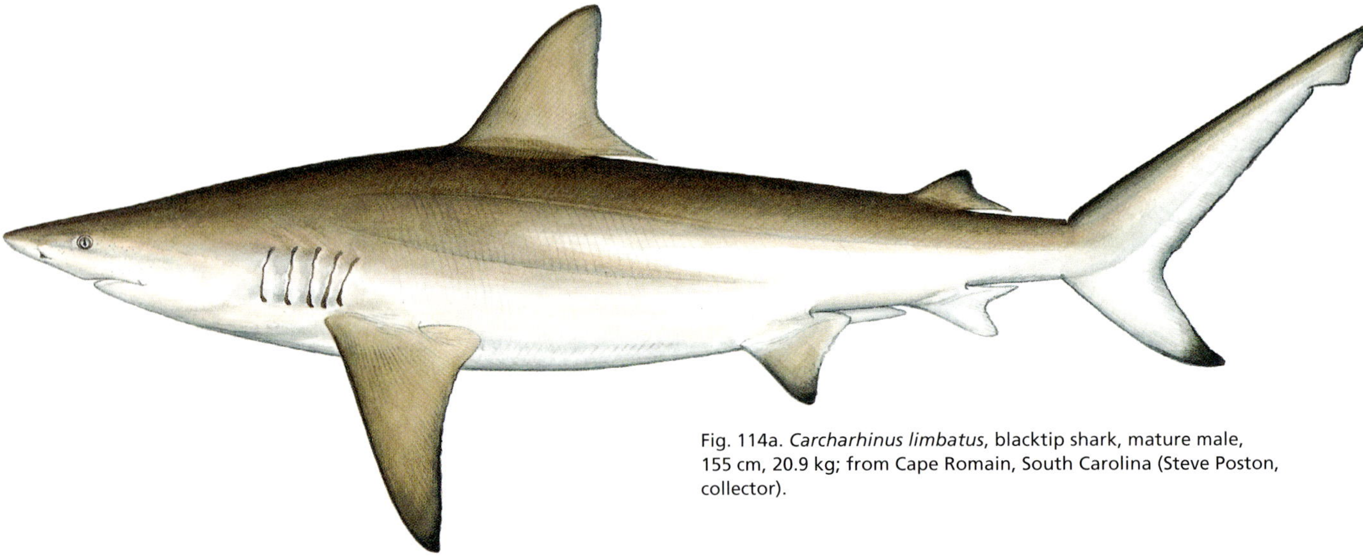

Fig. 114a. *Carcharhinus limbatus*, blacktip shark, mature male, 155 cm, 20.9 kg; from Cape Romain, South Carolina (Steve Poston, collector).

Common name Blacktip shark or "blacktip," reference to the black-tipped fins.

Spanish name *Tiburón de aletas negras, volador* (Mexico), *balicero, canalero* (Cuba), *volador* (A.F.S).

Scientific name *Carcharhinus limbatus* (Müller and Henle, 1839), originally described as *Carcharias* (*Prionodon*) *limbatus,* in *Systematische Beschreubung der Plagiostomen* 49, plate 19. *Carcharhinus:* from the Greek *karcharos,* sharp pointed or jagged, from *karcharias,* a species of dogfish having jagged teeth, and *rhine,* Greek file or rasp, also the name of a shark, originally given to *Squatina* due to its rasplike skin; *limbatus,* Latin for bordered, or having an area of one color bordered by another color, a reference to the black-tipped fins.

Synonyms A complete list of the scientific names given to this species over the last two centuries would take several pages. See Bigelow and Schroeder (1948) for a synonymy of the species. Some of the more frequently used names are *Carcharias aethlorus* Jordan and Gilbert 1882 [a], *Carcharias phorcys* Jordan and Evermann, 1902; *Carcharhinus natator* Meek and Hildebrand, 1923; and *Eulamia aethlorus* Beebe and Tee-Van, 1941.

Identification The blacktip shark is one of the *smooth-backed* carcharhinids. It has a *snout as long as the mouth width,* and a *first dorsal fin originating at about the midpoint of the inner margin of the pectoral fin.* The upper teeth have narrow, triangular, erect to slightly oblique cusps with *finely serrated edges,* and broad bases with coarser serrations; the lower teeth are erect, narrower than the upper teeth, and have finer serrations. Teeth number U: 14 to 16–1 to 3–14 to 16, L: 13 to 16–1 to 3–13 to 16. The dermal denticles are broad and overlapping; they have five parallel ridges and points, the central ones being longer. Coloration is dusky bronze in life, fading after death to pale or dark gray above with a *well-marked band of the upper gray extending down* along the lighter sides, from the level of the pectoral fins to the pelvic fins. The undersides are white or yellowish. The dorsal and pectoral fins, and the lower lobe of the caudal fin, are *black tipped,* although sometimes these markings fade with age. In Atlantic specimens, the *anal fin is white,* although rarely it can have a central dusky spot. Pacific specimens have a *black-tipped anal fin.*

Similar species The lack of an interdorsal ridge, the black-tipped fins, and symmetrical front teeth with finely serrated edges, distinguish the blacktip shark from other carcharhinid sharks in the area, except the spinner shark. The spinner shark is a more slender shark with a more posterior first dorsal fin, its origin over or posterior to the free rear tips of the pectoral fins; it has smooth-edged teeth in the lower jaw, and a black-tipped anal fin in specimens larger than 70 cm.

Range The blacktip shark is considered to be a cosmopolitan species found throughout tropical and subtropical waters (Bigelow and Schroeder 1948). It is likely that the "blacktip shark" consists of separate and distinct populations, some

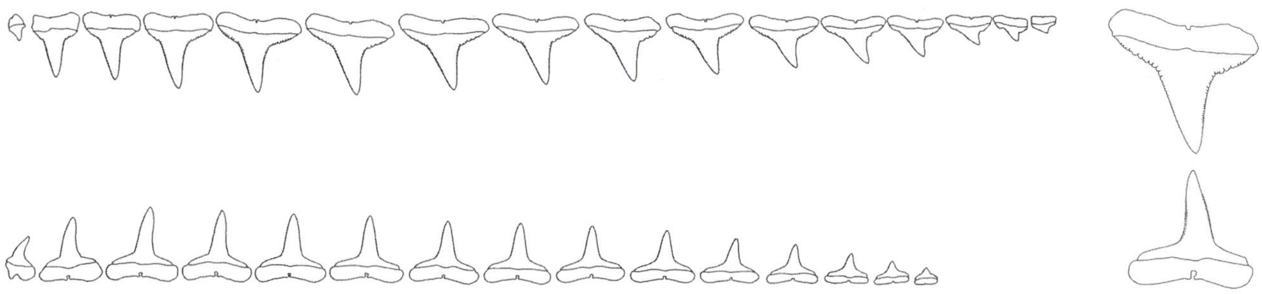

Fig. 114b. Teeth.

of which may eventually be recognized as different species (see discussion below). In the western Atlantic, the blacktip shark ranges from New England to Brazil. Along the eastern coast of North America, it ranges from New England to the Florida Keys and the Gulf of Mexico. It occurs only as a rare stray to the north of Cape Hatteras (Bigelow and Schroeder 1948) and is common from South Carolina southward to Florida and the Gulf of Mexico (Castro 1996). The blacktip shark that inhabits the eastern Pacific is a separate and distinct population, and probably a different species. It ranges along the western coast of North America, from California to the tropics. It is seasonally common in the Gulf of California. There is little published information on this population.

Size and weight The maximum size of the blacktip has been given as 240 cm by Bigelow and Schroeder (1948) and as 255 cm by Garrick (1982). None of the records of blacktip sharks off the southeastern United States that I was able to authenticate exceeded 202 cm (Castro 1996). Thus, it is likely that some of the early records refer to the spinner shark (*Carcharhinus brevipinna*) and more recent records may refer to blacktip sharks belonging to Indo-Pacific populations or to another species. Clark and von Schmidt (1965), working in the central Gulf coast of Florida, observed a maximum length of 175 cm for males and 191 cm for females. Killam (1987) reported a maximum size of 165 cm for males and 183.3 cm for females in Tampa Bay, Florida. Dodrill (1977) examined 104 specimens from Melbourne Beach, Florida, the largest of which measured 193 cm. The largest specimens seen by Gordon Hubbell (pers. comm., Dec. 1990), who fished extensively for this species in the 1970s, were a male 189.2 cm, weighing 38.55 kg, caught off Miami in July 1978, and a female 201.9 cm taken at the Port Salerno shark tournament in September 1987. The largest male I examined measured

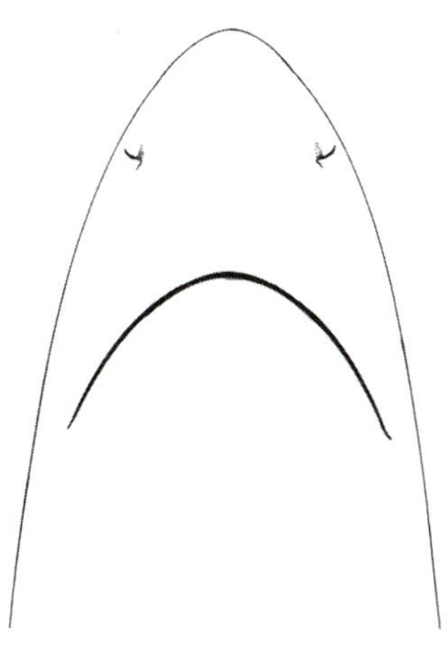

Fig. 114c. Snout.

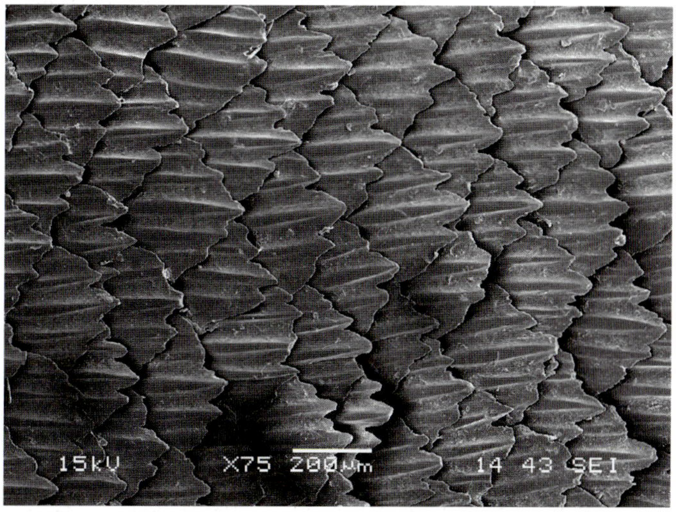

Fig. 114d. Dermal denticles.

174.3 cm and the largest female measured 193 cm and weighed 49.55 kg (Castro 1996). In summary, the blacktip shark seldom reaches 195 cm off the southeastern United States, and only exceptional specimens exceed 200 cm.

Representative sizes and weights for blacktip sharks taken off South Carolina and Florida, that I examined, are as follows:

Male: 74 cm, 2.5 kg; 131 cm, 13.4 kg; 143 cm, 14.9 kg; 151 cm, 20.0 kg; 165 cm, 30.5 kg; and 171 cm, 38.6 kg.

Female: 65 cm, 1.8 kg (neonate); 78 cm, 2.7 kg; 106 cm, 6.8 kg; 127 cm, 11.6 kg; 140 cm, 17.3 kg; 157 cm, 34.9 kg; 162 cm, 34.2 kg; 164 cm, 26 kg; 175 cm, 39.1 kg; 183.5 cm, 44.7 kg; and 193 cm, 46.7 kg (postpartum).

Blacktip sharks of the eastern Pacific are much larger than Atlantic specimens (probably reaching 240 cm), and have black-tipped anal fins. A male blacktip I examined at El Barril, Baja California, Mexico, measured 211 cm and weighed 65 kg, a size larger than that attained by Atlantic specimens. "Blacktip sharks" from the Indian Ocean reach an even larger size. I examined a gravid female taken off South Africa that measured 244 cm and weighed 120 kg. I believe these larger blacktip sharks to be a different species from *C. limbatus.*

Garrick (1982) examined a large number of museum specimens of blacktip sharks for his excellent review of the genus *Carcharhinus.* He believed the blacktip shark to be a single worldwide species predominating in tropical zones and seasonally abundant in temperate zones. Garrick found that, when viewed globally, anatomical and proportional differences, vertebral numbers, and color pattern overlap and intergrade to such an extent that it is not possible to significantly describe discrete intraspecific groupings from different localities. However, the blacktip of the western North Atlantic described here differs significantly from Pacific or Indian Ocean specimens in numerous traits such as size, markings, age at maturity, and brood size (Dudley and Cliff 1993; Garrick 1982). The question of whether these differences are due to intraspecific variation or whether there are several species included in *C. limbatus* has been a subject of debate for many years. Bigelow and Schroeder (1948: 351) stated that the "relationship to *limbatus* of the eastern Pacific and Atlantic can not be determined until specimens from the different ocean areas have been compared critically." Recent analysis by Keeny and Heist (2006) revealed that there are two geographically distinct lineages of blacktip sharks, a western Atlantic, Gulf of Mexico, and Caribbean clade, and an eastern Atlantic, Indian, and Pacific Ocean clade. As these different populations are studied it is likely that some will be found to be sufficiently distinct that they may be elevated to species status. Such was the case of the Australian blacktip shark, *Carcharhinus tilstoni,* a species first described by Whitley (1950a) and long considered to be synonymous with *C. limbatus.* It is now considered to be a separate species (Lavery and Shaklee 1991).

Biology The blacktip is a migratory species that inhabits shallow coastal waters and surface offshore waters. All the specimens I have seen in offshore waters have been males, suggesting that the sexes segregate by habitat, and that females remain in shallower coastal waters. The blacktip is a fast-swimming species, often seen leaping and spinning out of the water. It forms large schools at the time of migration. Unless otherwise noted, the information presented here refers to specimens from the eastern coast of North America.

Diet: The blacktip is primarily a predator of small bony fishes and small elasmobranchs. Off South Carolina, it preys mainly on menhaden (*Brevoortia* spp.) during the summer (Castro 1996). My analysis of stomach contents revealed shrimp trawl bycatch (unwanted fish discarded by shrimp trawlers) in 6% of the stomachs sampled in South Carolina (Castro 1996). I have often observed blacktip sharks following and congregating around shrimp boats off Georgia and the Carolinas. Perhaps these sharks are attracted by the disturbance of shrimp nets dragged on the bottom, or perhaps they are conditioned to wait for the large quantities of fish that are discarded overboard after the nets are retrieved. Dodrill (1977) found hardhead catfish (*Arius felis*) and other unidentified catfishes in 20% of the stomachs examined off Melbourne Beach, Florida. Other prey species included bluefish (*Pomatomus saltatrix*), crevalle jack (*Caranx hippos*), and ladyfish (*Elops saurus*). Wright (1981, see p. 184) noted hardhead catfish, mullet (*Mugil* spp.), white grunts (*Haemulon plumieri*), and stingrays (Dasyatidae) as the most common prey of blacktip sharks in the Florida Keys. In the Gulf of California, the blacktip sharks I examined all had large Humboldt squid in their stomachs.

Migrations: The blacktip is a warm water species that migrates into its nurseries and mating grounds in the temperate waters of Georgia and the Carolinas during the summer and moves southward to its wintering grounds off southern Florida and the Keys in winter (Castro 1996). The blacktip begins its northward migration from its wintering grounds off southern Florida and the Keys in early March. The schools pass Daytona Beach, Florida, from mid-March to late April when the surface water temperatures are 18–21°C. The first blacktip sharks arrive off South Carolina around the third week in April, although most adults do not arrive there until May. By late May or early June, the migration has reached North Carolina. Strays may continue to drift north in warm waters, but the main breeding population remains off the Carolinas and Georgia. Bigelow and Schroeder (1948) stated that although blacktips rarely reach the

New England coast, it was not unusual to see blacktips off New England drifting in the warm waters of the Gulf Stream. Breeding adults, neonates, and juveniles spend the summer in the shallow coastal waters of southern North Carolina, South Carolina and Georgia. The southern migration begins in late September, and by mid-October, when the surface water temperatures dip below 21°C, nearly all blacktips have left the summer grounds and nurseries. The blacktip schools pass Daytona Beach, Florida, from the last week in September to late November (Castro 1996). Dodrill (1977) stated that increased landings of blacktip sharks in Melbourne Beach, Florida, during November and December and a dominance of blacktips in the rod-and-reel catch in November suggested a run of this species at those times. Tabb and Manning (1961) stated that the blacktip was the most abundant shark in the shallow waters of Florida Bay, Florida, and associated brackish water areas, and that it was uncommon during the months of December through January. Wright (1981, see p. 184) indicated that the blacktip was one of the three most abundant species of sharks in the Florida Keys (the other two being the bull shark and the nurse shark) and that it occurred there at depths of less than 20 m throughout the year. He also stated that large seasonal runs occurred through the Florida Keys in late October and early November. A similar southern migration occurs in the Gulf of Mexico in fall. Clark and von Schmidt (1965), working off the central west coast of Florida, stated that the blacktip was fairly common most of the year, but that it was not found there in the winter (December–February).

The analysis of tag recovery data by Kohler et al. (1998) shows that blacktip sharks off eastern North America move primarily along the coast, supporting the idea that the eastern population of blacktip sharks winters off South Florida. Specimens tagged along the eastern coast of the United States moved north or south, with a few entering the Gulf of Mexico. Specimens tagged off Texas moved southward into Mexico. None of the sharks tagged in the United States appear to have crossed the Straits of Florida, given that there are no recovered tags from Cuba or the Caribbean.

Behavior: The ability of the blacktip to leap out of the water while spinning on its own axis has been known for many years. Bigelow and Schroeder (1948: 350) described this behavior as "frequently leaping clear of the water, sometimes revolving as many as three times around the horizontal axis in the air before falling back into the sea." This leaping and spinning behavior is usually associated with attacks on schooling fishes or after being hooked (Castro 1996). I observed this behavior several times when blacktips were feeding on schools of Spanish mackerel or menhaden. On one occasion, while a school of Spanish mackerel was passing under the boat, numerous sharks were seen jumping out of the water, creating a spectacular display. Some would leap vertically out of the water while rapidly spinning. Reaching a height of about two meters above the surface, a shark would appear to hang briefly in the air as it lost momentum and then, curling its body, would fall back into the water with a great splash (Castro 1996). Although it is obvious that a fast-moving shark may break the surface when attacking a school of fish from below, the reasons for and mechanics of the spinning are not well understood. The leaping behavior, however, is not confined to feeding situations, as many blacktips will jump one to three times when first hooked.

Most blacktip sharks in the coastal waters of the southeastern United States bear prominent circular erosion marks around the first dorsal fin base or between the two dorsal fins (Castro 1983). These lesions are caused by attachment of the parasitic leech *Stibarobdella macrothela* (formerly *Pontobdella*). This large, light green leech has been reported from numerous species of sharks throughout the world (Penner and Raj 1977).

Around divers, the blacktip is usually an extremely timid and nervous shark (when compared to other sharks such as the Caribbean reef shark). I have seen my friend Manny Puig attempting to coax blacktips to take bait from his hand (a practice I discourage, as it is very unsafe!). It took a very long time before the hungry sharks would make fast dashes to grab the bait and then swim away at great speed with the bait.

The fast movements and natural "nervousness" of the blacktip account for its being implicated in the slashing "attacks" on surfers off the eastern coast of central Florida. These highly localized interactions between sharks and humans are the result of people surfing in areas where small baitfish concentrate and attract large numbers of blacktips. In these turbid water areas, the sharks often mistake a splashing hand or foot for a small fish and sometimes cause a superficial slashing wound, which some surfers display proudly as a badge of courage. These highly localized areas are well known by reputation, but surfers continue to use them, accepting the risk. I doubt that this would be the case if bull sharks were involved.

Size at maturity

Male: In the male blacktip, as in other sharks, maturation involves the gradual development of the testes and the abrupt growth of the claspers, followed by their calcification and hardening (Castro 1996). The claspers begin to elongate abruptly when males reach a length of about 125 cm and become calcified by 145 cm. All the specimens under 142.5 cm that I examined had uncalcified claspers and were considered immature. Specimens 139–145 cm were in a transitional stage characterized by elongated claspers that were uncalcified or partly calcified, or when calcified they could barely fold. There was some individual variation as to whether the head of the clasper would open and whether sperm were present in the epididymis. Maturity is reached between 142.5 and 145 cm, and all male blacktips longer than 145 cm are mature; that is, they have hardened claspers that rotate anteriorly and whose heads open easily. The largest immature specimen I examined measured 145 cm and had partly calcified claspers and no sperm. The smallest mature male measured 142.5 cm and had calcified claspers, but sperm was not visible in the epididymis (Castro 1996).

The blacktip shows a clear seasonal change in testicular development. During the mating season, the testis appears as a sausagelike structure evaginating from the surface of the epigonal organ. After the mating season, the testis atrophies to a streaklike structure, just as in juveniles. Mature males appearing in Bulls Bay, South Carolina, in early May had fully developed testes and produced copious sperm. In specimens observed in August, the testes had begun to regress, while a male seen on 5 October had a streaklike testes that did not rise above the wall of the epigonal organ.

In early spring, males 139–145 cm are in a transitional stage, during which their claspers elongate and begin to calcify. These immature sharks with partly calcified claspers are found through early summer. By August, these sharks will have attained a length greater than 145 cm and will have hardened claspers. By the time such males mature, the mating season for the blacktip is past, and they will have to wait until the following spring to mate.

Female: Maturation in the female blacktip is marked by the growth of the oviducal gland, enlargement of the uteri, and ripening of the oocytes. The oviducal glands grow very slowly in juvenile blacktips. The glands measures 6–10 mm in juveniles less than 140 cm; after females reach 140 cm, the glands grows to a width of 26–34 mm. In recently mated or ovulating females, the oviducal glands swell and can reach 58 mm in width. The development of the uteri is also gradual. Immature females have narrow and symmetrical oviducts. With the onset of maturity, the oviducts posterior to the oviducal glands begin to enlarge to form the uteri. Enlargement begins at the caudal end and progresses anteriorly. Immature females 138.3–138.5 cm in length have uteri in which only the caudal third has expanded to a width of 20 mm. Females reach maturity at about 156 cm (Clark and von Schmidt 1965; Castro 1996). Vitellogenesis starts in late winter, and by late spring, nulliparous females are carrying ripe oocytes 30 mm in diameter. In this species, as in most carcharhinid sharks, only one ovary is developed. It lies at the anterior end of the left epigonal organ. Most adult females carry one or more atretic oocytes in the ovary. These are the remnants of oocytes that matured, but were not ovulated (Castro 1996).

Reproduction

Reproductive cycle: The blacktip shark is a placental viviparous species with a biennial reproductive cycle with a gestation period of about 340 days. Mating occurs in late May (time zero), while ovulation and the beginning of gestation follow in very early June. After a gestation period of almost one year, parturition occurs early the following May (end of year one). Parturition is followed by a "resting period" during which the females replace the energy lost by transfer to the developing offspring. Vitellogenesis and oogenesis are resumed very early in the spring of the second year, culminating in mating and ovulation in late May, at the end of the second year.

In late May and early June, the adult female population can be divided into two groups of equal size. One group consists of females that became pregnant the previous spring and are carrying term embryos. These females have inactive ovaries with oocytes with a mean diameter of 7 mm. After parturition, these females enter the resting period. They will not ovulate until the following spring, a year after having given birth. The other half of the adult population consists of ovulating females; those females that gave birth the previous spring and the nulliparous females that have just attained maturity. These nongravid females carry ripe oocytes measuring 30 mm in diameter. They mate in May and early June, gestate for almost one year, and give birth the following May or early June (Castro 1996).

Mating: In Bulls Bay, South Carolina, mating and ovulation occur from early May to early June, based on the presence of ovulating females with fresh mating bites on their flanks or with sperm in their uteri. Females bearing mating bites are seen from early May to early August, with the majority seen in May. Mating bites are often severe; usually, they are about 80–150 mm long and about 15–20 mm deep. Although these bites often cut through most of the body wall, I never saw one that had perforated the peritoneum. I examined a female on 7 June that had very fresh bites on its flanks and numerous teeth marks at the base of the right pectoral axil, indicating the animal had recently copulated and had been inseminated (Castro 1996). This female carried eight oocytes 30–31 mm in diameter, and it had two large masses of sperm (spermatozeugmata, or masses of active, dissociated sperm) in the uteri and cloaca. The spermatozeugmata were whitish gelatinous masses similar to those in finetooth sharks (Castro 1993a).

Development: Gestation usually begins in middle to late May and lasts slightly more than 11 months. In late May, females carry ellipsoidal, 50 × 30 mm, uterine eggs in the blastodisc stage. Embryos are nourished by stored yolk only (lecithotrophic) for the first eight weeks of gestation. After the eighth week of gestation, embryos measure 141–162 mm and their yolk sacs begin to implant into the uterine wall. The embryos are fully implanted by the 10th or 11th week of gestation, by which time they measure 178–194 mm. By the time of implantation, most of the stored yolk has been consumed, and the placental connection has been formed. After implantation, the embryos are nourished through the placenta, directly from the bloodstream of the mother.

Size at birth: The young are usually born when they measure about 55–60 cm (Castro 1996). Killam (1987) reported a mean size of 60.5 cm for 33 neonates seen in June in the Tampa Bay region, Florida.

Based on specimens seen in mid-July, in which the umbilical scars were already healed, it appears that the scars heal in about one month. Thus, the neonate stage lasts about one month (Castro 1996).

Brood size: Blacktip broods usually consist of four to six young, with exceptional broods having as many as nine (Bigelow and Schroeder 1948; Castro 1996). Clark and von Schmidt (1965) gave the brood size range as three to eight. Dodrill (1977) also gave a range of three to eight young and a mean of five young for 33 broods from specimens taken off Melbourne Beach, Florida. I reported broods ranging from two to six young (with a mean of four) in a sample of 39 females caught off South Carolina, that were known not to have aborted any young during capture (Castro 1996).

Nurseries In the southeastern United States, blacktip females enter shallow coastal areas in mid-April to mid-June to give birth to their young (Castro 1993b, 1996). I described a blacktip nursery in Bulls Bay, South Carolina, an estuarine system composed mainly of salt marshes and barrier islands (Castro 1993b). In Bulls Bay, I observed females carrying term embryos (58–74 cm) from mid-April to mid-June, and neonates (57–77 cm), juveniles, and adults from May to September. The latest date for a blacktip in Bulls Bay was 14 October. Their disappearance from the area usually coincided with the decline in surface water temperature below 21°C, and this usually occurred in early to mid-October (Castro 1996). This nursery area probably extends southward into Georgia and northern Florida. There are no reports of blacktip nurseries along the east central coast of Florida. Dodrill (1977) reported that he found no females with term young off Melbourne Beach after 15 May and that no neonates were caught along the beach there. Another nursery area may extend from the tip of Florida westward into the Gulf of Mexico. Killam (1987) reported 33 neonates with a mean size of 60.5 cm in the Tampa Bay area in June. There are blacktip nursery areas along the Yucatan peninsula, but little is known of these. Along western North America, the blacktip shark has nurseries in the Gulf of California. I have observed numerous 75- to 78-cm neonates, with open umbilical scars, along the Sonora coast, Mexico, in late May.

Age and growth Branstetter (1987a) used vertebral rings to estimate ages and growth rates for blacktips in the Gulf of Mexico, reporting that males mature at 130 cm and four to five years of age, and females at 150–155 cm and seven to eight years of age. Killam and Parsons (1989), working near Tampa Bay Florida, gave the size at maturity as 133–136 cm and four to five years of age for males, and 158–162 cm and six to seven years of age for females.

Relation to humans Commercially and recreationally, the blacktip is one of the two most important species of sharks off the southeastern United States (the other is the sandbar shark). In many areas, the blacktip is considered to be the one of the most palatable of sharks; consequently, other species are often sold under its name. It is usually caught on longline gear; however, migrating schools are also pursued with gillnets along the eastern coast of Florida. Seafood dealers favor the blacktip shark because the relatively small carcasses, usually 10–18 kg, are easily handled, and because consumers favor the pinkish color of its flesh. Blacktip sharks are also sought by many recreational anglers from the Carolinas to Florida and the Gulf of Mexico because of their fighting qualities and abundance near shore in the summer. Blacktip sharks have often been implicated in slashing attacks on surfers and swimmers along the east-central coast of Florida.

OCEANIC WHITETIP SHARK

Carcharhinus longimanus (Poey, 1861)

Fig. 115a. *Carcharhinus longimanus*, oceanic whitetip shark, female, 199 cm, from the Bahamas (Reinaldo "Rey" Pérez, collector).

Common name Oceanic whitetip shark, often used as "oceanic whitetip." The term "oceanic" distinguishes it from other whitetip sharks, such as the reef whitetip; "whitetip" refers to the conspicuous white fin tips.

Spanish name *Galano* (Cuba), *punta blanca* (Mexico), *tiburón oceánico* (A.F.S.).

Scientific name *Carcharhinus longimanus* (Poey, 1861), originally described as *Squalus longimanus* in *Memorias sobre la Historia Natural de la Isla de Cuba* 2: 338–339, plate 19, figs. 9 and 10. *Carcharhinus:* from the Greek *karcharos,* sharp pointed or jagged, from *karcharias,* a species of dogfish having jagged teeth, and *rhine,* Greek file or rasp, also the name of a shark, originally given to *Squatina* due to its rasplike skin; *longimanus,* from Latin *longus,* long, and *manus,* hand, an allusion to the long pectoral fins.

Synonyms The earliest name available for the species is *Squalus maou* Lesson, 1830, a name that was used for a few years in the recent past. However, Garrick (1982), in his revision of the genus, retained *Carcharhinus longimanus* (Poey, 1861) because of its long usage and because the name has been placed in the Official List of Specific Names in Zoology (name no. 2059). Other synonyms for the species include *Carcharias insularum* Snyder, 1904; *Pterolamiops magnipinnis* Smith, 1958; and *Pterolamiops budkeri* Fourmanoir, 1961, all based on differences in color, fin, or body shape that are now known to result from allometric growth and from comparing specimens of different sizes.

Identification The oceanic whitetip shark is characterized by a snout that is shorter than, or long as, the mouth width, a *first dorsal fin with a broad and evenly rounded apex, very long pectoral fins* (as long or longer than the distance from the tip of the snout to the last gill slit), the free rear tip of the anal fin that almost reaches to the base of the caudal fin, and a weak (occasionally absent) interdorsal ridge. The upper teeth are broadly triangular with serrated edges and mostly erect; the lower teeth have narrow, serrated cusps and broad bases. Teeth number U: 13 or 14–1 or 2–13 or 14, L: 13 to 15–1 to 3–13 to 15. The dermal denticles are broad with some overlapping. They have five parallel ridges and points. Color varies from olive gray to brown above and from dirty white to pale yellow below. Most specimens have one or more dark saddle blotches in front or behind the second dorsal fin. Usually, the first dorsal fin, pectorals, and caudal lobes are *white tipped,* hence the name.

Oceanic whitetip embryos and neonates have an interesting, but ephemeral, black fin pigmentation. The first dorsal fin has a light brown or tan tip with only a small area of black pigment at its apex. The second dorsal fin is strongly black tipped. The pectoral fin tips upper surfaces are light brown or tan, with a small area of black pigment at the

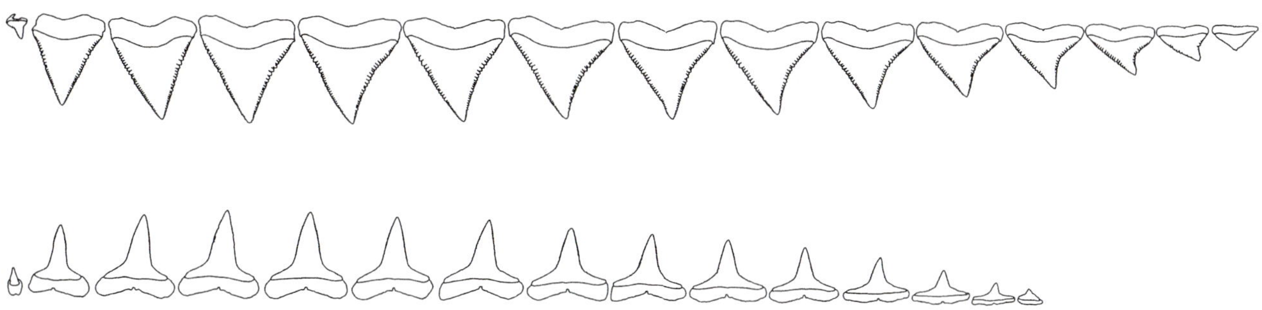

Fig. 115b. Teeth.

apex. The tips of their ventral surfaces are ink black. The pelvic fins are black tipped on both surfaces. The anal fin is black tipped. On the caudal fin, the upper lobe is usually white (rarely black tipped), and the lower lobe tip is black. There is usually one dark saddle blotch in front of the second dorsal fin, and one or two blotches on the caudal peduncle.

The dark pigmentation at the fin tips disappears gradually, and the remaining light tan coloration of the fin tips slowly changing to white or dirty white as the animal grows. The dark pigmentation of the pectoral fins disappears by the time the juveniles reach 80 cm. The pigmentation of the first dorsal fin disappears at about 95 cm. Some of the dark pigmentation of the lower lobe of the tail may last a long time. One 173-cm specimen that I examined retained some black pigment on its tail. The pigmentation of the pelvic fins usually disappears by 170 cm. The second dorsal and the anal fin may remain dark tipped into adulthood. The dark saddle blotches often persist to maturity.

Similar species The bull shark has a similarly rounded and short snout, but it lacks the interdorsal ridge and the very long pectoral fins and has a pointed (not evenly rounded) first dorsal fin.

Range The oceanic whitetip shark is circumtropical in warm oceanic waters. In North America, it ranges from Georges Bank to Florida and the Gulf of Mexico and from southern California to the tropics. It is common in the warm, deep water of the Gulf Stream, ranging far north in the current, but rarely straying into Canadian waters. According to Scott and Scott (1988) there are only two records of its presence in Canadian waters. One was a 199-cm female caught in the Gulf Stream south of Newfoundland in June 1950

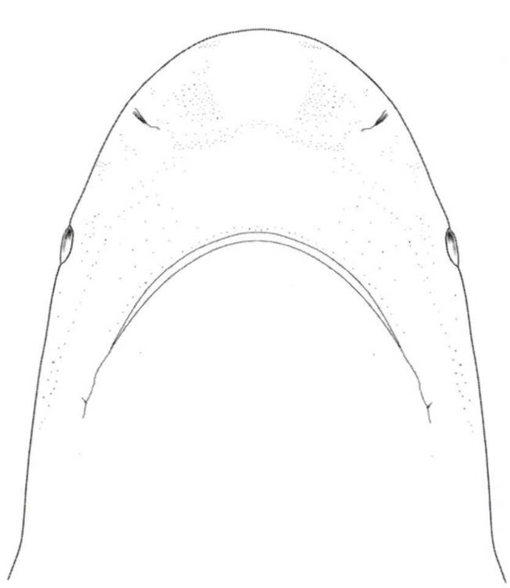

Fig. 115c. Snout.

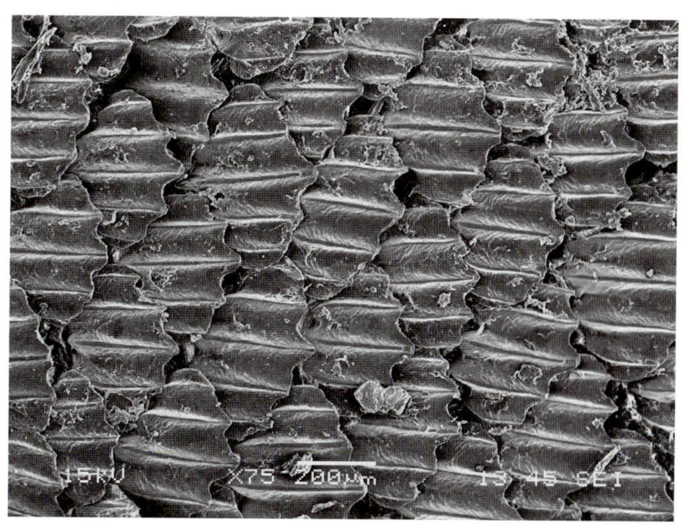

Fig. 115d. Dermal denticles.

(Schuck and Clark 1951). The other, a 221-cm male, was taken near the northeast edge of Georges Bank in October 1961, during an incursion of warm Gulf Stream water into the area (Tibbo 1962).

Hess (1962) reported the oceanic whitetip as the most frequently caught shark over depths in excess of 4,000 m from North Carolina to the Bahamas. He also reported the unusual capture of a specimen in shallow water (37 m) off the south coast of Great Abaco Island, British West Indies. Bigelow and Schroeder (1948: 361) reported that they could not find reliable records of the presence of the oceanic whitetip off the coast of Florida or anywhere else along the east coast of the United States, "in agreement with its oceanic nature." More recent work by Berkeley and Campos (1988) has demonstrated its presence off the Florida coast, although it is scarce there.

The oceanic whitetip is rare off western North America. Miller and Lea (1972) mention one specimen caught off Cortez Bank, California.

Size and weight Most oceanic whitetip sharks encountered are under 210 cm. Backus et al. (1956: 181) reported that "very few sharks over about 250 cm or under 150 cm in total length were caught" in a sample of 110 sharks, their largest specimen apparently being around 255 cm. Cadenat and Blache (1981) reported that the males they observed in the eastern Atlantic were 205–213 cm, while the females were 171–214 cm. I have recorded the following sizes and weights: 62 cm, 1.6 kg; 90 cm, 6 kg; 106 cm, 9.5 kg; 133 cm, 14 kg; 140 cm, 19.5 kg; 168 cm, 30 kg, and 173 cm, 30 kg.

Some authors give much larger measurements. Garrick (1982) says that they grow up to 300 cm, and Bigelow and Schroeder (1948) wrote that the largest oceanic whitetip for which they had exact measurements was 350 cm. Backus et al. (1956) mention Springer's observation that a small percentage of the population of some sharks is very much larger than the bulk of the adult population. This and other observations support the possibility that some of the extreme size records of oceanic whitetip sharks persistently referred to in the literature are valid but are based on abnormal individuals, apparently always females. However, the existence of such large oceanic whitetip sharks remains anecdotal.

Biology Bigelow and Schroeder (1948: 359) wrote: "Astonishingly little is known of the habits of *longimanus,* considering that it is one of the members of the genus that has been recognized the longest." Today, more than 60 years later, our knowledge of the species has increased little. Most biological knowledge comes from two studies, one by Backus et al. (1956) and one by Seki et al. (1998). There are several other studies that mention the oceanic whitetip, but most of these are fishery studies in which the authors merely reviewed catch data (often old data) without examining the animals. These studies, unfortunately, have added little to our biological knowledge.

The oceanic whitetip is a strictly pelagic species, found wherever the water temperature exceeds 22°C and the depth exceeds 180 m, very seldom straying into shallower areas. Most capture records are from areas where the surface water temperature ranges from 22 to 28°C. Shallow-water strays are usually found around islands, not far from deeper waters. Mather and Day (1954) found the oceanic whitetip to be the most abundant and widely distributed shark in the central Atlantic, where the surface water temperature ranged from 22.6 to 27.6°C. Tibbo (1962) caught a 221-cm male near the eastern tip of Georges Bank in October 1961, when the surface temperature was 22.9°C during a northward excursion of the Gulf Stream immediately following a hurricane. Off Natal, South Africa, Bass et al. (1973) caught oceanic whitetip sharks where the surface water temperature ranged from 24 to 25.3°C.

The oceanic whitetip used to be one of the most abundant sharks in warm oceanic waters. Authors writing in the 1950s called it the most abundant warm-water shark. D. W. Strasburg (1958: 339) wrote: "As implied by Bullis (1955) and pointed out by Mather and Day (1954) and Backus et al. (1956), the whitetip rather than the great blue is probably the most abundant warm-water pelagic shark." After more than five decades being taken as bycatch in the tuna and swordfish fisheries, the numbers of these sharks have declined severely. It has been estimated that, since the 1950s, oceanic whitetip sharks have declined by more than 99% (Baum and Myers 2004). Although other authors (Burgess et al. 2005) have questioned Baum and Meyers's numbers, the severe decline in oceanic whitetip numbers cannot be disputed.

Diet: Despite its abundance in the world's oceans, reports of actual stomach contents of oceanic whitetip sharks are few. Schuck and Clark (1951) reported finding a 90-cm dolphinfish (*Coryphaena hippurus*) in the stomach of an oceanic whitetip. Bullis (1961) reported blackfin tuna (*Thunnus atlanticus*) and skipjack tuna (*Katsuwonus pelamis*) in the stomachs of whitetip sharks in the Atlantic.

Bullis (1961: 195) reported an interesting and puzzling behavior. He observed a school of blackfin tuna feeding on numerous, but scattered, Spanish sardines (*Sardinella anchovia*). The tunas were very dense and feeding with

> fast darting rushes at the surface with frequent leaps clear out of the water. . . . Mixed with the tuna were many oceanic whitetip sharks ranging from 3 to 6 feet

[90–180 cm] in length. Each shark was swimming slowly in a rather erratic, sinuous course with its snout protruding from the water, its mouth wide open just about the surface level. At one point, the sharks were so numerous that they were spaced not over 10 feet apart in an area some 50 feet wide by 300 feet long. No attempt was made by the sharks to chase after or snap at the hundreds of tuna in the same area. It was an inescapable conclusion, however, that the whitetips were merely waiting and ready for those moments when tunas would accidentally swim or leap right into their mouths. The identical behavior of so many sharks, observed over a 30-minute interval, would indicate this to be a well-established feeding pattern. Only with the first hand observations, at close quarters, can one give complete credence to the likelihood that oceanic sharks obtain a segment of their food in the manner described.

Hess (1962) made a series of interesting observations on the hunting behavior of the oceanic whitetip shark while shark fishing in western North Atlantic and Bahamian waters. He used floating longlines with baited hooks suspended from a small chain some 50 cm beneath orange plastic seine floats. The oceanic whitetips were observed to approach the orange floats and nose them or play with them for a few minutes before diving and taking the bait. On four occasions, a series of four floats—one orange, one yellow, one blue-green, and one black—were baited as above and drifted from the vessel on separate lines. Of the 17 sharks taken by this method of fishing, all were caught on orange or yellow floats. Hess (1962: 655) suggested that the oceanic whitetip "is attracted by orange-yellow colors and search and find their food both visually and olfactorally. The use of high visibility orange colors in life-saving equipment, therefore, may be attractive to *C. longimanus.*"

The function of the huge, white-tipped fins of the oceanic whitetip shark has long been a subject of curiosity and debate. The late Arthur Myrberg, of the University of Miami, who did a great deal of open-ocean diving, provided a thought-provoking hypothesis to explain the function of the huge white-tipped fins (Myrberg 1991). He noted that the oceanic whitetip sharks arrived on the scene in an apparent glide, lacking the body flexibility and tail swaying of other species, such as the silky shark. Such gliding continued as the sharks patrolled the area around the observer. Their movements appeared not only effortless, but even lethargic. Myrberg also observed that as long as an oceanic whitetip remained nearby (i.e., several meters away), its silhouette was easily seen, but as it moved away, the silhouette faded against the background and only the bright fin tips remained visible. Often, oceanic whitetip sharks arrived and patrolled in pairs, the two sharks separated from one another by only a few meters. In such cases, as the pair moved

Fig. 115e. *Carcharhinus longimanus*, oceanic whitetip shark, term embryo, 58 cm; from a 213-cm female taken in the Windward Passage (Justin Grubish, collector).

Fig. 115f. *Carcharhinus longimanus*, oceanic whitetip shark, immature male, 86 cm, from 200 km northeast of the Bahamas (Reinaldo "Rey" Pérez, collector).

away, their silhouettes were rapidly replaced by a cluster of small white objects resembling a small school of fishes and gliding smoothly through the water. Thus, the white fin patches of the oceanic whitetip may act as lures for fast-moving fishes, attracting the would-be predators sufficiently close to the shark to be captured. Over time, selection pressure would increase the size of the "lures" to improve their effectiveness, and such an increase in size would also improve the shark's gliding ability. For the newborn and very small sharks, the white patches would present a potential problem by attracting predators. This problem is avoided by the blackish brown coloration of the fin tips of neonate and very small oceanic whitetips, which effectively mask or "hide" the lures. This dark pigmentation disappears by the time the specimens reach 80 cm and the fin tips get lighter as they grow.

The slow, gliding movements of the oceanic whitetip shark have engendered the common belief that it is a slow-moving, lethargic shark. In reality, these sharks can move as fast as any other sharks when they want to, as Myrberg (1991) has noted.

The migrations of the oceanic whitetip have not been described, but it is likely that they cover vast areas. There is evidence of sexual segregation, as with many other sharks. Backus et al. (1956) described an area in the northern Gulf of Mexico where females outnumbered males in roughly a 5:1 ratio, while in the Windward Passage, they reported taking a single female and 15 males. The same authors reported catching very few sharks longer than 250 cm or shorter than 150 cm in those areas. Meanwhile, in the area from the Gulf Stream off central Florida to the northern Bahamas, I have seen mainly specimens less than 130 cm.

Size at maturity The oceanic whitetip probably reaches maturity at about 180 cm, but the data are scanty.

Male: Seki et al. (1998) used the distribution of testis weight and clasper length to determine maturity in the male. These methods lack the precision of the standard of clasper calcification, as the testis weight usually varies with the season, and claspers elongate before they calcify. These authors estimated that males reached maturity between 154 and 196 cm (110 and 140 cm precaudal length), and that most males mature between 175 and 189 cm (given as 125 and 135 cm precaudal length).

Female: Backus et al. (1956) used the development of the oviduct as an indicator of maturity, although this character is a poor indicator of maturity. Nevertheless, their largest immature female was 198 cm, while the smallest gravid female measured 189 cm. Thus, given the data Backus et al. presented, females appear to reach maturity between 189 and 198 cm. Seki et al. (1998) stated that their smallest gravid female was 140 cm (98 cm precaudal length) but that the frequency of occurrence of pregnant females was low until 180 cm (130 cm precaudal length). No evidence of the gravid 140 cm (98 cm precaudal length) female was presented. This appears to be an error, a mistake in species identification or in coding the fishery data.

Reproduction Our knowledge of the reproductive processes of the oceanic whitetip shark is very limited. The oceanic whitetip shark is a viviparous, placental species. My examination of a female carrying term young revealed an inactive ovary, suggesting that the species has a biennial reproductive cycle, like most carcharhinid sharks. The length of gestation has not been determined.

Size at birth: Bigelow and Schroeder (1948) estimated that the young were born at 65–70 cm. The largest embryos that I have examined were the brood of six young 62–65 cm described below. Seki et al. (1998) gave the size at birth as between 55–77 cm for Pacific specimens. The 55-cm size was based on a free-swimming specimen those authors considered to be premature. Lessa et al. (1999b) reported a 71-cm neonate bearing a fresh umbilical scar from the equatorial southwestern Atlantic off Brazil. The smallest free-swimming specimen that I have seen measured 77 cm and was taken off the Bahamas on 9 September 1999.

Brood size: The broods of the oceanic whitetip are small, usually consisting of five to seven young. In the Atlantic Ocean, Backus et al. (1956) recorded 15 broods ranging from two to nine, and averaging six per brood. Bigelow and Schroeder (1948) reported a brood of six from a female caught off Cuba. I examined a brood of six young 61.5–65.0 cm, 1.59–1.87 kg, from a 213-cm female that was caught in the Windward Passage on 3 February 1994. Seki et al. (1998) presented data on 97 gravid females from the North Pacific. Broods ranged from 1 to 14 young, with a mode of 5 and an average of 6.2. Three gravid females examined by Strasburg (1958) in the tropical Pacific had five, six, and seven embryos. Five gravid females from Natal, South Africa, carried from six to eight embryos each and averaged seven (Bass et al. 1973).

Nurseries The nursery areas of the oceanic whitetip shark in the western North Atlantic have not been described. I have examined numerous small specimens 77–85 cm caught just north of the Bahamas, suggesting that birth had occurred near those islands. It is likely that the nursery areas, if they can be called that, are in the area north of the Antilles and the deep waters around the Bahamas. Obviously, further work is required to demonstrate this.

Seki et al. (1998) gave the distribution of catches of gravid females with large embryos longer than 63 cm and of "neonates" shorter than 84 cm in the Pacific Ocean. These "neonates" obviously included neonates and young of the year. Most of the young were found between 10° and 15° N latitude, and 10° E and 150° W longitudes.

Age and growth No data available.

Relation to humans The meat is often marketed locally in the Caribbean. The oceanic whitetip used to cause considerable damage to the catch in swordfish and tuna longline fisheries, and at the same time, these sharks used to be taken in these fisheries in great numbers. Nowadays its numbers are too reduced to be of economic importance.

The species is considered dangerous to humans and has been implicated in numerous attacks on shipwrecked sailors in warm waters, although some of these attacks are probably more accurately described as scavenging. While it is often difficult to identify species of attacking sharks in survivor accounts, the characteristic white-tipped dorsal fin of the oceanic whitetip is often sufficient for identification. In the famous case of the sinking of the cruiser USS *Indianapolis* in July 1945, it is likely that the oceanic whitetip shark, the most abundant oceanic shark in the tropical Pacific, was one of the species responsible for many of the deaths attributed to sharks. In one of the accounts of the sinking (*All the Drowned Sailors,* Lech 1982: 72), the characteristic dorsal fin provides a clue to the identity of the attackers: "The dorsal fin was 'almost as white as sheet of paper' while the body was of darker color. The shark could therefore always be seen because of the visibility of its white-tipped fin in the water."

DUSKY SHARK

Carcharhinus obscurus (Le Sueur, 1818)

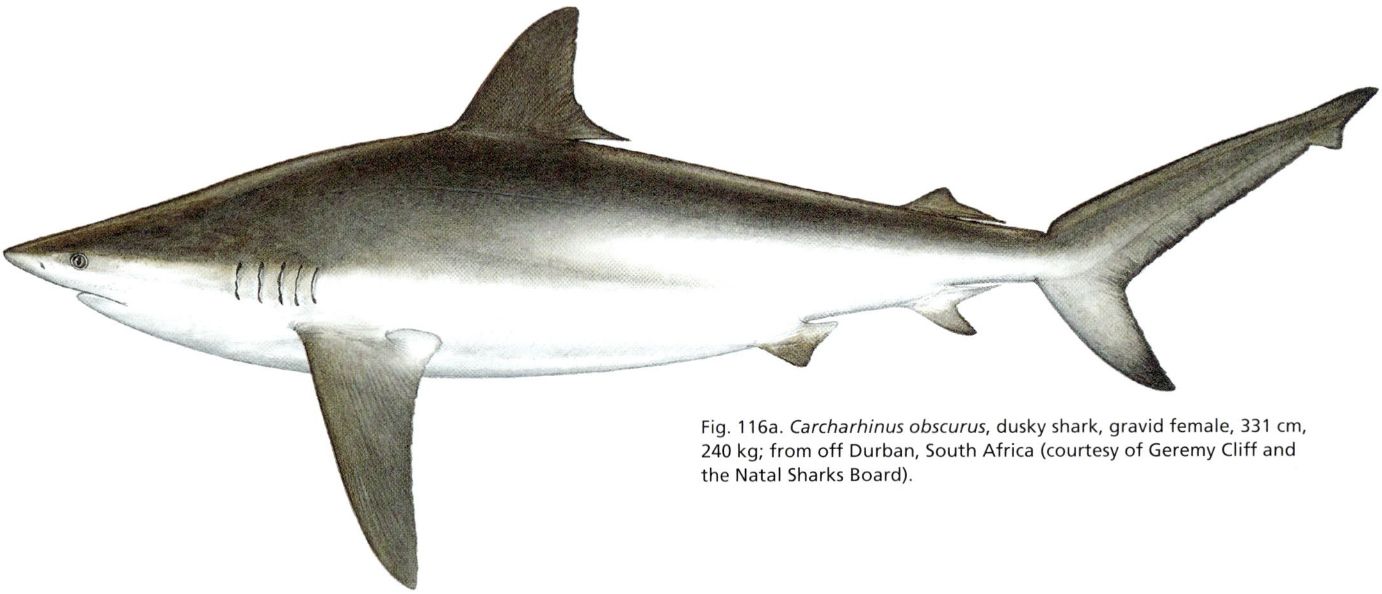

Fig. 116a. *Carcharhinus obscurus*, dusky shark, gravid female, 331 cm, 240 kg; from off Durban, South Africa (courtesy of Geremy Cliff and the Natal Sharks Board).

Common name Dusky shark or simply "dusky," a reference to its color.

Spanish name *Tiburón negro* (Mexico), *tiburón amarillo* (Cuba), *tiburón gambuso* (A.F.S.).

Scientific name *Carcharhinus obscurus* (Le Sueur, 1818), originally described as *Squalus obscurus* in *Journal of the Academy of Natural Sciences of Philadelphia* 1: 223–224. *Carcharhinus:* from the Greek *karcharos,* sharp pointed or jagged, from *karcharias,* a species of dogfish having jagged teeth, and *rhine,* Greek file or rasp, also the name of a shark, originally given to *Squatina* due to its rasplike skin; *obscurus,* Latin, dark or dusky.

Synonyms *Galeolamna eblis* Whitley, 1944; *Carcharhinus iranzae* Fourmanoir, 1961; *Carcharhinus obscurella* Deng, Xiong, and Zhan, 1981.

Identification The dusky shark has *a snout shorter than, or as long as, the mouth width,* a *sloping first dorsal fin originating over or slightly anterior to the free rear tips of the pectoral fins,* and a *marked interdorsal ridge.* The upper teeth are broadly triangular with serrated edges; the front teeth are nearly erect, and the others are slightly oblique. The lower teeth have narrow, nearly erect cusps with more finely serrated edges than the upper teeth. Teeth number U: 14 or 15–1 or 2–14 or 15, L: 13 to 15–1 or 2–13 to 15. The precaudal vertebrae number 86–97. The dermal denticles are broad and overlapping; they have five or six parallel ridges that do not terminate in distinct points. The denticle rear margins often lack distinct points and may have a scalloped look. Coloration is dark brownish gray to black above with metallic bronze or purple hues in life, white below.

Similar species The similar Galapagos shark is rarely reported from the continental waters of North America; it inhabits waters off oceanic islands, while the dusky inhabits continental coastlines. Large adults found in the open ocean should be examined carefully. The Galapagos shark has an erect, higher second dorsal fin (2.1–3.3% of the total length, compared with 1.5–2.3% of total length in the dusky shark) and a higher number of precaudal vertebrae (103–109). The sandbar shark is often confused with the dusky; it has a much higher and anteriorly placed first dorsal fin, with its origin over the pectoral fin axil, and its skin denticles are oval shaped and nonoverlapping. The Caribbean reef shark has much narrower upper teeth. The bignose shark has a more anterior origin of the first dorsal fin, over the pectoral axil, and its skin denticles are nonoverlapping. The silky shark has the origin of the first dorsal fin posterior to the free rear tips of the pectoral fins, and a second dorsal fin free rear tip twice as long as the fin height. The oceanic whitetip shark has a first dorsal fin and pectoral fins with broadly rounded tips; its second dorsal fin reaches almost to the base of the caudal fin. The bull shark lacks an interdorsal ridge.

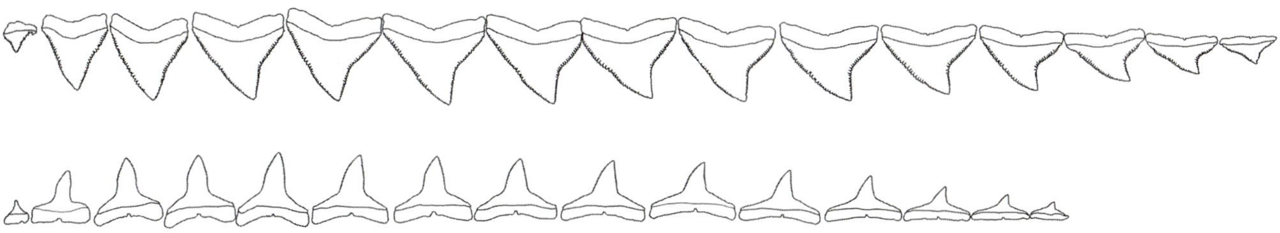

Fig.116b. Teeth.

Range The dusky shark has been reported from numerous localities throughout the world. It is probably cosmopolitan in warm and temperate continental water. In North America, it ranges from Georges Bank (Tibbo and McKenzie 1963), where it is rare, to Florida and the Gulf of Mexico (Springer 1938), and from southern California to the Gulf of California (Kato et al. 1967) and south to Chiapas (S. R. Soriano, pers. comm. October 2006). When dusky sharks were abundant in the 1960s and early 1970s, large females were regularly taken about 1–3 km off Midnight Pass, Sarasota (Clark and von Schmidt 1965; Cape Haze Marine Laboratory data). Today they are seldom encountered there.

Size and weight The dusky shark is the largest species of the genus *Carcharhinus.* The largest measured specimen that I have been able to find is a 384-cm, 372-kg female taken off Sarasota on 21 November 1970 (records of the Cape Haze Marine Laboratory). Springer (1938) reported that the largest female he examined measured 356 cm. Clark and von Schmidt (1965) reported females up to 342 cm and males up to 316 cm. Representative sizes and weights are as follows:

Male: 99 cm, 5.8–6.8 kg; 229 cm, 64.6 kg; and 328 cm, 196 kg.

Female: 83 cm, 5 kg; 267 cm, 99 kg; 318 cm, 200 kg; 334 cm, 238 kg; 345 cm, 331 kg (gravid); and 371 cm, 368 kg (term female).

Biology The dusky shark is one of the larger requiem sharks inhabiting continental waters of North America. Despite its former abundance, the life history of the dusky shark is poorly known, probably due to confusion with other large ridge-backed sharks.

The dusky shark is found from inshore waters to the outer reaches of the continental shelf. Although this large shark generally inhabits deeper water than the other large, coastal requiem sharks (e.g., bull, sandbar, and lemon sharks), it also comes close inshore. Dodrill (1977) reported that

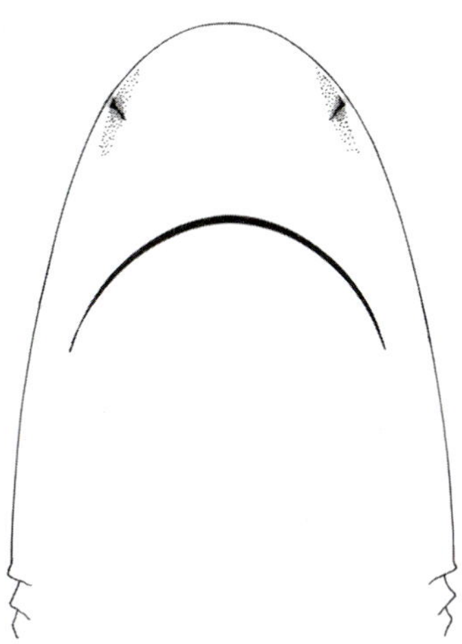

Fig. 116c. Snout.

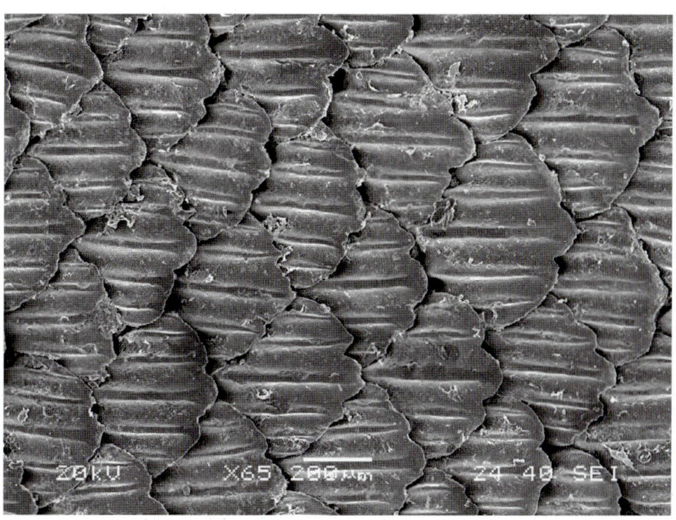

Fig. 116d. Dermal denticles.

off Melbourne Beach, Florida, the dusky shark was usually caught at depths of 5–7 m (some 300–500 m from the beach), but three had been caught at depths of 2–4 m (160–185 m from beach). Off Georgetown and McClellanville, South Carolina, I saw numerous dusky sharks caught in coastal waters 5–7 m deep during their spring and fall migrations in the 1980s. Some dusky sharks are found in the pelagic environment, although they are not as abundant there as they are over continental waters. Berkeley and Campos (1988) reported that the species constituted about 1.1% (7 out 613) of all sharks taken in their sampling of the commercial swordfish fishery operating in oceanic waters off Florida in 1981–1983.

The dusky shark is essentially a cool-water species and is usually found where the surface temperatures are 18–22°C; however, dusky sharks have been captured over a wider temperature range (15–30°C). In the northern part of the range, Tibbo and McKenzie (1963) reported catching adult males off Georges Bank in September, when the surface temperature was 19.3°C. Off South Carolina, I found dusky sharks are present during their spring and autumn migrations when surface waters were 19–21°C. Off Melbourne Beach, Florida, dusky sharks were found nearly year-round in temperatures of 15–30°C (Dodrill 1977). In the Gulf of Mexico, large female dusky sharks were found off Sarasota from November to February, when surface water temperatures ranged from 19 to 22°C; males were taken infrequently (Clark and von Schmidt 1965).

Dusky sharks have been reported to feed on numerous fishes and invertebrates: flatfishes, groupers, jacks, barracudas, other sharks, rays, mullet, eels, squid, octopuses, and many reef fishes. Springer (1946) reported finding 2.5–5 kg groupers and octopuses in the stomachs of dusky sharks caught off Florida.

Migrations: Although the dusky shark is known to move southward in winter and northward in summer along the East Coast and in the Gulf of Mexico, its migrations are poorly defined. Dusky sharks are found along the east coast of Florida throughout the year (Springer 1938); however, part of the population winters off south Florida, the Gulf of Mexico, and the northern Caribbean. Dodrill (1977) reported that the dusky shark was usually taken 300–500 m off Melbourne Beach, in central Florida, during all months except April, May and June. Springer (1946) reported taking 33 adults in four days (30 June to 4 July 1943), off Port Salerno, Florida, at the inshore edge of the Gulf Stream. Guitart Manday (1975) reported that dusky sharks were taken off Cuba only in the winter months and that the greatest numbers were taken in January. The gravid females travel farther south than the males, appearing off south Florida, the Keys, and Cuba from November to January. Large gravid females have been collected off the central Gulf coast of Florida (off Englewood and Sarasota) only from November to February, with the peak catches occurring from 10 December to 20 January (Springer 1938; Clark and von Schmidt 1965; Cape Haze Marine Laboratory records).

The dusky shark begins its northward migration earlier than other migratory species, probably due to its tolerance for cooler waters. The dusky shark is the first species to appear in South Carolina waters each spring. Neonates first appear off Georgetown, South Carolina, in March and April. I have seen dozens of neonates 100–104 cm caught there at depths of 5 m or less. Some of these sharks had healing umbilical scars and, in one case, remains of the umbilical cord. By summer, dusky sharks can be found off New England; Bigelow and Schroeder (1948) reported that all the captures in the northern part of the range have been during the warm months.

Size at maturity There are few reports of the dusky shark's size at maturity.

Male: Natanson et al. (1995) reported that males reach maturity at 279 cm.

Female: Dusky shark females appear to mature at a length of 270–280 cm. The smallest mature female in the Cape Haze Marine Laboratory records is 270.1 cm, taken on 1 December 1965, but most dusky sharks are just starting to develop at that size. The two smallest gravid females reported by Clark and von Schmidt (1965) measured 289 and 292 cm. Natanson el al. (1995) reported that females matured at 284 cm. The size at maturity for females (275 cm) is 72% of the maximum size of the species (384 cm).

Reproduction The dusky shark is a placental, viviparous species. Very little is known of its reproductive processes. Most of what we know about it is based on the observations of Eugenie Clark, Jon Dodrill, and Steve Branstetter, made in the 1960s and 1970s before the population of dusky sharks in the eastern United States declined severely due to commercial fishing. Unfortunately, little or no research was done on dusky sharks during the period when they were being fished out. Nowadays, the scarcity of adult dusky sharks and myriad regulations prevent research on the species. Hence, we are limited to studying the old data to determine the dusky shark's reproductive patterns.

Clark and von Schmidt (1965) reported that gravid dusky females taken off Sarasota, Florida, fell into two groups depending on the sizes of their embryos. One group, which the researchers referred to as group A, consisted of 14 gravid females caught during the winter months and carrying embryos 43–70 cm, and one exceptional female caught in August in offshore waters. Group B comprised four females carrying near- or full-term young 85–100 cm at the same

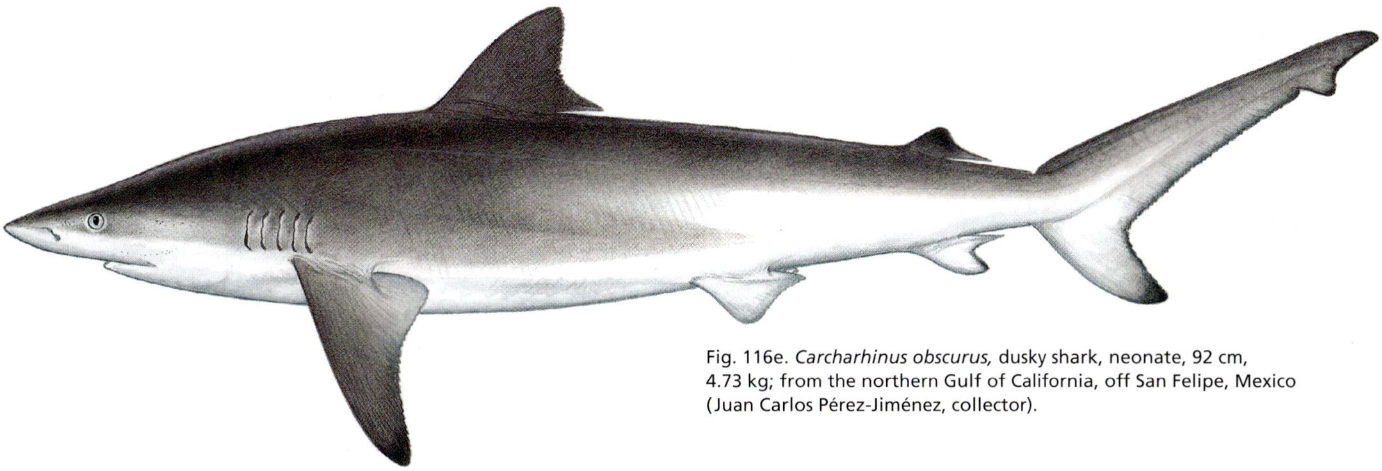

Fig. 116e. *Carcharhinus obscurus,* dusky shark, neonate, 92 cm, 4.73 kg; from the northern Gulf of California, off San Felipe, Mexico (Juan Carlos Pérez-Jiménez, collector).

time of the year. In addition, they showed some data from Springer (1938, 1940), which also showed the two groups. These researchers stated that the females carrying the larger embryos possibly gave birth in the Sarasota area, but they captured no free-swimming young in that area. Based on a single ripe male caught in April, the researchers postulated that group B would have a gestation period of eight or nine months, while group A would have a gestation period of about 16 months. Clark and von Schmidt (1965) concluded that the evidence for the 16-month gestation was stronger, when the birth size of 97 cm suggested by Bigelow and Schroeder (1948) was considered.

Dodrill (1977) reported the following mean sizes for embryos taken from females caught off Melbourne Beach, Florida: 25.5 cm on 11 September, 64.1 cm on 3 February, 59.2 cm on 27 February, 64.8 cm on 20 March, and 64.1 cm on 30 March. He also reported a female carrying 40 mm oocytes on 27 July. Dodrill then suggested that female dusky sharks gave birth every three years.

Later, Branstetter (1981), also working in the Gulf of Mexico, reported a gravid female carrying 10 embryos 79–86 cm on 21 July 1979, and a smaller embryo 77 cm from a brood of eight taken on 18 July. He examined two other females that had supposedly recently given birth and had visible placental scars. He also noted mating bites on all the mature females at that time (July).

My own review of the records of the Cape Haze Marine Laboratory revealed eight additional gravid females caught after the 1965 Clark and von Schmidt paper. Their embryos fell in the same two size groups mentioned above. When all the data (Gulf and Atlantic specimens) are plotted together (Clark and von Schmidt's 1965 data, Dodrill's 1977 data, Branstetter's 1981 data, and the additional Cape Haze Laboratory specimens), the data all fall on a curve that begins with mating in June and ends with term young in December of the following year, giving a gestation period of about 18 months. Dodrill's data on Atlantic specimens fits well with the Gulf data. The Cape Haze Laboratory data also reveal that term females carried only small, nonvitellogenic oocytes 5–10 mm in diameter. Therefore, the postpartum females would not be able to mate the next mating season in June or July and would probably mate the following year, some 18 months after parturition. In this model, the dusky shark would have an 18-month gestation period followed by an 18-month resting period. Thus, females would give birth every three years, as postulated by Dodrill (1977).

Size at birth: Dusky sharks are born at 90–100 cm. Embryos 94–95 cm weighed 4.1–4.6 kg, while a 98 cm embryo weighed 5.2 kg (Cape Haze Marine Laboratory data, and my own data).

Brood size: Dusky shark broods usually consist of 8–10 young, but larger females may carry up to 16. Analysis of 16 broods revealed a range of 5–10, and an average of eight per brood (Cape Haze Marine Laboratory data). Females carrying five, six, or seven young occurred only once in the sample, as most females carried eight or more young. A South African specimen that I examined carried 16 young 98–100 cm.

Nurseries The location of the nurseries of the dusky shark is not well defined. I have seen numerous neonate dusky sharks, 100–104 cm in length, off South Carolina in March and April. These animals had open umbilical scars with remains of the umbilical cord, or healing or partly-healed scars. These neonate sharks only spend a couple of months in the area and move northward as water warms in spring. I believe that these sharks were born off South Carolina or nearby. However, without seeing the gravid females, it is difficult to determine whether parturition had taken place there. Clark and von Schmidt (1965) reported catching numerous dusky shark females with term young (>90 cm) off Sarasota, Florida, in middle to late December. Although, several of these sharks gave birth in captivity in late December, these authors stated that no free-swimming young were caught in there.

Age and growth Natanson et al. (1995) used vertebral bands to age dusky sharks and reported that the males reach maturity at 279 cm and 19 years of age and females at 284 cm and 21 years. The oldest female in their sample was more than 33 years old.

Relation to humans Dusky sharks used to be abundant off the southeastern United States. Springer (1963: 112) stated that the dusky shark was taken in appreciable numbers in the Florida shark fishery of 1935–1950, and that the species was found "most abundantly in large schools at depths of 30 to 150 fathoms" along the east coast of Florida. The dusky shark was the fifth most abundant large shark[4] in the study carried out off Sarasota, Florida, by Clark and von Schmidt (1965). According to Dodrill (1977) the dusky shark was also the fifth most common shark longer than 100 cm in his survey of the sharks off Melbourne Beach, Florida, and represented 8.8% of the setline landings and 8.5% of all rod-and-reel landings of sharks larger than 100 cm.

The decline of the population of dusky sharks off the southeastern United States and in the Caribbean apparently started in the 1970s, probably as a consequence of the yellowfin tuna fisheries in the Gulf of Mexico. Guitart Manday (1975) was the first to report a marked decrease in the numbers of dusky sharks caught off Cuba compared to the abundance of the species 10 years before. According to Musick et al. (1993), the number of dusky sharks of all sizes taken in annual hook-and-line surveys in the Chesapeake Bight region declined rapidly after 1981, compared to the period 1974–1980. These authors noted that mature dusky sharks had been rare in their longline collections (9 since 1984) and that their most recent captures were in 1982. The development of the commercial shark fishery in the southeastern United States in the early 1980s further reduced the numbers of dusky sharks. Baum and Meyers (2004) have estimated that the abundance of dusky sharks in the Gulf of Mexico declined by 71–87% from 1950 to the late 1990s.

As a result of concerns over the disappearing dusky sharks, the species was given protected status in the federal waters of the eastern United States in 1999. The effectiveness of the protected status in restoring the dusky shark population is difficult to evaluate. Because dusky sharks cannot be prevented from taking hooks intended for sandbar sharks, and because dusky sharks can be confused with sandbar sharks, many dusky sharks are currently being taken. I have seen many juvenile dusky sharks still being caught in fishing operations off the southeastern United States, where their carcasses enter the markets as sandbar sharks. Given these facts, and the slow growth and reproductive rate of the dusky shark, it is unlikely that we will see a rapid recovery of the population.

4. The large sharks in the area were (in order of abundance): bull, sandbar, lemon, tiger, and dusky sharks.

CARIBBEAN REEF SHARK

Carcharhinus perezii (Poey, 1876)

Fig. 117a. *Carcharhinus perezii,* Caribbean reef shark, immature female, 178 cm, 39 kg; from off Key West, Florida (courtesy of Sean Reyngoudt and Chris Tanaka).

Common name Caribbean reef shark, both terms being obvious allusions to its habitat.

Spanish name *Cabeza dura* (Cuba).

Scientific name *Carcharhinus perezii* (Poey, 1876), originally described as *Platyodon perezii* in *Enumeratio Piscium Cubensium* 3: 194–196, plate 9, figs. 2 and 3. *Carcharhinus:* from the Greek *karcharos,* sharp pointed or jagged, from *karcharias,* a species of dogfish having jagged teeth, and *rhine,* Greek file or rasp, also the name of a shark, originally given to *Squatina* due to its rasplike skin; *perezii:* latinized form of Pérez, after Poey's friend Laureano Pérez Arcas, professor at the University of Madrid, whose textbook *Elementos de Zoología* was used at the University of Havana.

Synonyms *Carcharhinus springeri* Bigelow and Schroeder, 1944.

Identification The Caribbean reef shark has a snout shorter than the mouth width, a first dorsal fin originating over the free rear tips of the pectoral fins, a *second dorsal fin with a short free rear tip* (its inner margin is shorter than or equal to the height of the fin), and *an interdorsal ridge.* The upper teeth have *narrow cusps with serrated edges and broad bases;* the front two to four teeth on each side of the jaw are erect, and the others are increasingly oblique. The lower teeth have narrow, erect, triangular cusps with serrated edges and broad bases. Teeth number U: 12 or 13–1 or 2–12 or 13, L: 12 or 13–1–12 or 13. The dermal denticles are broad and closely overlapping; they have five to seven parallel ridges and corresponding points. Coloration is grayish brown to gray above and white below. The pectoral fins are black tipped, with the dark coloration extending along the ventral rear margin. The anal fin is black tipped, with the black coloration extending upward along the rear margin of the fin. The *lower caudal fin has a conspicuous black border.*

Similar species The dusky shark has broadly triangular, anterior upper teeth, and the free rear tip of its second dorsal fin is much longer (nearly twice the fin height). The silky shark has teeth with larger serrations at the bases than at the cusps, and the free rear tip of its second dorsal fin is much longer (about twice the height of the fin). The sandbar shark has the first dorsal fin originating more anteriorly, over the pectoral axil. The bignose shark has a longer snout (its length greater than the mouth width) and has broadly triangular teeth. The Galapagos shark has more erect first dorsal fin, a second dorsal fin with a longer free rear tip, and broadly triangular upper teeth. The bull shark lacks an interdorsal ridge.

Range The Caribbean reef shark inhabits the western Atlantic, from the southeast coast of Florida and the Caribbean Sea south to Brazil. The species has been reported from various localities in the western Atlantic: Islas Mujeres, Mexico (Clark 1975), Islas Los Hermanos, Venezuela (Cervigón 1964), and Fernando do Noronha Archipelago, Brazil (Sazima and Moura 2000). There are few reliable records from the Gulf of Mexico, perhaps due to confusion with similar

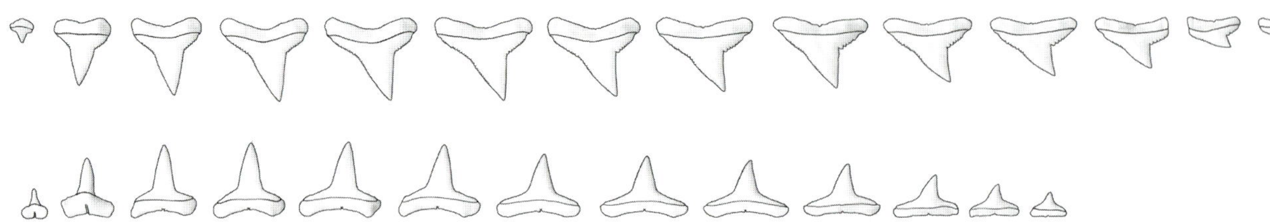

Fig.117b. Teeth.

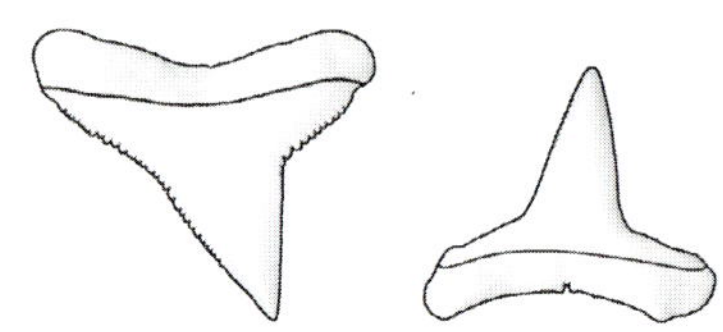

Fig.117c. Teeth (detail).

species. The Caribbean reef shark is one of the most abundant sharks around the coral reefs of the Bahamas and the Antilles.

The northernmost record for the Caribbean reef shark that I have is Port Salerno on the eastern coast of Florida, where I saw several at the Port Salerno Shark Fishing Tournaments in the late 1980s. Bigelow and Schroeder (1948: 404) mention a specimen "probably this species" from Englewood, on the west coast of Florida. I have not found any valid records of the species from that area, and neither Eugenie Clark, who did extensive sampling in the area in 1955–1970, nor I have ever heard of one being caught there. The Caribbean reef shark is locally common over coral reefs from Key Largo southward to Dry Tortugas and the Caribbean. It is abundant and easily encountered around Molasses Reef, Looe Key, Marquesas, and Dry Tortugas.

Size and weight The largest Caribbean reef shark that I have examined was a 249-cm female caught off Tobago in mid-March 1986. This specimen was so much larger than the reef sharks usually taken in Tobago that I had to examine its teeth to confirm the identification. Garrick (1982) mentions a record of a 295-cm specimen from the Bahamas. The specimen is listed as discarded, suggesting that Garrick did not actually examine the specimen. The reported size is so much larger than other Caribbean reef sharks that one

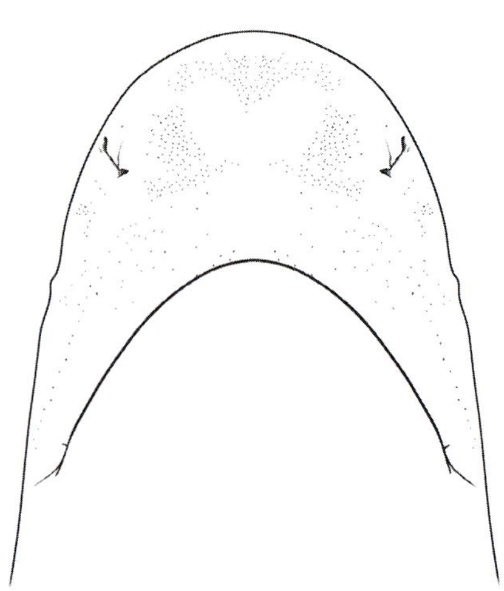

Fig. 117d. Snout.

Fig. 117e. Dermal denticles.

must suspect an identification error. The proportions and vertebral counts given for the specimen match those of the Galapagos shark, a large species of wide distribution around islands and occasionally found around the Bahamas.

Representative sizes and weights for the Caribbean reef shark are as follows:

Male: 69.8 cm, 1.9 kg; 73.3 cm, 2.3 kg; 109 cm, 8.6 kg; 115.9 cm, 8.9 kg; 143.3 cm, 20.5 kg; and 159 cm, 24.1 kg.

Female: 130.8 cm, 12.7 kg; 162 cm, 29.9 kg; 172 cm, 35.8 kg; and 215 cm, 70 kg.

Biology The Caribbean reef shark is abundant in the shallow coral reefs of extreme southern Florida and the Caribbean, but its distribution is highly localized, being confined to coral reef areas primarily in clear waters. It inhabits the outer edges of coral reefs, favoring rock ledges, caves, and drop-offs. The species often enters shallow waters less than 2–3 m deep, although it is most often seen at depths of 9–27 m. Despite its abundance in its habitat, the biology of the Caribbean reef shark is poorly known.

The Caribbean reef shark has the ability to remain motionless on the bottom, and individuals are often seen resting under a ledge or in caves. This behavior, sometimes erroneously referred to as "sleeping," was first reported by Randall (1968). Later, in a popular article of wide distribution, Clark (1975) reported Caribbean reef sharks resting on the bottom in caves off Isla Mujeres, Mexico. Some researchers have hypothesized that the sharks are able to remain motionless because of oxygen-rich currents in the area.

Little is known of the diet of the Caribbean reef shark. Specimens that I examined in Tobago had consumed flying fishes and small scombrid fishes (~30 cm).

Juvenile Caribbean reef sharks have been reported seeking out cleaning stations and posing on the bottom while being cleaned by the yellownose goby (*Elacatinus randalli*) at depths of 18–24 m in the Fernando do Noronha Archipelago, Brazil. Some sharks adopted a soliciting pose by lying on their sides, while others just simply took a position on the bottom (Sazima and Moura 2000).

Size at maturity

Male: Males mature at about 168–170 cm. I examined two males 159 and 166 cm that had fully elongated claspers that had not yet calcified. A 170 cm and 31 kg male had mature (calcified) claspers. All other males that I have seen larger than 170 cm were mature.

Female: Females appear to mature at about 190 cm, but this is based on limited data. The largest immature female that I have examined measured 182.4 cm, whereas a 209-cm female was gravid with four young.

Reproduction The Caribbean reef shark is a viviparous species, but little is known about its reproductive processes. A 249-cm female I examined on 16 May 1986 caught off Buccoo Reef, Tobago, carried 12 large oocytes 29 mm in diameter, suggesting that it was close to ovulation in early summer. The reproductive cycle appears to be biennial, as in most carcharhinid sharks. A female carrying 46.3- to 47.8-cm young on May 29 suggests birth in summer. This last specimen, in which pregnancy was well advanced, had an inactive ovary with 2 mm oocytes, suggesting a biennial cycle. My finding three neonates of similar sizes off Tobago in September suggests that the population is synchronous, with a definitive birth season in summer.

Size at birth: The young are probably born at 60–65 cm. The three neonates that I examined (69.8–74.5 cm) had healing umbilical scars. These last three were all caught off Tobago on 26 September 1985. Garrick (1982) mentions a 72.6-cm free-swimming specimen.

Brood size: I have examined only one gravid female and know of no others reported in the literature. It carried four young 46.3–47.8 cm when captured off Tobago on 29 May 1986.

Nurseries Little is known about the nurseries of the Caribbean reef shark, other than the young are found around shallow coral reefs.

Age and growth The Caribbean reef shark has not been aged.

Relation to humans The Caribbean reef shark is the most frequently seen shark in the "shark dives" of the Bahamas, in which dozens of sharks have been conditioned to be fed in front of divers. These enterprises have become important tourist attractions and booming businesses on some islands. The experience of seeing a couple of dozen sharks swimming all around you, waiting to be hand-fed, is truly exhilarating, as I can attest, and such dives have done much to promote shark conservation. First, to islanders, the live sharks are more valuable as permanent tourist attractions than are dead sharks used for meat and shark fin soup. Second, the people who participate in the dives suddenly become aware of the beauty and fragility of these great predators and often develop strongly positive attitudes about the conservation of sharks. The long-term effects of sharks becoming conditioned to be fed are unknown. Although shark-feeding activities have been occurring at some sites for more than 15 years, I am not aware of any reports of their effects on sharks. Some anecdotal evidence suggests that most sharks stay around feeding sites for only a few months and then disappear until the next season. Given that the main factors affecting the survival of sharks are the relentless pursuit and overfishing by commercial fleets and the tremendous bycatch in the tuna and swordfish fisheries, the feeding of a few dozen sharks, at specific localities, is likely to be of little ecological significance.

A shark-diving business similar to the Bahamian type flourished briefly in the early 1970s off Isla Mujeres, Mexico, where local divers took tourists to the "caves of the sleeping sharks." Unfortunately, the sharks were soon killed by local divers using spear guns with explosive heads or driven away, before the behavior could be studied further (E. Clark, pers. comm., Feb. 2004).

The Caribbean reef shark is of little commercial interest in the shark fisheries of the United States because of its limited range and relatively small size. The species received federal protection along the eastern coast of the United States in 1999. As there was no commercial interest in the species, and there were no documented declines in the species, it can be assumed that the reasons for placing it in the protected list were preventive in nature and reflected its lack of commercial importance in the shark fishery. It is difficult to assess the effects of protective legislation, as dressed carcasses are likely to be confused with those of blacktip sharks.

Caribbean reef sharks have been implicated in attacks on humans in Florida and the Bahamas. Most of the attacks of which I am aware have been on spear fishers who favor the clear-water reefs inhabited by this species. In the typical case, a spear fisher is handling or carrying a recently speared fish, when the usually unseen shark darts in to seize the catch, sometimes taking the person's fingers or hand along with the fish. There are anecdotal reports of skin divers being attacked while trailing behind boats.

SANDBAR SHARK

Carcharhinus plumbeus (Nardo, 1827)

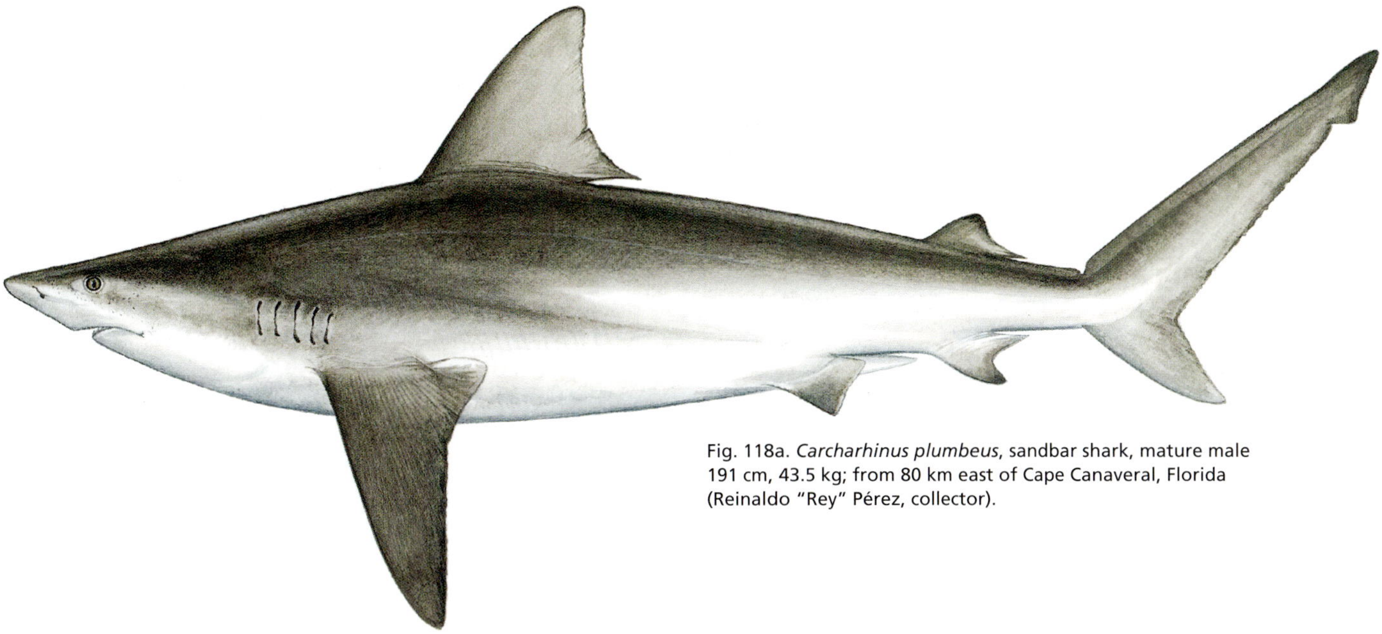

Fig. 118a. *Carcharhinus plumbeus*, sandbar shark, mature male 191 cm, 43.5 kg; from 80 km east of Cape Canaveral, Florida (Reinaldo "Rey" Pérez, collector).

Common name Sandbar shark; also brown shark or ground shark. In the early twentieth century, the species was usually known as the brown shark, ground shark (Thorne 1916), or New York ground shark (Nichols and Breder 1927). However, names such as blue shark (Smith, H. M. 1907; Hildebrand and Schroeder 1928) and small blue shark and white shark (Fowler 1906b) were also applied to it, demonstrating the confusion about requiem sharks that existed at the time. The name sandbar shark alludes to its shallow-water habits. The name sandbar appears to have been used first by Springer (1938); it appears to be a translation from the Cuban name *arenero,* which loosely translates as "sand dweller" (literally, "sand dealer"). The name ground shark refers to its being a bottom-dwelling shark.

Spanish name *Arenero* (Cuba), *tiburón aleta de cartón* (Mexico, A.F.S.).

Scientific name *Carcharhinus plumbeus* (Nardo, 1827), in *Isis* 20(6): 477, 483. *Carcharhinus:* from the Greek *karcharos,* sharp pointed or jagged, from *karcharias,* a species of dogfish having jagged teeth, and *rhine,* Greek file or rasp, also the name of a shark, originally given to *Squatina* due to its rasplike skin; *plumbeus:* Latin for leaden or lead colored, from the Latin *plumbum,* lead. Throughout most of the twentieth century, the sandbar shark was known as *Carcharhinus* or *Eulamia milberti* (Müller and Henle, 1839). So many scientific names have applied to the sandbar shark that Springer (1960:2) wrote: "A most extraordinary snarl has developed over the years in the determination of the scientific name to be applied to the sandbar shark." This confusion was engendered primarily by early authors, who described species without selecting truly diagnostic characteristics, who lacked sufficient specimens to see differences between young and adults, or who made identification errors.

Synonyms The names listed below are only a few of the more often encountered synonyms for the species. *Carcharias (Carcharhinus) milberti* Müller and Henle, 1839; *Carcharias ceruleus* DeKay, 1842; *Carcharhinus platyodon* Garman, 1913 (not *Squalus platyodon* Poey, 1861, which is *Carcharhinus leucas*). For more complete synonymy, see Bigelow and Schroeder (1948) and Compagno (1984b).

Identification The sandbar shark is characterized by a *snout shorter than the mouth width;* a *large first dorsal fin originating over the axil of the pectoral fin, its height equal to or greater than twice the snout length* (embryos and young specimens have slightly lower first dorsal fins); a *second dorsal fin origin over the anal fin origin;* and the presence of *an interdorsal ridge.* The upper teeth have broadly triangular cusps with finely serrated edges; the frontal teeth are erect, and the others are oblique with curved margins. The lower teeth have narrow, triangular, symmetrical, erect cusps with more finely serrated edges than the upper teeth. Teeth number

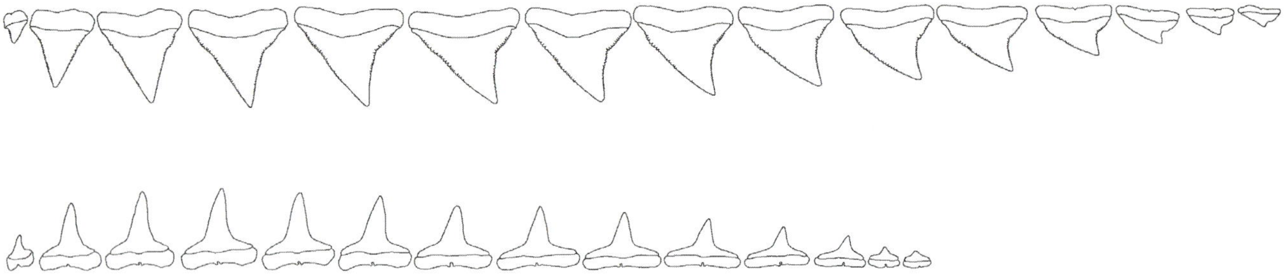

Fig. 118b. Teeth.

U: 14 to 16–1 or 2–14 to 16, L: 14 to 16–1 or 2–14 to 16. The *dermal denticles are of somewhat ovoid shape, wider than long, and barely overlap.* The denticles have three to five parallel ridges with weak points on their rear edges. The dorsal surfaces are brownish gray or brown, with bronze hues in life, fading to gray after death. The undersides are white or dirty white.

Similar species The dusky shark has the origin of the first dorsal fin placed farther back over the free rear tips of the pectoral fins; its denticles are overlapping and have ridges on their free edges. The bignose shark has a longer snout (its length is equal to or greater than the mouth width), and adults have a smaller first dorsal fin (its height is about 10% of the shark's total length). The bull shark has a shorter snout and dorsal fin and lacks an interdorsal ridge.

Range The sandbar shark is nearly cosmopolitan in warm temperate and subtropical waters. In the western Atlantic, it ranges from southern New England to Brazil (Bigelow and Schroeder 1948). In North America, it ranges from Cape Cod to Florida, the Gulf of Mexico, and eastern Mexico. It is abundant from New York to Key West and the Gulf of Mexico.

Size and weight The sandbar shark reaches a maximum length of 234 cm. According to Springer (1960), among the thousands of sandbar sharks he examined, the longest female was 234 cm and the longest male was 226 cm. Clark and von Schmidt (1965) reported that the largest of 91 adult females caught off Sarasota measured 230 cm. The largest that I have seen was a female that measured 220 cm. Representative weights are as follows:

Male: 95 cm, 5.2 kg; 112 cm, 10.4 kg; 127 cm, 12.2 kg, 133 cm, 13 kg; and 171 cm, 33 kg.

Female: 53 cm, 0.8 kg; 120 cm, 10 kg; 175 cm, 34 kg; 188 cm, 41 kg; 199 cm 54.4 kg; 203 cm, 61 kg; 212 cm, 58 kg; and 214 cm, 65 kg.

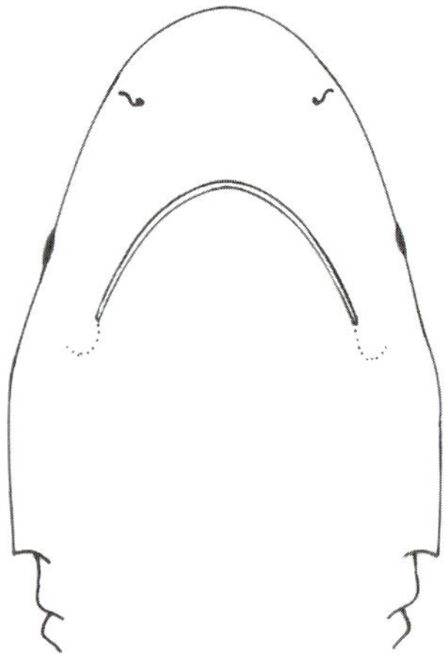

Fig. 118c. Snout.

Fig. 118d. Dermal denticles.

Biology The sandbar shark is a common, bottom-dwelling species that inhabits coastal waters and has wide depth and temperature ranges. Both juveniles and adults often enter shallow waters only 1–3 m deep. The adults are usually found at depths of 20–50 m and occasionally at depths of 175–250 m. Dodrill (1977) reported that sandbar sharks were often seen in daytime in the surf off Melbourne Beach, Florida, during the winter months, and that sometimes the sharks were in such shallow water as to become momentarily grounded.

The sandbar shark was once extremely abundant in eastern North America. Thorne (1916) stated that during a 15-year period he had seen no fewer than 2,500 from his boat in Great South Bay, New York, between Lindenhurst and Great River on the north side of the bay, and Cedar Island and Cherry Grove on the south side. He also mentioned that the greatest number seen in a day was 82, on 14 August 1916; of these, 42 were in sight at one time, and all were seen in the morning before noon. Bigelow and Schroeder (1948: 372) called it "undoubtedly the most abundant member of its genus along the middle Atlantic coast of the United States."

A relentless slaughter of sandbar sharks started in the early twentieth century. It probably started on a small scale with individuals harpooning them for sport. Thorne (1916), managing member of the New York Zoological Society, wrote that for the previous 15 years he had harpooned between 350 and 400 "ground sharks" and that the most he killed in one year was 60 in 1906. In the early 1920s, a commercial fishery developed in Florida primarily for skins and fins. Later, in the mid-1930s, the shark fishery expanded as the demand for shark liver oil skyrocketed. Springer (1960) estimated that from 1935 to 1950, between 5,000 and 15,000 sandbar sharks were taken yearly from the entire range of the fishery and that these were nearly all adults. This fishery lasted until 1950 (Springer 1952). In the mid-1980s, a new fishery developed from Key West to New York, engendered by the demand for shark fins and shark meat. The fishery produced peak catches of more than 197,000 sharks in 1988 (NOAA National Marine Fisheries Service 2002).

By the early 1990s, the population of sandbar sharks had been severely depleted. Musick et al. (1993) estimated that the adult sandbar stock had been reduced to only 15% of its numbers in 1980–1981. Despite such tremendous declines, the sandbar shark remains the most abundant large shark in the coastal waters of eastern North America.

Despite its abundance and economic importance, there are few studies on the sandbar shark in North America. Much of the present knowledge on the sandbar shark is the result of the work of Stewart Springer, who was first involved in the commercial shark fishery in Florida (1935–1950) and later joined the Bureau of Commercial Fisheries. Springer collected and analyzed records for some 100,000 adult sandbar sharks over 25 years and published a paper titled "Natural history of the sandbar shark *Eulamia milberti*" (Springer 1960). This paper remains the source for much of what we know about the species.

Diet: The sandbar shark feeds on whatever small bony fishes and small sharks are abundant in the area. Reported stomach contents include menhaden (*Brevoortia* spp.), gafftopsail catfish (*Bagre marinus*), blue runner (*Caranx crysos*), pinfish (*Lagodon rhomboides*), kingfish (*Menticirrhus* spp.), mullet (*Mugil* spp.), king mackerel (*Scomberomorus cavalla*), pipefishes (*Syngnathus* spp.), snappers (*Lutjanus* spp.), vermillion snappers (*Rhomboplites aurorubens*), barracuda (*Sphyraena barracuda*), bonnethead (*Sphyrna tiburo*), sharpnose shark (*Rhizoprionodon terraenovae*), scalloped hammerhead (*Sphyrna lewini*), octopods, squid, and crabs (Clark and von Schmidt 1965; J. I. Castro, unpublished observations). Some sandbar sharks caught off McClellanville, South Carolina, that I examined contained shrimp trawl bycatch (shrimp heads and a variety of small fishes), indicating that the species sometimes scavenges behind shrimp trawlers.

Neonates and juveniles feed heavily on blue crabs (*Callinectes sapidus*). When Medved et al. (1985) examined the stomach contents of some 414 juvenile sandbar sharks in Chincoteague Bay, Virginia, they found blue crabs in 67.4% of the stomachs. Menhaden (*Brevoortia tyrannus*) occurred in 13.3%. Other prey included summer flounder (*Paralichthys dentatus*), American eels (*Anguilla rostrata*), spot (*Leiostomus xanthurus*), and bluefish (*Pomatomus saltatrix*). Stillwell and Kohler (1993) reported that newborn sandbar sharks off Chincoteague fed mainly on crustaceans (82% by frequency of occurrence), and of these, 76% were soft-shell blue crabs.

Migrations: The migrations of the sandbar shark are complex. The adult females that mated in summer migrate southward in the fall and winter off Florida. In early spring, they begin their northward migration to the nursery areas from South Carolina to New York, where they will give birth. Dodrill (1977) reported that female sandbar sharks with large embryos were absent from the Melbourne Beach area, Florida, after early March, suggesting that they were the first to leave on their northward migration to the nursery areas. Most of the gravid females travel to the northern nurseries to give birth (Nichols and Murphy 1916; Thorne 1916; Springer 1960). Nichols and Murphy (1916) noted that numerous females enter Great South Bay, Long Island, New York, in midsummer to give birth and can be found there until September. Some females travel only as far as South Carolina to give birth (Castro 1993b). After giving birth, females winter off Georgia and the Carolinas with the bulk of the male population.

The adult males generally do not migrate northward with the females. Thorne (1928) mentions that of 300 sandbar sharks taken (as New York ground sharks) in Great South Bay, 287 were females and only three were males. Most males remain offshore from New York to Georgia year-round. Unlike the females, they never enter shallow water and remain offshore at depths of 20–40 m. Only a few of the males travel south with the gravid females in autumn. Springer (1960: 19) reported that "a very much larger number of adult female sandbar sharks were taken than adult males by the Florida fishermen." Dodrill (1977) reported that of 56 sandbar sharks caught off Melbourne Beach, Florida, 42 were females, 12 were males and the sex of two specimens was unknown. Of about 40 sandbar sharks that I have examined off south Florida, all were gravid females.

Neonates spend their first few months in the shallow water nurseries from New York to South Carolina. With the onset of cooler weather they migrate to deep water. According to Springer (1960), the only area in which young sandbar sharks are found in winter is off the Carolina coasts at depths of 140 m. They return to shallow waters in the spring. McCandless et al. (2007) reported that young sandbar sharks may return to their natal nurseries for up to five years after birth and potentially up to 12 years of age; afterward they begin to occupy offshore areas and undertake north–south migrations.

Size at maturity

Male: Males mature at about 180 cm. Springer (1960) reported that the smallest mature male he had seen measured 180 cm.

Female: Female sandbar sharks mature at about 185–190 cm. Springer (1960) stated that at least two female sandbar sharks that he had examined were mature at 183 cm and that five immature females of greater length had been collected. Springer (1960: 26) defined immature females if "none of the eggs [oocytes] in the ovary had begun to increase in size and if the oviducts were smaller in diameter than it is characteristic of the fully contracted oviducts in females following parturition." Since vitellogenesis begins long before the animal is functionally capable of reproduction (sometimes more than a year), this definition tends to underestimate the size at maturity. In this case, the error is minimal, because Springer's smallest gravid female measured 185 cm. Clark and von Schmidt (1965) reported that their smallest mature female measured 185 cm, but their smallest gravid female measured 190 cm. The smallest gravid female reported by Dodrill (1977) measured 193 cm. The largest immature sandbar shark that I have seen measured 190 cm and was caught off McClellanville, South Carolina, in June 1984. It had very thin oviducts and oocytes 7–8 mm. The smallest gravid female that I have seen measured 188.9 cm and carried nine young.

Reproduction The sandbar shark is a placental viviparous species. Off southeastern Florida, mating takes place primarily in June (Springer 1960). Oocytes reach a maximum diameter of about 35 mm at ovulation. A female that I examined was ovulating one 35-mm oocyte and carried embryos in blastodisc stage 1.2 mm in diameter when caught off Charleston, South Carolina, on 13 June. According to Springer (1960), fertilized eggs and the smallest detectable embryos were first observed in Salerno catches from early July to early August.

Springer (1960) estimated that the sandbar's gestation period lasts from 8 to 12 months, being nine months at the latitude of northern Florida. It appears unlikely that there could be such a wide variation of the gestation period in the population and that it would vary with latitude. My own estimate is that sandbar gestation lasts 11–12 months. This is based on finding ovulating females from Florida (Dodrill 1977) to South Carolina in June and on finding females giving birth in June off South Carolina. Most of the females carrying term young that I saw off South Carolina were caught from 18 May to 8 June, with the latest date for a term female being 20 June.

As in most requiem sharks, the sandbar shark has a biennial reproductive cycle. A female giving birth in June will not get pregnant again until June of the following year; it then gestates for a year and gives birth in June of the second year.

Size at birth: Sandbar embryos are generally born at 55–66 cm. Thorne (1916) gave the birth size as 55.8 cm. The "best estimate" by Springer's (1960) of the birth size in northern Florida waters was 60.9 cm. In South Carolina, I found that females carried young 48–66 cm in length from the last two weeks in May to the third week in June, with most young measuring 57–64 cm and the average being 60.5 cm. It has been suggested that embryos in the northern part of the range are somewhat smaller than those in the southern parts of the range (Springer 1960). A neonate caught on 29 July off South Carolina had an umbilical scar that was still open as a narrow slit, 8 mm long and 2 mm wide, while a second neonate, 52.5 cm in length, examined on 27 August, had a healed umbilical scar that was barely visible as a thin dark line 2 mm long, suggesting that in this species the umbilical scar disappears in about 2 months.

Brood size: Sandbar shark broods usually range from 6 to 10 young. Thorne (1916) reported broods of 8–11 in females caught off New York. According to Springer (1960), the number of young per brood in Florida ranged from 1 to 14, with a mode of 10 and an average of 9. I have found broods ranging from 6 to 10, with a mode of 9 and an average of 8, for females caught off South Carolina (Castro 1993b).

Nurseries Historically, the nursery areas of the sandbar shark have extended from Great South Bay, Long Island, New York, to just north of Cape Canaveral, Florida. Nichols and Murphy (1916) and Thorne (1916) reported that numerous females entered Great South Bay in midsummer to give birth. Although Springer (1960) stated that the sandbar nurseries extended to Cape Cod, I have not found any corroborating data or neonate specimens from the Cape Cod area, although juveniles may be common there (G. Skomal, Massachusetts Division of Marine Fisheries, pers. comm., Feb. 2004). Merson and Pratt (2007) suggest that the current northern limit of the sandbar nurseries is Great Bay, New Jersey, based on a contemporary survey that found no sandbar sharks in Great South Bay, Shinnecock Bay, or Peconic Bay, New York, or in Barnegat Bay, New Jersey. If this is correct, it would indicate a considerable recession of the nursery grounds in less than a century. Today, the major nurseries for the sandbar shark are in Delaware Bay (Merson and Pratt 2001) and Chesapeake Bay (Grubbs and Musick 2007). I found newborn sandbars abundant off Charleston and McClellanville (Bulls Bay), South Carolina (Castro 1993b). The southern boundary of the sandbar nurseries appears to be Cape Canaveral. Springer (1960) reported that young sandbar sharks are common from Cape Canaveral northwards, but he found no free-swimming young at Salerno or around southern Florida. Dodrill (1977) reported that he found no newborn sandbar sharks in the Melbourne Beach area, some 25 miles south of Cape Canaveral. Springer (1938) stated that no young sandbar sharks had been caught off Englewood, on the Gulf coast of Florida. Similarly, Clark and von Schmidt (1965) reported that although they had seen gravid female sandbar sharks in the area from November to March, they had not found any sandbar young off Sarasota.

Springer (1960) postulated the existence of a sandbar nursery area in the northern Gulf of Mexico, west of Mobile Bay and north of the 28th parallel, based on finding gravid females with term young in that area. Subsequent reports of neonate and gravid females with term young from diverse areas of the northern Gulf of Mexico suggest that sandbar sharks give birth throughout the northern Gulf. Branstetter (1981, 1987c) reported gravid females off Pensacola, Florida, and neonates from Galveston Bay, Texas. Carlson (1999) reported finding neonates and young-of-the year in Indian Pass and St. Andrew Sound, in the northeastern Gulf of Mexico. Neonates were captured in June and larger young were caught through October. According to Carlson, sandbar distribution in the northeastern Gulf of Mexico was limited to the mouth of the Apalachicola River, near St. Vincent Island; none were caught in other areas sampled.

Age and growth The sandbar shark is a slow-growing, long-lived species. Although the growth rate is known for juveniles, the rate for adults is uncertain. As in most species of sharks, the life span is also uncertain.

The first estimates of the age of sandbar sharks were made by Springer (1960: 28), who suggested "that it is probable that growth from birth to maturity takes about 2 years, occasionally 1 year, and rarely 3 years." These estimates were apparently based on the size frequency distributions he observed. By the 1970s, long-lasting tags and techniques for aging vertebral rings had appeared, and new age estimates for sandbar sharks were produced. Wass (1973), studying smaller sandbar sharks off Hawaii, estimated that both males and females reached maturity in three years. Lawler (1976) used vertebral centra to age specimens and calculated that 16 years was the minimum age at maturity in the sandbar shark. Casey et al. (1985) used vertebral ring counts, tagging data, and length frequency distribution to estimate the age of sandbar sharks from northeastern North America. They concluded that males matured at 13 years of age and females at 12 years. They also estimated that their oldest male, measuring 185 cm (154 cm fork length), was 15 years old, and that their oldest female, measuring 245 cm (204 cm fork length), was 21 years old. Their tag returns indicated that sandbar sharks may live for more than 30 years.

A few years later, Casey and Natanson (1992) revised the estimates of Casey et al. (1985), based on their long-term tag returns. They concluded that sandbar sharks grew much more slowly than previously estimated and may take 29–30 years to reach maturity. These authors then suggested that sandbar sharks could live for more than 50 years. Casey and Natanson also discussed the problems of using vertebral rings or other hard parts, which may not be reliable indicators for aging fish throughout all age classes. They also mentioned that it is likely that sharks may deposit annual rings on their vertebrae up to maturity and that further rings may not be related to time but may instead be deposited to strengthen the vertebral column. Interestingly, they did not consider the effects of the tags on small sharks. Some of their critical long-term recaptures had been tagged at lengths of 102, 112, and 185 cm fork length. Two of these sharks had been tagged with the metal dart-type "M" tag. When inserted on small sharks, these tags may penetrate the peritoneal cavity. I have seen two cases in which the tag had been applied near the base of the dorsal fin and the darts were embedded in the kidneys. It is possible that such injury could slow the growth rate of a shark. Manire and Gruber (1991) showed that dart-type tags inhibited the growth of lemon sharks and recommended that biologists refrain from marking young or small sharks with M-type dart tags, especially for age and growth studies.

The uncertainty about the age at maturity of the sandbar shark remains unresolved. Other recent or subsequent work on the sandbar shark has consisted of demographic models (Hoff 1990; Sminkey and Musick 1996), which use published and estimated life history parameters to estimate the effects of fishing on populations.

Relation to humans The sandbar shark has been the most economically important species in the shark fisheries of the United States because of its abundance and predictable migratory patterns. In the Florida shark fishery from 1935 to 1950, a special effort was made to catch sandbar sharks because of the size of their hides (Springer 1952) and the higher potency of its liver oil (Springer 1960). In the shark fisheries of the late twentieth century, sandbar sharks were (and are) pursued primarily for their fins (which are proportionally larger and more valuable than those of other species) and for their palatable flesh. Furthermore, fishermen and dealers preferred sandbar sharks because their medium size made them easier to handle (compared to larger species such as bull sharks and dusky sharks), while vendors preferred them because the size and color of the steaks were appealing to buyers.

The sandbar shark is a sturdy species that survives many years in captivity. According to Gary Violetta of Seaworld, Orlando, Florida (pers. comm., February 2004) many sandbar sharks survive 13–15 years in captivity, and one specimen had survived 19 years when it died while being transported to another tank. Its sturdiness is not the sole reason for its being the most often displayed requiem shark in American aquaria. Unlike the lemon shark, sandbar sharks can be kept with numerous other smaller fishes in the aquarium without consuming them.

SMALLTAIL SHARK

Carcharhinus porosus (Ranzani, 1840)

Fig. 119a. *Carcharhinus porosus*, smalltail shark, mature male, 86.3 cm, 4.1 kg; Trinidad, British West Indies (collected by the author and FAO team).

Common name Smalltail shark. In Trinidad it is known as the puppy shark.

Spanish name In Mexico, it is known as *cazón* due to its small size. The A.F.S. calls it *tiburón poroso.*

Scientific name *Carcharhinus porosus* (Ranzani 1840), originally described as *Carcharias porosus* in *De novis speciebus piscium, Novi Commentarii, Academiae Scientiarum Instituti Bononiensis* 4: 70–71, plate 9, figs. 1–3. *Carcharhinus:* from the Greek *karcharos,* sharp pointed or jagged, from *karcharias,* a species of dogfish having jagged teeth, and *rhine,* Greek file or rasp, also the name of a shark, originally given to *Squatina* due to its rasplike skin; *porosus,* having pores, from the Greek *porus,* pore or passage, probably a reference to the large and conspicuous pores located behind the eye.

Synonyms *Carcharias* (*Aprionodon*) *henlei* Müller and Henle, 1839. *Carcharhinus cerdale* Gilbert, 1898 is not a synonym but a different Pacific species (see p. XXX).

Identification This small shark (usually *less than 110 cm*) *lacks an interdorsal ridge.* It has a *broadly triangular first dorsal fin (the fin is shaped as an equilateral triangle in adults) originating over or behind the rear tip of the pectoral fin,* and a *second dorsal fin originating over the midpoint of the base of the anal fin.* There is a series of conspicuous pores located behind the eye. The *upper teeth have long triangular cusps that become increasingly oblique toward the corners of the mouth.* They have a weak notch on their medial edges (with the exception of the first large tooth near the symphysis that lacks the notch) and a strong notch on the outer edges. The edges have two or three coarse serrations near the bases up to the notch and finer serrations from the notch to the tip. The five or six outermost upper teeth have strongly oblique cusps with nearly straight inner margins and a strong notch on the outer margin. The lower teeth are erect and have cusps with a few coarse serrations at their bases and finer serrations toward the tips. The teeth number U: 14–1–14; L: 13–13. The dermal denticles are characteristic. They have a central strong ridge terminating in a long point and two shorter ridges on winglike sides. The surface has a characteristic microsculpture. Coloration is gray above and dirty white below. The *fins are unmarked.*

Range The smalltail shark ranges from the Gulf of Mexico to southern Brazil (Bigelow and Schroeder 1948). It is abundant along the northeastern coast of South America. It is the most abundant shallow water shark around Trinidad, where I have seen hundreds taken in a single bottom gillnet. According to McEachran and Fechhelm (1998), the smalltail shark is common southward of Corpus Christi, Texas. The species is seldom encountered in the northern Gulf of Mexico. I have examined two neonates taken in Bayou Goreau, Terrebonne Bay, Louisiana, on 9 May 1952 (Tulane University Fish Collection TU no. 1896). Bigelow and Schroeder (1948) reported that Stewart Springer had taken a series of specimens off Biloxi, Mississippi. It is interesting to note that most of the records of the species from the northern Gulf of Mexico or east of the Mississippi River that I have found are more than 50 years old, and almost none of present-day workers in those areas have ever seen the species. The presence of neonates in Louisiana demonstrates that the species once bred in that area. The reasons

Fig. 119b. Teeth.

for the species' scarcity today in the northern Gulf can only be guessed.

Similar species Throughout its range, the smalltail shark is most likely to be confused with the sharpnose sharks (both *Rhizoprionodon terraenovae* and *R. porosus*) because they are of similar size and inhabit the same habitat. These sharpnose sharks have conspicuous upper labial furrows and a series of large hyomandibular pores along each corner of the mouth, and adults usually have a few scattered white spots over the dorsal surface. The blacknose shark has a dusky spot on the nose and its second dorsal fin origin is over or just slightly behind the origin of the anal fin.

Size and weight The largest male that I examined in Trinidad measured 98.4 cm and weighed 4.4 kg, whereas a 98.0-cm male was the heaviest, weighing 5.7 kg; the largest females reached 110 cm and 8.7 kg. Representative weights are as follows:

Males: 57.1 cm, 1.05 kg; 73.1 cm, 2.0 kg; 85 cm, 3.3 kg; 86.8 cm, 4.0 kg; 92.1 cm, 4.0 kg; 92.7 cm, 5.0 kg; 95.2 cm, 4.6 kg; and 98.0 cm, 5.7 kg.

Females: 40.9 cm, 0.4 kg; 62.9 cm, 1.4 kg; 81.4 cm, 1.2 kg; 86.4 cm, 3.5 kg; 96.4 cm, 6.9 kg (term female); 100.0 cm, 7.1 kg; 109.7 cm, 8.7 kg; and 110 cm, 7.6 kg.

Biology The smalltail shark is a shallow-water tropical species abundant throughout the tropical Atlantic Americas. The notes that follow are based on my observations of the species in Trinidad, British West Indies, because I have never encountered it in the Gulf of Mexico, and there is no published information on the species in North America.

The smalltail shark inhabits shallow coastal waters, often over muddy bottoms. In Trinidad, large aggregations were found at depths of 18–36 m. Males and females were segregated during most of the year, with the males occupying the deeper parts of the range. It feeds on a wide variety of small bony fishes and shrimp.

Size at maturity Males mature over a wide range of 73–93 cm, based on clasper calcification, with most maturing at about 80 cm. Females mature at about 85 cm, based on the smallest gravid female (84.7 cm).

Reproduction The smalltail shark is a placental viviparous species as other requiem sharks. In Trinidad, I found females carrying term young throughout the year, indicating that the females are not synchronous in those tropical waters and that reproduction occurs throughout the year. However, there is a marked peak of birthing in late September. Term females had inactive ovaries, suggesting a biennial reproductive cycle, as in other species of the genus.

Brood size: Females carry four to seven young; most of the ones I observed carried four.

Size at birth: The largest embryos I saw in Trinidad measured 32–33 cm; the smallest free-swimming young that I encountered measured 31.8 cm, suggesting birth at 31–33 cm. The Bayou Goreau 34-cm neonates already had healed umbilical scars.

Nurseries The nurseries appear to be in shallow coastal waters. The presence of two 34-cm neonates in Bayou Goreau, Terrebonne Bay, Louisiana, suggests that the species may have once used that area as a nursery. In Trinidad,

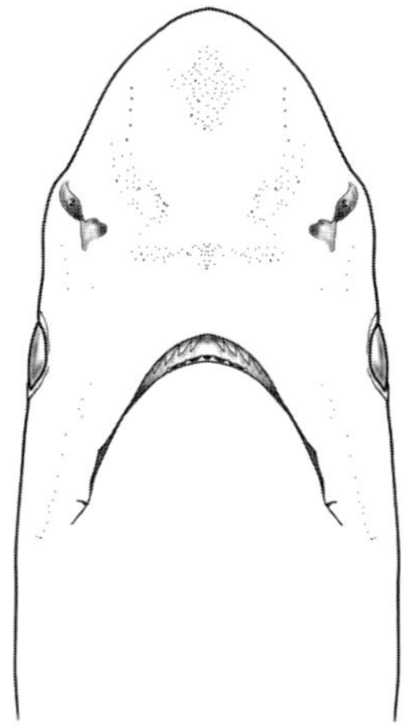

Fig. 119c. Snout.

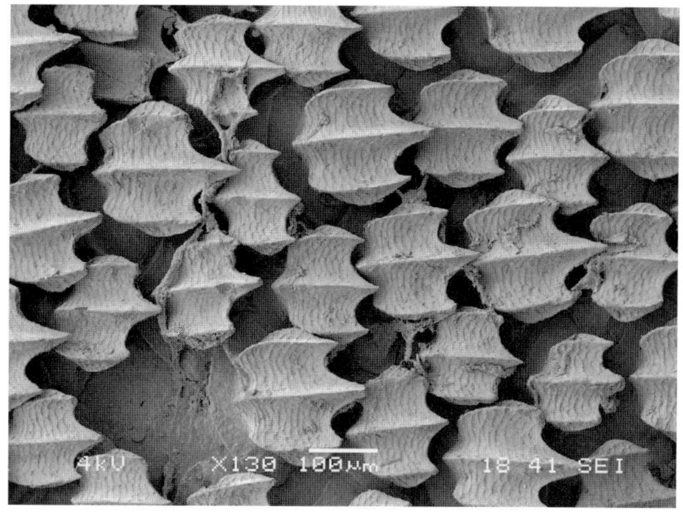

Fig. 119d. Dermal denticles (adult).

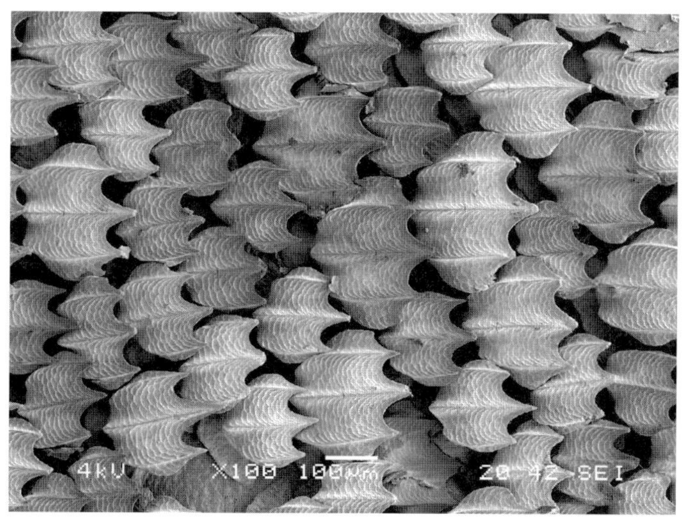

Fig. 119e. Dermal denticles (33-cm neonate).

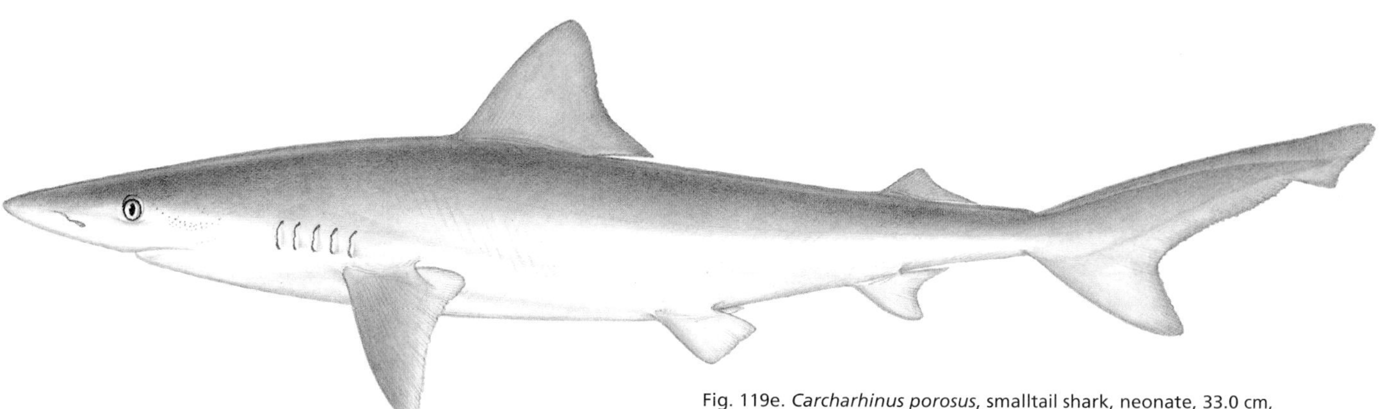

Fig. 119e. *Carcharhinus porosus*, smalltail shark, neonate, 33.0 cm, 0.2 kg; Bayou Goreau, Terrebonne Parish, Louisiana (TU# 1896, Courtesy of Hank Bart).

the nurseries were in shallow murky beaches; Manzanilla Bay, in eastern Trinidad, was an important nursery for the species.

Age and growth The species has not been aged.

Economic importance The species has no economic importance in North America. In Trinidad it is the most economically important shark because of its abundance. It is taken primarily as bycatch in gillnet and shrimp fisheries. The species should be vulnerable to overfishing because of its close-to-shore habitat and biennial reproductive cycle.

NIGHT SHARK

Carcharhinus signatus (Poey, 1876)

Fig. 120a. *Carcharhinus signatus*, night shark, mature male, 215 cm, 59.7 kg; caught 50 km east of Cape Canaveral, Florida (Reinaldo "Rey" Pérez, collector).

Common name Cuban night shark. The name of this species, first described from Cuba by Poey, is a translation of the Spanish name *tiburón de noche,* or "night shark," given to it by Cuban fishermen because it was caught only at night.

Spanish name *Tiburón de noche, jesuita* (Cuba); *tiburón nocturno, ojo de caballo, ojo verde* (Mexico). The A.F.S. name is *tiburón nocturno.*

Scientific name *Carcharhinus signatus* (Poey, 1876), originally described as *Hypoprion signatus* in *Repertorio Físico-Natural de la Isla de Cuba* 2: 452, plate 4, figs. 7 and 8. The description was based solely on a jaw, Poey having recognized that it came from an undescribed species. *Carcharhinus:* from the Greek *karcharos,* sharp pointed or jagged, from *karcharias,* a species of dogfish having jagged teeth, and *rhine,* Greek file or rasp, also the name of a shark, originally given to *Squatina* due to its rasplike skin; *signatus:* Latin marked or sealed.

Synonyms *Hypoprion longirostris* Poey, 1876; *Hypoprion bigelowi* Cadenat, 1956.

Identification The night shark is characterized by a *snout longer than the mouth width,* large *eyes with a noticeable green pupil,* a relatively small first dorsal fin (its height about 4% of the total length) originating behind the free rear tips of the pectoral fins, and a low interdorsal ridge. The *upper teeth have increasingly oblique cusps with a notch on the outer side at about one-third of the distance from the base to the tip, coarse serrations from the base to the notch* (or to the level of the notch on the median edge), and *smooth-edged tips.* The lower teeth are narrow and erect and have smooth-edged cusps. Teeth number U: 15–1 or 2–15, L: 15–1–15. The dermal denticles are small and greatly overlapping. They have three strong ridges terminating in three points, the central ones being much longer. The upper surfaces are brownish gray with silvery hues in life, fading to bluish gray after death; the undersides are grayish white.

Similar species The combination of long snout, the interdorsal ridge, the low dorsal fin originating behind the pectoral tips, and the characteristic teeth distinguish the night shark from other requiem sharks in the area. The teeth of small silky sharks are shaped like those of night sharks and have been confused with small night sharks by some. The silky shark has a first dorsal fin that begins behind the level of the pectoral fin free rear tip.

Range The night shark is abundant in the deep tropical waters of the western Atlantic, having been reported from Delaware (Mather and Gibbs 1957) to Brazil (Menni et al. 1995). It has also been reported from the eastern Atlantic (Ivory Coast, Cadenat and Blache 1981; Ghana to Angola, Compagno 1984b). Records from the eastern Atlantic are few, perhaps due to confusion with other species or due to its deep-water habitat. In eastern North America, the night shark is found in the deep waters of the Gulf Stream from off Virginia to Florida and in the Gulf of Mexico. It is very abundant in deep water off the northern coast of Cuba and the Straits of Florida.

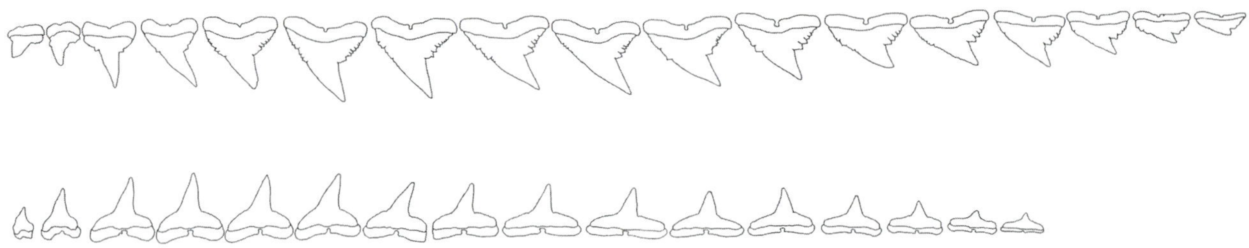

Fig. 120b. Teeth.

Size and weight The night shark reaches about 300 cm and 135 kg, although specimens larger than 275 cm are rare. Bigelow and Schroeder (1948) stated that night shark reached 277 cm, while Guitart Manday (1975) stated that the night shark reached 300 cm. Garrick (1985: 24) questioned those figures because he suspected errors or exaggerations, stating that "the largest recorded female *signatus*" is the 255-cm specimen mentioned by Raschi et al. (1982). I have examined the photographic archives of Ron Schatman of North Miami (see below), which include several specimens 240–260 cm and one exceptional female (gravid and the largest he ever encountered) caught in 1973, which measured 287 cm long and weighed 135 kg. Thus, the maximum sizes claimed by Bigelow and Schroeder and Guitart Manday are plausible, although it must be unusual for night sharks to reach such sizes in these days of great fishing mortality.

Representative weights are as follows:

Male: 66 cm, 1.7 kg; 75 cm, 2.3 kg; and 195 cm, 42 kg.

Female: 69 cm, 1.5 kg; 102 cm, 6.0 kg; 149 cm, 16.7 kg; 215 cm, 60 kg; 245 cm, 104 kg; 255 cm, 109 kg; 287 cm, 135 kg.

Biology The night shark is a tropical species that strays northward in the Gulf Stream. This is one the deeper dwelling requiem sharks, usually found over the continental slopes off Florida at depths of 150–250 m or deeper in daylight. It ascends in the water column at night and can be caught at depths of 30–50 m. Springer (1957a) reported that, in the deep-water hand-line fishery along the northwestern coast of Cuba in the 1950s, fishermen usually took the night shark at depths of 180 m at night and occasionally at 365 m in daytime. It feeds on fishes and shrimp.

The night shark is a migratory species, but its seasonal movements are poorly defined. Gravid females are common off the coast of the Carolinas and Georgia. Nongravid females generally travel farther south. Berkeley and Campos (1988) reported that of 142 females examined off Florida, only 2 (1.4%) were gravid. According to Martinez (1947),

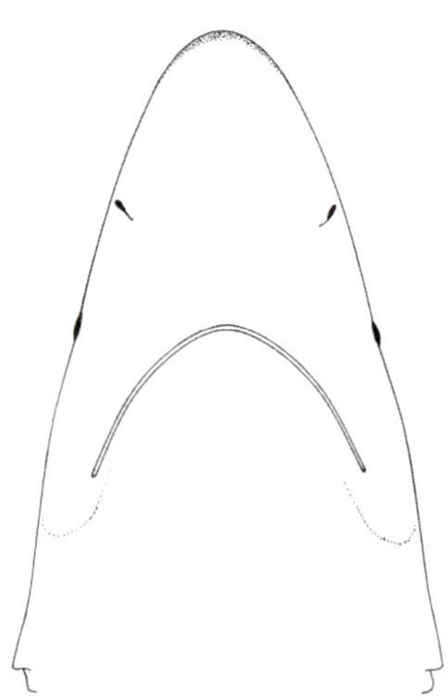

Fig. 120c. Snout.

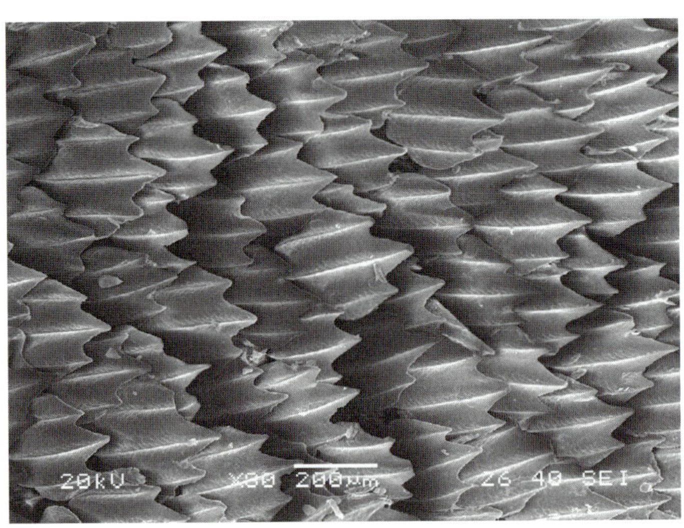

Fig. 120d. Dermal denticles.

male night sharks are most abundant along the northern coast of Cuba from February to July and during September and October, while females are most abundant in May and June. Adults used to be common off South Florida during the summer months.

Size at maturity Despite the abundance of night sharks off the United States coast, there are no estimates of the size at maturity for specimens in the area. Hazin et al. (2000) estimated that males mature at 185–190 cm and females mature at 200–205 cm.

Reproduction The night shark is a placental viviparous species. Its reproductive processes are very poorly known. Its reproductive cycle is probably biennial like the other requiem sharks. Whether the females in a population are reproductively synchronous is not known. Apparently, birth can take place in late summer, based on a report of a gravid female carrying 58- to 61-cm young in August and on my finding neonates with open umbilical scars in October.

Size at birth: The young are born at 60–70 cm (Branstetter 1990). This estimate agrees well with the largest embryos of 62 cm, and the smallest neonates of 68 cm, that I have observed.

Brood size: Broods of North American specimens usually consist of 10–20 young. Hazin et al. (2000) reported that females off Brazil carried 4–15 young.

Nurseries The nurseries appear to be in the shallower portions of the continental slopes. Raschi et al. (1982) reported capturing young 70–107 cm north of Cape Hatteras at depths of 135–175 m. I have caught neonates 68–75 cm, on the bottom, at depths of 260 m in the eastern Gulf of Mexico (26°27′ N, 84°41′ W) in late October.

Age and growth Santana and Lessa (2004), using specimens caught off northeastern Brazil, estimated that males mature at 180–190 cm and years of age, while females mature at 200–205 cm and 10 years of age.

Relation to humans During the heyday of the shark fishery of the 1940s, the night shark was the most important commercial species in the Cuban shark fishery (Martinez 1947), due to the high vitamin A content of its liver oil. The night shark made up 60–75% of the total catch of the fishery (Martinez 1947). Nowadays, it has no commercial use.

The night shark is often caught as bycatch in the tuna and swordfish longline fisheries. Berkeley and Campos (1988) reported that the night shark constituted 26% of the total catch in the swordfish fishery operating off the Florida eastern coast in the early 1980s. In the late 1970s and early 1980s, I had reports from commercial swordfish fishers catching very large numbers of night shark females during the summer off Georgia and South Carolina.

In the 1970s, the night shark was often taken by sports anglers from charter boats operating off hotels in Miami Beach, Florida. These anglers actually pursued sailfish in daytime and swordfish at night. However, when billfishes were not taken, the crews quickly rigged for night sharks, so that the customer would not go home empty handed. Unlike the elusive billfishes, the night sharks would always take the bait. Upon return to the docks, a commercial photographer would be waiting to record the anglers and their catches. Thus, a daily photographic record of the catches was created. I have examined such record for the Castaways Hotel in the 1970s, courtesy of Ron Schatman, who operated a sports fishing boat there. It contains dozens of photographs of large night sharks. These fishes are seldom encountered in the area now.

The night shark suffered tremendous mortality as bycatch in the swordfish fishery in the 1970s and 1980s. Large numbers of these sharks were routinely discarded dead. Consequently, after several decades of the swordfish and tuna fishery in our waters, the numbers and sizes of night sharks have declined severely. Unfortunately, most of the decline went unrecorded.

The night shark has been a protected species in the United States since 1997 and may not be landed. However, large numbers are still killed in the swordfish and tuna longline fisheries, because night sharks share the habitat with these fishes.

TIGER SHARK

Galeocerdo cuvier (Peron and Lesueur, 1822)

Fig. 121a. *Galeocerdo cuvier,* tiger shark, immature male 234 cm, 55.7 kg; from 270 km East of Savannah, Georgia (Reinaldo "Rey" Pérez, collector).

Common name Tiger shark, an allusion to the tigerlike stripes or markings of most specimens. In the past, it was also called a leopard shark because of the spotted markings of juveniles (Springer 1938).

Spanish name *Tiburón tigre, tigre, alecrín* (Cuba), *tintorera* (Mexico, A.F.S.).

Scientific name *Galeocerdo cuvier* (Peron and Lesueur, 1822), originally described as *Squalus cuvier* in *Journal of the Academy of Natural Sciences of Philadelphia* 2(2): 351–352. *Galeocerdo:* from the Greek *galeus,* a kind of shark, and *cerdo,* a wily one or a thief, the name of a fox; *cuvier:* after the famous French anatomist Georges Cuvier (1769–1832), who is considered the father of comparative anatomy and vertebrate paleontology. Formerly, the species name was written in Latinized form as *cuvieri.*

Synonyms *Squalus arcticus* Faber, 1829; *Galeus cepedianus* Agassiz, 1838; *Galeocerdo tigrinus* Müller and Henle, 1839.

Identification Of all the requiem sharks, the tiger shark is the easiest to identify because of its many diagnostic characteristics. It has a snout much shorter than the mouth width, with *long labial furrows* around the corners of the mouth, *small but visible spiracles* behind the eyes and *weak dermal ridges or keels on the caudal peduncle.* The shape of the teeth is the best diagnostic character; they have *curved cusps with finely serrated edges, a deep notch on the outer margin,* and are *similar in size and shape in both jaws.* Teeth number U: 9 to 12–0 or 1–9 to 12, L: 9 to 12–0 or 1–9 to 12. The dermal denticles are of different sizes and are well separated (or with minimal overlap). The denticles have one tapering central ridge that terminates on a point. Color varies from bluish or greenish gray to brown or black above and from light gray to dirty yellow or white below. Juveniles up to 150–180 cm have a mottled appearance; *their dorsal surfaces are covered with dark spots* on a lighter background. In larger sharks, the spots fuse together, forming vertical bars or *stripes producing a tigerlike appearance,* hence the name. In older sharks, the bars tend to fade but often persist on the flanks or in the caudal region. Some specimens from Tobago have been described as being shiny black above and immaculate white below (Springer 1950). However, all the tiger sharks I saw there in the mid-1980s were gray. I have seen one such specimen with a black back in the Ishigaki Islands. Rider et al. (2002) described a juvenile white tiger shark. This specimen, a 124-cm female, was caught 93 km east of Apalachicola, Florida, in September 1999. It had white dorsal and ventral surfaces. The dorsal darks spots, typical of juvenile tiger sharks, were faint but visible. The irises were light brown, indicating that it was not an albino.

Similar species Its characteristic teeth and markings distinguish the tiger shark from other sharks.

Range The tiger shark has a worldwide distribution in tropical and warm-temperate waters. Off North America, it is found from Cape Cod to Florida, the Gulf of Mexico, and the Caribbean Sea, and from southern California southward. It is common off the Carolinas, Georgia, and Florida and in the Gulf of Mexico. It is one of the most abundant large sharks in the Caribbean region. Reports of tiger sharks straying far from their normal range are common.

Bigelow and Welsh (1925: 27) wrote, "Every year a few young tiger sharks are taken in the fish traps in the Woods

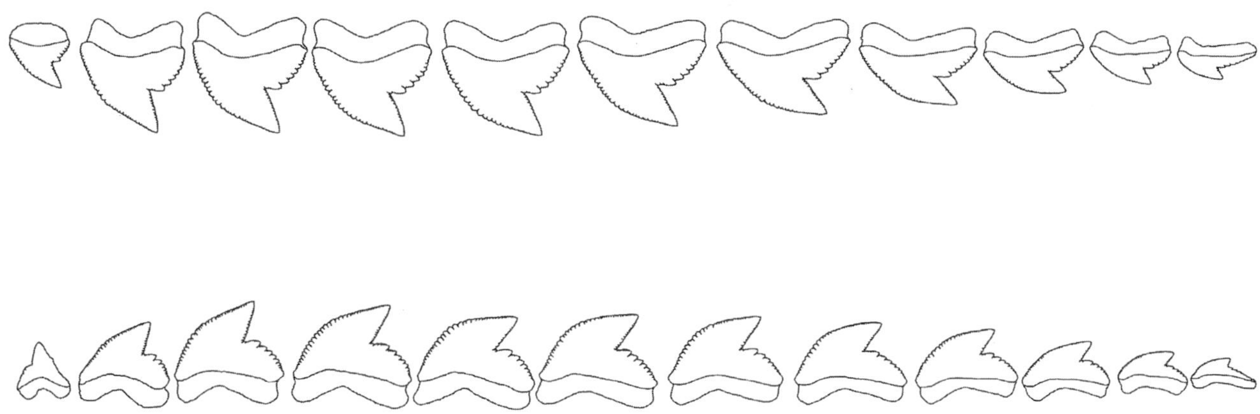

Fig. 121b. Teeth.

Hole region, where, according to the records of the Bureau of Fisheries, it is the latest shark to arrive, rarely being seen before August or after October." These specimens have been about 150 cm, at most 240 cm, and "very rarely does a full-grown tiger shark stray so far from its tropical home" (28).

Tiger sharks are occasional visitors in offshore waters from Cape Cod to Cape Hatteras during the warmer months. From Cape Hatteras southward, tiger sharks become quite common in coastal waters. It is a summer visitor in North Carolina, South Carolina, and Georgia. In the Gulf of Mexico, the species occurs in coastal waters from spring to fall and in deeper continental shelf and offshore regions throughout the year (Branstetter 1981; Branstetter et al. 1987). Tiger sharks are year-round residents throughout most of the Florida coast. Springer (1938) wrote that it was present throughout the year in peninsular Florida. Off Melbourne Beach, central Florida, Dodrill (1977) found tiger sharks from early February to mid-September. Dodrill also explained the absence of tiger sharks along the beach in November, December, and January as probably due to reduced fishing effort as a result of rough weather during those months. Small-scale commercial shark fishing operations off southern Brevard County in water 17–25 m deep landed tiger sharks throughout the year (Dodrill 1977). Clark and von Schmidt (1965) stated that tiger sharks were caught in the Gulf of Mexico, off Sarasota, Florida, in all months except September, when few lines were set due to stormy weather. Wright (1981, see p. 184) reported that tiger sharks were caught throughout the year in the Florida Keys, exhibiting no obvious seasonality. However, an examination of

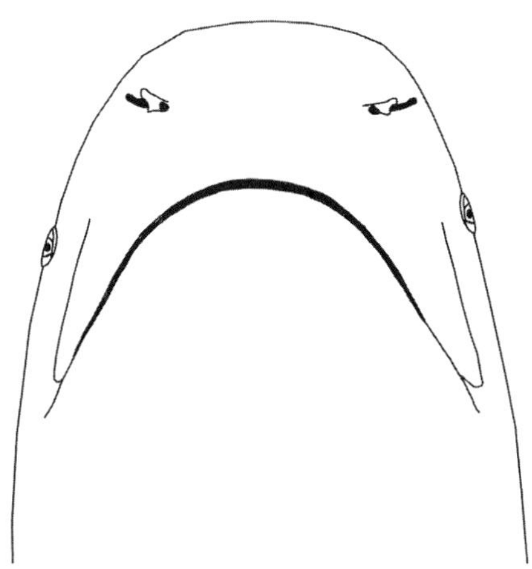

Fig. 121c. Snout.

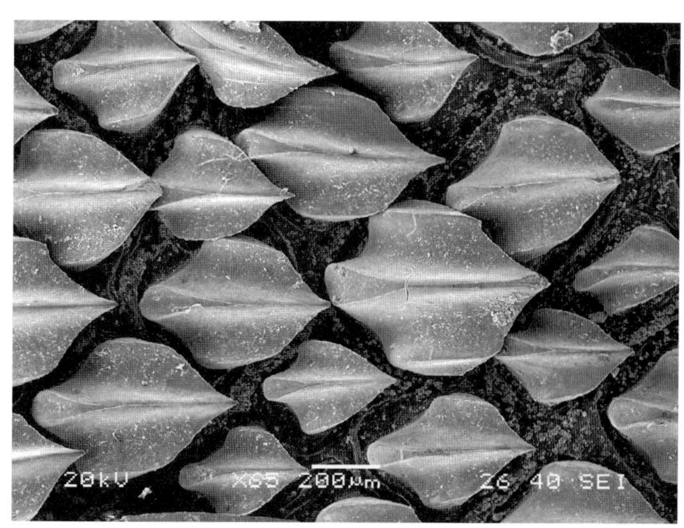

Fig. 121d. Dermal denticles.

his data reveals that only juveniles and males were caught in January and February. The large females larger than 300 cm were not found in the area during the coldest months. Today, tiger sharks are rare in Florida coastal waters due to overfishing.

Tiger sharks are rare along the western coast of the United States. Jordan and Evermann (1896) mentioned that it ranged from tropical waters north to San Diego. Starks (1917b) reported the tiger shark from southern California, based on a record of "some jaws of a specimen captured near San Diego" and fishermen's accounts. Scofield (1941) reported an anecdotal capture of a 274-cm female tiger shark off Las Flores Canyon, north of Oceanside, California. Seigel et al. (1995) reported an 88-cm juvenile that was caught on hook and line from a pier in Manhattan Beach, Los Angeles County, California, in September 1994. These authors also mentioned that two additional specimens were caught at the same location five weeks later. Presumably, tiger sharks only venture into California waters during the warm water expansions associated with El Niño.

Size and weight The tiger shark is one of the larger species of sharks. Tiger sharks 3–4 m are common, while specimens 4–4.5 m are unusual these days, and there are a couple of credible reports of tiger sharks larger than 4.5 m.

As has often happened with the larger species of sharks, the size attained by the tiger shark has been exaggerated in the literature. Bigelow and Schroeder (1948) mentioned that the longest tiger shark of which they had a positive, recent record of in the western Atlantic was a Cuban specimen of about 18 ft (5.49 m) based on a personal communication from Luis Howell-Rivero. No evidence supports this record. It must be noted that Howell-Rivero also reported the largest white shark in the world, a record that has not survived scrutiny (Randall 1987). In general, it is difficult to ascertain the sizes of these animals without any supporting evidence and after such a long time. One must choose between considering the reports to be exaggerations or that these larger animals once existed in Cuban waters. I choose to consider them errors or exaggerations because, in both cases, there is a wide gap between the size reported by Howell-Rivero and the second largest animal reported.

The second largest tiger shark reported in the western Atlantic was a 4.6-m male caught four miles off the Charleston jetties, South Carolina. Reportedly, the shark was tied along the vessel's gunwale and measured with a yardstick by the angler (Burton 1941). Thus, the accuracy of this measurement is questionable. Stewart Springer who had considerable experience as a commercial fisherman in the southeastern coast of the United States once wrote that he had "measured many over 12 feet [366 cm], one over 13 ft [396 cm], and none over 14 ft [427 cm]" (quoted in Gudger 1949: 49).

Reports from elsewhere confirm the abundance of tiger sharks up to 3.5 m and the scarcity of specimens larger than 4 m. Rivera-López (1970) measured 149 sharks caught off La Parguera, Puerto Rico. Of these, 56 measured 289–350 cm, 12 measured 351–378 cm, and one measured 381 cm. Cadenat and Blache (1981) reported measuring more than 800 tiger sharks caught in fisheries off Senegal, and most of them measured 300–375 cm. Whitley (1940) lists several tiger sharks larger than 4 m and only two larger specimens—a 442-cm, 537-kg shark caught off Sydney, Australia, in 1940, and a 472-cm tiger shark of unknown weight, taken off Maroubra, New South Wales, Australia, in 1936. Sarangdhar (1943) refers to a 399-cm female taken in India in January 1942 as "an outsize female" and a "monster." Simpfendorfer (1992) reported that off Townsville, Australia, most of the 835 tiger sharks caught in a beach protection program were between 170 and 350 cm, about 1% exceeded 400 cm, and the largest shark was 428 cm. Heithaus (2001b) reported four tiger sharks larger than 400 cm out of 252 sharks in Western Australia, the largest being 407 cm. According to Keiichi Sato (Okinawa Churaumi Aquarium, pers. comm., September 2008), the largest tiger sharks recorded in Ishigaki Island in 2007 were a 420-cm, 580-kg female, and two females, 391 and 393 cm respectively, both larger than 600 kg (scale limit).

Larger tiger sharks have been reported, but, invariably, these are second-hand reports, with no supporting evidence. Fourmanoir (1961) mentions receiving a photograph from an M. Teisseyre of a female tiger shark 7.4 m and 3,110 kg, carrying 75 young, caught off Indochina in 1957. The reported size of this animal and the great difference with the largest tiger sharks in the literature puts this specimen in the realm of fantasy (or whale sharks!) until evidence is produced. However, it must be stated that John Randall, a scientist of considerable experience with sharks and who has spent countless hours underwater, has expressed his belief that the existence of 6-m tiger sharks will eventually be demonstrated (Randall 1992).

In any case, the 4.6-m Charleston specimen (of questionable accuracy) and the 472-cm Maroubra specimen are the two largest tiger sharks that, in my opinion, are plausible in relation to other reports. If these sizes are correct, these two specimens, taken long ago, may represent the maximum sizes for tiger sharks at the end of their allotted lives. If larger tiger sharks exist today, the chances of encountering such large, old creatures are slim, because the chances of a tiger shark cruising the oceans for decades without a fatal encounter with the now ubiquitous longline hooks or gillnets must be very slim indeed. Given that the 423-cm female caught off North Myrtle Beach, South Carolina, taken in 1964 (see below), weighed 809 kg (the day after it was caught!), it is likely that the largest tigers may reach 1,000 kg.

I have recorded the following sizes and weights for tiger sharks taken along the southeastern coast of the United States: 125 cm, 5.6 kg: 147 cm, 13 kg; 185 cm, 59.4 kg; 246 cm, 66 kg; 285 cm, 123 kg; 295 cm, 145 kg; 312 cm, 177 kg (female); 323 cm, 210 kg; 362 cm, 326 kg (gravid female); 368 cm, 527 kg (gravid female); 376 cm, 460 kg (nongravid female); and 383 cm, 613 kg (gravid female).

Biology The tiger shark is a very interesting creature. It is essentially a tropical species that wanders into warm-temperate waters during the summer months. It is an adaptable species found in shallow continental waters, as well as in oceanic waters, or around oceanic islands. Although its dentition is highly specialized for preying on sea turtles and other large animals (see below), it is a voracious eater that may consume whatever is locally abundant from dolphins to conchs. It is also a scavenger and, on rare occasions, a man-eater. Reproductively, the tiger shark is an oceanic shark; its broods are very large, its young are relatively small, and its nurseries are widespread. Unlike many other sharks that show a clear diel cycle, approaching land by night and retreating to deeper water during the day, tiger sharks can be found in shallow inshore areas and beaches in the middle of the day. Dodrill (1977) reported catching tiger sharks both day and night off Melbourne Beach, Florida. The habit tiger sharks have of entering very shallow water is well documented. The largest tiger shark ever caught on rod and reel was caught from a beach pier in North Carolina during the middle of the day when other tiger sharks were being hooked (Millus 1980). All the gravid females collected by Rivera-López (1970) off La Parguera, Puerto Rico, were caught in water less than 2 m deep. A friend of mine, quite familiar with sharks and a reliable observer, was recently surprised to find a 300–320 cm tiger shark just a few meters away while snorkeling in water 1.2 m deep off Little Torch Key in the Florida Keys.

Although the tiger shark is a tropical species, it has a wide temperature tolerance, and most are caught in surface water temperatures of 17–28°C. Dodrill (1977) reported catching tiger sharks off Melbourne Beach, Florida, from February to September in a temperature range of 17.8–28.0°C. In Shark Bay, Western Australia, Heithaus (2001b) found a strong relationship between surface water temperature and tiger shark catches. Sharks were usually caught in water between 17 and 24°C, and occasionally in temperatures as low as 15°C. Catches appeared to be linked to both water temperature and prey availability.

Diet: The diet of the tiger shark has attracted considerable attention, and there are more papers about its diet than about the food habits of any other shark. Anatomically, with its unique teeth, large size, and a relatively large mouth, the tiger shark is specialized for preying on large sea turtles and other heavy-boned animals. Its teeth have relatively short crowns that are very wide at their bases and taper quickly to a point, and have coarsely serrated edges. The teeth are chisel-like in cross section and perfectly adapted for exerting tremendous pressure without breaking. Unlike most species of sharks, which have dissimilar teeth in the upper and lower jaws, tiger sharks have similar teeth in both jaws, so both jaws can exert a cutting action. Thus, tiger sharks can saw and cut through the very hard shell of large sea turtles and the bones of marine mammals with relative ease. The smaller sea turtles, of less than 50 cm carapace width, are swallowed whole by the larger tiger sharks.

Behaviorally, the tiger shark is an adaptable predator and a scavenger that takes whatever prey is locally abundant, easily caught, or easily consumed. Thus, reported stomach contents of tiger sharks include horseshoe crabs, blue crabs, horse conchs, skate eggs, smaller sharks, rays, all kinds of bony fishes, sea snakes, sea turtles, birds, marine mammals, offal of all kinds, and all sorts of drowned animals or people, regardless of their state of decomposition. In addition, tiger sharks often swallow indigestible items. As with most species of sharks, there is an ontogenetic shift in diet. Rancurel and Intes (1982), Simpfendorfer (1992), and Lowe et al. (1996) have all reported such dietary changes, with animals shifting to larger prey as they grow larger.

The anatomical adaptations for consuming large prey, the behavioral plasticity that allows it to pursue whatever slow-moving prey is locally abundant, and its scavenging habit make the tiger shark's diet the most extensive known among sharks. Listing all the animals that have been found inside tiger sharks would be a daunting task. A summary of the more common items follows.

Large invertebrates can form an important part of the tiger shark diet. These include horse conchs (*Pleuroploca gigantea*), spiny lobsters (*Palinurus argus*), horseshoe crabs (*Limulus polyphemus*), stone crabs (*Menippe mercenaria*), and blue crabs (*Callinectes sapidus*). Coles (1919) reported finding parts of three very large stone crabs in the stomach of a 381-cm tiger shark taken off Cape Lookout, North Carolina. According to Springer (1938), horseshoe crabs formed the largest part of the stomach contents examined in Florida, while spiny lobsters were consumed in the spring. A 351-cm female caught at Stump Pass off Englewood, Florida, in July 1959 contained 14 whole horseshoe crabs and pieces of approximately 12 others (Cape Haze Marine Laboratory, CHML no. 17). A 305-cm tiger shark caught off Morehead City, North Carolina, in July contained a large number of horseshoe crab shells, while another 353-cm specimen contained a large number of "soft-shell crabs" (Bell and Nichols 1921). At La Parguera, Puerto Rico, the spiny lobster was the most common invertebrate prey in tiger

Fig. 121e. *Galeocerdo cuvier*, tiger shark, immature male 91cm, 3 kg; from 53 km northeast of Sapelo Island, Georgia (Scott Bachman, collector).

sharks (Rivera-López 1970). Dodrill (1977) found horse conch remains or opercula in 6 of 16 stomachs (37.5%) of juvenile tiger sharks caught off Melbourne Beach, Florida. A very high proportion of the tiger sharks I have examined off South Carolina and of the specimens examined at the Cape Haze Marine Laboratory off Englewood, Florida, contained some crustacean remains.

Mark Rackley and Manny Puig have recently filmed an interesting feeding behavior of tiger sharks during an impromptu feeding session in offshore waters of the Gulf of Mexico. On several occasions, the sharks had taken and swallowed very large pieces of barracuda. Shortly thereafter, the sharks were observed regurgitating the food, catching it in the water column, cutting it into smaller pieces, and then swallowing the pieces. This behavior was filmed several times. This behavior may explain a puzzling observation by Springer (1963: 108), who reported that he had found "tiger shark stomachs full of conchs complete with shells and others full of conchs with neither whole calcareous shells nor any remnant of shells," but had not found partly dissolved conch shells in tiger shark stomachs. Springer speculated that "in some undetermined way the shells are removed in the shark's digestive tract." He also hypothesized that perhaps the shells are dissolved by the hydrochloric acid of the shark's stomach but failed to explain why the softer animal would not be digested also. It may be that the sharks regurgitate the conchs after the conch has relaxed out of the shells and died, and then pull the dead conch out of the shell and swallow it. Tiger sharks have also been seen everting their stomachs and swallowing them again (Eugenie Clark, pers. comm., October 1999). This may be a mechanism whereby tiger sharks eliminate indigestible materials.

Numerous species of bony fishes have been reported from tiger shark stomachs. Large tiger sharks seem to prefer other larger elasmobranchs such as hammerhead sharks, blacktip and blacknose sharks, rays, and skates (Coles 1919; Bell and Nichols 1921). Predation on skates may occur at various developmental stages of the prey, as the smaller tiger sharks 110–180 cm consume skate egg cases. I examined a 117-cm, 6.8-kg juvenile male taken in late May 1983 off Charleston, South Carolina, which had consumed nine large skate egg cases among other items. A 180-cm female caught off Englewood, Florida, on 6 May 1965 also contained skate egg cases (Eugenie Clark, Cape Haze Marine Laboratory, unpublished data).

Numerous studies show that tiger sharks prey on several species of sea turtles of all sizes. Witzell (1987) reported that of 404 tiger shark stomachs examined in eight previously published works, 21% contained the remains of large sea turtles. Along the southeastern United States, the loggerhead turtle (*Caretta caretta*) is the species most often found in tiger shark stomachs. Coles (1919) reported finding a freshly eaten 50-kg loggerhead turtle, which had been bitten through both shells in three places and the pieces of shell were badly crushed. Bell and Nichols (1921) reported finding numerous remains of loggerhead turtles in sharks examined off Morehead City, North Carolina. I found the remains of seven different large loggerhead turtles in the stomach of a 367-cm female tiger shark caught off Charleston, South Carolina, in August 1983. The bones and the turtle shells were well digested, but the horny beaks were still in good shape and allowed the turtles to be counted easily. Rivera-López (1970) also found turtle remains in 4% of the stomachs with food examined at La Parguera, Puerto Rico. Predation on sea turtles is not confined to large tiger sharks, nor is it confined to adult turtles. Dodrill (1977) reported finding 10 hatchling loggerhead turtles in the stomach of a 183-cm tiger shark caught off the Brevard County beaches, Florida, in August 1975.

Predation by tiger sharks on sea turtles is not limited to loggerhead turtles, but few studies identify the species taken. Heithaus et al. (2002) measured the frequency of tiger-shark–inflicted injuries on loggerheads and green turtles (*Chelonia mydas*) in Shark Bay, Western Australia. The injuries were more frequent on loggerheads than on green turtles, and many of the loggerhead turtles had one or both of the front flippers amputated by sharks. These authors suggested that the greater speed and maneuverability of the green turtle accounted for the fewer injuries inflicted upon it. Kauffman (1950) reported the hawksbill turtle (*Eretmochelys imbricata*) as the most common turtle found inside tiger sharks and their predominant source of food in the Philippines, being found in 50% of the 22 stomach contents examined. Green turtle parts were found infrequently, appearing on only two occasions. Bass et al. (1975c) mentions both green and hawksbill turtle remains being found in 13% of the stomachs of 39 sharks caught off the east coast of Africa. I encountered a small green turtle in the stomach of large tiger shark I examined in the Ishigaki Islands.

Tiger sharks prey heavily upon sea snakes in the Pacific and Indian oceans. Stevens and McLoughlin (1991) found sea snakes in 43% of the tiger shark stomachs examined off Northern Australia. Simpfendorfer (1992) surveyed reports on the stomach contents of 558 tiger sharks caught off Northern Australia between 1964 and 1986. Sea snakes were found in 50% of the stomachs examined, and tiger sharks of all sizes consumed them.

Tiger sharks often scavenge small migratory birds and take live seabirds at the surface. Small migrant birds often fall into the sea when they exhaust their metabolic reserves. Indeed, this must be a very common event because I have witnessed it at least 20 times off Florida, in the Caribbean, and off South America. Saunders and Clark (1962) reported that a 230-cm, 52-kg tiger shark caught off Sarasota, Florida, contained one yellow-billed cuckoo (*Coccyzus americanus*). A 182-cm female caught 250 m from the shore at Melbourne Beach, Florida, in May 1976 contained, among other animals, the remains of a yellow-billed cuckoo, a Bahama yellowthroat (*Geothlypis rostrata*), and a mourning dove (*Zenaida macroura*). Another 246-cm male tiger shark caught off Melbourne Beach, in September 1976, contained two wood thrushes (*Hylocichla mustelina*) (Dodrill and Gilmore 1978).

Tiger sharks also consume numerous species of seabirds. In the Northwestern Hawaiian Islands, seabirds were present in 75% of the tiger sharks sampled (de Crosta et al. 1984). Given the number of seagull and cormorant remains I have seen in tiger shark stomachs, I am convinced that these sharks actively take them on the surface and not just by scavenging. Nearly all the stomachs of large tiger sharks that I have examined in the southeastern United States contained at least a few seabird feathers or remains. Whitley (1940: 62) had already figured this out in the 1930s: "Judging from the stomach-contents of a number of Tiger Sharks, they frequently swallow mutton-birds (petrels) which they must snatch from the surface of the water." The consumption of birds is not limited to any one size of tiger sharks. I examined a 117-cm, 6.8-kg juvenile male taken in May 1983 off Charleston, South Carolina, which contained numerous gray/black feathers, presumed to be laughing gull (*Larus atricilla*) feathers. The records of the Cape Haze Marine Laboratory also show that cormorants, pelicans, and seagulls were often found in the stomachs of tiger sharks of all sizes caught off Englewood, Florida. One of the more interesting reports of birds in the stomachs of tiger sharks is by Carlson et al. (2002), who found remains of marsh-dwelling clapper rails (*Rallus longirostris*) in the stomachs of five juvenile tiger sharks (mean length, 102 cm) caught offshore of South Carolina. Apparently, these marsh birds had been driven offshore by storm winds or other factors.

Tiger sharks consume a variety of marine mammals found in tropical waters. Porpoises are the most often reported prey (Bell and Nichols 1921; Cape Haze Marine Laboratory records). In Shark Bay, Western Australia, Heithaus (2001a) found shark bite scars on 74.2% (95 of 128) of the noncalf bottlenose dolphins (*Tursiops aduncus*) examined, with most of the scars attributed to tiger sharks. Simpfendorfer et al. (2001) reported dugong (*Dugong dugon*) remains occurring in 16.5% of the stomachs sampled off Western Australia, with the occurrence increasing in the larger tiger sharks. Heithaus (2001b) also found dugong remains in 7 of 15 tiger shark stomachs he sampled off Western Australia. Most of the stomachs contained less than 1 kg of dugong flesh, and it was unclear whether they had obtained it by predation or scavenging. Tiger sharks are also known to feed on Hawaiian monk seals (*Monachus schauinslandi*) at French Frigate Shoals (Taylor and Naftel 1978; Balazs and Whittow 1979). Monk seal remains were found in three tiger sharks (280–403 cm) out of 23 caught at that location (Taylor and Naftel 1978). Again, whether the seals were taken by predation or scavenging is unclear.

Numerous land mammals have been found in the stomach's of tiger sharks. While some may have fallen overboard from ships, some are obviously dead domestic animals that have been jettisoned from ships or carried to sea by rivers. Thus, there are reports of rats, mongooses, cats, dogs, sheep, goats, and pigs being found in the stomachs of tiger sharks (Kauffman 1950; Randall 1992; Lowe et al. 1996). Tiger sharks congregate in areas where offal from coastal abattoirs is routinely discarded. Consequently, parts of cattle and horses have often been found in the stomachs of tiger sharks (Gudger 1949; Lowe et al. 1996). Tiger sharks consume even the most humble table scraps that are thrown overboard, from a frankfurter (in a 210-cm specimen, Cape Haze Marine Laboratoryno. 111, 16 August 1965) to steak bones wrapped in aluminum foil (CHML no. 63, 25 April 1964).

Human remains are found far more often inside tiger sharks than in any other shark species. Often, these are the remains of drowning victims. As scavengers, tiger sharks are more likely to consume decomposing cadavers than any other species of sharks.

Migrations: There are few data in the literature concerning the migrations of the tiger shark, other than statements of their presence at various localities at given times. Tagging data (Kohler et al. 1998) reveals south-to-north movements along the East Coast of the United States and that many tiger sharks move from the southeastern coast to the Caribbean and even to the northern coast of South America.

The migrations of the juveniles and adult males are not well understood; they appear to reside in the warmer waters of Florida and the Gulf of Mexico throughout the year. The migrations of the gravid females are clearer. Gravid females winter around the Caribbean Islands from September to April (Rivera-López 1970). In early spring, the gravid females migrate north to coastal waters of the southeastern United States to give birth in late May and early June. In late May and early June, one can find gravid females with large, near-term young, as well as females carrying very small embryos (less than 5 cm). The presence of females with very small young suggests that mating must also occur in that area.

Size at maturity Despite its being a common and ubiquitous species, there are few published data on size at maturity of tiger sharks in North America. According to Springer (1960), male tiger sharks mature at about 290 cm and females at about 300 cm. However, based on the specimens that I have observed, it appears that most specimens mature at about 310 cm.

Male: Specimens 295–305 cm are still immature. The largest immature male that I examined measured 295 cm and weighed 145 kg, caught off Charleston, South Carolina, in July 1983. It had elongated but uncalcified claspers. Branstetter (1981) reported that a 305-cm male had partially calcified claspers (i.e., was still immature).

Female: Clark and von Schmidt (1965) reported that the smallest mature female in their records was 297 cm. This size agrees well with the sizes given for tiger sharks in other areas. Simpfendorfer (1992) reported that the smallest pregnant female examined off Townsville, Australia, was 287 cm. The size at maturity (~300 cm) is 65% of the maximum size (460 cm) reported for North America.

Reproduction The tiger shark is an aplacental viviparous species, but there is little published information on its reproductive processes or development. Sarangdhar (1943) described 52-cm embryos removed from a female caught off Sassoon Dock, India, in 1942. These embryos had large yolk sacs with a considerable amount of yolk left in them and were enclosed in "water-filled" sacs formed in the shell membrane. The absence of any folding in their distal walls and the embryos being free inside the membranous egg case indicate that a placental connection does not form in the tiger shark (Sarangdhar 1943).

Two of the interesting aspects of tiger shark reproduction are the presence of a large quantity of a yellowish, clear fluid in each egg case (about 1 liter), and the fact that the 80- to 85-cm embryos appear to be too large to have been nourished solely by the yolk stored in the egg. Sarangdhar (1943: 110) was the first to suggest that "the liquid must probably also be having a nutritive function in the earlier stages, at least, of the embryonic development." Much later, unaware of Sarangdhar's paper, I wrote that term embryos that I had examined all had everted stomachs protruding slightly from the mouth and that it was possible that the abundant fluid may have nutritive properties ("uterine milk") and may be imbibed by the embryos through their everted stomachs (Castro 1983: 23). Eugenie Clark has also noted everted stomachs in 68-cm embryos examined 21 May 1961, but only 5 of the 37 embryos had everted stomachs. The function or functions of this fluid remain to be demonstrated.

Very little is known about the reproductive cycle of the tiger shark. In late May and early June, gravid females off the southern shores of the United States can be divided into two groups. One group consists of females that are carrying very large, near-term young. These females have inactive ovaries; they do not have ripe oocytes at the time of parturition. The other group consists of recently mated females that are carrying small embryos less than 10 cm. Most of these females have been taken in Florida waters from May to August (Clark and von Schmidt 1965; Dodrill 1977). Off La Parguera, Puerto Rico, Rivera-López (1970) found that 16 of 41 adult females were gravid. He observed embryos 30–65 mm in September with a progressive increase until April, when the largest embryos, 60.5–70.5 cm, were collected. These later embryos had no external yolk sac and were assumed to be "ready for birth." Rivera-López also remarked that free-swimming young were never encountered in the area. However, the assumption that embryos 60–70 cm are ready for birth because they have no external yolk sac is incorrect. Those embryos would not have been born until several months later, when they had reached 85 cm, with parturition probably occurring along the eastern coast of North America. An analysis of data presented by Rivera-López reveals that most of the gravid females caught at La Parguera were caught in January and February. One female was caught in March and one in April. No gravid females were caught at La Parguera from May to September, although numerous adult males were caught during those months. I believe that the reason for the absence of gravid females at La Parguera is that the Caribbean is only the wintering ground for the tiger shark, and females had departed for their nursery areas along the southern shores of the United States by April.

Based on the appearance of females carrying very small embryos off the southern United States from mid-May to mid September, and on females carrying large term embryos from late May to late July, it appears that the gestation period lasts just about a year (Castro 2009). This estimate agrees well with that of Clark and von Schmidt (1965), who concluded that gestation could be slightly over a year based on the presence of both early and late embryos in spring.

Since gravid females carrying term young do not have developed or ripe oocytes in the ovary at the time of parturition, the reproductive cycle of the tiger cannot be annual. The presence of two groups of gravid females in my data and that of Rivera-López suggests that the tiger shark has a biennial cycle. However, the data are quite limited and the estimate needs to be verified.

Size at birth: Bigelow and Schroeder (1948) and Baughman and Springer (1950) mentioned that free-swimming specimens measuring 46–48 cm had been recorded. This appears to be an error. Not only are there no other reports of neonates of such size, but broods of much larger embryos are usually reported. Clark and von Schmidt (1965) mention a brood of embryos 74.5–77 cm removed from a female caught off Sarasota, Florida, on 6 June. Eugenie Clark examined a female tiger shark caught off Isla Mujeres, Mexico, in June 1975 that was carrying embryos 81–89 cm (pers. comm., November 2000). The largest embryos that I have examined measured 85 cm, in a brood of 38 embryos 81–85 cm removed from a 368-cm, 527-kg female captured by M. Almond and M. Schultz off Charleston, South Carolina, on 31 May 1980. The smallest free-swimming specimen that I have seen was an 85-cm male caught on a longline set at 15–18 m depth off Charleston, South Carolina, in August 1983. Based on these observations, I estimate the tiger shark size at birth at 80–89 cm. This size range agrees well with the estimates for other oceans. Simpfendorfer (1992) reported that, off Townsville, Australia, the largest embryos were also 89 cm and that the smallest free-swimming individual was 84 cm, also indicating that the size at birth is 80–90 cm. Off Okinawa, the largest embryos examined by Makio Yanagisawa measured 84 cm, while the two smallest free-swimming young were 76 and 80 cm (pers. comm., December 2006).

Brood size: Tiger shark broods are large. I have examined or obtained data for 28 broods of North American tiger sharks. These ranged from 18 to 70 young (mean = 41, mode = 50). Dodrill (1977) reported catching a 312-cm female carrying 29 embryos 71–78.7 cm in early June 1977, south of Melbourne Beach, Florida. The 368-cm, 527-kg South Carolina female mentioned above contained 38 young measuring 81–85 cm, but it probably had aborted some during capture. A 359-cm female, taken in Winyah Bay, South Carolina, in June 1971, carried 57 young 75.8–80.8 cm (Grice Marine Biological Laboratory no. 71–100).

Bigelow and Schroeder (1948) mention a brood of 82 young from a 549-cm ("18 ft") female, based on a personal communication from Luis Howell-Rivero. To my knowledge, there is no documentation for this assertion, and both the size of the female and the number in the brood must be considered anecdotal or suspect because of the reasons stated above (see p. 259). The largest, reliably recorded, brood was reported by Dodrill (1977), who described a 364-cm female tiger shark caught off Melbourne Beach, Florida, in August 1975. This shark carried 70 embryos each about 17.6 cm long. In the Philippines, three females carried 41–62 young, averaging 52 per brood (Kauffman 1950).

Nurseries The tiger shark gives birth in June and July along a very large area ranging from central Florida to North Carolina, based on the presence of gravid females carrying term young 80–85 cm and of neonate tiger sharks of the same size range. The females do not come into very shallow water to deliver their offspring, like so many sharks do, but release their young at depths of 15–100 m. Natanson et al. (1999) noted that tagging and fishery data indicated the existence of a nursery ground for tiger sharks on the continental shelf of the southeastern United States, extending from about Augusta, Georgia, to Daytona Beach, Florida. Given that Augusta, Georgia, is some 200 km inland, one must assume that those authors are referring to the latitude of Augusta, Georgia, which would be off Georgetown, South Carolina.

Age and growth Branstetter et al. (1987) estimated that tiger shark males mature at 310 cm and females at 315–320 cm, but also said regional differences in juvenile growth rates result in different ages at maturity. Accordingly, these authors calculated that in the Gulf of Mexico, males mature in seven years, and females in eight years in the Atlantic, males and females mature in about 10 years. These estimates appear slightly high in comparison with the sizes at maturity described above, and it appears unlikely that there would be separate tiger shark populations in the Gulf of Mexico and the adjacent Atlantic Ocean. More recently, Natanson et al. (1999) used tag recapture data to estimate that the age at maturity is seven years for both sexes. Both Natanson et al. (1999) and Branstetter et al. (1987) have speculated that tiger sharks live at least 27–37 years.

Relation to humans The tiger shark is classed as a gamefish and affords the angler lively sport because of its great size and strength. The largest shark hooks and baits are usually used in this sport. The largest recorded tiger shark taken with rod and reel was caught from a pier in Cherry Grove, South Carolina, by Walter Maxwell in June 1964 (Millus 1980). This shark measured 4.23 m and weighed 807 kg (the shark was not weighed until the following morning). I have plotted the weight of this shark along with the weights of other large tiger sharks I have weighed myself, and it plots where it should be on the exponential line, suggesting that the measurements were accurate.

The tiger shark is of commercial value in the West Indies, Central America, and Mexico, where the flesh is eaten either fresh or salted. The skin can also be processed into leather of exceptional quality.

Tiger sharks are occasionally exhibited in aquaria, where they usually survive for only a few months. Tiger sharks generally feed poorly in captivity, and consequently usually lose weight after a few weeks of confinement. They then start swimming close to the surface with the head held at an abnormal upward angle. This exaggerated swimming posture usually presages death, and the animal must be released if it is to have a chance to survive. With special attention, the survival of tiger sharks in captivity may be prolonged. Crow and Hewitt (1988) have shown that with a special diet, daily feedings, and individual attention, a 200-cm shark survived 39 months at Sea World of Florida. Tiger sharks have survived at the aquarium in Veracruz, Mexico, for more than 4.5 years (Dehart 2004).

The tiger shark is one of the most dangerous sharks and is believed to be responsible for many casualties, especially in the West Indies. Baldridge (1974) reported that tiger sharks and white sharks were the species that attacked divers most often, with the tiger shark leading by a slight margin at the time. Although there is no doubt that tiger sharks have been responsible for many deaths worldwide, as noted above, some of the remains found in tiger sharks are due to the scavenging of drowned victims. Given the former abundance of large tiger sharks near popular beaches, it is remarkable that they were not implicated in more attacks on humans. Obviously, some factor inhibits these animals from attacking humans. Perhaps the search or attack patterns of tiger sharks are so strongly fixed that humans are perceived as food only in rare cases. In a recent case, a boy floating in an inner tube wandered into deeper water off Vero Beach, Florida, and was attacked and killed by what was believed to have been a tiger shark (R. Grant Gilmore, pers. comm., October 2004). In this case, the attack was probably released by either the person entering deeper water or by the image presented. A person in an inner tube must present an underwater image very similar to a turtle breathing on the surface, a round object with long appendages and a protruding head.

Extralimital Species

DAGGERNOSE SHARK

Isogomphodon oxyrhynchus (Müller and Henle, 1839)

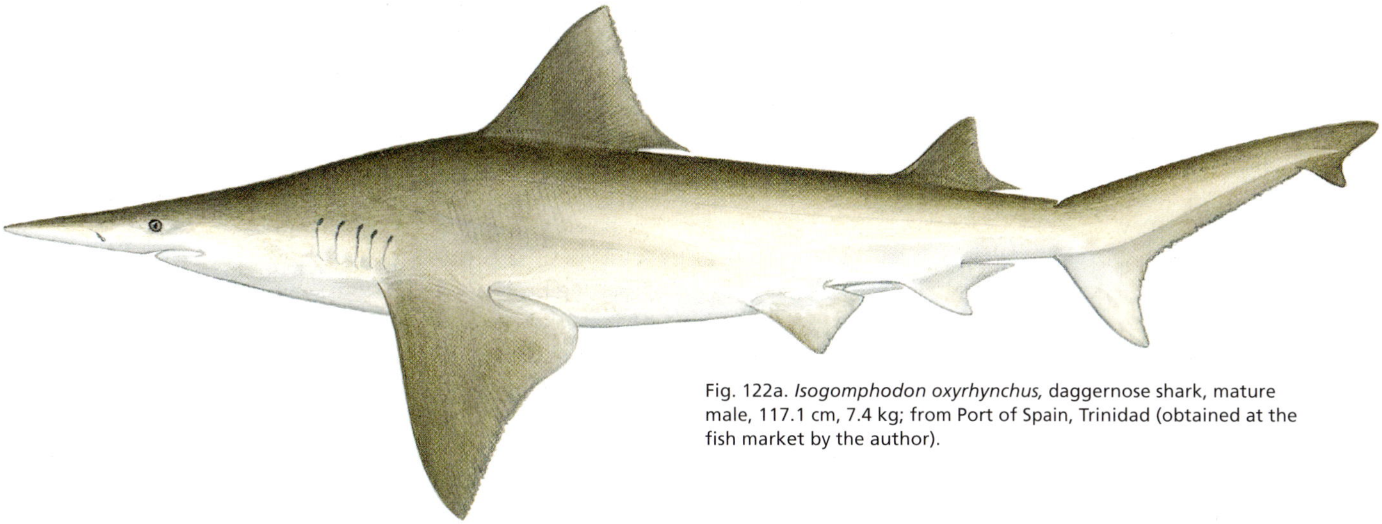

Fig. 122a. *Isogomphodon oxyrhynchus,* daggernose shark, mature male, 117.1 cm, 7.4 kg; from Port of Spain, Trinidad (obtained at the fish market by the author).

Common name Daggernose shark, a reference to its long and flat snout.

Spanish name None.

Scientific name *Isogomphodon oxyrhynchus* (Müller and Henle, 1839), originally described as *Carcharias* (*Prionodon*) *oxyrhynchus* in *Systematische Beschreubung der Plagiostomen* 2:41–42, plate 15. *Isogomphodon:* from the Greek *isos,* equal, or similar, and *gomph,* from the Greek *gomphos,* a bolt, a bond or fastening, plus *odon,* from the Greek *odontos,* tooth, thus "equally fastened teeth," a reference to its uniformly spaced teeth; *oxyrhynchus:* from the Greek *oxys,* sharp, and *rynchos,* snout, thus "sharp snouted."

Synonyms None.

Identification The daggernose shark is a smooth-backed species with an unusually long, flattened snout much longer than the mouth width; a first dorsal fin originating over the rear base of the pectoral fin; very large and broad pectoral fins (the width of the pectoral fin, from its origin to free rear tip, is about 75% of the length of the pectoral anterior margin), and minute round eyes about 0.8% of total length. Its teeth are proportionately small and have erect, symmetrical, narrow cusps and broad bases. The upper teeth have finely serrated edges; the lower teeth have smooth edges. The teeth are evenly spaced and are similar in both jaws. The dermal denticles have a strong central ridge ending in a long point, and two smaller ridges and points on winglike sides. Their surfaces have a characteristic micro sculpture. Coloration is yellowish gray above, with paler or whitish undersides. The fins are unmarked.

Similar species This is an unmistakable requiem shark that cannot be confused with any other species.

Range This small shark inhabits the northeastern coast of South America, from Venezuela and Trinidad south to Brazil. It has been reported from Trinidad, Surinam, Guyana, and French Guiana (Bigelow and Schroeder 1948; Uyeno et al. 1983). The limits of its distribution are not well known. I encountered a few in the fish market at Port of Spain, Trinidad. It is locally abundant along Brazil's northern coast (Lessa et al. 1999a).

Size and weight The daggernose shark reaches about 145 cm. Two 117-cm males I examined weighed 7.2 and 7.4 kg, while another of 126 cm weighed 9.2 kg. Daggernose shark females reach a maximum of 145 cm and 13 kg (Lessa et al. 1999a).

Biology There is little known about this peculiar shark of the tropical estuaries of northeastern South America. It inhabits turbid waters of mangrove coasts, estuaries, and river mouths. Its small eyes and large snout suggest that vision plays a minor role in the turbid or muddy waters where it lives. Its small teeth suggest that it preys on small fishes. Lessa et al. (1999a) reported catching the species at depths of 6–40 m.

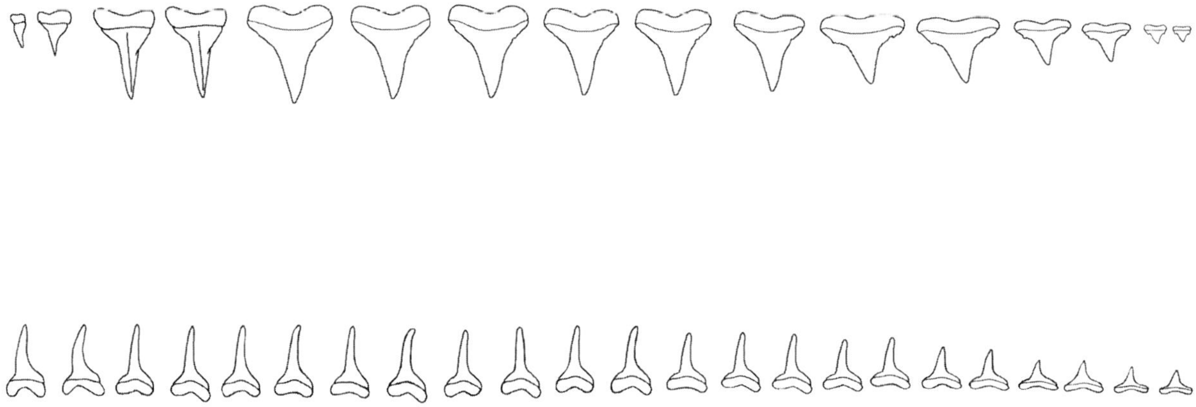

Fig. 122b. Teeth.

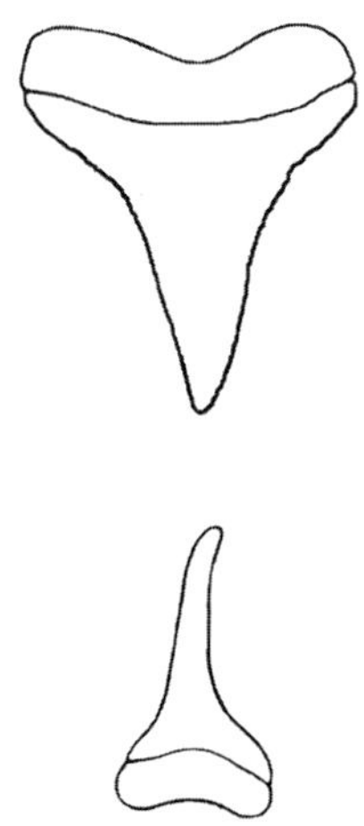

Fig. 122c. Teeth (detail).

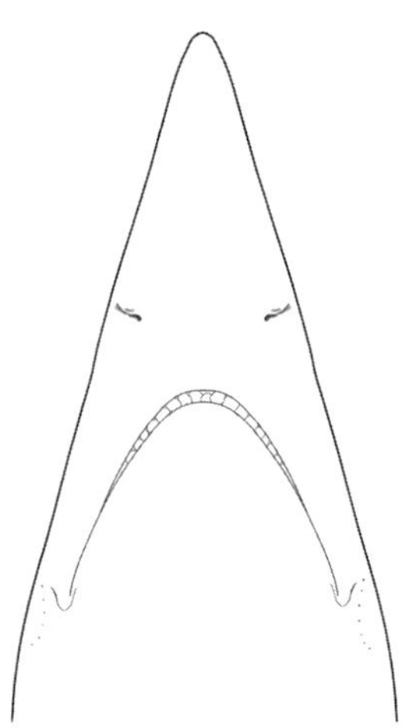

Fig. 122d. Snout.

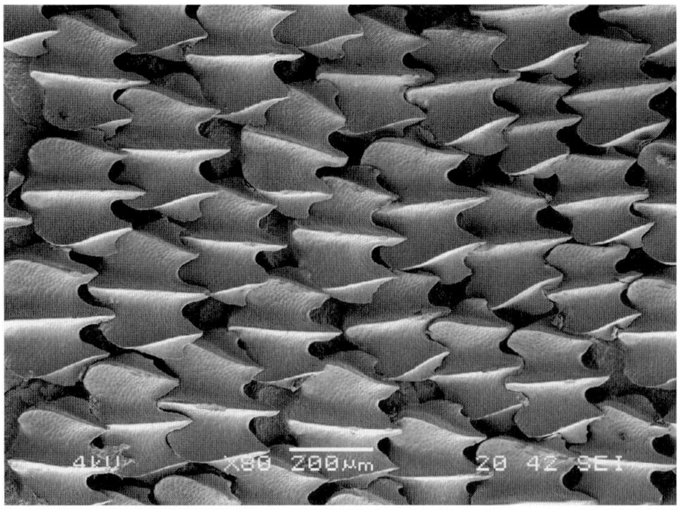

Fig. 122e. Dermal denticles.

Size at maturity Males mature at 93–103 cm, based on clasper calcification (Lessa et al. 1999a). The smallest gravid female reported measured 115 cm (Lessa et al. 1999a).

Reproduction The daggernose shark is a placental viviparous species. Both parturition and the beginning of gestation occur in December and January, although the reproductive cycle is apparently biennial (Lessa et al. 1999a).

Size at birth: Daggernose shark young are born at 38–43 cm (Lessa et al. 1999a).

Brood size: Broods usually consist of three to seven young, to a maximum of eight young (Lessa et al. 1999a).

Nurseries Unknown.

Age and growth Lessa et al. (2000) aged daggernose sharks from northern Brazil using stained vertebral sections. The oldest individuals in their sample were a 7-year-old male (116 cm) and a 12-year-old female (141 cm). These authors calculated that the largest males (144 cm) would be about 13 years old, while the largest females (160 cm) would be 20 years old.

Relation to humans None. The species is usually too small and too scarce to be of any economic importance. Small numbers of daggernose sharks are occasionally encountered in the Port of Spain fish market (Trinidad).

WHITENOSE SHARK

Nasolamia velox (Gilbert, 1898)

Fig. 123a. *Nasolamia velox*, whitenose shark, immature female, 95.7 cm, 3.6 kg, from off Guaymas, Mexico (Fernando Márquez, collector).

Common name Whitenose shark. The name appears to be a translation of the older Spanish name *pico blanco.* Why anyone would call this shark "*pico blanco*" or "whitenose shark" is a mystery to me. This shark has a conspicuous, squarish, *black* mark or spot at the very tip of the snout, and a bit of the white coloration of the underside extends into the upper surface along the sides of the black mark. This may have led some ill-natured troll to name it *pico blanco.* Given that we already have a blacknose shark, any name such as spotnose, black spotted, or even maculated shark would be better than whitenose shark.

Spanish name *Tiburón coyote, coyotito, zorrito, pico blanco, or trompa blanca* (Mexico); *Tiburón punta de zapato* (Guatemala).

Scientific name *Nasolamia velox* (Gilbert, 1898) (sometimes listed as *Nasolamia velox* Gilbert in Jordan and Evermann, 1898), originally described as *Carcharhinus velox* in *Bulletin of the United States National Museum* 47(3): 2747–2748. *Nasolamia:* from the Latin *nasus,* the nose, and *lamia,* from the Greek Lamia, mythical man-eating monster (see p. 278); *velox:* Latin swift or speedy.

Synonyms For nearly a century, the species was known as *Carcharhinus velox* Gilbert, 1898. In 1983, Compagno and Garrick erected a new genus, *Nasolamia,* for this species, based on its unique transversely oriented nasal capsules and cranial differences from other related sharks.

Identification This shark is characterized by a very long and narrow snout (much longer than the mouth width) with very *large, transverse, oval, closely set nostrils.* The nostrils are set at about half the distance from the tip of the snout to the front of the mouth. Its first dorsal fin originates over or behind the free rear tips of the pectoral fin, and a second dorsal fin originates over the origin of the anal fin. The upper teeth have strongly oblique cusps with a strong notch on their outer margins, finely serrated edges, and wide overlapping bases. The lower teeth have very narrow erect cusps with very fine serrations. Teeth number about U: 13–2–13, L: 12–1–12. The denticles are rhomboidal and closely overlapping, with six to eight ridges and tapering to a rear point. Color in life is brownish gray with metallic hues above, whitish below. The *tip of the snout has a small squarish black mark surrounded by a white area,* and the second dorsal fin tip is dusky.

Similar species The Pacific sharpnose shark has pronounced labial furrows, more widely separated nostrils, and a second dorsal that originates well behind the origin of the anal fin. The blue shark has a more posterior dorsal fin, originating well behind the free rear tips of the pectoral fins.

Range The whitenose shark ranges from the Pacific coast of Baja California (Roedel 1950) and the Gulf of California, southward to Peru (Chirichigno 1976). Being a tropical species, it is found off Baja California only occasionally. In the Gulf of California, I have seen a few specimens taken off Guaymas, Mexico, and only one taken off El Barril in

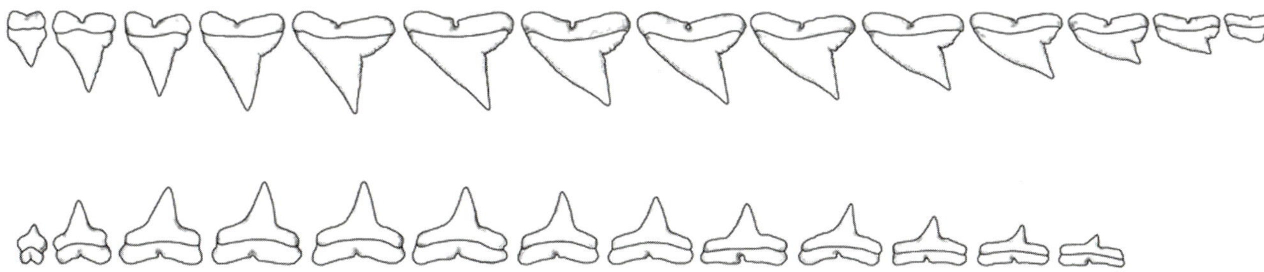

Fig. 123b. Teeth.

many trips to the sharks fishing camps in the area. The whitenose shark becomes abundant from the Gulf of Tehuantepec southward.

Size and weight The whitenose shark reaches about 165 cm. There are reports of white nose sharks reaching nearly 300 cm (Domínguez Arellano 2003, and other unpublished reports). These reports are apparently based on mistaken identifications, given that such reports also give a size at maturity of 115–130 cm for females and 125 cm for males. It is unlikely that a shark maturing at 125 cm would reach a length of 300 cm. The only weight data that I have are for females: 100 cm, 5 kg; 125 cm, 11 kg; and 161 cm, 22 kg.

Biology The whitenose shark is a tropical species inhabiting the eastern Pacific from southern Mexico and Guatemala southward. There are a few published or unpublished accounts of the whitenose shark (Porras Gándara 1997; Ruíz and Cabrera 2000; Domínguez Arellano 2003). These are obscure reports that must be used cautiously because they contain conflicting information, and one appears to contain misidentifications of a larger species (See above). These reports indicate that the whitenose shark is a migratory species seasonally common in the coastal waters of Guatemala at depths of 15–24 m, although occasionally it may be found as deep as 192 m (Porras Gándara 1997).

Size at maturity There is limited data on the size at maturity for the whitenose shark. According to Ruíz and Cabrera (2000), males mature at about 125 cm, while the smallest gravid female that they observed measured 134 cm.

Reproduction The whitenose is a placental viviparous species. Its reproductive processes are very poorly known. A female that I examined in June carried three embryos 1.5–2 mm in length.

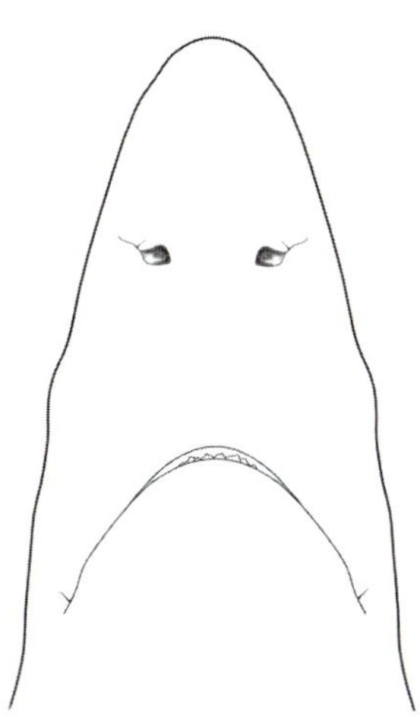

Fig. 123c. Snout.

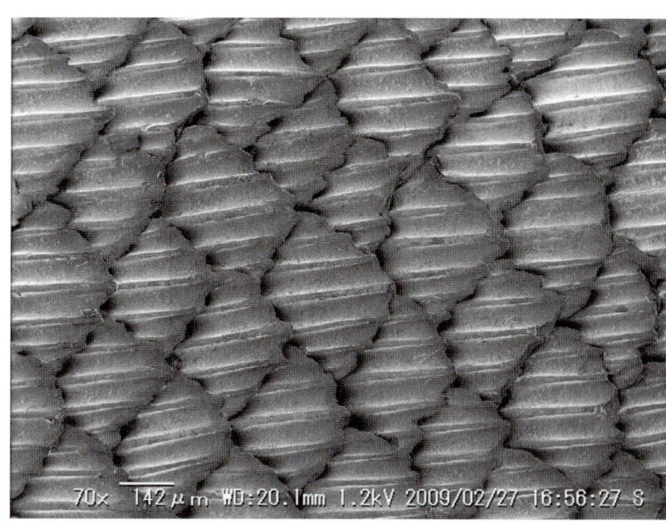

Fig. 123d. Dermal denticles.

Size at birth: The young appear to be born at about 45 cm.

Brood size: Brood size ranges from three to six young, with an average of five (Ruíz and Cabrera 2000).

Nurseries Unknown.

Age and growth The whitenose shark has not been aged.

Relation to humans The whitenose shark is of local economic importance in southern Mexico and Guatemala. It is the third most abundant species in the shark landings at Puerto Maderos, Chiapas, Mexico, and constitutes 10.4% of the total shark landings there (Domínguez Arellano 2003). In Guatemala, the whitenose shark is the second most abundant species (following the silky shark), and constitutes 16% of the catch (Ruíz and Cabrera 2000).

LEMON SHARK

Negaprion brevirostris (Poey, 1868)

Fig. 124a. *Negaprion brevirostris*, lemon shark, immature male, 180.6 cm, 32.9 kg; from off Cape Romain, South Carolina (Steve Poston, collector).

Common name Lemon shark, an allusion to the yellowish tinge of many adults.

Spanish name *Galano de ley* (Cuba); *tiburón amarillo, canchioc, tiburón limón* (Mexico). The A.F.S. name is *tiburón limón.* According to Applegate et al. (1979), the name *canchioc,* used in Yucatan, means *furia amarilla* (yellow fury), an apt name, or epithet, that anyone who has ever hooked a lemon shark will appreciate.

Scientific name *Negaprion brevirostris* (Poey, 1868), originally described as *Hypoprion brevirostris* in *Repertorio Físico-Natural de la Isla de Cuba* 2: 451, plate. 4, figs. 5, 6, 20. *Negaprion: nega,* from the Latin *nego,* to deny or refuse, and Greek *prion,* a saw, hence "not a saw," an allusion to the teeth not being serrated (the genus was erected by Whitley [1940] based on *Negaprion queenslandicus,* which has smooth-edged teeth); *brevirostris:* from the Latin *brevis,* short, and *rostris,* from the Latin *rostrum,* snout, thus "short-snouted."

Synonyms *Carcharias fronto* Jordan and Gilbert, 1882 [a].

Identification The lemon shark is characterized by a *snout shorter than the mouth width* and a *second dorsal fin that is nearly as large as the first.* The upper teeth have *narrow, triangular, smooth-edged, single cusps* and broad, finely serrated bases; the anterior teeth are erect, becoming progressively oblique toward the corners of the mouth. The lower teeth have narrow, triangular, erect, smooth-edged cusps. Teeth number U: 15–1 to 3–15, L: 13 or 14–3–13 or 14. The dermal denticles are much wider than long and are overlapping. The denticles have three to five parallel ridges and three well-separated points. The small young are bluish gray above, often with black fin edges, and have white undersides. Larger juveniles and adults are yellowish brown or yellowish olive gray above, occasionally dark brown; the undersides are usually yellowish.

Similar species The sand tiger has a similar yellowish olive gray color and two large dorsal fins, but its snout is conical, and it has long and slender teeth with one small cusplet on each side of the central cusp. The bull shark has a similarly short snout, but its second dorsal fin is much smaller than the first, and its upper teeth are broadly triangular with serrated edges.

Range The lemon shark inhabits the western Atlantic from New Jersey to Brazil (Bigelow and Schroeder 1948) and the eastern Pacific from the Gulf of California to Ecuador (Springer 1950b). Along the eastern coast of North America, the northernmost record of a lemon shark appears to be a large specimen from Beach Haven, New Jersey (Bigelow and Schroeder 1948). It is a summer visitor off North Carolina, where it enters Pamlico, Bogue, and Core sounds (Schwartz 2003). It is a common summer visitor from South Carolina to northern Florida. It is found on both coasts of central Florida from spring to fall. It is present and abundant all year in South Florida, the Florida Keys, and the Caribbean Sea. Along western North America, the lemon shark used to be common in the Gulf of California, but it became scarce during the last two decades of the twentieth century due to overfishing (E. de la Torre, pers. comm., Guaymas, January 2000). Jordan (1895) said that the species was

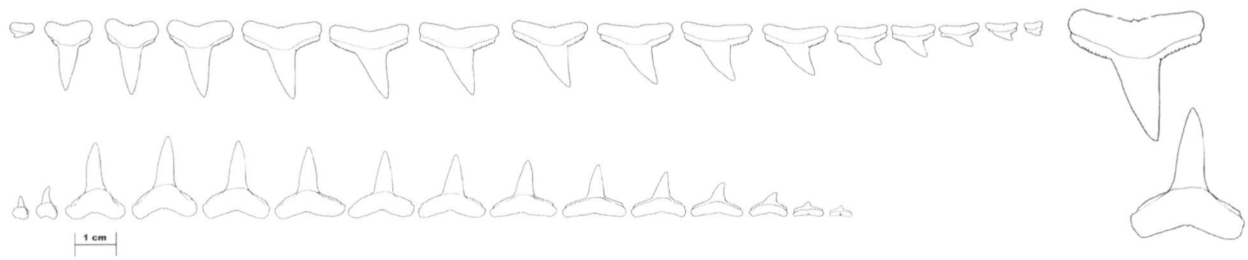

Fig. 124b. Teeth.

“not uncommon” about Mazatlan, but it is seldom seen there today. A few specimens are caught each year around Isla Tiburón and Isla Isabel (Mexico).

Size and weight Springer (1950b: 351) stated that the lemon shark “very rarely reaches a length of more than” 284 cm, and that the largest female that he had seen was 284 cm and the largest male was 279 cm. The largest specimen that I have seen was a female that measured 268 cm, caught off South Carolina. The largest reported by Dodrill (1977) off the central Atlantic coast of Florida measured 274 cm. The International Game Fish Association record-size lemon shark measured 290 cm and weighed 183.7 kg; it was caught off North Carolina in 1988 (International Game Fish Association 2006). Clark and von Schmidt (1965; and Cape Haze Marine Laboratory records) reported the largest specimens on record, a 291-cm female weighing 159 kg and a 296-cm male, both caught off the central gulf coast of Florida in May 1962. These two sharks were the only ones of some 250 lemon sharks listed in the capture records of the Cape Haze and Mote Marine Laboratories that exceeded 284 cm. Thus, Springer’s statement about the maximum length of the lemon shark remains true after more than half a century.

Specimens up to 250 cm are common from South Carolina to Florida and the Gulf of Mexico. Representative weights are as follows:

Male: 76 cm, 1.6 kg; 91 cm, 4.1 kg; 218 cm, 68 kg; 251 cm, 115 kg; and 291 cm, 159 kg.

Female: 249 cm, 94 kg; 269 cm, 104 kg; and 277 cm, 121 kg.

Biology The lemon shark is a common, large species of the tropical and temperate Americas that inhabits a wide variety

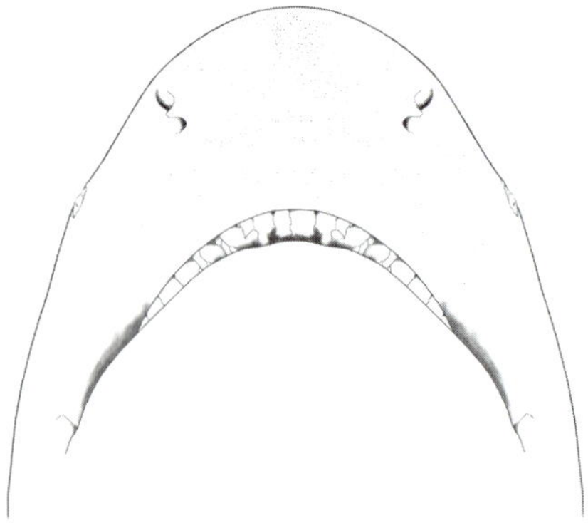

Fig. 124c. Snout.

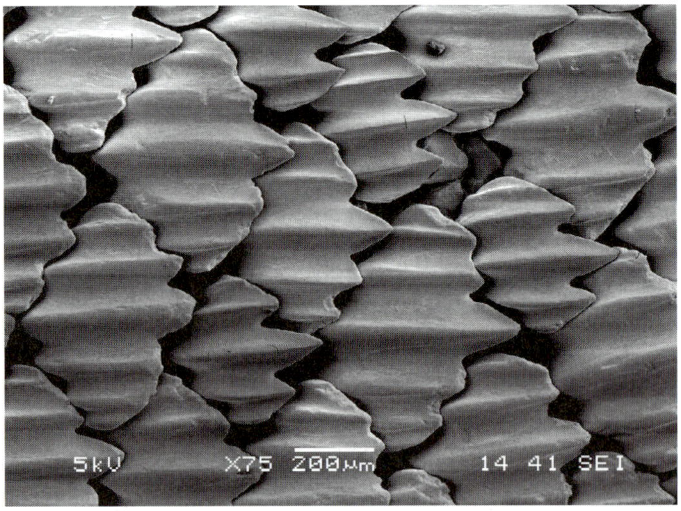

Fig. 124d. Dermal denticles.

of habitats in warm coastal waters. The lemon shark of eastern North America is one of the best-known requiem sharks. Springer (1950b) wrote the first comprehensive paper on the species, based on his experiences in the Florida shark fishery of the 1940s. Much of what we know about the lemon shark is the result of work by Samuel "Sonny" Gruber and his students at the University of Miami. Gruber, who made learning about the lemon shark a life-long quest, has published some 200 articles on the lemon shark on topics ranging from vision to aging.

Juvenile lemon sharks are common in clear water over sea-grass flats, lagoons, tidal creeks, and mangrove channels. They are social and form small groups (Springer 1950b, Gruber et al. 1988). They also establish overlapping home ranges and are highly site attached (Morrissey and Gruber 1993).

Multiple tracking studies performed by Gruber and his students in the Bahamas indicate that young lemon sharks do not migrate in their first year of life and remain associated with their natal nurseries for up to 10 years (Correia et al. 1995). The area of the home range grows as the shark increases in size. Juveniles (mean length = 73 cm) in the Bahamas occupied a home range of 0.68 km^2 (Morrissey and Gruber 1993), while a 168-cm shark had a home range of 18 km^2, and a 230-cm individual occupied 93 km^2 (Gruber et al. 1988). These studies suggest that the young sharks remain in shallow waters for foraging and protection, becoming more nomadic as they grow. Eventually, the sharks reach a size at which they can venture out of the shallow waters into deeper waters with reduced probability of predation by other sharks.

Adult lemon sharks tend to be solitary, spending the daylight hours offshore and approaching the shore at night, when they enter bays, sounds, saltwater creeks, inlets, and estuaries, often into very shallow water. Dodrill (1977) reported that all of his adult specimens were taken on setlines off Melbourne Beach between 4:00 P.M. and 8:30 A.M. Occasionally, adults congregate in large numbers, some of these aggregations consisting of sharks lying on the bottom (Cocking 2002). The reasons for these aggregations are unknown. Lemon sharks also congregate in dense swimming groups where there is abundant food. I have seen one such dense aggregation of swimming sharks in Dry Tortugas, Florida, where shrimp boats dumped their bycatch. It is likely that the sharks were attracted or conditioned to the dumping of bycatch at that location.

The lemon shark prefers surface water temperatures of 22–29°C, although they can tolerate much cooler temperatures. Clark (1963a) mentioned that lemon sharks, captive in pens open to the sea, and accustomed to feed regularly, fed less as the temperature fell to 20°C. When the temperature fell below 20°C in mid-December, the sharks stopped feeding. For the next 10 weeks, while the temperature in the pen remained below 20°C, the sharks showed practically no interest in food offered to them in any manner. The temperature remained below 20°C until 9 February, when it registered above 22°C. On 18 February, after the temperature had remained above 22° for 10 days, the sharks took some food again for the first time since mid-December. The sharks fed normally while the temperature remained above 22°C. When a sudden cold spell caused the temperature to drop to 19°C, the sharks again ignored food. Dodrill (1977) reported similar temperature preferences for the lemon shark, based on landings along Melbourne Beach, Florida, which occurred over a surf zone temperature range of 23–29°C. Dodrill also noted that the juveniles could tolerate temperatures as low as 16°C.

The lemon shark can remain motionless on the bottom for extended periods, apparently resting. This behavior was first observed in aquaria (Essapian 1962), and there are anecdotal reports of large adults seen resting on the bottom in deep water during the day. It is probable that this behavior is an energy-saving adaptation, resting during the day in deeper and cooler water may slow the metabolic rate and save energy for its crepuscular or nighttime foraging.

Due to their ability to survive in captivity for years, lemon sharks have been used in diverse experiments. Clark (1961, 1963a) conducted interesting experiments where lemon sharks were conditioned to press a target and ring a bell with the snout to receive food. This research showed that lemon sharks were capable of discriminating between shapes and colors and learning to push targets of the proper shape and color to receive food.

Diet: Young lemon sharks feed on small bony fishes and small crustacea, primarily shrimp. Springer (1950b) reported that the few young lemon sharks that had been examined at that time had large quantities of amphipods in their stomachs. Schmidt (1986) examined the stomach contents of 18 lemon sharks 58–100 cm from the shallow grass flats near Sandy Key in western Florida Bay. He found that they contained mainly small bony fishes (77% by number) including toadfish (*Opsanus beta*), pinfish (*Lagodon rhomboides*), silver jenny (*Eucinostomus gula*), grunts (*Haemulon* spp.), goldspotted killifish (*Floridichthys carpio*), striped anchovy (*Anchoa hepsetus*), and pink shrimp (*Farfantepenaeus duorarum,* 23% by number). Cortés and Gruber (1990) examined the stomach contents of young lemon sharks in the Florida Keys and reported that they fed mainly on penaeid shrimp and bony fishes such as snappers (Lutjanidae), porgies (Sparidae), halfbeaks (*Hemiramphus* spp.), and snake eels (Ophichthidae). The adults take whatever fish is locally available, including smaller sharks and rays, even smaller lemon sharks. I have found blacknose sharks and small nurse sharks in the stomachs of lemon sharks. There are also reports of lemon sharks feeding on octopus, blue crabs, and horseshoe crabs.

Size at maturity

Male: Males mature at about 245 cm on the basis of clasper calcification. A 232.5-cm male that I examined, caught off Charlotte Harbor, Florida, in mid-June had very soft claspers that protruded only 4 cm beyond the pelvic fins, and its epididymis was not developed. All the males that I have seen smaller than 245 cm were immature, while those larger than 245 cm were mature. Lemon sharks sampled by John Tyminski (pers. comm., June 2003) off Florida (Gulf of Mexico) were immature up to 241 cm and mature at 247 cm.

Female: Female lemon sharks usually mature at about 235–245 cm, although some may mature at 226–227 cm. Springer (1950b) reported that the lemon shark matures at 226 cm, but he did not explain the basis for this estimate. However, most females that I have seen or obtained records of seem to mature at a slightly larger size. Clark and von Schmidt (1965) reported that their smallest lemon shark carrying ripe oocytes measured 252 cm, and that the smallest gravid female measured 256 cm. Dodrill (1977) reported gravid females measuring 236 and 248 cm. Other than Springer's record, I have found only one record of a female maturing at less than 235 cm: Eugenie Clark recorded a 227-cm gravid female caught off Sarasota in May 1965 (Cape Haze Marine Laboratory records, Lemon shark no. 129, carrying 10 young 53–56 cm).

Reproduction The lemon shark is a placental viviparous species. Mating occurs off Sarasota, Florida, from the last week in April to mid-May, based on finding females carrying oocytes 3.5–4.5 cm in diameter, females with fresh mating bites, and ovulatory females with 4-cm oocytes and uterine eggs. Clark and von Schmidt (1965) reported mating behavior in captive sharks in May and June and sharks *in copula* on 1 May. According to Dodrill (1977), off Melbourne Beach, Florida, mating occurs between the third week in April and the third week of May, when the oocytes have reached 4–5 cm in diameter. Gestation lasts about 12 months. Gravid females carrying 55–65 cm term young enter shallow nursery areas (less than 1–2 m deep) to give birth from late April to the third week in May, with May being the month of peak births. The latest date, for which I have capture records, of a gravid female with term young is 18 May.

The lemon shark has a biennial reproductive cycle, and females give birth every two years. In early to mid-May, when some females are giving birth, other adult females are carrying 3.5- to 4.5-cm oocytes and may bear mating bites. The females that have just given birth will not have ripe oocytes and will not mate until the following May.

Size at birth: Lemon sharks are born at a length of 55–66 cm. According to Springer (1950b), the young measure 61–66 cm; however, I have collected free-swimming young in South Carolina as small as 55 cm with partly healed umbilical scars in June.

Brood size: The brood size of the lemon shark usually ranges from 7 to 18 young, with an average of 12. Springer (1950b) reported 11 young as the average size of 24 broods from the Florida Keys. Nineteen broods examined by Eugenie Clark off Sarasota ranged from 6 to 18 young, with an average of 12 (records of Cape Haze and Mote Marine Laboratories). Three broods that I examined off South Carolina ranged from 12 to 14 young, with an average of 13. Broods of more than 18 young are unusual. Garman (1913) mentions a second-hand report of a brood of 19 young taken in the Bahamas, while Dodrill (1977) also mentions a second-hand report of a brood of 21.

Nurseries The lemon shark has important nurseries along the shallow beaches and around the mangrove islands of South Florida and the Bahamas. Springer (1950b) first identified Florida Bay as one of the important nursery areas of the lemon shark and stated that, in May and June, adult females could be observed there in large numbers, in water so shallow that their fins broke the water surface. At this time, neonates were collected with dip nets. This nursery continues along the western coast of Florida, through the Ten Thousand Islands to at least Tampa Bay. The collection surveys of Mote Marine Laboratory from north of Tampa Bay to Yankeetown have not revealed any lemon shark neonates (J. Tyminski, pers. comm., January 2004). The shallow beaches and lagoons of the Bahamas are also important nurseries for the lemon shark. Morrissey and Gruber (1993) demonstrated that young lemon sharks in the Bahamas established home ranges in such shallow water that their 6-cm fins were exposed to air. Reyier et al. (2008) reported large aggregations of small lemon sharks (mean size 75-cm precaudal length) along the Atlantic beaches of Cape Canaveral. Although the sharks were present year-round, they were most abundant from November through March when the temperature 18–22°C. Reyier et al. suggested that the nearshore waters of Cape Canaveral are an important nursery for the lemon shark in South Florida.

Nurseries can be found as far north as South Carolina. Large gravid females are occasionally caught in May off Charleston and Bulls Bay (Castro 1993b), and neonates can be found in tidal creeks and around the barrier islands. I examined one 268-cm gravid female caught by Steve Poston off McClellanville on 7 May. This specimen carried

14 young 61.0–63.6 cm. I have also caught small neonates 55.0–55.9 cm, with partly healed umbilical scars, on 12 June off Wadmalaw Point, on the North Edisto River. In South Carolina, lemon shark neonates are not widely distributed and can only be found in certain tidal creeks and island channels. Neonates have also been reported from Georgia. Dahlberg and Heard (1969) reported catching a 58-cm lemon shark around Sapelo Island Beach in July.

The young lemon sharks remain along shallow beaches, sea-grass flats, tidal creeks, and island channels and around mangrove islands for a few years. In many of these areas the salinities vary widely, demonstrating that juvenile lemon sharks can tolerate a wide salinity range.

Age and growth Laboratory studies and tag–recapture studies have revealed the lemon shark to be a slow-growing species despite its tropical habitat. Early attempts at aging lemon sharks greatly overestimated their growth rate. Springer (1950b: 357) made the first attempt to determine the age at maturity for the lemon shark. He calculated that, if the rate of embryonic growth was constant, "maturity would not be reached for 650 or 700 days." Brown and Gruber (1988) used the vertebrae from tag-recaptured lemon sharks that had been injected with tetracycline hydrochloride at the time of tagging. The tetracycline formed a mark in the vertebral centra at the time of injection. These authors estimated that males mature at 11.6 years and females at 12.7 years and that the species lives for a minimum of 20 years.

Relation to humans In the Florida shark fishery of the 1940s, the lemon shark was an economically important species because of its size and its hide, which would produce high-quality leather (Springer 1938). It was also used for oil. In the Florida shark fishery of the late twentieth century, it was often of local economic importance.

The lemon shark is a hardy species that often feeds in captivity, where it survives up to four years (Clark 1963). Consequently, both small and large specimens have been used often for experimental work and aquarium displays. Juvenile lemon sharks are often exhibited in aquaria. The adults are not as popular for aquarium displays as sand tigers or sandbar sharks, because of their tendency to make a meal out of smaller aquarium dwellers.

In the heyday of shark fishing tournaments of the late 1970s and 1980s, anglers often sought lemon sharks because of their large size. Today, at some Florida beaches, juvenile lemon sharks are sought by fly-fishermen. Freshly hooked lemon sharks must be handled with great caution by anyone attempting to release them or bring them into a boat. These sharks often attempt to bite their tormentors, and sometimes they succeed in exacting retribution.

Lemon sharks have been implicated or suspected of a few attacks on bathers. Like any large shark that inhabits or enters shallow water, it must be considered potentially dangerous.

BLUE SHARK

Prionace glauca (Linnaeus, 1758)

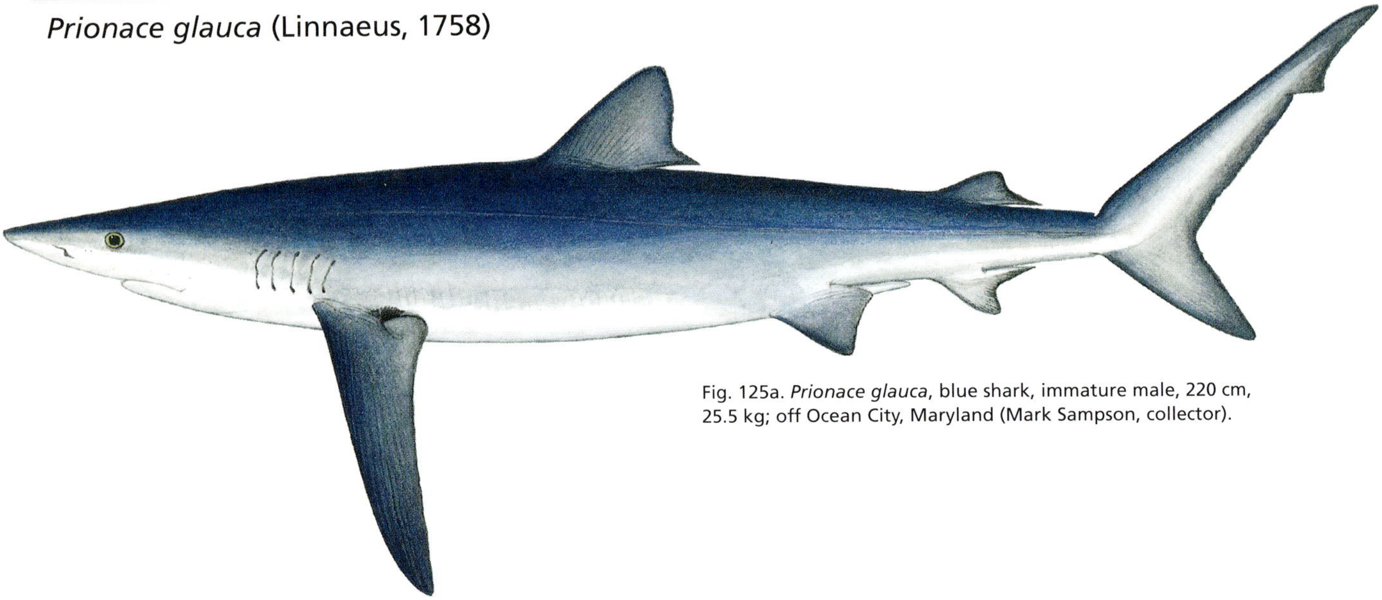

Fig. 125a. *Prionace glauca*, blue shark, immature male, 220 cm, 25.5 kg; off Ocean City, Maryland (Mark Sampson, collector).

Common name Blue shark, an obvious reference to the bright coloration of the live shark.

Spanish name *Tiburón azul.*

Scientific name *Prionace glauca* (Linneus, 1758), originally described as *Squalus glaucus* in *Systema Naturae,* 10th ed. I: 235. *Prionace:* Greek, saw-point, from *prion,* a saw, and *ake,* a point, an allusion to the shape of its teeth; *glauca:* from the Greek *glaukos,* bluish green or gray color.

Synonyms Although the blue shark has been clearly defined as a species since Linnaeus' day, it has been assigned to numerous genera (e.g., *Squalus, Carcharias, Prionodon, Galeus,* and *Glyphis*) over the last two centuries, depending on taxonomic whims or fashions of the time. Those desiring the specific details of the taxonomic history should consult the synonymy in Bigelow and Schroeder (1948).

Identification The blue shark has a snout longer than the width of its mouth and *very long pectoral fins* about as long as from the tip of the snout to the last gill slit; the *midpoint of the first dorsal fin base is closer to the pelvic fin origin than to the pectoral axil,* and it has *weak keels* on the caudal peduncle. The upper teeth have triangular, curved cusps with serrated edges and overlapping bases. The lower teeth have triangular, erect cusps with smooth or finely serrated edges. Teeth number U: 14–0 or 1–14, L: 13 to 15–1–13 to 15. The dermal denticles are broad and roundish, with three ridges and one main point flanked by smaller or indistinct points. Coloration is *dark indigo blue on top,* shading to *bright blue with metallic hues on the sides.* The undersides are white. The blue coloration changes to slate gray soon after death.

Similar species The deep indigo blue color on the back and bright blue of the sides, and the long pectoral fins, distinguish the blue shark from other requiem sharks. The shortfin mako has a similar blue coloration, but it differs in having a conical snout, long, smooth-edged teeth, pronounced caudal keels, and a lunate tail.

Range The blue shark is cosmopolitan in subtropical and temperate oceanic waters. In the western North Atlantic, the blue shark has been reported from Newfoundland and the Gulf of St. Lawrence (Scott and Scott 1988) to the Gulf of Mexico (Branstetter and McEachran 1983). Templeman (1963) noted that blue sharks are at times common in southern Newfoundland, although all of the records cited were of juveniles and were confined to dates from 15 August to 16 September. According to Templeman, the northernmost records are from the St. John's area, obviously due to the presence of a biological station there. The blue shark becomes very abundant in offshore waters from Georges Bank to offshore of Savannah, Georgia. McKenzie and Tibbo (1964a) found it to be abundant in the eastern Georges Bank and the Browns Bank off southwest Nova Scotia, but they did not specify the time of year. Bigelow and Schroeder (1953) reported that the blue shark is a regular summer visitor to the southern and western parts of the Gulf of Maine, appearing occasionally in July, and more often in August and September. Kenney et al. (1985) reported blue

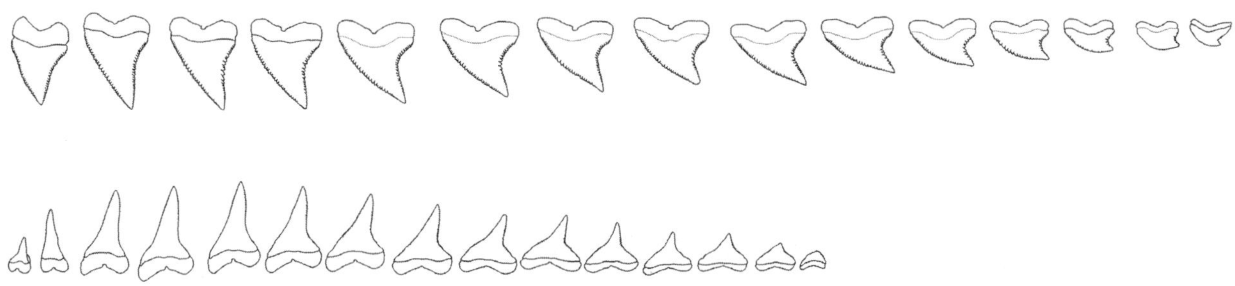

Fig. 125b. Teeth.

shark sightings, based on aerial and ship-borne surveys, over the shelf from the New York Bight to northern Georges Bank, and only scattered sightings in the northern parts of the Gulf of Maine. Bigelow and Schroeder (1948) reported counting 28 blue sharks in a one-hour run, 4–10 miles off Block Island, New York, in August 1943, estimating the number seen during the day at 150–200. Pratt (1979) stated that, based on offshore longline operations by the National Marine Fisheries Service between Cape Cod, Massachusetts, and Cape Hatteras, North Carolina, the blue shark is more abundant than any other large shark or big game fish in that area. Schwartz (1984b) reported that the species is found year-round in offshore waters of the Carolinas. The blue shark becomes scarce south of Savannah, Georgia. Along the Florida east coast, Berkeley and Campos (1988) found that blue sharks constituted only 1.5% of 613 sharks caught by swordfish boats sampled in 1981–1983. The blue shark is rare in the Gulf of Mexico (Baughman and Springer 1950; McEachran and Fechhelm 1998). Branstetter and McEachran (1983) reported four blue sharks taken by a commercial fisherman in the northwestern Gulf of Mexico, noting that the species was seldom caught or reported in the area. The species is also rare around the northwestern coast of Cuba and is found there only during winter. Guitart Manday (1975) reported that the blue shark composed only 1% of the catch in the Cuban shark fishery in 1971–1973.

Along the Pacific coast of North America, the blue shark ranges from the Gulf of Alaska to southern Mexico. The northernmost records come from Kodiak Island (Mecklenburg et al. 2002). Neave and Hanavan (1960) reported that the blue shark is found in the eastern Gulf of Alaska from July to September, while LeBrasseur (1964) found the species common in the southeastern Gulf during salmon migrations. In British Columbia, blue sharks are common off the west coast of Vancouver Island in summer, and off the

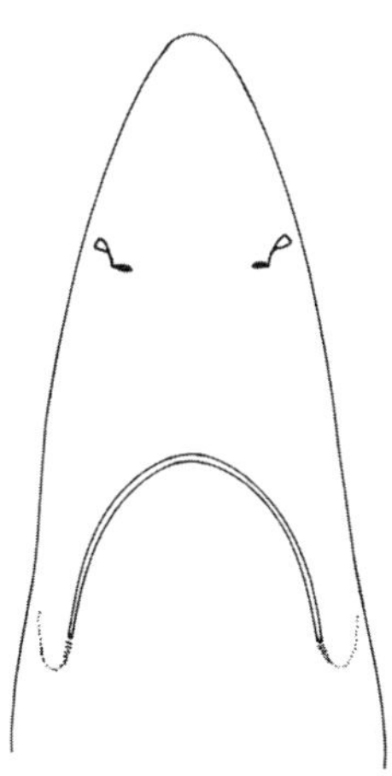

Fig. 125c. Snout.

Fig. 125d. Dermal denticles.

Queen Charlotte Islands (Hart 1973). They are common off California, having been reported from numerous localities: San Francisco (Starks 1917b; Berry and Perkins 1965), Monterey Bay (Harvey 1989), and San Diego (Eigenmann 1893; Berry and Perkins 1965), among others. The blue shark is common around the California Channel Islands nearly year-round. Sciarrotta and Nelson (1977) described the blue shark as abundant around Santa Catalina Island from March to December and scarce during January and February, while Tricas (1979) reported catching it there year-round.

Size and weight The blue shark may reach a length of 380 cm, although such specimens are exceptional. Bigelow and Schroeder (1948) stated that the largest blue shark that they had "positive record" of was 383 cm, but that the largest that they had handled was 335 cm. The largest specimen measured by McKenzie and Tibbo (1964a) was a 329-cm male caught in the Georges Bank area. The largest specimens reported by Skomal and Natanson (2003) were a 340-cm male and a 374-cm female. The largest Pacific Ocean specimen reported by Strasburg (1958) measured 304.5 cm. The largest specimen reported from the Indian Ocean was 348 cm and 210 kg (Gubanov and Grigor'yev 1975). Representative weights for blue sharks are as follows: 143 cm, 8.6 kg; 251 cm, 56 kg; 305 cm, 93 kg; 330 cm, 140 kg; and 340 cm, 142 kg (Kohler et al. 1996; N. Kohler, pers. comm. 2002).

Biology The blue shark is one of the most common and widest ranging sharks. It is a pelagic species that inhabits clear, deep, blue waters, usually in surface temperatures ranges of 10–20°C. It occasionally ventures close to land in clear, deep water around islands or in other areas when the temperatures are suitable. Although the blue shark has wide temperature and depth ranges, most are generally seen or caught from the surface to depths of 200 m, where the sea surface temperature ranges of 11–16°C. According to McKenzie and Tibbo (1964a), most blue sharks were caught close to the surface in temperatures of 12–16°C over the eastern Georges Bank. In the North Pacific, Strasburg (1958) reported that the best catches of blue sharks occurred in temperatures of 13–23°C in or just south of the transition zone between the relatively cold Aleutian current and the warmer North Pacific current. In subtropical shark fishing areas, the highest catches were in temperatures of 17–24°C. Strasburg (1958) gave evidence of "tropical submergence" in the blue shark, that is, that the species dwells in deeper water in the warmer tropics to stay within its thermal preferences.

Acoustic telemetry data gathered by Carey and Scharold (1990) showed that blue sharks made frequent excursions from the surface to a depth of several hundred meters and back up again. The oscillations, which were repeated every few hours, were largest in the daytime and were smaller in amplitude and confined to depths near the thermocline at night. Carey and Scharold interpreted these movements as a hunting tactic. For blue sharks, as for other sharks, olfaction is important for finding food. Olfactory cues exist in an ocean that is horizontally stratified in stabilized layers, and a chemical cue will spread horizontally as a widening ribbon on a horizontal layer. A predator moving vertically through the layers would have a greater chance of encountering an odor trail than if it searched horizontally in one or a few layers.

Diet The blue shark is an opportunistic predator that will take any locally abundant prey. Its diet consists primarily of squid and schooling bony fishes such as anchovies, sardines, herring, and bluefish. Kohler and Stillwell (1981) conducted a comprehensive analysis of blue shark diet off the northeast coast of the United States, examining the stomach contents of 863 specimens from 1972 to 1980. Bony fishes constituted the most important food items and a large number of species were recorded. Cods (Gadidae), bluefish (*Pomatomus saltatrix*), herrings (Clupeidae), and butterfish (*Peprilus triacanthus*) were the most common prey taken inshore. In offshore waters, other bony fishes and squid were the predominant prey items. Other prey included spiny dogfish, little skate (*Raja erinacea*), skate egg cases, and mammals. In a subsequent study, Kohler (1987) examined the stomach contents of some 1,199 blue sharks caught between Cape Hatteras and the Tail of the Grand Banks. In this study, squids were the primary component of the diet, and some 21 squid families were identified as prey.

In the Gulf of Alaska, LeBrasseur (1964) examined 29 blue shark stomachs and found salmon chunks (*Oncorhynchus* spp.), pomfret (*Brama raii*), saury (*Cololabis saira*), lanternfishes (Myctophidae), daggertooth (*Anotopterus pharao*), squid, shrimp, and salps (*Salpa fusiformis*). In Monterey Bay, California, Harvey (1989) examined the stomach contents of 150 blue sharks. He reported that the most important prey items were euphausiid shrimps (primarily *Thysanoessa spinifera*), northern anchovy (*Engraulis mordax*), Pacific hake (*Merluccius productus*), spiny dogfish, Pacific herring (*Clupea pallasii*), and five genera of cephalopods. Euphausids and northern anchovy were the most numerous items and constituted the most important portion of the diet. Off Santa Catalina Island, California, Tricas (1979) found that the northern anchovy was the major prey for the blue shark, while squid constituted the major prey in more distant offshore waters. Tricas also reported that blue sharks moved inshore in winter to feed on spawning schools of market squid (*Loligo opalescens*).

Migrations: The migrations of the blue shark are complex and poorly understood. These movements involve segregation by size and sex and may encompass vast areas, perhaps entire ocean basins. Much of what we know about the migrations of the blue shark along the eastern coast of North America is based on the work of John G. Casey and his coworkers, and hundreds of volunteer taggers, who tagged more than 60,000 blue sharks from 1965 to the early 1990s. According to Casey (1985), blue sharks are taken year-round in the Gulf Stream and the Sargasso Sea, and they winter eastward of the northern boundary of the Gulf Stream. Beginning in April or May, blue sharks move from the Gulf Stream toward North Carolina and northward to off Newfoundland. Blue sharks are common throughout summer on the continental shelf off southern New England, Georges Bank, Nova Scotia, and the Grand Banks. They are often found in inshore waters as shallow as 35 m off New York and southern New England.

Most of the blue sharks taken off eastern North America are juveniles of both sexes and adult males (Bigelow and Schroeder 1940, 1953). These juvenile blue sharks are usually larger than 122 cm. Of more than 3,500 blue sharks caught at fishing tournaments and from research vessels off the northeastern United States from 1975 to 1985, only six sharks measured less than 122 cm (Casey 1985). Adult males predominate in the continental waters of the eastern coast of North America. McKenzie and Tibbo (1964a) reported that of 306 blue sharks caught in the eastern Georges Bank, 86% were males, and none were gravid females. Mature female blue sharks are scarce in waters of the continental shelf, the continental slope, and the Gulf Stream, and only a few full-term gravid females or postpartum females have ever been taken anywhere off North America (Casey 1985). Pratt (1979) reported catching 16 gravid females off the coast of Long Island, New York, during June and July, all within 75 km of the shore. All of these females carried young 43–54 cm (37.1–46.6 cm fork length), so they could not have been too far from their birthing areas.

By contrast, in the eastern North Atlantic, catches consist primarily of adult females. Tucker (1958) reported that the blue shark catch off Looe, Cornwall, in 1957 totaled 4,170 sharks, "of which it is almost certain that less than 1 per cent were males" (Tucker 1958: 315). Also examining the Looe sport fishery in 1972–1973; Stevens (1974) reported that if the fishing data reflected the true sex ratio of blue sharks present, then only about 6% of the local population was male, and that most of the males appeared to be immature. At the same location, Vas (1990) reported that of 73 sharks brought to port in 1984–1985, only one (1.4%) was a male.

Blue sharks tagged off the northeastern coast of the United States have dispersed in all directions, with recaptures ranging from Europe to the northern coast of South America and the equatorial Atlantic (Kohler et al. 1998). The majority (97.6%) were recaptured at diverse locations in the western North Atlantic, with only 2.4% being recovered at localities in the eastern North Atlantic (Casey 1985). Stevens (1976) tagged 2,024 blue sharks off southwest England in the early 1970s. Most of these sharks migrated seasonally between 30° N and 50° N latitude in the eastern North Atlantic, although a few made transatlantic migrations. Two blue sharks tagged off Cornwall, England, were both recaptured off Long Island, New York, a straight line distance of more than 5,600 km. The small percentage of eastern Atlantic tagged sharks moving to the western Atlantic is probably explained by the small number of sharks tagged at a time when fishing effort was much less than in subsequent decades. In any case, it is evident that some mixing does occur between the eastern and western Atlantic locations.

The dispersal to all points of the compass by sharks tagged off the northeastern United States, and the transatlantic journeys by sharks from both sides of the Atlantic, suggest that blue sharks of the North Atlantic could constitute a single stock in the entire North Atlantic basin. Recent analysis of mitochondrial DNA reveals no genetic differences among blue sharks from the eastern and western North Atlantic (Skomal and Natanson 2003). However, the relationships of these two groups and their overall migration patterns remain unclear.

The migrations of blue sharks in the Pacific Ocean are very poorly known due to a lack of scientific effort in the area (although incredible numbers of blue sharks are caught in high seas fisheries in the Pacific). Strasburg (1958) simply stated that blue sharks appear to make pronounced northern migrations during the warmer months and speculated that these migrations were associated with reproductive activities.

Size at maturity

Male: According to Pratt (1979), the smallest mature male in his sample was 182 cm (153 cm fork length[5]) and 50% of males were mature by 218 cm (183 cm fork length). Pratt used the presence of spermatophores to determine sexual maturity, instead of using the standard of clasper calcification. Because sharks can start producing sperm and spermatophores before the claspers become calcified (Castro 1996), this method tends to underestimate the size at maturity. This probably accounts for his unusually wide size range for maturing blue sharks. It is likely that if clasper

5. Fork lengths are given here as reported by investigators. When the investigators have given a conversion formula, the total length is reported. Unless otherwise specified, the lengths given are total lengths.

calcification had been used as the criterion for maturity, it would have been found that the blue shark matures over a narrow size range like other carcharhinids.

Female: Most of the available evidence indicates that female blue sharks reach maturity at 203–220 cm. The wide size range is probably due to differences in research methods rather than population differences. Aasen (1966) stated that the smallest gravid female in his sample was 232 cm. Pratt (1979) examined 19 gravid females from the western North Atlantic. He did not give their sizes or the size of the smallest gravid female but stated that mature females ranged from 221 cm (given as 185 cm fork length, and converted to total length using his formula fork length = 1.73872 × total length + 0.82995). Bigelow and Schroeder (1948) estimated that females reached maturity at 213–244 cm. Other workers have given smaller estimates of size at maturity. J. A. Castro and Mejuto (1995) reported that, in the Gulf of Guinea, the smallest gravid females measured 203 cm (170 cm fork length) and that 50% were mature at 214 cm (180 cm fork length). Strasburg (1958) reported that the smallest gravid female in the equatorial Pacific Ocean measured 214 cm. Similarly, Gubanov and Grigor'yev (1975) stated that females in the Indian Ocean became mature at 215 cm (180 cm fork length) and 55 kg. Japanese researchers (Suda 1953; Nakano 1994) have stated that, in the North Pacific, pregnant females appear at about 140 cm (presumably) precaudal length, or at a total length of about 189 cm. This size estimate is much smaller than that of other observers. Although it is difficult to interpret this estimate, it may be the result of working with already cut market fish or with data collected by various fishery workers. Suda worked with market fish, and Nakano apparently interpreted data collected by others in the fishery. Pratt (1979: 457) wrote: "Defining sexual maturity in the female blue sharks is difficult because they pass through a distinct subadult phase in which the organs necessary for copulation are developed and those required for generation are dormant or developing. The subadult stage lasts for two summer seasons." According to this model, "subadults" carry ovarian follicles 2–6 mm in diameter in May and June, and sperm can be found in their oviducal glands during the summer. Pratt also stated that "subadults" were found inseminated at a minimum size of 161 cm total length (135 cm fork length). Presumably, these "subadults" (measuring 173–221 cm total length or 145–185 cm fork length) would store the sperm until the next spring (May and June), when they would ovulate and fertilization would then occur. This hypothesis deserves further investigation.

A core issue is the very concept of a "subadult," an animal that is neither juvenile nor adult. This notion, which is based solely on the size of the female (145–185 cm fork length) rather than on physiological or morphological features, is at odds with the observations of others. J. A. Castro and Mejuto (1995) showed that 20–50% of the females 170–185 cm fork length in the eastern tropical Atlantic are already gravid, at the sizes considered "subadults" by Pratt. Suda (1953) observed that mating occurred in June and embryo development proceeded immediately afterward and progressively for nine months. Brazilian workers claim that blue sharks off Brazil mate and embryo development follows directly (Alberto Amorim, pers. comm., 1998). To reconcile all of these observations requires a model where (1) juveniles mate and store sperm, and ovulation and fertilization occur a year or two later, and (2) adults do not store sperm, and development of their embryos follows mating. Clearly, many questions remain about this aspect of blue shark biology.

Reproduction The blue shark is a placental viviparous species. Despite the abundance and ubiquity of the blue shark, there are few studies on its reproductive processes. Much of the literature is confusing and difficult to interpret. The information presented here is an attempt to reconcile the data and conclusions of various authors through the last six decades. Some of the earlier works (e.g., Suda 1953; Aasen 1966) are sketchy. Aasen (1966) reported on blue sharks caught during the cruise of MS *Volstad Senior* in February–April 1965. Although Aasen recorded the capture date, location, size, and weight data for 647 blue sharks, there is insufficient information on the reproductive parameters.[6]

There are very few estimates of the length of gestation for the blue shark. Suda (1953) estimated the gestation period at nine months. According to Suda's data, the mating season begins in June, based on the appearance of "tooth cuts" (courtship and mating bites) on females. Small embryos less than 5 cm are seen in July, and mean embryo sizes continue to increase until March and April. By April, the embryos have reached their maximum length. Pratt (1979) estimated the gestation period by pooling embryo size data from different areas (North Atlantic and Mediterranean) and concluded that the gestation period lasted about 12 months. The length of the reproductive cycle is unknown.

Size at birth: Blue sharks are born at lengths of about 45–50 cm. Aasen (1966) calculated that the young are about 50 cm long at birth and stated that the largest term embryo

6. Interestingly, in the text, Aasen (1966) mentions that a total catch of 17,783 tons of blue shark (gutted weight) and that the total weight of blue shark catch was 32,041 tons. It appears that this is an error. Given that the largest blue sharks weigh some 100 kg, the total catch of 32,041 tons would translate as 320,041 sharks weighing 100 kg each, a ridiculously high number of blue sharks. If one assumes that the figure of 32,041 should be in kilograms, and given the 647 sharks listed in its appendix, the average blue shark weight comes out to be 49.5 kg, a more reasonable figure.

he examined was 53 cm. Aasen also gave the average length of a brood of 42 embryos as 49.3 cm in February. Strasburg (1958) reported that the largest blue shark embryos he saw in the nonequatorial Pacific Ocean were 34–48 cm and suggested they were very near term because their sizes approached the 45 cm size of the smallest longline-caught blue sharks. Pratt (1979) reported embryos 37.1–46.6 cm in females taken in June and July. Bigelow and Schroeder (1948) mentioned a free-living specimen 53.9 cm.

The birth size of the blue shark, 18–19% of the total length of the adult female, is rather small for a requiem shark, while its brood size is unusually large. These differences appear to be adaptations for life in the open ocean. Perhaps producing large numbers of small young enhances survival in the epipelagic environment because of the abundance of large predators there (Branstetter 1990). The tiger shark with its large broods and offshore nurseries might also exemplify this strategy.

Brood size: The blue shark is one of the most prolific sharks. Broods of 50–80 young are not unusual. Aasen (1966) reported that 12 specimens from the North Atlantic had 32–67 embryos each, the average being 45. Strasburg (1958) found the brood size from 4 to 38 in equatorial Pacific females and gave that of females taken between 24° and 35° N latitude as 23–40. Gubanov and Grigor'yev (1975) reported that the number of early stage embryos a female can carry was not correlated with her size and gave the following examples: 256 cm, 40 embryos; 275 cm, 84 embryos; 282 cm, 63 embryos; 286 cm, 57 embryos; 300 cm, 90 embryos; 310 cm, 13 embryos; 320 cm, 58 embryos; 334 cm, 135 embryos; and 345 cm, 60 embryos. J. A. Castro and Mejuto (1995) noted a mean of 37 embryos in a sample of 419 gravid females from the Gulf of Guinea. Nakano (1994) gave the brood range of females in the North Pacific Ocean as being from 1 to 62, averaging 25.6. As his sample consisted of 14,282 embryos, it is obvious that numerous individuals collected the data. They may have included females that had aborted embryos or were still in the process of fertilization, and thus, the low end of the brood range may be inaccurate. The largest brood reported consisted of 135 young, in a 334-cm female from the Indian Ocean (Gubanov and Grigor'yev 1975).

Diverse embryonic anomalies have been reported in the blue shark. Lozano Cabo (1969) reported a dicephalous specimen aborted on deck, along with 40 normal young, in a female caught in the Islas Chafarinas, off Morocco. Four dicephalous specimens were also reported from Japan (Goto et al. 1981). Abe (1972) reported a cyclopic embryo. I have an 18-cm embryo from Brazil (collected by Alberto Amorim) in which the nose is elongated into a conical snout and the jaw protrudes below, giving the embryo the appearance of a goblin shark.

Nurseries The location of blue shark nurseries off the eastern coast of North America is unknown. As Pratt wrote in 1979, small blue sharks in their first and second years must "inhabit an unknown niche." The apparent lack of neonates and young of the year blue sharks in the western North Atlantic suggests that the nurseries are far offshore or outside this area. In the southern California Bight, blue sharks with "late-term" young and small free-swimming young have been reported in June, suggesting that the bight may be part of nursery ground for the species (Hanan et al. 1993; Laughlin and Ugoretz 1999).

Age and growth The few available longevity estimates for the blue shark suggest a maximum age of 16–17 years. Cailliet and Bedford (1983) estimated a maximum age of 10 years for blue sharks of the Pacific Ocean. Skomal and Natanson (2003) derived age and growth estimates for the blue shark using 411 vertebral centra and 43 samples of tagged sharks recaptured in the North Atlantic. In this study the vertebrae of two oxytetracycline-injected recaptured sharks served to validate the hypothesis that one band pair (ring) is formed annually in the vertebrae in sharks up to 229 cm fork length. Based on their counts of vertebral band pairs, Skomal and Natanson posited the maximum ages for blue sharks of 16 years for males and 15 years for females. The maximum ages at liberty for tagged sharks also suggest a relatively short life span for blue sharks. Although nearly 5,000 blue sharks have been tagged in the 39-year history of the tagging program, the maximum times at liberty for blue sharks are 8.5 and 9.1 years (Skomal and Natanson 2003).

Relation to humans The meat and fins of the blue shark are of the lowest quality. Consequently, thousands of blue sharks are discarded at sea annually by the high seas fleets, usually after removing their fins.

Historical note: Some of the earliest descriptions of fishes are found in a Greek poem on fishing, *Halieutica,* attributed to Oppian of Cicilia, who lived during the reign of Marcus Aurelius (A.D. 161–180). In *Halieutica* (book I: 729–751), Oppian first describes how the dogfish and the angel shark protect their young and then continues:

> Others again protect their children by taking them in to the mouth as it were a house or nest; as, for example, the *Glaucus* [Greek blue-gray] which loves its children beyond all other fishes that are oviparous. For it both remains sitting by until the young come forth from the eggs and always swims beside them; and when it sees them afraid of a strange fish it opens its gape and takes them into its mouth until the terror has withdrawn, and then again ejects them from its throat. (Oppian 1963: 277)

The same fable, probably already old, appears in the writings of Aelian (A.D. 170–235) who collected animal stories rather indiscriminately: "Among the fishes the 'Blue-grey' [*glaucus*] is a model father. He maintains a strenuous watch over his mate's offspring, to ensure that they are not attacked or injured. . . . And if any of his young is afraid, he opens his mouth and takes the baby in. Later, when his fear has passed, he disgorges the one that took refuge exactly as he received it. And it resumes its swimming" (Scholfield 1971: 33–35). Later, Aelian writes: "The Shark brings forth its young through its mouth in the sea and takes them back again and then disgorges them by the same channel alive and unharmed" (153).

Some modern writers believed that the *glaucus* of ancient Greek and Roman writers is the blue shark. This is a problematic translation that includes, at least, two very different fishes. The details of this controversy are beyond the scope of this book (see D'Arcy Thompson's [1947] *A Glossary of Greek Fishes* for more details). The translation is accepted here simply to explain how the long-lived fable of parental care by the blue shark came about. The old fable continued to be improved into modern times; Pennant (1776, 3: 95) wrote:

> Aelian relates strange things of the affection this species bears to its young: among others, he says, that it will permit the small brood, when in danger, to swim down its mouth, and take shelter in its belly. This fact has been since confirmed by the observation of one of our best ichthyologists, and is no more incredible, than the young of the opossum should seek asylum in the ventral pouch of its parent, a fact too well known to be contested.

Nearly a century later, Couch (1868), while expressing skepticism about the fables of ancient writers, quoted a translation of Oppian's *Halieutica:*

> Of all oviparous kinds that throng the sea,
> The fond Blue Sharks in tender care surpass.
> They rear their fondlings, like some careful nurse,
> Observe their motions and restrain their course,
> Eye every wave, and shew the doubtful way,
> Teach where to hunt, and where to find their prey.
> When big with secret guilt the waters heave,
> They in their mouths their shelter'd young receive,
> But when waves at their own leisure roll,
> And no fierce robber drives the scatter'd shoal,
> Again the parent's pointed jaws compress'd,
> By force expel them from their pleasing rest.

Couch then noted (with some skepticism) that incredible as this strange proceeding may appear to us, it had received corroboration from the naturalist Rondeletius and other credible witnesses, including Mr. Darwin himself.

Amazingly, the fable about blue sharks caring for their offspring survived well into the twentieth century. In a chapter on blue sharks in *Nonsuch: Land of Water,* William Beebe (1932: 181) wrote: "There is no doubt of the attachment between parent and young sharks. In diving I have watched a mother shark and two young swimming together, week after week. The pair of youngsters sticking as close to her as pilotfishes." I presume that what Beebe saw were probably small cobias following the blue shark.

PACIFIC SHARPNOSE SHARK

Rhizoprionodon longurio Jordan and Gilbert, 1882

Fig. 126a. *Rhizoprionodon longurio*, Pacific sharpnose shark, mature male, 97.5 cm, 4.5 kg; off Guaymas, Mexico (Fernando Márquez, collector).

Common name Pacific sharpnose shark, references to its habitat and long snout. In common usage it is referred to as simply "sharpnose shark."

Spanish name *Cazón bironche* (Pacific Mexico, A.F.S.).

Scientific name *Rhizoprionodon longurio,* Jordan and Gilbert, 1882, originally described as *Carcharias longurio* in *Proceedings of the United States National Museum* 5: 106–107. *Rhizoprionodon: rhizo* from the Greek *rhiza* or *rhizoma,* a root, and *prion,* Greek, a saw, plus odont, from Greek *odontus,* tooth, thus "root of the sawtooth," an allusion to being the ancestor of the genus *Prionodon; longurio:* Latin, a slender youth, a stripling.

Synonyms None.

Identification The Pacific sharpnose shark has a pointed, long snout (longer than the mouth width), well-marked *labial furrows around the corners of the mouth,* a *second dorsal fin origin above or behind the middle of the anal fin base,* and a *series of conspicuous hyomandibular pores on the side of the head just behind the angle of the mouth. Teeth have oblique, smooth-edged cusps* (with very slightly serrated edges in large specimens) and a deep notch on their outer margins and are similar in both jaws. They number U: 13 to 15–1–13 to 15, L: 13 or 14–13 or 14. The dermal denticles have little overlap; they have three strong ridges terminating in points, the central ones being longer. Coloration is grayish brown above, with metallic hues in life, shading to white below.

Similar species The well-marked labial furrows and the position of the second dorsal fin, its origin well behind the anal fin origin, distinguish it from other requiem sharks in the area.

Range The Pacific sharpnose shark inhabits the tropical eastern Pacific from southern California to Peru (Kato et al. 1967). The northernmost record is for San Diego County (Hubbs and Mc Hugh 1950). It is common in the Gulf of California and along the Mexican Pacific (the type locality is Mazatlan). It has been described as the most abundant requiem shark in the in the Gulf of California by Thompson et al. (1979). It has been reported as seasonally very abundant around Isla Isabel, off Nayarit, Mexico, and as absent from the adjacent Islas Marias (Pérez-Jiménez et al. 2005b). Beebe and Tee-Van (1941) reported it from Panama.

Size and weight This is a small and slender species. The weights of males and nongravid females do not differ appreciably. Representative sizes and weights are as follows: 50 cm, 0.5 kg; 70 cm, 1.4 kg; 90 cm, 2.0 kg; 100 cm, 4.0 kg. Beebe and Tee-Van (1941) gave the maximum size as 107 cm and 4.1 kg. Castillo-Géniz (1990) recorded gravid females of 116 cm.

Biology This little shark inhabits warm coastal waters, usually over soft or muddy bottoms in less than 40 m. It is often mentioned in faunistic works, but there is only one dedicated work on the species (Márquez-Farias et al. 2005), much of it based on a thesis by Castillo-Géniz (1990). Nevertheless, its habits remain poorly known. Roedel (1950) reported capturing it at depths of 10–11 m and temperatures of 22°C. Its diet consists of small crustaceans and small bony fishes.

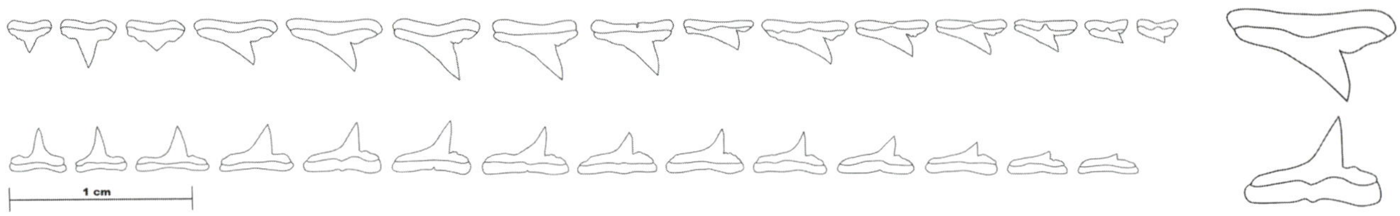

Fig. 126b. Teeth.

Migrations: The Pacific sharpnose shark is a migratory species that undertakes extensive, seasonal north–south movements along the Baja Peninsula. Kato and Hernandez Carvallo (1967:102) reported that one specimen "tagged in May at Cape San Lucas was recaptured 4 months later after it had traveled 600 miles northward on the western side of Baja California." According to Márquez-Farias et al. (2005), the Pacific sharpnose sharks in the Gulf of California move southward in winter, from southern Isla Tiburon in the central Gulf to the waters off Nayarit, Mexico. Pérez-Jiménez et al. (2005b) stated that the Pacific sharpnose shark appears in catches at Isla Isabel (off Nayarit) from mid-January on. Interestingly, these authors also reported that during their study a fisherman had pointed out that the Pacific sharpnose shark appears in large numbers off Isla Isabel only every two or three years, the year 2001 being one of the years of great abundance.

Size at maturity

Male: Males mature at about 93 cm based on clasper calcification (Márquez-Farias et al. 2005).

Female: Females usually mature at about 93 cm, although Márquez-Farias et al. (2005) cited two exceptionally small gravid females, 83 and 88 cm, respectively.

Reproduction The Pacific sharpnose shark is a placental viviparous species. Gestation is believed to last about 10–12 months (Márquez-Farias et al. 2005). The reproductive cycle is certainly annual, like other sharpnose sharks. Females give birth from April to June.

Size at birth: According to Márquez-Farias et al. (2005), the young are born at about 31 cm, although they also reported that the largest embryos reached 32 cm in February

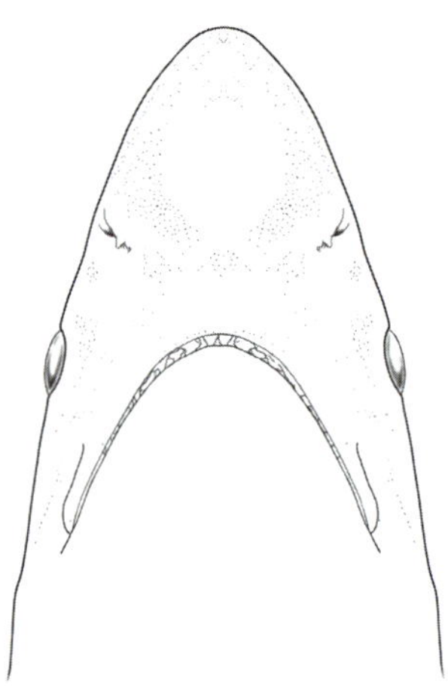

Fig. 126c. Snout.

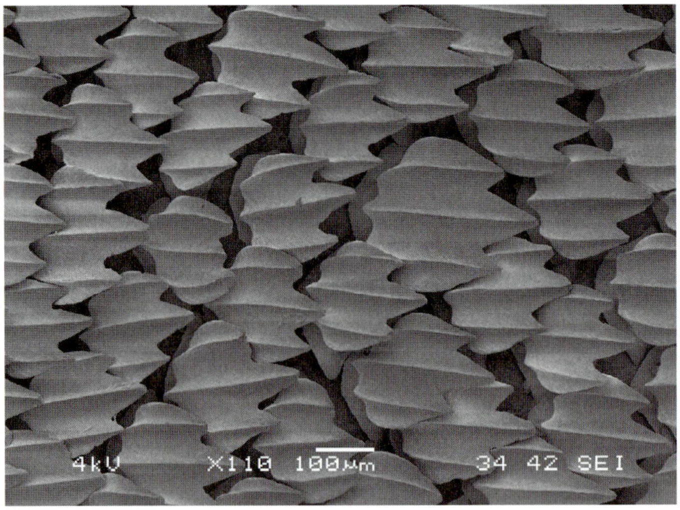

Fig. 126d. Dermal denticles.

and 37 cm in March. Pérez-Jiménez and Sosa-Nishisaki (2008) recorded neonate Pacific sharpnose sharks of 30–35 cm in the upper Gulf of California.

Brood size: Broods range from 1 to 12 embryos, with an average of 7.4 (Márquez-Farias et al. 2005).

Nurseries The nurseries of the Pacific sharpnose shark extend from the upper Gulf of California to the Nayarit coast, but, other than the report cited above (Pérez-Jiménez and Sosa-Nishisaki 2008), there is little documentation of this.

Age and growth There are no published age estimates for the Pacific sharpnose shark.

Relation to humans The Pacific sharpnose shark is one of the important species in the artisanal fisheries of western Mexico. It is an important component of the shark fishery in Mazatlan, Sinaloa, Mexico (Márquez-Farias et al. 2005). At Isla Isabel, Mexico, the Pacific sharpnose shark constituted 45% of the shark catch (Pérez-Jiménez et al. 2005b).

ATLANTIC SHARPNOSE SHARK

Rhizoprionodon terraenovae (Richardson, 1836)

Fig. 127a. *Rhizoprionodon terraenovae,* Atlantic sharpnose shark, mature male, 90.4 cm, 3.9 kg; off Saint Petersburg, Florida (Ernst Peebles, collector). This specimen had an unusual white-tipped lower caudal lobe.

Common name Atlantic sharpnose shark, sharpnose, a reference to its long snout. For many years the species was known simply as the sharp-nosed shark. When the Caribbean form was determined to be distinct from the Atlantic species, "Atlantic" was added to the name to distinguish it from the Caribbean form.

Spanish name *Cazón de ley* (Mexico, A.F.S.): from *cazón,* a small shark, and *de ley,* legal or lawful, meaning "the legal small shark," probably to distinguish it from the young of larger species; also *tripa, tutzun* (Mexico).

Scientific name *Rhizoprionodon terraenovae* (Richardson, 1836), originally described as *Squalus* (*Carcharias*) *terraenovae* in *Fauna Boreali Americana* 3: 289. *Rhizoprionodon: rhizo,* from the Greek *rhiza* or *rhizoma,* a root, and *prion,* a saw, plus odont, from *odontus,* tooth, thus "root of the sawtooth," an allusion to being the ancestor of the genus *Prionodon; terraenovae:* from Latin *terra,* earth, and *novae,* new, the name for Newfoundland, the supposed location of the type specimen. For many years the species was known as *Scoliodon terraenovae.* In 1964, V. G. Springer revised the genera *Scoliodon, Loxodon,* and *Rhizoprionodon* and placed the species in the latter genus.

Synonyms None.

Identification The Atlantic sharpnose shark is a small (less than 120 cm) and slender shark with snout longer than the mouth width, *long upper labial furrows,* a *conspicuous series of hyomandibular pores on the side of the head just behind the angle of the mouth, no spiracle visible,* and a *second dorsal fin originating over or behind the midpoint of the anal fin.* The *teeth are triangular, oblique, strongly notched on their outer margins,* smooth-edged (very weakly serrated in large specimens), and similar in both jaws. Teeth number U: 11 to 13–1–11 to 13, L: 12 or 13–2 or 13. The dermal denticles are much wider than long and are closely apposed with little overlapping. The denticles have five parallel ridges and points, the central ones being the longest. Coloration is brownish to olive gray, with metallic hues in life and a few *scattered white spots* above (adults only), shading to white below. The dorsal fins have black or dusky tips. Small specimens have black-edged dorsal and caudal fins.

Similar species The combination of a long snout, long labial furrows, and notched teeth distinguishes the Atlantic sharpnose shark from other carcharhinids in the area. The Atlantic sharpnose shark is replaced in the Caribbean by a very similar, closely related species, the Caribbean sharpnose shark. These two species can be separated only on the basis of vertebral counts; the Atlantic form has 58–65 precaudal vertebrae, while the Caribbean form has 65–75 precaudal vertebrae.

Range The Atlantic sharpnose shark is one of the most abundant sharks in eastern North American waters. It ranges from the Bay of Fundy to Yucatan. It is known from the Bay of Fundy by a single specimen collected in 1857, and it is a rare visitor to the Woods Hole area, Massachusetts (Bigelow and Schroeder 1948). It is only a straggler off the New Jersey coast (Breder 1929). The Atlantic sharpnose shark is most abundant from the Carolinas to Florida and the Gulf of Mexico. It is very abundant off North Carolina (Cape Lookout Shoals), except for a short period during winter and spring (Coles 1915). It is the most common summer shark along the southern part of the North Carolina coast and off South Carolina (Bigelow and Schroeder 1948). It is a year-round resident off South Carolina, being close to shore in the summer and migrating to deep waters in winter

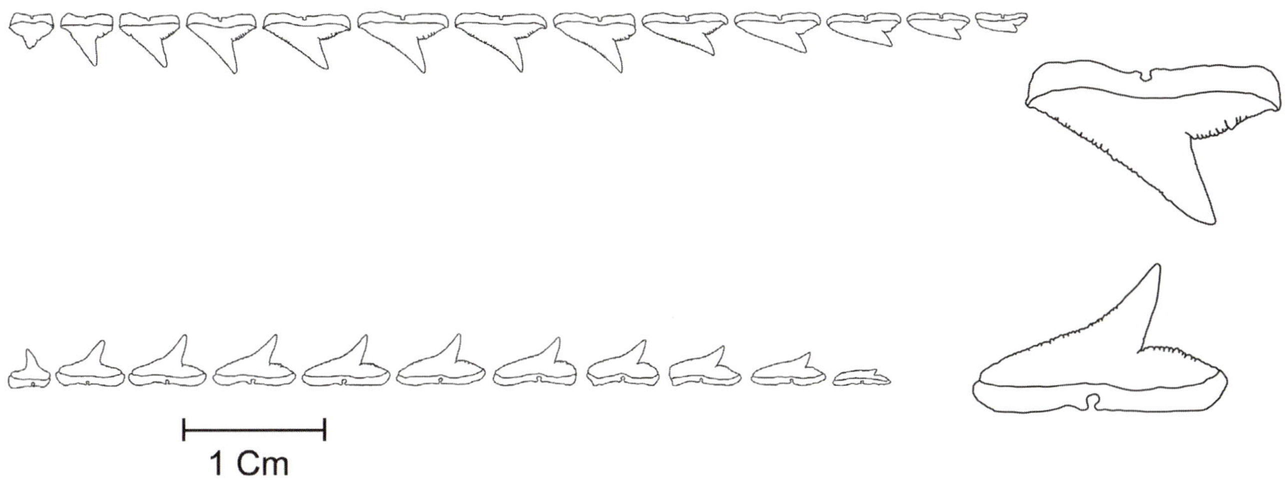

Fig. 127b. Teeth.

(Castro 1983). It is also present off Florida and the Keys throughout the year, and abundant in the Gulf of Mexico from June to September (Bigelow and Schroeder 1948). It has been reported from the Dry Tortugas (Gudger 1915b).

Size and weight The Atlantic sharpnose shark is one of the smaller coastal sharks in North America, the largest specimens reaching 104–108 cm. The weights of specimens I recorded in South Carolina are as follows:

Male: 49.5 cm, 0.37 kg; 61.0 cm, 0.77 kg; 70 cm, 1.3 kg; 79.6 cm, 1.68 kg; 90.5 cm, 2.24 kg; 100.3 cm, 3.73 kg; 101.6 and 5.1 kg; and 106.5 cm, 4.2 kg. This last male was the largest I examined.

Female: 48 cm, 0.6 kg; 53 cm, 0.8 kg; 67.2 cm, 1.2 kg; 86.2 cm, 2.9 kg; 90.1 cm, 3.4 kg; and 100 cm, 4.3 kg. Of 120 adult females that I examined off South Carolina, four measured 104 cm, and two reached 106 cm. The largest specimen reported from the Atlantic measured 108.8 cm (Loefer and Sedberry 2003).

Biology The Atlantic sharpnose shark is a common inhabitant of shallow coastal waters of the southeastern United

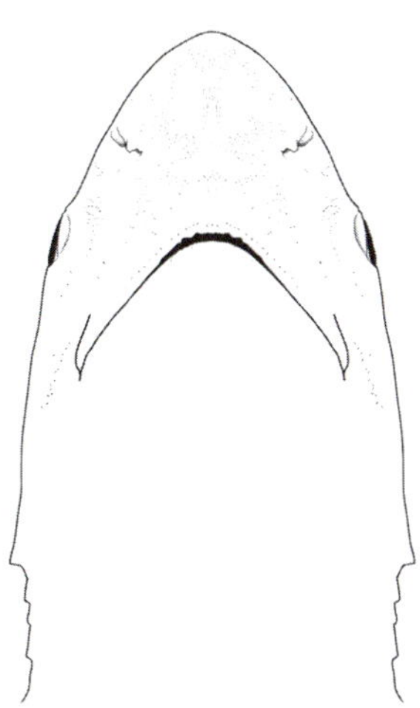

Fig. 127c. Snout.

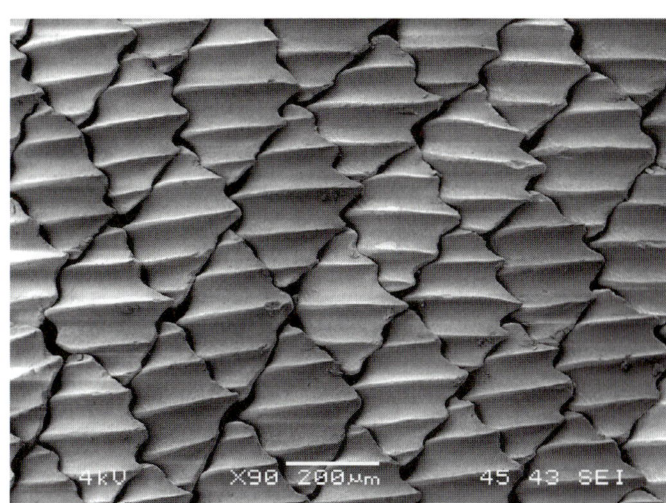

Fig. 127d. Dermal denticles.

States. The account given here is for the population found from North Carolina to southern Georgia in summer and comes from Castro and Wourms (1993). Large schools of males appear in the shallow waters of the Carolinas in depths of less than 2–4 m in mid-April, after the surface water temperature exceeds 20°C. Schools of females appear in shallow waters in mid-May. Juveniles born the previous summer appear in shallow waters about a month after the appearance of the females. Neonates and juveniles are abundant during the summer in energy-rich, shallow coastal and estuarine waters, where they feed on shrimp and small fishes.

During summer, adults are most abundant over "live bottoms" in 7–9 m, often occurring in very large schools. Except at mating time, adults are usually segregated by sex and size. Juveniles form schools composed of sharks of both sexes and similar size. Adults and juveniles remain in shallow waters less than 9 m deep until October. There is an interesting report by Backus (1957) of a 75-cm male taken at the surface in July 1956, about 100 km north of Barbuda, where the water depth is 6,035 m (I suspect an identification error).

Diet: The Atlantic sharpnose shark feeds on a wide variety of small bony fishes and crustaceans. Stomach content analysis of Atlantic sharpnose sharks I examined off South Carolina revealed bony fishes, penaeid shrimp, and stomatopods. Gelsleichter et al. (1999) examined the stomach contents of 129 Atlantic sharpnose sharks from the Chesapeake Bight region and reported that the diet consisted of bony fishes (57% by number) and crustaceans (34%). In this study, the smallmouth flounder (*Etropus microstomus*) occurred in 16% of the stomachs and was the most important bony fish prey.

Migrations The Atlantic sharpnose shark undertakes seasonal migrations, offshore and southward in autumn, and inshore and northward in spring, but these movements can only be described in broad terms. Although many Atlantic sharpnose sharks have been tagged along the eastern United States, few (1% or less) have been recovered, probably due to their small size or to the effects of tags on small sharks. The few tagging data available (Kohler et al. 1998) show north and south seasonal movements along the eastern coast of North America.

I have observed that both adults and juveniles remain in the summer grounds of Georgia and the Carolinas in waters less than 9 m deep until late October or early November, when they migrate to deeper water. Migration to deeper waters coincides with the lowering of the surface water temperature below 20°C. Schools are found in progressively deeper waters with the onset of winter. By January, schools of adults are encountered at depths of 27–36 m, although there is evidence that some Atlantic sharpnose sharks migrate southward (Kohler et al. 1998). As the waters warm in the spring, schools of Atlantic sharpnose sharks move gradually closer to shore and arrive at the 18 m depth contour in early April. By late May and early June, the Atlantic sharpnose shark is the most abundant shark off the Carolinas and Georgia.

Size at maturity

Male: Males mature at about 80 cm. In males I examined off South Carolina, the claspers begin to elongate when the animal reaches about 70 cm and are usually calcified by the time the animal reaches 80 cm (Castro and Wourms 1993). In rare cases, males as large as 91 cm had only partially calcified claspers. Other researchers have obtained similar results. Loefer and Sedberry (2003) determined that Atlantic males reached maturity at about 81.6 cm and three years of age. Parsons (1983) found that males in the northern Gulf of Mexico mature at about 80 cm.

Female: Females mature at about 81 cm. The smallest gravid female that I saw off South Carolina was an 82-cm specimen taken in May 1983 that was carrying two embryos, each 34.4 cm. Loefer and Sedberry (2003) examined 1093 Atlantic sharpnose sharks from the Atlantic coast and estimated that females mature at 81 cm and three years of age. There may be small differences between the Atlantic and the Gulf of Mexico populations. Parsons (1983) reported that, in the northern Gulf of Mexico, first ovulation occurred at a total length of 85–90 cm and that the smallest gravid females he observed were 87–89 cm; however, it appears that the number of specimens 80–85 cm he examined was small.

Reproduction The Atlantic sharpnose shark is a placental viviparous species. The reproductive tract of the female consists of a single ovary at the anterior end of the left epigonal organ, and two 12- to 15-cm-long oviducts that terminate anteriorly in a common ostium. Only the left ovary is functional.

Females bearing fresh mating bites are caught in coastal waters of South Carolina from late May to the end of July, in depths of 3.5–9 m. Recently mated females exhibit a series of single bite marks along one flank and multiple bite marks on the corresponding pectoral fin and axilla. Based on what is known about mating in sharks, these bite marks are the result of the male first grasping the female on the flanks and then moving its bite progressively toward the pectoral fin axilla. Once the male has grasped the female firmly at the pectoral fin axilla, copulation can ensue. In females with fresh bite marks, examination of simple sperm smear preparations revealed a great quantity of sperm in the oviducal gland and uterus.

The ovarian cycle begins in June, shortly after parturition, and proceeds concurrently with gestation. Undeveloped oocytes measure 1–2 mm in diameter. By late July, the developing oocytes measure 6 mm in diameter, attaining 10 mm by January at the midterm of gestation and reaching 18–19 mm by late May. Mating occurs in coastal waters from late May to mid July at depths of 6–9 m very, shortly after parturition. Ovulation occurs shortly after mating, at which time the oocytes have reached about 20 mm in diameter. For females just reaching maturity, ovulation occurs in late May or early June, and for postpartum females it occurs from mid-June to mid-July. Eggs are usually ovulated in pairs, one going down each oviduct and both moving through the oviducal glands simultaneously. At the time of ovulation, the oviducal gland enlarges to a width of 22 mm to allow egg passage (Castro 1988). The total number of eggs ovulated in a season usually ranges from two to seven, with the larger females having the largest broods. Ovulation may last anywhere from several days to perhaps several weeks, given that females carrying embryos 3–5 mm in the uteri have been observed ovulating other eggs. By middle to late July, nearly all mature females are carrying early stage embryos (Castro 1988).

During early gestation, the fertilized eggs are free in the uteri and embryos derive their nutrients from the large quantity of yolk stored in their yolk sacs. During early development, folds of the uterus wrap around each developing embryo, forming separate compartments or "crypts" for each embryo, first described by Gudger (1915b). Embryos as small as 3 mm can be found in well-formed, separate uterine compartments. However, in some cases, embryos 25–30 mm can be found still free in the uterus, indicating that the partitioning of the uterus into discrete compartments can sometimes be delayed (Castro and Wourms 1993).

In the Atlantic sharpnose shark, as in all placental sharks, an umbilical cord connects the embryo to the yolk-sac placenta. In most sharks the umbilical cord is smooth and unadorned. In the Atlantic sharpnose shark, mid- to late-term embryos possess an umbilical cord densely covered with elongated, flattened, dendritic filaments known as appendiculae (Alcock 1890). These structures were illustrated by Budker (1958). Development of the appendiculae begins at about 40 mm with the appearance of spiral folds on the umbilical cord. In 55-mm embryos, appendiculae appear as small knobby protuberances that cover the yolk stalk. In the 70- to 85-mm embryos, the appendiculae have developed into branched, filamentous structures. Their development continues through most of the gestation period. Wourms et al. (1988) reviewed appendicular function and concluded that their primary roles are nutrient absorption (of uterine milk or histotrophe), gas exchange, or both. Hamlett (1986) stated that the ultrastructural organization of the appendiculae and preliminary uptake experiments supported the hypothesis that they are nutrient-absorbing structures.

The development of the placenta begins with the vascularization of the yolk sac and the uterine wall at their point of contact, some five weeks after fertilization, when the embryos measure 40–55 mm. Implantation occurs when the embryos measure 70–85 mm, or at about 8–10 weeks of age (Castro and Wourms 1993).

Parsons (1983) estimated the length of the gestation period at 10–11 months for the Gulf of Mexico population, while Castro and Wourms (1993) estimated it at 11–12 months for the Atlantic population. This may be due to differences between the Gulf of Mexico and Atlantic populations of the species, or to Parsons not including very early stage embryos in his graph or calculations. There is also some uncertainty in estimates of the duration of gestation because the smaller, nongravid females, having just reached maturity, mate while the older gravid females are still carrying embryos. These smaller females mate two or three weeks before larger females give birth and subsequently mate. Presumably, those females that mate earlier will give birth earlier in the season the following year. Parturition takes place in shallow coastal waters from the third week in May to the third week in June; ovulation and mating follow closely thereafter. Thus, the Atlantic sharpnose shark has an annual reproductive cycle with concurrent ovarian and gestation periods.

Size at birth: Atlantic sharpnose sharks are usually born at 29–34 cm. Baughman and Springer (1950) stated that the young are born in the late spring and summer in the northern Gulf of Mexico at a size of 28–30 cm. Parsons (1983) stated that the young were born in the northern Gulf of Mexico at an average total length of 32 cm. Off South Carolina, the largest embryos that I examined measured 32–34.1 cm, while the smallest free-swimming young measured 29.2 cm. This last specimen was caught 12 June 1982 in a shrimp trawl in depths of 4.5–6 m off Folly Beach, South Carolina, being one of 13 newborn young 29.2–35.0 cm (average 31.8 cm) taken in that trawl. Loefer and Sedberry (2003) reported an unusually small free-swimming young 27.2 cm.

Brood size: Females usually carry four to seven young per brood. Term females stressed by capture often abort embryos; these embryos invariably emerging tail first. The average number of embryos in a sample of 60 gravid females was five embryos per female (Castro 1988). The maximum number of young reported in a brood is eight (Loefer and Sedberry 2003). Bigelow and Schroeder (1948) reported that pregnant females with as many as 12 embryos are taken around Cuba, based on a personal communication from L. Howell-Rivero. The sharpnose sharks in Cuban waters are considered to be a very closely related but different species, *Rhizoprionodon porosus,* the Caribbean sharpnose shark (Guitart Manday 1968), and their brood size is similar to those of the Atlantic sharpnose. Thus, it is likely that the specimens referred by Howell-Rivero belonged to another genus.

Nurseries The Atlantic sharpnose shark has its nurseries in the shallow coastal waters and estuarine waters from North Carolina to east central Florida during the summer months (Castro 1993b). The species is one of the most abundant sharks in North Carolina estuaries from May to October, and the young are born or encountered there starting in May (Schwartz and Burgess 1975). Females carrying embryos larger than 32 cm are found in shallow coastal waters of South Carolina from the third week in May to the third week in June. Neonates with fresh or healing umbilical scars can be found there from the last week in May to June (Castro 1993b). Gravid females carrying term young can be found off Daytona Beach, Florida, during the last week in May (Castro 1993b), and neonates still showing traces of an umbilical scar have been reported from Florida in July by Bigelow and Schroeder (1948). The location of the nursery areas in winter, or where the young of the year spend their first winter, is unknown.

Age and growth According to Parsons (1983), sharpnose sharks are born at about 32 cm in the northern Gulf of Mexico; in their first year they grow about 30 cm, and about 15 cm the following year. According to Branstetter (1987c), based on laboratory reared sharks injected with tetracycline and field studies, males reach maturity entering their third year of life, just prior to the mating season, while females reach maturity entering their fourth year of life, also just before the mating season. He estimated growth rates of 20–30 cm/year for neonates, 15 cm/year for one-year-old fish, 10 cm/year for two-year-old fish, and 5–7 cm/year for sharks that have reached maturity. According to these studies, Atlantic sharpnose sharks at their maximum size of 105–110 cm should be 8–10 years old.

Relation to humans Due to its abundance, the Atlantic sharpnose shark is economically important in the southeastern United States. Very large numbers of sharpnose sharks are caught in shrimp trawls from North Carolina to the Gulf of Mexico, yet the species appears to be one of the most resilient shark species. In the Gulf of Mexico, 1–2.5 million sharpnose sharks have been harvested each year from 1972 to 1994 (National Marine Fisheries Service 1995), apparently without causing the severe population declines evident in other shark species. The Atlantic sharpnose shark is of little interest to commercial shark fisheries because of its small size, but it is a favorite of sports anglers of the Carolinas who take many on very light tackle. Party boats operating from Texan ports used to catch hundreds daily. It is also a favorite dish of many (including the author). Adams and McMichael (1999) reported that total mercury levels for Atlantic sharpnose sharks larger than 50 cm precaudal length were always greater than 0.5 ppm.

WHITETIP REEF SHARK

Triaenodon obesus (Rüppell, 1837)

Fig. 128a. *Triaenodon obesus*, whitetip reef shark, female, 199.5 cm, 10 kg; from off Hawaii (Nick Whitney, collector).

Common name Whitetip reef shark, blunthead shark (South Africa).

Spanish name *Cazón coralero trompacorta* (A.F.S.).

Scientific name *Triaenodon obesus* (Rüppell, 1837), originally described as *Carcharias obesus* in *Neue Wirbelthiere zu der Fauna von Abyssinien gehorig:* 64–65, plate 18, fig. 2. *Triaenodon: triaen,* from the Greek *triaina,* a trident, and *odon,* Greek tooth, a reference to the three-pointed teeth of the species that resemble a trident; *obesus,* Latin, fat or plump—according to Rüppell, the blunt head gives the shark a plump appearance. For many years the genus *Triaenodon* was included among the Triakidae; however, today most authors include it in the Carcharhinidae.

Synonyms *Triaenodon apicalis* Whitley, 1939.

Identification The whitetip reef shark is a slender shark with a broad and flattened head with a *very short snout (its length is less than half of the mouth width)*; a *white-tipped first dorsal fin originating well behind the pectoral fin* (the midpoint of its base much closer to the pelvic fin origin than to the pectoral axil); a *large second dorsal fin* (its base is about three-fourths as long as the first dorsal fin base); and large conspicuous nasal flaps. The teeth are set in at least two functional rows. The teeth have a large, slender, smooth-edged, central cusp with one small cusp on each side. The frontal teeth are erect, but they become increasingly oblique toward the corners of the mouth. Teeth number about U: 22–1 or 2–22; L: 20 to 22–1 or 2–20 to 22. The teeth on both sides of the symphysis are much smaller than the others. The dermal denticles are broad and closely apposed with little overlapping. They have five parallel ridges with a single rounded point on the rear edge. Coloration is dark brownish or greenish gray above, with a few black blotches on the trunk, and light gray or whitish below. The *tips of the first dorsal fin and the upper lobe of the caudal fin are white.* The second dorsal fin is often white tipped also.

Similar species The short snout, the first dorsal fin set well behind the pectoral fins, and white-tipped first dorsal and upper caudal lobe differentiate the whitetip reef shark from the other carcharhinid sharks in the area.

Range The whitetip reef shark inhabits shallow coral reefs of the warm waters of the Indian and Pacific Oceans. This shark is so ubiquitous, that its range forms a wide belt from about 30° N latitude to 30° S latitude throughout the warm tropical waters of the Indo-Pacific region. It ranges from the eastern Africa coast (Northern Natal, southern Mozambique, Mauritius, Seychelles; Bass et al. 1975c), Comoros and Madagascar (Fourmanoir 1961), and the Red Sea (Rüppell, 1837), eastward through the warm tropical belt to northern Australia (Last and Stevens 1994) and southern Japan (Taniuchi 1975), through the numerous islands of the southwestern Pacific and Hawaii (Randall 1977), the Galapagos (Kato et al. 1967), and to the eastern Pacific. For detailed locality records of the species in the tropical Pacific, see Randall (1977).

There are few records of the whitetip reef shark being present in the eastern Pacific, where it has been reported from Islas Marias (off the Nayarit coast of Mexico) (Pérez-Jiménez et al. 2005b), the Revillagigedo Archipelago (Ochoa-López et al. 1997), and Panama (Kato et al. 1967). It is a rare catch at Islas Marias, its northernmost record in the eastern Pacific, but is apparently common around the Revillagigedo Islands.

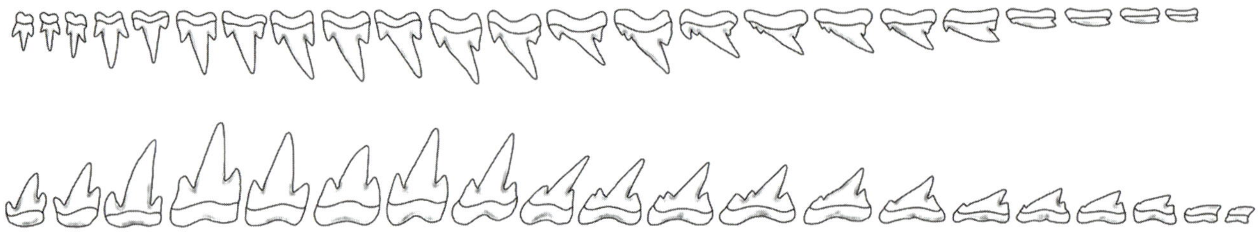

Fig. 128b. Teeth.

Size and weight The whitetip reef shark is a medium-size species reaching about 170 cm and 28 kg. Fourmanoir (1961) gave the maximum length of the species as 165 cm, based upon his examination of fishermen's catches in Madagascar. Randall (1977) wrote that the largest of 188 whitetip reef sharks examined by him was a 154.7-cm, 27.7-kg male, and that another specimen, taken off Johnston Island had measured 168 cm and weighed 23.6 kg.

Biology According to Randall (1977), who wrote the most comprehensive account of the species to date, the whitetip reef shark is a bottom-dwelling species usually found around shallow coral reefs. Although the species may be found at depths of a meter or less, it does not often enter shallower reefs and sand flats as often as the blacktip reef shark (*Carcharhinus melanopterus*). The whitetip reef shark often congregates under ledges or in caves in daytime (Taniuchi 197; Randall 1977). A given shark is often seen in the same cave, and Randall (1977) suggested that a shark has a home cave to which it may return after foraging. Observers have noted that this is a nimble and agile shark with a remarkable ability to enter small holes. Hobson (1963: 177) noted, "Large whitetips were seen disappearing into small holes from which they presently emerged, always head first."

The whitetip reef shark feeds on numerous reef fishes, octopuses, and crustaceans. Both Hobson (1963) and Randall (1977) have commented on the ability of this shark to capture prey hiding deep in a hole by inserting its head and wriggling into it. Hobson (1963: 177) noted that the species was "clumsy and ineffective in attempts at taking baits which were suspended in midwater. However, this same shark was remarkably effective in tracking down and capturing prey which had taken shelter deep in one of the many holes or crevices typical to a coral reef" and thus capturing prey that was not available to other species of sharks.

Size at maturity There are few reports of the size at maturity of the whitetip reef shark. Fourmanoir (1961) reported that the whitetip reef shark attains maturity at a length of one meter. However, in Randall's (1977) sample, the smallest

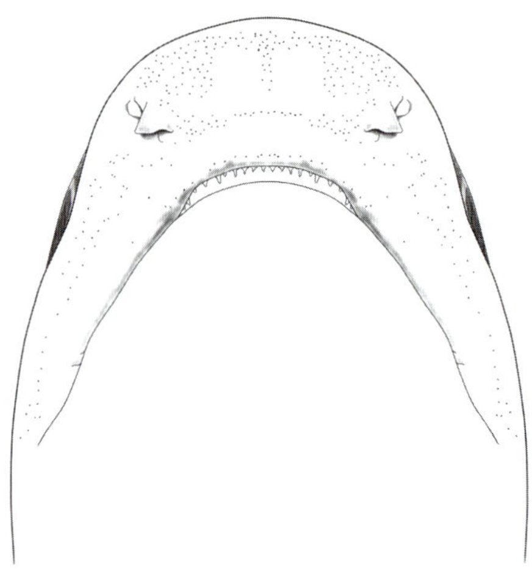

Fig. 128c. Snout.

Fig. 128d. Dermal denticles.

mature male measured 105 cm, while his smallest gravid female measured 129 cm (102 cm precaudal length).

Reproduction The whitetip reef shark is a placental viviparous species. Tricas and Le Feuvre (1985) observed copulating whitetip reef sharks in shallow waters of the Hawaiian Islands. The sharks copulated in a stationary position, with the male holding the female's pectoral fin in its mouth and the bodies angled upward at a 45-degree angle. Uchida et al. (1990) reported on the mating of captive whitetip reef sharks at the Okinawa Expo Aquarium. On 27 June 1983, two six-year-old, captive-born whitetip reef sharks from the same brood copulated with each other. Thirteen months after this copulation, the female gave birth to one young. Three years after that birth, the same female gave birth again, this time to two stillborn young and one that died the next day. Two hours after parturition, the female mated again (Uchida et al. 1990). Whitney et al. (2004) described a group courtship or mating behavior where several males competed to mate with a female.

Based on Randall's (1977) report of 16 out of 31 (adult) females containing developing young or ova, it is likely that the species has a biennial reproductive cycle, where half the adult females would be gravid and the other half would be postpartum and resting.

Size at birth: The young of the whitetip reef shark measure 52–60 cm at birth (Fourmanoir 1961). Randall (1977) reported that the largest embryos removed from females were 59–60 cm, and these sizes coincide with the size of the smallest free-swimming specimens taken and those of young born in captivity. Young born in captivity at the Okinawa Expo Aquarium were slightly larger, 59–69 cm, and averaged 63.4 cm (Uchida et al. 1990).

Brood size: Randall (1977) reported that 13 pregnant females carried from one to five embryos, the average being three.

Nurseries The nurseries of the whitetip reef shark have not been described. Small specimens and neonates can be found hidden in coral reefs, suggesting that they are dropped there by their mothers.

Age and growth The whitetip reef shark has not been aged.

Relation to humans Hobson (1963) and Randall (1977) cited a few cases of whitetip reef sharks being attracted to the powerful stimuli of speared fishes and then taking the wounded fish, or attempting to bite, or biting, the spear fisherman. Mike Heithaus (pers. comm., July 2006), reports that, at many Indo-Pacific locations, the whitetip reef shark is a popular attraction for divers who hand-feed the usually docile sharks. The whitetip reef shark is often exhibited in aquaria, where it survives for more than 10 years (G. Crow, Waikiki Aquarium, pers. comm., 21 February 2006), and successive generations of captive sharks have been produced (Uchida et al. 1990).

FAMILY SPHYRNIDAE

The Hammerhead Sharks

The flat, wide, hammer- or shovel-shaped heads of these sharks is so characteristic that they cannot be confused with any other sharks. These medium to large sharks of the coastal areas of the tropical and warm-temperate seas are considered to be the most advanced sharks, along with the closely related requiem sharks. They are characterized by a flat and wide head, eyes with a well-developed nictitating eyelid, and bladelike teeth with a single cusp, usually curved toward the corners of the mouth. Various reasons have been given for the evolution of the laterally expanded head, including increased agility when turning, an enlarged area for sensory perception, separation of the sensory organs, and prey handling (Kajiura 2001; Kajiura et al. 2005). Their development is viviparous. Hammerheads feed on rays, sharks, and many other fishes. Many species are migratory, often moving in large schools. They are economically important as food and game fishes, and some are considered dangerous to man. The family comprises at least nine species, seven or eight of which are found in North America. There are undescribed small hammerheads in the tropical Americas.

Several authors have suggested that hammerheads have a monophyletic origin (Gilbert 1967; Martin 1993). If they are indeed monophyletic, their reproductive adaptations suggest an early separation into two groups with different reproductive strategies. One group, comprising the golden hammerhead, the scoophead, the scalloped hammerhead, and the bonnethead, has an annual reproductive cycle and appendiculae on the umbilical cord (just like in sharpnose sharks *Rhizoprionodon*). The other group, which includes the great and smooth hammerheads, has a biennial reproductive cycle and smooth umbilical cord (just as in the genus *Carcharhinus*).

The taxonomy of hammerhead sharks has fluctuated considerably in recent times. The scientific names used here follow the latest revision of this family (Gilbert 1967); these names differ from those used in older literature. The identification of some species remains problematic, and confusion exists about the "scalloped hammerheads," all large species with scalloped anterior margins of the head and pelvic fins with straight rear margins. Springer (1940a) recognized two forms, *Sphyrna lewini* from the Pacific Ocean and *S. diplana* from the Atlantic. Later, these forms were synonymized (Fraser-Brunner 1950; Gilbert 1967). Cadenat (1950) described *S. couardi,* or whitefin hammerhead, from the eastern North Atlantic (Senegal). This large hammerhead also has a scalloped anterior margin of the head and pelvic fins with straight rear margins. Compagno (1984) recognized *S. lewini* as a cosmopolitan species and *S. couardi* as inhabiting the eastern Atlantic. Given the presence of the whitefin hammerhead in the eastern North Atlantic and wide ranges of the larger hammerheads, it is certainly possible, and even likely, that it is found in the western North Atlantic as well but has gone unrecognized. Recently, Quattro et al. (2006) reported that they found genetic evidence that there are two species of large scalloped hammerhead sharks in South Carolina waters, referring to one species as *S. lewini* and to the other as a "cryptic species." Whether the "cryptic species" is the whitefin hammerhead, *S. couardi,* or an undescribed species remains to be determined.

KEY TO THE SPHYRNIDAE

1a. Shovel-shaped head, anterior margin of head evenly rounded between the eyes, sometimes slightly pointed along the median line but lacking a marked central indentation 2

1b. Head laterally expanded but not shovel-shaped, anterior margin of the head undulated, curved, or nearly straight....... 4

2a. Head width 16–20% of total length (first dorsal fin free rear tip well ahead of a perpendicular over the origin of the pelvic fins, body with small scattered black spots)—see Fig. 133a........ **Bonnethead,** *Sphyrna tiburo* (p. 521).

2b. Head width 23–28% of total length 3

3a. Snout length about equal or only slightly longer than mouth width........ **Scoophead,** *Sphyrna media* (p. 514).

3b. Snout much longer than mouth width, usually more than 40% of head width (a small hammerhead of the tropical eastern Pacific) .... **Scalloped bonnethead,** *Sphyrna corona* (p. 507).

4a. Midpoint of first dorsal fin about equidistant from pectoral axil and origin of pelvic fins, coloration in life or when fresh bright yellow or orange on flanks, head, and fins .... **Smalleye or golden hammerhead,** *Sphyrna tudes* (p. 525).

4b. Midpoint of first dorsal fin much closer to pectoral axil than to pelvic fin origin; coloration in life grayish, brownish, or black 5

5a. Pelvic fins with strongly falcate rear margins, teeth with serrated edges (adults with anterior margin of head nearly straight with a central indentation) .... **Great hammerhead,** *Sphyrna mokarran* (p. 516).

5b. Pelvic fins with straight rear margins, not falcate; teeth smooth edged 6

6a. Anterior margin of head rounded, lacking a central indentation .... **Smooth hammerhead,** *Sphyrna zygaena* (p. 529).

6b. Anterior margin of head scalloped, with a central indentation 7

7a. Precaudal vertebrae 189–204 .... **Scalloped hammerhead,** *Sphyrna lewini* (p. 509).

7b. Precaudal vertebrae 162–179 .... **Cryptic hammerhead,** *Sphyrna* sp. (p. 505).

SCALLOPED BONNETHEAD

Sphyrna corona Springer, 1940

Fig. 129a. *Sphyrna corona,* Scalloped bonnethead shark, LACM # Golfo de Nicoya, Costa Rica. CK THIS

Common name Scalloped bonnethead (A.F.S.), a poorly chosen name, as the anterior margin of the head is not scalloped.

Spanish name *Cornuda coronada* (A.F.S.).

Scientific name *Sphyrna corona* Springer, 1940, in *Stanford Ichthyological Bulletin* 1(5): 163–164, fig. 4. *Sphyrna:* a corruption of the Greek name *sphyra,* a hammer or a mallet, an allusion to the shape of the head and, according to Jaeger (1962: 244), "one of Rafinesque's efforts at word-butchery"; *corona,* Latin for crown.

Synonyms None.

Identification This small hammerhead is characterized by a broadly rounded anterior margin of the head, occasionally with a shallow central indentation; *a very wide head,* its width 23–29% of the total length; *a snout length* (from the anterior margin of the head to the front of the mouth) more than 40% of the greatest head width, with *no inner narial groove;* and *a free rear tip of the first dorsal fin that reaches the level of the pelvic fins.* The upper teeth have oblique, smooth-edged cusps with a notch on their outer margins; the lower teeth are erect and smooth edged. Teeth number U: 14–14, L: 14–1–14. The dermal denticles are small and closely overlapping. The denticles have three to five points, the three central points being much larger. Coloration is gray or grayish brown above and paler below, with fins with a yellowish tinge but no dark markings. This is the smallest of the hammerheads. Most of the specimens examined have been less than 60 cm.

Similar species The scoophead has a shorter snout and anal fin with a more curved rear margin and more pointed apex.

Range Of all the hammerhead sharks, the scalloped bonnethead has one of the most restricted ranges. It occurs in the tropical eastern Pacific from southern Mexico (Chiapas) to at least the coast of Colombia (Gilbert 1967). Most of the museum records I have found are for Panama and Colombia. None of the Mexican biologists that I have contacted have ever seen a specimen. Whether this scarcity is because the species is not caught in the fisheries sampled by biologists, only rarely strays into Mexico, or has been extirpated from the area is unknown.

Size and weight The 67.2-cm holotype is the largest specimen reported. The species probably reaches about 90 cm.

Biology This is a poorly known small hammerhead that inhabits tropical coastal waters of the eastern Pacific. It is morphologically similar to the golden hammerhead (*Sphyrna tudes*) of the tropical Atlantic, suggesting that the two species may be the result of an ancestral species split into two groups by the appearance of the Isthmus of Panama. Nothing else is known of its habits.

Size at maturity No data available.

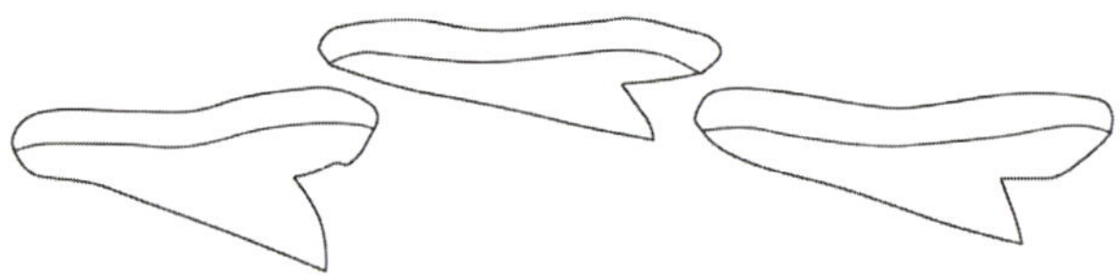

Fig. 129b. Teeth (upper).

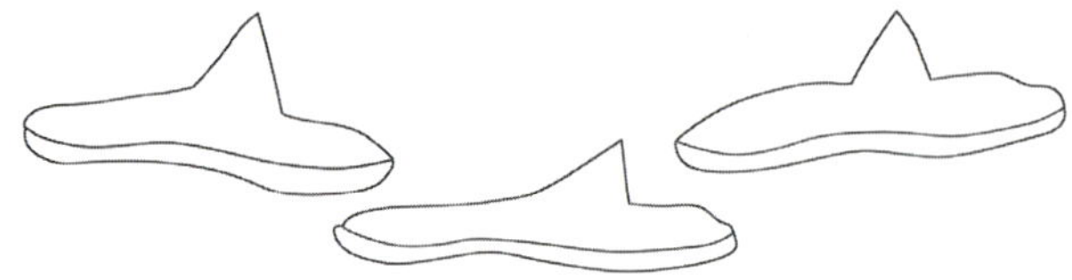

Fig. 129c. Teeth (lower).

Reproduction No data available.

Size at birth: Unknown. Gilbert (1967) mentions a 23-cm specimen in his study materials, presumably a free-swimming specimen. This suggests birth at a size smaller than 23 cm.

Brood size: Unknown.

Nurseries Unknown.

Age and growth The species has not been aged.

Relation to humans None.

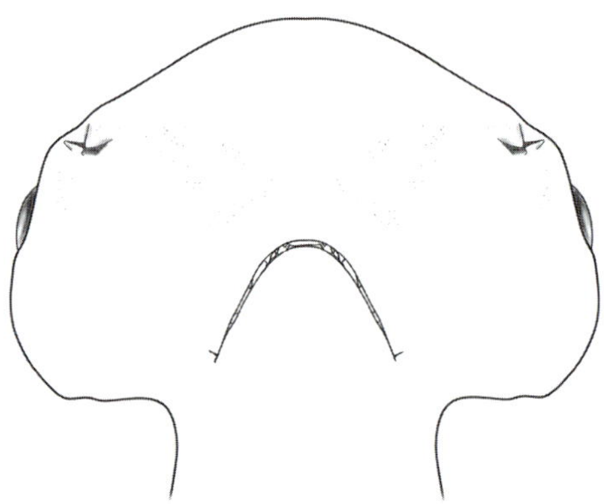

Fig. 129d. Snout.

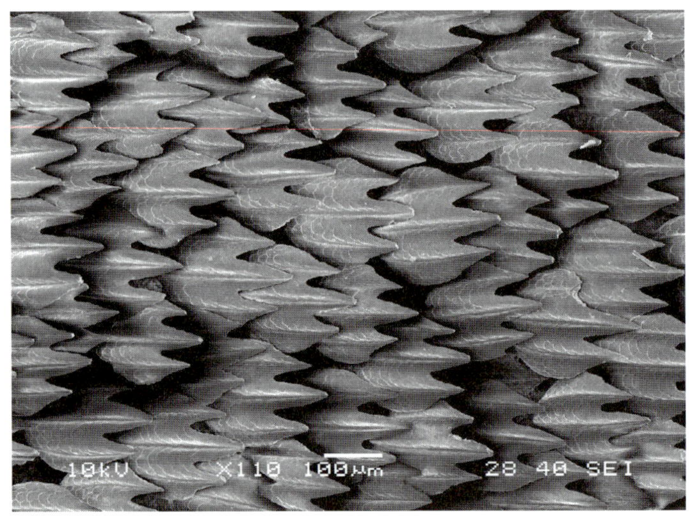

Fig. 129e. Dermal denticles.

SCALLOPED HAMMERHEAD

Sphyrna lewini (Griffith and Smith, 1834)

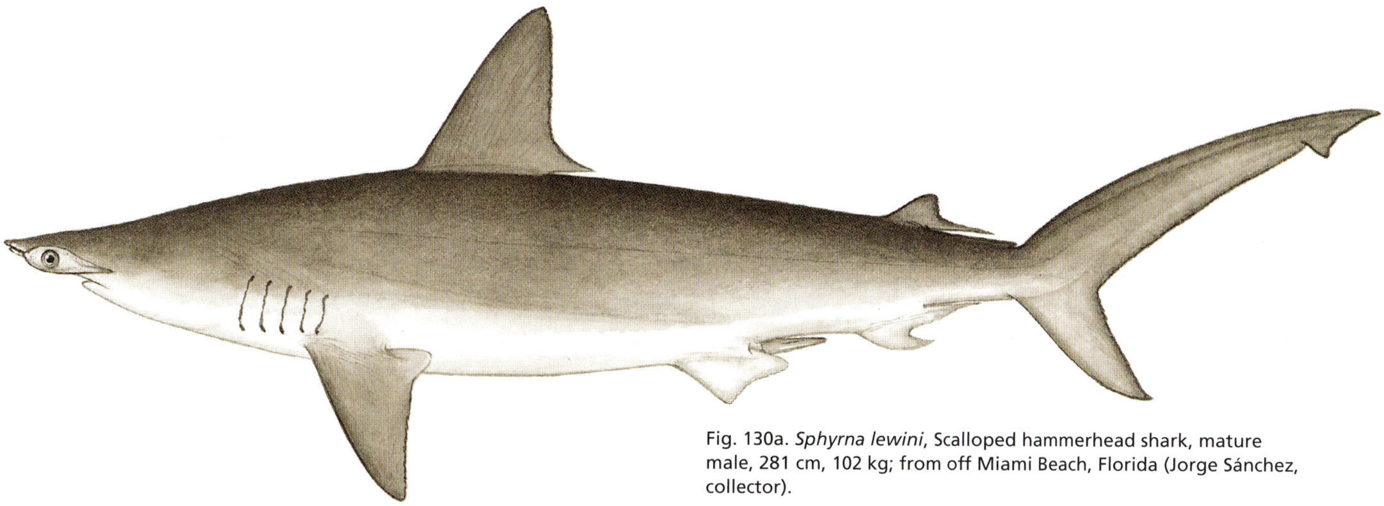

Fig. 130a. *Sphyrna lewini*, Scalloped hammerhead shark, mature male, 281 cm, 102 kg; from off Miami Beach, Florida (Jorge Sánchez, collector).

Common name Scalloped hammerhead, a reference to the highly scalloped anterior margin of the head.

Spanish name *Cornuda* (Cuba), *cornuda común* (Mexico, A.F.S.)

Scientific name *Sphyrna lewini* (Griffith and Smith, 1834), originally described as *Zygaena lewini* in Cuvier's *Règne Animal* 10: 640. *Sphyrna:* a corruption of the Greek name *sphyra,* a hammer or a mallet, an allusion to the shape of the head and, according to Jaeger (1962: 244), "one of Rafinesque's efforts at word-butchery"; *lewini:* latinized form of the name Lewin. Presumably named after Ludwig Lewin Jacobson (1783–1843), a brilliant Danish military surgeon and one of the foremost anatomists of his age, a time when the nascent comparative anatomy under Georges Cuvier was the main basis for the study of biology. Jacobson's anatomical researches led him to discover what is now known as the vomeronasal organ of Jacobson, an accessory olfactory organ of many tetrapods. Cuvier published a description of this organ in 1811.

Synonyms *Sphyrna diplana* Springer, 1940b. There are many doubtful or possible synonyms—for a comprehensive list, see C. R. Gilbert (1967).

Note: Until very recently, the scalloped hammerhead was considered to be the only large hammerhead with a scalloped anterior margin of the head in North America. Quattro et al. (2006) have indicated that it is likely there are two species of large scalloped hammerheads off the southeastern United States. Thus, it is possible that reports of the scalloped hammerhead may include data taken on another similar species, possibly the whitefin hammerhead (*S. couardi*).

Identification The scalloped hammerhead is characterized by a *marked central indentation on the anterior margin of the head,* which gives it a "scalloped" look; a first dorsal fin rear tip that does not reach the level of the pelvic fins; pelvic fins with nearly *straight rear margins;* and a *low and long second dorsal fin* (its length being about twice the height of the fin) reaching almost to the precaudal pit. The teeth are triangular, smooth edged or very finely serrated, and similar in both jaws. The central teeth are erect, while subsequent ones have oblique cusps. Teeth number U: 15 or 16–0 to 2–15 or 16, L: 15 or 16–1 or 2–15 or 16. The dermal denticles are very wide and overlap greatly. The denticles have eight parallel ridges and points of decreasing size. Coloration is deep olive to brownish gray above, shading to white below. The *ventral tips of the pectoral fins are dusky;* their coloration darkens with age, becoming black in large individuals. A 62-cm albino specimen has been reported from South Carolina waters (McKenzie 1970).

Similar species The whitefin hammerhead has unmarked pectoral fin tips. The great hammerhead has teeth with serrated edges and pelvic fins with curved rear margins. The smooth hammerhead lacks the central indentation on the anterior margin of the head. In the golden hammerhead,

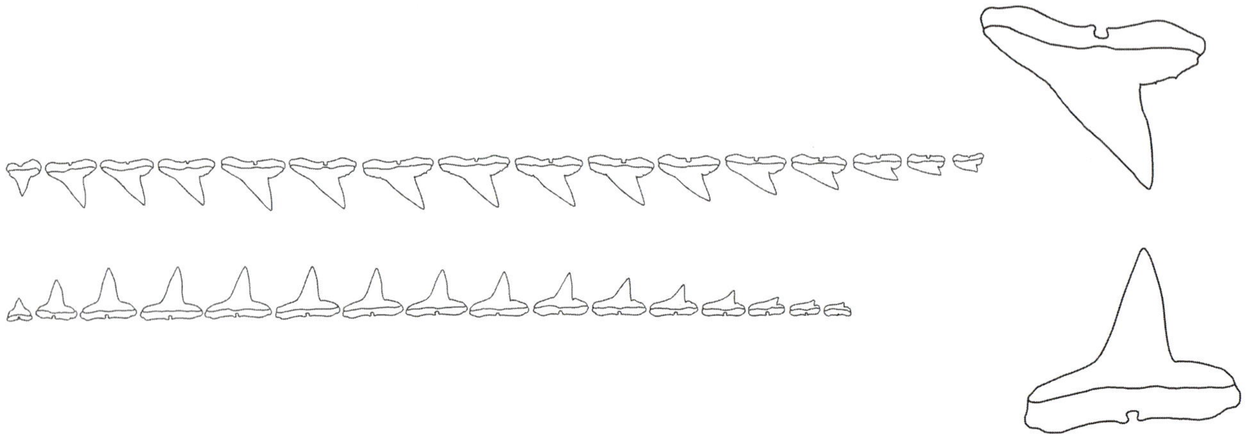

Fig. 130b. Teeth.

the scoophead, and scalloped bonnethead, the rear tip of the first dorsal fin reaches the level of the pelvic fins. The bonnethead has a shovel-shaped head.

Range The scalloped hammerhead is circumtropical in coastal and oceanic waters. In North American waters, it ranges from New Jersey to Florida and the Gulf of Mexico and from southern California southward. The northernmost record along the eastern coast of North America is a specimen caught about 90 miles off Cape May, New Jersey (Bigelow and Schroeder 1948). Older locality records north of Cape May probably refer to the smooth hammerhead. The scalloped hammerhead is common along the Virginia coast in summer, but it rarely enters the Chesapeake Bay (Murdy et al. 1997). It is one of the most common summer sharks in estuarine waters of the Carolinas and Georgia (Castro 1993b). It is also abundant year-round off coastal Florida, being one of the most abundant species off Melbourne Beach (Dodrill 1977). It is also one of the most common offshore sharks along Florida's eastern coast. Berkeley and Campos (1988) reported that the scalloped hammerhead was the most abundant species (32.5%) in the shark bycatch of the commercial swordfish fishery operating in waters deeper than 182 m (100 fathoms) along the eastern coast of Florida in 1981–1983.

Along the western coast of North America, the scalloped hammerhead is found from California (where it is rare) southward to the tropics. There are only a couple of records

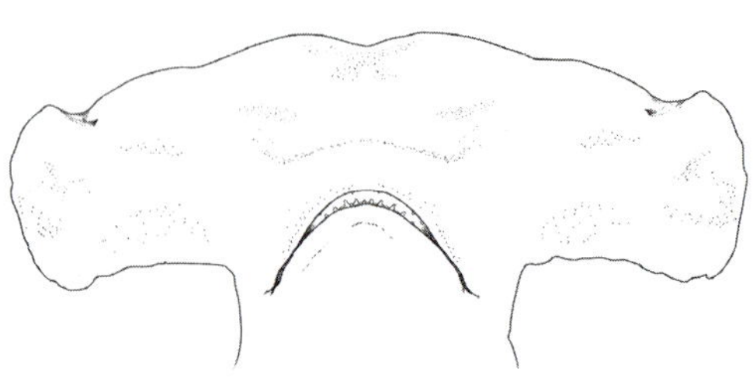

Fig. 130c. Snout.

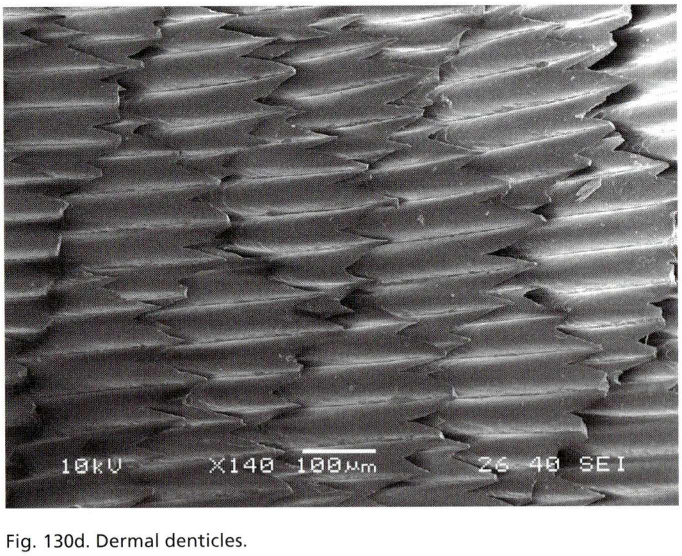

Fig. 130d. Dermal denticles.

of its presence off California. Fusaro and Anderson (1980) reported a 262-cm female caught some 2 km offshore near Santa Barbara, California. A photograph of this animal shows a peculiarly shaped head, and the identity of this animal must be considered uncertain. A few years later, Seigel (1985) reported a 54.5-cm juvenile caught in early October 1984 off Pacific Palisades, California, and mentioned two other unpublished records of scalloped hammerheads taken near Goleta Point. The scalloped hammerhead is common in the Gulf of California, and I have seen dozens captured by the migrant fishermen at El Barril, Baja California.

Size and weight The scalloped hammerhead is one of the larger hammerheads, reaching about 350 cm, and perhaps larger. The largest scalloped hammerhead that I have examined was a gravid female, 326 cm and 200 kg, caught off Miami by Mark Quartiano in January 2004. I have been informed of another larger specimen caught by the same angler, but I did not see it. According to Compagno (1984b), the scalloped hammerhead reaches lengths of 370–420 cm. Due to confusion with the larger great hammerhead, reports of scalloped hammerheads larger 330 cm must be taken with caution, until the species identity is well established. Representative sizes and weights are as follows:

Male: 141 cm, 13.5 kg; 195 cm, 29.4 kg; 218 cm, 54.6 kg; and 272 cm, 104 kg.

Female: 40 cm, 0.28 kg; 95 cm, 4 kg; 103 cm, 6.8 kg; 136.5 cm, 11.4 kg; 276 cm, 95 kg; and 277 cm, 109 kg.

Biology The scalloped hammerhead is a warm-water species seldom found in water temperatures below 22°C. It inhabits both turbid and clear waters. Juveniles are abundant year-round in shallow coastal and estuarine waters. The adults are found in offshore waters throughout the year but may enter shallow water during birthing or mating periods. The scalloped hammerhead feeds on stingrays, sharks, numerous bony fishes, shrimp, and crabs.

Some authors (Sadowsky 1965; Clarke, T. A. 1971; Dodrill 1977) have noted that females enter shallow water to give birth and depart shortly afterward. Sadowsky (1965) mentioned that although many juveniles were captured in the littoral waters of São Paulo, Brazil, the adult scalloped hammerhead (as *Sphyrna diplana*) approached the coast only during the reproductive period. T. A. Clarke (1971) observed that young and adult males were often caught inside Kanehoe Bay, Hawaii, but adult females were never taken in that area. According to Clarke, adult male scalloped hammerheads are absent, or at least very rare, in Kanehoe Bay and other breeding areas, except during the birthing season. He suggested that males entered bays only for reproductive purposes and that the females entered bays to give birth to their young and to mate with the males there. Dodrill (1977) mentioned the notable lack of hook-and-line landings of adult female scalloped hammerheads along the east coast of central Florida, while newborns were present in the surf. My observations in the shallow waters of South Carolina are similar to those above. Newborn and small scalloped hammerheads are very abundant in shallow waters, while the adults are abundant just a few miles offshore. I have seen only three gravid females in shallow waters, all carrying term young. Interestingly, on two occasions, a female with term young was caught in a shallow-water gillnet next to an adult male. Whether males were following the females, in response to exuded pheromones in anticipation of mating after parturition, or whether their presence was coincidental is unknown.

The scalloped hammerhead is a migratory species, but little is known of its movements. Schools of hundreds of scalloped hammerheads have been reported in the Gulf of California (Klimley and Nelson 1981). These sharks congregate around specific seamounts, swimming back and forth along the seamount ridge throughout the day. At night, the schools disperse, with sharks moving separately into the pelagic environment. Klimley et al. (1988) used ultrasonic transmitters to track sharks at the seamounts and found that individual sharks departed the seamount before dusk and returned to the same seamount at dawn the next day. The function of this behavior is not known. Much remains to be learned about the schooling behavior of scalloped hammerheads.

The hammerheads in these schools perform numerous ritualized behaviors, such as tilting the body laterally, accelerated swimming with head shaking, accelerated corkscrew swimming, rolling on the longitudinal axis, and many others that have been described by Klimley and Nelson (1981). Their observations of schooling hammerheads indicate that these fishes, in their long evolutionary history, have evolved complex social behaviors that are not yet understood.

Klimley and Nelson (1981: 359) stated, "The phenomenon of grouped hammerhead sharks presents a unique opportunity for the further understanding of shark social behavior, since the sharks are found in sufficiently clear water for observation, shallow enough for free diving, remain in sufficiently limited area for prolonged observations, and tolerate the approach of divers." Unfortunately, other than Peter Klimley's work, little progress has been made in the intervening decades. Generally, it is very difficult to carry out observations of large free-ranging sharks in the open ocean. Hammerheads are particularly difficult to study because they are easily spooked by the noises produced by scuba gear. Divers attempting to approach schools closely must use a silent rebreathing apparatus or confine themselves to free diving.

Size at maturity

Male: Scalloped hammerhead males can be mature at 169 cm, although some authors have reported that maturity is reached at about 185 cm. Springer (1940a), Bigelow and Schroeder (1948), and Dodrill (1977) have given the size at maturity for males as 180–185 cm. Branstetter (1981) reported that the smallest mature male he saw in the northern Gulf of Mexico was 197.5 cm. The smallest mature male that I examined measured 169.5 cm, and was caught on 4 June 1984 off Charleston, South Carolina.

Data from other populations indicate significant population differences in size at maturity. In Australian waters, Stevens and Lyle (1989) reported that males matured over an extended size range of 140–160 cm. Their smallest mature specimen measured 135 cm, while the largest immature male was 161 cm. In the southwestern Indian Ocean, Bass et al. (1975c) reported that scalloped hammerhead males mature at 140–165 cm. Off Taiwan, Chen et al. (1988) reported that, based on clasper calcification, scalloped hammerheads reached maturity at about 198 cm and that by 210 cm all were mature.

Female: Data on maturing females are few. Available data suggest that females mature at about 245 cm. A report by Branstetter (1981) noted that the smallest mature female scalloped hammerhead he examined measured 204 cm. Given that he saw no gravid females, it is likely that his criteria for maturity, which included a ruptured hymen and an ovary containing "developing eggs," allowed a large juvenile to be considered mature. Later, Branstetter (1987b) indicated that data he had collected agreed with the size at maturity of 250 cm suggested by Stevens (1984) and that a 249-cm female he examined had carried young previously. The smallest gravid female I have examined measured 246 cm (taken off Charleston, South Carolina, in June 1982).

Reproduction The scalloped hammerhead is a placental viviparous species with an annual reproductive cycle (producing a brood every year). A female that I examined in South Carolina carried 20 near-term young, while carrying about 30 large oocytes 2.8–3 cm in diameter and weighing 12.4–14.0 g. Presumably, these oocytes would have been ovulated shortly after parturition. Based on limited data on females giving birth in South Carolina in May and June and becoming pregnant shortly afterward, it appears that the gestation period lasts 10–11 months. This agrees with the gestation period of about 10 months estimated by Chen et al. (1988) for this species in Taiwan. The embryos have umbilical cords covered with flattened, short protrusions known as appendiculae (Alcock 1890).

Size at birth: Scalloped hammerhead young are born at 38–50 cm. A 274-cm female that I examined, taken in shallow water off McClellanville, South Carolina, on 31 May 1984, carried 20 young, 31–49 cm (average 42 cm). I have collected free-swimming young of 38 and 40 cm off Charleston, South Carolina, in mid-May. Dodrill (1977) also recorded a 38-cm free-swimming newborn.

Brood size: There are almost no reports of the brood size in the scalloped hammerhead in North America. Pérez-Jiménez and Venegas-Herrera (1997) examined five gravid females of 272–315 cm taken off Nayarit, Mexico, and reported broods of 29–44. I have examined four broods of 10–20 young. In Australia, Stevens and Lyle (1989) found that the mean brood size for four sharks was 16.5 with a range of 13–23. Off Taiwan, Chen et al. (1988) found that females carried 12–38 embryos or uterine eggs (mean of 25.8), but they also stated that females carried one to four nondeveloping uterine eggs.

Nurseries In North America, the scalloped hammerhead gives birth over a very wide area from southern North Carolina to Florida. However, the largest concentrations of neonates appear to be off South Carolina. Newborn young with open umbilical scars are found in shallow water (2–3 m) off South Carolina beaches in May and early June (Castro 1993b). I saw dozens of young 43–50 cm taken off Morris Island and Folly Beach, South Carolina, at depths of 2–3 m between 19 May and 22 June. By the third week in June most of the young I examined had healed umbilical scars. By the first week in August, the umbilical scar was not visible in young measuring 44–49 cm. Dodrill (1977) noted that a 38.1-cm newborn with an open umbilical scar was taken off Melbourne Beach, Florida, on 1 June, and mentioned that juveniles were found at depths less than 8.5 m throughout the year.

Age and growth Branstetter (1987b) analyzed the vertebral rings of 25 specimens, assuming that one vertebral ring formed annually, and concluded that males mature at about 180 cm and 10 years of age, while females mature at about 250 cm and 15 years of age. Piercy et al. (2007) also used vertebral rings to age 65 scalloped hammerheads of the northwestern Atlantic and Gulf of Mexico. Their oldest specimens were 30.5 years old, the male measuring 304 cm and the female measuring 313 cm.

Relation to humans There is little demand in U.S. markets for hammerhead flesh because consumers do not like its darker color, being conditioned to the pink tones of requiem shark flesh. I have eaten scalloped hammerheads (and other species) and I have found them edible.

Large numbers of scalloped hammerheads are caught in the southeastern waters of the United States. Large numbers of small juveniles are taken in the shrimp trawl fisheries operating in shallow coastal waters in summertime, while large adults are caught in coastal longline fisheries, as well as in tuna and swordfish fisheries. While the meat is essentially worthless, the fins are of high value in the fin trade. In the recent past, the usual practice in longline fisheries was to remove the fins from a hammerhead carcass and then dump the carcass at sea, a practice known as "finning" that is now illegal in United States waters. Accordingly, fishermen must bring carcasses with attached fins back to port. The intention of the legislation was to reduce hammerhead mortality, assuming that fishermen would release the hammerheads caught to avoid having to use vessel storage space for the worthless carcasses. Unfortunately, most species of hammerheads die quickly on longlines, and the high value of the fins precludes the release of live sharks. Consequently, in the directed shark fisheries of the southeastern United States, the carcasses are often brought pack to port to satisfy legal technicalities and then are discarded at a later time. Finning is very common on the high seas, and every year thousands of scalloped hammerheads meet this fate.

SCOOPHEAD

Sphyrna media Springer, 1940

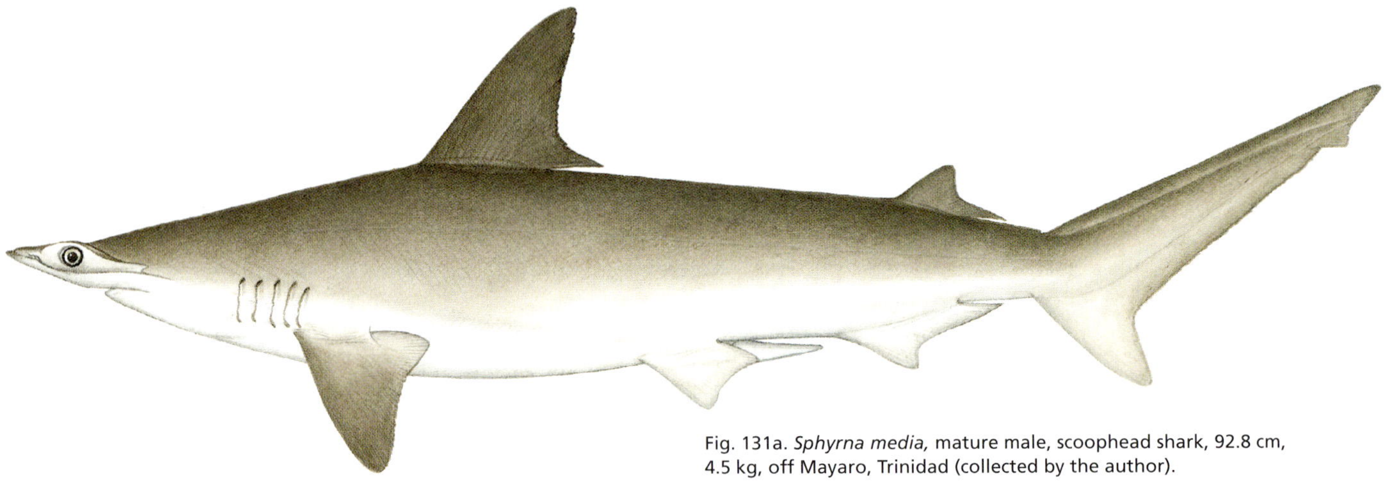

Fig. 131a. *Sphyrna media,* mature male, scoophead shark, 92.8 cm, 4.5 kg, off Mayaro, Trinidad (collected by the author).

Common name Scoophead, an allusion to the shape of its head.

Spanish name *Cornuda cuchara* (A.F.S.).

Scientific name *Sphyrna media* Springer, 1940, in *Stanford Ichthyological Bulletin* 1(5): 162–163, fig. 3. *Sphyrna:* a corruption of the Greek name *sphyra,* a hammer or a mallet, an obvious reference to the shape of the head and, according to Jaeger (1962: 244), "one of Rafinesque's efforts at word-butchery"; *media,* from the Latin *medius,* middle, an allusion to its being an intermediate form between other hammerheads.

Synonyms *Sphyrna tudes* Jordan and Evermann, 1896 (type of *Sphyrna media*), *Sphyrna nana* Sadowsky, 1965.

Identification This hammerhead is characterized by a slightly rounded anterior margin of the head without a central indentation, *a snout length* (from the anterior margin of the head to the front of the mouth) *less than 40% of the greatest head width, no inner narial groove,* and a free rear tip of the first dorsal fin that reaches to the level of the pelvic fins. The upper teeth have oblique, smooth-edged cusps with a notch on their outer margins; the lower teeth are erect and smooth edged. Teeth number U: 12 to 14–0 to 2–12 to 14, L: 13 or 14–1–13 or 14. The dermal denticles are small and overlap little. They have three to five ridges and points, the central ones being slightly longer than the others. Coloration is gray or grayish brown above, paler below, without fin markings.

Similar species The golden hammerhead has yellow fins and rear borders of its head, and it has a more pronounced notch on the anterior margin of the head. The yellow color of fresh specimens is sufficient to separate it. The scalloped bonnethead has a longer snout and an anal fin with a straighter rear margin and more rounded tip.

Range This small hammerhead has been reported from the eastern Pacific from the Gulf of California to Ecuador, the southern Caribbean, and the southwestern Atlantic (Compagno 1984b). The limits of its distribution are not well defined due to confusion with other hammerheads. I have never encountered one in the Gulf of California, despite extensive sampling. The type specimen is from Mazatlán, Mexico. The only specimens I have collected came from the northeastern coast of Trinidad, off Balandra and Mayaro, where it is far less abundant than other hammerheads.

Size and weight The scoophead reaches about 115 cm. Kato et al. (1967) give 150 cm as the maximum length for Pacific specimens. I doubt that Atlantic specimens reach that size. The following are the sizes and weights of specimens that I have examined:

Male: 60 cm, 0.8 kg; 73 cm, 1.4 kg; 83.2 cm, 2.6 kg; 90.3 cm, 3.4 kg; 97.0 cm, 4.6 kg; 100 cm, 4.5 kg; and 108.4 cm, 6.1 kg.

Female: 69.1 cm, 1.2 kg; 83.3 cm, 2.3 kg; 90.0 cm, 3.3 kg; 106.4 cm, 6.2 kg; and 111 cm, 7.2 kg. This last individual was the largest scoophead I have seen.

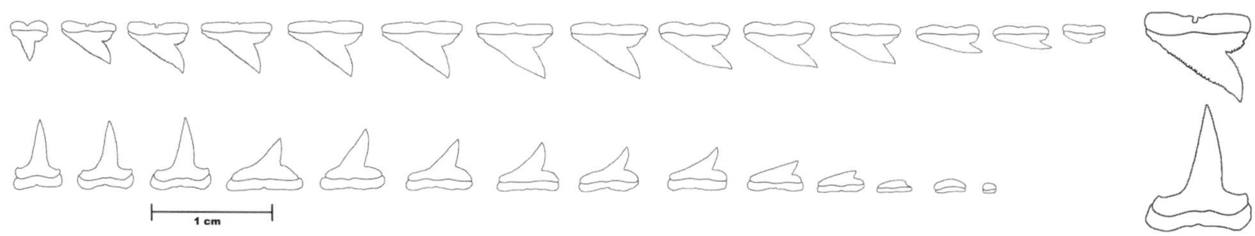

Fig. 131b. Teeth.

Biology This is one of the least known hammerheads of the Western Hemisphere. I encountered it in the murky waters of northeastern Trinidad, mostly at depths of 18–37 m. Analysis of stomach contents revealed that it feeds primarily on bony fishes, squid, octopuses, and shrimp. I also found a 12-cm ray (*Narcine*) and a small shark (*Mustelus*) in the stomachs of scoopheads.

Size at maturity Males mature at about 83 cm and females mature at about 90 cm.

Male: The largest immature male I have examined measured 83.2 cm and weighed 2.6 kg. It had long but uncalcified claspers 65 mm. The smallest mature male I examined measured 82.7 cm and weighed 2.38 kg. Eleven males larger than 86 cm were all mature.

Female: Females up to 88 cm had small oviducal glands less that 14 mm wide. The gravid females I examined had oviducal glands 14–17 mm. The smallest gravid female I examined measured 92 cm and weighed 3.5 kg. It carried one embryo 46 mm.

Reproduction The scoophead is a placental viviparous species. The embryos have long appendiculae on their umbilical cords. Females that I examined in Trinidad carried embryos 90–148 mm in early March, suggesting birth in summer.

Size at birth: No data available. The young are probably born at around 25–30 cm.

Brood size: I have only examined three gravid females. A small female 92.8 cm carried one embryo, while two larger females, 100.0 and 109.6 cm, carried seven and eight embryos respectively.

Nurseries Unknown.

Age and growth Unknown.

Relation to humans The scoophead is generally too scarce and too small to be of economic importance. Small numbers of scoophead are often seen in the fish market in the Port of Spain, Trinidad.

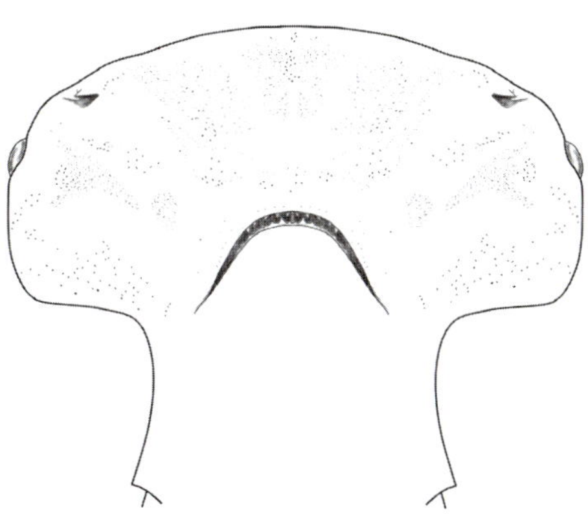

Fig. 131c. Snout.

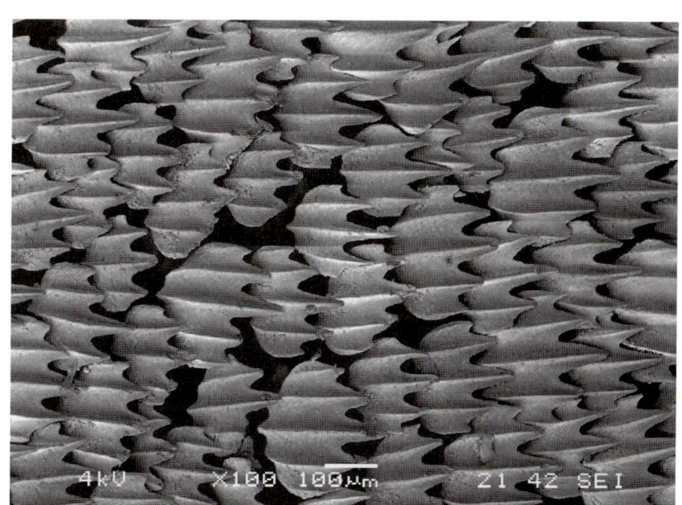

Fig. 131d. Dermal denticles.

GREAT HAMMERHEAD

Sphyrna mokarran (Rüppell, 1837)

Fig. 132a. *Sphyrna mokarran*, Great hammerhead shark, immature female, 273 cm, 97 kg; from Looe Key Reef, Florida (Mike Heithaus and Jeff Carrier, collectors).

Common name Great hammerhead, references to its great size and the shape of the head.

Spanish name *Cornuda de ley, cabeza de martillo* (Cuba), *tiburón martillo* (Mexico), *cornuda gigante* (Mexico, A.F.S.).

Scientific name *Sphyrna mokarran* (Rüppell, 1837), originally described as *Zygaena mokarran* in *Neue Wirbelthiere zu der Fauna von Abyssinien gehorig:* 66–67, plate 17, fig. 3. *Sphyrna:* a corruption of the Greek name *sphyra,* meaning hammer or mallet and, according to Jaeger (1962: 244), "one of Rafinesque's efforts at word-butchery"; *mokarran:* the Arabic common name for the species, from the Arabic *mogharn,* horned, probably an allusion to the head shape.

The scientific name of the great hammerhead has had a somewhat complex history. For many years, the great hammerhead was known as *Sphyrna tudes* based on Valenciennes's 1822 description. Tortonese (1950) pointed out that the three type specimens of *Zygaena tudes* (as Valenciennes named it) actually comprised two species, one being the golden hammerhead of the Americas, and one (a now missing type specimen) that could have been *Sphyrna mokarran.* Gilbert (1967), in an effort to correct this situation, assigned the name *S. tudes* to the golden hammerhead. Thus, the next available name (and description) for the great hammerhead is *S. mokarran* (Rüppell, 1837). For an excellent summary of this taxonomic confusion, see Compagno (1984b: 553).

Identification The great hammerhead is characterized by a *nearly straight anterior margin of the head with a deep central indentation,* a high second dorsal fin, and *pelvic fins with curved rear margins.* The shape of the head changes with age. Young specimens have slightly rounded anterior margins of the head. The anterior margin of the head becomes nearly straight in large adults. The teeth are triangular, with *strongly serrated edges,* and become increasingly oblique toward the corners of the mouth. Teeth number U: 17–2 or 3–17, L: 16 or 17–1 to 3–16 or 17, and are similar in both jaws. The dermal denticles are very broad and overlap greatly. The denticles have three to five ridges and points, the central ridge and point being much larger and wider than the others. Color varies from dark olive-green to brownish or light gray above, grading to white below.

Similar species All other hammerheads lack the pelvic fins with curved rear margins and teeth with strongly serrated edges.

Range The great hammerhead is circumtropical. In North American waters, it ranges from North Carolina southward to the tropics. It is a migratory species common off Georgia

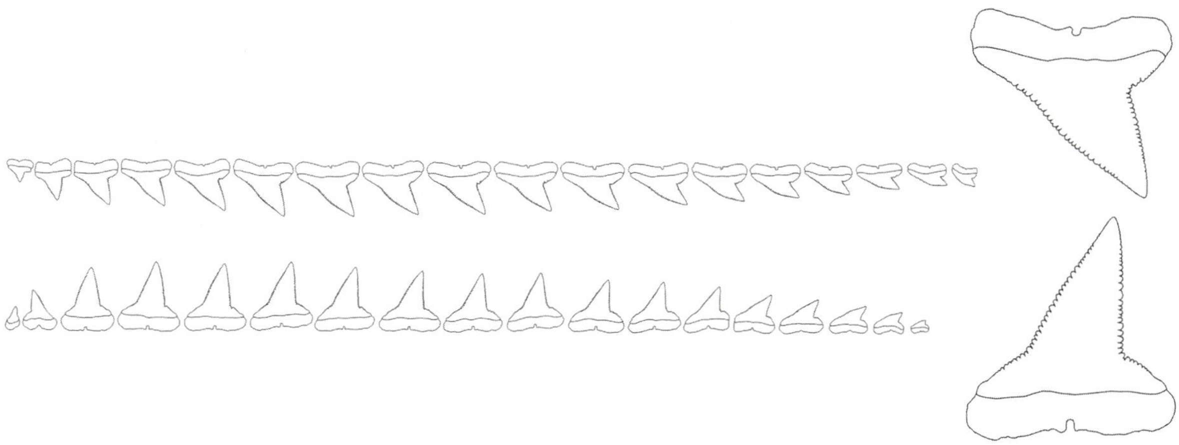

Fig. 132b. Teeth.

and the Carolinas during the summer months. It is found off northern and central Florida almost all year. In the Florida Keys, the great hammerhead can be found all year.

Size and weight This is one of the larger predatory sharks and the largest of the hammerhead sharks. Lengths reaching or exceeding 600 cm have been attributed to this species in the literature (e.g., Gilbert 1967; Compagno 1984b; Last and Stevens 1994) and anecdotally. However, a review of the literature reveals that few recorded specimens reach 480 cm, and there is only one first-hand report of a great hammerhead exceeding 500 cm. Gudger (1947) reviewed the evidence available at the time on the sizes attained by hammerheads and concluded that the largest known specimen was caught by Stewart Springer off Key West, Florida, and measured at 466 cm. Later, Springer (1960: 29) reported that one measured specimen was "18 feet long" (548 cm). Other hammerheads reported do not exceed 500 cm. The largest reported by Clark and von Schmidt (1965) off the western coast of Florida was 414 cm and weighed 397 kg (Cape Haze Marine Laboratory records). The largest specimen I found in the records of Mote Marine Laboratory (MML no. 215) was a 425-cm female taken off Sarasota, Florida, in June 1968. A female great hammerhead speared off Chub Key, Bahamas, in April 1986, was reported as measuring 472 cm and weighing 526 kg ("15 1/2 ft" and

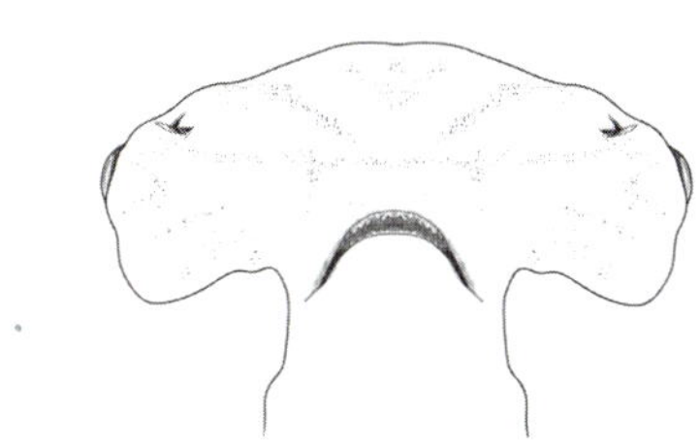

Fig. 132c. Snout (neonate).

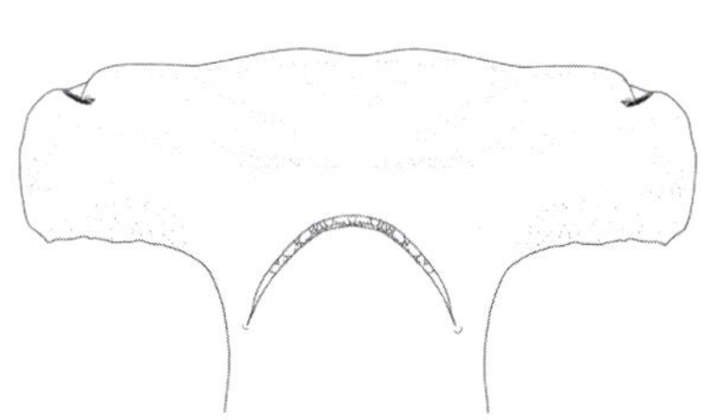

Fig. 132d. Snout (adult).

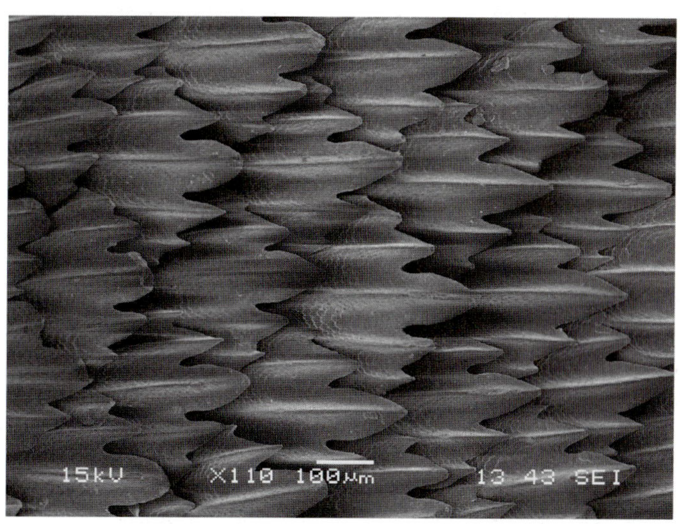

Fig. 132e. Dermal denticles.

1,160 lb, Hardie 1986). I examined a female taken off Sarasota, Florida, on 23 May 2006 that measured 430 cm and weighed 572 kg. This was the heaviest hammerhead I have seen.

The largest specimens reported from other areas are also smaller than 5 m. Cadenat and Blache (1981) reported that their largest specimen from the Mediterranean and the eastern Atlantic was a female measuring 443.5 cm and weighing 541 kg. Stevens and Lyle (1989) examined great hammerheads taken as incidental catch in the gillnet fishery in northern Australia and reported that largest male was 332 cm and the largest female was 390 cm. However, there is additional evidence of specimens over 5 m. Gordon Hubbell has in his collection of shark jaws a gigantic set from a great hammerhead obtained from Taiwan, estimated at about 548 cm (18 ft). Based on these jaws and on the specimen reported by Springer (1960), it appears that exceptional great hammerheads may reach 5.5 m.

Representative sizes and weights that I have recorded are as follows:

Male: 151 cm, 20.4 kg; 254 cm, 73 kg; 303 cm, 108 kg; 315 cm, 141 kg; and 318 cm, 138 kg.

Female: 200 cm, 37.6 kg; 258 cm, 73 kg; 319 cm, 152 kg; 319 cm, 170 kg; 322 cm, 164 kg; 328 cm, 189 kg; 351 cm, 241 kg; 414 cm, 397 kg; 437 cm, 318 kg; and 430 cm, 572 kg.

Biology The great hammerhead inhabits clear, warm waters in both oceanic and shallow coastal areas. This solitary fish is the top predator in many shallow-water, tropical environments, such as the coral reefs of the southern Florida coast.

The great hammerhead is usually seen where the surface water temperature is 22–30°C, although Dodrill (1977) reported a wider surface temperature range of 15.6–29.7°C. It wanders north in the summer and south in winter, but the extent of these migrations is not known. In summer, large juveniles and adults migrate northward as far as Cape Hatteras. Although large specimens are found off south Florida year around, gravid females depart the area in late October and are not seen again until April. This suggests that gravid females migrate much farther south, perhaps spending the winter off the southernmost Florida Keys or in Caribbean waters.

Diet: The great hammerhead consumes a wide variety of fishes. Analysis of stomach contents shows that they feed heavily on bottom dwelling fishes: stingrays, catfish, and flounders. Presumably, the enlarged surface area of the head allows for more efficient location of prey buried on the bottom. There are numerous reports of barbs (caudal spines or "stings") of southern stingrays (*Dasyatis americana*) being found in the mouth or throat area of great hammerheads caught off the southeastern states. Gudger (1947) reported finding 54 barbs in a 381-cm great hammerhead caught off Beaufort, North Carolina. More than 40 were embedded in the flesh adjacent to the jaw cartilage, and presumably there were more in the throat area. Wright (1981; see p. 184) mentions that 4 of 12 great hammerheads from the Florida Keys had stingray spines embedded in their jaws. Dodrill (1977) found that the most common great hammerhead prey off Melbourne, Florida, were catfishes (*Bagre marinus, Arius felis*), and stingrays (*Dasyatis* spp.) and reported that one specimen had 60 stingray barbs and catfish spines embedded in the jaws and flesh around the jaws. I once counted 23 stingray barbs in a 300-cm specimen from South Carolina. Dodrill (1977) also noted that great hammerheads had the highest percentage of stomachs containing food items (88.5%) of any large sharks collected. Whether this high percentage reflects an enhanced ability of the species to find and capture prey is not known.

The great hammerhead also feeds on numerous bony fishes such as tarpon (*Megalops atlantica*), jacks (*Caranx* spp.), and numerous sharks (Clark and von Schmidt 1965; Dodrill 1977). In Florida waters, the great hammerhead often enters very shallow waters along beaches in the pursuit of tarpon, its favorite prey. These great sharks have been observed, and even filmed, in pursuit of tarpon off Florida beaches, entering very shallow water. In the cases that I know of, the shark usually caught and consumed the tarpon. In Florida, nearly all the great hammerheads that I have examined in summer had in their stomachs the remains of one or more large tarpon. I examined one 351-cm female, caught off Marathon, Florida, that contained the remains of at least two large tarpons, one of which had scales 8 cm in diameter. Interestingly, great hammerheads seldom consume the tarpon's head. In nearly all the instances I have observed, the hammerheads had taken the tarpon from behind, severing off the body behind the gills or at midsection, and swallowing only the body. For these great fishes the hard, bony head of the tarpon is probably not worth eating.

Strong et al. (1990) reported an interesting behavior of the great hammerhead in the Bahamas. A great hammerhead estimated at 3 m was observed to overtake a rapidly fleeing southern stingray (*Dasyatis americana*). Upon reaching the ray, the hammerhead thrust its head down onto the stingray's dorsum, pushing the ray to the bottom with great force. The ray then struck the bottom and rebounded as the shark closed on it and delivered a second blow. This time, the shark maintained contact with the ray, pushing the ray to the bottom with its head. The shark then pivoted around and bit the anterior margin of the ray's left wing, removing a crescent-shaped piece of flesh. Later, the shark repeated

Fig. 132f. *Sphyrna mokarran*, great hammerhead shark, from Long Key Bridge, Florida (René de Dios, collector, courtesy of Gregg Poulakis).

the maneuver using its head to restrain the ray, pivoting, and taking a bite of the anterior margin of the right wing. The ray was left incapacitated on the bottom, although actively breathing and curling its fin margins. It did not move while the shark swam around it in diminishing circles for some 24 minutes. The shark then consumed the moribund ray, which did not resist. I have had other reports of great hammerheads using their heads to hit or pin down prey.

Size at maturity

Male: There are almost no published data on the size at maturity of male great hammerheads in North America. Springer (1938) stated that the species matures at about 300 cm, obviously referring to both sexes. The largest immature male that I examined (taken off Miami, Florida, in May) measured 300 cm. It had elongated but uncalcified claspers, suggesting that it was not far from maturity.

South African studies also support a length at maturity exceeding 300 cm. According to Cliff (1995), the length at 50% maturity for great hammerheads caught off South Africa is 309 cm (217 cm precaudal length); the largest immature male, based on its soft claspers, was 327 cm, and the smallest mature male was 275 cm. However, Stevens and Lyle (1989) reported that the smallest mature male seen off northern Australia was 210 cm and the largest immature specimen was 258 cm. It is likely that different species, or at least different populations, are involved.

Female: It appears that females mature at about 300 cm. Bigelow and Schroeder (1948) estimated maturity was reached at 300 cm. The smallest mature female examined by Clark and von Schmidt (1965) measured 300 cm. Gordon Hubbell (pers. comm., July 2003) examined a 300-cm female caught off Miami, Florida, that contained 28 young. The smallest gravid female I have seen measured 315 cm. This shark carried a total of 14 embryos 40–62 mm in September 1984. However, Dodrill (1977) mentions an unusual female, 256.2 cm, carrying 21 embryos 49.8–60.6 cm.

Reproduction The great hammerhead is a placental viviparous. Based on the sizes of embryos seen throughout the year, the gestation period appears to be 11 or 12 months long, beginning in July or August and ending the following May, June, or July. Analysis of females I examined off Florida and South Carolina indicates that females inhabiting the southeastern coast of the United States breed synchronously; that is, all females carry embryos of fairly similar sizes at a given time of the year. Great hammerheads off Australia are also synchronous breeders (Stevens and Lyle 1989). Term females I examined had inactive ovaries, suggesting a two-year reproductive cycle, similar to large carcharhinid sharks.

Size at birth: Based on the largest embryos and the smallest juveniles found during summer, great hammerhead sharks are born at lengths of 64–70 cm from late May to July. A female caught off Sarasota, Florida, carried embryos 61–70 cm on 16 May, while a second one carried 62- to 68-cm embryos on 23 May. Clark and von Schmidt (1965) examined another female from the Sarasota area carrying embryos 60–67.5 cm on 6 June, whereas Dodrill (1977) reported embryos 498–606 mm on 31 July. Free-swimming young 65–71 cm have been taken in the eastern Gulf of Mexico in June and July (see below).

Brood size: Great hammerhead broods are large, ranging from 13 to 56 young. Clark and von Schmidt (1965) examined gravid females in Florida carrying broods of 13, 21, and 41 young. Dodrill (1977) examined three gravid females caught off Melbourne Beach, Florida, that carried broods of 21, 23, and 36 young. I have collected data for 15 females caught off Florida and South Carolina; their mean brood size was 26, and the range was 13–56. A 430-cm female taken off Sarasota on 23 May 2006, which I examined, carried a brood of 56 young (53 viable young and three dead). This is the largest brood that I know of. Off Australia, Stevens and Lyle (1989) reported that the mean size of 30 broods was 15.4 young, with a range of 6–33, and that there was a significant relationship between increasing maternal size and brood size.

Nurseries The extent of the great hammerhead nurseries is not well defined. Although females with term young are often taken in the eastern Gulf of Mexico, the neonates and small young are seldom encountered and, consequently, the limits of the nurseries are unknown. Neonates and small juvenile great hammerheads are not found in the western Atlantic from North Carolina to eastern Florida. I have never seen a specimen less than 200 cm from that area, and Dodrill (1977) recorded only one specimen shorter than 200 cm (a male about 197 cm) in his three-year study off Melbourne Beach, Florida. Free-swimming young 65–71 cm with healed umbilical scars have been taken off Yankeetown, Florida, in June and July (J. Tyminski, MML, pers. comm., July 2003).

Age and growth Unknown.

Relation to humans In the Florida shark fisheries of the 1940s, the larger great hammerheads were especially sought for liver oil, which is high in vitamin A content. In today's fisheries, the great hammerhead is sought for its fins, which bring the highest prices in the shark fin trade. However, the great hammerhead is generally too scarce a catch to be of great economic importance. It is an incidental catch in the shark fisheries of the southeastern United States, as well as in offshore tuna and swordfish fisheries in tropical waters. Its flesh is not utilized in the United States because its dark color does not appeal to consumers, although it is eaten in other countries. Some years ago, a colleague of mine used the flesh of a huge, old great hammerhead in a consumer acceptance test. The flesh was smoked, thinly sliced, oiled, and dyed the color of salmon. It was avidly consumed by dozens of testers who asserted to me that it was salmon!

Attacks on humans have been attributed to the great hammerhead, and the species has long been considered dangerous to man because of its shallow-water habitat and its large size. The great hammerhead is potentially dangerous to spearfishers who may be carrying a fish that the shark desires. I know of one recent case in the Florida Keys where a commercial spearfisherman, Christopher Rackley, was threatened by an aggressive, large great hammerhead, forcing the diver to leave the water.

In the mid-1990s, two divers, Manny Puig and Mark Rackley, began using bait to attract great hammerheads for the purpose of filming them off Little Torch Key, Florida. After a few months of trials, they discovered that they could hand feed these great fishes without incident. They discovered that the hammerheads would focus entirely on the bait and ignore them. Some fascinating films resulted from these exhilarating encounters, several of which I observed. Incidentally, all of these underwater encounters with great hammerheads had to be carried out using only mask and snorkel. Great hammerheads, like other hammerhead sharks, are apparently disturbed by the sounds produced by scuba gear and cannot be approached easily by someone using such gear.

The great hammerhead is often sought as a trophy by fishers. The "record fish" are measured by weight; the heavier the fish, the greater its supposed merit. Unfortunately, this causes fishers to target the large gravid females because they are the heavier fish. Some "record size" great hammerheads have been taken off the Gulf coast of Florida, where females come to deliver their young. Most of these large sharks are simply dumped after being weighed. Great hammerheads have decreased in numbers in the last few decades due to the ubiquitous longline fisheries. The targeting of gravid females in the nursery areas further aggravates the status of this great fish, simply to satisfy the vanity of a few fishers. Given today's excellent tackle, there is little challenge or merit in landing these great sharks.

BONNETHEAD

Sphyrna tiburo (Linnaeus, 1758)

Fig. 133a. *Sphyrna tiburo*, bonnethead shark, mature male, 99 cm, 3.4 kg; from McClellanville, South Carolina (Steve Poston, collector).

Common name Bonnethead, bonnet shark, shovel nose or shovel head shark, all colorful references to the unusual shape of its head.

Spanish name *Cornuda de corona* (Cuba), *tiburón cabeza de pala* (Mexico), *cornuda cabeza de pala* (A.F.S.).

Scientific name *Sphyrna tiburo* (Linnaeus, 1758), originally described as *Squalus tiburo* in *Systema Naturae,* 10th ed. I: 234. *Sphyrna:* a corruption of the Greek name *sphyra,* a hammer or a mallet, an allusion to the shape of the head and, according to Jaeger (1962: 244), "one of Rafinesque's efforts at word-butchery"; *tiburo,* a corruption of *tiburón,* Spanish for shark.

Synonyms *Sphyrna vespertina* Springer, 1940.

Identification Its characteristic flattened, shovel-shaped head makes this hammerhead the easiest to identify. The *anterior margin of the head is evenly rounded between the eyes.* The frontal teeth have erect, smooth-edged cusps, while subsequent teeth have oblique cusps; the outermost teeth of the lower jaw are modified into flat crushers. Teeth number U: 12 to 14–0 or 1–12 to 14, L: 12 to 14–1–12 to 14. The dermal denticles are very broad and overlapping. The denticles have five parallel ridges ending in points. Coloration is brownish or greenish gray above and paler below.

Similar species All the other hammerheads have hammer-shaped (not shovel-shaped) heads.

Range The bonnethead is confined to the warm waters of the Western Hemisphere. Along the eastern coast of North America, it ranges from New England (Garman 1913), where it is rare, to the Gulf of Mexico and the South American coast down to Brazil (Bigelow and Schroeder 1948). According to Bigelow and Schroeder (1948), there are a couple of old records for New Jersey waters, one for Rhode Island and one for Massachusetts Bay. It is rare north of the Chesapeake Bay, although Murdy et al. (1977) stated that it was an occasional summer visitor to the lower Chesapeake Bay. The bonnethead is common in inshore waters of the Carolinas and Georgia in summer and off Florida and in the Gulf of Mexico in spring, summer, and fall. According to Jordan and Thompson (1905), it can be taken year-round in the Dry Tortugas. The bonnethead is one of the most abundant sharks in the southeastern North America. Baughman and Springer (1950) stated that the bonnethead was one of the most common sharks of the Texas coast, and Hoese and Moore (1958) stated that it was very common off Texas beaches during spring, summer, and fall. Márquez-Farias et al. (1998) called the bonnethead one of the most abundant species in the southern portion of the Gulf of Mexico (Bahia de Campeche).

Along the western coast of North America, it has been reported from southern California to Ecuador (Kato et al. 1967). These specimens were once considered a subspecies, *S. tiburo vespertina* (Springer 1940b). The bonnethead has been described as an occasional or uncommon visitor in southern California (Kato et al. 1967; Miller and Lea 1972). Bonnethead sharks used to be abundant along the Pacific coast of southern Mexico. Hernandez Carvallo (1967) described the bonnethead as "very abundant" and present throughout the year in the Pacific coast of southern Mexico. Jordan (1895) mentioned obtaining a specimen off Mazatlan. In the late twentieth century, the bonnethead be-

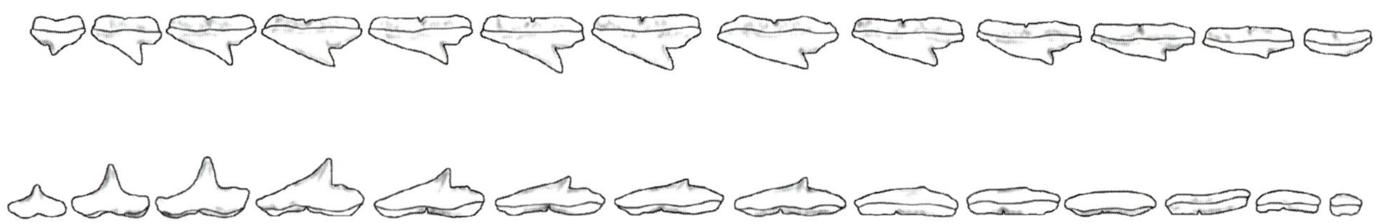

Fig. 133b. Teeth.

Fig. 133c. Teeth (detail).

came rare along the Mexican coast. I found a single report of a specimen taken in the late twentieth century. Biologist Leonardo Castillo (pers. comm., February 2009) encountered a single specimen off southern Sinaloa in 1994 and none off Chiapas in five years of sampling. All the museum specimens of bonnethead sharks from the western coast of Mexico that I have located were taken before 1960. In the Scripps Institution of Oceanography Marine Vertebrate Collection, there is a bonnethead specimen (SIO no. 47–53) taken in the upper Gulf of California (Bahia Ometepec) in 1947. However, Mexican biologist Juan Carlos Pérez Jiménez (pers. comm., February 2009) who did extensive sampling in the Gulf of California 1995–2005 never encountered one. I never encountered a bonnethead in my sampling of the central Gulf of California shark fisheries.

Size and weight Bonnetheads from different areas in the southeastern North America vary in size. Whether the observed size differences are due to clinal variations, genetic differences among different populations, or even different species is not known.

Male: Representative weights for the Florida population: 60 cm, 1.0 kg; 70 cm, 1.5 kg; 80 cm, 2.3 kg; and 90 cm, 3.1 kg.

Female: Representative weights for the Florida population: 60 cm, 0.9 kg; 70 cm, 1.6 kg; 80 cm, 2.4 kg; 90 cm, 3.2 kg; and 100 cm, 6.7 kg. Specimens caught off South Carolina in summer are much larger than those encountered in Florida. Representative weight for South Carolina females are

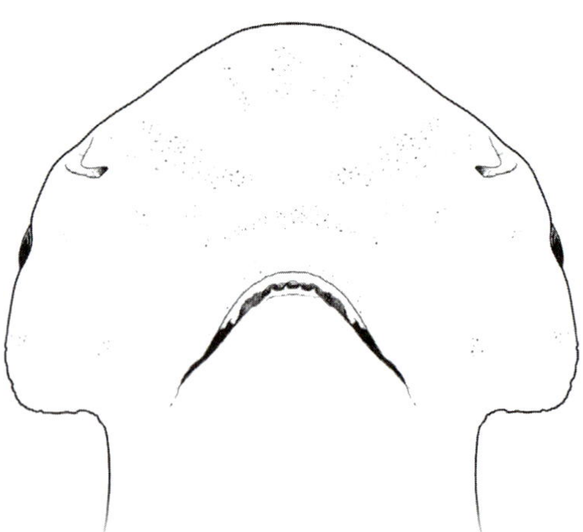

Fig. 133d. Snout.

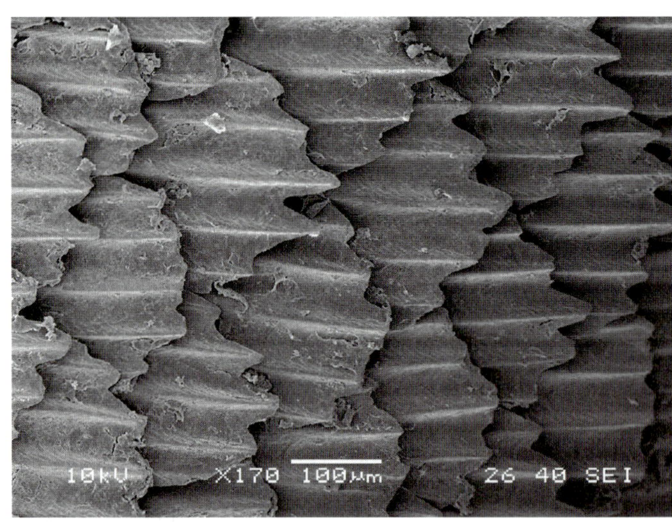

Fig. 133e. Dermal denticles.

115 cm, 6.4 kg; 118.5 cm, 9.07 kg; 122 cm, 10.43 kg; and 126 cm, 11.3 kg (gravid with 7 embryos 21–23 cm). The largest I have seen was a 130.8-cm female, caught off McClellanville, South Carolina, in mid-June 1990.

Biology The bonnethead inhabits warm, shallow coastal waters. Its habitat is difficult to define because it is a ubiquitous shark that can be found both in clear and turbid waters, along beaches and in river mouths, in sea grass beds as well as estuarine waters, and over both sandy and muddy bottoms. Juveniles are common over sea grass beds and in estuarine waters. I have often seen small juveniles swimming over very shallow (30 cm) sea grass beds in the Florida Keys and also at the water's edge in the islands of the South Carolina coast.

The bonnethead is a warm water shark. Manire and Rasmussen (1997) reported catching males in water temperatures of 19–31.8°C throughout the year off west central Florida. Most of the bonnetheads I caught off South Carolina during summer were in surface water temperatures of 21–25°C. Bonnetheads are believed to migrate southward in winter or to deeper offshore waters in Florida and the Gulf of Mexico, but little is known about such movements. Like many other hammerheads, bonnetheads often travel in schools. The schools generally consist of 5–15 individuals, but very large schools have been reported. E. Clark (1963b) reported that commercial fishermen had caught more than 700 bonnetheads from a school found off Lido Key, Sarasota, Florida, in November 1962.

In the early 1970s, Myrberg and Gruber (1974) studied a group of 11 captive bonnetheads held in a large pool at the Miami Seaquarium. These authors described some 18 different postures and movement patterns, assigning social relevance to some of these behaviors. The study revealed a size- and sex-dependent dominance hierarchy. Unfortunately, there has been no follow-up to this interesting ethological work in the three decades since it was carried out.

Diet: The bonnethead is an invertebrate predator that also feeds on small fishes. Gudger (1907) reported that that the food of the bonnethead consisted chiefly of blue crabs (*Callinectes sapidus*), shrimp (*Penaeus*), mantis shrimp (*Squilla*), and acorn barnacles.

Throughout its range in the eastern United States, the bonnethead preys heavily on blue crabs. Female bonnetheads prey on larger blue crabs. Hoese and Moore (1958) reported that stomach contents of adult females in Texas included mainly blue crabs and that all specimens had eaten one or two mature crabs. I observed the same dietary habit in bonnetheads taken off South Carolina and Florida. I have often been amazed by the large size of the crabs that bonnetheads could consume. For example, in the stomach of a 118.5-cm female bonnethead, taken off South Carolina, I found two blue crabs with well-calcified, intact carapaces, but minus legs and pincers. Their carapaces measured 17 cm wide from spine to spine. In a 95-cm female caught off Tampa Bay, I found a blue crab also with an intact carapace 13 cm wide. Obviously, the bonnethead locates and attacks blue crabs, violently tearing off their legs and pincers. It then maneuvers the crab sideways so it can be swallowed. How the shark can swallow the relatively large crab so that the sharp lateral spines of the carapace do not catch in its mouth, gill area, or esophagus, I do not know. Smaller crabs are usually crushed and swallowed whole. Other prey include swimming crabs (*Callinectes arenarius* and *Portunus* spp.), small stone crabs (*Menippe* spp.), octopuses, and numerous small fishes. I examined one 92.6-cm male, taken off Charleston, South Carolina, that contained one 6-cm-wide blue crab, one 14-cm mantis shrimp, one 16-cm spot (*Leiostomus xanthurus*), and a 17-cm porgy (*Pagrus pagrus*).

Size at maturity Despite its abundance in the shallow waters of the southeastern United States, there are few reports of bonnethead size at maturity.

Male: Clark and von Schmidt (1965) reported that off Sarasota, Florida, males were mature at 76 cm, except for one specimen that was mature at 68 cm. Having failed to notice clasper calcification in their early work, these authors attempted to use clasper length as a percentage of total length as an index for maturity in the bonnethead. They pointed out that, after maturity, the length of the clasper does not increase proportionally with the total length of the shark. Clark and von Schmidt also cited an example of how an exceptionally large adult male can have a clasper length falling into the "immature" category when expressed as a percent of total length. Parsons (1993a) used the relationship between total length and "clasper index" (the length of the clasper expressed as a percent of the total length of the shark) to determine maturity in the bonnethead. He did not relate this clasper index to usual standards of clasper calcification. In some species, the claspers elongate long before they become calcified (Castro 1996). This probably happens in the bonnethead also. Thus, when using clasper size alone as a criterion for maturity, it is possible to judge immature sharks with fully elongated, but uncalcified claspers as mature. Parsons (1993a) noted differences in sizes at maturity between the populations of Tampa Bay and those of Florida Bay. According to Parsons, in the Florida Bay population, development of the clasper reached its zenith, indicating maturation, when males reached about 68–70 cm. Parsons reported that the Tampa Bay males mature at about 80 cm, when the "clasper index" equals about 8%. If there is an error due to the usage of clasper length as an index of maturity, the error is likely to be small, and it appears certain that males are mature at 70–80 cm for the Florida Bay and Tampa Bay populations.

Female: Females in the Florida population of bonnetheads mature at about 85–90 cm. Clark and von Schmidt (1965) indicated that the smallest mature female they examined was 87 cm and the largest immature female measured 91 cm. Parsons (1993a) used ovulation as an indication of maturity in Florida bonnetheads and reported that female bonnetheads mature at about 80–85 cm in Florida Bay and at 85–90 cm in Tampa Bay.

Reproduction The bonnethead is a placental viviparous species. Its curious reproductive cycle has only recently been explained, mostly through the works by Glenn Parsons at the University of Mississippi and of Charles Manire at Mote Marine Laboratory. The bonnethead mates in the fall. Clark and von Schmidt (1965) reported males with enlarged testes and sperm oozing from the urogenital pore in early October. Manire et al. (1995), working in Tampa Bay, observed females with mating wounds on 21 October, having observed none with mating wounds a week earlier. By 4 November all mature females captured exhibited mating wounds. Interestingly, ovulation does not occur until four or five months later, in middle to late March, indicating that females store sperm over the winter. Parsons (1993a) reported that bonnetheads in Florida Bay carried ripe oocytes (about 2 cm in diameter) in early March and uterine eggs by late March, although no embryonic development could be discerned. The gestation period has been estimated at four to five months, with implantation occurring at about 2.5 months into the gestation period (Parsons 1993b; Manire et al. 1995). Off Florida, females give birth from early August to early September (Myrberg and Gruber 1974; Parsons 1993b).

The embryos have very long umbilical cords (sometimes as long as the embryo itself). The umbilical cord is covered along its entire length by long villi (Bigelow and Schroeder 1948). The function of these villi, known as appendiculae (Alcock 1890), is not entirely clear.

Both Parsons (1993a) and Manire et al. (1995) have stated that within a population, bonnethead females are reproductively well synchronized, although there are latitudinal differences in the timing of reproductive events in different Florida populations. Specimens that I examined, caught off McClellanville, South Carolina, in the first week of May, were carrying ripe oocytes 20–23 mm in diameter or eggs in blastodisc stage. By August, females carried embryos 21.2–23.6 cm. These few observations suggest that the reproductive processes of the East Coast population may lag behind the Gulf of Mexico population by about one to two months.

Size at birth: Bonnetheads are born at lengths of 24–34 cm. This relatively wide size range is probably due to population differences. Myrberg and Gruber (1974) reported that the mean size of two broods born in captivity was 24 cm. Parsons (1993a) observed that females in Florida Bay gave birth to young averaging 27 cm (range 24–29 cm), whereas females in Tampa Bay gave birth to young averaging 34.7 cm. Dodrill (1977) reported finding free-swimming specimens measuring 31 cm.

Brood size: Bonnethead broods range from 6 to 14. Hoese and Moore (1958) examined six females from the Texas coast and reported that the broods ranged from 11 to 14 young, with an average of 12. Myrberg and Gruber (1974) reported two Florida broods of 10 and 11 young both born in captivity. Six females that I examined off South Carolina had broods ranging from 10 to 12 young, whereas six Florida specimens carried broods of 6–10 young.

Nurseries Bonnethead nurseries are in the shallow beach areas throughout its range.

Age and growth The bonnethead is a rapidly growing shark. Captive young-of-the-year bonnetheads have been shown to grow 25–31 mm per month (Parsons 1993a; Carlson and Parsons 1997). Maximum age has been estimated, using vertebral ring counts, at 12 or more years for females and 8 or more years for males (Carlson and Parsons 1997).

Relation to humans Although it is occasionally marketed in South Carolina, it is of little economic importance. It is often exhibited in marine aquaria, where it usually survives for several months.

Extralimital Species

GOLDEN OR SMALLEYE HAMMERHEAD

Sphyrna tudes Valenciennes, 1822

Fig. 134a. *Sphyrna tudes,* golden or smalleye hammerhead shark, immature male, 66 cm, 1.26 kg; off Mayaro, Trinidad (Elizabeth Williams, Joan Williams and Dr. Roger Williams, collectors).

Common name Golden or smalleye hammerhead, obvious allusions to its striking yellow/orange color and its comparatively small eye. The name golden hammerhead is the obvious choice for anyone who has seen a live or fresh specimen. The name smalleye hammerhead, used by the A.F.S., was given to it by taxonomists who had seen only the preserved, faded specimens.

Spanish name The A.F. S. name is *cornuda ojichica.*

Scientific name *Sphyrna tudes* Valenciennes, 1822, in *Mémoires du Muséum d'Histoire Naturelle* (1822, 9: 225–226). *Sphyrna:* a corruption of the Greek name *sphyra,* a hammer or a mallet, an allusion to the shape of the head and, according to Jaeger (1962: 244), "one of Rafinesque's efforts at word-butchery"; *tudes:* Latin, hammer, another allusion to the shape of the head.

Synonyms *Sphyrna bigelowi* Springer, 1944.

Note: The name *Sphyrna tudes* was long applied to the great hammerhead, now *Sphyrna mokarran.* Apparently, in his description of the species, Valenciennes illustrated a specimen from Nice, France, and mentioned a late embryo from Cayenne, French Guiana. Over the years, this resulted in a great deal of taxonomic confusion, and the species was believed to exist in both tropical America and in the Mediterranean. Springer described the golden hammerhead as *Sphyrna bigelowi* in 1944. Later, Gilbert (1967) designated the Nice specimen as the lectotype[1] of *Sphyrna tudes* and the Cayenne specimen as the paralectotype, further confusing the issue. Ultimately, Cadenat and Blache (1981) realized that the two specimens represented two species, one American and one Mediterranean (*Sphyrna couardi*). To stabilize the nomenclature, the name *Sphyrna tudes* has been retained for the golden hammerhead of the American tropics. For a summary of this taxonomic confusion, see Compagno (1984b: 553).

Identification This hammerhead is characterized by a deep central indentation on the anterior margin of the head, a *well-developed inner narial groove,* and a first dorsal fin free rear tip of that reaches the level of the pelvic fins. The upper teeth have oblique, smooth-edged cusps; the lower teeth have erect, smooth-edged cusps. Teeth number U: 15 or 16–15 or 16, L: 15 or 16–15 or 16. The dermal denticles broad and overlapping. They have three parallel ridges and three points, the central ones being slightly longer than the others. The *bright yellow or orange color, with metallic or iridescent hues,* makes this shark one of the easiest to recognize, as well as one of the most beautiful sharks. The dorsal fins and the dorsal surface down to the level of the upper

1. When the type series of a species contains more than one specimen, a specimen can be designated as the lectotype, and it becomes the unique bearer of the species name. The other specimens of the type series then become paralectotypes.

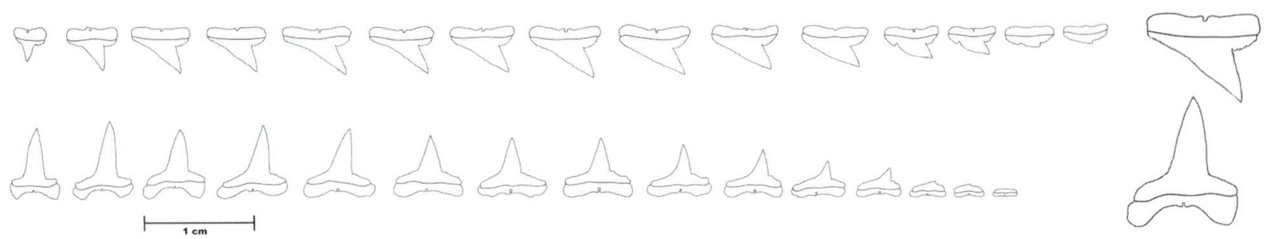

Fig. 134b. Teeth.

margins of the gill slits are yellowish gray. On the head, the anterior margin and the wide membranous rear margin are bright yellow or orange. The flanks, ventral surface, and pectoral, pelvic, anal, and caudal fins are also bright yellow or orange. In life, these colors have metallic or golden hues, which fade soon after death. Specimens preserved in alcohol turn gray with a yellowish cast because the yellow and orange pigments (carotenes) are soluble in alcohol. Live term fetuses and newborns are gray above with very pale yellow undersides (apparently acquiring the yellow pigment from the mother); the first dorsal fin and the upper caudal lobe have dusky or dark tips. Small juveniles up to 45 cm are yellowish gray above with pale to bright yellow undersides. At about 50 cm, juveniles change color to gray above with orange undersides. Specimens 55–70 cm have the brightest and most intense colors. The colors fade with the onset of sexual maturity and increasing age. Adults 90 cm and larger are gray with pale yellow undersides. Occasionally, large juveniles can have the very bright orange coloration of smaller (55–70 cm) specimens.

Similar species The coloration of live or fresh specimens is generally enough to separate the golden hammerhead from other hammerheads. The great hammerhead, the scalloped hammerhead, and the smooth hammerhead have the free rear tip of the first dorsal fin well ahead of the origin of the pelvic fins. The scoophead has a more rounded anterior margin of the head and lacks the deep central indentation.

Range The golden hammerhead has been reported from Venezuela to Uruguay (Cervigón 1966). Cervigón (1966) stated that the golden hammerhead was unknown to fishermen of Isla Margarita, off the northeastern coast of Venezuela, and that he found it only to the southeast of Trinidad, over muddy bottoms in the region of the Orinoco delta. Cervigón also stated that the species occurred down to 36 m

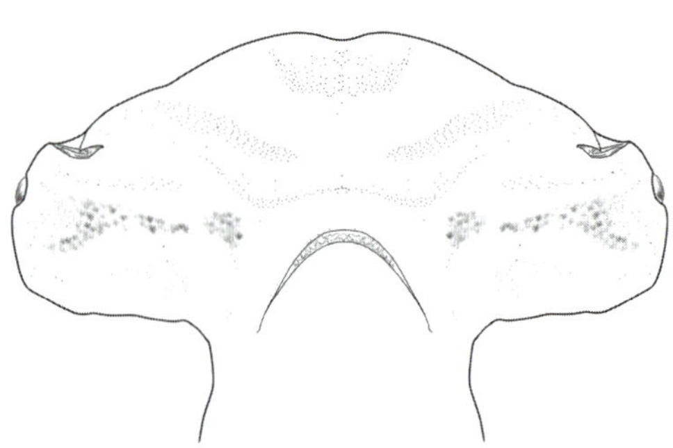

Fig. 134c. Snout.

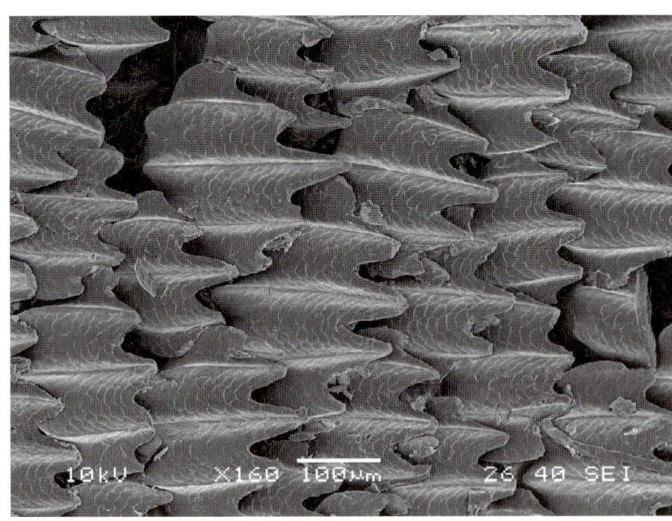

Fig. 134d. Dermal denticles.

and was not very abundant. Castro (1989) reported that the golden hammerhead was common off Trinidad, from Matelot, on the northern coast, to Mayaro Point on the eastern coast. Sadowsky (1965) reported the golden hammerhead as abundant in littoral waters off Cananeia, Brazil.

A record of this species off Mississippi in the northern Gulf of Mexico based on Gilbert (1961) is incorrect because of erroneous locality data (Robins et al. 1980). The golden hammerhead is included in this book as an extralimital species and to complete the account of the hammerhead of the Americas.

Size and weight The golden hammerhead is a small shark. Average size for adults off Trinidad was 108 cm and 6.4 kg. The largest male I examined was 121 cm and weighed 8.7 kg, whereas the largest female was a nongravid specimen 120 cm and 9.1 kg (Castro 1989). Sadowsky (1965) reported 10 females 122–148 cm from Brazil.

Biology The golden hammerhead inhabits coastal waters of the northwestern coast of South America to a depth of 40 m. Other than brief mentions in faunistic works, the species is known only from my work in Trinidad (Castro 1989). Off Trinidad, it is confined to the muddy waters from Matelot to Mayaro Point; it was not found elsewhere around the island, despite extensive sampling. Adult males and juveniles of both sexes form schools of individuals of uniform size that are most abundant at depths of 27–36 m. Mature females do not appear to school and were caught in shallower water 9–18 m deep. Newborns and juveniles less than 40 cm are found in shallower beach waters and move to deeper waters a few months after birth.

The most unusual characteristic of the golden hammerhead is its color, which results from ingestion of a carotenoid pigment in the food. Juveniles develop the coloration after they start feeding on shrimp. Their color intensifies to orange as they reach 50–60 cm and begins to fade after they change to a fish diet at about 67 cm. Large juveniles and adults appear to pick up pigments from catfish and their eggs. Whether the catfish pigment is also acquired from shrimp, or whether it is produced by the catfish is unknown (Castro 1989).

There are few references in the literature to the strikingly bright and unusual coloration of the golden hammerhead. This is obviously because the pigment fades quickly from preserved specimens, being soluble in alcohol, and museum-bound naturalists saw only faded specimens. The earliest reference to the unusual coloration that I found is by Springer (1944), who attributed the yellowish cast of museum specimens to their preservation in alcohol (Springer was certainly not a museum-confined naturalist!). Much later, probably after seeing live specimens, Springer (1979b) stated that: "The species is usually the ordinary shark-gray colour but an occasional shark of the species is a vivid golden yellow. Probably this is a colour imparted by some organism that the shark has eaten. A carotene like substance might be suspected." All the golden hammerheads I have seen off Trinidad were yellow or orange. A carotenoid pigment is responsible for the coloration, and the shark acquires the pigment and a metabolite through the food chain (Castro 1989). A sympatric species, the yellow smoothhound (*Mustelus higmani*), a species of the eastern coast (Atlantic) of South America, also feeds on shrimp and has an overall yellowish tinge, with bright yellow dorsal and caudal fins. However, its colors do not approach the intensity and brightness of the golden hammerhead. It is likely that the golden yellow coloration of the golden hammerhead and the yellow smoothhound serves to camouflage the sharks in the muddy brownish waters where they live. The few times that I observed gillnet-caught golden hammerheads under water, in the brownish waters of northern Trinidad, they were practically invisible at even at close ranges.

Diet: Juveniles less than 67 cm feed mainly on penaeid shrimp (usually *Xiphopenaeus kroyeri*). Stomach contents often consist of an orange paste of partially digested shrimp that closely matched the shark skin color (Castro 1989). Large juveniles and adults feed on small bony fishes, primarily sea catfish (*Ariidae*), and their eggs. Of 116 specimens larger than 67 cm I examined, 103 (89%) had the remains of bony fishes in the stomach, 21 (18%) had shrimp, and 11 (9%) had catfish eggs (some sharks had more than one type of food). Many adult sharks had stomachs full of intact or partially digested catfish eggs (Castro 1989).

The golden hammerhead provides an interesting study in niche partitioning. There are five species of hammerheads found in the coastal waters of Trinidad. Three of these, the golden hammerhead, the scoophead, and the bonnethead, were commonly taken together over the shallow muddy bottoms of eastern Trinidad, whereas two species, the scalloped hammerhead and the great hammerhead, were caught only occasionally in that habitat. These various hammerheads avoid interspecific competition by dietary and habitat differences. The golden hammerhead is the most abundant hammerhead over muddy coastal bottoms of the Orinoco delta region, where the depth is less than 40 m. The bonnethead is the dominant hammerhead over the shallow sandy beaches of southern Trinidad. It feeds primarily on crabs (*Calappa*) and stomatopods (*Squilla*). The scoophead is found in deeper water over rocky bottoms and feeds on small elasmobranchs (*Narcine*), octopuses, squids, and flounders. The scalloped hammerhead is predominantly a clear-water species that occasionally wanders into murky waters; it is most abundant in the clearer waters of the northern coast and feeds on a wide range of batoids and bony fishes. The great hammerhead is an oceanic-coastal species found around reefs and clear coastal waters. It seldom enters muddy waters, and it feeds on batoids and numerous bony fishes. Unlike its sympatric congeners, the golden hammerhead has a proportionally small eye. This may be an adaptation to the murky waters of its habitat, where vision must play only a small role in prey location. All the other sympatric hammerheads have proportionally larger eyes and prefer clearer waters.

Size at maturity

Male: Golden hammerhead males reach maturity at about 80 cm. The claspers develop abruptly after the animal reaches 75 cm and become calcified at about 80 cm. Of 54 males under 80 cm, 53 (98%) had soft, immature claspers, whereas of 95 males longer than 80 cm, 93 (98%) had calcified claspers (Castro 1989).

Female: Females reach maturity at about 98 cm (Castro 1989).

Reproduction The golden hammerhead is a placental viviparous species. The embryos are enclosed within large egg envelopes that contain a considerable amount of fluid (histotrophe). The placenta is similar to that of the bonnethead (Schlernitzauer and Gilbert 1966). The embryos have umbilical cords bearing numerous appendiculae, which are short, spatulate, and broad projections growing out of the umbilical cord (Alcock 1890). Ovulation and the beginning of gestation occur in August and September, presumably right after mating. Parturition occurs in late May and June, after a gestation period of 10–11 months (Castro 1989).

Female golden hammerheads with near-term embryos (about 30 cm) carry ripe oocytes 10 mm in diameter, indicating that gestation is concurrent with the ovarian cycle. Thus, it appears that females conceive again shortly after parturition and that the species reproduces yearly. An annual reproductive cycle would be consistent with the pattern of the scalloped hammerhead and the bonnethead, which also reproduce annually.

Size at birth: The young measure about 30–31 cm at birth. The smallest free-swimming young I captured off Trinidad were 32 and 33.5 cm (Castro 1989).

Brood size: Broods I examined ranged from 5 to 12; the average of 12 broods was 8 (Castro 1989).

Nurseries Parturition occurs in shallow coastal waters at depths of 9–18 m. In Trinidad, the waters from Balandra to Mayaro Point, with their shallow bays and abundant shrimp and catfish, are the golden hammerhead nurseries (Castro 1989).

Age and growth The species has not been aged.

Relation to humans When I lived in Port of Spain, Trinidad, in 1985–1986, the golden hammerhead used to be common in the fish market. However, when I visited the island in 1996, few specimens were to be seen in the market. Similarly, in Mayaro, the fishermen used to land dozens of golden hammerheads daily in the mid-1980s, but Isaw few in 1996. This stark change suggested that the introduction of monofilament gillnets in the 1980s severely affected the golden hammerhead populations in the area. Subsequent reports from Trinidad have confirmed my observation.

SMOOTH HAMMERHEAD

Sphyrna zygaena (Linnaeus, 1758)

Fig. 135a. *Sphyrna zygaena*, smooth hammerhead shark, immature male, 156 cm, 14 kg; off Mazatlan, Mexico (Humberto Ortega and Fernando Márquez, collectors).

Common name Smooth hammerhead, a reference to the rounded anterior margin of the head, which lacks the central indentation seen in the other large hammerheads.

Spanish name *Cabeza de martillo* (Cuba), *cornuda prieta* (Mexico, A.F.S.).

Scientific name *Sphyrna zygaena* (Linnaeus, 1758), originally described as *Squalus zygaena* in *Systema Naturae,* 10th ed. I: 234). *Sphyrna:* a corruption of the Greek name *sphyra,* a hammer or a mallet, an obvious allusion to the shape of the head and, according to Jaeger (1962: 244), "one of Rafinesque's efforts at word-butchery"; *zygaena:* from the Greek *zygaina,* the ancient name for a hammerhead, derived from *zygon* meaning yoke, an allusion to the shape of the head.

Synonyms *Zygaena subarcuata* Storer, 1848.

Note: This species is one of the two hammerheads originally named by Linnaeus. Thus, its scientific name has been stable for centuries, unlike those of many other hammerheads, and synonyms are seldom encountered in the scientific literature.

Identification The smooth hammerhead is characterized by the *rounded anterior margin of the head that lacks a central indentation,* and by a first dorsal fin free rear tip well anterior to the pelvic fin origin. The teeth have oblique, smooth-edged or weakly serrated cusps with a notch on their outer edges; they are similar in both jaws and number U: 13 to 15–0 or 1–13 to 15, L: 12 to 14–1–12 to 14. The dermal denticles are very broad and overlap greatly. They have five to seven diverging ridges and an equal number of points. Coloration is variable. All the specimens I have seen in the Gulf of California have been dark gray or black above and dirty white to white below, with dusky or black-tipped pectoral fins. Most of the East Coast specimens that I have seen also had this black coloration, although one Florida specimen had light olive brown dorsal surfaces.

Similar species The combination of an anterior margin of the head without a central indentation and the relative position of its first dorsal fin well anterior of the pelvic fins distinguishes it from the other hammerhead sharks, except for the bonnethead, which has a shovel-shaped head. However, confusion among species is common in the older literature.

Range The smooth hammerhead inhabits temperate waters throughout the world and is rare or absent in tropical waters. In eastern North America, it ranges from the Bay Fundy and the eastern coast of Nova Scotia to the Florida Keys. The northernmost records are from Halifax Harbor (Vladykov 1935) and Sambro Island, Nova Scotia (McKenzie 1939). There is also a record from Brier Island, at the entrance to the Bay of Fundy (McKenzie 1939). Most Canadian records are of small specimens taken in late summer. The southernmost record I know of is a specimen caught off Islamorada, in the Florida Keys, in October 1978 and examined by Gordon Hubbell. The species is rare off the southeastern United States. Murdy et al. (1997) called the smooth hammerhead a rare to occasional visitor to the Chesapeake Bay. I only know of few specimens taken off eastern

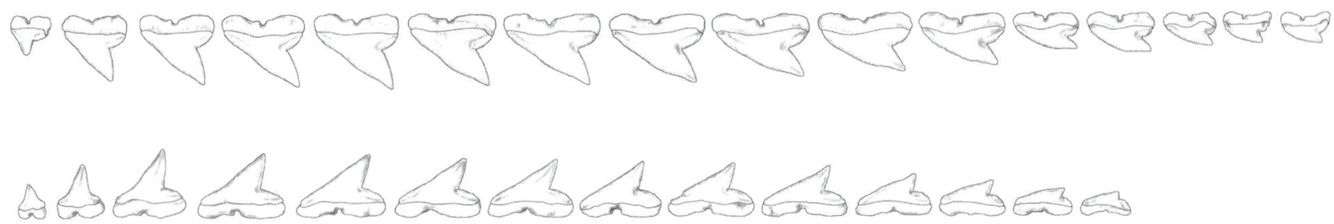

Fig. 135b. Teeth.

Florida. In 2009 three specimens were taken off Miami Beach by Mark Quartiano. The smooth hammerhead has not been reported from the Gulf of Mexico. Along the western coast of North America, it is found from central California (Miller and Lea 1972) to Chile. It is locally abundant in the Gulf of California, where I have seen dozens of specimens of all sizes taken by migrant fishermen on both sides of the Gulf. It is seasonally abundant along the coasts of Sinaloa and Nayarit, Mexico.

Size and weight According to Bigelow and Schroeder (1948), the smooth hammerhead reaches 380 cm and more than 400 kg. C. R. Gilbert (1967: 37) noted, "Maximum total length probably between 12 and 13 feet" (between 366 cm and 396 cm) but did not cite any specimens. I have not observed such large sharks, the largest I have examined measured 266 cm. Representative sizes and weights that I have recorded are as follows:

Male: 61 cm, 0.8 kg; 75 cm, 1.5 kg; 140 cm, 10.4 kg; 152 cm, 13 kg; 208 cm, 43 kg; 232 cm, 60 kg; 238 cm, 61 kg; and 266 cm, 101 kg.

Female: 60 cm, 0.9 kg; 120 cm, 7.7 kg; 180 cm, 30 kg; 234 cm, 70 kg; and 246 cm, 66 kg.

Biology The smooth hammerhead is a nearly cosmopolitan species in temperate seas. It inhabits much cooler waters than other hammerhead species; thus, it ranges much farther north along the North American coasts than any other hammerhead. Its habits are poorly known.

Along the eastern coast of North America, the smooth hammerhead is a scarce fish. Neonates and juveniles have been reported from Delaware Bay to the Bay of Fundy. All the specimens caught south of Cape Hatteras have been large juveniles or adults as far as I know. I examined only three Florida specimens: a 246-cm female taken off Port Salerno

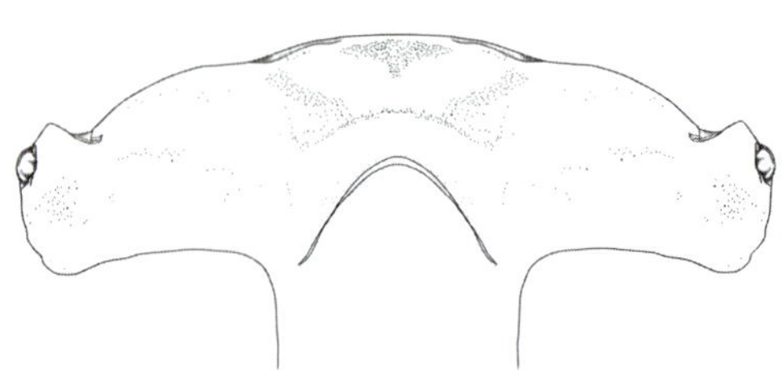

Fig. 135c. Snout.

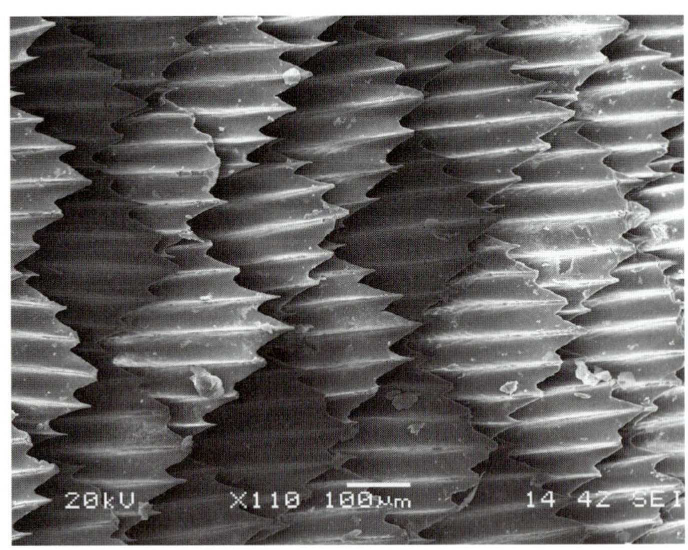

Fig. 135d. Dermal denticles.

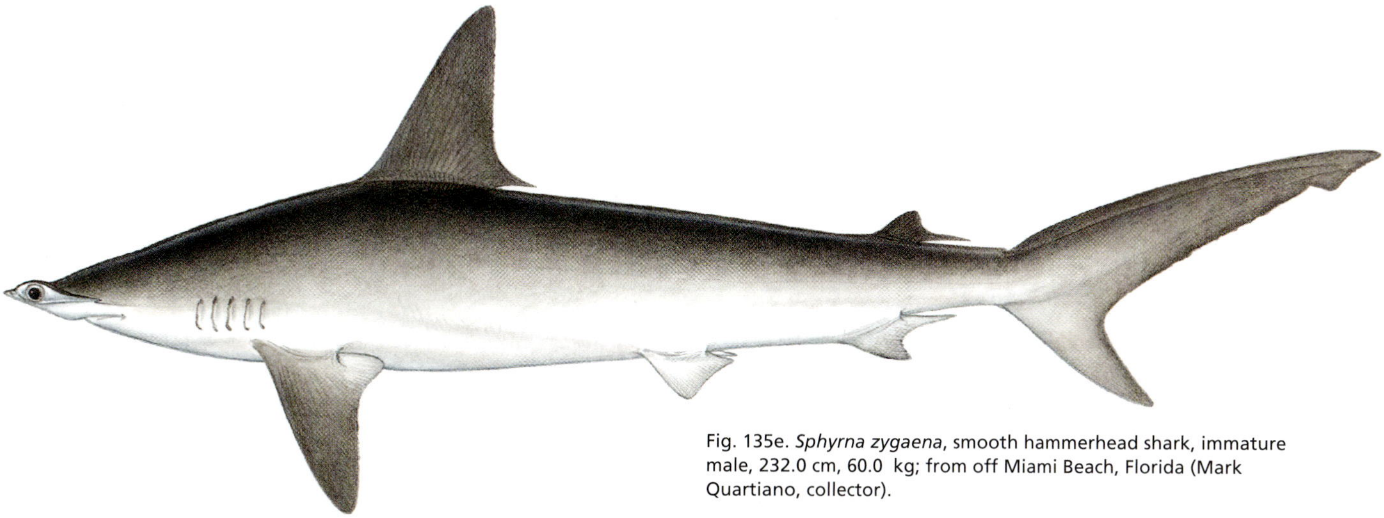

Fig. 135e. *Sphyrna zygaena*, smooth hammerhead shark, immature male, 232.0 cm, 60.0 kg; from off Miami Beach, Florida (Mark Quartiano, collector).

on 11 September 1983; a 208-cm male caught off St. Augustine in early May 1991, taken in water 37-m deep and surface temperature of 25.6°C; and a 232-cm male taken, at a depth of 55 m and a surface water temperature of 26.6°C off Miami Beach on 20 November 2008. It is likely that the smooth hammerhead is highly migratory, like most other hammerhead sharks, but nothing has been written about its migrations in North American waters.

Size at maturity There are few data on the size at maturity. I have seen only small juveniles and adults. The largest juvenile male I examined measured 232 cm and has 15-cm immature claspers. According to Cadenat and Blache (1981), the species matures between 240 and 270 cm.

Reproduction The smooth hammerhead is a placental viviparous species. The umbilical cords of the young are smooth; they lack the appendiculae that cover the umbilical cords of the embryos of some hammerhead species. The two gravid females carrying term young that I have examined had inactive ovaries; that is, they did not have ripe oocytes. This indicates that the reproductive cycle is probably biennial, like that of the great hammerhead.

Size at birth: The young measure 53–65 cm at birth. One 273-cm gravid female that I examined carried 22 embryos measuring 56–65 cm. Another female, measuring 263 cm total length, carried 28 embryos. I examined numerous neonates measuring 52–60 cm in the last week of March in Bahia Kino and at the fishing camp of El Choyudo, Sonora, Mexico.

Brood size: Based on very limited data, it appears that smooth hammerhead broods range from 22 to 28 young.

Nurseries The locations of smooth hammerhead nurseries along the eastern coast of North America are not well defined. Small numbers of neonates and small juveniles have been found from Delaware Bay to the Bay of Fundy. Along the western coast of North America, the smooth hammerhead has nurseries in the coastal waters of the upper Gulf of California and along the Sonora coast, Mexico. I found numerous neonates with unhealed umbilical scars along the Sonora coast from late March to May; small juveniles are caught there almost year-round. It is likely that the nursery extends farther southward along the Mexican Pacific.

Age and growth The species has not been aged.

Relation to humans The smooth hammerhead is of no economic importance along the eastern coast of North America due to its scarcity. Large numbers are caught by migrant fishermen in the Gulf of California, and the species may be of local economic importance for food and fins.

APPENDIX 1

A Primer on Shark Reproduction

This appendix provides a simplified summary of the reproductive processes of North American sharks, to enable the casual reader to understand the section on reproduction of the various species. Readers desiring more information can consult the excellent reviews by Wourms (1977, 1981). Although several classifications of the reproductive systems of sharks have emerged in recent years, these reviews remain useful.

The survival and evolutionary success of sharks are partly due to their reproductive adaptations that depart from the egg laying typical of bony fish (oviparity). These adaptations are internal fertilization, the production of fully developed young, and viviparity (giving birth to live young).

Oviparity, or egg laying, is the primitive mode of reproduction in bony fishes. Oviparous fishes produce large numbers of eggs and sperm that are released into the water. Fertilization occurs in the water, so it is said to be external. Most bony fishes retain this mode of reproduction. Their eggs are usually small and have only a minute amount of stored yolk. With limited available nutrients, the embryos of oviparous fishes must hatch in an undeveloped or larval condition and continue their development after hatching. The minute larvae must quickly find food or perish. The larvae often require weeks or months to complete their development. Both eggs and larvae are highly vulnerable to predators and environmental factors for prolonged periods and suffer heavy mortality.

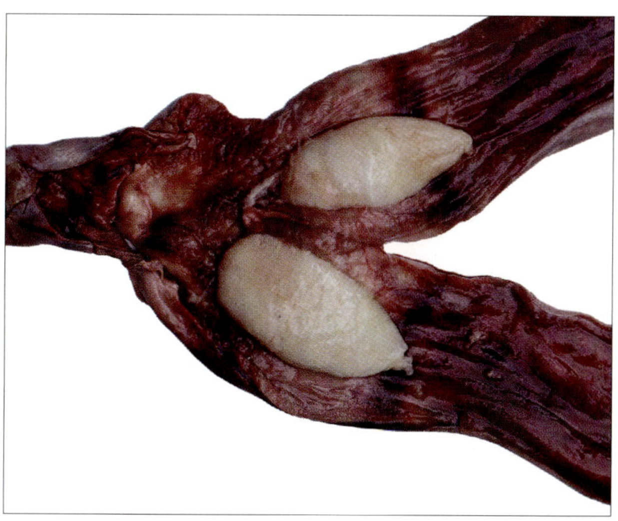

Figure A1. Masses of sperm (spermozeugmata) deposited by the male at the base of the uterus.

Early in their evolutionary history, sharks evolved several adaptations that reduce the mortality of their young and enhance their chances of survival. The first of these is internal fertilization. In animals with internal fertilization, the male transfers its sperm to the female, and fertilization occurs inside the female's body. In male sharks, the pelvic fins are modified into stout, tubelike copulatory organs known as claspers, which are inserted into the female during copulation to transfer sperm (figure A1).

The clasper's tip is armed with hooks or spurs, and it expands when inserted into the female, effectively attaching the male, hence the name "claspers." All sharks have internal fertilization, so their embryos spend their most vulnerable stages protected inside their mother's uterus, and their survival is enhanced.

The second survival-enhancing reproductive adaptation of sharks is the production of large young that hatch, or are born, as fully developed, active, miniature sharks that are relatively large, often reaching 10–25% of the length of the mother. To achieve, this sharks have evolved several mechanisms to nurture their young to large size, and these mechanisms are discussed below. A large size at hatching or birth reduces the number of potential predators and competitors, while increasing the number of potential prey, thus enhancing survival.

The third survival-enhancing adaptation of sharks is the evolution of viviparity, or giving birth to live young. Most sharks are viviparous: they retain their embryos through their entire developmental period and then give birth to fully formed young. Most species have evolved diverse the means of quickly transferring nutrients to their embryos, such as egg eating or through a yolk-sac placenta, so the embryos are not limited to the nutrients that can be packed in an egg. By one method or another, these embryos are fed directly and continuously by the mother, so they can

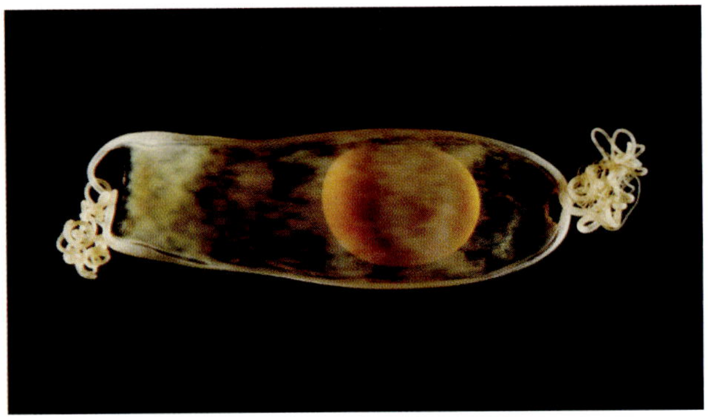

Figure A2. Trans-illuminated egg case of the chain dogfish, showing the fertilized egg inside.

Figure A3. Trans-illuminated egg case of the chain dogfish. The slit vents of the posterior egg case are open and the embryo constantly fans its tail to ventilate the case. The case is the only protection of the embryo has during the many months of development in its dark, cold environment.

be brought faster to a larger size, which enhances their survival.

The production of large, fully developed young requires that large amounts of nutrients be available to the embryos to nurture them to large size. Thus, because of the great energy demands on the mother, the number of young in a brood is usually small, and most species of sharks produce broods of a dozen or less, although some large species may carry broods of a hundred or more. Sharks have evolved several modes of nourishing their young to large size. Traditionally these modes have been divided into three broad functional categories: oviparity or egg laying, aplacental viviparity, and placental viviparity.[1] These are simply functional groups and not phylogenetic categories, as some families of sharks have both aplacental and placental species.

Oviparity in sharks differs from that of bony fishes. Oviparous sharks produce large eggs that contain sufficient yolk to nourish the embryo throughout development and allow it to emerge as a fully developed miniature shark. The eggs are enclosed in horny egg cases or capsules (figure A2) that are often attached to upright bottom structures (e.g., corals or sponges) by means of adhesive tendrils, or are deposited on the bottom. Once deposited, eggs receive no further care from the parents; there are no known cases of parental care in sharks. The horny egg case is the sole protection of the embryo (figure A3). The egg cases are generally of two types. They can be rectangular with adhesive tendrils at four corners for their attachment, or they can be conical spirals that can bury on the bottom. After the first weeks of development, embryonic enzymes cause small slits to open on the sides of the egg cases; these slits allow for the ventilation of the egg case. The embryo aids in exchanging the water in the case by fanning its tail, thus renewing the oxygen and voiding metabolites. Development can last from several months to longer than a year (perhaps lasting as much as two years), depending on water temperature. The hatchlings of oviparous sharks are proportionally small in comparison to the mother's size, because these embryos have only the small amount of yolk in the egg case to nurture them through development.

Aplacental viviparity, also known as ovoviviparity, is the most common mode of reproduction in sharks. These are viviparous sharks that nourish their young to large size by various adaptations that do not include a placental connection. This is a heterogeneous group that includes sharks of different lineages. One group, represented by the squaloid sharks, produces large eggs that contain sufficient yolk to sustain embryos for periods of about two years (figures A4 and A5). No additional nutrients are shunted to the em-

1. The modes of reproduction in sharks have been the subject of several classifications in recent years. Most of these classifications are simply new combinations of old terms and add little to our understanding. Here I retain the traditional terms as this is intended to be a simple primer on reproduction for the reader.

Figure A4. Spiny dogfish embryo during the first year of development. All the necessary nutrients for its development are stored in the yolk sac.

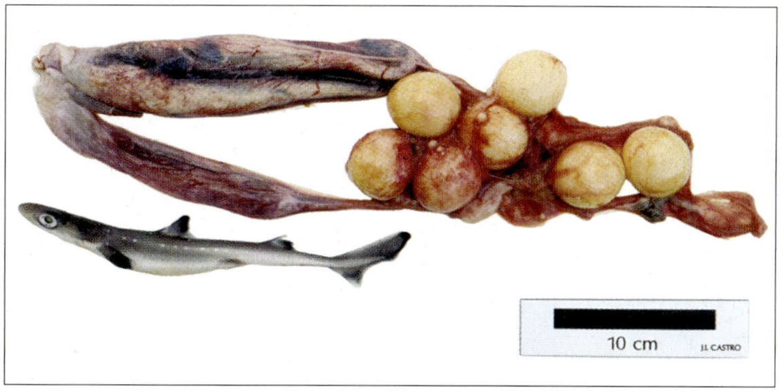

Figure A5. Reproductive tract of the term spiny dogfish. One uterus has been opened to reveal the fully formed embryo inside. The other uterus still holds two other embryos. The ovary has ripe oocytes that will be ovulated during mating shortly after parturition.

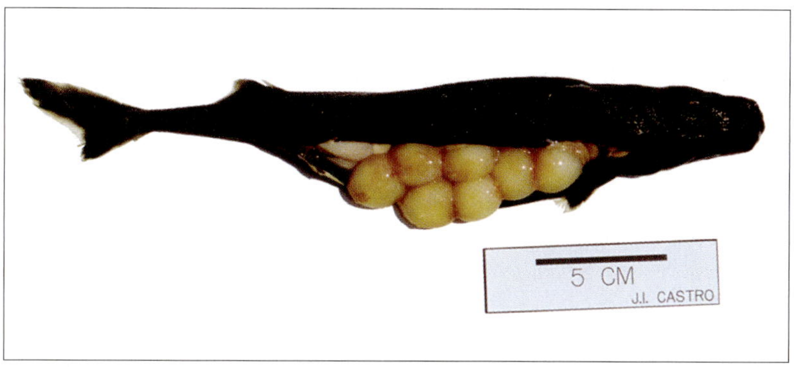

Figure A6. Ovulatory female of the broadband lanternshark. The abdomen has been opened to reveal the huge oocytes. These oocytes contain sufficient yolk to nourish the embryo during its long gestation.

bryos. Their eggs can be very large in proportion to the size of the female (figure A6). Another group, which includes the nurse shark and the whale shark, produces large numbers of eggs of medium size that produce small young (figure A7). These species produce unusually large broods (for sharks); nurse sharks usually have broods of 30–50 young, while whale sharks may produce 300 or more offspring. Because their eggs are proportionally small and their broods so large, their young are relatively small at birth, about 12% of the mother's total length in the case of the nurse shark and about 6% for the whale shark. Due to the large broods, it may take several weeks for all the eggs to be ovulated and fertilized. Thus, the embryos of the first eggs fertilized may be several weeks older than the last fertilized eggs, and consequently, there is great disparity in the developmental stages of the embryos. Some embryos may still be in their egg cases while the older embryos may be ready for birth (figure A8).

The tiger shark demonstrates another more interesting type of aplacental viviparity. The tiger shark produces relatively small eggs and large young. Apparently, the young are nourished to a large size by a uterine fluid (histotrophe). Tiger shark embryos are found in huge fluid-filled egg cases (figure A9), and the young apparently imbibe this fluid. Little is known about this type of embryonic nourishment.

Figure A7. Nurse shark embryo about midterm, shortly after emerging from the egg case.

Figure A8; Nurse shark uterus, opened to reveal embryos in different stages of development. Three embryos are still in their egg cases at the base of the uterus. The other embryos are in progressive stages of yolk sac absorption. Some have fully absorbed yolk sacs.

The lamnid and thresher sharks constitute yet another group of aplacental viviparous sharks characterized by large embryos that feed directly on eggs in the uterus: oophagy ("egg eating"). Oophagy is an ancient adaptation of lamnoid embryos. In their early stages oophagous embryos are lecithotrophic (nourished by yolk stored in their yolk sacs), as are the embryos of most species of sharks. Their yolk sacs are minute, and they usually contain less than 1 cc of yolk; their contents are quickly consumed during the first few weeks of development. These embryos develop a precocious dentition, and by the time they reach 5 cm they have well-defined teeth. After consuming the contents of their yolk sacs, these embryos feed on ovulated eggs in the uterus. In these sharks, ovulation does not stop after a few eggs are fertilized but continues through much of gestation, producing a nearly continuous supply of food for the embryos. Their precocious teeth allow them to pierce the soft egg cases and to consume their contents.

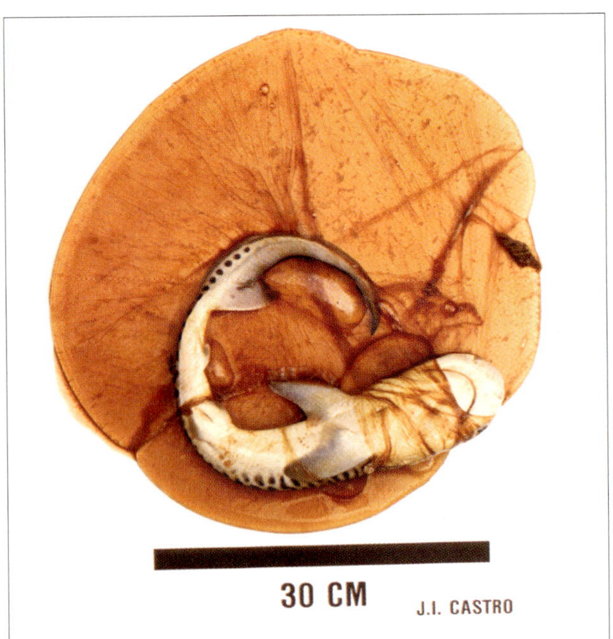

Figure A9. Near term tiger shark embryo in its fluid-filled egg case.

Several variations of oophagy are known. In the some sharks (the makos, porbeagle, salmon shark, sand tiger, and probably the white shark), several egg cases containing a single egg are ovulated into each uterus. The embryos hatch out of their egg cases when only 3–4 cm long and quickly consume the contents of their yolk sacs. Thereafter, the embryos feed on subsequent egg cases containing several eggs each, ovulated for their feeding. In these sharks the female ovulates eggs only during the earlier months of gestation. Consequently, these embryos consume great quantities of egg cases in early gestation, acquiring greatly distended yolk stomachs (figure A10). Several embryos may occupy each uterus, and in some species, as many as a dozen

Figure A10. 30-cm shortfin mako embryo, showing a well-developed yolk stomach.

Figure A12. A 5-cm sand tiger embryo showing its precocious dentition that it will soon use to kill its competing siblings in the uterus.

embryos may coexist in a uterus (figure A11). Having released most of its vitellogenic oocytes, the ovary ceases ovulating during the last few months of gestation, and the embryos are nourished by the nutrients stored in their huge yolk stomachs. They will digest the yolk over the following months, and their stomachs will slim down by the time they are born.

A peculiar type of oophagy exists in the sand tiger. In this species, when the largest embryo reaches 4–5 cm, it seeks the smaller embryos in the uterus and kills them (figures A12–A16). It will consume them later along with the egg cases. This behavior is known as uterine cannibalism, and only one embryo survives in each uterus, having consumed its smaller siblings. By late term, the ovary has released most of its eggs, and embryos have consumed and stored the yolk in their huge yolk stomachs (figure A17).

In other sharks, such as the threshers, the females ovulate continuously through gestation, providing the embryos with a continuous supply of nourishing eggs. Consequently, these embryos do not acquire distended yolk stomachs (figure A18). In the smalltooth thresher, the first egg case produced in each oviduct receives a single large, fertilized egg about 1 cm in diameter (figure A19). Subsequent egg cases enclose multiple small eggs 5–8 mm in diameter (figure A20). The embryos quickly exhaust the yolk in their yolk sac and then begin eating the eggs in the uterus. In early gestation, the uterus is quickly filled with egg cases as the cases are ovulated faster than the small embryo can consume

Figure A11. Shortfin mako brood. These embryos have already consumed all the egg cases necessary for their development and store the yolk in their huge yolk stomachs.

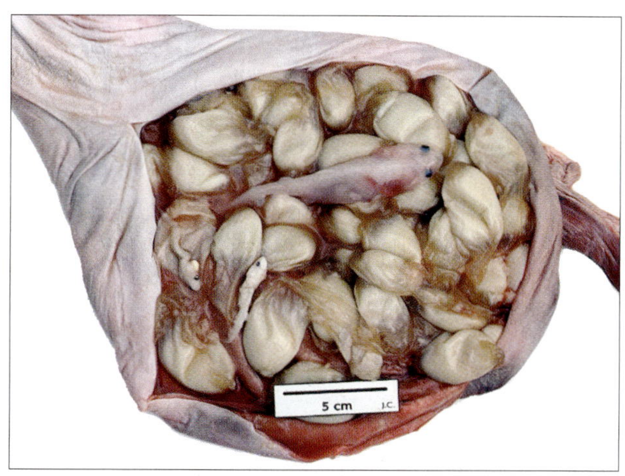

Figure A13. A sand tiger uterus during early gestation (June). The largest embryo (8-cm) has already killed all its siblings in that uterus. Two of the smaller dead embryos can be seen to the left. The embryos swim in a sea of egg cases that will soon be consumed by the largest survivor.

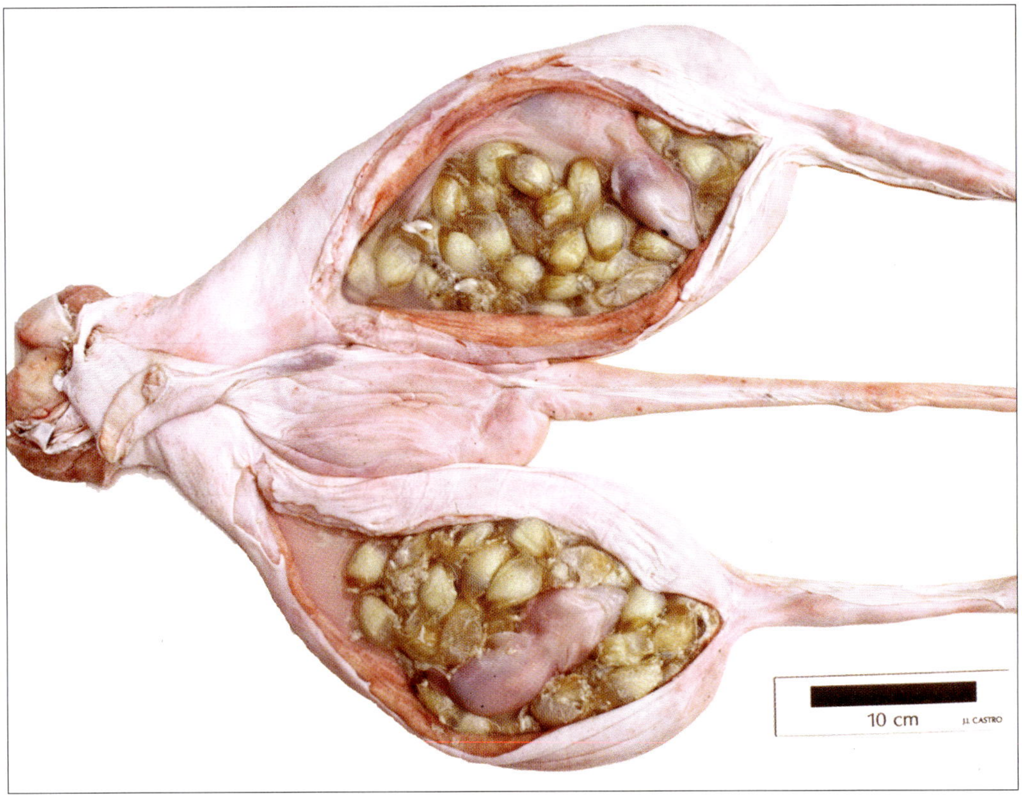

Figure A14. Panoramic view of the sand tiger uteri in early gestation. Each uterus is occupied by one 28-cm embryo that has killed its competing siblings and is surrounded by egg cases.

them (figure A21). As the embryo grows, the number of uneaten egg cases in the uterus decreases, eventually reaching the point when the embryos consume each egg case shortly after it reaches the uterus (figure A22). It is likely that one egg case is ovulated into each oviduct each day or every other day to feed the sole embryo. The crocodile shark exhibits one more type of oophagy, where the yolk is fed almost continuously to the four embryos (figure A23). All of these variations of the oophagy theme demonstrate that in the long evolution of the lamnoid sharks, multiple variations have occurred.

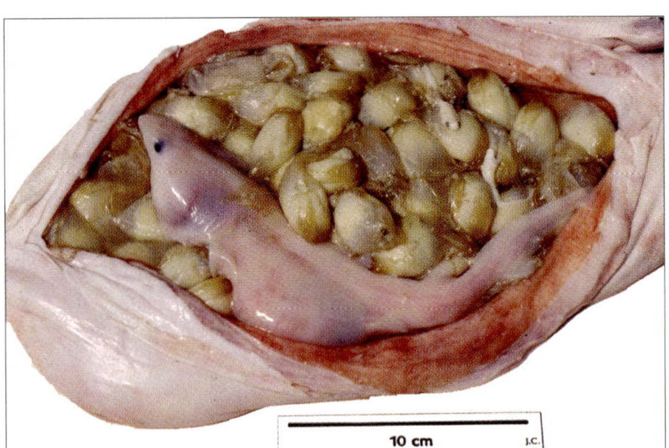

Figure A15. Detail of uterus in Fig. A14. The surviving embryo has already consumed many egg cases and it is starting to store yolk in its developing yolk stomach. It will eventually consume its dead siblings along with egg cases.

Placental viviparity is the most advanced mode of reproduction in sharks. The eggs of placental sharks are relatively small. The egg is surrounded by a very thin, almost transparent egg case. The egg case is large enough to encompass the term embryo, but at first it is highly pleated and folded over itself at the posterior end of the egg (figure A24). It will expand as the embryo grows into it (figure A25). During their first weeks of development, the embryos of placental sharks derive their nourishment from yolk stored in the yolk sac (figures A26–A28). The amount of stored yolk is modest and is consumed in just a few weeks. As the embryo grows and consumes the yolk and the yolk sac is emptied of its contents, the yolk stalk that connects the yolk sac to the embryo's gut becomes much elongated (figure A29). Where the flaccid, nearly empty yolk sac touches the uterine wall, the tissues of both the yolk sac and the uterine wall become

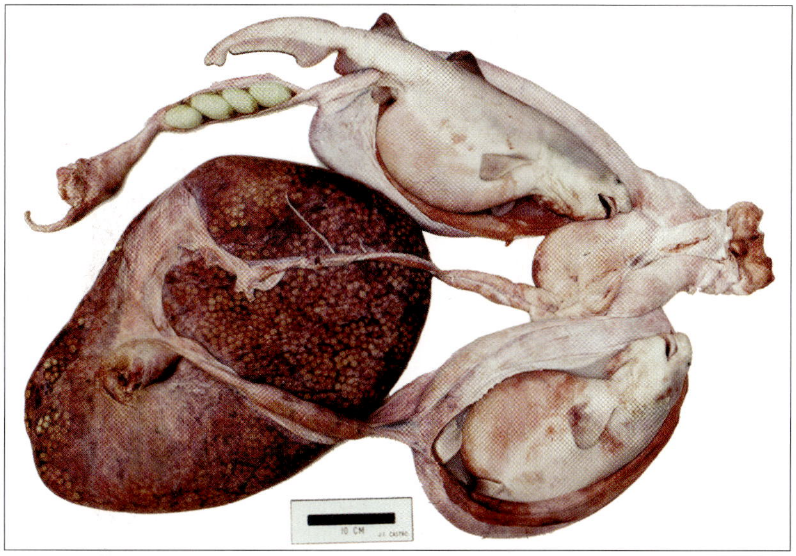

Figure A16. Sand tiger reproductive tract at the end of the fist trimester. The ovary is still huge and ovulates its oocytes continuously. The 5–6 mm oocytes can be seen on the surface of the ovary, and 15–20 oocytes are packed into each egg case. Four egg cases can be seen in the upper oviduct en route to the embryo; three other egg cases can be discerned through the skin of the other oviduct.

Figure A17. Two 80-cm embryos and flaccid, spent ovary at the end of October. The embryos will spend the last three months consuming the yolk stored in their yolk stomachs. They will be born when their yolk stomachs have been absorbed.

highly vascularized and grow into each other, becoming intimately connected and forming the placenta (sometimes referred as the yolk sac placenta; figure A30). Once the placenta is formed, creating a connection between the blood stream of mother and embryo, the embryo can be nourished directly from the mother's blood stream (figure A31). With a constant supply of nutrients, the embryos of placental sharks can grow to large size, usually in about a year (figure A32). In most placental sharks, the embryos are each enclosed in separate compartments, or crypts, in the uterus. This adaptation probably serves to prevent interactions between embryos and to keep their umbilical cords

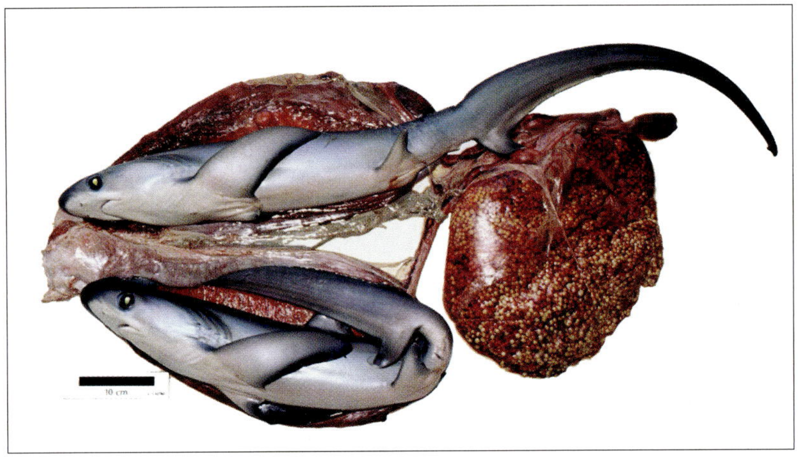

Figure A18. Pelagic thresher embryos (55-cm) at the end of their first trimester. Note that these oophagous embryos do not develop the distended yolk stomachs of other lamnoid sharks. These embryos are fed during their entire gestation period.

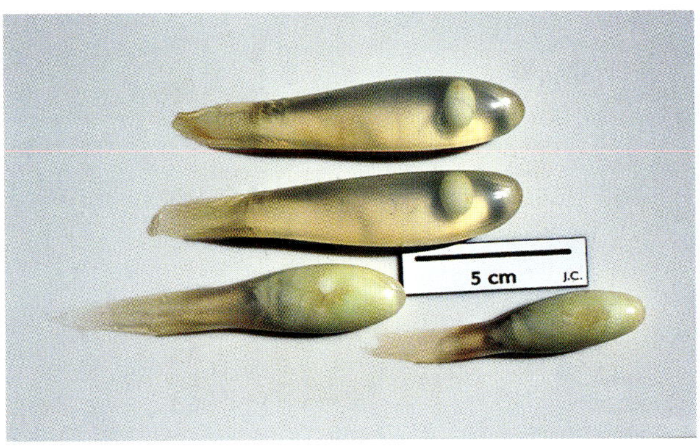

Figure A19. The first egg cases produced by a pelagic thresher. Each of the first two egg cases contains a single 1-cm egg; one such egg case goes into each uterus. Subsequent egg cases contain many unfertilized smaller eggs and are for consumption by the embryos.

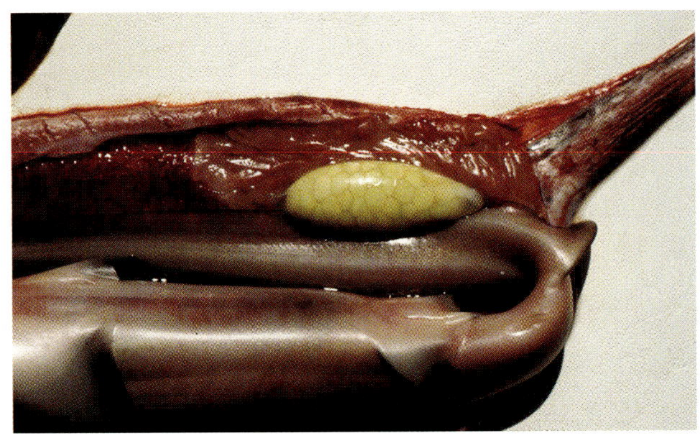

Figure A20. Pelagic thresher egg case en route to the embryo. Notice the many eggs packed into it.

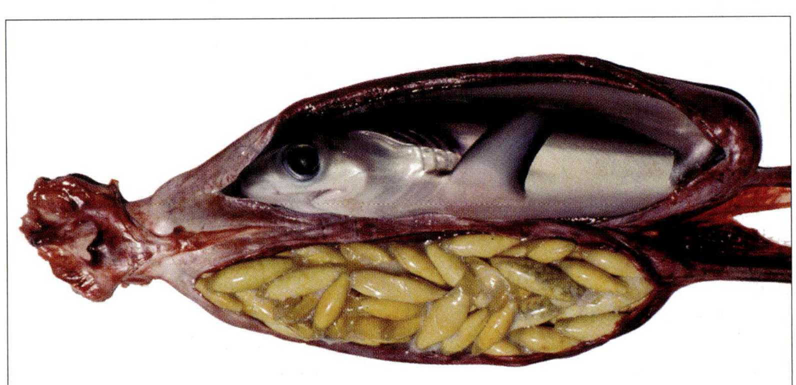

Figure A21. An interesting case when an embryo died in one uterus, and the uterus became filled with the egg cases that the embryo would have consumed. The embryo in the other uterus survived and was developing normally.

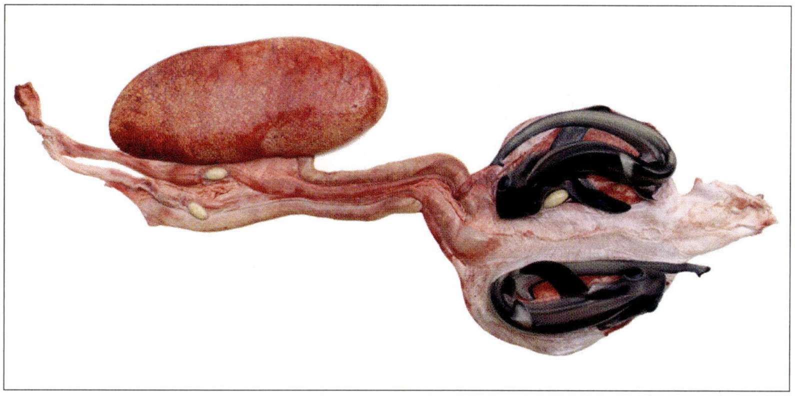

Figure A22. Bigeye thresher reproductive tract at about the end of the first trimester. Two egg cases are en route to the embryos. One embryo had already consumed one egg case; the other embryo had not yet consumed its egg case.

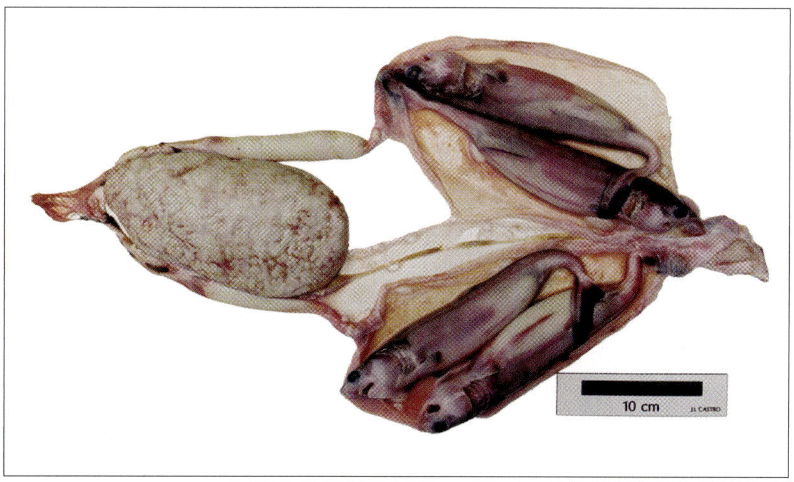

Figure A23. Reproductive tract of the crocodile shark, with the usual four embryos. Note size of the ovary and nearly continuous yolk shunt.

Figure A24. Early egg of the Atlantic sharpnose shark. The developing embryo in blastodisc stage can be discerned as a whitish round spot in the anterior (lower) side of the egg. The much folded egg case can be seen at the other end.

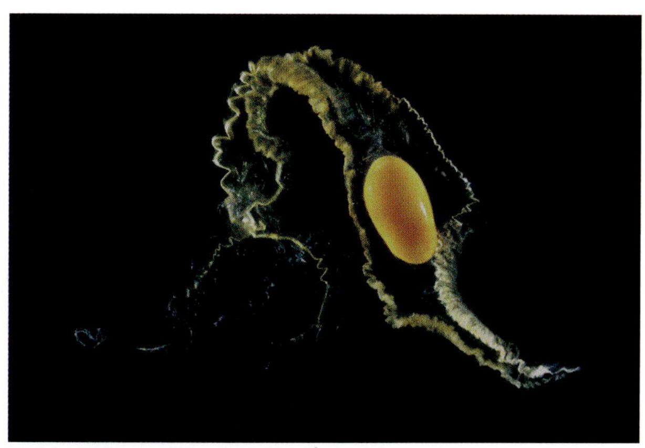

Figure A25. The same egg as in Fig. A24, showing the expanded egg case. The embryo can be seen as the whitish round spot on the lower part of the egg. During development the embryo grows into the egg case.

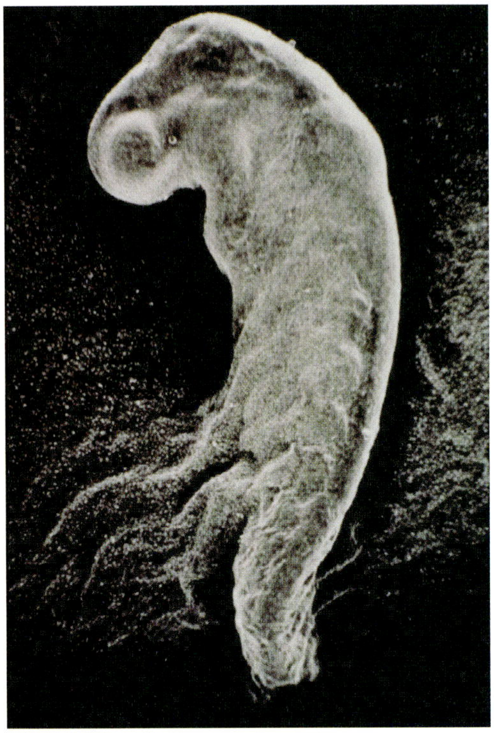

Figure A26. A 1.8-mm Atlantic sharpnose shark embryo rises from the surface of the egg.

Figure A27. A 12-mm Atlantic sharpnose shark embryo. At this stage the embryo is fully dependent on yolk stored in the yolk sac. The thin egg case surrounds the embryo and yolk sac.

from tangling with each other. In some placental sharks, such as the sharpnose sharks and some hammerheads, the umbilical cord is covered with long fingerlike protrusions known as *appendiculae*. Placental viviparity is confined to the more advanced sharks: the smoothhounds, the requiem sharks, and the hammerheads.

Sharks have very long reproductive cycles, which determine how often an animal is capable of reproducing. In males it is how often the animal is capable of producing ripe sperm once they have reached adult size. This usually occurs yearly for a few weeks in animals living in the temperate zone and for much longer periods or year-round in some tropical sharks. In females, the reproductive cycle includes the time it takes to develop the yolky eggs (vitel-

Figure A28. A 40-mm embryo inside its egg case. At this stage the embryo is still dependent on yolk reserves.

Figure A29. A 80-mm Atlantic sharpnose shark embryo has consumed most of the stored yolk in its yolk sac. The yolk stalk has grown long and the surface of the flaccid yolk sac is highly folded.

logenesis), mate, ovulate, lay the eggs, and gestate (nurture) the embryos to birth. The female reproductive cycle can be divided into two phases: the vitellogenic period, and the gestation period. These two periods can run concurrently or consecutively. For example, in the spiny dogfish the two periods run concurrently. A newly matured female spiny dogfish that has just mated for the first time begins to gestate and almost simultaneously begins vitellogenesis to develop the next batch of oocytes to be ovulated after her gestation ends. Thus, she nourishes both embryos and oocytes, and gestation and vitellogenesis can be said to occur concurrently. In the spiny dogfish the gestation period lasts about 23 months (Ketchen 1972), and the vitellogenic period appears to be just as long

In the carcharhinid sharks, the vitellogenic period and gestation run consecutively. For example, a female mating and ovulating for the first time in June will gestate for about a year and will give birth the following June (a year later). During gestation her ovaries are inactive and the oocytes do not develop much (figures 33 and 34). After birth the female enters a "resting period" of several months when there ovary remains inactive. In early winter, its vitellogenic cycle starts and the oocytes begin to develop and grow in size. The oocytes be become mature or ripe by the next June (the second year), when the animal will again mate and ovulate, and the cycle starts over. Thus, the cycle lasts two years and the shark reproduces every two years. Such cycles are called biennial.

Some sharks such as the Atlantic sharpnose shark and some hammerhead sharks have annual cycles. In these sharks the two periods run concurrently, and females carry developing young and developing oocytes (figure 35), both periods lasting about a year. The female mates and conceives again almost immediately after birth. Thus, these sharks reproduce annually. Reproductive cycles lasting more than three years have been suggested for some species of sharks, but the evidence is very limited. In general, the length of the reproductive cycle of most sharks is unknown, because it is difficult subject to study, and large numbers of specimens must be examined to arrive to meaningful conclusions.

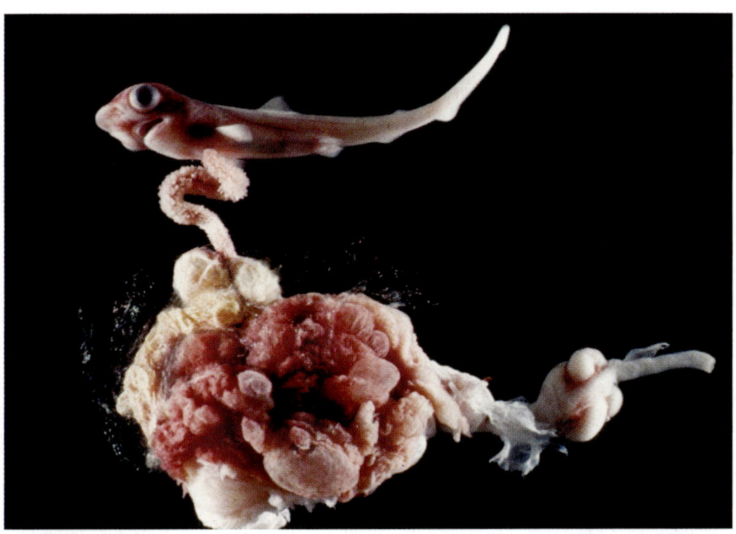

Figure A30. Atlantic sharpnose shark embryo at implantation. The much distended yolk sac surface comes into intimate contact with the much-folded, highly-vascularized uterine lining, forming the yolk sac placenta.

Figure A31. Implanted Atlantic sharpnose embryos (120–130 mm). Once the placenta is formed, nutrients can be shunted to the embryos directly from the blood stream of the mother and they can grow rapidly. Note that their umbilical cords are covered with long villi-like structures know as appendiculae.

Figure A32. Atlantic sharpnose shark at term. Note the long umbilical cord terminating in the fetal portion of the placenta. The umbilical cord is covered with appendiculae.

Figure A33. Biennial reproductive cycle. A blacktip shark term uterus with fully developed young and an inactive ovary.

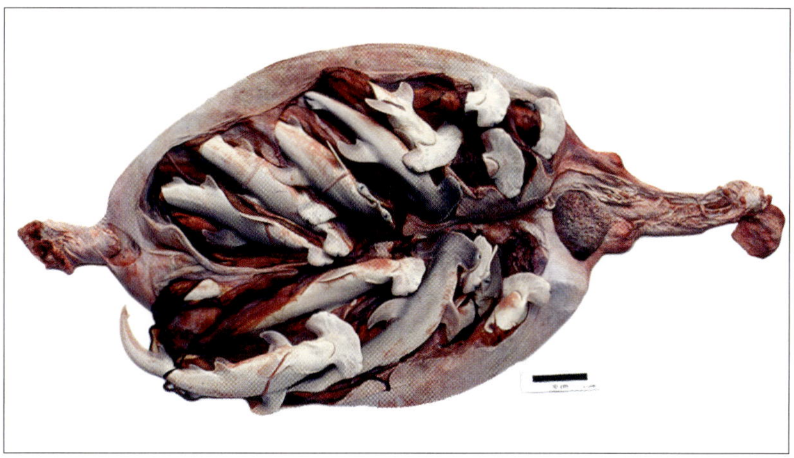

Figure A34. Biennial reproductive cycle. A great hammerhead uterus with term young and an inactive ovary (on midline, above size label). Note smooth umbilical cords.

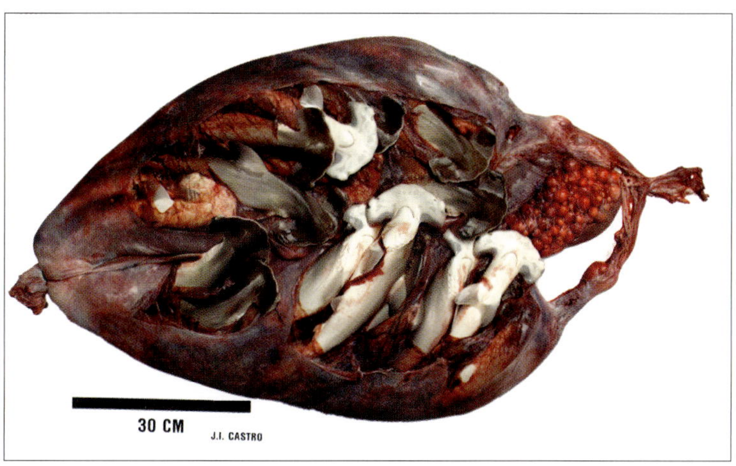

Figure A35: Annual reproductive cycle. A scalloped hammerhead uterus about two months from parturition. The embryos are well developed and the ovary is active. In the last two months the oocytes will double in size. Note umbilical cords with appendiculae.

APPENDIX 2

An Essay on the Conservation of Sharks

Man has been systematically hunting large animals to extinction for eons. In North America, primitive man hunted the huge mammals first: the mammoths, mastodons, ground sloths, giant bison, and so forth. By about 10,000 years ago, primitive man had hunted all the large land animals to extinction in what has often been called the "Pleistocene megafauna overkill hypothesis." Although the slaughter of the megafauna is well correlated with the proliferation of humans on the continent, some authors doubt that primitive humans, with only spears as their main weapons, could have killed, perhaps in two or three millennia, the immense herds of animals that existed in that epoch. These authors suggest that the disappearance of the megafauna instead may have been caused by severe climatic changes or a meteor impact phenomenon similar to that which caused the disappearance of dinosaurs. The critics of the Pleistocene overkill hypothesis also question the similar destruction of the megafauna that occurred in Australia around the time humans first entered that continent some 45,000–50,000 years ago. However, spears and high intelligence are effective weapons against unwary animals that have no fear of humans, as rifles are against much more wary animals in much more recent times.

Although we can debate for a long time what actually occurred in prehistoric times, there is little doubt of what has occurred in historical times. The slaughter of the large animals continued in North America and Africa when Europeans entered those continents, with the flintlock rifle in the former continent and with the modern rifle in the latter. In North America, the bison, the elk, the deer, the pronghorn antelope, and many other smaller land animals survived in large numbers until Europeans arrived in the sixteenth and seventeenth centuries, and colonists started hunting the animals heavily in the eighteenth century. Unrestricted hunting continued for two centuries, with subsistence hunting often developing into market hunting[1] by the late nineteenth century. Even the larger aquatic mammals, the baleen whales, the sea elephants, the fur seals, the walrus, and the manatee, among many others, often living far at sea, in remote areas, or around inaccessible islands at the highest latitudes, were hunted relentlessly. By the beginning of the twentieth century, most of the large animals had been severely reduced in range and numbers, and many were severely threatened or faced extinction. It is likely that, in a not too distant future, when the few remaining large animals are confined to a few parks, zoos, or aquariums, the slaughter of all the large animals in North America and Africa in the nineteenth and twentieth centuries will be referred to as the *modern megafauna overkill.* The modern overkill will include the slaughter of the sharks that we are witnessing now.

A similar pattern of destruction has occurred throughout the globe; most of the large populations of large wild animals were destroyed by the first half of the twentieth century. By then, most of the surviving large land, shore, or sea animals had ceased to be economically attractive because of their scarcity, or were protected by legislation because of their scarcity. Mankind then turned to large fishes for prey. The maritime conflicts of the two world wars of the twentieth century inhibited the development of high seas fishing fleets for a few decades. Large-scale ocean fisheries developed rapidly after the 1950s, with many of the large bony fishes, such as tunas and swordfish, being the primary prey. The sizes of the fleets, fishing knowledge, improvements on the efficiency of the gear, and landings all mounted steadily throughout the latter half of the century; by the end of the century, most of the stocks of large oceanic bony fishes were severely diminished or depleted. The sharks were among the last of the large vertebrates to be targeted and hunted in a systematic and relentless manner by large-scale fleets.

Although sharks have been utilized for food for hundreds of years in many parts of the world, most shark fisheries in the past were small and localized. In North America

1. Market hunting is hunting to supply markets in towns. The demand of modern markets is so great that the naturally occurring animal populations, regardless of taxon or reproductive potential, cannot support the unregulated hunting and fishing brought on by market demand. Historical examples of this include the destruction of the passenger pigeon, the bison, the sturgeons, and the codfish, among many others.

there were no shark fisheries until the early twentieth century. Small shark fisheries developed in the eastern United States and the West Indies in the early 1920s, their principal product being shark leather. In the late 1930s, it was discovered that shark liver oil contained high quantities of vitamin A, which was normally obtained from cod liver oil, often imported from Europe. When World War II interrupted the European supply of vitamin A in 1939, the American shark leather fishery grew into the shark liver oil fishery of the early 1940s, encompassing both coasts. On the West Coast, the fishery extended from California to Washington and involved many boats. On the East Coast, the shark fishery of the 1940s was based in Florida and was small, seldom involving more than five boats in Florida and a total of 16 in the southeastern states (Springer, S. 1952). In these fisheries the livers were removed from the sharks and the carcasses discarded into the ocean, although in some cases small quantities of dried fins and shark leather were also processed. By the late 1940s, these fisheries were depleted because of overfishing and fishing in the nursery areas. The appearance of synthetic vitamin A on the market in 1950 caused the fishery to be discontinued (Springer, S. 1952).

The economic changes of the 1970s brought great changes in the utilization of sharks in North America. Until the early 1970s there were poor political relations between the United States and China, and almost no economic trade. The political relations began to improve around 1972 and have continued to do so into the present. Improving political climate fostered economic trade, and in the 1980s business channels developed between the two countries. A demand for dried shark fins, used for making the traditional dish of shark fin soup, has existed in China and the Far East for many decades. In the late 1970s and early 1980s, improving economic conditions in China and other countries of the Far East resulted in high prices being paid for the fins (averaging over US$50/kg) for export to those countries. The high prices, in turn, engendered commercial fisheries for sharks throughout the world. A new commercial shark fishery sprang along the southeastern United States, fueled by the high price of the fins, the cheap shark meat being only of secondary importance (the carcasses were often discarded at sea). In most fisheries the fins were primarily exported to Hong Kong and the meat was consumed locally, if at all. Although the price of the fins would fluctuate, it was generally around US$50/kg in the late 1990s, whereas the price of shark meat was about $0.50–2.00/kg in the United States and much lower elsewhere. The fins were ideal objects of commerce for they could be cut from the shark easily and required little storage space in a dry place (they were often put to dry on the vessel's rigging). By contrast, the meat required prompt attention, dressing the carcass, and storage in salt, cold brine, or ice and took considerable storage space.

At about the same time the commercial shark fishery of the southeastern United States was developing in the mid-1970s, the movie *Jaws* created an awareness of sharks and stimulated a recreational fishery for sharks, which were perceived as malignant man-eaters. In this recreational fishery, thousands of sharks would be killed simply to be displayed dead for a brief time or to display their dried jaws as trophies. This national shark killing frenzy would last nearly two decades.

The killing of the sharks, in both the recreational fishery and the commercial fisheries, increased exponentially in the late 1980s and early 1990s. Although concern for the destruction of sharks had been expressed as early as 1979,[2] lack of consensus or concern by regulatory agencies prevented the regulation of the shark fishery at that time. By the early 1990s, public concern and conservation organizations prompted the National Marine Fisheries Service to enact a fishery management plan by 1992. Although the initial plan was severely limited because of lack of fishery and species data, the management of the shark fisheries would continue to improve over the next decade.

At the international level, concern for shark stocks was first exhibited in only a handful of countries. These happened to be, primarily, English-speaking countries, such as Australia, the United States, Canada, the United Kingdom, and South Africa. Some of these countries have instituted fairly successful programs intended to stop the decline of the shark stocks. In most cases the overfishing continues, but at least there are programs in place and there is public interest on the subject. The reasons why these English-speaking countries have been the leaders at shark conservation are probably their high level of public education and awareness, the presence in these cultures of empowered constituencies that can shape the political will, the existence of political processes whereby citizens can efficiently create and enact conservation laws, and traditions of law enforcement.

Although other countries have demonstrated some concern for sharks, there is little ecological awareness in most countries, or the process to create and enforce conservation laws does not exist. Simply stated, there are no conservation laws in most countries or in the high seas. Even where conservation laws exist, usually there is no enforcement of such laws. If such enforcement exists, is it often weak or sporadic, and often there is deep-rooted corruption that thwarts any efforts at enforcement. In some countries permits or exemp-

2. In 1979, the Gulf of Mexico Fishery Management Council published a "Draft Management Plan for Sharks and Other Elasmobranchs." This was one of the early attempts to regulate the shark fisheries.

tions to break the law can be bought at such high national levels that local officials are intimidated into ignoring the most flagrant law violators.

The conservation of wildlife is always a complex problem that depends on animal behavior as well as various economic and social variables. The conservation of sharks is even more complex. Many species of sharks are seasonally migratory, traveling great distances in their life cycles, and often crossing national boundaries or even oceans. Even where sharks are not targeted by fisheries, many sharks are taken as bycatch in lines or nets intended for bony fishes and are often dead when such gear is retrieved. The high price of the fins and their ease of transportation make shark fishing an attractive and lucrative commercial enterprise; shark fins can be an ephemeral bonanza for industrialist fishermen as well as for impoverished artisanal fishermen. People's fear of sharks as man-eaters and old attitudes of sharks being "trash fish" may prevent fishery managers from addressing the conservation problem.

A few organizations concerned with the conservation and management of species of sharks have emerged, most of them based in English-speaking countries. Most of these organizations have focused their efforts on passing legislation limiting catches or protecting species, establishing codes of conduct for fishermen, increasing public awareness, and so on. These efforts have often been successful at the national level in the United States, the United Kingdom, Australia, and South Africa. However, on a worldwide scale, the results of efforts to conserve shark populations have been poor. Organizations spearheading efforts to conserve sharks, often nongovernmental organizations, usually focus on the passing of legislation designed to achieve those purposes and not on the measurable effects of such legislation. When the legislation is enacted, conservationists consider most of their work done, and the responsibility passes to the government and the law enforcement agencies. These government bureaucracies and enforcement agencies usually have other priorities, little interest on sharks, and few resources, and they are often ineffective at conservation efforts. In many cases the bureaucrats charged with the protection of species are ignorant or have minimal interest on the animals they are charged to protect. Thus, the passage of legislation protecting sharks should not be confused with the effective conservation of these animals. Conservation success and progress cannot be measured by legislation approved, guidelines issued by international organizations, manifestos signed, and so forth. The effectiveness of conservation efforts must be periodically evaluated by their *measurable* positive effects on the populations or by the decreases in mortality.

In the last decade, despite the efforts of conservation organizations, the worldwide catches of sharks continue to increase despite the great reduction in the stocks. Much of the slaughter of sharks often occurs in coastal areas where there is no law or little law enforcement, or in the high seas. Given all the problems outlined above, it appears unlikely that protective legislation can be enacted in enough countries, or that countries will agree on legislation governing fishing for sharks in the high seas, or that such laws could be enforced in the high seas, in time to avoid the destruction of the shark stocks. Furthermore, many of the sharks killed are taken as bycatch in tuna and swordfish fisheries, and at present, there is no interest or political will to terminate the bycatch kill in these lucrative, politically powerful fisheries.

The shark fin trade, with the fins destined for Asian markets, drives the relentless hunt for sharks today. In today's world, with its immense population and global markets, the impact of trade on the shark populations is unsustainable. It is people's demands for these products and the availability of the products that create the markets. As long as the market demand exists, the fishermen, whether impoverished artisanal fishermen in a pirogue or industrialized fishermen in a factory ship, will seek and kill sharks, and countless middlemen and merchants will sell them.

From the paragraphs above, it is clear that if any conservation efforts directed at sharks are to succeed, they must be directed *at the markets.* However, few conservation organizations direct their efforts toward markets. There are many reasons for this. First, people's traditional wants or values are difficult to change. Second, the markets are large and politically powerful. Third, fishermen, merchants, and other business people are not going to listen to any arguments by conservationists that may affect their profits, and they could not care less about populations of sharks, or tigers, or rhinos or any other animal, for that matter. Thus, many conservation organizations prefer dealing with government officials and pursue legislative solutions, despite their poor record, because it easier to pass legislation than to attempt to change markets and businesses.

People's traditional attitudes and wants are difficult to change indeed. They are, however, changeable, and markets change accordingly. For example, from the late 1970s to the early 1990s, a shark killing frenzy occurred in the United States, and both the recreational and commercial fisheries seemed determined to catch every shark in U.S. waters. However, in a relatively short period, people became sufficiently educated and enlightened to demand that the wanton killing of these magnificent predators cease. In less than a decade, a conservation ethic for sharks developed, and regulatory legislation was enacted to control both fisheries. An unregulated, exponentially growing commercial fishery was replaced by a relatively well-regulated fishery, and strict bag limits and a list of prohibited species were imposed on the recreational fishery. The attitudes toward sharks had certainly changed! A similar change in attitudes occurred toward the slaughter of dolphins that occurred in the high seas for decades. Thousands of dolphins used to be killed yearly in the tuna fisheries, and this was tolerated by the ill-informed public. In the 1990s, consumers became aware of the slaughter of dolphins and demanded "dolphin-safe tuna." The industry responded by changing its procedures to ensure that the dolphin kill was minimized.

For conservation efforts to be effective, they must be directed to people and their markets. The targets for such efforts must be education, stemming the demand that fuels the markets, and the markets themselves. Education is usually a long-term goal, and it works best when directed to children. Educational efforts must be rooted the understanding that most species of sharks and other large, long-lived fish species cannot support intensive fisheries, and certainly not fisheries of the scale that exist today. The development of a strong conservation ethic must be developed in children, and many organizations can contribute to this achievable and relatively inexpensive goal. The ethic of leaving the world to future generations in the same or better shape must be developed, and it must include the animal legacy. A more complex problem is the education of the policy makers, business leaders, and money lenders, whose judgment is already tainted by ignorance, greed, or short-term gains. In addition to educational efforts, time and money must be spent on influencing the markets. Organizations can start grass roots efforts to coordinate use of their persuasive economic power to influence businesses and markets. A coordinated, strategic campaign could do much to influence people and markets. Only when a modern ethic of enlightened conservation has replaced the irrational utilization practices and greed of the past, can we hope for a better world.

REFERENCES CITED

Aasen, O. 1961. Some observations on the biology of the porbeagle shark, (*Lamna nasus* L.). *International Council for the Exploration of the Sea* CM 1961 (109). 7 pp.

Aasen, O. 1963. Length and growth of the porbeagle (*Lamna nasus,* Bonnaterre) in the northwest Atlantic. *Fiskeridirektoratets Skrifter Serie Havundersøkelser, Reports on Norwegian Fishery and Marine Investigations* 13 (6): 20–37.

Aasen, O. 1966. Blåhaien, *Prionace glauca* (Linneus), 1758. *Fiskeridirektøratets Skrifter Serie Havundersøkelser, Reports on Norwegian Fishery and Marine Investigations* 1: 1–15. Reprinted as: Blue shark, *Prionace glauca* (Linnaeus), 1758. Translation by the Foreign Fisheries (Translations), International Activities Staff, National Marine Fisheries Service, NOAA. Washington, D.C.: U.S. Department of Commerce. Not dated.

Abe, T. 1962. A record of a little squaloid shark, *Squaliolus laticaudus,* from Suruga Bay. *Japanese Journal of Ichthyology* 8 (5/6): 147–151.

Abe, T. 1972. On a cyclopic embryo of the blue shark. *U O* (Japanese Society of Ichthyologists) 9 (9): 1–3.

Abe, T. 1973. The white area behind the mouth of *Pseudocarcharias kamoharai* (Matsubara). *U O* (Japanese Society of Ichthyologists) no. 15: 1–2 (no volume number).

Abe, T., S. Isokawa, T. Misu, T. Kishimoto, Y. Shimma, and H. Shimma. 1968. Notes on some members of Osteodonti (class Chondrichthyes). 1. *Bulletin of Tokai Regional Fisheries Research Laboratory* 56: 1–6.

Abe, T., S. Isokawa, K. Aoki, T. Kishimoto, Y. Shimma, and H. Shimma. 1969. Notes on some members of Osteodonti (class Chondrichthyes). 2. *Bulletin of Tokai Regional Fisheries Research Laboratory* 60: 1–3.

Able, K. W., and D. Flescher. 1991. Distribution and habitat of chain dogfish, *Scyliorhinus retifer,* in the Mid-Atlantic Bight. *Copeia* 1991 (1): 231–234.

Abella, A. 1972. Hallazgo de una nueva especie de *Carcharhinus,* en las costas de Rocha, Uruguay. *Boletín de la Sociedad Zoológica del Uruguay* 2: 102–106.

Ackerman, J. T., M. C. Kondratieff, S. A. Matern, and J. J. Cech, Jr. 2000. Tidal influence on spatial dynamics of leopard sharks, *Triakis semifasciata,* in Tomales Bay, California. *Environmental Biology of Fishes* 58: 33–43.

Ackerman, L. T. 1971. Contributions to the Biology of the Leopard Shark, *Triakis semifasciata* (Girard) in Elkhorn Slough, Monterey Bay, California. Master's thesis, Sacramento State College. 54 pp.

Adams, D. H., and R. H. McMichael, Jr. 1999. Mercury levels in four species of sharks from the Atlantic coast of Florida. *Fishery Bulletin* 97: 372–379.

Aguirre, H., V. J. Madrid, and J. A. Virgen. 2002. Presence of *Echinorhinus cookei* off central Pacific Mexico. *Journal of Fish Biology* 61: 1403–1409.

Ainley, D. G., C. S. Strong, H. R. Huber, T. J. Lewis, and S. H. Morrell. 1981. Predation by sharks on pinnipeds at the Farallon Islands. *Fishery Bulletin* 78: 941–945.

Aires da Silva, A., P. C. Duarte, A. Giga, and G. Menezes. 1998. First record of the spined pygmy shark, *Squaliolus laticaudus* (Smith & Radcliffe, 1912) in the Azores, extending its distribution in the northeastern Atlantic. *Arquipélago* (Bulletin of the University of the Azores) 16A: 57–61.

Alcock, A. 1890. Observations on the gestation of some sharks and rays. *Journal of the Asiatic Society of Bengal* 59 (pt. 2, no. 1): 51–57.

Allen, W. E. 1923. Behavior of the thresher shark. *Science* 58 (1489): 81–82.

Allis, E. P., Jr. 1923. The cranial anatomy of *Chlamydoselachus anguineus. Acta Zoologica* 4: 123–221.

Alluchon-Gerard, M. J., and J. Mellinger. 1971. Mise en évidence de plusieurs phases distinctes dans la croissance embryonnaire de la roussette (*Scyllium canicula* CUV.). *Annales d'Embryologie et de Morphogenèse* 4 (1): 19–35.

Alverson, D. L., and M. E. Stansby. 1963. The Spiny Dogfish (*Squalus acanthias*) in the Northeastern Pacific. U.S. Fish and Wildlife Service, Special Scientific Report—Fisheries, no. 447. Washington, D.C.: U.S. Department of the Interior. 25 pp.

Ames, J. A., J. J. Geibel, F. E. Wendell, and C. A. Pattison. 1996. White shark-inflicted wounds of sea otters in California, 1968–1992. In *Great White Sharks: The Biology of Carcharodon carcharias,* edited by A. P. Klimley and D. G. Ainley, pp. 309–316. San Diego: Academic Press.

Ames, J. A., and G. V. Morejohn. 1980. Evidence of white shark, *Carcharodon carcharias,* attacks on sea otters, *Enhydra lutris. California Fish and Game* 66 (4): 196–209.

Amorim, A. F., C. A. Arfelli, and J. I. Castro. 2000. Description of a juvenile megamouth shark, *Megachasma pelagios,* caught off Brazil. *Environmental Biology of Fishes* 59: 117–123.

Anderson, M. E., G. M. Cailliet, and B. S. Antrim. 1979. Notes on some uncommon deep-sea fishes from the Monterey Bay area, California. *California Fish and Game* 65 (4): 256–264.

Anderson, R. C., and J. D. Stevens. 1996. Review of information on diurnal vertical migration in the bignose shark (*Carcharhinus altimus*). *Journal of Marine and Freshwater Research* 47: 605–608.

Anderson, S. D., and K. J. Goldman. 2001. Temperature measurements from salmon sharks, *Lamna ditropis,* in Alaskan waters. *Copeia* 2001 (3): 794–796.

Anonymous. 1992. Reproducción del tiburón gato (*Ginglymostoma cirratum*). *Boletín CEINER* (Boletín del Centro de Investigaciones, Educación y Recreación; Cartagena, Colombia) 1 (10): 4.

Anthoni, U., C. Christophersen, L. Gram, N. H. Nielsen, and P. Nielsen. 1991. Poisonings from flesh of the Greenland shark *Somniosus microcephalus* may be due to trimethylamine. *Toxicon* 29 (10): 1205–1212.

Antunes, M. T. 1970. Présence de *Alopias superciliosus* (Lowe) dans les mers du Portugal remarques sur les *Alopias* (Selachii) recents et fossiles. *Arquivos do Museu Bocage, 2ª sér.,* 2 (19): 363–380.

Applegate, S. P. 1966. A possible record-sized bonito shark, *Isurus oxyrinchus* Rafinesque, from Southern California. *California Fish and Game* 52 (3): 204–209.

Applegate, S. P. 1977. A new record-size bonito shark, *Isurus oxyrinchus* Rafinesque, from Southern California. *California Fish and Game* 63 (2): 126–127.

Applegate, S. P., L. Espinosa Arrubarrena, L. B. Menchaca Lopez, and F. Sotelo Macias. 1979, *Tiburones Mexicanos.* Mexico: Dirección General de Ciencia y Tecnología del Mar. 146 pp.

Aristotle. 1970. *Historia Animalium,* Vol. 2. Translated by A. L. Peck. Cambridge, Mass.: Harvard University Press.

Arnold, P. W. 1972. Predation on harbour porpoise, *Phocoena phocoena,* by a white shark, *Carcharodon carcharias. Journal of the Fisheries Research Board of Canada* 29 (8): 1213–1214.

Ayres, W. O. 1854. Description of new species of California fishes. *Proceedings of the California Academy of Natural Sciences,* ser. 1, 1: 16.

Ayres, W. O. 1855. Descriptions of new species of California fishes. *Proceedings of the California Academy of Natural Sciences,* ser. 1, 1: 76–77.

Ayres, W. O. 1859. On new fishes of the California coast. *Proceedings of the California Academy of Natural Sciences* 2: 25–32.

Backus, R. H. 1957. Notes on western North Atlantic sharks. *Copeia* 1957 (3): 246–248.

Backus, R. H. 1960. Notes on western North Atlantic sharks, no. 2. *Copeia* 1960 (3): 243–245.

Backus, R. H., S. Springer, and E. L. Arnold, Jr. 1956. A contribution to the natural history of the white-tip shark, *Pterolamiops longimanus* (Poey). *Deep-Sea Research* 3: 178–188.

Baker, A. N. 1985. Pygmy right whale *Caperea marginata* (Gray, 1846). In *Handbook of Marine Mammal,* Vol. 3, *The Sirenians and Baleen Whales,* edited by S. H. Ridgway and R. Harrison, pp. 345–354. Orlando: Academic Press.

Balart, E. F., J. González-García, and C. Villavicencio-Garayzar. 2000. Notes on the biology of *Cephalurus cephalus* and *Parmaturus xaniurus* (Chondrichthyes: Scyliorhinidae) from the west coast of Baja California Sur, México. *Fishery Bulletin* 98: 219–221.

Balazs, G. H., and G. C. Whittow. 1979. First record of a tiger shark observed feeding on a Hawaiian monk seal. *Elepaio (Hawaii Audubon Society)* 39 (9): 107–109.

Baldridge, H. D., Jr. 1974. *Shark Attack: A Program of Data Reduction and Analysis.* Contributions from the Mote Marine Laboratory 1 (2). Sarasota, Fla.: Mote Marine Laboratory. 98 pp.

Baldridge, H. D. 1970. Sinking factors and average densities of Florida sharks as functions of liver buoyancy. *Copeia* 1970 (4): 744–754.

Bane, G. W. 1966. Observations on the silky shark, *Carcharhinus falciformis,* in the Gulf of Guinea. *Copeia* 1966 (2): 354–356.

Bane, G. W., and A. W. Bane. 1971. *Bay Fishes of Northern California* Southampton: Mariscos Publications. 143 pp.

Baranes, A., and A. Ben-Tuvia. 1978. Note on *Carcharhinus altimus* (Springer, 1950) from the northern Red Sea. *Cybium,* 3e sér. (4): 61–64.

Barber, R. 1992. *Bestiary: An English Version of the Bodleian Library, Oxford M.S. Bodley 764 with All the Original Miniatures Reproduced in Facsimile.* London: Folio Society. 205 pp.

Barboza du Bocage, J. V., and F. de Brito Capello. 1864. Sur quelques espèces inédites de Squalidae de la tribu *Acanthiana,* Gray, qui fréquentent les côtes du Portugal. *Proceedings of the Zoological Society of London* 1864: 260–263.

Barnhart, P. S. 1936. *Marine Fishes of Southern California.* Berkeley: University of California Press. 209 pp.

Barraclough, W. E. 1948. The decline of the soupfin shark fishery in British Columbia. Fisheries Research Board of Canada. *Progress Reports of the Pacific Coast Stations* 77: 91–94.

Barraclough, W. E. 1953. The development of the dogfish fishery in British Columbia. *Proceedings of the Seventh Pacific Science Congress of the Pacific Science Association* (4): 513–519.

Bass, A. J. 1979. Records of little-known sharks from Australian waters. *Proceedings of the Linnean Society of New South Wales* 103 (4): 247–254.

Bass, A. J., and J. A. Ballard. 1972. Buoyancy control in the shark *Odontaspis taurus* (Rafinesque). *Copeia* 1972 (3): 594.

Bass, A. J., J. D. D'Aubrey, and N. Kistnasamy. 1973. Sharks of the East Coast of Southern Africa. I. The Genus *Carcharhinus* (Carcharhinidae). Investigational Report no. 33. Durban: Oceanographic Research Institute. 168 pp.

Bass, A. J., J. D. D'Aubrey, and N. Kistnasamy. 1975a. Sharks of the East Coast of Southern Africa. 5. The Families Hexanchidae, Chlamydoselachidae, Heterodontidae, Pristiophoridae and Squatinidae. Investigational Report no. 43. Durban: Oceanographic Research Institute. 50 pp.

Bass, A. J., J. D. D'Aubrey, and N. Kistnasamy. 1975b. Sharks of the East Coast of Southern Africa. 4. The Families Odontaspididae, Scapanorhynchidae, Isuridae, Cetorhinidae, Alopiidae, Orectolobidae and Rhiniodontidae. Investigational Report no. 39. Durban: Oceanographic Research Institute. 102 pp.

Bass, A. J., J. D. D'Aubrey, and N. Kistnasamy. 1975c. Sharks of the East Coast of Southern Africa. 3. The Families Carcharhinidae (excluding *Mustelus* and *Carcharhinus*) and Sphyrnidae. Investigational Report no. 38. Durban: Oceanographic Research Institute. 100 pp.

Bass, A. J., J. D. D'Aubrey, and N. Kistnasamy. 1976. Sharks of the East Coast of Southern Africa. 6. The Families Oxynotidae, Squalidae, Dalatiidae and Echinorhinidae. Investigational Report no. 45. Durban: Oceanographic Research Institute. 103 pp.

Bauchot, M. L., and A. Pras. 1980. *Guide des poisons marins de'Europe.* Paris: Delachaux & Niestlé, Editeurs. 427 pp.

Baughman, J. L. 1955. The oviparity of the whale shark, *Rhineodon typus,* with records of this and other fishes in Texas waters. *Copeia* 1955 (1): 54–55.

Baughman, J. L., and S. Springer. 1950. Biological and economic notes on the sharks of the Gulf of Mexico, with especial reference to those of Texas, and with a key for their identification. *American Midland Naturalist* 44 (1): 96–152.

Baum, J. K., and R. A. Myers. 2004. Shifting baselines and the decline of pelagic sharks in the Gulf of Mexico. *Ecology Letters* 7: 135–145.

Beamish, R. J., and M. S. Smith. 1976. A preliminary Report on the Distribution, Abundance, and Biology of Juvenile Spiny Dogfish (*Squalus acanthias*) in the Strait of Georgia and Their Relationship with Other Fishes. Technical Report no. 629. Ottawa: Fisheries and Marine Service, Environment Canada. 44 pp.

Bean, B. A. 1905. Fishes of the Bahama Islands. In *The Bahama Islands,* edited by G. B. Shattuck, pp. 293–296. New York: Macmillan.

Bean, T. H. 1881. Check-list of duplicates of North American fishes distributed by the Smithsonian Institution in behalf of the United States National Museum, 1877–1880. *Proceedings of the United States National Museum* 3: 75–116.

Bean, T. H. 1884. The first occurrence of *Pseudotriacis microdon,* Capello, on the coast of the United States. *Proceedings of the United States National Museum* 6: 147–150.

Benchley, P. 1974. *Jaws.* Garden City: Doubleday. 311 pp.

Bearden, C. M. 1965a. Occurrence of spiny dogfish, *Squalus acanthias,* and other elasmobranchs in South Carolina coastal waters. *Copeia* 1965 (3): 378.

Bearden, C. M. 1965b. *Elasmobranch Fishes of South Carolina.* Contributions from Bears Bluff Laboratories, no. 42. Wadmalaw Island, South Carolina: Bears Bluff Laboratories. 22 pp.

Beck, B., and A. W. Mansfield. 1969. Observations on the Greenland shark, *Somniosus microcephalus,* in Northern Baffin Island. *Journal of the Fisheries Research Board of Canada* 26 (1): 143–145.

Beckley, L. E., G. Cliff, M. J. Smale, and L. J. V. Compagno. 1997. Recent strandings and sightings of whale sharks in South Africa. *Environmental Biology of Fishes* 50: 343–348.

Bedford, D. 1985. Pelagic Shark/Swordfish Drift Gill Net Fishery. Management Information Document. Long Beach: California Department of Fish and Game, State Fisheries Library. 74 pp.

Bedford, D. 1992. Thresher shark. In *California's Living Marine Resources and Their Utilization,* edited by W. S. Leet, C. M. Dewees, and C. W. Haugen, pp. 49–51. California Sea Grant, Sea Grant Extension

Publication UCSGEP-92–12. Davis: University of California.

Beebe, W. 1932. *Nonsuch: Land of Water.* New York: Brewer, Warren & Putnam. 259 pp.

Beebe, W. 1941. External characters of six embryo nurse sharks, *Ginglymostoma cirratum* (Gmelin). *Zoologica* 26: 9–12.

Beebe, W., and J. Tee-Van. 1941. Eastern Pacific expeditions of the New York Zoological Society. 25. Fishes from the tropical Eastern Pacific [from Cedros Island, lower California, south to the Galápagos Islands and northern Peru]. Part 2. Sharks. *Zoologica* 26 (15): 93–122.

Behnke, R. J. 2002. *Trout and Salmon of North America.* 2002. New York: Chanticleer Press. 359 pp.

Bell, J. C., and J. T. Nichols. 1921. Notes on the food of Carolina sharks. *Copeia* 1921 (92): 17–20.

Belloc, G. 1937. Sur la capture dans le Bassin D'Arcachon d'un squale nouveau pour la faune de France: *Euprotomicrus sarmenti* Noronha. *Bulletin du Muséum National d'Histoire Naturelle,* n.s. 2ª, 9: 370–372.

Bennett, F. D. 1840. *Narrative of a Whaling Voyage Round the Globe, From the Year 1833 to 1836,* Vol. 2. Facsimile 1970, Amsterdam: N. Israel. 395 pp.

Benz, G. W., Z. Lucas, and L. F. Lowry. 1998. New host and ocean records for the copepod *Ommatokoita elongata* (Siphonostomatoida: Lernaeopodidae), a parasite of the eyes of sleeper sharks. *Journal of Parasitology* 84 (6): 1271–1274.

Benz, G. W., E. R. Hoffmayer, W. B. Driggers III, D. Allen, L. E. Bishop, and D. A. Brown. 2007. First record of a sleeper shark in the western Gulf of Mexico and comments on taxonomic uncertainty within *Somniosus* (*Somniosus*). *Bulletin of Marine Science* 80 (2): 343–351.

Berkeley, S. A., and W. L. Campos. 1988. Relative abundance and fishery potential of pelagic sharks along Florida's east coast. *Marine Fisheries Review* 50 (1): 9–16.

Berland, B. 1961. Copepod *Ommatokoita elongata* (Grant) in the eyes of the Greenland shark–a possible cause of mutual dependence. *Nature* 191 (4790): 829–830.

Bernal, D., and C. A. Sepulveda. 2005. Evidence for temperature elevation in the aerobic swimming musculature of the common thresher shark, *Alopias vulpinus. Copeia* 2005 (1): 146–151.

Berra, T. M., and J. B. Hutchins. 1990. A specimen of megamouth shark, *Megachasma pelagios* (Megachasmidae) from Western Australia. *Records of the Western Australian Museum* 14 (4): 651–656.

Berry, F. H., and H. C. Perkins. 1965. Survey of pelagic fishes of the California current area. *Fishery Bulletin* 65: 625–682.

Berrow, S. D. 1994. Incidental capture of elasmobranchs in the bottom-set gill-net fishery off the south coast of Ireland. *Journal of the Marine Biological Association of the United Kingdom* 74: 837–847.

Best, E. 1929. *Fishing Methods and Devices of the Maori.* Wellington: V. R. Ward Government Printer. 264 pp.

Bethea, D. M., J. A. Buckel, and J. K. Carlson. 2004. Foraging ecology of the early life stages of four sympatric shark species. *Marine Ecology Progress Series* 268: 245–264.

Bigelow, H. B., and W. C. Schroeder. 1940. Notes on New England fishes. *Copeia* 1940 (2): 139.

Bigelow, H. B., and W. C. Schroeder. 1941. *Cephalurus,* a new genus of Scyliorhinid shark with redescription of the genotype, *Catulus cephalus* Gilbert. *Copeia* 1941 (2): 73–76.

Bigelow, H. B., and W. C. Schroeder. 1944. New sharks from the western North Atlantic. *Proceedings of the New England Zoölogical Club* 23: 21–36.

Bigelow, H. B., and W. C. Schroeder. 1945. *Guide to Commercial Shark Fishing in the Caribbean Area.* Fishery Leaflet 135, Fish and Wildlife Service. Washington, D.C.: U.S. Department of the Interior. 149 pp.

Bigelow, H. B., and W. C. Schroeder. 1948. *Fishes of the Western North Atlantic, Lancelets, Cyclostomes, and Sharks.* Memoir no. 1, Part 1. New Haven, Conn.: Sears Foundation for Marine Research, Yale University. 576 pp.

Bigelow, H. B., and W. C. Schroeder. 1953. *Fishes of the Gulf of Maine.* First Revision. Fishery Bulletin 74. *Fishery Bulletin of the Fish and Wildlife Service* 53: 1–577.

Bigelow, H. B., and W. C. Schroeder. 1954. Deep water elasmobranchs and chimaeroids from the northwestern Atlantic slope. *Bulletin of the Museum of Comparative Zoology at Harvard College* 112 (2): 37–87.

Bigelow, H. B., and W. C. Schroeder. 1957. A study of the sharks of the suborder Squaloidea. *Bulletin of the Museum of Comparative Zoology at Harvard College* 117 (1): 1–150.

Bigelow, H. B., and W. C. Schroeder. 1961. *Carcharhinus nicaraguensis,* a synonym of the bull shark, *C. leucas. Copeia* 1961 (3): 359.

Bigelow, H. B., and W. W. Welsh. 1925. Fishes of the Gulf of Maine. *Bulletin of the United States Bureau of Fisheries* 40 (pt. 1): 1–567.

Bigelow, H. B., W. C. Schroeder, and S. Springer. 1943. A new species of *Carcharhinus* from the western Atlantic. *Proceedings of the New England Zoölogical Club* 22: 69–74.

Bigelow, H. B., W. C. Schroeder, and S. Springer. 1953. New and little known sharks from the Atlantic and from the Gulf of Mexico. *Bulletin of the Museum of Comparative Zoology at Harvard College* 109 (3): 213–276.

Bigelow, H. B., W. C. Schroeder, and S. Springer. 1955. Three new shark records from the Gulf of Mexico. *Breviora* (Museum of Comparative Zoology) 49: 1–12.

Bini, G. 1967. *Atlante dei pesci delle Coste Italiane,* Vol. 1. Italy: Mondo Sommerso Editrice. 206 pp.

Bishop, S. D. H., M. P. Francis, C. Duffy, and J. C. Montgomery. 2006. Age, growth, maturity, longevity, and natural mortality of the shortfin mako shark, (*Isurus oxyrinchus*) in New Zealand waters. *Marine and Freshwater Research* 57:143–154.

Blagoderov, A. I. 1994. Seasonal distribution and some notes on the biology of salmon shark (*Lamna ditropis*) in the Northwest Pacific Ocean. *Journal of Ichthyology* 34 (2): 115–121.

Blainville, D. M. P. 1810. Note sur plusieurs espèces de Squales; confondues sous le nom de *Squalus maximus* de Linnée. *Nouveau Bulletin des Sciences, par la Société Philomatique* 2 (58): 169–171.

Blasco, M., and R. Muñoz-Chapuli. 1981. Presencia de *Alopias superciliosus* en las costas Ibéricas y datos sobre su morfología. *Arquivos do Museu Bocage,* sér. B, 1 (6): 53–59.

Bloch, M. E., and J. G. Schneider. 1801. *Systema Ichthyologiae.* Historiae Naturalis Classica, Vol. 55. Edited by J. Cramer and H. K. Swann. Reprint, New York: Wheldon & Wesley, Ltd., 1967. 584 pp.

Block, B. A., J. R. Finnerty, A. F. R. Stewart, and J. Kidd. 1993. Evolution of endothermy in fish: Mapping physiological traits on a molecular phylogeny. *Science* 260: 210–214.

Bøje, O. 1939. Toxin in the flesh of the Greenland shark. *Meddelelser om Grønland, Udgivner af Kommissionen for Videnskabelige Undersøgelser I Grønland* 125 (5): 1–16.

Bolívar, I. 1907. Indicación de algunos peces notables de La Coruña. *Boletín de la Real Sociedad Española de Historia Natural* 7: 206–209.

Bonaparte, C. L. P. 1846. *Catalogo Metodico dei Pesci Europei.* Napoli: Stamperia E. Cartiere del Fibreno. 97 pp.

Bone, Q., and A. D. Chubb. 1983. The retial system of the locomotor muscles in the thresher shark. *Journal of the Marine Biological Association of the United Kingdom* 63: 239–241.

Bonfil-Sanders, R. 1987. *Composición por especies de la pesquería de tiburón y cazón en Yucatán y relaciones morfométricas para las principales especies.* Contribuciones de Investigación Pesquera, Technical Document no. 1: 1–10. Yucalpeten, Yucatan, Mexico: Instituto Nacional de la Pesca, Secretaría de Pesca.

Bonfil, R. 1995. Is the ragged-tooth shark cosmopolitan? First record from the western North Atlantic. *Journal of Fish Biology* 47: 341–344.

Bonfil, R., R. Mena, and D. de Anda. 1993. Biological parameters of commercially exploited silky sharks, *Carcharhinus falciformis,* from the Campeche Bank, Mexico. In *Conservation Biology of Elasmobranchs,* edited by S. Branstetter, pp. 73–86. NOAA Technical Report NMFS 115. U.S. Department of Commerce.

Bonfil, R., M. Meyer, M. C. Scholl, R. Johnson, S. O'Brien, H. Oosthuizen, S. Swanson, D. Kotze, and M. Paterson. 2005. Transoceanic migration, spatial dynamics, and population linkages of white sharks. *Science* 310: 100–103.

Bonham, K. 1942. Records of three sharks from the Washington coast. *Copeia* 1942 (4): 264–266.

Bonham, K. 1954. Food of the dogfish *Squalus acanthias. Fisheries Research Papers,* pt. 3 (State of Washington, Department of Fisheries) 1 (2): 25–36.

Bonham, K., and F. B. Sanford. 1949. Preliminary report on the soupfin shark fishery in Washington. *State of Washington, Department of Fisheries Biological Report* 49A: 203–211.

Bonham, K., F. B. Sanford, W. Clegg, and G. C. Bucher. 1949. Biological and vitamin A studies of dogfish landed in the state of Washington. *State of Washington, Department of Fisheries Biological Report* 49A: 83–114.

Bonnaterre, L. 1788. *Tableau Encyclopédique et Méthodique des Trois Règnes de la Nature, Ichthyologie.* Paris: Panckoucke. 215 pp.

Bordalo Machado, P. B., and I. Figueiredo. 2000. A technique for ageing the birdbeak dogfish (*Deania calcea* Lowe, 1839) from dorsal spines. *Fisheries Research* 45: 93–98.

Borets, L. A. 1986. Ichthyofauna of the Northwestern and Hawaiian submarine ranges. *Journal of Ichthyology* 26 (3): 1–13.

Borucinska, J. D., G. W. Benz, and H. E. Whiteley. 1998. Ocular lesions associated with attachment of the parasitic copepod *Ommatokoita elongata* (Grant) to corneas of Greenland sharks, *Somniosus microcephalus* (Bloch & Schneider). *Journal of Fish Diseases* 21: 415–422.

Boustany, A. M., S. F. Davis, P. Pyle, S. D. Anderson, B. J. Le Boeuf, and B. A. Block. 2002. Satellite tagging: Expanded niche for white sharks. *Nature* 415: 35–36.

Branstetter, S. 1981. Biological notes on the sharks of the north central Gulf of Mexico. *Contributions in Marine Science* 24: 13–34.

Branstetter, S. 1987a. Age and growth estimates for blacktip, *Carcharhinus limbatus* and spinner,

C. brevipinna, sharks from the northwestern Gulf of Mexico. *Copeia* 1987 (4): 964–974.

Branstetter, S., 1987b. Age, growth and reproductive biology of the silky shark, *Carcharhinus falciformis*, and the scalloped hammerhead, *Sphyrna lewini*, from the northwestern Gulf of Mexico. *Environmental Biology of Fishes* 19: 161–173.

Branstetter, S. 1987c. Age and growth validation of newborn sharks held in laboratory aquaria, with comments on the life history of the Atlantic sharpnose shark, *Rhizoprionodon terraenovae*. *Copeia* 1987 (2): 291–300.

Branstetter, S. 1990. Early life-history implications of selected carcharhinoid and lamnoid sharks of the northwest Atlantic. In *Elasmobranchs as Living Resources: Advances in the Biology, Ecology, Systematics, and the Status of the Fisheries*, edited by H. L. Pratt, Jr., S. H. Gruber and T. Taniuchi, pp. 17–28. NOAA Technical Report NMFS 90. Springfield, Va.: U.S. Department of Commerce.

Branstetter, S., and J. D. McEachran. 1983. A first record of the bigeye thresher, *Alopias superciliosus*, the blue shark, *Prionace glauca*, and the pelagic stingray, *Dasyatis violacea*, from the Gulf of Mexico. *Northeast Gulf Science* 6 (1): 59–61.

Branstetter, S., and J. D. McEachran. 1986a. A first record of *Odontaspis noronhai* (Lamniformes: Odontaspididae) for the western North Atlantic, with notes on two uncommon sharks from the Gulf of Mexico. *Northeast Gulf Science* 8 (2): 153–160.

Branstetter, S., and J. D. McEachran. 1986b. Age and growth of four carcharhinid sharks common to the Gulf of Mexico: In *Indo-Pacific Fish Biology: Proceedings of the Second International Conference on Indo-Pacific Fishes*, edited by T. Uyeno, R. Arai, T. Taniuchi, and K. Matsuura, pp. 361–371. Tokyo: Ichthyological Society of Japan.

Branstetter, S., and J. A. Musick. 1994. Age and growth estimates for the sand tiger in the northwestern Atlantic Ocean. *Transactions of the American Fisheries Society* 123: 242–254.

Branstetter, S., J. A. Musick, and J. A. Colvocoresses. 1987. A comparison of the age and growth of the tiger shark, *Galeocerdo cuvieri*, from off Virginia and from the northwestern Gulf of Mexico. *Fishery Bulletin* 85: 269–279.

Branstetter, S., and R. Stiles. 1987. Age and growth estimates of the bull shark, *Carcharhinus leucas*, from the northern Gulf of Mexico. *Environmental Biology of Fishes* 20: 169–181.

Breder, C. M, Jr. 1929. *Field Book of Marine Fishes of the Atlantic Coast from Labrador to Texas*. New York: G. P. Putnam's Sons. 332 pp.

Bright, D. B. 1959. The occurrence and food of the sleeper shark, *Somniosus pacificus*, in a central Alaska Bay. *Copeia* 1959 (1): 76–77.

Bright, D. B. 1960. A record of the porbeagle, *Lamna nasus*, from Cook Inlet, Alaska. *Copeia* 1960 (2): 145–146.

Broussonet, P. M. A. 1780. Mémoire sur les différentes espèces de chiens de mer. *Histoire de l'Academie Royale des Sciences* (Paris) 1780: 641–680.

Brown, C. A., and S. H. Gruber. 1988. Age assessment of the lemon shark, *Negaprion brevirostris*, using tetracycline validated vertebral centra. *Copeia* 1988 (3): 747–753.

Bruce, B. D. 1992. Preliminary observations on the biology of the white shark, *Carcharodon carcharias*, in South Australian waters. *Australian Journal of Marine and Freshwater Research* 43: 1–11.

Bryan, P. G. 1973. Three new shark records from Guam, Mariana Islands. *Micronesica* (Journal of the University of Guam): 9: 159–160.

Budker, P. 1958. La viviparité chez les sélaciens. In *Traité de Zoologie: Anatomie, Systématique, Biologie*. 2nd ed., tome 13, edited by P. P. Grassé, pp. 1755–1790. Paris: Libraires de l'Académie de Médecine.

Buencuerpo, V., S. Rios, and J. Morón. 1998. Pelagic sharks associated with the swordfish, *Xiphias gladius*, fishery in the eastern North Atlantic Ocean and the Strait of Gibraltar. *Fishery Bulletin* 96: 667–685.

Bullis, H. R., Jr. 1955. Preliminary report on exploratory long-line fishing for tuna in the Gulf of Mexico and the Caribbean Sea. Part 1: Exploratory tuna fishing by the *Oregon*. *Commercial Fisheries Review* 17 (10): 1–15.

Bullis, H. R., Jr. 1961. Observations on the feeding behavior of white-tip sharks on schooling fishes. *Ecology* 42 (1): 194–195.

Bullis, H. R., Jr. 1967. Depth segregations and distribution of sex-maturity groups in the marbled catshark, *Galeus arae*. In *Sharks, Skates, and Rays*, edited byP. W. Gilbert, R. F. Mathewson, and D. P. Rall, pp. 141–148. Baltimore: Johns Hopkins University Press.

Bullis, H. R., Jr., and J. R. Thompson. 1965. Collections by the exploratory fishing vessels *Oregon, Silver Bay, Combat*, and *Pelican* made during 1956–1960 in the southwestern North Atlantic. U.S. Fish and Wildlife Service, Special Scientific Report—Fisheries no. 510. Washington, D.C.: U.S. Department of the Interior. 130 pp.

Burckhardt, R. 1900. On the luminous organs of selachian fishes. *Annals and Magazine of Natural History*, ser. 6, 7: 558–568.

Burgess, G. H., G. W. Link, Jr., and S. W. Ross. 1979. Additional marine fishes new or rare to Carolina waters. *Northeast Gulf Science* 3 (2): 74–87.

Burgess, G. H., and M. Callahan. 1996. Worldwide patterns of white shark attacks on humans. In *Great White Sharks: The Biology of Carcharodon carcharias,* edited by A. P. Klimley and D. G. Ainley, pp. 457–469. San Diego: Academic Press.

Burgess, G. H., L. R. Beerkircher, G. M. Cailliet, J. K. Carlson, E. Cortés, K. J. Goldman, R. D. Grubbs, J. A. Musick, M. K. Musyl, and C. A. Simpfendorfer. 2005. Is the collapse of shark populations in the Northwest Atlantic Ocean and Gulf of Mexico real? *Fisheries* 30 (10): 19–26.

Burne, R. H. 1923. Some peculiarities of the blood-vascular system of the porbeagle shark (*Lamna cornubica*). *Proceedings of the Royal Society of London,* ser. B, 212: 209–257.

Burton, E. M. 1935. Shark attacks along the South Carolina coast. *Scientific Monthly* 40: 279–283.

Burton, E. M. 1940. *Aprionodon isodon* from South Carolina. *Copeia* 1940 (2): 140.

Burton, E. M. 1941. A record tiger shark from South Carolina. *Copeia* 1941 (1): 40.

Byers, R. D. 1940. The California shark fishery. *California Fish and Game* 26: 23–38.

Cadenat, J. 1950. *Poissons de mer du Sénégal. Initiations Africaines,* Vol. 3. Dakar: Institut Francais D'Afrique Noire. 345 pp.

Cadenat, J. 1956. Note d'ichtyologie ouest-africaine. 15. Sur un requin-renard nouveau pour la faune d'Afrique Occidentale francaise *Alopias superciliosus* (Lowe) 1840? *Bulletin de l'Institut Français d'Afrique Noire* 18, ser. A (4): 1257–1266.

Cadenat, J. 1960. Notes d'ichtyologie ouest-africaine. 28. *Deania cremouxi,* sp. nov. des côtes du Sénégal. *Bulletin de l'Institut Français d'Afrique Noire* 22, ser. A (1): 312–324.

Cadenat, J. 1961. Notes d'ichtyologie ouest-africaine. 34. Liste complémentaire des espèces de poissons de mer (provenant des côtes de l'Afrique occidentale) en collection à la section de biologie marine de l'I.F.A.N. à Gorée. *Bulletin de l'Institut Français d'Afrique Noire* 23, ser. A (1): 231–245.

Cadenat, J. 1963. Notes d'ichtyologie ouest-africaine. 39. Notes sur les requins de la famille des *Carchariidae* et formes apparentées de l'Atlantique Ouest-Africain (avec la description d'une espèce nouvelle: *Pseudocarcharias pelagicus,* classée dans un sous-genre nouveau). *Bulletin de l'Institut Français d'Afrique Noire* 25, ser. A (2): 526–537.

Cadeant, J., and G.-E. Maul. 1966. Note d'Ichthyologie ouest-africaine. 43. Description d'une espèce nouvelle du genre *Apristurus, Apristurus maderensis* [Sélaciens Scyliorhinidae]. *Bulletin de l'Institut Français d'Afrique Noire* 28, ser. A (2): 769–782.

Cadenat, J., and J. Blache. 1981. *Requins de Mediterranée et de d'Atlantique. Faune Tropicale,* Vol. 21. Paris: ORSTOM. 330 pp.

Caille, G., M., and E. K. Olsen. 2000. A bramble shark, *Echinorhinus brucus,* caught near the Patagonian coast, Argentina. *Revista de Biología Marina y Oceanografía* 35 (1): 117–119.

Cailliet, G. M., and D. W. Bedford. 1983. The biology of three pelagic sharks from California waters, and their emerging fisheries: A review. *California Cooperative Oceanic Fisheries Investigations, Progress Report* 24: 57–69.

Cailliet, G. M., L. J. Natanson, B. A. Welden, and D. A. Ebert. 1985. Preliminary studies on the age and growth of the white shark, *Carcharodon carcharias,* using vertebral bands. *Memoirs of the Southern California Academy of Sciences* 9: 49–60.

Calderwood, W. L. 1888. Notes on an intra-uterine specimen of the porbeagle (*Lamna cornubica*). In *Sixth Annual Report of the Fishery Board for Scotland being for the Year 1887,* part 2, pp. 263–264. Edinburgh: Neill & Co.

Caloyianis, N. 2000. Arctic sharks: Adventures with the Greenland shark. *Ocean Realm* Winter 2000: 42–51.

Campana, S. E., and W. N. Joyce. 2004. Temperature and depth associations of porbeagle shark (*Lamna nasus*) in the northwest Atlantic. *Fisheries Oceanography* 13 (1): 52–64.

Campana, S., L. Marks, W. Joyce, and S. Harley. 2001. Analytical assessment of the porbeagle shark (*Lamna nasus*) population in the northwest Atlantic, with estimates of long-term sustainable yield. Research Document 2001/067. Ottawa: Canadian Science Advisory Secretariat. 59 pp.

Campana, S. E., L. J. Natanson, and S. Myklevoll. 2002. Bomb dating and age determination of large pelagic sharks. *Canadian Journal of Fisheries and Aquatic Sciences* 59: 450–455.

Capapé, C. 1980. Nouvelle description de *Heptranchias perlo* (Bonnaterre, 1788) (Pisces, Pleurotremata, Hexanchidae). Données sur la biologie de la reproduction et le régime alimentaire des spécimens des côtes Tunisiennes. *Bulletin de l'Office National des Pêches de Tunisie* 4 (2): 231–264.

Cappetta, H. 1980. Les sélaciens du Crétace supérieur du Liban. 1: Requins. *Paleontographica* Abt. A, 168 (1–4): 69–148.

Cappetta, H. 1985. Capture de cinq espèces de poissons en dehors de leur aire de distribution connue. *Cybium* 9 (4): 401–403.

Cappetta, H. 1987. *Chondrichthyes II; Mesozoic and Cenozoic Elasmobranchii.* Handbook of Paleoichthyology 3B. Stuttgart: Gustav Fischer Verlag. 193 pp.

Carey, F. G. 1973. Fishes with warm bodies. *Scientific American* 228 (2): 36–44.

Carey, F. G., and E. Clark. 1995. Depth telemetry from the sixgill shark, *Hexanchus griseus,* at Bermuda. *Environmental Biology of Fishes* 42: 7–14.

Carey, F. G., and J. V. Scharold. 1990. Movements of blue sharks (*Prionace glauca*) in depth and course. *Marine Biology* 106: 329–342.

Carey, F. G., and J. M. Teal. 1969. Mako and porbeagle: Warm-bodied sharks. *Comparative Biochemistry and Physiology* 28: 199–204.

Carey, F. G., J. G. Casey, H. L. Pratt, D. Urquhart, and J. E. McCosker. 1985. Temperature, heat production and heat exchange in lamnid sharks. *Memoirs of the Southern California Academy of Sciences* 9: 92–108.

Carey, F. G., L. Middleton, C. Stillwell, W. Pratt, N. Kohler, and C. Cavin. 1978. Mako Shark Experiment on "*WIECZNO,*" March 1978. Laboratory Reference no. 78–27. Woods Hole, Mass.: National Marine Fisheries Service, Northeast Fisheries Center. 12 pp.

Carlson, J. K. 1999. Occurrence of neonate and juvenile sandbar sharks, *Carcharhinus plumbeus,* in the northeastern Gulf of Mexico. *Fishery Bulletin* 97: 387–391.

Carlson, J. K., and I. E. Baremore. 2005. Growth dynamics of the spinner shark (*Carcharhinus brevipinna*) off the United States southeast and Gulf of Mexico coasts: A comparison of methods. *Fishery Bulletin* 103: 280–291.

Carlson, J. K., and G. R. Parsons. 1997. Age and growth of the bonnethead shark, *Sphyrna tiburo,* from northwest Florida, with comments on clinal variation. *Environmental Biology of Fishes* 50: 331–341.

Carlson, J. K., E. Cortés, and D. M. Bethea. 2003. Life history and population dynamics of the finetooth shark (*Carcharhinus isodon*) in the northeastern Gulf of Mexico. *Fishery Bulletin* 101 (2): 281–292.

Carlson, J. K., E. Cortés, and A. G. Johnson. 1999. Age and growth of the blacknose shark, *Carcharhinus acronotus,* in the eastern Gulf of Mexico. *Copeia* 1999 (3): 684–691.

Carlson, J. K., M. A. Grace, and P. K. Lago. 2002. An observation of juvenile tiger sharks feeding on clapper rails off the southeastern coast of the United States. *Southeastern Naturalist* 1 (3): 307–310.

Carrassón, M., C. Stefanescu, and J. E. Cartes. 1992. Diets and bathymetric distributions of two bathyal sharks of the Catalan deep sea (western Mediterranean). *Marine Ecology Progress Series* 82: 21–30.

Carrier, J. C., and C. A. Luer. 1990. Growth rates in the nurse shark, *Ginglymostoma cirratum. Copeia* 1990 (3): 686–692.

Carrier, J. C., H. L. Pratt, Jr., and L. K. Martin. 1994. Group reproductive behaviors in free-living nurse sharks, *Ginglymostoma cirratum. Copeia* 1994 (3): 646–656.

Carroll, L. (C. L. Dodgson). 1936. *Through the Looking Glass.* Reprinted in *The Complete Works of Lewis Carroll.* pp. 133–272. New York: Vintage Books, 1976.

Casey, J. G. 1985. Transatlantic migrations of the blue shark; a case history of cooperative shark tagging. In *World Angling Resources and Challenges: Proceedings of the First World Angling Conference,* edited by R. H. Stroud, pp. 253–268. Fort Lauderdale, Fla.: International Game Fish Association.

Casey, J. G. 1989. (Gamefish profile) *Isurus oxyrinchus:* The shortfin mako shark. *The Edge Canyon Fishing Annual,* pp. 90–94. Point Pleasant, N.J.: Offshore Informal Publications.

Casey, J. G., and N. E. Kohler. 1992. Tagging studies on the shortfin mako shark (*Isurus oxyrinchus*) from the western North Atlantic. *Australian Journal of Marine and Freshwater Research* 43: 45–60.

Casey, J. G., and L. J. Natanson. 1992. Revised estimates of age and growth of the sandbar shark (*Carcharhinus plumbeus*) from the western North Atlantic. *Canadian Journal of Fisheries and Aquatic Sciences* 49 (7): 1474–1477.

Casey, J. G., and H. L. Pratt, Jr. 1985. Distribution of the white shark, *Carcharodon carcharias,* in the western North Atlantic. *Memoirs of the Southern California Academy of Sciences* 9: 2–14.

Casey, J. G., H. L. Pratt, Jr., and C. E. Stillwell. 1985. Age and growth of the sandbar shark (*Carcharhinus plumbeus*) from the western North Atlantic. *Canadian Journal of Fisheries and Aquatic Sciences* 42 (5): 963–975.

Casterlin, M. E., and W. W. Reynolds. 1979. Diel activity patterns of the smooth dogfish shark, *Mustelus canis. Bulletin of Marine Science* 29 (3): 440–442.

Castillo Géniz, J. L. 1990. Contribución al conocimiento de la biología y pesquería del cazón bironche, *Rhizoprionodon longurio* (Jordan y Gilbert, 1882) (Elasmobranchii, Carcharhinidae) del sur de Sinaloa. México. Thesis, Universidad Nacional Autónoma de México. 128 pp.

Castro, A. L. F., and R. S. Rosa. 2005. Use of natural marks on population estimates of the nurse shark, *Ginglymostoma cirratum,* at Atol das Rocas biological reserve, Brazil. *Environmental Biology of Fishes* 72: 213–221.

Castro, J. A., and J. Mejuto. 1995. Reproductive parameters of blue shark, *Prionace glauca,* and other sharks in the Gulf of Guinea. *Marine and Freshwater Research* 46: 967–973.

Castro, J. I. 1983. *The Sharks of North American Waters.* College Station: Texas A&M University Press. 180 pp.

Castro, J. I. 1988. Investigations in the Reproductive Biology of Sharks. Ph.D. diss., Clemson University. 115 pp.

Castro, J. I. 1989. Biology of the golden hammerhead, *Sphyrna tudes,* off Trinidad. *Environmental Biology of Fishes* 24: 3–11.

Castro, J. I. 1993a. The biology of the finetooth shark, *Carcharhinus isodon. Environmental Biology of Fishes* 36: 219–232.

Castro, J. I. 1993b. The shark nursery of Bulls Bay, South Carolina, with a review of the shark nurseries of the southeastern coast of the United States. *Environmental Biology of Fishes* 38: 37–48.

Castro, J. I. 1996. The biology of the blacktip shark, *Carcharhinus limbatus,* off the southeastern United States. *Bulletin of Marine Science* 59 (3): 508–522.

Castro, J. I. 2000. The biology of the nurse shark, *Ginglymostoma cirratum,* off the Florida east coast and the Bahama Islands. *Environmental Biology of Fishes* 58: 1–22.

Castro, J. I. 2001a. *Guía para la identificación de las especies de tiburones de importancia comercial del Golfo de México.* Mexico City: Secretaría de Agricultura, Ganadería, Desarrollo Rural, Pesca y Alimentación, Instituto Nacional de la Pesca. 4 pp.

Castro, J. I. 2001b. *Guía para la identificación de las especies de tiburones de importancia comercial del Océano Pacífico.* Mexico City: Secretaría de Agricultura, Ganadería, Desarrollo Rural, Pesca y Alimentación, Instituto Nacional de la Pesca. 4 pp.

Castro, J. I. 2009. Observations on the reproductive cycles of some viviparous North American sharks. *Aqua* 15 (4): 205–222.

Castro, J. I., and J. P. Wourms. 1993. Reproduction, placentation, and embryonic development of the Atlantic sharpnose shark, *Rhizoprionodon terraenovae. Journal of Morphology* 218: 257–280.

Castro, J. I., P. M. Bubucis, and N. A. Overstrom. 1988. The reproductive biology of the chain dogfish, *Scyliorhinus retifer. Copeia* 1988 (3): 740–746.

Castro, J. I., C. M. Woodley, and R. L. Brudek. 1999. A Preliminary Evaluation of the Status of Shark Species. FAO Fisheries Technical Paper, no. 380. Rome: Food and Agriculture Organization of the United Nations. 72 pp.

Castro, J. I., E. Clark, K. Yano, and K. Nakaya. 1997. The gross anatomy of the female reproductive tract and associated organs of the Fukuoka megamouth shark (*Megachasma pelagios*). In *Biology of the Megamouth Shark,* edited by K. Yano, J. F. Morrissey, Y. Yabumoto, and K. Nakaya, pp. 115–119. Tokyo: Tokai University Press.

Castro-Aguirre, J. L. 1981. Especies de la familia Scyliorhinidae (Elasmobranchii, Galeoidea), de la costa occidental de México, con especial referencia a *Cephalurus cephalus* (Gilbert). *Anales de la Escuela Nacional de Ciencias Biologicas de Mexico* 24: 71–93.

Castro-Aguirre, J. L., A. Antuna-Mendiola, A. F. González-Acosta, and J. De la Cruz-Agüero. 2005. *Mustelus albipinnis* sp. nov. (Chondrichthyes: Carcharhiniformes: Triakidae) de la costa suroccidental de Baja California Sur, México. *Hidrobiológica* 15 (2 especial): 123–130.

Castro-Aguirre, J. L., and F. de Lachica-Bonilla. 1973. Nuevos registros de peces marinos en la costa del Pacífico mexicano. *Revista de la Sociedad Mexicana de Historia Natural* 34: 147–181.

Castro-Aguirre, J. L., and F. García-Domínguez. 1988. Sobre la presencia de *Isistius brasiliensis* (Quoy y Gaimard) (Squaliformes: Squalidae: Dalatiinae) en el Golfo de México. *Anales de la Escuela Nacional de Ciencias Biologicas de México* 32 (1–4): 91–108.

Castro-Aguirre, J. L., H. Espinosa Pérez, and L. Huidobro Campos. 2006. Dos nuevas especies del género *Squatina* (Chondrichthyes: Squatinidae) del Golfo de México. *Revista de Biología Tropical* 54 (3): 1031–1040.

Cervigón, F. 1961. Una nueva especie de *Oxynotus* de las costas de Venezuela. *Novedades Cientificas, Contribuciones Ocasionales del Museo de Historia Natural La Salle, Serie Zoológica* 27: 3–10.

Cervigón, F. 1964. Nuevas citas de peces para Venezuela y datos sobre algunas especies poco conocidas. *Novedades Cientiíficas, Contribuciones Ocasionales del Museo de Historia Natural La Salle, serie Zoológica* 31: 1–18.

Cervigón, F. 1966. *Los Peces Marinos de Venezuela,* Vol. 1. Caracas: Estación de Investigaciones Marinas de Margarita, Fundación La Salle de Ciencias Naturales. 436 pp.

Chang, W.-B., M.-Y. Leu, and L.-S. Fang. 1997. Embryos of the whale shark, *Rhincodon typus:* Early growth and size distribution. *Copeia* 1997 (2): 444–446.

Chatwin, B. M., and C. R. Forrester. 1953. Feeding habits of dogfish (*Squalus suckleyi* (Girard). Fisheries Research Board of Cananda, *Progress Reports of the Pacific Coast Stations* no. 95: 35–38.

Chávez-Ramos, H., and J. L. Castro-Aguirre. 1974. Notas y observaciones sobre la presencia de *Echinorhinus cookei* Pietschmann, 1928, en el Golfo de California,

México. *Anales de la Escuela Nacional de Ciencias Biológicas de México* 21: 155–164.

Chen, C., T. Taniuchi, and Y. Nose. 1979. Blainville's dogfish, *Squalus blainville,* form Japan, with notes on *S. mitsukurii* and S. japonicus. *Japanese Journal of Ichthyology* 26: 26–42.

Chen, C.-T., T.-C. Liu, and S.-J. Joung. 1988. Notes on reproduction in the scalloped hammerhead, *Sphyrna lewini,* in northeastern Taiwan waters. *Fishery Bulletin* 86: 389–393.

Chen, C.-T., K.-M. Liu, and Y.-C. Chang. 1997a. Reproductive biology of the bigeye thresher shark, *Alopias superciliosus* (Lowe, 1839) (Chondrichthyes: Alopiidae), in the northwestern Pacific. *Ichthyological Research* 44 (3): 227–235.

Chen, C.-T., K.-M. Liu, and S.-J. Joung. 1997b. Preliminary report on Taiwan's whale shark fishery. *Traffic Bulletin* 17 (1): 1–5.

Chen, J. T. F. 1963. A review of the sharks of Taiwan. *Biological Bulletin 19. Ichthyological Series* no. 1: 1–102. Tunghai University, Taichung, Taiwan.

Cherel, Y., and G. Duhamel. 2004. Antarctic jaws: Cephalopod prey of sharks in Kerguelen waters. *Deep-Sea Research,* pt. 1, 51: 17–31.

Chiaramonte, G. E. 1998. The shark genus *Carcharhinus* Blainville, 1816 (Chondrichthyes: Carcharhinidae) in Argentine waters. *Marine and Freshwater Research* 49: 747–752.

Chirichigno F. N. 1963. *Nuevos tiburones para la fauna del Perú.* Serie de Divulgación Científica no. 19. Callao, Peru: Servicio de Pesquería del Perú. 20 pp.

Chirichigno F. N. 1976. *Nuevas adiciones a la ictiofauna marina del Perú.* Serie de Informes Especiales no. IM 174. Callao, Peru: Instituto del Mar del Perú. 155 pp.

Chirichigno F. N. 1998. *Claves para identificar los peces marinos del Perú.* Callao, Peru: Instituto del Mar del Perú.

Choy, B. K., and D. H. Adams. 1995. An observation of a basking shark, *Cetorhinus maximus,* feeding along a thermal front off the east central coast of Florida. *Florida Scientist* 58 (4): 313–319.

Chu, Y.-T., C.-W. Meng, and J.-X. Liu. 1981. Description of a new genus and a new species of Squalidae of China. *Acta Zootaxonomica Sinica* 6 (1): 100–103.

Chute, G. R. 1930. The lily-iron returns to Monterey Bay: Shark fishing recommences on a harpoon basis. *California Fish and Game* 16 (2): 143–152.

Cigala Fulgosi, F., and G. Gandolfi. 1983. Re-description of the external morphology of *Somniosus rostratus* (Risso, 1826), with special reference to its squamation and cutaneous sensory organs, and aspects of their functional morphology (Pisces Selachii Squalidae). *Monitore Zoologique Italiano,* n.s., 17: 27–70.

Cione, A. L. 1998. Record of a sleeper shark, *Somniosus* (Squaliformes), on the Argentinean coast. *Neotrópica* 44 (111–112): 125–127.

Clark, A. H. 1887. Part 15. The whale-fishery. 1. History and present condition of the fishery. In *The Fisheries and Fishery Industries of the United States,* Sec. 5, Vol. 2, edited by G. B. Goode, pp. 3–293. Washington, D.C.: Government Printing Office.

Clark, E. 1947. Notes on the inflating power of the swell shark, *Cephaloscyllium uter. Copeia* 1947 (4): 278–280.

Clark, E. 1961. Visual discrimination in lemon sharks. In *Abstracts of Symposium Papers,* pp. 175–176. Honolulu, Hawaii: Tenth Pacific Science Congress.

Clark, E. 1963a. The maintenance of sharks in captivity, with a report on their instrumental conditioning. In *Sharks and Survival,* edited by P. W. Gilbert, pp. 115–149. Boston: American Institute of Biological Sciences.

Clark, E. 1963b. Massive aggregations of large rays and sharks in and near Sarasota, Florida. *Zoologica* 48 (2): 61–64.

Clark, E. 1975. Into the lairs of "sleeping" sharks. *National Geographic* 147 (4): 570–584.

Clark, E. 1992. Gentle monsters of the deep; whale sharks. *National Geographic* 182 (6): 122–138.

Clark, E., and E. Kristof. 1990. Deep-sea Elasmobranchs observed from submersibles off Bermuda, Grand Cayman, and Freeport, Bahamas. In *Elasmobranchs as Living Resources: Advances in the Biology, Ecology, Systematics, and the Status of the Fisheries,* edited by H. L. Pratt, Jr., S. H. Gruber, and T. Taniuchi, pp. 269–284. NOAA Technical Report NMFS 90. Springfield, Va.: U.S. Department of Commerce.

Clark, E., and D. R. Nelson. 1997. Young whale sharks, *Rhincodon typus,* feeding on a copepod bloom near La Paz, Mexico. *Environmental Biology of Fishes* 50: 63–73.

Clark, E., and K. von Schmidt. 1965. Sharks of the central Gulf coast of Florida. *Bulletin of Marine Science* 15 (1): 13–83.

Clarke, M. R. 1980. Cephalopoda in the diet of sperm whales of the southern hemisphere and their bearing on sperm whale biology. *Discovery Reports* 37: 1–324.

Clarke, T. A. 1971. The ecology of the scalloped hammerhead shark, *Sphyrna lewini,* in Hawaii. *Pacific Science* 25 (2): 133–144.

Clarke, W. D. 1963. Function of bioluminescence in mesopelagic organisms. *Nature* 196: 1244–1246.

Clemens, W. A., and G. V. Wilby. 1946. *Fishes of the Pacific Coast of Canada.* Bulletin 68. Ottawa: Fisheries Research Board of Canada. 368 pp.

Cliff, G. 1995. Sharks caught in the protective gill nets off Kwazulu-Natal, South Africa. 8. The great hammerhead shark *Sphyrna mokarran* (Rüppell). *South African Journal of Marine Science* 15: 105–114.

Cliff, G., and S. F. J. Dudley. 1992. Sharks caught in the protective gill nets off Natal, South Africa. 6. The copper shark *Carcharhinus brachyurus* (Günther). *South African Journal of Marine Science* 12: 663–674.

Cliff, G., S. F. J. Dudley, and B. Davis. 1989. Sharks caught in the protective gill nets off Natal, South Africa. 2. The great white shark, *Carcharodon carcharias* (Linnaeus). *South African Journal of Marine Science* 8: 131–144.

Cliff, G., S. F. J. Dudley, and B. Davis. 1990. Sharks caught in the protective gill nets off Natal, South African. 3. The shortfin mako *Isurus oxyrinchus* (Rafinesque). *South African Journal of Marine Science* 9: 115–126.

Cliff, G., R. P. Van der Elst, A. Govender, T. K. Witthuhn, and E. M. Bullen. 1996. First estimates of mortality and population size of white sharks on the South African coast. In *Great White Sharks: The Biology of Carcharodon carcharias,* edited by A. P. Klimley and D. G. Ainley, pp. 393–400. San Diego: Academic Press.

Coad, B. W., and J. Gilhen. 2002. A rare leucistic spiny dogfish, *Squalus acanthias,* from the Bay of Fundy, Nova Scotia. *Canadian Field-Naturalist* 116: 120–121.

Cocking, S. 2002. Looking for the meaning behind shark clustering. *The Herald* (Miami, Florida), 10 March: 18C.

Cohen, J. L. 1973. An albino grey smoothhound *Mustelus californicus* Gill. *California Fish and Game* 59 (3): 210–211.

Coles, R. J. 1915. Notes on the sharks and rays of Cape Lookout, N.C. *Proceedings of the Biological Society of Washington* 28: 89–94.

Coles, R. J. 1919. The large sharks of Cape Lookout, North Carolina. The white shark or maneater, tiger shark and hammerhead. *Copeia* 1919 (69): 34–43.

Collett, R. 1890. Sur quelques poissons rapportés de Madère par le Prince de Monaco. *Bulletin de la Societe Zoologique de France,* Séance du 25 Novembre 1890: 218–224.

Collett, R. 1897. On *Chlamydoselachus anguineus,* Garm., a remarkable shark found in Norway 1896. *Festskrift til Hs. Maj. Kong Oscar II. ved regjerings jubilet den 18 de september, 1897, fra det K Norske Frederiks universitet* 2: 1–17.

Collett, R. 1904. Diagnosis of four hitherto undescribed fishes from the depths south of the Faroe Islands. Paper read at Meeting of the Videnskabsselskab, May 3rd, 1904. *Christiania Videnskabs-Selskabs Forhandlinger for 1904,* no. 9: 3–7.

Collett, R. 1905. *Meddelelser om Norges Fiske i Aarene 1884–1901. Christiania Videnskabs-Selskabs Forhandlinger* 3 (7): 79–91.

Collier, R. S. 1992. Recurring attacks by white sharks on divers at two Pacific sites off Mexico and California. *Environmental Biology of Fishes* 33: 319–325.

Collyer, R. D. 1953. The bramble shark (*Echinorhinus brucus*) at Guadalupe Island, Mexico. *California Fish and Game* 39 (2): 266.

Commercial Fisheries Review. 1960. Frozen salmon shark exported to Italy. *Commercial Fisheries Review* October: 67.

Compagno, L. J. V. 1984a. *FAO Species Catalogue. Vol. 4. Sharks of the World. An Annotated and Illustrated Catalogue of Shark Species Known to Date. Part 1. Hexanchiformes to Lamniformes.* FAO Fisheries Synopsis 125 (4, pt. 1): 1–249. Rome: Food and Agriculture Organization of the United Nations. 249 pp.

Compagno, L. J. V. 1984b. *FAO Species Catalogue. Vol. 4. Sharks of the World. An Annotated and Illustrated Catalogue of Shark Species Known to Date. Part 2. Carcharhiniformes.* FAO Fisheries Synopsis 125 (4, pt. 2): 251–655. Rome: Food and Agriculture Organization of the United Nations. 405 pp.

Compagno, L. J. V. 1988. *Sharks of the Order Carcharhiniformes.* Princeton, N.J.: Princeton University Press. 486 pp.

Compagno, L. J. V. 2001. *Sharks of the World. An Annotated and Illustrated Catalogue of Shark Species Known to Date. Vol. 2. Bullhead, Mackerel and Carpet Sharks (Heterodontiformes, Lamniformes and Orectolobiformes).* FAO Species Catalogue for Fishery Purposes no. 1, Vol. 2. Rome: Food and Agriculture Organization of the United Nations. 269 pp.

Compagno, L. J. V., and J. A. F. Garrick. 1983. *Nasolamia,* new genus, for the shark *Carcharhinus velox* Gilbert, 1898 (Elasmobranchii: Carcharhinidae). *Zoology Publications from Victoria University of Wellington* 76: 1–16.

Compagno, L. J. V., M. Dando, and S. Fowler. 2005. *A Field Guide to the Sharks of the World.* London: HarperCollins. 368 pp.

Conrath, C. L., J. Gelsleichter, and J. A. Musick. 2002. Age and growth of the smooth dogfish (*Mustelus canis*) in the northwest Atlantic Ocean. *Fishery Bulletin* 100: 674–682.

Correia (Correa [*sic*]), J., J. De Marignac, and S. Gruber. 1995. Young lemon shark behaviour in Bimini lagoon. *Bahamas Journal of Science* 3 (1): 7 pp.

Correia da Costa, M. E. 1994. Tubarões de profundidade: Estudo da biologia e pescaria de *Centrophorus granulosus* (Bloch & Schneider, 1801), *Dalatias licha*

(Bonnaterre, 1788) e *Deania calceus* (Lowe, 1839) no Atlântico Central. Thesis, Universidade do Algarve, Portugal. 77 pp.

Cortés, E. 1999. Standardized diet compositions and trophic levels of sharks. *Journal of Marine Science* 56: 707–717.

Cortés, E., and S. H. Gruber. 1990. Diet, feeding habits and estimates of daily ration of young lemon sharks, *Negaprion brevirostris* (Poey). *Copeia* 1990 (1): 204–218.

Costa, F. E. S., F. M. S. Braga, A. F. Amorim, and C. A. Arfelli. 1995. Reproductive biology of shortfin mako, *Isurus oxyrinchus,* Rafinesque 1809. Program and Abstracts, p. 30.7[th] Reunião do Grupo de Trabalho Sobre Pesca e Pesquisa de Raias e Tubarões no Brasil. 20–24 November 1995. Rio Grande, Brazil. Pelotas, Brazil: Universidade Federal de Pelotas.

Couch, J. 1846. Account of the capture, in Cornwall, of the six-branchial gray shark; but lately recognized in the British fauna. *Zoologist: A Monthly Journal of Natural History* 4: 1337–1340.

Couch, J. 1868. *A History of the Fishes of the British Islands.* 2nd ed. 4 vols. London: Groombridge and Sons.

Cowan, I. M. 1938. Some fish records from the coast of British Columbia. *Copeia* 1938 (2): 97.

Cox, K. W. 1963. Egg-cases of some elasmobranchs and a cyclostome from Californian waters. *California Fish and Game* 49 (4): 271–289.

Crane, N. L., and J. N. Heine. 1992. Observations of the prickly shark (*Echinorhinus cookei*) in Monterey Bay, California. *California Fish and Game* 78 (4): 166–168.

Croker, R. S. 1942. Mackerel shark (*Lamna nasus*) taken in California. *California Fish and Game* 28 (2): 124–125.

Cross, J. N. 1988. Aspects of the biology of two scyliorhinid sharks, *Apristurus brunneus* and *Parmaturus xaniurus,* from the upper continental slope off southern California. *Fishery Bulletin* 86: 691–702.

Crow, G. L., and J. D. Hewitt, IV. 1988. Longevity records for captive tiger sharks *Galeocerdo cuvier* with notes on behaviour and management. *International Zoo Yearbook* 27: 237–240.

Crow, G. L., C. G. Lowe, and B. M. Wetherbee. 1996. Shark records from longline fishing programs in Hawaii with comments on Pacific Ocean distributions. *Pacific Science* 50 (4): 382–392.

Crozier, P., M. Zuffa, and T. Storai. 2002. On the capture of a giant specimen of Greenland shark *Somniosus microcephalus* (Bloch & Schneider, 1801) with some considerations about maximum sizes collected for the genus *Somniosus* compared with the genus *Carcharodon.* Abstracts, p. 23. 6th Annual European Elasmobranch Association Conference, Cardiff, U.K., 6–8 September 2002. National Museums and Galleries of Wales.

Cruz, R., R. Brito, E. Díaz, and R. Lalana. 1986. Ecología de la langosta (*Panulirus argus*) al SE de la Isla de la Juventud. I. Colonización de arrecifes artificiales. *Revista de Investigaciones Marinas* 7 (3): 3–17.

Cruz-Martínez, A., X. Chiappa-Carrara, and V. Arenas-Fuentes. 2005. Age and growth of the bull shark, *Carcharhinus leucas,* from southern Gulf of Mexico. *Journal of Northwest Atlantic Fishery Science* 35: 367–374.

Cuvier, G. 1817. *Le Règne Animal Distribué D'après son Organisation, pour Servir de Base a l'Histoire Naturelle des Animaux et d'Introduction a l'Anatomie Comparee. Tome 2. Les reptiles, les Possions, les Mollusques et les Annélides.* Poissons. Paris: Deterville. pp. 104–142.

Cuvier, G. 1995. *Historical Portrait of the Progress of Ichthyology from Its Origins to Our Own Time.* Edited by T. Pietsch. Baltimore: Johns Hopkins Univeristy Press. 366 pp.

D'Aubrey, J. D. 1964. Preliminary Guide to the Sharks Found Off the East Coast of South Africa. Investigational Report no. 8. Durban: Oceanographic Research Institute. 95 pp.

Dahlberg, M. D. 1975. *Guide to Coastal Fishes of Georgia and Nearby States.* Athens: University of Georgia Press. 186 pp.

Dahlberg, M. D., and R. W. Heard, III. 1969. Observations on elasmobranchs from Georgia. *Quarterly Journal of the Florida Academy of Sciences* 32 (1): 21–25.

Daniel, J. F. 1914. The anatomy of *Heterodontus francisci.* 1. The exoskeleton. *University of California Publications in Zoology* 13 (6): 147–166.

Daniel, J. F. 1915. The anatomy of *Heterodontus francisci.* 2. The endoskeleton. *Journal of Morphology* 26 (3): 447–493.

Daniel, J. F. 1934. *The Elasmobranch Fishes.* 3rd ed., rev. Berkeley: University of California Press. 332 pp.

Daugherty, A. E. 1964. The sand shark, *Carcharias ferox* (Risso), in California. *California Fish and Game* 50 (1): 4–10.

Day, F. 1880–1884. *The Fishes of Great Britain and Ireland,* Vol. 2. London: Williams and Norgate. 388 pp.

de Brito Capello, F. 1868. Description de trois nouveaux poissons des mers du Portugal. *Jornal de Sciencias Mathematicas, Physicas, e Naturaes* 1: 318–323.

de Brito Capello, F. 1870. Catalogo dos peixes de Portugal que existem no museu de Lisboa. *Jornal de Sciencias Mathematicas, Physicas, e Naturaes* 2: 131–153.

de Buen, F. 1926. *Catálogo Ictiológico del Mediterráneo Español y de Marruecos.* Instituto Español de Oceanografía. Madrid. No. 2. 221 pp.

de Buen, F. 1950. El tiburón vitamínico de la costa uruguaya, *Galeorhinus vitaminicus,* nov. sp., y algunas consideraciones generales sobre su biología. *Publicaciones Científicas* (Montevideo) 4: 153–162.

de Buen, F. 1952. El tiburón vitamínico. *Revista de la Facultad de Humanidades y Ciencias* (Montevideo) 7: 87–116.

de Buen, F. 1960. Tiburones, rayas y quimeras en la estación de biología marina de Montemar, Chile. *Revista de Biología Marina* (Universidad de Chile, Valparaiso) 10 (1, 2, and 3): 1–50.

De Crosta, M. A., L. R. Taylor, Jr., and J. D. Parrish. 1984. Age determination, growth, and energetics of three species of carcharhinid sharks in Hawaii. In *Proceedings of the Second Symposium on Resource Investigations in the Northwestern Hawaiian Islands,* Vol. 2, edited by R. W. Grigg and K. Y. Tanoue, pp. 75–95. Sea Grant Miscellaneous Report, UNIHI-SEAGRANT-MR-84–01. Honolulu: University of Hawaii.

De Groot, S. J., and H. Nijssen. 1976. The shark *Dalatias licha* (Bonnaterre, 1788) caught alive in the coastal waters of the Netherlands (Pisces, Selachii, Dalatiidae). *Bulletin Zoologisch Museum Universiteit van Amsterdam* 5 (9): 73–76.

De Kay, J. E. 1842. *Zoology of New-York or the New-York Fauna. Part 4, Fishes.* Albany, N.Y.: W. & A. White & J. Visscher. 415 pp.

de Wit, L. A. 1975. Changes in the species composition of sharks in south San Francisco Bay. *California Fish and Game* 61 (2): 106–111.

de Wit, L. A., and T. B. Scanland. 1980. Notes on seasonal collections of sharks near Point Conception, California. *Bulletin of the Southern California Academy of Sciences* 79 (3): 112–118.

Dean, B. 1907. Notes on Acanthodian sharks. *The American Journal of Anatomy* 7 (2): 209–226.

Dehart, A. 2004. Husbandry of tiger sharks, *Galeocerdo cuvier.* In *Elasmobranch Husbandry Manual: Captive Care of Sharks, Rays, and Their Relatives,* edited by M. Smith, D. Warmolts, D. Thoney, and R. Hueter, pp. 483–486. Columbus: Ohio Biological Survey, Inc.

DeLacy, A. C., and W. M. Chapman. 1935. Notes on some elasmobranchs of Puget Sound, with descriptions of their egg cases. *Copeia* 1935 (2): 63–67.

DeLacy, A. C., B. S. Miller, and S. F. Borton. 1972. *Checklist of Puget Sound Fishes.* Washington Seagrant Program 72–3. Seattle: University of Washington. 43 pp.

Dempster, R. P., and E. S. Herald. 1961. Notes on the hornshark, *Heterodontus francisci,* with observations on mating activities. *Occasional Papers of the California Academy of Sciences* 33: 1–7.

Deng, S. M., G. Q. Xiong, and H. X. Zhan. 1981. On three new species of sharks of the genus *Carcharhinus* from China. *Acta Zootaxonomica Sinica* 6 (2): 216–220.

Depperman, B. 1953. Maternity model mako. *Sports Afield* March: 86–87.

Deraniyagala, P. E. P. 1936. Oceanic big-game fishing off Ceylon. *Country Life* 2 May: 31.

Desbrosses, P. 1938. Croissance et migrations du requin griset *Hexanchus griseus* (Bonnaterre 1788) Rafinesque 1810. *Revue des Travaux de l'Office des Pêches Maritimes* 11 (1, no. 41): 53–57.

DeWitt, J. W. 1955. A record of an attack by a leopard shark (*Triakis semifasciata* Girard). *California Fish and Game* 41 (4): 348.

Di Noronha, A. C. 1926. XIII. A new species of deep water shark (*Squaliolus sarmenti*) from Madeira. *Annals of the Carnegie Museum* 16 (3–4): 385–389.

Díaz de Astarloa, J. M., D. E. Figueroa, L. Lucífora, R. C. Menni, B. L. Prenski, and G. Chiaramonte. 1999. New records of the Pacific sleeper shark, *Somniosus pacificus* (Chondrichthyes: Squalidae), from the southwest Atlantic. *Ichthyological Research* 46 (3): 303–308.

Dickens, D. A. G. 1956. Deep sea shark (Indian Ocean). *Marine Observers' Log, London* 26: 73–74.

Dodrill, J. W. 1977. A Hook and Line Survey of the Sharks of Melbourne Beach, Brevard County, Florida. Master's thesis, Florida Institute of Technology. 304 pp.

Dodrill, J. W., and R. G. Gilmore. 1978. Land birds in the stomachs of tiger sharks *Galeocerdo cuvieri* (Peron and Lesueur). *Auk* 95: 585–586.

Dodrill, J. W., and R. G. Gilmore. 1979. First North American continental record of the longfin mako (*Isurus paucus* Guitart Manday). *Florida Scientist* 42 (1): 52–58.

Doflein, F. 1906. *Ostasienfahrt. Ergebnisse und Beobachtungen Eines Naturforschers in China, Japan und Celyon.* Leipzig: Teubner. 511 pp.

Dollfus, R. P. 1955. *Premièer contribution à l'établissement d'un fichier ichthyologique du Maroc atlantique de Tanger à l'embouchure de l'Oued Dra.* Travaux de L'institute Scientifique Chérifien, serie Zoologie, no. 6. Tangier: Editions Internationales. 227 pp.

Domínguez Arellano, A. P. 2003. Algunos aspectos reproductivos del tiburón coyotito (*Nasolamia velox*) en el Golfo de Tehuantepec del periodo 1996–2001. Thesis, Universidad Nacional Autónoma de México. 70 pp.

Domanevskiy, L. N. 1975. The frill shark, *Chlamidoselachus [sic] anguineus,* from the Cape Blanc area (central eastern Atlantic). *Journal of Ichthyology* 15 (6): 1000–1002.

Dooley, J. K. 1972. Fishes associated with the pelagic *Sargassum* complex, with a discussion of the *Sargassum* community. *Contributions in Marine Science* 16: 1–32.

Drainville, G., and L. Brassard. 1960. Le requin *Somniosus microcephalus* dans la rivière Saguenay. *Le Naturaliste Canadien* 87 (12): 269–277.

Driggers, W. B., III. 2002. Age, Growth, Reproduction and Population Genetics of the Blacknose Shark, *Carharhinus acronotus,* in the Western North Atlantic. Ph.D. diss., University of South Carolina, Columbia. 104 pp.

Driggers, W. B., III., J. K. Carlson, B. Cullum, J. M. Dean, D. Oakley, and G. Ulrich. 2004. Age and growth of the blacknose shark, *Carcharhinus acronotus,* in the western North Atlantic Ocean with comments on regional variation in growth rates. *Environmental Biology of Fishes* 71: 171–178.

Driggers, W. B., III, G. H. Burgess, A. N. Hamilton Jr., N. M. Hopkins, and C. M. Schobernd. *Squaliolus laticaudus* in the western North Atlantic Ocean: Distributional and life history observations. Unpublished manuscript.

Dudley, S. F. J., and G. Cliff. 1993. Sharks caught in the protective gill nets off Natal, South Africa. 7. The blacktip shark *Carcharhinus limbatus* (Valenciennes). *South African Journal of Marine Science* 13: 237–254.

Duffy, C. A. J. 1997. Further records of the goblin shark, *Mitsukurina owstoni* (Lamniformes: Mitsukurinidae), from New Zealand. *New Zealand Journal of Zoology* 24: 167–171.

Duhamel, G., and C. Ozouf-Costaz. 1982. Presence de *Lamna nasus* (Bonnaterre, 1788) aux Iles Kerguelen. *Cybium* 6 (4): 15–18.

Duméril, A. H. A. 1865. *Histoire Naturelle des Poissons ou Ichthyologie Générale.* Tome 1. *Elasmobranchés.* Plagiostomes et Holocéphales ou Chimères, pt. 2. Paris: Roret. 720 pp.

Dunbar, M. J., and H. H. Hildebrand. 1952. Contribution to the study of fishes of Ungava Bay. *Journal of the Fisheries Research Board of Canada* 9 (2): 83–128.

Dunbrack, R., and R. Zielinski. 2003. Seasonal and diurnal activity of sixgill sharks (*Hexanchus griseus*) on a shallow water reef in the Strait of Georgia, British Columbia. *Canadian Journal of Zoology* 81: 1107–1111.

Duncker, G., and E. Mohr. 1929. Die fische der Südsee-expedition der Hamburgischen Wissenschaftlichen Stiftung 1908–1909. 3 Teil. *Mitteilungen aus dem Zoologischen Staatsinstitut und Zoologischer Museum in Hamburg* (44): 57–84.

Dyer, W. J., D. I. Fraser, and S. N. Tibbo. 1963. Composition and palatability of porbeagle flesh. *Journal of the Fisheries Research Board of Canada* 20 (5): 1153–1158.

Ebert, D. A. 1985. Color variation in the sevengill shark, *Notorynchus maculatus* Ayres, along the California coast. *California Fish and Game* 71 (1): 58–59.

Ebert, D. A. 1986a. Aspects on the biology of hexanchid sharks along the California coast. In *Indo-Pacific Fish Biology: Proceedings of the Second International Conference on Indo-Pacific Fishes,* edited by T. Uyeno, R. Arai, T. Taniuchi, and K. Matsuura, pp. 437–449. Tokyo: Ichthyological Society of Japan.

Ebert, D. A. 1986b. Biological aspects of the sixgill shark, *Hexanchus griseus. Copeia* 1986 (1): 131–135.

Ebert, D. A. 1989. Life history of the sevengill shark, *Notorynchus cepedianus* Peron, in two northern California bays. *California Fish and Game* 75 (2): 102–112.

Ebert, D. A. 1991a. Diet of the sevengill shark *Notorhynchus cepedianus* in the temperate coastal waters of southern Africa. *South African Journal of Marine Science 11: 565–572.*

Ebert, D. A. 1991b. Observations on the predatory behaviour of the sevengill shark *Notorynchus cepedianus. South African Journal of Marine Science* 11: 455–465.

Ebert, D. A. 1996. Biology of the sevengill shark *Notorynchus cepedianus* (Peron, 1807) in the temperate coastal waters of southern Africa. *South African Journal of Marine Science* 17: 93–103.

Ebert, D. A. 2003. *Sharks, Rays, and Chimaeras of California.* Berkeley: University of California Press. 284 pp.

Ebert, D. A., L. J. V. Compagno, and L. J. Natanson. 1987. Biological notes on the Pacific sleeper shark, *Somniosus pacificus* (Chondrichthyes: Squalidae). *California Fish and Game* 73 (2): 117–123.

Ebert, D. A., and T. B. Ebert. 2005. Reproduction, diet and habitat use of leopard sharks, *Triakis semifasciata* (Girard), in Humboldt Bay, California, USA. *Marine and Freshwater Research* 56: 1089–1098.

Ebert, D. A., H. F. Mollet, A. Baldridge, T. Thomas, K. A. Forney, and W. E. Ripley. 2004. Occurrence of the whale shark, *Rhincodon typus* Smith 1828, in California waters. *Northwestern Naturalist* 85: 26–28.

Eckert, S. A., and B. S. Stewart. 2001. Telemetry and satellite tracking of whale sharks, *Rhincodon typus,* in the Sea of Cortez, Mexico, and the north Pacific Ocean. *Environmental Biology of Fishes* 60: 299–308.

Eckert, S. A., L. L. Dolar, G. L. Kooyman, W. Perrin, and R. A. Rahman. 2002. Movements of whale sharks

(*Rhincodon typus*) in South-east Asian waters as determined by satellite telemetry. *Journal of Zoology (London)* 257: 111–115.

Edwards, H. M. 1920. The growth of the swell shark within the egg case. *California Fish and Game* 6 (4): 153–157.

Eigenmann, R. S. 1890. Description of a new species of *Euprotomicrus. Proceedings of the California Academy of Sciences,* 2nd ser., 3: 35.

Eigenmann, C. H. 1893. The fishes of San Diego. *Proceedings of the United States National Museum* 15: 123–178.

Eisert, C. F. 1969. Reproduction in the ovoviviparous sharks, *Etmopterus virens* and *Squalus acanthias.* Master's thesis, Cornell University, Ithaca, N.Y. 90 pp.

Ellis, R. 1975. *The Book of Sharks.* New York: Grosset & Dunlap. 320 pp.

Engelhardt, R. 1912. Über einige neue Selachier-Formen. *Zoologischer Anzeiger* (Leipzig) 39: 643–648.

Eschricht, H., and H. Müller. 1835. Über die arteriösen und venösen wundernetze an der leber und einen merkwürdigen bau dieses organes beim thunfische, *Thynnus vulgaris.* Berlin: Abhandlungen der Koniglichen Akademie der Wissenschaften zu Berlin. 32 pp.

Espinosa Pérez, H., J. L. Castro-Aguirre, and L. Huidobro Campos. 2004. *Listados Faunísticos de México. 9. Catálogo Sistemático de Tiburones (Elasmobranchii: Selachimorpha).* Mexico: Universidad Nacional Autónoma de México. 134 pp.

Essapian, F. S. 1962. Notes on the behavior of sharks in captivity. *Copeia* 1962 (2): 457–459.

Evermann, B. W. 1919. Note on the basking shark. *Copeia* 1919 (74): 77–78.

Evermann, B. W., and B. A. Bean. 1898. Indian River and its fishes. *Report of U.S. Commissioner of Fish and Fisheries for the year 1896* 22: 227–262.

Faber, F. 1829. *Naturgeschichte der Fische Islands.* Frankfurt-Main: Druck und Verlag von Heinrich Ludwig Brönner. 206 pp.

Fabricius, O. 1780. *Fauna Groenlandica systematice sistens, animalia Groenlandiae occidentalis.* Copenhagen: Ioannis Gottlob Rothe. 452 pp.

Fairfax, D. 1998. *The Basking Shark in Scotland: Natural History, Fishery and Conservation.* East Linton, Scotland: Tuckwell Press. 206 pp.

F.A.O. 1994. *FAO Yearbook, Fishery Statistics, Catches and Landings,* Vol. 78. FAO Fisheries Series no. 46, FAO Statistics Series no. 128. Rome: Food and Agriculture Organization of the United Nations. 699 pp.

Feder, H. M., C. H. Turner, and C. Limbaugh. 1974. *Observations on Fishes Associated with Kelp Beds in Southern California.* Fish Bulletin 160. Sacramento, Ca.: The Resources Agency of California, Department of Fish and Game. 144 pp.

Finstad, W. O., and D. R. Nelson. 1975. Circadian activity rhythm in the horn shark, *Heterodontus francisci:* Effect of light intensity. *Bulletin of the Southern California Academy of Sciences* 74 (1): 20–26.

Fisk, A. T., S. A. Tittlemier, J. L. Pranschke, and R. J. Norstrom. 2002. Using anthropogenic contaminants and stable isotopes to assess the feeding ecology of Greenland sharks. *Ecology* 83 (8): 2162–2172.

Fitch, J. E. 1948. Use of DUKW's in the fishery for basking sharks, *Cetorhinus maximus. California Fish and Game* 34 (4): 219–220.

Fitch, J. E. 1951 The whale shark, *Rhineodon typus,* off northern Baja California. *California Fish and Game* 37: 351.

Fitch, J. E. 1977. News Release. California Department of Fish and Game, Sacramento, 21 March.

Fitch, J. E., and W. L. Craig. 1964. First records for the bigeye thresher (*Alopias superciliosus*) and slender tuna (*Allothunnus fallai*) from California, with notes on eastern Pacific scombrid otoliths. *California Fish and Game* 50 (3): 195–206.

Flammang, B. E. 2005. Distribution and Reproductive Ecology of Deep-Sea Catsharks (Chondrichthyes: Scyliorhinidae) of the Eastern North Pacific. Master's thesis, California State University, Monterey Bay. 85 pp.

Flammang, B. E., D. A. Ebert, and G. M. Cailliet. 2008. Reproductive biology of deep-sea catsharks (Chondrichthyes: Scyliorhinidae) in the eastern north Pacific. *Environmental Biology of Fishes* 81: 35–49.

Flemming, K. 1999. Elasmobranchs. In *Report on the 1980–1995 Fish, Shrimp, and Crab Sampling in the San Francisco Estuary, California,* edited by J. J. Orsi, pp. 143–149. Report 63. Sacramento: California Bay-Delta Authority Science Program.

Flores G., H., and P. Rojas Z. 1979. Presencia de *Echinorhinus cookei* Pietschmann, 1928 frente a las costas de Chile (Pisces: Squalidae, Echinorhinidae). *Investigaciones Marinas* (Valparaíso) 7 (1): 41–45.

Follett, W. I. 1976. First record of albinism in the leopard shark (*Triakis semifasciata* Girard). *California Fish and Game* 62 (2): 163–164.

Ford, E. 1921. A contribution to our knowledge of the life-histories of the dogfishes landed at Plymouth. *Journal of the Marine Biological Association of the United Kingdom* 12: 468–505.

Forster, G. R., J. R. Badcock, M. R. Longbottom, N. R. Merrett, and K. S. Thomson. 1970. Results of the Royal Society Indian Ocean deep slope fishing expedition, 1969. *Proceedings of the Royal Society of London* B175: 367–404.

Fourmanoir, P. 1961. Requins de la côte ouest de Madagascar. *Mémoirs de l'Institut Scientifique de Madagascar,* sér. F, 4: 1–81.

Fouts, W. R., and D. R. Nelson. 1999. Prey capture by the Pacific angel shark, *Squatina californica:* Visually mediated strikes and ambush-site characteristics. *Copeia* 1999 (2): 304–312.

Fowler, H. W. 1906a. Some cold-blooded vertebrates of the Florida Keys. *Proceedings of the Academy of Natural Sciences of Philadelphia* 58 (pt. 1): 77–113.

Fowler, H. W. 1906b. *The Fishes of New Jersey.* Annual Report of the New Jersey State Museum, pt. 2: 35–477. Trenton: MacCrellish & Quigley, State Printers.

Fowler, H. W. 1925. Fishes from Natal, Zululand, and Portuguese East Africa. *Proceedings of the Academy of Natural Sciences of Philadelphia* 77: 187–268.

Fowler, H. W. 1941. *The Fishes of the Groups Elasmobranchii, Holocephali, Isospondyli, and Ostarophysi Obtained by the Unites States Bureau of Fisheries Steamer "Albatross" in 1907 to 1910, Chiefly in the Philippine Islands and Adjacent Seas.* Smithsonian Institution, United States National Museum, Bulletin 100, Vol. 13. Washington, D.C.: U.S. Government Printing Office. 879 pp.

Fowler, H. W. 1944. Results of the Fifth George Vanderbilt expedition (1941) (Bahamas, Caribbean Sea, Panama, Galapagos Archipelago and Mexican Pacific Islands). *Academy of Natural Sciences of Philadelphia, Monograph* 6: 57–529.

Fowler, H. W. 1945. *A Study of the Fishes of the Southern Piedmont and Coastal Plain.* Monograph 7. Philadelphia: Academy of Natural Sciences of Philadelphia, 408 pp.

Fox, D. L. 1976. *Animal Biochromes and Structural Colors.* Berkeley: University of California Press. 433 pp.

Fox, H. M., and G. Vevers. 1960. *The Nature of Animal Colours.* New York: Mcmillan. 246 pp.

Francis, M. P. 1996. Observations on a pregnant white shark with a review of reproductive biology. In *Great White Sharks: The Biology of Carcharodon carcharias,* edited by A. P. Klimley and D. G. Ainley, pp. 157–172. San Diego: Academic Press.

Francis, M. P., and J. D. Stevens. 2000. Reproduction, embryonic development, and growth of the porbeagle shark, *Lamna nasus,* in the southwest Pacific Ocean. *Fishery Bulletin* 98: 41–63.

Francis, M. P., J. D. Stevens, and P. R. Last. 1988. New records of *Somniosus* (Elasmobranchii: Squalidae) from Australasia, with comments on the taxonomy of the genus. *New Zealand Journal of Marine and Freshwater Research* 22: 401–409.

Franke, R., and A. Acero P. 1991. Registros nuevos y comentarios adicionales sobre peces cartilaginosos del Parque Nacional Natural Gorgona (Pacífico Colombiano). I. Tiburones. *Trianea (Acta Científica y Tecnológica INDERENA)* 4: 527–540.

Fraser-Brunner, A. 1950. A synopsis of the hammerhead sharks (*Sphyrna*), with description of a new species. *Records of the Australian Museum* 22 (3): 213–219.

Frentzel-Beyme, B. Z., and F. W. Köster. 2002. On the biology of the sharpnose sevengill shark, *Heptranchias perlo,* from the Great Meteor Seamount (Central Eastern Atlantic). In *Proceedings of the 4th European Elasmobranch Association Meeting,* edited by M. Vacchi, G. La Mesa, F. Serena, and B. Séret, pp. 77–96. Livorno, Italy: ICRAM, ARPAT, and SFI.

Frøiland, Ø. 1975. Albinisme hos hai. *Fauna* (Oslo) 28 (3): 170–173.

Fudge, D. S., and E. D. Stevens. 1996. The visceral *retia mirabilia* of tuna and sharks: An annotated translation and discussion of the Eschricht & Müller 1835 paper and related papers. *Guelph Ichthyology Reviews* 4: 1–92.

Fujita, K. 1981. Oviphagous embryos of the Pseudocarchariid shark, *Pseudocarcharias kamoharai,* from the Central Pacific. *Japanese Journal of Ichthyology* 28 (1): 37–44.

Furnestin, J., J. Dardogmac, C. Maurin, A. Vincent, R. Coupe, and H. Boutiere. 1958. Données nouvelles sur les poissions du Maroc Atlantique. *Revue des Travaux de l'Institut des Pêches Maritimes* 22 (4): 381–493.

Fusaro, C., and S. Anderson. 1980. First California record: The scalloped hammerhead shark, *Sphyrna lewini,* in coastal Santa Barbara waters. *California Fish and Game* 66 (2): 121–123.

Gadig, O. B. F., and U. L. Gomes. 2002. First report on embryos of *Isistius brasiliensis. Journal of Fish Biology* 60: 1322–1325.

Gaida, I. H. 1997. Population structure of the Pacific angel shark, *Squatina californica* (Squatiniformes: Squatinidae), around the California Channel Islands. *Copeia* 1997 (4): 738–744.

Galloway, L. 1994. *The Forty Fathom Bank.* San Francisco: Chronicle Books. 108 pp.

Garman, S. 1880. New species of selachians in the Museum collection. *Bulletin of the Museum of Comparative Zoology at Harvard College* 6 (1): 167–172.

Garman, S. 1884. An extraordinary shark. *Essex Institute Historical Collection* 16: 47–55.

Garman, S. 1885. I.-*Chlamydoselachus anguineus* Garm.—a living species of cladodont shark. *Bulletin of the Museum of Comparative Zoölogy at Harvard College* 12 (1): 1–35.

Garman, S. 1888. On the lateral canal system of the Selachia and Holocephala. *Bulletin of the Museum of Comparative Zoölogy at Harvard College* 17 (2): 57–119.

Garman, S. 1899. Report on an exploration off the west coasts of Mexico, Central and South America, and off the Galapagos Islands. XXVI. The Fishes. *Memoirs of the Museum of Comparative Zoölogy at Harvard College* 24, 2 vols. 431 pp.

Garman, S. 1906. New plagiostomia. *Bulletin of the Museum of Comparative Zoölogy at Harvard College* 46 (11): 203–208.

Garman, S. 1913. *The Plagiostoma (Sharks, Skates, and Rays).* Memoirs of the Museum of Comparative Zoölogy at Harvard College 36, 2 vols. Cambridge, Mass.: Harvard College. 515 pp.

Garrick, J. A. F. 1959. Studies on New Zealand Elasmobranchii. Part 8. Two northern hemisphere species of *Centroscymnus* in New Zealand waters. *Transactions of the Royal Society of New Zealand* 87 (pts. 1 and 2): 75–89.

Garrick, J. A. F. 1960a. Studies on New Zealand Elasmobranchii. Part 10. The genus *Echinorhinus,* with an account of a second species, *E. cookei* Pietschmann, 1928, from New Zealand waters. *Transactions of the Royal Society of New Zealand* 88 (pt. 1): 105–117.

Garrick, J. A. F. 1960b. Studies on New Zealand Elasmobranchii. Part 11. Squaloids of the genera *Deania, Etmopterus, Oxynotus* and *Dalatias* in New Zealand waters. *Transactions of the Royal Society of New Zealand* 88 (pt. 3): 489–517, plate 26.

Garrick, J. A. F. 1964. Additional information on the morphology of an embryo whale shark. *Proceedings of the United States National Museum* 115 (3476): 1–8.

Garrick, J. A. F. 1967. Revision of sharks of the genus *Isurus* with description of a new species (Galeoidea, Lamnidae). *Proceedings of the United States National Museum* 118 (3537): 663–690.

Garrick, J. A. F. 1974. First record of an odontaspidid shark in New Zealand waters. *New Zealand Journal of Marine and Freshwater Research* 8 (4): 621–630.

Garrick, J. A. F. 1982. Sharks of the genus *Carcharhinus.* NOAA Technical Report NMFS Circular 445. Rockville, MD. U.S. Department of Commerce. 194 pp.

Garrick, J. A. F. 1985. Additions to a revision of the shark genus *Carcharhinus:* Synonimy of *Aprionodon* and *Hypoprion,* and description of a new species of *Carcharhinus* (Carcharhinidae). NOAA Technical Report NMFS 34. Springfield, Virginia.U.S. Department of Commerce. 26 pp.

Garrick, J. A. F., and J. M. Moreland. 1968. Notes on a bramble shark, *Echinorhinus cookei,* from Cook Strait, New Zealand. *Records of the Dominion Museum* 6 (10): 133–139.

Garrick, J. A. F., and L. J. Paul. 1971. *Heptranchias dakini* Whitley, 1931, a synonym of *H. perlo* (Bonnaterre, 1788), the sharpsnouted sevengill or perlon shark, with notes on sexual dimorphism in this species. *Zoology Publications* (Victoria University) 54: 1–14.

Garrick, J. A. F., and S. Springer. 1964. *Isistius plutodus,* a new squaloid shark from the Gulf of Mexico. *Copeia* 1964 (4): 678–682.

Garrick, J. A. F., and L. P. Schultz. 1963. A guide to the kinds of potentially dangerous sharks. In *Sharks and Survival,* edited by P. W. Gilbert, pp. 3–60. Boston: D. C. Heath.

Garrick, J. A. F., R. Backus, R. H. Gibbs, Jr. 1964. *Carcharhinus floridanus,* the silky shark, a synonym of *C. falciformis. Copeia* 1964 (2): 369–375.

Gauld, J. A. 1989. Records of porbeagles landed in Scotland, with observations on the biology, distribution and exploitation of the species. Scottish Fisheries Research Report no. 45. Aberdeen, Scotland. Department of Agriculture and Fisheries for Scotland. 15 pp.

Geddes, T. 1960. *Hebridean Sharker.* London: Herbert Jenkins Ltd. 183 pp.

Gelsleichter, J., J. A. Musick, and S. Nichols. 1999. Food habits of the smooth dogfish, *Mustelus canis,* dusky shark, *Carcharhinus obscurus,* Atlantic sharpnose shark, *Rhizoprionodon terraenovae,* and the sand tiger, *Carcharias taurus,* from the northwest Atlantic ocean. *Environmental Biology of Fishes* 54: 205–217.

Gilbert, C. H. 1892. Scientific results of explorations by the U.S. Fish Commission steamer *Albatross.* no. 22. Descriptions of thirty-four new species of fishes collected in 1888 and 1889, principally among the Santa Barbara Islands and in the Gulf of California. *Proceedings of the United States National Museum* 14 (880): 539–566.

Gilbert, C. H. 1905. The deep-sea fishes of the Hawaiian Islands. *Bulletin of the United States Fish Commission* 23 (pt. 2, sec. 2): 575–713.

Gilbert, C. H., and E. C. Starks. 1904. The fishes of Panama Bay. *Memoirs of the California Academy of Sciences* 4: 1–304.

Gilbert, C. R. 1961. First record for the hammerhead shark, *Sphyrna tudes,* in United States waters. *Copeia* 1961 (4): 480.

Gilbert, C. R. 1967. A revision of the hammerhead sharks (family Sphyrnidae). *Proceedings of the United States National Museum* 119 (3539): 1–88.

Gilbert, P. W. 1959. The ability of yolk-sac stage dogfish pups to survive outside the uterus. *Bulletin of the Mount Desert Island Biological Laboratory:* 68.

Gilbert, P. W., and C. Gilbert. 1973. Sharks and shark deterrents. *Underwater Journal* 5 (2): 69–79.

Gilbert, P. W., and D. A. Schlernitzauer. 1966. The placenta and gravid uterus of *Carcharhinus falciformis. Copeia* 1966 (3): 451–457.

Gilchrist, J. D. F. 1922. Deep-sea fishes procured by the S.S. "*Pickle*" (part 1). *South African Fisheries and Marine Biological Survey,* special report nos. 2 and 3: 41–79.

Gilhen, J. 1989. The bluntnose sixgill shark *Hexanchus griseus* (Bonnaterre, 1788), new to the fish fauna of Atlantic Canada. *Proceedings of the Nova Scotian Institute of Science* 39: 75–77.

Gilhen, J., and B. W. Coad. 1999. The false catshark, *Pseudotriakis microdon* Brito Capello, 1867, new to the fish fauna of Atlantic Canada. *Canadian Field-Naturalist* 113 (3): 514–516.

Gilhen, J., B. W. Coad, and A. Hebda. 2003. The chain dogfish, *Scyliorhinus retifer* (Garman, 1881), new to the Canadian Atlantic ichthyofauna. *Canadian Field-Naturalist* 117 (3): 475–477.

Gill, T. 1862 (1863). On the classification of the families and genera of the Squali of California. *Proceedings of the Academy of Natural Sciences of Philadelphia,* nos. 10, 11, 12: 483–501.

Gill, T. 1864 (1865). Second contribution to the Selachology of California. *Proceedings of the Academy of Natural Sciences of Philadelphia* 16: 147–151.

Gill, T., and J. F. Bransford. 1877. Synopsis of the fishes of Lake Nicaragua. *Proceedings of the Academy of Natural Sciences of Philadelphia,* 3rd ser.: 175–191.

Gilmore, R. G. 1983. Observations on the embryos of the longfin mako, *Isurus paucus,* and the bigeye thresher, *Alopias superciliosus. Copeia* 1983 (2): 375–382.

Gilmore, R. G. 1993. Reproductive biology of lamnoid sharks. *Environmental Biology of Fishes* 38: 95–114.

Gilmore, R. G., J. W. Dodrill, and P. A. Linley. 1983. Reproduction and embryonic development of the sand tiger shark, *Odontaspis taurus* (Rafinesque). *Fishery Bulletin* 81: 201–225.

Girard, C. 1856. Characteristics of some cartilaginous fishes of the Pacific coast of North America. *Proceedings of the Academy of Natural Sciences of Philadelphia* 7: 196–197.

Girard, M., and M.-H. Du Buit. 1999. Reproductive biology of two deep-water sharks from the British Isles, *Centroscymnus coelolepis* and *Centrophorus squamosus,* (Chondrichthyes: Squalidae). *Journal of the Marine Biological Association* 79: 923–931.

Glover, C. J. M. 1976. The goblin shark *Scapanorhynchus owstoni* (Jordan, 1898): Confirmation of the first Australian record. *South Australian Naturalist* 50 (4): 69–72.

Glukhov, A. A., and A. P. Kuz'michev. 1984. New record of *Squaliolus laticaudus* (Squalidae) and *Neocyttus helgae* (Zeidae) in the northeast Atlantic. *Journal of Ichthyology* 24 (3): 122–124.

Gmelin, J. F. 1788 [1789]. Pisces Chondropterygii. Squalua. In *Caroli a Linné, Systema Naturae per Regna Tria Naturae* Vol. 1, pt. 3. Leipzig. 1516 pp.

Goldman, K. J. 1997. Regulation of body temperature in the white shark, *Carcharodon carcharias. Journal of Comparative Physiology* B 167: 423–429.

Goldman, K. J., and S. D. Anderson. 1999. Space utilization and swimming depth of white sharks, *Carcharodon carcharias,* at the South Farallon Islands, central California. *Environmental Biology of Fishes* 56: 351–364.

Goldman, K. J., and J. A. Musick. 2006. Growth and maturity of salmon sharks (*Lamna ditropis*) in the eastern and western North Pacific, and comments on back-calculation methods. *Fishery Bulletin* 104: 278–292.

Goldman, K. J., S. D. Anderson, J. E. McCosker, and A. P. Klimley. 1996. Temperature, swimming depth, and movements of a white shark at the South Farallon Islands, California. In *Great White Sharks: The Biology of Carcharodon carcharias,* edited by A. P. Klimley and D. G. Ainley, pp. 111–120. San Diego: Academic Press.

Goldman, K. J., S. D. Anderson, R. J. Latour, and J. A. Musick. 2004. Homeothermy in adult salmon sharks, *Lamna ditropis. Environmental Biology of Fishes* 71: 403–411.

Golovan, G. A. 1976. (Translation) Rare and firstly recorded Chondrostean and Teleostean fishes of the continental slope of West Africa. *Trudy Instituta Okeanologii im P.P. Shirshova* 104: 277–317.

Gómez, G. O, L. A. Zapata, R. Franke A., and G. E. Ramos. 2003. Hábitos alimentarios de *Mustelus lunulatus y M. henlei* (pisces:Triakidae) colectados en el Parque Nacional Natural Gorgona, Pacífico Colombiano. *Boletin de Investigaciones Marinas y Costeras* (Santa Marta, Colombia) 32: 219–230.

Goode, G. B. 1884. Part 3. The food fishes of the United States. In *The Fisheries and Fishery Industries of the United States,* Sec. 1, Natural History of Useful Aquatic Animals, pp. 163–682. Washington, D.C.: U.S. Government Printing Office.

Goode, G. B., and T. H. Bean. 1879. A catalogue of the fishes of Essex County, Massachusetts, including the fauna of Massachusetts Bay and the contiguous deep waters. *Bulletin of the Essex Institute* 11 (1, 2, 3): 1–38.

Goode, G. B., and T. H. Bean. 1896. *Oceanic ichthyology.* 2 vols. Washington, D.C.: Smithsonian Institution.

Gordon, B. L. 1956. The amazing angel shark. *Bulletin of the International Oceanographic Foundation* 2 (2): 108–111.

Gordon, J. D. M., and S. C. Swan. 1997. The distribution and abundance of deep-water sharks on the continental slope to the west of the British Isles. CM 1997/ BB:11. Copenhagen, Denmark. International Council for the Exploration of the Sea. 23 pp.

Gore, M. A., D. Rowat, J. Hall, F. R. Gell, and R. F. Ormond. 2008. Transatlantic migrations and deep mid-ocean diving by basking shark. *Biology Letters* 4 (4): 395–398.

Gosse, P. H. 1851. *A Naturalist's Sojourn in Jamaica.* London: Longman, Brown, Green, and Longmans. 508 pp.

Goto, M., T. Taniuchi, N. Kuga, and M. Iwata. 1981. Four dicephalous specimens of the blue shark, *Prionace glauca,* from Japan. *Japanese Journal of Ichthyology* 28 (2): 157–165.

Gotshall, D. W., and T. Jow. 1965. Sleeper sharks (*Somniosus pacificus*) off Trinidad, California, with life history notes. *California Fish and Game* 51 (4): 294–298.

Graham, R. T., and C. M. Roberts. 2007. Assesing the size, growth rate and structure of a seasonal population of whale sharks (*Rhincodon typus* Smith 1828) using conventional tagging and photo identification. *Fisheries Research* 84: 71–80.

Graham, R. T., C. M. Roberts, and J. C. R. Smart. 2005. Diving behavior of whale sharks in relation to a predictable food pulse. *Journal of the Royal Society Interface.* doi: 10.1098/rsif.2005.0082.

Grant, C. J., R. L. Sandland, and A. M. Olsen. 1979. Estimation of growth, mortality and yield per recruit of the Australian school shark, *Galeorhinus australis* (Macleay), from tag recoveries. *Australian Journal of Marine and Freshwater Research* 30: 625–637.

Grant, R. E. 1827. On the structure and characters of the *Lernaea elongata,* Gr. A new species from the Arctic Seas. *The Edinburgh Journal of Science* 7: 147–154.

Graves, R. 1955. *The Greek Myths,* Vol. 1. Baltimore: Penguin Books. 370 pp.

Gray, S., and C. S. Johnson. 1979. *Solution to An/BQR-19 Hydrophone Boot Problem.* (Raytheon Company, Submarine Signal Division). San Diego: Naval Ocean Systems Center.

Gray, J. E. 1851. *List of the Specimens of Fish of the Collection of the British Museum. Part 1—Chondropterygii.* London: Trustees of the British Museum of Natural History. 160 pp.

Grey, Z. 1934. The great mako. *Natural History* 34 (3): 221–233.

Griffith, E. 1834. Supplementary additions. In *The Animal Kingdom,* Vol. 10, by B. Cuvier, p. 640. London: Whittaker and Co.

Grover, C. A. 1972a. Predation on egg-cases of the swell shark, *Cephaloscyllium ventriosum. Copeia* 1972 (4): 871–872.

Grover, C. A. 1972b. Population differences in the swell shark *Cephaloscyllium ventriosum. California Fish and Game* 58 (3): 191–197.

Grover, C. A. 1974. Juvenile denticles of the swell shark *Cephaloscyllium ventriosum:* Function in hatching. *Canadian Journal of Zoology* 52: 359–363.

Grubbs, R. D., and J. A. Musick. 2007. Spatial delineation of summer nursery areas for juvenile sandbar sharks in Chesapeake Bay, Viriginia. In *Shark Nursery Grounds of the Gulf of Mexico and the East Coast Waters of the United States,* edited by C. T. McCandless, N. E. Kohler, and H. L. Pratt, Jr., pp. 63–85. Bethesda: American Fisheries Society, Symposium 50.

Gruber, S. H., and L. J. V. Compagno. 1981. Taxonomic status and biology of the bigeye thresher, *Alopias superciliosus. Fishery Bulletin* 79: 617–640.

Gruber, S. H., D. I. Hamasaki, and B. L. Davis. 1975. Window to the epiphysis in sharks. *Copeia* 1975 (2): 378–380.

Gruber, S. H., D. R. Nelson, and J. F. Morrissey. 1988. Patterns of activity and space utilization of lemon sharks, *Negaprion brevirostris,* in a shallow Bahamian lagoon. *Bulletin of Marine Science* 43 (1): 61–76.

Gubanov, Y. P. 1972. On the biology of the thresher shark (*Alopias vulpinus* (Bonnaterre)) in the northwest Indian Ocean. *Journal of Ichthyology* 12 (4): 591–600.

Gubanov, Y. P. 1974. The capture of a giant specimen of the mako shark (*Isurus glaucus*) in the Indian Ocean. *Journal of Ichthyology* 14 (4): 589–591.

Gubanov, Y. P. 1978. The reproduction of some species of pelagic sharks from the equatorial zone of the Indian Ocean. *Journal of Ichthyology* 18 (5): 781–792.

Gubanov, Y. P., and V. N. Grigor'yev. 1975. Observations on the distribution and biology of the blue shark, *Prionace glauca* (Carcharhinidae) of the Indian Ocean. *Journal of Ichthyology* 15 (1): 37–43.

Gudger, E. W. 1907. A note on the hammerhead shark (*Sphyrna zygaena*) and its food. *Science,* n.s., 25: 1005.

Gudger, E. W. 1912. Summary of work done on the fishes of Tortugas. *Yearbook of the Carnegie Institute of Washington* 11: 148–150.

Gudger, E. W. 1915a. Natural history of the whale shark *Rhineodon typus* Smith. *Zoologica* 1 (19): 349–389.

Gudger, E. W. 1915b. Utero-gestation in the sharp-nosed shark, *Scoliodon terranovae* [*sic*]. *Science,* n.s., 41 (1055): 439.

Gudger, E. W. 1923. An extraordinary capture of the giant shark, *Rhineodon typus. Natural History,* 23: 62–63.

Gudger, E. W. 1927. A second whale shark, *Rhineodon typus,* impaled on the bow of a steamship. *Zoological Society Bulletin* 30 (3): 76–77.

Gudger, E. W. 1934. The geographical distribution of the whale-shark (*Rhineodon typus*). *Proceedings of the Zoological Society* (4): 863–893.

Gudger, E. W. 1935. Maxillary breathing valves in the sharks *Chlamydoselachus* and *Cetorhinus* with notes on breathing valves in thirteen marine teleosts. *Journal of Morphology* 57 (1): 91–104.

Gudger, E. W. 1936. *Rhineodon* at New York's front door. *Natural History* 37 (2): 159–169.

Gudger, E. W. 1937. A whale shark impaled on the bow of a steamer near the Tuamotus, South Seas. *Science* 85 (2204): 314.

Gudger, E. W. 1939. A school of whale sharks in the Bahama Islands. *Nature* 143: 79–80.

Gudger, E. W. 1940. The breeding habits, reproductive organs and external embryonic development of *Chlamydoselachus,* based on notes and drawings by Bashford Dean. In *The Bashford Dean Memorial Volume Archaic Fishes,* pt. 2, art. 7, edited by E. W. Gudger, pp. 521–633 [646]. New York: American Museum of Natural History.

Gudger, E. W. 1941. The feeding organs of the whale shark, *Rhineodon typus. Journal of Morphology* 68 (1): 81–99.

Gudger, E. W. 1947. Sizes attained by the large hammerhead sharks. *Copeia* 1947 (4): 228–236.

Gudger, E. W. 1948. The basking shark, *Cetorhinus maximus,* on the North Carolina coast. *Journal of the Elisha Mitchell Scientific Society* 64 (1): 41–44.

Gudger, E. W. 1949. Natural history notes on tiger sharks, *Galeocerdo tigrinus,* caught at Key West, Florida, with emphasis on food and feeding habits. *Copeia* 1949 (1): 39–47.

Gudger, E. W. 1952. Oviparity—the mode of reproduction of the whale shark, *Rhineodon typus. Copeia* 1952 (4): 266–267.

Gudger, E. W., and W. H. Hoffmann. 1931. The whale shark off Havana harbor. *Scientific Monthly* 32 (1): 33–36.

Gudger, E. W., and B. G. Smith. 1933. The natural history of the frilled shark *Chlamydoselachus anguineus.* In *The Bashford Dean Memorial Volume Archaic Fishes,* pt. 1, art. 5, edited by E. W. Gudger, pp. 245–319. New York: American Museum of Natural History.

Guitart, D. J. 1972. Un nuevo género y especie de tiburón de la familia *Triakidae. Serie Poeyana* (Instituto de Zoología, Academia de Ciencias de Cuba) 99: 1–4.

Guitart D. 1974. *Sinopsis de los Peces Marinos de Cuba.* Vol 1. Havana: Academia de Ciencias de Cuba. 121 pp.

Guitart, D. [J.], and J. Fernandez Milera. 1974. El monstruo marino de Cojímar. *Mar y Pesca* 104: 10–11.

Guitart Manday, D. 1966. Nuevo nombre para una especie de tiburón del género *Isurus* (Elasmobranchii: Isuridae) de aguas cubanas. *Poeyana* (Instituto de Biología), ser. A, 15: 1–9.

Guitart Manday, D. 1968. Guía para los tiburones de aguas cubanas. *Serie Oceanológica* (Academia de Ciencias de Cuba) 1: 1–61.

Guitart Manday, D. 1975. Las pesquerías pelágico-oceánicas de corto radio de acción en la región noroccidental de Cuba. *Serie Oceanológica* (Academia de Ciencias de Cuba) 31: 1–26.

Gunnerus, J. E. 1765. Brugden, (*Squalus maximus*). *Skrifter-Det Kongelige Norske Videnskabers Selskab. p.* 33, plate 2.

Gunter, G. 1942. A list of the fishes of the mainland of North and Middle America recorded from both freshwater and sea water. *American Midland Naturalist* 28 (2): 305–326.

Günther, A. 1870. *Catalogue of the Fishes of the British Museum. Catalogue of the Physostomi, Containing the Families Gymnotidae, Symbranchidae, Muraenidae, Pegasidae, and of the Lophobranchii, Plectognathi, Dipnoi, Ganoidei, Chondropterygii, Cyclostomata, Leptocardii, in the British Museum,* Vol. 8. London: Trustees of the British Museum. 549 pp.

Günther, A. 1877. Preliminary notes on new fishes collected in Japan during the expedition of H.M.S. 'Challenger.' *Annals and Magazine of Natural History,* 4th ser., 20: 433–445.

Gushchin, A. V., V. V. Sukhovershin, I. I. Konovalenko, and V. S. Sukhorukova. 1987. On the capture of the polar shark genus *Somniosus* (Squalidae) in the Southern Hemisphere. *Journal of Ichthyology* 27 (1): 115–117.

Guzmán M., L., and I. Campodonico G. 1976. *Notorynchus cepedianus* (Peron, 1807) en la entrada oriental del Estrecho de Magallanes (Elasmobranchii, Hexanchidae). *Anales Instituto de la Patagonia* (Punta Arenas) 7: 207–210.

Haeseker, S. L., and J. J. Cech, Jr. 1993. Food habits of the brown smoothhound shark (*Mustelus henlei*) from two sites in Tomales Bay. *California Fish and Game* 79 (3): 89–95.

Hakluyt, R. 1589 *Voyages,* Vol. 1, *The Principall Navigations of the English Nation.* Reprint, London: J. M. Dent & Sons Ltd., 1939. 468 pp.

Halstead, B. W. 1970. Class Chondrichthyes. Venomous sharks, rays, chimaeras. In *Poisonous and Venomous Marine Animals of the World,* Vol. 3, pp. 1–12. Washington, D.C.: U.S. Government Printing Office.

Halstead, B. W., P. S. Auerbach, and D. R. Campbell. 1990. *A Color Atlas of Dangerous Marine Animals.* Boca Raton, Fla.: CRC Press. 192 pp.

Hamasaki, D. I., and P. Streck. 1971. Properties of the *epiphysis cerebri* of the small-spotted dogfish shark, *Scyliorhinus caniculus* L. *Vision Research* 11: 189–198.

Hamlett, W. C. 1986. Prenatal nutrient absorptive structures in selachians. In *Indo-Pacific Fish Biology:*

Proceedings of the Second International Conference on Indo-Pacific Fishes, edited by T. Uyeno, R. Arai, T. Taniuchi, and K. Matsuura, pp. 333–344. Tokyo: Ichthyological Society of Japan.

Hanan, D. A., D. B. Holts, and A. L. Coan, Jr. 1993. *The California Drift Gill Net Fishery for Sharks and Swordfish, 1981–1982 through 1990–1991.* Fish Bulletin 175. Sacramento: Department of Fish and Game, Resources Agency, State of California. 95 pp.

Hansen, P. M. 1949. Studies on the biology of the cod in Greenland waters. In *Conseil Permanent International Pour L'Exploration de la Mer. Rapports et Procès-Verbaux des Réunions,* Vol. 123, pp. 5–77. Copenhagen: Andr. Fred. Høst & Fils.

Hansen, P. M. 1953. Havkalen, den arktiske dybhavshaj. *Grønland* 5: 182–188.

Hansen, P. M. 1963. Tagging experiments with the Greenland shark (*Somniosus microcephalus* (Bloch and Schneider)) in Subarea 1. In Special Publication no. 4, pp. 172–175. Dartmouth, Nova Scotia: International Commission for the Northwest Atlantic Fisheries.

Hardie, J. 1986. Diver kills huge shark with spear. *Miami Herald,* 24 May: 1E, 5E.

Hart, J. L. 1973. Pacific fishes of Canada. Bulletin 180. Ottawa: Fisheries Research Board of Canada. 740 pp.

Harvey, J. T. 1989. Food habits, seasonal abundance, size, and sex of the blue shark, *Prionace glauca,* in Monterey Bay, California. *California Fish and Game* 75 (1): 33–44.

Harvey-Clark, C. 1995. Protection of sixgill sharks. In *Marine Protected Areas and Sustainable Fisheries. Proceedings of the Symposium on Marine Protected Areas and Sustainable Fisheries Conducted at the Second International Conference on Science and the Management of Protected Areas,* edited by N. L. Shackell and J. H. M. Willison, pp. 286–289. Wolfville, Nova Scotia: Science and Management of Protected Areas Association.

Hawkins, R. 1622 *The Observations of Sir Richard Hawkins.* Reprint, edited by J. A. Williamson. London: The Argonaut Press, 1933. 190 pp.

Hazin, F. H. V., F. M. Lucena, T. S. A. L. Souza, C. E. Boeckman, M. K. Broadhurst, and R. C. Menni. 2000. Maturation of the night shark, *Carcharhinus signatus,* in the southwestern equatorial Atlantic Ocean. *Bulletin of Marine Science* 66 (1): 173–185.

Hazin, F. H. V., P. G. Oliveira, and M. K. Broadhurst. 2002. Reproduction of the blacknose shark (*Carcharhinus acronotus*) in coastal waters off northeastern Brazil. *Fishery Bulletin* 100: 143–148.

Heemstra, P. C. 1973. A Revision of the Shark Genus *Mustelus* (Squaliformes: Carcharhinidae). Ph.D. diss., University of Miami, Coral Gables. 187 pp.

Heemstra, P. C. 1997. A review of the smooth-hound sharks (genus *Mustelus,* family Triakidae) of the western Atlantic Ocean, with descriptions of two new species and a new subspecies. *Bulletin of Marine Science* 60 (3): 894–928.

Heithaus, M. R. 2001a. Shark attacks on bottlenose dolphins (*Tursiops aduncus*) in Shark Bay, western Australia: Attack rate, bite scar frequencies, and attack seasonality. *Marine Mammal Science* 17 (3): 526–539.

Heithaus, M. R. 2001b. The biology of tiger sharks, *Galeocerdo cuvier,* in Shark Bay, Western Australia: Sex ratio, size distribution, diet, and seasonal changes in catch rates. *Environmental Biology of Fishes* 61: 25–36.

Heithaus, M. R., A. Frid, and L. M. Dill. 2002. Shark-inflicted injury frequencies, escape ability, and habitat use of green and loggerhead turtles. *Marine Biology* 140: 229–236.

Helbing, H. 1904. *Beiträge zur Anatomie und Systematik der Laemargiden.* Mit. 2, Tafeln no. 8 and 9, Nova Acta. Kaiserl. Leop.-Carol. Deutschen Akademie der Naturforscher 82 (4). Halle: Druck von Ehrhardt Karras. 530 pp.

Heller, J. H., M. S. Heller, S. Springer, and E. Clark. 1957. Squalene content of various shark livers. *Nature* 179: 919–920.

Henningsen, A. D,, M. J. Smale, I. Gordon, R. Gardner, R. Marin-Osorno, and N. Kinnunen. 2004. Captive breeding and sexual conflict in elasmobranchs. In *The elasmobranch husbandry manual: Captive care of sharks, rays, and their relatives.* Edited by M. Smith, D. Warmolts, D. Thoney, and R. Hueter, pages 237–248. Columbus, Ohio: Ohio Biological Survey.

Herald, E. S. 1953. The 1952 shark derbies at Elkhorn Slough, Monterey Bay, and at Coyote Point, San Francisco Bay. *California Fish and Game* 39 (2): 237–243.

Herald, E. S. 1968. Size and aggressiveness of the sevengill shark (*Notorynchus maculatus*). *Copeia* 1968 (2): 412–414.

Herald, E. S., and W. E. Ripley. 1951. The relative abundance of sharks and bat stingrays in San Francisco Bay. *California Fish and Game* 37 (3): 315–329.

Herald, E. S., W. Schneebeli, N. Green, and K. Innes. 1960. Catch records for seventeen shark derbies held at Elkhorn Slough, Monterey Bay, California. *California Fish and Game* 46 (1): 59–67.

Herdendorf, C. E., and T. M. Berra. 1995. A Greenland shark from the wreck of the *SS Central America* at 2,200 meters. *Transactions of the American Fisheries Society* 124: 950–953.

Hernández Carvallo, A. Observations on the hammerhead sharks (*Sphyrna*) in waters near Mazatlan, Sinaloa, Mexico. In *Sharks, Skates and Rays,* edited by P. W.

Gilbert, R. F. Mathewson, and D. P. Rall, pp. 79–83. Baltimore: Johns Hopkins Press.

Herre, A. W. 1935. Notes on fishes in the Zoological Museum of Stanford University. 2. Two new genera and species of Japanese sharks and a Japanese species of *Narcetes. Copeia* 1935 (3): 122–127.

Herre, A. W. 1936. *Fishes of the Crane Pacific Expedition.* Zoological Series 21, no. 353. Chicago: Field Museum of Natural History. 472 pp.

Hess, P. W. 1962. Notes on some sharks in the western North Atlantic and Bahama areas. *Copeia* 1962 (3): 653–656.

Heyman, W. D., R. T. Graham, B. Kjerfve, and R. E. Johannes. 2001. Whale sharks *Rhincodon typus* aggregate to feed on fish spawn in Belize. *Marine Ecology Progress Series* 215: 275–282.

Hickling, C. F. 1930. A contribution towards the life-history of the spur-dog. *Journal of the Marine Biological Association of the United Kingdom* 16: 529–570.

Hildebrand, S. F., and W. C. Schroeder. 1928. Fishes of Chesapeake Bay. *Bulletin of the United States Bureau of Fisheries* 43 (pt. I): 1–388.

Hildebrand, S. F. 1946. *A Descriptive Catalog of the Shore Fishes of Peru.* United States National Museum, Bulletin 189. Washington, D.C.: Smithsonian Institution. 530 pp.

Hinton, S. 1962. Longevity of fishes in captivity, as of September, 1956. *Zoologica* 47 (pt. 2, no. 9): 105–116.

Hisaw, F. L., and A. Albert. 1947. Observations on the reproduction of the spiny dogfish, *Squalus acanthias. Biological Bulletin* 92 (3): 187–199.

Hixon, M. A. 1979. Term fetuses from a large common thresher shark, *Alopias vulpinus. California Fish and Game* 65 (3): 191–192.

Hiyama, Y. 1951. *Systematic list of fishes of the Ryukyu Islands.* Tokyo: General Headquarters, Supreme Commander for the Allied Powers, Natural Resources Section, Report no. 150. 47 pp.

Hobson, E. S. 1963. Feeding behavior in three species of sharks. *Pacific Science* 17 (2): 171–194.

Hoese, H. D., and R. B. Moore. 1958. Notes on the life history of the bonnetnose shark, *Sphyrna tiburo. Texas Journal of Science* 10 (1): 69–71.

Hoese, H. D., and R. H. Moore. 1977. *Fishes of the Gulf of Mexico, Texas, Louisiana, and Adjacent Waters.* College Station: Texas A.& M. University Press. 327 pp.

Hoey, J. J. 1983. Analysis of Longline Fishing Effort for Apex Predators (Swordfish, Shark, and Tuna) in the Western North Atlantic and Gulf of Mexico. Ph.D. diss., University of Rhode Island. 288 pp.

Hoff, T. B. 1990. Conservation and Management of the Western North Atlantic Shark Resource Based on the Life History Strategy Limitations of Sandbar Sharks. Ph.D. diss., University of Delaware. 282 pp.

Hoffman, W. 1981. Whale sharks associated with fish schools off south Texas. *Northeast Gulf Science* 5 (1): 55–57.

Holden, M. J. 1974. Problems in the rational exploitation of Elasmobranch populations and some suggested solutions. In *Sea Fisheries Research,* edited by F. R. Harden Jones, pp. 117–137. New York: John Wiley & Sons.

Holland, G. A. 1957. Migration and growth of the dogfish shark, *Squalus acanthias* (Linnaeus), of the eastern North Pacific. *Fisheries Research Papers* (State of Washington, Department of Fisheries) 2 (1): 43–59.

Holts, D. B. 1988. Review of U.S. west coast commercial shark fisheries. *Marine Fisheries Review* 50 (1): 1–8.

Holts, D. B., and D. Bedford. 1989. Report of the assessment methods workshop for sharks; March 14–15, 1989. Administrative Report LJ-89-11. La Jolla, Calif.: National Marine Fisheries Service, Southwest Fisheries Center. 20 pp.

Holts, D. B., and D. W. Bedford. 1993. Horizontal and vertical movements of the shortfin mako shark, *Isurus oxyrinchus,* in the southern California Bight. *Australian Journal of Marine and Freshwater Research* 44: 901–909.

Howell Rivero, L. 1936. Some new, rare and little-known fishes from Cuba. *Proceedings of the Boston Society of Natural History* 41 (4): 41–76.

Hubbell, G. 1996. Using tooth structure to determine the evolutionary history of the white shark. In *Great White Sharks: The Biology of Carcharodon carcharias,* edited by A. P. Klimley and D. G. Ainley, pp. 9–18. San Diego: Academic Press.

Hubbs, C. L. 1917. A note on the food of *Squalus sucklii,* the California dogfish. *Copeia* 43: 37–38.

Hubbs, C. L. 1923. Notes on a small collection of fishes from Monhegan Island, Maine. *Copeia* 123: 101–103.

Hubbs, C. L. 1938. The scientific names of the American "smooth dogfish," *Mustelus canis* (Mitchill), and of the related European species. *Occasional Papers of the Museum of Zoology* (University of Michigan), no. 374: 1–19.

Hubbs, C. L. 1952. Antitropical distribution of fishes and other organisms. Symposium on Problems of Bipolarity and of Pan-Temperate Faunas. Pacific Science Congress 1949, at Auckland, New Zealand and Christchurch 3: 324–329.

Hubbs, C. L., and F. N. Clark. 1945. Occurrence of the bramble shark (*Echinorhinus brucus*) in California. *California Fish and Game* 31 (2): 64–67.

Hubbs, C. L., and W. I. Follett. 1947. *Lamna ditropis,* new species, the salmon shark of the north Pacific. *Copeia* 1947 (3): 194.

Hubbs, C. L., and J. L. McHugh. 1950. Pacific sharpnose shark (*Scoliodon longurio*) in California and Baja California. *California Fish and Game* 36 (1): 7–11.

Hubbs, C. L., and J. L. McHugh. 1951. Relationships of the pelagic shark *Euprotomicrus bispinatus,* with description of a specimen from off California. *Proceedings of the California Academy of Sciences,* 4th ser., 27 (6): 159–176.

Hubbs, C. L., and L. R. Taylor, Jr. 1969. Data on the life history and characters of *Galeus piperatus,* a dwarf shark of Golfo de California. *Fiskeridirektoratets Skrifter Serie Havundersokelser* 15: 310–330.

Hubbs, C. L., T. Iwai, and K. Matsubara. 1967. External and internal characters, horizontal and vertical distribution, luminescence, and food of the dwarf pelagic shark, *Euprotomicrus bispinatus. Bulletin of the Scripps Institution of Oceanography* (University of California) 10: 1–81.

Huber, D. R., T. G. Eason, R. E. Hueter, and P. J. Motta. 2005. Analysis of the bite force and mechanical design of the feeding mechanism of the durophagous horn shark *Heterodontus francisci. Journal of Experimental Biology* 208: 3553–3571.

Hulbert, L. B., M. F. Sigler, and C. R. Lunsford. 2006. Depth and movement behaviour of the Pacific sleeper shark in the north-east Pacific ocean. *Journal of Fish Biology* 69: 406–425.

Huntsman, A. G. 1922. The fishes of the Bay of Fundy. *Contributions to Canadian Biology* 3: 49–72.

Hussain, S. M. 1989. Buoyancy mechanism, and density of sand tiger shark (*Eugomphodus taurus*). *Indian Journal of Fisheries* 36 (3): 266–268.

Hussakof, L. 1909. A new goblin shark, *Scapanorhynchus jordani,* from Japan. *Bulletin of the American Museum of Natural History* 26 (art. 19): 257–262.

Hussakof, L. 1915. The capture of a basking shark on Long Island. *Copeia* 1915 (21): 25–27.

Hsu, H.-H., S.-J. Joung, Y.-Y. Liao, and K.-M. Liu. 2007. Satellite tracking of juvenile whale sharks, *Rhincodon typus,* in the northwestern Pacific. *Fisheries Research* 84: 25–31.

Ifft, J. D., and D. J. Zinn. 1948. Tooth succession in the smooth dogfish, *Mustelus canis. Biological Bulletin* 95 (1): 100–106.

Iglésias, S. P., K. Nakaya, and M. Stehmann. 2004. *Apristurus melanoasper,* a new species of deep-water catshark from the North Atlantic (Chondrichthyes: Carcharhiniformes: Scyliorhinidae). *Cybium* 28 (4): 345–356.

International Commission on Zoological Nomenclature. 1984. Official list of generic names in zoology. *Bulletin of Zoological Nomenclature* 41 (4): 215–217.

International Council for the Exploration of the Sea. 1995. *Report of the Study Group on Elasmobranch Fishes.* ICES CM 1995/G: 3. Copenhagen, Denmark: Demersal Fish Committee. 88 pp.

International Game Fish Association. 2006. *2006 World record game fishes.* Dania Beach: International Game Fish Association. 384 pp.

Isouchi, T. 1970. A cigar shark, *Isistius brasiliensis* from tropical water of the eastern Pacific. *Japanese Journal of Ichthyology* 17 (3): 124–125.

Ivanov, O. A. 1986. On the distribution of the bigeye thresher shark, *Alopias superciliosus,* in the Pacific Ocean. *Journal of Ichthyology* 26 (5): 121–122.

Iversen, E. S. 1959. Pelagic puzzle. *Sea Frontiers* 5 (3): 175–178.

Iwasaki, Y. 1970. On the distribution and environment of the whale shark, *Rhincodon typus,* in skipjack fishing grounds in the western Pacific Ocean. *Journal of the College of Marine Science and Technology* (Tokai University) 4: 37–51.

Izawa, K., and T. Shibata. 1993. A young basking shark, *Cetorhinus maximus,* from Japan. *Japanese Journal of Ichthyology* 40 (2): 237–245.

Jacobson, H. (trans.) 1885. Havkalfangsten fra Island. [The Iceland shark-fisheries]. From *Fiskeritidende,* no. 46. Tranlated from Danish in *Bulletin of the United States Fish Commission* 5 (19): 301–304.

Jaeger, E. C. 1962. *A Source-Book of Biological Names and Terms.* 3rd ed. Springfield, Ill.: Charles C. Thomas. 323 pp.

Jahn, A. E., and R. L. Haedrich. 1987. Notes on the pelagic squaloid shark *Isistius brasiliensis. Biological Oceanography* 5: 297–309.

Jakobsdóttir, K. B. 2001. Biological aspects of two deep-water squalid sharks: *Centroscyllium fabricii* (Reinhardt, 1825) and *Etmopterus princeps* (Collett, 1904) in Icelandic waters. *Fisheries Research* 51: 247–265.

JAMARC (Japan Marine Fishery Resource Research Center). 1981. The development of the shark resources in the North Pacific Ocean, fiscal years 1979 and 1980. Translated by Tamio Otsu. Translation no. 63, from *Japan Marine Fishery Resource Research Center (JAMARC) Kaihatsu News (Development News)* no. 11, pp. 17–25, August 1980, and no. 17, pp. 12–22, June 1981. National Marine Fisheries Service, National Oceanic and Atmospheric Administration. Honolulu, Hawaii: U.S. Department of Commerce. 21 pp.

Jaquet, M. 1905. Description de quelques parties du squelette du *Pseudotriacis microdon* Capello. *Bulletin du Musée Océanographique de Monaco* 36: 1–28.

Jeffers, G. W. 1933. Fishes observed in the Strait of Belle Isle. *Contributions to Canadian Biology and Fisheries* (ser. A, general no. 13), n.s., 7 (16): 203–211.

Jenkins, J. T. 1925. *Fishes of the British Isles.* London: Frederick Warne & Co. Ltd. 376 pp.

Jensen, A. C. 1966. Life history of the spiny dogfish. *Fishery Bulletin* 65: 527–554.

Jensen, A. C. 1969. Spiny dogfish tagging and migration in North America and Europe. *International Commission for the Northwest Atlantic Fisheries, Research Bulletin* 6: 72–78.

Jensen, A. S. 1914. *The Selachians of Greenland.* Copenhagen: Bianco Lunos Bogtrykkeri. 40 pp.

Jensen, A. S. 1948. Contributions to the ichthyofauna of Greenland 8–24. *Spolia Zoologica Musei Hauniensis* 9. *Skrifter undgivet af Universitetets Zoologiske Museum* 9. Copehhagen: I Kommission Hos. E. Munksgaard. 182 pp.

Jensen, C. F., L. J. Natanson, H. L. J. Pratt, Jr., N. E. Kohler, and S. E. Campana. 2002. The reproductive biology of the porbeagle shark (*Lamna nasus*) in the western north Atlantic Ocean. *Fishery Bulletin* 100: 727–738.

Jensen, N. H. 1976. Reproduction of the bull shark, *Carcharhinus leucas,* in the Lake Nicaragua-Rio San Juan System. *Investigations of the ichthyofauna of Nicaraguan Lakes,* edited by T. B. Thorson, pp. 539–559. Lincoln: University of Nebraska Press.

Johnson, C. S. 1978. Sea creatures and the problem of equipment damage. *U.S. Naval Institute Proceedings* 104–8-906: 106–107.

Jones, B. C., and G. H. Geen. 1976a. Bibliography of spiny dogfish (*Squalus acanthias*) and related species. Technical Report no. 655. Canada Fisheries and Marine Service. 84 pp.

Jones, B. C., and G. H. Geen. 1976b. Observations on the brown cat shark, *Apristurus brunneus* (Gilbert), in British Columbia coastal waters. *Syesis* 10: 169–170.

Jones, B. C., and G. H. Geen. 1976c. Taxonomic reevaluation of the spiny dogfish (*Squalus acanthias* L.) in the northeastern Pacific Ocean. *Journal of the Fisheries Research Board of Canada* 33 (11): 2500–2506.

Jones, B. C., and G. H. Geen. 1977a. Age and growth of spiny dogfish (*Squalus acanthias*) in the Strait of Georgia British Columbia. Technical Report no. 699. Canada Fisheries and Marine Service. 16 pp.

Jones, B. C., and G. H. Geen. 1977b. Food and feeding of spiny dogfish (*Squalus acanthias*) in British Columbia waters. *Journal of the Fisheries Research Board of Canada* 34 (11): 2067–2078.

Jones, B. C., and G. H. Geen. 1977c. Reproduction and embryonic development of spiny dogfish *Squalus acanthias* in the Strait of Georgia, British Columbia. *Journal of the Fisheries Research Board of Canada* 34 (9): 1286–1292.

Jones, E. C. 1971. *Isistius brasiliensis,* a squaloid shark, the probable cause of crater wounds on fishes and cetaceans. *Fishery Bulletin* 69: 791–798.

Jones, L. 2003. *Encounters with Sharks, Dolphins and Big Fish.* Durban: L. Jones. 240 pp.

Jones, L. 2005. *Hooks, Spears and Spanners.* Durban: L. Jones. 254 pp.

Jones, L. M., and R. L. Shipp. 2006. Oculocutaneous albinism in a finetooth shark, *Carcharhinus isodon,* from Mobile Bay, Alabama. *Gulf of Mexico Science* (1/2): 81–82.

Jones, R. T., E. A. Hudson, and J. C. Andrews. 1983. Methods for transport and long-term maintenance of spiny dogfish sharks. *Laboratory Animal Science* 1983: 388–389.

Jordan, D. S. 1895. The fishes of Sinaloa. *Contributions to Biology from the Hopkins Laboratory of Biology* 1: 377–514.

Jordan, D. S. 1898. Description of a species of fish (*Mitsukurina owstoni*) from Japan, the type of a distinct family of lamnoid sharks. *Proceedings of the California Academy of Sciences,* 3rd ser., 1 (6): 199–204.

Jordan, D. S. 1907. *Fishes.* New York: Henry Holt. 789 pp.

Jordan, D. S., and B. W. Evermann. 1898 (1896–1898). *The Fishes of North and Middle America: A Descriptive Catalogue of the Species of Fish-like Vertebrates Found in the Waters of North America, North of the Isthmus of Panama.* Bulletin of the United States National Museum no. 47. Washington, D.C.: Government Printing Office. 4 vols.

Jordan, D. S., and B. W. Evermann. 1902 (1903). Descriptions of new genera and species of fishes from the Hawaiian Islands. In *U.S. Fish Commission Bulletin for 1902,* pp. 161–208.

Jordan, D. S., and H. W. Fowler. 1903. A review of the Elasmobranchiate fishes of Japan. *Proceedings of the United States National Museum* 26 (1324): 593–674.

Jordan, D. S., and C. H. Gilbert. 1880 (1881). Description of a new species of notidanoid shark (*Hexanchus corinus*), from the Pacific coast of the United States. *Proceedings of the United States National Museum* 3: 352–355.

Jordan, D. S., and C. H. Gilbert. 1882a (1883). Description of a new shark (*Carcharias lamiella*) from San Diego, California. *Proceedings of United States National Museum* 5: 110–111.

Jordan, D. S., and C. H. Gilbert. 1882b (1883). Description of four new species of sharks from Mazatlan, Mexico. *Proceedings of the United States National Museum* 5: 102–110.

Jordan, D. S., and C. H. Gilbert. 1883. *Synopsis of the Fishes of North America.* Bulletin of the United States National Museum, no. 16. 1018 pp.

Jordan, D. S., and J. O. Snyder. 1902. Descriptions of two new species of squaloid sharks from Japan. *Proceedings of the United States National Museum* 24 (1279): 79–81.

Jordan, D. S., and J. O. Snyder. 1904. On a collection of fishes made by Mr. Alan Owston in the deep waters of Japan. *Smithsonian Miscellaneous Collections* 45 (1447): 230–240.

Jordan, D. S., and E. C. Starks. 1901. Description of three new species of fishes from Japan. *Proceedings of the California Academy of Sciences,* 3rd ser., 2: 381–386.

Jordan, D. S., and J. C. Thompson. 1905. The fish fauna of the Tortugas Archipelago. *Bulletin of the Bureau of Fisheries* (for 1904) 24: 229–256.

Joseph, D. C. 1954. A record-size thresher from southern California. *California Fish and Game* 40 (4): 433–435.

Joung, S.-J., C.-T. Chen, E. Clark, S. Uchida, and W. Y. P. Huang. 1996. The whale shark, *Rhincodon typus,* is a livebearer: 300 embryos found in one 'megamamma' supreme. *Environmental Biology of Fishes* 46: 219–223.

Joung, S.-J., and H.-H. Hsu. 2005. Reproduction and embryonic development of the shortfin mako, *Isurus oxyrinchus* Rafinesque 1810, in the northwestern Pacific. *Zoological Studies* 44 (4): 487–496.

Joyce, W. N., S. E Campana, L. J. Natanson, N. E. Kohler, H. L. Pratt, Jr., and C. F. Jensen. 2002. Analysis of stomach contents of the porbeagle shark (*Lamna nasus* Bonaterre) in the northwest Atlantic. *ICES Journal of Marine Science* 59:1263–1269.

Kaikini, A. S., V. Ramamohana Rao, and M. H. Dhulkhed. 1959. A note on the whale shark *Rhincodon typus* Smith, stranded off Mangalore. *Journal of the Marine Biological Association of India* 1: 92–93.

Kajiura, S. M. 2001. Head morphology and electrosensory pore distribution of carcharhinid and sphyrnid sharks. *Environmental Biology of Fishes* 61: 125–133.

Kajiura, S. M., J. B. Forni, and A. P. Summers. 2005. Olfactory morphology of carcharhinid and sphyrnid sharks: Does the cephalofoil confer a sensory advantage? *Journal of Morphology* 264: 253–263.

Karbhari, J. P., and C. J. Josekutty. 1986. On the largest whale shark *Rhincodon typus* Smith landed alive at Cuffe Parade, Bombay. *Marine Fisheries Information Service, Technical and Extension Series* (Central Marine Fisheries Research Institute, Cochin, India, Indian Council of Agricultural Research) 66: 31–35.

Kato, S., and A. H. Carvallo. 1967. Shark tagging in the eastern Pacific Ocean, 1962–1965. In *Sharks, Skates and Rays,* edited by P. W. Gilbert, R. F. Mathewson, and D. P. Rall, pp. 93–109. Baltimore: Johns Hopkins Press.

Kato, S., S. Springer, and M. H. Wagner. 1967. *Field Guide to Eastern Pacific and Hawaiian Sharks.* Circular 271. Bureau of Commercial Fisheries, Fish and Wildlife Service. Washington, D.C.: U.S. Department of the Interior. 47 pp.

Kauffman, D. E. 1950. Notes on the Biology of the Tiger Shark (*Galeocerdo articus*) from Philippine Waters. U.S. Fish and Wildlife Service, Research Report 16. Washington, D.C.: U.S. Department of the Interior. 10 pp.

Kauffman, D. E. 1955. Noteworthy recoveries of tagged dogfish. *Fisheries Research Papers, Washington Department of Fisheries* 1 (3): 39–40.

Keeny, D. B., and E. J. Heist. 2006. Worldwide phylogeography of the blacktip shark (*Carcharhinus limbatus*) inferred from mitochondrial DNA reveals isolation of western Atlantic populations coupled with recent Pacific dispersal. *Molecular Ecology* 15: 3669–3679.

Kenney, R. D., R. E. Owen, and H. E. Winn. 1985. Shark distributions off the northeast United States from marine mammal surveys. *Copeia* 1985 (1): 220–223.

Kerstetter, D. W., and M. L. Taylor. 2008. Live release of a bigeye sand tiger *Odontaspis noronhai* (Elasmobranchii: Lamniformes) in the western North Atlantic Ocean. *Bulletin of Marine Science* 83 (3): 465–469.

Ketchen, K. S. 1972. Size at maturity, fecundity, and embryonic growth of the spiny dogfish (*Squalus acanthias*) in British Columbia waters. *Journal of the Fisheries Research Board of Canada* 29 (12): 1717–1723.

Ketchen, K. S. 1975. Age and growth of dogfish *Squalus acanthias* in British Columbia waters. *Journal of the Fisheries Research Board of Canada* 32 (1): 43–59.

Ketchen, K. S. 1986. *The Spiny Dogfish (Squalus acanthias) in the Northeast Pacific and a History of Its Utilization.* Canadian Special Publication of Fisheries and Aquatic Science 88. Ottawa: Department of Fisheries and Oceans. 78 pp.

Killam, K. A. 1987. The Reproductive Biology, Age, and Growth of the Blacktip Shark, *Carcharhinus limbatus* (Valenciennes) near Tampa Bay, Florida. Master's thesis, University of South Florida. 109 pp.

Killam, K. A., and G. R. Parsons. 1989. Age and growth of the blacktip shark, *Carcharhinus limbatus,* near Tampa Bay, Florida. *Fishery Bulletin* 87: 845–857.

King, J. E., and I. I. Ikehara. 1956. Some unusual fishes from the Central Pacific. *Pacific Science* 10: 17–24.

Kitafuji, M., and K. Yamamoto. 1998. Rearing of the whale shark, *Rhincodon typus,* in the Osaka aquarium "Kaiyukan." *Journal of Japanese Association of Zoological Aquariums* 39 (2): 47–54.

Klawe, W. L. 1966. Observations on the opah, *Lampris regius* (Bonnaterre). *Nature* 210 (5039): 965–966.

Klimley, A. P. 1980. Observations of courtship and copulation in the nurse shark, *Ginglymostoma cirratum.* *Copeia* 1980 (4): 878–882.

Klimley, A. P. 1985. The areal distribution and autoecology of the white shark, *Carcharodon carcharias,* off the west coast of North America. *Memoirs of the Southern California Academy of Sciences* 9: 15–40.

Klimley, A. P. 1994. The predatory behavior of the white shark. *American Scientist* 82 (2): 122–133.

Klimley, A. P., and S. D. Anderson. 1996. Residency patterns of white sharks at the South Farallon Islands, California. In *Great White Sharks: The Biology of Carcharodon carcharias,* edited by A. P. Klimley and D. G. Ainley, pp. 365–373. San Diego: Academic Press.

Klimley, A. P., S. B. Butler, D. R. Nelson, and A. T. Stull. 1988. Diel movements of scalloped hammerhead sharks, *Sphyrna lewini* Griffith and Smith, to and from a seamount in the Gulf of California. *Journal of Fish Biology* 33: 751–761.

Klimley, A. P., and D. R. Nelson. 1981. Schooling of the scalloped hammerhead shark, *Sphyrna lewini,* in the Gulf of California. *Fishery Bulletin* 79: 356–360.

Knaggs, E. H., J. S. Sunada, and R. N. Lea. 1975. Notes on some fishes collected off the outer coast of Baja California. *California Fish and Game* 61: 56–59.

Kneeland, S., Jr. 1847. Art. 36.—Dissection of *Scymnus brevipinna* (Lesueur). *Boston Journal of Natural History* 5: 479–485.

Kobayashi, H. 1986. Studies on deep-sea sharks in Kumano-nada region. *Bulletin of the Faculty of Fisheries* (Mie University) 13: 25–133.

Kobayashi, H., Y. Yamaguchi, T. Nonoda, K. Izawa, and H. Ban. 1982. The sharks caught on the continental shelf and slope in Kumano Nada region along the Pacific coast of Japan. *Bulletin of the Faculty of Fisheries* (Mie University) 9: 101–123.

Koefoed, E. 1927. Fishes from the Sea-Bottom. Report of the Scientific Results of the "*Michael Sars*" North Atlantic Deep-Sea Expedition. 1910: 4 (pt. 1): 1–148.

Koefoed, E. 1957. A uterine foetus and the uterus from a Greenland shark. In *Notes on the Greenland Shark Acanthorhinus carcharias (Gunn). Fiskeridirektoratets Skrifter, Serie Havundersøkelser (Report on Norwegian Fishery and Marine Investigations)* 11 (10): 8–12.

Kohler, N. E. 1987. Aspects of the Feeding Ecology of the Blue Shark, *Prionace glauca* in the Western North Atlantic. Ph.D. diss., University of Rhode Island. 163 pp.

Kohler, N. E., J. G. Casey, and P. A. Turner. 1995. Length-weight relationships for 13 species of sharks from the western North Atlantic. *Fishery Bulletin* 93: 412–418.

Kohler, N. E., J. G. Casey, and P. A. Turner. 1996. Length-length and length-weight relationships for 13 shark species from the western North Atlantic. NOAA Technical Memorandum NMFS-NE-110. U.S. Department of Commerce. 22 pp.

Kohler, N. E., J. G. Casey, and P. A. Turner. 1998. NMFS Cooperative Shark Tagging Program, 1962–93: An atlas of shark tag and recapture data. *Marine Fisheries Review* 60 (2): 1–87.

Kohler, N. E., and C. E. Stillwell. 1981. Food habits of the blue shark (*Prionace glauca*) in the northwest Atlantic. C.M.1981/H: 61. International Council for the Exploration of the Sea. 7 pp.

Kondyurin, V. V., and N. A. Myagkov. 1983. Catches of newborn Greenland shark, *Somniosus microcephalus* (Bloch and Schneider) (Dalatiidae). *Journal of Ichthyology* 23 (6): 140–141.

Konstantinou, H., and J. R. Cozzi. 1998. *Galeus springeri,* a new species of sawtail catshark from the Caribbean Sea (Chondrichthyes, Scyliorhinidae). *Copeia* 1998 (1): 151–158.

Konstantinov, K. G., and G. P. Nizovtsev. 1979. The basking shark, *Cetorhinus maximus,* in Kandalaksha Bay of the White Sea. *Journal of Ichthyology* 19 (1): 155–156.

Kormanik, G. A. 1989. Nitrogen budget in developing embryos of the spiny dogfish *Squalus acanthias. Journal of Experimental Biology* 144: 583–587.

Krefft, G. 1968. Neue und erstmalig nachgewiesene Knorpelfische aus dem Archibenthal des Südwest-atlantiks, einschließlich einer Diskussion einiger *Etmpoterus*-Arten südlicher Meere. *Archiv fuer Fischerei Wissenschaft* 19 (1): 1–42.

Krefft, G. 1980. Results of the research cruises of FRV "*Walther Herwig*" to South America. 53. Sharks from the pelagic trawl catches obtained during Atlantic transects, including some specimens from other cruises. *Archiv fuer Fischerei Wissenschaft* 30 (1): 1–16.

Krefft, G., and E. Tortonese. 1973. In *Check-list of the Fishes of North-Eastern Atlantic and of the Mediterranean* (CLOFNAM), edited by J. C. Hureau and T. Monod, Vol. 1. Paris: UNESCO. 683 pp.

Kruska, D. C. T. 1988. The brain of the basking shark (*Cetorhinus maximus*). *Brain, Behavior and Evolution* 32: 353–363.

Kubota, T., Y. Shiobara, and T. Kubodera. 1991. Food habits of the frilled shark *Chlamydoselachus anguineus* collected from Suruga Bay, central Japan. *Nippon Suisan Gakkaishi* 57 (1): 15–20.

Kukuyev, E. I. 1996. The new finds in recently born individuals of the whale shark, *Rhiniodon typus,* (Rhiniodontidae) in the Atlantic Ocean. *Journal of Ichthyology* 36 (2): 203–205.

Kunzlik, P. A. 1988. *The Basking Shark.* Scottish Fisheries Information Pamphlet Number 14. Aberdeen,

Scotland: Department of Agriculture and Fisheries for Scotland. 21 pp.

Kusher, D. I., S. E. Smith, and G. M. Cailliet. 1992. Validated age and growth of the leopard shark, *Triakis semifasciata,* with comments on reproduction. *Environmental Biology of Fishes* 35: 187–203.

Kyne, P. M., J. W. Johnson, W. T. White, and M. B. Bennet. 2005. First records of the false catshark, *Pseudotriakis microdon* Capello, 1868, from the waters of eastern Australia and Indonesia. *Memoirs of the Queensland Museum* 51 (2): 525–530.

Kyushin, K., K. Amaoka, K. Nakaya, H. Ida, Y. Tanino, and T. Senta. 1985. *Fishes of the South China Sea.* Tokyo: Japan Marine Fishery Resource Research Center. 333 pp.

La Gorce, J. O. 1939. *The Book of Fishes.* Washington, D.C.: National Geographic Society. 367 pp.

Lahille, F. 1928. Nota sobre unos peces elasmobranquios. *Anales del Museo Nacional de Historia Natural "Bernardino Rivadavia"* (Buenos Aires) 34: 299–339.

Last, P. R., and J. D. Stevens. 1994. *Sharks and Rays of Australia.* Victoria, Australia: CSIRO. 513 pp.

Last, P. R., W. T. White, and J. J. Pogonoski (eds.). 2008. Descritions of new Australian condrichthyans. CSIRO Marine and Atmospheric Research Paper no. 22. Hobart, Australia: CSIRO. 358 pp.

Laughlin, L., and J. Ugoretz. 1999. An observation of blue shark parturition in the southern California Bight. *California Fish and Game* 85 (2): 83–84.

Lavenberg, R. J., and J. A. Seigel. 1985. The Pacific's megamystery—megamouth. *Terra* 23 (4): 29–31.

Lavery, S., and J. B. Shaklee. 1991. Genetic evidence for separation of two sharks, *Carcharhinus limbatus* and *C. tilstoni,* from Northern Australia. *Marine Biology* 108: 1–4.

Lawler, E. F., Jr. 1976. The Biology of the Sandbar Shark *Carcharhinus plumbeus* (Nardo, 1827) in the Lower Chesapeake Bay and Adjacent Waters. Master's thesis, College of William and Mary. 49 pp.

Lawson, J. 1709. *A New Voyage to Carolina.* Facsimile, London: Readex Microprint, 1966. 258 pp.

Le Boeuf, B. J., J. E. McCosker, and J. Hewitt. 1987. Crater wounds on northern elephant seals: The cookiecutter shark strikes again. *Fishery Bulletin* 85: 387–392.

Le Boeuf, B. J., M. Riedman, and R. S. Keyes. 1982. White shark predation on pinnipeds in California coastal waters. *Fishery Bulletin* 80: 891–895.

Lea, R. N., and D. J. Miller. 1985. Shark attacks off the California and Oregon coasts: An update, 1980–84. *Memoirs of the Southern California Academy of Sciences* 9: 136–150.

LeBrasseur, R. J. 1964. Stomach contents of blue shark (*Prionace glauca* L.) taken in the Gulf of Alaska. *Journal of the Fisheries Research Board of Canada* 21 (4): 861–862.

Lech, R. L. 1982. *All the Drowned Sailors.* New York: Military Heritage Press.

Lee, R. S. 1969. The filetail catshark, *Parmaturus xaniurus,* in midwater in the Santa Barbara basin off California. *California Fish and Game* 55: 88–90.

Leet, W. S., C. M. Dewees, and C. W. Haugen. 1992. California's living marine resources and their utilization. In *California's Living Marine Resources and Their Utilization,* edited by W. S. Leet, C. M. Dewees, and C. W. Haugen, pp. 48–49. Sea Grant Extension Publication UCSGEP-92–12. Davis: University of California.

Lessa, R., V. Batista, and Z. Almeida. 1999a. Occurrence and biology of the daggernose shark *Isogomphodon oxyrhynchus* (Chondrichthyes: Carcharhinidae) off the Maranhão coast (Brazil). *Bulletin of Marine Science* 64 (1): 115–128.

Lessa, R., F. M. Santana, and R. Paglerani. 1999b. Age, growth and stock structure of the oceanic whitetip shark, *Carcharhinus longimanus,* from the southwestern equatorial Atlantic. *Fisheries Research* 42: 21–30.

Lessa, R., F. M. Santana, V. Batista, and Z. Almeida. 2000. Age and growth of the daggernose shark, *Isogomphodon oxyrhynchus,* from northern Brazil. *Marine and Freshwater Research* 51: 339–347.

Le Sueur, C. A. 1818. Descriptions of several new species of North American fishes. *Journal of the Academy of Natural Sciences of Philadelphia,* n.s., 1: 222–235.

Le Sueur, C. A. 1822. Description of a *Squalus,* of a very large size, which was taken on the coast of New Jersey. *Journal of the Academy of Natural Sciences of Philadelphia* 2 (pt. 2): 343–352.

Lewis, R. W. 1969. Studies on the stomach oils of marine animals—I. Oils of the black shark *Dalatias licha* (Bonnaterre). *Comparative Biochemistry and Physiology* 31: 715–724.

Lien, J., and L. Fawcett. 1986. Distribution of basking sharks, *Cetorhinus maximus,* incidentally caught in inshore fishing gear in Newfoundland. *The Canadian Field-Naturalist* 100: 246–252.

Limbaugh, C. 1955. Fish life in the kelp beds and the effects of kelp harvesting. University of California, Institute of Marine Resources, Reference 55–9: 50–51.

Limbaugh, C. 1963. Field notes on sharks. In *Sharks and Survival,* edited by P. W. Gilbert, pp. 63–94. Boston: D. C. Heath and Company.

Lindberg, G. U., and M. I. Legeza. 1967. *Fishes of the Sea of Japan and the Adjacent Areas of the Sea of Okhotsk and the Yellow sea.* Part 1. *Amphioxi Petromyzones Myxini Elasmobranchii Holocephali.* Edited by A. P.

Andriyashev. Jerusalem: Israel Program for Scientific Translations. 207 pp.

Lineweaver, T. H., III, and R. H. Backus. 1970. *The Natural History of Sharks.* Philadelphia: J. B. Lippincott. 256 pp.

Linnaeus [Linneus], C. 1758. *Systema Naturae. Regnum Animale. A Photographic Facsimile of the First Volume of the Tenth Edition (1758).* Reprint, Surrey: British Museum (Natural History), 1956. 823 pp.

Linné [Linneus], C. 1806. *A General System of Nature.* London: Lackington, Allen, and Co. 7 vols.

Liu, K.-M., C.-T. Chen, T.-H. Liao, and S.-J. Joung. 1999. Age, growth, and reproduction of the pelagic thresher shark, *Alopias pelagicus* in the northwestern Pacific. *Copeia* 1999 (1): 68–74.

Llano, G. A. 1955. *Airmen against the Sea.* ADTIC Publication G-104 ed. Research Studies Institute, Air University. Alabama: Maxwell Air Force Base. 114 pp.

Loefer, J. K., and G. R. Sedberry. 2003. Life history of the Atlantic sharpnose shark (*Rhizoprionodon terraenovae*) (Richardson, 1836) off the southeastern United States. *Fishery Bulletin* 101: 75–88.

Loefer, J. K., G. R. Sedberry, and J. C. McGovern. 2005. Vertical movements of a shortfin mako in the western North Atlantic as determined by pop-up satellite tagging. *Southeastern Naturalist* 4 (2): 237–246.

Lohberger, J. 1910. Über zwei riesige embryonen von *Lamna. Abhandlungen der math.- phys. Klasse der K. Bayer (*Akademie der Wissenschaften) 4 (suppl. bd. 2 Abhandlg): 1–45.

Long, D. J. 1994. First records of the combtooth dogfish shark, *Centroscyllium nigrum* (Chondrichthyes: Squalidae) from the Pacific subantarctic of Chile. *Japanese Journal of Ichthyology* 40 (4): 478–481.

Long, D. J. 1996a. First confirmed record of teleost predation on a shark egg case. *California Fish and Game* 82 (2): 103–104.

Long, D. J. 1996b. Records of white shark-bitten leatherback sea turtles along the central California coast. In *Great White Sharks: The Biology of Carcharodon carcharias,* edited by A. P. Klimley and D. G. Ainley, pp. 317–319. San Diego: Academic Press.

Long, D. J., and R. E. Jones. 1996. White shark predation and scavenging on cetaceans in the eastern North Pacific Ocean. In *Great White Sharks: The Biology of Carcharodon carcharias,* edited by A. P. Klimley and D. G. Ainley, pp. 293–307. San Diego: Academic Press.

Lord, J. K. 1866. *The Naturalist in Vancouver Island and British Columbia,* Vol. 1. London: Richard Bentley. 375 pp.

Lowe, C. G., B. M. Wetherbee, G. L. Crow, and A. L. Tester. 1996. Ontogenetic dietary shifts and feeding behavior of the tiger shark, *Galeocerdo cuvier,* in Hawaiian waters. *Environmental Biology of Fishes* 47: 203–211.

Lowe, R. H. 1962. The fishes of the British Guiana Continental Shelf, Atlantic coast of South America, with notes on their natural history. *Journal of the Linnean Society of London, Zoology* 44 (301): 669–700.

Lowe, R. T. 1839. Supplement to "A synopsis of the fishes of Maderia." *Transactions of the Zoological Society of London* 3 (1): 1–20.

Lozano Cabo, F. 1969. Nota sobre un caso de bicefalismo en la tintorera (*Prionace glauca* (L.)), capturada en las inmediaciones de las islas Chafarinas. *Boletín de la Real Sociedad Española de Historia Natural, Sección Biológica* 67: 27–28.

Lucifora, L. O., R. C. Menni, and A. H. Escalante. 2002. Reproductive ecology and abundance of the sand tiger shark, *Carcharias taurus,* from the southwestern Atlantic. *ICES Journal of Marine Science* 59: 553–561.

MacCoy, C. V. 1931. A record of the basking shark for York Harbor, Maine. *Bulletin of the Boston Society of Natural History* 61: 17–20.

MacCoy, C. V. 1939. Greenland shark from Jeffreys. *New England Naturalist* 25 (2): 24–26.

MacGinitie, G. E. 1931. Basking sharks on the Pacific coast. *Science* 73 (1897): 496.

Mackintosh, N. A., and J. F. G. Wheeler. 1929. Southern blue and fin whales. *Discovery Reports* 1: 257–540.

MacLeay, W. J. 1881. *Descriptive Catalogue of Australian Fishes,* Vol. 2. Sydney: F.W. White. 324 pp.

Magenta Cunha, C., and M. Bueno Gonzalez. 2006. Pregnancy in *Squaliolus laticaudus* (Elasmobranch: Dalatiidae) from Brazil. *Environmental Biology of Fishes* 75: 465–469.

Makihara, M. 1980. Toward effective utilization of the abundant shark resources. Translation no. 52, from *The Fishing and Food Industry Weekly* (898): 22–26. Translated by Tamio Otsu, Southwest Fisheries Center Honolulu Laboratory, National Marine Fisheries Service, NOAA. Honolulu, Hawaii: U.S. Department of Commerce.

Manire, C. A., and S. H. Gruber. 1991. Effect of M-type dart tags on field growth of juvenile lemon sharks. *Transactions of the American Fisheries Society* 120: 776–780.

Manire, C. A., and L. E. L. Rasmussen. 1997. Serum concentrations of steroid hormones in the mature male bonnethead shark, *Sphyrna tiburo. General and Comparative Endocrinology* 107: 414–420.

Manire, C. A., L. E. L. Rasmussen, D. L. Hess, and R. E. Hueter. 1995. Serum steroid hormones and the reproductive cycle of the female bonnethead shark,

Sphyrna tiburo. General and Comparative Endocrinology 97: 366–376.
Mansfield, A. W. 1963. *Seals of Arctic and Eastern Canada.* Bulletin 137. Ottawa. Fisheries Research Board of Canada. 30 pp.
Mansueti, R. J. 1957. Eight-foot shark blunders into Patuxent River seine. *Maryland Tidewater News* 13 (8): 1, 4.
Manzer, J. I. 1946. *Interesting Movements as Shown by the Recoveries of Certain Species of Tagged Fish.* Progress Reports of the Pacific Coast Stations no. 67. Vancouver, B. C.: Fisheries Research Board of Canada. 31 pp.
Markle, D. F., W. B. Scott, and A. C. Kohler. 1980. New and rare records of Canadian fishes and the influence of hydrography on resident and nonresident Scotian Shelf ichthyofauna. *Canadian Journal of Fisheries and Aquatic Sciences* 37: 49–65.
Marques da Silva, H. M. 1983. *Preliminary Studies of the Exploited Stock of Kitefin Shark Scymnorhinus licha* (Bonnaterre, 1788) in the Azores. CM 1983/G: 18. Copenhagen, Denmark International Council for the Exploration of the Sea. 13 pp.
Marques da Silva, H. M. 1988. *Growth and Reproduction of Kitefin Shark Dalatias licha (Bonn, 1788) in Azorean Waters.* C.M. 1988/G: 21. Copenhagen, Denmark International Council for the Exploration of the Sea. 15 pp.
Márquez-Farias, J. F., J. L. Castillo-Géniz, and C. Rodríguez de la Cruz. 1998. Demografía del cazón pech, *Sphyrna tiburo* (Linnaeus, 1758), en el sureste del Golfo de México. *Ciencias Marinas* 24 (1): 13–34.
Márquez-Farias, J. F., D. Corro-Espinosa, and J. L. Castillo-Géniz. 2005. Observations on the biology of the Pacific sharpnose shark (*Rhizoprionodon longurio,* Jordan and Gilbert, 1882), captured in southern Sinaloa, México. *Journal of Northwest Atlantic Fisheries Science* 35: 107–114.
Marshall, N. B. 1956. Deep sea shark (Indian ocean). *Marine Observer's Log, London* 26: 73–74.
Marshall, N. B., and N. R. Merrett. 1977. The existence of a benthopelagic fauna in the deep-sea. In *A Voyage of Discovery/ George Deacon 70th Anniversary Volume,* pp. 483–497. Oxford: Pergamon Press.
Martin, A. 1993. Hammerhead shark origins. *Nature* 364: 494.
Martinez, J. L. 1947. *Part I—the Cuban shark industry.* Fishery Leaflet no. 250, Fish and Wildlife Service. Washington, D. C. U.S. Department of the Interior. 18 pp.
Masai, H., Y. Sato, and M. Aoki. 1973. The brain of *Mitsukurina owstoni. Journal für Hirnforschung* 14 (6): 493–500.
Massachusetts Commissioners on Fisheries and Game. 1906. *Report of the Commissioners on Fisheries and Game upon the Damage Caused to the Fisheries of Massachusetts by Dogfish during the Year 1905.* Public Document no. 25:95–169. Boston: Wright and Potter Printing Co.
Matallanas, J. 1982. Feeding habits of *Scymnorhinus licha* in Catalan waters. *Journal of Fish Biology* 20: 155–163.
Mather, F. J., III, and C. G. Day. 1954. Observations of pelagic fishes of the tropical Atlantic. *Copeia* 1954 (3): 179–188.
Mather, F. J., III, and R. H. Gibbs, Jr. 1957. Distributional records of fishes from waters off New England and the middle Atlantic states. *Copeia* 1957 (3): 242–244.
Mathews, C. P. 1975. Some observations on the ecology of *Galeus piperatus* Springer and Wagner, a little known shark endemic to the northern Gulf of California. *Journal of Fish Biology* 7: 77–82.
Mathews, C. P., and M. F. Ruiz. 1974. *Cephalurus cephalus,* a small shark, taken in the northern Gulf of California, with a description. *Copeia* 1974 (2): 556–560.
Matsubara, K. 1936. A new carcharoid shark found in Japan. *Zoological Magazine* (Japan) 48 (7): 380–382.
Matthews, L. H. 1950. Reproduction in the basking shark, *Cetorhinus maximus. Philosophical Transactions of the Royal Society of London* B 234 (612): 247–316.
Matthews, L. H. 1962. The shark that hibernates. Washington, D.C.: *Smithsonian Report for 1962,* pp. 415–421.
Matthews, L. H., and H. W. Parker. 1950. Notes on the anatomy and biology of the basking shark (*Cetorhinus maximus* (Gunner)). *Proceedings of the Zoological Society of London* 120: 535–576.
Matthews, L. H., and H. W. Parker. 1951. Basking sharks leaping. *Proceedings of the Zoological Society of London* 121 (pt. 2): 461–462.
Mauchline, J., and J. D. M. Gordon. 1983. Diets of the sharks and chimaeroids of the Rockall Trough, northeastern Atlantic Ocean. *Marine Biology* 75: 269–278.
Maul, G. E. 1955. Five species of rare sharks new for Madeira including two new to science. *Notulae Naturae of the Academy of Natural Sciences of Philadelphia* 279: 1–13.
Maxwell, G. 1952. *Harpoon Venture.* New York: Viking Press. 304 pp.
McCandless, C. T., H. L. Pratt, Jr., N. E. Kohler, R. R. Merson, and C. W. Recksiek. 2007. Distribution, localized abundance, movements, and migrations of juvenile sandbar sharks tagged in Delaware Bay. In *Shark Nursery Grounds of the Gulf of Mexico and the*

East Coast of the United States, edited by C. T. McCandless, N. E. Kohler, and H. L. Pratt, Jr., pp. 45–62. Bethesda: American Fisheries Society.

McCormick, H. W., T. Allen, and W. E. Young. 1963. *Shadows in the Sea; The Sharks, Skates, and Rays.* Philadelphia: Chilton Company. 415 pp.

McCosker, J. E. 1981. Great white shark: An ichthyologist takes a closer look at the fish behind the legend. *Science 81* July/August: 42–51.

McCosker, J. E. 1985. White shark attack behavior: Observations of and speculations about predator and prey strategies. *Memoirs of the Southern California Academy of Sciences* 9: 123–135.

McCosker, J. E., and R. N. Lea. 1996. White shark attacks in the eastern Pacific Ocean: An update and analysis. In *Great White Sharks: The Biology of Carcharodon carcharias,* edited by A. P. Klimley and D. G. Ainley, pp. 419–434. San Diego: Academic Press.

McEachran, J. D., and S. Branstetter. 1984. Squalidae. In *Poissons de l'Atlantique du nord-est et de la Méditerranée,* edited by P. J. P. Whitehead, M.- L. Bauchot, J.- C. Hureau, J. Nielsen, and E. Tortonese, Vol. 1, pp. 128–147. Paris: UNESCO.

McEachran, J. D., and J. D. Fechhelm. 1998. *Fishes of the Gulf of Mexico,* Vol. 1. Austin: University of Texas Press. 1112 pp.

McGrouther, M. A. 2001. First record of the large-tooth cookie-cutter shark *Isistius plutodus* from Australian waters. *Memoirs of the Queensland Museum* 46 (2): 442.

McKenzie, M. D. 1970. First record of albinism in the hammerhead shark, *Sphyrna lewini* (Pisces: Sphyrnidae). *The Journal of the Elisha Mitchell Scientific Society* 86 (1): 35–37.

McKenzie, R. A. 1939. Some marine fish and salp records. *Proceedings of the Nova Scotian Institute of Science* 20 (pt. 1): 13–20.

McKenzie, R. A. 1963. Studies of the porbeagle, *Lamna nasus* (Bonnaterre), 1788. Fisheries Research Board of Canada, original manuscript of the Biological Station, St. Andrews, New Brunswick, no. 967. 6 pp.

McKenzie, R. A., and S. N. Tibbo. 1964a. A morphometric description of blue shark (*Prionace glauca*) from Canadian Atlantic waters. *Journal Fisheries Research Board of Canada* 21 (4): 865–866.

McKenzie, R. A., and S. N. Tibbo. 1964b. A morphometric description of porbeagle (*Lamna nasus*) from Canadian Atlantic waters. *Journal of the Fisheries Research Board of Canada* 21 (4): 863–864.

McKinnell, S., and M. P. Seki. 1998. Shark bycatch in the Japanese high seas squid driftnet fishery in the North Pacific Ocean. *Fisheries Research* 39: 127–138.

McLaughlin, R. H., and A. K. O'Gower. 1971. Life history and underwater studies of a heterodont shark. *Ecological Monographs* 41 (4): 271–289.

McManus, M. W., C. S. Johnson, and M. M. Jeffries. 1984. Training Nurse Sharks using Operant Conditioning. Technical Report 977. San Diego: Naval Ocean Systems Center. 12 pp.

McMillan, D. G., and W. W. Morse. 1999. Essential Fish Habitat Source Document: Spiny Dogfish, *Squalus acanthias,* Life History and Habitat Characteristics. NOAA Technical Memorandum NMFS-NE-150. Woods Hole, Massachusetts: U.S. Department of Commerce.16 pp.

McNally, K. 1976. *The Sun-Fish Hunt.* Belfast: Blackstaff Press. 80 pp.

Mecklenburg, C. W., T. A. Mecklenburg, and L. K. Thorsteinson. 2002. *Fishes of Alaska.* Bethesda: American Fisheries Society.1037 pp.

Medved, R. J., C. E. Stillwell, and J. J. Casey. 1985. Stomach contents of young sandbar sharks, *Carcharhinus plumbeus,* in Chincoteague Bay, Virginia. *Fishery Bulletin* 83: 395–402.

Meek, S. E., and S. F. Hildebrand. 1923. *The Marine Fishes of Panama. Part 1.* Field Museum of Natural History Publication no. 215, Zoölogical Seriesvol. 15, part 1.,. Chicago: Field Museum of Natural History. 330 pp.

Mendez Loeza, I. 2004. Hábitos alimentarios de los cazones *Mustelus californicus* y *Mustelus lunulatus* (Triakidae) en el alto Golfo de California. Thesis, Universidad Autónoma de Baja California, Ensenada, Mexico. 59 pp.

Menni, R. C. 1986. Shark biology in Argentina: A review. In *Indo-Pacific Fish Biology: Proceedings of the Second International Conference on Indo-Pacific Fishes,* edited by T. Uyeno, R. Arai, T. Taniuchi, and K. Matsuura, pp. 425–436. Tokyo: Ichthyological Society of Japan.

Menni, R. C., F. H. V. Hazín, and R. P. Lessa. 1995. Occurrence of the night shark *Carcharhinus signatus* and the pelagic stingray *Dasyatis violacea* off northeastern Brazil. *Neotropica* 41: 105–110.

Merrett, N. R. 1973. A new shark of the genus *Squalus* (Squalidae: Squaloidea) from the equatorial western Indian Ocean; with notes on *Squalus blainvillei. Journal of Zoology* 171: 93–110.

Merson R. R., and H. L. Pratt, Jr. 2007. Sandbar shark nurseries in New Jersey and New York: Evidence of northern pupping grounds along the United States East Coast. In *Shark Nursery Grounds of the Gulf of Mexico and the East Coast Waters of the United States,* edited by C. T. McCandless, N. E. Kohler, and H. L.

Pratt, Jr., pp. 35–43. Bethesda: American Fisheries Society, Symposium 50.

Merson R. R., and H. L. Pratt, Jr. 2001. Distribution, movements, and growth of young sandbar sharks, *Carcharhinus plumbeus,* in the nursery grounds of Delaware Bay. *Environmental Biology of Fishes* 61: 13–24.

Michel, M. 2002. Environmental Factors Affecting the Distribution and Abundance of Sharks in the Mangrove Estuary of the Ten Thousand Islands, Florida, USA. Ph.D. diss., University of Basel, Basel. 134 pp.

Miller, B., and D. W. Greenfield. 1965. A juvenile sixgilled shark (*Hexanchus corinus*) from the San Juan Islands, Washington. *Journal of the Fisheries Research Board of Canada* 22 (3): 857–859.

Miller, D. J., and R. N. Lea. 1972. Guide to the coastal marine fishes of California. Fish Bulletin 157. Sacramento, Ca.: California Department of Fish and Game. 235 pp.

Millus, D. 1980. Big-game fishing's greatest feat. *South Carolina Wildlife* 27 (3): 10–14.

Milstein, C. B. 1978. Sharks taken by longline in inshore New Jersey waters during 1973 and 1974. *Bulletin of the New Jersey Academy of Science* 23 (1): 34–38.

Mitchill, S. L. 1815. The fishes of New York. *Transactions of the Literary and Philosophical Society* 1: 355–492.

Molin, R. 1860. *Sullo Scheletro Degli Squali.* Venice: Presso la Segretaria dell' I.R. Istituto. 93 pp.

Mollet, H. F., G. Cliff, H. L. Pratt, Jr., and J. D. Stevens. 2000. Reproductive biology of the female shortfin mako, *Isurus oxyrinchus* Rafinesque, 1810, with comments on the embryonic development of lamnoids. *Fishery Bulletin* 98: 299–318.

Moore, J. A., K. E. Hartel, J. E. Craddock, and J. K. Galbraith. 2003. An annotated list of deepwater fishes from off the New England region, with new area records. *Northeastern Naturalist* 10 (2): 159–248.

Moreno, J. A. 1995. *Guía de los Tiburones de Aguas Ibéricas, Atlántico Nororiental y Mediterráneo.* Madrid: Ediciones Pirámide, SA. 310 pp.

Moreno, J. A., and A. Hoyos. 1983. *Carcharhinus acarenatus,* nov. sp., nouveau requin carcharhinide de L'Atlantique nord-oriental et de la Mediterranée occidentale. *Cybium* 7 (1): 57–64.

Moreno, J. A., and J. Morón. 1992. Comparative study of the genus *Isurus* (Rafinesque, 1810), and description of a form ('Marrajo Criollo') apparently endemic to the Azores. *Australian Journal of Marine and Freshwater Research* 43: 109–122.

Moreno, J. A., J. I. Parajúa, and J. Morón. 1989. Biología reproductiva y fenología de *Alopias vulpinus* (Bonnaterre, 1788) (Squaliformes: Alopiidae) en el Atlántico nor-oriental y Mediterráneo occidental. *Scientia Marina* 53 (1): 37–46.

Moreno, S., and J. Pol. 1992. *Somniosus pacificus,* adición a la ictiofauna cubana de elasmobranquios. *Revista de Investigaciones Marinas* 13 (3): 215–221.

Mori, T. 1952. Check list of the fishes of Korea. *Memoirs of the Hyogo University of Agriculture* B no. 1, 1 (3): 1–228.

Morrissey, J. F., and S. H. Gruber. 1993. Habitat selection by juvenile lemon sharks (*Negaprion brevirostris*). *Environmental Biology of Fishes* 38: 311–319.

Morrow, J. E. 1955. A thresher shark from Long Island Sound. *Postilla* (Yale Peabody Museum of Natural History) 24: 1–3.

Moulton, J. M. 1954. The sand shark, *Carcharias taurus,* from Casco Bay, Maine. *Copeia* 1954 (4): 296.

Müller, J., and J. Henle. 1838–1841. *Systematische Beschereibung der Plagiostomen.* Berlin: Verlag von Veit und Comp. 200 pp.

Muñoz-Chapuli, R., and F. Ramos. 1989. Review of the *Centrophorus* sharks (Elasmobranchii, Squalidae) of the eastern Atlantic. *Cybium* 13 (1): 65–81.

Murdy, E. O., R. S. Birdsong, and J. A. Musick. 1997. *Fishes of Chesapeake Bay.* Washington, D.C.: Smithsonian Institution Press. 324 pp.

Murphy, R. C. 1919. Notes on a mackerel shark, *Isurus tigris* (Atwood), from New York. *Copeia* 1919 (69): 32–34.

Murray, J. Sir., and J. Hjort. 1912. *The Depths of the Ocean.* London: Macmillan and Co., 821 pp.

Musick, J. A., and J. D. McEachran. 1969. The squaloid shark *Echinorhinus brucus* off Virginia. *Copeia* 1969 (1): 205–206.

Musick, J. A., S. Branstetter, and J. A. Colvocoresses. 1993. Trends in shark abundance from 1974 to 1991 for the Chesapeake Bight region of the U.S. Mid-Atlantic coast. In *Conservation Biology of Elasmobranchs,* edited by S. Branstetter. NOAA Technical Report NMFS 115, pp. 1–18. Springfield, Va.: U.S. Department of Commerce.

Myers, G. S. 1952. Sharks and sawfishes in the Amazon. *Copeia* 1952 (4): 268–269.

Myrberg, A. A., Jr. 1991. Distinctive markings of sharks: Ethological considerations of visual function. *The Journal of Experimental Zoology Supplement* 5: 156–166.

Myrberg, A. A., Jr., and S. H. Gruber. 1974. The behavior of the bonnethead shark, *Sphyrna tiburo. Copeia* 1974 (2): 358–374.

Nair, R. V., and R. S. Lal Mohan. 1971. On the occurrence of the spiny shark *Echinorhinus brucus* (Bonnaterre) from the east coast of India with a note on its distribution. *Indian Journal of Animal Science* 41 (10): 1011–1014.

Nakamura, H. 1935. On the two species of the thresher shark from Formosan waters. *Memoirs of the Faculty of Sciences and Agriculture, Taihoku Imperial University* 14: 1–6.

Nakamura, H. 1936. Reports of the investigation of the sharks of Taiwan. *Reports of the Taiwan Fisheries Experimental Station* 7 (1): 1–54.

Nakano, H. 1994. Age, reproduction and migration of blue shark in the North Pacific Ocean. *Bulletin of the National Research Institute of Far Seas Fisheries* 31: 141–144.

Nakano, H., H. Matsunaga, H. Okamoto, and M. Okazaki. 2003. Acoustic tracking of bigeye thresher shark *Alopias superciliosus* in the eastern Pacific Ocean. *Marine Ecology Progress Series* 265: 255–261.

Nakano, H., and M. Tabuchi. 1990. Occurrence of the cookiecutter shark *Isistius brasiliensis* in surface waters of the North Pacific Ocean. *Japanese Journal of Ichthyology* 37 (1): 60–63.

Nakatsu, L. M. 1957. A review of the soupfin shark fishery of the Pacific Coast. *Commercial Fisheries Review* 19 (12): 5–8.

Nakaya, K. 1971. Descriptive notes on a porbeagle, *Lamna nasus,* from Argentine waters, compared with the North Pacific salmon shark, *Lamna ditropis. Bulletin of the Faculty of Fisheries* (Hokkaido University) 21 (4): 269–279.

Nakaya, K. 1989. Discovery of a megamouth shark from Japan. *Report of the Japanese Society for Elasmobranch Studies* 26: 36–39.

Nakaya, K. 1996. A fatal shark attack in Aichi Prefecture, Japan, with other confirmed attack cases in Japanese waters. *Fisheries Science* 62 (5): 830–831.

Nakaya, K., and A. J. Bass. 1978. The frill shark *Chlamydoselachus anguineus* in New Zealand seas. *New Zealand Journal of Marine and Freshwater Research* 12 (4): 397–398.

Nakaya, K., and K. Sato. 1998. Taxonomic review of *Apristurus laurussonii* (Saemundsson, 1922) from the eastern North Atlantic (Elasmobranchii, Scyliorhinidae). *Cybium* 22 (2): 149–157.

Nammack, M. F., J. A. Musick, and J. A. Colvocoresses. 1985. Life history of spiny dogfish off the northeastern United States. *Transactions of the American Fisheries Society* 114: 367–376.

Nardo, D. 1826. Prodromus observationum et disquisitionum ichthyologiae Adriaticae. *Isis* 20 (6): 473–488.

Natanson, L. J., and G. M. Cailliet. 1986. Reproduction and development of the Pacific angel shark, *Squatina californica,* off Santa Barbara, California. *Copeia* 1986 (4): 987–994.

Natanson, L. J., and G. M. Cailliet. 1990. Vertebral growth zone deposition in Pacific angel shark. Copeia 1990 (4): 1133–1145.

Natanson, L. J., J. G. Casey, and N. E. Kohler. 1995. Age and growth estimates for the dusky shark, *Carcharhinus obscurus,* in the western North Atlantic Ocean. *Fishery Bulletin* 93: 116–126.

Natanson, L. J., J. J. Mello, and S. E. Campana. 2002. Validated age and growth of the porbeagle shark (*Lamna nasus*) in the western North Atlantic Ocean. *Fishery Bulletin* 100: 266–278.

Natanson, L. J., J. G. Casey, N. E. Kohler, and T. Colket, IV. 1999. Growth of the tiger shark, *Galeocerdo cuvier,* in the western North Atlantic based on tag returns and length frequencies; and a note on the effects of tagging. *Fishery Bulletin* 97: 944–953.

Natanson, L. J., N. E. Kohler, D. Ardizzone, G. M. Cailliet, S. P. Wintner, and H. F. Mollet. 2006. Validated age and growth estimates for the shortfin mako, *Isurus oxyrinchus,* in the North Atlantic Ocean. *Environmental Biology of Fishes* 77: 367–383.

Natanson, L. J., S. P. Wintner, F. Johansson, A. Piercy, P. Campbell, A. De Maddalena, S. J. B. Gulak, B Human, F. Cigala Fulgosi, D. A. Ebert, F. Hemida, F. H. Mollen, S. Vanni, G. H. Burgess, L. J. V. Compagno, and A. Wedderburn-Maxwell. 2008. Ontogenetic vertebral growth patterns in the basking shark *Cetorhinus maximus. Marine Ecology Progress Series* 361: 267–278.

National Marine Fisheries Service. 1992. *Fishery Management Plan for Sharks of the Atlantic Coast.* Washington, D.C.: U.S. Department of Commerce. 160 pp.

National Marine Fisheries Service. 1995. *By-catch Estimates and Estimates of Relative Abundance for Sharks.* Document SB-3–23, March 8, 1995. Pascagoula, Miss.: National Oceanic and Atmospheric Administration (NOAA). Unpaginated.

National Marine Fisheries Service. 1996. Atlantic Shark Fisheries: Quotas, Bag Limits, Prohibitions, and Requirements. 50 CFR §678. *Federal Register:* 61 (no. 246): 67295–67299.

National Marine Fisheries Service. 1997. 50 CFR Part 678. Atlantic Shark Fisheries: Quotas, Bag Limits, Prohibitions, and Requirements. *Federal Register* 62 (no. 66): 16648–16656.

National Marine Fisheries Service. 1999. 50 CFR, Parts 285, 600, 630, 635, 644, and 678Atlantic Highly Migratory Species (HMS) Fisheries; Fishery Management Plan, Plan Amendment, and Consolidation of Regulations. *Federal Register* 64 (no. 12): 3154–3194.

Neal, H. V. 1898. The segmentation of the nervous system in *Squalus acanthias. Bulletin of the Museum of*

Comparative Zoölogy at Harvard College 31 (7): 147–294.

Neave, F., and M. G. Hanavan. 1960. Seasonal distribution of some epipelagic fishes in the Gulf of Alaska region. *Journal of the Fisheries Research Board of Canada* 17 (2): 221–233.

Needham, J. 1942. *Biochemistry and Morphogenesis.* Cambridge: Cambridge University Press. 785 pp.

Nelson, D. R., and R. H. Johnson. 1970. Diel activity rhythms in the nocturnal, bottom-dwelling sharks, *Heterodontus francisci* and *Cephaloscyllium ventriosum. Copeia* 1970 (4): 732–739.

Nelson, D. R., J. N. McKibben, W. R. Strong, Jr., C. G. Lowe, J. A. Sisneros, D. M. Schroeder, and R. J. Lavenberg. 1997. An acoustic tracking of a megamouth shark, *Megachasma pelagios:* A crepuscular vertical migrator. *Environmental Biology of Fishes* 49: 389–399.

Nelson, J. D., and S. A. Eckert. 2007. Foraging ecology of whale sharks (*Rhincodon typus*) within Bahía de Los Angeles, Baja California Norte, México. *Fisheries Research* 84: 47–64.

Nelson, J. S., E. J. Crossman, H. Espinosa-Pérez, L. T. Findley, C. R. Gilbert, R. N. Lea, and J. D. Williams. 2004. *Common and Scientific Names of Fishes from the United States, Canada, and Mexico.* 6th ed. Special Publication 29. Bethesda: American Fisheries Society. 386 pp.

Nemoto, T. 1955. White scars on whales (1) Lamprey marks. *Scientific Reports of the Whales Research Institute* (Tokyo) 10: 69–77.

Nichols, J. T. 1927. A new shark from the continental slope off Florida. *American Museum Novitates* 256: 1–2.

Nichols, J. T., and C. M. Breder, Jr. 1927. *The Marine Fishes of New York and Southern New England.* New York Aquarium Nature Series. New York: New York Zoological Society. 192 pp.

Nichols, J. T., and F. E. Firth. 1939. Rare fishes off the Atlantic Coast including a new grammicolepid. *Proceedings of the Biological Society of Washington* 52: 85–88.

Nichols, J. T., and R. C. Murphy. 1916. Long Island fauna—4: The sharks (order Selachii). *Science Bulletin* (Brooklyn Institute of Arts and Sciences, Brooklyn Museum) 3 (1): 1–34.

Nishikawa, T. 1898. Notes on some embryos of *Chlamydoselachus anguineus* Garm. *Annotationes Zoologicae Japonenses* 2: 95–102.

Noble, E. R. 1948. On the recent frilled shark catch. *Science* 108 (2806): 380.

Norris, H. W. 1923. The occurrence of *Mustelus lunulatus* on the California coast. *Copeia* 1923 (114): 1–2.

Norris, H. W., and S. P. Hughes. 1920. The cranial, occipital, and anterior spinal nerves of the dogfish, *Squalus acanthias. Journal of Comparative Neurology* 31 (5): 293–395.

Norris, K. S., and T. P. Dohl. 1980. Behavior of the Hawaiian spinner dolphin, *Stenella longirostris. Fishery Bulletin* 77: 821–849.

Ochoa-López, E., C. Villavicencio-Garayzar, and A. Ruiz-Gaytán. 1997. First record of the whitetip reef shark, *Triaenodon obesus* in the west coast of México (Revillagigedo Archipelago). *Revista de Biología Tropical* 44 (3)/45 (1) pt. B: 687–688.

O'Connor, P. F. 1953. *Shark!* New York: Norton. 255 pp.

Okamura, O., K. Amaoka, M. Takeda, K.Yano, K. Okada, and S. Chinuki. 1995. Fishes collected by the R/V *Shinkai Maru* around Greenland. Tokyo: Japan Marine Fishery Research Center. 304 pp.

Olsen, A. M. 1953. Tagging of school shark, *Galeorhinus australis* (Macleay) (Carcharhanidae [sic]) in south-eastern Australian waters. *Australian Journal of Marine and Freshwater Research* 4 (1): 95–104.

Olsen, A. M. 1954. The biology, migration, and growth rate of the school shark, *Galeorhinus australis* (Macleay) (Carcharhanidae [*sic*]) in south-eastern Australian waters. *Australian Journal of Marine and Freshwater Research* 5 (3): 353–410.

Olsen, A. M. 1984. Synopsis of biological data on the school shark *Galeorhinus australis* (Macleay 1881). *FAO Fisheries Synopsis* no. 139. Rome: Food and Agriculture Organization of the United Nations. 42 pp.

Oppian. 1963. *Cygenetica and Halieutica.* In *Oppian, Colluthus, Tryphiodorus,* translated by A. W. Mair, pp. 1–531. London: William Heinemann Ltd.

Orlov, A. M. 1999. Capture of especially large sleeper shark *Somniosus pacificus* (Squalidae) with some notes on its ecology in northwestern Pacific. *Journal of Ichthyology* 39 (7): 548–553.

Oshitani, S., H. Nakano, and S. Tanaka. 2003. Age growth of the silky shark *Carcharhinus falciformis* from the Pacific Ocean. *Fisheries Science* 69: 456–464.

Otake, T., and K. Mizue. 1981. Direct evidence for oophagy in thresher shark, *Alopias pelagicus. Japanese Journal of Ichthyology* 28 (2): 171–172.

Otsu, T., and H. O. Yoshida. 1957. Another record of the rare shark, *Euprotomicrus bispinatus,* from the Pacific Ocean. *Copeia* 1957 (1): 58.

Parin, N. V. 1966. Data on the biology and distribution of the pelagic sharks *Euprotomicrus bispinatus* and *Isistius brasiliensis* (Squalidae, Pisces). In *Fishes of the Pacific and Indian Oceans Biology and Distribution,* edited by T. S. Rass, pp. 173–195. Translated from Russian. Jerusalem: Israel Program for Scientific Translations.

Parin, N. V. 1975. First Pacific Ocean record of the dalatiid shark, *Isistius plutodus* Garrick and Springer, collected near Okinawa, Japan. *U O* (Japanese Society of Ichthyologists) 25: 1–3.
Parker, G. H. 1910. Influence of the eyes, ears, and other allied sense organs on the movements of the dogfish, *Mustelus canis* (Mitchill). *Bulletin of the Bureau of Fisheries* 29 (738): 43–57.
Parker, G. H. 1914. The directive influence of the sense of smell in the dogfish. *Bulletin of the United States Bureau of Fisheries* 33 (798): 61–68.
Parker, G. H. 1938. Melanophore responses in the young of *Mustelus canis. Proceedings of the American Academy of Arts and Sciences* 72 (7): 269–282.
Parker, H. W., and M. Boeseman. 1955. The basking shark, *Cetorhinus maximus,* in winter. *Proceedings of the Zoological Society of London* 124: 185–194.
Parker, H. W., and F. C. Stott. 1965. Age, size and vertebral calcification in the basking shark, *Cetorhinus maximus* (Gunnerus). *Zoologische Mededelingen* 40 (34): 305–319.
Parker, T. J. 1882. Art. 29–Notes on the anatomy and embryology of *Scymnus lichia. Transactions and Proceedings of the Royal Society of New Zealand* (Wellington) 15: 222–234.
Parra, A. 1787. *Descripción de Diferentes Piezas de Historia Natural, las Mas del Ramo Marítimo, Representadas en Setenta y Cinco Laminas.* Havana: Imprenta de la Capitania General. 195 pp.
Parsons, G. R. 1983. The reproductive biology of the Atlantic sharpnose shark, *Rhizoprionodon terraenovae* (Richardson). *Fishery Bulletin* 81: 61–73.
Parsons, G. R. 1986. Observations on the reproductive biology of the marbled catshark, *Galeus arae antillensis. Northeast Gulf Science* 8 (2): 149–150.
Parsons, G. R. 1993a. Age determination and growth of the bonnethead shark *Sphyrna tiburo:* A comparison of two populations. *Marine Biology* 112: 23–31.
Parsons, G. R. 1993b. Geographic variation in reproduction between two populations of the bonnethead shark, *Sphyrna tiburo. Environmental Biology of Fishes* 38: 25–35.
Parsons, G. R., G. W. Ingram, Jr., and R. Havard. 2002. First record of the goblin shark, *Mitsukurina owstoni,* Jordan (family Mitsukurinidae) in the Gulf of Mexico. *Southeastern Naturalist* 1 (2): 189–192.
Pascoe, P. L. 1986. Fish otoliths from the stomach of a thresher shark, *Alopias vulpinus. Journal of the Marine Biological Association of the United Kingdom* 66: 315–317.
Paterson, R. A., C. J. Quayle, and S. M. Van Dyck. 1993. A humpback whale calf and two subadult dense-beaked whales recently stranded in southern Queensland. *Memoirs of the Queensland Museum* 33 (1): 291–297.
Paust, B., and R. Smith. 1986. Salmon Shark Manual: The Development of a Commercial Salmon Shark, *Lamna ditropis,* Fishery in the North Pacific. Alaska Sea Grant Report 86–01, edited by T. Frady. Fairbanks: University of Alaska. 430 pp.
Peck, A. L. (trans.). 1979. *Aristotle Historia Animalium,* Vol. 2. Cambridge, Mass.: Harvard University Press.
Peillard, L. 1963. *The Laconia Affair.* New York: G. P. Putnam's Sons. 270 pp.
Pennant, T. 1776. *British Zoology.* Vol. 3. London: Benjamin White. 371 pp.
Penner, L. R., and P. J. S. Raj. 1977. Concerning the marine leech, *Pontobdella macrothela* Schmarda, 1861 (Piscicolidae: Hirudinea). *Publicaciones especiales* (Instituto de Biología, Mexico) 4: 519–530.
Penrith, M. J. 1969. New records of deep-water fishes from South West Africa. *Cimbebasia,* ser. A, 1 (3): 59–75.
Penrith, M. J. 1972. Earliest description and name for the whale shark. *Copeia* 1972 (2): 362.
Pequeño, R. G. 1979. El género *Notorhyncus* en Chile (Elasmobranchii: Hexanchidae). *Revista de Biología Marina* (University of Chile) 16 (3): 248–254.
Pequeño R. G., J. Lamilla, and A. Crovetto. 1991. Capture of *Somniosus cf. pacificus* Bigelow and Schroeder, 1944, off Valdivia, Chile, with notes on the gut contents (Chondrichthyes, Squalidae). *Estudios Oceanológicos* 10: 117–122.
Pérez Canto, C. 1886. Historia Natural—Estudio sobre algunos escualos de la costa de Chile. *Anales de la Universidad de Chile* 69 (1): 1003–1016.
Pérez Jiménez, J. C. 2006. Biología y taxonomía de los tiburones del género *Mustelus* (Elasmobranchii) de la región norte del Golfo de California. Ph.D. diss., Centro de Investigación Científica y de Educación Superior de Ensenada, Ensenada, México. 188 pp.
Pérez Jiménez, J. C., O. Sosa Nishizaki, and J. L. Castillo Geniz. 2005a. A new eastern North Pacific smooth-hound shark (genus *Mustelus,* family Triakidae) from the Gulf of California. *Copeia* 2005 (4): 834–845.
Pérez-Jiménez,[1] J. C., O. Sosa-Nishizaki, E. Furlong-Estrada, D. Corro-Espinosa, A. Venegas-Herrera, and O. V. Barragán-Cuencas. 2005b. Artisanal shark fishery at "Tres Marias" Islands and Isabel Island in the Central Mexican Pacific. *Journal of Northwest Atlantic Fishery Science* 35: 333–343.

1. Author's note: These authors' last names are hyphenated in some papers and not in others. Here I follow the usage in the different publications.

Pérez-Jiménez, J. C., and O. Sosa-Nishizaki. 2008. Reproductive biology of the brown smoothhound shark *Mustelus henlei,* in the northern Gulf of California, México. *Journal of Fish Biology* 73: 782–792.

Perley, M. H. 1852. Descriptive catalogue [in part] of the fishes of New Brunswick and Nova Scotia. In *Reports on the Sea and River Fisheries of New Brunswick,* 2nd ed., pp. 179–228. Fredericton: J. Simpson.

Péron, F. 1807. *Voyage de Découvertes aux Terres Australes, exécuté par ordre de sa majesté l'Empereur et Roi, sur les Corvettes la Géographe, la Naturaliste et la Goulette le Casuarina, pendant les années 1800, 1801, 1803 et 1804,* Vol. 1. Paris. 496 pp.

Peters, W. 1852. Diagnosen von neuen flussfischen aus Mossambique. *Bericht über die zur Bekanntmachung geeigneten Verhandlungen der Königl. Preuß. Akademie der Wissenschaften zu Berlin* 1852: 275–276, 681–686.

Philippi, R. A. 1887. Sobre los tiburones y algunos otros peces de Chile. *Anales de la Universidad de Chile* 71: 535–574.

Phillipps, W. J. 1935. Sharks of New Zealand: No 4. *New Zealand Journal of Science and Technology* 16: 236–241.

Phillips, J. B. 1948. Basking shark fishery revived in California. *California Fish and Game* 34 (1): 11–23.

Phillips, J. B. 1952. Sleeper shark, *Somniosus pacificus,* caught off Fort Bragg, California. *California Fish and Game* 39 (1): 147–149.

Piercy, A. N., J. K. Carlson, J. A. Sulikowski, and G. H. Burgess. 2007. Age and growth of the scalloped hammerhead shark, *Sphyrna lewini,* in the north-west Atlantic Ocean and Gulf of Mexico. *Marine and Freshwater Research* 58: 34–40.

Pietschmann, V. 1907. Zwei neue Selachier aus Japan. *Anzeiger der Kaiserlichen Akademie der Wissenschaften* 44 (1–27): 394–396.

Pietschmann, V. 1928. Sitzung der mathematisch-naturwissenschaftlichen klasse vom 13. Dezember 1928. *Anzeiger der Akademie der Wissenschaften* 65 (1–27): 297–298.

Pietschmann, V. 1930. Remarks on Pacific fishes. *Bernice P. Bishop Museum Bulletin* 73: 23 pp.

Pike, G. C. 1951. Lamprey marks on whales. *Journal of the Fisheries Research Board of Canada* 8 (4): 275–280.

Piotrovskiy, A. S., and V. G. Prut'ko. 1980. The occurrence of the goblin shark, *Scapanorhynchus owstoni* (Chondrichthyes, Scapanorhynchidae), in the Indian Ocean. *Journal of Ichthyology* 20 (1): 124–125.

Pittenger, G. G. 1984. Movements, distribution, feeding, and growth of the Pacific angel shark, *Squatina californica,* at Catalina Island, California. Master's thesis, California State University, Long Beach. 83 pp.

Poey, F. 1858–1861. *Memorias Sobre la Historia Natural de la Isla de Cuba,* Vol. 2. Havana: Imprenta de la Viuda de Barcina. 442 pp.

Poey, F. 1876. Enumeratio piscium cubensium, part 3. *Anales de la Sociedad Española de Historia Natural* 5: 373–404.

Poey, F. 1868. *Peces cubanos. Especies nuevas. Repertorio Físico-Natural de la Isla de Cuba,* Vol. 2. Havana: La Viuda de Barcina y Comp. pp. 229–245.

Poll, M. 1951. Poissons. 1. Généralités. 2. Sélaciens et chimères. *Résultats Scientifiques. Expédition Océanographique Belge dans les Eaux Côtières Africaines de L'Atlantique Sud (1948–1949),* Vol. 4, Fasc. 1. Bruxelles: Institut Royal Des Sciences Naturelles de Belgique. 154 pp.

Pollard, D. A., M. P. L. Smith, and A. K. Smith. 1996. The biology and conservation status of the grey nurse shark (*Carcharias taurus* Rafinesque 1810) in New South Wales, Australia. *Aquatic Conservation: Marine and Freshwater Ecosystems* 6: 1–20.

Porras Gándara, G. 1997. Contribución al conocimiento de la biología del tiburón Punta de Zapato (*Nasolamia velox*) capturado por la pesca artesanal en el Pacífico de Guatemala. Thesis, Centro de Estudios del Mar y Acuicultura, Universidad de San Carlos de Guatemala, Guatemala. 38 pp.

Pratt, H. L., Jr. 1979. Reproduction in the blue shark, *Prionace glauca. Fishery Bulletin* 77: 445–470.

Pratt, H. L., Jr. 1996. Reproduction in the male white shark. In *Great White Sharks: The Biology of Carcharodon carcharias,* edited by P. A. Klimley and D. G. Ainley, pp. 131–138. San Diego: Academic Press.

Pratt, H. L., Jr., and J. C. Carrier. 1995. Wild mating of the nurse sharks. *National Geographic* 187 (5): 44–53.

Pratt, H. L., Jr., and J. G. Casey. 1983. Age and growth of the shortfin mako, *Isurus oxyrinchus,* using four methods. *Canadian Journal of Fisheries and Aquatic Sciences* 40 (11): 1944–1957.

Pratt, H. L., Jr., J. G. Casey, and R. B. Conklin. 1982. Observations on large white sharks, *Carcharodon carcharias,* off Long Island, New York. *Fishery Bulletin* 80: 153–156.

Putnam, F. W. 1874. [A large shark tooth exhibited at the regular meeting, Monday, April 20, 1874]. *Bulletin of the Essex Institute* 6 (4): 72.

Pyle, P., M. J. Schramm, C. Keiper, and S. D. Anderson. 1999. Predation on a white shark (*Carcharodon carcharias*) by a killer whale (*Orcinus orca*) and a possible case of competitive displacement. *Marine Mammal Science* 15 (2): 563–568.

Quattro, J. M., D. S. Stoner, W. B. Driggers, C. A. Anderson, K. A. Priede, E. C. Hoppmann, N. H. Campbell, K. M. Duncan, and J. M. Grady. 2006. Genetic evidence of cryptic speciation within hammerhead sharks (genus *Sphyrna*). *Marine Biology* 148: 1143–1155.

Quéro, J.-C. 1972. Capture de deux *Scapanorhynchus owstoni* (Jordan, 1898) (Selachii, Scapanorhynchidae) par des chalutiers de la Rochelle. *Annales de la Société des Sciences Naturelles de la Charente-Maritime* 1972: 168–170.

Quéro, J.-C. 1975. *Somniosus bauchotae* sp. nov. (Selachii, Squalidae, Scymnorhininae) espece nouvelle de l'Atlantique N.O. *Revue Des Travaux De L'Institut Des Pêches Maritimes* 39 (4): 455–469.

Quéro, J.-C., R. Verron, Y. Cattin.1976. Observations ichthyologiques effectuées au port de la Rochelle en 1975. Annales de la Societé, R. Verron, and Y. Cattin. *Anales de la Societé des Sciences Naturelles de la Charente-Maritime* 1976/01 6 (3): 177–190.

Quéro, J.-C., and R. Emonnet. 1993. Disparition ou raréfaction d'espèces marines au large d'Arcachon. In Actes du 3me Colloque International "*Oceanographie du Golfe de Gascogne*" pp. 221–225. Nieuil-sûr-Mer, France: IFREMER.

Quoy, J. R. C., and P. Gaimard. 1824. Zoologie. In *Voyage Autour du Monde . . . Exécuté sur les Corvettes de S. M. L'Uranie et La Physicienne, Pendant les Années 1817, 1818, 1819 et 1820,* pp. 197–198. Paris: Chez Pillet Aîné, Imprimeur-Libraire.

Radcliffe, L. 1916. The sharks and rays of Beaufort, North Carolina. *Bulletin of the Bureau of Fisheries* 34: 239–284.

Rae, B. B. 1962. Porbeagle sharks. *Scottish Fisheries Bulletin* 18: 17–19.

Rafinesque, C. S. 1810. *Caratteri di Alcuni Nuovi Generi e Nuove Specie di Animali e Piante Della Sicilia.* Palermo: Le Stampe di Sanfilippo. 105 pp.

Rago, P. J., K. A. Sosebee, J. K. T. Brodziak, S. A. Murawski, and E. D. Anderson. 1998. Implications of recent increases in catches on the dynamics of Northwest Atlantic spiny dogfish (*Squalus acanthias*). *Fisheries Research* 39: 165–181.

Ramachandran, A., and T. V. Sankar. 1990. Fins and fin rays from whale shark (*Rhiniodon typus* Smith). *Fishery Technology* 27 (2): 138–140.

Rancurel, P. and A. Intes. 1982. Le requin tigre, *Galeocerdo cuvieri* Lacépède, des eaux néocaledoniennes examen des contenus stomacaux. *Tethys* 10 (3): 195–199.

Randall, B. M., R. M. Randall, and L. J. V. Compagno. 1988. Injuries to jackass penguins (*Spheniscus demersus*): Evidence for shark involvement. *Journal of Zoology (London)* 214: 589–599.

Randall, J. E. 1963. A fatal attack by the shark *Carcharhinus galapagensis* at St. Thomas, Virgin Islands. *Caribbean Journal of Science* 3 (4): 201–205.

Randall, J. E. 1967. Food habits of reef fishes of the West Indies. *Studies in Tropical Oceanography* 5: 665–847.

Randall, J. E. 1968. *Caribbean Reef Fishes.* Hong Kong: T. F. H. Publications, Inc. 318 pp.

Randall, J. E. 1973. Size of the great white shark (*Carcharodon*). *Science* 181: 169–170.

Randall, J. E. 1977. Contribution to the biology of the whitetip reef shark (*Triaenodon obesus*). *Pacific Science* 31 (2): 143–164.

Randall, J. E. 1986. *Sharks of Arabia.* London: IMMEL Publishing. 148 pp.

Randall, J. E. 1987. Refutation of lengths of 11.3, 9.0 and 6.4 m attributed to the white shark, *Carcharodon carcharias. California Fish and Game* 73 (3): 163–168.

Randall, J. E. 1992. Review of the biology of the tiger shark (*Galeocerdo cuvier*). *Australian Journal of Marine and Freshwater Research* 43: 21–31.

Randall, J. E., and M. F. Levy. 1976. A near-fatal shark attack by a mako in the northern Red Sea. *Israel Journal of Zoology* 25: 61–70.

Ranzani, C. 1840. De novis speciebus piscium. Dissertatio prima. *Novi Commentarii Academiae Scientiarum, Instituti Bononiensis* 4: 65–83.

Ranzi, S. 1934. Le basi fisio-morfologiche dello sviluppo embrionale dei selaci.-parti 2 e 3. *Publicazioni della Stazione Zoologica di Napoli* 13 (3): 332–437.

Rappé, G. 1984. An unrecorded 19th-century capture of *Echinorhinus brucus* (Bonnaterre, 1788), a former rare visitor to the North Sea. *Naturalist* 109 (970): 113–114.

Raschi, W., J. A. Musick, and L. J. V. Compagno. 1982. *Hypoprion bigelowi,* a synonym of *Carcharhinus signatus* (Pisces: Carcharhinidae), with a description of ontogenetic heterodonty in this species and notes on its natural history. *Copeia* 1982 (1): 102–109.

Rathbun, R. 1887. The sponge fishery and trade. In *The Fisheries and Fishery Industries of the United States. Section 5, History and Methods of the Fisheries,* Vol. 2, edited by G. B. Goode, pp. 819–841. Washington, D.C.: Government Printing Office.

Raven, H. C. 1926. The Greenland shark. *Natural History* 26 (6): 654–655.

Regan, C. T. 1908. A synopsis of the sharks of the family Scyliorhinidae. *Annals and Magazine of Natural History,* 8 ser., 1(6): 453–465.

Reid, G. K. 1957. External morphology of an embryo whale shark, *Rhincodon typus* Smith. *Copeia* 1957 (2): 157–158.

Reinhardt, J. C. 1825. Physisk Classe. In *Oversigt det Kongelige Danske Videnskabernes Selskabs Forhand-*

linger og dets Medlemmers Arbeider fra 31 Mai 1824 til 31 Mai 1825, pp. 2–3. Copenhagen.

Reyier, E. A., D. H. Adams, and R. H. Lowers. 2008. First evidence of a high density nursery ground for the lemon shark, *Negaprion brevirostris,* near Cape Canaveral, Florida. *Florida Scientist* 71: 134–148.

Richards, J. B. 1987. Developing a localized fishery: The pacific angel shark. In *Sharks: An inquiry into Biology, Behavior, Fisheries, and Use,* edited by S. Cook, pp. 147–160. Portland: Oregon State University Extension Service.

Richards, J. B. 1992. Pacific Angel Shark. In *California's Living Marine Resources and Their Utilization,* edited by W. S. Leet, C. M. Dewees, and C. W. Haugen, pp. 46–47. Sea Grant Extension Publication UCSGEP-92–12. Davis: University of California.

Richards, W. J. 1999. Problems with unofficial and inaccurate geographical names in the fisheries literature. *Marine Fisheries Review* 61 (3): 56–57.

Richardson, J. 1836. *Fauna Boreali-Americana or the Zoology of the Northern Parts of British America,* Part 3, *The Fish.* Reprint, New York: Arno Press, 1978, pp. 278–313.

Rider, S. J., M. Athorn, and G. O. Bailey. 2002. First record of a white tiger shark, *Galeocerdo cuvier,* from the northeastern Gulf of Mexico. *Florida Scientist* 65 (1): 13–15.

Ripley, W. E. 1946. The soupfin shark and the fishery. In *The Biology of the Soupfin Galeorhinus zyopterus and Biochemical Studies of the Liver,* pp. 7–37. Department of Natural Resources Division of Fish and Game, Bulletin no. 64. Eureka: California State Printing Office.

Ripley, W. E. 1949. Shark and shark liver. In *The Commercial Fish catch of California for the Year 1947 with an Historical Review 1916–1947,* pp. 129–134. Department of Natural Resources Division of Fish and Game, Fish Bulletin no. 74. Eureka: California State Printing Office.

Risso, A. 1810 (Facsimile 1966). *Icthyologie de Nice.* Amsterdam: A. Asher & Co. 388 pp.

Risso, A. 1826. *Histoire Naturelle des Principales Productions de l'Europe Méridionale et Particulièrement de Celles des Environs de Nice et des Alpes Maritimes,* Vol. 3, pp. 115–141. Paris.

Rivera-López, J. 1970. Studies on the Biology of the Nurse Shark, *Ginglymostoma cirratum* Bonnaterre and the Tiger Shark, *Galeocerdo cuvieri* Peron and Le Sueur. Master's thesis, University of Puerto Rico, Mayaguez. 60 pp.

Robins, C. R., and R. N. Lea. 1975. Proposed suppression of *Rhiniodon* Smith, 1828 (Pisces) in favour of *Rhincodon* Smith, 1829 as the generic name of the whale shark Z.N. (S.) 2090. *Bulletin of Zoological Nomenclature* 32 (pt. 3): 163–167.

Robins, C. R., R. M. Bailey, C. E. Bond, J. R. Brooker, E. A. Lachner, R. N. Lea, and W. B. Scott. 1980. *A List of Common and Scientific Names of Fishes from the United States and Canada.* 4th ed., Special publication no. 12. Bethesda: American Fisheries Society. 174 pp.

Rodríguez, A., E. Valdés-Muñoz, and R. Valdés. 1984. *Lista de nombres científicos y comunes de peces marinos cubanos (Nomenclator).* Cuba: Centro de Investigaciones Pesqueras, Ministerio de la Industria Pesquera. 82 pp.

Roedel, P. M. 1950. Notes on two species of sharks from Baja California. *California Fish and Game* 36 (3): 330–332.

Roedel, P. M. 1951. The brown shark, *Apristurus brunneus,* in California. *California Fish and Game* 37 (1): 61–63.

Roedel, P. M. 1953. Common ocean fishes of the California Coast. *California Department of Fish and Game, Fish Bulletin* 91: 1–184.

Roedel, P. M., and W. E. Ripley. 1950. California sharks and rays. *California Department of Natural Resources, Division of Fish and Game, Fish Bulletin* 75: 1–88.

Rojas, M. J. R. 2006. Reproducción y alimentación del tiburón enano *Mustelus dorsalis* (Pisces: Triakidae) en el Golfo de Nicoya, Costa Rica: Elementos para un manejo sostenible. *Revista de Biología Tropical* 54 (3): 861–871.

Romanov, E. V., and V. V. Samorov. 1994. On discoveries of the crocodile shark, *Pseudocarcharias kamoharai* (Pseudocarchariidae), in the equatorial Indian Ocean. *Journal of Ichthyology* 34 (4): 155–157.

Rosenblatt, R. H., and W. J. Baldwin. 1958. A review of the eastern Pacific sharks of the genus *Carcharhinus,* with a redescription of *C.* malpeloensi*s* (Fowler) and California records of *C. remotus* (Duméril). *California Fish and Game* 44 (2): 137–159.

Roule, L. 1912. Notice sur les Sélaciens conservés dans les collections du Musée Océanographique. *Bulletin de l'Institut Océanographique, Monaco* 243: 1–36.

Rountree, R. A., and K. W. Able. 1996. Seasonal abundance, growth, and foraging habits of juvenile smooth dogfish, *Mustelus canis,* in a New Jersey estuary. *Fishery Bulletin* 94: 522–534.

Rowat, D., and M. Gore. 2007. Regional scale horizontal and local scale vertical movements of whale sharks in the Indian Ocean off Seychelles. *Fisheries Research* 84: 32–40.

Ruíz A., C. L., and M. Ixquiac-Cabrera. 2000. *Evaluación del Potencial de Explotación del Recurso Tiburón en las Costas del Pacífico de Guatemala.* Guatemala: Fondo Nacional de Ciencia y Tecnología, Centro de Estudios del Mar y Acuicultura, Unidad Especial de Pesca y Acuicultura. 55 pp.

Rüppell, E. 1835–1840. *Neue Wirbelthiere zu der Fauna von Abyssinien Gehörig, Entdeckt und Beschrieben. Fishes des Rothen Meeres.* Frankfurt-am-Main: Commission bei Siegmund Schmerber. 148 pp.

Russo, R. A. 1975. Observations on the food habits of leopard sharks (*Triakis semifasciata*) and brown smoothhounds (*Mustelus henlei*). *California Fish and Game* 61 (2): 95–103.

Sadowsky, V. 1965. The hammerhead sharks of the littoral zone of São Paulo, Brazil, with the description of a new species. *Bulletin of Marine Science* 15 (1): 1–15.

Sadowsky, V. 1967a. Selachier aus dem Litoral von Sao Paulo, Brasilien. *Beiträge zur Neotropischen Fauna* 5 (2): 71–88.

Sadowsky, V. 1967b. The adult stage of the shark *Carcharhinus remotus* (Duméril, 1865). *Senckenbergiana Biologica* 48 (5/6): 327–334.

Sadowsky, V. 1970. First record of broad-snouted sevengilled shark from Cananéia, coast of Brazil. *Boletim do Instituto Oceanográfico,* São Paulo 18 (1): 33–35.

Sadowsky, V., A. F. de Amorim, and C. A. Arfelli. 1984. Second occurrence of *Odontaspis noronhai* (Maul, 1955). *Boletim do Instituto de Pesca* (São Paulo) 11: 69–79.

Sadowsky, V., A. F. de Amorim, and C. A. Arfelli. 1985. Record of unusual number of dwarf shark, *Squaliolus laticaudus,* off the south coast of Brazil. *Boletim do Instituto de Pesca,* São Paulo 12 (4): 45–50.

Sadowsky, V., C. A. Arfelli, A. F. Amorim, and O. B. E. Gadig. 1987. Primeiro registro e ocorrência de *Carcharhinus altimus* (Chondrichtyes, Selachii) no Atlântico Su1.14th Sociedade Brasiliera de Zoologia Congress, Juiz de Fora. Resumos, Abstract 83.

Saemundsson, B. 1922. Zoologiske Meddelelser fra Island. *Videnskabelige meddelelser fra Dansk Naturhistorisk Forening i Kjøbenhavn* 74: 159–201.

Saemundsson, B. 1949. Marine pisces. In *The Zoology of Iceland,* Vol. 4, Part 72, edited by Á. Friõriksson and S. L. Tuxen, pp. 1–150. Copenhagen: Ejnar Munksgaard.

Sánchez Roig, M. 1931. Peces Cubanos. *Revista de Agricultura, Comercio y Trabajo:* Year 13 (9), 11: 17–28.

Sano, O. 1966. Predation of shark on salmon. Symposium of Biological Studies on Tunas, Sharks and other Oceanic Fishes in the Pacific, Eleventh Pacific Science Congress, Tokyo. 7 pp.

Santana, F. M., and R. Lessa. 2004. Age determination and growth of the night shark (*Carcharhinus signatus*) off the northeastern Brazilian coast. *Fishery Bulletin* 102: 156–167.

Sarangdhar, P. N. 1943. Tiger shark—*Galeocerdo tigrinus* Muller and Henle. Feeding and breeding habits. *Journal of the Bombay Natural History Society* 44 (1): 102–110.

Sasaki, K., and T. Uyeno. 1987. *Squaliolus aliae,* a dalatiid shark distinct from *S. laticaudus. Japanese Journal of Ichthyology* 34 (3): 373–376.

Saunders, G. B., and E. Clark. 1962. Yellow-billed cuckoo in stomach of tiger shark. *Auk* 79: 118.

Saville, K. J., A. M. Lindley, E. G. Maries, J. C. Carrier, and H. L. Pratt, Jr. 2002. Multiple paternity in the nurse shark, *Ginglymostoma cirratum. Environmental Biology of Fishes* 63: 347–351.

Sazima, I., and R. L. Moura. 2000. Shark (*Carcharhinus perezi*), cleaned by the goby (*Elacatinus randalli*), at Fernando de Noronha Archipelago, western South Atlantic. *Copeia* 2000 (1): 297–299.

Scammon, R. E. 1911. *Normal Plates of the Development of Squalus acanthias.* Jena: Verlag Von Gustav Fischer. 140 pp.

Scattergood, L. W. 1948. Notes on some Gulf of Maine fishes. *Copeia* 1948 (2): 142–144.

Scattergood, L. W. 1949. Notes on the Maine shark fishery. *Copeia* 1949 (1): 69–71.

Scattergood, L. W. 1952. Notes on Gulf of Maine fishes in 1950 and 1951. *Copeia* 1952 (3): 205–206.

Scattergood, L. W. 1962a. White sharks, *Carcharodon carcharias,* in Maine, 1959–1960. *Copeia* 1962 (2): 446–447.

Scattergood, L. W. 1962b. First record of mako, *Isurus oxyrinchus,* in Maine waters. *Copeia* 1962 (2): 462.

Schlernitzauer, D. A., and P. W. Gilbert. 1966. Placentation and associated aspects of gestation in the bonnethead shark, *Sphyrna tiburo. Journal of Morphology* 120 (3): 219–230.

Schmidt, T. W. 1986. Food of young juvenile lemon sharks, *Negaprion brevirostris* (Poey), near Sandy Key, western Florida Bay. *Florida Scientist* 49 (1): 7–10.

Scholander, P. F. 1958. Counter current exchange a principle in biology. *Hvalrådets Skrifter, Scientific Results of Marine Biological Research* (Oslo) 44: 1–24.

Scholfield, A. F. 1971. *Aelian on the Characteristics of Animals.* Cambridge: Harvard University Press. 3 vols.

Scholl, J. P. 1983. Skull fragments of the California sea lion (*Zalophus californianus*) in stomach of a white shark (*Carcharodon carcharias*). *Journal of Mammology* 64 (2): 332.

Schott, J. W. 1964. Chromatic patterns of the leopard shark, *Triakis semifasciata* Girard. *California Fish and Game* 50 (3): 207–214.

Schroeder, W. C. 1938. Records of *Carcharodon carcharias* (Linnaeus) and *Pseudopriacanthus altus* (Gill) from the Gulf of Maine, summer of 1937. *Copeia* 1938 (1): 46.

Schuck, H. A., and J. R. Clark. 1951. Record of a white-tipped shark, *Carcharhinus longimanus,* from the northwestern Atlantic. *Copeia* 1951 (2): 172.

Schwartz, F. J. 1984a. Occurrence, abundance, and biology of the blacknose shark, *Carcharhinus acronotus* in North Carolina. *Northeast Gulf Science* 7 (1): 29–47.

Schwartz, F. J. 1984b. *Sharks, Sawfish, Skates, and Rays of the Carolinas.* Morehead City, North Carolina: Institute of Marine Sciences. 101 pp.

Schwartz, F. J. 1993. A North Carolina capture of the bramble shark, *Echinorhinus brucus,* family Echinorhinidae, the fourth in the western Atlantic. *Journal of the Elisha Mitchell Scientific Society* 109 (3): 158–162.

Schwartz, F. J., and G. H. Burgess. 1975. *Sharks of North Carolina and Adjacent Waters.* Raleigh, North Carolina Department of Natural and Economic Resources. 57 pp.

Schwartz, F. J. 2003. *Sharks, Skates, and Rays of the Carolinas.* Chapel Hill: University of North Carolina Press. 161 pp.

Schultz, L., E. S. Herald, E. A. Lachner, A. D. Welander, L. P. Woods. 1953. Fishes of the Marshall and Marianas Islands, Vol. 1. *United States National Museum Bulletin* 202: 1–685.

Sciarrotta, T. C., and D. R. Nelson. 1977. Diel behavior of the blue shark, *Prionace glauca,* near Santa Catalina Island, California. *Fishery Bulletin* 75: 519–528.

Scofield, N. B. 1920. Commercial fishery notes. Sleeper shark captured. *California Fish and Game* 6 (2): 80.

Scofield, W. L. 1941. Occurrence of the tiger shark in California. *California Fish and Game* 27 (4): 271–272.

Scoresby, W. 1820 (Reprint 1969). *An Account of the Arctic Regions with a History and Description of the Northern Whale-Fishery,* Vol. 2. New York: Augustus M. Kelley. 574 pp.

Scott, W. B., and M. G. Scott. 1988. *Atlantic Fishes of Canada.* Toronto: University of Toronto Press. 731 pp.

Scott, W. B., and S. N. Tibbo. 1968. Food and feeding habits of swordfish, *Xiphias gladius,* in the western North Atlantic. *Journal of the Fisheries Research Board of Canada* 25 (5): 903–919.

Sedberry, G. R., H. S. Meister, and J. K. Loefer. 2007. First *in-situ* observation of a frill shark, *Chlamydoselachus anguineus,* and record for the western North Atlantic. *Journal of the North Carolina Academy of Sciences* 123 (3): 127–132.

Seigel, J. A. 1978. Revision of the dalatiid sharks genus *Squaliolus:* Anatomy systematics, ecology. *Copeia* 1978 (4): 602–614.

Seigel, J. A. 1985. The scalloped hammerhead, *Sphyrna lewini,* in coastal southern California waters: Three records including the first reported juvenile. *California Fish and Game* 71 (3): 189–190.

Seigel, J. A., and L. J. V. Compagno. 1986. New records of the ragged-tooth shark, *Odontaspis ferox,* from California waters. *California Fish and Game* 72 (3): 172–176.

Seigel, J. A., D. J. Long, J. M. Rounds, and J. Hernandez. 1995. The tiger shark, *Galeocerdo cuvier,* in coastal southern California waters. *California Fish and Game* 81 (4): 163–166.

Seigel, J. A., T. W. Pietsch, B. H. Robison, and T. Abe. 1977. *Squaliolus sarmenti* and *S. alii,* synonyms of the dwarf deepsea shark, *Squaliolus laticaudus. Copeia* 1977 (4): 788–791.

Seki, T., T. Taniuchi, H. Nakano, and M. Shimizu. 1998. Age, growth and reproduction of the oceanic whitetip shark from the Pacific Ocean. *Fisheries Science* 64 (1): 14–20.

Séret, B. 1995. Première capture d'un requin grande gueule (Chondrichthyes, Megachasmidae) dans l'Atlantique, au large du Sénégal. *Cybium* 19 (4): 425–427.

Sport Fishing Institute. 1963. Rare cow shark. *SFI Bulletin* (Sport Fishing Institute, Washington, D.C.) 137: 7.

Shann, E. W. 1911. A description of the advanced embryonic stage of *Lamna cornubica.* Part 3. In *28th Annual Report of the Fishery Board for Scotland,* pp. 73–79.

Shann, E. W. 1923. The embryonic development of the porbeagle shark, *Lamna cornubica. Proceedings of the Zoological Society of London* 11: 161–171.

Shark News. 1998. Closure of Philippines whale shark fishery. 1998: 11.

Sheehan, T. F. 1998. First record of the ragged-tooth shark, *Odontaspis ferox,* off the U.S. Atlantic Coast. *Marine Fisheries Review* 60 (1): 33–34.

Sheldon, R. E. 1911. The sense of smell in selachians. *Journal of Experimental Zoölogy* 10 (1): 51–62.

Shen, S.-C., and W.-H. Ting. 1972. Ecological and morphological study on fish-fauna from the waters around Taiwan and its adjacent Islands. *Bulletin Institute of Zoology, Academia Sinica* 11 (1): 13–31.

Sherburne, S. W. 1973. Cell types, differential cell counts and blood cell measurements of a Portuguese shark, *Centroscymnus coelolepis,* captured at 700 fathoms. *Fishery Bulletin* 71: 435–439.

Shirai, S. 1992. *Squalean Phylogeny a New Framework of "Squaloid" Sharks and Related Taxa.* Sapporo: Hokkaido University Press. 151 pp.

Shirai, S., and K. Nakaya. 1992. Functional morphology of feeding apparatus of the cookie-cutter shark, *Isistius brasiliensis* (Elasmobranchii, Dalatiidae). *Zoological Science* 9 (4): 811–821.

Shirai, S., and H. Tachikawa. 1993. Taxonomic resolution of the *Etmopterus pusillus* species group (Elasmobranchii, Etmopteridae), with description of *E. bigelowi,* n. sp. *Copeia* 1993 (2): 483–495.

Sigalas, R. 1939–1940. Sur *Euprotomicrus sarmenti* Noronha. *Actes de la Société Linnéenne de Bordeaux* 91: 70–71.

Signeux, J. 1949. Notes paléoichthyologiques. *Bulletin du Muséum National d'Histoire Naturelle* (Paris), ser. 2, 21 (5): 633–635.

Silas, E. G. 1986. The whale shark (*Rhiniodon typus* Smith) in Indian coastal waters: Is the species endangered or vulnerable? *Marine Fisheries Information Service, Technical and Extension Series* (Cochin, India) 66: 1–19.

Silas, E. G., and M. S. Rajagopalan. 1963. On a recent capture of a whale shark (*Rhincodon typus* Smith) at Tuticorin, with a note on information to be obtained on whale sharks from Indian waters. *Journal of the Marine Biological Association of India* 5 (1): 153–157.

Silas, E. G., and G. S. D. Selvaraj. 1972. Descriptions of the adult and embryo of the bramble shark *Echinorhinus brucus* (Bonnaterre) obtained from the continental slope of India. *Journal of the Marine Biological Association of India* 14 (1): 395–401.

Simpfendorfer, C. 1992. Biology of tiger sharks (*Galeocerdo cuvier*) caught by the Queensland shark meshing program off Townsville, Australia. *Australian Journal of Marine and Freshwater Research* 43: 33–43.

Simpfendorfer, C. A., G. G. Freitas, T. R. Wiley, and M. R. Heupel. 2005. Distribution and habitat partitioning of immature bull sharks (*Carcharhinus leucas*) in a southwest Florida estuary. *Estuaries* 28 (1): 78–85.

Simpfendorfer, C. A., A. B. Goodreid, and R. B. McAuley. 2001. Size, sex and geographic variation in the diet of the tiger shark, *Galeocerdo cuvier,* from Western Australian waters. *Environmental Biology of Fishes* 61: 37–46.

Sims, D. W., A. M. Fox, and D. A. Merrett. 1997. Basking shark occurrence off south-west England in relation to zooplankton abundance. *Journal of Fish Biology* 51: 436–440.

Sims, D. W., and D. A. Merrett. 1997. Determination of zooplankton characteristics in the presence of surface feeding basking sharks *Cetorhinus maximus. Marine Ecology Progress Series* 158: 297–302.

Sims, D. W., and V. A. Quayle. 1998. Selective foraging behaviour of basking sharks on zooplankton in a small-scale front. *Nature* 393: 460–464.

Sims, D. W., E. J. Southall, V. A. Quayle, and A. M. Fox. 2000. Annual social behaviour of basking sharks associated with coastal front areas. *Proceedings of the Royal Society of London B* 267: 1897–1904.

Skomal, G. B., and L. J. Natanson. 2003. Age and growth of the blue shark (*Prionace glauca*) in the North Atlantic Ocean. *Fishery Bulletin* 101: 627–639.

Skomal, G. B., S. I. Zeeman, J. H. Chisholm, E. L. Summers, H. J. Walsh, K. W. McMahon, and S. R. Thorrold. 2009. Transequatorial migrations by basking sharks in the western Atlantic Ocean. *Current Biology* 19: 1–4; doi 10.1016/j.cub.2009.04/019.

Smale, M. J. 1991. Occurrence and feeding of three shark species, *Carcharhinus brachyurus, C. obscurus, Sphyrna zygaena,* on the eastern Cape Coast of South Africa. *South African Journal of Marine Science* 11: 31–42.

Sminkey, T. R., and J. A. Musick. 1996. Demographic analysis of the sandbar shark, *Carcharhinus plumbeus,* in the western North Atlantic. *Fishery Bulletin* 94: 341–347.

Sminkey, T. R., and C. R. Tabit. 1992. Reproductive biology of the chain dogfish, *Scyliorhinus retifer,* from the Mid-Atlantic Bight. *Copeia* 1992 (1): 251–253.

Smith, A. 1828. Description of new, or imperfectly known objects of the animal kingdom found in the south of Africa. *South African Commercial Advertiser* 3 (145): 2.

Smith, A. 1849. *Illustrations of the Zoology of South Africa; Consisting Chiefly of Figures and Descriptions of the Objects of Natural History Collected During an Expedition into the Interior of South Africa, in the Years 1834, 1835, and 1836. Pisces.* London: Smith, Elder and Co. 77 unnumbered pages.

Smith, B. G. 1937. The anatomy of the frilled shark *Chamydoselachus anguineus* Garman. In *The Bashford Dean Memorial Volume Archaic Fishes.* Part 2, article 6, edited by E. W. Gudger, pp. 333–505. New York: American Museum of Natural History.

Smith, B. G. 1942. The Heterodontid sharks: Their natural history and the external development of *Heterodontus japonicus* based on notes and drawings by Bashford Dean. In *The Bashford Dean Memorial Volume Archaic Fishes,* Part 2, Article 8, edited by E. W. Gudger, pp. 647–770. New York: American Museum of Natural History.

Smith, C. R., and A. R. Baco. 2003. Ecology of whale falls at the deep-sea floor. *Oceanography and Marine Biology: An Annual Review* 41: 311–354.

Smith, H. M. 1898. The fishes found in the vicinity of Woods Hole. *Bulletin of the United States Fish Commission* 17: 85–111.

Smith, H. M. 1907. *The Fishes of North Carolina.* Raleigh: E. M. Uzzell & Co. 449 pp.

Smith, H. M. 1912. The squaloid sharks of the Philippine Archipelago, with descriptions of new genera and species. *Proceedings of the United States National Museum* 41 (1877): 677–685.

Smith, H. M. 1922. Second occurrence of *Squatina* in New England. *Copeia* 1922 (106): 33.

Smith, J. L. B. 1936. Two interesting new fishes from South Africa. *Transactions of the Royal Society of South Africa* 24 (pt. 1, September 18, 1935): 1–6.

Smith, J. L. B. 1949. *The Sea Fishes of Southern Africa.* South Africa: Central News Agency, Ltd. 550 pp.

Smith, J. L. B. 1958. White-tip sharks. *Veld & Vlei* July: 16–25.

Smith, J. L. B. 1967. The lizard shark, *Chlamydoselachus anguineus* Garman, in South Africa. Occasional Paper no. 10, pp. 105–114. Grahamstown, South Africa: Department of Ichthyology, Rhodes University.

Smith, J. V. C. 1833. *Natural History of the Fishes of Massachusetts, Embracing a Practical Essay on Angling.* Facsimile, New York: Freshet Press, 1970. 399 pp.

Smith, R. L., and D. Rhodes. 1983. Body temperature of the salmon shark, *Lamna ditropis. Journal of the Marine Biological Association of the United Kingdom* 63: 243–244.

Smith, S. E. 2005. Leopard shark mating observed off La Jolla, California. *California Fish and Game* 91 (1): 128–135.

Smith, S. E., R. A. Mitchell, and D. Fuller. 2003. Age-validation of a leopard shark (*Triakis semifasciata*) recaptured after 20 years. *Fishery Bulletin* 101: 194–198.

Smith-Vaniz, W. F., B. B. Collette, and B. E. Luckhurst. 1999. *Fishes of Bermuda: History, Zoogeography, Annotated Checklist and Identification Keys.* American Society of Ichthyologists and Herpetologists Special Publication no. 4. Lawrence, Kansas: Allen Press. 424 pp.

Snelson, F. F., Jr., T. J. Mulligan, and S. E. Williams. 1984. Food habits, occurrence, and population structure of the bull shark, *Carcharhinus leucas,* in Florida coastal lagoons. *Bulletin of Marine Science* 34 (1): 71–80.

Snodgrass, R. E., and E. Heller. 1905. Papers from the Hopkins-Stanford Galapagos expedition, 1898–1899. 17. Shore fishes of the Revillagigedo, Clipperton, Cocos and Galapagos Islands. *Proceedings of the Washington Academy of Sciences* 6: 333–427.

Snyder, J. O. 1904. A catalogue of the shore fishes collected by the steamer *Albatross* about the Hawaiian Islands in 1902. In *U.S. Fish Commission Bulletin for 1902,* pp. 513–538. Washington, D.C.: Government Printing Office.

Soldatov, V. K., and G. J. Lindberg. 1930. A review of the fishes of the seas of the Far East. *Bulletins of the Pacific Scientific Fisheries Institute* (Vladivostock) 5 (5): 1–576.

Soto, J. M. R. 2000. Segundo registro do tubarão-dente-de-agulha *Carcharhinus isodon* (Valenciennes, 1839) (Chondrichthyes: Carcharhinidae) no Atlântico Sul ocidental. *Comunicacoes do Museu de Ciencias e Tecnologia da PUCRS* (Porto Alegre), Série Zoologia, 13: 105–108.

Soto, J. M. R., W. Nisa-Castro-Neto, and R. C. Mazzoleni. 1995. Novos registros de *Echinorhinus brucus, Centroscymnus cryptacanthus e Somniosus microcephalus* (Squaliformes: Echinorhinidae, Squalidae), com a ampliação de suas distribuições. p. 66. In *Resumos da 7a Reunião do Grupo de Trabalho sobre Pesca e Pesquisa de Tubarões e Raias no Brasil.* Rio Grande: Fundação Universidade do Rio Grande. 79 pp.

Southwell. T. 1913. Notes on the fauna of the Ceylon pearl banks. *Ceylon Administration Reports,* 1912–1913. Part 4. Education, Science, and Art. Report of the Marine Biologist, pp. E 32–E 49. Colombo: Government Record Office.

Spalding, D. J. 1964. *Comparative Feeding Habits of the Fur Seal, Sea Lion, and Harbour Seal on the British Columbia Coast.* Bulletin 146. Ottawa: Fisheries Research Board of Canada. 52 pp.

Springer, S. 1938. Notes on the sharks of Florida. *Proceedings of the Florida Academy of Sciences* 3: 9–41.

Springer, S. 1939a. The great white shark, *Carcharodon carcharias* (Linnaeus), in Florida waters. *Copeia* 1939 (2): 114–115.

Springer, S. 1939b. Two new Atlantic species of dog sharks, with a key to the species of *Mustelus. Proceedings of the United States National Museum* 86 (3058): 461–468.

Springer, S. 1940a. A new species of hammerhead shark of the genus *Sphyrna. Proceedings of the Florida Academy of Sciences* 5: 46–52.

Springer, S. 1940b. Three new sharks of the genus *Sphyrna* from the Pacific coast of tropical America. *Stanford Ichthyological Bulletin* 1 (5): 161–169.

Springer, S. 1943. A second species of thresher shark from Florida. *Copeia* 1943 (1): 54–55.

Springer, S. 1944. *Sphyrna bigelowi,* a new hammerhead shark from off the Atlantic coast of South America, with notes on *Sphyrna mokarran* from New South Wales. *Journal of the Washington Academy of Sciences* 34 (8): 274–276.

Springer, S. 1946. A collection of fishes from the stomachs of sharks taken off Salerno, Florida. *Copeia* 1946 (3): 174–175.

Springer, S. 1948. Oviphagous embryos of the sand shark, *Carcharias taurus. Copeia* 1948 (3): 153–157.

Springer, S. 1950a. A revision of North American sharks allied to the genus *Carcharhinus. American Museum Novitates* 1451: 1–13.

Springer, S. 1950b. Natural history notes on the lemon shark, *Negaprion brevirostris. Texas Journal of Science* 3: 349–359.

Springer, S. 1952. The effect of fluctuations in the availability of sharks on a shark fishery. In *Proceedings of the Gulf and Caribbean Fisheries Institute, Fourth Annual Session, Miami Beach, November 1951,* pp. 140–145. Coral Gables, Fl.: Univeristy of Miami.

Springer, S. 1957a. Night sharks, *Hypoprion,* from the Gulf of Mexico and the Straits of Florida. *Copeia* 1957 (2): 160.

Springer, S. 1957b. Some observations on the behavior of schools of fishes in the Gulf of Mexico and adjacent waters. *Ecology* 38 (1): 166–171.

Springer, S. 1959. A new shark of the family Squalidae from the Carolina continental slope. *Copeia* 1959 (1): 30–33.

Springer, S. 1960. Natural history of the sandbar shark *Eulamia milberti.* Fishery Bulletin 178. *Fishery Bulletin of the Fish and Wildlife Service* 61: 1–38.

Springer, S. 1963. Field observations on large sharks of the Florida-Caribbean region. In *Sharks and Survival,* edited by P. W. Gilbert, pp. 95–113. Boston: D. C. Heath.

Springer, S. 1966. A review of western Atlantic cat sharks, Scyliorhinidae, with descriptions of a new genus and five new species. *Fishery Bulletin* 65: 581–624.

Springer, S. 1967. Social organization of shark populations. In *Sharks, Skates and Rays,* edited by P. W. Gilbert, R. F. Mathewson, and D. P. Rall, pp. 149–174. Baltimore: Johns Hopkins Press.

Springer, S. 1979a. A Revision of the Catsharks, Family Scyliorhinidae. NOAA Technical Report NMFS Circular 422. U.S. Department of Commerce. 152 pp.

Springer, S. 1979b. Report on shark fishing in the western central Atlantic. WECAF Reports no. 3. United Nations Development Programme. Panama: Food and Agriculture Organization of the United Nations. 39 pp.

Springer, S., and H. R. Bullis, Jr. 1956. *Collections by the* Oregon *in the Gulf of Mexico. U.S. Fish and Wildlife Service.* Special Scientific Report-Fisheries no. 196. Washington, D.C.: U.S. Department of the Interior. 134 pp.

Springer, S., and H. R. Bullis, Jr. 1960. A new species of sawshark, *Pristiophorus schroederi,* from the Bahamas. *Bulletin of Marine Science of the Gulf and Caribbean* 10 (2): 241–254.

Springer, S., and P. W. Gilbert. 1976. The basking shark, *Cetorhinus maximus,* from Florida and California, with comments on its biology and systematics. *Copeia* 1976 (1): 47–54.

Springer, S., and V. Sadowsky. 1970. Subspecies of the western Atlantic cat shark, *Scyliorhinus retifer. Proceedings of the Biological Society of Washington* 83 (7): 83–98.

Springer, S., and M. H. Wagner. 1966. *Galeus piperatus,* a new shark of the family Scyliorhinidae from the Gulf of California. *Contributions in Science* (Los Angeles County Museum) 110: 1–9.

Springer, S., and R. A. Waller. 1969. *Hexanchus vitulus,* a new sixgill shark from the Bahamas. *Bulletin of Marine Science* 19 (1): 159–174.

Springer, V. G. 1964. A revision of the carcharhinid shark genera *Scoliodon, Loxodon,* and *Rhizoprionodon. Proceedings of the United States National Museum* 115 (3493): 559–632.

Springer, V. G., and J. A. F. Garrick. 1964. A survey of vertebral numbers in sharks. *Proceedings of the United States National Museum* 116 (3496): 73–96.

Springer, V. G., and K. D. Woodburn. 1960. An Ecological Study of the Fishes of the Tampa Bay Area. Professional Papers Series, no. 1. St. Petersburg: Florida State Board of Conservation, Marine Laboratory. 15 pp.

Squire, J. L., Jr. 1967. Observations of basking sharks and great white sharks in Monterey Bay, 1948–50. *Copeia* 1967 (1): 247–250.

Squire, J. L., Jr. 1990. Distribution and apparent abundance of the basking shark, *Cetorhinus maximus,* off the central and southern California coast, 1962–85. *Marine Fisheries Review* 52 (2): 8–11.

Standora, E. A., and D. R. Nelson. 1977. A telemetric study of the behavior of free-swimming Pacific angel sharks, *Squatina californica. Bulletin of the Southern California Academy of Sciences* 76 (3): 193–201.

Starks, E. C. 1917a. On the differential characters between *Mustelus henlei* and *Mustelus californicus. Copeia* 1917 (46): 61–63.

Starks, E. C. 1917b. The sharks of California. *California Fish and Game* 3 (4): 145–153.

Starks, E. C., and E. L. Morris. 1907. The marine fishes of southern California. *University of California Publications, Zoology* 3 (11): 159–251.

Stead, D. G. 1938. The "bronze whaler." An undescribed Australian shark. *Australian Naturalist* 10 (3): 98–105.

Steiner, P. A. 2002. A Comparative Study on the Activity Patterns of the Bull Shark, *Carcharhinus leucas,* and the Blacktip Shark, *Carcharhinus limbatus,* in the Ten Thousand Islands Estuary, Florida. Ph.D. diss. University of Basel. 115 pp.

Stevens, J. D. 1974. The occurrence and significance of tooth cuts on the blue shark (*Prionace glauca* L.) from British waters. *Journal of the Marine Biological Association of the United Kingdom* 54: 373–378.

Stevens, J. D. 1976. First results of shark tagging in the north-east Atlantic, 1972–1975. *Journal of the Marine Biological Association of the United Kingdom* 56: 929–937.

Stevens, J. D. 1983. Observations on reproduction in the shortfin mako *Isurus oxyrinchus. Copeia* 1983 (1): 126–130.

Stevens, J. D. 1984. Biological observations on sharks caught by sport fisherman off New South Wales. *Australian Journal of Marine and Freshwater Research* 35: 573–590.

Stevens, J. D. 1990. Further results from a tagging study of pelagic sharks in the north-east Atlantic. *Journal of the Marine Biological Association of the United Kingdom* 70: 707–720.

Stevens, J. D. 1993. The Status of Chondrichthyan Resources in the South West Pacific. Report of the South West Pacific Subgroup of the IUCN Shark Specialist Group. Tasmania, Australia: CSIRO Division of Fisheries. 39 pp.

Stevens, J. D., M. C. Dunning, and S. Machida. 1983. Occurrence of the porbeagle shark, *Lamna nasus,* in the Tasman Sea. *Japanese Journal of Ichthyology* 30 (3): 301–307.

Stevens, J. D., and J. M. Lyle. 1989. Biology of three hammerhead sharks (*Eusphyra blochii, Sphyrna mokarran* and *S. lewini*) from northern Australia. *Australian Journal of Marine and Freshwater Research* 40: 129–146.

Stevens, J. D., and K. J. McLoughlin. 1991. Distribution, size and sex composition, reproductive biology and diet of sharks from northern Australia. *Australian Journal of Marine and Freshwater Research* 42: 151–199.

Stevens, J. D., B. M. Norman, J. S. Gunn, and T. L. O. Davis. 1998. Movement and Behavioural Patterns of Whale Sharks at Ningaloo Reef: The Implications for tourism. National Ecotourism Final Report: Application ref. no. 30051. Australia: CSIRO Marine Research. 35 pp.

Stevens, J. D., and J. R. Paxton. 1985. A new record of the goblin shark, *Mitsukurina owstoni* (family Mitsukurinidae), from eastern Australia. *Proceedings of the Linnean Society of New South Wales* (1984) 108 (1): 37–45.

Stick, K. C., and L. Hreha. 1988. Summary of the 1986 and 1987 Washington/Oregon experimental thresher shark gill net fishery. Progress Report no. 266. Olympia, Wa.: State of Washington, Department of Fisheries. 57 pp.

Stillwell, C. E., and J. G. Casey. 1976. Observations on the bigeye thresher shark, *Alopias superciliosus,* in the western North Atlantic. *Fishery Bulletin* 74: 221–225.

Stillwell, C. E., and N. E. Kohler. 1982. Food, feeding habits and estimates of daily ration of the shortfin mako (*Isurus oxyrinchus*) in the northwest Atlantic. *Canadian Journal of Fisheries and Aquatic Sciences* 39 (3): 407–414.

Stillwell, C. E., and N. E. Kohler. 1993. Food habits of the sandbar shark, *Carcharhinus plumbeus,* off the U.S. northeast coast, with estimates of daily ration. *Fishery Bulletin* 91: 138–150.

Stokesbury, M. J. W., C. Harvey-Clark, J. Gallant, B. A. Block, and R. A. Myers. 2005. Movement and environmental preferences of Greenland sharks (*Somniosus microcephalus*) electronically tagged in the St. Lawrence Estuary, Canada. *Marine Biology* 148: 159–165.

Storer, D. H. 1839. Reports on the fishes, reptiles and birds of Massachusetts. In *Reports of the Commissioner on the Zoological and Botanical Survey of the State of Massachusetts* 1: 184–186. Boston: Dutton and Wentworth, State Printers.

Storer, D. H. 1846. *A Synopsis of the Fishes of North America.* Cambridge, Mass.: Metcalf and Company. 298 pp.

Stott, F. C. 1982. A note on catches of basking sharks, *Cetorhinus maximus* (Gunnerus), off Norway and their relation to possible migration paths. *Journal of Fish Biology* 21: 227–230.

Stott, F. C. 1984. The Growth of *Cetorhinus maximus* (Gunnerus)—A Reply to Criticism. CM 1984/H: 2. *Copenhagen:* International Council for the Exploration of the Sea. 4 pp.

Strasburg, D. W. 1958. Distribution, abundance, and habits of pelagic sharks in the central Pacific Ocean. Fishery Bulletin 138. *Fishery Bulletin of the Fish and Wildlife Service* 58: 335–361.

Strasburg, D. W. 1963. The diet and dentition of *Isistius brasiliensis,* with remarks on tooth replacement in other sharks. *Copeia* 1963 (1): 33–40.

Strong, W. R., Jr. 1989. Behavioral Ecology of Horn Sharks, *Heterodontus francisci,* at Santa Catalina Island, California, with Emphasis on Patterns of Space Utilization. Master's thesis, California State University, Long Beach. 252 pp.

Strong, W. R., Jr., F. F. Snelson, Jr., and S. H. Gruber. 1990. Hammerhead shark predation on stingrays: An observation of prey handling by *Sphyrna mokarran. Copeia* 1990 (3): 836–840.

Suda, A. 1953. Ecological study on the blue shark (*Prionace glauca* Linne). Translation of a research report in *South Seas Area Fisheries Research Laboratory,* report no. 26, supplement 1: 1–15.

Sulikowski, J. A., W. B. Driggers, III, T. S. Ford, R. K. Boonstra, and J. K. Carlson. 2007. Reproductive cycle of the blacknose shark *Carcharhinus acronotus* in the Gulf of Mexico. *Journal of Fish Biology* 70: 1–13.

Sund, O. 1943. Småstykker. Et brugdebarsel. *Naturen* 67: 285–286.

Svetlov, M. F. 1978. The porbeagle, *Lamna nasus,* in Antarctic waters. *Journal of Ichthyology* 18 (5): 850–851.

Swenander, G. 1907. Über die ernährung des embryos der *Lamna cornubica.* In *Zoologiska Studier,* Tollgnade Tullberg, pp. 283–288. Uppsala: Almqvist & Wiksells Boktryckeri-A.-B.

Tabb, D. C., and R. B. Manning. 1961. A checklist of the flora and fauna of northern Florida Bay and adjacent brackish waters of the Florida mainland collected during the period July, 1957 through September, 1960. *Bulletin of Marine Science of the Gulf and Caribbean* 11 (4): 552–649.

Takada, K., H. Hiruda, S. Wakisaka, S. Kudo, and K. Yano. 1997. A summary of the freezing, thawing, preservation, and display methodology used on a large megamouth shark, *Megachasma pelagios.* In *Biology of the Megamouth Shark,* edited by K. Yano, J. F. Morrissey, Y. Yabumoto, and K. Nakaya, pp. 11–19. Tokyo: Tokai University Press.

Talent, L. G. 1973. Albinism in embryo gray smooth-hound sharks, *Mustelus californicus,* from Elkhorn Slough, Monterey Bay, California. *Copeia* 1973 (3): 595–597.

Talent, L. G. 1976. Food habits of the leopard shark, *Triakis semifasciata,* in Elkhorn Slough, Monterey Bay, California. *California Fish and Game* 62 (4): 286–298.

Talent, L. G. 1985. The occurrence, seasonal distribution, and reproductive condition of elasmobranch fishes in Elkhorn Slough, California. *California Fish and Game* 71 (4): 210–219.

Tanaka, S. 1912. *Figures and Descriptions of the Fishes of Japan Including Riukiu Islands, Bonin Islands, Formosa, Kurile Islands, Korea, and Southern Sakhalin.* Tokyo: Kazamashobo. 557 pp.

Tanaka, S. 1998. A little sleeper shark, *Somniosus rostratus* (Risso, 1826), from Suruga Bay, Japan. *Report of Japanese Society for Elasmobranch Studies* 34: 20–23.

Tanaka, S., Y. Shiobara, S. Hioki, H. Abe, G. Nishi, K. Yano, and K. Suzuki. 1990. The reproductive biology of the frilled shark, *Chlamydoselachus anguineus,* from Suruga Bay, Japan. *Japanese Journal of Ichthyology* 37 (3): 273–291.

Tanaka, S., K. Teshima, and K. Mizue. 1975. Studies on sharks-10. Morphological and ecological study on the reproductive organs in male *Heptranchias perlo. Bulletin of the Faculty of Fisheries* (Nagasaki University) 40: 15–22.

Tanaka, S., K. Yano, and T. Ichihara. 1982. Notes on a Pacific sleeper shark, *Somniosus pacificus,* from Suruga Bay, Japan. *Bulletin of the Faculty of Fisheries* (Nagasaki University) 15: 345–358.

Taniuchi, T. 1975. Reef whitetip shark, *Triaenodon obesus,* from Japan. *Japanese Journal of Ichthyology* 22 (3): 167–170.

Taniuchi, T., and J. A. F. Garrick. 1986. A new species of *Scymnodalatias* from the southern oceans, and comments on other squaliform sharks. *Japanese Journal of Ichthyology* 33 (2): 119–134.

Taniuchi, T., H. Kobayashi, and T. Otake. 1984. Occurrence and reproductive mode of the false cat shark, *Pseudotriakis microdon,* in Japan. *Japanese Journal of Ichthyology* 31 (1): 88–92.

Taniuchi, T., and F. Yanagisawa. 1983. Occurrence of the prickly shark, *Echinorhinus cookei,* at Kumanonada, Japan. *Japanese Journal of Ichthyology* 29 (4): 465–468.

Taylor, G. 1994. *Whale Sharks the Giants of Ningaloo Reef.* Australia: Angus & Robertson. 176 pp.

Taylor, L. R., Jr. 1972a. A Revision of the Shark Family Heterodontidae (Heterodontiformes, Selachii). Ph.D. diss., University of California, San Diego. 176 pp.

Taylor, L. R., Jr. 1972b. *Apristurus kampae,* a new species of Scyliorhinid shark from the eastern Pacific Ocean. *Copeia* 1972 (1): 71–78.

Taylor, L. R., Jr., and J. L. Castro-Aguirre. 1972. *Heterodontus mexicanus,* a new horn shark from the Golfo de California. *Anales de la Escuela Nacional de Ciencias Biológicas de Mexico* 19: 123–143.

Taylor, L. R., Jr., L. J. V. Compagno, and P. J. Struhsaker. 1983. Megamouth-a new species, genus, and family of lamnoid shark (*Megachasma pelagios,* family Mega-chasmidae) from the Hawaiian Islands. *Proceedings of the California Academy of Sciences* 43 (8): 87–110.

Taylor, L. R., and G. Naftel. 1978. Preliminary investigations of shark predation on the Hawaiian monk seal at Pearl and Hermes Reef and French Frigate Shoals, Report no. MMC-77/07, PB-285 626. Washington, D.C.: U.S. Marine Mammal Commission. 34 pp.

Templeman, W. 1944. The life-history of the spiny dogfish (*Squalus acanthias*) and the vitamin A values of dogfish liver oil. Research Bulletin no. 15 (Fisheries). St. John's, Newfoundland: Newfoundland Department of Natural Resources. 102 pp.

Templeman, W. 1954. Migrations of spiny dogfish tagged in Newfoundland waters. *Journal of the Fisheries Research Board of Canada* 11 (4): 351–354.

Templeman, W. 1958. Grand bank tagged dogfish moves to Iceland. Fisheries Research Board of Canada. *Atlantic Progress Report* 70: 28–30.

Templeman, W. 1963. Distribution of sharks in the Canadian Atlantic (with special reference to Newfoundland waters). Bulletin 140. Ottawa: Fisheries Research Board of Canada. 77 pp.

Templeman, W. 1966. Marine Resources of Newfoundland. Bulletin 154. Ottawa: Fisheries Research Board of Canada. 170 pp.

Templeman, W. 1976. Transatlantic migrations of spiny dogfish (*Squalus acanthias*). *Journal of the Fisheries Research Board of Canada* 33 (11): 2605–2609.

Teng, H.-T. 1959a. Studeis [*sic*] on the elasmobranch fishes from Formosa. Part 4—*Squaliolus alii,* a new species of deep sea squaloid shark from Tung-Kang, Formosa. *Taiwan Fisheries Research Institute (Keelung), Report-Laboratory of Fishery Biology, Taiwan* 8: 1–6.

Teng, H.-T. 1959b. Studies on the elasmobranch fishes from Formosa. Part 6—a new species of deep-sea shark (*Centrophorus niaukang*) from Formosa. *Taiwan Fisheries Research Institute (Chi-lung), Report-Laboratory of Fishery Biology, Taiwan* 9: 1–6.

Teng, H.-T. 1962. *Classification and Distribution of the Chondrichthyes of Taiwan* (in Japanese). Kyoto: Ogawa Press. 304 pp. Seen in translation by K. Nakaya and H.-C. Yang (only the descriptions of new species and subspecies are supplied). Laboratory of Marine Zoology, Faculty of Fisheries, Hokkaido University. 16 pp.

Tester, A. L. 1969. Cooperative Shark Research and Control Program: Final Report 1967–69. Honolulu: University of Hawaii. 47 pp.

TeWinkel, L. E. 1950. Notes on ovulation, ova, and early development in the smooth dogfish, *Mustelus canis. Biological Bulletin* 99 (3): 474–486.

Thomerson, J. E., T. B. Thorson, and R. L. Hempel. 1977. The bull shark, *Carcharhinus leucas,* from the upper Mississippi River near Alton, Illinois. *Copeia* 1977 (1): 166–168.

Thomson, D. A., L. T. Findley, and A. N. Kerstitch. 1979. *Reef Fishes of the Sea of Cortez.* New York: John Wiley & Sons. 302 pp.

Thompson, D. A. W. 1947. *A Glossary of Greek Fishes.* Oxford: Oxford University Press. 302 pp.

Thompson, W. 1856. Order 8. Plagiostomi. In *The Natural History of Ireland,* Vol. 4, pp. 247–264. London: Henry G. Bohn.

Thorne, E. 1916. Occurrence of ground sharks, *Carcharhinus,* in Great South Bay. *Copeia* 1916 (35): 69–71.

Thorne, E. 1928. Great South Bay as a shark nursery. *Bulletin of the New York Zoological Society* 21 (4): 114–115.

Thorson, T. B. 1971. Movement of bull sharks, *Carcharhinus leucas,* between Caribbean Sea and Lake Nicaragua demonstrated by tagging. *Copeia* 1971 (2): 336–338.

Thorson, T. B. 1972. The status of the bull shark, *Carcharhinus leucas,* in the Amazon River. *Copeia* 1972 (3): 601–605.

Thorson, T. B., ed. 1976. *Investigations of the Ichthyofauna of Nicaraguan Lakes.* Lincoln, Nebr.: School of Life Sciences, University of Nebraska. 663 pp.

Thorson, T. B., and E. J. Lacy, Jr. 1982. Age, growth rate and longevity of *Carcharhinus leucas* estimated from tagging and vertebral rings. *Copeia* 1982 (1): 110–116.

Tibbo, S. N. 1962. New records for occurrence of the white-tip shark, *Pterolamiops longimanus* (Poey), and the dolphin, *Coryphaena hippurus* L., in the Northwest Atlantic. *Journal of the Fisheries Research Board of Canada* 19 (3): 517–518.

Tibbo, S. N., and R. A. McKenzie. 1963. An occurrence of dusky sharks, *Carcharhinus obscurus,* (Lesueur) 1818, in the Northwest Atlantic. *Journal of the Fisheries Research Board of Canada* 20 (4): 1101–1102.

Tibbo, S. N., R. A. McKenzie, and W. B. Scott. 1963. An occurrence of mako sharks, *Isurus oxyrinchus,* Rafinesque 1810, in the Canadian Atlantic. *Journal of the Fisheries Research Board of Canada* 20 (5): 1352–1355.

Topsell, E. 1658 (Reprint 1967). *The History of Four-Footed Beasts and Serpents and Insects,* Vol. 1. London: Frank Cass & Co. 586 pp.

Torchio, M., and M. Michelangeli. 1971. Prima segnalazione in acque italiane di uno squalide del genere *Centroscymnus. Natura* 62 (3): 241–245.

Tortonese, E. 1937–1938. Intorno agli Squali del genere *Alopias. Bollettino dei Musei di Zoologia e di Anatomia Comparata della R. Università di Torino* 46: ser. 3, no. 78: 1–7.

Tortonese, E. 1950. A note on the hammerhead shark, *Sphyrna tudes Val.,* after a study of the types. *Annals and Magazine of Natural History,* ser. 12, 3: 1030–1033.

Tortonese, E. 1956. *Leptocardia, Ciclostomata, Selachii. Fauna d'Italia.* Bologna: Calderini. 331 pp.

Tortonese, E. 1969. The squaliforms of the Ligurian Sea: A revised list with notes. *Israel Journal of Zoology* 18: 233–236.

Tricas, T. C. 1979. Relationships of the blue shark, *Prionace glauca,* and its prey species near Santa Catalina Island, California. *Fishery Bulletin* 77: 175–182.

Tricas, T. C., and E. M. LeFeuvre. 1985. Mating in the reef white-tip shark *Triaenodon obesus. Marine Biology* 84 (3): 233–237.

Trono, R. B. 1996. Report on the Preliminary Investigation of the whale shark (*Rhincodon typus*) fishery in Bohol Sea, Philippines. Quezon City: WWF-Philippine Program.

Trunov, J. A. 1968. The whalefish (*Barbourisia rufa,* fam. Barbourisiidae) and the frilled shark (*Chlamydoselachus anguineus,* fam. Chlamydoselachidae) from southwest African coastal waters. *Journal of Ichthyology* 8 (1): 135–138.

Tucker, D. W. 1958. Supplementary note on the breeding of the blue shark *Prionace glauca* (L.). *Annals and Magazine of Natural History,* ser. 13, 1 (5): 315–316.

Tumokhin, I. G. 1980. Discovery of the frill shark, *Chlamydoselachus anguineus,* in the south-west Indian Ocean. *Journal of Ichthyology* 20 (1): 125–126.

Turnbull, S. D., and J. E. Randell. 2006. Rare occurrence of a *Rhincodon typus* (whale shark) in the Bay of Fundy, Canada. *Northeastern Naturalist* 13 (1): 57–58.

Turner, P. W. 1875. Observations on the spiny shark (*Echinorhinus spinosus*). *Journal of Anatomy and Physiology* 9: 297–301.

Uchida, S., M. Toda, and Y. Kamei. 1990. Reproduction of elasmobranchs in captivity. In *Elasmobranchs as Living Resources: Advances in the Biology, Ecology, Systematics, and the Status of the Fisheries,* edited by H. L. Pratt, Jr., S. H. Gruber, and T. Taniuchi, pp. 211–237. NOAA Technical Report NMFS 90. Springfield, Virginia: U.S. Department of Commerce.

Uchida, S., M. Toda, K. Teshima, and K. Yano. 1996. Pregnant white sharks and full-term embryos from Japan. In *Great White Sharks: The Biology of Carcharodon carcharias,* edited by A. P. Klimley and D. G. Ainley, pp. 139–155. San Diego: Academic Press.

Ugoretz, J. K., and J. A. Seigel. 1999. First eastern Pacific record of the goblin shark, *Mitsukurina owstoni* (Lamniformes: Mitsukurinidae). *California Fish and Game* 85 (3): 118–120.

Uribe Martínez, J. A. 1990. *Guía de campo para la identificación de especies de tiburones y cazones en la Sonda de Campeche.* Serie Documentos de Trabajo Año II no. 23. Mexico City: Instituto Nacional de la Pesca. 48 pp.

Urquhart, D. L. 1981. The North Pacific salmon shark. *Sea Frontiers* 27 (6): 361–363.

Uyeno, T., K. Matsuura, and E. Fujii, eds. 1983. *Fishes trawled off Suriname and French Guiana.* Tokyo: Japan Marine Fishery Resource Research Center. 519 pp.

Vaillant, L. 1888. Poissons. In *Expéditions Scientifiques du Travailleur et du Talisman Pendant les Années 1880, 1881, 1882, 1883.* Paris: G. Masson, Editeur. 406 pp.

Vaillant, L. 1901. Sur un griset (*Hexanchus griseus* L. Gm.) du Golfe de Gascogne. *Bulletin du Muséum d'Histoire Naturelle* 7: 202–204.

Valenciennes, M. A. 1822. Sur le sous-genre Marteau, *Zygaena. Mémoires du Muséum d'Histoire Naturelle* 9: 222–228.

Van Deinse, A. B., and M. J. Adriani. 1953. On the absence of gill rakers in specimens of the basking shark, *Cetorhinus maximus* (Gunner). *Zoologische Mededelingen* 31 (27): 307–310.

Van Dykhuizen, G., H. F. Mollet, and J. M. Ezcurra. 1998. Homing behavior of a sevengill shark released from the Monterey Bay Aquarium. *California Fish and Game* 84 (4): 180–181.

Vas, P. 1990. The abundance of the blue shark, *Prionace glauca,* in the western English Channel. *Environmental Biology of Fishes* 29: 209–225.

Veríssimo, A., L. Gordo, and I. Figueiredo. 2003. Reproductive biology and embryonic development of *Centroscymnus coelolepis* in Portuguese mainland waters. *ICES Journal of Marine Science* 60: 1335–1341.

Villavicencio-Garayzar, C. J. 1996. The ragged-tooth shark, *Odontaspis ferox* (Risso, 1810), in the Gulf of California. *California Fish and Game* 82 (4): 195–196.

Vladykov, V. D. 1933. Biological and oceanographic conditions in Hudson Bay. 9. Fishes from the Hudson Bay region (except the Coregonidae). *Biological Board of Canada,* ser. A, general, 29 (2): 13–61.

Vladykov, V. D. 1935. Some unreported and rare fishes for the coast of Nova Scotia. *Proceedings of the Nova Scotian Institute of Science* 19 (pt. 1): 1–8.

Vovcsko, J. 1995. West Coast dogfish find new respect. *National Fisherman* 76 (6): 34–36.

Walford, L. A. 1935. The sharks and rays of California. Fish Bulletin no. 45. Sacramento, Ca.: Division of Fish and Game of California, Bureau of Commercial Fisheries. 66 pp.

Walter, J. P., and D. A. Ebert. 1991. Preliminary estimates of age of the bronze whaler *Carcharhinus brachyurus* (Chondrichthyes: Carcharhinidae) from southern Africa, with a review of some life history parameters. *South African Journal of Marine Science* 10: 37–44.

Wass, R. C. 1973. Size, growth, and reproduction of the sandbar shark, *Carcharhinus milberti,* in Hawaii. *Pacific Science* 27 (4): 305–318.

Watkins, A. 1958. *The Sea My Hunting Ground.* New York: St. Martin's Press. 249 pp.

Webber, J. D., and J. J. Cech, Jr. 1998. Nondestructive diet analysis of the leopard shark from two sites in Tomales Bay, California. *California Fish and Game* 84 (1): 18–24.

Weng, K. C., and B. A. Block. 2004. Diel vertical migration of the bigeye thresher shark (*Alopias superciliosus*), a species possessing orbital *retia mirabilia. Fishery Bulletin* 102: 221–229.

Weng, K. C., P. C. Castilho, J. M. Morrissette, A. M. Landeira-Fernandez, D. B. Holts, R. J. Schallert, K. J. Goldman, and B. A. Block. 2005. Satellite tagging and cardiac physiology reveal niche expansion in salmon sharks. *Science* 310: 104–106.

Went, A. E. J., and S. Ó Súilleabháin. 1967. Fishing for the sun-fish or basking shark in Irish waters. *Proceedings of the Royal Irish Academy* 65 (C, 5): 91–115.

Westrheim, S. J. 1950. The 1949 soupfin shark fishery of Oregon. *Fish Commission Research Briefs* (Portland, Oregon) 3 (1): 39–49.

Wetherbee, B. M., G. L. Crow, and C. G. Lowe. 1996. Biology of the Galapagos shark, *Carcharhinus galapagensis,* in Hawai'i. *Environmental Biology of Fishes* 45: 299–310.

Wheeler, A. 1962. New records for distribution of the frilled shark. *Nature* 196 (4855): 689–690.

Wheeler, A. 1969. *The Fishes of the British Isles and North-West Europe.* East Lansing: Michigan State University Press. 613 pp.

Wheeler, A., and R. W. Blacker. 1972. Rare and little-known fishes in British seas in 1968 and 1969. *Journal of Fish Biology* 4: 141–170.

Wheeler, J. F. G. 1962. Notes on the three common species of sharks in the Mauritius-Seychelles area. *Proceedings of the Royal Society of Arts and Sciences of Mauritius* 2 (2): 146–160.

Whitley, G. P. 1929. Studies in ichthyology, no. 3. *Records of the Australian Museum* 17 (3): 101–143.

Whitley, G. P. 1930. The teeth of fishes. *Australian Museum Magazine* 4 (3): 92–99.

Whitley, G. P. 1931a. New names for Australian fishes. *Australian Zoologist* 6: 310–334.

Whitley, G. P. 1931b. Studies in ichthyology, no. 5. *Records of the Australian Museum* 18 (4): 138–158.

Whitley, G. P. 1934. Notes on some Australian sharks. *Memoirs of the Queensland Museum* 10 (4): 180–200.

Whitley, G. P. 1937. Studies in ichthyology, no. 10. *Records of the Australian Museum* 20 (1): 3–24.

Whitley, G. P. 1940. *The Fishes of Australia, Part 1. The Sharks, Rays, Devil-Fish, and other Primitive Fishes of Australia and New Zealand.* Sydney: Royal Zoological Society of New South Wales. 280 pp.

Whitley, G. P. 1943. Ichthyological descriptions and notes. *Proceedings of the Linnean Society of New South Wales* 68 (3, 4): 114–144.

Whitley, G. P. 1944. New sharks and fishes from western Australia. *Australian Zoologist* 10 (pt. 3): 252–273.

Whitley, G. P. 1950a. A new shark from north-western Australia. *Western Australian Naturalist* 2 (5): 100–105.

Whitley, G. P. 1950b. Studies in ichthyology, no. 14. *Records of the Australian Museum* 22 (3): 234–245.

Whitley, G. P. 1968. A check-list of the fishes recorded from the New Zealand region. *Australian Zoologist* 15 (1): 1–102.

Whitney, N. M., H. L. Pratt Jr., and J. C. Carrier. 2004. Group courtship, mating behaviour and siphon sac function in the whitetip reef shark, *Triaenodon obesus. Animal Behavior* 68 (6): 1435–1442.

Widder, E. A. 1998. A predatory use of counterillumination in the squaloid shark, *Isistius brasiliensis. Environmental Biology of Fishes* 53: 267–273.

Wiley, T. R. and C. A. Simpfendorfer. 2007. Occurrence of *Carcharhinus isodon* (Finetooth shark) in Florida Bay. *Southeastern Naturalist* 6 (1): 183–186.

Wilimovsky, N. J. 1954. List of the fishes of Alaska. *Stanford Ichthyological Bulletin* 4 (5): 279–294.

Williamson, G. R. 1963. Common porpoise from the stomach of a Greenland shark. *Journal of the Fisheries Research Board of Canada* 20 (4): 1085–1086.

Wilson, D. E., and G. W. Hughes. 1978. The first record of the brown cat shark, *Apristurus brunneus* (Gilbert, 1891) from Alaskan waters. *Syesis* 11: 283.

Wintner, S. P. 2000. Preliminary study of vertebral growth rings in the whale shark, *Rhincodon typus,* from the east coast of South Africa. *Environmental Biology of Fishes* 59: 441–451.

Wintner, S. P., and G. Cliff. 1999. Age and growth determination of the white shark, *Carcharodon carcharias,* from the east coast of South Africa. *Fishery Bulletin* 97: 153–169.

Wintner, S. P., S. F. J. Dudley, N. Kistnasamy, and B. Everett. 2002. Age and growth estimates for the Zambezi shark, *Carcharhinus leucas,* from the east coast of South Africa. *Marine and Freshwater Research* 53: 557–566.

Witzell, W. N. 1987. Selective predation on large cheloniid sea turtles by tiger sharks (*Galeocerdo cuvier*). *Japanese Journal of Herpetology* 12 (1): 22–29.

Wolfson, F. H. 1983. Records of seven juveniles of the whale shark, *Rhiniodon typus. Journal of Fish Biology* 22: 647–655.

Wolfson, F. H. 1986. Occurrences of the whale shark, *Rhincodon typus* Smith. In *Indo-Pacific Fish Biology: Proceedings of the Second International Conference on Indo-Pacific Fishes,* edited by T. Uyeno, R. Arai, T. Taniuchi, and K. Matsuura, pp. 208–226. Tokyo: Ichthyological Society of Japan.

Wolfson, F. H., and G. N. di Sciara. 1981. The whale shark, *Rhiniodon typus* Smith, 1828: An annotated bibliography. *Atti della Societá Italiana di Scienze Naturali e del Museo Civico di Storia Naturale en Milano* 122 (3–4): 171–203.

Woodward, A. S. 1899. Note on *Scapanorhynchus,* a Cretaceous shark apparently surviving in Japanese seas. *The Annals and Magazine of Natural History,* Zoology, Botany and Geology Series 7, 3: 487–489.

Woodward, S. A. S. 1942. Some new and little-known upper Cretaceous fishes from Mount Lebanon. *The Annals and Magazine of Natural History* 11 (56): 537–568.

Wourms, J. P. 1977. Reproduction and development in chondrichichthyan fishes. *American Zoologist* 17: 379–410.

Wourms, J. P. 1981. Viviparity: The maternal-fetal relationship in fishes. *American Zoologist* 21: 473–515.

Wourms, J. P., B. D. Grove, and J. Lombardi. 1988. The maternal-embryonic relationship in viviparous fishes. In *Fish Physiology,* Vol. 11B, edited by W. S. Hoar and D. J. Randall, pp. 1–134. New York: Academic Press.

Wright, E. P. 1870. Six months at the Seychelles: A letter written to A. Searle Hart . . . for the information of the Board of Trinity College, Dublin. In *Spicilegia Biologica: Or Papers on Zoological and Botanical Subjects,* pp. 63–65.

Yabumoto, Y., M. Goto, K. Yano, and T. Uyeno. 1997. Dentition of a female megamouth, *Megachasma pelagios,* collected from Hakata Bay, Japan. In *Biology of the Megamouth Shark,* edited by K. Yano, J. F. Morrissey, Y. Yabumoto, and K. Nakaya, pp. 63–75. Tokyo: Tokai University Press.

Yamaguchi, R., and K. Nakaya. 1997. Fukuoka megamouth, a probable victim of the cookie-cutter shark. In *Biology of the Megamouth Shark,* edited by K. Yano, J. F. Morrissey, Y. Yabumoto, and K. Nakaya, pp. 171–175. Tokyo: Tokai University Press.

Yang, H.-C., and H. Horike. 1998. Note on the death of a baby whale shark from Taiwan, reared in a Japanese aquarium. *Fisherman Magazine* (Taiwan Provincial Fisherman's Association) 252: 44.

Yang, M.-S., and B. N. Page. 1999. Diet of the Pacific sleeper shark, *Somniosus pacificus,* in the Gulf of Alaska. *Fishery Bulletin* 97: 406–409.

Yano, K. 1991. Catch distribution, stomach contents and size at maturity of two squaloid sharks, *Deania calceus* and *D. crepidalbus,* from the southeast Atlantic off Namibia. *Bulletin of the Japanese Society of Fisheries Oceanography* 55 (3): 189–196.

Yano, K. 1992. Comments on the reproductive mode of the false cat shark *Pseudotriakis microdon. Copeia* 1992 (2): 460–468.

Yano, K. 1995a. *Fishes Collected by the R/V Shinkai Maru around Greenland.* Edited by O. Okamura, K. Amaoka, M. Takeda, K. Yano, K. Okada, and S. Chinuki. Tokyo, Japan: Marine Fishery Resources Research Center. 304 pp.

Yano, K. 1995b. Reproductive biology of the black dogfish, *Centroscyllium fabricii,* collected from waters off western Greenland. *Journal of the Marine Biological Association of the United Kingdom* 75: 285–310.

Yano, K., J. F. Morrissey, Y. Yabumoto, and K. Nakaya (eds.). 1997. *Biology of the Megamouth Shark.* Tokyo: Tokai University Press. 203 pp.

Yano, K., and J. A. Musick. 1992. Comparison of morphometrics of Atlantic and Pacific specimens of the false catshark, *Pseudotriakis microdon,* with notes on stomach contents. *Copeia* 1992 (3): 877–886.

Yano, K., J. D. Stevens, and L. J. V. Compagno. 2004. A review of the systematics of the sleeper shark genus *Somniosus with* redescriptions of *Somniosus (Somniosus) antarcticus* and *Somniosus* (*Rhinoscymnus*) *longus* (Squaliformes: Somniosidae). *Ichthyological Research* 51: 360–373.

Yano, K., and S. Tanaka. 1983b. Portuguese shark, *Centroscymnus coelolepis* from Japan, with notes on *C. owstoni. Japanese Journal of Ichthyology* 30 (3): 208–216.

Yano, K., and S. Tanaka. 1984a. Review of the deep sea squaloid shark genus *Scymnodon* of Japan, with a description of a new species. *Japanese Journal of Ichthyology* 30 (4): 341–360.

Yano, K., and S. Tanaka. 1984b. Some biological aspects of the deep sea squaloid shark *Centroscymnus* from Suruga Bay, Japan. *Bulletin of the Japanese Society of Scientific Fisheries* 50 (2): 249–256.

Yano, K., and S. Tanaka. 1987. Reproductive organs of deep sea sharks, *Centroscymnus owstoni* and *C. coelolepis. Journal of the Faculty of Marine Science and Technology* (Tokai University) 25: 57–67.

Yano, K., and S. Tanaka. 1988. Size at maturity, reproductive cycle, fecundity, and depth segregation of the deep sea squaloid sharks *Centroscymnus owstoni* and *C. coelolepis* in Suruga Bay, Japan. *Nippon Suisan Gakkaishi* 54 (2): 167–174.

Yano, K., Y. Yabumoto, S. Tanaka, O. Tsukada, and M. Furuta. 1999. Capture of a mature female megamouth shark, *Megachasma pelagios,* from Mie, Japan. In *Proceedings of the 5th Indo-Pacific Fish Conference, Nouméa, 1977,* edited by B. Séret and J.-Y. Sire, pp. 335–349. Paris: Societé Française d'Ichtyologie.

Yano, K., M. Miya, M. Aizawa, T. Noichi. 2003. Some aspects of the biology of the goblin shark, *Mitsukurina owstoni,* collected from the Tokyo Submarine Canyon and adjacent waters, Japan. *Ichthyological Research* 54: 388–398.

Yano, K., A. Ali, A. C. Gambang, I. A. Hamid, S. A. Razak, and A. Zainal. 2005. *Sharks and Rays of Malaysia and Brunei Darussalam.* KualaTerngganu, Malaysia: Southeast Asian Fisheries Development Center. 557 pp.

Yarrell, W. 1841. *A History of British Fishes,* 2nd ed., Vol. 2. London: John Van Voorst. 628 pp.

Yasuhara, T., M. Murofushi, and H. Suguro. 1983. On the frilled shark caught in the offing of the coast of Heda, Suruga Bay, Japan. Contribution 257. *Report of Mishima Research Institute of Science for Living* (Nihon University) 6: 25–30.

Young, W. E., and H. S. Mazet. 1934. *Shark! Shark! The Thirty-Year Odyssey of a Pioneer Shark Hunter.* New York: Gotham House. 287 pp.

Yudin, K. G., and G. M. Cailliet. 1990. Age and growth of the gray smoothhound, *Mustelus californicus,* and the brown smoothhound, *M. henlei,* sharks from central California. *Copeia* 1990 (1): 191–204.

Ziegler, H. E. 1908. Ein embryo von *Chlamydoselachus anguineus* Garm. *Anatomischer Anzeiger* 33 (22/23): 561–574.

Zirkle, C. 1936. Animals impregnated by the wind. *Isis* 25 (1): 95–130.

GLOSSARY

Abdominal length: The length of the abdomen, measured from the pectoral fin insertion or axil to the pelvic fin origin.

Abdominal ridges: Ridges along the ventral sides of the abdomen of some sharks.

Abyss: Also abyssal region; once defined as that depth beyond the reach of fishermen, today it is generally defined as that portion of the oceans between 2,000 and 6,000 m and a temperature of <4°C. The abyssal region occupies most of the volume of the oceans and is the largest ecological unit of the oceans.

Adelphophagy: Literally, "brother-eating," a type of embryonic nutrition of some lamnoid sharks where one embryo consumes its smaller siblings in the uterus. Also known as embryonic cannibalism.

Allometric growth: The growth rate of one part of the body being different from the growth rate of the whole, a change in proportions with growth.

Ampullae of Lorenzini: Minute electroreceptor organs that cover the head of shark, particularly the snout area. These organs enable these fishes to detect weak electric fields at short ranges, allowing them to detect animals in the dark or hidden out of view.

Annuli: Concentric rings in vertebrae produced by the periodic deposition of salts or minerals. These rings can sometimes used to age animals.

Anterior nasal flap: A flap covering the anterior part of the nostril.

Antitropical: Refers to organisms that inhabit the temperate zones of both hemispheres and avoid the tropics, a distribution that has also been called "bipolar." The term antitropical was coined by Hubbs (1952).

Apex predator: A predator at the top of the food chain, one that feeds on other predators.

Aplacental viviparous: Nourishment of embryos to a large size by various adaptations that do not include a placental connection, (see appendix 1).

Appendiculae: Small, dendritic, fingerlike or spatulate projections of the umbilical cord of some elasmobranchs, such as sharpnose and hammerhead sharks.

Arcuate: Curved or shaped like a bow.

Artisanal fishery: A fishery composed of small boats, usually in third-world countries. This favorite neologism of international agencies cannot be found in dictionaries but is included here because it is widely used in the literature of those agencies.

Axil: The insertion of the pectoral fin, the posterior base of that fin.

Barbel: Elongated nasal flap of some sharks. Also, nasal barbel.

Bathypelagic: Said of animals living in the water column in the bathyal zone, that part of the continental slopes between 300 and 1,000 m.

Biennial: A reproductive cycle lasting two years.

Brood: The total offspring of a shark produced in one reproductive cycle.

Bycatch: Fishes or organisms taken in a fishery directed at another species.

Caudal crest: A band of raised or enlarged dermal denticles on the upper edge of the tail's caudal lobe.

Ceratotrichia: The horny rays inside a shark's fin.

Claspers: Rodlike processes of the pelvic fins of male elasmobranchs that serve as intromittent copulatory organs.

Cloaca: A common chamber into which intestines and the genital and urinary canals open in most vertebrates, except most mammals.

Congener: Belonging to the same genus.

Counterillumination: A type of camouflage found in fishes living in the "twilight zone" (200–1,000 m), where some light penetrates, its color varying with depth and time of day. These fishes would be detected by deeper swimming predators as dark silhouettes against the light filtering from above. To avoid being detected thus, some fishes (e.g., the lanternsharks) have ventral photophores (light-producing organs) that emit a bioluminescence that matches the environmental light at the depth where the species dwells. By matching the environmental light, the outline of the fish is very hard to detect visually from below. In addition, these fishes may have a metallic colored band along the flanks that serves to reflect light so as to be invisible to predators looking from the side.

Countershading: A type of camouflage common in fishes, where the dorsal coloration is dark and the ventral coloration is light, and the sides can be of silvery or metallic color. When the fish is seen from above, the dark coloration matches the bottom or the environmental light and the fish blends with its surroundings. When looked at from below against the surface glare, the fish silhouette blends in with the diffused bright light from above. When looked at from the sides light, the fish may reflect environmental light and blend in with the diffused light. All these adaptations make the fish hard to see. (This is an oversimplified description of a complex adaptation.)

Crepuscular: Describes animals that become active in early evening twilight or before sunrise.

Crown: The upper portion of tooth or a dermal denticle that protrudes above the base.

Cusp: The point of the crown on a tooth.

Cusplet: Small lateral cusps at the base of the crown in the teeth of some sharks. Also called basal denticles.

Demersal: Living on or near the bottom of the sea.

Dermal denticles: The minute, hard scales that cover the skin of sharks, also known as placoid scales or skin denticles. These structures give shark skin the feel of sandpaper.

Dicephalous: Having two heads. Sometimes seen as "bicephalous," which is a bit of a barbarism.

Egg case: The egg shell of an elasmobranch. It is keratinous and tough in oviparous elasmobranchs and thin and diaphanous in viviparous forms.

Elasmobranch: A subclass (Elasmobranchii) of the cartilaginous fishes (Chondrichthyes) that includes the sharks and rays.

Electroreceptor: *See* ampullae of Lorenzini.

Embryonic cannibalism: *See* adelphophagy.

Endemic: Prevalent in or restricted to a particular region.

Endogenous: Originating within an organism.

Endolymphatic pores: Two tiny pores or openings (less than 1 mm in diameter) on the median line of the dorsal chondrocranium of sharks. These pores lead to the endolymphatic canals and the inner ear, the organ of equilibrium and hearing.

Endothermy: In warm-blooded sharks, the ability to control the body temperature at a higher temperature than the surrounding environment through metabolism and the use of countercurrent heat exchangers.

Epididymis: A duct of the male urogenital system of sharks, straight in juveniles and highly coiled in mature sharks, an indicator of the reproductive state of sharks when full of sperm.

Epigonal organs: In some sharks, large paired, usually straplike organs found along the dorsal viscera and below the retroperitoneal kidneys, which produce various types of blood cells. The anterior dorsal portion of the epigonal organs engulfs the gonads.

Epipelagic: Living in the upper 300 m of the oceans, usually far from shore.

Exogenous: Originating outside an organism.

Extralimital species: Species not found in the area covered by this book (North America).

Gestation period: The period of intrauterine development of an embryo, from fertilization to birth.

Gill arches: The cartilaginous skeletal support of the gills.

Gill filament: The filaments of the gills where gas exchange occurs.

Gillnet: A large, wall-like net employed in fisheries for entangling fishes.

Gill rakers: Black, keratinous, bristlelike structures about 10 cm in length that protrude from the gill arches toward the pharynx of basking sharks for straining plankton from the water.

Heterocercal tail: A tail with unequal upper and lower lobes, with vertebrae extending in to the upper lobe.

Histotrophe: Nutritive substances supplied to the embryos of viviparous animals from sources other than the maternal blood stream (e.g., uterine milk from uterine glands). The word was defined by Needham (1942).

Holotype: The single specimen used as basis for the original description of a species.

Hybodont: A group of extinct sharks that lived from the Permian to the Cretaceous periods, that constitute an intermediate level between the primitive sharks and the modern sharks. They were characterized by large dorsal fin spines with a triangular plate behind the spine, and generally had two types of teeth, cusped frontal teeth and rear molariform teeth.

Hyomandibular pore: Large external pores extending along the line of the mandible and below the eye.

Ichthyologist: A person who studies fishes; a harmless drudge.

Imbricated: Having parts overlapping each other like roof tiles, such as dermal denticles or scales.

Incubation period: The period of development of an embryo developing in an egg case outside the mother. Note that sharks do not incubate their eggs.

Inner narial groove: A groove originating in the nares and extending along the front of the rostrum toward the tip of the snout.

Insertion: The point of attachment of a fin at the rear of the base.

Interdorsal ridge: An elevated dermal ridge running between the dorsal fins of some sharks.

Interdorsal space: The space between the dorsal fins, from the insertion of the first dorsal fin to the origin of the second.

Intermediate teeth: Smaller teeth between the frontal and lateral teeth.

Internarial distance: The distance separating the inner margins of the nostrils. Also internostril distance.

Junior synonym: Of two synonyms, the later established. In a case when two names are used for the same species or taxon, the name that is published or established first (the senior synonym) has priority and is generally the scientific name used.

Juvenile: An organism that has not attained sexual maturity, with some exceptions, such as neotenic amphibians.

Keel: usually used as caudal keels; a flattened area of the caudal peduncle of lamnoid sharks that serves to strengthen the tail.

Labial fold: A small fold at the mouth corners, usually perpendicular to the mouth axis.

Labial furrow: Deep folds or grooves around the corners of the mouth.

Lanceolate: Slightly broad at the base and tapering to a point, like a lance.

Larva: An embryo that is independent and self-sustaining before it assumes the characteristic features of the adult.

Lateral line: In fishes, a longitudinal line along each side of the body, made up of sensory cells.

Lecithotrophic: Feeding on yolk; often used to describe embryos nourished solely by yolk stored in the yolk sac.

Longline: A fishing line often several kilometers long, from which lines (known as gangions) hang vertically with hooks at their ends.

Lunate: Said of a tail with nearly equal upper and lower lobes, somewhat crescent shaped.

Mantle length: In squids, the length of the body proper, excluding the tentacles.

Mesopelagic: Said of fishes dwelling in the water column at depths of 200–1,000 m. The mesopelagic zone is an aphotic zone, although some light filters down to the upper boundary of the zone.

Molariform: Molarlike; teeth with blunt or flat crowns used for crushing hard or shelled prey; they are usually located at the rear of the jaw.

Monophyletic: A group descending from a single parental form or taxon.

Monotypic: Having only one species.

Multicuspid: A tooth having several cusps or cusplets.

Neonate: A recently born animal.

Nictitating eyelid: Also third eyelid or nictitating membrane; a movable eyelid of some sharks. In some cases, it can be drawn entirely over the eye for protection while passing close to objects or when biting.

Nursery area: Areas where females give birth and the neonates and young sharks spend their first few weeks, months, or years. These areas are usually in shallow waters of high productivity.

Oophagy: The mode of embryonic nourishment of lamnoid sharks; the embryos feed on eggs ovulated into the uterus for their nourishment.

Ovarian cycle: How often a shark enters the reproductive cycle and ovulates a batch of eggs.

Oviducal gland: A gland that appears like a swelling of the oviduct where the egg shell is produced; also known as nidamental or shell gland.

Oviparity: Egg laying (see appendix 1).

Oviparous: An animal that lays eggs.

Ovoviviparous: *See* aplacental viviparity (see also appendix 1).

Ovulation: The release of an ovum (egg) from the ovary.

Paratype: When a series of animals is used to describe a species, one specimen is designated as the holotype (the "type specimen"); the rest of the type series are known as paratypes.

Pedicel: In some sharks, a stalk that forms the base of a dermal denticle.

Pelagic: Living in the water column.

Photophore: A light-emitting organ, a bioluminescent structure.

Pineal window: A slightly depressed, unpigmented spot in the center of the chondrocranium of some sharks, believed to be a vestigial light receptor.

Planktotrophic: An animal that feeds on plankton.

Precaudal pit: A small pit, often lunate in shape, on the caudal peduncle at the base of the tail.

Precaudal vertebrae: The vertebrae from the base of the skull to the caudal fin origin.

Rete mirabile: Also known as a "countercurrent heat exchanger," a network of closely apposed, parallel, small veins and arteries that supply or drain the viscera and red muscle of some fishes, minimizing the heat being dissipated at the gills and resulting in a net temperature increase (see footnote on p. 256). Plural *retia mirabilia.*

Ridge-backed: Describes a carcharhinid shark with an interdorsal ridge.

Rostrum: The snout area.

Saddle marks: Dark markings or blotches on the dorsal surface of some sharks, such as catsharks or the leopard shark.

Selachia: Members of the Selachii, one of the two orders of elasmobrachs, which includes all the sharks. Selachia are characterized by pectoral fins well demarcated from the body, gill slits lateral in position, and a tail used for locomotion.

Smooth-backed: Describes a carcharhinid shark lacking an interdorsal ridge.

Spermatogenesis: The process of sperm formation from spermatogonium, to primary and secondary spermatocytes, to spermatid, to spermatozoa.

Spermatophores: A capsule of albuminous substance containing a number of spermatozoa.

Spermozeugma: A large mass of active spermatozoa in a matrix ready for delivery to the female. Plural *spermozeugmata.*

Spiracle: An opening usually posterior to the eyes on top of the head of some sharks; it represents a vestigial first gill pouch and may have rudimentary gill lamellae on its anterior surface. The oxygenated blood from these lamellae is diverted to the eye and brain. Generally, the spiracle in sharks is not involved in respiration, but it is likely that in some sharks it serves a respiratory function when it takes a long time to bite off a chunk of meat and swallow it.

Squalene: A hydrocarbon (C_{30}) and tripentene, originally isolated from shark live oil, but now known to occur in many other animals and plants. It has many uses, but these days it is mainly used (as squalane) in cosmetics as a moisturizer because it penetrates the skin and does not feel oily.

Subadults: A poorly defined term, presumably for an animal that has not attained sexual maturity, a large juvenile. One problem with the term is that it cannot easily be separated from a juvenile.

Subterminal mouth: A mouth located on the ventral side of the head and away from the tip of the snout.

Supraorbital shelf: Some catshark genera have the chondrocranium cartilage extending outward over the eye orbits, forming narrow shelves that can be detected by touch without dissection. See S. Springer (1979a: 5) for a diagram of this structure. Also supraorbital crest.

Sympatric: Occupying the same, or overlapping, areas of geographical distribution.

Symphysis: The center of each jaw, the line of union where the two halves of the jaws meet and fuse together. In many species there are one or more tiny teeth at the symphysis.

Synonym: Each of two or more names used to denote the same species or taxon.

Tendrils: Long, stringlike, coiling, adhesive protrusions at the ends of the egg cases of some elasmobranchs for attaching the egg cases to bottom structures.

Terminal mouth: A mouth at the tip of the snout, like in the whale shark or the frill shark.

Type specimen: The single specimen used as basis for the original description of the species; the holotype.

Umbilical scar: A scar between the pectoral fins caused by the closing of the umbilicus in placental sharks. The umbilicus closes and heals in a few weeks. An open (unhealed) wound indicates a recent birth.

Vitellogenesis: The production of yolk.

Viviparous: Giving birth of live offspring.

Wind eggs: An archaic term for unfertile eggs. The idea that animals could be fertilized by the wind (anaemophily) dates back to the earliest civilizations. The idea was old in the time of Aristotle, who wrote about it. For a charming account of anaemophily, see Zirkle (1936).

Xanthic: A yellow or yellowish individual.

Yolk sac: Elasmobranch eggs are proportionally very large and have large quantities of yolk to nourish the embryo through development. In development the embryo rises away from the egg and remains attached to it by a yolk stalk. The egg then becomes known as the yolk sac. The connecting yolk stalk carries the yolk to the digestive tract.

Yolk-sac placenta: The placenta of sharks formed by the union of the yolk sac and the uterine wall.

Yolk-sac scar: A small scar on the ventral side between the two pectoral fins in neonate aplacental sharks where the yolk sac has been absorbed into the body, an indication of recent birth.

Yolk stomach: A greatly distended stomach that occurs in some oophagous shark embryos from ingestion and storage of large quantities of yolk, eggs, or egg cases.

INDICES

SCIENTIFIC NAMES

T

Z

COMMON NAMES

S

T

V

W